Rational (Reciprocal) Function

$f(x) = \dfrac{1}{x}$

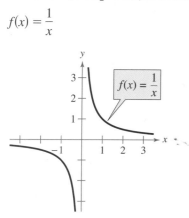

Domain: $(-\infty, 0) \cup (0, \infty)$
Range: $(-\infty, 0) \cup (0, \infty)$
No intercepts
Decreasing on $(-\infty, 0)$ and $(0, \infty)$
Odd function
Origin symmetry
Vertical asymptote: y-axis
Horizontal asymptote: x-axis

Exponential Function

$f(x) = a^x, \ a > 0, \ a \neq 1$

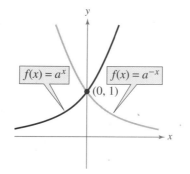

Domain: $(-\infty, \infty)$
Range: $(0, \infty)$
Intercept: $(0, 1)$
Increasing on $(-\infty, \infty)$
 for $f(x) = a^x$
Decreasing on $(-\infty, \infty)$
 for $f(x) = a^{-x}$
x-axis is a horizontal asymptote
Continuous

Logarithmic Function

$f(x) = \log_a x, \ a > 0, \ a \neq 1$

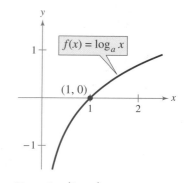

Domain: $(0, \infty)$
Range: $(-\infty, \infty)$
Intercept: $(1, 0)$
Increasing on $(0, \infty)$
y-axis is a vertical asymptote
Continuous
Reflection of graph of $f(x) = a^x$
 in the line $y = x$

Sine Function

$f(x) = \sin x$

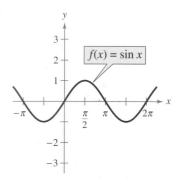

Domain: $(-\infty, \infty)$
Range: $[-1, 1]$
Period: 2π
x-intercepts: $(n\pi, 0)$
y-intercept: $(0, 0)$
Odd function
Origin symmetry

Cosine Function

$f(x) = \cos x$

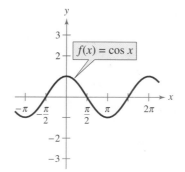

Domain: $(-\infty, \infty)$
Range: $[-1, 1]$
Period: 2π
x-intercepts: $\left(\dfrac{\pi}{2} + n\pi, 0\right)$
y-intercept: $(0, 1)$
Even function
y-axis symmetry

Tangent Function

$f(x) = \tan x$

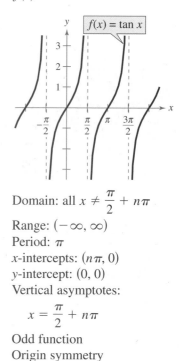

Domain: all $x \neq \dfrac{\pi}{2} + n\pi$

Range: $(-\infty, \infty)$
Period: π
x-intercepts: $(n\pi, 0)$
y-intercept: $(0, 0)$
Vertical asymptotes:

$$x = \dfrac{\pi}{2} + n\pi$$

Odd function
Origin symmetry

Cosecant Function

$f(x) = \csc x$

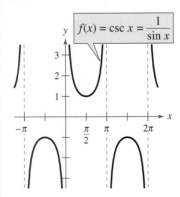

Domain: all $x \neq n\pi$
Range: $(-\infty, -1] \cup [1, \infty)$
Period: 2π
No intercepts
Vertical asymptotes: $x = n\pi$
Odd function
Origin symmetry

Sécant Function

$f(x) = \sec x$

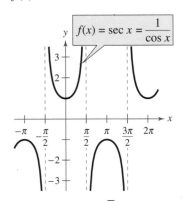

Domain: all $x \neq \dfrac{\pi}{2} + n\pi$

Range: $(-\infty, -1] \cup [1, \infty)$
Period: 2π
y-intercept: $(0, 1)$
Vertical asymptotes:

$$x = \frac{\pi}{2} + n\pi$$

Even function
y-axis symmetry

Cotangent Function

$f(x) = \cot x$

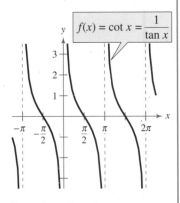

Domain: all $x \neq n\pi$
Range: $(-\infty, \infty)$
Period: π

x-intercepts: $\left(\dfrac{\pi}{2} + n\pi, 0\right)$

Vertical asymptotes: $x = n\pi$
Odd function
Origin symmetry

Inverse Sine Function

$f(x) = \arcsin x$

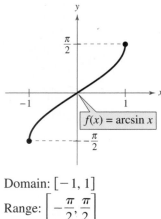

Domain: $[-1, 1]$

Range: $\left[-\dfrac{\pi}{2}, \dfrac{\pi}{2}\right]$

Intercept: $(0, 0)$
Odd function
Origin symmetry

Inverse Cosine Function

$f(x) = \arccos x$

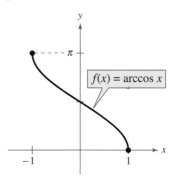

Domain: $[-1, 1]$
Range: $[0, \pi]$

y-intercept: $\left(0, \dfrac{\pi}{2}\right)$

Inverse Tangent Function

$f(x) = \arctan x$

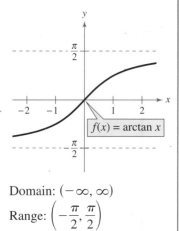

Domain: $(-\infty, \infty)$

Range: $\left(-\dfrac{\pi}{2}, \dfrac{\pi}{2}\right)$

Intercept: $(0, 0)$
Horizontal asymptotes:

$$y = \pm\frac{\pi}{2}$$

Odd function
Origin symmetry

Algebra and Trigonometry
A Graphing Approach

Fourth Edition

Ron Larson

Robert P. Hostetler
The Pennsylvania State University
The Behrend College

Bruce H. Edwards
The University of Florida

With the assistance of David C. Falvo
The Pennsylvania State University
The Behrend College

Houghton Mifflin Company   Boston   New York

Vice President and Publisher: Jack Shira
Associate Sponsoring Editor: Cathy Cantin
Development Manager: Maureen Ross
Assistant Editor: Lisa Pettinato
Assistant Editor: James Cohen
Supervising Editor: Karen Carter
Senior Project Editor: Patty Bergin
Editorial Assistant: Allison Seymour
Production Technology Supervisor: Gary Crespo
Executive Marketing Manager: Michael Busnach
Senior Marketing Manager: Danielle Potvin
Marketing Associate: Nicole Mollica
Senior Manufacturing Coordinator: Priscilla Bailey
Composition and Art: Meridian Creative Group
Cover Design Manager: Diana Coe

Cover photograph of The Reichstag © Photonica

We have included examples and exercises that use real-life data as well as technology output from a variety of software. This would not have been possible without the help of many people and organizations. Our wholehearted thanks go to all their time and effort.

Printed in the U.S.A.

Library of Congress Catalog Card Number: 2003113990

ISBN: 0-618-39455-9

23456789–DOW–08 07 06 05 04

Contents

iv   Contents

CONTENTS

A Word from the Authors

Welcome to *Algebra and Trigonometry: A Graphing Approach*, Fourth Edition. We are pleased to present this new edition of our textbook in which we focus on making the mathematics accessible, supporting student success, and offering instructors flexibility in how the course can be taught.

Accessible to Students

Over the years we have taken care to write this text with the student in mind. Paying careful attention to the presentation, we use precise mathematical language and a clear writing style to develop an effective learning tool. We believe that every student can learn mathematics, and we are committed to providing a text that makes the mathematics of the college algebra course accessible to all students. For the Fourth Edition, we have revised and improved many text features designed for this purpose.

Throughout the text, we present solutions to many examples from multiple perspectives—algebraic, graphic, and numeric. The side-by-side format of this pedagogical feature helps students to see that a problem can be solved in more than one way and to see that different methods yield the same result. The side-by-side format also addresses many different learning styles.

We have found that many students taking an algebra and trigonometry course, grasp mathematical concepts more easily when they work with them in the context of real-life situations. Students have numerous opportunities to do this throughout the Fourth Edition, in examples and exercises, including developing models to fit current real data. To reinforce the concept of functions, we have compiled all the elementary functions as a *Library of Functions*. Each function is introduced at the first point of use in the text with a definition and description of basic characteristics; all elementary functions are also presented in a summary on the front endpapers of the text for convenient reference.

We have carefully written and designed each page to make the book more readable and accessible to students. For example, to avoid unnecessary page turning and disruptions to students' thought processes, each example and corresponding solution begins and ends on the same page.

Supports Student Success

During more than thirty years of teaching and writing, we have learned many things about the teaching and learning of mathematics. We have found that students are most successful when they know what they are expected to learn and why it is important to learn it. With that in mind, we have enhanced the thematic study thread throughout the Fourth Edition.

Each chapter begins with a list of section references and a study guide, *What You Should Learn*, which is a comprehensive overview of the chapter concepts. This study guide helps students prepare to study and learn the material in the chapter.

Using the same pedagogical theme, each section begins with a set of section learning objectives—*What You Should Learn*. These are followed by an engaging real-life application—*Why You Should Learn It*—that motivates students and illustrates an area where the mathematical concepts will be applied in an example or exercise in the section. The *Chapter Summary—What Did You Learn?*—at the end of each chapter is a section-by-section overview that ties the learning objectives from the chapter to sets of *Review Exercises* at the end of each chapter.

Throughout the text, other features further improve accessibility. *Study Tips* are provided throughout the text at point-of-use to reinforce concepts and to help students learn how to study mathematics. *Explorations* have been expanded in order to reinforce mathematical concepts. Each Example with worked-out solution is followed by a *Checkpoint*, which directs the student to work a similar exercise from the exercise set. The *Section Exercises* now begin with a *Vocabulary Check*, which gives the students an opportunity to test their understanding of the important terms in the section. *Synthesis Exercises* check students' conceptual understanding of the topics in each section and *Review Exercises* provide additional practice with the concepts in the chapter or previous chapters. *Chapter Tests*, at the end of each chapter, and periodic *Cumulative Tests* offer students frequent opportunities for self-assessment and to develop strong study- and test-taking skills.

The use of technology also supports students with different learning styles, and graphing calculators are fully integrated into the text presentation. In the Fourth Edition, a robust *Technology Support Appendix* has been added to make it easier for students to use technology. *Technology Support* notes are provided throughout the text at point-of-use. These notes guide students to the *Technology Support Appendix*, where they can learn how to use specific graphing calculator features to enhance their understanding of the concepts presented in the text. These notes also direct students to the *Graphing Technology Guide*, on the textbook website, for keystroke support that is available for numerous calculator models. *Technology Tips* are provided in the text at point-of-use to call attention to the strengths and weaknesses of graphing technology, as well as to offer alternative methods for solving or checking a problem using technology. Because students are often misled by the limitations of graphing calculators, we have, where appropriate, used color to enhance the graphing calculator displays in the textbook. This enables students to visualize the mathematical concepts clearly and accurately and avoid common misunderstandings.

Numerous additional text-specific resources are available to help students succeed in the algebra and trigonometry course. These include "live" online tutoring, instructional DVDs and videos, and a variety of other resources, such as tutorial support and self-assessment, which are available on CD-ROM and the Web. In addition, the *Student Success Organizer* is a note-taking guide that helps students organize their class notes and create an effective study and review tool.

Flexible Options for Instructors

From the time we first began writing textbooks in the early 1970s, we have always considered it a critical part of our role as authors to provide instructors with flexible programs. In addition to addressing a variety of learning styles, the optional features within the text allow instructors to design their courses to meet their instructional needs and the needs of their students. For example, the

Explorations throughout the text can be used as a quick introduction to concepts or as a way to reinforce student understanding.

Our goal when developing the exercise sets was to address a wide variety of learning styles and teaching preferences. New to this edition are the *Vocabulary Check* questions, which are provided at the beginning of every exercise set to help students learn proper mathematical terminology. In each exercise set we have included a variety of exercise types, including questions requiring writing and critical thinking, as well as real-data applications. The problems are carefully graded in difficulty from mastery of basic skills to more challenging exercises. Some of the more challenging exercises include the *Synthesis Exercises* that combine skills and are used to check for conceptual understanding. *Review Exercises*, placed at the end of each exercise set, reinforce previously learned skills in preparation for the next lesson. In addition, Houghton Mifflin's Eduspace® website offers instructors the option to assign homework and tests online—and also includes the ability to grade these assignments automatically.

Several other print and media resources are also available to support instructors. The *Instructor Success Organizer* includes suggested lesson plans and is an especially useful tool for larger departments that want all sections of a course to follow the same outline. The *Instructor's Edition* of the *Student Success Organizer* can be used as a lecture outline for every section of the text and includes additional examples for classroom discussion and important definitions. This is another valuable resource for schools trying to have consistent instruction and it can be used as a resource to support less experienced instructors. When used in conjunction with the *Student Success Organizer* these resources can save instructors preparation time and help students concentrate on important concepts. For a complete list of resources available with this text, see page xv.

We hope you enjoy the Fourth Edition!

Ron Larson

Robert P. Hostetler

Bruce H. Edwards

Acknowledgments

We would like to thank the many people who have helped us prepare the text and the supplements package. Their encouragement, criticisms, and suggestions have been invaluable to us.

Fourth Edition Reviewers

Tony Homayoon Akhlaghi, Bellevue Community College; Kimberly Bennekin, Georgia Perimeter College; Charles M. Biles, Humboldt State University; Phyllis Barsch Bolin, Oklahoma Christian University; Khristo Boyadzheiv, Ohio Northern University; Jennifer Dollar, Grand Rapids Community College; Susan E. Enyart, Otterbein College; Patricia K. Gramling, Trident Technical College; Rodney Holke-Farnam, Hawkeye Community College; Deborah Johnson, Cambridge South Dorchester High School; Susan Kellicut, Seminole Community College; Richard J. Maher, Loyola University; Rupa M. Patel, University of Portland; Lila F. Roberts, Georgia Southern University; Keith Schwingendorf, Purdue University North Central; Pamela K. M. Smith, Fort Lewis College; Hayat Weiss, Middlesex Community College; Fred Worth, Henderson State University.

We would like to thank the staff of Larson Texts, Inc. and the staff of Meridian Creative Group, who assisted in proofreading the manuscript, preparing and proofreading the art package, and typesetting the supplements.

On a personal level, we are grateful to our wives, Deanna Gilbert Larson, Eloise Hostetler, and Consuelo Edwards for their love, patience, and support. Also, a special thanks goes to R. Scott O'Neil.

If you have suggestions for improving this text, please feel free to write us. Over the past two decades we have received many useful comments from both instructors and students, and we value these very much.

Ron Larson
Robert P. Hostetler
Bruce H. Edwards

Features Highlights

Colleges and universities track enrollment figures in order to determine the financial outlook of the institution. The growth in student enrollment at a college or university can be modeled by a linear equation.

1 Functions and Their Graphs

What You Should Learn

1.1 Graphs of Equations
1.2 Lines in the Plane
1.3 Functions
1.4 Graphs of Functions
1.5 Shifting, Reflecting, and Stretching Graphs
1.6 Combinations of Functions
1.7 Inverse Functions

In this chapter, you will learn how to:

■ Sketch graphs of equations by point plotting or by using a graphing utility.

■ Find and use the slope of a line to write and graph linear equations.

■ Evaluate functions and find their domains.

■ Analyze graphs of functions.

■ Identify and graph shifts, reflections, and nonrigid transformations of functions.

■ Find arithmetic combinations and compositions of functions.

■ Find inverse functions graphically and algebraically.

73

● "What You Should Learn"

Each chapter begins with *What You Should Learn*, a comprehensive overview of the chapter concepts. The photograph and caption illustrate a real-life application of a key concept. Section references help students prepare for the chapter.

● "What You Should Learn" and "Why You Should Learn It"

Sections begin with *What You Should Learn*, an outline of the main concepts covered in the section, and *Why You Should Learn It*, a real-life application or mathematical reference that illustrates the relevance of the section content.

Section 1.3   Functions   99

1.3 Functions

Introduction to Functions

Many everyday phenomena involve pairs of quantities that are related to each other by some rule of correspondence. The mathematical term for such a rule of correspondence is a **relation**. Here are two examples.

1. The simple interest I earned on an investment of $1000 for 1 year is related to the annual interest rate r by the formula $I = 1000r$.

2. The area A of a circle is related to its radius r by the formula $A = \pi r^2$.

Not all relations have simple mathematical formulas. For instance, people commonly match up NFL starting quarterbacks with touchdown passes, and hours of the day with temperature. In each of these cases, there is some relation that matches each item from one set with exactly one item from a different set. Such a relation is called a **function.**

Definition of a Function

A **function** f from a set A to a set B is a relation that assigns to each element x in the set A exactly one element y in the set B. The set A is the **domain** (or set of inputs) of the function f, and the set B contains the **range** (or set of outputs).

To help understand this definition, look at the function that relates the time of day to the temperature in Figure 1.29.

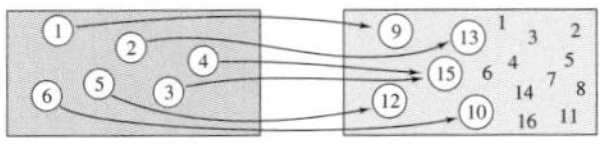

Time of day (P.M.)   Temperature (in degrees C)

Set *A* is the domain.
Inputs: 1, 2, 3, 4, 5, 6

Set *B* contains the range.
Outputs: 9, 10, 12, 13, 15

Figure 1.29

This function can be represented by the ordered pairs $\{(1, 9°), (2, 13°), (3, 15°), (4, 15°), (5, 12°), (6, 10°)\}$. In each ordered pair, the first coordinate (*x*-value) is the **input** and the second coordinate (*y*-value) is the **output.**

Characteristics of a Function from Set A to Set B

1. Each element of *A* must be matched with an element of *B*.
2. Some elements of *B* may not be matched with any element of *A*.
3. Two or more elements of *A* may be matched with the same element of *B*.
4. An element of *A* (the domain) cannot be matched with two different elements of *B*.

What you should learn
● Decide whether relations between two variables represent a function.
● Use function notation and evaluate functions.
● Find the domains of functions.
● Use functions to model and solve real-life problems.
● Evaluate difference quotients.

Why you should learn it
Many natural phenomena can be modeled by functions, such as the force of water against the face of a dam, explored in Exercise 81 on page 111.

Kunio Owaki/Corbis

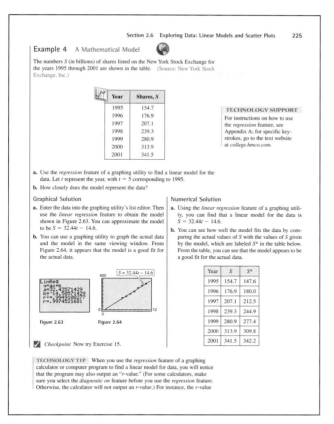

Examples

Many examples present side-by-side solutions from multiple approaches—algebraic, graphical, and numerical. This format addresses a variety of learning styles and shows students that different solution methods yield the same result.

Checkpoint

The *Checkpoint* directs students to work a similar problem in the exercise set for extra practice.

Library of Functions

The *Library of Functions* feature defines each elementary function and its characteristics at first point of use.

Explorations

The *Exploration* engages students in active discovery of mathematical concepts, strengthens critical thinking skills, and helps them to develop an intuitive understanding of theoretical concepts.

Study Tips

Study Tips reinforce concepts and help students learn how to study mathematics.

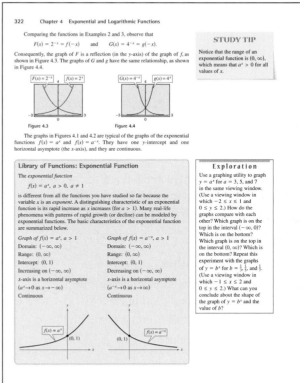

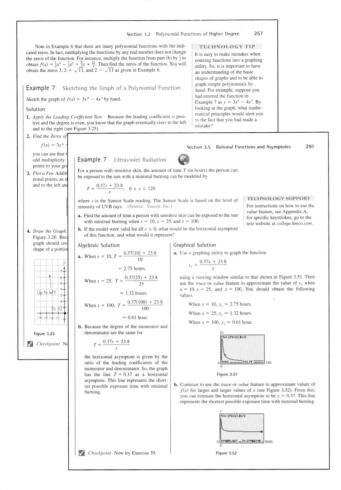

● Technology Tip

Technology Tips point out the pros and cons of technology use in certain mathematical situations. *Technology Tips* also provide alternative methods of solving or checking a problem by the use of a graphing calculator.

● Technology Support

The *Technology Support* feature guides students to the *Technology Support Appendix* if they need to reference a specific calculator feature. These notes also direct students to the *Graphing Technology Guide*, on the textbook website, for keystroke support that is available for numerous calculator models.

● Real-Life Applications

A wide variety of real-life applications, many using current real data, are integrated throughout the examples and exercises. The 🌐 indicates an example that involves a real-life application.

● Algebra of Calculus

Throughout the text, special emphasis is given to the algebraic techniques used in calculus. 𝑓 indicates an example or exercise in which the algebra of calculus is featured.

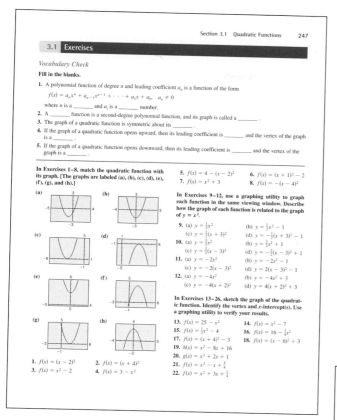

- ## Vocabulary Check

Section exercises begin with a *Vocabulary Check* that serves as a review of the important mathematical terms in each section.

- ## Section Exercises

The section exercise sets consist of a variety of computational, conceptual, and applied problems.

- ## Synthesis and Review Exercises

Each exercise set concludes with two types of exercises.

Synthesis exercises promote further exploration of mathematical concepts, critical thinking skills, and writing about mathematics. The exercises require students to show their understanding of the relationships between many concepts in the section.

Review Exercises reinforce previously learned skills and concepts.

FEATURES

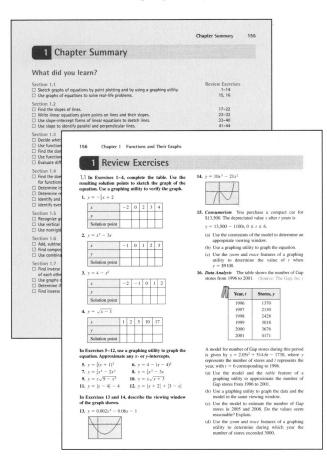

● Chapter Summary

The *Chapter Summary*, "*What Did You Learn?*" is a section-by-section overview that ties the learning objectives from the chapter to sets of Review Exercises for extra practice.

● Review Exercises

The chapter *Review Exercises* provide additional practice with the concepts in the chapter.

● Chapter Tests and Cumulative Tests

Chapter Tests, at the end of each chapter, and periodic *Cumulative Tests* offer students frequent opportunities for self-assessment and to develop strong study- and test-taking skills.

Supplements

Resources

Text website (college.hmco.com)

Many text-specific resources for students and instructors can be found at the Houghton Mifflin website. They include, but are not limited to, the following features for the student and instructor.

Student Website

- Student Success Organizer
- Digital Lessons
- Graphing Technology Guide
- Graphing Calculator Programs
- Chapter Projects
- Historical Notes
- Ace Quizzes

Instructor Website

- Instructor Success Organizer
- Digital Art and Tables
- Digital Lessons
- Graphing Technology Guide
- Graphing Calculator Programs
- Chapter Projects
- Answers to Chapter Projects
- Link to Student website

Additional Resources for the Student

Study and Solutions Guide by Bruce H. Edwards (University of Florida)

HM mathSpace® Tutorial CD-ROM: This new tutorial CD-ROM allows students to practice skills and review concepts as many times as necessary by using algorithmically generated exercises and step-by-step solutions for practice. The CD-ROM contains a variety of other student resources as well.

Instructional Videotapes by Dana Mosely

Instructional Videotapes for Graphing Calculators by Dana Mosely

SMARTTHINKING™ Live, On-Line Tutoring: Houghton Mifflin has partnered with SMARTTHINKING™ to provide an easy-to-use, effective, on-line tutorial service. Through state-of-the-art tools and a two-way whiteboard, students communicate in real-time with qualified e-structors who can help the students understand difficult concepts and guide them through the problem solving process while studying or completing homework. Live online tutoring support, Question submission, Pre-scheduled tutoring time, and Reviews of past online sessions are four levels of service offered to the students.

Eduspace®: Eduspace® is a text-specific online learning environment that combines algorithmic tutorials with homework capabilities. Text-specific content is available to help you understand the mathematics covered in this textbook.

Eduspace® with eSolutions: Eduspace® with eSolutions combines all the features of Eduspace® with an electronic version of the textbook exercises and the complete solutions to the odd-numbered exercises. The result is a convenient and comprehensive way to do homework and view your course materials.

Additional Resources for the Instructor

Instructor's Annotated Edition (IAE)

Instructor's Solutions Guide and Test Item File by Bruce H. Edwards (University of Florida)

HM ClassPrep with HM Testing CD-ROM: This CD-ROM is a combination of two course management tools.

- HM Testing 6.0 computerized testing software provides instructors with an array of algorithmic test items, allowing for the creation of an unlimited number of tests for each chapter, including cumulative tests and final exams. HM Testing also offers online testing via a Local Area Network (LAN) or the Internet, as well as a grade book function.
- HM ClassPrep features supplements and text-specific resources.

Eduspace®: Eduspace® is a text-specific online learning environment that combines algorithmic tutorials with homework capabilities and classroom management functions. Electronic grading and Course Management are two levels of service provided for instructors. Please contact your Houghton Mifflin sales representative for detailed information about the course content available for this text.

Eduspace® with eSolutions: Eduspace® with eSolutions combines all the features of Eduspace® with an electronic version of the textbook exercises and the complete solutions to the odd-numbered exercises, providing students with a convenient and comprehensive way to do homework and view course materials.

The stopping distance of an automobile depends on the distance traveled during the driver's reaction time and the distance traveled after the brakes are applied. The total stopping distance can be modeled by a polynomial.

David Young-Wolff/PhotoEdit

P Prerequisites

What You Should Learn

In this chapter, you will learn how to:

■ Represent, classify, and order real numbers and use inequalities.

■ Evaluate algebraic expressions using the basic rules of algebra.

■ Use properties of exponents and radicals to simplify and evaluate expressions.

■ Add, subtract, and multiply polynomials.

■ Factor expressions completely.

■ Determine the domains of algebraic expressions and simplify rational expressions.

■ Use algebraic techniques common in calculus.

■ Plot points in the coordinate plane and use the Distance and Midpoint Formulas.

■ Organize data and represent data graphically.

1

P.1 Real Numbers

Real Numbers

Real numbers are used in everyday life to describe quantities such as age, miles per gallon, and population. Real numbers are represented by symbols such as

$$-5, 9, 0, \tfrac{4}{3}, 0.666\ldots, 28.21, \sqrt{2}, \pi, \text{ and } \sqrt[3]{-32}.$$

Here are some important **subsets** (each member of subset B is also a member of set A) of the set of real numbers.

$$\{1, 2, 3, 4, \ldots\} \qquad \text{Set of natural numbers}$$

$$\{0, 1, 2, 3, 4, \ldots\} \qquad \text{Set of whole numbers}$$

$$\{\ldots, -3, -2, -1, 0, 1, 2, 3, \ldots\} \qquad \text{Set of integers}$$

A real number is **rational** if it can be written as the ratio p/q of two integers, where $q \neq 0$. For instance, the numbers

$$\frac{1}{3} = 0.3333\ldots = 0.\overline{3}, \quad \frac{1}{8} = 0.125, \text{ and } \frac{125}{111} = 1.126126\ldots = 1.\overline{126}$$

are rational. The decimal representation of a rational number either *repeats* (as in $\frac{173}{55} = 3.1\overline{45}$) or *terminates* (as in $\frac{1}{2} = 0.5$). A real number that cannot be written as the ratio of two integers is called **irrational**. Irrational numbers have infinite nonrepeating decimal representations. For instance, the numbers

$$\sqrt{2} = 1.4142135\ldots \approx 1.41 \quad \text{and} \quad \pi = 3.1415925\ldots \approx 3.14$$

are irrational. (The symbol $\approx$ means "is approximately equal to.") Figure P.1 shows subsets of real numbers and their relationships to each other.

Real numbers are represented graphically by a **real number line.** The point 0 on the real number line is the **origin.** Numbers to the right of 0 are positive and numbers to the left of 0 are negative, as shown in Figure P.2. The term **nonnegative** describes a number that is either positive or zero.

Figure P.2 The Real Number Line

There is a *one-to-one correspondence* between real numbers and points on the real number line. That is, every point on the real number line corresponds to exactly one real number, called its **coordinate,** and every real number corresponds to exactly one point on the real number line, as shown in Figure P.3.

Every point on the real number line corresponds to exactly one real number.

Every real number corresponds to exactly one point on the real number line.

Figure P.3 One-to-One Correspondence

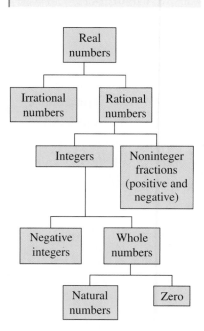

Figure P.1 Subsets of Real Numbers

Ordering Real Numbers

One important property of real numbers is that they are **ordered**.

> ### Definition of Order on the Real Number Line
>
> If a and b are real numbers, a is **less than** b if $b - a$ is positive. This order is denoted by the **inequality** $a < b$. This relationship can also be described by saying that b is **greater than** a and writing $b > a$. The inequality $a \leq b$ means that a is **less than or equal to** b, and the inequality $b \geq a$ means that b is **greater than or equal to** a. The symbols $<$, $>$, $\leq$, and $\geq$ are **inequality symbols.**

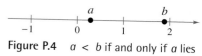

Figure P.4 $a < b$ if and only if a lies to the left of b.

Geometrically, this definition implies that $a < b$ if and only if a lies to the *left* of b on the real number line, as shown in Figure P.4.

Example 1   Interpreting Inequalities

Describe the subset of real numbers represented by each inequality.

a. $x \leq 2$ **b.** $x > -1$ **c.** $-2 \leq x < 3$

Solution

a. The inequality $x \leq 2$ denotes all real numbers less than or equal to 2, as shown in Figure P.5.

b. The inequality $x > -1$ denotes all real numbers greater than -1, as shown in Figure P.6.

c. The inequality $-2 \leq x < 3$ means that $x \geq -2$ *and* $x < 3$. The "double inequality" denotes all real numbers between -2 and 3, including -2 but not including 3, as shown in Figure P.7.

✓ *Checkpoint* Now try Exercise 31.

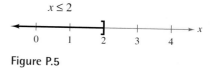

Figure P.5

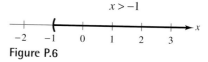

Figure P.6

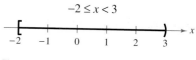

Figure P.7

Inequalities can be used to describe subsets of real numbers called **intervals.** In the bounded intervals below, the real numbers a and b are the **endpoints** of each interval.

Bounded Intervals on the Real Number Line

Notation	Interval Type	Inequality	Graph
$[a, b]$	Closed	$a \leq x \leq b$	
(a, b)	Open	$a < x < b$	
$[a, b)$		$a \leq x < b$	
$(a, b]$		$a < x \leq b$	

The symbols ∞, **positive infinity,** and −∞, **negative infinity,** do not represent real numbers. They are simply convenient symbols used to describe the unboundedness of an interval such as $(1, \infty)$ or $(-\infty, 3]$.

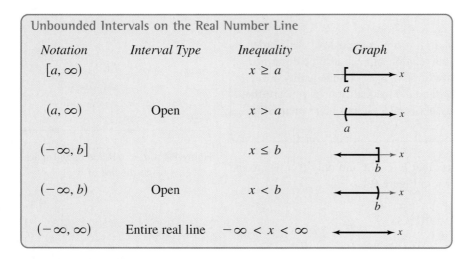

Unbounded Intervals on the Real Number Line

Notation	Interval Type	Inequality	Graph
$[a, \infty)$		$x \geq a$	
(a, ∞)	Open	$x > a$	
$(-\infty, b]$		$x \leq b$	
$(-\infty, b)$	Open	$x < b$	
$(-\infty, \infty)$	Entire real line	$-\infty < x < \infty$	

Example 2   Using Inequalities to Represent Intervals

Use inequality notation to describe each of the following.

a. c is at most 2.   **b.** All x in the interval $(-3, 5]$

Solution

a. The statement "c is at most 2" can be represented by $c \leq 2$.

b. "All x in the interval $(-3, 5]$" can be represented by $-3 < x \leq 5$.

✓ *Checkpoint* Now try Exercise 35.

Example 3   Interpreting Intervals

Give a verbal description of each interval.

a. $(-1, 0)$ **b.** $[2, \infty)$ **c.** $(-\infty, 0)$

Solution

a. This interval consists of all real numbers that are greater than -1 and less than 0.

b. This interval consists of all real numbers that are greater than or equal to 2.

c. This interval consists of all negative real numbers.

✓ *Checkpoint* Now try Exercise 41.

The **Law of Trichotomy** states that for any two real numbers a and b, precisely one of three relationships is possible:

$a = b$, $a < b$, or $a > b$.   Law of Trichotomy

Absolute Value and Distance

The **absolute value** of a real number is its *magnitude*, or the distance between the origin and the point representing the real number on the real number line.

> **Definition of Absolute Value**
>
> If a is a real number, the **absolute value** of a is
>
> $$|a| = \begin{cases} a, & \text{if } a \geq 0 \\ -a, & \text{if } a < 0 \end{cases}.$$

Notice from this definition that the absolute value of a real number is never negative. For instance, if $a = -5$, then $|-5| = -(-5) = 5$. The absolute value of a real number is either positive or zero. Moreover, 0 is the only real number whose absolute value is 0. So, $|0| = 0$.

Example 4   Evaluating the Absolute Value of a Number

Evaluate $\dfrac{|x|}{x}$ for (a) $x > 0$ and (b) $x < 0$.

Solution

a. If $x > 0$, then $|x| = x$ and $\dfrac{|x|}{x} = \dfrac{x}{x} = 1$.

b. If $x < 0$, then $|x| = -x$ and $\dfrac{|x|}{x} = \dfrac{-x}{x} = -1$.

✓ *Checkpoint* Now try Exercise 47.

> **Properties of Absolute Value**
>
> **1.** $|a| \geq 0$ **2.** $|-a| = |a|$
>
> **3.** $|ab| = |a||b|$ **4.** $\left|\dfrac{a}{b}\right| = \dfrac{|a|}{|b|}, \quad b \neq 0$

Absolute value can be used to define the distance between two points on the real number line. For instance, the distance between -3 and 4 is

$$|-3 - 4| = |-7| = 7$$

as shown in Figure P.8.

> **Distance Between Two Points on the Real Line**
>
> Let a and b be real numbers. The **distance between a and b** is
>
> $$d(a, b) = |b - a| = |a - b|.$$

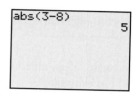
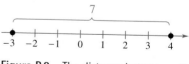

Figure P.8 The distance between -3 and 4 is 7.

Algebraic Expressions

One characteristic of algebra is the use of letters to represent numbers. The letters are **variables,** and combinations of letters and numbers are **algebraic expressions.** Here are a few examples of algebraic expressions.

$$5x, \qquad 2x - 3, \qquad \frac{4}{x^2 + 2}, \qquad 7x + y$$

Definition of an Algebraic Expression

An **algebraic expression** is a combination of letters (**variables**) and real numbers (**constants**) combined using the operations of addition, subtraction, multiplication, division, and exponentiation.

The **terms** of an algebraic expression are those parts that are separated by *addition.* For example,

$$x^2 - 5x + 8 = x^2 + (-5x) + 8$$

has three terms: x^2 and $-5x$ are the **variable terms** and 8 is the **constant term.** The numerical factor of a variable term is the **coefficient** of the variable term. For instance, the coefficient of $-5x$ is -5, and the coefficient of x^2 is 1.

To **evaluate** an algebraic expression, substitute numerical values for each of the variables in the expression. Here are two examples.

Expression	Value of Variable	Substitute	Value of Expression
$-3x + 5$	$x = 3$	$-3(3) + 5$	$-9 + 5 = -4$
$3x^2 + 2x - 1$	$x = -1$	$3(-1)^2 + 2(-1) - 1$	$3 - 2 - 1 = 0$

When an algebraic expression is evaluated, the **Substitution Principle** is used. It states, "If $a = b$, then a can be replaced by b in any expression involving a." In the first evaluation shown above, for instance, 3 is *substituted* for x in the expression $-3x + 5$.

Basic Rules of Algebra

There are four arithmetic operations with real numbers: *addition, multiplication, subtraction,* and *division,* denoted by the symbols $+$, $\times$ or $\cdot$, $-$, and $\div$ or $/$. Of these, addition and multiplication are the two primary operations. Subtraction and division are the inverse operations of addition and multiplication, respectively.

Subtraction: Add the opposite of b.   *Division: Multiply by the reciprocal of b.*

$$a - b = a + (-b) \qquad\qquad \text{If } b \neq 0, \text{ then } a/b = a\left(\frac{1}{b}\right) = \frac{a}{b}.$$

In these definitions, $-b$ is the **additive inverse** (or opposite) of b, and $1/b$ is the **multiplicative inverse** (or reciprocal) of b. In the fractional form a/b, a is the **numerator** of the fraction and b is the **denominator.**

Because the properties of real numbers below are true for variables and algebraic expressions, as well as for real numbers, they are often called the **Basic Rules of Algebra.** Try to formulate a verbal description of each property. For instance, the first property states that *the order in which two real numbers are added does not affect their sum.*

Basic Rules of Algebra

Let a, b, and c be real numbers, variables, or algebraic expressions.

	Property	*Example*
Commutative Property of Addition:	$a + b = b + a$	$4x + x^2 = x^2 + 4x$
Commutative Property of Multiplication:	$ab = ba$	$(1 - x)x^2 = x^2(1 - x)$
Associative Property of Addition:	$(a + b) + c = a + (b + c)$	$(x + 5) + x^2 = x + (5 + x^2)$
Associative Property of Multiplication:	$(ab)c = a(bc)$	$(2x \cdot 3y)(8) = (2x)(3y \cdot 8)$
Distributive Properties:	$a(b + c) = ab + ac$	$3x(5 + 2x) = 3x \cdot 5 + 3x \cdot 2x$
	$(a + b)c = ac + bc$	$(y + 8)y = y \cdot y + 8 \cdot y$
Additive Identity Property:	$a + 0 = a$	$5y^2 + 0 = 5y^2$
Multiplicative Identity Property:	$a \cdot 1 = a$	$(4x^2)(1) = 4x^2$
Additive Inverse Property:	$a + (-a) = 0$	$6x^3 + (-6x^3) = 0$
Multiplicative Inverse Property:	$a \cdot \dfrac{1}{a} = 1, \quad a \neq 0$	$(x^2 + 4)\left(\dfrac{1}{x^2 + 4}\right) = 1$

Because subtraction is defined as "adding the opposite," the Distributive Properties are also true for subtraction. For instance, the "subtraction form" of $a(b + c) = ab + ac$ is $a(b - c) = ab - ac$.

Properties of Negation and Equality

Let a, b, and c be real numbers, variables, or algebraic expressions.

Property	*Example*
1. $(-1)a = -a$	$(-1)7 = -7$
2. $-(-a) = a$	$-(-6) = 6$
3. $(-a)b = -(ab) = a(-b)$	$(-5)3 = -(5 \cdot 3) = 5(-3)$
4. $(-a)(-b) = ab$	$(-2)(-x) = 2x$
5. $-(a + b) = (-a) + (-b)$	$-(x + 8) = (-x) + (-8) = -x - 8$
6. If $a = b$, then $a + c = b + c$.	$\frac{1}{2} + 3 = 0.5 + 3$
7. If $a = b$, then $ac = bc$.	$4^2(2) = 16(2)$
8. If $a + c = b + c$, then $a = b$.	$1.4 - 1 = \frac{7}{5} - 1$
9. If $ac = bc$ and $c \neq 0$, then $a = b$.	$\dfrac{3}{4} = \dfrac{\sqrt{9}}{4}$

STUDY TIP

Be sure you see the difference between the *opposite of a number* and a *negative number*. If a is already negative, then its opposite, $-a$, is positive. For instance, if $a = -2$, then $-a = -(-2) = 2$.

Properties of Zero

Let a and b be real numbers, variables, or algebraic expressions.

1. $a + 0 = a$ and $a - 0 = a$
2. $a \cdot 0 = 0$

3. $\dfrac{0}{a} = 0, \quad a \neq 0$
4. $\dfrac{a}{0}$ is undefined.

5. Zero-Factor Property: If $ab = 0$, then $a = 0$ or $b = 0$.

STUDY TIP

The "or" in the Zero-Factor Property includes the possibility that either or both factors may be zero. This is an **inclusive or,** and it is the way the word "or" is generally used in mathematics.

Properties and Operations of Fractions

Let a, b, c, and d be real numbers, variables, or algebraic expressions such that $b \neq 0$ and $d \neq 0$.

1. Equivalent Fractions: $\dfrac{a}{b} = \dfrac{c}{d}$ if and only if $ad = bc$.

2. Rules of Signs: $-\dfrac{a}{b} = \dfrac{-a}{b} = \dfrac{a}{-b}$ and $\dfrac{-a}{-b} = \dfrac{a}{b}$

3. Generate Equivalent Fractions: $\dfrac{a}{b} = \dfrac{ac}{bc}, \quad c \neq 0$

4. Add or Subtract with Like Denominators: $\dfrac{a}{b} \pm \dfrac{c}{b} = \dfrac{a \pm c}{b}$

5. Add or Subtract with Unlike Denominators: $\dfrac{a}{b} \pm \dfrac{c}{d} = \dfrac{ad \pm bc}{bd}$

6. Multiply Fractions: $\dfrac{a}{b} \cdot \dfrac{c}{d} = \dfrac{ac}{bd}$

7. Divide Fractions: $\dfrac{a}{b} \div \dfrac{c}{d} = \dfrac{a}{b} \cdot \dfrac{d}{c} = \dfrac{ad}{bc}, \quad c \neq 0$

Example 5 Properties and Operations of Fractions

a. $\dfrac{x}{3} + \dfrac{2x}{5} = \dfrac{5 \cdot x + 3 \cdot 2x}{15} = \dfrac{11x}{15}$
Add fractions with unlike denominators.

b. $\dfrac{7}{x} \div \dfrac{3}{2} = \dfrac{7}{x} \cdot \dfrac{2}{3} = \dfrac{14}{3x}$
Divide fractions.

✓ *Checkpoint* Now try Exercise 101.

STUDY TIP

In Property 1 of fractions, the phrase "if and only if" implies two statements. One statement is: If $a/b = c/d$, then $ad = bc$. The other statement is: If $ad = bc$, where $b \neq 0$ and $d \neq 0$, then $a/b = c/d$.

If a, b, and c are integers such that $ab = c$, then a and b are **factors** or **divisors** of c. A **prime number** is an integer that has exactly two positive factors: itself and 1. For example, 2, 3, 5, 7, and 11 are prime numbers. The numbers 4, 6, 8, 9, and 10 are **composite** because they can be written as the product of two or more prime numbers. The number 1 is neither prime nor composite. The **Fundamental Theorem of Arithmetic** states that every positive integer greater than 1 can be written as the product of prime numbers. For instance, the **prime factorization** of 24 is $24 = 2 \cdot 2 \cdot 2 \cdot 3$.

P.1   Exercises

Vocabulary Check

Fill in the blanks.

1. A real number is _____ if it can be written as the ratio $\frac{p}{q}$ of two integers, where $q \neq 0$.

2. _____ numbers have infinite nonrepeating decimal representations.

3. The distance between a point on the real number line and the origin is the _____ of the real number.

4. Numbers that can be written as the product of two or more prime numbers are called _____ numbers.

5. Integers that have exactly two positive factors, the integer itself and 1, are called _____ numbers.

6. An algebraic expression is a combination of letters called _____ and real numbers called _____ .

7. The _____ of an algebraic expression are those parts separated by addition.

8. The numerical factor of a variable term is the _____ of the variable term.

9. The _____ states: If $ab = 0$, then $a = 0$ or $b = 0$.

In Exercises 1–6, determine which numbers are (a) natural numbers, (b) whole numbers, (c) integers, (d) rational numbers, and (e) irrational numbers.

1. $\left\{ -9, -\frac{7}{2}, 5, \frac{2}{3}, \sqrt{2}, 0, 1, -4, -1 \right\}$

2. $\left\{ \sqrt{5}, -7, -\frac{7}{3}, 0, 3.12, \frac{5}{4}, -2, -8, 3 \right\}$

3. $\{ 2.01, 0.666 \ldots, -13, 0.010110111 \ldots, 1, -10, 20 \}$

4. $\{ 2.3030030003 \ldots, 0.7575, -4.63, \sqrt{10}, -2, 0.03, -10 \}$

5. $\left\{ -\pi, -\frac{1}{3}, \frac{6}{3}, \frac{1}{2}\sqrt{2}, -7.5, -2, 3, -3 \right\}$

6. $\left\{ 25, -17, -\frac{12}{5}, \sqrt{9}, 3.12, \frac{1}{2}\pi, 6, -4, 18 \right\}$

In Exercises 7–12, use a calculator to find the decimal form of the rational number. If it is a nonterminating decimal, write the repeating pattern.

7. $\frac{5}{8}$

8. $\frac{17}{4}$

9. $\frac{41}{333}$

10. $\frac{6}{11}$

11. $-\frac{100}{11}$

12. $-\frac{218}{33}$

In Exercises 13–16, use a graphing utility to rewrite the rational number as the ratio of two integers.

13. 4.6

14. 12.3

15. 6.5

16. −1.83

In Exercises 17 and 18, approximate the numbers and place the correct inequality symbol (< or >) between them.

17.

18.

In Exercises 19–24, plot the two real numbers on the real number line. Then place the correct inequality symbol (< or >) between them.

19. $-4, -8$

20. $-3.5, 1$

21. $\frac{3}{2}, 7$

22. $1, \frac{16}{3}$

23. $\frac{5}{6}, \frac{2}{3}$

24. $-\frac{8}{7}, -\frac{3}{7}$

In Exercises 25–32, (a) verbally describe the subset of real numbers represented by the inequality, (b) sketch the subset on the real number line, and (c) state whether the interval is bounded or unbounded.

25. $x \leq 5$

26. $x > 3$

27. $x < 0$

28. $x \geq 4$

29. $-2 < x < 2$

30. $0 \leq x \leq 5$

31. $-1 \leq x < 0$

32. $0 < x \leq 6$

In Exercises 33–38, use inequality and interval notation to describe the set.

33. x is negative.

34. z is at least 10.

35. y is nonnegative.

36. y is no more than 25.

37. p is less than 9 but no less than -1.

38. The annual rate of inflation r is expected to be at least 2.5%, but no more than 5%.

In Exercises 39–42, give a verbal description of the interval.

39. $(-6, \infty)$

40. $(-\infty, 4]$

41. $(-\infty, 2]$

42. $[1, \infty)$

In Exercises 43–48, evaluate the expression.

43. $|-10|$

44. $|0|$

45. $-3|-3|$

46. $|-1| - |-2|$

47. $\dfrac{|x + 2|}{x + 2}$

48. $\dfrac{|x - 1|}{x - 1}$

In Exercises 49–54, place the correct symbol ($<$, $>$, or $=$) between the pair of real numbers.

49. $|-3| \quad -|-3|$

50. $|-4| \quad |4|$

51. $-5 \quad -|5|$

52. $-|-6| \quad |-6|$

53. $-|-2| \quad -|2|$

54. $-(-2) \quad -2$

In Exercises 55–60, find the distance between a and b.

55. $a = 126, b = 75$

56. $a = -126, b = -75$

57. $a = -\frac{5}{2}, b = 0$

58. $a = \frac{1}{4}, b = \frac{11}{4}$

59. $a = \frac{16}{5}, b = \frac{112}{75}$

60. $a = 9.34, b = -5.65$

In Exercises 61–66, use absolute value notation to describe the situation.

61. The distance between x and 5 is no more than 3.

62. The distance between x and -10 is at least 6.

63. y is at least six units from 0.

64. y is at most two units from a.

65. While traveling on the Pennsylvania Turnpike, you pass milepost 57 near Pittsburgh, then milepost 236 near Gettysburg. How many miles do you travel during that time period?

66. The temperature in Bismarck, North Dakota was 60° at noon, then 23° at midnight. What was the change in temperature over the 12-hour period?

Budget Variance **In Exercises 67–70, the accounting department of a company is checking to determine whether the actual expenses of a department differ from the budgeted expenses by more than $500 or by more than 5%. Fill in the missing parts of the table, and determine whether the actual expense passes the "budget variance test."**

		Budgeted Expense, b	Actual Expense, a	$|a - b|$	$0.05b$
67.	Wages	$112,700	$113,356		
68.	Utilities	$9400	$9772		
69.	Taxes	$37,640	$37,335		
70.	Insurance	$2575	$2613		

Federal Deficit **In Exercises 71–76, use the bar graph, which shows the receipts of the federal government (in billions of dollars) for selected years from 1960 through 2002. In each exercise you are given the expenditures of the federal government. Find the magnitude of the surplus or deficit for the year.**
(Source: U.S. Office of Management and Budget)

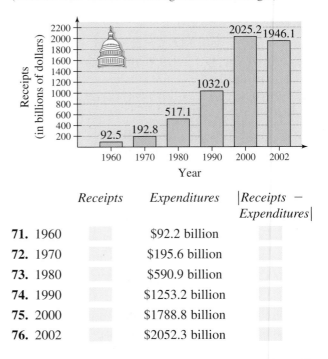

		Receipts	Expenditures	$|Receipts -$ Expenditures$|$
71.	1960		$92.2 billion	
72.	1970		$195.6 billion	
73.	1980		$590.9 billion	
74.	1990		$1253.2 billion	
75.	2000		$1788.8 billion	
76.	2002		$2052.3 billion	

In Exercises 77–82, identify the terms. Then identify the coefficients of the variable terms of the expression.

77. $7x + 4$

78. $2x - 9$

79. $\sqrt{3}x^2 - 8x - 11$

80. $7\sqrt{5}x^2 + 3$

81. $4x^3 + \dfrac{x}{2} - 5$ **82.** $3x^4 + \dfrac{2x^3}{5}$

In Exercises 83–86, evaluate the expression for each value of x. (If not possible, state the reason.)

	Expression		Values	
83.	$4x - 6$	(a) $x = -1$	(b) $x = 0$
84.	$9 - 7x$	(a) $x = -3$	(b) $x = 3$
85.	$-x^2 + 5x - 4$	(a) $x = -1$	(b) $x = 1$
86.	$\dfrac{x}{x + 2}$	(a) $x = 2$	(b) $x = -2$

In Exercises 87–94, identify the rule(s) of algebra illustrated by the statement.

87. $x + 9 = 9 + x$

88. $2\left(\dfrac{1}{2}\right) = 1$

89. $\dfrac{1}{h + 6}(h + 6) = 1, \quad h \neq -6$

90. $(x + 3) - (x + 3) = 0$

91. $2(x + 3) = 2x + 6$

92. $(z - 2) + 0 = z - 2$

93. $x + (y + 10) = (x + y) + 10$

94. $\dfrac{1}{7}(7 \cdot 12) = \left(\dfrac{1}{7} \cdot 7\right)12 = 1 \cdot 12 = 12$

In Exercises 95–104, perform the operations. (Write fractional answers in simplest form.)

95. $\dfrac{3}{16} + \dfrac{5}{16}$ **96.** $\dfrac{6}{7} - \dfrac{4}{7}$

97. $\dfrac{5}{8} - \dfrac{5}{12} + \dfrac{1}{6}$ **98.** $\dfrac{10}{11} + \dfrac{6}{33} - \dfrac{13}{66}$

99. $\dfrac{x}{6} + \dfrac{3x}{4}$ **100.** $\dfrac{2x}{5} + \dfrac{x}{10}$

101. $\dfrac{12}{x} \div \dfrac{1}{8}$ **102.** $\dfrac{11}{x} \div \dfrac{3}{4}$

103. $\left(\dfrac{2}{5} \div 4\right) - \left(4 \cdot \dfrac{3}{8}\right)$ **104.** $\left(\dfrac{3}{5} \div 3\right) - \left(6 \cdot \dfrac{4}{8}\right)$

In Exercises 105–110, use a calculator to evaluate the expression. (Round your answer to two decimal places.)

105. $14\left(-3 + \dfrac{3}{7}\right)$ **106.** $3\left(-\dfrac{5}{12} + \dfrac{3}{8}\right)$

107. $\dfrac{11.46 - 5.37}{3.91}$ **108.** $\dfrac{12.24 - 8.4}{2.5}$

109. $\dfrac{\frac{2}{3}(-2 - 6)}{-\frac{2}{5}}$ **110.** $\dfrac{\frac{1}{5}(-8 - 9)}{-\frac{1}{3}}$

111. (a) Use a calculator to complete the table.

n	1	0.5	0.01	0.0001	0.000001
$5/n$					

    (b) Use the result from part (a) to make a conjecture about the value of $5/n$ as n approaches 0.

112. (a) Use a calculator to complete the table.

n	1	10	100	10,000	100,000
$5/n$					

    (b) Use the result from part (a) to make a conjecture about the value of $5/n$ as n increases without bound.

Synthesis

True or False? **In Exercises 113 and 114, determine whether the statement is true or false. Justify your answer.**

113. Let $a > b$, then $\dfrac{1}{a} > \dfrac{1}{b}$, where $a \neq 0$ and $b \neq 0$.

114. Because $\dfrac{a + b}{c} = \dfrac{a}{c} + \dfrac{b}{c}$, then $\dfrac{c}{a + b} = \dfrac{c}{a} + \dfrac{c}{b}$.

In Exercises 115 and 116, use the real numbers A, B, and C shown on the number line. Determine the sign of each expression.

115. (a) $-A$ **116.** (a) $-C$

    (b) $B - A$ (b) $A - C$

117. ***Exploration***   Consider $|u + v|$ and $|u| + |v|$.

    (a) Are the values of the expressions always equal? If not, under what conditions are they unequal?

    (b) If the two expressions are not equal for certain values of u and v, is one of the expressions always greater than the other? Explain.

118. ***Think About It***   Is there a difference between saying that a real number is positive and saying that a real number is nonnegative? Explain.

119. ***Writing***   Describe the differences among the sets of whole numbers, natural numbers, integers, rational numbers, and irrational numbers.

P.2   Exponents and Radicals

Integer Exponents

Repeated *multiplication* can be written in **exponential form.**

Repeated Multiplication	Exponential Form
$a \cdot a \cdot a \cdot a \cdot a$	a^5
$(-4)(-4)(-4)$	$(-4)^3$
$(2x)(2x)(2x)(2x)$	$(2x)^4$

In general, if a is a real number, variable, or algebraic expression and n is a positive integer, then

$$a^n = \underbrace{a \cdot a \cdot a \cdots a}_{n \text{ factors}}$$

where n is the **exponent** and a is the **base.** The expression a^n is read "a to the nth **power.**" An exponent can be negative as well. Property 3 below shows how to use a negative exponent.

Let a and b be real numbers, variables, or algebraic expressions, and let m and n be integers. (All denominators and bases are nonzero.)

Property	Example
1. $a^m a^n = a^{m+n}$	$3^2 \cdot 3^4 = 3^{2+4} = 3^6 = 729$
2. $\dfrac{a^m}{a^n} = a^{m-n}$	$\dfrac{x^7}{x^4} = x^{7-4} = x^3$
3. $a^{-n} = \dfrac{1}{a^n} = \left(\dfrac{1}{a}\right)^n$	$y^{-4} = \dfrac{1}{y^4} = \left(\dfrac{1}{y}\right)^4$
4. $a^0 = 1, \quad a \neq 0$	$(x^2 + 1)^0 = 1$
5. $(ab)^m = a^m b^m$	$(5x)^3 = 5^3 x^3 = 125x^3$
6. $(a^m)^n = a^{mn}$	$(y^3)^{-4} = y^{3(-4)} = y^{-12} = \dfrac{1}{y^{12}}$
7. $\left(\dfrac{a}{b}\right)^m = \dfrac{a^m}{b^m}$	$\left(\dfrac{2}{x}\right)^3 = \dfrac{2^3}{x^3} = \dfrac{8}{x^3}$
8. $|a^2| = |a|^2 = a^2$	$|(-2)^2| = |-2|^2 = 2^2 = 4$

It is important to recognize the difference between expressions such as $(-2)^4$ and -2^4. In $(-2)^4$, the parentheses indicate that the exponent applies to the negative sign as well as to the 2, but in $-2^4 = -(2^4)$, the exponent applies only to the 2. So, $(-2)^4 = 16$, whereas $-2^4 = -16$. It is also important to know when to use parentheses when evaluating exponential expressions using a graphing calculator. Figure P.9 shows that a graphing calculator follows the order of operations.

What you should learn

- Use properties of exponents.
- Use scientific notation to represent real numbers.
- Use properties of radicals.
- Simplify and combine radicals.
- Rationalize denominators and numerators.
- Use properties of rational exponents.

Why you should learn it

Real numbers and algebraic expressions are often written with exponents and radicals. For instance, in Exercise 93 on page 23, you will use an expression involving a radical to find the size of a particle that can be carried by a stream moving at a certain velocity.

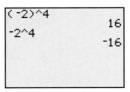

SuperStock

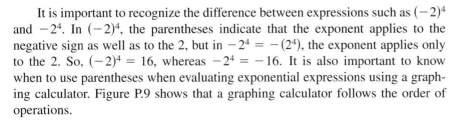

Figure P.9

The properties of exponents listed on the previous page apply to *all* integers *m* and *n*, not just positive integers. For instance, by Property 2, you can write

$$\frac{3^4}{3^{-5}} = 3^{4-(-5)} = 3^{4+5} = 3^9.$$

Example 1   Using Properties of Exponents

a. $(-3ab^4)(4ab^{-3}) = -12(a)(a)(b^4)(b^{-3}) = -12a^2b$

b. $(2xy^2)^3 = 2^3(x)^3(y^2)^3 = 8x^3y^6$

c. $3a(-4a^2)^0 = 3a(1) = 3a, \qquad a \neq 0$

✓ *Checkpoint* Now try Exercise 15.

Example 2   Rewriting with Positive Exponents

a. $x^{-1} = \dfrac{1}{x}$ Property 3

b. $\dfrac{1}{3x^{-2}} = \dfrac{1(x^2)}{3} = \dfrac{x^2}{3}$ The exponent -2 does not apply to 3.

c. $\dfrac{1}{(3x)^{-2}} = (3x)^2 = 9x^2$ The exponent -2 does apply to 3.

d. $\dfrac{12a^3b^{-4}}{4a^{-2}b} = \dfrac{12a^3 \cdot a^2}{4b \cdot b^4} = \dfrac{3a^5}{b^5}$ Properties 3 and 1

e. $\left(\dfrac{3x^2}{y}\right)^{-2} = \dfrac{3^{-2}(x^2)^{-2}}{y^{-2}}$ Properties 5 and 7

$\qquad = \dfrac{3^{-2}x^{-4}}{y^{-2}}$ Property 6

$\qquad = \dfrac{y^2}{3^2x^4} = \dfrac{y^2}{9x^4}$ Property 3, and simplify.

✓ *Checkpoint* Now try Exercise 19.

Example 3   Calculators and Exponents

Expression	Graphing Calculator Keystrokes	Display
a. $3^{-2} + 4^{-1}$	3 [^] [(-)] 2 [+] 4 [^] [(-)] 1 [ENTER]	.3611111111
b. $\dfrac{3^5 + 1}{3^5 - 1}$	[(] 3 [^] 5 [+] 1 [)] [÷]	
	[(] 3 [^] 5 [-] 1 [)] [ENTER]	1.008264463

✓ *Checkpoint* Now try Exercise 23.

TECHNOLOGY TIP   The graphing calculator keystrokes given in this text may not be the same as the keystrokes for your graphing calculator. Be sure you are familiar with the use of the keys on your own calculator.

STUDY TIP

Rarely in algebra is there only one way to solve a problem. Don't be concerned if the steps you use to solve a problem are not exactly the same as the steps presented in this text. The important thing is to use steps that you understand *and*, of course, that are justified by the rules of algebra. For instance, you might prefer the following steps for Example 2(e).

$$\left(\frac{3x^2}{y}\right)^{-2} = \left(\frac{y}{3x^2}\right)^2 = \frac{y^2}{9x^4}$$

Scientific Notation

Exponents provide an efficient way of writing and computing with very large (or very small) numbers. For instance, there are about 359 billion billion gallons of water on Earth—that is, 359 followed by 18 zeros.

359,000,000,000,000,000,000

It is convenient to write such numbers in **scientific notation.** This notation has the form $\pm c \times 10^n$, where $1 \leq c < 10$ and n is an integer. So, the number of gallons of water on Earth can be written in scientific notation as

$3.59 \times 100,000,000,000,000,000,000 = 3.59 \times 10^{20}$.

The *positive* exponent 20 indicates that the number is *large* (10 or more) and that the decimal point has been moved 20 places. A *negative* exponent indicates that the number is *small* (less than 1). For instance, the mass (in grams) of one electron is approximately

$9.0 \times 10^{-28} = 0.00000000000000000000000000009$.

$$\underbrace{\hspace{5cm}}_{\text{28 decimal places}}$$

Example 4   Scientific Notation

a. $1.345 \times 10^2 = 134.5$ **b.** $0.0000782 = 7.82 \times 10^{-5}$

c. $-9.36 \times 10^{-6} = -0.00000936$ **d.** $836,100,000 = 8.361 \times 10^8$

✓ *Checkpoint* Now try Exercise 27.

TECHNOLOGY TIP Most calculators automatically switch to scientific notation when they are showing large or small numbers that exceed the display range. Try evaluating $86,500,000 \times 6000$. If your calculator follows standard conventions, its display should be

$\boxed{5.19 \ 11}$ or $\boxed{5.19 \ \text{E} \ 11}$

which is 5.19×10^{11}.

Example 5   Using Scientific Notation with a Calculator

Use a calculator to evaluate $65,000 \times 3,400,000,000$.

Solution

Because $65,000 = 6.5 \times 10^4$ and $3,400,000,000 = 3.4 \times 10^9$, you can multiply the two numbers using the following graphing calculator keystrokes.

6.5 $\boxed{\text{EE}}$ 4 $\boxed{\times}$ 3.4 $\boxed{\text{EE}}$ 9 $\boxed{\text{ENTER}}$

After entering these keystrokes, the calculator display should read $\boxed{2.21 \ \text{E} \ 14}$. So, the product of the two numbers is

$(6.5 \times 10^4)(3.4 \times 10^9) = 2.21 \times 10^{14} = 221,000,000,000,000$.

✓ *Checkpoint* Now try Exercise 35.

Radicals and Their Properties

A **square root** of a number is one of its two equal factors. For example, 5 is a square root of 25 because 5 is one of the two equal factors of $25 = 5 \cdot 5$. In a similar way, a **cube root** of a number is one of its three equal factors, as in $125 = 5^3$.

Definition of the *n*th Root of a Number

Let a and b be real numbers and let $n \geq 2$ be a positive integer. If

$$a = b^n$$

then b is an ***n*th root of *a*.** If $n = 2$, the root is a **square root.** If $n = 3$, the root is a **cube root.**

Some numbers have more than one *n*th root. For example, both 5 and -5 are square roots of 25. The *principal square root* of 25, written as $\sqrt{25}$, is the positive root, 5. The **principal *n*th root** of a number is defined as follows.

Principal *n*th Root of a Number

Let a be a real number that has at least one *n*th root. The **principal *n*th root of *a*** is the *n*th root that has the same sign as a. It is denoted by a **radical symbol**

$$\sqrt[n]{a}. \qquad \text{Principal } n\text{th root}$$

The positive integer n is the **index** of the radical, and the number a is the **radicand.** If $n = 2$, omit the index and write $\sqrt{a}$ rather than $\sqrt[2]{a}$. (The plural of index is *indices*.)

A common misunderstanding is that the square root sign implies both negative and positive roots. This is not correct. The square root sign implies only a positive root. When a negative root is needed, you must use the negative sign with the square root sign.

Incorrect: $\sqrt{4} = \pm 2$ Correct: $-\sqrt{4} = -2$ and $\sqrt{4} = 2$

Example 6   Evaluating Expressions Involving Radicals

a. $\sqrt{36} = 6$ because $6^2 = 36$.

b. $-\sqrt{36} = -6$ because $-\left(\sqrt{36}\right) = -\left(\sqrt{6^2}\right) = -(6) = -6$.

c. $\sqrt[3]{\dfrac{125}{64}} = \dfrac{5}{4}$ because $\left(\dfrac{5}{4}\right)^3 = \dfrac{5^3}{4^3} = \dfrac{125}{64}$.

d. $\sqrt[5]{-32} = -2$ because $(-2)^5 = -32$.

e. $\sqrt[4]{-81}$ is not a real number because there is no real number that can be raised to the fourth power to produce -81.

✓ *Checkpoint* Now try Exercise 41.

Here are some generalizations about the *n*th roots of a real number.

Generalizations About *n*th Roots of Real Numbers

Real Number a	Integer n	Root(s) of a	Example
$a > 0$	$n > 0$, n is even.	$\sqrt[n]{a}, -\sqrt[n]{a}$	$\sqrt[4]{81} = 3, -\sqrt[4]{81} = -3$
$a > 0$ or $a < 0$	n is odd.	$\sqrt[n]{a}$	$\sqrt[3]{-8} = -2$
$a < 0$	n is even.	No real roots	$\sqrt{-4}$ is not a real number.
$a = 0$	n is even or odd.	$\sqrt[n]{0} = 0$	$\sqrt[5]{0} = 0$

Integers such as 1, 4, 9, 16, 25, and 36 are called **perfect squares** because they have integer square roots. Similarly, integers such as 1, 8, 27, 64, and 125 are called **perfect cubes** because they have integer cube roots.

Properties of Radicals

Let a and b be real numbers, variables, or algebraic expressions such that the indicated roots are real numbers, and let m and n be positive integers.

Property	*Example*
1. $\sqrt[n]{a^m} = \left(\sqrt[n]{a}\right)^m$	$\sqrt[3]{8^2} = \left(\sqrt[3]{8}\right)^2 = (2)^2 = 4$
2. $\sqrt[n]{a} \cdot \sqrt[n]{b} = \sqrt[n]{ab}$	$\sqrt{5} \cdot \sqrt{7} = \sqrt{5 \cdot 7} = \sqrt{35}$
3. $\dfrac{\sqrt[n]{a}}{\sqrt[n]{b}} = \sqrt[n]{\dfrac{a}{b}}$, $b \neq 0$	$\dfrac{\sqrt[4]{27}}{\sqrt[4]{9}} = \sqrt[4]{\dfrac{27}{9}} = \sqrt[4]{3}$
4. $\sqrt[m]{\sqrt[n]{a}} = \sqrt[mn]{a}$	$\sqrt[3]{\sqrt{10}} = \sqrt[6]{10}$
5. $\left(\sqrt[n]{a}\right)^n = a$	$\left(\sqrt{3}\right)^2 = 3$
6. For n even, $\sqrt[n]{a^n} = |a|$.	$\sqrt{(-12)^2} = |-12| = 12$
For n odd, $\sqrt[n]{a^n} = a$.	$\sqrt[3]{(-12)^3} = -12$

Example 7   Using Properties of Radicals

Use the properties of radicals to simplify each expression.

a. $\sqrt{8} \cdot \sqrt{2}$ **b.** $\left(\sqrt[3]{5}\right)^3$ **c.** $\sqrt[3]{x^3}$ **d.** $\sqrt[6]{y^6}$

Solution

a. $\sqrt{8} \cdot \sqrt{2} = \sqrt{8 \cdot 2} = \sqrt{16} = 4$

b. $\left(\sqrt[3]{5}\right)^3 = 5$

c. $\sqrt[3]{x^3} = x$

d. $\sqrt[6]{y^6} = |y|$

✓ *Checkpoint* Now try Exercise 55.

TECHNOLOGY TIP

There are four methods of evaluating radicals on most graphing calculators. For square roots, you can use the *square root key* 〔√〕. For cube roots, you can use the *cube root key* 〔³√〕 (or menu choice). For other roots, you can first convert the radical to exponential form and then use the *exponential key* 〔^〕 or you can use the *x*th *root key* 〔ˣ√〕 (or menu choice). For example, the screens below show you how to evaluate $\sqrt{36}$, $\sqrt[3]{-8}$, $\sqrt{16}$, and $\sqrt[5]{32}$ using one of the four methods described.

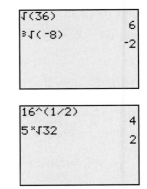

Simplifying Radicals

An expression involving radicals is in **simplest form** when the following conditions are satisfied.

1. All possible factors have been removed from the radical.
2. All fractions have radical-free denominators (accomplished by a process called *rationalizing the denominator*).
3. The index of the radical is reduced.

To simplify a radical, factor the radicand into factors whose exponents are multiples of the index. The roots of these factors are written outside the radical, and the "leftover" factors make up the new radicand.

Example 8   Simplifying Even Roots

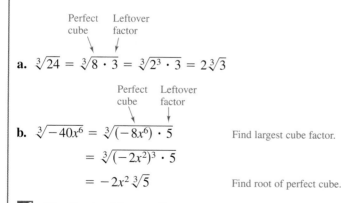

a. $\sqrt[4]{48} = \sqrt[4]{16 \cdot 3} = \sqrt[4]{2^4 \cdot 3} = 2\sqrt[4]{3}$

b. $\sqrt{75x^3} = \sqrt{25x^2 \cdot 3x}$ Find largest square factor.

$\quad\quad\quad\; = \sqrt{(5x)^2 \cdot 3x}$

$\quad\quad\quad\; = 5x\sqrt{3x}$ Find root of perfect square.

c. $\sqrt[4]{(5x)^4} = |5x| = 5|x|$

✓ *Checkpoint* Now try Exercise 57(a).

Example 9   Simplifying Odd Roots

a. $\sqrt[3]{24} = \sqrt[3]{8 \cdot 3} = \sqrt[3]{2^3 \cdot 3} = 2\sqrt[3]{3}$

b. $\sqrt[3]{-40x^6} = \sqrt[3]{(-8x^6) \cdot 5}$ Find largest cube factor.

$\quad\quad\quad\quad\; = \sqrt[3]{(-2x^2)^3 \cdot 5}$

$\quad\quad\quad\quad\; = -2x^2\sqrt[3]{5}$ Find root of perfect cube.

✓ *Checkpoint* Now try Exercise 57(b).

Radical expressions can be combined (added or subtracted) if they are **like radicals**—that is, if they have the same index and radicand. For instance, $\sqrt{2}$, $3\sqrt{2}$, and $\frac{1}{2}\sqrt{2}$ are like radicals, but $\sqrt{3}$ and $\sqrt{2}$ are unlike radicals. To determine whether two radicals can be combined, you should first simplify each radical.

STUDY TIP

When you simplify a radical, it is important that both expressions are defined for the same values of the variable. For instance, in Example 8(b), $\sqrt{75x^3}$ and $5x\sqrt{3x}$ are both defined only for nonnegative values of x. Similarly, in Example 8(c), $\sqrt[4]{(5x)^4}$ and $5|x|$ are both defined for all real values of x.

Example 10   Combining Radicals

a. $2\sqrt{48} - 3\sqrt{27} = 2\sqrt{16 \cdot 3} - 3\sqrt{9 \cdot 3}$ Find square factors.

$\qquad\qquad\quad = 8\sqrt{3} - 9\sqrt{3}$ Find square roots and multiply by coefficients.

$\qquad\qquad\quad = (8 - 9)\sqrt{3}$ Combine like terms.

$\qquad\qquad\quad = -\sqrt{3}$ Simplify.

b. $\sqrt[3]{16x} - \sqrt[3]{54x^4} = \sqrt[3]{8 \cdot 2x} - \sqrt[3]{27 \cdot x^3 \cdot 2x}$ Find cube factors.

$\qquad\qquad\quad = 2\sqrt[3]{2x} - 3x\sqrt[3]{2x}$ Find cube roots.

$\qquad\qquad\quad = (2 - 3x)\sqrt[3]{2x}$ Combine like terms.

 Checkpoint Now try Exercise 61.

Try using your calculator to check the result of Example 10(a). You should obtain -1.732050808, which is the same as the calculator's approximation for $-\sqrt{3}$.

Rationalizing Denominators and Numerators

To rationalize a denominator or numerator of the form $a - b\sqrt{m}$ or $a + b\sqrt{m}$, multiply both numerator and denominator by a **conjugate**: $a + b\sqrt{m}$ and $a - b\sqrt{m}$ are conjugates of each other. If $a = 0$, then the rationalizing factor for $\sqrt{m}$ is itself, $\sqrt{m}$.

Example 11   Rationalizing Denominators

Rationalize the denominator of each expression.

a. $\dfrac{5}{2\sqrt{3}}$ **b.** $\dfrac{2}{\sqrt[3]{5}}$

Solution

a. $\dfrac{5}{2\sqrt{3}} = \dfrac{5}{2\sqrt{3}} \cdot \dfrac{\sqrt{3}}{\sqrt{3}}$ $\sqrt{3}$ is rationalizing factor.

$\qquad = \dfrac{5\sqrt{3}}{2(3)}$ Multiply.

$\qquad = \dfrac{5\sqrt{3}}{6}$ Simplify.

b. $\dfrac{2}{\sqrt[3]{5}} = \dfrac{2}{\sqrt[3]{5}} \cdot \dfrac{\sqrt[3]{5^2}}{\sqrt[3]{5^2}}$ $\sqrt[3]{5^2}$ is rationalizing factor.

$\qquad = \dfrac{2\sqrt[3]{5^2}}{\sqrt[3]{5^3}} = \dfrac{2\sqrt[3]{25}}{5}$ Multiply and simplify.

 Checkpoint Now try Exercise 67.

STUDY TIP

Notice in Example 11(b) that the numerator and denominator are multiplied by $\sqrt[3]{5^2}$ to produce a perfect cube radicand.

Example 12   Rationalizing a Denominator with Two Terms

Rationalize the denominator of $\dfrac{2}{3 + \sqrt{7}}$.

Solution

$$\frac{2}{3 + \sqrt{7}} = \frac{2}{3 + \sqrt{7}} \cdot \frac{3 - \sqrt{7}}{3 - \sqrt{7}}$$

Multiply numerator and denominator by conjugate of denominator.

$$= \frac{2(3 - \sqrt{7})}{(3)^2 - (\sqrt{7})^2}$$

Find products. In denominator, $(a + b)(a - b) = a^2 - ab + ab - b^2$
$= a^2 - b^2$.

$$= \frac{2(3 - \sqrt{7})}{2} = 3 - \sqrt{7}$$

Simplify and divide out common factors.

✓ *Checkpoint* Now try Exercise 69.

Sometimes it is necessary to rationalize the numerator of expressions from calculus.

Example 13   Rationalizing a Numerator   ∫ algebra of calculus

Rationalize the numerator of $\dfrac{\sqrt{5} - \sqrt{7}}{2}$.

Solution

$$\frac{\sqrt{5} - \sqrt{7}}{2} = \frac{\sqrt{5} - \sqrt{7}}{2} \cdot \frac{\sqrt{5} + \sqrt{7}}{\sqrt{5} + \sqrt{7}}$$

Multiply numerator and denominator by conjugate of numerator.

$$= \frac{(\sqrt{5})^2 - (\sqrt{7})^2}{2(\sqrt{5} + \sqrt{7})}$$

Find products. In numerator, $(a + b)(a - b) = a^2 - ab + ab - b^2$
$= a^2 - b^2$.

$$= \frac{-2}{2(\sqrt{5} + \sqrt{7})} = \frac{-1}{\sqrt{5} + \sqrt{7}}$$

Simplify and divide out common factors.

✓ *Checkpoint* Now try Exercise 73.

> **STUDY TIP**
>
> Do not confuse the expression $\sqrt{5} + \sqrt{7}$ with the expression $\sqrt{5 + 7}$. In general, $\sqrt{x + y}$ does not equal $\sqrt{x} + \sqrt{y}$. Similarly, $\sqrt{x^2 + y^2}$ does not equal $x + y$.

Rational Exponents

Definition of Rational Exponents

If a is a real number and n is a positive integer such that the principal nth root of a exists, then $a^{1/n}$ is defined as

$$a^{1/n} = \sqrt[n]{a} \text{ where } 1/n \text{ is the } \textbf{rational exponent} \text{ of } a.$$

Moreover, if m is a positive integer that has no common factor with n, then

$$a^{m/n} = (a^{1/n})^m = (\sqrt[n]{a})^m \quad \text{and} \quad a^{m/n} = (a^m)^{1/n} = \sqrt[n]{a^m}.$$

The symbol ∫ indicates an example or exercise that highlights algebraic techniques specifically used in calculus.

The numerator of a rational exponent denotes the *power* to which the base is raised, and the denominator denotes the *index* or the *root* to be taken.

$$b^{m/n} = \left(\sqrt[n]{b}\right)^m = \sqrt[n]{b^m}$$

When you are working with rational exponents, the properties of integer exponents still apply. For instance,

$$2^{1/2}2^{1/3} = 2^{(1/2)+(1/3)} = 2^{5/6}.$$

STUDY TIP

Rational exponents can be tricky, and you must remember that the expression $b^{m/n}$ is not defined unless $\sqrt[n]{b}$ is a real number. This restriction produces some unusual-looking results. For instance, the number $(-8)^{1/3}$ is defined because $\sqrt[3]{-8} = -2$, but the number $(-8)^{2/6}$ is undefined because $\sqrt[6]{-8}$ is not a real number.

Example 14 Changing from Radical to Exponential Form

a. $\sqrt{3} = 3^{1/2}$

b. $\sqrt{(3xy)^5} = \sqrt[2]{(3xy)^5} = (3xy)^{(5/2)}$

c. $2x\sqrt[4]{x^3} = (2x)(x^{3/4}) = 2x^{1+(3/4)} = 2x^{7/4}$

✓ *Checkpoint* Now try Exercise 75.

Example 15 Changing from Exponential to Radical Form

a. $(x^2 + y^2)^{3/2} = \left(\sqrt{x^2 + y^2}\right)^3 = \sqrt{(x^2 + y^2)^3}$

b. $2y^{3/4}z^{1/4} = 2(y^3z)^{1/4} = 2\sqrt[4]{y^3z}$

c. $a^{-3/2} = \dfrac{1}{a^{3/2}} = \dfrac{1}{\sqrt{a^3}}$

d. $x^{0.2} = x^{1/5} = \sqrt[5]{x}$

✓ *Checkpoint* Now try Exercise 77.

Rational exponents are useful for evaluating roots of numbers on a calculator, reducing the index of a radical, and simplifying calculus expressions.

Example 16 Simplifying with Rational Exponents

a. $(-32)^{-4/5} = \left(\sqrt[5]{-32}\right)^{-4} = (-2)^{-4} = \dfrac{1}{(-2)^4} = \dfrac{1}{16}$

b. $(-5x^{5/3})(3x^{-3/4}) = -15x^{(5/3)-(3/4)} = -15x^{11/12}, \qquad x \neq 0$

c. $\sqrt[9]{a^3} = a^{3/9} = a^{1/3} = \sqrt[3]{a}$ Reduce index.

d. $\sqrt[3]{\sqrt{125}} = \sqrt[6]{125} = \sqrt[6]{(5)^3} = 5^{3/6} = 5^{1/2} = \sqrt{5}$

e. $(2x - 1)^{4/3}(2x - 1)^{-1/3} = (2x - 1)^{(4/3)-(1/3)} = 2x - 1, \qquad x \neq \dfrac{1}{2}$

✓ *Checkpoint* Now try Exercise 83.

STUDY TIP

The expression in Example 16(e) is not defined when $x = \frac{1}{2}$ because

$$\left(2 \cdot \tfrac{1}{2} - 1\right)^{-1/3} = (0)^{-1/3}$$

is not a real number.

P.2   Exercises

Vocabulary Check

Fill in the blanks.

1. In the exponential form a^n, n is the _____ and a is the _____ .

2. A convenient way of writing very large or very small numbers is called _____ .

3. One of the two equal factors of a number is called a _____ of the number.

4. The _____ of a number is the nth root that has the same sign as a, and is denoted by $\sqrt[n]{a}$.

5. In the radical form $\sqrt[n]{a}$, the positive integer n is called the _____ of the radical and the number a is called the _____ .

6. When an expression involving radicals has all possible factors removed, radical-free denominators, and a reduced index, it is in _____.

7. The expressions $a + b\sqrt{m}$ and $a - b\sqrt{m}$ are _____ of each other.

8. The process used to create a radical-free denominator is known as _____ the denominator.

9. In the expression $b^{m/n}$, m denotes the _____ to which the base is raised and n denotes the _____ or root to be taken.

In Exercises 1–8, evaluate each expression.

1. (a) $4^2 \cdot 3$ (b) $3 \cdot 3^3$

2. (a) $\dfrac{5^5}{5^2}$ (b) $\dfrac{3^2}{3^4}$

3. (a) $(3^3)^2$ (b) -3^2

4. (a) $(2^3 \cdot 3^2)^2$ (b) $\left(-\frac{3}{5}\right)^3\left(\frac{5}{3}\right)^2$

5. (a) $\dfrac{3}{3^{-4}}$ (b) $24(-2)^{-5}$

6. (a) $\dfrac{4 \cdot 3^{-2}}{2^{-2} \cdot 3^{-1}}$ (b) $(-2)^0$

7. (a) $2^{-1} + 3^{-1}$ (b) $(2^{-1})^{-2}$

8. (a) $3^{-1} + 2^{-2}$ (b) $(3^{-2})^2$

In Exercises 9–14, evaluate the expression for the value of x.

Expression	Value
9. $7x^{-2}$	2
10. $6x^0 - (6x)^0$	7
11. $2x^3$	-3
12. $-3x^4$	-2
13. $4x^2$	$-\frac{1}{2}$
14. $5(-x)^3$	$\frac{1}{3}$

In Exercises 15–20, simplify each expression.

15. (a) $(-5z)^3$ (b) $5x^4(x^2)$

16. (a) $(3x)^2$ (b) $(4x^3)^2$

17. (a) $\dfrac{7x^2}{x^3}$ (b) $\dfrac{12(x + y)^3}{9(x + y)}$

18. (a) $\dfrac{r^4}{r^6}$ (b) $\left(\dfrac{4}{y}\right)^3\left(\dfrac{3}{y}\right)^4$

19. (a) $[(x^2y^{-2})^{-1}]^{-1}$ (b) $\left(\dfrac{a^{-2}}{b^{-2}}\right)\left(\dfrac{b}{a}\right)^3$

20. (a) $(2x^5)^0$, $x \neq 0$ (b) $(5x^2z^6)^3(5x^2z^6)^{-3}$

In Exercises 21–24, use a calculator to evaluate the expression. (Round your answer to three decimal places.)

21. $(-4)^3(5^2)$

22. $(8^{-4})(10^3)$

23. $\dfrac{3^6}{7^3}$

24. $\dfrac{4^3}{3^{-4}}$

In Exercises 25–28, write the number in scientific notation.

25. Land area of Earth: 57,300,000 square miles

26. Light year: 9,460,000,000,000 kilometers

27. Relative density of hydrogen: 0.0000899 gram per cubic centimeter

28. One micron (millionth of a meter): 0.00003937 inch

In Exercises 29–32, write the number in decimal notation.

29. Worldwide Coca-Cola products daily consumption: 5.64×10^8 drinks   (Source: The Coca-Cola Company)

30. Interior temperature of sun: 1.5×10^7 degrees Celsius

31. Charge of electron: 1.6022×10^{-19} coulomb

32. Width of human hair: 9.0×10^{-5} meter

In Exercises 33 and 34, evaluate the expression without using a calculator.

33. $\sqrt{25 \times 10^8}$

34. $\sqrt[3]{8 \times 10^{15}}$

In Exercises 35–38, use a calculator to evaluate each expression. (Round your answer to three decimal places.)

35. (a) $(9.3 \times 10^6)^3(6.1 \times 10^{-4})$

(b) $\dfrac{(2.414 \times 10^4)^6}{(1.68 \times 10^5)^5}$

36. (a) $750\left(1 + \dfrac{0.11}{365}\right)^{800}$

(b) $\dfrac{67{,}000{,}000 + 93{,}000{,}000}{0.0052}$

37. (a) $\sqrt{4.5 \times 10^9}$

(b) $\sqrt[3]{6.3 \times 10^4}$

38. (a) $(2.65 \times 10^{-4})^{1/3}$

(b) $\sqrt{9 \times 10^{-4}}$

In Exercises 39–48, evaluate the expression without using a calculator.

39. $\sqrt{121}$

40. $\sqrt{16}$

41. $-\sqrt[3]{-27}$

42. $\dfrac{\sqrt[4]{81}}{3}$

43. $\left(\sqrt[3]{-125}\right)^3$

44. $\sqrt[4]{562^4}$

45. $32^{-3/5}$

46. $\left(\tfrac{9}{4}\right)^{-1/2}$

47. $\left(-\dfrac{1}{64}\right)^{-1/3}$

48. $-\left(\dfrac{1}{125}\right)^{-4/3}$

In Exercises 49–54, use a calculator to approximate the number. (Round your answer to three decimal places.)

49. $\sqrt[5]{-27^3}$

50. $\sqrt[3]{45^2}$

51. $(3.4)^{2.5}$

52. $(6.1)^{-2.9}$

53. $(1.2^{-2})\sqrt{75} + 3\sqrt{8}$

54. $\dfrac{-5 + \sqrt{33}}{5}$

In Exercises 55 and 56, use the properties of radicals to simplify each expression.

55. (a) $\left(\sqrt[4]{3}\right)^4$ (b) $\sqrt[5]{96x^5}$

56. (a) $\sqrt{12} \cdot \sqrt{3}$ (b) $\sqrt[4]{x^4}$

In Exercises 57–62, simplify each expression.

57. (a) $\sqrt{54xy^4}$

(b) $\sqrt[3]{\dfrac{32a^2}{b^2}}$

58. (a) $\sqrt[3]{54}$

(b) $\sqrt{32x^3y^4}$

59. (a) $2\sqrt{50} + 12\sqrt{8}$

(b) $10\sqrt{32} - 6\sqrt{18}$

60. (a) $5\sqrt{x} - 3\sqrt{x}$

(b) $-2\sqrt{9y} + 10\sqrt{y}$

61. (a) $3\sqrt{x+1} + 10\sqrt{x+1}$

(b) $7\sqrt{80x} - 2\sqrt{125x}$

62. (a) $5\sqrt{10x^2} - \sqrt{90x^2}$

(b) $8\sqrt[3]{27x} - \tfrac{1}{2}\sqrt[3]{64x}$

In Exercises 63–66, complete the statement with <, =, or >.

63. $\sqrt{5} + \sqrt{3}$ ⬚ $\sqrt{5+3}$

64. $\sqrt{\dfrac{3}{11}}$ ⬚ $\dfrac{\sqrt{3}}{\sqrt{11}}$

65. 5 ⬚ $\sqrt{3^2 + 2^2}$

66. 5 ⬚ $\sqrt{3^2 + 4^2}$

In Exercises 67–70, rationalize the denominator of the expression. Then simplify your answer.

67. $\dfrac{1}{\sqrt{3}}$

68. $\dfrac{8}{\sqrt[3]{2}}$

69. $\dfrac{5}{\sqrt{14} - 2}$

70. $\dfrac{3}{\sqrt{5} + \sqrt{6}}$

In Exercises 71–74, rationalize the numerator of the expression and simplify your answer.

71. $\dfrac{\sqrt{8}}{2}$

72. $\dfrac{\sqrt{2}}{3}$

73. $\dfrac{\sqrt{5}+\sqrt{3}}{3}$

74. $\dfrac{\sqrt{7}-3}{4}$

In Exercises 75–82, fill in the missing form of the expression.

Radical Form	Rational Exponent Form
75. $\sqrt[3]{64}$	
76.	$-(144^{1/2})$
77.	$32^{1/5}$
78. $\sqrt[3]{614.125}$	
79. $\sqrt[3]{-216}$	
80.	$(-243)^{1/5}$
81. $\sqrt[4]{81^3}$	
82.	$16^{5/4}$

In Exercises 83–86, perform the operations and simplify.

83. $\dfrac{(2x^2)^{3/2}}{2^{1/2}x^4}$

84. $\dfrac{x^{4/3}y^{2/3}}{(xy)^{1/3}}$

85. $\dfrac{x^{-3}\cdot x^{1/2}}{x^{3/2}\cdot x^{-1}}$

86. $\dfrac{5^{-1/2}\cdot 5x^{5/2}}{(5x)^{3/2}}$

In Exercises 87 and 88, reduce the index of each radical and rewrite in radical form.

87. (a) $\sqrt[4]{3^2}$ (b) $\sqrt[6]{(x+1)^4}$

88. (a) $\sqrt[6]{x^3}$ (b) $\sqrt[4]{(3x^2)^4}$

In Exercises 89 and 90, write each expression as a single radical. Then simplify your answer.

89. (a) $\sqrt{\sqrt{32}}$ (b) $\sqrt{\sqrt[4]{2x}}$

90. (a) $\sqrt{\sqrt{243(x+1)}}$ (b) $\sqrt{\sqrt[3]{10a^7b}}$

91. **Period of a Pendulum** The period T (in seconds) of a pendulum is given by $T = 2\pi\sqrt{L/32}$, where L is the length of the pendulum (in feet). Find the period of a pendulum whose length is 2 feet.

92. **Mathematical Modeling** A funnel is filled with water to a height of h centimeters. The formula

$$t = 0.03[12^{5/2} - (12 - h)^{5/2}], \quad 0 \le h \le 12$$

represents the amount of time t (in seconds) it will take for the funnel to empty. Find t for $h = 7$ centimeters.

93. **Erosion** A stream of water moving at the rate of v feet per second can carry particles of size $0.03\sqrt{v}$ inches. Find the size of the particle that can be carried by a stream flowing at the rate of $\frac{3}{4}$ foot per second.

94. **Environment** There were 2.319×10^8 tons of municipal waste generated in 2000. Find the number of tons for each of the categories in the graph. (Source: Franklin Associates, Ltd.)

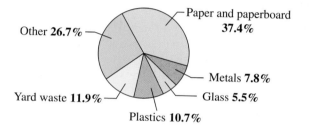

Paper and paperboard 37.4%
Other 26.7%
Metals 7.8%
Yard waste 11.9%
Glass 5.5%
Plastics 10.7%

Synthesis

True or False? In Exercises 95 and 96, determine whether the statement is true or false. Justify your answer.

95. $\dfrac{x^{k+1}}{x} = x^k$

96. $(a^n)^k = a^{(n^k)}$

97. **Think About It** Verify that $a^0 = 1$, $a \ne 0$. (*Hint:* Use the property of exponents $a^m/a^n = a^{m-n}$.)

98. **Think About It** Is the real number 52.7×10^5 written in scientific notation? Explain.

99. **Exploration** List all possible digits that occur in the units place of the square of a positive integer. Use that list to determine whether $\sqrt{5233}$ is an integer.

100. **Think About It** Square the real number $2/\sqrt{5}$ and note that the radical is eliminated from the denominator. Is this equivalent to rationalizing the denominator? Why or why not?

The symbol **∫** indicates an example or exercise that highlights algebraic techniques specifically used in calculus.

P.3 Polynomials and Factoring

Polynomials

An algebraic expression is a collection of variables and real numbers. The most common type of algebraic expression is the **polynomial.** Some examples are

$$2x + 5, \quad 3x^4 - 7x^2 + 2x + 4, \quad \text{and} \quad 5x^2y^2 - xy + 3.$$

The first two are *polynomials in x* and the third is a *polynomial in x and y*. The terms of a polynomial in x have the form ax^k, where a is the **coefficient** and k is the **degree** of the term. For instance, the polynomial

$$2x^3 - 5x^2 + 1 = 2x^3 + (-5)x^2 + (0)x + 1$$

has coefficients $2, -5, 0,$ and 1.

Definition of a Polynomial in x

Let $a_0, a_1, a_2, \ldots, a_n$ be *real numbers* and let n be a *nonnegative integer.* A **polynomial in x** is an expression of the form

$$a_n x^n + a_{n-1} x^{n-1} + \cdots + a_1 x + a_0$$

where $a_n \neq 0$. The polynomial is of **degree** n, a_n is the **leading coefficient,** and a_0 is the **constant term.**

In **standard form,** a polynomial in x is written with descending powers of x. Polynomials with one, two, and three terms are called **monomials, binomials,** and **trinomials,** respectively.

A polynomial that has all zero coefficients is called the **zero polynomial,** denoted by 0. No degree is assigned to this particular polynomial. For polynomials in more than one variable, the degree of a *term* is the sum of the exponents of the variables in the term. The degree of the *polynomial* is the highest degree of its terms. For instance, the degree of the polynomial $-2x^3y^6 + 4xy - x^7y^4$ is 11 because the sum of the exponents in the last term is the greatest. Expressions such as the following are not polynomials.

$$x^3 - \sqrt{3x} = x^3 - (3x)^{1/2} \qquad \text{The exponent 1/2 is not an integer.}$$

$$x^2 + 5x^{-1} \qquad \text{The exponent } -1 \text{ is not a nonnegative integer.}$$

Example 1   Writing Polynomials in Standard Form

Polynomial	Standard Form	Degree
a. $4x^2 - 5x^7 - 2 + 3x$	$-5x^7 + 4x^2 + 3x - 2$	7
b. $4 - 9x^2$	$-9x^2 + 4$	2
c. 8	$8 \ (8 = 8x^0)$	0

✓ *Checkpoint* Now try Exercise 15.

What you should learn

- Write polynomials in standard form.
- Add, subtract, and multiply polynomials.
- Use special products to multiply polynomials.
- Remove common factors from polynomials.
- Factor special polynomial forms.
- Factor trinomials as the product of two binomials.
- Factor by grouping.

Why you should learn it

Polynomials can be used to model and solve real-life problems. For instance, in Exercise 178 on page 36, a polynomial is used to model the rate of change of a chemical reaction.

Sean Brady/24th Street Group

STUDY TIP

Expressions are not polynomials if:

1. A variable is underneath a radical.

2. A polynomial expression (with degree greater than 0) is in the denominator of a term.

Operations with Polynomials

You can add and subtract polynomials in much the same way you add and subtract real numbers. Simply add or subtract the *like terms* (terms having the same variables to the same powers) by adding their coefficients. For instance, $-3xy^2$ and $5xy^2$ are like terms and their sum is

$$-3xy^2 + 5xy^2 = (-3 + 5)xy^2 = 2xy^2.$$

Example 2 Sums and Differences of Polynomials

Perform the indicated operation.

a. $(5x^3 - 7x^2 - 3) + (x^3 + 2x^2 - x + 8)$

b. $(7x^4 - x^2 - 4x + 2) - (3x^4 - 4x^2 + 3x)$

Solution

a. $(5x^3 - 7x^2 - 3) + (x^3 + 2x^2 - x + 8)$

$\quad = (5x^3 + x^3) + (-7x^2 + 2x^2) - x + (-3 + 8)$ Group like terms.

$\quad = 6x^3 - 5x^2 - x + 5$ Combine like terms.

b. $(7x^4 - x^2 - 4x + 2) - (3x^4 - 4x^2 + 3x)$

$\quad = 7x^4 - x^2 - 4x + 2 - 3x^4 + 4x^2 - 3x$ Distributive Property

$\quad = (7x^4 - 3x^4) + (-x^2 + 4x^2) + (-4x - 3x) + 2$ Group like terms.

$\quad = 4x^4 + 3x^2 - 7x + 2$ Combine like terms.

✓ *Checkpoint* Now try Exercise 23.

> **STUDY TIP**
>
> When a negative sign precedes an expression within parentheses, remember to distribute the negative sign to each term inside the parentheses.
>
> $-(x^2 - x + 3) = -x^2 + x - 3$

To find the product of two polynomials, use the left and right Distributive Properties.

Example 3 Multiplying Polynomials: The FOIL Method

$$(3x - 2)(5x + 7) = 3x(5x + 7) - 2(5x + 7)$$

$$= (3x)(5x) + (3x)(7) - (2)(5x) - (2)(7)$$

$$= 15x^2 + 21x - 10x - 14$$

Product of First terms	Product of Outer terms	Product of Inner terms	Product of Last terms

$$= 15x^2 + 11x - 14$$

Note that when using the **FOIL Method** (which can only be used to multiply two binomials), the outer (O) and inner (I) terms are like terms and can be combined into one term.

✓ *Checkpoint* Now try Exercise 39.

Example 4   The Product of Two Trinomials

Find the product of $4x^2 + x - 2$ and $-x^2 + 3x + 5$.

Solution

When multiplying two polynomials, be sure to multiply *each* term of one polynomial by *each* term of the other. A vertical format is helpful.

$$
\begin{array}{rl}
4x^2 + x - 2 & \quad \text{Write in standard form.} \\
\times -x^2 + 3x + 5 & \quad \text{Write in standard form.} \\
\hline
20x^2 + 5x - 10 & \quad 5(4x^2 + x - 2) \\
12x^3 + 3x^2 - 6x & \quad 3x(4x^2 + x - 2) \\
-4x^4 - x^3 + 2x^2 & \quad -x^2(4x^2 + x - 2) \\
\hline
-4x^4 + 11x^3 + 25x^2 - x - 10 & \quad \text{Combine like terms.}
\end{array}
$$

✓ *Checkpoint* Now try Exercise 59.

Special Products

Special Products

Let u and v be real numbers, variables, or algebraic expressions.

Special Product	*Example*
Sum and Difference of Same Terms	
$(u + v)(u - v) = u^2 - v^2$	$(x + 4)(x - 4) = x^2 - 4^2 = x^2 - 16$
Square of a Binomial	
$(u + v)^2 = u^2 + 2uv + v^2$	$(x + 3)^2 = x^2 + 2(x)(3) + 3^2 = x^2 + 6x + 9$
$(u - v)^2 = u^2 - 2uv + v^2$	$(3x - 2)^2 = (3x)^2 - 2(3x)(2) + 2^2 = 9x^2 - 12x + 4$
Cube of a Binomial	
$(u + v)^3 = u^3 + 3u^2v + 3uv^2 + v^3$	$(x + 2)^3 = x^3 + 3x^2(2) + 3x(2^2) + 2^3 = x^3 + 6x^2 + 12x + 8$
$(u - v)^3 = u^3 - 3u^2v + 3uv^2 - v^3$	$(x - 1)^3 = x^3 - 3x^2(1) + 3x(1^2) - 1^3 = x^3 - 3x^2 + 3x - 1$

Example 5   The Product of Two Trinomials

Find the product of $x + y - 2$ and $x + y + 2$.

Solution

By grouping $x + y$ in parentheses, you can write the product of the trinomials as a special product.

$$
\begin{aligned}
(x + y - 2)(x + y + 2) &= [(x + y) - 2][(x + y) + 2] \\
&= (x + y)^2 - 2^2 = x^2 + 2xy + y^2 - 4
\end{aligned}
$$

✓ *Checkpoint* Now try Exercise 61.

Factoring

The process of writing a polynomial as a product is called **factoring.** It is an important tool for solving equations and for simplifying rational expressions.

Unless noted otherwise, when you are asked to factor a polynomial, you can assume that you are looking for factors with integer coefficients. If a polynomial cannot be factored using integer coefficients, it is **prime** or **irreducible over the integers.** For instance, the polynomial $x^2 - 3$ is irreducible over the integers. Over the real numbers, this polynomial can be factored as

$$x^2 - 3 = (x + \sqrt{3})(x - \sqrt{3}).$$

A polynomial is **completely factored** when each of its factors is prime. So,

$$x^3 - x^2 + 4x - 4 = (x - 1)(x^2 + 4) \qquad \text{Completely factored}$$

is completely factored, but

$$x^3 - x^2 - 4x + 4 = (x - 1)(x^2 - 4) \qquad \text{Not completely factored}$$

is not completely factored. Its complete factorization is

$$x^3 - x^2 - 4x + 4 = (x - 1)(x + 2)(x - 2).$$

The simplest type of factoring involves a polynomial that can be written as the product of a monomial and another polynomial. The technique used here is the Distributive Property, $a(b + c) = ab + ac$, in the *reverse* direction. For instance, the polynomial $5x^2 + 15x$ can be factored as follows.

$$5x^2 + 15x = 5x(x) + 5x(3) \qquad \text{5x is a common factor.}$$
$$= 5x(x + 3)$$

The first step in completely factoring a polynomial is to remove (factor out) any common factors, as shown in the next example.

Example 6   Removing Common Factors

Factor each expression.

a. $6x^3 - 4x$ **b.** $3x^4 + 9x^3 + 6x^2$ **c.** $(x - 2)(2x) + (x - 2)(3)$

Solution

a. $6x^3 - 4x = 2x(3x^2) - 2x(2) = 2x(3x^2 - 2)$ $\qquad$ 2x is a common factor.

b. $3x^4 + 9x^3 + 6x^2 = 3x^2(x^2) + 3x^2(3x) + 3x^2(2)$ $\qquad$ 3x² is a common factor.
$$= 3x^2(x^2 + 3x + 2)$$

c. $(x - 2)(2x) + (x - 2)(3) = (x - 2)(2x + 3)$ $\qquad$ x − 2 is a common factor.

✓ *Checkpoint* Now try Exercise 73.

Factoring Special Polynomial Forms

Some polynomials have special forms that arise from the special product forms on page 26. You should learn to recognize these forms so that you can factor such polynomials easily.

Factoring Special Polynomial Forms

Factored Form	Example

Difference of Two Squares

$$u^2 - v^2 = (u + v)(u - v) \qquad\qquad 9x^2 - 4 = (3x)^2 - 2^2 = (3x + 2)(3x - 2)$$

Perfect Square Trinomial

$$u^2 + 2uv + v^2 = (u + v)^2 \qquad\qquad x^2 + 6x + 9 = x^2 + 2(x)(3) + 3^2 = (x + 3)^2$$
$$u^2 - 2uv + v^2 = (u - v)^2 \qquad\qquad x^2 - 6x + 9 = x^2 - 2(x)(3) + 3^2 = (x - 3)^2$$

Sum or Difference of Two Cubes

$$u^3 + v^3 = (u + v)(u^2 - uv + v^2) \qquad x^3 + 8 = x^3 + 2^3 = (x + 2)(x^2 - 2x + 4)$$
$$u^3 - v^3 = (u - v)(u^2 + uv + v^2) \qquad 27x^3 - 1 = (3x)^3 - 1^3 = (3x - 1)(9x^2 + 3x + 1)$$

One of the easiest special polynomial forms to factor is the difference of two squares. Think of this form as follows.

$$u^2 - v^2 = (u + v)(u - v)$$

Difference   Opposite signs

To recognize perfect square terms, look for coefficients that are squares of integers and variables raised to *even powers*.

Example 7   Removing a Common Factor First

$$3 - 12x^2 = 3(1 - 4x^2) \qquad\qquad \text{3 is a common factor.}$$
$$= 3[1^2 - (2x)^2] \qquad\qquad \text{Difference of two squares}$$
$$= 3(1 + 2x)(1 - 2x) \qquad\qquad \text{Factored form}$$

✓ *Checkpoint* Now try Exercise 77.

> **STUDY TIP**
>
> In Example 7, note that the first step in factoring a polynomial is to check for a common factor. Once the common factor is removed, it is often possible to recognize patterns that were not immediately obvious.

Example 8   Factoring the Difference of Two Squares

a. $(x + 2)^2 - y^2 = [(x + 2) + y][(x + 2) - y]$
$$= (x + 2 + y)(x + 2 - y)$$

b. $16x^4 - 81 = (4x^2)^2 - 9^2 \qquad\qquad \text{Difference of two squares}$
$$= (4x^2 + 9)(4x^2 - 9)$$
$$= (4x^2 + 9)[(2x)^2 - 3^2] \qquad\qquad \text{Difference of two squares}$$
$$= (4x^2 + 9)(2x + 3)(2x - 3) \qquad\qquad \text{Factored form}$$

✓ *Checkpoint* Now try Exercise 81.

A perfect square trinomial is the square of a binomial, as shown below.

$$u^2 + 2uv + v^2 = (u + v)^2 \qquad \text{or} \qquad u^2 - 2uv + v^2 = (u - v)^2$$

Like signs   Like signs

Note that the first and last terms are squares and the middle term is twice the product of u and v.

Example 9   Factoring Perfect Square Trinomials

Factor each trinomial.

a. $x^2 - 10x + 25$ **b.** $16x^2 + 8x + 1$

Solution

a. $x^2 - 10x + 25 = x^2 - 2(x)(5) + 5^2$ Rewrite in $u^2 - 2uv + v^2$ form.

$\qquad = (x - 5)^2$

b. $16x^2 + 8x + 1 = (4x)^2 + 2(4x)(1) + 1^2$ Rewrite in $u^2 + 2uv + v^2$ form.

$\qquad = (4x + 1)^2$

✓ *Checkpoint* Now try Exercise 87.

The next two formulas show the sums and differences of cubes. Pay special attention to the signs of the terms.

Like signs   Like signs

$$u^3 + v^3 = (u + v)(u^2 - uv + v^2) \quad u^3 - v^3 = (u - v)(u^2 + uv + v^2)$$

Unlike signs   Unlike signs

> **Exploration**
>
> Rewrite $u^6 - v^6$ as the difference of two squares. Then find a formula for completely factoring $u^6 - v^6$. Use your formula to completely factor $x^6 - 1$ and $x^6 - 64$.

Example 10   Factoring the Difference of Cubes

Factor $x^3 - 27$.

Solution

$\qquad x^3 - 27 = x^3 - 3^3$ Rewrite 27 as 3^3.

$\qquad = (x - 3)(x^2 + 3x + 9)$ Factor.

✓ *Checkpoint* Now try Exercise 91.

Example 11   Factoring the Sum of Cubes

$3x^3 + 192 = 3(x^3 + 64)$ 3 is a common factor.

$\qquad = 3(x^3 + 4^3)$ Rewrite 64 as 4^3.

$\qquad = 3(x + 4)(x^2 - 4x + 16)$ Factor.

✓ *Checkpoint* Now try Exercise 93.

Trinomials with Binomial Factors

To factor a trinomial of the form $ax^2 + bx + c$, use the following pattern.

Factors of a

$$ax^2 + bx + c = (\boxed{}x + \boxed{})(\boxed{}x + \boxed{})$$

Factors of c

The goal is to find a combination of factors of a and c so that the outer and inner products add up to the middle term bx. For instance, in the trinomial $6x^2 + 17x + 5$, you can write all possible factorizations and determine which one has outer and inner products that add up to $17x$.

$$(6x + 5)(x + 1), (6x + 1)(x + 5), (2x + 1)(3x + 5), (2x + 5)(3x + 1)$$

You can see that $(2x + 5)(3x + 1)$ is the correct factorization because the outer (O) and inner (I) products add up to $17x$.

$$\overset{\text{F}}{} \quad \overset{\text{O}}{} \quad \overset{\text{I}}{} \quad \overset{\text{L}}{} \qquad \overset{\text{O + I}}{}$$
$$(2x + 5)(3x + 1) = 6x^2 + 2x + 15x + 5 = 6x^2 + 17x + 5.$$

Example 12   Factoring a Trinomial: Leading Coefficient Is 1

Factor $x^2 - 7x + 12$.

Solution

The possible factorizations are

$$(x - 2)(x - 6), \quad (x - 1)(x - 12), \quad \text{and} \quad (x - 3)(x - 4).$$

Testing the middle term, you will find the correct factorization to be

$$x^2 - 7x + 12 = (x - 3)(x - 4). \qquad \text{O + I} = -4x + (-3x) = -7x$$

✓ **Checkpoint** Now try Exercise 103.

Example 13   Factoring a Trinomial: Leading Coefficient Is Not 1

Factor $2x^2 + x - 15$.

Solution

The eight possible factorizations are as follows.

$$(2x - 1)(x + 15), (2x + 1)(x - 15), (2x - 3)(x + 5), (2x + 3)(x - 5),$$

$$(2x - 5)(x + 3), (2x + 5)(x - 3), (2x - 15)(x + 1), (2x + 15)(x - 1)$$

Testing the middle term, you will find the correct factorization to be

$$2x^2 + x - 15 = (2x - 5)(x + 3). \qquad \text{O + I} = 6x - 5x = x$$

✓ **Checkpoint** Now try Exercise 111.

Factoring by Grouping

Sometimes polynomials with more than three terms can be factored by a method called **factoring by grouping.**

Example 14   Factoring by Grouping

Use factoring by grouping to factor $x^3 - 2x^2 - 3x + 6$.

Solution

$$\begin{aligned} x^3 - 2x^2 - 3x + 6 &= (x^3 - 2x^2) - (3x - 6) &&\text{Group terms.} \\ &= x^2(x - 2) - 3(x - 2) &&\text{Factor groups.} \\ &= (x - 2)(x^2 - 3) &&(x - 2) \text{ is a common factor.} \end{aligned}$$

✓ *Checkpoint* Now try Exercise 115.

Factoring a trinomial can involve quite a bit of trial and error. Some of this trial and error can be lessened by using factoring by grouping. The key to this method of factoring is knowing how to rewrite the middle term. In general, to factor a trinomial $ax^2 + bx + c$ by grouping, choose factors of the product ac that add up to b and use these factors to rewrite the middle term.

Example 15   Factoring a Trinomial by Grouping

Use factoring by grouping to factor $2x^2 + 5x - 3$.

Solution

In the trinomial $2x^2 + 5x - 3$, $a = 2$ and $c = -3$, which implies that the product ac is -6. Now, because -6 factors as $(6)(-1)$ and $6 - 1 = 5 = b$, rewrite the middle term as $5x = 6x - x$. This produces the following.

$$\begin{aligned} 2x^2 + 5x - 3 &= 2x^2 + 6x - x - 3 &&\text{Rewrite middle term.} \\ &= (2x^2 + 6x) - (x + 3) &&\text{Group terms.} \\ &= 2x(x + 3) - (x + 3) &&\text{Factor groups.} \\ &= (x + 3)(2x - 1) &&(x + 3) \text{ is a common factor.} \end{aligned}$$

So, the trinomial factors as $2x^2 + 5x - 3 = (x + 3)(2x - 1)$.

✓ *Checkpoint* Now try Exercise 117.

Guidelines for Factoring Polynomials

  1. Factor out any common factors using the Distributive Property.

  2. Factor according to one of the special polynomial forms.

  3. Factor as $ax^2 + bx + c = (mx + r)(nx + s)$.

  4. Factor by grouping.

P.3   Exercises

Vocabulary Check

Fill in the blanks.

1. For the polynomial $a_nx^n + a_{n-1}x^{n-1} + \cdots + a_1x + a_0$, the degree is _____ and the leading coefficient is _____ .

2. A polynomial that has all zero coefficients is called the _____ .

3. A polynomial with one term is called a _____ .

4. The letters in "FOIL" stand for the following.
   F _____   O _____   I _____   L _____

5. If a polynomial cannot be factored using integer coefficients, it is called _____ .

6. The polynomial $u^2 + 2uv + v^2$ is called a _____ .

In Exercises 1–6, match the polynomial with its description. [The polynomials are labeled (a), (b), (c), (d), (e), and (f).]

(a) $6x$ (b) $1 - 4x^3$

(c) $x^3 + 2x^2 - 4x + 1$ (d) 7

(e) $-3x^5 + 2x^3 + x$ (f) $\frac{3}{4}x^4 + x^2 + 14$

1. A polynomial of degree zero

2. A trinomial of degree five

3. A binomial with leading coefficient -4

4. A monomial of positive degree

5. A trinomial with leading coefficient $\frac{3}{4}$

6. A third-degree polynomial with leading coefficient 1

In Exercises 7–10, write a polynomial that fits the description. (There are many correct answers.)

7. A third-degree polynomial with leading coefficient -2

8. A fifth-degree polynomial with leading coefficient 8

9. A fourth-degree polynomial with a negative leading coefficient

10. A third-degree trinomial with an even leading coefficient

In Exercises 11–16, write the polynomial in standard form. Then identify the degree and leading coefficient of the polynomial.

11. $3x + 4x^2 + 2$

12. $x^2 - 4 - 3x^4$

13. $1 + x^7$

14. $-21x$

15. $1 - x + 6x^4 - 2x^5$

16. $7 + 8x$

In Exercises 17–20, determine whether the expression is a polynomial. If so, write the polynomial in standard form.

17. $7x - 2x^3 + 10$

18. $4x^3 + x - x^{-1}$

19. $\sqrt{x^2 - x^4}$

20. $\dfrac{x^2 + 2x - 3}{6}$

In Exercises 21–36, perform the operations and write the result in standard form.

21. $(6x + 5) - (8x + 15)$

22. $(2x^2 + 1) - (x^2 - 2x + 1)$

23. $-(x^3 - 2) + (4x^3 - 2x)$

24. $-(5x^2 - 1) - (-3x^2 + 5)$

25. $(15x^2 - 6) - (-8.1x^3 - 14.7x^2 - 17)$

26. $(15.6x^4 - 18x - 19.4) - (13.9x^4 - 9.2x + 15)$

27. $3x(x^2 - 2x + 1)$ 28. $y^2(4y^2 + 2y - 3)$

29. $-5z(3z - 1)$ 30. $(-3x)(5x + 2)$

31. $(1 - x^3)(4x)$ 32. $-4x(3 - x^3)$

33. $(2.5x^2 + 5)(-3x)$ 34. $(2 - 3.5y)(4y^3)$

35. $-2x\left(\frac{1}{8}x + 3\right)$ 36. $6y\left(4 - \frac{3}{8}y\right)$

In Exercises 37–68, multiply or find the special product.

37. $(x + 3)(x + 4)$ 38. $(x - 5)(x + 10)$

39. $(3x - 5)(2x + 1)$ 40. $(7x - 2)(4x - 3)$

41. $(2x - 5y)^2$ 42. $(5 - 8x)^2$

43. $(x + 10)(x - 10)$ 44. $(2x + 3)(2x - 3)$

45. $(x + 2y)(x - 2y)$ 46. $(2x + 3y)(2x - 3y)$

47. $(2r^2 - 5)(2r^2 + 5)$

48. $(3a^3 - 4b^2)(3a^3 + 4b^2)$

49. $(x + 1)^3$ **50.** $(x - 2)^3$

51. $(2x - y)^3$ **52.** $(3x + 2y)^3$

53. $\left(\frac{1}{2}x - 5\right)^2$ **54.** $\left(\frac{3}{5}t + 4\right)^2$

55. $\left(\frac{1}{4}x - 3\right)\left(\frac{1}{4}x + 3\right)$ **56.** $\left(2x + \frac{1}{6}\right)\left(2x - \frac{1}{6}\right)$

57. $(2.4x + 3)^2$ **58.** $(1.8y - 5)^2$

59. $(-x^2 + x - 5)(3x^2 + 4x + 1)$

60. $(x^2 + 3x + 2)(2x^2 - x + 4)$

61. $[(m - 3) + n][(m - 3) - n]$

62. $[(x + y) + 1][(x + y) - 1]$

63. $[(x - 3) + y]^2$ **64.** $[(x + 1) - y]^2$

65. $5x(x + 1) - 3x(x + 1)$

66. $(2x - 1)(x + 3) + 3(x + 3)$

67. $(u + 2)(u - 2)(u^2 + 4)$

68. $(x + y)(x - y)(x^2 + y^2)$

In Exercises 69–74, factor out the common factor.

69. $2x + 8$ **70.** $5y - 30$

71. $2x^3 - 6x$ **72.** $4x^3 - 6x^2 + 12x$

73. $3x(x - 5) + 8(x - 5)$

74. $(5x - 4)^2 + (5x - 4)$

In Exercises 75–82, factor the difference of two squares.

75. $x^2 - 64$ **76.** $x^2 - 81$

77. $32y^2 - 18$ **78.** $4 - 36y^2$

79. $4x^2 - \frac{1}{9}$ **80.** $\frac{25}{36}y^2 - 49$

81. $(x - 1)^2 - 4$ **82.** $25 - (z + 5)^2$

In Exercises 83–90, factor the perfect square trinomial.

83. $x^2 - 4x + 4$ **84.** $x^2 + 10x + 25$

85. $x^2 + x + \frac{1}{4}$ **86.** $x^2 - \frac{4}{3}x + \frac{4}{9}$

87. $4t^2 + 4t + 1$ **88.** $9x^2 - 12x + 4$

89. $9t^2 + \frac{3}{2}t + \frac{1}{16}$ **90.** $4t^2 + \frac{8}{5}t + \frac{4}{25}$

In Exercises 91–100, factor the sum or difference of cubes.

91. $x^3 - 8$ **92.** $x^3 + 27$

93. $y^3 + 216$ **94.** $z^3 - 125$

95. $x^3 - \frac{8}{27}$ **96.** $x^3 + \frac{8}{125}$

97. $8x^3 - 1$ **98.** $27x^3 + 8$

99. $\frac{1}{8}x^3 + 1$ **100.** $\frac{27}{64}x^3 - 1$

In Exercises 101–114, factor the trinomial.

101. $x^2 + x - 2$ **102.** $x^2 + 5x + 6$

103. $s^2 - 5s + 6$ **104.** $t^2 - t - 6$

105. $20 - y - y^2$ **106.** $24 + 5z - z^2$

107. $3x^2 - 5x + 2$ **108.** $3x^2 + 13x - 10$

109. $2x^2 - x - 1$ **110.** $2x^2 - x - 21$

111. $5x^2 + 26x + 5$ **112.** $8x^2 - 45x - 18$

113. $-5u^2 - 13u + 6$ **114.** $-6x^2 + 23x + 4$

In Exercises 115–118, factor by grouping.

115. $x^3 - x^2 + 2x - 2$

116. $x^3 + 5x^2 - 5x - 25$

117. $6x^2 + x - 2$

118. $3x^2 + 10x + 8$

In Exercises 119–150, completely factor the expression.

119. $x^3 - 16x$ **120.** $12x^2 - 48$

121. $x^3 - x^2$ **122.** $6x^2 - 54$

123. $x^2 - 2x + 1$ **124.** $9x^2 - 6x + 1$

125. $1 - 4x + 4x^2$ **126.** $16 - 6x - x^2$

127. $2x^2 + 4x - 2x^3$ **128.** $7y^2 + 15y - 2y^3$

129. $9x^2 + 10x + 1$ **130.** $13x + 6 + 5x^2$

131. $\frac{1}{8}x^2 - \frac{1}{96}x - \frac{1}{16}$ **132.** $\frac{1}{81}x^2 + \frac{2}{9}x - 8$

133. $3x^3 + x^2 + 15x + 5$

134. $5 - x + 5x^2 - x^3$

135. $3u - 2u^2 + 6 - u^3$

136. $x^4 - 4x^3 + x^2 - 4x$

137. $25 - (z + 5)^2$ **138.** $(t - 1)^2 - 49$

139. $(x^2 + 1)^2 - 4x^2$ **140.** $(x^2 + 8)^2 - 36x^2$

141. $2t^3 - 16$ **142.** $5x^3 + 40$

143. $4x(2x - 1) + 2(2x - 1)^2$

144. $5(3 - 4x)^2 - 8(3 - 4x)(5x - 1)$

145. $2(x + 1)(x - 3)^2 - 3(x + 1)^2(x - 3)$

146. $7(3x + 2)^2(1 - x)^2 + (3x + 2)(1 - x)^3$

147. $7x(2)(x^2 + 1)(2x) - (x^2 + 1)^2(7)$

148. $3(x - 2)^2(x + 1)^4 + (x - 2)^3(4)(x + 1)^3$

149. $2x(x - 5)^4 - x^2(4)(x - 5)^3$

150. $5(x^6 + 1)^4(6x^5)(3x + 2)^3 + 3(3x + 2)^2(3)(x^6 + 1)^5$

151. *Compound Interest* After 2 years, an investment of \$500 compounded annually at an interest rate r will yield an amount of $500(1 + r)^2$.

(a) Write this polynomial in standard form.

(b) Use a calculator to evaluate the polynomial for the values of r shown in the table.

r	$2\frac{1}{2}\%$	3%	4%	$4\frac{1}{2}\%$	5%
$500(1 + r)^2$					

(c) What conclusion can you make from the table?

152. *Compound Interest* After 3 years, an investment of \$1200 compounded annually at an interest rate r will yield an amount of $1200(1 + r)^3$.

(a) Write this polynomial in standard form.

(b) Use a calculator to evaluate the polynomial for the values of r shown in the table.

r	2%	3%	$3\frac{1}{2}\%$	4%	$4\frac{1}{2}\%$
$1200(1 + r)^3$					

(c) What conclusion can you make from the table?

153. *Geometry* An overnight shipping company is designing a closed box by cutting along the solid lines and folding along the broken lines on the rectangular piece of corrugated cardboard shown in the figure. The length and width of the rectangle are 45 centimeters and 15 centimeters, respectively. Find the volume of the box in terms of x. Find the volume when $x = 3$, $x = 5$, and $x = 7$.

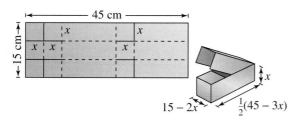

154. *Geometry* A take-out fast food restaurant is constructing an open box made by cutting squares out of the corners of a piece of cardboard that is 18 centimeters by 26 centimeters. The edge of each cut-out square is x inches. Find the volume of the box in terms of x. Find the volume when $x = 1$, $x = 2$, and $x = 3$.

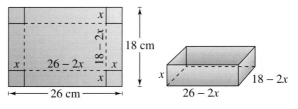

Figure for 154

155. *Stopping Distance* The stopping distance of an automobile is the distance traveled during the driver's reaction time plus the distance traveled after the brakes are applied. In an experiment, these distances were measured (in feet) when the automobile was traveling at a speed of x miles per hour on dry, level pavement, as shown in the bar graph. The distance traveled during the reaction time R was $R = 1.1x$, and the braking distance B was

$B = 0.0475x^2 - 0.001x + 0.23$.

(a) Determine the polynomial that represents the total stopping distance T.

(b) Use the result of part (a) to estimate the total stopping distance when $x = 30$, $x = 40$, and $x = 55$.

(c) Use the bar graph to make a statement about the total stopping distance required for increasing speeds.

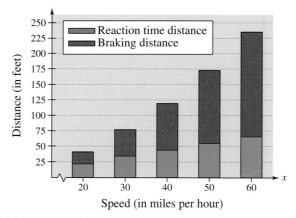

156. *Engineering* A uniformly distributed load is placed on a one-inch-wide steel beam. When the span of the beam is x feet and its depth is 6 inches, the safe load S (in pounds) is approximated by

$S_6 = (0.06x^2 - 2.42x + 38.71)^2$.

When the depth is 8 inches, the safe load is approximated by

$S_8 = (0.08x^2 - 3.30x + 51.93)^2$.

(a) Use the bar graph to estimate the difference in the safe loads for these two beams when the span is 12 feet.

(b) How does the difference in safe load change as the span increases?

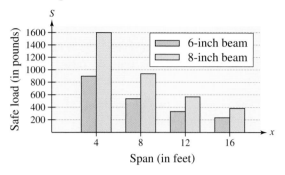

Geometric Modeling In Exercises 157–160, match the factoring formula with the correct geometric factoring model. [The models are labeled (a), (b), (c), and (d).] For instance, a factoring model for

$$2x^2 + 3x + 1 = (2x + 1)(x + 1)$$

is shown in the figure.

(a)

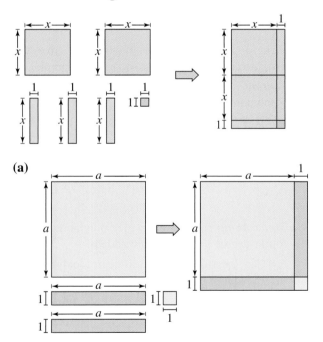

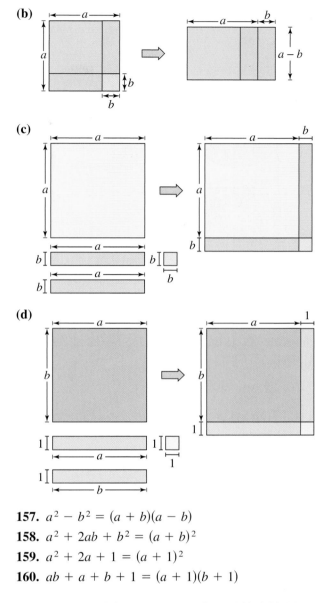

157. $a^2 - b^2 = (a + b)(a - b)$

158. $a^2 + 2ab + b^2 = (a + b)^2$

159. $a^2 + 2a + 1 = (a + 1)^2$

160. $ab + a + b + 1 = (a + 1)(b + 1)$

Geometric Modeling In Exercises 161–164, draw a geometric factoring model to represent the factorization.

161. $3x^2 + 7x + 2 = (3x + 1)(x + 2)$

162. $x^2 + 4x + 3 = (x + 3)(x + 1)$

163. $2x^2 + 7x + 3 = (2x + 1)(x + 3)$

164. $x^2 + 3x + 2 = (x + 2)(x + 1)$

Geometry   **In Exercises 165–168, write an expression in factored form for the area of the shaded portion of the figure.**

165.   **166.**

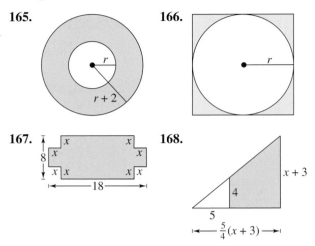

167.   **168.**

In Exercises 169–172, find all values of b for which the trinomial can be factored with integer coefficients.

169. $x^2 + bx - 15$ **170.** $x^2 + bx - 12$

171. $x^2 + bx + 50$ **172.** $x^2 + bx + 24$

In Exercises 173–176, find two integer values of c such that the trinomial can be factored. (There are many correct answers.)

173. $2x^2 + 5x + c$ **174.** $3x^2 - x + c$

175. $3x^2 - 10x + c$ **176.** $2x^2 + 9x + c$

177. *Geometry*   The cylindrical shell shown in the figure has a volume of $V = \pi R^2 h - \pi r^2 h$.

(a) Factor the expression for the volume.

(b) From the result of part (a), show that the volume is

$$2\pi \text{ (average radius)(thickness of the shell)}h.$$

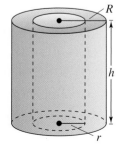

178. *Chemical Reaction*   The rate of change of an autocatalytic chemical reaction is $kQx - kx^2$, where Q is the amount of the original substance, x is the amount of substance formed, and k is a constant of proportionality. Factor the expression.

Synthesis

True or False?   **In Exercises 179–181, determine whether the statement is true or false. Justify your answer.**

179. The product of two binomials is always a second-degree polynomial.

180. The difference of two perfect squares can be factored as the product of conjugate pairs.

181. The sum of two perfect squares can be factored as the binomial sum squared.

182. *Exploration*   Find the degree of the product of two polynomials of degrees m and n.

183. *Exploration*   Find the degree of the sum of two polynomials of degrees m and n if $m < n$.

184. *Writing*   A student's homework paper included the following.

$$(x - 3)^2 = x^2 + 9$$

Write a paragraph fully explaining the error and give the correct method for squaring a binomial.

185. *Writing*   Explain what is meant when it is said that a polynomial is in factored form.

186. *Think About It*   Is $(3x - 6)(x + 1)$ completely factored? Explain.

187. *Error Analysis*   Describe the error.

$$9x^2 - 9x - 54 = (3x + 6)(3x - 9)$$
$$= 3(x + 2)(x - 3)$$

188. *Think About It*   A third-degree polynomial and a fourth-degree polynomial are added.

(a) Can the sum be a fourth-degree polynomial? Explain or give an example.

(b) Can the sum be a second-degree polynomial? Explain or give an example.

(c) Can the sum be a seventh-degree polynomial? Explain or give an example.

189. *Think About It*   Must the sum of two second-degree polynomials be a second-degree polynomial? If not, give an example.

P.4   Rational Expressions

Domain of an Algebraic Expression

The set of real numbers for which an algebraic expression is defined is the **domain** of the expression. Two algebraic expressions are **equivalent** if they have the same domain and yield the same values for all numbers in their domain. For instance, the expressions $(x + 1) + (x + 2)$ and $2x + 3$ are equivalent because

$$(x + 1) + (x + 2) = x + 1 + x + 2 = x + x + 1 + 2 = 2x + 3.$$

Example 1   Finding the Domain of an Algebraic Expression

a. The domain of the polynomial

$$2x^3 + 3x + 4$$

is the set of all real numbers. In fact, the domain of any polynomial is the set of all real numbers, unless the domain is specifically restricted.

b. The domain of the radical expression

$$\sqrt{x - 2}$$

is the set of real numbers greater than or equal to 2, because the square root of a negative number is not a real number.

c. The domain of the expression

$$\frac{x + 2}{x - 3}$$

is the set of all real numbers except $x = 3$, which would result in division by zero, which is undefined.

✓ *Checkpoint* Now try Exercise 5.

The quotient of two algebraic expressions is a **fractional expression.** Moreover, the quotient of two *polynomials* such as

$$\frac{1}{x}, \qquad \frac{2x - 1}{x + 1}, \qquad \text{or} \qquad \frac{x^2 - 1}{x^2 + 1}$$

is a **rational expression.**

Simplifying Rational Expressions

Recall that a fraction is in simplest form if its numerator and denominator have no factors in common aside from ± 1. To write a fraction in simplest form, divide out common factors.

$$\frac{a \cdot \cancel{c}}{b \cdot \cancel{c}} = \frac{a}{b}, \qquad c \neq 0.$$

The key to success in simplifying rational expressions lies in your ability to *factor* polynomials. When simplifying rational expressions, be sure to factor each polynomial completely before concluding that the numerator and denominator have no factors in common.

Example 2   Simplifying a Rational Expression

Write $\dfrac{x^2 + 4x - 12}{3x - 6}$ in simplest form.

Solution

$$\frac{x^2 + 4x - 12}{3x - 6} = \frac{(x + 6)(x - 2)}{3(x - 2)} \qquad \text{Factor completely.}$$

$$= \frac{x + 6}{3}, \qquad x \neq 2 \qquad \text{Divide out common factors.}$$

Note that the original expression is undefined when $x = 2$ (because division by zero is undefined). To make sure that the simplified expression is *equivalent* to the original expression, you must restrict the domain of the simplified expression by excluding the value $x = 2$.

✓ *Checkpoint* Now try Exercise 15.

It may sometimes be necessary to change the sign of a factor to simplify a rational expression, as shown in Example 3.

Example 3   Simplifying Rational Expressions

Write $\dfrac{12 + x - x^2}{2x^2 - 9x + 4}$ in simplest form.

Solution

$$\frac{12 + x - x^2}{2x^2 - 9x + 4} = \frac{(4 - x)(3 + x)}{(2x - 1)(x - 4)} \qquad \text{Factor completely.}$$

$$= \frac{-(x - 4)(3 + x)}{(2x - 1)(x - 4)} \qquad (4 - x) = -(x - 4)$$

$$= -\frac{3 + x}{2x - 1}, \qquad x \neq 4 \qquad \text{Divide out common factors.}$$

✓ *Checkpoint* Now try Exercise 23.

Operations with Rational Expressions

To multiply or divide rational expressions, you can use the properties of fractions discussed in Section P.1. Recall that to divide fractions you invert the divisor and multiply.

Example 4   Multiplying Rational Expressions

$$\frac{2x^2 + x - 6}{x^2 + 4x - 5} \cdot \frac{x^3 - 3x^2 + 2x}{4x^2 - 6x} = \frac{(2x - 3)(x + 2)}{(x + 5)(x - 1)} \cdot \frac{x(x - 2)(x - 1)}{2x(2x - 3)}$$

$$= \frac{(x + 2)(x - 2)}{2(x + 5)}, \quad x \neq 0, x \neq 1, x \neq \frac{3}{2}$$

✓ *Checkpoint* Now try Exercise 39.

Example 5   Dividing Rational Expressions

Divide $\dfrac{x^3 - 8}{x^2 - 4}$ by $\dfrac{x^2 + 2x + 4}{x^3 + 8}$.

Solution

$$\frac{x^3 - 8}{x^2 - 4} \div \frac{x^2 + 2x + 4}{x^3 + 8} = \frac{x^3 - 8}{x^2 - 4} \cdot \frac{x^3 + 8}{x^2 + 2x + 4} \qquad \text{Invert and multiply.}$$

$$= \frac{(x - 2)(x^2 + 2x + 4)}{(x + 2)(x - 2)} \cdot \frac{(x + 2)(x^2 - 2x + 4)}{(x^2 + 2x + 4)}$$

$$= x^2 - 2x + 4, \quad x \neq \pm 2 \qquad \begin{array}{l}\text{Divide out common} \\ \text{factors.}\end{array}$$

✓ *Checkpoint* Now try Exercise 41.

To add or subtract rational expressions, you can use the LCD (least common denominator) method or the basic definition

$$\frac{a}{b} \pm \frac{c}{d} = \frac{ad \pm bc}{bd}, \qquad b \neq 0 \text{ and } d \neq 0. \qquad \text{Basic definition}$$

This definition provides an efficient way of adding or subtracting *two* fractions that have no common factors in their denominators.

Example 6   Subtracting Rational Expressions

Subtract $\dfrac{2}{3x + 4}$ from $\dfrac{x}{x - 3}$.

Solution

$$\frac{x}{x - 3} - \frac{2}{3x + 4} = \frac{x(3x + 4) - 2(x - 3)}{(x - 3)(3x + 4)} \qquad \text{Basic definition}$$

$$= \frac{3x^2 + 4x - 2x + 6}{(x - 3)(3x + 4)} \qquad \text{Distributive Property}$$

$$= \frac{3x^2 + 2x + 6}{(x - 3)(3x + 4)} \qquad \text{Combine like terms.}$$

✓ *Checkpoint* Now try Exercise 45.

> **STUDY TIP**
>
> When subtracting rational expressions, remember to distribute the negative sign to *all* the terms in the quantity that is being subtracted.

For three or more fractions, or for fractions with a repeated factor in the denominators, the LCD method works well. Recall that the least common denominator of several fractions consists of the product of all prime factors in the denominators, with each factor given the highest power of its occurrence in any denominator. Here is a numerical example.

$$\frac{1}{6} + \frac{3}{4} - \frac{2}{3} = \frac{1 \cdot 2}{6 \cdot 2} + \frac{3 \cdot 3}{4 \cdot 3} - \frac{2 \cdot 4}{3 \cdot 4} \qquad \text{The LCD is 12.}$$

$$= \frac{2}{12} + \frac{9}{12} - \frac{8}{12}$$

$$= \frac{3}{12} = \frac{1}{4}$$

Sometimes the numerator of the answer has a factor in common with the denominator. In such cases the answer should be simplified. For instance, in the example above, $\frac{3}{12}$ was simplified to $\frac{1}{4}$.

Example 7   Combining Rational Expressions: The LCD Method

Perform the operations and simplify.

$$\frac{3}{x - 1} - \frac{2}{x} + \frac{x + 3}{x^2 - 1}$$

Solution

Using the factored denominators $(x - 1)$, x, and $(x + 1)(x - 1)$, you can see that the LCD is $x(x + 1)(x - 1)$.

$$\frac{3}{x - 1} - \frac{2}{x} + \frac{x + 3}{(x + 1)(x - 1)}$$

$$= \frac{3(x)(x + 1)}{x(x + 1)(x - 1)} - \frac{2(x + 1)(x - 1)}{x(x + 1)(x - 1)} + \frac{(x + 3)(x)}{x(x + 1)(x - 1)}$$

$$= \frac{3(x)(x + 1) - 2(x + 1)(x - 1) + (x + 3)(x)}{x(x + 1)(x - 1)}$$

$$= \frac{3x^2 + 3x - 2x^2 + 2 + x^2 + 3x}{x(x + 1)(x - 1)} \qquad \text{Distributive Property}$$

$$= \frac{(3x^2 - 2x^2 + x^2) + (3x + 3x) + 2}{x(x + 1)(x - 1)} \qquad \text{Group like terms.}$$

$$= \frac{2x^2 + 6x + 2}{x(x + 1)(x - 1)} \qquad \text{Combine like terms.}$$

$$= \frac{2(x^2 + 3x + 1)}{x(x + 1)(x - 1)} \qquad \text{Factor.}$$

☑ *Checkpoint* Now try Exercise 51.

Complex Fractions

Fractional expressions with separate fractions in the numerator, denominator, or both are called **complex fractions.** Here are two examples.

$$\frac{\left(\dfrac{1}{x}\right)}{x^2 + 1} \quad \text{and} \quad \frac{\left(\dfrac{1}{x}\right)}{\left(\dfrac{1}{x^2 + 1}\right)}$$

A complex fraction can be simplified by combining the fractions in its numerator into a single fraction and then combining the fractions in its denominator into a single fraction. Then invert the denominator and multiply.

Example 8   Simplifying a Complex Fraction

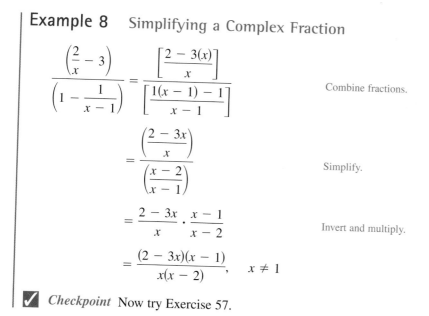

$$\frac{\left(\dfrac{2}{x} - 3\right)}{\left(1 - \dfrac{1}{x - 1}\right)} = \frac{\left[\dfrac{2 - 3(x)}{x}\right]}{\left[\dfrac{1(x - 1) - 1}{x - 1}\right]} \qquad \text{Combine fractions.}$$

$$= \frac{\left(\dfrac{2 - 3x}{x}\right)}{\left(\dfrac{x - 2}{x - 1}\right)} \qquad \text{Simplify.}$$

$$= \frac{2 - 3x}{x} \cdot \frac{x - 1}{x - 2} \qquad \text{Invert and multiply.}$$

$$= \frac{(2 - 3x)(x - 1)}{x(x - 2)}, \qquad x \neq 1$$

✓ *Checkpoint* Now try Exercise 57.

In Example 8, the restriction $x \neq 1$ is added to the final expression to make its domain agree with the domain of the original expression.

Another way to simplify a complex fraction is to multiply each term in its numerator and denominator by the LCD of all fractions in its numerator and denominator. This method is applied to the fraction in Example 8 as follows.

$$\frac{\left(\dfrac{2}{x} - 3\right)}{\left(1 - \dfrac{1}{x - 1}\right)} = \frac{\left(\dfrac{2}{x} - 3\right)}{\left(1 - \dfrac{1}{x - 1}\right)} \cdot \frac{x(x - 1)}{x(x - 1)} \qquad \text{LCD is } x(x - 1).$$

$$= \frac{\left(\dfrac{2 - 3x}{x}\right) \cdot x(x - 1)}{\left(\dfrac{x - 2}{x - 1}\right) \cdot x(x - 1)} \qquad \text{Combine fractions.}$$

$$= \frac{(2 - 3x)(x - 1)}{x(x - 2)}, \qquad x \neq 1 \qquad \text{Simplify.}$$

The next four examples illustrate some methods for simplifying rational expressions involving negative exponents and radicals. These types of expressions occur frequently in calculus.

To simplify an expression with negative exponents, one method is to begin by factoring out the common factor with the smaller exponent. Remember that when factoring, you subtract exponents. For instance, in $3x^{-5/2} + 2x^{-3/2}$ the smaller exponent is $-\frac{5}{2}$ and the common factor is $x^{-5/2}$.

$$3x^{-5/2} + 2x^{-3/2} = x^{-5/2}[3(1) + 2x^{-3/2-(-5/2)}]$$

$$= x^{-5/2}(3 + 2x^1) = \frac{3 + 2x}{x^{5/2}}$$

Example 9   Simplifying an Expression with Negative Exponents

algebra of calculus

Simplify $x(1 - 2x)^{-3/2} + (1 - 2x)^{-1/2}$.

Solution

Begin by factoring out the common factor with the smaller exponent.

$$x(1 - 2x)^{-3/2} + (1 - 2x)^{-1/2} = (1 - 2x)^{-3/2}[x + (1 - 2x)^{(-1/2)-(-3/2)}]$$

$$= (1 - 2x)^{-3/2}[x + (1 - 2x)^1]$$

$$= \frac{1 - x}{(1 - 2x)^{3/2}}$$

✓ *Checkpoint* Now try Exercise 63.

A second method for simplifying this type of expression involves multiplying the numerator and denominator by a term to eliminate the negative exponent.

Example 10   Simplifying an Expression with Negative Exponents

algebra of calculus

Simplify $\dfrac{(4 - x^2)^{1/2} + x^2(4 - x^2)^{-1/2}}{4 - x^2}$.

Solution

$$\frac{(4 - x^2)^{1/2} + x^2(4 - x^2)^{-1/2}}{4 - x^2}$$

$$= \frac{(4 - x^2)^{1/2} + x^2(4 - x^2)^{-1/2}}{4 - x^2} \cdot \frac{(4 - x^2)^{1/2}}{(4 - x^2)^{1/2}}$$

$$= \frac{(4 - x^2)^1 + x^2(4 - x^2)^0}{(4 - x^2)^{3/2}}$$

$$= \frac{4 - x^2 + x^2}{(4 - x^2)^{3/2}} = \frac{4}{(4 - x^2)^{3/2}}$$

✓ *Checkpoint* Now try Exercise 67.

Example 11   Rewriting a Difference Quotient

The following expression from calculus is an example of a *difference quotient*.

$$\frac{\sqrt{x+h}-\sqrt{x}}{h}$$

Rewrite this expression by rationalizing its numerator.

Solution

$$\frac{\sqrt{x+h}-\sqrt{x}}{h} = \frac{\sqrt{x+h}-\sqrt{x}}{h} \cdot \frac{\sqrt{x+h}+\sqrt{x}}{\sqrt{x+h}+\sqrt{x}}$$

$$= \frac{\left(\sqrt{x+h}\right)^2 - \left(\sqrt{x}\right)^2}{h\left(\sqrt{x+h}+\sqrt{x}\right)}$$

$$= \frac{h}{h\left(\sqrt{x+h}+\sqrt{x}\right)}$$

$$= \frac{1}{\sqrt{x+h}+\sqrt{x}}, \qquad h \neq 0$$

Notice that the original expression is undefined when $h = 0$. So, you must exclude $h = 0$ from the domain of the simplified expression so that the expressions are equivalent.

 Checkpoint Now try Exercise 69.

Difference quotients, like that in Example 11, occur frequently in calculus. Often, they need to be rewritten in an equivalent form that can be evaluated when $h = 0$. Note that the equivalent form is not simpler than the original form, but it has the advantage in that it is defined when $h = 0$.

Example 12   Rewriting a Difference Quotient

Rewrite the expression by rationalizing its numerator.

$$\frac{\sqrt{x-4}-\sqrt{x}}{4}$$

Solution

$$\frac{\sqrt{x-4}-\sqrt{x}}{4} = \frac{\sqrt{x-4}-\sqrt{x}}{4} \cdot \frac{\sqrt{x-4}+\sqrt{x}}{\sqrt{x-4}+\sqrt{x}}$$

$$= \frac{\left(\sqrt{x-4}\right)^2 - \left(\sqrt{x}\right)^2}{4\left(\sqrt{x-4}+\sqrt{x}\right)}$$

$$= \frac{-4}{4\left(\sqrt{x-4}+\sqrt{x}\right)}$$

$$= -\frac{1}{\sqrt{x-4}+\sqrt{x}}$$

 Checkpoint Now try Exercise 70.

P.4   Exercises

Vocabulary Check

Fill in the blanks.

1. The set of real numbers for which an algebraic expression is defined is the _____ of the expression.

2. The quotient of two algebraic expressions is a fractional expression and the quotient of two polynomials is a _____ .

3. Fractional expressions with separate fractions in the numerator, denominator, or both are called _____ .

4. To simplify an expression with negative exponents, it is possible to begin by factoring out the common factor with the _____ exponent.

5. Two algebraic expressions that have the same domain and yield the same values for all numbers in their domains are called _____ .

In Exercises 1–8, find the domain of the expression.

1. $3x^2 - 4x + 7$

2. $2x^2 + 5x - 2$

3. $4x^3 + 3, \quad x \geq 0$

4. $6x^2 - 9, \quad x > 0$

5. $\dfrac{1}{3 - x}$

6. $\dfrac{x + 6}{3x + 2}$

7. $\sqrt{x + 7}$

8. $\sqrt{4 - x}$

In Exercises 9 and 10, find the missing factor in the numerator so that the two fractions are equivalent.

9. $\dfrac{5}{2x} = \dfrac{5()}{6x^2}$

10. $\dfrac{3}{4} = \dfrac{3()}{4(x + 1)}$

In Exercises 11–28, write the rational expression in simplest form.

11. $\dfrac{15x^2}{10x}$

12. $\dfrac{18y^2}{60y^5}$

13. $\dfrac{3xy}{xy + x}$

14. $\dfrac{2x^2y}{xy - y}$

15. $\dfrac{4y - 8y^2}{10y - 5}$

16. $\dfrac{9x^2 + 9x}{2x + 2}$

17. $\dfrac{x - 5}{10 - 2x}$

18. $\dfrac{12 - 4x}{x - 3}$

19. $\dfrac{y^2 - 16}{y + 4}$

20. $\dfrac{x^2 - 25}{5 - x}$

21. $\dfrac{x^3 + 5x^2 + 6x}{x^2 - 4}$

22. $\dfrac{x^2 + 8x - 20}{x^2 + 11x + 10}$

23. $\dfrac{y^2 - 7y + 12}{y^2 + 3y - 18}$

24. $\dfrac{3 - x}{x^2 + 11x + 10}$

25. $\dfrac{2 - x + 2x^2 - x^3}{x - 2}$

26. $\dfrac{x^2 - 9}{x^3 + x^2 - 9x - 9}$

27. $\dfrac{z^3 - 8}{z^2 + 2z + 4}$

28. $\dfrac{y^3 - 2y^2 - 3y}{y^3 + 1}$

In Exercises 29 and 30, complete the table. What can you conclude?

29.

x	0	1	2	3	4	5	6
$\dfrac{x^2 - 2x - 3}{x - 3}$							
$x + 1$							

30.

x	0	1	2	3	4	5	6
$\dfrac{x - 3}{x^2 - x - 6}$							
$\dfrac{1}{x + 2}$							

31. **Error Analysis**   Describe the error.

$$\frac{5x^3}{2x^3 + 4} = \frac{5x^3}{2x^3 + 4} = \frac{5}{2 + 4} = \frac{5}{6}$$

32. **Error Analysis**   Describe the error.

$$\frac{x^3 + 25x}{x^2 - 2x - 15} = \frac{x(x^2 + 25)}{(x - 5)(x + 3)}$$

$$= \frac{x(x - 5)(x + 5)}{(x - 5)(x + 3)} = \frac{x(x + 5)}{x + 3}$$

Geometry   In Exercises 33 and 34, find the ratio of the area of the shaded portion of the figure to the total area of the figure.

33.

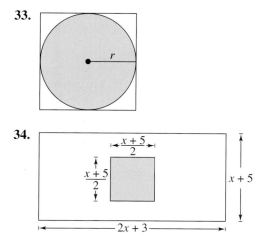

34.

In Exercises 53–60, simplify the complex fraction.

53. $\dfrac{\left(\dfrac{x}{2} - 1\right)}{(x - 2)}$

54. $\dfrac{(x - 4)}{\left(\dfrac{x}{4} - \dfrac{4}{x}\right)}$

55. $\dfrac{\left[\dfrac{x^2}{(x + 1)^2}\right]}{\left[\dfrac{x}{(x + 1)^3}\right]}$

56. $\dfrac{\left(\dfrac{x^2 - 1}{x}\right)}{\left[\dfrac{(x - 1)^2}{x}\right]}$

57. $\dfrac{\left[\dfrac{1}{(x + h)^2} - \dfrac{1}{x^2}\right]}{h}$

58. $\dfrac{\left(\dfrac{x + h}{x + h + 1} - \dfrac{x}{x + 1}\right)}{h}$

59. $\dfrac{\left(\sqrt{x} - \dfrac{1}{2\sqrt{x}}\right)}{\sqrt{x}}$

60. $\dfrac{\left(\dfrac{t^2}{\sqrt{t^2 + 1}} - \sqrt{t^2 + 1}\right)}{t^2}$

In Exercises 35–42, perform the multiplication or division and simplify.

35. $\dfrac{5}{x - 1} \cdot \dfrac{x - 1}{25(x - 2)}$

36. $\dfrac{x + 13}{x^3(3 - x)} \cdot \dfrac{x(x - 3)}{5}$

37. $\dfrac{r}{r - 1} \div \dfrac{r^2}{r^2 - 1}$

38. $\dfrac{4y - 16}{5y + 15} \div \dfrac{4 - y}{2y + 6}$

39. $\dfrac{t^2 - t - 6}{t^2 + 6t + 9} \cdot \dfrac{t + 3}{t^2 - 4}$

40. $\dfrac{y^3 - 8}{2y^3} \cdot \dfrac{4y}{y^2 - 5y + 6}$

41. $\dfrac{3(x + y)}{4} \div \dfrac{x + y}{2}$

42. $\dfrac{x + 2}{5(x - 3)} \div \dfrac{x - 2}{5(x - 3)}$

In Exercises 43–52, perform the addition or subtraction and simplify.

43. $\dfrac{5}{x - 1} + \dfrac{x}{x - 1}$

44. $\dfrac{2x - 1}{x + 3} - \dfrac{1 - x}{x + 3}$

45. $\dfrac{6}{2x + 1} - \dfrac{x}{x + 3}$

46. $\dfrac{3}{x - 1} + \dfrac{5x}{3x + 4}$

47. $\dfrac{3}{x - 2} + \dfrac{5}{2 - x}$

48. $\dfrac{2x}{x - 5} - \dfrac{5}{5 - x}$

49. $\dfrac{1}{x^2 - x - 2} - \dfrac{x}{x^2 - 5x + 6}$

50. $\dfrac{2}{x^2 - x - 2} + \dfrac{10}{x^2 + 2x - 8}$

51. $-\dfrac{1}{x} + \dfrac{2}{x^2 + 1} - \dfrac{1}{x^3 + x}$

52. $\dfrac{2}{x + 1} + \dfrac{2}{x - 1} + \dfrac{1}{x^2 - 1}$

In Exercises 61–66, simplify the expression by removing the common factor with the smaller exponent.

61. $x^5 - 2x^{-2}$

62. $x^5 - 5x^{-3}$

63. $x^2(x^2 + 1)^{-5} - (x^2 + 1)^{-4}$

64. $2x(x - 5)^{-3} - 4x^2(x - 5)^{-4}$

65. $2x^2(x - 1)^{1/2} - 5(x - 1)^{-1/2}$

66. $4x^3(2x - 1)^{3/2} - 2x(2x - 1)^{-1/2}$

In Exercises 67 and 68, simplify the expression.

67. $\dfrac{2x^{3/2} - x^{-1/2}}{x^2}$

68. $\dfrac{-x^2(x^2 + 1)^{-1/2} + 2x(x^2 + 1)^{-3/2}}{x^3}$

In Exercises 69 and 70, rationalize the numerator of the expression.

69. $\dfrac{\sqrt{x + 2} - \sqrt{x}}{2}$

70. $\dfrac{\sqrt{z - 3} - \sqrt{z}}{3}$

71. ***Rate***   A photocopier copies at a rate of 16 pages per minute.

(a) Find the time required to copy 1 page.

(b) Find the time required to copy x pages.

(c) Find the time required to copy 60 pages.

72. *Monthly Payment*   The formula that approximates the annual interest rate r of a monthly installment loan is given by

$$r = \frac{\left[\dfrac{24(NM - P)}{N}\right]}{\left(P + \dfrac{NM}{12}\right)}$$

where N is the total number of payments, M is the monthly payment, and P is the amount financed.

(a) Approximate the annual interest rate for a five-year car loan of $20,000 that has monthly payments of $400.

(b) Simplify the expression for the annual interest rate r, and then rework part (a).

Probability   **In Exercises 73 and 74, consider an experiment in which a marble is tossed into a box whose base is shown in the figure. The probability that the marble will come to rest in the shaded portion of the box is equal to the ratio of the shaded area to the total area of the figure. Find the probability.**

73.   **74.**

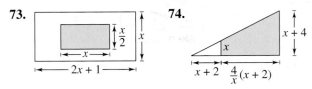

75. *Refrigeration*   When food (at room temperature) is placed in a refrigerator, the time required for the food to cool depends on the amount of food, the air circulation in the refrigerator, the original temperature of the food, and the temperature of the refrigerator. Consider the model that gives the temperature of food that is at 75°F and is placed in a 40°F refrigerator as

$$T = 10\left(\frac{4t^2 + 16t + 75}{t^2 + 4t + 10}\right)$$

where T is the temperature (in degrees Fahrenheit) and t is the time (in hours).

(a) Complete the table.

t	0	2	4	6	8	10
T						

t	12	14	16	18	20	22
T						

(b) What value of T does the mathematical model appear to be approaching?

76. *Plants*   The table shows the numbers of endangered and threatened plant species in the United States for the years 1996 through 2002.   (Source: U.S. Fish and Wildlife Service)

Year	Endangered	Threatened
1996	513	101
1997	553	115
1998	567	135
1999	581	140
2000	593	142
2001	595	145
2002	598	147

Mathematical models for this data are

$$\text{Endangered plants} = \frac{-141.341t + 663.9}{-0.227t + 1.0}$$

and

$$\text{Threatened plants} = -1.80t^2 + 39.7t - 72$$

where t represents the year, with $t = 6$ corresponding to 1996.

(a) Using the models, create a table to estimate the number of endangered plant species and the number of threatened plant species for the given years. Compare these estimates with the actual data.

(b) Determine a model for the ratio of the number of threatened plant species to the number of endangered plant species. Use the model to find this ratio for the given years.

Synthesis

True or False?   **In Exercises 77 and 78, determine whether the statement is true or false. Justify your answer.**

77. $\dfrac{x^{2n} - 1^{2n}}{x^n - 1^n} = x^n + 1^n$ **78.** $\dfrac{x^{2n} - n^2}{x^n - n} = x^n + n$

79. *Think About It*   How do you determine whether a rational expression is in simplest form?

80. *Think About It*   Is the following statement true for all nonzero real numbers a and b? Explain.

$$\frac{ax - b}{b - ax} = -1$$

P.5 | The Cartesian Plane

The Cartesian Plane

Just as you can represent real numbers by points on a real number line, you can represent ordered pairs of real numbers by points in a plane called the **rectangular coordinate system,** or the **Cartesian plane,** after the French mathematician René Descartes (1596–1650).

The Cartesian plane is formed by using two real number lines intersecting at right angles, as shown in Figure P.10. The horizontal real number line is usually called the **x-axis,** and the vertical real number line is usually called the **y-axis.** The point of intersection of these two axes is the **origin,** and the two axes divide the plane into four parts called **quadrants.**

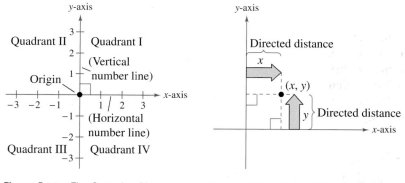

Figure P.10   The Cartesian Plane

Figure P.11   Ordered Pair (x, y)

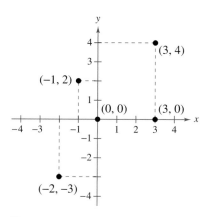
Each point in the plane corresponds to an **ordered pair** (x, y) of real numbers x and y, called **coordinates** of the point. The **x-coordinate** represents the directed distance from the y-axis to the point, and the **y-coordinate** represents the directed distance from the x-axis to the point, as shown in Figure P.11.

$$\underset{\substack{\text{Directed distance} \\ \text{from } y\text{-axis}}}{\qquad} (x, y) \underset{\substack{\text{Directed distance} \\ \text{from } x\text{-axis}}}{\qquad}$$

The notation (x, y) denotes both a point in the plane and an open interval on the real number line. The context will tell you which meaning is intended.

Example 1   Plotting Points in the Cartesian Plane

Plot the points $(-1, 2)$, $(3, 4)$, $(0, 0)$, $(3, 0)$, and $(-2, -3)$.

Solution

To plot the point $(-1, 2)$, imagine a vertical line through -1 on the x-axis and a horizontal line through 2 on the y-axis. The intersection of these two lines is the point $(-1, 2)$. This point is one unit to the left of the y-axis and two units up from the x-axis. The other four points can be plotted in a similar way (see Figure P.12).

✓ *Checkpoint* Now try Exercise 3.

Figure P.12

The beauty of a rectangular coordinate system is that it enables you to see relationships between two variables. It would be difficult to overestimate the importance of Descartes's introduction of coordinates to the plane. Today, his ideas are in common use in virtually every scientific and business-related field.

In the next example, data is represented graphically by points plotted on a rectangular coordinate system. This type of graph is called a **scatter plot.**

Example 2   Sketching a Scatter Plot

From 1996 through 2001, the amount A (in millions of dollars) spent on archery equipment in the United States is shown in the table, where t represents the year. Sketch a scatter plot of the data by hand.   (Source: National Sporting Goods Association)

Year, t	Amount, A
1996	276
1997	270
1998	255
1999	262
2000	254
2001	262

Solution

Before you sketch the scatter plot, it is helpful to represent each pair of values by an ordered pair (t, A), as follows.

(1996, 276), (1997, 270), (1998, 255), (1999, 262), (2000, 254), (2001, 262)

To sketch a scatter plot of the data shown in the table, first draw a vertical axis to represent the amount (in millions of dollars) and a horizontal axis to represent the year. Then plot the resulting points, as shown in Figure P.13. Note that the break in the t-axis indicates that the numbers between 0 and 1996 have been omitted.

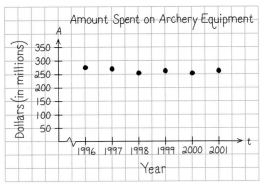

Figure P.13

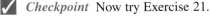 *Checkpoint* Now try Exercise 21.

TECHNOLOGY SUPPORT

For instructions on how to use the *list editor*, see Appendix A; for specific keystrokes, go to the text website at *college.hmco.com*.

TECHNOLOGY TIP   You can use a graphing utility to graph the scatter plot in Example 2. First, enter the data into the graphing utility's *list editor* as shown in Figure P.14. Then use the *statistical plotting* feature to set up the scatter plot, as shown in Figure P.15. Finally, display the scatter plot (use a viewing window in which $1995 \leq x \leq 2002$ and $0 \leq y \leq 300$) as shown in Figure P.16.

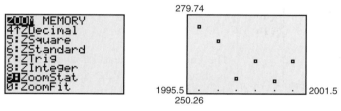

Figure P.14   Figure P.15   Figure P.16

Some graphing utilities have a *ZoomStat* feature, as shown in Figure P.17. This feature automatically selects an appropriate viewing window that displays all the data in the list editor, as shown in Figure P.18.

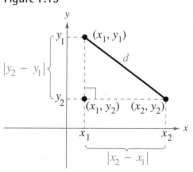

Figure P.17   Figure P.18

The Distance Formula

Recall from the Pythagorean Theorem that, for a right triangle with hypotenuse of length c and sides of lengths a and b, you have $a^2 + b^2 = c^2$ as shown in Figure P.19. (The converse is also true. That is, if $a^2 + b^2 = c^2$, then the triangle is a right triangle.)

Suppose you want to determine the distance d between two points (x_1, y_1) and (x_2, y_2) in the plane. With these two points, a right triangle can be formed, as shown in Figure P.20. The length of the vertical side of the triangle is $|y_2 - y_1|$, and the length of the horizontal side is $|x_2 - x_1|$. By the Pythagorean Theorem,

$$d^2 = |x_2 - x_1|^2 + |y_2 - y_1|^2$$
$$d = \sqrt{|x_2 - x_1|^2 + |y_2 - y_1|^2}$$
$$d = \sqrt{(x_2 - x_1)^2 + (y_2 - y_1)^2}.$$

This result is called the **Distance Formula.**

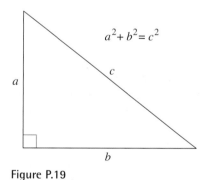

Figure P.19

The Distance Formula

The distance d between the points (x_1, y_1) and (x_2, y_2) in the plane is

$$d = \sqrt{(x_2 - x_1)^2 + (y_2 - y_1)^2}.$$

Figure P.20

Example 3   Finding a Distance

Find the distance between the points $(-2, 1)$ and $(3, 4)$.

Algebraic Solution

Let $(x_1, y_1) = (-2, 1)$ and $(x_2, y_2) = (3, 4)$. Then apply the Distance Formula as follows.

$$d = \sqrt{(x_2 - x_1)^2 + (y_2 - y_1)^2} \qquad \text{Distance Formula}$$

$$= \sqrt{[3 - (-2)]^2 + (4 - 1)^2} \qquad \begin{array}{l}\text{Substitute for}\\ x_1, y_1, x_2, \text{ and } y_2.\end{array}$$

$$= \sqrt{(5)^2 + (3)^2} \qquad \text{Simplify.}$$

$$= \sqrt{34} \approx 5.83 \qquad \text{Simplify.}$$

So, the distance between the points is about 5.83 units.

You can use the Pythagorean Theorem to check that the distance is correct.

$$d^2 \overset{?}{=} 3^2 + 5^2 \qquad \text{Pythagorean Theorem}$$

$$\left(\sqrt{34}\right)^2 \overset{?}{=} 3^2 + 5^2 \qquad \text{Substitute for } d.$$

$$34 = 34 \qquad \text{Distance checks. } \checkmark$$

 *Checkpoint* Now try Exercise 23.

Graphical Solution

Use centimeter graph paper to plot the points $A(-2, 1)$ and $B(3, 4)$. Carefully sketch the line segment from A to B. Then use a centimeter ruler to measure the length of the segment.

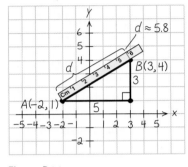

Figure P.21

The line segment measures about 5.8 centimeters, as shown in Figure P.21. So, the distance between the points is about 5.8 units.

Example 4   Verifying a Right Triangle

Show that the points $(2, 1)$, $(4, 0)$, and $(5, 7)$ are the vertices of a right triangle.

Solution

The three points are plotted in Figure P.22. Using the Distance Formula, you can find the lengths of the three sides as follows.

$$d_1 = \sqrt{(5 - 2)^2 + (7 - 1)^2} = \sqrt{9 + 36} = \sqrt{45}$$

$$d_2 = \sqrt{(4 - 2)^2 + (0 - 1)^2} = \sqrt{4 + 1} = \sqrt{5}$$

$$d_3 = \sqrt{(5 - 4)^2 + (7 - 0)^2} = \sqrt{1 + 49} = \sqrt{50}$$

Because $(d_1)^2 + (d_2)^2 = 45 + 5 = 50 = (d_3)^2$, you can conclude that the triangle must be a right triangle.

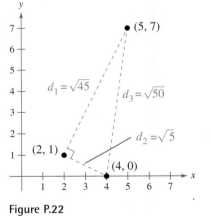

Figure P.22

 *Checkpoint* Now try Exercise 37.

The Midpoint Formula

To find the **midpoint** of the line segment that joins two points in a coordinate plane, find the average values of the respective coordinates of the two endpoints using the **Midpoint Formula.** See Appendix B for a proof of the Midpoint Formula.

> **The Midpoint Formula**
>
> The midpoint of the line segment joining the points (x_1, y_1) and (x_2, y_2) is given by the Midpoint Formula
>
> $$\text{Midpoint} = \left(\frac{x_1 + x_2}{2}, \frac{y_1 + y_2}{2}\right).$$

Example 5 Finding a Line Segment's Midpoint

Find the midpoint of the line segment joining the points $(-5, -3)$ and $(9, 3)$.

Solution

Let $(x_1, y_1) = (-5, -3)$ and $(x_2, y_2) = (9, 3)$.

$$\text{Midpoint} = \left(\frac{x_1 + x_2}{2}, \frac{y_1 + y_2}{2}\right) \qquad \text{Midpoint Formula}$$

$$= \left(\frac{-5 + 9}{2}, \frac{-3 + 3}{2}\right) \qquad \text{Substitute for } x_1, y_1, x_2, \text{ and } y_2.$$

$$= (2, 0) \qquad \text{Simplify.}$$

The midpoint of the line segment is $(2, 0)$, as shown in Figure P.23.

✓ *Checkpoint* Now try Exercise 43.

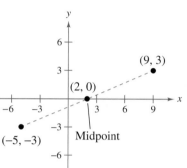

Figure P.23

Example 6 Estimating Annual Sales

The Wm. Wrigley Jr. Company had annual sales of $2.15 billion in 2000 and $2.75 billion in 2002. Without knowing any additional information, what would you estimate the 2001 sales to have been?   (Source: Wm. Wrigley Jr. Company)

Solution

One solution to the problem is to assume that sales followed a *linear* pattern. With this assumption, you can estimate the 2001 sales by finding the midpoint of the line segment connecting the points $(2000, 2.15)$ and $(2002, 2.75)$.

$$\text{Midpoint} = \left(\frac{2000 + 2002}{2}, \frac{2.15 + 2.75}{2}\right)$$

$$= (2001, 2.45)$$

So, you would estimate the 2001 sales to have been about $2.45 billion, as shown in Figure P.24. (The actual 2001 sales were $2.43 billion.)

✓ *Checkpoint* Now try Exercise 51.

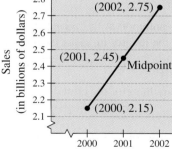

Figure P.24

The Equation of a Circle

The Distance Formula provides a convenient way to define circles. A **circle of radius** r with center at the point (h, k) is shown in Figure P.25. The point (x, y) is on this circle if and only if its distance from the center (h, k) is r. This means that a **circle** in the plane consists of all points (x, y) that are a given positive distance r from a fixed point (h, k). Using the Distance Formula, you can express this relationship by saying that the point (x, y) lies on the circle if and only if

$$\sqrt{(x - h)^2 + (y - k)^2} = r.$$

By squaring each side of this equation, you obtain the **standard form of the equation of a circle.**

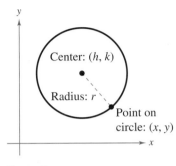

Figure P.25

Standard Form of the Equation of a Circle

The **standard form of the equation of a circle** is

$$(x - h)^2 + (y - k)^2 = r^2.$$

The point (h, k) is the **center** of the circle, and the positive number r is the **radius** of the circle. The standard form of the equation of a circle whose center is the origin, $(h, k) = (0, 0)$, is $x^2 + y^2 = r^2$.

Example 7   Writing the Equation of a Circle

The point $(3, 4)$ lies on a circle whose center is at $(-1, 2)$, as shown in Figure P.26. Write the standard form of the equation of this circle.

Solution

The radius r of the circle is the distance between $(-1, 2)$ and $(3, 4)$.

$$r = \sqrt{[3 - (-1)]^2 + (4 - 2)^2} \qquad \text{Substitute for } x, y, h, \text{ and } k.$$

$$= \sqrt{16 + 4} \qquad \text{Simplify.}$$

$$= \sqrt{20} \qquad \text{Radius.}$$

Using $(h, k) = (-1, 2)$ and $r = \sqrt{20}$, the equation of the circle is

$$(x - h)^2 + (y - k)^2 = r^2 \qquad \text{Equation of circle}$$

$$[x - (-1)]^2 + (y - 2)^2 = \left(\sqrt{20}\right)^2 \qquad \text{Substitute for } h, k, \text{ and } r.$$

$$(x + 1)^2 + (y - 2)^2 = 20. \qquad \text{Standard form}$$

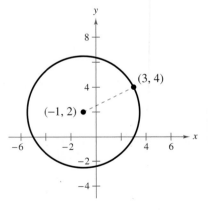

Figure P.26

 Checkpoint Now try Exercise 57.

Example 8   Translating Points in the Plane

The triangle in Figure P.27 has vertices at the points $(-1, 2)$, $(1, -4)$, and $(2, 3)$. Shift the triangle three units to the right and two units upward and find the vertices of the shifted triangle, as shown in Figure P.28.

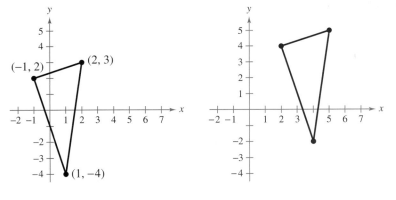

Figure P.27   Figure P.28

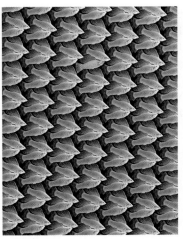

Paul Morrell

Much of computer graphics, including this computer-generated goldfish tessellation, consists of transformations of points in a coordinate plane. One type of transformation, a translation, is illustrated in Example 8. Other types of transformations include reflections, rotations, and stretches.

Solution

To shift the vertices three units to the right, add 3 to each of the x-coordinates. To shift the vertices two units upward, add 2 to each of the y-coordinates.

Original Point	*Translated Point*
$(-1, 2)$	$(-1 + 3, 2 + 2) = (2, 4)$
$(1, -4)$	$(1 + 3, -4 + 2) = (4, -2)$
$(2, 3)$	$(2 + 3, 3 + 2) = (5, 5)$

Plotting the translated points and sketching the line segments between them produces the shifted triangle shown in Figure P.28.

✓ *Checkpoint*   Now try Exercise 69.

Example 8 shows how to translate points in a coordinate plane. The following transformed points are related to the original points as follows.

Original Point	*Transformed Point*	
(x, y)	$(-x, y)$	$(-x, y)$ is a reflection of the original point in the y-axis.
(x, y)	$(x, -y)$	$(x, -y)$ is a reflection of the original point in the x-axis.
(x, y)	$(-x, -y)$	$(-x, -y)$ is a reflection of the original point through the origin.

The figure provided with Example 8 was not really essential to the solution. Nevertheless, it is strongly recommended that you develop the habit of including sketches with your solutions, even if they are not required, because they serve as useful problem-solving tools.

P.5 Exercises

Vocabulary Check

1. Match each term with its definition.

   (a) x-axis
   (b) y-axis
   (c) origin
   (d) quadrants
   (e) x-coordinate
   (f) y-coordinate

   (i) point of intersection of vertical axis and horizontal axis
   (ii) directed distance from the x-axis
   (iii) horizontal real number line
   (iv) four regions of the coordinate plane
   (v) directed distance from the y-axis
   (vi) vertical real number line

In Exercises 2–5, fill in the blanks.

2. An ordered pair of real numbers can be represented in a plane called the rectangular coordinate system or the _____ plane.

3. The _____ is a result derived from the Pythagorean Theorem.

4. Finding the average values of the respective coordinates of the two endpoints of a line segment in a coordinate plane is also known as using the _____ .

5. The standard form of the equation of a circle is _____ , where the point (h, k) is the _____ of the circle and the positive number r is the _____ of the circle.

In Exercises 1 and 2, approximate the coordinates of the points.

1.   2.

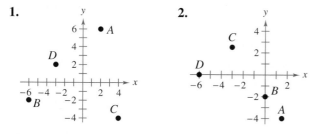

In Exercises 3–6, plot the points in the Cartesian plane.

3. $(-4, 2), (-3, -6), (0, 5), (1, -4)$
4. $(4, -2), (0, 0), (-4, 0), (-5, -5)$
5. $(3, 8), (0.5, -1), (5, -6), (-2, -2.5)$
6. $\left(1, -\frac{1}{2}\right), \left(-\frac{3}{4}, 2\right), (3, -3), \left(\frac{3}{2}, \frac{4}{3}\right)$

In Exercises 7–10, find the coordinates of the point.

7. The point is located five units to the left of the y-axis and four units above the x-axis.

8. The point is located three units below the x-axis and two units to the right of the y-axis.

9. The point is located six units below the x-axis and the coordinates of the point are equal.

10. The point is on the x-axis and 10 units to the left of the y-axis.

In Exercises 11–20, determine the quadrant(s) in which (x, y) is located so that the condition(s) is (are) satisfied.

11. $x > 0$ and $y < 0$
12. $x < 0$ and $y < 0$
13. $x = -4$ and $y > 0$
14. $x > 2$ and $y = 3$
15. $y < -5$
16. $x > 4$
17. $x < 0$ and $-y > 0$
18. $-x > 0$ and $y < 0$
19. $xy > 0$
20. $xy < 0$

In Exercises 21 and 22, sketch a scatter plot of the data shown in the table.

21. *Meteorology*   The table shows the lowest temperature on record y (in degrees Fahrenheit) in Duluth, Minnesota, for each month x, where $x = 1$ represents January.   (Source: NOAA)

Month, x	Temperature, y
1	−39
2	−39
3	−29
4	−5
5	17
6	27
7	35
8	32
9	22
10	8
11	−23
12	−34

22. *Number of Stores*   The table shows the number y of Wal-Mart stores for each year x from 1994 through 2001.   (Source: Wal-Mart Stores, Inc.)

Year	Number of stores, y
1994	2759
1995	2943
1996	3054
1997	3406
1998	3599
1999	3985
2000	4189
2001	4414

In Exercises 23–32, find the distance between the points algebraically and verify graphically by using centimeter graph paper and a centimeter ruler.

23. $(6, -3), (6, 5)$ 24. $(1, 4), (8, 4)$

25. $(-3, -1), (2, -1)$ 26. $(-3, -4), (-3, 6)$

27. $(-2, 6), (3, -6)$ 28. $(8, 5), (0, 20)$

29. $\left(\frac{1}{2}, \frac{4}{3}\right), (2, -1)$

30. $\left(-\frac{2}{3}, 3\right), \left(-1, \frac{5}{4}\right)$

31. $(-4.2, 3.1), (-12.5, 4.8)$

32. $(9.5, -2.6), (-3.9, 8.2)$

In Exercises 33–36, (a) find the length of each side of the right triangle and (b) show that these lengths satisfy the Pythagorean Theorem.

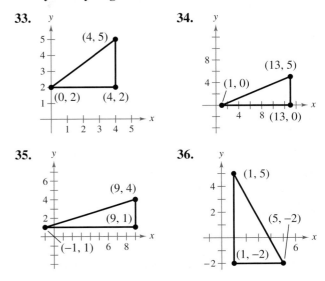

33.

34.

35.

36.

In Exercises 37–40, show that the points form the vertices of the polygon.

37. Right triangle: $(4, 0), (2, 1), (-1, -5)$

38. Isosceles triangle: $(1, -3), (3, 2), (-2, 4)$

39. Parallelogram: $(2, 5), (0, 9), (-2, 0), (0, -4)$

40. Parallelogram: $(0, 1), (3, 7), (4, 4), (1, -2)$

In Exercises 41–50, (a) plot the points, (b) find the distance between the points, and (c) find the midpoint of the line segment joining the points.

41. $(1, 1), (9, 7)$

42. $(1, 12), (6, 0)$

43. $(-4, 10), (4, -5)$

44. $(-7, -4), (2, 8)$

45. $(-1, 2), (5, 4)$

46. $(2, 10), (10, 2)$

47. $\left(\frac{1}{2}, 1\right), \left(-\frac{5}{2}, \frac{4}{3}\right)$

48. $\left(-\frac{1}{3}, -\frac{1}{3}\right), \left(-\frac{1}{6}, -\frac{1}{2}\right)$

49. $(6.2, 5.4), (-3.7, 1.8)$

50. $(-16.8, 12.3), (5.6, 4.9)$

Sales **In Exercises 51 and 52, use the Midpoint Formula to estimate the sales of PETCO Animal Supplies, Inc. and PetsMART, Inc. in 2000. The sales for the two companies in 1998 and 2002 are shown in the tables. Assume that the sales followed a linear pattern.**

51. PETCO

Year	Sales (in millions)
1998	$839.6
2002	$1480.0

(Source: PETCO Animal Supplies, Inc.)

52. PetsMART

Year	Sales (in millions)
1998	$2109.3
2002	$2750.0

(Source: PetsMART, Inc.)

53. *Exploration*   A line segment has (x_1, y_1) as one endpoint and (x_m, y_m) as its midpoint. Find the other endpoint (x_2, y_2) of the line segment in terms of $x_1, y_1, x_m,$ and $y_m.$ Use the result to find the coordinates of the endpoint of a line segment if the coordinates of the other endpoint and midpoint are, respectively,

(a) $(1, -2), (4, -1)$

(b) $(-5, 11), (2, 4)$

54. *Exploration*   Use the Midpoint Formula three times to find the three points that divide the line segment joining (x_1, y_1) and (x_2, y_2) into four parts. Use the result to find the points that divide the line segment joining the given points into four equal parts.

(a) $(1, -2), (4, -1)$

(b) $(-2, -3), (0, 0)$

In Exercises 55–62, write the standard form of the equation of the specified circle.

55. Center: $(0, 0)$; radius: 3

56. Center: $(0, 0)$; radius: 6

57. Center: $(2, -1)$; radius: 4

58. Center: $\left(0, \frac{1}{3}\right)$; radius: $\frac{1}{3}$

59. Center: $(-1, 2)$; solution point: $(0, 0)$

60. Center: $(3, -2)$; solution point: $(-1, 1)$

61. Endpoints of a diameter: $(0, 0), (6, 8)$

62. Endpoints of a diameter: $(-4, -1), (4, 1)$

In Exercises 63–68, find the center and radius, and sketch the circle.

63. $x^2 + y^2 = 25$

64. $x^2 + y^2 = 16$

65. $(x - 1)^2 + (y + 3)^2 = 4$

66. $x^2 + (y - 1)^2 = 49$

67. $\left(x - \frac{1}{2}\right)^2 + \left(y - \frac{1}{2}\right)^2 = \frac{9}{4}$

68. $\left(x - \frac{2}{3}\right)^2 + \left(y + \frac{1}{4}\right)^2 = \frac{25}{9}$

In Exercises 69–72, the polygon is shifted to a new position in the plane. Find the coordinates of the vertices of the polygon in the new position.

69.

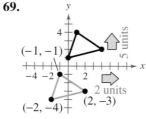

70.

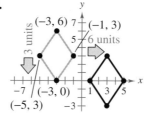

71. Original coordinates of vertices:

$(0, 2), (3, 5), (5, 2), (2, -1)$

Shift: three units upward, one unit to the left

72. Original coordinates of vertices:

$(1, -1), (3, 2), (1, -2)$

Shift: two units downward, three units to the left

Analyzing Data   **In Exercises 73 and 74, refer to the scatter plot, which shows the mathematics entrance test scores *x* and the final examination scores *y* in an algebra course for a sample of 10 students.**

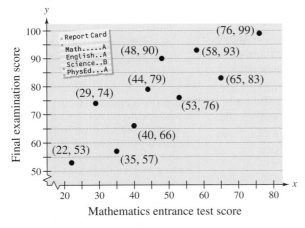

73. Find the entrance exam score of any student with a final exam score in the 80s.

74. Does a higher entrance exam score necessarily imply a higher final exam score? Explain.

75. *Rock and Roll Hall of Fame*   The graph shows the numbers of recording artists inducted to the Rock and Roll Hall of Fame from 1986 to 2003.

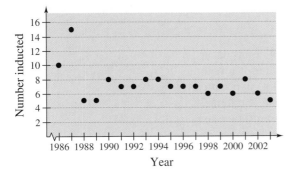

(a) Describe any trends in the data. From these trends, predict the number of artists that will be elected in 2005.

(b) Why do you think the numbers elected in 1986 and 1987 were greater than in other years?

76. *Flying Distance*   A jet plane flies from Naples, Italy in a straight line to Rome, Italy, which is 120 kilometers west and 150 kilometers north of Naples. How far does the plane fly?

77. *Sports*   In a football game, a quarterback throws a pass from the 15-yard line, 10 yards from the sideline as shown in the figure. The pass is caught on the 40-yard line, 45 yards from the same sideline. How long is the pass?

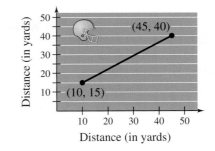

78. *Make a Conjecture*   Plot the points $(2, 1)$, $(-3, 5)$, and $(7, -3)$ on a rectangular coordinate system. Then change the sign of the indicated coordinate(s) of each point and plot the three new points on the same rectangular coordinate system. Make a conjecture about the location of a point when each of the following occurs.

(a) The sign of the *x*-coordinate is changed.

(b) The sign of the *y*-coordinate is changed.

(c) The signs of both the *x*- and *y*-coordinates are changed.

Synthesis

True or False?   **In Exercises 79–81, determine whether the statement is true or false. Justify your answer.**

79. In order to divide a line segment into 16 equal parts, you would have to use the Midpoint Formula 16 times.

80. The points $(-8, 4)$, $(2, 11)$, and $(-5, 1)$ represent the vertices of an isosceles triangle.

81. If four points represent the vertices of a polygon, and the four sides are equal, then the polygon must be a square.

82. *Think About It*   What is the *y*-coordinate of any point on the *x*-axis? What is the *x*-coordinate of any point on the *y*-axis?

83. *Think About It*   When plotting points on the rectangular coordinate system, is it true that the scales on the *x*- and *y*-axes must be the same? Explain.

P.6   Exploring Data: Representing Data Graphically

Line Plots

Statistics is the branch of mathematics that studies techniques for collecting, organizing, and interpreting data. In this section, you will study several ways to organize data. The first is a **line plot,** which uses a portion of a real number line to order numbers. Line plots are especially useful for ordering small sets of numbers (about 50 or less) by hand.

Many statistical measures can be obtained from a line plot. Two such measures are the *frequency* and *range* of the data. The **frequency** measures the number of times a value occurs in a data set. The **range** is the difference between the greatest and least data values. For example, consider the data values

20, 21, 21, 25, 32.

The frequency of 21 in the data set is 2 because 21 occurs twice. The range is 12 because the difference between the greatest and least data values is $32 - 20 = 12$.

Example 1   Constructing a Line Plot

Use a line plot to organize the following test scores. Which score occurs with the greatest frequency? What is the range of scores?

93, 70, 76, 67, 86, 93, 82, 78, 83, 86, 64, 78, 76, 66, 83
83, 96, 74, 69, 76, 64, 74, 79, 76, 88, 76, 81, 82, 74, 70

Solution

Begin by scanning the data to find the smallest and largest numbers. For this data, the smallest number is 64 and the largest is 96. Next, draw a portion of a real number line that includes the interval $[64, 96]$. To create the line plot, start with the first number, 93, and enter an $\times$ above 93 on the number line. Continue recording $\times$'s for each number in the list until you obtain the line plot shown in Figure P.29. From the line plot, you can see that 76 occurs with the greatest frequency. Because the range is the difference between the greatest and least data values, the range of scores is $96 - 64 = 32$.

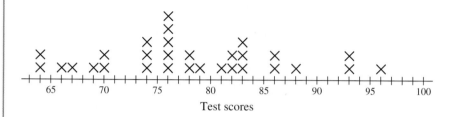

Test scores

Figure P.29

 Checkpoint Now try Exercise 1.

Histograms and Frequency Distributions

When you want to organize large sets of data, it is useful to group the data into intervals and plot the frequency of the data in each interval. A **frequency distribution** can be used to construct a **histogram.** A histogram uses a portion of a real number line as its horizontal axis. The bars of a histogram are not separated by spaces.

Example 2   Constructing a Histogram

The table at the right shows the percent of the resident population of each state and the District of Columbia that was at least 65 years old in 2000. Construct a frequency distribution and a histogram for the data.   (Source: U.S. Census Bureau)

Solution

To begin constructing a frequency distribution, you must first decide on the number of intervals. There are several ways to group this data. However, because the smallest number is 5.7 and the largest is 17.6, it seems that seven intervals would be appropriate. The first would be the interval $[5, 7)$, the second would be $[7, 9)$, and so on. By tallying the data into the seven intervals, you obtain the frequency distribution shown below. You can construct the histogram by drawing a vertical axis to represent the number of states and a horizontal axis to represent the percent of the population 65 and older. Then, for each interval, draw a vertical bar whose height is the total tally, as shown in Figure P.30.

Interval	Tally
$[5, 7)$	I
$[7, 9)$	I
$[9, 11)$	IIII
$[11, 13)$	HHT HHT HHT HHT II
$[13, 15)$	HHT HHT HHT HHT
$[15, 17)$	II
$[17, 19)$	I

AK	5.7	MT	13.4
AL	13.0	NC	12.0
AR	14.0	ND	14.7
AZ	13.0	NE	13.6
CA	10.6	NH	12.0
CO	9.7	NJ	13.2
CT	13.8	NM	11.7
DC	12.2	NV	11.0
DE	13.0	NY	12.9
FL	17.6	OH	13.3
GA	9.6	OK	13.2
HI	13.3	OR	12.8
IA	14.9	PA	15.6
ID	11.3	RI	14.5
IL	12.1	SC	12.1
IN	12.4	SD	14.3
KS	13.3	TN	12.4
KY	12.5	TX	9.9
LA	11.6	UT	8.5
MA	13.5	VA	11.2
MD	11.3	VT	12.7
ME	14.4	WA	11.2
MI	12.3	WI	13.1
MN	12.1	WV	15.3
MO	13.5	WY	11.7
MS	12.1		

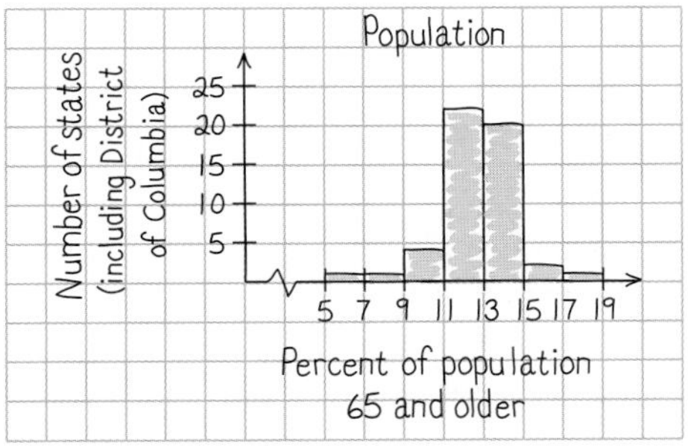

Figure P.30

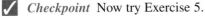

✓ *Checkpoint* Now try Exercise 5.

Example 3   Constructing a Histogram

A company has 48 sales representatives who sold the following numbers of units during the first quarter of 2005. Construct a frequency distribution for this data.

107	162	184	170	177	102	145	141
105	193	167	149	195	127	193	191
150	153	164	167	171	163	141	129
109	171	150	138	100	164	147	153
171	163	118	142	107	144	100	132
153	107	124	162	192	134	187	177

Interval	Tally
100–109	✚✚✚ |||
110–119		
120–129	|||
130–139	|||
140–149	✚✚✚ ||
150–159	✚✚✚
160–169	✚✚✚ |||
170–179	✚✚✚	
180–189	||
190–199	✚✚✚

Solution

To begin constructing a frequency distribution, you must first decide on the number of intervals. There are several ways to group this data. However, because the smallest number is 100 and the largest is 195, it seems that 10 intervals would be appropriate. The first interval would be 100–109, the second would be 110–119, and so on. By tallying the data into the 10 intervals, you obtain the distribution shown at the right above. A histogram for the distribution is shown in Figure P.31.

✓ *Checkpoint* Now try Exercise 6.

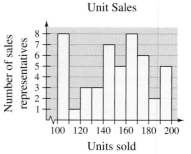

Figure P.31

Bar Graphs

A **bar graph** is similar to a histogram, except that the bars can be either horizontal or vertical and the labels of the bars are not necessarily numbers. Another difference between a bar graph and a histogram is that the bars in a bar graph are usually separated by spaces.

Example 4   Constructing a Bar Graph

The data below shows the monthly normal precipitation (in inches) in Houston, Texas. Construct a bar graph for this data. What can you conclude? (Source: National Climatic Data Center)

January	3.7	February	3.0	March	3.4
April	3.6	May	5.2	June	5.4
July	3.2	August	3.8	September	4.3
October	4.5	November	4.2	December	3.7

Solution

To create a bar graph, begin by drawing a vertical axis to represent the precipitation and a horizontal axis to represent the month. The bar graph is shown in Figure P.32. From the graph, you can see that Houston receives a fairly consistent amount of rain throughout the year—the driest month tends to be February and the wettest month tends to be June.

✓ *Checkpoint* Now try Exercise 9.

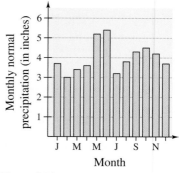

Figure P.32

12. Approximate the difference in the price of one pound of chicken breast from the highest price shown in the graph to the price in 1994.

Advertising **In Exercises 13 and 14, use the line graph, which shows the cost of a 30-second television spot (in thousands of dollars) during the Super Bowl from 1995 to 2002.** (Source: The Associated Press)

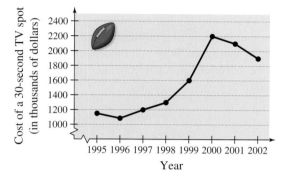

13. Approximate the percent increase in the cost of a 30-second spot from Super Bowl XXX in 1996 to Super Bowl XXXVI in 2002.

14. Estimate the increase or decrease in the cost of a 30-second spot from (a) Super Bowl XXIV in 1995 to Super Bowl XXXIII in 1999, and (b) Super Bowl XXXIV in 2000 to Super Bowl XXXVI in 2002.

15. *Oil Imports* The table shows the amount of crude oil imported into the United States (in millions of barrels) for the years 1995 through 2001. Construct a line graph for the data and state what information the graph reveals. (Source: U.S. Energy Information Administration)

Year	Imports
1995	2693
1996	2740
1997	3002
1998	3178
1999	3187
2000	3260
2001	3405

16. *Entertainment* The table shows the percent of U.S. households owning televisions that owned more than one television set for selected years from 1960 to 2000. Construct a line graph for the data and state what information the graph reveals. (Source: Nielsen Media Research)

Year	Percent
1960	12
1965	22
1970	35
1975	43
1980	50
1985	57
1990	65
1995	71
2000	76

17. *Government* The table shows the number of U.S. representatives from the state of New York for selected years from 1930 to 2000. Use a graphing utility to construct a line graph for the data. (Source: U.S. Census Bureau)

Year	Representatives
1930	45
1940	45
1950	43
1960	41
1970	39
1980	34
1990	31
2000	29

18. *Personal Savings*   The table shows the amount (in billions of dollars) of personal savings in the United States from 1995 to 2001. Use a graphing utility to construct a line graph for the data.   (Source: Bureau of Economic Analysis)

Year	Personal savings
1995	179.8
1996	158.5
1997	121.0
1998	265.4
1999	160.9
2000	201.5
2001	169.7

19. *Travel*   The table shows the places of origin and numbers of travelers (in millions) to the United States in 2000. Choose an appropriate display to organize the data.   (Source: U.S. Department of Commerce)

Place of origin	Travelers
Canada	14.6
Caribbean	1.3
Europe	11.6
Far East	7.6
Mexico	10.3
South America	2.9

20. *Education*   The table shows the number of college degrees (in thousands) awarded in the United States from 1996 to 2002. Choose an appropriate display to organize the data.   (Source: U.S. Department of Education)

Year	Degrees
1996	1692
1997	1717
1998	1739
1999	1763
2000	1820
2001	1766
2002	1786

Synthesis

21. *Writing*   Describe the differences between a bar graph and a histogram.

22. *Think About It*   How can you decide which type of graph to use when you are organizing data?

23. *Graphical Interpretation*   The graphs shown below represent the same data points. Which of the two graphs is misleading, and why? Discuss other ways in which graphs can be misleading. Try to find another example of a misleading graph in a newspaper or magazine. Why is it misleading? Why would it be beneficial for someone to use a misleading graph?

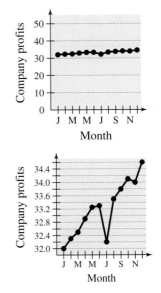

P Chapter Summary

What did you learn?

	Review Exercises
Section P.1	
☐ Represent and classify real numbers.	1, 2
☐ Order real numbers and use inequalities.	3–6
☐ Find the absolute values of real numbers and the distance between two real numbers.	7–12
☐ Evaluate algebraic expressions.	13–16
☐ Use the basic rules and properties of algebra.	17–26
Section P.2	
☐ Use properties of exponents.	27–30
☐ Use scientific notation to represent real numbers.	31–34
☐ Use properties of radicals.	35, 36
☐ Simplify and combine radicals.	37–50
☐ Rationalize denominators and numerators.	51–54
☐ Use properties of rational exponents.	55–58
Section P.3	
☐ Write polynomials in standard form.	59, 60
☐ Add, subtract, and multiply polynomials.	61–68
☐ Use special products to multiply polynomials.	69–76
☐ Remove common factors from polynomials.	77–84
☐ Factor special polynomial forms.	85–88
☐ Factor trinomials as the product of two binomials.	89–92
☐ Factor by grouping.	93–96
Section P.4	
☐ Find domains of algebraic expressions.	97–100
☐ Simplify rational expressions.	101–104
☐ Add, subtract, multiply, and divide rational expressions.	105–112
☐ Simplify complex fractions.	113, 114
Section P.5	
☐ Plot points in the Cartesian plane and sketch scatter plots.	115–122
☐ Use the Distance Formula to find the distance between two points.	123, 124
☐ Use the Midpoint Formula to find the midpoint of a line segment.	125, 126
☐ Find the equation of a circle.	127, 128
☐ Translate points in the plane.	129, 130
Section P.6	
☐ Use line plots to order and analyze data.	131
☐ Use histograms to represent frequency distributions.	132
☐ Use bar graphs and line graphs to represent and analyze data.	133, 134

P Review Exercises

P.1 In Exercises 1 and 2, determine which numbers are (a) natural numbers, (b) whole numbers, (c) integers, (d) rational numbers, and (e) irrational numbers.

1. $\left\{11, -14, -\frac{8}{9}, \frac{5}{2}, \sqrt{6}, 0.4\right\}$

2. $\left\{\sqrt{15}, -22, -\frac{10}{3}, 0, 5.2, \frac{3}{7}\right\}$

In Exercises 3 and 4, use a calculator to find the decimal form of each rational number. If it is a nonterminating decimal, write the repeating pattern. Then plot the numbers on a real number line and place the correct inequality symbol ($<$ or $>$) between them.

3. (a) $\frac{5}{6}$ (b) $\frac{7}{8}$ 4. (a) $\frac{1}{3}$ (b) $\frac{9}{25}$

In Exercises 5 and 6, verbally describe the subset of real numbers represented by the inequality. Then sketch the subset on the real number line.

5. $x \leq 7$ 6. $x > 1$

In Exercises 7 and 8, find the distance between a and b.

7. $a = -74$, $b = 48$ 8. $a = -123$, $b = -9$

In Exercises 9–12, use absolute value notation to describe the situation.

9. The distance between x and 8 is at least 3.

10. The distance between x and 25 is no more than 10.

11. The distance between y and -30 is less than 5.

12. The distance between y and -16 is greater than 8.

In Exercises 13–16, evaluate the expression for each value of x. (If not possible, state the reason.)

Expression	*Values*	
13. $10x - 3$	(a) $x = -1$	(b) $x = 3$
14. $x^2 - 11x + 24$	(a) $x = -2$	(b) $x = 2$
15. $-2x^2 - x + 3$	(a) $x = 3$	(b) $x = -3$
16. $\dfrac{4x}{x - 1}$	(a) $x = -1$	(b) $x = 1$

In Exercises 17–20, identify the rule of algebra illustrated by the statement.

17. $2x + (3x - 10) = (2x + 3x) - 10$

18. $\dfrac{2}{y + 4} \cdot \dfrac{y + 4}{2} = 1$, $y \neq -4$

19. $(t + 4)(2t) = (2t)(t + 4)$

20. $0 + (a - 5) = a - 5$

In Exercises 21–26, perform the operations. (Write fractional answers in simplest form.)

21. $\frac{2}{3} + \frac{8}{9}$ 22. $\frac{3}{4} - \frac{1}{6} + \frac{1}{8}$

23. $\frac{3}{16} \div \frac{9}{2}$ 24. $\frac{5}{8} \cdot \frac{2}{3}$

25. $\frac{x}{5} + \frac{7x}{12}$ 26. $\frac{9}{x} \div \frac{1}{6}$

P.2 In Exercises 27–30, simplify each expression.

27. (a) $(-2z)^3$ (b) $(a^2b^4)(3ab^{-2})$

28. (a) $\dfrac{(8y)^0}{y^2}$ (b) $\dfrac{40(b - 3)^5}{75(b - 3)^2}$

29. (a) $\dfrac{6^2u^3v^{-3}}{12u^{-2}v}$ (b) $\dfrac{3^{-4}m^{-1}n^{-3}}{9^{-2}mn^{-3}}$

30. (a) $(x + y^{-1})^{-1}$ (b) $\left(\dfrac{x^{-3}}{y}\right)\left(\dfrac{x}{y}\right)^{-1}$

In Exercises 31 and 32, write the number in scientific notation.

31. *Revenues of Target Corporation in 2002:*
    $43,800,000,000   (Source: Target Corporation)

32. *Number of meters in one foot:* 0.3048

In Exercises 33 and 34, write the number in decimal notation.

33. *Distance between the Sun and Jupiter:*
    4.836×10^8 miles

34. *Ratio of day to year:* 2.74×10^{-3}

In Exercises 35 and 36, use the properties of radicals to simplify the expression.

35. $\left(\sqrt[4]{78}\right)^4$ 36. $\sqrt[3]{9} \cdot \sqrt[3]{3}$

In Exercises 37–42, simplify by removing all possible factors from the radical.

37. $\sqrt{4x^4}$ 38. $\sqrt[5]{64x^6}$

39. $\sqrt{\frac{81}{144}}$ 40. $\sqrt[3]{\frac{125}{216}}$

41. $\sqrt[3]{\dfrac{2x^3}{27}}$ **42.** $\sqrt{\dfrac{75x^2}{y^4}}$

In Exercises 43–48, simplify the expression.

43. $\sqrt{50} - \sqrt{18}$ **44.** $3\sqrt{32} + 4\sqrt{98}$

45. $8\sqrt{3x} - 5\sqrt{3x}$ **46.** $-11\sqrt{36y} - 6\sqrt{y}$

47. $\sqrt{8x^3} + \sqrt{2x}$ **48.** $3\sqrt{14x^2} - \sqrt{56x^2}$

Strength of a Wooden Beam **In Exercises 49 and 50, use the figure, which shows the rectangular cross section of a wooden beam cut from a log of diameter 24 inches.**

49. Find the area of the cross section when $w = 12\sqrt{2}$ inches and $h = \sqrt{24^2 - \left(12\sqrt{2}\right)^2}$ inches. What is the shape of the cross section? Explain.

50. The rectangular cross section will have a maximum strength when $w = 8\sqrt{3}$ inches and $h = \sqrt{24^2 - \left(8\sqrt{3}\right)^2}$ inches. Find the area of the cross section.

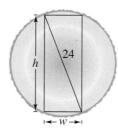

In Exercises 51 and 52, rationalize the denominator of the expression. Then simplify your answer.

51. $\dfrac{1}{2 - \sqrt{3}}$ **52.** $\dfrac{1}{\sqrt{x} - 1}$

∫ **In Exercises 53 and 54, rationalize the numerator of the expression. Then simplify your answer.**

53. $\dfrac{\sqrt{20}}{4}$ **54.** $\dfrac{\sqrt{2} - \sqrt{11}}{3}$

In Exercises 55–58, simplify the expression.

55. $81^{3/2}$ **56.** $64^{-2/3}$

57. $\left(-3x^{2/5}\right)\left(-2x^{1/2}\right)$ **58.** $(x - 1)^{1/3}(x - 1)^{-1/4}$

P.3 In Exercises 59 and 60, write the polynomial in standard form. Then identify the degree and leading coefficient of the polynomial.

59. $15x^2 - 2x^5 + 3x^3 + 5 - x^4$

60. $-2x^4 + x^2 - 10 - x + x^3$

In Exercises 61–68, perform the operations and write the result in standard form.

61. $-(3x^2 + 2x) + (1 - 5x)$

62. $8y - [2y^2 - (3y - 8)]$

63. $(2x^3 - 5x^2 + 10x - 7) + (4x^2 - 7x - 2)$

64. $(6x^4 - 4x^3 - x + 3 - 20x^2) - (16 + 9x^4 - 11x^2)$

65. $(x^2 - 2x + 1)(x^3 - 1)$

66. $(x^3 - 3x)(2x^2 + 3x + 5)$

67. $(y^2 - y)(y^2 + 1)(y^2 + y + 1)$

68. $\left(x - \dfrac{1}{x}\right)(x + 2)$

In Exercises 69–74, find the special product.

69. $(x + 8)(x - 8)$ **70.** $(7x + 4)(7x - 4)$

71. $(x - 4)^3$ **72.** $(2x - 1)^3$

73. $(m - 4 + n)(m - 4 - n)$

74. $(x - y - 6)(x - y + 6)$

75. *Geometry* Use the area model to write two different expressions for the area. Then equate the two expressions and name the algebraic property that is illustrated.

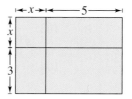

76. *Compound Interest* After 2 years, an investment of $2500 compounded annually at an interest rate r will yield an amount of $2500(1 + r)^2$. Write this polynomial in standard form.

In Exercises 77–82, factor out the common factor.

77. $7x + 35$ **78.** $10x - 2$

79. $x^3 - x$ **80.** $x(x - 3) + 4(x - 3)$

81. $2x^3 + 18x^2 - 4x$ **82.** $-6x^4 - 3x^3 + 12x$

83. *Geometry* The surface area of a right circular cylinder is $S = 2\pi r^2 + 2\pi rh$.

(a) Draw a right circular cylinder of radius r and height h. Use the figure to explain how the surface area formula is obtained.

(b) Factor the expression for surface area.

84. *Business* The revenue for selling x units of a product at a price of p dollars per unit is $R = xp$. For a flat panel television the revenue is

$$R = 1600x - 0.50x^2.$$

Factor the expression and determine an expression that gives the price in terms of x.

In Exercises 85–92, factor the expression.

85. $x^2 - 169$

86. $9x^2 - \frac{1}{25}$

87. $x^3 + 216$

88. $64x^3 - 27$

89. $x^2 - 6x - 27$

90. $x^2 - 9x + 14$

91. $2x^2 + 21x + 10$

92. $3x^2 + 14x + 8$

In Exercises 93–96, factor by grouping.

93. $x^3 - 4x^2 - 3x + 12$

94. $x^3 - 6x^2 - x + 6$

95. $2x^2 - x - 15$

96. $6x^2 + x - 12$

P.4 In Exercises 97–100, find the domain of the expression.

97. $-5x^2 - x - 1$

98. $9x^4 + 7, \quad x > 0$

99. $\dfrac{4}{2x - 3}$

100. $\sqrt{x + 12}$

In Exercises 101–104, write the rational expression in simplest form.

101. $\dfrac{4x^2}{4x^3 + 28x}$

102. $\dfrac{6xy}{xy + 2x}$

103. $\dfrac{x^2 - x - 30}{x^2 - 25}$

104. $\dfrac{x^2 - 9x + 18}{8x - 48}$

In Exercises 105–112, perform the operation and simplify your answer.

105. $\dfrac{x^2 - 4}{x^4 - 2x^2 - 8} \cdot \dfrac{x^2 + 2}{x^2}$

106. $\dfrac{2x - 1}{x + 1} \cdot \dfrac{x^2 - 1}{2x^2 - 7x + 3}$

107. $\dfrac{x^2(5x - 6)}{2x + 3} \div \dfrac{5x}{2x + 3}$

108. $\dfrac{4x - 6}{(x - 1)^2} \div \dfrac{2x^2 - 3x}{x^2 + 2x - 3}$

109. $x - 1 + \dfrac{1}{x + 2} + \dfrac{1}{x - 1}$

110. $2x + \dfrac{3}{2(x - 4)} - \dfrac{1}{2(x + 2)}$

111. $\dfrac{1}{x} - \dfrac{x - 1}{x^2 + 1}$

112. $\dfrac{1}{x - 1} + \dfrac{1 - x}{x^2 + x + 1}$

In Exercises 113 and 114, simplify the complex fraction.

113. $\dfrac{\left(\dfrac{1}{x} - \dfrac{1}{y}\right)}{(x^2 - y^2)}$

114. $\dfrac{\left(\dfrac{1}{2x - 3} - \dfrac{1}{2x + 3}\right)}{\left(\dfrac{1}{2x} - \dfrac{1}{2x + 3}\right)}$

P.5 In Exercises 115–118, plot the point in the Cartesian plane and determine the quadrant in which it is located.

115. $(8, -3)$

116. $(-4, -9)$

117. $\left(-\frac{5}{2}, 10\right)$

118. $(-6.5, -0.5)$

In Exercises 119 and 120, determine the quadrant(s) in which (x, y) is located so that the conditions are satisfied.

119. $x > 0$ and $y = -2$

120. $xy = 4$

Patents In Exercises 121 and 122, use the table, which shows the number of patents P (in thousands) issued in the United States from 1994 through 2001. (Source: U.S. Patent and Trademark Office)

Year	Patents, P
1994	113.6
1995	113.8
1996	121.7
1997	124.1
1998	163.1
1999	169.1
2000	176.0
2001	184.0

121. Sketch a scatter plot of the data.

122. What statement can be made about the number of patents issued in the United States?

In Exercises 123 and 124, plot the points and find the distance between the points.

123. $(-3, 8), (1, 5)$ **124.** $(5.6, 0), (0, 8.2)$

In Exercises 125 and 126, plot the points and find the midpoint of the line segment joining the points.

125. $(-12, 5), (4, -7)$

126. $(1.8, 7.4), (-0.6, -14.5)$

In Exercises 127 and 128, write the standard form of the equation of the specified circle.

127. Center: $(3, -1)$; solution point: $(-5, 1)$

128. Endpoints of a diameter: $(-4, 6), (10, -2)$

In Exercises 129 and 130, the polygon is shifted to a new position in the plane. Find the coordinates of the vertices of the polygon in the new position.

129. Original coordinates of vertices:

$(4, 8), (6, 8), (4, 3), (6, 3)$

Shift: three units downward, two units to the left

130. Original coordinates of vertices:

$(0, 1), (3, 3), (0, 5), (-3, 3)$

Shift: five units upward, four units to the right

P.6

131. ***Consumer Awareness*** Use a line plot to organize the following sample of prices (in dollars) of running shoes. Which price occurred with the greatest frequency?

100, 65, 67, 88, 69, 60, 100, 100, 88, 79, 99, 75, 65, 89, 68, 74, 100, 66, 81, 95, 75, 69, 85, 91, 71

132. ***Sports*** The list shows the free-throw percentages for the players in the 2002 WNBA playoffs. Use a frequency distribution and a histogram to organize the data. (Source: WNBA)

82, 50, 60, 100, 67, 71, 100, 50, 50, 17, 100, 100, 70, 71, 75, 88, 100, 83, 40, 86, 75, 50, 50, 73, 60, 93, 100, 67, 100, 80, 50, 80, 70, 88, 88, 100, 73, 69, 94, 90, 84, 36, 75, 100, 100, 68, 71, 68, 87, 88, 50, 50, 100, 91, 71, 100, 50

133. ***Meteorology*** The normal daily maximum and minimum temperatures (in °F) for each month for the city of Chicago are shown in the table. Construct a double bar graph for the data. (Source: National Climatic Data Center)

Month	Max.	Min.
Jan.	29.6	14.3
Feb.	34.7	19.2
Mar.	46.1	28.5
Apr.	58.0	37.6
May	69.9	47.5
Jun.	79.2	57.2
Jul.	83.5	63.2
Aug.	81.2	62.2
Sep.	73.9	53.7
Oct.	62.1	42.1
Nov.	47.1	31.6
Dec.	34.4	20.4

Table for 133

134. ***Travel*** The table shows the numbers (in millions) of automobile trips taken by U.S. residents from 1995 to 2001. Construct a line graph for the data and state what information the graph reveals. (Source: Travel Industry Association of America)

Year	Trips
1995	396.2
1996	400.7
1997	402.7
1998	410.5
1999	387.7
2000	386.3
2001	396.1

Synthesis

True or False? **In Exercises 135 and 136, determine whether the statement is true or false. Justify your answer.**

135. $\dfrac{x^3 - 1}{x - 1} = x^2 + x + 1$ for all values of x.

136. A binomial sum squared is equal to the sum of the terms squared.

Error Analysis **In Exercises 137 and 138, describe the error.**

137. $(2x)^4 = 2x^4$ **138.** $\sqrt{3^2 + 4^2} = 3 + 4$

139. ***Writing*** Explain why $\sqrt{5u} + \sqrt{3u} \neq 2\sqrt{2u}$.

P Chapter Test

Take this test as you would take a test in class. After you are finished, check your work against the answers in the back of the book.

1. Place the correct symbol ($<$ or $>$) between $-\frac{10}{3}$ and $-|-4|$.

2. Find the distance between the real numbers -17 and 39.

3. Identify the rule of algebra illustrated by $(5 - x) + 0 = 5 - x$.

In Exercises 4 and 5, evaluate each expression without using a calculator.

4. (a) $27\left(-\dfrac{2}{3}\right)$ (b) $\dfrac{5}{18} \div \dfrac{15}{8}$ (c) $\left(-\dfrac{2}{7}\right)^3$ (d) $\left(\dfrac{3^2}{2}\right)^{-3}$

5. (a) $\sqrt{5} \cdot \sqrt{125}$ (b) $\dfrac{\sqrt{72}}{\sqrt{2}}$ (c) $\dfrac{5.4 \times 10^8}{3 \times 10^3}$ (d) $(3 \times 10^4)^3$

In Exercises 6 and 7, simplify each expression.

6. (a) $3z^2(2z^3)^2$ (b) $(u - 2)^{-4}(u - 2)^{-3}$ (c) $\left(\dfrac{x^{-2}y^2}{3}\right)^{-1}$

7. (a) $9z\sqrt{8z} - 3\sqrt{2z^3}$ (b) $-5\sqrt{16y} + 10\sqrt{y}$ (c) $\sqrt[3]{\dfrac{16}{v^5}}$

8. Write the polynomial $3 - 2x^5 + 3x^3 - x^4$ in standard form. Identify the degree and leading coefficient.

In Exercises 9–12, perform the operations and simplify.

9. $(x^2 + 3) - [3x + (8 - x^2)]$ 10. $\left(x + \sqrt{5}\right)\left(x - \sqrt{5}\right)$

11. $\dfrac{8x}{x - 3} + \dfrac{24}{3 - x}$ 12. $\dfrac{\left(\dfrac{2}{x} - \dfrac{2}{x + 1}\right)}{\left(\dfrac{4}{x^2 - 1}\right)}$

In Exercises 13–15, factor the expression completely.

13. $2x^4 - 3x^3 - 2x^2$ 14. $x^3 + 2x^2 - 4x - 8$ 15. $8x^3 - 27$

16. Rationalize each denominator: (a) $\dfrac{16}{\sqrt[3]{16}}$ and (b) $\dfrac{6}{1 - \sqrt{3}}$.

17. Write an expression for the area of the shaded region in the figure at the right and simplify the result.

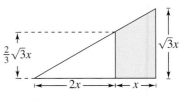

Figure for 17

18. Plot the points $(-2, 5)$ and $(6, 0)$. Find the coordinates of the midpoint of the line segment joining the points and the distance between the points.

19. The numbers (in millions) of votes cast for the Democratic candidates for president in 1980, 1984, 1988, 1992, 1996, and 2000 were 35.5, 37.6, 41.8, 44.9, 47.4, and 51.0, respectively. Construct a bar graph for this data. (Source: Congressional Quarterly, Inc.)

Colleges and universities track enrollment figures in order to determine the financial outlook of the institution. The growth in student enrollment at a college or university can be modeled by a linear equation.

David Young-Wolff/PhotoEdit

1 Functions and Their Graphs

What You Should Learn

In this chapter, you will learn how to:

■ Sketch graphs of equations by point plotting or by using a graphing utility.

■ Find and use the slope of a line to write and graph linear equations.

■ Evaluate functions and find their domains.

■ Analyze graphs of functions.

■ Identify and graph shifts, reflections, and nonrigid transformations of functions.

■ Find arithmetic combinations and compositions of functions.

■ Find inverse functions graphically and algebraically.

Introduction to Library of Functions

In Chapter 1, you will be introduced to the concept of a *function*. As you proceed through the text, you will see that functions play a primary role in modeling real-life situations.

There are three basic types of functions that have proven to be the most important in modeling real-life situations. These functions are algebraic functions, exponential and logarithmic functions, and trigonometric and inverse trigonometric functions. These three types of functions are referred to as the *elementary functions*, though they are often placed in the two categories of *algebraic functions* and *transcendental functions*. Each time a new type of function is studied in detail in this text, it will be highlighted in a box similar to this one. The graphs of many of these functions are shown on the inside front cover of this text.

Algebraic Functions

These functions are formed by applying algebraic operations to the identity function $f(x) = x$.

Name	*Function*	*Location*
Linear	$f(x) = ax + b$	Section 1.2
Quadratic	$f(x) = ax^2 + bx + c$	Section 3.1
Cubic	$f(x) = ax^3 + bx^2 + cx + d$	Section 3.2
Polynomial	$P(x) = a_n x^n + a_{n-1} x^{n-1} + \cdots + a_2 x^2 + a_1 x + a_0$	Section 3.2
Rational	$f(x) = \dfrac{N(x)}{D(x)}$, $N(x)$ and $D(x)$ are polynomial functions	Section 3.5
Radical	$f(x) = \sqrt[n]{P(x)}$	Section 1.3

Transcendental Functions

These functions cannot be formed from the identity function by using algebraic operations.

Name	*Function*	*Location*
Exponential	$f(x) = a^x, a > 0, a \neq 1$	Section 4.1
Logarithmic	$f(x) = \log_a x, x > 0, a > 0, a \neq 1$	Section 4.2
Trigonometric	$f(x) = \sin x, f(x) = \cos x, f(x) = \tan x,$ $f(x) = \csc x, f(x) = \sec x, f(x) = \cot x$	Section 5.3
Inverse Trigonometric	$f(x) = \arcsin x, f(x) = \arccos x, f(x) = \arctan x$	Section 5.7

Nonelementary Functions

Some useful nonelementary functions include the following.

Name	*Function*	*Location*
Absolute value	$f(x) = |g(x)|$, $g(x)$ is an elementary function	Section 1.3
Piecewise-defined	$f(x) = \begin{cases} 3x + 2, & x \geq 1 \\ -2x + 4, & x < 1 \end{cases}$	Section 1.3
Greatest integer	$f(x) = [\![g(x)]\!]$, $g(x)$ is an elementary function	Section 1.4
Data defined	Formula for temperature: $F = \dfrac{9}{5}C + 32$	Section 1.3

1.1 Graphs of Equations

The Graph of an Equation

News magazines often show graphs comparing the rate of inflation, the federal deficit, or the unemployment rate to the time of year. Businesses use graphs to report monthly sales statistics. Such graphs provide geometric pictures of the way one quantity changes with respect to another. Frequently, the relationship between two quantities is expressed as an **equation.** This section introduces the basic procedure for determining the geometric picture associated with an equation.

For an equation in the variables x and y, a point (a, b) is a **solution point** if the substitution of $x = a$ and $y = b$ satisfies the equation. Most equations have *infinitely many* solution points. For example, the equation $3x + y = 5$ has solution points $(0, 5)$, $(1, 2)$, $(2, -1)$, $(3, -4)$, and so on. The set of all solution points of an equation is the **graph of the equation.**

Example 1   Determining Solution Points

Determine whether (a) $(2, 13)$ and (b) $(-1, -3)$ lie on the graph of $y = 10x - 7$.

Solution

a. $y = 10x - 7$ Write original equation.

$13 \overset{?}{=} 10(2) - 7$ Substitute 2 for x and 13 for y.

$13 = 13$ $(2, 13)$ is a solution. ✓

The point $(2, 13)$ *does* lie on the graph of $y = 10x - 7$ because it is a solution point of the equation.

b. $y = 10x - 7$ Write original equation.

$-3 \overset{?}{=} 10(-1) - 7$ Substitute -1 for x and -3 for y.

$-3 \neq -17$ $(-1, -3)$ is not a solution.

The point $(-1, -3)$ *does not* lie on the graph of $y = 10x - 7$ because it is not a solution point of the equation.

✓ *Checkpoint* Now try Exercise 3.

The basic technique used for sketching the graph of an equation is the point-plotting method.

Sketching the Graph of an Equation by Point Plotting

1. If possible, rewrite the equation so that one of the variables is isolated on one side of the equation.

2. Make a table of values showing several solution points.

3. Plot these points on a rectangular coordinate system.

4. Connect the points with a smooth curve or line.

What you should learn

- Sketch graphs of equations by point plotting.
- Graph equations using a graphing utility.
- Use graphs of equations to solve real-life problems.

Why you should learn it

The graph of an equation can help you see relationships between real-life quantities. For example, Exercise 71 on page 85 shows how a graph can be used to understand the relationship between life expectancy and the year a child is born.

Bruce Avres/Getty Images

Example 2   Sketching a Graph by Point Plotting

Use point plotting and graph paper to sketch the graph of $3x + y = 6$.

Solution

In this case you can isolate the variable y.

$$y = 6 - 3x \qquad \text{Solve equation for } y.$$

Using negative, zero, and positive values for x, you can obtain the following table of values (solution points).

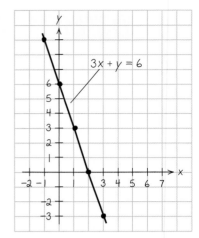

Figure 1.1

x	-1	0	1	2	3
$y = 6 - 3x$	9	6	3	0	-3
Solution point	$(-1, 9)$	$(0, 6)$	$(1, 3)$	$(2, 0)$	$(3, -3)$

Next, plot these points and connect them, as shown in Figure 1.1. It appears that the graph is a straight line. You will study lines extensively in Section 1.2.

 Checkpoint Now try Exercise 7.

The points at which a graph touches or crosses an axis are called the **intercepts** of the graph. For instance, in Example 2 the point $(0, 6)$ is the y-intercept of the graph because the graph crosses the y-axis at that point. The point $(2, 0)$ is the x-intercept of the graph because the graph crosses the x-axis at that point.

Example 3   Sketching a Graph by Point Plotting

Use point plotting and graph paper to sketch the graph of $y = x^2 - 2$.

Solution

Because the equation is already solved for y, make a table of values by choosing several convenient values of x and calculating the corresponding values of y.

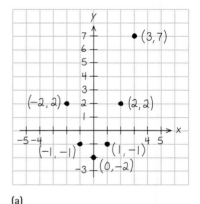

(a)

x	-2	-1	0	1	2	3
$y = x^2 - 2$	2	-1	-2	-1	2	7
Solution point	$(-2, 2)$	$(-1, -1)$	$(0, -2)$	$(1, -1)$	$(2, 2)$	$(3, 7)$

Next, plot the corresponding solution points, as shown in Figure 1.2(a). Finally, connect the points with a smooth curve, as shown in Figure 1.2(b). This graph is called a *parabola*. You will study parabolas in Section 3.1.

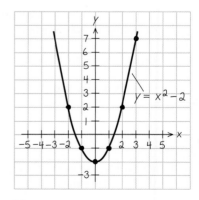

(b)
Figure 1.2

 Checkpoint Now try Exercise 9.

In this text, you will study two basic ways to create graphs: *by hand* and *using a graphing utility*. For instance, the graphs in Figures 1.1 and 1.2 were sketched by hand and the graph in Figure 1.6 was created using a graphing utility.

Using a Graphing Utility

One of the disadvantages of the point-plotting method is that to get a good idea about the shape of a graph, you need to plot *many* points. With only a few points, you could badly misrepresent the graph. For instance, consider the equation

$$y = \frac{1}{30}x(x^4 - 10x^2 + 39).$$

Suppose you plotted only five points: $(-3, -3), (-1, -1), (0, 0), (1, 1),$ and $(3, 3)$, as shown in Figure 1.3(a). From these five points, you might assume that the graph of the equation is a line. That, however, is not correct. By plotting several more points and connecting the points with a smooth curve, you can see that the actual graph is not a line at all, as shown in Figure 1.3(b).

TECHNOLOGY TIP

This section presents a brief overview of how to use a graphing utility to graph an equation. For more extensive coverage of this topic, see Appendix A and the *Graphing Technology Guide* on the text website at *college.hmco.com*.

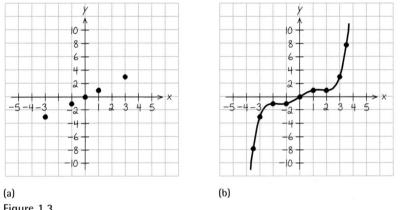

(a)   (b)

Figure 1.3

From this, you can see that the point-plotting method leaves you with a dilemma. The method can be very inaccurate if only a few points are plotted and it is very time-consuming to plot a dozen (or more) points. Technology can help solve this dilemma. Plotting several (even several hundred) points on a rectangular coordinate system is something that a computer or calculator can do easily.

TECHNOLOGY TIP   The point-plotting method is the method used by *all* graphing utilities. Each computer or calculator screen is made up of a grid of hundreds or thousands of small areas called *pixels*. Screens that have many pixels per square inch are said to have a higher *resolution* than screens with fewer pixels.

Using a Graphing Utility to Graph an Equation

To graph an equation involving x and y on a graphing utility, use the following procedure.

1. Rewrite the equation so that y is isolated on the left side.

2. Enter the equation into the graphing utility.

3. Determine a *viewing window* that shows all important features of the graph.

4. Graph the equation.

Example 4   Using a Graphing Utility to Graph an Equation

Use a graphing utility to graph $2y + x^3 = 4x$.

Solution

To begin, solve the equation for y in terms of x.

$$2y + x^3 = 4x \qquad \text{Write original equation.}$$

$$2y = -x^3 + 4x \qquad \text{Subtract } x^3 \text{ from each side.}$$

$$y = -\frac{1}{2}x^3 + 2x \qquad \text{Divide each side by 2.}$$

Enter this equation into a graphing utility (see Figure 1.4). Using a standard viewing window (see Figure 1.5), you can obtain the graph shown in Figure 1.6.

TECHNOLOGY TIP

Many graphing utilities are capable of creating a table of values such as the following, which shows some points of the graph in Figure 1.6. For instructions on how to use the *table* feature, see Appendix A; for specific keystrokes, go to the text website at *college.hmco.com*.

Figure 1.4

Figure 1.5

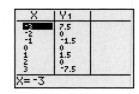

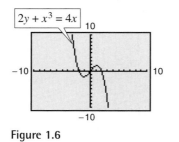

Figure 1.6

✓ *Checkpoint* Now try Exercise 39.

TECHNOLOGY TIP   By choosing different viewing windows for a graph, it is possible to obtain very different impressions of the graph's shape. For instance, Figure 1.7 shows three different viewing windows for the graph of the equation in Example 4. However, none of these views show *all* of the important features of the graph as does Figure 1.6. For instructions on how to set up a viewing window, see Appendix A; for specific keystrokes, go to the text website at *college.hmco.com*.

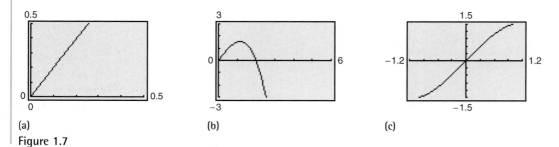

(a)   (b)   (c)

Figure 1.7

TECHNOLOGY TIP   The standard viewing window on many graphing utilities does not give a true geometric perspective because the screen is rectangular, which distorts the image. That is, perpendicular lines will not appear to be perpendicular and circles will not appear to be circular. To overcome this, you can use a *square setting*, as demonstrated in Example 5.

Example 5   Using a Graphing Utility to Graph a Circle

Use a graphing utility to graph $x^2 + y^2 = 9$.

Solution

The graph of $x^2 + y^2 = 9$ is a circle whose center is the origin and whose radius is 3. (See Section P.5.) To graph the equation, begin by solving the equation for y.

$$x^2 + y^2 = 9 \qquad \text{Write original equation.}$$

$$y^2 = 9 - x^2 \qquad \text{Subtract } x^2 \text{ from each side.}$$

$$y = \pm\sqrt{9 - x^2} \qquad \text{Take square root of each side.}$$

Remember that when you take the square root of a variable expression, you must account for both the positive and negative solutions. The graph of

$$y = \sqrt{9 - x^2} \qquad \text{Upper semicircle}$$

is the upper semicircle. The graph of

$$y = -\sqrt{9 - x^2} \qquad \text{Lower semicircle}$$

is the lower semicircle. Enter *both* equations in your graphing utility and generate the resulting graphs. In Figure 1.8, note that if you use a standard viewing window, the two graphs do not appear to form a circle. You can overcome this problem by using a *square setting*, in which the horizontal and vertical tick marks have equal spacing, as shown in Figure 1.9. On many graphing utilities, a square setting can be obtained by using a y to x ratio of 2 to 3. For instance, in Figure 1.9, the y to x ratio is

$$\frac{Y_{\max} - Y_{\min}}{X_{\max} - X_{\min}} = \frac{4 - (-4)}{6 - (-6)} = \frac{8}{12} = \frac{2}{3}.$$

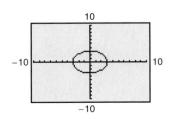

Figure 1.8

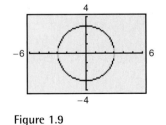

Figure 1.9

✓ *Checkpoint* Now try Exercise 55.

TECHNOLOGY TIP

Notice that when you graph a circle by graphing two separate equations for y, your graphing utility may not connect the two semicircles. This is because some graphing utilities are limited in their resolution. So, in this text, a blue curve is placed behind the graphing utility's display to indicate where the graph should appear.

Applications

Throughout this course, you will learn that there are many ways to approach a problem. Two of the three common approaches are illustrated in Example 6.

    A Numerical Approach: Construct and use a table.

    An Algebraic Approach: Use the rules of algebra.

    A Graphical Approach: Draw and use a graph.

You should develop the habit of using at least two approaches to solve every problem in order to build your intuition and to check that your answer is reasonable.

    The following two applications show how to develop mathematical models to represent real-world situations. You will see that both a graphing utility and algebra can be used to understand and solve the problems posed.

Example 6   Running a Marathon

A runner runs at a constant rate of 4.9 miles per hour. The verbal model and algebraic equation relating distance run and elapsed time are as follows.

$$\text{Verbal Model:} \quad \boxed{\text{Distance}} = \boxed{\text{Rate}} \cdot \boxed{\text{Time}} \qquad \text{Equation: } d = 4.9t$$

a. Determine how far the runner can run in 3.1 hours.

b. Determine how long it will take to run a 26.2-mile marathon.

TECHNOLOGY SUPPORT

For instructions on how to use the *value* feature, the *zoom* and *trace* features, and the *table* feature, see Appendix A; for specific keystrokes, go to the text website at *college.hmco.com*.

Algebraic Solution

a. To begin, find how far the runner can run in 3.1 hours by substituting 3.1 for t in the equation.

$d = 4.9t$	Write original equation.
$= 4.9(3.1)$	Substitute 3.1 for t.
≈ 15.2	Use a calculator.

So, the runner can run about 15.2 miles in 3.1 hours. Use estimation to check your answer. Because 4.9 is about 5 and 3.1 is about 3, the distance is about $5(3) = 15$. So, 15.2 is reasonable.

b. You can find how long it will take to run a 26.2-mile marathon as follows. (For help with solving linear equations, see Appendix D.)

$d = 4.9t$	Write original equation.
$26.2 = 4.9t$	Substitute 26.2 for d.
$\dfrac{26.2}{4.9} = t$	Divide each side by 4.9.
$5.3 \approx t$	Use a calculator.

So, it will take about 5.3 hours to run 26.2 miles.

✓ *Checkpoint* Now try Exercise 67.

Graphical Solution

a. Use a graphing utility to graph the equation $d = 4.9t$. (Represent d by y and t by x.) Be sure to use a viewing window that shows the graph when $x = 3.1$. Then use the *value* feature or the *zoom* and *trace* features of the graphing utility to estimate that when $x = 3.1$, the distance is $y \approx 15.2$ miles, as shown in Figure 1.10(a).

b. Adjust the viewing window so that it shows the graph when $y = 26.2$. Use the *zoom* and *trace* features to estimate that when $y = 26.2$, the time is $x \approx 5.4$ hours, as shown in Figure 1.10(b).

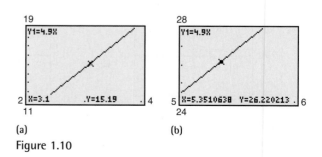

(a)   (b)

Figure 1.10

Note that the viewing window on your graphing utility may differ slightly from those shown in Figure 1.10.

Example 7   Monthly Wage

You receive a monthly salary of $2000 plus a commission of 10% of sales. The verbal model and algebraic equations relating the wages, the salary, and the commission are as follows.

Verbal Model: Wages = Salary + Commission on sales

Equation: $y = 2000 + 0.1x$

a. Sales are $x = 1480$ in August. What are your wages for that month?

b. You receive $2225 for September. What are your sales for that month?

Numerical Solution

a. To find the wages in August, evaluate the equation when $x = 1480$.

$$y = 2000 + 0.1x \qquad \text{Write original equation.}$$

$$= 2000 + 0.1(1480) \qquad \text{Substitute 1480 for } x.$$

$$= 2148 \qquad \text{Simplify.}$$

So, the wages in August are $2148.

b. You can use the *table* feature of a graphing utility to create a table that shows the wages for different sales amounts. First enter the equation in the graphing utility. Then set up a table, as shown in Figure 1.11. The graphing utility produces the table shown in Figure 1.12.

Figure 1.11

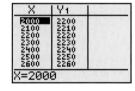

Figure 1.12

From the table, you can see that wages of $2225 result from sales between $2200 and $2300. You can improve this estimate by setting up the table shown in Figure 1.13. The graphing utility produces the table shown in Figure 1.14.

Figure 1.13

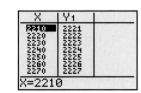

Figure 1.14

From the table, you can see that wages of $2225 result from sales of $2250.

✓ *Checkpoint* Now try Exercise 72.

Graphical Solution

a. You can use a graphing utility to graph $y = 2000 + 0.1x$ and then estimate the wages when $x = 1480$. Be sure to use a viewing window that shows the graph when $x \geq 0$ and $y > 2000$. Then, by using the *value* feature or the *zoom* and *trace* features near $x = 1480$, you can estimate that the wages are about $2148, as shown in Figure 1.15(a).

b. Use the graphing utility to find the value along the x-axis (sales) that corresponds to a y-value of 2225 (wages). Using the *zoom* and *trace* features, you can estimate the sales to be about $2250, as shown in Figure 1.15(b).

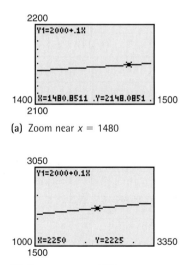

(a) Zoom near $x = 1480$

(b) Zoom near $y = 2225$

Figure 1.15

1.1   Exercises

Vocabulary Check

Fill in the blanks.

1. For an equation in x and y, if the substitution of $x = a$ and $y = b$ satisfies the equation, then the point (a, b) is a _____ .

2. The set of all solution points of an equation is the _____ of the equation.

3. The points at which a graph touches or crosses an axis are called the _____ of the graph.

In Exercises 1–6, determine whether each point lies on the graph of the equation.

Equation	Points	
1. $y = \sqrt{x + 4}$	(a) $(0, 2)$	(b) $(5, 3)$
2. $y = x^2 - 3x + 2$	(a) $(2, 0)$	(b) $(-2, 8)$
3. $y = 4 - \lvert x - 2 \rvert$	(a) $(1, 5)$	(b) $(1.2, 3.2)$
4. $2x - y - 3 = 0$	(a) $(1, 2)$	(b) $(1, -1)$
5. $x^2 + y^2 = 20$	(a) $(3, -2)$	(b) $(-4, 2)$
6. $y = \frac{1}{3}x^3 - 2x^2$	(a) $\left(2, -\frac{16}{3}\right)$	(b) $(-3, 9)$

In Exercises 7–10, complete the table. Use the resulting solution points to sketch the graph of the equation. Use a graphing utility to verify the graph.

7. $y = -2x + 3$

x	-1	0	1	$\frac{3}{2}$	2
y					
Solution point					

8. $y = \frac{3}{2}x - 1$

x	-2	0	$\frac{2}{3}$	1	2
y					
Solution point					

9. $y = x^2 - 2x$

x	-1	0	1	2	3
y					
Solution point					

10. $y = 3 - \lvert x - 2 \rvert$

x	0	1	2	3	4
y					
Solution point					

11. *Exploration*

(a) Complete the table for the equation $y = \frac{1}{4}x - 3$.

x	-2	-1	0	1	2
y					

(b) Use the solution points to sketch the graph. Then use a graphing utility to verify the graph.

(c) Repeat parts (a) and (b) for the equation $y = -\frac{1}{4}x - 3$. Use the result to describe any differences between the graphs.

12. *Exploration*

(a) Complete the table for the equation

$$y = \frac{6x}{x^{-2} + 1}.$$

x	-2	-1	0	1	2
y					

(b) Use the solution points to sketch the graph. Then use a graphing utility to verify the graph.

(c) Continue the table in part (a) for x-values of 5, 10, 20, and 40. What is the value of y approaching? Can y be negative for positive values of x? Explain.

In Exercises 13–18, match the equation with its graph. [The graphs are labeled (a), (b), (c), (d), (e), and (f).]

(a)

(b)

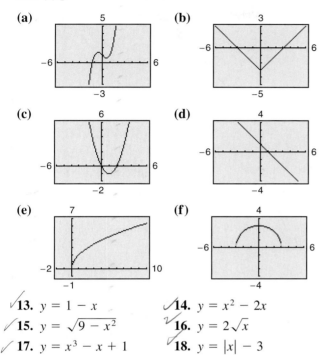

(c)

(d)

(e)

(f)

13. $y = 1 - x$

14. $y = x^2 - 2x$

15. $y = \sqrt{9 - x^2}$

16. $y = 2\sqrt{x}$

17. $y = x^3 - x + 1$

18. $y = |x| - 3$

In Exercises 19–32, sketch the graph of the equation.

19. $y = -4x + 1$

20. $y = 2x - 3$

21. $y = 2 - x^2$

22. $y = x^2 - 1$

23. $y = x^2 - 3x$

24. $y = -x^2 - 4x$

25. $y = x^3 + 2$

26. $y = x^3 - 3$

27. $y = \sqrt{x - 3}$

28. $y = \sqrt{1 - x}$

29. $y = |x - 2|$

30. $y = 5 - |x|$

31. $x = y^2 - 1$

32. $x = y^2 - 4$

In Exercises 33–46, use a graphing utility to graph the equation. Use a standard viewing window. Approximate any x- or y-intercepts of the graph.

33. $y = x - 7$

34. $y = x + 1$

35. $y = 3 - \frac{1}{2}x$

36. $y = \frac{2}{3}x - 1$

37. $y = x^2 - 4x + 3$

38. $y = \frac{1}{2}(x + 4)(x - 2)$

39. $y = x(x - 2)^2$

40. $y = -x^3 + 1$

41. $y = \dfrac{2x}{x - 1}$

42. $y = \dfrac{4}{x}$

43. $y = x\sqrt{x + 3}$

44. $y = (6 - x)\sqrt{x}$

45. $y = \sqrt[3]{x}$

46. $y = \sqrt[3]{x + 1}$

In Exercises 47–50, use a graphing utility to graph the equation. Begin by using a standard viewing window. Then graph the equation a second time using the specified viewing window. Which viewing window is better? Explain.

47. $y = \frac{5}{2}x + 5$

48. $y = -3x + 50$

| Xmin = 0 |
| Xmax = 6 |
| Xscl = 1 |
| Ymin = 0 |
| Ymax = 10 |
| Yscl = 1 |

| Xmin = -1 |
| Xmax = 4 |
| Xscl = 1 |
| Ymin = -5 |
| Ymax = 60 |
| Yscl = 5 |

49. $y = -x^2 + 10x - 5$

50. $y = 4(x + 5)\sqrt{4 - x}$

| Xmin = -1 |
| Xmax = 11 |
| Xscl = 1 |
| Ymin = -5 |
| Ymax = 25 |
| Yscl = 5 |

| Xmin = -6 |
| Xmax = 6 |
| Xscl = 1 |
| Ymin = -5 |
| Ymax = 50 |
| Yscl = 5 |

In Exercises 51–54, describe the viewing window of the graph shown.

51. $y = 4x^2 - 25$

52. $y = x^3 - 3x^2 + 4$

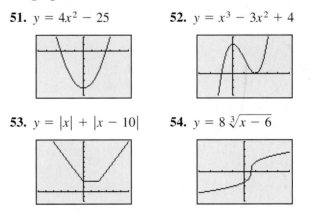

53. $y = |x| + |x - 10|$

54. $y = 8\sqrt[3]{x - 6}$

In Exercises 55–58, solve for y and use a graphing utility to graph each of the resulting equations in the same viewing window. (Adjust the viewing window so that the circle appears circular.)

55. $x^2 + y^2 = 64$

56. $x^2 + y^2 = 49$

57. $(x - 1)^2 + (y - 2)^2 = 16$

58. $(x - 3)^2 + (y - 1)^2 = 25$

In Exercises 59–62, explain how to use a graphing utility to verify that $y_1 = y_2$. Identify the rule of algebra that is illustrated.

59. $y_1 = \frac{1}{4}(x^2 - 8)$

   $y_2 = \frac{1}{4}x^2 - 2$

60. $y_1 = \frac{1}{2}x + (x + 1)$

   $y_2 = \frac{3}{2}x + 1$

61. $y_1 = \frac{1}{5}[10(x^2 - 1)]$

   $y_2 = 2(x^2 - 1)$

62. $y_1 = (x - 3) \cdot \dfrac{1}{x - 3}$

   $y_2 = 1$

In Exercises 63–66, use a graphing utility to graph the equation. Use the *trace* feature of the graphing utility to approximate the unknown coordinate of each solution point accurate to two decimal places. (*Hint:* You may need to use the *zoom* feature of the graphing utility to obtain the required accuracy.)

63. $y = \sqrt{5 - x}$

   (a) $(2, y)$

   (b) $(x, 3)$

64. $y = x^3(x - 3)$

   (a) $(2.25, y)$

   (b) $(x, 20)$

65. $y = x^5 - 5x$

   (a) $(-0.5, y)$

   (b) $(x, -4)$

66. $y = |x^2 - 6x + 5|$

   (a) $(2, y)$

   (b) $(x, 1.5)$

67. *Depreciation*   A manufacturing plant purchases a new molding machine for \$225,000. The depreciated value (drop in value) y after t years is

$$y = 225{,}000 - 20{,}000t, \qquad 0 \le t \le 8.$$

(a) Use the constraints of the model to determine an appropriate viewing window.

(b) Use a graphing utility to graph the equation.

(c) Use the *value* feature or the *zoom* and *trace* features of a graphing utility to determine the value of y when $t = 5.8$. Verify your answer algebraically.

(d) Use the *value* feature or the *zoom* and *trace* features of a graphing utility to determine the value of y when $t = 2.35$. Verify your answer algebraically.

68. *Consumerism*   You purchase a personal watercraft for \$8100. The depreciated value y after t years is

$$y = 8100 - 929t, \qquad 0 \le t \le 6.$$

(a) Use the constraints of the model to determine an appropriate viewing window.

(b) Use a graphing utility to graph the equation.

(c) Use the *zoom* and *trace* features of a graphing utility to determine the value of t when $y = 5545.25$. Verify your answer algebraically.

(d) Use the *value* feature or the *zoom* and *trace* features of a graphing utility to determine the value of y when $t = 5.5$. Verify your answer algebraically.

69. *Geometry*   A rectangle of length x and width w has a perimeter of 12 meters.

(a) Draw a diagram to represent the rectangle. Use the specified variables to label its sides.

(b) Show that the width of the rectangle is $w = 6 - x$ and its area is $A = x(6 - x)$.

(c) Use a graphing utility to graph the area equation.

(d) Use the *zoom* and *trace* features of a graphing utility to determine the value of A when $w = 4.9$ meters. Verify your answer algebraically.

(e) From the graph in part (c), estimate the dimensions of the rectangle that yield a maximum area.

70. *Data Analysis*   The table shows the median (middle) sales prices (in thousands of dollars) of new one-family homes in the United States from 1996 to 2001.   (Sources: U.S. Census Bureau and U.S. Department of Housing and Urban Development)

Year	Median sales price, y
1996	140
1997	146
1998	153
1999	161
2000	169
2001	175

A model for the median sales price during this period is given by

$$y = -0.167t^3 + 4.32t^2 - 29.3t + 196, \quad 6 \le t \le 11$$

where y represents the sales price and t represents the year, with $t = 6$ corresponding to 1996.

(a) Use the model and the *table* feature of a graphing utility to find the median sales prices from 1996 to 2001.

(b) Use a graphing utility to graph the data from the table above and the model in the same viewing window.

(c) Use the model to estimate the median sales prices in 2005 and 2010. Do the values seem reasonable? Explain.

(d) Use the *zoom* and *trace* features of a graphing utility to determine during which year(s) the median sales price exceeded $160,000.

71. *Population Statistics* The table shows the life expectancy of a child (at birth) in the United States for selected years from 1930 to 2000. (Source: U.S. National Center for Health Statistics)

Year	Life expectancy, y
1930	59.7
1940	62.9
1950	68.2
1960	69.7
1970	70.8
1980	73.7
1990	75.4
2000	76.9

A model for the life expectancy during this period is given by

$$y = \frac{59.97 + 0.98t}{1 + 0.01t}, \quad 0 \le t \le 70$$

where y represents the life expectancy and t is the time in years, with $t = 0$ corresponding to 1930.

(a) What does the y-intercept of the graph of the model represent?

(b) Use the *zoom* and *trace* features of a graphing utility to determine the year when the life expectancy was 73.2. Verify your answer algebraically.

(c) Determine the life expectancy in 1948 both graphically and algebraically.

(d) Use the model to estimate the life expectancy of a child born in 2010.

72. *Electronics* The resistance y (in ohms) of 1000 feet of solid copper wire at 68 degrees Fahrenheit can be approximated by the mathematical model

$$y = \frac{10,770}{x^2} - 0.37, \quad 5 \le x \le 100$$

where x is the diameter of the wire in mils (0.001 inch). (Source: American Wire Gage)

(a) Complete the table.

x	10	20	30	40	50
y					

x	60	70	80	90	100
y					

(b) Use your table to approximate the value of x when the resistance is 4.8 ohms. Then determine the answer algebraically.

(c) Use the *value* feature or the *zoom* and *trace* features of a graphing utility to determine the resistance when $x = 85.5$.

(d) What can you conclude in general about the relationship between the diameter of the copper wire and the resistance?

Synthesis

True or False? In Exercises 73 and 74, determine whether the statement is true or false. Justify your answer.

73. A parabola can have only one x-intercept.

74. The graph of a linear equation can have either no x-intercepts or only one x-intercept.

75. *Writing* Explain how to find an appropriate viewing window for the graph of an equation.

76. *Writing* Your employer offers you a choice of wage scales: a monthly salary of $3000 plus commission of 7% of sales or a salary of $3400 plus a 5% commission. Write a short paragraph discussing how you would choose your option. At what sales level would the options yield the same salary?

Review

In Exercises 77–80, perform the operations and simplify.

77. $7\sqrt{72} - 5\sqrt{18}$

78. $-10\sqrt{25y} - \sqrt{y}$

79. $7^{3/2} \cdot 7^{11/2}$

80. $\dfrac{10^{17/4}}{10^{5/4}}$

In Exercises 81 and 82, perform the operation and write the result in standard form.

81. $(9x - 4) + (2x^2 - x + 15)$

82. $(3x^2 - 5)(-x^2 + 1)$

1.2 Lines in the Plane

The Slope of a Line

In this section, you will study lines and their equations. The **slope** of a nonvertical line represents the number of units the line rises or falls vertically for each unit of horizontal change from left to right. For instance, consider the two points (x_1, y_1) and (x_2, y_2) on the line shown in Figure 1.16. As you move from left to right along this line, a change of $(y_2 - y_1)$ units in the vertical direction corresponds to a change of $(x_2 - x_1)$ units in the horizontal direction. That is,

$$y_2 - y_1 = \text{the change in } y$$

and

$$x_2 - x_1 = \text{the change in } x.$$

The slope of the line is given by the ratio of these two changes.

Dwayne Newton/PhotoEdit

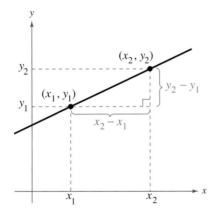

Figure 1.16

Definition of the Slope of a Line

The **slope** m of the nonvertical line through (x_1, y_1) and (x_2, y_2) is

$$m = \frac{y_2 - y_1}{x_2 - x_1} = \frac{\text{change in } y}{\text{change in } x}$$

where $x_1 \neq x_2$.

When this formula for slope is used, the *order of subtraction* is important. Given two points on a line, you are free to label either one of them as (x_1, y_1) and the other as (x_2, y_2). However, once you have done this, you must form the numerator and denominator using the same order of subtraction.

$$m = \frac{y_2 - y_1}{x_2 - x_1} \qquad m = \frac{y_1 - y_2}{x_1 - x_2} \qquad m = \frac{y_2 - y_1}{x_1 - x_2}$$

Correct   Correct   Incorrect

Throughout this text, the term *line* always means a *straight* line.

Example 1   Finding the Slope of a Line

Find the slope of the line passing through each pair of points.

a. $(-2, 0)$ and $(3, 1)$ **b.** $(-1, 2)$ and $(2, 2)$ **c.** $(0, 4)$ and $(1, -1)$

Solution

Difference in *y*-values

a. $m = \dfrac{\overbrace{y_2 - y_1}}{\underbrace{x_2 - x_1}} = \dfrac{1 - 0}{3 - (-2)} = \dfrac{1}{3 + 2} = \dfrac{1}{5}$

Difference in *x*-values

b. $m = \dfrac{2 - 2}{2 - (-1)} = \dfrac{0}{3} = 0$

c. $m = \dfrac{-1 - 4}{1 - 0} = \dfrac{-5}{1} = -5$

The graphs of the three lines are shown in Figure 1.17. Note that the square setting gives the correct "steepness" of the lines.

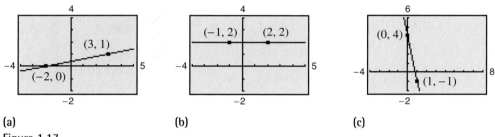

(a)   (b)   (c)

Figure 1.17

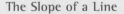 *Checkpoint* Now try Exercise 9.

The definition of slope does not apply to vertical lines. For instance, consider the points $(3, 4)$ and $(3, 1)$ on the vertical line shown in Figure 1.18. Applying the formula for slope, you obtain

$$m = \frac{4 - 1}{3 - 3} = \frac{3}{0}. \qquad \text{Undefined}$$

Because division by zero is undefined, the slope of a vertical line is undefined.

From the slopes of the lines shown in Figures 1.17 and 1.18, you can make the following generalizations about the slope of a line.

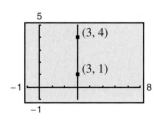

Figure 1.18

The Slope of a Line

1. A line with positive slope ($m > 0$) *rises* from left to right.

2. A line with negative slope ($m < 0$) *falls* from left to right.

3. A line with zero slope ($m = 0$) is *horizontal*.

4. A line with undefined slope is *vertical*.

Exploration

Use a graphing utility to compare the slopes of the lines $y = 0.5x$, $y = x$, $y = 2x$, and $y = 4x$. What do you observe about these lines? Compare the slopes of the lines $y = -0.5x$, $y = -x$, $y = -2x$, and $y = -4x$. What do you observe about these lines? (*Hint:* Use a square setting to guarantee a true geometric perspective.)

The Point-Slope Form of the Equation of a Line

If you know the slope of a line *and* you also know the coordinates of one point on the line, you can find an equation for the line. For instance, in Figure 1.19, let (x_1, y_1) be a point on the line whose slope is m. If (x, y) is any *other* point on the line, it follows that

$$\frac{y - y_1}{x - x_1} = m.$$

This equation in the variables x and y can be rewritten in the **point-slope form** of the equation of a line.

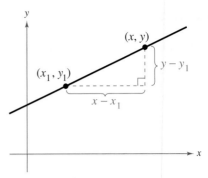

Figure 1.19

Point-Slope Form of the Equation of a Line

The **point-slope form** of the equation of the line that passes through the point (x_1, y_1) and has a slope of m is

$$y - y_1 = m(x - x_1).$$

The point-slope form is most useful for finding the equation of a line if you know at least one point that the line passes through and the slope of the line. You should remember this form of the equation of a line.

Example 2   The Point-Slope Form of the Equation of a Line

Find an equation of the line that passes through the point $(1, -2)$ and has a slope of 3.

Solution

$y - y_1 = m(x - x_1)$	Point-slope form
$y - (-2) = 3(x - 1)$	Substitute for y_1, m, and x_1.
$y + 2 = 3x - 3$	Simplify.
$y = 3x - 5$	Solve for y.

The line is shown in Figure 1.20.

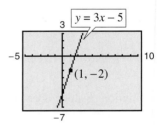

Figure 1.20

✓ *Checkpoint* Now try Exercise 25.

The point-slope form can be used to find an equation of a nonvertical line passing through two points (x_1, y_1) and (x_2, y_2). First, find the slope of the line.

$$m = \frac{y_2 - y_1}{x_2 - x_1}, \quad x_1 \neq x_2$$

Then use the point-slope form to obtain the equation

$$y - y_1 = \frac{y_2 - y_1}{x_2 - x_1}(x - x_1).$$

This is sometimes called the **two-point form** of the equation of a line.

STUDY TIP

When you find an equation of the line that passes through two given points, you need to substitute the coordinates of only one of the points into the point-slope form. It does not matter which point you choose because both points will yield the same result.

Example 3   A Linear Model for Sales Prediction

During 2000, Nike's net sales were \$9.0 billion, and in 2001 net sales were \$9.5 billion. Write a linear equation giving the net sales y in terms of the year x. Then use the equation to predict the net sales for 2002.   (Source: Nike, Inc.)

Solution

Let $x = 0$ represent 2000. In Figure 1.21, let $(0, 9.0)$ and $(1, 9.5)$ be two points on the line representing the net sales. The slope of this line is

$$m = \frac{9.5 - 9.0}{1 - 0} = 0.5. \qquad m = \frac{y_2 - y_1}{x_2 - x_1}$$

By the point-slope form, the equation of the line is as follows.

$$y - 9.0 = 0.5(x - 0) \qquad \text{Write in point-slope form.}$$

$$y = 0.5x + 9.0 \qquad \text{Simplify.}$$

Now, using this equation, you can predict the 2002 net sales $(x = 2)$ to be

$$y = 0.5(2) + 9.0 = 1 + 9.0 = \$10.0 \text{ billion.}$$

✓ *Checkpoint* Now try Exercise 43.

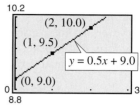

Figure 1.21

Library of Functions: Linear Function

In the next section, you will be introduced to the precise meaning of the term *function*. The simplest type of function is a *linear function* of the form

$$f(x) = mx + b.$$

As its name implies, the graph of a linear function is a line that has a slope of m and a y-intercept at $(0, b)$. The basic characteristics of a linear function are summarized below. (Note that some of the terms below will be defined later in the text.)

Graph of $f(x) = mx + b, m > 0$ *Graph of* $f(x) = mx + b, m < 0$

Domain: $(-\infty, \infty)$ Domain: $(-\infty, \infty)$

Range: $(-\infty, \infty)$ Range: $(-\infty, \infty)$

x-intercept: $(-b/m, 0)$ x-intercept: $(-b/m, 0)$

y-intercept: $(0, b)$ y-intercept: $(0, b)$

Increasing   Decreasing

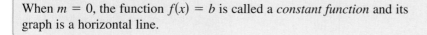

When $m = 0$, the function $f(x) = b$ is called a *constant function* and its graph is a horizontal line.

STUDY TIP

The prediction method illustrated in Example 3 is called **linear extrapolation**. Note in the top figure below that an extrapolated point does not lie between the given points. When the estimated point lies between two given points, as shown in the bottom figure, the procedure used to predict the point is called **linear interpolation**.

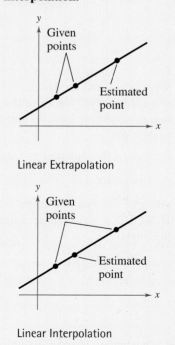

Linear Extrapolation

Linear Interpolation

Sketching Graphs of Lines

Many problems in coordinate geometry can be classified as follows.

1. Given a graph (or parts of it), find its equation.

2. Given an equation, sketch its graph.

For lines, the first problem is solved easily by using the point-slope form. This formula, however, is not particularly useful for solving the second type of problem. The form that is better suited to graphing linear equations is the **slope-intercept form** of the equation of a line, $y = mx + b$.

Slope-Intercept Form of the Equation of a Line

The graph of the equation

$$y = mx + b$$

is a line whose slope is m and whose y-intercept is $(0, b)$.

Example 4   Using the Slope-Intercept Form

Determine the slope and y-intercept of each linear equation. Then describe its graph.

a. $x + y = 2$ **b.** $y = 2$

Algebraic Solution

a. Begin by writing the equation in slope-intercept form.

$$x + y = 2 \qquad \text{Write original equation.}$$

$$y = 2 - x \qquad \text{Subtract } x \text{ from each side.}$$

$$y = -x + 2 \qquad \text{Write in slope-intercept form.}$$

From the slope-intercept form of the equation, the slope is -1 and the y-intercept is $(0, 2)$. Because the slope is negative, you know that the graph of the equation is a line that falls one unit for every unit it moves to the right.

b. By writing the equation $y = 2$ in slope-intercept form

$$y = (0)x + 2$$

you can see that the slope is 0 and the y-intercept is $(0, 2)$. A zero slope implies that the line is horizontal.

✓ *Checkpoint* Now try Exercise 45.

Graphical Solution

a. Solve the equation for y to obtain $y = 2 - x$. Enter this equation in your graphing utility. Use a decimal viewing window to graph the equation. To find the y-intercept, use the *value* or *trace* feature. When $x = 0$, $y = 2$, as shown in Figure 1.22(a). So, the y-intercept is $(0, 2)$. To find the slope, continue to use the *trace* feature. Move the cursor along the line until $x = 1$. At this point, $y = 1$. So the graph falls 1 unit for every unit it moves to the right, and the slope is -1.

b. Enter the equation $y = 2$ in your graphing utility and graph the equation. Use the *trace* feature to verify the y-intercept $(0, 2)$ as shown in Figure 1.22(b), and to see that the value of y is the same for all values of x. So, the slope of the horizontal line is 0.

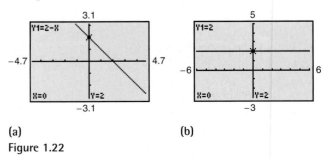

(a)   (b)

Figure 1.22

From the slope-intercept form of the equation of a line, you can see that a horizontal line ($m = 0$) has an equation of the form $y = b$. This is consistent with the fact that each point on a horizontal line through $(0, b)$ has a y-coordinate of b. Similarly, each point on a vertical line through $(a, 0)$ has an x-coordinate of a. So, a vertical line has an equation of the form $x = a$. This equation cannot be written in slope-intercept form because the slope of a vertical line is undefined. However, *every* line has an equation that can be written in the **general form**

$$Ax + By + C = 0 \qquad \text{General form of the equation of a line}$$

where A and B are not *both* zero.

> ### Exploration
>
> Graph the lines $y_1 = 2x + 1$, $y_2 = \frac{1}{2}x + 1$, and $y_3 = -2x + 1$ in the same viewing window. What do you observe?
>
> Graph the lines $y_1 = 2x + 1$, $y_2 = 2x$, and $y_3 = 2x - 1$ in the same viewing window. What do you observe?

Summary of Equations of Lines

1. General form: $Ax + By + C = 0$

2. Vertical line: $x = a$

3. Horizontal line: $y = b$

4. Slope-intercept form: $y = mx + b$

5. Point-slope form: $y - y_1 = m(x - x_1)$

Example 5   Different Viewing Windows

The graphs of the two lines

$$y = -x - 1 \qquad \text{and} \qquad y = -10x - 1$$

are shown in Figure 1.23. Even though the slopes of these lines are quite different (-1 and -10, respectively), the graphs seem misleadingly similar because the viewing windows are different.

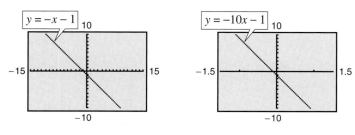

Figure 1.23

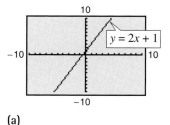

(a)

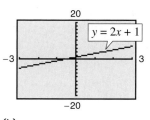

(b)

 Checkpoint Now try Exercise 49.

TECHNOLOGY TIP   When a graphing utility is used to graph a line, it is important to realize that the graph of the line may not visually appear to have the slope indicated by its equation. This occurs because of the viewing window used for the graph. For instance, Figure 1.24 shows graphs of $y = 2x + 1$ produced on a graphing utility using three different viewing windows. Notice that the slopes in Figures 1.24(a) and (b) do not visually appear to be equal to 2. However, if you use a *square setting*, as in Figure 1.24(c), the slope visually appears to be 2.

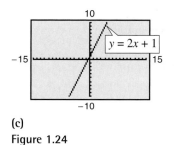

(c)

Figure 1.24

Parallel and Perpendicular Lines

The slope of a line is a convenient tool for determining whether two lines are parallel or perpendicular.

> **Parallel Lines**
>
> Two distinct nonvertical lines are **parallel** if and only if their slopes are equal. That is,
>
> $m_1 = m_2.$

Example 6   Equations of Parallel Lines

Find the slope-intercept form of the equation of the line that passes through the point $(2, -1)$ and is parallel to the line $2x - 3y = 5$.

Solution

Begin by writing the equation of the given line in slope-intercept form.

$$2x - 3y = 5 \qquad \text{Write original equation.}$$

$$-2x + 3y = -5 \qquad \text{Multiply by } -1.$$

$$3y = 2x - 5 \qquad \text{Add } 2x \text{ to each side.}$$

$$y = \frac{2}{3}x - \frac{5}{3} \qquad \text{Write in slope-intercept form.}$$

Therefore, the given line has a slope of $m = \frac{2}{3}$. Any line parallel to the given line must also have a slope of $\frac{2}{3}$. So, the line through $(2, -1)$ has the following equation.

$$y - (-1) = \frac{2}{3}(x - 2) \qquad \text{Write in point-slope form.}$$

$$y + 1 = \frac{2}{3}x - \frac{4}{3} \qquad \text{Simplify.}$$

$$y = \frac{2}{3}x - \frac{7}{3} \qquad \text{Write in slope-intercept form.}$$

Notice the similarity between the slope-intercept form of the original equation and the slope-intercept form of the parallel equation. The graphs of both equations are shown in Figure 1.25.

✓ *Checkpoint*   Now try Exercise 55(a).

> **Perpendicular Lines**
>
> Two nonvertical lines are **perpendicular** if and only if their slopes are negative reciprocals of each other. That is,
>
> $m_1 = -\dfrac{1}{m_2}.$

TECHNOLOGY TIP

Be careful when you graph equations such as $y = \frac{2}{3}x - \frac{7}{3}$ on your graphing utility. A common mistake is to type in the equation as

$$Y1 = 2/3X - 7/3,$$

which may not be interpreted by your graphing utility as the original equation. You should use one of the following formulas.

$$Y1 = 2X/3 - 7/3$$

$$Y1 = (2/3)X - 7/3$$

Do you see why?

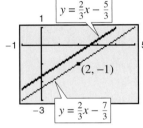

Figure 1.25

Example 7   Equations of Perpendicular Lines

Find the slope-intercept form of the equation of the line that passes through the point $(2, -1)$ and is perpendicular to the line $2x - 3y = 5$.

Solution

From Example 6, you know that the equation can be written in the slope-intercept form $y = \frac{2}{3}x - \frac{5}{3}$. You can see that the line has a slope of $\frac{2}{3}$. So, any line perpendicular to this line must have a slope of $-\frac{3}{2}$ $\left(\text{because } -\frac{3}{2} \text{ is the negative reciprocal of } \frac{2}{3}\right)$. So, the line through the point $(2, -1)$ has the following equation.

$$y - (-1) = -\frac{3}{2}(x - 2) \qquad \text{Write in point-slope form.}$$

$$y + 1 = -\frac{3}{2}x + 3 \qquad \text{Simplify.}$$

$$y = -\frac{3}{2}x + 2 \qquad \text{Write in slope-intercept form.}$$

The graphs of both equations are shown in Figure 1.26.

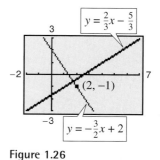

Figure 1.26

✓ *Checkpoint* Now try Exercise 55(b).

Example 8   Graphs of Perpendicular Lines

Use a graphing utility to graph the lines

$$y = x + 1$$

and

$$y = -x + 3$$

in the same viewing window. The lines are supposed to be perpendicular (they have slopes of $m_1 = 1$ and $m_2 = -1$). Do they appear to be perpendicular on the display?

Solution

If the viewing window is nonsquare, as in Figure 1.27, the two lines will not appear perpendicular. If, however, the viewing window is square, as in Figure 1.28, the lines will appear perpendicular.

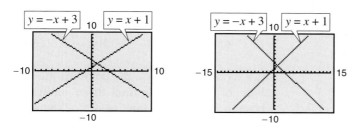

Figure 1.27   Figure 1.28

 Checkpoint Now try Exercise 61.

1.2   Exercises

Vocabulary Check

1. Match each equation with its form.

   (a) $Ax + By + C = 0$ (i) vertical line

   (b) $x = a$ (ii) slope-intercept form

   (c) $y = b$ (iii) general form

   (d) $y = mx + b$ (iv) point-slope form

   (e) $y - y_1 = m(x - x_1)$ (v) horizontal line

In Exercises 2–5, fill in the blanks.

2. For a line, the ratio of the change in y to the change in x is called the _____ of the line.

3. Two lines are _____ if and only if their slopes are equal.

4. Two lines are _____ if and only if their slopes are negative reciprocals of each other.

5. The prediction method _____ is the method used to estimate a point on a line that does not lie between the given points.

In Exercises 1 and 2, identify the line that has the indicated slope.

1. (a) $m = \frac{2}{3}$ (b) m is undefined.   (c) $m = -2$

2. (a) $m = 0$ (b) $m = -\frac{3}{4}$ (c) $m = 1$

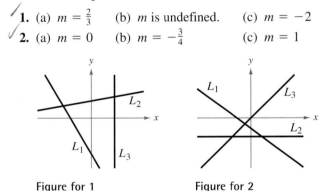

Figure for 1   Figure for 2

In Exercises 3 and 4, sketch the lines through the point with the indicated slopes on the same set of coordinate axes.

	Point		Slopes		
3.	$(2, 3)$	(a)	0	(b)	1
		(c)	2	(d)	-3
4.	$(-4, 1)$	(a)	3	(b)	-3
		(c)	$\frac{1}{2}$	(d)	Undefined

In Exercises 5 and 6, estimate the slope of the line.

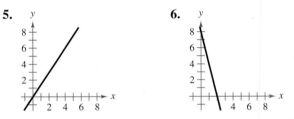

In Exercises 7–10, use a graphing utility to plot the points and use the *draw* feature to graph the line segment connecting the two points. (Use a square setting.) Then find the slope of the line passing through the pair of points.

7. $(0, -10), (-4, 0)$ 8. $(2, 4), (4, -4)$

9. $(-6, -1), (-6, 4)$ 10. $(-3, -2), (1, 6)$

In Exercises 11–18, use the point on the line and the slope of the line to find three additional points through which the line passes. (There are many correct answers.)

	Point	Slope
11.	$(2, 1)$	$m = 0$
12.	$(3, -2)$	$m = 0$
13.	$(1, 5)$	m is undefined.
14.	$(-4, 1)$	m is undefined.

Point	Slope
15. $(0, -9)$	$m = -2$
16. $(-5, 4)$	$m = 2$
17. $(7, -2)$	$m = \frac{1}{2}$
18. $(-1, -6)$	$m = -\frac{1}{2}$

In Exercises 19–24, (a) find the slope and *y*-intercept (if possible) of the equation of the line algebraically, (b) sketch the line by hand, and (c) use a graphing utility to verify your answers to parts (a) and (b).

19. $5x - y + 3 = 0$ **20.** $2x + 3y - 9 = 0$

21. $5x - 2 = 0$ **22.** $3x + 7 = 0$

23. $3y + 5 = 0$ **24.** $-11 - 8y = 0$

In Exercises 25–32, find the general form of the equation of the line that passes through the given point and has the indicated slope. Sketch the line by hand. Use a graphing utility to verify your sketch, if possible.

Point	Slope
25. $(0, -2)$	$m = 3$
26. $(-3, 6)$	$m = -2$
27. $(0, 0)$	$m = 4$
28. $(-2, -5)$	$m = \frac{3}{4}$
29. $(6, -1)$	m is undefined.
30. $(-10, 4)$	m is undefined.
31. $\left(-\frac{1}{2}, \frac{3}{2}\right)$	$m = 0$
32. $(2.3, -8.5)$	$m = 0$

In Exercises 33–42, find the slope-intercept form of the equation of the line that passes through the points. Use a graphing utility to graph the line.

33. $(5, -1), (-5, 5)$ **34.** $(4, 3), (-4, -4)$

35. $(-8, 1), (-8, 7)$ **36.** $(-1, 4), (6, 4)$

37. $\left(2, \frac{1}{2}\right), \left(\frac{1}{2}, \frac{5}{4}\right)$ **38.** $(1, 1), \left(6, -\frac{2}{3}\right)$

39. $\left(-\frac{1}{10}, -\frac{3}{5}\right), \left(\frac{9}{10}, -\frac{9}{5}\right)$ **40.** $\left(\frac{3}{4}, \frac{3}{2}\right), \left(-\frac{4}{3}, \frac{7}{4}\right)$

41. $(1, 0.6), (-2, -0.6)$ **42.** $(-8, 0.6), (2, -2.4)$

43. *Annual Salary* A jeweler's salary was $28,500 in 2000 and $32,900 in 2002. The jeweler's salary follows a linear growth pattern. What will the jeweler's salary be in 2006?

44. *Annual Salary* A librarian's salary was $25,000 in 2000 and $27,500 in 2002. The librarian's salary follows a linear growth pattern. What will the librarian's salary be in 2006?

In Exercises 45–48, determine the slope and *y*-intercept of the linear equation. Then describe its graph.

45. $x - 2y = 4$ **46.** $3x + 4y = 1$

47. $x = -6$ **48.** $y = 12$

In Exercises 49 and 50, use a graphing utility to graph the equation using each of the suggested viewing windows. Describe the difference between the two graphs.

49. $y = 0.5x - 3$

Xmin = -5 Xmax = 10 Xscl = 1 Ymin = -1 Ymax = 10 Yscl = 1	Xmin = -2 Xmax = 10 Xscl = 1 Ymin = -4 Ymax = 1 Yscl = 1

50. $y = -8x + 5$

Xmin = -5 Xmax = 5 Xscl = 1 Ymin = -10 Ymax = 10 Yscl = 1	Xmin = -5 Xmax = 10 Xscl = 1 Ymin = -80 Ymax = 80 Yscl = 20

In Exercises 51–54, determine whether the lines L_1 and L_2 passing through the pairs of points are parallel, perpendicular, or neither.

51. L_1: $(0, -1), (5, 9)$ **52.** L_1: $(-2, -1), (1, 5)$
      L_2: $(0, 3), (4, 1)$ L_2: $(1, 3), (5, -5)$

53. L_1: $(3, 6), (-6, 0)$ **54.** L_1: $(4, 8), (-4, 2)$
      L_2: $(0, -1), \left(5, \frac{7}{3}\right)$ L_2: $(3, -5), \left(-1, \frac{1}{3}\right)$

In Exercises 55–60, write the slope-intercept forms of the equations of the lines through the given point (a) parallel to the given line and (b) perpendicular to the given line.

Point	Line
55. $(2, 1)$	$4x - 2y = 3$
56. $(-3, 2)$	$x + y = 7$
57. $\left(-\frac{2}{3}, \frac{7}{8}\right)$	$3x + 4y = 7$
58. $(-3.9, -1.4)$	$6x + 2y = 9$
59. $(3, -2)$	$x - 4 = 0$
60. $(-4, 1)$	$y + 2 = 0$

Graphical Analysis   In Exercises 61–64, identify any relationships that exist among the lines, and then use a graphing utility to graph the three equations in the same viewing window. Adjust the viewing window so that each slope appears visually correct. Use the slopes of the lines to verify your results.

61. (a) $y = 2x$ (b) $y = -2x$ (c) $y = \frac{1}{2}x$

62. (a) $y = \frac{2}{3}x$ (b) $y = -\frac{3}{2}x$ (c) $y = \frac{2}{3}x + 2$

63. (a) $y = -\frac{1}{2}x$ (b) $y = -\frac{1}{2}x + 3$ (c) $y = 2x - 4$

64. (a) $y = x - 8$ (b) $y = x + 1$ (c) $y = -x + 3$

65. *Sales*   The following are the slopes of lines representing annual sales y in terms of time x in years. Use each slope to interpret any change in annual sales for a one-year increase in time.

(a) The line has a slope of $m = 135$.

(b) The line has a slope of $m = 0$.

(c) The line has a slope of $m = -40$.

66. *Revenue*   The following are the slopes of lines representing daily revenues y in terms of time x in days. Use each slope to interpret any change in daily revenues for a one-day increase in time.

(a) The line has a slope of $m = 400$.

(b) The line has a slope of $m = 100$.

(c) The line has a slope of $m = 0$.

67. *Earnings per Share*   The graph shows the earnings per share of stock for Circuit City for the years 1992 through 2002.   (Source: Circuit City Stores, Inc.)

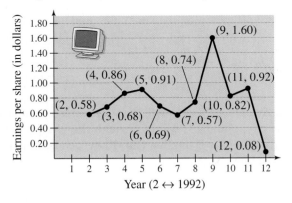

(a) Use the slopes to determine the year(s) in which the earnings per share of stock showed the greatest increase and decrease.

(b) Find the equation of the line between the years 1992 and 2002.

(c) Interpret the meaning of the slope of the equation from part (b) in the context of the problem.

(d) Use the equation from part (b) to estimate the earnings per share of stock for the year 2006. Do you think this is an accurate estimation? Explain.

68. *Earnings per Share*   The graph shows the earnings per share of stock for Harley-Davidson, Inc. for the years 1992 through 2002.   (Source: Harley-Davidson, Inc.)

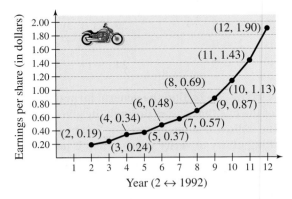

(a) Use the slopes to determine the years in which the earnings per share of stock showed the greatest increase and the smallest increase.

(b) Find the equation of the line between the years 1992 and 2002.

(c) Interpret the meaning of the slope of the equation from part (b) in the context of the problem.

(d) Use the equation from part (b) to estimate the earnings per share of stock for the year 2006. Do you think this is an accurate estimation? Explain.

69. *Height*   The "rise to run" ratio of the roof of a house determines the steepness of the roof. The rise to run ratio of a roof is 3 to 4. Determine the maximum height in the attic of the house if the house is 32 feet wide.

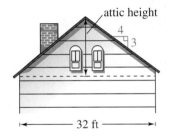

70. Road Grade   When driving down a mountain road, you notice warning signs indicating that it is a "12% grade." This means that the slope of the road is $-\frac{12}{100}$. Approximate the amount of horizontal change in your position if you note from elevation markers that you have descended 2000 feet vertically.

Rate of Change   In Exercises 71–74, you are given the dollar value of a product in 2004 and the rate at which the value of the product is expected to change during the next 5 years. Write a linear equation that gives the dollar value V of the product in terms of the year t. (Let $t = 4$ represent 2004.)

	2004 Value	*Rate*
71.	$2540	$125 increase per year
72.	$156	$4.50 increase per year
73.	$20,400	$2000 decrease per year
74.	$245,000	$5600 decrease per year

Graphical Interpretation   In Exercises 75–78, match the description with its graph. Determine the slope of each graph and how it is interpreted in the given context. [The graphs are labeled (a), (b), (c), and (d).]

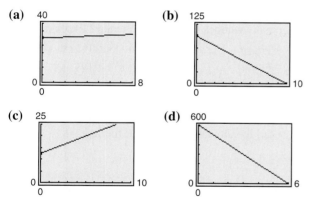

75. You are paying $10 per week to repay a $100 loan.

76. An employee is paid $12.50 per hour plus $1.50 for each unit produced per hour.

77. A sales representative receives $30 per day for food plus $0.35 for each mile traveled.

78. A word processor that was purchased for $600 depreciates $100 per year.

79. Meteorology   Find the equation of the line that shows the relationship between the temperature in degrees Celsius C and degrees Fahrenheit F. Remember that water freezes at $0°C$ ($32°F$) and boils at $100°C$ ($212°F$).

80. Meteorology   Use the result of Exercise 79 to complete the table.

C		$-10°$	$10°$			$177°$
F	$0°$			$68°$	$90°$	

81. Depreciation   A pizza shop purchases a used pizza oven for $875. After 5 years, the oven will have to be replaced.

(a) Write a linear equation giving the value V of the oven during the 5 years it will be used.

(b) Use a graphing utility to graph the linear equation representing the depreciation of the oven, and use the *value* or *trace* feature to complete the table.

t	0	1	2	3	4	5
V						

(c) Verify your answers in part (b) algebraically by using the equation you found in part (a).

82. Depreciation   A school district purchases a high-volume printer, copier, and scanner for $25,000. After 10 years, the equipment will have to be replaced. Its value at that time is expected to be $2000.

(a) Write a linear equation giving the value V of the equipment during the 10 years it will be used.

(b) Use a graphing utility to graph the linear equation representing the depreciation of the equipment, and use the *value* or *trace* feature to complete the table.

t	0	1	2	3	4	5	6	7	8	9	10
V											

(c) Verify your answers in part (b) algebraically by using the equation you found in part (a).

83. Cost, Revenue, and Profit   A contractor purchases a bulldozer for $36,500. The bulldozer requires an average expenditure of $5.25 per hour for fuel and maintenance, and the operator is paid $11.50 per hour.

(a) Write a linear equation giving the total cost C of operating the bulldozer for t hours. (Include the purchase cost of the bulldozer.)

(b) Assuming that customers are charged $27 per hour of bulldozer use, write an equation for the revenue R derived from t hours of use.

(c) Use the profit formula $(P = R - C)$ to write an equation for the profit derived from t hours of use.

(d) Use the result of part (c) to find the break-even point (the number of hours the bulldozer must be used to yield a profit of 0 dollars).

84. *Rental Demand*   A real estate office handles an apartment complex with 50 units. When the rent per unit is $580 per month, all 50 units are occupied. However, when the rent is $625 per month, the average number of occupied units drops to 47. Assume that the relationship between the monthly rent p and the demand x is linear.

(a) Write the equation of the line giving the demand x in terms of the rent p.

(b) Use a graphing utility to graph the demand equation and use the *trace* feature to estimate the number of units occupied when the rent is $655. Verify your answer algebraically.

(c) Use the demand equation to predict the number of units occupied when the rent is lowered to $595. Verify your answer graphically.

85. *Education*   In 1990, Penn State University had an enrollment of 75,365 students. By 2002, the enrollment had increased to 83,038.   (Source: Penn State Fact Book)

(a) What was the average annual change in enrollment from 1990 to 2002?

(b) Use the average annual change in enrollment to estimate the enrollments in 1984, 1997, and 2000.

(c) Write the equation of a line that represents the given data. What is its slope? Interpret the slope in the context of the problem.

86. *Writing*   Using the results from Exercise 85, write a short paragraph discussing the concepts of *slope* and *average rate of change*.

Synthesis

True or False?   **In Exercises 87 and 88, determine whether the statement is true or false. Justify your answer.**

87. The line through $(-8, 2)$ and $(-1, 4)$ and the line through $(0, -4)$ and $(-7, 7)$ are parallel.

88. If the points $(10, -3)$ and $(2, -9)$ lie on the same line, then the point $\left(-12, -\frac{37}{2}\right)$ also lies on that line.

Exploration   **In Exercises 89 and 90, use the values of a and b and a graphing utility to graph the equation of the line**

$$\frac{x}{a} + \frac{y}{b} = 1, \qquad a \neq 0, b \neq 0.$$

Use the graphs to make a conjecture about what a and b represent. Verify your conjecture.

89. $a = 5$, $b = -3$ 90. $a = -6$, $b = 2$

In Exercises 91–94, use the results of Exercises 89 and 90 to write an equation of the line that passes through the points.

91. x-intercept: $(2, 0)$ 92. x-intercept: $(-5, 0)$
    y-intercept: $(0, 3)$ y-intercept: $(0, -4)$

93. x-intercept: $\left(-\frac{1}{6}, 0\right)$ 94. x-intercept: $\left(\frac{3}{4}, 0\right)$
    y-intercept: $\left(0, -\frac{2}{3}\right)$ y-intercept: $\left(0, \frac{4}{5}\right)$

95. *Think About It*   The slopes of two lines are -3 and $\frac{5}{2}$. Which is steeper?

96. *Think About It*   Is it possible for two lines with positive slopes to be perpendicular? Explain.

97. *Writing*   Explain how you could show that the points $A(2, 3)$, $B(2, 9)$, and $C(7, 3)$ are the vertices of a right triangle.

98. *Writing*   Write a brief paragraph explaining whether or not any pair of points on a line can be used to calculate the slope of the line.

Review

In Exercises 99–104, determine whether the expression is a polynomial. If it is, write the polynomial in standard form.

99. $x + 20$ 100. $3x - 10x^2 + 1$

101. $4x^2 + x^{-1} - 3$ 102. $2x^2 - 2x^4 - x^3 + 2$

103. $\dfrac{x^2 + 3x + 4}{x^2 - 9}$ 104. $\sqrt{x^2 + 7x + 6}$

In Exercises 105–108, factor the trinomial.

105. $x^2 - 6x - 27$ 106. $x^2 - 11x + 28$

107. $2x^2 + 11x - 40$ 108. $3x^2 - 16x + 5$

1.3 Functions

Introduction to Functions

Many everyday phenomena involve pairs of quantities that are related to each other by some rule of correspondence. The mathematical term for such a rule of correspondence is a **relation.** Here are two examples.

1. The simple interest I earned on an investment of $1000 for 1 year is related to the annual interest rate r by the formula $I = 1000r$.

2. The area A of a circle is related to its radius r by the formula $A = \pi r^2$.

Not all relations have simple mathematical formulas. For instance, people commonly match up NFL starting quarterbacks with touchdown passes, and hours of the day with temperature. In each of these cases, there is some relation that matches each item from one set with exactly one item from a different set. Such a relation is called a **function.**

> #### Definition of a Function
>
> A **function** f from a set A to a set B is a relation that assigns to each element x in the set A exactly one element y in the set B. The set A is the **domain** (or set of inputs) of the function f, and the set B contains the **range** (or set of outputs).

To help understand this definition, look at the function that relates the time of day to the temperature in Figure 1.29.

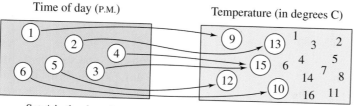

Time of day (P.M.)   Temperature (in degrees C)

Set A is the domain.
Inputs: 1, 2, 3, 4, 5, 6

Set B contains the range.
Outputs: 9, 10, 12, 13, 15

Figure 1.29

This function can be represented by the ordered pairs $\{(1, 9°), (2, 13°), (3, 15°),$ $(4, 15°), (5, 12°), (6, 10°)\}$. In each ordered pair, the first coordinate (x-value) is the **input** and the second coordinate (y-value) is the **output.**

> #### Characteristics of a Function from Set A to Set B
>
> 1. Each element of A must be matched with an element of B.
> 2. Some elements of B may not be matched with any element of A.
> 3. Two or more elements of A may be matched with the same element of B.
> 4. An element of A (the domain) cannot be matched with two different elements of B.

What you should learn

- Decide whether relations between two variables represent a function.
- Use function notation and evaluate functions.
- Find the domains of functions.
- Use functions to model and solve real-life problems.
- Evaluate difference quotients.

Why you should learn it

Many natural phenomena can be modeled by functions, such as the force of water against the face of a dam, explored in Exercise 81 on page 111.

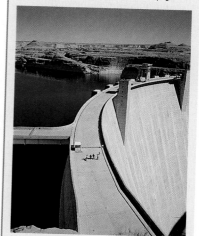

Kunio Owaki/Corbis

Library of Functions: Data Defined Function

Many functions do not have simple mathematical formulas, but are defined by real-life data. Such functions arise when you are using collections of data to model real-life applications. Functions can be represented in four ways.

1. *Verbally* by a sentence that describes how the input variable is related to the output variable

   Example: The input value x is the election year from 1952 to 2004 and the output value y is the elected president of the United States.

2. *Numerically* by a table or a list of ordered pairs that matches input values with output values

   Example: In the set of ordered pairs $\{(2, 34), (4, 40), (6, 45), (8, 50), (10, 54)\}$, the input value is the age of a male child in years and the output value is the height of the child in inches.

3. *Graphically* by points on a graph in a coordinate plane in which the input values are represented by the horizontal axis and the output values are represented by the vertical axis

   Example: See Figure 1.30.

4. *Algebraically* by an equation in two variables

   Example: The formula for temperature, $F = \frac{9}{5}C + 32$, where F is the temperature in degrees Fahrenheit and C is the temperature in degrees Celsius, is an equation that represents a function. You will see that it is often convenient to approximate data using a mathematical model or formula.

STUDY TIP

To determine whether or not a relation is a function, you must decide whether each input value is matched with exactly one output value. If any input value is matched with two or more output values, the relation is not a function.

Example 1 Testing for Functions

Decide whether the relation represents y as a function of x.

STUDY TIP

Be sure you see that the *range* of a function is not the same as the use of *range* relating to the viewing window of a graphing utility.

a.

Input x	2	2	3	4	5
Output y	11	10	8	5	1

b.

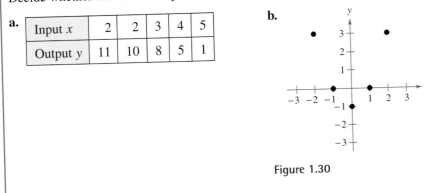

Figure 1.30

Solution

a. This table *does not* describe y as a function of x. The input value 2 is matched with two different y-values.

b. The graph in Figure 1.30 *does* describe y as a function of x. Each input value is matched with exactly one output value.

✓ *Checkpoint* Now try Exercise 5.

In algebra, it is common to represent functions by equations or formulas involving two variables. For instance, the equation $y = x^2$ represents the variable y as a function of the variable x. In this equation, x is the **independent variable** and y is the **dependent variable.** The domain of the function is the set of all values taken on by the independent variable x, and the range of the function is the set of all values taken on by the dependent variable y.

Example 2   Testing for Functions Represented Algebraically

Which of the equations represent(s) y as a function of x?

a. $x^2 + y = 1$ **b.** $-x + y^2 = 1$

Solution

To determine whether y is a function of x, try to solve for y in terms of x.

a. Solving for y yields

$$x^2 + y = 1 \qquad \text{Write original equation.}$$

$$y = 1 - x^2. \qquad \text{Solve for } y.$$

Each value of x corresponds to exactly one value of y. So, y is a function of x.

b. Solving for y yields

$$-x + y^2 = 1 \qquad \text{Write original equation.}$$

$$y^2 = 1 + x \qquad \text{Add } x \text{ to each side.}$$

$$y = \pm\sqrt{1 + x}. \qquad \text{Solve for } y.$$

The $\pm$ indicates that for a given value of x there correspond two values of y. For instance, when $x = 3$, $y = 2$ or $y = -2$. So, y is not a function of x.

✓ *Checkpoint* Now try Exercise 19.

Function Notation

When an equation is used to represent a function, it is convenient to name the function so that it can be referenced easily. For example, you know that the equation $y = 1 - x^2$ describes y as a function of x. Suppose you give this function the name "f." Then you can use the following **function notation.**

Input	Output	Equation
x	$f(x)$	$f(x) = 1 - x^2$

The symbol $f(x)$ is read as the *value of f at x* or simply *f of x*. The symbol $f(x)$ corresponds to the y-value for a given x. So, you can write $y = f(x)$. Keep in mind that f is the *name* of the function, whereas $f(x)$ is the *output value* of the function at the *input value* x. In function notation, the *input* is the independent variable and the *output* is the dependent variable. For instance, the function $f(x) = 3 - 2x$ has *function values* denoted by $f(-1), f(0)$, and so on. To find these values, substitute the specified input values into the given equation.

For $x = -1$, $f(-1) = 3 - 2(-1) = 3 + 2 = 5$.

For $x = 0$, $f(0) = 3 - 2(0) = 3 - 0 = 3$.

Exploration

Use a graphing utility to graph $x^2 + y = 1$. Then use the graph to write a convincing argument that each x-value has at most one y-value.

Use a graphing utility to graph $-x + y^2 = 1$. (*Hint:* You will need to use two equations.) Does the graph represent y as a function of x? Explain.

TECHNOLOGY TIP

You can use a graphing utility to evaluate a function. Use the Evaluating an Algebraic Expression Program found on the website *college.hmco.com.* The program will prompt you for a value of x, and then evaluate the expression in the equation editor for that value of x. Try using the program to evaluate several different functions of x.

Although f is often used as a convenient function name and x is often used as the independent variable, you can use other letters. For instance,

$$f(x) = x^2 - 4x + 7, \quad f(t) = t^2 - 4t + 7, \quad \text{and} \quad g(s) = s^2 - 4s + 7$$

all define the same function. In fact, the role of the independent variable is that of a "placeholder." Consequently, the function could be described by

$$f() = ()^2 - 4() + 7.$$

Example 3   Evaluating a Function

Let $g(x) = -x^2 + 4x + 1$. Find (a) $g(2)$, (b) $g(t)$, and (c) $g(x + 2)$.

Solution

a. Replacing x with 2 in $g(x) = -x^2 + 4x + 1$ yields the following.

$$g(2) = -(2)^2 + 4(2) + 1 = -4 + 8 + 1 = 5$$

b. Replacing x with t yields the following.

$$g(t) = -(t)^2 + 4(t) + 1 = -t^2 + 4t + 1$$

c. Replacing x with $x + 2$ yields the following.

$$
\begin{aligned}
g(x + 2) &= -(x + 2)^2 + 4(x + 2) + 1 & &\text{Substitute } x + 2 \text{ for } x. \\
&= -(x^2 + 4x + 4) + 4x + 8 + 1 & &\text{Multiply.} \\
&= -x^2 - 4x - 4 + 4x + 8 + 1 & &\text{Distributive Property} \\
&= -x^2 + 5 & &\text{Simplify.}
\end{aligned}
$$

✓ *Checkpoint* Now try Exercise 33.

In Example 3, note that $g(x + 2)$ is not equal to $g(x) + g(2)$. In general, $g(u + v) \neq g(u) + g(v)$.

Library of Functions: Piecewise-Defined Function

A *piecewise-defined function* is a function that is defined by two or more equations over a specified domain. The *absolute value function* given by $f(x) = |x|$ can be written as a piecewise-defined function. The basic characteristics of the absolute value function are summarized below.

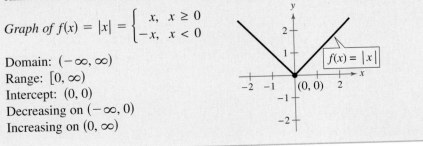

Graph of $f(x) = |x| = \begin{cases} x, & x \geq 0 \\ -x, & x < 0 \end{cases}$

Domain: $(-\infty, \infty)$
Range: $[0, \infty)$
Intercept: $(0, 0)$
Decreasing on $(-\infty, 0)$
Increasing on $(0, \infty)$

$f(x) = |x|$

An illustration of a piecewise-defined function is given in Example 4.

Example 4   A Piecewise–Defined Function

Evaluate the function when $x = -1$ and 0.

$$f(x) = \begin{cases} x^2 + 1, & x < 0 \\ x - 1, & x \geq 0 \end{cases}$$

Solution

Because $x = -1$ is less than 0, use $f(x) = x^2 + 1$ to obtain

$$f(-1) = (-1)^2 + 1 = 2.$$

For $x = 0$, use $f(x) = x - 1$ to obtain

$$f(0) = (0) - 1 = -1.$$

✓ *Checkpoint* Now try Exercise 37.

TECHNOLOGY TIP

Most graphing utilities can graph piecewise-defined functions. For instructions on how to enter a piecewise-defined function into your graphing utility, consult your user's manual. You may find it helpful to set your graphing utility to *dot mode* before graphing.

The Domain of a Function

The domain of a function can be described explicitly or it can be *implied* by the expression used to define the function. The **implied domain** is the set of all real numbers for which the expression is defined. For instance, the function

$$f(x) = \frac{1}{x^2 - 4} \qquad \text{Domain excludes } x\text{-values that result in division by zero.}$$

has an implied domain that consists of all real x other than $x = \pm 2$. These two values are excluded from the domain because division by zero is undefined. Another common type of implied domain is that used to avoid even roots of negative numbers. For example, the function

$$f(x) = \sqrt{x} \qquad \text{Domain excludes } x\text{-values that result in even roots of negative numbers.}$$

is defined only for $x \geq 0$. So, its implied domain is the interval $[0, \infty)$. In general, the domain of a function *excludes* values that would cause division by zero *or* result in the even root of a negative number.

Exploration

Use a graphing utility to graph $y = \sqrt{4 - x^2}$. What is the domain of this function? Then graph $y = \sqrt{x^2 - 4}$. What is the domain of this function? Do the domains of these two functions overlap? If so, for what values?

Library of Functions: Radical Function

Radical functions arise from the use of rational exponents. The most common radical function is the *square root function* given by $f(x) = \sqrt{x}$. The basic characteristics of the square root function are summarized below.

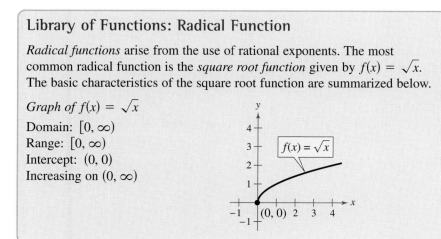

Graph of $f(x) = \sqrt{x}$

Domain: $[0, \infty)$
Range: $[0, \infty)$
Intercept: $(0, 0)$
Increasing on $(0, \infty)$

STUDY TIP

Because the square root function is not defined for $x < 0$, you must be careful when analyzing the domains of complicated functions involving the square root symbol.

Example 5   Finding the Domain of a Function

Find the domain of each function.

a. $f: \{(-3, 0), (-1, 4), (0, 2), (2, 2), (4, -1)\}$

b. $g(x) = -3x^2 + 4x + 5$ **c.** $h(x) = \dfrac{1}{x + 5}$

d. Volume of a sphere: $V = \frac{4}{3}\pi r^3$ **e.** $k(x) = \sqrt{4 - 3x}$

Solution

a. The domain of f consists of all first coordinates in the set of ordered pairs.

   Domain $= \{-3, -1, 0, 2, 4\}$

b. The domain of g is the set of all *real* numbers.

c. Excluding x-values that yield zero in the denominator, the domain of h is the set of all real numbers $x \neq -5$.

d. Because this function represents the volume of a sphere, the values of the radius r must be positive. So, the domain is the set of all real numbers r such that $r > 0$.

e. This function is defined only for x-values for which $4 - 3x \geq 0$. By solving this inequality, you will find that the domain of k is all real numbers that are less than or equal to $\frac{4}{3}$.

 Checkpoint   Now try Exercise 51.

STUDY TIP

In Example 5(e), $4 - 3x \geq 0$ is a *linear inequality*. For help with solving linear inequalities, see Appendix D. You will study more about inequalities in Section 2.5.

   In Example 5(d), note that the *domain of a function may be implied by the physical context*. For instance, from the equation $V = \frac{4}{3}\pi r^3$, you would have no reason to restrict r to positive values, but the physical context implies that a sphere cannot have a negative or zero radius.

   For some functions, it may be easier to find the domain and range of the function by examining its graph.

Example 6   Finding the Domain and Range of a Function

Use a graphing utility to find the domain and range of the function

$$f(x) = \sqrt{9 - x^2}.$$

Solution

Graph the function as $y = \sqrt{9 - x^2}$, as shown in Figure 1.31. Using the *trace* feature of a graphing utility, you can determine that the x-values extend from -3 to 3 and the y-values extend from 0 to 3. So, the domain of the function f is all real numbers such that $-3 \leq x \leq 3$ and the range of f is all real numbers such that $0 \leq y \leq 3$.

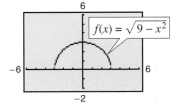

Figure 1.31

 Checkpoint   Now try Exercise 61.

Applications

Example 7   Cellular Phone Subscribers

The number N (in millions) of cellular phone subscribers in the United States increased in a linear pattern from 1995 to 1997, as shown in Figure 1.32. Then, in 1998, the number of subscribers took a jump, and until 2001, increased in a *different* linear pattern. These two patterns can be approximated by the function

$$N(t) = \begin{cases} 10.75t - 20.1, & 5 \le t \le 7 \\ 20.11t - 92.8, & 8 \le t \le 11 \end{cases}$$

where t represents the year, with $t = 5$ corresponding to 1995. Use this function to approximate the number of cellular phone subscribers for each year from 1995 to 2001.   (Source: Cellular Telecommunications & Internet Association)

Solution

From 1995 to 1997, use $N(t) = 10.75t - 20.1$

$$\underbrace{33.7,}_{1995} \quad \underbrace{44.4,}_{1996} \quad \underbrace{55.2}_{1997}$$

From 1998 to 2001, use $N(t) = 20.11t - 92.8$.

$$\underbrace{68.1,}_{1998} \quad \underbrace{88.2,}_{1999} \quad \underbrace{108.3,}_{2000} \quad \underbrace{128.4}_{2001}$$

✔ *Checkpoint* Now try Exercise 79.

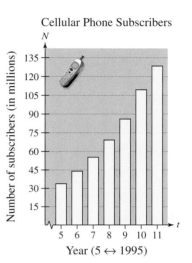

Cellular Phone Subscribers

Figure 1.32

Example 8   The Path of a Baseball

A baseball is hit at a point 3 feet above the ground at a velocity of 100 feet per second and an angle of 45°. The path of the baseball is given by the function

$$f(x) = -0.0032x^2 + x + 3$$

where y and x are measured in feet. Will the baseball clear a 10-foot fence located 300 feet from home plate?

Algebraic Solution

The height of the baseball is a function of the horizontal distance from home plate. When $x = 300$, you can find the height of the baseball as follows.

$$f(x) = -0.0032x^2 + x + 3 \qquad \text{Write original function.}$$

$$f(300) = -0.0032(300)^2 + 300 + 3 \qquad \text{Substitute 300 for } x.$$

$$= 15 \qquad \text{Simplify.}$$

When $x = 300$, the height of the baseball is 15 feet, so the baseball will clear a 10-foot fence.

✔ *Checkpoint* Now try Exercise 81.

Graphical Solution

Use a graphing utility to graph the function $y = -0.0032x^2 + x + 3$. Use the *value* feature or the *zoom* and *trace* features of the graphing utility to estimate that $y = 15$ when $x = 300$, as shown in Figure 1.33. So, the ball will clear a 10-foot fence.

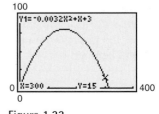

Figure 1.33

Difference Quotients

One of the basic definitions in calculus employs the ratio

$$\frac{f(x + h) - f(x)}{h}, \qquad h \neq 0.$$

This ratio is called a **difference quotient,** as illustrated in Example 9.

Example 9   Evaluating a Difference Quotient

For $f(x) = x^2 - 4x + 7$, find $\dfrac{f(x + h) - f(x)}{h}$.

Solution

$$\frac{f(x + h) - f(x)}{h} = \frac{[(x + h)^2 - 4(x + h) + 7] - (x^2 - 4x + 7)}{h}$$

$$= \frac{x^2 + 2xh + h^2 - 4x - 4h + 7 - x^2 + 4x - 7}{h}$$

$$= \frac{2xh + h^2 - 4h}{h}$$

$$= \frac{h(2x + h - 4)}{h} = 2x + h - 4, \ h \neq 0$$

✓ *Checkpoint* Now try Exercise 85.

> **STUDY TIP**
>
> Notice in Example 9 that h cannot be zero in the original expression. Therefore, you must restrict the domain of the simplified expression by adding $h \neq 0$ so that the simplified expression is equivalent to the original expression.

Summary of Function Terminology

Function: A **function** is a relationship between two variables such that to each value of the independent variable there corresponds exactly one value of the dependent variable.

Function Notation: $y = f(x)$

  f is the *name* of the function.
  y is the **dependent variable,** or output value.
  x is the **independent variable,** or input value.
  $f(x)$ is the *value of the function at x.*

Domain: The **domain** of a function is the set of all values (inputs) of the independent variable for which the function is defined. If x is in the domain of f, f is said to be *defined* at x. If x is not in the domain of f, f is said to be *undefined* at x.

Range: The **range** of a function is the set of all values (outputs) assumed by the dependent variable (that is, the set of all function values).

Implied Domain: If f is defined by an algebraic expression and the domain is not specified, the **implied domain** consists of all real numbers for which the expression is defined.

The symbol ∫ indicates an example or exercise that highlights algebraic techniques specifically used in calculus.

1.3   Exercises

Vocabulary Check

Fill in the blanks.

1. A relation that assigns to each element x from a set of inputs, or _____ , exactly one element y in a set of outputs, or _____ , is called a _____ .

2. For an equation that represents y as a function of x, the _____ variable is the set of all x in the domain, and the _____ variable is the set of all y in the range.

3. The function $f(x) = \begin{cases} x^2 - 4, x \le 0 \\ 2x + 1, x > 0 \end{cases}$ is an example of a _____ function.

4. If the domain of the function f is not given, then the set of values of the independent variable for which the expression is defined is called the _____ .

5. In calculus, one of the basic definitions is that of a _____ , given by $\dfrac{f(x + h) - f(x)}{h}, h \neq 0$.

In Exercises 1–4, does the relationship describe a function? Explain your reasoning.

1. *Domain*   *Range*   2. *Domain*   *Range*

$$-2 \longrightarrow 5$$
$$-1 \quad 6$$
$$0 \quad 7$$
$$1 \quad 8$$
$$2$$

$$-2 \longrightarrow 3$$
$$-1 \longrightarrow 4$$
$$0 \quad 5$$
$$1$$
$$2$$

3. *Domain*   *Range*   4. *Domain*   *Range*

National League → Cubs, Pirates, Dodgers

American League → Orioles, Yankees, Twins

(Year)   (Number of North Atlantic tropical storms and hurricanes)

1994 → 7
1995 → 12
1996 → 13
1997 → 14
1998 → 15
1999 → 19
2000
2001

In Exercises 5–8, does the table describe a function? Explain your reasoning.

5.

Input Value	-2	-1	0	1	2
Output Value	-8	-1	0	1	8

6.

Input Value	0	1	2	1	0
Output Value	-4	-2	0	2	4

7.

Input Value	10	7	4	7	10
Output Value	3	6	9	12	15

8.

Input Value	0	3	9	12	15
Output Value	3	3	3	3	3

In Exercises 9 and 10, which sets of ordered pairs represent functions from A to B? Explain.

9. $A = \{0, 1, 2, 3\}$ and $B = \{-2, -1, 0, 1, 2\}$
   (a) $\{(0, 1), (1, -2), (2, 0), (3, 2)\}$
   (b) $\{(0, -1), (2, 2), (1, -2), (3, 0), (1, 1)\}$
   (c) $\{(0, 0), (1, 0), (2, 0), (3, 0)\}$
   (d) $\{(0, 2), (3, 0), (1, 1)\}$

10. $A = \{a, b, c\}$ and $B = \{0, 1, 2, 3\}$
    (a) $\{(a, 1), (c, 2), (c, 3), (b, 3)\}$
    (b) $\{(a, 1), (b, 2), (c, 3)\}$
    (c) $\{(1, a), (0, a), (2, c), (3, b)\}$
    (d) $\{(c, 0), (b, 0), (a, 3)\}$

Circulation of Newspapers   **In Exercises 11 and 12, use the graph, which shows the circulation (in millions) of daily newspapers in the United States.** (Source: Editor & Publisher Company)

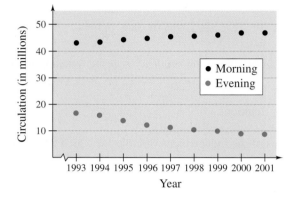

Year

11. Is the circulation of morning newspapers a function of the year? Is the circulation of evening newspapers a function of the year? Explain.

12. Let $f(x)$ represent the circulation of evening newspapers in year x. Find $f(2000)$.

In Exercises 13–24, determine whether the equation represents y as a function of x.

13. $x^2 + y^2 = 4$ 14. $x = y^2$

15. $x^2 + y = -1$ 16. $y = \sqrt{x + 5}$

17. $2x + 3y = 4$ 18. $x = -y + 5$

19. $y^2 = x^2 - 1$ 20. $x + y^2 = 3$

21. $y = |4 - x|$ 22. $|y| = 4 - x$

23. $x = -7$ 24. $y = 8$

In Exercises 25 and 26, fill in the blanks using the specified function and the given values of the independent variable. Simplify the result.

25. $f(x) = \dfrac{1}{x + 1}$

   (a) $f(4) = \dfrac{1}{() + 1}$

   (b) $f(0) = \dfrac{1}{() + 1}$

   (c) $f(4t) = \dfrac{1}{() + 1}$

   (d) $f(x + c) = \dfrac{1}{() + 1}$

26. $g(x) = x^2 - 2x$

   (a) $g(2) = ()^2 - 2()$

   (b) $g(-3) = ()^2 - 2()$

   (c) $g(t + 1) = ()^2 - 2()$

   (d) $g(x + c) = ()^2 - 2()$

In Exercises 27–38, evaluate the function at each specified value of the independent variable and simplify.

27. $f(x) = 2x - 3$
   (a) $f(1)$ (b) $f(-3)$ (c) $f(x - 1)$

28. $g(y) = 7 - 3y$
   (a) $g(0)$ (b) $g(\frac{7}{3})$ (c) $g(s + 2)$

29. $h(t) = t^2 - 2t$
   (a) $h(2)$ (b) $h(1.5)$ (c) $h(x + 2)$

30. $V(r) = \frac{4}{3}\pi r^3$
   (a) $V(3)$ (b) $V(\frac{3}{2})$ (c) $V(2r)$

31. $f(y) = 3 - \sqrt{y}$
   (a) $f(4)$ (b) $f(0.25)$ (c) $f(4x^2)$

32. $f(x) = \sqrt{x + 8} + 2$
   (a) $f(-8)$ (b) $f(1)$ (c) $f(x - 8)$

33. $q(x) = \dfrac{1}{x^2 - 9}$
   (a) $q(0)$ (b) $q(3)$ (c) $q(y + 3)$

34. $q(t) = \dfrac{2t^2 + 3}{t^2}$
   (a) $q(2)$ (b) $q(0)$ (c) $q(-x)$

35. $f(x) = \dfrac{|x|}{x}$
   (a) $f(2)$ (b) $f(-2)$ (c) $f(x^2)$

36. $f(x) = |x| + 4$
   (a) $f(2)$ (b) $f(-2)$ (c) $f(x^2)$

37. $f(x) = \begin{cases} 2x + 1, & x < 0 \\ 2x + 2, & x \ge 0 \end{cases}$
   (a) $f(-1)$ (b) $f(0)$ (c) $f(2)$

38. $f(x) = \begin{cases} x^2 + 2, & x \le 1 \\ 2x^2 + 2, & x > 1 \end{cases}$
   (a) $f(-2)$ (b) $f(1)$ (c) $f(2)$

In Exercises 39–42, complete the table.

39. $h(t) = \frac{1}{2}|t + 3|$

t	-5	-4	-3	-2	-1
$h(t)$					

40. $f(s) = \dfrac{|s - 2|}{s - 2}$

s	0	1	$\frac{3}{2}$	$\frac{5}{2}$	4
$f(s)$					

41. $f(x) = \begin{cases} -\frac{1}{2}x + 4, & x \le 0 \\ (x - 2)^2, & x > 0 \end{cases}$

x	-2	-1	0	1	2
$f(x)$					

42. $h(x) = \begin{cases} 9 - x^2, & x < 3 \\ x - 3, & x \ge 3 \end{cases}$

x	1	2	3	4	5
$h(x)$					

In Exercises 43–46, find all real values of x such that $f(x) = 0$.

43. $f(x) = 15 - 3x$

44. $f(x) = 5x + 1$

45. $f(x) = \dfrac{3x - 4}{5}$

46. $f(x) = \dfrac{12 - x^2}{5}$

In Exercises 47 and 48, find the value(s) of x for which $f(x) = g(x)$.

47. $f(x) = x^2, \quad g(x) = x + 2$

48. $f(x) = x^2 + 2x + 1, \quad g(x) = 3x + 3$

In Exercises 49–58, find the domain of the function.

49. $f(x) = 5x^2 + 2x - 1$

50. $g(x) = 1 - 2x^2$

51. $h(t) = \dfrac{4}{t}$

52. $s(y) = \dfrac{3y}{y + 5}$

53. $f(x) = \sqrt[3]{x - 4}$

54. $f(x) = \sqrt[4]{x^2 + 3x}$

55. $g(x) = \dfrac{1}{x} - \dfrac{3}{x + 2}$

56. $h(x) = \dfrac{10}{x^2 - 2x}$

57. $g(y) = \dfrac{y + 2}{\sqrt{y - 10}}$

58. $f(x) = \dfrac{\sqrt{x + 6}}{6 + x}$

In Exercises 59–62, use a graphing utility to graph the function. Find the domain and range of the function.

59. $f(x) = \sqrt{4 - x^2}$

60. $f(x) = \sqrt{x^2 + 1}$

61. $g(x) = |2x + 3|$

62. $g(x) = |x - 5|$

In Exercises 63–66, assume that the domain of f is the set $A = \{-2, -1, 0, 1, 2\}$. Determine the set of ordered pairs representing the function f.

63. $f(x) = x^2$

64. $f(x) = x^2 - 3$

65. $f(x) = |x| + 2$

66. $f(x) = |x + 1|$

67. *Geometry*   Write the area A of a circle as a function of its circumference C.

68. *Geometry*   Write the area A of an equilateral triangle as a function of the length s of its sides.

69. *Exploration*   The cost per unit to produce a radio model is $60. The manufacturer charges $90 per unit for orders of 100 or less. To encourage large orders, the manufacturer reduces the charge by $0.15 per radio for each unit ordered in excess of 100 (for example, there would be a charge of $87 per radio for an order size of 120).

(a) The table shows the profit P (in dollars) for various numbers of units ordered, x. Use the table to estimate the maximum profit.

Units, x	Profit, P
110	3135
120	3240
130	3315
140	3360
150	3375
160	3360
170	3315

(b) Plot the points (x, P) from the table in part (a). Does the relation defined by the ordered pairs represent P as a function of x?

(c) If P is a function of x, write the function and determine its domain.

70. *Exploration* An open box of maximum volume is to be made from a square piece of material, 24 centimeters on a side, by cutting equal squares from the corners and turning up the sides (see figure).

(a) The table shows the volume V (in cubic centimeters) of the box for various heights x (in centimeters). Use the table to estimate the maximum volume.

Height, x	Volume, V
1	484
2	800
3	972
4	1024
5	980
6	864

(b) Plot the points (x, V) from the table in part (a). Does the relation defined by the ordered pairs represent V as a function of x?

(c) If V is a function of x, write the function and determine its domain.

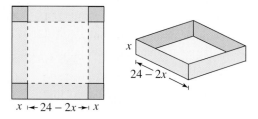

$x \mid\leftarrow 24 - 2x \rightarrow\mid x$

71. *Geometry* A right triangle is formed in the first quadrant by the x- and y-axes and a line through the point $(2, 1)$ (see figure). Write the area A of the triangle as a function of x, and determine the domain of the function.

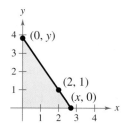

72. *Geometry* A rectangle is bounded by the x-axis and the semicircle $y = \sqrt{36 - x^2}$ (see figure). Write the area A of the rectangle as a function of x, and determine the domain of the function.

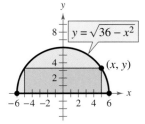

73. *Postal Regulations* A rectangular package to be sent by the U.S. Postal Service can have a maximum combined length and girth (perimeter of a cross section) of 108 inches (see figure).

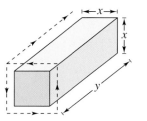

(a) Write the volume V of the package as a function of x.

(b) What is the domain of the function?

(c) Use a graphing utility to graph the function. Be sure to use the appropriate viewing window.

(d) What dimensions will maximize the volume of the package? Explain.

74. *Cost, Revenue, and Profit* A company produces a toy for which the variable cost is $12.30 per unit and the fixed costs are $98,000. The toy sells for $17.98. Let x be the number of units produced and sold.

(a) The total cost for a business is the sum of the variable cost and the fixed costs. Write the total cost C as a function of the number of units produced.

(b) Write the revenue R as a function of the number of units sold.

(c) Write the profit P as a function of the number of units sold. (*Note:* $P = R - C$.)

Revenue In Exercises 75–78, use the table, which shows the monthly revenue y (in thousands of dollars) of a landscaping business for each month of 2003, with $x = 1$ representing January.

Month, x	Revenue, y
1	5.2
2	5.6
3	6.6
4	8.3
5	11.5
6	15.8
7	12.8
8	10.1
9	8.6
10	6.9
11	4.5
12	2.7

A mathematical model that represents this data is

$$f(x) = \begin{cases} -1.97x + 26.3 \\ 0.505x^2 - 1.47x + 6.3 \end{cases}$$

75. What is the domain of each part of the piecewise-defined function? Explain your reasoning.

76. Use the mathematical model to find $f(5)$. Interpret your results in the context of the problem.

77. Use the mathematical model to find $f(11)$. Interpret your results in the context of the problem.

78. How do the values obtained from the model in Exercises 76 and 77 compare with the actual data values?

79. *Motor Vehicles*   The number n (in billions) of miles traveled by vans, pickup trucks, and sport utility vehicles in the United States from 1990 to 2000 can be approximated by the model

$$n(t) = \begin{cases} -9.2t^2 + 84.5t + 575, & 0 \le t \le 4 \\ 26.8t + 657, & 5 \le t \le 10 \end{cases}$$

where t represents the year, with $t = 0$ corresponding to 1990. Use the *table* feature of a graphing utility to approximate the number of miles traveled by vans, pickup trucks, and sport utility vehicles for each year from 1990 to 2000. (Source: U.S. Federal Highway Administration)

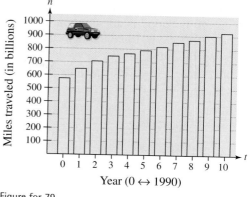

Figure for 79

80. *Transportation*   For groups of 80 or more people, a charter bus company determines the rate per person according to the formula

$$\text{Rate} = 8 - 0.05(n - 80), \quad n \ge 80$$

where the rate is given in dollars and n is the number of people.

(a) Write the revenue R of the bus company as a function of n.

(b) Use the function from part (a) to complete the table. What can you conclude?

n	90	100	110	120	130	140	150
$R(n)$							

(c) Use a graphing utility to graph R and determine the number of people that will produce a maximum revenue. Compare the result with your conclusion from part (b).

81. *Physics*   The force F (in tons) of water against the face of a dam is estimated by the function $F(y) = 149.76\sqrt{10}y^{5/2}$, where y is the depth of the water in feet.

(a) Complete the table. What can you conclude from the table?

y	5	10	20	30	40
$F(y)$					

(b) Use a graphing utility to graph the function. Describe your viewing window.

(c) Use the table to approximate the depth at which the force against the dam is 1,000,000 tons. How could you find a better estimate?

(d) Verify your answer in part (c) graphically.

82. Data Analysis The graph shows the retail sales (in billions of dollars) of prescription drugs in the United States from 1995 through 2001. Let $f(x)$ represent the retail sales in year x. (Source: National Association of Chain Drug Stores)

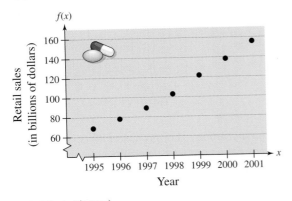

(a) Find $f(1998)$.

(b) Find $\dfrac{f(2001) - f(1995)}{2001 - 1995}$

and interpret the result in the context of the problem.

(c) An approximate model for the function is

$$P(t) = -0.1556t^3 + 4.657t^2 - 28.75t + 115.7,$$
$$5 \le t \le 11$$

where P is the retail sales (in billions of dollars) and t represents the year, with $t = 5$ corresponding to 1995. Complete the table and compare the results with the data.

t	5	6	7	8	9	10	11
$P(t)$							

(d) Use a graphing utility to graph the model and data in the same viewing window. Comment on the validity of the model.

ʃ In Exercises 83–88, find the difference quotient and simplify your answer.

83. $f(x) = 2x, \quad \dfrac{f(x + c) - f(x)}{c}, \quad c \neq 0$

84. $g(x) = 3x - 1, \quad \dfrac{g(x + h) - g(x)}{h}, \quad h \neq 0$

85. $f(x) = x^2 - x + 1, \quad \dfrac{f(2 + h) - f(2)}{h}, \quad h \neq 0$

86. $f(x) = x^3 + x, \quad \dfrac{f(x + h) - f(x)}{h}, \quad h \neq 0$

87. $f(t) = \dfrac{1}{t}, \quad \dfrac{f(t) - f(1)}{t - 1}, \quad t \neq 1$

88. $f(x) = \dfrac{4}{x + 1}, \quad \dfrac{f(x) - f(7)}{x - 7}, \quad x \neq 7$

Synthesis

True or False? In Exercises 89 and 90, determine whether the statement is true or false. Justify your answer.

89. The domain of the function $f(x) = x^4 - 1$ is $(-\infty, \infty)$, and the range of $f(x)$ is $(0, \infty)$.

90. The set of ordered pairs $\{(-8, -2), (-6, 0), (-4, 0), (-2, 2), (0, 4), (2, -2)\}$ represents a function.

Exploration In Exercises 91 and 92, match the data with one of the functions $g(x) = cx^2$ or $r(x) = c/x$ and determine the value of the constant c such that the function fits the data given in the table.

91.

x	-4	-1	0	1	4
y	-8	-32	Undef.	32	8

92.

x	-4	-1	0	1	4
y	-32	-2	0	-2	-32

93. Writing In your own words, explain the meanings of *domain* and *range*.

94. Think About It Describe an advantage of function notation.

Review

In Exercises 95–98, perform the operations and simplify.

95. $12 - \dfrac{4}{x + 2}$

96. $\dfrac{3}{x^2 + x - 20} + \dfrac{x}{x^2 + 4x - 5}$

97. $\dfrac{2x^3 + 11x^2 - 6x}{5x} \cdot \dfrac{x + 10}{2x^2 + 5x - 3}$

98. $\dfrac{x + 7}{2(x - 9)} \div \dfrac{x - 7}{2(x - 9)}$

The symbol ʃ indicates an example or exercise that highlights algebraic techniques specifically used in calculus.

1.4 Graphs of Functions

The Graph of a Function

In Section 1.3, functions were represented graphically by points on a graph in a coordinate plane in which the input values are represented by the horizontal axis and the output values are represented by the vertical axis. The **graph of a function** f is the collection of ordered pairs $(x, f(x))$ such that x is in the domain of f. As you study this section, remember the geometric interpretations of x and $f(x)$.

x = the directed distance from the y-axis

$f(x)$ = the directed distance from the x-axis

Example 1 shows how to use the graph of a function to find the domain and range of the function.

What you should learn

- Find the domains and ranges of functions and use the Vertical Line Test for functions.
- Determine intervals on which functions are increasing, decreasing, or constant.
- Determine relative maximum and relative minimum values of functions.
- Identify and graph step functions and other piecewise-defined functions.
- Identify even and odd functions.

Why you should learn it

Graphs of functions provide a visual relationship between two variables. Exercise 81 on page 123 shows how the graph of a step function can represent the cost of a telephone call.

Example 1   Finding the Domain and Range of a Function

Use the graph of the function f shown in Figure 1.34 to find (a) the domain of f, (b) the function values $f(-1)$ and $f(2)$, and (c) the range of f.

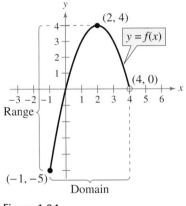

Figure 1.34

Jeff Greenberg/Peter Arnold, Inc.

Solution

a. The closed dot at $(-1, -5)$ indicates that $x = -1$ is in the domain of f, whereas the open dot at $(4, 0)$ indicates that $x = 4$ is not in the domain. So, the domain of f is all x in the interval $[-1, 4)$.

b. Because $(-1, -5)$ is a point on the graph of f, it follows that

$$f(-1) = -5.$$

Similarly, because $(2, 4)$ is a point on the graph of f, it follows that

$$f(2) = 4.$$

c. Because the graph does not extend below $f(-1) = -5$ or above $f(2) = 4$, the range of f is the interval $[-5, 4]$.

✓ *Checkpoint* Now try Exercise 3.

STUDY TIP

The use of dots (open or closed) at the extreme left and right points of a graph indicates that the graph does not extend beyond these points. If no such dots are shown, assume that the graph extends beyond these points.

Example 2 Finding the Domain and Range of a Function

Find the domain and range of

$$f(x) = \sqrt{x - 4}.$$

Algebraic Solution

Because the expression under a radical cannot be negative, the domain of $f(x) = \sqrt{x - 4}$ is the set of all real numbers such that $x - 4 \geq 0$. Solve this linear inequality for x as follows. (For help with solving linear inequalities, see Appendix D.)

$$x - 4 \geq 0 \qquad \text{Write original inequality.}$$

$$x \geq 4 \qquad \text{Add 4 to each side.}$$

So, the domain is the set of all real numbers greater than or equal to 4. Because the value of a radical expression is never negative, the range of $f(x) = \sqrt{x - 4}$ is the set of all nonnegative real numbers.

 Checkpoint Now try Exercise 7.

Graphical Solution

Use a graphing utility to graph the equation $y = \sqrt{x - 4}$, as shown in Figure 1.35. Use the *trace* feature to determine that the x-coordinates of points on the graph extend from 4 to the right. When x is greater than or equal to 4, the expression under the radical is nonnegative. So, you can conclude that the domain is the set of all real numbers greater than or equal to 4. From the graph, you can see that the y-coordinates of points on the graph extend from 0 upwards. So you can estimate the range to be the set of all nonnegative real numbers.

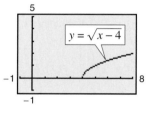

Figure 1.35

By the definition of a function, at most one y-value corresponds to a given x-value. It follows, then, that a vertical line can intersect the graph of a function at most once. This leads to the **Vertical Line Test** for functions.

> **Vertical Line Test for Functions**
>
> A set of points in a coordinate plane is the graph of y as a function of x if and only if no vertical line intersects the graph at more than one point.

Example 3 Vertical Line Test for Functions

Use the Vertical Line Test to decide whether the graphs in Figure 1.36 represent y as a function of x.

Solution

a. This is *not* a graph of y as a function of x because you can find a vertical line that intersects the graph twice.

b. This *is* a graph of y as a function of x because every vertical line intersects the graph at most once.

 Checkpoint Now try Exercise 13.

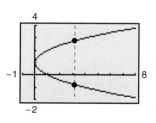

(a)

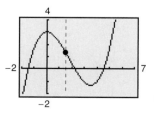

(b)

Figure 1.36

Increasing and Decreasing Functions

The more you know about the graph of a function, the more you know about the function itself. Consider the graph shown in Figure 1.37. Moving from *left to right*, this graph falls from $x = -2$ to $x = 0$, is constant from $x = 0$ to $x = 2$, and rises from $x = 2$ to $x = 4$.

Increasing, Decreasing, and Constant Functions

A function f is **increasing** on an interval if, for any x_1 and x_2 in the interval,

$$x_1 < x_2 \text{ implies } f(x_1) < f(x_2).$$

A function f is **decreasing** on an interval if, for any x_1 and x_2 in the interval,

$$x_1 < x_2 \text{ implies } f(x_1) > f(x_2).$$

A function f is **constant** on an interval if, for any x_1 and x_2 in the interval,

$$f(x_1) = f(x_2).$$

TECHNOLOGY TIP

Most graphing utilities are designed to graph functions of x more easily than other types of equations. For instance, the graph shown in Figure 1.36(a) represents the equation $x - (y - 1)^2 = 0$. To use a graphing utility to duplicate this graph you must first solve the equation for y to obtain $y = 1 \pm \sqrt{x}$, and then graph the two equations $y_1 = 1 + \sqrt{x}$ and $y_2 = 1 - \sqrt{x}$ in the same viewing window.

Example 4   Increasing and Decreasing Functions

In Figure 1.38, determine the open intervals on which each function is increasing, decreasing, or constant.

Solution

a. Although it might appear that there is an interval in which this function is constant, you can see that if $x_1 < x_2$, then $(x_1)^3 < (x_2)^3$, which implies that $f(x_1) < f(x_2)$. So, the function is increasing over the entire real line.

b. This function is increasing on the interval $(-\infty, -1)$, decreasing on the interval $(-1, 1)$, and increasing on the interval $(1, \infty)$.

c. This function is increasing on the interval $(-\infty, 0)$, constant on the interval $[0, 2]$, and decreasing on the interval $(2, \infty)$.

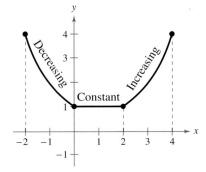

Figure 1.37

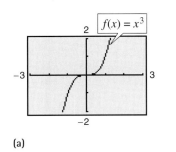

(a)

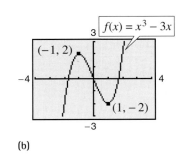

(b)

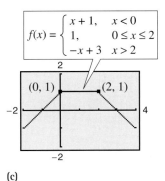

$$f(x) = \begin{cases} x + 1, & x < 0 \\ 1, & 0 \le x \le 2 \\ -x + 3 & x > 2 \end{cases}$$

(c)

Figure 1.38

 Checkpoint Now try Exercise 19.

Relative Minimum and Maximum Values

The points at which a function changes its increasing, decreasing, or constant behavior are helpful in determining the relative maximum or relative minimum values of the function.

Definition of Relative Minimum and Relative Maximum

A function value $f(a)$ is called a **relative minimum** of f if there exists an interval (x_1, x_2) that contains a such that

$$x_1 < x < x_2 \quad \text{implies} \quad f(a) \leq f(x).$$

A function value $f(a)$ is called a **relative maximum** of f if there exists an interval (x_1, x_2) that contains a such that

$$x_1 < x < x_2 \quad \text{implies} \quad f(a) \geq f(x).$$

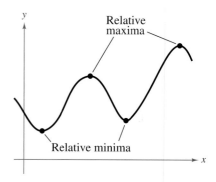

Figure 1.39

Figure 1.39 shows several different examples of relative minima and relative maxima. In Section 3.1, you will study a technique for finding the *exact points* at which a second-degree polynomial function has a relative minimum or relative maximum. For the time being, however, you can use a graphing utility to find reasonable approximations of these points.

Example 5   Approximating a Relative Minimum

Use a graphing utility to approximate the relative minimum of the function given by $f(x) = 3x^2 - 4x - 2$.

Solution

The graph of f is shown in Figure 1.40. By using the *zoom* and *trace* features of a graphing utility, you can estimate that the function has a relative minimum at the point

$(0.67, -3.33)$.   See Figure 1.41.

Later, in Section 3.1, you will be able to determine that the exact point at which the relative minimum occurs is $\left(\frac{2}{3}, -\frac{10}{3}\right)$.

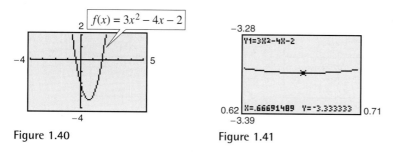

Figure 1.40

Figure 1.41

✓ *Checkpoint* Now try Exercise 29.

TECHNOLOGY TIP

When you use a graphing utility to estimate the *x*- and *y*-values of a relative minimum or relative maximum, the *zoom* feature will often produce graphs that are nearly flat, as shown in Figure 1.41. To overcome this problem, you can manually change the vertical setting of the viewing window. The graph will vertically stretch if the values of Ymin and Ymax are closer together.

TECHNOLOGY TIP Some graphing utilities have built-in programs that will find minimum or maximum values. These features are demonstrated in Example 6.

Example 6   Approximating Relative Minima and Maxima

Use a graphing utility to approximate the relative minimum and relative maximum of the function given by $f(x) = -x^3 + x$.

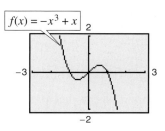

$f(x) = -x^3 + x$

Solution

The graph of f is shown in Figure 1.42. By using the *zoom* and *trace* features or the *minimum* and *maximum* features of the graphing utility, you can estimate that the function has a relative minimum at the point

Figure 1.42

$$(-0.58, -0.38) \qquad \text{See Figure 1.43.}$$

and a relative maximum at the point

$f(x) = -x^3 + x$

$$(0.58, 0.38). \qquad \text{See Figure 1.44.}$$

If you take a course in calculus, you will learn a technique for finding the exact points at which this function has a relative minimum and a relative maximum.

Minimum
X=-.5773489 Y=-.3849002

✓ *Checkpoint*   Now try Exercise 31.

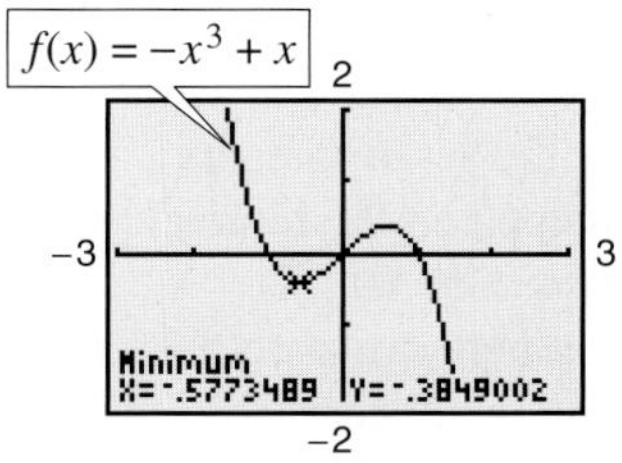

Figure 1.43

Example 7   Temperature

During a 24-hour period, the temperature y (in degrees Fahrenheit) of a certain city can be approximated by the model

$$y = 0.026x^3 - 1.03x^2 + 10.2x + 34, \qquad 0 \le x \le 24$$

where x represents the time of day, with $x = 0$ corresponding to 6 A.M. Approximate the maximum and minimum temperatures during this 24-hour period.

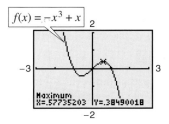

$f(x) = -x^3 + x$

Maximum
X=.57735203 Y=.3849001B

Figure 1.44

Solution

To solve this problem, graph the function as shown in Figure 1.45. Using the *zoom* and *trace* features or the *maximum* feature of a graphing utility, you can determine that the maximum temperature during the 24-hour period was approximately 64°F. This temperature occurred at about 12:36 P.M. ($x \approx 6.6$), as shown in Figure 1.46. Using the *zoom* and *trace* features or the *minimum* feature, you can determine that the minimum temperature during the 24-hour period was approximately 34°F, which occurred at about 1:48 A.M. ($x \approx 19.8$), as shown in Figure 1.47.

TECHNOLOGY SUPPORT

For instructions on how to use the *minimum* and *maximum* features, see Appendix A; for specific keystrokes, go to the text website at *college.hmco.com*.

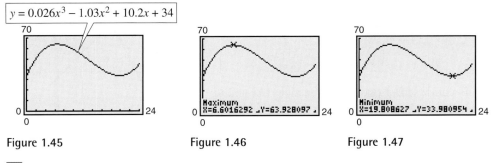

$y = 0.026x^3 - 1.03x^2 + 10.2x + 34$

Figure 1.45

Figure 1.46

Figure 1.47

✓ *Checkpoint*   Now try Exercise 87.

Graphing Step Functions and Piecewise-Defined Functions

Library of Functions: Greatest Integer Function

The *greatest integer function*, denoted by $[\![x]\!]$ and defined as the greatest integer less than or equal to x, has an infinite number of breaks or steps—one at each integer value in its domain. The basic characteristics of the greatest integer function are summarized below.

Graph of $f(x) = [\![x]\!]$

Domain: $(-\infty, \infty)$

Range: the set of integers

x-intercepts: in the interval $[0, 1)$

y-intercept: $(0, 0)$

Constant between each pair of consecutive integers

Jumps vertically one unit at each integer value

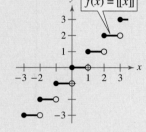

Could you describe the greatest integer function using a piecewise-defined function? How does the graph of the greatest integer function differ from the graph of a line with a slope of zero?

TECHNOLOGY TIP

Most graphing utilities display graphs in *connected mode*, which means that the graph has no breaks. When you are sketching graphs that do have breaks, it is better to use *dot mode*. Graph the greatest integer function [often called Int (x)] in connected and dot modes, and compare the two results.

Because of the vertical jumps described above, the greatest integer function is an example of a **step function** whose graph resembles a set of stairsteps. Some values of the greatest integer function are as follows.

$$[\![-1]\!] = (\text{greatest integer} \le -1) = -1$$

$$\left[\!\left[\tfrac{1}{10}\right]\!\right] = \left(\text{greatest integer} \le \tfrac{1}{10}\right) = 0$$

$$[\![1.5]\!] = (\text{greatest integer} \le 1.5) = 1$$

In Section 1.3, you learned that a piecewise-defined function is a function that is defined by two or more equations over a specified domain. To sketch the graph of a piecewise-defined function, you need to sketch the graph of each equation on the appropriate portion of the domain.

Example 8   Graphing a Piecewise-Defined Function

Sketch the graph of $f(x) = \begin{cases} 2x + 3, & x \le 1 \\ -x + 4, & x > 1 \end{cases}$ by hand.

Solution

This piecewise-defined function is composed of two linear functions. At and to the left of $x = 1$, the graph is the line given by $y = 2x + 3$. To the right of $x = 1$, the graph is the line given by $y = -x + 4$ (see Figure 1.48). Notice that the point $(1, 5)$ is a solid dot and the point $(1, 3)$ is an open dot. This is because $f(1) = 5$.

 Checkpoint Now try Exercise 41.

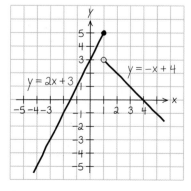

Figure 1.48

Even and Odd Functions

A graph has *symmetry with respect to the y-axis* if whenever (x, y) is on the graph, so is the point $(-x, y)$. A graph has *symmetry with respect to the origin* if whenever (x, y) is on the graph, so is the point $(-x, -y)$. A graph has *symmetry with respect to the x-axis* if whenever (x, y) is on the graph, so is the point $(x, -y)$. A function whose graph is symmetric with respect to the y-axis is an **even** function. A function whose graph is symmetric with respect to the origin is an **odd** function. A graph that is symmetric with respect to the x-axis is not the graph of a function (except for the graph of $y = 0$). These three types of symmetry are illustrated in Figure 1.49.

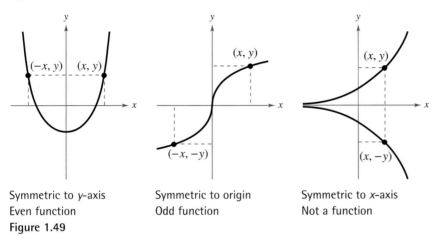

Symmetric to y-axis   Symmetric to origin   Symmetric to x-axis
Even function   Odd function   Not a function

Figure 1.49

Test for Even and Odd Functions

A function f is **even** if, for each x in the domain of f, $f(-x) = f(x)$.
A function f is **odd** if, for each x in the domain of f, $f(-x) = -f(x)$.

Example 9   Testing for Evenness and Oddness

Is the function given by $f(x) = |x|$ even, odd, or neither?

Algebraic Solution

This function is even because

$$f(-x) = |-x|$$
$$= |x|$$
$$= f(x).$$

Graphical Solution

Use a graphing utility to enter $y = |x|$ in the equation editor, as shown in Figure 1.50. Then graph the function using a standard viewing window, as shown in Figure 1.51. You can see that the graph appears to be symmetric about the y-axis. So, the function is even.

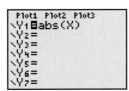

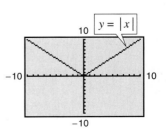

✓ *Checkpoint* Now try Exercise 49.

Figure 1.50

Figure 1.51

Example 10   Even and Odd Functions

Determine whether each function is even, odd, or neither.

a. $g(x) = x^3 - x$

b. $h(x) = x^2 + 1$

c. $f(x) = x^3 - 1$

Algebraic Solution

a. This function is odd because

$$g(-x) = (-x)^3 - (-x)$$
$$= -x^3 + x$$
$$= -(x^3 - x)$$
$$= -g(x).$$

b. This function is even because

$$h(-x) = (-x)^2 + 1$$
$$= x^2 + 1$$
$$= h(x).$$

c. Substituting $-x$ for x produces

$$f(-x) = (-x)^3 - 1$$
$$= -x^3 - 1.$$

Because $f(x) = x^3 - 1$ and $-f(x) = -x^3 + 1$, you can conclude that $f(-x) \neq f(x)$ and $f(-x) \neq -f(x)$. So, the function is neither even nor odd.

✓ *Checkpoint* Now try Exercise 51.

Graphical Solution

a. In Figure 1.52, the graph is symmetric with respect to the origin. So, this function is odd.

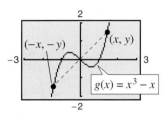

Figure 1.52

b. In Figure 1.53, the graph is symmetric with respect to the y-axis. So, this function is even.

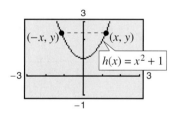

Figure 1.53

c. In Figure 1.54, the graph is neither symmetric with respect to the origin nor with respect to the y-axis. So, this function is neither even nor odd.

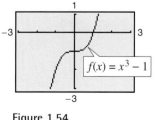

Figure 1.54

To help visualize symmetry with respect to the origin, place a pin at the origin of a graph and rotate the graph 180°. If the result after rotation coincides with the original graph, the graph is symmetric with respect to the origin.

1.4 Exercises

Vocabulary Check

Fill in the blanks.

1. The graph of a function f is a collection of _____ (x, y) such that x is in the domain of f.

2. The _____ is used to determine whether the graph of an equation is a function of y in terms of x.

3. A function f is _____ on an interval if, for any x_1 and x_2 in the interval, $x_1 < x_2$ implies $f(x_1) > f(x_2)$.

4. A function value $f(a)$ is a relative _____ of f if there exists an interval (x_1, x_2) containing a such that $x_1 < x < x_2$ implies $f(a) \leq f(x)$.

5. The function $f(x) = [\![x]\!]$ is called the _____ function, and is an example of a step function.

6. A function f is _____ if, for each x in the domain of f, $f(-x) = f(x)$.

In Exercises 1–4, use the graph of the function to find the domain and range of f. Then find $f(0)$.

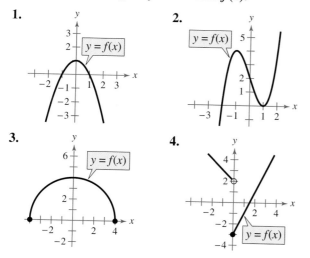

In Exercises 5–10, use a graphing utility to graph the function and estimate its domain and range. Then find the domain and range algebraically.

5. $f(x) = 2x^2 + 3$

6. $f(x) = -x^2 - 1$

7. $f(x) = \sqrt{x - 1}$

8. $h(t) = \sqrt{4 - t^2}$

9. $f(x) = |x + 3|$

10. $f(x) = -\frac{1}{4}|x - 5|$

In Exercises 11–16, use the Vertical Line Test to determine whether y is a function of x. Describe how you can use a graphing utility to produce the given graph.

11. $y = \frac{1}{2}x^2$

12. $y = \frac{1}{4}x^3$

13. $x - y^2 = 1$

14. $x^2 + y^2 = 25$

15. $x^2 = 2xy - 1$

16. $x = |y + 2|$

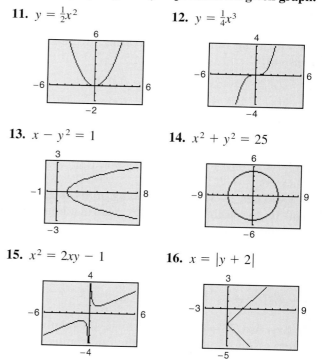

In Exercises 17–20, determine the intervals over which the function is increasing, decreasing, or constant.

17. $f(x) = \frac{3}{2}x$

18. $f(x) = x^2 - 4x$

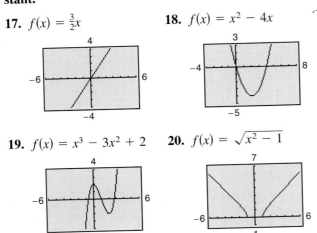

19. $f(x) = x^3 - 3x^2 + 2$

20. $f(x) = \sqrt{x^2 - 1}$

In Exercises 21–28, (a) use a graphing utility to graph the function and (b) determine the open intervals on which the function is increasing, decreasing, or constant.

21. $f(x) = 3$

22. $f(x) = x$

23. $f(x) = x^{2/3}$

24. $f(x) = -x^{3/4}$

25. $f(x) = x\sqrt{x + 3}$

26. $f(x) = \sqrt{1 - x}$

27. $f(x) = |x + 1| + |x - 1|$

28. $f(x) = -|x + 4| - |x + 1|$

In Exercises 29–34, use a graphing utility to approximate (to two decimal places) any relative minimum or maximum values of the function.

29. $f(x) = x^2 - 6x$

30. $f(x) = 3x^2 - 2x - 5$

31. $y = 2x^3 + 3x^2 - 12x$

32. $y = x^3 - 6x^2 + 15$

33. $h(x) = (x - 1)\sqrt{x}$

34. $g(x) = x\sqrt{4 - x}$

In Exercises 35–40, (a) approximate the relative minimum or maximum values of the function by sketching its graph using the point-plotting method, (b) use a graphing utility to approximate (to two decimal places) any relative minimum or maximum values, and (c) compare your answers from parts (a) and (b).

35. $f(x) = x^2 - 4x - 5$

36. $f(x) = 3x^2 - 12$

37. $f(x) = x^3 - 8x$

38. $f(x) = -x^3 + 7x$

39. $f(x) = (x - 4)^{2/3}$

40. $f(x) = \sqrt{4x^2 + 1}$

In Exercises 41–48, sketch the graph of the piecewise-defined function by hand.

41. $f(x) = \begin{cases} 2x + 3, & x < 0 \\ 3 - x, & x \geq 0 \end{cases}$

42. $f(x) = \begin{cases} x + 6, & x \leq -4 \\ 2x - 4, & x > -4 \end{cases}$

43. $f(x) = \begin{cases} \sqrt{4 + x}, & x < 0 \\ \sqrt{4 - x}, & x \geq 0 \end{cases}$

44. $f(x) = \begin{cases} 1 - (x - 1)^2, & x \leq 2 \\ \sqrt{x - 2}, & x > 2 \end{cases}$

45. $f(x) = \begin{cases} x + 3, & x \leq 0 \\ 3, & 0 < x \leq 2 \\ 2x - 1, & x > 2 \end{cases}$

46. $g(x) = \begin{cases} x + 5, & x \leq -3 \\ -2, & -3 < x < 1 \\ 5x - 4, & x \geq 1 \end{cases}$

47. $f(x) = \begin{cases} 2x + 1, & x \leq -1 \\ x^2 - 2, & x > -1 \end{cases}$

48. $h(x) = \begin{cases} 3 + x, & x < 0 \\ x^2 + 1, & x \geq 0 \end{cases}$

In Exercises 49–56, algebraically determine whether the function is even, odd, or neither. Verify your answer using a graphing utility.

49. $f(t) = t^2 + 2t - 3$

50. $f(x) = x^6 - 2x^2 + 3$

51. $g(x) = x^3 - 5x$

52. $h(x) = x^3 - 5$

53. $f(x) = x\sqrt{1 - x^2}$

54. $f(x) = x\sqrt{x + 5}$

55. $g(s) = 4s^{2/3}$

56. $f(s) = 4s^{3/2}$

Think About It   In Exercises 57–62, find the coordinates of a second point on the graph of a function f if the given point is on the graph and the function is (a) even and (b) odd.

57. $\left(-\frac{3}{2}, 4\right)$

58. $\left(-\frac{5}{3}, -7\right)$

59. $(4, 9)$

60. $(5, -1)$

61. $(x, -y)$

62. $(2a, 2c)$

In Exercises 63–72, use a graphing utility to graph the function and determine whether it is even, odd, or neither. Verify your answer algebraically.

63. $f(x) = 5$

64. $f(x) = -9$

65. $f(x) = 3x - 2$

66. $f(x) = 5 - 3x$

67. $h(x) = x^2 - 4$

68. $f(x) = -x^2 - 8$

69. $f(x) = \sqrt{1 - x}$ **70.** $g(t) = \sqrt[3]{t - 1}$

71. $f(x) = |x + 2|$ **72.** $f(x) = -|x - 5|$

In Exercises 73–76, graph the function and determine the interval(s) (if any) on the real axis for which $f(x) \geq 0$. Use a graphing utility to verify your results.

73. $f(x) = 4 - x$ **74.** $f(x) = 4x + 2$

75. $f(x) = x^2 - 9$ **76.** $f(x) = x^2 - 4x$

In Exercises 77 and 78, use a graphing utility to graph the function. State the domain and range of the function. Describe the pattern of the graph.

77. $s(x) = 2\left(\frac{1}{4}x - \left[\!\left[\frac{1}{4}x\right]\!\right]\right)$ **78.** $g(x) = 2\left(\frac{1}{4}x - \left[\!\left[\frac{1}{4}x\right]\!\right]\right)^2$

79. Geometry   The perimeter of a rectangle is 100 meters.

   (a) Show that the area of the rectangle is given by $A = x(50 - x)$, where x is its length.

   (b) Use a graphing utility to graph the area function.

   (c) Use a graphing utility to approximate the maximum area of the rectangle and the dimensions that yield the maximum area.

80. Cost, Revenue, and Profit   The marketing department of a company estimates that the demand for a color scanner is $p = 100 - 0.0001x$, where p is the price per scanner and x is the number of scanners. The cost of producing x scanners is $C = 350,000 + 30x$ and the profit for producing and selling x scanners is

$$P = R - C = xp - C.$$

Use a graphing utility to graph the profit function and estimate the number of scanners that would produce a maximum profit.

81. Communications   The cost of using a telephone calling card is \$1.05 for the first minute and \$0.38 for each additional minute or portion of a minute.

   (a) A customer needs a model for the cost C of using the calling card for a call lasting t minutes. Which of the following is the appropriate model?

$$C_1(t) = 1.05 + 0.38[\![t - 1]\!]$$

$$C_2(t) = 1.05 - 0.38[\![-(t - 1)]\!]$$

   (b) Use a graphing utility to graph the appropriate model. Use the *value* feature or the *zoom* and *trace* features to estimate the cost of a call lasting 18 minutes and 45 seconds.

82. Delivery Charges   The cost of sending an overnight package from New York to Atlanta is \$9.80 for a package weighing up to but not including 1 pound and \$2.50 for each additional pound or portion of a pound. Use the greatest integer function to create a model for the cost C of overnight delivery of a package weighing x pounds, where $x > 0$. Sketch the graph of the function.

In Exercises 83 and 84, write the height h of the rectangle as a function of x.

83.   **84.**

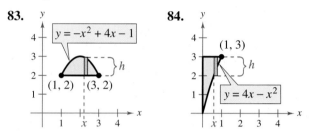

In Exercises 85 and 86, write the length L of the rectangle as a function of y.

85.   **86.**

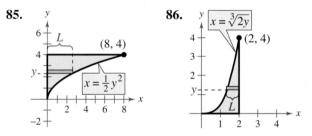

87. Population   During a seven-year period, the population P (in thousands) of North Dakota increased and then decreased according to the model

$$P = -0.76t^2 + 9.9t + 618, \quad 5 \leq t \leq 11$$

where t represents the year, with $t = 5$ corresponding to 1995. (Source: U.S. Census Bureau)

   (a) Use a graphing utility to graph the model over the appropriate domain.

   (b) Use the graph from part (a) to determine during which years the population was increasing. During which years was the population decreasing?

   (c) Use the *zoom* and *trace* features or the *maximum* feature of a graphing utility to approximate the maximum population between 1995 and 2001.

88. *Fluid Flow* The intake pipe of a 100-gallon tank has a flow rate of 10 gallons per minute, and two drain pipes have a flow rate of 5 gallons per minute each. The graph shows the volume V of fluid in the tank as a function of time t. Determine in which pipes the fluid is flowing in specific subintervals of the one-hour interval of time shown on the graph. (There are many correct answers.)

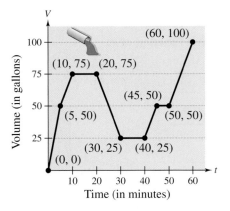

Synthesis

True or False? **In Exercises 89 and 90, determine whether the statement is true or false. Justify your answer.**

89. A function with a square root cannot have a domain that is the set of all real numbers.

90. It is possible for an odd function to have the interval $[0, \infty)$ as its domain.

91. *Proof* Prove that a function of the following form is odd.

$$y = a_{2n+1}x^{2n+1} + a_{2n-1}x^{2n-1} + \cdots + a_3x^3 + a_1x$$

92. *Proof* Prove that a function of the following form is even.

$$y = a_{2n}x^{2n} + a_{2n-2}x^{2n-2} + \cdots + a_2x^2 + a_0$$

93. If f is an even function, determine if g is even, odd, or neither. Explain.

(a) $g(x) = -f(x)$ (b) $g(x) = f(-x)$

(c) $g(x) = f(x) - 2$ (d) $g(x) = -f(x - 2)$

94. *Think About It* Does the graph in Exercise 13 represent x as a function of y? Explain.

95. *Think About It* Does the graph in Exercise 14 represent x as a function of y? Explain.

96. *Writing* Write a short paragraph describing three different functions that represent the behaviors of quantities between 1990 and 2004. Describe one quantity that decreased during this time, one that increased, and one that was constant. Present your results graphically.

Review

In Exercises 97–100, identify the terms. Then identify the coefficients of the variable terms of the expression.

97. $-2x^2 + 8x$

98. $10 + 3x$

99. $\dfrac{x}{3} - 5x^2 + x^3$

100. $7x^4 + \sqrt{2}x^2$

In Exercises 101–104, find (a) the distance between the two points and (b) the midpoint of the line segment joining the points.

101. $(-2, 7), (6, 3)$

102. $(-5, 0), (3, 6)$

103. $\left(\frac{5}{2}, -1\right), \left(-\frac{3}{2}, 4\right)$

104. $\left(-6, \frac{2}{3}\right), \left(\frac{3}{4}, \frac{1}{6}\right)$

In Exercises 105–108, evaluate the function at each specified value of the independent variable and simplify.

105. $f(x) = 5x - 1$

(a) $f(6)$ (b) $f(-1)$ (c) $f(x - 3)$

106. $f(x) = -x^2 - x + 3$

(a) $f(4)$ (b) $f(-2)$ (c) $f(x - 2)$

107. $f(x) = x\sqrt{x - 3}$

(a) $f(3)$ (b) $f(12)$ (c) $f(6)$

108. $f(x) = -\frac{1}{2}x|x + 1|$

(a) $f(-4)$ (b) $f(10)$ (c) $f\left(-\frac{2}{3}\right)$

In Exercises 109 and 110, find the difference quotient and simplify your answer.

109. $f(x) = x^2 - 2x + 9$, $\dfrac{f(3 + h) - f(3)}{h}$, $h \neq 0$

110. $f(x) = 5 + 6x - x^2$, $\dfrac{f(6 + h) - f(6)}{h}$, $h \neq 0$

1.5 Shifting, Reflecting, and Stretching Graphs

Summary of Graphs of Common Functions

One of the goals of this text is to enable you to build your intuition for the basic shapes of the graphs of different types of functions. For instance, from your study of lines in Section 1.2, you can determine the basic shape of the graph of the linear function $f(x) = mx + b$. Specifically, you know that the graph of this function is a line whose slope is m and whose y-intercept is $(0, b)$.

The six graphs shown in Figure 1.55 represent the most commonly used functions in algebra. Familiarity with the basic characteristics of these simple graphs will help you analyze the shapes of more complicated graphs.

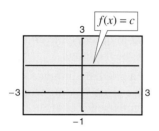

(a) Constant Function

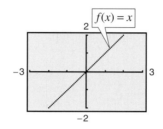

(b) Identity Function

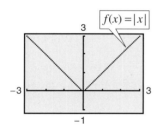

(c) Absolute Value Function

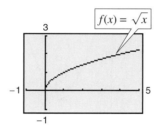

(d) Square Root Function

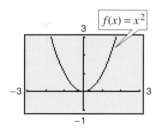

(e) Quadratic Function

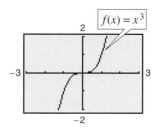

(f) Cubic Function

Figure 1.55

Throughout this section, you will discover how many complicated graphs are derived by shifting, stretching, shrinking, or reflecting the common graphs shown above. Shifts, stretches, shrinks, and reflections are called *transformations*. Many graphs of functions can be created from a combination of these transformations.

Vertical and Horizontal Shifts

Many functions have graphs that are simple transformations of the common graphs summarized in Figure 1.55. For example, you can obtain the graph of

$$h(x) = x^2 + 2$$

by shifting the graph of $f(x) = x^2$ *upward* two units, as shown in Figure 1.56. In function notation, h and f are related as follows.

$$h(x) = x^2 + 2$$
$$= f(x) + 2 \qquad \text{Upward shift of two units}$$

Similarly, you can obtain the graph of

$$g(x) = (x - 2)^2$$

by shifting the graph of $f(x) = x^2$ to the *right* two units, as shown in Figure 1.57. In this case, the functions g and f have the following relationship.

$$g(x) = (x - 2)^2$$
$$= f(x - 2) \qquad \text{Right shift of two units}$$

<table>
<tr><td>

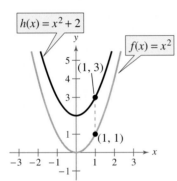

Figure 1.56   Vertical shift upward: two units

</td><td>

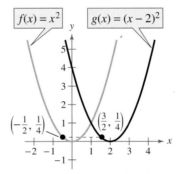

Figure 1.57   Horizontal shift to the right: two units

</td></tr>
</table>

> **Exploration**
>
> Use a graphing utility to display (in the same viewing window) the graphs of $y = x^2 + c$, where $c = -2, 0, 2,$ and 4. Use the result to describe the effect that c has on the graph.
>
> Use a graphing utility to display (in the same viewing window) the graphs of $y = (x + c)^2$, where $c = -2, 0, 2,$ and 4. Use the result to describe the effect that c has on the graph.

The following list summarizes horizontal and vertical shifts.

Vertical and Horizontal Shifts

Let c be a positive real number. **Vertical and horizontal shifts** in the graph of $y = f(x)$ are represented as follows.

1. Vertical shift c units *upward*: $h(x) = f(x) + c$

2. Vertical shift c units *downward*: $h(x) = f(x) - c$

3. Horizontal shift c units to the *right*: $h(x) = f(x - c)$

4. Horizontal shift c units to the *left*: $h(x) = f(x + c)$

In items 3 and 4, be sure you see that $h(x) = f(x - c)$ corresponds to a *right* shift and $h(x) = f(x + c)$ corresponds to a *left* shift for $c > 0$.

Example 1   Shifts in the Graph of a Function

Compare the graph of each function with the graph of $f(x) = x^3$.

a. $g(x) = x^3 - 1$ **b.** $h(x) = (x - 1)^3$ **c.** $k(x) = (x + 2)^3 + 1$

Solution

a. Graph $f(x) = x^3$ and $g(x) = x^3 - 1$ [see Figure 1.58(a)]. You can obtain the graph of g by shifting the graph of f one unit downward.

b. Graph $f(x) = x^3$ and $h(x) = (x - 1)^3$ [see Figure 1.58(b)]. You can obtain the graph of h by shifting the graph of f one unit to the right.

c. Graph $f(x) = x^3$ and $k(x) = (x + 2)^3 + 1$ [see Figure 1.58(c)]. You can obtain the graph of k by shifting the graph of f two units to the left and then one unit upward.

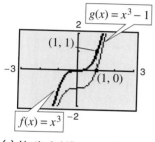
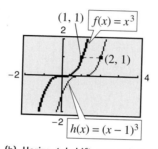
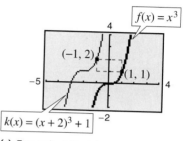

(a) Vertical shift: one unit downward

Figure 1.58

(b) Horizontal shift: one unit right

(c) Two units left and one unit upward

✓ *Checkpoint* Now try Exercise 3.

Example 2   Finding Equations from Graphs

The graph of $f(x) = x^2$ is shown in Figure 1.59. Each of the graphs in Figure 1.60 is a transformation of the graph of f. Find an equation for each function.

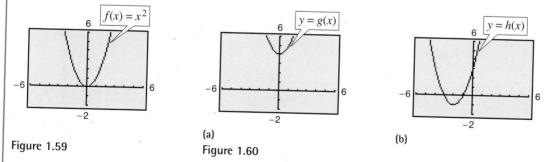

Figure 1.59

(a)

Figure 1.60

(b)

Solution

a. The graph of g is a vertical shift of four units upward of the graph of $f(x) = x^2$. So, the equation for g is $g(x) = x^2 + 4$.

b. The graph of h is a horizontal shift of two units to the left, and a vertical shift of one unit downward, of the graph of $f(x) = x^2$. So, the equation for h is $h(x) = (x + 2)^2 - 1$.

✓ *Checkpoint* Now try Exercise 21.

Reflecting Graphs

The second common type of transformation is called a **reflection**. For instance, if you consider the x-axis to be a mirror, the graph of $h(x) = -x^2$ is the mirror image (or reflection) of the graph of $f(x) = x^2$ (see Figure 1.61).

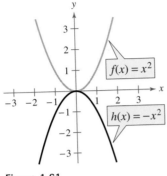

Figure 1.61

Reflections in the Coordinate Axes

Reflections in the coordinate axes of the graph of $y = f(x)$ are represented as follows.

    1. Reflection in the x-axis: $h(x) = -f(x)$

    2. Reflection in the y-axis: $h(x) = f(-x)$

Example 3   Finding Equations from Graphs

The graph of $f(x) = x^4$ is shown in Figure 1.62. Each of the graphs in Figure 1.63 is a transformation of the graph of f. Find an equation for each function.

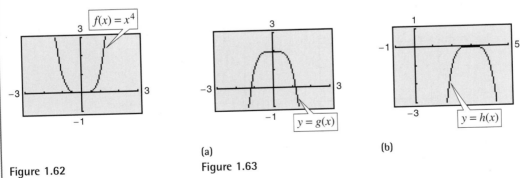

Figure 1.62

(a)   (b)

Figure 1.63

Solution

a. The graph of g is a reflection in the x-axis *followed by* an upward shift of two units of the graph of $f(x) = x^4$. So, the equation for g is $g(x) = -x^4 + 2$.

b. The graph of h is a horizontal shift of three units to the right *followed by* a reflection in the x-axis of the graph of $f(x) = x^4$. So, the equation for h is $h(x) = -(x - 3)^4$.

✓ *Checkpoint* Now try Exercise 25.

Example 4   Reflections and Shifts

Compare the graph of each function with the graph of $f(x) = \sqrt{x}$.

a. $g(x) = -\sqrt{x}$ **b.** $h(x) = \sqrt{-x}$ **c.** $k(x) = -\sqrt{x+2}$

Algebraic Solution

a. Relative to the graph of $f(x) = \sqrt{x}$, the graph of g is a reflection in the x-axis because

$$g(x) = -\sqrt{x}$$
$$= -f(x).$$

b. The graph of h is a reflection of the graph of $f(x) = \sqrt{x}$ in the y-axis because

$$h(x) = \sqrt{-x}$$
$$= f(-x).$$

c. From the equation

$$k(x) = -\sqrt{x+2}$$
$$= -f(x+2)$$

you can conclude that the graph of k is a left shift of two units, followed by a reflection in the x-axis, of the graph of $f(x) = \sqrt{x}$.

✔ *Checkpoint* Now try Exercise 27.

Graphical Solution

a. Use a graphing utility to graph f and g in the same viewing window. From the graph in Figure 1.64, you can see that the graph of g is a reflection of the graph of f in the x-axis.

b. Use a graphing utility to graph f and h in the same viewing window. From the graph in Figure 1.65, you can see that the graph of h is a reflection of the graph of f in the y-axis.

c. Use a graphing utility to graph f and k in the same viewing window. From the graph in Figure 1.66, you can see that the graph of k is a left shift of two units of the graph of f, followed by a reflection in the x-axis.

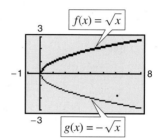

Figure 1.64

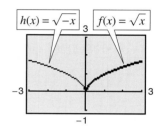

Figure 1.65

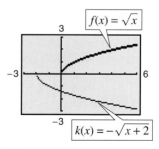

Figure 1.66

When graphing functions involving square roots, remember that the domain must be restricted to exclude negative numbers inside the radical. For instance, here are the domains of the functions in Example 4.

Domain of $g(x) = -\sqrt{x}$: $x \geq 0$

Domain of $h(x) = \sqrt{-x}$: $x \leq 0$

Domain of $k(x) = -\sqrt{x+2}$: $x \geq -2$

Nonrigid Transformations

Horizontal shifts, vertical shifts, and reflections are called **rigid transformations** because the basic shape of the graph is unchanged. These transformations change only the *position* of the graph in the coordinate plane. **Nonrigid transformations** are those that cause a *distortion*—a change in the shape of the original graph. For instance, a nonrigid transformation of the graph of $y = f(x)$ is represented by $y = cf(x)$ (each y-value is multiplied by c), where the transformation is a **vertical stretch** if $c > 1$ and a **vertical shrink** if $0 < c < 1$. Another nonrigid transformation of the graph of $y = f(x)$ is represented by $h(x) = f(cx)$ (each x-value is multiplied by $1/c$), where the transformation is a **horizontal shrink** if $c > 1$ and a **horizontal stretch** if $0 < c < 1$.

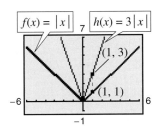

Example 5 Nonrigid Transformations

Compare the graph of each function with the graph of $f(x) = |x|$.

a. $h(x) = 3|x|$ **b.** $g(x) = \frac{1}{3}|x|$

Figure 1.67

Solution

a. Relative to the graph of $f(x) = |x|$, the graph of

$$h(x) = 3|x|$$
$$= 3f(x)$$

is a vertical stretch (each y-value is multiplied by 3) of the graph of f. (See Figure 1.67.)

b. Similarly, the graph of

$$g(x) = \frac{1}{3}|x|$$
$$= \frac{1}{3}f(x)$$

is a vertical shrink $\left(\text{each } y\text{-value is multiplied by } \frac{1}{3}\right)$ of the graph of f. (See Figure 1.68.)

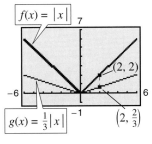

Figure 1.68

✓ *Checkpoint* Now try Exercise 37.

Example 6 Nonrigid Transformations

Compare the graph of $h(x) = f\left(\frac{1}{2}x\right)$ with the graph of $f(x) = 2 - x^3$.

Solution

Relative to the graph of $f(x) = 2 - x^3$, the graph of

$$h(x) = f\left(\tfrac{1}{2}x\right) = 2 - \left(\tfrac{1}{2}x\right)^3 = 2 - \tfrac{1}{8}x^3$$

is a horizontal stretch (each x-value is multiplied by 2) of the graph of f. (See Figure 1.69.)

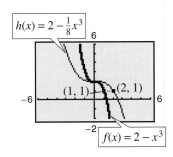

Figure 1.69

✓ *Checkpoint* Now try Exercise 43.

1.5 Exercises

Vocabulary Check

In Exercises 1–5, fill in the blanks.

1. The graph of a _____ is U-shaped.

2. The graph of an _____ is V-shaped.

3. Horizontal shifts, vertical shifts, and reflections are called _____ .

4. A reflection in the x-axis of $y = f(x)$ is represented by $h(x) =$ _____ , while a reflection in the y-axis of $y = f(x)$ is represented by $h(x) =$ _____ .

5. A nonrigid transformation of $y = f(x)$ represented by $cf(x)$ is a vertical stretch if _____ and a vertical shrink if _____ .

6. Match the rigid transformation of $y = f(x)$ with the correct representation, where $c > 0$.
   (a) $h(x) = f(x) + c$ (i) horizontal shift c units to the left
   (b) $h(x) = f(x) - c$ (ii) vertical shift c units upward
   (c) $h(x) = f(x - c)$ (iii) horizontal shift c units to the right
   (d) $h(x) = f(x + c)$ (iv) vertical shift c units downward

In Exercises 1–12, sketch the graphs of the three functions by hand on the same rectangular coordinate system. Verify your result with a graphing utility.

1. $f(x) = x$
   $g(x) = x - 4$
   $h(x) = 3x$

2. $f(x) = \frac{1}{2}x$
   $g(x) = \frac{1}{2}x + 2$
   $h(x) = \frac{1}{2}(x - 2)$

3. $f(x) = x^2$
   $g(x) = x^2 + 2$
   $h(x) = (x - 2)^2$

4. $f(x) = x^2$
   $g(x) = x^2 - 4$
   $h(x) = (x + 2)^2 + 1$

5. $f(x) = -x^2$
   $g(x) = -x^2 + 1$
   $h(x) = -(x - 2)^2$

6. $f(x) = (x - 2)^2$
   $g(x) = (x - 2)^2 + 2$
   $h(x) = -(x - 2)^2 + 4$

7. $f(x) = x^2$
   $g(x) = \frac{1}{2}x^2$
   $h(x) = (2x)^2$

8. $f(x) = x^2$
   $g(x) = \frac{1}{4}x^2 + 2$
   $h(x) = -\frac{1}{4}x^2$

9. $f(x) = |x|$
   $g(x) = |x| - 1$
   $h(x) = |x - 3|$

10. $f(x) = |x|$
    $g(x) = |2x|$
    $h(x) = -2|x + 2| - 1$

11. $f(x) = \sqrt{x}$
    $g(x) = \sqrt{x + 1}$
    $h(x) = \sqrt{x - 2} + 1$

12. $f(x) = \sqrt{x}$
    $g(x) = \frac{1}{2}\sqrt{x}$
    $h(x) = -\frac{1}{2}\sqrt{x + 4}$

13. Use the graph of f to sketch each graph. To print an enlarged copy of the graph, go to the website *www.mathgraphs.com.*
    (a) $y = f(x) + 2$
    (b) $y = -f(x)$
    (c) $y = f(x - 2)$
    (d) $y = f(x + 3)$
    (e) $y = 2f(x)$
    (f) $y = f(-x)$
    (g) $y = f\left(\frac{1}{2}x\right)$

14. Use the graph of f to sketch each graph. To print an enlarged copy of the graph, go to the website *www.mathgraphs.com.*
    (a) $y = f(x) - 1$
    (b) $y = f(x + 1)$
    (c) $y = f(x - 1)$
    (d) $y = -f(x - 2)$
    (e) $y = f(-x)$
    (f) $y = \frac{1}{2}f(x)$
    (g) $y = f(2x)$

In Exercises 15–26, identify the common function and describe the transformation shown in the graph. Write an equation for the graphed function.

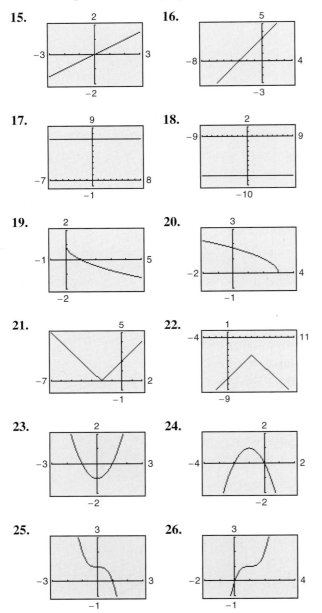

15.

16.

17.

18.

19.

20.

21.

22.

23.

24.

25.

26.

In Exercises 27–32, compare the graph of the function with the graph of $f(x) = \sqrt{x}$.

27. $y = -\sqrt{x} - 1$

28. $y = \sqrt{x} + 2$

29. $y = \sqrt{x - 2}$

30. $y = \sqrt{x + 4}$

31. $y = \sqrt{2x}$

32. $y = \sqrt{-x + 3}$

In Exercises 33–38, compare the graph of the function with the graph of $f(x) = |x|$.

33. $y = |x + 5|$

34. $y = |x| - 3$

35. $y = -|x|$

36. $y = |-x|$

37. $y = 4|x|$

38. $y = |\frac{1}{2}x|$

In Exercises 39–44, compare the graph of the function with the graph of $f(x) = x^3$.

39. $g(x) = 4 - x^3$

40. $g(x) = -(x - 1)^3$

41. $h(x) = \frac{1}{4}(x + 2)^3$

42. $h(x) = -2(x - 1)^3 + 3$

43. $p(x) = \left(\frac{1}{3}x\right)^3 + 2$

44. $p(x) = [3(x - 2)]^3$

In Exercises 45–48, use a graphing utility to graph the three functions in the same viewing window. Describe the graphs of g and h relative to the graph of f.

45. $f(x) = x^3 - 3x^2$

   $g(x) = f(x + 2)$

   $h(x) = \frac{1}{2}f(x)$

46. $f(x) = x^3 - 3x^2 + 2$

   $g(x) = f(x - 1)$

   $h(x) = f(3x)$

47. $f(x) = x^3 - 3x^2$

   $g(x) = -\frac{1}{3}f(x)$

   $h(x) = f(-x)$

48. $f(x) = x^3 - 3x^2 + 2$

   $g(x) = -f(x)$

   $h(x) = f(2x)$

In Exercises 49 and 50, use the graph of $f(x) = x^3 - 3x^2$ (see Exercise 45) to write a formula for the function g shown in the graph.

49.

50.

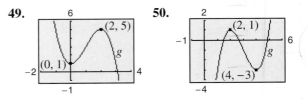

In Exercises 51–64, g is related to one of the six common functions on page 125. (a) Identify the common function f. (b) Describe the sequence of transformations from f to g. (c) Sketch the graph of g by hand. (d) Use function notation to write g in terms of the common function f.

51. $g(x) = 2 - (x + 5)^2$

52. $g(x) = -(x + 10)^2 + 5$

53. $g(x) = 3 + 2(x - 4)^2$

54. $g(x) = -\frac{1}{4}(x + 2)^2 - 2$

55. $g(x) = 3(x - 2)^3$ **56.** $g(x) = -\frac{1}{2}(x + 1)^3$

57. $g(x) = (x - 1)^3 + 2$

58. $g(x) = -(x + 3)^3 - 10$

59. $g(x) = |x + 4| + 8$

60. $g(x) = |x + 3| + 9$

61. $g(x) = -2|x - 1| - 4$

62. $g(x) = \frac{1}{2}|x - 2| - 3$

63. $g(x) = -\frac{1}{2}\sqrt{x + 3} - 1$

64. $g(x) = -\sqrt{x + 1} - 6$

65. *Profit*   The profit P per week on a case of soda pop is given by the model

$$P(x) = 80 + 20x - 0.5x^2, \qquad 0 \le x \le 20$$

where x is the amount spent on advertising. In this model, x and P are both measured in hundreds of dollars.

(a) Use a graphing utility to graph the profit function.

(b) The business estimates that taxes and operating costs will increase by an average of $2500 per week during the next year. Rewrite the profit function to reflect this expected decrease in profits. Describe the transformation applied to the graph of the function.

(c) Rewrite the profit function so that x measures advertising expenditures in dollars. $\left[\text{Find } P\!\left(\frac{x}{100}\right).\right]$ Describe the transformation applied to the graph of the profit function.

66. *Automobile Aerodynamics*   The number of horse-power H required to overcome wind drag on an automobile is approximated by the model

$$H(x) = 0.002x^2 + 0.005x - 0.029, \quad 10 \le x \le 100$$

where x is the speed of the car in miles per hour.

(a) Use a graphing utility to graph the function.

(b) Rewrite the function so that x represents the speed in kilometers per hour. [Find $H(x/1.6)$.] Describe the transformation applied to the graph of the function.

67. *Fuel Use*   The amount of fuel F (in billions of gallons) used by trucks from 1980 through 2000 can be approximated by the function $F(t) = 0.036t^2 + 20.1$, where $t = 0$ represents 1980. (Source: U.S. Federal Highway Administration)

(a) Describe the transformation of the common function $f(t) = t^2$. Then sketch the graph over the interval $0 \le t \le 20$.

(b) Rewrite the function so that $t = 0$ represents 1990. Explain how you got your answer.

68. *Finance*   The amount M (in billions of dollars) of mortgage debt outstanding in the United States from 1990 through 2001 can be approximated by the function $M(t) = 29.9t^2 + 3892$, where $t = 0$ represents 1990.   (Source: Board of Governors of the Federal Reserve System)

(a) Describe the transformation of the common function $f(t) = t^2$. Then sketch the graph over the interval $0 \le t \le 11$.

(b) Rewrite the function so that $t = 0$ represents 2000. Explain how you got your answer.

Synthesis

True or False?   **In Exercises 69 and 70, determine whether the statement is true or false. Justify your answer.**

69. The graphs of $f(x) = |x| - 5$ and $g(x) = |-x| - 5$ are identical.

70. Relative to the graph of $f(x) = \sqrt{x}$, the graph of the function $h(x) = -\sqrt{x + 9} - 13$ is shifted 9 units to the left and 13 units downward, then reflected in the x-axis.

71. *Exploration*   Use a graphing utility to graph each function. Describe any similarities and differences you observe among the graphs.

(a) $y = x$ (b) $y = x^2$ (c) $y = x^3$

(d) $y = x^4$ (e) $y = x^5$ (f) $y = x^6$

72. *Conjecture*   Use the results of Exercise 71.

(a) Make a conjecture about the shapes of the graphs of $y = x^7$ and $y = x^8$. Use a graphing utility to verify your conjecture.

(b) Sketch the graphs of $y = (x - 3)^3$ and $y = (x + 1)^2$ by hand. Use a graphing utility to verify your graphs.

Review

In Exercises 73 and 74, determine whether the lines L_1 and L_2 passing through the pairs of points are parallel, perpendicular, or neither.

73. L_1: $(-2, -2), (2, 10)$ **74.** L_1: $(-1, -7), (4, 3)$
 L_2: $(-1, 3), (3, 9)$ L_2: $(1, 5), (-2, -7)$

In Exercises 75–78, find the domain of the function.

75. $f(x) = \dfrac{4}{9 - x}$ **76.** $f(x) = \dfrac{\sqrt{x - 5}}{x - 7}$

77. $f(x) = \sqrt{100 - x^2}$ **78.** $f(x) = \sqrt[3]{16 - x^2}$

1.6 Combinations of Functions

Arithmetic Combinations of Functions

Just as two real numbers can be combined by the operations of addition, subtraction, multiplication, and division to form other real numbers, two *functions* can be combined to create new functions. If $f(x) = 2x - 3$ and $g(x) = x^2 - 1$, you can form the sum, difference, product, and quotient of f and g as follows.

$$f(x) + g(x) = (2x - 3) + (x^2 - 1)$$
$$= x^2 + 2x - 4 \qquad \text{Sum}$$

$$f(x) - g(x) = (2x - 3) - (x^2 - 1)$$
$$= -x^2 + 2x - 2 \qquad \text{Difference}$$

$$f(x) \cdot g(x) = (2x - 3)(x^2 - 1)$$
$$= 2x^3 - 3x^2 - 2x + 3 \qquad \text{Product}$$

$$\frac{f(x)}{g(x)} = \frac{2x - 3}{x^2 - 1}, \qquad x \neq \pm 1 \qquad \text{Quotient}$$

The domain of an **arithmetic combination** of functions f and g consists of all real numbers that are common to the domains of f and g. In the case of the quotient $f(x)/g(x)$, there is the further restriction that $g(x) \neq 0$.

What you should learn

● Add, subtract, multiply, and divide functions.
● Find compositions of one function with another function.
● Use combinations of functions to model and solve real-life problems.

Why you should learn it

Combining functions can sometimes help you better understand the big picture. For instance, Exercises 75 and 76 on page 143 illustrate how to use combinations of functions to analyze U.S. health expenditures.

SuperStock

Sum, Difference, Product, and Quotient of Functions

Let f and g be two functions with overlapping domains. Then, for all x common to both domains, the sum, difference, product, and quotient of f and g are defined as follows.

1. Sum: $\qquad (f + g)(x) = f(x) + g(x)$

2. Difference: $\qquad (f - g)(x) = f(x) - g(x)$

3. Product: $\qquad (fg)(x) = f(x) \cdot g(x)$

4. Quotient: $\qquad \left(\dfrac{f}{g}\right)(x) = \dfrac{f(x)}{g(x)}, \quad g(x) \neq 0$

Example 1 Finding the Sum of Two Functions

Given $f(x) = 2x + 1$ and $g(x) = x^2 + 2x - 1$, find $(f + g)(x)$. Then evaluate the sum when $x = 2$.

Solution

$$(f + g)(x) = f(x) + g(x) = (2x + 1) + (x^2 + 2x - 1) = x^2 + 4x$$

When $x = 2$, the value of this sum is $(f + g)(2) = 2^2 + 4(2) = 12$.

✓ *Checkpoint* Now try Exercise 13.

Example 2   Finding the Difference of Two Functions

Given $f(x) = 2x + 1$ and $g(x) = x^2 + 2x - 1$, find $(f - g)(x)$. Then evaluate the difference when $x = 2$.

Algebraic Solution

The difference of the functions f and g is

$$(f - g)(x) = f(x) - g(x)$$

$$= (2x + 1) - (x^2 + 2x - 1)$$

$$= -x^2 + 2.$$

When $x = 2$, the value of this difference is

$$(f - g)(2) = -(2)^2 + 2$$

$$= -2.$$

Note that $(f - g)(2)$ can also be evaluated as follows.

$$(f - g)(2) = f(2) - g(2)$$

$$= [2(2) + 1] - [2^2 + 2(2) - 1]$$

$$= 5 - 7$$

$$= -2$$

✓ *Checkpoint* Now try Exercise 15.

Graphical Solution

You can use a graphing utility to graph the difference of two functions. Enter the functions as follows (see Figure 1.70).

$$y_1 = 2x + 1$$

$$y_2 = x^2 + 2x - 1$$

$$y_3 = y_1 - y_2$$

Graph y_3 as shown in Figure 1.71. Then use the *value* feature or the *zoom* and *trace* features to estimate that the value of the difference when $x = 2$ is -2.

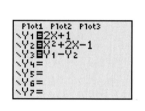

Figure 1.70

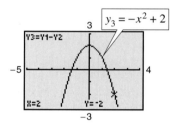

Figure 1.71

In Examples 1 and 2, both f and g have domains that consist of all real numbers. So, the domain of both $(f + g)$ and $(f - g)$ is also the set of all real numbers. Remember that any restrictions on the domains of f or g must be considered when forming the sum, difference, product, or quotient of f and g. For instance, the domain of $f(x) = 1/x$ is all $x \neq 0$, and the domain of $g(x) = \sqrt{x}$ is $[0, \infty)$. This implies that the domain of $(f + g)$ is $(0, \infty)$.

Example 3   Finding the Product of Two Functions

Given $f(x) = x^2$ and $g(x) = x - 3$, find $(fg)(x)$. Then evaluate the product when $x = 4$.

Solution

$$(fg)(x) = f(x)g(x)$$

$$= (x^2)(x - 3)$$

$$= x^3 - 3x^2$$

When $x = 4$, the value of this product is

$$(fg)(4) = 4^3 - 3(4)^2 = 16.$$

✓ *Checkpoint* Now try Exercise 17.

Example 4   Finding the Quotient of Two Functions

Find $(f/g)(x)$ and $(g/f)(x)$ for the functions given by $f(x) = \sqrt{x}$ and $g(x) = \sqrt{4 - x^2}$. Then find the domains of f/g and g/f.

Solution

The quotient of f and g is

$$\left(\frac{f}{g}\right)(x) = \frac{f(x)}{g(x)} = \frac{\sqrt{x}}{\sqrt{4 - x^2}},$$

and the quotient of g and f is

$$\left(\frac{g}{f}\right)(x) = \frac{g(x)}{f(x)} = \frac{\sqrt{4 - x^2}}{\sqrt{x}}.$$

The domain of f is $[0, \infty)$ and the domain of g is $[-2, 2]$. The intersection of these domains is $[0, 2]$. So, the domains for f/g and g/f are as follows.

Domain of (f/g): $[0, 2)$ Domain of (g/f): $(0, 2]$

✓ *Checkpoint* Now try Exercise 19.

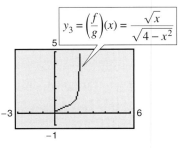

Figure 1.72

TECHNOLOGY TIP You can confirm the domain of f/g in Example 4 with your graphing utility by entering the three functions $y_1 = \sqrt{x}$, $y_2 = \sqrt{4 - x^2}$, and $y_3 = y_1/y_2$, and graphing y_3 as shown in Figure 1.72. Use the *trace* feature to determine that the x-coordinates of points on the graph extend from 0 to 2 but do not include 2. So, you can estimate the domain of f/g to be $[0, 2)$. You can confirm the domain of g/f in Example 4 by entering $y_4 = y_2/y_1$ and graphing y_4 as shown in Figure 1.73. Use the *trace* feature to determine that the x-coordinates of points on the graph extend from 0 to 2 but do not include 0. So, you can estimate the domain of g/f to be $(0, 2]$.

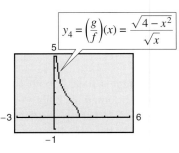

Figure 1.73

Compositions of Functions

Another way of combining two functions is to form the **composition** of one with the other. For instance, if $f(x) = x^2$ and $g(x) = x + 1$, the composition of f with g is

$$f(g(x)) = f(x + 1) = (x + 1)^2.$$

This composition is denoted as $f \circ g$ and read as "f of g."

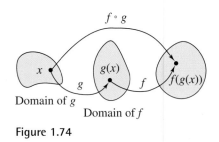

Figure 1.74

Definition of Composition of Two Functions

The **composition** of the function f with the function g is

$$(f \circ g)(x) = f(g(x)).$$

The domain of $f \circ g$ is the set of all x in the domain of g such that $g(x)$ is in the domain of f. (See Figure 1.74.)

Example 5   Forming the Composition of f with g

Find $(f \circ g)(x)$ for $f(x) = \sqrt{x}, x \geq 0$, and $g(x) = x - 1, x \geq 1$. If possible, find $(f \circ g)(2)$ and $(f \circ g)(0)$.

Solution

$$
\begin{aligned}
(f \circ g)(x) &= f(g(x)) && \text{Definition of } f \circ g \\
&= f(x - 1) && \text{Definition of } g(x) \\
&= \sqrt{x - 1}, \quad x \geq 1 && \text{Definition of } f(x)
\end{aligned}
$$

The domain of $f \circ g$ is $[1, \infty)$. So, $(f \circ g)(2) = \sqrt{2 - 1} = 1$ is defined, but $(f \circ g)(0)$ is not defined because 0 is not in the domain of $f \circ g$.

✓ *Checkpoint* Now try Exercise 35.

The composition of f with g is generally not the same as the composition of g with f. This is illustrated in Example 6.

> **Exploration**
>
> Let $f(x) = x + 2$ and $g(x) = 4 - x^2$. Are the compositions $f \circ g$ and $g \circ f$ equal? You can use your graphing utility to answer this question by entering and graphing the following functions.
>
> $$y_1 = (4 - x^2) + 2$$
>
> $$y_2 = 4 - (x + 2)^2$$
>
> What do you observe? Which function represents $f \circ g$ and which represents $g \circ f$?

Example 6   Compositions of Functions

Given $f(x) = x + 2$ and $g(x) = 4 - x^2$, evaluate (a) $(f \circ g)(x)$ and (b) $(g \circ f)(x)$ when $x = 0, 1, 2$, and 3.

Algebraic Solution

a.
$$
\begin{aligned}
(f \circ g)(x) &= f(g(x)) && \text{Definition of } f \circ g \\
&= f(4 - x^2) && \text{Definition of } g(x) \\
&= (4 - x^2) + 2 && \text{Definition of } f(x) \\
&= -x^2 + 6
\end{aligned}
$$
$$
\begin{aligned}
(f \circ g)(0) &= -0^2 + 6 = 6 \\
(f \circ g)(1) &= -1^2 + 6 = 5 \\
(f \circ g)(2) &= -2^2 + 6 = 2 \\
(f \circ g)(3) &= -3^2 + 6 = -3
\end{aligned}
$$

b.
$$
\begin{aligned}
(g \circ f)(x) &= g(f(x)) && \text{Definition of } g \circ f \\
&= g(x + 2) && \text{Definition of } f(x) \\
&= 4 - (x + 2)^2 && \text{Definition of } g(x) \\
&= 4 - (x^2 + 4x + 4) \\
&= -x^2 - 4x
\end{aligned}
$$
$$
\begin{aligned}
(g \circ f)(0) &= -0^2 - 4(0) = 0 \\
(g \circ f)(1) &= -1^2 - 4(1) = -5 \\
(g \circ f)(2) &= -2^2 - 4(2) = -12 \\
(g \circ f)(3) &= -3^2 - 4(3) = -21
\end{aligned}
$$

Note that $f \circ g \neq g \circ f$.

✓ *Checkpoint* Now try Exercise 37.

Numerical Solution

a. You can use the *table* feature of a graphing utility to evaluate $f \circ g$ when $x = 0, 1, 2$, and 3. Enter $y_1 = g(x)$ and $y_2 = f(g(x))$ in the equation editor (see Figure 1.75). Then set the table to *ask* mode to find the desired function values (see Figure 1.76). Finally, display the table, as shown in Figure 1.77.

b. You can evaluate $g \circ f$ when $x = 0, 1, 2$, and 3 by using a procedure similar to that of part (a). You should obtain the table shown in Figure 1.78.

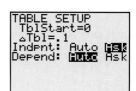

Figure 1.75   **Figure 1.76**

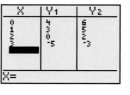

Figure 1.77   **Figure 1.78**

From the tables you can see that $f \circ g \neq g \circ f$.

To determine the domain of a composite function $f \circ g$, you need to restrict the outputs of g so that they are in the domain of f. For instance, to find the domain of $f \circ g$ given that $f(x) = 1/x$ and $g(x) = x + 1$, consider the outputs of g. These can be any real number. However, the domain of f is restricted to all real numbers except 0. So, the outputs of g must be restricted to all real numbers except 0. This means that $g(x) \neq 0$, or $x \neq -1$. So, the domain of $f \circ g$ is all real numbers except $x = -1$.

Example 7   Finding the Domain of a Composite Function

Find the domain of the composition $(f \circ g)(x)$ for the functions given by

$$f(x) = x^2 - 9 \quad \text{and} \quad g(x) = \sqrt{9 - x^2}.$$

Algebraic Solution

The composition of the functions is as follows.

$$(f \circ g)(x) = f(g(x))$$
$$= f\left(\sqrt{9 - x^2}\right)$$
$$= \left(\sqrt{9 - x^2}\right)^2 - 9$$
$$= 9 - x^2 - 9$$
$$= -x^2$$

From this, it might appear that the domain of the composition is the set of all real numbers. This, however, is not true. Because the domain of f is the set of all real numbers and the domain of g is $[-3, 3]$, the domain of $(f \circ g)$ is $[-3, 3]$.

✓ *Checkpoint* Now try Exercise 39.

Graphical Solution

You can use a graphing utility to graph the composition of the functions $(f \circ g)(x)$ as $y = \left(\sqrt{9 - x^2}\right)^2 - 9$. Enter the functions as follows.

$$y_1 = \sqrt{9 - x^2} \qquad y_2 = y_1^2 - 9$$

Graph y_2 as shown in Figure 1.79. Use the *trace* feature to determine that the x-coordinates of points on the graph extend from -3 to 3. So, you can graphically estimate the domain of $(f \circ g)(x)$ to be $[-3, 3]$.

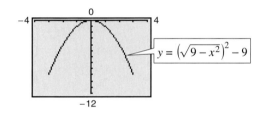

$$y = \left(\sqrt{9 - x^2}\right)^2 - 9$$

Figure 1.79

Example 8   A Case in Which $f \circ g = g \circ f$

Given $f(x) = 2x + 3$ and $g(x) = \frac{1}{2}(x - 3)$, find each composition.

a. $(f \circ g)(x)$ **b.** $(g \circ f)(x)$

Solution

a. $(f \circ g)(x) = f(g(x))$

$$= f\left(\frac{1}{2}(x - 3)\right)$$
$$= 2\left[\frac{1}{2}(x - 3)\right] + 3$$
$$= x - 3 + 3 = x$$

b. $(g \circ f)(x) = g(f(x))$

$$= g(2x + 3)$$
$$= \frac{1}{2}\left[(2x + 3) - 3\right]$$
$$= \frac{1}{2}(2x) = x$$

✓ *Checkpoint* Now try Exercise 43.

STUDY TIP

In Example 8, note that the two composite functions $f \circ g$ and $g \circ f$ are equal, and both represent the identity function. That is, $(f \circ g)(x) = x$ and $(g \circ f)(x) = x$. You will study this special case in the next section.

In Examples 5, 6, 7, and 8 you formed the composition of two given functions. In calculus, it is also important to be able to identify two functions that make up a given composite function. Basically, to "decompose" a composite function, look for an "inner" and an "outer" function.

Example 9   Identifying a Composite Function

Write the function $h(x) = (3x - 5)^3$ as a composition of two functions.

Solution

One way to write h as a composition of two functions is to take the inner function to be $g(x) = 3x - 5$ and the outer function to be $f(x) = x^3$. Then you can write

$$h(x) = (3x - 5)^3$$
$$= f(3x - 5)$$
$$= f(g(x)).$$

✓ *Checkpoint* Now try Exercise 55.

Example 10   Identifying a Composite Function

Write the function

$$h(x) = \frac{1}{(x - 2)^2}$$

as a composition of two functions.

Solution

One way to write h as a composition of two functions is to take the inner function to be $g(x) = x - 2$ and the outer function to be

$$f(x) = \frac{1}{x^2}$$
$$= x^{-2}.$$

Then you can write

$$h(x) = \frac{1}{(x - 2)^2}$$
$$= (x - 2)^{-2}$$
$$= f(x - 2)$$
$$= f(g(x)).$$

✓ *Checkpoint* Now try Exercise 59.

Exploration

The function in Example 10 can be decomposed in other ways. For which of the following pairs of functions is $h(x)$ equal to $f(g(x))$?

a. $g(x) = \dfrac{1}{x - 2}$ and

$f(x) = x^2$

b. $g(x) = x^2$ and

$f(x) = \dfrac{1}{x - 2}$

c. $g(x) = \dfrac{1}{x}$ and

$f(x) = (x - 2)^2$

Example 11   Bacteria Count

The number N of bacteria in a refrigerated food is given by

$$N(T) = 20T^2 - 80T + 500, \qquad 2 \le T \le 14$$

where T is the temperature of the food in degrees Celsius. When the food is removed from refrigeration, the temperature of the food is given by

$$T(t) = 4t + 2, \qquad 0 \le t \le 3$$

where t is the time (in hours).

a. Find the composition $N(T(t))$ and interpret its meaning in context.

b. Find the number of bacteria in the food when $t = 2$ hours.

c. Find the time when the bacterial count reaches 2000.

Solution

a. $N(T(t)) = 20(4t + 2)^2 - 80(4t + 2) + 500$

$\qquad = 20(16t^2 + 16t + 4) - 320t - 160 + 500$

$\qquad = 320t^2 + 320t + 80 - 320t - 160 + 500$

$\qquad = 320t^2 + 420$

The composite function $N(T(t))$ represents the number of bacteria as a function of the amount of time the food has been out of refrigeration.

b. When $t = 2$, the number of bacteria is

$N = 320(2)^2 + 420$

$\quad = 1280 + 420$

$\quad = 1700.$

c. The bacterial count will reach $N = 2000$ when $320t^2 + 420 = 2000$. You can solve this equation for t algebraically as follows.

$$320t^2 + 420 = 2000$$

$$320t^2 = 1580$$

$$t^2 = \frac{79}{16}$$

$$t = \frac{\sqrt{79}}{4}$$

$$t \approx 2.22 \text{ hours}$$

So, the count will reach 2000 when $t \approx 2.22$ hours. When you solve this equation, note that the negative value is rejected because it is not in the domain of the composite function. You can use a graphing utility to confirm your solution. First graph the equation $N = 320t^2 + 420$, as shown in Figure 1.80. Then use the *zoom* and *trace* features to approximate $N = 2000$ when $t \approx 2.22$, as shown in Figure 1.81.

 Checkpoint Now try Exercise 79.

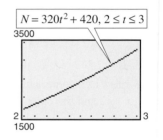

Figure 1.80

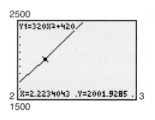

Figure 1.81

1.6   Exercises

Vocabulary Check

Fill in the blanks.

1. Two functions f and g can be combined by the arithmetic operations of _____ , _____ , _____ , and _____ to create new functions.

2. The _____ of the function f with g is $(f \circ g)(x) = f(g(x))$.

3. The domain of $f \circ g$ is the set of all x in the domain of g such that _____ is in the domain of f.

4. To decompose a composite function, look for an _____ and _____ function.

In Exercises 1–4, use the graphs of f and g to graph $h(x) = (f + g)(x)$. To print an enlarged copy of the graph, go to the website *www.mathgraphs.com*.

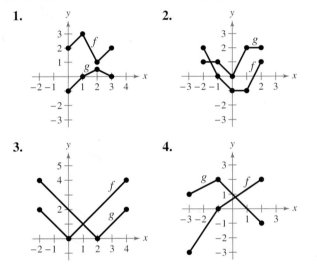

1.   **2.**

3.   **4.**

In Exercises 5–12, find (a) $(f + g)(x)$, (b) $(f - g)(x)$, (c) $(fg)(x)$, and (d) $(f/g)(x)$. What is the domain of f/g?

5. $f(x) = x + 3$, $g(x) = x - 3$

6. $f(x) = 2x - 5$, $g(x) = 1 - x$

7. $f(x) = x^2$, $g(x) = 1 - x$

8. $f(x) = 2x - 5$, $g(x) = 4$

9. $f(x) = x^2 + 5$, $g(x) = \sqrt{1 - x}$

10. $f(x) = \sqrt{x^2 - 4}$, $g(x) = \dfrac{x^2}{x^2 + 1}$

11. $f(x) = \dfrac{1}{x}$, $g(x) = \dfrac{1}{x^2}$

12. $f(x) = \dfrac{x}{x + 1}$, $g(x) = x^3$

In Exercises 13–26, evaluate the indicated function for $f(x) = x^2 + 1$ and $g(x) = x - 4$ algebraically. If possible, use a graphing utility to verify your answer.

13. $(f + g)(3)$ 14. $(f - g)(-2)$

15. $(f - g)(0)$ 16. $(f + g)(1)$

17. $(fg)(4)$ 18. $(fg)(-6)$

19. $\left(\dfrac{f}{g}\right)(-5)$ 20. $\left(\dfrac{f}{g}\right)(0)$

21. $(f - g)(2t)$ 22. $(f + g)(t - 4)$

23. $(fg)(-5t)$ 24. $(fg)(3t^2)$

25. $\left(\dfrac{f}{g}\right)(-t)$ 26. $\left(\dfrac{f}{g}\right)(t + 2)$

In Exercises 27–30, use a graphing utility to graph the functions f, g, and $f + g$ in the same viewing window.

27. $f(x) = \frac{1}{2}x$, $g(x) = x - 1$

28. $f(x) = \frac{1}{3}x$, $g(x) = -x + 4$

29. $f(x) = x^2$, $g(x) = -2x$

30. $f(x) = 4 - x^2$, $g(x) = x$

In Exercises 31–34, use a graphing utility to graph f, g, and $f + g$ in the same viewing window. Which function contributes most to the magnitude of the sum when $0 \le x \le 2$? Which function contributes most to the magnitude of the sum when $x > 6$?

31. $f(x) = 3x$, $g(x) = -\dfrac{x^3}{10}$

32. $f(x) = \dfrac{x}{2}$, $g(x) = \sqrt{x}$

33. $f(x) = 3x + 2$, $g(x) = -\sqrt{x + 5}$

34. $f(x) = x^2 - \frac{1}{2}$, $g(x) = -3x^2 - 1$

In Exercises 35–38, find (a) $f \circ g$, (b) $g \circ f$, and, if possible, (c) $(f \circ g)(0)$.

35. $f(x) = x^2, \quad g(x) = x - 1$

36. $f(x) = \sqrt[3]{x - 1}, \quad g(x) = x^3 + 1$

37. $f(x) = 3x + 5, \quad g(x) = 5 - x$

38. $f(x) = x^3, \quad g(x) = \dfrac{1}{x}$

In Exercises 39–44, (a) find $f \circ g$, $g \circ f$, and the domain of $f \circ g$. (b) Use a graphing utility to graph $f \circ g$ and $g \circ f$. Determine whether $f \circ g = g \circ f$.

39. $f(x) = \sqrt{x + 4}, \quad g(x) = x^2$

40. $f(x) = \sqrt[3]{x + 1}, \quad g(x) = x^3 - 1$

41. $f(x) = \frac{1}{3}x - 3, \quad g(x) = 3x + 1$

42. $f(x) = \sqrt{x}, \quad g(x) = \sqrt{x}$

43. $f(x) = x^{2/3}, \quad g(x) = x^6$

44. $f(x) = |x|, \quad g(x) = x + 6$

In Exercises 45–50, (a) find $(f \circ g)(x)$ and $(g \circ f)(x)$, (b) determine algebraically whether $(f \circ g)(x) = (g \circ f)(x)$, and (c) verify your answer to part (b) by comparing a table of values for each composition.

45. $f(x) = 5x + 4, \quad g(x) = 4 - x$

46. $f(x) = \frac{1}{4}(x - 1), \quad g(x) = 4x + 1$

47. $f(x) = \sqrt{x + 6}, \quad g(x) = x^2 - 5$

48. $f(x) = x^3 - 4, \quad g(x) = \sqrt[3]{x + 10}$

49. $f(x) = |x + 3|, \quad g(x) = 2x - 1$

50. $f(x) = \dfrac{6}{3x - 5}, \quad g(x) = -x$

In Exercises 51–54, use the graphs of f and g to evaluate the functions.

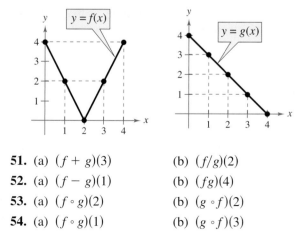

51. (a) $(f + g)(3)$ (b) $(f/g)(2)$

52. (a) $(f - g)(1)$ (b) $(fg)(4)$

53. (a) $(f \circ g)(2)$ (b) $(g \circ f)(2)$

54. (a) $(f \circ g)(1)$ (b) $(g \circ f)(3)$

In Exercises 55–62, find two functions f and g such that $(f \circ g)(x) = h(x)$. (There are many correct answers.)

55. $h(x) = (2x + 1)^2$

56. $h(x) = (1 - x)^3$

57. $h(x) = \sqrt[3]{x^2 - 4}$

58. $h(x) = \sqrt{9 - x}$

59. $h(x) = \dfrac{1}{x + 2}$

60. $h(x) = \dfrac{4}{(5x + 2)^2}$

61. $h(x) = (x + 4)^2 + 2(x + 4)$

62. $h(x) = (x + 3)^{3/2} + 4(x + 3)^{1/2}$

In Exercises 63–72, determine the domains of (a) f, (b) g, and (c) $f \circ g$. Use a graphing utility to verify your results.

63. $f(x) = \sqrt{x + 4}, \quad g(x) = x^2$

64. $f(x) = \sqrt{x + 3}, \quad g(x) = \dfrac{x}{2}$

65. $f(x) = x^2 + 1, \quad g(x) = \sqrt{x}$

66. $f(x) = x^{1/4}, \quad g(x) = x^4$

67. $f(x) = \dfrac{1}{x}, \quad g(x) = x + 3$

68. $f(x) = \dfrac{1}{x}, \quad g(x) = \dfrac{1}{2x}$

69. $f(x) = |x - 4|, \quad g(x) = 3 - x$

70. $f(x) = \dfrac{2}{|x|}, \quad g(x) = x - 1$

71. $f(x) = x + 2, \quad g(x) = \dfrac{1}{x^2 - 4}$

72. $f(x) = \dfrac{3}{x^2 - 1}, \quad g(x) = x + 1$

73. *Stopping Distance*   The research and development department of an automobile manufacturer has determined that when required to stop quickly to avoid an accident, the distance (in feet) a car travels during the driver's reaction time is given by $R(x) = \frac{3}{4}x$, where x is the speed of the car in miles per hour. The distance (in feet) traveled while the driver is braking is given by $B(x) = \frac{1}{15}x^2$.

(a) Find the function that represents the total stopping distance T.

(b) Use a graphing utility to graph the functions R, B, and T in the same viewing window for $0 \le x \le 60$.

(c) Which function contributes most to the magnitude of the sum at higher speeds? Explain.

74. *Sales*   From 2000 to 2005, the sales R_1 (in thousands of dollars) for one of two restaurants owned by the same parent company can be modeled by

$$R_1 = 480 - 8t - 0.8t^2, \qquad t = 0, 1, 2, 3, 4, 5$$

where $t = 0$ represents 2000. During the same six-year period, the sales R_2 (in thousands of dollars) for the second restaurant can be modeled by

$$R_2 = 254 + 0.78t, \qquad t = 0, 1, 2, 3, 4, 5.$$

(a) Write a function R_3 that represents the total sales for the two restaurants.

(b) Use a graphing utility to graph R_1, R_2, and R_3 (the total sales function) in the same viewing window.

Data Analysis   In Exercises 75 and 76, use the table, which shows the total amount spent (in billions of dollars) on health services and supplies in the United States and Puerto Rico for the years 1994 through 2000. The variables y_1, y_2, and y_3 represent out-of-pocket payments, insurance premiums, and other types of payments, respectively. (Source: U.S. Centers for Medicare and Medicaid Services)

Year	y_1	y_2	y_3
1994	143.9	312.1	40.7
1995	146.5	330.1	44.9
1996	152.1	344.8	48.2
1997	162.3	359.4	52.1
1998	174.5	383.2	55.6
1999	184.4	409.4	57.3
2000	194.5	443.9	57.2

Models for the data are $y_1 = 8.93t + 103.0$, $y_2 = 1.886t^2 - 5.24t + 305.7$, and $y_3 = -0.361t^2 + 7.97t + 14.2$, where t represents the year, with $t = 4$ corresponding to 1994.

75. Use the models and the *table* feature of a graphing utility to create tables showing the values for y_1, y_2, and y_3 for each year from 1994 to 2000. Compare these values with the original data.

76. Use a graphing utility to graph y_1, y_2, y_3, and $y_1 + y_2 + y_3$ in the same viewing window. Use the model $y_1 + y_2 + y_3$ to estimate the total amount spent on health services and supplies for the years 2005 and 2010.

77. *Ripples*   A pebble is dropped into a calm pond, causing ripples in the form of concentric circles (see figure). The radius (in feet) of the outer ripple is given by $r(t) = 0.6t$, where t is the time (in seconds) after the pebble strikes the water. The area of the circle is given by $A(r) = \pi r^2$. Find and interpret $(A \circ r)(t)$.

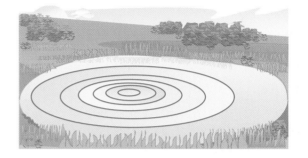

78. *Geometry*   A square concrete foundation was prepared as a base for a large cylindrical gasoline tank (see figure).

(a) Write the radius r of the tank as a function of the length x of the sides of the square.

(b) Write the area A of the circular base of the tank as a function of the radius r.

(c) Find and interpret $(A \circ r)(x)$.

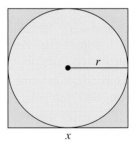

79. *Cost*   The weekly cost C of producing x units in a manufacturing process is given by

$$C(x) = 60x + 750.$$

The number of units x produced in t hours is $x(t) = 50t$.

(a) Find and interpret $(C \circ x)(t)$.

(b) Use a graphing utility to graph the cost as a function of time. Use the *trace* feature to estimate (to two-decimal-place accuracy) the time that must elapse until the cost increases to $15,000.

80. Air Traffic Control An air traffic controller spots two planes at the same altitude flying toward each other. Their flight paths form a right angle at point P. One plane is 150 miles from point P and is moving at 450 miles per hour. The other plane is 200 miles from point P and is moving at 450 miles per hour. Write the distance s between the planes as a function of time t.

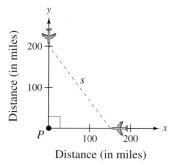

Distance (in miles)

81. Salary You are a sales representative for an automobile manufacturer. You are paid an annual salary plus a bonus of 3% of your sales over $500,000. Consider the two functions

$$f(x) = x - 500,000 \quad \text{and} \quad g(x) = 0.03x.$$

If x is greater than $500,000, which of the following represents your bonus? Explain.

(a) $f(g(x))$ (b) $g(f(x))$

82. Consumer Awareness The suggested retail price of a new car is p dollars. The dealership advertised a factory rebate of $1200 and an 8% discount.

(a) Write a function R in terms of p giving the cost of the car after receiving the rebate from the factory.

(b) Write a function S in terms of p giving the cost of the car after receiving the dealership discount.

(c) Form the composite functions $(R \circ S)(p)$ and $(S \circ R)(p)$ and interpret each.

(d) Find $(R \circ S)(18,400)$ and $(S \circ R)(18,400)$. Which yields the lower cost for the car? Explain.

Synthesis

True or False? **In Exercises 83 and 84, determine whether the statement is true or false. Justify your answer.**

83. If $f(x) = x + 1$ and $g(x) = 6x$, then

$$(f \circ g)(x) = (g \circ f)(x).$$

84. If you are given two functions $f(x)$ and $g(x)$, you can calculate $(f \circ g)(x)$ if and only if the range of g is a subset of the domain of f.

85. Proof Prove that the product of two odd functions is an even function, and that the product of two even functions is an even function.

86. Conjecture Use examples to hypothesize whether the product of an odd function and an even function is even or odd. Then prove your hypothesis.

87. Proof Given a function f, prove that $g(x)$ is even and $h(x)$ is odd, where $g(x) = \frac{1}{2}[f(x) + f(-x)]$ and $h(x) = \frac{1}{2}[f(x) - f(-x)]$.

88. (a) Use the result of Exercise 87 to prove that any function can be written as a sum of even and odd functions. (*Hint:* Add the two equations in Exercise 87.)

(b) Use the result of part (a) to write each function as a sum of even and odd functions.

$$f(x) = x^2 - 2x + 1, \quad g(x) = \frac{1}{x + 1}$$

Review

In Exercises 89–92, find three points that lie on the graph of the equation.

89. $y = -x^2 + x - 5$ **90.** $y = \frac{1}{5}x^3 - 4x^2 + 1$

91. $x^2 + y^2 = 24$ **92.** $y = \dfrac{x}{x^2 - 5}$

In Exercises 93–96, find an equation of the line that passes through the two points.

93. $(-4, -2), (-3, 8)$ **94.** $(1, 5), (-8, 2)$

95. $\left(\frac{3}{2}, -1\right), \left(-\frac{1}{3}, 4\right)$ **96.** $(0, 1.1), (-4, 3.1)$

In Exercises 97–102, use the graph of f to sketch the graph of the specified function. To print an enlarged copy of the graph, go to the website *www.mathgraphs.com.*

97. $f(x - 4)$

98. $f(x + 2)$

99. $f(x) + 4$

100. $f(x) - 1$

101. $2f(x)$

102. $f\left(\frac{1}{2}x\right)$

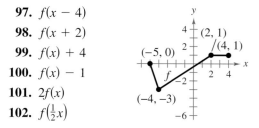

1.7 Inverse Functions

Inverse Functions

Recall from Section 1.3 that a function can be represented by a set of ordered pairs. For instance, the function $f(x) = x + 4$ from the set $A = \{1, 2, 3, 4\}$ to the set $B = \{5, 6, 7, 8\}$ can be written as follows.

$$f(x) = x + 4: \ \{(1, 5), (2, 6), (3, 7), (4, 8)\}$$

In this case, by interchanging the first and second coordinates of each of these ordered pairs, you can form the **inverse function** of f, which is denoted by f^{-1}. It is a function from the set B to the set A, and can be written as follows.

$$f^{-1}(x) = x - 4: \ \{(5, 1), (6, 2), (7, 3), (8, 4)\}$$

Note that the domain of f is equal to the range of f^{-1}, and vice versa, as shown in Figure 1.82. Also note that the functions f and f^{-1} have the effect of "undoing" each other. In other words, when you form the composition of f with f^{-1} or the composition of f^{-1} with f, you obtain the identity function.

$$f(f^{-1}(x)) = f(x - 4) = (x - 4) + 4 = x$$

$$f^{-1}(f(x)) = f^{-1}(x + 4) = (x + 4) - 4 = x$$

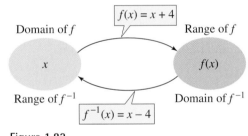

Domain of f $\boxed{f(x) = x + 4}$ Range of f

x $f(x)$

Range of f^{-1} $\boxed{f^{-1}(x) = x - 4}$ Domain of f^{-1}

Figure 1.82

Example 1   Finding Inverse Functions Informally

Find the inverse function of $f(x) = 4x$. Then verify that both $f(f^{-1}(x))$ and $f^{-1}(f(x))$ are equal to the identity function.

Solution

The function f *multiplies* each input by 4. To "undo" this function, you need to *divide* each input by 4. So, the inverse function of $f(x) = 4x$ is given by

$$f^{-1}(x) = \frac{x}{4}.$$

You can verify that both $f(f^{-1}(x))$ and $f^{-1}(f(x))$ are equal to the identity function as follows.

$$f(f^{-1}(x)) = f\left(\frac{x}{4}\right) = 4\left(\frac{x}{4}\right) = x \qquad f^{-1}(f(x)) = f^{-1}(4x) = \frac{4x}{4} = x$$

 Checkpoint Now try Exercise 1.

Example 2   Finding Inverse Functions Informally

Find the inverse function of $f(x) = x - 6$. Then verify that both $f(f^{-1}(x))$ and $f^{-1}(f(x))$ are equal to the identity function.

Solution

The function f *subtracts* 6 from each input. To "undo" this function, you need to *add* 6 to each input. So, the inverse function of $f(x) = x - 6$ is given by

$$f^{-1}(x) = x + 6.$$

You can verify that both $f(f^{-1}(x))$ and $f^{-1}(f(x))$ are equal to the identity function as follows.

$$f(f^{-1}(x)) = f(x + 6) = (x + 6) - 6 = x$$
$$f^{-1}(f(x)) = f^{-1}(x - 6) = (x - 6) + 6 = x$$

 Checkpoint Now try Exercise 3.

A table of values can help you understand inverse functions. For instance, the following table shows several values of the function in Example 2. Interchange the rows of this table to obtain values of the inverse function.

x	-2	-1	0	1	2
$f(x)$	-8	-7	-6	-5	-4

x	-8	-7	-6	-5	-4
$f^{-1}(x)$	-2	-1	0	1	2

In the table at the left, each output is 6 less than the input, and in the table at the right, each output is 6 more than the input.

The formal definition of an inverse function is as follows.

Definition of Inverse Function

Let f and g be two functions such that

$$f(g(x)) = x \qquad \text{for every } x \text{ in the domain of } g$$

and

$$g(f(x)) = x \qquad \text{for every } x \text{ in the domain of } f.$$

Under these conditions, the function g is the **inverse function** of the function f. The function g is denoted by f^{-1} (read "f-inverse"). So,

$$f(f^{-1}(x)) = x \qquad \text{and} \qquad f^{-1}(f(x)) = x.$$

The domain of f must be equal to the range of f^{-1}, and the range of f must be equal to the domain of f^{-1}.

If the function g is the inverse function of the function f, it must also be true that the function f is the inverse function of the function g. For this reason, you can say that the functions f and g are *inverse functions of each other*.

Example 3   Verifying Inverse Functions Algebraically

Show that the functions are inverse functions of each other.

$$f(x) = 2x^3 - 1 \quad \text{and} \quad g(x) = \sqrt[3]{\frac{x+1}{2}}$$

Solution

$$f(g(x)) = f\left(\sqrt[3]{\frac{x+1}{2}}\right) = 2\left(\sqrt[3]{\frac{x+1}{2}}\right)^3 - 1$$

$$= 2\left(\frac{x+1}{2}\right) - 1$$

$$= x + 1 - 1$$

$$= x$$

$$g(f(x)) = g(2x^3 - 1) = \sqrt[3]{\frac{(2x^3 - 1) + 1}{2}}$$

$$= \sqrt[3]{\frac{2x^3}{2}}$$

$$= \sqrt[3]{x^3}$$

$$= x$$

✓ *Checkpoint* Now try Exercise 15.

Example 4   Verifying Inverse Functions Algebraically

Which of the functions is the inverse function of $f(x) = \dfrac{5}{x-2}$?

$$g(x) = \frac{x-2}{5} \quad \text{or} \quad h(x) = \frac{5}{x} + 2$$

Solution

By forming the composition of f with g, you have

$$f(g(x)) = f\left(\frac{x-2}{5}\right) = \frac{5}{\left(\dfrac{x-2}{5}\right) - 2} = \frac{25}{x - 12} \neq x.$$

Because this composition is not equal to the identity function x, it follows that g *is not* the inverse function of f. By forming the composition of f with h, you have

$$f(h(x)) = f\left(\frac{5}{x} + 2\right) = \frac{5}{\left(\dfrac{5}{x} + 2\right) - 2} = \frac{5}{\left(\dfrac{5}{x}\right)} = x.$$

So, it appears that h is the inverse function of f. You can confirm this by showing that the composition of h with f is also equal to the identity function.

✓ *Checkpoint* Now try Exercise 19.

TECHNOLOGY TIP

Most graphing utilities can graph $y = x^{1/3}$ in two ways:

$$y_1 = x \wedge (1/3) \text{ or}$$

$$y_1 = \sqrt[3]{x}.$$

However, you may not be able to obtain the complete graph of $y = x^{2/3}$ by entering $y_1 = x \wedge (2/3)$. If not, you should use

$$y_1 = (x \wedge (1/3))^2 \text{ or}$$

$$y_1 = \sqrt[3]{x^2}.$$

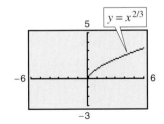

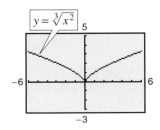

The Graph of an Inverse Function

The graphs of a function f and its inverse function f^{-1} are related to each other in the following way. If the point (a, b) lies on the graph of f, then the point (b, a) must lie on the graph of f^{-1}, and vice versa. This means that the graph of f^{-1} is a *reflection* of the graph of f in the line $y = x$, as shown in Figure 1.83.

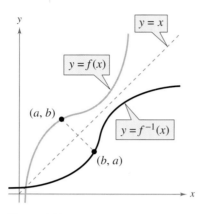

Figure 1.83

TECHNOLOGY TIP

In Examples 3 and 4, inverse functions were verified algebraically. A graphing utility can also be helpful in checking whether one function is the inverse function of another function. Use the Graph Reflection Program found on the website *college.hmco.com* to verify Example 4 graphically.

Example 5   Verifying Inverse Functions Graphically and Numerically

Verify that the functions f and g from Example 3 are inverse functions of each other graphically and numerically.

Graphical Solution

You can *graphically* verify that f and g are inverse functions of each other by using a graphing utility to graph f and g in the same viewing window. (Be sure to use a square setting.) From the graph in Figure 1.84, you can verify that the graph of g is the reflection of the graph of f in the line $y = x$.

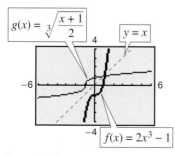

Figure 1.84

Numerical Solution

You can *numerically* verify that f and g are inverse functions of each other. Begin by entering the compositions $f(g(x))$ and $g(f(x))$ into a graphing utility as follows.

$$y_1 = f(g(x)) = 2\left(\sqrt[3]{\frac{x+1}{2}}\right)^3 - 1$$

$$y_2 = g(f(x)) = \sqrt[3]{\frac{(2x^3 - 1) + 1}{2}}$$

Then use the *table* feature of the graphing utility to create a table, as shown in Figure 1.85. Note that the entries for x, y_1, and y_2 are the same. So, $f(g(x)) = x$ and $g(f(x)) = x$. You can now conclude that f and g are inverse functions of each other.

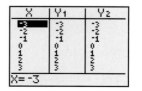

Figure 1.85

 Checkpoint Now try Exercise 25.

The Existence of an Inverse Function

Consider the function $f(x) = x^2$. The first table at the right is a table of values for $f(x) = x^2$. The second table was created by interchanging the rows of the first table. The second table does not represent a function because the input $x = 4$ is matched with two different outputs: $y = -2$ and $y = 2$. So, $f(x) = x^2$ does not have an inverse function.

x	-2	-1	0	1	2
$f(x)$	4	1	0	1	4

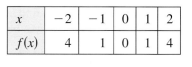

x	4	1	0	1	4
$g(x)$	-2	-1	0	1	2

To have an inverse function, a function must be **one-to-one,** which means that no two elements in the domain of f correspond to the same element in the range of f.

> ### Definition of a One-to-One Function
>
> A function f is **one-to-one** if, for a and b in its domain, $f(a) = f(b)$ implies that $a = b$.

> ### Existence of an Inverse Function
>
> A function f has an inverse function f^{-1} if and only if f is one-to-one.

From its graph, it is easy to tell whether a function of x is one-to-one. Simply check to see that every horizontal line intersects the graph of the function at most once. This is called the **Horizontal Line Test.** For instance, Figure 1.86 shows the graph of $y = x^4$. On the graph, you can find a horizontal line that intersects the graph twice.

Two special types of functions that pass the Horizontal Line Test are those that are increasing or decreasing on their entire domains.

1. If f is *increasing* on its entire domain, then f is one-to-one.

2. If f is *decreasing* on its entire domain, then f is one-to-one.

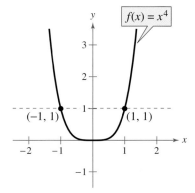

Figure 1.86 $f(x) = x^4$ is not one-to-one.

Example 6   Testing for One-to-One Functions

Is the function $f(x) = \sqrt{x} + 1$ one-to-one?

Algebraic Solution

Let a and b be nonnegative real numbers with $f(a) = f(b)$.

$$\sqrt{a} + 1 = \sqrt{b} + 1 \qquad \text{Set } f(a) = f(b).$$
$$\sqrt{a} = \sqrt{b}$$
$$a = b$$

So, $f(a) = f(b)$ implies that $a = b$. We can conclude that f is one-to-one and *does* have an inverse function.

Graphical Solution

Use a graphing utility to graph the function $y = \sqrt{x} + 1$. From Figure 1.87, you can see that a horizontal line will intersect the graph at most once and the function is increasing. So, f is one-to-one and *does* have an inverse function.

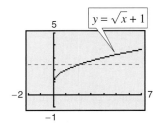

Figure 1.87

✓ *Checkpoint* Now try Exercise 33.

Finding Inverse Functions Algebraically

For simple functions you can find inverse functions by inspection. For more complicated functions, however, it is best to use the following guidelines.

> **Finding an Inverse Function**
>
> 1. Use the Horizontal Line Test to decide whether f has an inverse function.
> 2. In the equation for $f(x)$, replace $f(x)$ by y.
> 3. Interchange the roles of x and y, and solve for y.
> 4. Replace y by $f^{-1}(x)$ in the new equation.
> 5. Verify that f and f^{-1} are inverse functions of each other by showing that the domain of f is equal to the range of f^{-1}, the range of f is equal to the domain of f^{-1}, and $f(f^{-1}(x)) = x$ and $f^{-1}(f(x)) = x$.

It is important to note that in Step 1 above, the domain of f is assumed to be the entire real line. However, the domain of f may be restricted so that f does have an inverse function. For instance, if the domain of $f(x) = x^2$ is restricted to the nonnegative real numbers, then f does have an inverse function.

Example 7   Finding an Inverse Function Algebraically

Find the inverse function of $f(x) = \dfrac{5 - 3x}{2}$.

Solution

The graph of f in Figure 1.88 passes the Horizontal Line Test. So you know that f is one-to-one and has an inverse function.

$$f(x) = \frac{5 - 3x}{2} \qquad \text{Write original equation.}$$

$$y = \frac{5 - 3x}{2} \qquad \text{Replace } f(x) \text{ by } y.$$

$$x = \frac{5 - 3y}{2} \qquad \text{Interchange } x \text{ and } y.$$

$$2x = 5 - 3y \qquad \text{Multiply each side by 2.}$$

$$3y = 5 - 2x \qquad \text{Isolate the } y\text{-term.}$$

$$y = \frac{5 - 2x}{3} \qquad \text{Solve for } y.$$

$$f^{-1}(x) = \frac{5 - 2x}{3} \qquad \text{Replace } y \text{ by } f^{-1}(x).$$

The domain and range of both f and f^{-1} consist of all real numbers. Verify that $f(f^{-1}(x)) = x$ and $f^{-1}(f(x)) = x$.

✓ *Checkpoint* Now try Exercise 53.

TECHNOLOGY TIP

Many graphing utilities have a built-in feature to draw an inverse function. To see how this works, consider the function $f(x) = \sqrt{x}$. The inverse function of f is given by $f^{-1}(x) = x^2$, $x \geq 0$. Enter the function $y_1 = \sqrt{x}$. Then graph it in the standard viewing window and use the *draw inverse* feature. You should obtain the figure below, which shows both f and its inverse function f^{-1}. For instructions on how to use the *draw inverse* feature, see Appendix A; for specific keystrokes, go to the text website at *college.hmco.com*.

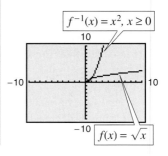

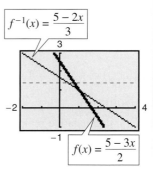

Figure 1.88

Example 8 Finding an Inverse Function Algebraically

Find the inverse function of $f(x) = x^3 - 4$ and use a graphing utility to graph f and f^{-1} in the same viewing window.

Solution

$f(x) = x^3 - 4$	Write original function.
$y = x^3 - 4$	Replace $f(x)$ by y.
$x = y^3 - 4$	Interchange x and y.
$y^3 = x + 4$	Isolate y.
$y = \sqrt[3]{x + 4}$	Solve for y.
$f^{-1}(x) = \sqrt[3]{x + 4}$	Replace y by $f^{-1}(x)$.

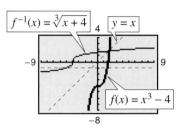

The graph of f in Figure 1.89 passes the Horizontal Line Test. So, you know that f is one-to-one and has an inverse function. The graph of f^{-1} in Figure 1.89 is the reflection of the graph of f in the line $y = x$.

Figure 1.89

✓ *Checkpoint* Now try Exercise 55.

Example 9 Finding an Inverse Function Algebraically

Find the inverse function of $f(x) = \sqrt{2x - 3}$ and use a graphing utility to graph f and f^{-1} in the same viewing window.

Solution

$f(x) = \sqrt{2x - 3}$	Write original equation.
$y = \sqrt{2x - 3}$	Replace $f(x)$ by y.
$x = \sqrt{2y - 3}$	Interchange x and y.
$x^2 = 2y - 3$	Square each side.
$2y = x^2 + 3$	Isolate y.
$y = \dfrac{x^2 + 3}{2}$	Solve for y.
$f^{-1}(x) = \dfrac{x^2 + 3}{2}, \quad x \geq 0$	Replace y by $f^{-1}(x)$.

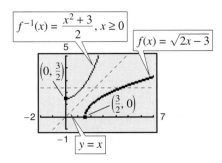

The graph of f in Figure 1.90 passes the Horizontal Line Test. So you know that f is one-to-one and has an inverse function. The graph of f^{-1} in Figure 1.90 is the reflection of the graph of f in the line $y = x$. Note that the range of f is the interval $[0, \infty)$, which implies that the domain of f^{-1} is the interval $[0, \infty)$. Moreover, the domain of f is the interval $[\frac{3}{2}, \infty)$, which implies that the range of f^{-1} is the interval $[\frac{3}{2}, \infty)$.

✓ *Checkpoint* Now try Exercise 59.

Figure 1.90

1.7   Exercises

Vocabulary Check

Fill in the blanks.

1. If the composite functions $f(g(x)) = x$ and $g(f(x)) = x$, then the function g is the _____ function of f, and is denoted by _____ .

2. The domain of f is the _____ of f^{-1}, and the _____ of f^{-1} is the range of f.

3. The graphs of f and f^{-1} are reflections of each other in the line _____ .

4. To have an inverse function, a function f must be _____ ; that is, $f(a) = f(b)$ implies $a = b$.

5. A graphical test for the existence of an inverse function is called the _____ Line Test.

In Exercises 1–8, find the inverse function of f informally. Verify that $f(f^{-1}(x)) = x$ and $f^{-1}(f(x)) = x$.

1. $f(x) = 6x$ 2. $f(x) = \frac{1}{3}x$

3. $f(x) = x + 7$ 4. $f(x) = x - 3$

5. $f(x) = 2x + 1$ 6. $f(x) = \dfrac{x - 1}{4}$

7. $f(x) = \sqrt[3]{x}$ 8. $f(x) = x^5$

In Exercises 9–14, (a) show that f and g are inverse functions algebraically and (b) verify that f and g are inverse functions numerically by creating a table of values for each function.

9. $f(x) = -\dfrac{7}{2}x - 3$, $g(x) = -\dfrac{2x + 6}{7}$

10. $f(x) = \dfrac{x - 9}{4}$, $g(x) = 4x + 9$

11. $f(x) = x^3 + 5$, $g(x) = \sqrt[3]{x - 5}$

12. $f(x) = \dfrac{x^3}{2}$, $g(x) = \sqrt[3]{2x}$

13. $f(x) = -\sqrt{x - 8}$; $g(x) = 8 + x^2$, $x \leq 0$

14. $f(x) = \sqrt[3]{3x - 10}$, $g(x) = \dfrac{x^3 + 10}{3}$

In Exercises 15–20, show that f and g are inverse functions algebraically. Use a graphing utility to graph f and g in the same viewing window. Describe the relationship between the graphs.

15. $f(x) = x^3$, $g(x) = \sqrt[3]{x}$

16. $f(x) = \dfrac{1}{x}$, $g(x) = \dfrac{1}{x}$

17. $f(x) = \sqrt{x - 4}$; $g(x) = x^2 + 4$, $x \geq 0$

18. $f(x) = 9 - x^2$, $x \geq 0$; $g(x) = \sqrt{9 - x}$

19. $f(x) = 1 - x^3$, $g(x) = \sqrt[3]{1 - x}$

20. $f(x) = \dfrac{1}{1 + x}$, $x \geq 0$; $g(x) = \dfrac{1 - x}{x}$, $0 < x \leq 1$

In Exercises 21–24, match the graph of the function with the graph of its inverse function. [The graphs of the inverse functions are labeled (a), (b), (c), and (d).]

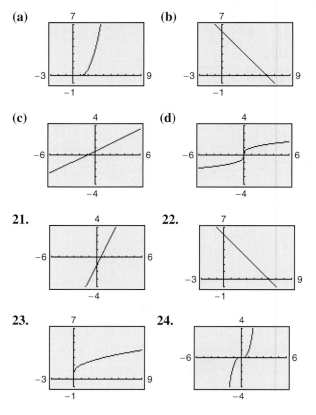

In Exercises 25–28, show that f and g are inverse functions (a) graphically and (b) numerically.

25. $f(x) = 2x, \quad g(x) = \dfrac{x}{2}$

26. $f(x) = x - 5, \quad g(x) = x + 5$

27. $f(x) = \dfrac{x - 1}{x + 5}, \quad g(x) = -\dfrac{5x + 1}{x - 1}$

28. $f(x) = \dfrac{x + 3}{x - 2}, \quad g(x) = \dfrac{2x + 3}{x - 1}$

In Exercises 29–40, use a graphing utility to graph the function and use the Horizontal Line Test to determine whether the function is one-to-one and so has an inverse function.

29. $f(x) = 3 - \tfrac{1}{2}x$

30. $f(x) = \tfrac{1}{4}(x + 2)^2 - 1$

31. $h(x) = \dfrac{x^2}{x^2 + 1}$

32. $g(x) = \dfrac{4 - x}{6x^2}$

33. $h(x) = \sqrt{16 - x^2}$

34. $f(x) = -2x\sqrt{16 - x^2}$

35. $f(x) = 10$

36. $f(x) = -0.65$

37. $g(x) = (x + 5)^3$

38. $f(x) = x^5 - 7$

39. $h(x) = |x + 4| - |x - 4|$

40. $f(x) = -\dfrac{|x - 6|}{|x + 6|}$

In Exercises 41–52, determine algebraically whether the function is one-to-one. If it is, find its inverse function. Verify your answer graphically.

41. $f(x) = x^4$

42. $g(x) = x^2 - x^4$

43. $f(x) = \dfrac{3x + 4}{5}$

44. $f(x) = 3x + 5$

45. $f(x) = \dfrac{1}{x^2}$

46. $h(x) = \dfrac{4}{x^2}$

47. $f(x) = (x + 3)^2, \quad x \geq -3$

48. $q(x) = (x - 5)^2, \quad x \leq 5$

49. $f(x) = \sqrt{2x + 3}$

50. $f(x) = \sqrt{x - 2}$

51. $f(x) = |x - 2|, \quad x \leq 2$

52. $f(x) = \dfrac{x^2}{x^2 + 1}$

In Exercises 53–62, find the inverse function of f. Use a graphing utility to graph both f and f^{-1} in the same viewing window. Describe the relationship between the graphs.

53. $f(x) = 2x - 3$

54. $f(x) = 3x$

55. $f(x) = x^5$

56. $f(x) = x^3 + 1$

57. $f(x) = x^{3/5}$

58. $f(x) = x^2, \quad x \geq 0$

59. $f(x) = \sqrt{4 - x^2}, \quad 0 \leq x \leq 2$

60. $f(x) = \sqrt{16 - x^2}, \quad -4 \leq x \leq 0$

61. $f(x) = \dfrac{4}{x}$

62. $f(x) = \dfrac{6}{\sqrt{x}}$

Think About It   In Exercises 63–66, delete part of the graph of the function so that the part that remains is one-to-one. Find the inverse function of the remaining part and give the domain of the inverse function. (There are many correct answers.)

63. $f(x) = (x - 2)^2$

64. $f(x) = 1 - x^4$

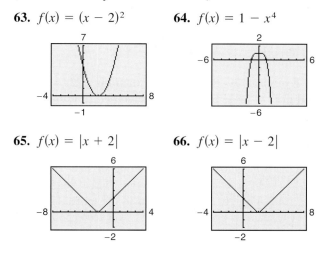

65. $f(x) = |x + 2|$

66. $f(x) = |x - 2|$

In Exercises 67 and 68, use the graph of the function f to complete the table and sketch the graph of f^{-1}.

67.

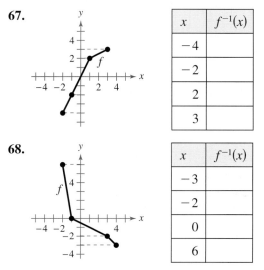

x	$f^{-1}(x)$
-4	
-2	
2	
3	

68.

x	$f^{-1}(x)$
-3	
-2	
0	
6	

Graphical Reasoning In Exercises 69–72, (a) use a graphing utility to graph the function, (b) use the *draw inverse* feature of the graphing utility to draw the inverse of the function, and (c) determine whether the graph of the inverse relation is an inverse function, explaining your reasoning.

69. $f(x) = x^3 + x + 1$

70. $h(x) = x\sqrt{4 - x^2}$

71. $g(x) = \dfrac{3x^2}{x^2 + 1}$

72. $f(x) = \dfrac{4x}{\sqrt{x^2 + 15}}$

In Exercises 73–78, use the functions $f(x) = \frac{1}{8}x - 3$ and $g(x) = x^3$ to find the indicated value or function.

73. $(f^{-1} \circ g^{-1})(1)$

74. $(g^{-1} \circ f^{-1})(-3)$

75. $(f^{-1} \circ f^{-1})(6)$

76. $(g^{-1} \circ g^{-1})(-4)$

77. $(f \circ g)^{-1}$

78. $g^{-1} \circ f^{-1}$

In Exercises 79–82, use the functions $f(x) = x + 4$ and $g(x) = 2x - 5$ to find the specified function.

79. $g^{-1} \circ f^{-1}$

80. $f^{-1} \circ g^{-1}$

81. $(f \circ g)^{-1}$

82. $(g \circ f)^{-1}$

83. *Transportation* The total value of new car sales f (in billions of dollars) in the United States from 1995 through 2001 is shown in the table. The time (in years) is given by t, with $t = 5$ corresponding to 1995. (Source: National Automobile Dealers Association)

Year, t	Sales, $f(t)$
5	456.2
6	490.0
7	507.5
8	546.3
9	606.5
10	650.3
11	690.4

(a) Does f^{-1} exist?

(b) If f^{-1} exists, what does it mean in the context of the problem?

(c) If f^{-1} exists, find $f^{-1}(650.3)$.

(d) If the table above were extended to 2002 and if the total value of new car sales for that year were $546.3 billion, would f^{-1} exist? Explain.

84. *Hourly Wage* Your wage is $8.00 per hour plus $0.75 for each unit produced per hour. So, your hourly wage y in terms of the number of units produced is $y = 8 + 0.75x$.

(a) Find the inverse function. What does each variable in the inverse function represent?

(b) Use a graphing utility to graph the function and its inverse function.

(c) Use the *trace* feature of a graphing utility to find the hourly wage when 10 units are produced per hour.

(d) Use the *trace* feature of a graphing utility to find the number of units produced when your hourly wage is $22.25.

Synthesis

True or False? In Exercises 85 and 86, determine whether the statement is true or false. Justify your answer.

85. If f is an even function, f^{-1} exists.

86. If the inverse function of f exists, and the graph of f has a y-intercept, the y-intercept of f is an x-intercept of f^{-1}.

87. *Proof* Prove that if f and g are one-to-one functions, $(f \circ g)^{-1}(x) = (g^{-1} \circ f^{-1})(x)$.

88. *Proof* Prove that if f is a one-to-one odd function, f^{-1} is an odd function.

Review

In Exercises 89–92, write the rational expression in simplest form.

89. $\dfrac{27x^3}{3x^2}$

90. $\dfrac{5x^2y}{xy + 5x}$

91. $\dfrac{x^2 - 36}{6 - x}$

92. $\dfrac{x^2 + 3x - 40}{x^2 - 3x - 10}$

In Exercises 93–98, determine whether the equation represents y as a function of x.

93. $4x - y = 3$

94. $x = 5$

95. $x^2 + y^2 = 9$

96. $x^2 + y = 8$

97. $y = \sqrt{x + 2}$

98. $x - y^2 = 0$

1 Chapter Summary

What did you learn?

1 Review Exercises

1.1 In Exercises 1–4, complete the table. Use the resulting solution points to sketch the graph of the equation. Use a graphing utility to verify the graph.

1. $y = -\frac{1}{2}x + 2$

x	-2	0	2	3	4
y					
Solution point					

2. $y = x^2 - 3x$

x	-1	0	1	2	3
y					
Solution point					

3. $y = 4 - x^2$

x	-2	-1	0	1	2
y					
Solution point					

4. $y = \sqrt{x - 1}$

x	1	2	5	10	17
y					
Solution point					

In Exercises 5–12, use a graphing utility to graph the equation. Approximate any x- or y-intercepts.

5. $y = \frac{1}{4}(x + 1)^3$ **6.** $y = 4 - (x - 4)^2$

7. $y = \frac{1}{4}x^4 - 2x^2$ **8.** $y = \frac{1}{4}x^3 - 3x$

9. $y = x\sqrt{9 - x^2}$ **10.** $y = x\sqrt{x + 3}$

11. $y = |x - 4| - 4$ **12.** $y = |x + 2| + |3 - x|$

In Exercises 13 and 14, describe the viewing window of the graph shown.

13. $y = 0.002x^2 - 0.06x - 1$

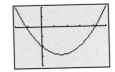

14. $y = 10x^3 - 21x^2$

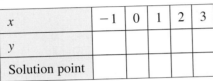

15. *Consumerism* You purchase a compact car for $13,500. The depreciated value y after t years is

$$y = 13,500 - 1100t, \quad 0 \le t \le 6.$$

(a) Use the constraints of the model to determine an appropriate viewing window.

(b) Use a graphing utility to graph the equation.

(c) Use the *zoom* and *trace* features of a graphing utility to determine the value of t when $y = \$9100$.

16. *Data Analysis* The table shows the number of Gap stores from 1996 to 2001.   (Source: The Gap, Inc.)

Year, t	Stores, y
1996	1370
1997	2130
1998	2428
1999	3018
2000	3676
2001	4171

A model for number of Gap stores during this period is given by $y = 2.05t^2 + 514.6t - 1730$, where y represents the number of stores and t represents the year, with $t = 6$ corresponding to 1996.

(a) Use the model and the *table* feature of a graphing utility to approximate the number of Gap stores from 1996 to 2001.

(b) Use a graphing utility to graph the data and the model in the same viewing window.

(c) Use the model to estimate the number of Gap stores in 2005 and 2008. Do the values seem reasonable? Explain.

(d) Use the *zoom* and *trace* features of a graphing utility to determine during which year the number of stores exceeded 3000.

1.2 **In Exercises 17–22, plot the two points and find the slope of the line passing through the pair of points.**

17. $(-3, 2), (8, 2)$

18. $(7, -1), (7, 12)$

19. $\left(\frac{3}{2}, 1\right), \left(5, \frac{5}{2}\right)$

20. $\left(-\frac{3}{4}, \frac{5}{6}\right), \left(\frac{1}{2}, -\frac{5}{2}\right)$

21. $(-4.5, 6), (2.1, 3)$

22. $(-2.7, -6.3), (-1, -1.2)$

In Exercises 23–32, use the point on the line and the slope of the line to find the general form of the equation of the line, and find three additional points through which the line passes. (There are many correct answers.)

	Point	Slope
23.	$(2, -1)$	$m = \frac{1}{4}$
24.	$(-3, 5)$	$m = -\frac{3}{2}$
25.	$(0, -5)$	$m = \frac{3}{2}$
26.	$(3, 0)$	$m = -\frac{2}{3}$
27.	$\left(\frac{1}{5}, -5\right)$	$m = -1$
28.	$\left(0, \frac{7}{8}\right)$	$m = -\frac{4}{5}$
29.	$(-2, 6)$	$m = 0$
30.	$(-8, 8)$	$m = 0$
31.	$(10, -6)$	m is undefined.
32.	$(5, 4)$	m is undefined.

In Exercises 33–36, find the slope-intercept form of the equation of the line that passes through the points. Use a graphing utility to graph the line.

33. $(2, -1), (4, -1)$ **34.** $(0, 0), (0, 10)$

35. $(-1, 0), (6, 2)$ **36.** $(1, 6), (4, 2)$

Rate of Change **In Exercises 37 and 38, you are given the dollar value of a product in 2005 *and* the rate at which the value of the item is expected to change during the 5 years following. Use this information to write a linear equation that gives the dollar value V of the product in terms of the year t. (Let $t = 5$ represent 2005.)**

	2005 Value	Rate
37.	$12,500	$850 increase per year
38.	$72.95	$5.15 decrease per year

39. *Sales*   During the second and third quarters of the year, an e-commerce business had sales of $160,000 and $185,000, respectively. The growth of sales follows a linear pattern. Estimate sales during the fourth quarter.

40. *Depreciation*   The dollar value of a VCR in 2004 is $85, and the product will decrease in value at an expected rate of $10.75 per year.

(a) Write a linear equation that gives the dollar value V of the VCR in terms of the year t. (Let $t = 4$ represent 2004.)

(b) Use a graphing utility to graph the equation found in part (a).

(c) Use the *value* or *trace* feature of your graphing utility to estimate the dollar value of the VCR in 2008.

In Exercises 41–44, write the slope-intercept forms of the equations of the lines through the given point (a) parallel to the given line and (b) perpendicular to the given line. Verify your result with a graphing utility (use a square setting).

	Point	Line
41.	$(3, -2)$	$5x - 4y = 8$
42.	$(-8, 3)$	$2x + 3y = 5$
43.	$(-6, 2)$	$x = 4$
44.	$(3, -4)$	$y = 2$

1.3 **In Exercises 45 and 46, which sets of ordered pairs represent functions from A to B? Explain.**

45. $A = \{10, 20, 30, 40\}$ and $B = \{0, 2, 4, 6\}$

(a) $\{(20, 4), (40, 0), (20, 6), (30, 2)\}$

(b) $\{(10, 4), (20, 4), (30, 4), (40, 4)\}$

(c) $\{(40, 0), (30, 2), (20, 4), (10, 6)\}$

(d) $\{(20, 2), (10, 0), (40, 4)\}$

46. $A = \{u, v, w\}$ and $B = \{-2, -1, 0, 1, 2\}$

(a) $\{(v, -1), (u, 2), (w, 0), (u, -2)\}$

(b) $\{(u, -2), (v, 2), (w, 1)\}$

(c) $\{(u, 2), (v, 2), (w, 1), (w, 1)\}$

(d) $\{(w, -2), (v, 0), (w, 2)\}$

In Exercises 47–50, determine whether the equation represents y as a function of x.

47. $16x - y^4 = 0$ **48.** $2x - y - 3 = 0$

49. $y = \sqrt{1 - x}$ **50.** $|y| = x + 2$

In Exercises 51–54, evaluate the function at each value of the independent variable and simplify.

51. $f(x) = x^2 + 1$

    (a) $f(2)$ (b) $f(-4)$

    (c) $f(t^2)$ (d) $-f(x)$

52. $g(x) = x^{4/3}$

    (a) $g(8)$ (b) $g(t + 1)$

    (c) $g(-27)$ (d) $g(-x)$

53. $h(x) = \begin{cases} 2x + 1, & x \le -1 \\ x^2 + 2, & x > -1 \end{cases}$

    (a) $h(-2)$ (b) $h(-1)$

    (c) $h(0)$ (d) $h(2)$

54. $f(x) = \dfrac{3}{2x - 5}$

    (a) $f(1)$ (b) $f(-2)$

    (c) $f(t)$ (d) $f(10)$

In Exercises 55–60, find the domain of the function.

55. $f(x) = (x - 1)(x + 2)$

56. $f(x) = x^2 - 4x - 32$

57. $f(x) = \sqrt{25 - x^2}$ **58.** $f(x) = \sqrt{x^2 + 8x}$

59. $g(s) = \dfrac{5}{3s - 9}$ **60.** $f(x) = \dfrac{2}{3x + 4}$

61. *Cost*   A hand tool manufacturer produces a product for which the variable cost is $5.35 per unit and the fixed costs are $16,000. The company sells the product for $8.20 and can sell all that it produces.

    (a) Write the total cost C as a function of x, the number of units produced.

    (b) Write the profit P as a function of x.

62. *Consumerism*   The retail sales R (in billions of dollars) of lawn care products and services in the United States from 1994 to 2001 can be approximated by the model

$$R(t) = \begin{cases} -0.67t + 11.0, & 4 \le t \le 7 \\ 0.600t^2 - 10.06t + 50.7, & 8 \le t \le 11 \end{cases}$$

where t represents the year, with $t = 4$ corresponding to 1994. Use the *table* feature of a graphing utility to approximate the retail sales of lawn care products and services for each year from 1994 to 2001. (Source: The National Gardening Association)

In Exercises 63 and 64, find the difference quotient and simplify your answer.

63. $f(x) = 2x^2 + 3x - 1$, $\dfrac{f(x + h) - f(x)}{h}$, $h \ne 0$

64. $f(x) = x^3 - 5x^2 + x$, $\dfrac{f(x + h) - f(x)}{h}$, $h \ne 0$

1.4 **In Exercises 65–68, use a graphing utility to graph the function and estimate its domain and range. Then find the domain and range algebraically.**

65. $f(x) = 3 - 2x^2$ **66.** $f(x) = \sqrt{2x^2 - 1}$

67. $h(x) = \sqrt{36 - x^2}$ **68.** $g(x) = |x + 5|$

In Exercises 69–72, (a) use a graphing utility to graph the equation and (b) use the Vertical Line Test to determine whether y is a function of x.

69. $y = \dfrac{x^2 + 3x}{6}$ **70.** $y = -\dfrac{2}{3}|x + 5|$

71. $3x + y^2 = 2$ **72.** $x^2 + y^2 = 49$

In Exercises 73–76, (a) use a graphing utility to graph the function and (b) determine the open intervals on which the function is increasing, decreasing, or constant.

73. $f(x) = x^3 - 3x$ **74.** $f(x) = \sqrt{x^2 - 9}$

75. $f(x) = x\sqrt{x - 6}$ **76.** $f(x) = \dfrac{|x + 8|}{2}$

In Exercises 77–80, use a graphing utility to approximate (to two decimal places) any relative minimum or maximum values of the function.

77. $f(x) = (x^2 - 4)^2$ **78.** $f(x) = x^2 - x - 1$

79. $h(x) = 4x^3 - x^4$ **80.** $f(x) = x^3 - 4x^2 - 1$

In Exercises 81 and 82, sketch the graph of the piecewise-defined function by hand.

81. $f(x) = \begin{cases} 3x + 5, & x < 0 \\ x - 4, & x \ge 0 \end{cases}$

82. $f(x) = \begin{cases} x^2 + 7, & x < 1 \\ x^2 - 5x + 6, & x \ge 1 \end{cases}$

In Exercises 83 and 84, algebraically determine whether the function is even, odd, or neither. Verify your answer using a graphing utility.

83. $f(x) = (x^2 - 8)^2$ **84.** $f(x) = 2x^3 - x^2$

1.5 In Exercises 85–88, identify the common function and describe the transformation shown in the graph. Write an equation for the graphed function.

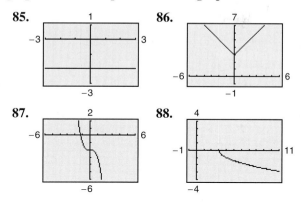

85.

86.

87.

88.

In Exercises 89–100, *h* is related to one of the six common functions on page 125. (a) Identify the common function *f*. (b) Describe the sequence of transformations from *f* to *h*. (c) Sketch the graph of *h* by hand. (d) Use function notation to write *h* in terms of the common function *f*.

89. $h(x) = x^2 - 6$ 90. $h(x) = (x - 3)^2 - 2$

91. $h(x) = (x - 1)^3 + 7$ 92. $h(x) = (x + 2)^3 + 5$

93. $h(x) = \sqrt{x} - 5$ 94. $h(x) = |x + 8| - 1$

95. $h(x) = -x^2 - 3$ 96. $h(x) = -(x - 2)^2 - 8$

97. $h(x) = -2x^2 + 3$ 98. $h(x) = \frac{1}{2}(x - 3)^2 + 6$

99. $h(x) = -\frac{1}{2}|x| + 9$ 100. $h(x) = \sqrt{3x} - 5$

1.6 In Exercises 101–110, let $f(x) = 3 - 2x$, $g(x) = \sqrt{x}$, and $h(x) = 3x^2 + 2$, and find the indicated values.

101. $(f - g)(4)$ 102. $(f + h)(5)$

103. $(f + g)(25)$ 104. $(g - h)(1)$

105. $(fh)(1)$ 106. $\left(\dfrac{g}{h}\right)(1)$

107. $(h \circ g)(7)$ 108. $(g \circ f)(-2)$

109. $(f \circ h)(-4)$ 110. $(g \circ h)(6)$

Data Analysis   In Exercises 111 and 112, the numbers (in thousands) of students taking the SAT (y_1) and ACT (y_2) for the years 1996 through 2001 can be modeled by $y_1 = -2.75t^2 + 86.8t + 659$ and $y_2 = -1.88t^2 + 62.4t + 616$, where *t* represents the year, with $t = 6$ corresponding to 1996. (Source: College Entrance Examination Board and ACT, Inc.)

111. Use a graphing utility to graph y_1, y_2, and $y_1 + y_2$ in the same viewing window.

112. Use the model $y_1 + y_2$ to estimate the total number of students taking the SAT and ACT in 2006.

1.7 In Exercises 113 and 114, find the inverse function of *f* informally. Verify that $f(f^{-1}(x)) = x$ and $f^{-1}(f(x)) = x$.

113. $f(x) = 6x$ 114. $f(x) = x + 5$

In Exercises 115 and 116, show that *f* and *g* are inverse functions (a) graphically and (b) numerically.

115. $f(x) = 3 - 4x$, $g(x) = \dfrac{3 - x}{4}$

116. $f(x) = \sqrt{x + 1}$, $g(x) = x^2 - 1, x \geq 0$

In Exercises 117–120, use a graphing utility to graph the function and use the Horizontal Line Test to determine whether the function is one-to-one and so has an inverse function.

117. $f(x) = \frac{1}{2}x - 3$ 118. $f(x) = (x - 1)^2$

119. $h(t) = \dfrac{2}{t - 3}$ 120. $g(x) = \sqrt{x + 6}$

In Exercises 121–126, find the inverse function of *f* algebraically.

121. $f(x) = \dfrac{x}{12}$ 122. $f(x) = \dfrac{7x + 3}{8}$

123. $f(x) = 4x^3 - 3$ 124. $f(x) = x^3 - 2$

125. $f(x) = \sqrt{x + 10}$ 126. $f(x) = 4\sqrt{6 - x}$

Synthesis

True or False?   In Exercises 127–129, determine whether the statement is true or false. Justify your answer.

127. If the graph of the common function $f(x) = x^2$ is moved six units to the right, moved three units upward, and reflected in the *x*-axis, then the point $(-1, 28)$ will lie on the graph of the transformation.

128. If $f(x) = x^n$ where *n* is odd, f^{-1} exists.

129. There exists no function *f* such that $f = f^{-1}$.

1 Chapter Test

Take this test as you would take a test in class. After you are finished, check your work against the answers in the back of the book.

In Exercises 1–6, use the point-plotting method to graph the equation by hand and identify any x- and y-intercepts. Verify your results using a graphing utility.

1. $y = 4 - \frac{3}{4}|x|$

2. $y = 4 - (x - 2)^2$

3. $y = x - x^3$

4. $y = -x^3 + 2x - 4$

5. $y = \sqrt{3 - x}$

6. $y = \frac{1}{2}x\sqrt{x + 3}$

7. A line with slope $m = \frac{3}{2}$ passes through the point $(3, -1)$. List three additional points on the line. Then sketch the line.

8. Find an equation of the line that passes through the point $(0, 4)$ and is (a) parallel to and (b) perpendicular to the line $5x + 2y = 3$.

9. Does the graph at the right represent y as a function of x? Explain.

10. Evaluate $f(x) = |x + 2| - 15$ at each value of the independent variable and simplify.

   (a) $f(-8)$ (b) $f(14)$ (c) $f(t - 6)$

11. Find the domain of $f(x) = 10 - \sqrt{3 - x}$.

12. An electronics company produces a car stereo for which the variable cost is $5.60 and the fixed costs are $24,000. The product sells for $99.50. Write the total cost C as a function of x. Write the profit P as a function of x.

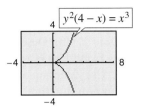

Figure for 9

In Exercises 13 and 14, determine the open intervals on which the function is increasing, decreasing, or constant.

13. $h(x) = \frac{1}{4}x^4 - 2x^2$

14. $g(t) = |t + 2| - |t - 2|$

In Exercises 15 and 16, use a graphing utility to approximate (to two decimal places) any relative minimum or maximum values of the function.

15. $f(x) = -x^3 - 5x^2 + 12$

16. $f(x) = x^5 - x^3 + 2$

In Exercises 17–19, (a) identify the common function f, (b) describe the sequence of transformations from f to g, and (c) sketch the graph of g.

17. $g(x) = -2(x - 5)^3 + 3$

18. $g(x) = \sqrt{-x - 7}$

19. $g(x) = 4|-x| - 7$

20. Use the functions $f(x) = x^2$ and $g(x) = \sqrt{2 - x}$ to find the specified function and its domain.

   (a) $(f - g)(x)$ (b) $\left(\dfrac{f}{g}\right)(x)$ (c) $(f \circ g)(x)$ (d) $(g \circ f)(x)$

In Exercises 21–23, determine whether the function has an inverse function, and if so, find the inverse function.

21. $f(x) = x^3 + 8$

22. $f(x) = x^2 + 6$

23. $f(x) = \dfrac{3x\sqrt{x}}{8}$

A quadratic equation can be used to model the number of hours spent annually per person using the internet in the United States.

Royalty Free/Corbis

2 Solving Equations and Inequalities

What You Should Learn

In this chapter, you will learn how to:

■ Solve and use linear equations, including those involving fractions.

■ Write and use mathematical models to solve real-life problems.

■ Find intercepts, zeros, and solutions of equations graphically.

■ Perform operations with complex numbers and plot complex numbers in the complex plane.

■ Solve quadratic equations, polynomial equations, and equations involving radicals, fractions, and absolute values.

■ Solve linear inequalities, inequalities involving absolute values, polynomial inequalities, and rational inequalities.

■ Use scatter plots and a graphing utility to find linear models for data.

2.1 Linear Equations and Problem Solving

Equations and Solutions of Equations

An **equation** in x is a statement that two algebraic expressions are equal. For example, $3x - 5 = 7$, $x^2 - x - 6 = 0$, and $\sqrt{2x} = 4$ are equations. To **solve** an equation in x means to find all values of x for which the equation is true. Such values are **solutions**. For instance, $x = 4$ is a solution of the equation $3x - 5 = 7$, because $3(4) - 5 = 7$ is a true statement.

The solutions of an equation depend on the kinds of numbers being considered. For instance, in the set of rational numbers, $x^2 = 10$ has no solution because there is no rational number whose square is 10. However, in the set of real numbers the equation has the two solutions $\sqrt{10}$ and $-\sqrt{10}$.

An equation that is true for *every* real number in the domain of the variable is called an **identity**. For example, $x^2 - 9 = (x + 3)(x - 3)$ is an identity because it is a true statement for any real value of x, and $x/(3x^2) = 1/(3x)$, where $x \neq 0$, is an identity because it is true for any nonzero real value of x.

An equation that is true for just *some* (or even none) of the real numbers in the domain of the variable is called a **conditional equation**. For example, the equation $x^2 - 9 = 0$ is conditional because $x = 3$ and $x = -3$ are the only values in the domain that satisfy the equation. The equation $2x + 1 = 2x - 3$ is also conditional because there are no real values of x for which the equation is true. Learning to solve conditional equations is the primary focus of this chapter.

A **linear equation in one variable x** is an equation that can be written in the standard form $ax + b = 0$, where a and b are real numbers, with $a \neq 0$. For a review of solving one- and two-step linear equations, see Appendix D.

To solve an equation involving fractional expressions, find the least common denominator (LCD) of all terms in the equation and multiply every term by this LCD. This procedure clears the equation of fractions as demonstrated in Example 1.

Example 1 Solving an Equation Involving Fractions

Solve $\dfrac{x}{3} + \dfrac{3x}{4} = 2$.

Solution

$$\frac{x}{3} + \frac{3x}{4} = 2 \qquad \text{Write original equation.}$$

$$(12)\frac{x}{3} + (12)\frac{3x}{4} = (12)2 \qquad \text{Multiply each term by the LCD of 12.}$$

$$4x + 9x = 24 \qquad \text{Divide out and multiply.}$$

$$13x = 24 \qquad \text{Combine like terms.}$$

$$x = \frac{24}{13} \qquad \text{Divide each side by 13.}$$

✓ *Checkpoint* Now try Exercise 15.

What you should learn

- Solve equations involving fractional expressions.
- Write and use mathematical models to solve real-life problems.
- Use common formulas to solve real-life problems.

Why you should learn it

Linear equations are useful for modeling situations in which you need to find missing information. For instance, Exercise 43 on page 169 shows how to use a linear equation to determine the score you must get on a test in order to get an A for the course.

PhotoEdit

STUDY TIP

After solving an equation, you should check each solution in the original equation. For instance, you can check the solution to Example 1 as follows.

$$\frac{x}{3} + \frac{3x}{4} = 2$$

$$\frac{\frac{24}{13}}{3} + \frac{3\left(\frac{24}{13}\right)}{4} \stackrel{?}{=} 2$$

$$\frac{8}{13} + \frac{18}{13} \stackrel{?}{=} 2$$

$$2 = 2 \checkmark$$

When multiplying or dividing an equation by a *variable* expression, it is possible to introduce an **extraneous solution**—one that does not satisfy the original equation. The next example demonstrates the importance of checking your solution when you have multiplied or divided by a variable expression.

Example 2   An Equation with an Extraneous Solution

Solve $\dfrac{1}{x-2} = \dfrac{3}{x+2} - \dfrac{6x}{x^2-4}$.

Algebraic Solution

The LCD is

$$x^2 - 4 = (x+2)(x-2).$$

Multiplying each term by the LCD and simplifying produces the following.

$$\frac{1}{x-2}(x+2)(x-2)$$

$$= \frac{3}{x+2}(x+2)(x-2) - \frac{6x}{x^2-4}(x+2)(x-2)$$

$$x + 2 = 3(x-2) - 6x, \quad x \neq \pm 2$$

$$x + 2 = 3x - 6 - 6x$$

$$4x = -8$$

$$x = -2 \qquad \text{Extraneous solution}$$

A check of $x = -2$ in the original equation shows that it yields a denominator of zero. So, $x = -2$ is an extraneous solution, and the original equation has *no solution*.

✓ *Checkpoint* Now try Exercise 29.

Graphical Solution

Use a graphing utility (in *dot* mode) to graph the left and right sides of the equation

$$y_1 = \frac{1}{x-2} \quad \text{and} \quad y_2 = \frac{3}{x+2} - \frac{6x}{x^2-4}$$

in the same viewing window, as shown in Figure 2.1. The graphs of the equations do not appear to intersect. This means that there is no point for which the left side of the equation y_1 is equal to the right side of the equation y_2. So, the equation appears to have *no solution*.

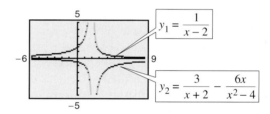

Figure 2.1

Using Mathematical Models to Solve Problems

One of the primary goals of this text is to learn how algebra can be used to solve problems that occur in real-life situations. This procedure is called **mathematical modeling.**

A good approach to mathematical modeling is to use two stages. Begin by using the verbal description of the problem to form a *verbal model*. Then, after assigning labels to the quantities in the verbal model, form a *mathematical model* or an *algebraic equation*.

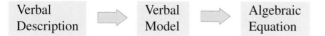

When you are trying to construct a verbal model, it is helpful to look for a *hidden equality*—a statement that two algebraic expressions are equal. These two expressions might be explicitly stated as being equal, or they might be known to be equal (based on prior knowledge or experience).

TECHNOLOGY TIP

Notice in Figure 2.1 that the equations were graphed using the *dot* mode of a graphing utility. In this text, a blue or light red curve is placed behind the graphing utility's display to indicate where the graph should appear. You will learn more about how graphing utilities graph these types of equations in Section 3.6.

Example 3 Finding the Dimensions of a Room

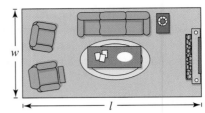

Figure 2.2

A rectangular family room is twice as long as it is wide, and its perimeter is 84 feet. Find the dimensions of the family room.

Solution

For this problem, it helps to draw a diagram, as shown in Figure 2.2.

Verbal Model: 2 · ⟨ Length ⟩ + 2 · ⟨ Width ⟩ = ⟨ Perimeter ⟩

Labels:
Perimeter = 84	(feet)
Width = w	(feet)
Length = $l = 2w$	(feet)

Equation:
$$2(2w) + 2w = 84 \qquad \text{Original equation}$$
$$6w = 84 \qquad \text{Group like terms.}$$
$$w = 14 \qquad \text{Divide each side by 6.}$$

Because the length is twice the width, you have

$$l = 2w \qquad \text{Length is twice width.}$$
$$= 2(14) \qquad \text{Substitute 14 for } w.$$
$$= 28. \qquad \text{Simplify.}$$

So, the dimensions of the room are 14 feet by 28 feet.

✓ *Checkpoint* Now try Exercise 41.

Example 4 A Distance Problem

A plane is flying nonstop from New York to San Francisco, a distance of about 2900 miles, as shown in Figure 2.3. After $1\frac{1}{2}$ hours in the air, the plane flies over Chicago (a distance of about 800 miles from New York). Estimate the time it will take the plane to fly from New York to San Francisco.

Solution

Verbal Model: ⟨ Distance ⟩ = ⟨ Rate ⟩ · ⟨ Time ⟩

Labels:
Distance = 2900	(miles)
Time = t	(hours)
Rate = $\dfrac{\text{Distance to Chicago}}{\text{Time to Chicago}} = \dfrac{800}{1.5}$	(miles per hour)

Equation:
$$2900 = \frac{800}{1.5}t$$
$$5.44 \approx t$$

The trip will take about 5.44 hours, or about 5 hours and 27 minutes.

✓ *Checkpoint* Now try Exercise 45.

STUDY TIP

Students sometimes say that although a solution looks easy when it is worked out in class, they don't see where to begin when solving a problem alone. Keep in mind that no one—not even great mathematicians—can expect to look at every mathematical problem and know immediately where to begin. Many problems involve some trial and error before a solution is found. To make algebra work for you, put in a lot of time, expect to try solution methods that end up not working, and learn from both your successes and your failures.

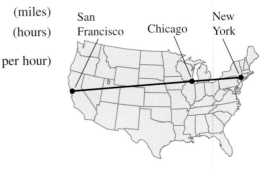

San Francisco   Chicago   New York

Figure 2.3

Example 5   Height of a Building

To determine the height of the Aon Center Building (in Chicago), you measure the shadow cast by the building and find it to be 142 feet long, as shown in Figure 2.4. Then you measure the shadow cast by a 48-inch post and find it to be 6 inches long. Estimate the building's height.

Solution

To solve this problem, you use a result from geometry that states that the ratios of corresponding sides of similar triangles are equal.

Verbal Model: $\dfrac{\text{Height of building}}{\text{Length of building's shadow}} = \dfrac{\text{Height of post}}{\text{Length of post's shadow}}$

Labels:

Height of building = x	(feet)
Length of building's shadow = 142	(feet)
Height of post = 48	(inches)
Length of post's shadow = 6	(inches)

Equation: $\dfrac{x}{142} = \dfrac{48}{6}$ $x = 1136$

So, the Aon Center Building is about 1136 feet high.

✓ *Checkpoint* Now try Exercise 51.

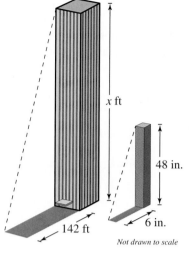

x ft

48 in.

142 ft   6 in.

Not drawn to scale

Figure 2.4

Example 6   An Inventory Problem

A store has $30,000 of inventory in 13-inch and 19-inch color televisions. The profit on a 13-inch set is 22% and the profit on a 19-inch set is 40%. The profit for the entire stock is 35%. How much was invested in each type of television?

Solution

Verbal Model: $\boxed{\text{Profit from 13-inch sets}} + \boxed{\text{Profit from 19-inch sets}} = \boxed{\text{Total profit}}$

Labels:

Inventory of 13-inch sets = x	(dollars)
Inventory of 19-inch sets = $30{,}000 - x$	(dollars)
Profit from 13-inch sets = $0.22x$	(dollars)
Profit from 19-inch sets = $0.40(30{,}000 - x)$	(dollars)
Total profit = $0.35(30{,}000) = 10{,}500$	(dollars)

Equation: $0.22x + 0.40(30{,}000 - x) = 10{,}500$

$$-0.18x = -1500$$

$$x \approx 8333.33$$

So, $8333.33 was invested in 13-inch sets and $30{,}000 - x$, or $21,666.67, was invested in 19-inch sets.

✓ *Checkpoint* Now try Exercise 55.

Common Formulas

Many common types of geometric, scientific, and investment problems use ready-made equations called **formulas.** Knowing these formulas will help you translate and solve a wide variety of real-life applications.

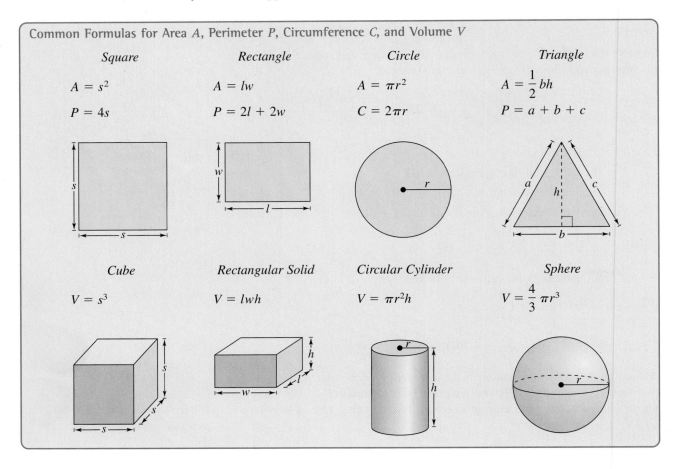

Common Formulas for Area A, Perimeter P, Circumference C, and Volume V

Square

$A = s^2$

$P = 4s$

Rectangle

$A = lw$

$P = 2l + 2w$

Circle

$A = \pi r^2$

$C = 2\pi r$

Triangle

$A = \dfrac{1}{2} bh$

$P = a + b + c$

Cube

$V = s^3$

Rectangular Solid

$V = lwh$

Circular Cylinder

$V = \pi r^2 h$

Sphere

$V = \dfrac{4}{3} \pi r^3$

Miscellaneous Common Formulas

Temperature:	$F = \dfrac{9}{5} C + 32$	F = degrees Fahrenheit, C = degrees Celsius
Simple Interest:	$I = Prt$	I = interest, P = principal (original deposit), r = annual interest rate, t = time in years
Compound Interest:	$A = P\left(1 + \dfrac{r}{n}\right)^{nt}$	A = balance, P = principal (original deposit), r = annual interest rate, n = compoundings (number of times interest is calculated) per year, t = time in years
Distance:	$d = rt$	d = distance traveled, r = rate, t = time

When working with applied problems, you often need to rewrite one of the common formulas. For instance, the formula for the perimeter of a rectangle, $P = 2l + 2w$, can be solved for w as $w = \frac{1}{2}(P - 2l)$.

|←— 4 cm —→|

Example 7   Using a Formula

A cylindrical can has a volume of 600 cubic centimeters and a radius of 4 centimeters, as shown in Figure 2.5. Find the height of the can.

Solution

The formula for the volume of a cylinder is $V = \pi r^2 h$. To find the height of the can, solve for h as

$$h = \frac{V}{\pi r^2}.$$

Then, using $V = 600$ and $r = 4$, find the height.

$$h = \frac{600}{\pi(4)^2} = \frac{600}{16\pi} \approx 11.94$$

You can use unit analysis to check that your answer is reasonable.

$$\frac{600 \text{ cm}^3}{16\pi \text{ cm}^2} \approx 11.94 \text{ cm}$$

 Checkpoint Now try Exercise 57.

Figure 2.5

Example 8   Using a Formula

The average daily temperature in San Diego, California is 64.4°F. What is San Diego's average daily temperature in degrees Celsius?   (Source: U.S. National Oceanic and Atmospheric Administration)

Solution

Use $F = 64.4$ in the formula for temperature to find the temperature in degrees Celsius.

$$F = \frac{9}{5}C + 32 \qquad \text{Formula for temperature}$$

$$64.4 = \frac{9}{5}C + 32 \qquad \text{Substitute 64.4 for } F.$$

$$32.4 = \frac{9}{5}C \qquad \text{Subtract 32 from each side.}$$

$$18 = C \qquad \text{Simplify.}$$

The average daily temperature in San Diego is 18°C.

 Checkpoint Now try Exercise 60.

2.1   Exercises

Vocabulary Check

Fill in the blanks.

1. A(n) _____ is a statement that equates two algebraic expressions.
2. To find all values that satisfy an equation is to _____ the equation.
3. There are two types of equations, _____ and _____ .
4. A linear equation in one variable is an equation that can be written in the standard form _____ .
5. When solving an equation, it is possible to introduce an _____ solution, which is a value that does not satisfy the original equation.
6. _____ is a procedure used in algebra to solve problems that occur in real-life situations.
7. Many real-life problems can be solved using ready-made equations called _____ .

In Exercises 1–6, determine whether each value of x is a solution of the equation.

Equation	Values

1. $\dfrac{5}{2x} - \dfrac{4}{x} = 3$ (a) $x = -\frac{1}{2}$ (b) $x = 4$

                       (c) $x = 0$ (d) $x = \frac{1}{4}$

2. $\dfrac{x}{2} + \dfrac{6x}{7} = \dfrac{19}{14}$ (a) $x = -2$ (b) $x = 1$

                       (c) $x = \dfrac{1}{2}$ (d) $x = 7$

3. $3 + \dfrac{1}{x + 2} = 4$ (a) $x = -1$ (b) $x = -2$

                       (c) $x = 0$ (d) $x = 5$

4. $\dfrac{(x + 5)(x - 3)}{2} = 24$ (a) $x = -3$ (b) $x = -2$

                       (c) $x = 7$ (d) $x = 9$

5. $\dfrac{\sqrt{x + 4}}{6} + 3 = 4$ (a) $x = -3$ (b) $x = 0$

                       (c) $x = 21$ (d) $x = 32$

6. $\dfrac{\sqrt[3]{x - 8}}{3} = -\dfrac{2}{3}$ (a) $x = -16$ (b) $x = 0$

                       (c) $x = 9$ (d) $x = 16$

In Exercises 7–12, determine whether the equation is an identity or a conditional equation.

7. $2(x - 1) = 2x - 2$
8. $-7(x - 3) + 4x = 3(7 - x)$
9. $x^2 - 8x + 5 = (x - 4)^2 - 11$
10. $x^2 + 2(3x - 2) = x^2 + 6x - 4$
11. $3 + \dfrac{1}{x + 1} = \dfrac{4x}{x + 1}$ 12. $\dfrac{5}{x} + \dfrac{3}{x} = 24$

In Exercises 13 and 14, solve the equation using two methods. Then explain which method is easier.

13. $\dfrac{3x}{8} - \dfrac{4x}{3} = 4$ 14. $\dfrac{3z}{8} - \dfrac{z}{10} = 6$

In Exercises 15–30, solve the equation (if possible). Then use a graphing utility to verify your solution.

15. $\dfrac{x}{5} - \dfrac{x}{2} = 3$ 16. $\dfrac{5x}{4} + \dfrac{1}{2} = x - \dfrac{1}{2}$

17. $\frac{3}{2}(z + 5) - \frac{1}{4}(z + 24) = 0$

18. $\dfrac{3x}{2} + \dfrac{1}{4}(x - 2) = 10$

19. $\dfrac{100 - 4u}{3} = \dfrac{5u + 6}{4} + 6$

20. $\dfrac{17 + y}{y} + \dfrac{32 + y}{y} = 100$

21. $\dfrac{5x - 4}{5x + 4} = \dfrac{2}{3}$ 22. $\dfrac{10x + 3}{5x + 6} = \dfrac{1}{2}$

23. $\dfrac{1}{x - 3} + \dfrac{1}{x + 3} = \dfrac{10}{x^2 - 9}$

24. $\dfrac{1}{x - 2} + \dfrac{3}{x + 3} = \dfrac{4}{x^2 + x - 6}$

25. $\dfrac{7}{2x + 1} - \dfrac{8x}{2x - 1} = -4$

26. $\dfrac{x}{x + 4} + \dfrac{4}{x + 4} + 2 = 0$

27. $\dfrac{1}{x} + \dfrac{2}{x - 5} = 0$

28. $3 = 2 + \dfrac{2}{z + 2}$

29. $\dfrac{3}{x^2 - 3x} + \dfrac{4}{x} = \dfrac{1}{x - 3}$

30. $\dfrac{6}{x} - \dfrac{2}{x + 3} = \dfrac{3(x + 5)}{x(x + 3)}$

In Exercises 31–38, solve for the indicated variable.

31. *Area of a Triangle*

Solve for h: $A = \frac{1}{2}bh$

32. *Area of a Trapezoid*

Solve for b: $A = \frac{1}{2}(a + b)h$

33. *Investment at Compound Interest*

Solve for P: $A = P\left(1 + \dfrac{r}{n}\right)^{nt}$

34. *Investment at Simple Interest*

Solve for r: $A = P + Prt$

35. *Geometric Progression*

Solve for r: $S = \dfrac{rL - a}{r - 1}$

36. *Arithmetic Progression*

Solve for n: $L = a + (n - 1)d$

37. *Volume of an Oblate Spheroid*

Solve for b: $V = \dfrac{4}{3}\pi a^2 b$

38. *Volume of a Spherical Segment*

Solve for r: $V = \dfrac{1}{3}\pi h^2(3r - h)$

Anthropology **In Exercises 39 and 40, use the following information. The relationship between the length of an adult's femur (thigh bone) and the height of the adult can be approximated by the linear equations**

$y = 0.432x - 10.44$ **Female**

$y = 0.449x - 12.15$ **Male**

where y is the length of the femur in inches and x is the height of the adult in inches (see figure).

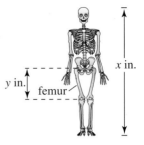

Figure for 39 and 40

39. An anthropologist discovers a femur belonging to an adult human female. The bone is 16 inches long. Estimate the height of the female.

40. From the foot bones of an adult human male, an anthropologist estimates that the person's height was 69 inches. A few feet away from the site where the foot bones were discovered, the anthropologist discovers a male adult femur that is 19 inches long. Is it likely that both the foot bones and the thigh bone came from the same person?

41. *Geometry*   A room is 1.5 times as long as it is wide, and its perimeter is 25 meters.

   (a) Draw a diagram that gives a visual representation of the problem. Identify the length as l and the width as w.

   (b) Write l in terms of w and write an equation for the perimeter in terms of w.

   (c) Find the dimensions of the room.

42. *Geometry*   A picture frame has a total perimeter of 3 meters. The height of the frame is $\frac{2}{3}$ times its width.

   (a) Draw a diagram that gives a visual representation of the problem. Identify the width as w and the height as h.

   (b) Write h in terms of w and write an equation for the perimeter in terms of w.

   (c) Find the dimensions of the picture frame.

43. *Course Grade*   To get an A in a course, you must have an average of at least 90 on four tests of 100 points each. The scores on your first three tests were 87, 92, and 84.

   (a) Write a verbal model for the test average for the course.

   (b) What must you score on the fourth test to get an A for the course?

44. Course Grade You are taking a course that has four tests. The first three tests are 100 points each and the fourth test is 200 points. To get an A in the course, you must have an average of at least 90% on the four tests. Your scores on the first three tests were 87, 92, and 84. What must you score on the fourth test to get an A for the course?

45. Travel Time You are driving on a Canadian freeway to a town that is 300 kilometers from your home. After 30 minutes you pass a freeway exit that you know is 50 kilometers from your home. Assuming that you continue at the same constant speed, how long will it take for the entire trip?

46. Travel Time On the first part of a 317-mile trip, a salesperson averaged 58 miles per hour. The salesperson averaged only 52 miles per hour on the last part of the trip because of an increased volume of traffic. The total time of the trip was 5 hours and 45 minutes. Find the amount of time at each of the two speeds.

47. Average Speed A truck driver traveled at an average speed of 55 miles per hour on a 200-mile trip to pick up a load of freight. On the return trip (with the truck fully loaded), the average speed was 40 miles per hour. Find the average speed for the round trip.

48. Wind Speed An executive flew in the corporate jet to a meeting in a city 1500 kilometers away. After traveling the same amount of time on the return flight, the pilot mentioned that they still had 300 kilometers to go. The air speed of the plane was 600 kilometers per hour. How fast was the wind blowing? (Assume that the wind direction was parallel to the flight path and constant all day.)

49. Speed of Light Light travels at the speed of 3.0×10^8 meters per second. Find the time in minutes required for light to travel from the sun to Earth (a distance of 1.5×10^{11} meters).

50. Radio Waves Radio waves travel at the same speed as light, 3.0×10^8 meters per second. Find the time required for a radio wave to travel from Mission Control in Houston to NASA astronauts on the surface of the moon 3.84×10^8 meters away.

51. Height To obtain the height of a barn silo, you measure the silo's shadow and find that it is 80 feet long. You also measure the shadow of a four-foot stake and find that it is $3\frac{1}{2}$ feet long.

(a) Draw a diagram that illustrates the problem. Let h represent the height of the silo.

(b) Find the height of the silo.

52. Height A person who is 6 feet tall walks away from a flagpole toward the tip of the shadow of the flagpole. When the person is 30 feet from the flagpole, the tips of the person's shadow and the shadow cast by the flagpole coincide at a point 5 feet in front of the person.

(a) Draw a diagram that illustrates the problem. Let h represent the height of the flagpole.

(b) Find the height of the flagpole.

53. Investment You plan to invest $12,000 in two funds paying $4\frac{1}{2}\%$ and 5% simple interest. (There is more risk in the 5% fund.) Your goal is to obtain a total annual interest income of $560 from the investments. What is the smallest amount you can invest in the 5% fund in order to meet your objective?

54. Investment You plan to invest $25,000 in two funds paying 3% and $4\frac{1}{2}\%$ simple interest. (There is more risk in the $4\frac{1}{2}\%$ fund.) Your goal is to obtain a total annual interest income of $1000 from the investments. What is the smallest amount you can invest in the $4\frac{1}{2}\%$ fund in order to meet your objective?

55. Mixture Problem A grocer mixes peanuts that cost $2.49 per pound and walnuts that cost $3.89 per pound to make 100 pounds of a mixture that costs $3.19 per pound. How much of each kind of nut is put into the mixture?

56. Mixture Problem A forester mixes gasoline and oil to make 2 gallons of mixture for his two-cycle chainsaw engine. This mixture is 32 parts gasoline and 1 part two-cycle oil. How much gasoline must be added to bring the mixture to 40 parts gasoline and 1 part oil?

57. Height A triangular sail has an area of 182.25 square feet. The sail has a base of 13.5 feet. Find the height of the sail.

58. Geometry The volume of a rectangular package is 2304 cubic inches. The length of the package is 3 times its width, and the height is one and a half times its width.

(a) Draw a diagram that illustrates the problem. Label the height, width, and length accordingly.

(b) Find the dimensions of the package.

59. Geometry The volume of a globe is about 47,712.94 cubic centimeters. Use a graphing utility to find the radius of the globe. Round your result to two decimal places.

60. *Meteorology* The line graph shows the temperatures (in degrees Fahrenheit) on a summer day in Buffalo, New York from 10:00 A.M. to 6:00 P.M. Create a new line graph showing the temperatures throughout the day in degrees Celsius.

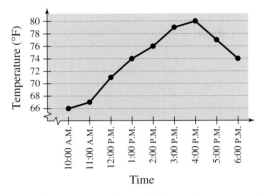

Statics **In Exercises 61 and 62, you have a uniform beam of length L with a fulcrum x feet from one end (see figure). Objects with weights W_1 and W_2 are placed at opposite ends of the beam. The beam will balance when**

$$W_1 x = W_2(L - x).$$

Find x such that the beam will balance.

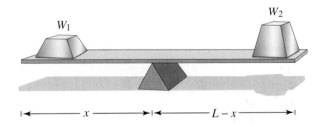

61. Two children weighing 50 pounds and 75 pounds are going to play on a seesaw that is 10 feet long.

62. A person weighing 200 pounds is attempting to move a 550-pound rock with a bar that is 5 feet long.

Synthesis

True or False? **In Exercises 63 and 64, determine whether the statement is true or false. Justify your answer.**

63. The equation

$$x(3 - x) = 10$$

is a linear equation.

64. The volume of a cube with a side length of 9.5 inches is greater than the volume of a sphere with a radius of 5.9 inches.

In Exercises 65 and 66, write a linear equation that has the given solution. (There are many correct answers.)

65. $x = -3$

66. $x = \frac{1}{4}$

67. *Think About It* What is meant by *equivalent equations*? Give an example of two equivalent equations.

68. *Writing* In your own words, describe how to clear an equation of fractions.

Review

In Exercises 69–74, sketch the graph of the equation by hand. Verify using a graphing utility.

69. $y = \frac{5}{8}x - 2$

70. $y = \dfrac{3x - 5}{2} + 2$

71. $y = (x - 3)^2 + 7$

72. $y = \frac{1}{3}x^2 - 4$

73. $y = -\frac{1}{2}|x + 4| - 1$

74. $y = |x - 2| + 10$

In Exercises 75–80, evaluate the combination of functions for $f(x) = -x^2 + 4$ and $g(x) = 6x - 5$.

75. $(f + g)(-3)$

76. $(g - f)(-1)$

77. $(fg)(8)$

78. $\left(\dfrac{f}{g}\right)\left(\dfrac{1}{2}\right)$

79. $(f \circ g)(4)$

80. $(g \circ f)(2)$

2.2 Solving Equations Graphically

Intercepts, Zeros, and Solutions

In Section 1.1, you learned that the intercepts of a graph are the points at which the graph intersects the x- or y-axis.

What you should learn

- Find x- and y-intercepts of graphs of equations.
- Find solutions of equations graphically.
- Find the points of intersection of two graphs.

Why you should learn it

Because some real-life problems involve equations that are difficult to solve algebraically, it is helpful to use a graphing utility to approximate the solutions of such equations. For instance, you can use a graphing utility to find the intersection point of the equations in Example 7 on page 178 to determine the year during which the number of morning newspapers in the United States exceeded the number of evening newspapers.

> **Definition of Intercepts**
>
> 1. The point $(a, 0)$ is called an **x-intercept** of the graph of an equation if it is a solution point of the equation. To find the x-intercept(s), set y equal to 0 and solve the equation for x.
>
> 2. The point $(0, b)$ is called a **y-intercept** of the graph of an equation if it is a solution point of the equation. To find the y-intercept(s), set x equal to 0 and solve the equation for y.

Sometimes it is convenient to denote the x-intercept as simply the x-coordinate of the point $(a, 0)$ rather than the point itself. Unless it is necessary to make a distinction, "intercept" will be used to mean either the point or the coordinate.

It is possible for a graph to have no intercepts, one intercept, or several intercepts. For instance, consider the four graphs shown in Figure 2.6.

Simon Bottomley/Getty Images

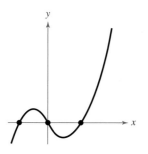

Three x-Intercepts
One y-Intercept

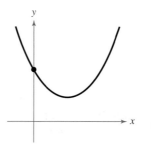

No x-Intercepts
One y-Intercept

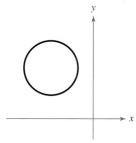

No Intercepts

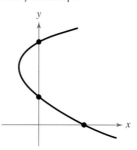

One x-Intercept
Two y-Intercepts

Figure 2.6

As you study this section, you will see the connection between intercepts, zeros, and solutions of functions.

Example 1   Finding x- and y-Intercepts

Find the x- and y-intercepts of the graph of $2x + 3y = 5$.

Solution

To find the x-intercept, let $y = 0$ and solve for x. This produces

$$2x = 5 \quad \Longrightarrow \quad x = \tfrac{5}{2}$$

which implies that the graph has one x-intercept: $\left(\tfrac{5}{2}, 0\right)$. To find the y-intercept, let $x = 0$ and solve for y. This produces

$$3y = 5 \quad \Longrightarrow \quad y = \tfrac{5}{3}$$

which implies that the graph has one y-intercept: $\left(0, \tfrac{5}{3}\right)$. See Figure 2.7.

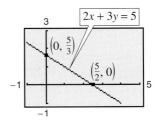

Figure 2.7

 Checkpoint   Now try Exercise 1.

A **zero** of a function $y = f(x)$ is a number a such that $f(a) = 0$. So, to find the zeros of a function, you must solve the equation $f(x) = 0$.

The concepts of x-intercepts, zeros of functions, and solutions of equations are closely related. In fact, the following statements are equivalent.

1. The point $(a, 0)$ is an x-*intercept* of the graph of $y = f(x)$.
2. The number a is a *zero* of the function f.
3. The number a is a *solution* of the equation $f(x) = 0$.

Example 2   Verifying Zeros of Functions

Verify that the real numbers -2 and 3 are zeros of the function $f(x) = x^2 - x - 6$.

Algebraic Solution

To verify that -2 is a zero of f, check that $f(-2) = 0$.

$$f(x) = x^2 - x - 6 \qquad \text{Write original function.}$$

$$f(-2) = (-2)^2 - (-2) - 6 \qquad \text{Substitute } -2 \text{ for } x.$$

$$= 4 + 2 - 6 = 0 \qquad -2 \text{ is a solution. } ✓$$

To verify that 3 is a zero of f, check that $f(3) = 0$.

$$f(x) = x^2 - x - 6 \qquad \text{Write original function.}$$

$$f(3) = (3)^2 - (3) - 6 \qquad \text{Substitute } 3 \text{ for } x.$$

$$= 9 - 3 - 6 = 0 \qquad 3 \text{ is a solution. } ✓$$

Graphical Solution

Use a graphing utility to graph $y = x^2 - x - 6$. From the graph in Figure 2.8, it appears that the function has x-intercepts (where y is zero) at $x = -2$ and at $x = 3$. Use the *zero* or *root* feature to confirm this. So, you can approximate the zeros of the function to be -2 and 3.

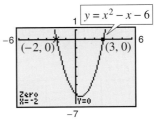

Figure 2.8

 Checkpoint   Now try Exercise 15.

Note that the graph of $y = x^2 - x - 6$ in Figure 2.8 has x-*intercepts* of $(3, 0)$ and $(-2, 0)$ because the equation $0 = x^2 - x - 6$ has the *solutions* $x = 3$ and $x = -2$.

The close connection among x-intercepts, zeros, and solutions is crucial to your study of algebra. You can take advantage of this connection in two ways. Use your algebraic "equation-solving skills" to find the x-intercepts of a graph and your "graphing skills" to approximate the solutions of an equation.

Finding Solutions Graphically

Polynomial equations of degree 1 or 2 can be solved in relatively straightforward ways. Solving polynomial equations of a higher degree can, however, be quite difficult, especially if you rely only on algebraic techniques. For such equations, a graphing utility can be very helpful.

Graphical Approximations of Solutions of an Equation

1. Write the equation in *general form*, $f(x) = 0$, with the nonzero terms on one side of the equation and zero on the other side.

2. Use a graphing utility to graph the function $y = f(x)$. Be sure the viewing window shows all the relevant features of the graph.

3. Use the *zero* or *root* feature or the *zoom* and *trace* features of the graphing utility to approximate the x-intercepts of the graph of f.

In Chapter 3 you will learn techniques for determining the number of solutions of a polynomial equation. For now, you should know that a polynomial equation of degree n cannot have more than n different solutions.

Example 3   Finding Solutions of an Equation Graphically

Use a graphing utility to approximate the solutions of $2x^3 - 3x + 2 = 0$.

Solution

Graph the function $y = 2x^3 - 3x + 2$. You can see from the graph that there is one x-intercept. It lies between -2 and -1 and is approximately -1.5. By using the *zero* or *root* feature of a graphing utility, you can improve the approximation. Choose a left bound of $x = -2$ (see Figure 2.9) and a right bound of $x = -1$ (see Figure 2.10). To three-decimal-place accuracy, the solution is $x \approx -1.476$, as shown in Figure 2.11. Check this approximation on your calculator. You will find that the value of y is $y = 2(-1.476)^3 - 3(-1.476) + 2 \approx -0.003$.

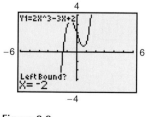

Figure 2.9

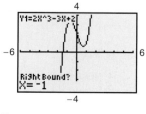

Figure 2.10

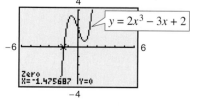

Figure 2.11

☑ *Checkpoint* Now try Exercise 39.

Exploration

In Chapter 3 you will learn that a cubic equation such as

$$24x^3 - 36x + 17 = 0$$

can have up to three real solutions. Use a graphing utility to graph

$$y = 24x^3 - 36x + 17.$$

Describe a viewing window that enables you to determine the number of real solutions of the equation

$$24x^3 - 36x + 17 = 0.$$

Use the same technique to determine the number of real solutions of

$$97x^3 - 102x^2 - 200x - 63 = 0.$$

TECHNOLOGY TIP   You can also use a graphing calculator's *zoom* and *trace* features to approximate the solution of an equation. Here are some suggestions for using the *zoom-in* feature of a graphing utility.

1. With each successive zoom-in, adjust the *x*-scale (if necessary) so that the resulting viewing window shows at least the two scale marks between which the solution lies.

2. The accuracy of the approximation will always be such that the error is less than the distance between two scale marks.

3. If you have a *trace* feature on your graphing utility, you can generally add one more decimal place of accuracy without changing the viewing window.

Unless stated otherwise, this book will approximate all real solutions with an error of *at most* 0.01.

Example 4   Approximating Solutions of an Equation Graphically

Use a graphing utility to approximate the solutions of $x^2 + 3 = 5x$.

Solution

In general form, this equation is

$$x^2 - 5x + 3 = 0. \qquad \text{Equation in general form}$$

So, you can begin by graphing

$$y = x^2 - 5x + 3 \qquad \text{Function to be graphed}$$

as shown in Figure 2.12. This graph has two *x*-intercepts, and by using the *zoom* and *trace* features you can approximate the corresponding solutions to be $x \approx 0.70$ and $x \approx 4.30$, as shown in Figures 2.13 and 2.14.

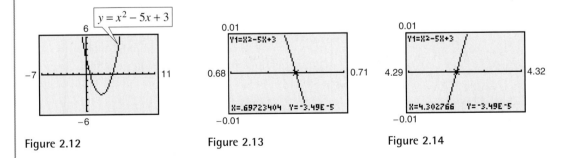

Figure 2.12   Figure 2.13   Figure 2.14

✓ *Checkpoint*   Now try Exercise 43.

TECHNOLOGY TIP   Remember from Example 3 that the built-in *zero* or *root* features of a graphing utility will approximate solutions of equations or *x*-intercepts of graphs. If your graphing utility has such features, try using them to approximate the solutions in Example 4.

TECHNOLOGY TIP

Remember that the more decimal places in the solution, the more accurate the solution is. You can reach the desired accuracy when zooming in as follows.

- To approximate the zero to the nearest hundredth, set the *x*-scale to 0.01.
- To approximate the zero to the nearest thousandth, set the *x*-scale to 0.001.

Points of Intersection of Two Graphs

An ordered pair that is a solution of two different equations is called a **point of intersection** of the graphs of the two equations. For instance, in Figure 2.15 you can see that the graphs of the following equations have two points of intersection.

$$y = x + 2 \qquad \text{Equation 1}$$

$$y = x^2 - 2x - 2 \qquad \text{Equation 2}$$

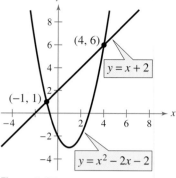

Figure 2.15

The point $(-1, 1)$ is a solution of both equations, and the point $(4, 6)$ is a solution of both equations. To check this algebraically, substitute $x = -1$ and $x = 4$ into each equation.

Check that $(-1, 1)$ *is a solution.*

Equation 1: $y = -1 + 2 = 1$ Solution checks. ✓

Equation 2: $y = (-1)^2 - 2(-1) - 2 = 1$ Solution checks. ✓

Check that $(4, 6)$ *is a solution.*

Equation 1: $y = 4 + 2 = 6$ Solution checks. ✓

Equation 2: $y = (4)^2 - 2(4) - 2 = 6$ Solution checks. ✓

To find the points of intersection of the graphs of two equations, solve each equation for y (or x) and set the two results equal to each other. The resulting equation will be an equation in one variable, which can be solved using standard procedures, as shown in Example 5.

Example 5   Finding Points of Intersection

Find the points of intersection of the graphs of $2x - 3y = -2$ and $4x - y = 6$.

Algebraic Solution

To begin, solve each equation for y to obtain

$$y = \frac{2}{3}x + \frac{2}{3} \quad \text{and} \quad y = 4x - 6.$$

Next, set the two expressions for y equal to each other and solve the resulting equation for x, as follows.

$\frac{2}{3}x + \frac{2}{3} = 4x - 6$ Equate expressions for y.

$2x + 2 = 12x - 18$ Multiply each side by 3.

$-10x = -20$ Subtract $12x$ and 2 from each side.

$x = 2$ Divide each side by -10.

When $x = 2$, the y-value of each of the original equations is 2. So, the point of intersection is $(2, 2)$.

✓ *Checkpoint* Now try Exercise 59.

Graphical Solution

To begin, solve each equation for y to obtain $y_1 = \frac{2}{3}x + \frac{2}{3}$ and $y_2 = 4x - 6$. Then use a graphing utility to graph both equations in the same viewing window. In Figure 2.16, the graphs appear to have one point of intersection. Use the *intersect* feature of the graphing utility to approximate the point of intersection to be $(2, 2)$.

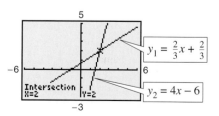

Figure 2.16

TECHNOLOGY TIP   Another way to approximate the points of intersection of two graphs is to graph both equations with a graphing utility and use the *zoom* and *trace* features to find the point or points at which the two graphs intersect.

Example 6   Approximating Points of Intersection Graphically

Approximate the point(s) of intersection of the graphs of the following equations.

$$y = x^2 - 3x - 4 \qquad \text{Equation 1 (quadratic function)}$$

$$y = x^3 + 3x^2 - 2x - 1 \qquad \text{Equation 2 (cubic function)}$$

Solution

Begin by using a graphing utility to graph both functions, as shown in Figure 2.17. From this display, you can see that the two graphs have only one point of intersection. Then, using the *zoom* and *trace* features, approximate the point of intersection to be $(-2.17, 7.25)$, as shown in Figure 2.18.

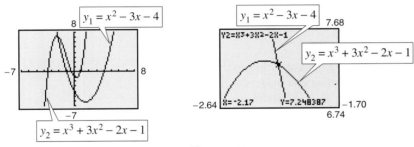

Figure 2.17   **Figure 2.18**

To test the reasonableness of this approximation, you can evaluate both functions at $x = -2.17$.

Quadratic Function:

$$y = (-2.17)^2 - 3(-2.17) - 4$$

$$\approx 7.22$$

Cubic Function:

$$y = (-2.17)^3 + 3(-2.17)^2 - 2(-2.17) - 1$$

$$\approx 7.25$$

Because both functions yield approximately the same y-value, you can conclude that the approximate coordinates of the point of intersection are $x \approx -2.17$ and $y \approx 7.25$.

✓ *Checkpoint*   Now try Exercise 63.

TECHNOLOGY TIP   If you choose to use the *intersect* feature of your graphing utility to find the point of intersection of the graphs in Example 6, you will see that it yields the same result.

TECHNOLOGY TIP

The table shows some points on the graphs of the equations in Example 5. Find the points of intersection of the graphs by finding the value(s) of x for which y_1 and y_2 are equal.

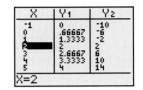

The method shown in Example 6 gives a nice graphical picture of the points of intersection of two graphs. However, for actual approximation purposes, it is better to use the algebraic procedure described in Example 5. That is, the point of intersection of $y = x^2 - 3x - 4$ and $y = x^3 + 3x^2 - 2x - 1$ coincides with the solution of the equation

$$x^3 + 3x^2 - 2x - 1 = x^2 - 3x - 4 \qquad \text{Equate } y\text{-values.}$$

$$x^3 + 2x^2 + x + 3 = 0. \qquad \text{Write in general form.}$$

By graphing $y = x^3 + 2x^2 + x + 3$ on a graphing utility and using the *zoom* and *trace* features (or the *zero* or *root* feature), you can approximate the solution of this equation to be $x \approx -2.17$. The corresponding y-value for *both* of the functions given in Example 6 is $y \approx 7.25$.

Example 7   A Historical Look at Newspapers

Between 1990 and 2001, the number of morning newspapers in the United States was *increasing* and the number of evening newspapers was *decreasing*. Two models that approximate the numbers of newspapers are

$$M = 20.5t + 557, \quad 0 \le t \le 11 \qquad \text{Morning newspapers}$$

$$E = -34.4t + 1070, \quad 0 \le t \le 11 \qquad \text{Evening newspapers}$$

where t represents the year, with $t = 0$ corresponding to 1990. According to these two models, when would you expect the number of morning newspapers to have exceeded the number of evening newspapers?   (Source: Editor & Publisher Co.)

Algebraic Solution

Set the two expressions equal to each other and solve the resulting equation for t, as follows.

$$20.5t + 557 = -34.4t + 1070 \qquad \text{Equate expressions.}$$

$$54.9t + 557 = 1070 \qquad \text{Add } 34.4t \text{ to each side.}$$

$$54.9t = 513 \qquad \text{Subtract 557 from each side.}$$

$$t = \frac{513}{54.9} \qquad \text{Divide each side by 54.9.}$$

$$t \approx 9.34 \qquad \text{Use a calculator.}$$

So, from the given models, you would expect that the number of morning newspapers exceeded the number of evening newspapers sometime during 1999.

✓ *Checkpoint* Now try Exercise 76.

Graphical Solution

Use a graphing utility to graph both equations in the same viewing window. From Figure 2.19, the graphs appear to have one point of intersection. Use the *intersect* feature of the graphing utility to approximate the point of intersection to be (9.34, 748.56). So, you would expect that the number of morning newspapers exceeded the number of evening newspapers sometime during 1999.

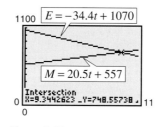

Figure 2.19

TECHNOLOGY TIP   If you choose to use the *zoom* and *trace* features of your graphing utility to find the point of intersection of the graphs in Example 7, you will see that these features yield the same result.

2.2 Exercises

Vocabulary Check

Fill in the blanks.

1. The points $(a, 0)$ and $(0, b)$ are called the _____ and _____ , respectively, of the graph of an equation.

2. A _____ of a function is a number a such that $f(a) = 0$.

3. An ordered pair that is a solution of two different equations is called a _____ of the graphs of the two equations.

In Exercises 1–12, find the x- and y-intercepts of the graph of the equation.

1. $y = x - 5$

2. $y = -\frac{3}{4}x - 3$

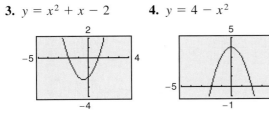

3. $y = x^2 + x - 2$

4. $y = 4 - x^2$

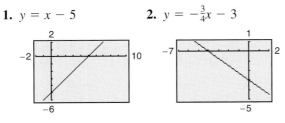

5. $y = x\sqrt{x + 2}$

6. $y = -\frac{1}{2}x\sqrt{x + 3} + 1$

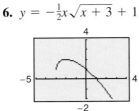

7. $xy = 4$

8. $4xy = 3x - 1$

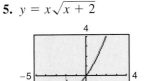
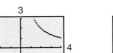
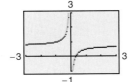

9. $y = |x - 2| - 4$

10. $y = 3 - \frac{1}{2}|x + 1|$

11. $xy - 2y - x + 1 = 0$

12. $x^2y - x^2 + 4y = 0$

In Exercises 13–18, the zero(s) of the function are given. Verify the zero(s) both algebraically and graphically.

Function	Zero(s)
13. $f(x) = 5(4 - x)$	$x = 4$
14. $f(x) = 3(x - 5) + 9$	$x = 2$
15. $f(x) = x^3 - 6x^2 + 5x$	$x = 0, 5, 1$
16. $f(x) = x^3 - 9x^2 + 18x$	$x = 0, 3, 6$
17. $f(x) = \dfrac{x + 2}{3} - \dfrac{x - 1}{5} - 1$	$x = 1$
18. $f(x) = x - 3 - \dfrac{10}{x}$	$x = -2, 5$

Graphical Analysis   **In Exercises 19–22, use a graphing utility to graph the equation and approximate any x-intercepts. Set $y = 0$ and solve the resulting equation. Compare the results with the x-intercepts of the graph.**

19. $y = 2(x - 1) - 4$

20. $y = 4(x + 3) - 2$

21. $y = 20 - (3x - 10)$

22. $y = 10 + 2(x - 2)$

In Exercises 23–36, solve the equation algebraically. Then write the equation in the form $f(x) = 0$ and use a graphing utility to verify the algebraic solution.

23. $2.7x - 0.4x = 1.2$

24. $3.5x - 8 = 0.5x$

25. $25(x - 3) = 12(x + 2) - 10$

26. $1200 = 300 + 2(x - 500)$

27. $\dfrac{3x}{2} + \dfrac{1}{4}(x - 2) = 10$

28. $\dfrac{2x}{3} + \dfrac{1}{2}(x - 5) = 6$

29. $0.60x + 0.40(100 - x) = 1.2$

30. $0.75x + 0.2(80 - x) = 20$

31. $\dfrac{2x}{3} = 10 - \dfrac{24}{x}$

32. $\dfrac{x-3}{25} = \dfrac{x-5}{12}$

33. $\dfrac{3}{x+2} - \dfrac{4}{x-2} = 5$

34. $\dfrac{6}{x} + \dfrac{8}{x+5} = 3$

35. $(x+2)^2 = x^2 - 6x + 1$

36. $(x+1)^2 + 2(x-2) = (x+1)(x-2)$

In Exercises 37–50, use a graphing utility to approximate any solutions (accurate to three decimal places) of the equation. [Remember to write the equation in the form $f(x) = 0$.]

37. $\frac{1}{4}(x^2 - 10x + 17) = 0$

38. $-\frac{1}{2}(x^2 - 6x + 6) = 0$

39. $x^3 + x + 4 = 0$

40. $\frac{1}{9}x^3 + x + 4 = 0$

41. $2x^3 - x^2 - 18x + 9 = 0$

42. $4x^3 + 12x^2 - 26x - 24 = 0$

43. $x^4 = 2x^3 + 1$

44. $x^5 = 3 + 2x^3$

45. $\dfrac{2}{x+2} = 3$

46. $\dfrac{5}{x} = 1 + \dfrac{3}{x+2}$

47. $|x - 3| = 4$

48. $|x + 1| = 6$

49. $\sqrt{x - 2} = 3$

50. $\sqrt{x - 4} = 8$

51. *Exploration*

(a) Use a graphing utility to complete the table.

x	-1	0	1	2	3	4
$3.2x - 5.8$						

(b) Use the table in part (a) to determine the interval in which the solution to the equation $3.2x - 5.8 = 0$ is located. Explain your reasoning.

(c) Use a graphing utility to complete the table.

x	1.5	1.6	1.7	1.8	1.9	2
$3.2x - 5.8$						

(d) Use the table in part (c) to determine the interval in which the solution to the equation $3.2x - 5.8 = 0$ is located. Explain how this process can be used to approximate the solution to any desired degree of accuracy.

(e) Use a graphing utility to verify graphically the solution to $3.2x - 5.8 = 0$ found in part (d).

52. *Exploration* Use the procedure from Exercise 51 to approximate the solution of the equation $0.3(x - 1.5) - 2 = 0$ accurate to two decimal places.

In Exercises 53–58, determine any point(s) of intersection algebraically. Then verify your result numerically by creating a table of values for each function.

53. $y = 2 - x$
$y = 2x - 1$

54. $y = 7 - x$
$y = \frac{3}{2} - \frac{11}{2}x$

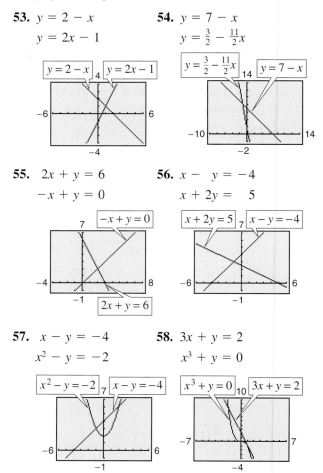

55. $2x + y = 6$
$-x + y = 0$

56. $x - y = -4$
$x + 2y = 5$

57. $x - y = -4$
$x^2 - y = -2$

58. $3x + y = 2$
$x^3 + y = 0$

In Exercises 59–64, use a graphing utility to approximate any points of intersection (accurate to three decimal places) of the graphs of the equations. Verify your results algebraically.

59. $y = 9 - 2x$
$y = x - 3$

60. $y = \frac{1}{3}x + 2$
$y = \frac{5}{2}x - 11$

61. $y = 4 - x^2$
$y = 2x - 1$

62. $y = x^3 - 3$
$y = 5 - 2x$

63. $y = 2x^2$
$y = x^4 - 2x^2$

64. $y = -x$
$y = 2x - x^2$

In Exercises 65 and 66, evaluate the expression in two ways. (a) Calculate entirely on your calculator by storing intermediate results and then rounding the final answer to two decimal places. (b) Round both the numerator and denominator to two decimal places before dividing, and then round the final answer to two decimal places. Does the method in part (b) decrease the accuracy? Explain.

65. $\dfrac{1 + 0.73205}{1 - 0.73205}$ **66.** $\dfrac{1 + 0.86603}{1 - 0.86603}$

67. *Travel Time* On the first part of a 280-mile trip, a salesperson averaged 63 miles per hour. The salesperson averaged only 54 miles per hour on the last part of the trip because of an increased volume of traffic.

(a) Write the total time t for the trip as a function of the distance x traveled at an average speed of 63 miles per hour.

(b) Use a graphing utility to graph the time function. What is the domain of the function?

(c) Approximate the number of miles traveled at 63 miles per hour when the total time is 4 hours and 45 minutes.

68. *Production* An electronics company has fixed costs of $25,000 per month and a variable cost of $18.65 per 13-inch TV/VCR combination manufactured. (*Fixed costs* are those that occur regardless of the level of production.)

(a) Write the total monthly costs C as a function of the number of units x produced.

(b) Use a graphing utility to graph the cost function.

(c) Use the graph from part (b) to approximate the number of units that can be produced per month if total costs cannot exceed $200,000. Verify algebraically. Is this problem better solved algebraically or graphically? Explain.

69. *Mixture Problem* A 55-gallon barrel contains a mixture with a concentration of 33% sodium chloride. You remove x gallons of this mixture and replace it with 100% sodium chloride.

(a) Write the amount A of sodium chloride in the final mixture as a function of x.

(b) Use a graphing utility to graph the concentration function. What is the domain of the function?

(c) Approximate (accurate to one decimal place) the value of x when the final mixture is 60% sodium chloride.

70. *Geometry* A rectangular horse corral with a perimeter of 230 meters has a length of x.

(a) Draw a diagram that gives a visual representation of the problem.

(b) Write the corral's area A as a function of x.

(c) Use a graphing utility to graph the area function. What is the domain of the function?

(d) Approximate (accurate to one decimal place) the dimensions of the corral when its area is 2000 square meters.

Geometry **In Exercises 71 and 72, (a) write a function for the area of the region, (b) use a graphing utility to graph the function, and (c) approximate the value of x when the area of the region is 200 square units.**

71.   **72.**

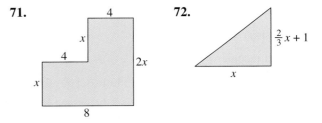

73. *Geometry* Consider the swimming pool in the figure. (When finding its volume, use the fact that the volume is the area of the region on the vertical sidewall times the width of the pool.)

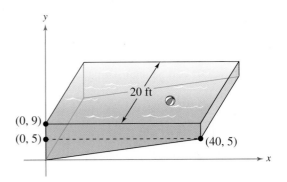

(a) Find the volume of the pool.

(b) Find an equation of the line representing the base of the pool.

(c) The depth of the water at the deep end of the pool is d feet. Show that the volume of water is

$$V(d) = \begin{cases} 80d^2, & 0 \le d \le 5 \\ 800d - 2000, & 5 < d \le 9 \end{cases}.$$

(d) Graph the volume function.

(e) Use a graphing utility to complete the table.

d	3	5	7	9
V				

(f) Approximate the depth of the water at the deep end when the volume is 4800 cubic feet.

(g) How many gallons of water are in the pool? (There are 7.48 gallons of water in 1 cubic foot.)

74. **Income Tax** The following information describes a possible negative income tax for a family consisting of two adults and two children. The plan would guarantee the poor a minimum income while encouraging a family to increase its private income ($0 \leq x \leq 20{,}000$). (A *subsidy* is a grant of money.)

Family's earned income: $I = x$

Subsidy: $S = 10{,}000 - \frac{1}{2}x$

Total income: $T = I + S$

(a) Write the total income T in terms of x.

(b) Use a graphing utility to find the earned income x when the subsidy is $6600. Verify your answer algebraically.

(c) Use a graphing utility to find the earned income x when the total income is $13,800. Verify your answer algebraically.

(d) Find the subsidy S graphically when the total income is $12,500.

75. **Labor Force** The number y (in millions) of women in the civilian labor force in the United States from 1990 to 2001 can be approximated by the model

$$y = 0.91t + 56.4, \quad 0 \leq t \leq 11$$

where t represents the year, with $t = 0$ corresponding to 1990. According to this model, during which year did the number of women in the civilian labor force reach 60 million? Explain how to answer the question graphically and algebraically. (Source: U.S. Bureau of Labor Statistics)

76. **Population** The population (in thousands) of Arizona A, and Minnesota M, from 1995 to 2001 can be modeled by

$$A = 142.9t + 3729, \quad 5 \leq t \leq 11$$

$$M = 52.1t + 4400, \quad 5 \leq t \leq 11$$

where t represents the year, with $t = 5$ corresponding to 1995. (Source: U.S. Census Bureau)

(a) Use a graphing utility to graph each model in the same viewing window over the appropriate domain. Approximate the point of intersection.

(b) Find the point of intersection algebraically. What does the point of intersection represent?

(c) Use the models to estimate the population of each state in 2006.

Synthesis

True or False? **In Exercises 77–79, determine whether the statement is true or false. Justify your answer.**

77. To find the y-intercept of a graph, let $x = 0$ and solve the equation for y.

78. Every linear equation has at least one y-intercept or x-intercept.

79. Two linear equations can have either one point of intersection or no points of intersection.

80. **Writing** You are solving the equation

$$\frac{x}{x-1} - \frac{99}{100} = 0$$

for x, and you obtain $x = -99.1$ as your solution. Substituting this value back into the equation produces

$$\frac{-99.1}{-99.1 - 1} - \frac{99}{100} = 0.00000999$$

$$= 9.99 \times 10^{-6} \approx 0.$$

Does this mean that -99.1 is a good approximation of the solution? Write a short paragraph explaining why or why not.

Review

In Exercises 81–84, rationalize the denominator.

81. $\dfrac{12}{5\sqrt{3}}$

82. $\dfrac{4}{\sqrt{10} - 2}$

83. $\dfrac{3}{8 + \sqrt{11}}$

84. $\dfrac{14}{3\sqrt{10} - 1}$

In Exercises 85–88, find the product.

85. $(x + 6)(3x - 5)$

86. $(3x + 13)(4x - 7)$

87. $(2x - 9)(2x + 9)$

88. $(4x + 1)^2$

2.3 Complex Numbers

The Imaginary Unit *i*

Some quadratic equations have no real solutions. For instance, the quadratic equation $x^2 + 1 = 0$ has no real solution because there is no real number x that can be squared to produce -1. To overcome this deficiency, mathematicians created an expanded system of numbers using the **imaginary unit *i*,** defined as

$$i = \sqrt{-1} \qquad \text{Imaginary unit}$$

where $i^2 = -1$. By adding real numbers to real multiples of this imaginary unit, you obtain the set of **complex numbers.** Each complex number can be written in the **standard form $a + bi$.** For instance, the standard form of the complex number $\sqrt{-9} - 5$ is $-5 + 3i$ because

$$\sqrt{-9} - 5 = \sqrt{3^2(-1)} - 5 = 3\sqrt{-1} - 5 = 3i - 5 = -5 + 3i.$$

In the standard form $a + bi$, the real number a is called the **real part** of the **complex number $a + bi$,** and the number bi (where b is a real number) is called the **imaginary part** of the complex number.

> **Definition of a Complex Number**
>
> If a and b are real numbers, the number $a + bi$ is a **complex number,** and it is said to be written in **standard form.** If $b = 0$, the number $a + bi = a$ is a real number. If $b \neq 0$, the number $a + bi$ is called an **imaginary number.** A number of the form bi, where $b \neq 0$, is called a **pure imaginary number.**

The set of real numbers is a subset of the set of complex numbers, as shown in Figure 2.20. This is true because every real number a can be written as a complex number using $b = 0$. That is, for every real number a, you can write $a = a + 0i$.

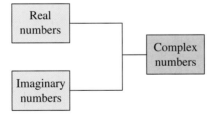

Figure 2.20

> **Equality of Complex Numbers**
>
> Two complex numbers $a + bi$ and $c + di$, written in standard form, are equal to each other
>
> $$a + bi = c + di \qquad \text{Equality of two complex numbers}$$
>
> if and only if $a = c$ and $b = d$.

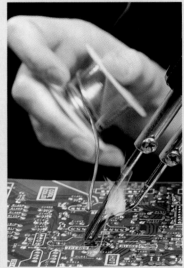

Operations with Complex Numbers

To add (or subtract) two complex numbers, you add (or subtract) the real and imaginary parts of the numbers separately.

Addition and Subtraction of Complex Numbers

If $a + bi$ and $c + di$ are two complex numbers written in standard form, their sum and difference are defined as follows.

Sum: $(a + bi) + (c + di) = (a + c) + (b + d)i$

Difference: $(a + bi) - (c + di) = (a - c) + (b - d)i$

The **additive identity** in the complex number system is zero (the same as in the real number system). Furthermore, the **additive inverse** of the complex number $a + bi$ is

$-(a + bi) = -a - bi.$ Additive inverse

So, you have $(a + bi) + (-a - bi) = 0 + 0i = 0$.

Example 1   Adding and Subtracting Complex Numbers

a. $(3 - i) + (2 + 3i) = 3 - i + 2 + 3i$ Remove parentheses.

$= 3 + 2 - i + 3i$ Group like terms.

$= (3 + 2) + (-1 + 3)i$

$= 5 + 2i$ Write in standard form.

b. $2i + (-4 - 2i) = 2i - 4 - 2i$ Remove parentheses.

$= -4 + 2i - 2i$ Group like terms.

$= -4$ Write in standard form.

c. $3 - (-2 + 3i) + (-5 + i) = 3 + 2 - 3i - 5 + i$

$= 3 + 2 - 5 - 3i + i$

$= 0 - 2i$

$= -2i$

d. $(3 + 2i) + (4 - i) - (7 + i) = 3 + 2i + 4 - i - 7 - i$

$= 3 + 4 - 7 + 2i - i - i$

$= 0 + 0i$

$= 0$

✓ *Checkpoint* Now try Exercise 19.

In Examples 1(b) and 1(d), note that the sum of complex numbers can be a real number.

Many of the properties of real numbers are valid for complex numbers as well. Here are some examples.

Associative Property of Addition and Multiplication
Commutative Property of Addition and Multiplication
Distributive Property of Multiplication over Addition

Notice how these properties are used when two complex numbers are multiplied.

$$(a + bi)(c + di) = a(c + di) + bi(c + di)$$ Distributive Property

$$= ac + (ad)i + (bc)i + (bd)i^2$$ Distributive Property

$$= ac + (ad)i + (bc)i + (bd)(-1)$$ $i^2 = -1$

$$= ac - bd + (ad)i + (bc)i$$ Commutative Property

$$= (ac - bd) + (ad + bc)i$$ Associative Property

The procedure above is similar to multiplying two polynomials and combining like terms, as in the FOIL method shown in Section P.3.

> **Exploration**
>
> Complete the following:
>
> $i^1 = i$ $i^7 = $ ▨
>
> $i^2 = -1$ $i^8 = $ ▨
>
> $i^3 = -i$ $i^9 = $ ▨
>
> $i^4 = 1$ $i^{10} = $ ▨
>
> $i^5 = $ ▨ $i^{11} = $ ▨
>
> $i^6 = $ ▨ $i^{12} = $ ▨
>
> What pattern do you see? Write a brief description of how you would find i raised to any positive integer power.

Example 2   Multiplying Complex Numbers

a. $\sqrt{-4} \cdot \sqrt{-16} = (2i)(4i)$ Write each factor in i-form.

$$= 8i^2$$ Multiply.

$$= 8(-1)$$ $i^2 = -1$

$$= -8$$ Simplify.

b. $(2 - i)(4 + 3i) = 8 + 6i - 4i - 3i^2$ Product of binomials

$$= 8 + 6i - 4i - 3(-1)$$ $i^2 = -1$

$$= 8 + 3 + 6i - 4i$$ Group like terms.

$$= 11 + 2i$$ Write in standard form.

c. $(3 + 2i)(3 - 2i) = 9 - 6i + 6i - 4i^2$ Product of binomials

$$= 9 - 4(-1)$$ $i^2 = -1$

$$= 9 + 4$$ Simplify.

$$= 13$$ Write in standard form.

d. $4i(-1 + 5i) = 4i(-1) + 4i(5i)$ Distributive Property

$$= -4i + 20i^2$$ Simplify.

$$= -4i + 20(-1)$$ $i^2 = -1$

$$= -20 - 4i$$ Write in standard form.

e. $(3 + 2i)^2 = 9 + 6i + 6i + 4i^2$ Product of binomials

$$= 9 + 12i + 4(-1)$$ $i^2 = -1$

$$= 9 - 4 + 12i$$ Group like terms.

$$= 5 + 12i$$ Write in standard form.

> **STUDY TIP**
>
> Before you perform operations with complex numbers, be sure to rewrite the terms or factors in i-form first and then procede with the operations, as shown in Example 2(a).

✓ *Checkpoint* Now try Exercise 29.

Complex Conjugates

Notice in Example 2(c) that the product of two complex numbers can be a real number. This occurs with pairs of complex numbers of the forms $a + bi$ and $a - bi$, called **complex conjugates.**

$$(a + bi)(a - bi) = a^2 - abi + abi - b^2i^2 = a^2 - b^2(-1) = a^2 + b^2$$

Example 3   Multiplying Conjugates

Multiply $3 - 5i$ by its complex conjugate.

Solution

The complex conjugate of $3 - 5i$ is $3 + 5i$.

$$(3 - 5i)(3 + 5i) = 3^2 - (5i)^2 = 9 - 25i^2 = 9 - 25(-1) = 34$$

✓ *Checkpoint* Now try Exercise 37.

To write the quotient of $a + bi$ and $c + di$ in standard form, where c and d are not both zero, multiply the numerator and denominator by the complex conjugate of the *denominator* to obtain

$$\frac{a + bi}{c + di} = \frac{a + bi}{c + di}\left(\frac{c - di}{c - di}\right)$$ Multiply numerator and denominator by complex conjugate of denominator.

$$= \frac{(ac + bd) + (bc - ad)i}{c^2 + d^2}.$$ Standard form

Example 4   Writing a Quotient of Complex Numbers in Standard Form

Write the quotient $\dfrac{2 + 3i}{4 - 2i}$ in standard form.

Solution

$$\frac{2 + 3i}{4 - 2i} = \frac{2 + 3i}{4 - 2i}\left(\frac{4 + 2i}{4 + 2i}\right)$$ Multiply numerator and denominator by complex conjugate of denominator.

$$= \frac{8 + 4i + 12i + 6i^2}{16 - 4i^2}$$ Expand.

$$= \frac{8 - 6 + 16i}{16 + 4}$$ $i^2 = -1$

$$= \frac{2 + 16i}{20}$$ Simplify.

$$= \frac{1}{10} + \frac{4}{5}i$$ Write in standard form.

✓ *Checkpoint* Now try Exercise 49.

TECHNOLOGY TIP

Some graphing utilities can perform operations with complex numbers. For instance, on some graphing utilities, to divide $2 + 3i$ by $4 - 2i$, use the following keystrokes.

$$\boxed{(}\;2\;\boxed{+}\;3\;\boxed{i}\;\boxed{)}\;\boxed{\div}$$

$$\boxed{(}\;4\;\boxed{-}\;2\;\boxed{i}\;\boxed{)}\;\boxed{\text{ENTER}}$$

The display will be as follows.

$$.1 + .8i \quad \text{or} \quad \frac{1}{10} + \frac{4}{5}i$$

Fractals and the Mandelbrot Set

Most applications involving complex numbers are either theoretical or very technical, and are therefore not appropriate for inclusion in this text. However, to give you some idea of how complex numbers can be used in applications, a general description of their use in **fractal geometry** is presented.

To begin, consider a coordinate system called the **complex plane.** Just as every real number corresponds to a point on the real number line, every complex number corresponds to a point in the complex plane, as shown in Figure 2.21. In this figure, note that the vertical axis is called the **imaginary axis** and the horizontal axis is called the **real axis.** The point that corresponds to the complex number $a + bi$ is (a, b).

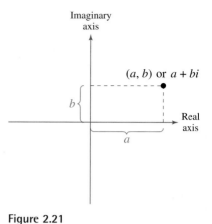

Figure 2.21

Example 5   Plotting Complex Numbers

Plot each complex number in the complex plane.

a. $2 + 3i$ **b.** $-1 + 2i$ **c.** 4 **d.** $-3i$

Solution

a. To plot the complex number $2 + 3i$, move (from the origin) two units to the right on the real axis and then three units upward, as shown in Figure 2.22. In other words, plotting the complex number $2 + 3i$ in the complex plane is comparable to plotting the point $(2, 3)$ in the Cartesian plane. (Note that in Figure 2.22, i is called the imaginary unit because it is located one unit from the origin on the imaginary axis of the complex plane.)

b. The complex number $-1 + 2i$ corresponds to the point $(-1, 2)$, as shown in Figure 2.22.

c. The complex number 4 corresponds to the point $(4, 0)$, as shown in Figure 2.22.

d. The complex number $-3i$ corresponds to the point $(0, -3)$, as shown in Figure 2.22.

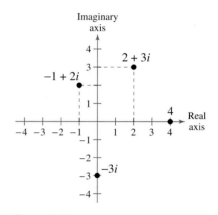

Figure 2.22

✓ *Checkpoint* Now try Exercise 67.

In the hands of a person who understands fractal geometry, the complex plane can become an easel on which stunning pictures, called **fractals,** can be drawn. The most famous such picture is called the **Mandelbrot Set,** named after the Polish-born mathematician Benoit Mandelbrot. To draw the Mandelbrot Set, consider the following sequence of numbers.

$$c, \, c^2 + c, \, (c^2 + c)^2 + c, \, [(c^2 + c)^2 + c]^2 + c, \ldots$$

The behavior of this sequence depends on the value of the complex number c. For some values of c this sequence is **bounded,** which means that the absolute value of each number $\left(|a + bi| = \sqrt{a^2 + b^2} \right)$ in the sequence is less than some fixed number N. For other values of c, the sequence is **unbounded,** which means that the absolute values of the terms of the sequence become infinitely large. If the sequence is bounded, the complex number c is in the Mandelbrot Set; if the sequence is unbounded, the complex number c is not in the Mandelbrot Set.

Example 6   Members of the Mandelbrot Set

a. The complex number -2 is in the Mandelbrot Set because for $c = -2$, the corresponding Mandelbrot sequence is $-2, 2, 2, 2, 2, 2, \ldots$, which is bounded.

b. The complex number i is also in the Mandelbrot Set because for $c = i$, the corresponding Mandelbrot sequence is

$$i, \quad -1 + i, \quad -i, \quad -1 + i, \quad -i, \quad -1 + i, \quad \ldots$$

which is bounded.

c. The complex number $1 + i$ is not in the Mandelbrot Set because for $c = 1 + i$, the corresponding Mandelbrot sequence is

$$1 + i, \quad 1 + 3i, \quad -7 + 7i, \quad 1 - 97i, \quad -9407 - 193i,$$

$$88,454,401 + 3,631,103i, \quad \ldots$$

which is unbounded.

✓ *Checkpoint* Now try Exercise 73.

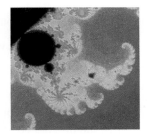

With this definition, a picture of the Mandelbrot Set would have only two colors: one color for points that are in the set (the sequence is bounded), and one for points that are outside the set (the sequence is unbounded). Figure 2.23 shows a black and yellow picture of the Mandelbrot Set. The points that are black are in the Mandelbrot Set and the points that are yellow are not.

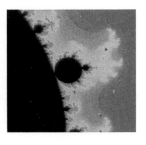

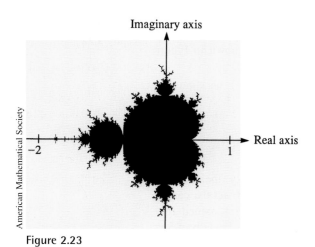

Figure 2.23

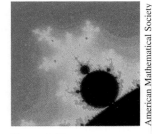

Figure 2.24

To add more interest to the picture, computer scientists discovered that the points that are not in the Mandelbrot Set can be assigned a variety of colors, depending on "how quickly" their sequences diverge (become infinitely large). Figure 2.24 shows three different appendages of the Mandelbrot Set. (The black portions of the picture represent points that are in the Mandelbrot Set.)

Figure 2.25 shows another type of fractal. From this picture, you can see why fractals have fascinated people since their discovery (around 1980). The fractal shown was produced on a graphing calculator. The problem for creating the fractal fern is available on the website *college.hmco.com*.

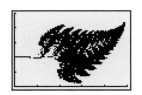

Figure 2.25   A Fractal Fern

2.3 Exercises

Vocabulary Check

1. Match the type of complex number with its definition.

   (a) real number   (i) $a + bi, a = 0, b \neq 0$

   (b) imaginary number   (ii) $a + bi, b = 0$

   (c) pure imaginary number   (iii) $a + bi, a \neq 0, b \neq 0$

In Exercises 2–5, fill in the blanks.

2. The imaginary unit i is defined as $i =$ _____ , where $i^2 =$ _____ .

3. The set of real multiples of the imaginary unit i combined with the set of real numbers is called the set of _____ numbers, which are written in the standard form _____ .

4. Complex numbers can be plotted in the complex plane, where the horizontal axis is the _____ axis and the vertical axis is the _____ axis.

5. The most famous fractal is called the _____ .

In Exercises 1–4, find real numbers a and b such that the equation is true.

1. $a + bi = -9 + 4i$

2. $a + bi = 12 + 5i$

3. $(a - 1) + (b + 3)i = 5 + 8i$

4. $(a + 6) + 2bi = 6 - 5i$

In Exercises 5–14, write the complex number in standard form.

5. $4 + \sqrt{-25}$ 6. $3 + \sqrt{-9}$

7. 7 8. 42

9. $-5i + i^2$ 10. $-3i^2 + i$

11. $\left(\sqrt{-75}\right)^2$ 12. $\left(\sqrt{-4}\right)^2 - 7$

13. $\sqrt{-0.09}$ 14. $\sqrt{-0.0004}$

In Exercises 15–24, perform the addition or subtraction and write the result in standard form.

15. $(4 + i) + (7 - 2i)$ 16. $(11 - 2i) + (-3 + 6i)$

17. $\left(-1 + \sqrt{-8}\right) + \left(8 - \sqrt{-50}\right)$

18. $\left(7 + \sqrt{-18}\right) - \left(3 + 3\sqrt{2}i\right)$

19. $13i - (14 - 7i)$ 20. $22 + (-5 + 8i) + 10i$

21. $\left(\frac{3}{2} + \frac{5}{2}i\right) + \left(\frac{5}{3} + \frac{11}{3}i\right)$ 22. $-\left(\frac{3}{4} + \frac{7}{5}i\right) - \left(\frac{5}{6} - \frac{1}{6}i\right)$

23. $(1.6 + 3.2i) + (-5.8 + 4.3i)$

24. $-(-3.7 - 12.8i) - \left(6.1 - \sqrt{-24.5}\right)$

In Exercises 25–36, perform the operation and write the result in standard form.

25. $\sqrt{-6} \cdot \sqrt{-2}$ 26. $\sqrt{-5} \cdot \sqrt{-10}$

27. $\left(\sqrt{-10}\right)^2$ 28. $\left(\sqrt{-75}\right)^2$

29. $(1 + i)(3 - 2i)$ 30. $(6 - 2i)(2 - 3i)$

31. $4i(8 + 5i)$ 32. $-3i(6 - i)$

33. $\left(\sqrt{14} + \sqrt{10}i\right)\left(\sqrt{14} - \sqrt{10}i\right)$

34. $\left(3 + \sqrt{-5}\right)\left(7 - \sqrt{-10}\right)$

35. $(4 + 5i)^2 - (4 - 5i)^2$ 36. $(1 - 2i)^2 - (1 + 2i)^2$

In Exercises 37–44, write the complex conjugate of the complex number. Then multiply the number by its complex conjugate.

37. $4 + 3i$ 38. $7 - 5i$

39. $-6 - \sqrt{5}i$ 40. $-3 + \sqrt{2}i$

41. $\sqrt{-20}$ 42. $\sqrt{-13}$

43. $3 - \sqrt{-2}$ 44. $1 + \sqrt{-8}$

In Exercises 45–52, write the quotient in standard form.

45. $\dfrac{6}{i}$ 46. $-\dfrac{5}{2i}$

47. $\dfrac{2}{4 - 5i}$ 48. $\dfrac{3}{1 - i}$

49. $\dfrac{2+i}{2-i}$

50. $\dfrac{8-7i}{1-2i}$

51. $\dfrac{i}{(4-5i)^2}$

52. $\dfrac{5i}{(2+3i)^2}$

In Exercises 53–56, perform the operation and write the result in standard form.

53. $\dfrac{2}{1+i} - \dfrac{3}{1-i}$

54. $\dfrac{2i}{2+i} + \dfrac{5}{2-i}$

55. $\dfrac{i}{3-2i} + \dfrac{2i}{3+8i}$

56. $\dfrac{1+i}{i} - \dfrac{3}{4-i}$

In Exercises 57–62, simplify the complex number and write it in standard form.

57. $-6i^3 + i^2$

58. $4i^2 - 2i^3$

59. $\left(\sqrt{-75}\right)^3$

60. $\left(\sqrt{-2}\right)^6$

61. $\dfrac{1}{i^3}$

62. $\dfrac{1}{(2i)^3}$

63. Cube each complex number. What do you notice?

(a) 2 (b) $-1+\sqrt{3}\,i$ (c) $-1-\sqrt{3}\,i$

64. Raise each complex number to the fourth power and simplify.

(a) 2 (b) -2 (c) $2i$ (d) $-2i$

In Exercises 65 and 66, determine the complex number shown in the complex plane.

65.

66.

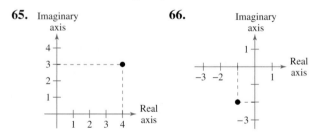

In Exercises 67–72, plot the complex number in the complex plane.

67. $4-5i$

68. $-7+2i$

69. $3i$

70. $-5i$

71. 1

72. -6

Fractals   In Exercises 73–76, find the first six terms of the sequence given on page 187. From the terms, do you think the given complex number is in the Mandelbrot Set? Explain your reasoning.

73. $c = \frac{1}{2}i$

74. $c = 2$

75. $c = 1$

76. $c = -i$

Impedance   In Exercises 77 and 78, use the following information. The opposition to current in an electrical circuit is called its impedance. The impedance z in a parallel circuit with two pathways satisfies the equation $1/z = 1/z_1 + 1/z_2$, where z_1 is the impedance (in ohms) of pathway 1 and z_2 is the impedance (in ohms) of pathway 2. Use the table to determine the impedance of the parallel circuit. (*Hint:* You can find the impedance of each pathway in a parallel circuit by adding the impedances of all components in the pathway.)

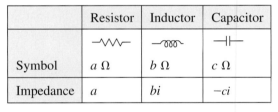

	Resistor	Inductor	Capacitor
Symbol	$a\,\Omega$	$b\,\Omega$	$c\,\Omega$
Impedance	a	bi	$-ci$

77.

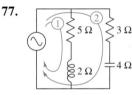

78.

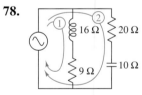

Synthesis

True or False?   In Exercises 79 and 80, determine whether the statement is true or false. Justify your answer.

79. There is no complex number that is equal to its conjugate.

80. $i^{44} + i^{150} - i^{74} - i^{109} + i^{61} = -1$

Review

In Exercises 81–84, perform the operation and write the result in standard form.

81. $(4x-5)(4x+5)$

82. $(x+2)^3$

83. $\left(3x-\frac{1}{2}\right)(x+4)$

84. $(2x-5)^2$

2.4 Solving Equations Algebraically

Quadratic Equations

A **quadratic equation in x** is an equation that can be written in the general form

$$ax^2 + bx + c = 0$$

where a, b, and c are real numbers with $a \neq 0$. A quadratic equation in x is also known as a **second-degree polynomial equation in x.** You should be familiar with the following four methods for solving quadratic equations.

Paul Souders/Getty Images

Solving a Quadratic Equation

Factoring: If $ab = 0$, then $a = 0$ or $b = 0$.   Zero-Factor Property

Example: $x^2 - x - 6 = 0$

$$(x - 3)(x + 2) = 0$$

$$x - 3 = 0 \implies x = 3$$

$$x + 2 = 0 \implies x = -2$$

Extracting Square Roots: If $u^2 = c$, where $c > 0$, then $u = \pm\sqrt{c}$.

Example: $(x + 3)^2 = 16$

$$x + 3 = \pm 4$$

$$x = -3 \pm 4$$

$$x = 1 \quad \text{or} \quad x = -7$$

Completing the Square: If $x^2 + bx = c$, then

$$x^2 + bx + \left(\frac{b}{2}\right)^2 = c + \left(\frac{b}{2}\right)^2$$

$$\left(x + \frac{b}{2}\right)^2 = c + \frac{b^2}{4}.$$

Example: $x^2 + 6x = 5$

$$x^2 + 6x + 3^2 = 5 + 3^2$$

$$(x + 3)^2 = 14$$

$$x + 3 = \pm\sqrt{14}$$

$$x = -3 \pm \sqrt{14}$$

Quadratic Formula: If $ax^2 + bx + c = 0$, then $x = \dfrac{-b \pm \sqrt{b^2 - 4ac}}{2a}$.

Example: $2x^2 + 3x - 1 = 0$

$$x = \frac{-3 \pm \sqrt{3^2 - 4(2)(-1)}}{2(2)} = \frac{-3 \pm \sqrt{17}}{4}$$

Example 1   Solving a Quadratic Equation by Factoring

Solve each quadratic equation by factoring.

a. $6x^2 = 3x$ **b.** $9x^2 - 6x + 1 = 0$

Solution

a.

$6x^2 = 3x$	Write original equation.
$6x^2 - 3x = 0$	Write in general form.
$3x(2x - 1) = 0$	Factor.
$3x = 0 \implies x = 0$	Set 1st factor equal to 0.
$2x - 1 = 0 \implies x = \frac{1}{2}$	Set 2nd factor equal to 0.

b.

$9x^2 - 6x + 1 = 0$	Write original equation.
$(3x - 1)^2 = 0$	Factor.
$3x - 1 = 0 \implies x = \frac{1}{3}$	Set repeated factor equal to 0.

Throughout the text, when solving equations, be sure to check your solutions either *algebraically* by substituting in the original equation or *graphically*.

Check

a.

$6x^2 = 3x$	Write original equation.
$6(0)^2 \stackrel{?}{=} 3(0)$	Substitute 0 for x.
$0 = 0$	Solution checks. ✓
$6\left(\frac{1}{2}\right)^2 \stackrel{?}{=} 3\left(\frac{1}{2}\right)$	Substitute $\frac{1}{2}$ for x.
$\frac{6}{4} = \frac{3}{2}$	Solution checks. ✓

b.

$9x^2 - 6x + 1 = 0$	Write original equation.
$9\left(\frac{1}{3}\right)^2 - 6\left(\frac{1}{3}\right) + 1 \stackrel{?}{=} 0$	Substitute $\frac{1}{3}$ for x.
$1 - 2 + 1 \stackrel{?}{=} 0$	Simplify.
$0 = 0$	Solution checks. ✓

Similarly, you can graphically check your solutions using the graphs in Figure 2.26.

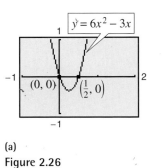

(a)

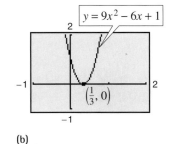

(b)

Figure 2.26

✓ *Checkpoint* Now try Exercise 7.

STUDY TIP

Quadratic equations always have two solutions. From the graph in Figure 2.26(b), it looks like there is only one solution to the equation $9x^2 - 6x + 1 = 0$.

Because the equation is a perfect square trinomial, its two factors are identical. As a result, the equation has two *repeated* solutions.

Solving a quadratic equation by extracting square roots is an efficient method to use when the quadratic equation can be written in the form $ax^2 + c = 0$, as shown in Example 2.

Example 2   Extracting Square Roots

Solve each quadratic equation.

a. $4x^2 = 12$ **b.** $(x - 3)^2 = 7$

Solution

a. $4x^2 = 12$ Write original equation.

$x^2 = 3$ Divide each side by 4.

$x = \pm\sqrt{3}$ Take square root of each side.

This equation has two solutions: $x = \sqrt{3}$ and $x = -\sqrt{3}$.

b. $(x - 3)^2 = 7$ Write original equation.

$x - 3 = \pm\sqrt{7}$ Take square root of each side.

$x = 3 \pm\sqrt{7}$ Add 3 to each side.

This equation has two solutions: $x = 3 + \sqrt{7}$ and $x = 3 - \sqrt{7}$.

The graphs of $y = 4x^2 - 12$ and $y = (x - 3)^2 - 7$, shown in Figure 2.27, verify the solutions.

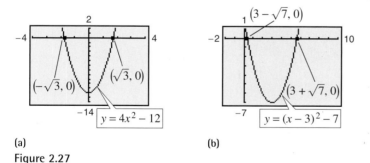

(a)   (b)

Figure 2.27

☑ *Checkpoint* Now try Exercise 19.

TECHNOLOGY TIP   Note that the solutions shown in Example 2 are listed in *exact* form. Most graphing utilities produce decimal approximations of solutions rather than exact forms. For instance, if you solve the equations in Example 2 using a graphing utility, you will obtain $x \approx \pm1.732$ in part (a) and $x \approx 5.646$ and $x \approx 0.354$ in part (b). Some graphing utilities have symbolic algebra programs that *can* list the exact form of a solution.

Completing the square can be used to solve any quadratic equation, but it is best suited for quadratic equations in general form $ax^2 + bx + c = 0$ with $a = 1$ and b an even number (see page 191). If the leading coefficient of the quadratic is not 1, divide each side of the equation by this coefficient *before* completing the square, as shown in Example 4.

Example 3   Completing the Square: Leading Coefficient Is 1

Solve $x^2 + 2x - 6 = 0$ by completing the square.

Solution

$x^2 + 2x - 6 = 0$	Write original equation.
$x^2 + 2x = 6$	Add 6 to each side.
$x^2 + 2x + 1^2 = 6 + 1^2$	Add 1^2 to each side.

(Half of 2)^2

$(x + 1)^2 = 7$	Simplify.
$x + 1 = \pm\sqrt{7}$	Take square root of each side.
$x = -1 \pm \sqrt{7}$	Solutions

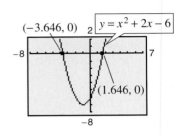

Figure 2.28

Using a calculator, the two solutions are $x \approx 1.646$ and $x \approx -3.646$, which agree with the graphical solutions shown in Figure 2.28.

 Checkpoint Now try Exercise 23.

Example 4   Completing the Square: Leading Coefficient Is Not 1

Solve $2x^2 + 8x + 3 = 0$ by completing the square.

Solution

$2x^2 + 8x + 3 = 0$	Write original equation.
$2x^2 + 8x = -3$	Subtract 3 from each side.
$x^2 + 4x = -\dfrac{3}{2}$	Divide each side by 2.
$x^2 + 4x + 2^2 = -\dfrac{3}{2} + 2^2$	Add 2^2 to each side.

(Half of 4)^2

$(x + 2)^2 = \dfrac{5}{2}$	Simplify.
$x + 2 = \pm\sqrt{\dfrac{5}{2}}$	Take square root of each side.
$x + 2 = \pm\dfrac{\sqrt{10}}{2}$	Rationalize denominator.
$x = -2 \pm \dfrac{\sqrt{10}}{2}$	Solutions

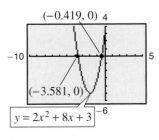

Figure 2.29

Using a calculator, the two solutions are $x \approx -0.419$ and $x \approx -3.581$, which agree with the graphical solutions shown in Figure 2.29.

 Checkpoint Now try Exercise 27.

Example 5   Completing the Square: Leading Coefficient Is Not 1

Solve $3x^2 - 4x - 5 = 0$ by completing the square.

Solution

$$3x^2 - 4x - 5 = 0 \qquad \text{Write original equation.}$$

$$3x^2 - 4x = 5 \qquad \text{Add 5 to each side.}$$

$$x^2 - \frac{4}{3}x = \frac{5}{3} \qquad \text{Divide each side by 3.}$$

$$x^2 - \frac{4}{3}x + \left(-\frac{2}{3}\right)^2 = \frac{5}{3} + \left(-\frac{2}{3}\right)^2 \qquad \text{Add } \left(-\frac{2}{3}\right)^2 \text{ to each side.}$$

$$\left(\text{Half of } -\frac{4}{3}\right)^2$$

$$\left(x - \frac{2}{3}\right)^2 = \frac{19}{9} \qquad \text{Simplify.}$$

$$x - \frac{2}{3} = \pm \frac{\sqrt{19}}{3} \qquad \text{Take square root of each side.}$$

$$x = \frac{2}{3} \pm \frac{\sqrt{19}}{3} \qquad \text{Solutions}$$

Using a calculator, the two solutions are $x \approx 2.120$ and $x \approx -0.786$, which agree with the graphical solutions shown in Figure 2.30.

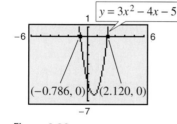

Figure 2.30

✓ *Checkpoint* Now try Exercise 31.

Often in mathematics you are taught the long way of solving a problem first. Then, the longer method is used to develop shorter techniques. The long way stresses understanding and the short way stresses efficiency.

For instance, you can think of completing the square as a "long way" of solving a quadratic equation. When you use the method of completing the square to solve a quadratic equation, you must complete the square for *each* equation separately. In the derivation on the following page, you complete the square *once* in a general setting to obtain the Quadratic Formula, which is a shortcut for solving a quadratic equation.

$$ax^2 + bx + c = 0$$ Quadratic equation in general form, $a \neq 0$

$$ax^2 + bx = -c$$ Subtract c from each side.

$$x^2 + \frac{b}{a}x = -\frac{c}{a}$$ Divide each side by a.

$$x^2 + \frac{b}{a}x + \left(\frac{b}{2a}\right)^2 = -\frac{c}{a} + \left(\frac{b}{2a}\right)^2$$ Complete the square.

$$\left(\text{half of } \frac{b}{a}\right)^2$$

$$\left(x + \frac{b}{2a}\right)^2 = \frac{b^2 - 4ac}{4a^2}$$ Simplify.

$$x + \frac{b}{2a} = \pm\sqrt{\frac{b^2 - 4ac}{4a^2}}$$ Extract square roots.

$$x = -\frac{b}{2a} \pm \frac{\sqrt{b^2 - 4ac}}{2|a|}$$ Solutions

Note that because $\pm 2|a|$ represents the same numbers as $\pm 2a$, you can omit the absolute value sign. So, the formula simplifies to

$$x = \frac{-b \pm \sqrt{b^2 - 4ac}}{2a}.$$

> **Exploration**
>
> Use a graphing utility to graph the three quadratic equations
>
> $$y_1 = x^2 - 2x$$
> $$y_2 = x^2 - 2x + 1$$
> $$y_3 = x^2 - 2x + 2$$
>
> in the same viewing window. Compute the *discriminant* $\sqrt{b^2 - 4ac}$ for each and discuss the relationship between the discriminant and the number of zeros of the quadratic function.

Example 6   Quadratic Formula: Two Distinct Solutions

Solve $x^2 + 3x = 9$ using the Quadratic Formula.

Algebraic Solution

$$x^2 + 3x = 9$$ Write original equation.

$$x^2 + 3x - 9 = 0$$ Write in general form.

$$x = \frac{-b \pm \sqrt{b^2 - 4ac}}{2a}$$ Quadratic Formula

$$x = \frac{-3 \pm \sqrt{3^2 - 4(1)(-9)}}{2(1)}$$ Substitute 3 for b, 1 for a, and -9 for c.

$$x = \frac{-3 \pm \sqrt{45}}{2}$$ Simplify.

$$x = \frac{-3 \pm 3\sqrt{5}}{2}$$ Simplify radical.

$$x \approx 1.85 \text{ or } -4.85$$ Solutions

The equation has two solutions: $x \approx 1.85$ and $x \approx -4.85$. Check these solutions in the original equation.

✓ *Checkpoint* Now try Exercise 47.

Graphical Solution

Use a graphing utility to graph $y_1 = x^2 + 3x$ and $y_2 = 9$ in the same viewing window. Use the *intersect* feature of the graphing utility to approximate the points where the graphs intersect. From Figure 2.31, it appears that the graphs intersect at $x \approx 1.85$ and $x \approx -4.85$. These x-coordinates of the intersection points are the solutions of the equation $x^2 + 3x = 9$.

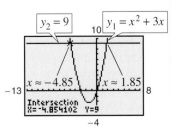

Figure 2.31

Example 7   Quadratic Formula: One Repeated Solution

Solve $8x^2 - 24x + 18 = 0$.

Algebraic Solution

This equation has a common factor of 2. You can simplify the equation by dividing each side of the equation by 2.

$$8x^2 - 24x + 18 = 0 \qquad \text{Write original equation.}$$

$$4x^2 - 12x + 9 = 0 \qquad \text{Divide each side by 2.}$$

$$x = \frac{-b \pm \sqrt{b^2 - 4ac}}{2a} \qquad \text{Quadratic Formula}$$

$$x = \frac{-(-12) \pm \sqrt{(-12)^2 - 4(4)(9)}}{2(4)}$$

$$x = \frac{12 \pm \sqrt{0}}{8} = \frac{3}{2} \qquad \text{Repeated solution}$$

This quadratic equation has only one solution: $x = \frac{3}{2}$. Check this solution in the original equation.

✓ *Checkpoint* Now try Exercise 49.

Graphical Solution

Use a graphing utility to graph

$$y = 8x^2 - 24x + 18.$$

Use the *zero* feature of the graphing utility to approximate the value(s) of x for which the function is equal to zero. From the graph in Figure 2.32, it appears that the function is equal to zero when $x = 1.5 = \frac{3}{2}$. This is the only solution of the equation $8x^2 - 24x + 18 = 0$.

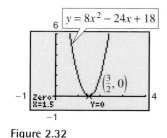

Figure 2.32

Example 8   Complex Solutions of a Quadratic Equation

Solve $3x^2 - 2x + 5 = 0$.

Algebraic Solution

By the Quadratic Formula, you can write the solutions as follows.

$$3x^2 - 2x + 5 = 0 \qquad \text{Write original equation.}$$

$$x = \frac{-b \pm \sqrt{b^2 - 4ac}}{2a} \qquad \text{Quadratic Formula}$$

$$= \frac{-(-2) \pm \sqrt{(-2)^2 - 4(3)(5)}}{2(3)} \qquad \begin{array}{l}\text{Substitute } -2 \text{ for } b, \\ 3 \text{ for } a, \text{ and } 5 \text{ for } c.\end{array}$$

$$= \frac{2 \pm \sqrt{-56}}{6} \qquad \text{Simplify.}$$

$$= \frac{2 \pm 2\sqrt{14}\,i}{6} \qquad \text{Simplify radical.}$$

$$= \frac{1}{3} \pm \frac{\sqrt{14}}{3}\,i \qquad \text{Solutions}$$

The equation has no real solution, but it has two complex solutions: $x = \frac{1}{3}\left(1 + \sqrt{14}\,i\right)$ and $x = \frac{1}{3}\left(1 - \sqrt{14}\,i\right)$.

✓ *Checkpoint* Now try Exercise 51.

Graphical Solution

Use a graphing utility to graph

$$y = 3x^2 - 2x + 5.$$

Note in Figure 2.33 that the graph of the function appears to have no x-intercepts. From this you can conclude that the equation $3x^2 - 2x + 5 = 0$ has no real solution. You can solve the equation algebraically to find the complex solutions.

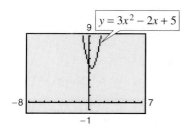

Figure 2.33

Polynomial Equations of Higher Degree

The methods used to solve quadratic equations can sometimes be extended to polynomial equations of higher degree, as shown in the next two examples.

Example 9   Solving an Equation of Quadratic Type

Solve $x^4 - 3x^2 + 2 = 0$.

Solution

The expression $x^4 - 3x^2 + 2$ is said to be in *quadratic form* because it is written in the form $au^2 + bu + c$, where u is any expression in x, namely x^2. You can use factoring to solve the equation as follows.

$$x^4 - 3x^2 + 2 = 0 \qquad \text{Write original equation.}$$
$$(x^2)^2 - 3(x^2) + 2 = 0 \qquad \text{Write in quadratic form.}$$
$$(x^2 - 1)(x^2 - 2) = 0 \qquad \text{Partially factor.}$$
$$(x + 1)(x - 1)(x^2 - 2) = 0 \qquad \text{Factor completely.}$$
$$x + 1 = 0 \implies x = -1 \qquad \text{Set 1st factor equal to 0.}$$
$$x - 1 = 0 \implies x = 1 \qquad \text{Set 2nd factor equal to 0.}$$
$$x^2 - 2 = 0 \implies x = \pm\sqrt{2} \qquad \text{Set 3rd factor equal to 0.}$$

The equation has four solutions: $x = -1$, $x = 1$, $x = \sqrt{2}$, and $x = -\sqrt{2}$. Check these solutions in the original equation. Figure 2.34 verifies the solutions graphically.

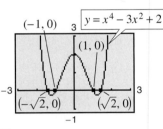

Figure 2.34

✓ *Checkpoint* Now try Exercise 63.

Example 10   Solving a Polynomial Equation by Factoring

Solve $2x^3 - 6x^2 - 6x + 18 = 0$.

Solution

This equation has a common factor of 2. You can simplify the equation by first dividing each side of the equation by 2.

$$2x^3 - 6x^2 - 6x + 18 = 0 \qquad \text{Write original equation.}$$
$$x^3 - 3x^2 - 3x + 9 = 0 \qquad \text{Divide each side by 2.}$$
$$x^2(x - 3) - 3(x - 3) = 0 \qquad \text{Group terms.}$$
$$(x - 3)(x^2 - 3) = 0 \qquad \text{Factor by grouping.}$$
$$x - 3 = 0 \implies x = 3 \qquad \text{Set 1st factor equal to 0.}$$
$$x^2 - 3 = 0 \implies x = \pm\sqrt{3} \qquad \text{Set 2nd factor equal to 0.}$$

The equation has three solutions: $x = 3$, $x = \sqrt{3}$, and $x = -\sqrt{3}$. Check these solutions in the original equation. Figure 2.35 verifies the solutions graphically.

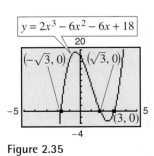

Figure 2.35

✓ *Checkpoint* Now try Exercise 67.

Equations Involving Radicals

An equation involving a radical expression can often be cleared of radicals by raising each side of the equation to an appropriate power. When using this procedure, remember to check for extraneous solutions.

Example 11   Solving an Equation Involving a Radical

Solve $\sqrt{2x + 7} - x = 2$.

Algebraic Solution

$\sqrt{2x + 7} - x = 2$	Write original equation.
$\sqrt{2x + 7} = x + 2$	Isolate radical.
$2x + 7 = x^2 + 4x + 4$	Square each side.
$x^2 + 2x - 3 = 0$	Write in general form.
$(x + 3)(x - 1) = 0$	Factor.
$x + 3 = 0 \implies x = -3$	Set 1st factor equal to 0.
$x - 1 = 0 \implies x = 1$	Set 2nd factor equal to 0.

By substituting into the original equation, you can determine that $x = -3$ is extraneous, whereas $x = 1$ is valid. So, the equation has only one real solution: $x = 1$.

✓ *Checkpoint* Now try Exercise 83.

Graphical Solution

First rewrite the equation as $\sqrt{2x + 7} - x - 2 = 0$. Then use a graphing utility to graph $y = \sqrt{2x + 7} - x - 2$, as shown in Figure 2.36. Notice that the domain is $x \geq -\frac{7}{2}$ because the expression under the radical cannot be negative. There appears to be one solution near $x = 1$. Use the *zoom* and *trace* features, as shown in Figure 2.37, to approximate the only solution to be $x = 1$.

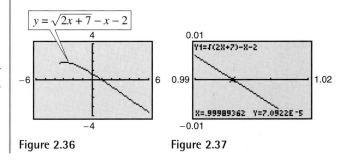

Figure 2.36   Figure 2.37

Example 12   Solving an Equation Involving Two Radicals

$\sqrt{2x + 6} - \sqrt{x + 4} = 1$	Original equation
$\sqrt{2x + 6} = 1 + \sqrt{x + 4}$	Isolate $\sqrt{2x + 6}$.
$2x + 6 = 1 + 2\sqrt{x + 4} + (x + 4)$	Square each side.
$x + 1 = 2\sqrt{x + 4}$	Isolate $2\sqrt{x + 4}$.
$x^2 + 2x + 1 = 4(x + 4)$	Square each side.
$x^2 - 2x - 15 = 0$	Write in general form.
$(x - 5)(x + 3) = 0$	Factor.
$x - 5 = 0 \implies x = 5$	Set 1st factor equal to 0.
$x + 3 = 0 \implies x = -3$	Set 2nd factor equal to 0.

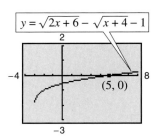

Figure 2.38

By substituting into the original equation, you can determine that $x = -3$ is extraneous, whereas $x = 5$ is valid. Figure 2.38 verifies that $x = 5$ is the only solution.

✓ *Checkpoint* Now try Exercise 89.

Example 13   Solving an Equation with Rational Exponents

Solve $(x + 1)^{2/3} = 4$.

Algebraic Solution

$$(x + 1)^{2/3} = 4 \qquad \text{Write original equation.}$$

$$\sqrt[3]{(x + 1)^2} = 4 \qquad \text{Rewrite in radical form.}$$

$$(x + 1)^2 = 64 \qquad \text{Cube each side.}$$

$$x + 1 = \pm 8 \qquad \text{Take square root of each side.}$$

$$x = 7, x = -9 \qquad \text{Subtract 1 from each side.}$$

Substitute $x = 7$ and $x = -9$ into the original equation to determine that both are valid solutions.

✓ *Checkpoint*   Now try Exercise 91.

Graphical Solution

Use a graphing utility to graph $y_1 = \sqrt[3]{(x + 1)^2}$ and $y_2 = 4$ in the same viewing window. Use the *intersect* feature of the graphing utility to approximate the solutions to be $x = -9$ and $x = 7$, as shown in Figure 2.39.

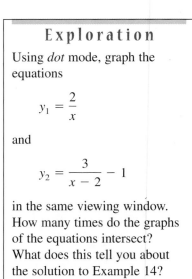

Figure 2.39

Equations Involving Fractions or Absolute Values

As demonstrated in Section 2.1, you can algebraically solve an equation involving fractions by multiplying each side of the equation by the least common denominator of all terms in the equation to clear the equation of fractions.

Example 14   Solving an Equation Involving Fractions

Solve $\dfrac{2}{x} = \dfrac{3}{x - 2} - 1$.

Solution

For this equation, the least common denominator of the three terms is $x(x - 2)$, so you can begin by multiplying each term of the equation by this expression.

$$\frac{2}{x} = \frac{3}{x - 2} - 1 \qquad \text{Write original equation.}$$

$$x(x - 2)\frac{2}{x} = x(x - 2)\frac{3}{x - 2} - x(x - 2)(1) \qquad \text{Multiply each term by the LCD.}$$

$$2(x - 2) = 3x - x(x - 2), \qquad x \neq 0, 2 \qquad \text{Simplify.}$$

$$x^2 - 3x - 4 = 0 \qquad \text{Write in general form.}$$

$$(x - 4)(x + 1) = 0 \qquad \text{Factor.}$$

$$x - 4 = 0 \implies x = 4 \qquad \text{Set 1st factor equal to 0.}$$

$$x + 1 = 0 \implies x = -1 \qquad \text{Set 2nd factor equal to 0.}$$

The equation has two solutions: $x = 4$ and $x = -1$. Check these solutions in the original equation. Use a graphing utility to verify these solutions graphically.

✓ *Checkpoint*   Now try Exercise 101.

Exploration

Using *dot* mode, graph the equations

$$y_1 = \frac{2}{x}$$

and

$$y_2 = \frac{3}{x - 2} - 1$$

in the same viewing window. How many times do the graphs of the equations intersect? What does this tell you about the solution to Example 14?

TECHNOLOGY TIP

Graphs of functions involving variable denominators can be tricky because of the way graphing utilities skip over points at which the denominator is zero. You will study graphs of such functions in Sections 3.5 and 3.6.

Example 15   Solving an Equation Involving Absolute Value

Solve $|x^2 - 3x| = -4x + 6$.

Solution

Begin by writing the equation as $|x^2 - 3x| + 4x - 6 = 0$. From the graph of $y = |x^2 - 3x| + 4x - 6$ in Figure 2.40, you can estimate the solutions to be $x = -3$ and $x = 1$. These can be verified by substitution into the equation. To solve *algebraically* an equation involving an absolute value, you must consider the fact that the expression inside the absolute value symbols can be positive or negative. This results in *two* separate equations, each of which must be solved.

First Equation:

$$x^2 - 3x = -4x + 6 \qquad \text{Use positive expression.}$$
$$x^2 + x - 6 = 0 \qquad \text{Write in general form.}$$
$$(x + 3)(x - 2) = 0 \qquad \text{Factor.}$$
$$x + 3 = 0 \implies x = -3 \qquad \text{Set 1st factor equal to 0.}$$
$$x - 2 = 0 \implies x = 2 \qquad \text{Set 2nd factor equal to 0.}$$

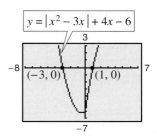

Second Equation:

$$-(x^2 - 3x) = -4x + 6 \qquad \text{Use negative expression.}$$
$$x^2 - 7x + 6 = 0 \qquad \text{Write in general form.}$$
$$(x - 1)(x - 6) = 0 \qquad \text{Factor.}$$
$$x - 1 = 0 \implies x = 1 \qquad \text{Set 1st factor equal to 0.}$$
$$x - 6 = 0 \implies x = 6 \qquad \text{Set 2nd factor equal to 0.}$$

Figure 2.40

Check

$$|(-3)^2 - 3(-3)| \overset{?}{=} -4(-3) + 6 \qquad \text{Substitute } -3 \text{ for } x.$$
$$18 = 18 \qquad -3 \text{ checks. } \checkmark$$
$$|2^2 - 3(2)| \overset{?}{=} -4(2) + 6 \qquad \text{Substitute 2 for } x.$$
$$2 \neq -2 \qquad 2 \text{ does not check.}$$
$$|1^2 - 3(1)| \overset{?}{=} -4(1) + 6 \qquad \text{Substitute 1 for } x.$$
$$2 = 2 \qquad 1 \text{ checks. } \checkmark$$
$$|6^2 - 3(6)| \overset{?}{=} -4(6) + 6 \qquad \text{Substitute 6 for } x.$$
$$18 \neq -18 \qquad 6 \text{ does not check.}$$

The equation has only two solutions: $x = -3$ and $x = 1$, just as you obtained by graphing.

✓ *Checkpoint* Now try Exercise 107.

Exploration

In Figure 2.40, the graph of $y = |x^2 - 3x| + 4x - 6$ appears to be a straight line to the right of the y-axis. Is it? Explain how you decided.

Applications

A common application of quadratic equations involves an object that is falling (or projected into the air). The general equation that gives the height of such an object is called a **position equation,** and on *Earth's* surface it has the form

$$s = -16t^2 + v_0 t + s_0.$$

In this equation, s represents the height of the object (in feet), v_0 represents the initial velocity of the object (in feet per second), s_0 represents the initial height of the object (in feet), and t represents the time (in seconds). Note that this position equation ignores air resistance.

Example 16   Falling Time

A construction worker on the 24th floor of a building project (see Figure 2.41) accidentally drops a wrench and yells, "Look out below!" Could a person at ground level hear this warning in time to get out of the way?

Solution

Assume that each floor of the building is 10 feet high, so that the wrench is dropped from a height of 235 feet (the construction worker's hand is 5 feet below the ceiling of the 24th floor). Because sound travels at about 1100 feet per second, it follows that a person at ground level hears the warning within 1 second of the time the wrench is dropped. To set up a mathematical model for the height of the wrench, use the position equation

$$s = -16t^2 + v_0 t + s_0. \qquad \text{Position equation}$$

Because the object is dropped rather than thrown, the initial velocity is $v_0 = 0$ feet per second. So, with an initial height of $s_0 = 235$ feet, you have the model

$$s = -16t^2 + (0)t + 235 = -16t^2 + 235.$$

After falling for 1 second, the height of the wrench is $-16(1)^2 + 235 = 219$. After falling for 2 seconds, the height of the wrench is $-16(2)^2 + 235 = 171$. To find the number of seconds it takes the wrench to hit the ground, let the height s be zero and solve the equation for t.

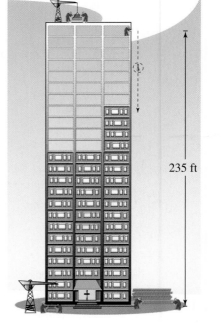

235 ft

$$s = -16t^2 + 235 \qquad \text{Write position equation.}$$

$$0 = -16t^2 + 235 \qquad \text{Substitute 0 for } s.$$

$$16t^2 = 235 \qquad \text{Add } 16t^2 \text{ to each side.}$$

$$t^2 = \frac{235}{16} \qquad \text{Divide each side by 16.}$$

$$t = \frac{\sqrt{235}}{4} \approx 3.83 \qquad \text{Extract positive square root.}$$

Figure 2.41

The wrench will take about 3.83 seconds to hit the ground. If the person hears the warning 1 second after the wrench is dropped, the person still has almost 3 more seconds to get out of the way.

 Checkpoint Now try Exercise 125.

Example 17   Quadratic Modeling: Internet Use

From 1996 to 2001, the number of hours h spent annually per person using the Internet in the United States closely followed the quadratic model

$$h = -0.05t^2 + 29.6t - 168$$

where t represents the year, with $t = 6$ corresponding to 1996. The number of hours per year is shown graphically in Figure 2.42. According to this model, in which year will the number of hours spent per person reach or surpass 300? (Source: Veronis Suhler Stevenson)

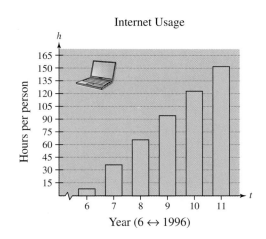

Figure 2.42

Solution

To find when the number of hours spent per person will reach 300, you need to solve the equation

$$-0.05t^2 + 29.6t - 168 = 300.$$

To begin, write the equation in general form.

$$-0.05t^2 + 29.6t - 468 = 0$$

Then apply the Quadratic Formula.

$$t = \frac{-29.6 \pm \sqrt{(29.6)^2 - 4(-0.05)(-468)}}{2(-0.05)}$$

$$\approx 16.3 \text{ or } 575.7$$

Choose the smaller value $t = 16.3$. Because $t = 6$ corresponds to 1996, it follows that $t = 16.3$ must correspond to some time in 2006. So, the number of hours spent annually per person using the Internet should reach 300 during 2006.

 Checkpoint Now try Exercise 129.

> **TECHNOLOGY TIP**   You can solve Example 17 with your graphing utility by graphing the two functions $y_1 = -0.05t^2 + 29.6t - 168$ and $y_2 = 300$ in the same viewing window and finding their point of intersection. You should obtain $x \approx 16.3$, which verifies the answer obtained algebraically.

Another type of application that often involves a quadratic equation is one dealing with the hypotenuse of a right triangle. These types of applications often use the Pythagorean Theorem, which states that

$$a^2 + b^2 = c^2 \qquad \text{Pythagorean Theorem}$$

where a and b are the legs of a right triangle and c is the hypotenuse, as indicated in Figure 2.43.

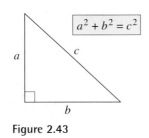

Figure 2.43

Example 18   An Application Involving the Pythagorean Theorem

An L-shaped sidewalk from the athletic center to the library on a college campus is shown in Figure 2.44. The sidewalk was constructed so that the length of one sidewalk forming the L is twice as long as the other. The length of the diagonal sidewalk that cuts across the grounds between the two buildings is 102 feet. How many feet does a person save by walking on the diagonal sidewalk?

Figure 2.44

Solution

Using the Pythagorean Theorem, you have

$a^2 + b^2 = c^2$	Pythagorean Theorem
$x^2 + (2x)^2 = 102^2$	Substitute for a, b, and c.
$5x^2 = 10{,}404$	Combine like terms.
$x^2 = 2080.8$	Divide each side by 5.
$x = \pm\sqrt{2080.8}$	Take the square root of each side.
$x = \sqrt{2080.8}$	Extract positive square root.

The total distance covered by walking on the L-shaped sidewalk is

$$x + 2x = 3x$$
$$= 3\sqrt{2080.8}$$
$$\approx 136.8 \text{ feet.}$$

Walking on the diagonal sidewalk saves a person about $136.8 - 102 = 34.8$ feet.

✓ *Checkpoint* Now try Exercise 135.

2.4   Exercises

Vocabulary Check

Fill in the blanks.

1. An equation of the form $ax^2 + bx + c = 0$, where a, b, and c are real numbers and $a \neq 0$, is a _____ , or a second-degree polynomial equation in x.

2. The four methods that can be used to solve a quadratic equation are _____ , _____ , _____ , and the _____ .

3. The part of the Quadratic Formula $\sqrt{b^2 - 4ac}$, known as the _____ , determines the type of solutions of a quadratic equation.

4. The general equation that gives the height of an object (in feet) in terms of the time t (in seconds) is called the _____ equation, and has the form $s =$ _____ , where v_0 represents the _____ and s_0 represents the _____ .

In Exercises 1–4, write the quadratic equation in general form. Do not solve the equation.

1. $2x^2 = 3 - 5x$
2. $x^2 = 25x + 26$
3. $\frac{1}{5}(3x^2 - 10) = 12x$
4. $x(x + 2) = 3x^2 + 1$

In Exercises 5–14, solve the quadratic equation by factoring. Check your solutions in the original equation.

5. $6x^2 + 3x = 0$
6. $9x^2 - 1 = 0$
7. $x^2 - 2x - 8 = 0$
8. $x^2 - 10x + 9 = 0$
9. $3 + 5x - 2x^2 = 0$
10. $2x^2 = 19x + 33$
11. $x^2 + 4x = 12$
12. $-x^2 + 8x = 12$
13. $(x + a)^2 - b^2 = 0$
14. $x^2 + 2ax + a^2 = 0$

In Exercises 15–22, solve the equation by extracting square roots. List both the exact solutions and the decimal solutions rounded to two decimal places.

15. $x^2 = 49$
16. $x^2 = 144$
17. $(x - 12)^2 = 16$
18. $(x - 5)^2 = 25$
19. $(2x - 1)^2 = 12$
20. $(4x + 7)^2 = 44$
21. $(x - 7)^2 = (x + 3)^2$
22. $(x + 5)^2 = (x + 4)^2$

In Exercises 23–32, solve the quadratic equation by completing the square. Verify your answer graphically.

23. $x^2 + 4x - 32 = 0$
24. $x^2 - 2x - 3 = 0$
25. $x^2 + 6x + 2 = 0$
26. $x^2 + 8x + 14 = 0$
27. $9x^2 - 18x + 3 = 0$
28. $4x^2 - 4x - 99 = 0$
29. $8 + 4x - x^2 = 0$
30. $-x^2 + x - 1 = 0$
31. $2x^2 + 5x - 8 = 0$
32. $9x^2 - 12x - 14 = 0$

Graphical Reasoning **In Exercises 33–38, (a) use a graphing utility to graph the equation, (b) use the graph to approximate any x-intercepts of the graph, (c) set $y = 0$ and solve the resulting equation, and (d) compare the result of part (c) with the x-intercepts of the graph.**

33. $y = (x + 3)^2 - 4$
34. $y = 1 - (x - 2)^2$
35. $y = -4x^2 + 4x + 3$
36. $y = x^2 + 3x - 4$
37. $y = \frac{1}{4}(4x^2 - 20x + 25)$
38. $y = -\frac{1}{4}(x^2 - 2x + 9)$

In Exercises 39–44, use a graphing utility to determine the number of real solutions of the quadratic equation.

39. $2x^2 - 5x + 5 = 0$
40. $2x^2 - x - 1 = 0$
41. $\frac{4}{7}x^2 - 8x + 28 = 0$
42. $\frac{1}{3}x^2 - 5x + 25 = 0$
43. $-0.2x^2 + 1.2x - 8 = 0$
44. $9 + 2.4x - 8.3x^2 = 0$

In Exercises 45–52, use the Quadratic Formula to solve the equation. Use a graphing utility to verify your solutions graphically.

45. $2 + 2x - x^2 = 0$
46. $x^2 - 10x + 22 = 0$
47. $x^2 + 8x - 4 = 0$
48. $4x^2 - 4x - 4 = 0$
49. $28x - 49x^2 = 4$
50. $9x^2 + 24x + 16 = 0$
51. $4x^2 + 16x + 17 = 0$
52. $9x^2 - 6x + 37 = 0$

In Exercises 53–60, solve the equation using any convenient method.

53. $x^2 - 2x - 1 = 0$

54. $11x^2 + 33x = 0$

55. $(x + 3)^2 = 81$

56. $x^2 - 14x + 49 = 0$

57. $x^2 - x - \frac{11}{4} = 0$

58. $x^2 + 3x - \frac{3}{4} = 0$

59. $(x + 1)^2 = x^2$

60. $a^2x^2 - b^2 = 0,\ a \neq 0$

In Exercises 61–78, find all solutions of the equation algebraically. Use a graphing utility to verify the solutions graphically.

61. $4x^4 - 18x^2 = 0$

62. $20x^3 - 125x = 0$

63. $x^4 - 4x^2 + 3 = 0$

64. $x^4 + 5x^2 - 36 = 0$

65. $5x^3 + 30x^2 + 45x = 0$

66. $9x^4 - 24x^3 + 16x^2 = 0$

67. $x^3 - 3x^2 - x + 3 = 0$

68. $x^4 + 2x^3 - 8x - 16 = 0$

69. $4x^4 - 65x^2 + 16 = 0$ **70.** $36t^4 + 29t^2 - 7 = 0$

71. $\frac{1}{t^2} + \frac{8}{t} + 15 = 0$ **72.** $6 - \frac{1}{x} - \frac{1}{x^2} = 0$

73. $6\left(\frac{s}{s+1}\right)^2 + 5\left(\frac{s}{s+1}\right) - 6 = 0$

74. $8\left(\frac{t}{t-1}\right)^2 - 2\left(\frac{t}{t-1}\right) - 3 = 0$

75. $2x + 9\sqrt{x} - 5 = 0$ **76.** $6x - 7\sqrt{x} - 3 = 0$

77. $3x^{1/3} + 2x^{2/3} = 5$

78. $9t^{2/3} + 24t^{1/3} + 16 = 0$

Graphical Analysis **In Exercises 79–82, (a) use a graphing utility to graph the equation, (b) use the graph to approximate any x-intercepts of the graph, (c) set y = 0 and solve the resulting equation, and (d) compare the result of part (c) with the x-intercepts of the graph.**

79. $y = x^3 - 2x^2 - 3x$

80. $y = 2x^4 - 15x^3 + 18x^2$

81. $y = x^4 - 10x^2 + 9$

82. $y = x^4 - 29x^2 + 100$

In Exercises 83–94, find all solutions of the equation algebraically. Check your solutions both algebraically and graphically.

83. $\sqrt{x - 10} - 4 = 0$ **84.** $\sqrt{2x + 5} + 3 = 0$

85. $\sqrt{x + 1} - 3x = 1$ **86.** $\sqrt{x + 5} - 2x = 3$

87. $\sqrt[3]{2x + 1} + 8 = 0$ **88.** $\sqrt[3]{4x - 3} + 2 = 0$

89. $\sqrt{x} - \sqrt{x - 5} = 1$ **90.** $\sqrt{x} + \sqrt{x - 20} = 10$

91. $(x - 5)^{2/3} = 16$

92. $(x^2 - x - 22)^{4/3} = 16$

93. $3x(x - 1)^{1/2} + 2(x - 1)^{3/2} = 0$

94. $4x^2(x - 1)^{1/3} + 6x(x - 1)^{4/3} = 0$

Graphical Analysis **In Exercises 95–98, (a) use a graphing utility to graph the equation, (b) use the graph to approximate any x-intercepts of the graph, (c) set y = 0 and solve the resulting equation, and (d) compare the result of part (c) with the x-intercepts of the graph.**

95. $y = \sqrt{11x - 30} - x$ **96.** $y = 2x - \sqrt{15 - 4x}$

97. $y = \sqrt{7x + 36} - \sqrt{5x + 16} - 2$

98. $y = 3\sqrt{x} - \dfrac{4}{\sqrt{x}} - 4$

In Exercises 99–108, find all solutions of the equation. Use a graphing utility to verify your solutions graphically.

99. $\dfrac{20 - x}{x} = x$ **100.** $\dfrac{4}{x} - \dfrac{5}{3} = \dfrac{x}{6}$

101. $\dfrac{1}{x} - \dfrac{1}{x + 1} = 3$ **102.** $\dfrac{x}{x^2 - 4} + \dfrac{1}{x + 2} = 3$

103. $x = \dfrac{3}{x} + \dfrac{1}{2}$ **104.** $4x + 1 = \dfrac{3}{x}$

105. $|2x - 1| = 5$ **106.** $|3x + 2| = 7$

107. $|x| = x^2 + x - 3$ **108.** $|x - 10| = x^2 - 10x$

Graphical Analysis **In Exercises 109–112, (a) use a graphing utility to graph the equation, (b) use the graph to approximate any x-intercepts of the graph, (c) set y = 0 and solve the resulting equation, and (d) compare the result of part (c) with the x-intercepts of the graph.**

109. $y = \dfrac{1}{x} - \dfrac{4}{x - 1} - 1$ **110.** $y = x + \dfrac{9}{x + 1} - 5$

111. $y = |x + 1| - 2$ **112.** $y = |x - 2| - 3$

Think About It **In Exercises 113–118, find an equation having the given solutions. (There are many correct answers.)**

113. $-6, 5$ **114.** $-\frac{7}{3}, \frac{6}{7}$

115. $\sqrt{2}, -\sqrt{2}, 4$

116. $2, \sqrt{5}, -\sqrt{5}$

117. $-2, 2, i, -i$

118. $4i, -4i, 6, -6$

Think About It **In Exercises 119 and 120, find x such that the distance between the points is 13.**

119. $(1, 2), (x, -10)$

120. $(-8, 0), (x, 5)$

121. *Geometry*   The floor of a one-story building is 14 feet longer than it is wide. The building has 1632 square feet of floor space.

   (a) Draw a diagram that gives a visual representation of the floor space. Represent the width as w and show the length in terms of w.

   (b) Write a quadratic equation in terms of w.

   (c) Find the length and width of the building floor.

122. *Geometry*   An above-ground swimming pool with a square base is to be constructed such that the surface area of the pool is 576 square feet. The height of the pool is to be 4 feet. What should the dimensions of the base be? (*Hint:* The surface area is $S = x^2 + 4xh$.)

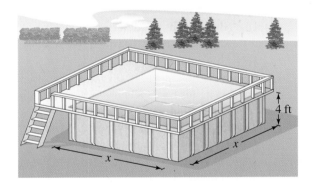

123. *Packaging*   An open gift box is to be made from a square piece of material by cutting two-centimeter squares from each corner and turning up the sides (see figure). The volume of the finished gift box is to be 200 cubic centimeters. Find the size of the original piece of material.

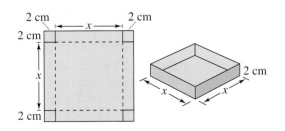

124. *Exploration*   A rancher has 100 meters of fencing to enclose two adjacent rectangular corrals as shown in the figure.

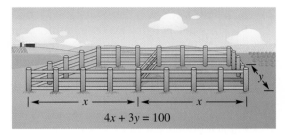

$$4x + 3y = 100$$

   (a) Write the area of the enclosed region as a function of x.

   (b) Use a graphing utility to generate additional rows of the table. Use the table to estimate the dimensions that will produce a maximum area.

x	y	Area
2	$\frac{92}{3}$	$\frac{368}{3} \approx 123$
4	28	224

   (c) Use a graphing utility to graph the area function, and use the graph to estimate the dimensions that will produce a maximum area.

   (d) Use the graph to approximate the dimensions such that the enclosed area is 350 square meters.

   (e) Find the required dimensions of part (d) algebraically.

In Exercises 125–127, use the position equation given on page 202 as the model for the problem.

125. *CN Tower*   At 1815 feet tall, the CN Tower in Toronto, Ontario is the world's tallest self-supporting structure. An object is dropped from the top of the tower.

   (a) Find the position equation

$$s = -16t^2 + v_0 t + s_0.$$

   (b) Complete the table.

t	0	2	4	6	8	10	12
s							

   (c) From the table in part (b), determine the time interval during which the object reaches the ground. Find the time algebraically.

126. *Military* A cargo plane flying at 8000 feet over level terrain drops a 500-pound supply package.

(a) How long will it take the package to strike the ground?

(b) The plane is flying at 600 miles per hour. How far will the package travel horizontally during its descent?

127. *Sports* You throw a baseball straight up into the air at a velocity of 45 feet per second. You release the baseball at a height of 5.5 feet and catch it when it falls back to a height of 6 feet.

(a) Use the position equation to write a mathematical model for the height of the baseball.

(b) Find the height of the baseball after 0.5 second.

(c) How many seconds is the baseball in the air?

(d) Use a graphing utility to verify your answer in part (c).

128. *Transportation* The total number y of electric-powered vehicles in the United States from 1992 through 2001 can be approximated by the model

$$y = 75.76t^2 + 912, \quad 2 \leq t \leq 11$$

where t represents the year, with $t = 2$ corresponding to 1992. (Source: Energy Information Administration)

(a) Determine algebraically when the number of electric-powered vehicles reached 7000.

(b) Verify your answer to part (a) by creating a table of values for the model.

(c) Use a graphing utility to graph the model.

(d) Use the *zoom* and *trace* features of a graphing utility to find the year in which the total number of electric-powered vehicles reached 9000.

(e) Verify your answer to part (d) algebraically.

129. *Agriculture* The total number S (in millions) of sheep and lambs on farms in the United States from 1995 through 2002 can be approximated by the model $S = 0.032t^2 - 0.87t + 12.6$, $5 \leq t \leq 12$, where t represents the year, with $t = 5$ corresponding to 1995. (Source: U.S. Department of Agriculture)

(a) Use a graphing utility to graph the model.

(b) Extend the model past 2002. Does the model predict that the number of sheep and lambs will eventually increase? If so, estimate when the number of sheep and lambs will once again reach 8 million.

130. *Biology* The metabolic rate of an ectothermic organism increases with increasing temperature within a certain range. Experimental data for oxygen consumption C (in microliters per gram per hour) of a beetle at certain temperatures yielded the model $C = 0.45x^2 - 1.65x + 50.75$, $10 \leq x \leq 25$, where x is the air temperature in degrees Celsius.

(a) Use a graphing utility to graph the consumption model over the specified domain.

(b) Use the graph to approximate the air temperature resulting in oxygen consumption of 150 microliters per gram per hour.

(c) The temperature is increased from 10°C to 20°C. The oxygen consumption is increased by approximately what factor?

131. *Fuel Efficiency* The distance d (in miles) a car can travel on one tank of fuel is approximated by $d = -0.024s^2 + 1.455s + 431.5$, $0 < s \leq 75$, where s is the average speed of the car in miles per hour.

(a) Use a graphing utility to graph the function over the specified domain.

(b) Use the graph to determine the greatest distance that can be traveled on a tank of fuel. How long will the trip take?

(c) Determine the greatest distance that can be traveled in this car in 8 hours with no refueling. How fast should the car be driven? [*Hint:* The distance traveled in 8 hours is $8s$. Graph this expression in the same viewing window as the graph in part (a) and approximate the point of intersection.]

132. *Saturated Steam* The temperature T (in degrees Fahrenheit) of saturated steam increases as pressure increases. This relationship is approximated by the model

$$T = 75.82 - 2.11x + 43.51\sqrt{x}, \quad 5 \leq x \leq 40$$

where x is the absolute pressure in pounds per square inch.

(a) Use a graphing utility to graph the function over the specified domain.

(b) The temperature of steam at sea level ($x = 14.696$) is 212°F. Evaluate the model at this pressure and verify the result graphically.

(c) Use the model to approximate the pressure for a steam temperature of 240°F.

133. *Meteorology*   A meteorologist is positioned 100 feet from the point at which a weather balloon is launched. When the balloon is at height h, the distance d (in feet) between the meteorologist and the balloon is $d = \sqrt{100^2 + h^2}$.

(a) Use a graphing utility to graph the equation. Use the *trace* feature to approximate the value of h when $d = 200$.

(b) Complete the table. Use the table to approximate the value of h when $d = 200$.

h	160	165	170	175	180	185
d						

(c) Find h algebraically when $d = 200$.

(d) Compare the results of each method. In each case, what information did you gain that wasn't revealed by another solution method?

134. *Geometry*   An equilateral triangle has a height of 10 inches. How long is each of its sides? (*Hint:* Use the height of the triangle to partition the triangle into two congruent right triangles.)

135. *Flying Speed*   Two planes leave simultaneously from Chicago's O'Hare Airport, one flying due north and the other due east. The northbound plane is flying 50 miles per hour faster than the eastbound plane. After 3 hours the planes are 2440 miles apart. Find the speed of each plane. (*Hint:* draw a diagram.)

136. *Flying Distance*   A chartered airplane flies to three cities whose locations form the vertices of a right triangle (see figure). The total flight distance (from Indianapolis to Peoria to Springfield and back to Indianapolis) is approximately 448 miles. It is 195 miles between Indianapolis and Peoria. Approximate the other two distances.

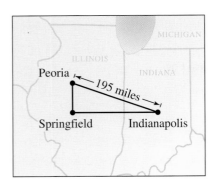

Synthesis

True or False?   **In Exercises 137–139, determine whether the statement is true or false. Justify your answer.**

137. The quadratic equation $-3x^2 - x = 10$ has two real solutions.

138. If $(2x - 3)(x + 5) = 8$, then $2x - 3 = 8$ or $x + 5 = 8$.

139. An equation can never have more than one extraneous solution.

140. *Exploration*   Solve $3(x + 4)^2 + (x + 4) - 2 = 0$ in two ways.

(a) Let $u = x + 4$, and solve the resulting equation for u. Then find the corresponding values of x that are the solutions of the original equation.

(b) Expand and collect like terms in the original equation, and solve the resulting equation for x.

(c) Which method is easier? Explain.

141. *Exploration*   Given that a and b are nonzero real numbers, determine the solutions of the equations.

(a) $ax^2 + bx = 0$ (b) $ax^2 - ax = 0$

142. *Writing*   On a graphing utility, store the value 5 in A, -2 in B, and 1 in C. Use the graphing utility to graph $y = C(x - A)(x - B)$. Explain how the values of A and B can be determined from the graph. Now store any other nonzero value in C. Does the value of C affect the x-intercepts of the graph? Explain. Find values of A, B, and C such that the graph opens downward and has x-intercepts at $(-5, 0)$ and $(0, 0)$. Summarize your findings.

Review

In Exercises 143–146, completely factor the expression over the real numbers.

143. $x^5 - 27x^2$ **144.** $x^3 - 5x^2 - 14x$

145. $x^3 + 5x^2 - 2x - 10$ **146.** $5(x + 5)x^{1/3} + 4x^{4/3}$

In Exercises 147–152, determine whether y is a function of x.

147. $5x + 8y = -1$ **148.** $-x^2 + y^2 = 2$

149. $x + y^2 = 10$ **150.** $-2y = \sqrt{x + 6}$

151. $y = |x - 3|$ **152.** $|y| = 1 - x$

2.5 Solving Inequalities Algebraically and Graphically

Properties of Inequalities

Simple inequalities were reviewed in Section P.1. There, the inequality symbols $<$, $\leq$, $>$, and $\geq$ were used to compare two numbers and to denote subsets of real numbers. For instance, the simple inequality $x \geq 3$ denotes all real numbers x that are greater than or equal to 3.

In this section you will study inequalities that contain more involved statements such as

$$5x - 7 > 3x + 9 \qquad \text{and} \qquad -3 \leq 6x - 1 < 3.$$

As with an equation, you **solve an inequality** in the variable x by finding all values of x for which the inequality is true. These values are **solutions** of the inequality and are said to **satisfy** the inequality. For instance, the number 9 is a solution of the first inequality listed above because

$$5(9) - 7 > 3(9) + 9$$

$$38 > 36.$$

On the other hand, the number 7 is not a solution because

$$5(7) - 7 \not> 3(7) + 9$$

$$28 \not> 30.$$

The set of all real numbers that are solutions of an inequality is the **solution set** of the inequality.

The set of all points on the real number line that represent the solution set is the **graph of the inequality**. Graphs of many types of inequalities consist of intervals on the real number line.

The procedures for solving linear inequalities in one variable are much like those for solving linear equations. To isolate the variable, you can make use of the **properties of inequalities**. These properties are similar to the properties of equality, but there are two important exceptions. When each side of an inequality is multiplied or divided by a negative number, *the direction of the inequality symbol must be reversed* in order to maintain a true statement. Here is an example.

$$-2 < 5 \qquad \text{Original inequality}$$

$$(-3)(-2) > (-3)(5) \qquad \text{Multiply each side by } -3 \text{ and reverse inequality.}$$

$$6 > -15 \qquad \text{Simplify.}$$

Two inequalities that have the same solution set are **equivalent inequalities**. For instance, the inequalities

$$x + 2 < 5 \qquad \text{and} \qquad x < 3$$

are equivalent. To obtain the second inequality from the first, you can subtract 2 from each side of the inequality. The properties listed at the top of the next page describe operations that can be used to create equivalent inequalities.

What you should learn

- Use properties of inequalities to solve linear inequalities.
- Solve inequalities involving absolute values.
- Solve polynomial inequalities.
- Solve rational inequalities.
- Use inequalities to model and solve real-life problems.

Why you should learn it

An inequality can be used to determine when a real-life quantity exceeds a given level. For instance, Exercise 66 on page 221 shows how to use a quadratic inequality to determine when the total number of bachelor's degrees conferred in the United States will exceed 1.4 million.

Cliff Hollis/Liaison/Getty Images

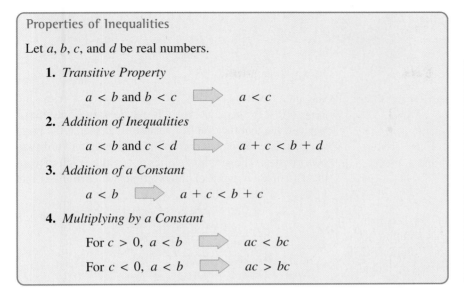

Properties of Inequalities

Let a, b, c, and d be real numbers.

1. *Transitive Property*

$$a < b \text{ and } b < c \implies a < c$$

2. *Addition of Inequalities*

$$a < b \text{ and } c < d \implies a + c < b + d$$

3. *Addition of a Constant*

$$a < b \implies a + c < b + c$$

4. *Multiplying by a Constant*

$$\text{For } c > 0, \ a < b \implies ac < bc$$

$$\text{For } c < 0, \ a < b \implies ac > bc$$

Exploration

Use a graphing utility to graph $f(x) = 5x - 7$ and $g(x) = 3x + 9$ in the same viewing window. (Use $-1 \le x \le 15$ and $-5 \le y \le 50$.) For which values of x does the graph of f lie above the graph of g? Explain how the answer to this question can be used to solve the inequality in Example 1.

Each of the properties above is true if the symbol $<$ is replaced by $\le$ and $>$ is replaced by $\ge$. For instance, another form of Property 3 is as follows.

$$a \le b \implies a + c \le b + c$$

Solving a Linear Inequality

The simplest type of inequality to solve is a **linear inequality** in one variable, such as $2x + 3 > 4$. (See Appendix D for help with solving one-step linear inequalities.)

Example 1   Solving a Linear Inequality

Solve $5x - 7 > 3x + 9$.

Solution

$5x - 7 > 3x + 9$	Write original inequality.
$2x - 7 > 9$	Subtract $3x$ from each side.
$2x > 16$	Add 7 to each side.
$x > 8$	Divide each side by 2.

So, the solution set is all real numbers that are greater than 8. The interval notation for this solution set is $(8, \infty)$. The number line graph of this solution set is shown in Figure 2.45. Note that a parenthesis at 8 on the number line indicates that 8 *is not* part of the solution set.

 Checkpoint   Now try Exercise 11.

Checking the solution set of an inequality is not as simple as checking the solution(s) of an equation because there are simply too many x-values to substitute into the original inequality. However, you can get an indication of the validity of the solution set by substituting a few convenient values of x. For instance, in Example 1, try substituting $x = 5$ and $x = 10$ into the original inequality.

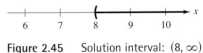

Figure 2.45   Solution interval: $(8, \infty)$

Note that the four inequalities forming the solution steps of Example 1 are all *equivalent* in the sense that each has the same solution set.

Example 2 Solving an Inequality

Solve $1 - \frac{3}{2}x \geq x - 4$.

Algebraic Solution

$$1 - \frac{3}{2}x \geq x - 4 \qquad \text{Write original inequality.}$$

$$2 - 3x \geq 2x - 8 \qquad \text{Multiply each side by the LCD.}$$

$$2 - 5x \geq -8 \qquad \text{Subtract } 2x \text{ from each side.}$$

$$-5x \geq -10 \qquad \text{Subtract 2 from each side.}$$

$$x \leq 2 \qquad \text{Divide each side by } -5 \text{ and reverse inequality.}$$

The solution set is all real numbers that are less than or equal to 2. The interval notation for this solution set is $(-\infty, 2]$. The number line graph of this solution set is shown in Figure 2.46. Note that a bracket at 2 on the number line indicates that 2 *is* part of the solution set.

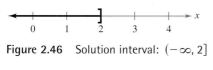

Figure 2.46 Solution interval: $(-\infty, 2]$

✓ *Checkpoint* Now try Exercise 13.

Graphical Solution

Use a graphing utility to graph $y_1 = 1 - \frac{3}{2}x$ and $y_2 = x - 4$ in the same viewing window. In Figure 2.47, you can see that the graphs appear to intersect at the point $(2, -2)$. Use the *intersect* feature of the graphing utility to confirm this. The graph of y_1 lies above the graph of y_2 to the left of their point of intersection, which implies that $y_1 \geq y_2$ for all $x \leq 2$.

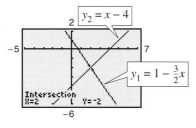

Figure 2.47

Sometimes it is possible to write two inequalities as a **double inequality,** as demonstrated in Example 3.

Example 3 Solving a Double Inequality

Solve $-3 \leq 6x - 1$ and $6x - 1 < 3$.

Algebraic Solution

$$-3 \leq 6x - 1 < 3 \qquad \text{Write as a double inequality.}$$

$$-2 \leq 6x < 4 \qquad \text{Add 1 to each part.}$$

$$-\frac{1}{3} \leq x < \frac{2}{3} \qquad \text{Divide by 6 and simplify.}$$

The solution set is all real numbers that are greater than or equal to $-\frac{1}{3}$ *and* less than $\frac{2}{3}$. The interval notation for this solution set is $\left[-\frac{1}{3}, \frac{2}{3}\right)$. The number line graph of this solution set is shown in Figure 2.48.

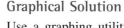

Figure 2.48 Solution interval $\left[-\frac{1}{3}, \frac{2}{3}\right)$

✓ *Checkpoint* Now try Exercise 15.

Graphical Solution

Use a graphing utility to graph $y_1 = 6x - 1$, $y_2 = -3$, and $y_3 = 3$ in the same viewing window. In Figure 2.49, you can see that the graphs appear to intersect at the points $\left(-\frac{1}{3}, -3\right)$ and $\left(\frac{2}{3}, 3\right)$. Use the *intersect* feature of the graphing utility to confirm this. The graph of y_1 lies above the graph of y_2 to the right of $\left(-\frac{1}{3}, -3\right)$ *and* the graph of y_1 lies below the graph of y_3 to the left of $\left(\frac{2}{3}, 3\right)$. This implies that $y_2 \leq y_1 < y_3$ when $-\frac{1}{3} \leq x < \frac{2}{3}$.

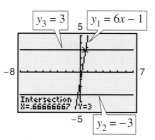

Figure 2.49

Inequalities Involving Absolute Value

Solving an Absolute Value Inequality

Let x be a variable or an algebraic expression and let a be a real number such that $a \geq 0$.

1. The solutions of $|x| < a$ are all values of x that lie between $-a$ and a.

$$|x| < a \quad \text{if and only if} \quad -a < x < a. \qquad \text{Double inequality}$$

2. The solutions of $|x| > a$ are all values of x that are less than $-a$ or greater than a.

$$|x| > a \quad \text{if and only if} \quad x < -a \quad \text{or} \quad x > a. \qquad \text{Compound inequality}$$

These rules are also valid if $<$ is replaced by $\leq$ and $>$ is replaced by $\geq$.

Example 4   Solving Absolute Value Inequalities

Solve each inequality.

a. $|x - 5| < 2$ **b.** $|x - 5| > 2$

Algebraic Solution

a. $|x - 5| < 2$ Write original inequality.

     $-2 < x - 5 < 2$ Write double inequality.

       $3 < x < 7$ Add 5 to each part.

The solution set is all real numbers that are greater than 3 *and* less than 7. The interval notation for this solution set is $(3, 7)$. The number line graph of this solution set is shown in Figure 2.50.

b. The absolute value inequality $|x - 5| > 2$ is equivalent to the following compound inequality: $x - 5 < -2$ *or* $x - 5 > 2$.

Solve first inequality: $x - 5 < -2$ Write first inequality.

               $x < 3$ Add 5 to each side.

Solve second inequality: $x - 5 > 2$ Write second inequality.

               $x > 7$ Add 5 to each side.

The solution set is all real numbers that are less than 3 *or* greater than 7. The interval notation for this solution set is $(-\infty, 3) \cup (7, \infty)$. The symbol $\cup$ is called a *union* symbol and is used to denote the combining of two sets. The number line graph of this solution set is shown in Figure 2.51.

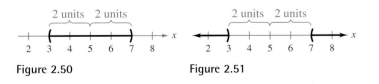

Figure 2.50   Figure 2.51

✓ *Checkpoint* Now try Exercise 29.

Graphical Solution

a. Use a graphing utility to graph $y_1 = |x - 5|$ and $y_2 = 2$ in the same viewing window. In Figure 2.52, you can see that the graphs appear to intersect at the points $(3, 2)$ and $(7, 2)$. Use the *intersect* feature of the graphing utility to confirm this. The graph of y_1 lies below the graph of y_2 when $3 < x < 7$. So, you can approximate the solution set to be all real numbers greater than 3 *and* less than 7.

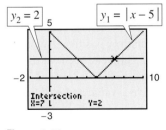

Figure 2.52

b. In Figure 2.52, you can see that the graph of y_1 lies above the graph of y_2 when $x < 3$ *or* when $x > 7$. So, you can approximate the solution set to be all real numbers that are less than 3 *or* greater than 7.

Polynomial Inequalities

To solve a polynomial inequality such as $x^2 - 2x - 3 < 0$, use the fact that a polynomial can change signs only at its zeros (the x-values that make the polynomial equal to zero). Between two consecutive zeros, a polynomial must be entirely positive or entirely negative. This means that when the real zeros of a polynomial are put in order, they divide the real number line into intervals in which the polynomial has no sign changes. These zeros are the **critical numbers** of the inequality, and the resulting open intervals are the **test intervals** for the inequality. For instance, the polynomial above factors as

$$x^2 - 2x - 3 = (x + 1)(x - 3)$$

and has two zeros, $x = -1$ and $x = 3$, which divide the real number line into three test intervals: $(-\infty, -1)$, $(-1, 3)$, and $(3, \infty)$. To solve the inequality $x^2 - 2x - 3 < 0$, you need to test only one value from each test interval.

TECHNOLOGY TIP

Some graphing utilities will produce graphs of inequalities. For instance, you can graph $2x^2 + 5x > 12$ by setting the graphing utility to *dot* mode and entering $y = 2x^2 + 5x > 12$. Using the settings $-10 \leq x \leq 10$ and $-4 \leq y \leq 4$, your graph should look like the graph shown below. Solve the problem algebraically to verify that the solution is $(-\infty, -4) \cup \left(\frac{3}{2}, \infty\right)$.

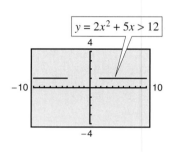

Finding Test Intervals for a Polynomial

To determine the intervals on which the values of a polynomial are entirely negative or entirely positive, use the following steps.

1. Find all real zeros of the polynomial, and arrange the zeros in increasing order. The zeros of a polynomial are its critical numbers.

2. Use the critical numbers to determine the test intervals.

3. Choose one representative x-value in each test interval and evaluate the polynomial at that value. If the value of the polynomial is negative, the polynomial will have negative values for *every* x-value in the interval. If the value of the polynomial is positive, the polynomial will have positive values for every x-value in the interval.

Example 5   Investigating Polynomial Behavior

To determine the intervals on which $x^2 - x - 6$ is entirely negative and those on which it is entirely positive, factor the quadratic as $x^2 - x - 6 = (x + 2)(x - 3)$. The critical numbers occur at $x = -2$ and $x = 3$. So, the test intervals for the quadratic are $(-\infty, -2)$, $(-2, 3)$, and $(3, \infty)$. In each test interval, choose a representative x-value and evaluate the polynomial, as shown in the table.

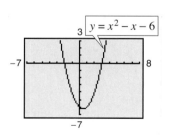

Interval	x-Value	Value of Polynomial	Sign of Polynomial
$(-\infty, -2)$	$x = -3$	$(-3)^2 - (-3) - 6 = 6$	Positive
$(-2, 3)$	$x = 0$	$(0)^2 - (0) - 6 = -6$	Negative
$(3, \infty)$	$x = 5$	$(5)^2 - (5) - 6 = 14$	Positive

Figure 2.53

The polynomial has negative values for every x in the interval $(-2, 3)$ and positive values for every x in the intervals $(-\infty, -2)$ and $(3, \infty)$. This result is shown graphically in Figure 2.53.

 Checkpoint Now try Exercise 43.

To determine the test intervals for a polynomial inequality, the inequality must first be written in general form with the polynomial on one side.

Example 6 Solving a Polynomial Inequality

Solve $2x^2 + 5x > 12$.

Algebraic Solution

$$2x^2 + 5x - 12 > 0 \qquad \text{Write inequality in general form.}$$

$$(x + 4)(2x - 3) > 0 \qquad \text{Factor.}$$

Critical Numbers: $x = -4, x = \frac{3}{2}$

Test Intervals: $(-\infty, -4), \left(-4, \frac{3}{2}\right), \left(\frac{3}{2}, \infty\right)$

Test: Is $(x + 4)(2x - 3) > 0$?

After testing these intervals, you can see that the polynomial $2x^2 + 5x - 12$ is positive on the open intervals $(-\infty, -4)$ and $\left(\frac{3}{2}, \infty\right)$. Therefore, the solution set of the inequality is

$$(-\infty, -4) \cup \left(\frac{3}{2}, \infty\right).$$

✓ *Checkpoint* Now try Exercise 47.

Graphical Solution

First write the polynomial inequality $2x^2 + 5x > 12$ as $2x^2 + 5x - 12 > 0$. Then use a graphing utility to graph $y = 2x^2 + 5x - 12$. In Figure 2.54, you can see that the graph is *above* the x-axis when x is less than -4 *or* when x is greater than $\frac{3}{2}$. So, you can graphically approximate the solution set to be $(-\infty, -4) \cup \left(\frac{3}{2}, \infty\right)$.

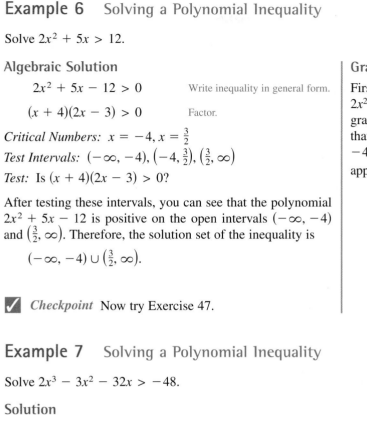

Figure 2.54

Example 7 Solving a Polynomial Inequality

Solve $2x^3 - 3x^2 - 32x > -48$.

Solution

$$2x^3 - 3x^2 - 32x + 48 > 0 \qquad \text{Write inequality in general form.}$$

$$x^2(2x - 3) - 16(2x - 3) > 0 \qquad \text{Factor by grouping.}$$

$$(x^2 - 16)(2x - 3) > 0 \qquad \text{Distributive Property}$$

$$(x - 4)(x + 4)(2x - 3) > 0 \qquad \text{Factor difference of two squares.}$$

The critical numbers are $x = -4$, $x = \frac{3}{2}$, and $x = 4$; and the test intervals are $(-\infty, -4), \left(-4, \frac{3}{2}\right), \left(\frac{3}{2}, 4\right)$, and $(4, \infty)$.

Interval	x-Value	Polynomial Value	Conclusion
$(-\infty, -4)$	$x = -5$	$2(-5)^3 - 3(-5)^2 - 32(-5) + 48 = -117$	Negative
$\left(-4, \frac{3}{2}\right)$	$x = 0$	$2(0)^3 - 3(0)^2 - 32(0) + 48 = 48$	Positive
$\left(\frac{3}{2}, 4\right)$	$x = 2$	$2(2)^3 - 3(2)^2 - 32(2) + 48 = -12$	Negative
$(4, \infty)$	$x = 5$	$2(5)^3 - 3(5)^2 - 32(5) + 48 = 63$	Positive

From this you can conclude that the polynomial is positive on the open intervals $\left(-4, \frac{3}{2}\right)$ and $(4, \infty)$. So, the solution set is $\left(-4, \frac{3}{2}\right) \cup (4, \infty)$.

✓ *Checkpoint* Now try Exercise 49.

STUDY TIP

When solving a quadratic inequality, be sure you have accounted for the particular type of inequality symbol given in the inequality. For instance, in Example 7, note that the original inequality contained a "greater than" symbol and the solution consisted of two open intervals. If the original inequality had been

$$2x^3 - 3x^2 - 32x \geq -48,$$

the solution would have consisted of the closed interval $\left[-4, \frac{3}{2}\right]$ and the interval $[4, \infty)$.

Example 8 Unusual Solution Sets

a. The solution set of

$$x^2 + 2x + 4 > 0$$

consists of the entire set of real numbers, $(-\infty, \infty)$. In other words, the value of the quadratic $x^2 + 2x + 4$ is positive for every real value of x, as indicated in Figure 2.55(a). (Note that this quadratic inequality has *no* critical numbers. In such a case, there is only one test interval—the entire real number line.)

b. The solution set of

$$x^2 + 2x + 1 \le 0$$

consists of the single real number $\{-1\}$, because the quadratic $x^2 + 2x + 1$ has one critical number, $x = -1$, and it is the only value that satisfies the inequality, as indicated in Figure 2.55(b).

c. The solution set of

$$x^2 + 3x + 5 < 0$$

is empty. In other words, the quadratic $x^2 + 3x + 5$ is not less than zero for any value of x, as indicated in Figure 2.55(c).

d. The solution set of

$$x^2 - 4x + 4 > 0$$

consists of all real numbers *except* the number 2. In interval notation, this solution set can be written as $(-\infty, 2) \cup (2, \infty)$. The graph of $x^2 - 4x + 4$ lies above the x-axis except at $x = 2$, where it touches it, as indicated in Figure 2.55(d).

TECHNOLOGY TIP

One of the advantages of technology is that you can solve complicated polynomial inequalities that might be difficult, or even impossible, to factor. For instance, you could use a graphing utility to approximate the solution to the inequality

$$x^3 - 0.2x^2 - 3.16x + 1.4 < 0.$$

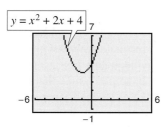

(a)

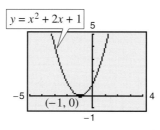

(b)

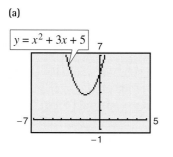

(c)

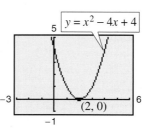

(d)

Figure 2.55

Rational Inequalities

The concepts of critical numbers and test intervals can be extended to inequalities involving rational expressions. To do this, use the fact that the value of a rational expression can change sign only at its *zeros* (the *x*-values for which its numerator is zero) and its *undefined values* (the *x*-values for which its denominator is zero). These two types of numbers make up the *critical numbers* of a rational inequality.

Example 9   Solving a Rational Inequality

Solve $\dfrac{2x - 7}{x - 5} \leq 3$.

Algebraic Solution

$$\frac{2x - 7}{x - 5} \leq 3 \qquad \text{Write original inequality.}$$

$$\frac{2x - 7}{x - 5} - 3 \leq 0 \qquad \text{Write in general form.}$$

$$\frac{2x - 7 - 3x + 15}{x - 5} \leq 0 \qquad \text{Write as single fraction.}$$

$$\frac{-x + 8}{x - 5} \leq 0 \qquad \text{Simplify.}$$

Now, in standard form you can see that the critical numbers are $x = 5$ and $x = 8$, and you can proceed as follows.

Critical Numbers: $x = 5, x = 8$
Test Intervals: $(-\infty, 5), (5, 8), (8, \infty)$
Test: Is $\dfrac{-x + 8}{x - 5} \leq 0$?

Interval	x-Value	Polynomial Value	Conclusion
$(-\infty, 5)$	$x = 0$	$\dfrac{-0 + 8}{0 - 5} = -\dfrac{8}{5}$	Negative
$(5, 8)$	$x = 6$	$\dfrac{-6 + 8}{6 - 5} = 2$	Positive
$(8, \infty)$	$x = 9$	$\dfrac{-9 + 8}{9 - 5} = -\dfrac{1}{4}$	Negative

By testing these intervals, you can determine that the rational expression $(-x + 8)/(x - 5)$ is negative in the open intervals $(-\infty, 5)$ and $(8, \infty)$. Moreover, because $(-x + 8)/(x - 5) = 0$ when $x = 8$, you can conclude that the solution set of the inequality is $(-\infty, 5) \cup [8, \infty)$.

✓ *Checkpoint* Now try Exercise 55.

Graphical Solution

Use a graphing utility to graph

$$y_1 = \frac{2x - 7}{x - 5} \quad \text{and} \quad y_2 = 3$$

in the same viewing window. In Figure 2.56, you can see that the graphs appear to intersect at the point $(8, 3)$. Use the *intersect* feature of the graphing utility to confirm this. The graph of y_1 lies below the graph of y_2 in the intervals $(-\infty, 5)$ and $[8, \infty)$. So, you can graphically approximate the solution set to be all real numbers less than 5 *or* greater than or equal to 8.

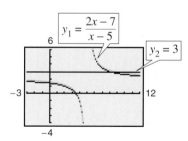

Figure 2.56

Note in Example 9 that $x = 5$ is not included in the solution set because the inequality is undefined when $x = 5$.

Application

In Section 1.3 you studied the *implied domain* of a function, the set of all x-values for which the function is defined. A common type of implied domain is used to avoid even roots of negative numbers, as shown in Example 10.

Example 10   Finding the Domain of an Expression

Find the domain of $\sqrt{64 - 4x^2}$.

Solution

Because $\sqrt{64 - 4x^2}$ is defined only if $64 - 4x^2$ is nonnegative, the domain is given by $64 - 4x^2 \geq 0$.

$$64 - 4x^2 \geq 0 \qquad \text{Write in general form.}$$

$$16 - x^2 \geq 0 \qquad \text{Divide each side by 4.}$$

$$(4 - x)(4 + x) \geq 0 \qquad \text{Factor.}$$

The inequality has two critical numbers: $x = -4$ and $x = 4$. A test shows that $64 - 4x^2 \geq 0$ in the *closed interval* $[-4, 4]$. The graph of $y = \sqrt{64 - 4x^2}$, shown in Figure 2.57, confirms that the domain is $[-4, 4]$.

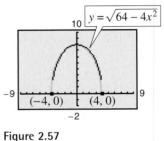

Figure 2.57

 Checkpoint Now try Exercise 63.

Example 11   Height of a Projectile

A projectile is fired straight upward from ground level with an initial velocity of 384 feet per second. During what time period will its height exceed 2000 feet?

Solution

In Section 2.4 you saw that the position of an object moving vertically can be modeled by the *position equation*

$$s = -16t^2 + v_0 t + s_0$$

where s is the height in feet and t is the time in seconds. In this case, $s_0 = 0$ and $v_0 = 384$. So, you need to solve the inequality $-16t^2 + 384t > 2000$. Using a graphing utility, graph $y_1 = -16t^2 + 384t$ and $y_2 = 2000$, as shown in Figure 2.58. From the graph, you can determine that $-16t^2 + 384t > 2000$ for t between approximately 7.6 and 16.4. You can verify this result algebraically.

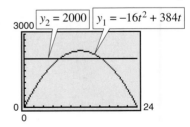

$$-16t^2 + 384t > 2000 \qquad \text{Write original inequality.}$$

$$t^2 - 24t < -125 \qquad \text{Divide by } -16 \text{ and reverse inequality.}$$

$$t^2 - 24t + 125 < 0 \qquad \text{Write in general form.}$$

Figure 2.58

By the Quadratic Formula the critical numbers are $t = 12 - \sqrt{19}$ and $t = 12 + \sqrt{19}$, or approximately 7.64 and 16.36. A test will verify that the height of the projectile will exceed 2000 feet when $7.64 < t < 16.36$; that is, during the time interval $(7.64, 16.36)$ seconds.

 Checkpoint Now try Exercise 65.

2.5 Exercises

Vocabulary Check

Fill in the blanks.

1. To solve a linear inequality in one variable, you can use the properties of inequalities, which are identical to those used to solve an equation, with the exception of multiplying or dividing each side by a _____ constant.

2. It is sometimes possible to write two inequalities as one inequality, called a _____ inequality.

3. The solutions to $|x| \leq a$ are those values of x such that _____ .

4. The solutions to $|x| \geq a$ are those values of x such that _____ or _____ .

5. The critical numbers of a rational expression are its _____ and its _____ .

In Exercises 1–4, match the inequality with its graph. [The graphs are labeled (a), (b), (c), and (d).]

(a)

(b)

(c)

(d)

1. $x < 3$

2. $x \geq 5$

3. $-3 < x \leq 4$

4. $0 \leq x \leq \frac{9}{2}$

In Exercises 5–8, determine whether each value of x is a solution of the inequality.

Inequality	Values

5. $5x - 12 > 0$ (a) $x = 3$ (b) $x = -3$

                        (c) $x = \frac{5}{2}$ (d) $x = \frac{3}{2}$

6. $-5 < 2x - 1 \leq 1$ (a) $x = -\frac{1}{2}$ (b) $x = -\frac{5}{2}$

                        (c) $x = \frac{4}{3}$ (d) $x = 0$

7. $-1 < \dfrac{3 - x}{2} \leq 1$ (a) $x = 0$ (b) $x = \sqrt{5}$

                        (c) $x = 1$ (d) $x = 5$

8. $|x - 10| \geq 3$ (a) $x = 13$ (b) $x = -1$

                        (c) $x = 14$ (d) $x = 9$

In Exercises 9–18, solve the inequality and sketch the solution on the real number line. Use a graphing utility to verify your solution graphically.

9. $-10x < 40$

10. $6x > 15$

11. $4(x + 1) < 2x + 3$

12. $2x + 7 < 3(x - 4)$

13. $\frac{3}{4}x - 6 \leq x - 7$

14. $3 + \frac{2}{7}x > x - 2$

15. $-8 \leq 1 - 3(x - 2) < 13$

16. $0 \leq 2 - 3(x + 1) < 20$

17. $-4 < \dfrac{2x - 3}{3} < 4$

18. $0 \leq \dfrac{x + 3}{2} < 5$

Graphical Analysis **In Exercises 19–22, use a graphing utility to approximate the solution.**

19. $5 - 2x \geq 1$

20. $20 < 6x - 1$

21. $3(x + 1) < x + 7$

22. $4(x - 3) \leq 8 - x$

In Exercises 23–26, use a graphing utility to graph the equation and graphically approximate the values of x that satisfy the specified inequalities. Then solve each inequality algebraically.

Equation	Inequalities	

23. $y = 2x - 3$ (a) $y \geq 1$ (b) $y \leq 0$

24. $y = -3x + 8$ (a) $-1 \leq y \leq 3$ (b) $y \leq 0$

25. $y = -\frac{1}{2}x + 2$ (a) $0 \leq y \leq 3$ (b) $y \geq 0$

26. $y = \frac{2}{3}x + 1$ (a) $y \leq 5$ (b) $y \geq 0$

In Exercises 27–34, solve the inequality and sketch the solution on the real number line.

27. $|5x| > 10$

28. $\left|\dfrac{x}{2}\right| \leq 1$

29. $|x - 7| < 6$

30. $|x - 20| \geq 4$

31. $|x + 14| + 3 > 17$

32. $\left|\dfrac{x - 3}{2}\right| \geq 5$

33. $10|1 - 2x| < 5$

34. $3|4 - 5x| \leq 9$

In Exercises 35 and 36, use a graphing utility to graph the equation and graphically approximate the values of x that satisfy the specified inequalities. Then solve each inequality algebraically.

Equation	*Inequalities*
35. $y = |x - 3|$	(a) $y \le 2$ (b) $y \ge 4$
36. $y = \left|\frac{1}{2}x + 1\right|$	(a) $y \le 4$ (b) $y \ge 1$

In Exercises 37–42, use absolute value notation to define the interval (or pair of intervals) on the real number line.

37.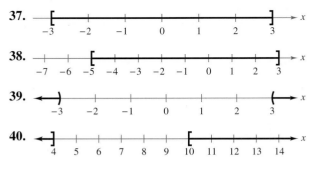

38.

39.

40.

41. All real numbers within 10 units of 7

42. All real numbers no more than 8 units from -5

In Exercises 43 and 44, determine the intervals on which the polynomial is entirely negative and those on which it is entirely positive.

43. $x^2 - 4x - 5$ **44.** $2x^2 - 4x - 3$

In Exercises 45–50, solve the inequality and graph the solution on the real number line. Use a graphing utility to verify your solution graphically.

45. $(x + 2)^2 < 25$ **46.** $(x - 3)^2 \ge 1$

47. $x^2 + 4x + 4 \ge 9$ **48.** $x^2 - 6x + 9 < 16$

49. $x^3 - 4x \ge 0$ **50.** $x^4(x - 3) \le 0$

In Exercises 51 and 52, use a graphing utility to graph the equation and graphically approximate the values of x that satisfy the specified inequalities. Then solve each inequality algebraically.

Equation	*Inequalities*
51. $y = -x^2 + 2x + 3$	(a) $y \le 0$ (b) $y \ge 3$
52. $y = x^3 - x^2 - 16x + 16$	(a) $y \le 0$ (b) $y \ge 36$

In Exercises 53–56, solve the inequality and graph the solution on the real number line. Use a graphing utility to verify your solution graphically.

53. $\dfrac{1}{x} - x > 0$ **54.** $\dfrac{1}{x} - 4 < 0$

55. $\dfrac{x + 6}{x + 1} - 2 < 0$ **56.** $\dfrac{x + 12}{x + 2} - 3 \ge 0$

In Exercises 57 and 58, use a graphing utility to graph the equation and graphically approximate the values of x that satisfy the specified inequalities. Then solve each inequality algebraically.

Equation	*Inequalities*
57. $y = \dfrac{3x}{x - 2}$	(a) $y \le 0$ (b) $y \ge 6$
58. $y = \dfrac{5x}{x^2 + 4}$	(a) $y \ge 1$ (b) $y \le 0$

In Exercises 59–64, find the domain of x in the expression.

59. $\sqrt{x - 5}$ **60.** $\sqrt[4]{6x + 15}$

61. $\sqrt[3]{6 - x}$ **62.** $\sqrt[3]{2x^2 - 8}$

63. $\sqrt{x^2 - 4}$ **64.** $\sqrt[4]{4 - x^2}$

65. *Data Analysis*   You want to determine whether there is a relationship between an athlete's weight x (in pounds) and the athlete's maximum bench-press weight y (in pounds). The table shows a sample of data from 12 athletes.

Athlete's weight, x	Bench-press weight, y
165	170
184	185
150	200
210	255
196	205
240	295
202	190
170	175
185	195
190	185
230	250
160	150

(a) Use a graphing utility to plot the data.

(b) A model for this data is $y = 1.3x - 36$. Use a graphing utility to graph the equation in the same viewing window used in part (a).

(c) Use the graph to estimate the values of x that predict a maximum bench-press weight of at least 200 pounds.

(d) Use the graph to write a statement about the accuracy of the model. If you think the graph indicates that an athlete's weight is not a good indicator of the athlete's maximum bench-press weight, list other factors that might influence an individual's maximum bench-press weight.

66. **Education** The number D (in thousands) of earned bachelor's degrees conferred annually in the United States for selected years from 1975 to 2000 is approximated by the model $D = 0.42t^2 - 1.3t + 911$, where t represents the year, with $t = 5$ corresponding to 1975. (Source: U.S. National Center for Education Statistics)

(a) Use a graphing utility to graph the model.

(b) According to this model, estimate when the number of degrees will exceed 1,400,000.

Music In Exercises 67–70, use the following information. Michael Kasha of Florida State University used physics and mathematics to design a new classical guitar. He used the model for the frequency of the vibrations on a circular plate

$$v = \frac{2.6t}{d^2}\sqrt{\frac{E}{\rho}}$$

where v is the frequency (in vibrations per second), t is the plate thickness (in millimeters), d is the diameter of the plate, E is the elasticity of the plate material, and ρ is the density of the plate material. For fixed values of d, E, and ρ, the graph of the equation is a line, as shown in the figure.

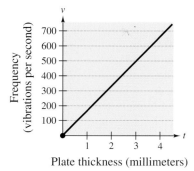

Plate thickness (millimeters)

67. Estimate the frequency when the plate thickness is 2 millimeters.

68. Estimate the plate thickness when the frequency is 600 vibrations per second.

69. Approximate the interval for the plate thickness when the frequency is between 200 and 400 vibrations per second.

70. Approximate the interval for the frequency when the plate thickness is less than 3 millimeters.

Synthesis

True or False? In Exercises 71 and 72, determine whether the statement is true or false. Justify your answer.

71. If $-10 \le x \le 8$, then $-10 \ge -x$ and $-x \ge -8$.

72. The solution set of the inequality $\frac{3}{2}x^2 + 3x + 6 \ge 0$ is the entire set of real numbers.

In Exercises 73 and 74, consider the polynomial $(x - a)(x - b)$ and the real number line (see figure).

73. Identify the points on the line where the polynomial is zero.

74. In each of the three subintervals of the line, write the sign of each factor and the sign of the product. For which x-values does the polynomial possibly change signs?

Review

In Exercises 75–78, find the distance between each pair of points. Then find the midpoint of the line segment joining the points.

75. $(-4, 2), (1, 12)$

76. $(1, -2), (10, 3)$

77. $(3, 6), (-5, -8)$

78. $(0, -3), (-6, 9)$

In Exercises 79–82, sketch a graph of the function.

79. $f(x) = -x^2 + 6$

80. $f(x) = \frac{1}{3}(x - 5)^2$

81. $f(x) = -|x + 5| - 6$

82. $f(x) = \frac{1}{2}|-x| - 4$

In Exercises 83–86, find the inverse function.

83. $y = 12x$

84. $y = 5x + 8$

85. $y = x^3 + 7$

86. $y = \sqrt[3]{x - 7}$

Scatter Plots and Correlation

Many real-life situations involve finding relationships between two variables, such as the year and the number of people in the labor force. In a typical situation, data is collected and written as a set of ordered pairs. The graph of such a set, called a *scatter plot*, was discussed briefly in Section P.5.

Example 1 Constructing a Scatter Plot

The data in the table shows the number *P* (in millions) of people in the United States who were part of the labor force from 1995 through 2001. Construct a scatter plot of the data. (Source: U.S. Bureau of Labor Statistics)

Year	People, *P*
1995	132
1996	134
1997	136
1998	138
1999	139
2000	141
2001	142

Solution

Begin by representing the data with a set of ordered pairs. Let *t* represent the year, with $t = 5$ corresponding to 1995.

$$(5, 132), (6, 134), (7, 136), (8, 138), (9, 139), (10, 141), (11, 142)$$

Then plot each point in a coordinate plane, as shown in Figure 2.59.

✓ *Checkpoint* Now try Exercise 1.

From the scatter plot in Figure 2.59, it appears that the points describe a relationship that is nearly linear. The relationship is not *exactly* linear because the labor force did not increase by precisely the same amount each year.

A mathematical equation that approximates the relationship between *t* and *P* is a *mathematical model*. When developing a mathematical model to describe a set of data, you strive for two (often conflicting) goals—accuracy and simplicity. For the data above, a linear model of the form $P = at + b$ appears to be best. It is simple and relatively accurate.

What you should learn

- Construct scatter plots and interpret correlation.
- Use scatter plots and a graphing utility to find linear models for data.

Why you should learn it

Many real-life data follow a linear pattern. For instance, in Exercise 17 on page 229, you will find a linear model for the winning times in the women's 400-meter freestyle swimming Olympic event.

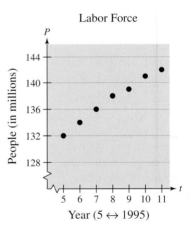

Nick Wilson/Getty Images

Figure 2.59

Consider a collection of ordered pairs of the form (x, y). If y tends to increase as x increases, the collection is said to have a **positive correlation.** If y tends to decrease as x increases, the collection is said to have a **negative correlation.** Figure 2.60 shows three examples: one with a positive correlation, one with a negative correlation, and one with no (discernible) correlation.

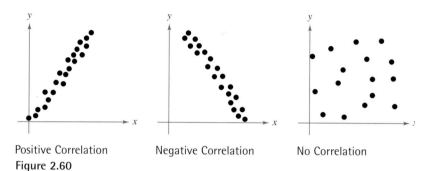

Positive Correlation   Negative Correlation   No Correlation

Figure 2.60

Example 2   Interpreting Correlation

On a Friday, 22 students in a class were asked to record the number of hours they spent studying for a test on Monday and the number of hours they spent watching television. The results are shown below. (The first coordinate is the number of hours and the second coordinate is the score obtained on the test.)

Study Hours: (0, 40), (1, 41), (2, 51), (3, 58), (3, 49), (4, 48), (4, 64), (5, 55), (5, 69), (5, 58), (5, 75), (6, 68), (6, 63), (6, 93), (7, 84), (7, 67), (8, 90), (8, 76), (9, 95), (9, 72), (9, 85), (10, 98)

TV Hours: (0, 98), (1, 85), (2, 72), (2, 90), (3, 67), (3, 93), (3, 95), (4, 68), (4, 84), (5, 76), (7, 75), (7, 58), (9, 63), (9, 69), (11, 55), (12, 58), (14, 64), (16, 48), (17, 51), (18, 41), (19, 49), (20, 40)

a. Construct a scatter plot for each set of data.

b. Determine whether the points are positively correlated, are negatively correlated, or have no discernable correlation. What can you conclude?

Solution

a. Scatter plots for the two sets of data are shown in Figure 2.61.

b. The scatter plot relating study hours and test scores has a positive correlation. This means that the more a student studied, the higher his or her score tended to be. The scatter plot relating television hours and test scores has a negative correlation. This means that the more time a student spent watching television, the lower his or her score tended to be.

 Checkpoint Now try Exercise 3.

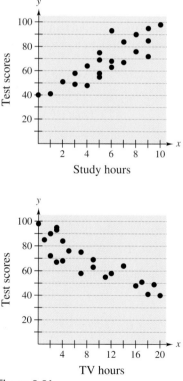

Figure 2.61

Fitting a Line to Data

Finding a linear model to represent the relationship described by a scatter plot is called **fitting a line to data.** You can do this graphically by simply sketching the line that appears to fit the points, finding two points on the line, and then finding the equation of the line that passes through the two points.

Example 3   Fitting a Line to Data

Find a linear model that relates the year to the number of people in the United States labor force. (See Example 1.)

Year	People, P
1995	132
1996	134
1997	136
1998	138
1999	139
2000	141
2001	142

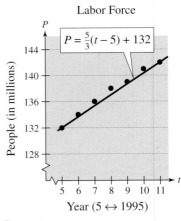

Figure 2.62

Solution

Let t represent the year, with $t = 5$ corresponding to 1995. After plotting the data in the table, draw the line that you think best represents the data, as shown in Figure 2.62. Two points that lie on this line are $(5, 132)$ and $(11, 142)$. Using the point-slope form, you can find the equation of the line to be

$$P = \frac{5}{3}(t - 5) + 132. \qquad \text{Linear model}$$

✓ *Checkpoint*   Now try Exercise 11(a) and (b).

Once you have found a model, you can measure how well the model fits the data by comparing the actual values with the values given by the model, as shown in the following table.

	t	5	6	7	8	9	10	11
Actual ⇒	P	132	134	136	138	139	141	142
Model ⇒	P	132	133.7	135.3	137	138.7	140.3	142

The sum of the squares of the differences between the actual values and the model values is the **sum of the squared differences.** The model that has the least sum is the **least squares regression line** for the data. For the model in Example 3, the sum of the squared differences is 2.16. The least squares regression line for the data is

$$P = 1.7t + 124. \qquad \text{Best-fitting linear model}$$

Its sum of squared differences is 1.04. See Appendix C for more on the least squares regression line.

STUDY TIP

The model in Example 3 is based on the two data points chosen. If different points are chosen, the model may change somewhat. For instance, if you choose $(8, 138)$ and $(10, 141)$, the new model is $P = \frac{3}{2}(t - 8) + 138$.

Example 4 A Mathematical Model

The numbers S (in billions) of shares listed on the New York Stock Exchange for the years 1995 through 2001 are shown in the table. (Source: New York Stock Exchange, Inc.)

Year	Shares, S
1995	154.7
1996	176.9
1997	207.1
1998	239.3
1999	280.9
2000	313.9
2001	341.5

TECHNOLOGY SUPPORT

For instructions on how to use the *regression* feature, see Appendix A; for specific keystrokes, go to the text website at *college.hmco.com*.

a. Use the *regression* feature of a graphing utility to find a linear model for the data. Let t represent the year, with $t = 5$ corresponding to 1995.

b. How closely does the model represent the data?

Graphical Solution

a. Enter the data into the graphing utility's list editor. Then use the *linear regression* feature to obtain the model shown in Figure 2.63. You can approximate the model to be $S = 32.44t - 14.6$.

b. You can use a graphing utility to graph the actual data and the model in the same viewing window. From Figure 2.64, it appears that the model is a good fit for the actual data.

Numerical Solution

a. Using the *linear regression* feature of a graphing utility, you can find that a linear model for the data is $S = 32.44t - 14.6$.

b. You can see how well the model fits the data by comparing the actual values of S with the values of S given by the model, which are labeled S^* in the table below. From the table, you can see that the model appears to be a good fit for the actual data.

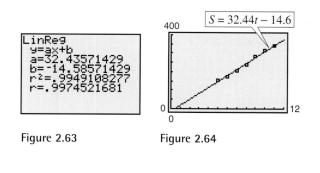

Figure 2.63   Figure 2.64

Year	S	S^*
1995	154.7	147.6
1996	176.9	180.0
1997	207.1	212.5
1998	239.3	244.9
1999	280.9	277.4
2000	313.9	309.8
2001	341.5	342.2

✓ *Checkpoint* Now try Exercise 15.

TECHNOLOGY TIP   When you use the *regression* feature of a graphing calculator or computer program to find a linear model for data, you will notice that the program may also output an "*r*-value." (For some calculators, make sure you select the *diagnostic on* feature before you use the *regression* feature. Otherwise, the calculator will not output an *r*-value.) For instance, the *r*-value from Example 4 was $r \approx 0.997$. This r-value is the **correlation coefficient** of the data and gives a measure of how well the model fits the data. Correlation coefficients vary between -1 and 1. Basically, the closer $|r|$ is to 1, the better the points can be described by a line. Three examples are shown in Figure 2.65.

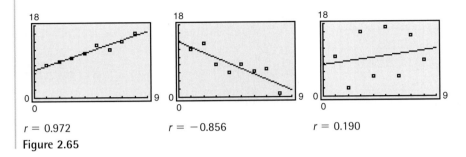

$r = 0.972$ $r = -0.856$ $r = 0.190$

Figure 2.65

Example 5   Finding a Least Squares Regression Line

The following ordered pairs (w, h) represent the shoe sizes w and the heights h (in inches) of 25 men. Use the *regression* feature of a graphing utility to find the least squares regression line for the data.

(10.0, 70.5)	(10.5, 71.0)	(9.5, 69.0)	(11.0, 72.0)	(12.0, 74.0)
(8.5, 67.0)	(9.0, 68.5)	(13.0, 76.0)	(10.5, 71.5)	(10.5, 70.5)
(10.0, 71.0)	(9.5, 70.0)	(10.0, 71.0)	(10.5, 71.0)	(11.0, 71.5)
(12.0, 73.5)	(12.5, 75.0)	(11.0, 72.0)	(9.0, 68.0)	(10.0, 70.0)
(13.0, 75.5)	(10.5, 72.0)	(10.5, 71.0)	(11.0, 73.0)	(8.5, 67.5)

Solution

After entering the data into a graphing utility (see Figure 2.66), you obtain the model shown in Figure 2.67. So, the least squares regression line for the data is

$$h = 1.84w + 51.9.$$

In Figure 2.68, this line is plotted with the data. Note that the plot does not have 25 points because some of the ordered pairs graph as the same point. The correlation coefficient for this model is $r \approx 0.981$, which implies that the model is a good fit for the data.

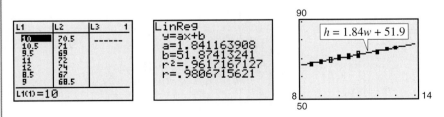

Figure 2.66   **Figure 2.67**   **Figure 2.68**

 Checkpoint Now try Exercise 17.

2.6 Exercises

Vocabulary Check

Fill in the blanks.

1. Consider a collection of ordered pairs of the form (x, y). If y tends to increase as x increases, then the collection is said to have a _____ correlation.

2. Consider a collection of ordered pairs of the form (x, y). If y tends to decrease as x increases, then the collection is said to have a _____ correlation.

3. The process of finding a linear model for a set of data is called _____ .

4. Correlation coefficients vary between _____ and _____ .

---

1. *Sales* The following ordered pairs give the years of experience x for 15 sales representatives and the monthly sales y (in thousands of dollars).

   $(1.5, 41.7)$, $(1.0, 32.4)$, $(0.3, 19.2)$, $(3.0, 48.4)$,
   $(4.0, 51.2)$, $(0.5, 28.5)$, $(2.5, 50.4)$, $(1.8, 35.5)$,
   $(2.0, 36.0)$, $(1.5, 40.0)$, $(3.5, 50.3)$, $(4.0, 55.2)$,
   $(0.5, 29.1)$, $(2.2, 43.2)$, $(2.0, 41.6)$

   (a) Create a scatter plot of the data.

   (b) Does the relationship between x and y appear to be approximately linear? Explain.

2. *Quiz Scores* The following ordered pairs give the scores on two consecutive 15-point quizzes for a class of 18 students.

   $(7, 13)$, $(9, 7)$, $(14, 14)$, $(15, 15)$, $(10, 15)$, $(9, 7)$,
   $(14, 11)$, $(14, 15)$, $(8, 10)$, $(9, 10)$, $(15, 9)$, $(10, 11)$,
   $(11, 14)$, $(7, 14)$, $(11, 10)$, $(14, 11)$, $(10, 15)$, $(9, 6)$

   (a) Create a scatter plot for the data.

   (b) Does the relationship between consecutive quiz scores appear to be approximately linear? If not, give some possible explanations.

In Exercises 3–6, the scatter plots of sets of data are shown. Determine whether there is positive correlation, negative correlation, or no discernable correlation between the variables.

3.   4.

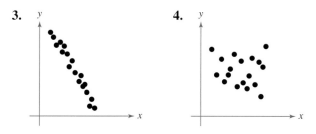

5.   6.

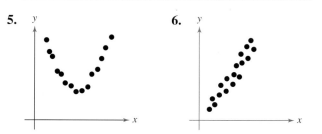

In Exercises 7–10, (a) for the data points given, draw a line of best fit through two of the points and find the equation of the line through the points, (b) use the *regression* **feature of a graphing utility to find a linear model for the data, (c) graph the data points and the lines obtained in parts (a) and (b) in the same viewing window, and (d) comment on the validity of both models. To print an enlarged copy of the graph, go to the website** *www.mathgraphs.com.*

7.   8.

9.   10.

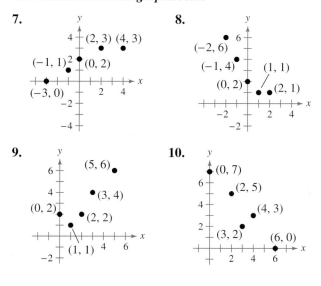

11. *Hooke's Law*   Hooke's Law states that the force F required to compress or stretch a spring (within its elastic limits) is proportional to the distance d that the spring is compressed or stretched from its original length. That is, $F = kd$, where k is the measure of the stiffness of the spring and is called the *spring constant*. The table shows the elongation d in centimeters of a spring when a force of F kilograms is applied.

Force, F	Elongation, d
20	1.4
40	2.5
60	4.0
80	5.3
100	6.6

(a) Sketch a scatter plot of the data.

(b) Find the equation of the line that seems to best fit the data.

(c) Use the *regression* feature of a graphing utility to find a linear model for the data. Compare this model with the model from part (b).

(d) Use the model from part (c) to estimate the elongation of the spring when a force of 55 kilograms is applied.

12. *Radio*   The number R of U.S. radio stations for selected years from 1970 through 2000 is shown in the table.   (Source: M Street Corporation)

Year	Radio stations, R
1970	6,760
1975	7,744
1980	8,566
1985	10,359
1990	10,788
1995	11,834
2000	13,058

(a) Use the *regression* feature of a graphing utility to find a linear model for the data. Let t represent the year, with $t = 0$ corresponding to 1970.

(b) Use a graphing utility to plot the data and graph the model in the same viewing window.

(c) Interpret the slope of the model in the context of the problem.

(d) Use the model to predict the number of radio stations in 2010.

13. *Sports*   The average salary S (in millions of dollars) for professional baseball players from 1996 through 2002 is shown in the table.   (Source: Associated Press and Major League Baseball)

Year	Salary, S
1996	1.1
1997	1.3
1998	1.4
1999	1.6
2000	1.8
2001	2.1
2002	2.3

(a) Use the *regression* feature of a graphing utility to find a linear model for the data. Let t represent the year, with $t = 6$ corresponding to 1996.

(b) Use a graphing utility to plot the data and graph the model in the same viewing window.

(c) Interpret the slope of the model in the context of the problem.

(d) Use the model to predict the average salary for a professional baseball player in 2006.

14. *Number of Stores*   The table shows the number T of Target stores from 1997 to 2002.   (Source: Target Corp.)

Year	Number of stores, T
1997	1130
1998	1182
1999	1243
2000	1307
2001	1381
2002	1476

(a) Use the *regression* feature of a graphing utility to find a linear model for the data. Let t represent the year, with $t = 7$ corresponding to 1997.

(b) Use a graphing utility to plot the data and graph the model in the same viewing window.

(c) Interpret the slope of the model in the context of the problem.

(d) Use the model to find the year in which the number of Target stores will exceed 1800.

(e) Create a table showing the actual values of T and the values of T given by the model. How closely does the model represent the data?

15. Communications   The table shows the average monthly spending S (in dollars) on paging and messaging services in the United States from 1997 to 2002.   (Source: The Strategis Group)

Year	Spending, S
1997	8.30
1998	8.50
1999	8.65
2000	8.80
2001	9.00
2002	9.25

(a) Use the *regression* feature of a graphing utility to find a linear model for the data. Let t represent the year, with $t = 7$ corresponding to 1997.

(b) Use a graphing utility to plot the data and graph the model in the same viewing window.

(c) Interpret the slope of the model in the context of the problem.

(d) Use the model to estimate the average monthly spending on paging and messaging services in 2008.

(e) Create a table showing the actual values of S and the values of S given by the model. How closely does the model represent the data?

16. Advertising and Sales   The table shows the advertising expenditures x and sales volume y for a company for seven randomly selected months. Both are measured in thousands of dollars.

Month	Advertising expenditures, x	Sales volume, y
1	2.4	202
2	1.6	184
3	2.0	220
4	2.6	240
5	1.4	180
6	1.6	164
7	2.0	186

Table for 16

(a) Use the *regression* feature of a graphing utility to find a linear model for the data.

(b) Use a graphing utility to plot the data and graph the model in the same viewing window.

(c) Interpret the slope of the model in the context of the problem.

(d) Use the model to estimate sales for advertising expenditures of $1500.

17. Sports   The following ordered pairs (x, y) represent the Olympic year x and the winning time y (in minutes) in the women's 400-meter freestyle swimming event.   (Source: The New York Times Almanac 2003)

(1948, 5.30)	(1976, 4.16)
(1952, 5.20)	(1980, 4.15)
(1956, 4.91)	(1984, 4.12)
(1960, 4.84)	(1988, 4.06)
(1964, 4.72)	(1992, 4.12)
(1968, 4.53)	(1996, 4.12)
(1972, 4.32)	(2000, 4.10)

(a) Use the *regression* feature of a graphing utility to find a linear model for the data. Let x represent the year, with $x = 0$ corresponding to 1950.

(b) What information is given by the sign of the slope of the model?

(c) Use a graphing utility to plot the data and graph the model in the same viewing window.

(d) How closely does the model fit the data?

(e) Can the model be used to estimate the winning times in the future? Explain.

18. *Elections*   The data shows the percent x of the voting-age population that was registered to vote and the percent y that actually voted by state in 2000. (Source: U.S. Census Bureau)

AK (72.5, 65.6)   AL (73.6, 59.6)   AR (59.4, 49.4)
AZ (53.3, 46.7)   CA (52.8, 46.4)   CO (64.1, 53.6)
CT (62.5, 55.2)   D.C. (72.4, 65.6)   DE (67.9, 62.2)
FL (60.5, 51.6)   GA (61.1, 49.0)   HI (47.0, 39.7)
IA (72.2, 64.1)   ID (61.4, 53.9)   IL (66.7, 56.8)
IN (68.5, 58.5)   KS (67.7, 60.2)   KY (69.7, 54.9)
LA (75.4, 64.6)   MA (70.3, 60.1)   MD (65.6, 57.1)
ME (80.3, 69.2)   MI (69.1, 60.1)   MN (76.7, 67.8)
MO (74.3, 65.4)   MS (72.2, 59.8)   MT (70.0, 62.2)
NC (66.1, 53.2)   ND (91.1, 69.8)   NE (71.8, 58.9)
NH (69.6, 63.3)   NJ (63.2, 55.2)   NM (59.5, 51.3)
NV (52.3, 46.5)   NY (58.6, 51.0)   OH (67.0, 58.1)
OK (68.3, 58.3)   OR (68.2, 60.8)   PA (65.3, 55.7)
RI (69.7, 60.1)   SC (68.0, 58.9)   SD (70.9, 58.7)
TN (62.1, 52.3)   TX (61.4, 48.2)   UT (64.7, 56.3)
VA (64.1, 57.2)   VT (72.0, 63.3)   WA (66.1, 58.6)
WI (76.5, 67.8)   WV (63.1, 52.1)   WY (68.6, 62.5)

(a) Use the *regression* feature of a graphing utility to find a linear model for the data.

(b) Use a graphing utility to plot the data and graph the model in the same viewing window.

(c) Interpret the graph in part (b). Use the graph to identify any states that appear to differ substantially from most of the others.

(d) Interpret the slope of the model in the context of the problem.

Synthesis

True or False?   **In Exercises 19 and 20, determine whether the statement is true or false. Justify your answer.**

19. A linear regression model with a positive correlation will have a slope that is greater than 0.

20. If the correlation coefficient for a linear regression model is close to -1, the regression line cannot be used to describe the data.

21. *Writing*   A linear mathematical model for predicting prize winnings at a race is based on data for 3 years. Write a paragraph discussing the potential accuracy or inaccuracy of such a model.

22. *Research Project*   Use your school's library, the Internet, or some other reference source to locate data that you think describes a linear relationship. Create a scatter plot of the data and find the least squares regression line that represents the points. Interpret the slope and y-intercept in the context of the data. Write a summary of your findings.

Review

In Exercises 23–26, use inequality and interval notation to describe the set.

23. P is no more than 2.

24. x is positive.

25. z is at least -3 and at most 10.

26. W is less than 7 but no less than -6.

In Exercises 27 and 28, simplify the complex fraction.

27. $\dfrac{x^2 - 4}{\left(\dfrac{x + 2}{5}\right)}$

28. 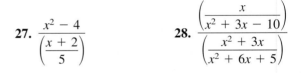 $\dfrac{\left(\dfrac{x}{x^2 + 3x - 10}\right)}{\left(\dfrac{x^2 + 3x}{x^2 + 6x + 5}\right)}$

In Exercises 29–32, evaluate the function at each value of the independent variable and simplify.

29. $f(x) = 2x^2 - 3x + 5$
  (a) $f(-1)$ (b) $f(w + 2)$

30. $g(x) = 5x^2 - 6x + 1$
  (a) $g(-2)$ (b) $g(z - 2)$

31. $h(x) = \begin{cases} 1 - x^2, & x \le 0 \\ 2x + 3, & x > 0 \end{cases}$
  (a) $h(1)$ (b) $h(0)$

32. $k(x) = \begin{cases} 5 - 2x, & x < -1 \\ x^2 + 4, & x \ge -1 \end{cases}$
  (a) $k(-3)$ (b) $k(-1)$

In Exercises 33–38, solve the equation algebraically. Check your solution graphically.

33. $6x + 1 = -9x - 8$ **34.** $3(x - 3) = 7x + 2$

35. $8x^2 - 10x - 3 = 0$ **36.** $10x^2 - 23x - 5 = 0$

37. $2x^2 - 7x + 4 = 0$ **38.** $2x^2 - 8x + 5 = 0$

2 Chapter Summary

What did you learn?

	Review Exercises
Section 2.1	
☐ Solve equations involving fractional expressions.	1–8
☐ Write and use mathematical models to solve real-life problems.	9–14
☐ Use common formulas to solve real-life problems.	15, 16
Section 2.2	
☐ Find x- and y-intercepts of graphs of equations.	17–22
☐ Find solutions of equations graphically.	23–28
☐ Find the points of intersection of two graphs.	29–32
Section 2.3	
☐ Use the imaginary unit i to write complex numbers.	33–36
☐ Add, subtract, and multiply complex numbers.	37–44
☐ Use complex conjugates to write the quotient of two complex numbers in standard form.	45–48
☐ Plot complex numbers in the complex plane.	49–54
Section 2.4	
☐ Solve quadratic equations by factoring, extracting square roots, completing the square, and using the Quadratic Formula.	55–64
☐ Solve polynomial equations of degree three or greater.	65–68
☐ Solve equations involving radicals.	69–78
☐ Solve equations involving fractions or absolute values.	79–86
☐ Use quadratic equations to model and solve real-life problems.	87, 88
Section 2.5	
☐ Use properties of inequalities to solve linear inequalities.	89–94
☐ Solve inequalities involving absolute values.	95–100
☐ Solve polynomial inequalities.	101–106
☐ Solve rational inequalities.	107–110
☐ Use inequalities to model and solve real-life problems.	111, 112
Section 2.6	
☐ Construct scatter plots and interpret correlation.	113, 114
☐ Use scatter plots and a graphing utility to find linear models for data.	115–118

2 Review Exercises

2.1 In Exercises 1 and 2, determine whether each value of x is a solution of the equation.

Equation	Values

1. $6 + \dfrac{3}{x - 4} = 5$ (a) $x = 5$ (b) $x = 0$
                           (c) $x = -2$ (d) $x = 1$

2. $6 + \dfrac{2}{x + 3} = \dfrac{6x + 1}{3}$ (a) $x = -3$ (b) $x = 3$
                           (c) $x = 0$ (d) $x = -\frac{2}{3}$

In Exercises 3–8, solve the equation (if possible). Then use a graphing utility to verify your solution.

3. $\dfrac{18}{x} = \dfrac{10}{x - 4}$

4. $\dfrac{5}{x - 2} = \dfrac{13}{2x - 3}$

5. $14 + \dfrac{2}{x - 1} = 10$

6. $6 - \dfrac{11}{x} = 3 + \dfrac{7}{x}$

7. $\dfrac{9x}{3x - 1} - \dfrac{4}{3x + 1} = 3$

8. $\dfrac{5}{x - 5} + \dfrac{1}{x + 5} = \dfrac{2}{x^2 - 25}$

9. *Profit* In October, a greeting card company's total profit was 12% more than it was in September. The total profit for the two months was $689,000. Find the profit for each month.

10. *Cost Sharing* A group of farmers agree to share equally in the cost of a $48,000 piece of machinery. If they can find two more farmers to join the group, each person's share of the cost will decrease by $4000. How many farmers are presently in the group?

11. *Mixture Problem* A car radiator contains 10 liters of a 30% antifreeze solution. How many liters will have to be replaced with pure antifreeze if the resulting solution is to be 50% antifreeze?

12. *Average Speed* You drove 56 miles one way on a service call. On the return trip, your average speed was 8 miles per hour greater and the trip took 10 fewer minutes. What was your average speed on the return trip?

13. *Height* To obtain the height of a tree, you measure the tree's shadow and find that it is 8 meters long. You also measure the shadow of a two-meter lamp-post and find that it is 75 centimeters long.

(a) Draw a diagram that illustrates the problem. Let h represent the height of the tree.

(b) Find the height of the tree in meters.

14. *Investment* You invest $12,000 in a fund paying $2\frac{1}{2}\%$ simple interest and $10,000 in a fund with a variable interest rate. At the end of the year, you were notified that the total interest for both funds was $870. Find the equivalent simple interest rate on the variable–rate fund.

15. *Meteorology* The average daily temperature for the month of January in Juneau, Alaska is 25.7° F. What is Juneau's average daily temperature for the month of January in degrees Celsius? (Source: U.S. National Oceanic and Atmospheric Administration)

16. *Geometry* A basketball and a baseball have circumferences of 30 inches and $9\frac{1}{4}$ inches, respectively. Find the volume of each.

2.2 In Exercises 17–22, find the x- and y-intercepts of the graph of the equation.

17. $-x + y = 3$ **18.** $x - 5y = 20$

19. $y = x^2 - 9x + 8$ **20.** $y = 25 - x^2$

21. $y = -|x + 5| - 2$ **22.** $y = 6 - 2|x - 3|$

In Exercises 23–28, use a graphing utility to approximate any solutions (accurate to three decimal places) of the equation. [Remember to write the equation in the form $f(x) = 0$.]

23. $5(x - 2) - 1 = 0$ **24.** $12 - 5(x - 7) = 0$

25. $3x^3 - 2x + 4 = 0$ **26.** $\frac{1}{3}x^3 - x + 4 = 0$

27. $x^4 - 3x + 1 = 0$ **28.** $6 - \frac{1}{2}x^2 + \frac{5}{6}x^4 = 0$

In Exercises 29–32, determine any point(s) of intersection algebraically. Use a graphing utility to verify your answer(s).

29. $3x + 5y = -7$ **30.** $x - y = 3$
     $-x - 2y = 3$ $2x + y = 12$

31. $x^2 + 2y = 14$ **32.** $y = -x + 7$
     $3x + 4y = 1$ $y = 2x^3 - x + 9$

2.3 In Exercises 33–36, write the complex number in standard form.

33. $6 + \sqrt{-25}$

34. $-\sqrt{-12} + 3$

35. $-2i^2 + 7i$

36. $-i^2 - 4i$

In Exercises 37–44, perform the operations and write the result in standard form.

37. $(7 + 5i) + (-4 + 2i)$

38. $\left(\dfrac{\sqrt{2}}{2} - \dfrac{\sqrt{2}}{2}i\right) - \left(\dfrac{\sqrt{2}}{2} + \dfrac{\sqrt{2}}{2}i\right)$

39. $5i(13 - 8i)$

40. $(1 + 6i)(5 - 2i)$

41. $(10 - 8i)(2 - 3i)$

42. $i(6 + i)(3 - 2i)$

43. $(3 + 7i)^2 + (3 - 7i)^2$

44. $(4 - i)^2 - (4 + i)^2$

In Exercises 45–48, write the quotient in standard form.

45. $\dfrac{6 + i}{i}$

46. $\dfrac{4}{-3i}$

47. $\dfrac{3 + 2i}{5 + i}$

48. $\dfrac{1 - 7i}{2 + 3i}$

In Exercises 49–54, plot the complex number in the complex plane.

49. $2 - 5i$

50. $-1 + 4i$

51. $-6i$

52. $7i$

53. 3

54. -2

2.4 In Exercises 55–64, solve the equation using any convenient method. Use a graphing utility to verify your solution(s).

55. $6x = 3x^2$

56. $15 + x - 2x^2 = 0$

57. $(x + 4)^2 = 18$

58. $16x^2 = 25$

59. $x^2 - 12x + 30 = 0$

60. $x^2 + 6x - 3 = 0$

61. $2x^2 + 9x - 5 = 0$

62. $-x^2 - x + 15 = 0$

63. $x^2 + 4x + 10 = 0$

64. $2x^2 - 6x + 21 = 0$

In Exercises 65–86, find all real solutions of the equation algebraically. Use a graphing utility to verify the solutions graphically.

65. $3x^3 - 26x^2 + 16x = 0$

66. $216x^4 - x = 0$

67. $5x^4 - 12x^3 = 0$

68. $4x^3 - 6x^2 = 0$

69. $\sqrt{x + 4} = 3$

70. $\sqrt{x - 2} - 8 = 0$

71. $2\sqrt{x} - 5 = 0$

72. $\sqrt{3x - 2} = 4 - x$

73. $\sqrt{2x + 3} + \sqrt{x - 2} = 2$

74. $5\sqrt{x} - \sqrt{x - 1} = 6$

75. $(x - 1)^{2/3} - 25 = 0$

76. $(x + 2)^{3/4} = 27$

77. $(x + 4)^{1/2} + 5x(x + 4)^{3/2} = 0$

78. $8x^2(x^2 - 4)^{1/3} + (x^2 - 4)^{4/3} = 0$

79. $3\left(1 - \dfrac{1}{5t}\right) = 0$

80. $\dfrac{1}{x - 2} = 3$

81. $\dfrac{4}{(x - 4)^2} = 1$

82. $\dfrac{1}{(t + 1)^2} = 1$

83. $|x - 5| = 10$

84. $|2x + 3| = 7$

85. $|x^2 - 3| = 2x$

86. $|x^2 - 6| = x$

87. *Population*   The population P of South Dakota (in thousands) from 1995 through 2001 can be approximated by the model

$$P = 0.11t^2 + 1.5t + 728, \quad 5 \le t \le 11$$

where t represents the year, with $t = 5$ corresponding to 1995.   (Source: U.S. Census Bureau)

(a) Use the model to approximate algebraically when the population reached 750,000.

(b) Verify your answer to part (a) by creating a table of values for the model.

(c) Use a graphing utility to graph the model.

(d) Use the *zoom* and *trace* features of a graphing utility to determine when the population exceeded 740,000.

(e) Verify your answer to part (d) algebraically.

88. *Life Insurance*   The value y (in trillions of dollars) of life insurance policies in the United States from 1992 through 2000 can be approximated by the model

$$y = 0.045t^2 + 0.20t + 9.8, \quad 2 \le t \le 10$$

where t represents the year, with $t = 2$ corresponding to 1992.   (Source: American Council of Life Insurers)

(a) Use a graphing utility to graph the model.

(b) Use the graph to determine the year in which the value of life insurance policies was $15 trillion.

(c) Is this model accurate for predicting the value of life insurance policies in the future? Explain.

2.5 In Exercises 89–110, solve the inequality and sketch the solution on the real number line. Use a graphing utility to verify your solution graphically.

89. $8x - 3 < 6x + 15$

90. $9x - 8 \le 7x + 16$

91. $\frac{1}{2}(3 - x) > \frac{1}{3}(2 - 3x)$

92. $4(5 - 2x) \ge \frac{1}{2}(8 - x)$

93. $-2 < -x + 7 \le 10$

94. $-6 \le 3 - 2(x - 5) < 14$

95. $|x - 2| < 1$

96. $|x| \le 4$

97. $\left|x - \frac{3}{2}\right| \ge \frac{3}{2}$

98. $|x - 3| > 4$

99. $4|3 - 2x| \le 16$

100. $|x + 9| + 7 > 19$

101. $x^2 - 2x \ge 3$

102. $x^2 - 6x - 27 < 0$

103. $4x^2 - 23x \le 6$

104. $6x^2 + 5x < 4$

105. $x^3 - 16x \ge 0$

106. $12x^3 - 20x^2 < 0$

107. $\dfrac{x - 5}{3 - x} < 0$

108. $\dfrac{2}{x + 1} \le \dfrac{3}{x - 1}$

109. $\dfrac{3x + 8}{x - 3} \le 4$

110. $\dfrac{x + 8}{x + 5} - 2 < 0$

111. *Accuracy of Measurement*   The side of a square is measured as 20.8 inches with a possible error of $\frac{1}{16}$ inch. Using these measurements, determine the interval containing the area of the square.

112. *Meteorology*   An electronic device is to be operated in an environment with relative humidity h in the interval defined by

$$|h - 50| \le 30.$$

What are the minimum and maximum relative humidities for the operation of this device?

2.6

113. *Education*   The following ordered pairs give the entrance exam scores x and the grade-point averages y after 1 year of college for 10 students.

(75, 2.3), (82, 3.0), (90, 3.6), (65, 2.0), (70, 2.1), (88, 3.5), (93, 3.9), (69, 2.0), (80, 2.8), (85, 3.3)

(a) Create a scatter plot for the data.

(b) Does the relationship between x and y appear to be approximately linear? Explain.

114. *Stress Test*   A machine part was tested by bending it x centimeters 10 times per minute until it failed (y equals the time to failure in hours). The results

are given as the following ordered pairs.

(3, 61), (6, 56), (9, 53), (12, 55), (15, 48), (18, 35), (21, 36), (24, 33), (27, 44), (30, 23)

(a) Create a scatter plot for the data.

(b) Does the relationship between x and y appear to be approximately linear? If not, give some possible explanations.

115. *Falling Object*   In an experiment, students measured the speed s (in meters per second) of a ball t seconds after it was released. The results are shown in the table.

Time, t	Speed, s
0	0
1	11.0
2	19.4
3	29.2
4	39.4

(a) Sketch a scatter plot of the data.

(b) Find the equation of the line that seems to best fit the data.

(c) Use the *regression* feature of a graphing utility to find a linear model for the data. Compare with the model from part (b).

(d) Use the model from part (c) to estimate the speed of the ball after 2.5 seconds.

116. *Sales*   The table shows the sales S (in millions of dollars) for Timberland from 1995 to 2002. (Source: The Timberland Co.)

Year	Sales, S
1995	655.1
1996	690.0
1997	796.5
1998	862.2
1999	917.2
2000	1091.5
2001	1183.6
2002	1190.9

(a) Use the *regression* feature of a graphing utility to find a linear model for the data. Let t represent the year, with $t = 5$ corresponding to 1995.

(b) Use a graphing utility to plot the data and graph the model in the same viewing window.

(c) Interpret the slope of the model in the context of the problem.

(d) Use the model to find the year in which the sales will exceed \$1300 million.

(e) Create a table showing the actual values of S and the values of S given by the model. How closely does the model represent the data?

117. *Height* The following ordered pairs (x, y) represent the percent y of women between the ages of 20 and 29 who are under a certain height x (in feet). (Source: U.S. National Center for Health Statistics)

(4.67, 0.6)	(5.42, 62.7)
(4.75, 0.7)	(5.50, 74.0)
(4.83, 1.2)	(5.58, 84.7)
(4.92, 3.1)	(5.67, 92.4)
(5.00, 6.0)	(5.75, 96.2)
(5.08, 11.5)	(5.83, 98.6)
(5.17, 21.8)	(5.92, 99.5)
(5.25, 34.3)	(6.00, 100.0)
(5.33, 48.9)	

(a) Use the *regression* feature of a graphing utility to find a linear model for the data.

(b) Use a graphing utility to plot the data and graph the model in the same viewing window.

(c) How closely does the model fit the data?

(d) Can the model be used to estimate the percent of women who are under a height of greater than 6 feet?

118. *Sports* The following ordered pairs (x, y) represent the Olympic year x and the winning time y (in minutes) in the men's 1500-meter speed skating event. (Source: The New York Times Almanac 2003)

(1964, 2.17)	(1988, 1.87)
(1968, 2.06)	(1992, 1.91)
(1972, 2.05)	(1994, 1.85)
(1976, 1.99)	(1998, 1.80)
(1980, 1.92)	(2002, 1.73)
(1984, 1.97)	

(a) Use the *regression* feature of a graphing utility to find a linear model for the data. Let x represent the year, with $x = 4$ corresponding to 1964.

(b) What information is given by the sign of the slope of the model?

(c) Use a graphing utility to plot the data and graph the model in the same viewing window.

(d) How closely does the model fit the data?

(e) Can the model be used to estimate the winning times in the future? Explain.

Synthesis

True or False? In Exercises 119–121, determine whether the statement is true or false. Justify your answer.

119. The graph of a function may have two distinct y-intercepts.

120. The sum of two complex numbers cannot be a real number.

121. The sign of the slope of a regression line is always positive.

122. *Writing* In your own words, explain the difference between an identity and a conditional equation.

123. *Writing* Describe the relationship among the x-intercepts of a graph, the zeros of a function, and the solutions of an equation.

124. Consider the linear equation $ax + b = 0$.

(a) What is the sign of the solution if $ab > 0$?

(b) What is the sign of the solution if $ab < 0$?

125. *Error Analysis* Describe the error.
$$\sqrt{-6}\sqrt{-6} = \sqrt{(-6)(-6)} = \sqrt{36} = 6$$

126. *Error Analysis* Describe the error.
$$-i(\sqrt{-4} - 1) = -i(4i - 1)$$
$$= -4i^2 - i$$
$$= 4 - i$$

127. Write each of the powers of i as i, $-i$, 1, or -1.

(a) i^{40} (b) i^{25} (c) i^{50} (d) i^{67}

2 | Chapter Test

Take this test as you would take a test in class. After you are finished, check your work against the answers given in the back of the book.

In Exercises 1 and 2, solve the equation (if possible). Then use a graphing utility to verify your solution.

1. $\dfrac{12}{x} - 7 = -\dfrac{27}{x} + 6$

2. $\dfrac{4}{3x - 2} - \dfrac{9x}{3x + 2} = -3$

In Exercises 3–8, perform the operations and write the result in standard form.

3. $(-8 - 3i) + (-1 - 15i)$

4. $\left(10 + \sqrt{-20}\right) - \left(4 - \sqrt{-14}\right)$

5. $(2 + i)(6 - i)$

6. $(4 + 3i)^2 - (5 + i)^2$

In Exercises 7 and 8, write the quotient in standard form.

7. $\dfrac{8 + 5i}{6 - i}$

8. $\dfrac{5i}{2 + i}$

In Exercises 9–12, use a graphing utility to graph the equation and approximate any x-intercepts. Set $y = 0$ and solve the resulting equation. Compare the results with the x-intercepts of the graph.

9. $y = 3x^2 + 1$

10. $y = 2 + 8x^{-2}$

11. $y = x^3 - 4x^2 + 5x$

12. $y = x^3 + x$

In Exercises 13–16, solve the equation using any convenient method. Use a graphing utility to verify the solutions graphically.

13. $x^2 - 10x + 9 = 0$

14. $x^2 + 12x - 2 = 0$

15. $4x^2 - 81 = 0$

16. $5x^2 + 14x - 3 = 0$

In Exercises 17–20, find all solutions of the equation algebraically. Use a graphing utility to verify the solutions graphically.

17. $3x^3 - 4x^2 - 12x + 16 = 0$

18. $x + \sqrt{22 - 3x} = 6$

19. $(x^2 + 6)^{2/3} = 16$

20. $|8x - 1| = 21$

In Exercises 21–23, solve the inequality and sketch the solution on the real number line. Use a graphing utility to verify your solution graphically.

21. $-\frac{5}{6} < x - 2 < \frac{1}{8}$

22. $2|x - 8| < 10$

23. $\dfrac{3 - 5x}{2 + 3x} < -2$

24. The table shows the number of local telephone access lines L (in millions) in the United States from 1994 through 2000, where t represents the year, with $t = 4$ corresponding to 1994. Use the *regression* feature of a graphing utility to find a linear model for the data. Use the model to find the year in which the number of local telephone access lines will exceed 300 million. (Source: U.S. Federal Communications Commission)

Year, t	Lines, L
4	157
5	166
6	178
7	194
8	205
9	228
10	245

Table for 24

P–2 Cumulative Test

Take this test to review the material from earlier chapters. After you are finished, check your work against the answers in the back of the book.

In Exercises 1–3, simplify the expression.

1. $\dfrac{14x^2y^{-3}}{32x^{-1}y^2}$ **2.** $8\sqrt{60} - 2\sqrt{135} - \sqrt{15}$ **3.** $\sqrt{28x^4y^3}$

In Exercises 4–6, perform the operation and simplify the result.

4. $4x - [2x + 5(2 - x)]$ **5.** $(x - 2)(x^2 + x - 3)$ **6.** $\dfrac{2}{x + 3} - \dfrac{1}{x + 1}$

In Exercises 7–9, factor the expression completely.

7. $25 - (x - 2)^2$ **8.** $x - 5x^2 - 6x^3$ **9.** $54 - 16x^3$

10. Find the midpoint of the line segment connecting the points $\left(-\frac{7}{2}, 4\right)$ and $(6.5, -8)$. Then find the distance between the points.

11. Write the standard form of the equation of a circle with center $\left(-\frac{1}{2}, -8\right)$ and a radius of $\frac{5}{4}$.

In Exercises 12–14, use point plotting to sketch the graph of the equation.

12. $x - 3y + 12 = 0$ **13.** $y = x^2 - 9$ **14.** $y = \sqrt{4 - x}$

In Exercises 15–17, (a) write the general form of the equation of the line that satisfies the given conditions and (b) find three additional points through which the line passes.

15. The line contains the points $(-5, 8)$ and $\left(\frac{1}{2}, -6\right)$.

16. The line contains the point $\left(-\frac{1}{2}, 1\right)$ and has a slope of -2.

17. The line has an undefined slope and contains the point $\left(-\frac{3}{7}, \frac{1}{8}\right)$.

In Exercises 18 and 19, evaluate the function at each value of the independent variable and simplify.

18. $f(x) = \dfrac{x}{x - 2}$

   (a) $f(6)$ (b) $f(2)$ (c) $f(s + 2)$

19. $f(x) = \begin{cases} 3x - 8, & x \le -\frac{5}{3} \\ 3x^2 + 9x - 8, & x > -\frac{5}{3} \end{cases}$

   (a) $f\left(-\frac{5}{3}\right)$ (b) $f(-1)$ (c) $f(0)$

20. Does the graph at the right represent y as a function of x? Explain.

21. Use a graphing utility to graph the function $f(x) = 2|x - 5| - |x + 5|$. Then determine the open intervals over which the function is increasing, decreasing, or constant.

22. Compare the graph of each function with the graph of $f(x) = \sqrt[3]{x}$.

   (a) $r(x) = \dfrac{1}{2}\sqrt[3]{x}$ (b) $h(x) = \sqrt[3]{x} + 2$ (c) $g(x) = \sqrt[3]{x + 2}$

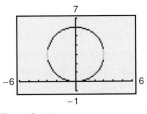

Figure for 20

In Exercises 23–26, evaluate the indicated function for

$$f(x) = -x^2 + 3x - 10 \quad \text{and} \quad g(x) = 4x + 1.$$

23. $(f + g)(-4)$ **24.** $(g - f)\left(\frac{3}{4}\right)$ **25.** $(g \circ f)(-2)$ **26.** $(fg)(-1)$

27. Determine whether $h(x) = 5x - 2$ has an inverse function. If so, find it.

28. Plot the complex number $-5 + 4i$ in the complex plane.

In Exercises 29–32, use a graphing utility to graph the equation and approximate any *x*-intercepts of the graph. Set $y = 0$ and solve the resulting equation. Compare the results with the *x*-intercepts of the graph.

29. $y = 4x^3 - 12x^2 + 8x$ **30.** $y = 12x^3 - 84x^2 + 120x$

31. $y = |2x - 3| - 5$ **32.** $y = \sqrt{x^2 + 1} + x - 9$

In Exercises 33 and 34, solve the equation for the indicated variable.

33. Solve for X: $Z = \sqrt{R^2 - X^2}$ **34.** Solve for p: $L = \dfrac{k}{3\pi r^2 p}$

In Exercises 35–38, solve the inequality and graph the solution on the real number line. Use a graphing utility to verify your solution graphically.

35. $\dfrac{x}{5} - 6 \le -\dfrac{x}{2} + 6$ **36.** $2x^2 + x \ge 15$

37. $|7 + 8x| > 5$ **38.** $\dfrac{2(x - 2)}{x + 1} \le 0$

39. A soccer ball has a volume of about 370.7 cubic inches. Find the radius of the soccer ball (accurate to three decimal places).

40. A rectangular plot of land with a perimeter of 546 feet has a width of x.

    (a) Write the area A of the plot as a function of x.

    (b) Use a graphing utility to graph the area function. What is the domain of the function?

    (c) Approximate the dimensions of the plot when the area is 15,000 square feet.

41. The total revenues R (in millions of dollars) for Papa John's from 1995 through 2001 are shown in the table.   (Source: Papa John's International)

    (a) Use the *regression* feature of a graphing utility to find a linear model for the data. Let t represent the year, with $t = 5$ corresponding to 1995.

    (b) Use a graphing utility to plot the data and graph the model in the same viewing window.

    (c) Interpret the slope of the model in the context of the problem.

    (d) Use the model to estimate the revenues for Papa John's in 2007.

    (e) Create a table showing the actual values of R and the values of R given by the model. How closely does the model represent the data?

Year	Revenues, R
1995	253.4
1996	360.1
1997	508.8
1998	669.8
1999	805.3
2000	944.7
2001	971.2

Table for 41

The average monthly rate for basic cable television service in the United States has increased from 1995 to 2001. You can use a cubic polynomial to model this growth and predict future cable rates.

Jose Luis Pelaez, Inc./Corbis

3

Polynomial and Rational Functions

What You Should Learn

In this chapter, you will learn how to:

■ Sketch and analyze graphs of quadratic and polynomial functions.

■ Use long division and synthetic division to divide polynomials by other polynomials.

■ Determine the numbers of rational and real zeros of polynomial functions, and find the zeros.

■ Determine the domains, find the asymptotes, and sketch the graphs of rational functions.

■ Classify scatter plots and use a graphing utility to find quadratic models for data.

3.1 Quadratic Functions

The Graph of a Quadratic Function

In this and the next section, you will study the graphs of polynomial functions.

Definition of Polynomial Function

Let n be a nonnegative integer and let $a_n, a_{n-1}, \ldots, a_2, a_1, a_0$ be real numbers with $a_n \neq 0$. The function given by

$$f(x) = a_n x^n + a_{n-1} x^{n-1} + \cdots + a_2 x^2 + a_1 x + a_0$$

is called a **polynomial function of x with degree n.**

Polynomial functions are classified by degree. For instance, the polynomial function

$$f(x) = a, \quad a \neq 0 \qquad \text{Constant function}$$

has degree 0 and is called a **constant function.** In Chapter 1, you learned that the graph of this type of function is a horizontal line. The polynomial function

$$f(x) = mx + b, \quad m \neq 0 \qquad \text{Linear function}$$

has degree 1 and is called a **linear function.** You also learned in Chapter 1 that the graph of the linear function $f(x) = mx + b$ is a line whose slope is m and whose y-intercept is $(0, b)$. In this section you will study second-degree polynomial functions, which are called **quadratic functions.**

Definition of Quadratic Function

Let a, b, and c be real numbers with $a \neq 0$. The function given by

$$f(x) = ax^2 + bx + c \qquad \text{Quadratic function}$$

is called a **quadratic function.**

Often real-life data can be modeled by quadratic functions. For instance, the table at the right shows the height h (in feet) of a projectile fired from a height of 6 feet with an initial velocity of 256 feet per second at any time t (in seconds). A quadratic model for the data in the table is $h(t) = -16t^2 + 256t + 6$ for $0 \leq t \leq 16$.

The graph of a quadratic function is a special type of U-shaped curve called a **parabola.** Parabolas occur in many real-life applications, especially those involving reflective properties, such as satellite dishes or flashlight reflectors. You will study these properties in a later chapter.

All parabolas are symmetric with respect to a line called the **axis of symmetry,** or simply the **axis** of the parabola. The point where the axis intersects the parabola is called the **vertex** of the parabola.

t	h
0	6
2	454
4	774
6	966
8	1030
10	966
12	774
13	454
16	6

Library of Functions: Quadratic Function

The simplest type of *quadratic function* is $f(x) = ax^2$, also known as the *squaring function*. The basic characteristics of the squaring function are summarized below.

Graph of $f(x) = ax^2$, $a > 0$

Domain: $(-\infty, \infty)$
Range: $[0, \infty)$
Intercept: $(0, 0)$
Decreasing on $(-\infty, 0)$
Increasing on $(0, \infty)$
Even function
y-Axis symmetry
Relative minimum or vertex: $(0, 0)$

Graph of $f(x) = ax^2$, $a < 0$

Domain: $(-\infty, \infty)$
Range: $(-\infty, 0]$
Intercept: $(0, 0)$
Increasing on $(-\infty, 0)$
Decreasing on $(0, \infty)$
Even function
y-Axis symmetry
Relative maximum or vertex: $(0, 0)$

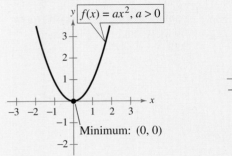

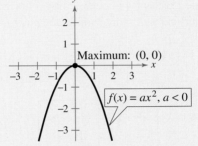

For the general quadratic form $f(x) = ax^2 + bx + c$, if the leading coefficient a is positive, the parabola opens upward; and if the leading coefficient a is negative, the parabola opens downward. Later in this section you will learn ways to find the coordinates of the vertex of a parabola.

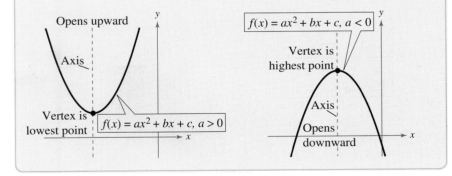

Exploration

Use a graphing utility to graph the parabola

$$y = x^2 + c$$

for $c = -3, -2, -1, 1, 2,$ and 3. What can you conclude about the parabola when $c < 0$? When $c > 0$?

When sketching the graph of $f(x) = ax^2$, it is helpful to use the graph of $y = x^2$ as a reference, as discussed in Section 1.5. There you saw that when $a > 1$, the graph of $y = af(x)$ is a vertical stretch of the graph of $y = f(x)$. When $0 < a < 1$, the graph of $y = af(x)$ is a vertical shrink of the graph of $y = f(x)$. This is demonstrated again in Example 1.

Example 1   Graphing Simple Quadratic Functions

Describe how the graph of each function is related to the graph of $y = x^2$.

a. $f(x) = \dfrac{1}{3}x^2$

b. $g(x) = 2x^2$

c. $h(x) = -x^2 + 1$

d. $k(x) = (x + 2)^2 - 3$

Solution

a. Compared with $y = x^2$, each output of f "shrinks" by a factor of $\frac{1}{3}$. The result is a parabola that opens upward and is broader than the parabola represented by $y = x^2$, as shown in Figure 3.1.

b. Compared with $y = x^2$, each output of g "stretches" by a factor of 2, creating a narrower parabola, as shown in Figure 3.2.

c. With respect to the graph of $y = x^2$, the graph of h is obtained by a *reflection* in the x-axis and a vertical shift one unit *upward*, as shown in Figure 3.3.

d. With respect to the graph of $y = x^2$, the graph of k is obtained by a horizontal shift two units *to the left* and a vertical shift three units *downward*, as shown in Figure 3.4.

> **STUDY TIP**
>
> In Example 1, note that the coefficient a determines how widely the parabola given by $f(x) = ax^2$ opens. If $|a|$ is small, the parabola opens more widely than if $|a|$ is large.

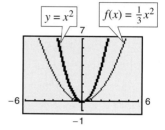

Figure 3.1

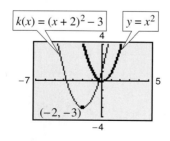

Figure 3.2

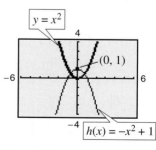

Figure 3.3

Figure 3.4

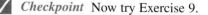

✓ *Checkpoint* Now try Exercise 9.

Recall from Section 1.5 that the graphs of $y = f(x \pm c)$, $y = f(x) \pm c$, $y = -f(x)$, and $y = f(-x)$ are rigid transformations of the graph of $y = f(x)$.

$y = f(x \pm c)$ Horizontal shift $y = -f(x)$ Reflection in x-axis

$y = f(x) \pm c$ Vertical shift $y = f(-x)$ Reflection in y-axis

The Standard Form of a Quadratic Function

The equation in Example 1(d) is written in the **standard form**

$$f(x) = a(x - h)^2 + k.$$

This form is especially convenient for sketching a parabola because it identifies the vertex of the parabola as (h, k).

Standard Form of a Quadratic Equation

The quadratic function given by

$$f(x) = a(x - h)^2 + k, \qquad a \neq 0$$

is in **standard form**. The graph of f is a parabola whose axis is the vertical line $x = h$ and whose vertex is the point (h, k). If $a > 0$, the parabola opens upward, and if $a < 0$, the parabola opens downward.

> **Exploration**
>
> Use a graphing utility to graph $y = ax^2$ with $a = -2, -1, -0.5, 0.5, 1,$ and 2. How does changing the value of a affect the graph?
>
> Use a graphing utility to graph $y = (x - h)^2$ with $h = -4, -2, 2,$ and 4. How does changing the value of h affect the graph?
>
> Use a graphing utility to graph $y = x^2 + k$ with $k = -4, -2, 2,$ and 4. How does changing the value of k affect the graph?

Example 2   Identifying the Vertex of a Quadratic Function

Describe the graph of $f(x) = 2x^2 + 8x + 7$ and identify the vertex.

Solution

Write the quadratic function in standard form by completing the square. Recall that the first step is to factor out any coefficient of x^2 that is not 1.

$f(x) = 2x^2 + 8x + 7$	Write original function.
$= 2(x^2 + 4x) + 7$	Factor 2 out of x-terms.
$= 2(x^2 + 4x + 4 - 4) + 7$	Because $b = 4$, add and subtract $(4/2)^2 = 4$ within parentheses.

$$\left(\frac{4}{2}\right)^2$$

$= 2(x^2 + 4x + 4) - 2(4) + 7$	Regroup terms.
$= 2(x + 2)^2 - 1$	Write in standard form.

From the standard form, you can see that the graph of f is a parabola that opens upward with vertex $(-2, -1)$, as shown in Figure 3.5. This corresponds to a left shift of two units and a downward shift of one unit relative to the graph of $y = 2x^2$.

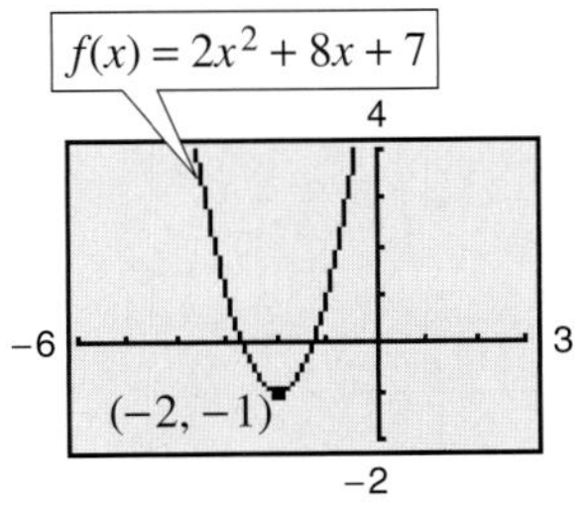

$$f(x) = 2x^2 + 8x + 7$$

Figure 3.5

☑ *Checkpoint* Now try Exercise 19.

To find the x-intercepts of the graph of $f(x) = ax^2 + bx + c$, solve the equation $ax^2 + bx + c = 0$. If $ax^2 + bx + c$ does not factor, you can use the Quadratic Formula to find the x-intercepts, or a graphing utility to approximate the x-intercepts. Remember, however, that a parabola may not have x-intercepts.

Example 3 Identifying x-Intercepts of a Quadratic Function

Describe the graph of $f(x) = -x^2 + 6x - 8$ and identify any x-intercepts.

Solution

$$f(x) = -x^2 + 6x - 8 \qquad \text{Write original function.}$$

$$= -(x^2 - 6x) - 8 \qquad \text{Factor } -1 \text{ out of } x\text{-terms.}$$

$$= -(x^2 - 6x + 9 - 9) - 8 \qquad \begin{array}{l}\text{Because } b = 6, \text{ add and subtract}\\ (6/2)^2 = 9 \text{ within parentheses.}\end{array}$$

$$\left(\frac{6}{2}\right)^2$$

$$= -(x^2 - 6x + 9) - (-9) - 8 \qquad \text{Regroup terms.}$$

$$= -(x - 3)^2 + 1 \qquad \text{Write in standard form.}$$

The graph of f is a parabola that opens downward with vertex $(3, 1)$, as shown in Figure 3.6. The x-intercepts are determined as follows.

$$-(x^2 - 6x + 8) = 0 \qquad \text{Factor out } -1.$$

$$-(x - 2)(x - 4) = 0 \qquad \text{Factor.}$$

$$x - 2 = 0 \quad \Longrightarrow \quad x = 2 \qquad \text{Set 1st factor equal to 0.}$$

$$x - 4 = 0 \quad \Longrightarrow \quad x = 4 \qquad \text{Set 2nd factor equal to 0.}$$

So, the x-intercepts are $(2, 0)$ and $(4, 0)$, as shown in Figure 3.6.

✓ *Checkpoint* Now try Exercise 23.

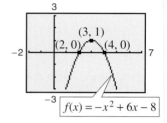

Figure 3.6

Example 4 Writing the Equation of a Parabola in Standard Form

Write the standard form of the equation of the parabola whose vertex is $(1, 2)$ and that passes through the point $(3, -6)$, as shown in Figure 3.7.

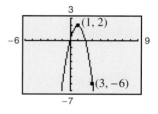

Figure 3.7

Solution

Because the vertex of the parabola is $(h, k) = (1, 2)$, the equation has the form

$$f(x) = a(x - 1)^2 + 2. \qquad \text{Substitute for } h \text{ and } k \text{ in standard form.}$$

Because the parabola passes through the point $(3, -6)$, it follows that $f(3) = -6$. So, you obtain

$$-6 = a(3 - 1)^2 + 2$$

$$-6 = 4a + 2$$

$$-2 = a.$$

The equation in standard form is $f(x) = -2(x - 1)^2 + 2$. Try graphing $f(x) = -2(x - 1)^2 + 2$ with a graphing utility to confirm that its vertex is $(1, 2)$ and that it passes through the point $(3, -6)$.

✓ *Checkpoint* Now try Exercise 35.

STUDY TIP

In Example 4, there are infinitely many different parabolas that have a vertex at $(1, 2)$. Of these, however, the only one that passes through the point $(3, -6)$ is the one given by

$$f(x) = -2(x - 1)^2 + 2.$$

Finding Minimum and Maximum Values

Many applications involve finding the maximum or minimum value of a quadratic function. By writing the quadratic function $f(x) = ax^2 + bx + c$ in standard form,

$$f(x) = a\left(x + \frac{b}{2a}\right)^2 + \left(c - \frac{b^2}{4a}\right)$$

you can see that the vertex occurs at $x = -b/(2a)$, which implies the following.

> **Minimum and Maximum Values of Quadratic Functions**
>
> **1.** If $a > 0$, f has a *minimum* at $x = -\dfrac{b}{2a}$.
>
> **2.** If $a < 0$, f has a *maximum* at $x = -\dfrac{b}{2a}$.

Example 5   The Maximum Height of a Baseball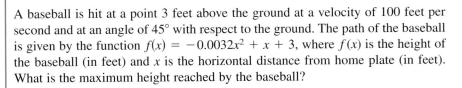

A baseball is hit at a point 3 feet above the ground at a velocity of 100 feet per second and at an angle of 45° with respect to the ground. The path of the baseball is given by the function $f(x) = -0.0032x^2 + x + 3$, where $f(x)$ is the height of the baseball (in feet) and x is the horizontal distance from home plate (in feet). What is the maximum height reached by the baseball?

Algebraic Solution

For this quadratic function, you have

$$f(x) = ax^2 + bx + c = -0.0032x^2 + x + 3$$

which implies that $a = -0.0032$ and $b = 1$. Because the function has a maximum when $x = -b/(2a)$, you can conclude that the baseball reaches its maximum height when it is x feet from home plate, where x is

$$x = -\frac{b}{2a}$$

$$= -\frac{1}{2(-0.0032)} = 156.25 \text{ feet.}$$

At this distance, the maximum height is

$$f(156.25) = -0.0032(156.25)^2 + 156.25 + 3$$

$$= 81.125 \text{ feet.}$$

✓ *Checkpoint* Now try Exercise 63.

Graphical Solution

Use a graphing utility to graph $y = -0.0032x^2 + x + 3$ so that you can see the important features of the parabola. Use the *maximum* feature (see Figure 3.8) or the *zoom* and *trace* features (see Figure 3.9) of the graphing utility to approximate the maximum height on the graph to be $y \approx 81.125$ feet at $x \approx 156.25$. Note that when using the *zoom* and *trace* features, you might have to change the y-scale in order to avoid a graph that is "too flat."

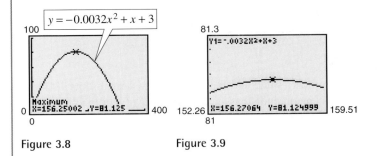

Figure 3.8   Figure 3.9

Example 6   Cost

A soft-drink manufacturer has daily production costs of

$$C = 70,000 - 120x + 0.055x^2$$

where C is the total cost (in dollars) and x is the number of units produced. Estimate numerically the number of units that should be produced each day to yield a minimum cost.

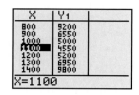

Figure 3.10

Solution

Enter the function $y = 70,000 - 120x + 0.055x^2$ into your graphing utility. Then use the *table* feature of the graphing utility to create a table. Set the table to start at $x = 0$ and set the table step to 100. By scrolling through the table you can see that the minimum cost is between 1000 units and 1200 units, as shown in Figure 3.10. You can improve this estimate by starting the table at $x = 1000$ and setting the table step to 10. From the table in Figure 3.11, you can see that approximately 1090 units should be produced to yield a minimum cost of \$4545.50.

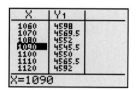

Figure 3.11

 Checkpoint Now try Exercise 65.

Example 7   Hairdressers and Cosmetologists

The number h (in thousands) of hairdressers and cosmetologists in the United States from 1994 to 2001 can be approximated by the model

$$h = 4.17t^2 - 48.1t + 881, \qquad 4 \leq t \leq 11$$

where t represents the year, with $t = 4$ corresponding to 1994. Using this model, determine the year in which the number of hairdressers and cosmetologists was the least.   (Source: U.S. Bureau of Labor Statistics)

Algebraic Solution

Use the fact that the minimum point of the parabola occurs when $t = -b/(2a)$. For this function, you have $a = 4.17$ and $b = -48.1$. So,

$$t = -\frac{b}{2a}$$

$$= -\frac{-48.1}{2(4.17)}$$

$$\approx 5.8$$

From this t-value and the fact that $t = 4$ represents 1994, you can conclude that the least number of hairdressers and cosmetologists occurred sometime during 1995.

Graphical Solution

Use a graphing utility to graph

$$y = 4.17x^2 - 48.1x + 881$$

for $4 \leq x \leq 11$, as shown in Figure 3.12. Use the *minimum* feature (see Figure 3.12) or the *zoom* and *trace* features (see Figure 3.13) of the graphing utility to approximate the minimum point of the parabola to be $x \approx 5.8$. So, you can conclude that the least number of hairdressers and cosmetologists occurred sometime during 1995.

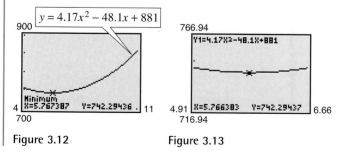

Figure 3.12   Figure 3.13

 Checkpoint Now try Exercise 67.

3.1 Exercises

Vocabulary Check

Fill in the blanks.

1. A polynomial function of degree n and leading coefficient a_n is a function of the form

   $$f(x) = a_n x^n + a_{n-1} x^{n-1} + \cdots + a_1 x + a_0, \quad a_n \neq 0$$

   where n is a _____ and a_i is a _____ number.

2. A _____ function is a second-degree polynomial function, and its graph is called a _____ .

3. The graph of a quadratic function is symmetric about its _____ .

4. If the graph of a quadratic function opens upward, then its leading coefficient is _____ and the vertex of the graph is a _____ .

5. If the graph of a quadratic function opens downward, then its leading coefficient is _____ and the vertex of the graph is a _____ .

In Exercises 1–8, match the quadratic function with its graph. [The graphs are labeled (a), (b), (c), (d), (e), (f), (g), and (h).]

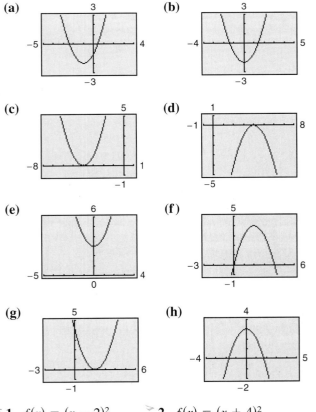

(a) **(b)** **(c)** **(d)** **(e)** **(f)** **(g)** **(h)**

1. $f(x) = (x - 2)^2$
2. $f(x) = (x + 4)^2$
3. $f(x) = x^2 - 2$
4. $f(x) = 3 - x^2$

5. $f(x) = 4 - (x - 2)^2$
6. $f(x) = (x + 1)^2 - 2$
7. $f(x) = x^2 + 3$
8. $f(x) = -(x - 4)^2$

In Exercises 9–12, use a graphing utility to graph each function in the same viewing window. Describe how the graph of each function is related to the graph of $y = x^2$.

9. (a) $y = \frac{1}{2}x^2$ (b) $y = \frac{1}{2}x^2 - 1$
    (c) $y = \frac{1}{2}(x + 3)^2$ (d) $y = -\frac{1}{2}(x + 3)^2 - 1$

10. (a) $y = \frac{3}{2}x^2$ (b) $y = \frac{3}{2}x^2 + 1$
     (c) $y = \frac{3}{2}(x - 3)^2$ (d) $y = -\frac{3}{2}(x - 3)^2 + 1$

11. (a) $y = -2x^2$ (b) $y = -2x^2 - 1$
     (c) $y = -2(x - 3)^2$ (d) $y = 2(x - 3)^2 - 1$

12. (a) $y = -4x^2$ (b) $y = -4x^2 + 3$
     (c) $y = -4(x + 2)^2$ (d) $y = 4(x + 2)^2 + 3$

In Exercises 13–26, sketch the graph of the quadratic function. Identify the vertex and x-intercept(s). Use a graphing utility to verify your results.

13. $f(x) = 25 - x^2$
14. $f(x) = x^2 - 7$
15. $f(x) = \frac{1}{2}x^2 - 4$
16. $f(x) = 16 - \frac{1}{4}x^2$
17. $f(x) = (x + 4)^2 - 3$
18. $f(x) = (x - 6)^2 + 3$
19. $h(x) = x^2 - 8x + 16$
20. $g(x) = x^2 + 2x + 1$
21. $f(x) = x^2 - x + \frac{5}{4}$
22. $f(x) = x^2 + 3x + \frac{1}{4}$

23. $f(x) = -x^2 + 2x + 5$

24. $f(x) = -x^2 - 4x + 1$

25. $h(x) = 4x^2 - 4x + 21$

26. $f(x) = 2x^2 - x + 1$

In Exercises 27–34, use a graphing utility to graph the quadratic function. Identify the vertex and x-intercept(s). Then check your results algebraically by writing the quadratic function in standard form.

27. $f(x) = -(x^2 + 2x - 3)$

28. $f(x) = -(x^2 + x - 30)$

29. $g(x) = x^2 + 8x + 11$

30. $f(x) = x^2 + 10x + 14$

31. $f(x) = -2x^2 + 16x - 31$

32. $f(x) = -4x^2 + 24x - 41$

33. $g(x) = \frac{1}{2}(x^2 + 4x - 2)$

34. $f(x) = \frac{3}{5}(x^2 + 6x - 5)$

In Exercises 35–38, write an equation for the parabola in standard form. Use a graphing utility to graph the equation and verify your result.

35.   **36.**

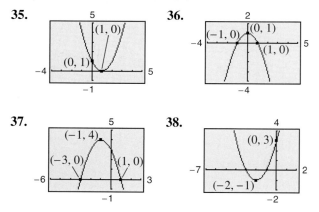

37.   **38.**

In Exercises 39–42, write the standard form of the quadratic function that has the indicated vertex and whose graph passes through the given point. Verify your result with a graphing utility.

39. Vertex: $(-2, 5)$;   Point: $(0, 9)$

40. Vertex: $(4, -1)$;   Point: $(2, 3)$

41. Vertex: $\left(\frac{5}{2}, -\frac{3}{4}\right)$;   Point: $(-2, 4)$

42. Vertex: $\left(-\frac{5}{2}, 0\right)$;   Point: $\left(-\frac{7}{2}, -\frac{16}{3}\right)$

Graphical Reasoning   In Exercises 43–46, determine the x-intercept(s) of the graph visually. How do the x-intercepts correspond to the solutions of the quadratic equation when $y = 0$?

43.   **44.**

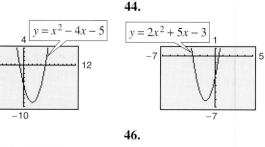

45.   **46.**

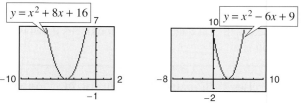

In Exercises 47–52, use a graphing utility to graph the quadratic function. Find the x-intercepts of the graph and compare them with the solutions of the corresponding quadratic equation when $y = 0$.

47. $y = x^2 - 4x$

48. $y = -2x^2 + 10x$

49. $y = 2x^2 - 7x - 30$

50. $y = 4x^2 + 25x - 21$

51. $y = -\frac{1}{2}(x^2 - 6x - 7)$

52. $y = \frac{7}{10}(x^2 + 12x - 45)$

In Exercises 53–56, find two quadratic functions, one that opens upward and one that opens downward, whose graphs have the given x-intercepts. (There are many correct answers.)

53. $(-1, 0), (3, 0)$ **54.** $(0, 0), (10, 0)$

55. $(-3, 0), \left(-\frac{1}{2}, 0\right)$ **56.** $\left(-\frac{5}{2}, 0\right), (2, 0)$

In Exercises 57–60, find two positive real numbers whose product is a maximum.

57. The sum is 110.

58. The sum is S.

59. The sum of the first and twice the second is 24.

60. The sum of the first and three times the second is 42.

61. Geometry An indoor physical fitness room consists of a rectangular region with a semicircle on each end. The perimeter of the room is to be a 200-meter single-lane running track.

(a) Draw a diagram that illustrates the problem. Let x and y represent the length and width of the rectangular region, respectively.

(b) Determine the radius of the semicircular ends of the track. Determine the distance, in terms of y, around the inside edge of the two semicircular parts of the track.

(c) Use the result of part (b) to write an equation, in terms of x and y, for the distance traveled in one lap around the track. Solve for y.

(d) Use the result of part (c) to write the area A of the rectangular region as a function of x.

(e) Use a graphing utility to graph the area function from part (d). Use the graph to approximate the dimensions that will produce a rectangle of maximum area.

62. Numerical, Graphical, and Analytical Analysis A rancher has 200 feet of fencing to enclose two adjacent rectangular corrals (see figure). Use the following methods to determine the dimensions that will produce a maximum enclosed area.

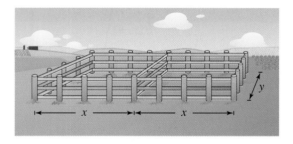

(a) Write the area A of the corral as a function of x.

(b) Use the *table* feature of a graphing utility to create a table showing possible values of x and the corresponding areas of the corral. Use the table to estimate the dimensions that will produce the maximum enclosed area.

(c) Use a graphing utility to graph the area function. Use the graph to approximate the dimensions that will produce the maximum enclosed area.

(d) Write the area function in standard form to find algebraically the dimensions that will produce the maximum area.

(e) Compare your results from parts (b), (c), and (d).

63. Height of a Ball The height y (in feet) of a ball thrown by a child is given by

$$y = -\frac{1}{12}x^2 + 2x + 4$$

where x is the horizontal distance (in feet) from where the ball is thrown (see figure).

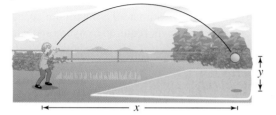

(a) Use a graphing utility to graph the path of the ball.

(b) How high is the ball when it leaves the child's hand? (*Hint:* Find y when $x = 0$.)

(c) What is the maximum height of the ball?

(d) How far from the child does the ball strike the ground?

64. Path of a Diver The path of a diver is given by

$$y = -\frac{4}{9}x^2 + \frac{24}{9}x + 12$$

where y is the height (in feet) and x is the horizontal distance (in feet) from the end of the diving board (see figure). What is the maximum height of the diver? Verify your answer using a graphing utility.

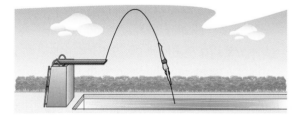

65. Cost A manufacturer of lighting fixtures has daily production costs of

$$C = 800 - 10x + 0.25x^2$$

where C is the total cost (in dollars) and x is the number of units produced. Use the *table* feature of a graphing utility to determine how many fixtures should be produced each day to yield a minimum cost.

66. *Automobile Aerodynamics*   The number of horsepower y required to overcome wind drag on a certain automobile is approximated by

$$y = 0.002s^2 + 0.005s - 0.029, \quad 0 \le s \le 100$$

where s is the speed of the car (in miles per hour).

(a) Use a graphing utility to graph the function.

(b) Graphically estimate the maximum speed of the car if the power required to overcome wind drag is not to exceed 10 horsepower. Verify your result algebraically.

67. *Graphical Analysis*   From 1960 to 2001, the average annual per capita consumption C of cigarettes by Americans (age 18 and older) can be modeled by $C = 4274 + 3.4t - 1.52t^2$, $0 \le t \le 41$, where t is the year, with $t = 0$ corresponding to 1960. (Source: Tobacco Situation and Outlook Yearbook)

(a) Use a graphing utility to graph the model.

(b) Use the graph of the model to approximate the maximum average annual consumption. Beginning in 1966, all cigarette packages were required by law to carry a health warning. Do you think the warning had any effect? Explain.

(c) In 2000, the U.S. population (age 18 and over) was 209,128,000. Of these, about 48,300,000 were smokers. What was the average annual cigarette consumption *per smoker* in 2000? What was the average daily cigarette consumption *per smoker*?

68. *Data Analysis*   The number y (in millions) of VCRs in use in the United States for the years 1994 through 2000 can be modeled by

$$y = -0.17t^2 + 4.3t + 60, \quad 4 \le t \le 10$$

where t represents the year, with $t = 4$ corresponding to 1994. (Source: Television Bureau of Advertising, Inc.)

(a) Use a graphing utility to graph the model.

(b) Do you think the model can be used to estimate VCR use in the year 2008? Explain.

Synthesis

True or False?   **In Exercises 69 and 70, determine whether the statement is true or false. Justify your answer.**

69. The function $f(x) = -12x^2 - 1$ has no x-intercepts.

70. The graphs of $f(x) = -4x^2 - 10x + 7$ and $g(x) = 12x^2 + 30x + 1$ have the same axis of symmetry.

71. *Profit*   The profit P (in millions of dollars) for a recreational vehicle retailer is modeled by a quadratic function of the form $P = at^2 + bt + c$, where t represents the year. If you were president of the company, which of the following models would you prefer? Explain your reasoning.

(a) a is positive and $t \ge -b/(2a)$.

(b) a is positive and $t \le -b/(2a)$.

(c) a is negative and $t \ge -b/(2a)$.

(d) a is negative and $t \le -b/(2a)$.

72. *Writing*   The parabola in the figure below has an equation of the form

$$y = ax^2 + bx - 4.$$

Find the equation of this parabola in two different ways, by hand and with technology (graphing utility or computer software). Write a paragraph describing the methods you used and comparing the results of the two methods.

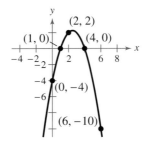

Review

In Exercises 73–76, determine algebraically any points of intersection of the graphs of the equations. Verify your results using the *intersect* feature of a graphing utility.

73. $x + y = 8$
$-\frac{2}{3}x + y = 6$

74. $y = 3x - 10$
$y = \frac{1}{4}x + 1$

75. $y = 9 - x^2$
$y = x + 3$

76. $y = x^3 + 2x - 1$
$y = -2x + 15$

In Exercises 77–80, perform the operation and write the result in standard form.

77. $(6 - i) - (2i + 11)$

78. $(2i + 5)^2 - 21$

79. $(3i + 7)(-4i + 1)$

80. $(4 - i)^3$

3.2 Polynomial Functions of Higher Degree

Graphs of Polynomial Functions

You should be able to sketch accurate graphs of polynomial functions of degrees 0, 1, and 2. The graphs of polynomial functions of degree greater than 2 are more difficult to sketch by hand. However, in this section you will learn how to recognize some of the basic features of the graphs of polynomial functions. Using these features along with point plotting, intercepts, and symmetry, you should be able to make reasonably accurate sketches *by hand*.

The graph of a polynomial function is **continuous.** Essentially, this means that the graph of a polynomial function has no breaks, holes, or gaps, as shown in Figure 3.14.

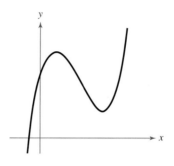

(a) Polynomial functions have continuous graphs.

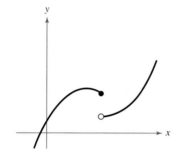

(b) Functions with graphs that are not continuous are not polynomial functions.

Figure 3.14

Another feature of the graph of a polynomial function is that it has only smooth, rounded turns, as shown in Figure 3.15(a). It cannot have a sharp turn such as the one shown in Figure 3.15(b).

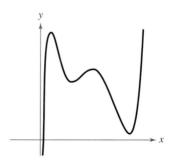

(a) Polynomial functions have graphs with smooth, rounded turns.

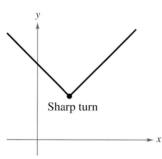

Sharp turn

(b) Graphs of polynomial functions cannot have sharp turns.

Figure 3.15

Informally, you can say that a function is continuous if its graph can be drawn with a pencil without lifting the pencil from the paper.

What you should learn

● Use transformations to sketch graphs of polynomial functions.

● Use the Leading Coefficient Test to determine the end behavior of graphs of polynomial functions.

● Find and use zeros of polynomial functions as sketching aids.

● Use the Intermediate Value Theorem to help locate zeros of polynomial functions.

Why you should learn it

You can use polynomial functions to model various aspects of nature, such as the growth of a red oak tree, as shown in Exercise 88 on page 262.

Leonard Lee Rue III/Earth Scenes

Library of Functions: Polynomial Function

The graphs of polynomial functions of degree 1 are lines, and those of functions of degree 2 are parabolas. The graphs of polynomial functions of higher degree are smooth and continuous. A polynomial function of degree n has the form

$$f(x) = a_n x^n + a_{n-1} x^{n-1} + \cdots + a_2 x^2 + a_1 x + a_0$$

where n is a positive integer and $a_n \neq 0$. The polynomial functions that have the simplest graphs are monomials of the form $f(x) = x^n$, where n is an integer greater than zero. If n is even, the graph is similar to the graph of $f(x) = x^2$ and touches the axis at the x-intercept. If n is odd, the graph is similar to the graph of $f(x) = x^3$ and crosses the axis at the x-intercept. The greater the value of n, the flatter the graph near the origin. The basic characteristics of the *cubic function* $f(x) = x^3$ are summarized below.

Graph of $f(x) = x^3$

Domain: $(-\infty, \infty)$
Range: $(-\infty, \infty)$
Intercept: $(0, 0)$
Increasing on $(-\infty, \infty)$
Odd function
Origin symmetry

Example 1   Transformations of Monomial Functions

Sketch the graph of each function.

a. $f(x) = -x^5$ **b.** $g(x) = x^4 + 1$ **c.** $h(x) = (x + 1)^4$

Solution

a. Because the degree of $f(x) = -x^5$ is odd, the graph is similar to the graph of $y = x^3$. Moreover, the negative coefficient reflects the graph in the x-axis, as shown in Figure 3.16.

b. The graph of $g(x) = x^4 + 1$ is an upward shift of one unit of the graph of $y = x^4$, as shown in Figure 3.17.

c. The graph of $h(x) = (x + 1)^4$ is a left shift of one unit of the graph of $y = x^4$, as shown in Figure 3.18.

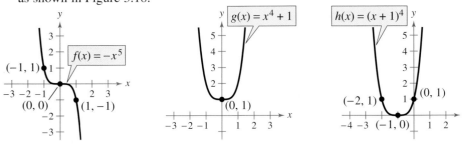

Figure 3.16   Figure 3.17   Figure 3.18

✓ *Checkpoint* Now try Exercise 9.

The Leading Coefficient Test

In Example 1, note that all three graphs eventually rise or fall without bound as x moves to the right. Whether the graph of a polynomial eventually rises or falls can be determined by the function's degree (even or odd) and by its leading coefficient, as indicated in the **Leading Coefficient Test.**

Leading Coefficient Test

As x moves without bound to the left or to the right, the graph of the polynomial function $f(x) = a_n x^n + \cdots + a_1 x + a_0, \, a_n \neq 0$, eventually rises or falls in the following manner.

1. When n is odd:

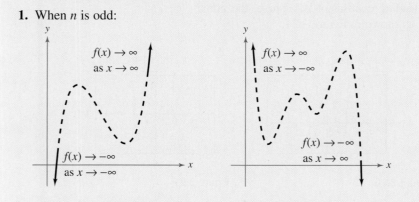

If the leading coefficient is positive $(a_n > 0)$, the graph falls to the left and rises to the right.

If the leading coefficient is negative $(a_n < 0)$, the graph rises to the left and falls to the right.

2. When n is even:

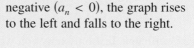
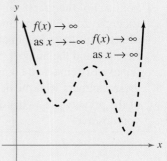
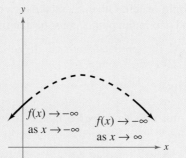

If the leading coefficient is positive $(a_n > 0)$, the graph rises to the left and right.

If the leading coefficient is negative $(a_n < 0)$, the graph falls to the left and right.

Note that the dashed portions of the graphs indicate that the test determines only the right-hand and left-hand behavior of the graph.

As you continue to study polynomial functions and their graphs, you will notice that the degree of a polynomial plays an important role in determining other characteristics of the polynomial and its graph.

Example 2   Applying the Leading Coefficient Test

Use the Leading Coefficient Test to describe the right-hand and left-hand behavior of the graph of each polynomial function.

a. $f(x) = -x^3 + 4x$ **b.** $f(x) = x^4 - 5x^2 + 4$ **c.** $f(x) = x^5 - x$

Solution

a. Because the degree is odd and the leading coefficient is negative, the graph rises to the left and falls to the right, as shown in Figure 3.19.

b. Because the degree is even and the leading coefficient is positive, the graph rises to the left and right, as shown in Figure 3.20.

c. Because the degree is odd and the leading coefficient is positive, the graph falls to the left and rises to the right, as shown in Figure 3.21.

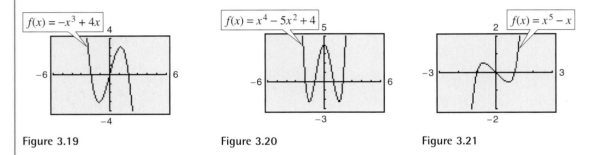

Figure 3.19   Figure 3.20   Figure 3.21

☑ *Checkpoint* Now try Exercise 17.

In Example 2, note that the Leading Coefficient Test only tells you whether the graph *eventually* rises or falls to the right or left. Other characteristics of the graph, such as intercepts and minimum and maximum points, must be determined by other tests.

Zeros of Polynomial Functions

It can be shown that for a polynomial function f of degree n, the following statements are true.

1. The function f has at most n real zeros. (You will study this result in detail in Section 3.4 on the Fundamental Theorem of Algebra.)

2. The graph of f has at most $n - 1$ relative **extrema** (relative **minima** or **maxima**).

Recall that a **zero** of a function f is a number x for which $f(x) = 0$. Finding the zeros of polynomial functions is one of the most important problems in algebra. You have already seen that there is a strong interplay between graphical and algebraic approaches to this problem. Sometimes you can use information about the graph of a function to help find its zeros. In other cases, you can use information about the zeros of a function to find a good viewing window.

> **Exploration**
>
> For each of the graphs in Example 2, count the number of zeros of the polynomial function and the number of relative extrema, and compare these numbers with the degree of the polynomial. What do you observe?

Real Zeros of Polynomial Functions

If f is a polynomial function and a is a real number, the following statements are equivalent.

    1. $x = a$ is a *zero* of the function f.

    2. $x = a$ is a *solution* of the polynomial equation $f(x) = 0$.

    3. $(x - a)$ is a *factor* of the polynomial $f(x)$.

    4. $(a, 0)$ is an *x*-intercept of the graph of f.

TECHNOLOGY SUPPORT

For instructions on how to use the *zero* or *root* feature, see Appendix A; for specific keystrokes, go the text website at *college.hmco.com*.

Finding zeros of polynomial functions is closely related to factoring and finding *x*-intercepts, as demonstrated in Examples 3, 4, and 5.

Example 3   Finding Zeros of a Polynomial Function

Find all real zeros of $f(x) = x^3 - x^2 - 2x$.

Algebraic Solution

$$f(x) = x^3 - x^2 - 2x \qquad \text{Write original function.}$$
$$0 = x^3 - x^2 - 2x \qquad \text{Substitute 0 for } f(x).$$
$$0 = x(x^2 - x - 2) \qquad \text{Remove common monomial factor.}$$
$$0 = x(x - 2)(x + 1) \qquad \text{Factor completely.}$$

So, the real zeros are $x = 0$, $x = 2$, and $x = -1$, and the corresponding *x*-intercepts are $(0, 0)$, $(2, 0)$, and $(-1, 0)$.

Check

$$(0)^3 - (0)^2 - 2(0) = 0 \qquad x = 0 \text{ is a zero. } ✓$$
$$(2)^3 - (2)^2 - 2(2) = 0 \qquad x = 2 \text{ is a zero. } ✓$$
$$(-1)^3 - (-1)^2 - 2(-1) = 0 \qquad x = -1 \text{ is a zero. } ✓$$

✓ *Checkpoint*   Now try Exercise 35.

Graphical Solution

Use a graphing utility to graph $y = x^3 - x^2 - 2x$. In Figure 3.22, the graph appears to have the *x*-intercepts $(0, 0)$, $(2, 0)$, and $(-1, 0)$. Use the *zero* or *root* feature, or the *zoom* and *trace* features, of the graphing utility to verify these intercepts. Note that this third-degree polynomial has two relative extrema, at $(-0.5486, 0.6311)$ and $(1.2152, -2.1126)$.

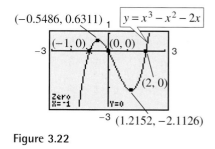

Figure 3.22

Example 4   Analyzing a Polynomial Function

Find all real zeros and relative extrema of $f(x) = -2x^4 + 2x^2$.

Solution

$$0 = -2x^4 + 2x^2 \qquad \text{Substitute 0 for } f(x).$$
$$0 = -2x^2(x^2 - 1) \qquad \text{Remove common monomial factor.}$$
$$0 = -2x^2(x - 1)(x + 1) \qquad \text{Factor completely.}$$

So, the real zeros are $x = 0$, $x = 1$, and $x = -1$, and the corresponding *x*-intercepts are $(0, 0)$, $(1, 0)$, and $(-1, 0)$, as shown in Figure 3.23. Using the *minimum* and *maximum* features of a graphing utility, you can approximate the three relative extrema to be $(-0.7071, 0.5)$, $(0, 0)$, and $(0.7071, 0.5)$.

✓ *Checkpoint*   Now try Exercise 47.

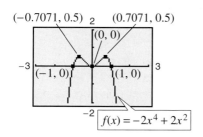

Figure 3.23

> **Repeated Zeros**
>
> For a polynomial function, a factor of $(x - a)^k$, $k > 1$, yields a **repeated zero** $x = a$ of **multiplicity** k.
>
> 1. If k is odd, the graph *crosses* the x-axis at $x = a$.
> 2. If k is even, the graph *touches* the x-axis (but does not cross the x-axis) at $x = a$.

Example 5   Finding Zeros of a Polynomial Function

Find all real zeros of $f(x) = x^5 - 3x^3 - x^2 - 4x - 1$.

Solution

Use a graphing utility to obtain the graph shown in Figure 3.24. From the graph, you can see that there are three zeros. Using the *zero* or *root* feature, you can determine that the zeros are approximately $x \approx -1.861$, $x \approx -0.254$, and $x \approx 2.115$. It should be noted that this fifth-degree polynomial factors as

$$f(x) = x^5 - 3x^3 - x^2 - 4x - 1 = (x^2 + 1)(x^3 - 4x - 1).$$

The three zeros obtained above are the zeros of the cubic factor $x^3 - 4x - 1$ (the quadratic factor $x^2 + 1$ has two complex zeros and so no *real* zeros).

✓ *Checkpoint* Now try Exercise 49.

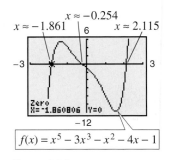

Figure 3.24

Example 6   Finding a Polynomial Function with Given Zeros

Find polynomial functions with the following zeros. (There are many correct solutions.)

a. $-\dfrac{1}{2}, 3, 3$ **b.** $3, 2 + \sqrt{11}, 2 - \sqrt{11}$

Solution

a. Note that the zero $x = -\frac{1}{2}$ corresponds to either $\left(x + \frac{1}{2}\right)$ or $(2x + 1)$. To avoid fractions, choose the second factor and write

$$f(x) = (2x + 1)(x - 3)^2$$
$$= (2x + 1)(x^2 - 6x + 9) = 2x^3 - 11x^2 + 12x + 9.$$

b. For each of the given zeros, form a corresponding factor and write

$$f(x) = (x - 3)\left[x - \left(2 + \sqrt{11}\right)\right]\left[x - \left(2 - \sqrt{11}\right)\right]$$
$$= (x - 3)\left[(x - 2) - \sqrt{11}\right]\left[(x - 2) + \sqrt{11}\right]$$
$$= (x - 3)\left[(x - 2)^2 - \left(\sqrt{11}\right)^2\right]$$
$$= (x - 3)(x^2 - 4x + 4 - 11)$$
$$= (x - 3)(x^2 - 4x - 7) = x^3 - 7x^2 + 5x + 21.$$

✓ *Checkpoint* Now try Exercise 57.

Note in Example 6 that there are many polynomial functions with the indicated zeros. In fact, multiplying the functions by any real number does not change the zeros of the function. For instance, multiply the function from part (b) by $\frac{1}{2}$ to obtain $f(x) = \frac{1}{2}x^3 - \frac{7}{2}x^2 + \frac{5}{2}x + \frac{21}{2}$. Then find the zeros of the function. You will obtain the zeros 3, $2 + \sqrt{11}$, and $2 - \sqrt{11}$ as given in Example 6.

Example 7 Sketching the Graph of a Polynomial Function

Sketch the graph of $f(x) = 3x^4 - 4x^3$ by hand.

Solution

1. *Apply the Leading Coefficient Test.* Because the leading coefficient is positive and the degree is even, you know that the graph eventually rises to the left and to the right (see Figure 3.25).
2. *Find the Zeros of the Polynomial.* By factoring

$$f(x) = 3x^4 - 4x^3 = x^3(3x - 4)$$

you can see that the zeros of f are $x = 0$ (of odd multiplicity 3) and $x = \frac{4}{3}$ (of odd multiplicity 1). So, the x-intercepts occur at $(0, 0)$ and $\left(\frac{4}{3}, 0\right)$. Add these points to your graph, as shown in Figure 3.25.

3. *Plot a Few Additional Points.* To sketch the graph by hand, find a few additional points, as shown in the table. Be sure to choose points between the zeros and to the left and right of the zeros. Then plot the points (see Figure 3.26).

x	-1	0.5	1	1.5
$f(x)$	7	-0.3125	-1	1.6875

4. *Draw the Graph.* Draw a continuous curve through the points, as shown in Figure 3.26. Because both zeros are of odd multiplicity, you know that the graph should cross the x-axis at $x = 0$ and $x = \frac{4}{3}$. If you are unsure of the shape of a portion of the graph, plot some additional points.

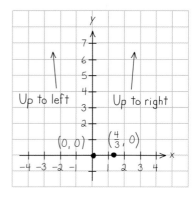

Figure 3.25

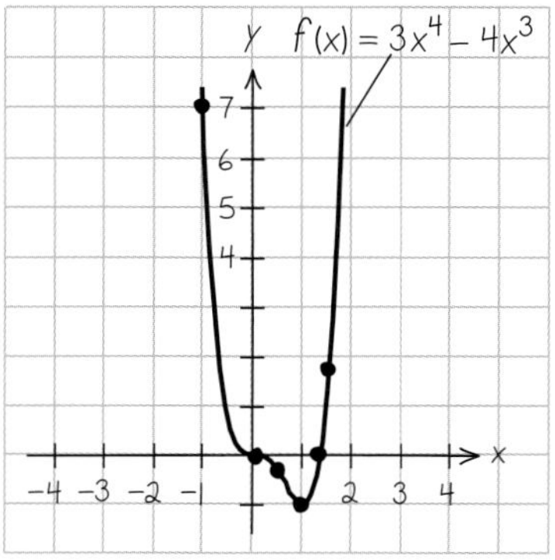

Figure 3.26

✓ *Checkpoint* Now try Exercise 65.

TECHNOLOGY TIP

It is easy to make mistakes when entering functions into a graphing utility. So, it is important to have an understanding of the basic shapes of graphs and to be able to graph simple polynomials *by hand*. For example, suppose you had entered the function in Example 7 as $y = 3x^5 - 4x^3$. By looking at the graph, what mathematical principles would alert you to the fact that you had made a mistake?

Exploration

Partner Activity Multiply three, four, or five distinct linear factors to obtain the equation of a polynomial function of degree 3, 4, or 5. Exchange equations with your partner and sketch, *by hand*, the graph of the equation that your partner wrote. When you are finished, use a graphing utility to check each other's work.

Example 8   Sketching the Graph of a Polynomial Function

Sketch the graph of $f(x) = -2x^3 + 6x^2 - \frac{9}{2}x$.

Solution

1. *Apply the Leading Coefficient Test.*   Because the leading coefficient is negative and the degree is odd, you know that the graph eventually rises to the left and falls to the right (see Figure 3.27).

2. *Find the Zeros of the Polynomial.*   By factoring

$$f(x) = -2x^3 + 6x^2 - \frac{9}{2}x$$
$$= -\frac{1}{2}x(4x^2 - 12x + 9)$$
$$= -\frac{1}{2}x(2x - 3)^2$$

   you can see that the zeros of f are $x = 0$ (of odd multiplicity 1) and $x = \frac{3}{2}$ (of even multiplicity 2). So, the x-intercepts occur at $(0, 0)$ and $\left(\frac{3}{2}, 0\right)$. Add these points to your graph, as shown in Figure 3.27.

3. *Plot a Few Additional Points.*   To sketch the graph by hand, find a few additional points, as shown in the table. Then plot the points (see Figure 3.28.)

x	-0.5	0.5	1	2
$f(x)$	4	-1	-0.5	-1

4. *Draw the Graph.*   Draw a continuous curve through the points, as shown in Figure 3.28. As indicated by the multiplicities of the zeros, the graph crosses the x-axis at $(0, 0)$ and touches (but does not cross) the x-axis at $\left(\frac{3}{2}, 0\right)$.

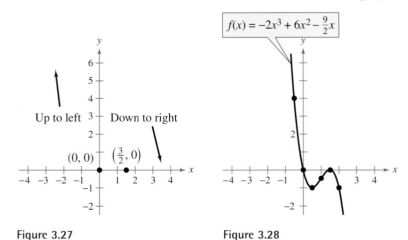

Figure 3.27   Figure 3.28

✓ *Checkpoint* Now try Exercise 67.

TECHNOLOGY TIP   Remember that when using a graphing utility to verify your graphs, you may need to adjust your viewing window in order to see all the features of the graph.

The Intermediate Value Theorem

The **Intermediate Value Theorem** concerns the existence of real zeros of polynomial functions. The theorem states that if $(a, f(a))$ and $(b, f(b))$ are two points on the graph of a polynomial function such that $f(a) \neq f(b)$, then for any number d between $f(a)$ and $f(b)$ there must be a number c between a and b such that $f(c) = d$. (See Figure 3.29.)

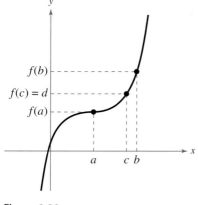

> ### Intermediate Value Theorem
>
> Let a and b be real numbers such that $a < b$. If f is a polynomial function such that $f(a) \neq f(b)$, then in the interval $[a, b]$, f takes on every value between $f(a)$ and $f(b)$.

Figure 3.29

This theorem helps you locate the real zeros of a polynomial function in the following way. If you can find a value $x = a$ at which a polynomial function is positive, and another value $x = b$ at which it is negative, you can conclude that the function has at least one real zero between these two values. For example, the function $f(x) = x^3 + x^2 + 1$ is negative when $x = -2$ and positive when $x = -1$. Therefore, it follows from the Intermediate Value Theorem that f must have a real zero somewhere between -2 and -1.

Example 9   Approximating the Zeros of a Function

Find three intervals of length 1 in which the polynomial $f(x) = 12x^3 - 32x^2 + 3x + 5$ is guaranteed to have a zero.

Graphical Solution

Use a graphing utility to graph

$$y = 12x^3 - 32x^2 + 3x + 5$$

as shown in Figure 3.30.

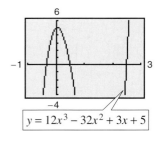

Figure 3.30

From the figure, you can see that the graph crosses the x-axis three times—between -1 and 0, between 0 and 1, and between 2 and 3. So, you can conclude that the function has zeros in the intervals $(-1, 0)$, $(0, 1)$, and $(2, 3)$.

✓ *Checkpoint* Now try Exercise 73.

Numerical Solution

Use the *table* feature of a graphing utility to create a table of function values. Scroll through the table looking for consecutive function values that differ in sign. For instance, from the table in Figure 3.31 you can see that $f(-1)$ and $f(0)$ differ in sign. So, you can conclude from the Intermediate Value Theorem that the function has a zero between -1 and 0. Similarly, $f(0)$ and $f(1)$ differ in sign, so the function has a zero between 0 and 1. Likewise, $f(2)$ and $f(3)$ differ in sign, so the function has a zero between 2 and 3. So, you can conclude that the function has zeros in the intervals $(-1, 0)$, $(0, 1)$, and $(2, 3)$.

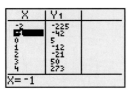

Figure 3.31

3.2   Exercises

Vocabulary Check

Fill in the blanks.

1. The graphs of all polynomial functions are _____ , which means that the graphs have no breaks, holes, or gaps.

2. The _____ is used to determine the left-hand and right-hand behavior of the graph of a polynomial function.

3. A polynomial function of degree n has at most _____ real zeros and at most _____ turning points, called _____ .

4. If $x = a$ is a zero of a polynomial function f, then the following statements are true.
   (a) $x = a$ is a _____ of the polynomial equation $f(x) = 0$.
   (b) _____ is a factor of the polynomial $f(x)$.
   (c) $(a, 0)$ is an _____ of the graph of f.

5. If a zero of a polynomial function is of even multiplicity, then the graph of f _____ the x-axis, and if the zero is of odd multiplicity, then the graph of f _____ the x-axis.

6. The _____ Theorem states that if f is a polynomial function such that $f(a) \neq f(b)$, then in the interval $[a, b]$, f takes on every value between $f(a)$ and $f(b)$.

In Exercises 1–8, match the polynomial function with its graph. [The graphs are labeled (a) through (h).]

1. $f(x) = -2x + 3$

2. $f(x) = x^2 - 4x$

3. $f(x) = -2x^2 - 5x$

4. $f(x) = 2x^3 - 3x + 1$

5. $f(x) = -\frac{1}{4}x^4 + 3x^2$

6. $f(x) = -\frac{1}{3}x^3 + x^2 - \frac{4}{3}$

7. $f(x) = x^4 + 2x^3$

8. $f(x) = \frac{1}{5}x^5 - 2x^3 + \frac{9}{5}x$

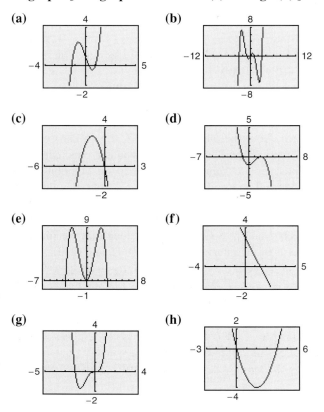

(a) (b) (c) (d) (e) (f) (g) (h)

In Exercises 9–12, sketch the graph of $y = x^n$ and each specified transformation.

9. $y = x^3$
   (a) $f(x) = (x - 2)^3$
   (b) $f(x) = x^3 - 2$
   (c) $f(x) = -\frac{1}{2}x^3$
   (d) $f(x) = (x - 2)^3 - 2$

10. $y = x^5$
   (a) $f(x) = (x + 3)^5$
   (b) $f(x) = x^5 + 3$
   (c) $f(x) = 1 - \frac{1}{2}x^5$
   (d) $f(x) = -\frac{1}{2}(x + 1)^5$

11. $y = x^4$
   (a) $f(x) = (x + 5)^4$
   (b) $f(x) = x^4 - 5$
   (c) $f(x) = 4 - x^4$
   (d) $f(x) = \frac{1}{2}(x - 1)^4$

12. $y = x^6$
   (a) $f(x) = -\frac{1}{8}x^6$
   (b) $f(x) = x^6 - 4$
   (c) $f(x) = -\frac{1}{4}x^6 + 1$
   (d) $f(x) = (x + 2)^6 - 4$

Graphical Analysis   In Exercises 13–16, use a graphing utility to graph the functions *f* and *g* in the same viewing window. Zoom out far enough so that the right-hand and left-hand behaviors of *f* and *g* appear identical. Show both graphs.

13. $f(x) = 3x^3 - 9x + 1, \quad g(x) = 3x^3$

14. $f(x) = -\frac{1}{3}(x^3 - 3x + 2), \quad g(x) = -\frac{1}{3}x^3$

15. $f(x) = -(x^4 - 4x^3 + 16x), \quad g(x) = -x^4$

16. $f(x) = 3x^4 - 6x^2, \quad g(x) = 3x^4$

In Exercises 17–24, use the Leading Coefficient Test to determine the right-hand and left-hand behavior of the graph of the polynomial function. Use a graphing utility to verify your result.

17. $f(x) = 2x^4 - 3x + 1$ 18. $f(x) = \frac{1}{3}x^3 + 5x$

19. $g(x) = 5 - \frac{7}{2}x - 3x^2$ 20. $h(x) = 1 - x^6$

21. $f(x) = \dfrac{6 - 2x + 4x^2 - 5x^3}{3}$

22. $f(x) = \dfrac{3x^4 - 2x + 5}{4}$

23. $h(t) = -\frac{2}{3}(t^2 - 5t + 3)$

24. $f(s) = -\frac{7}{8}(s^3 + 5s^2 - 7s + 1)$

In Exercises 25–34, find all the real zeros of the polynomial function. Determine the multiplicity of each zero. Use a graphing utility to verify your result.

25. $f(x) = x^2 - 25$ 26. $f(x) = 49 - x^2$

27. $h(t) = t^2 - 6t + 9$ 28. $f(x) = x^2 + 10x + 25$

29. $f(x) = x^2 + x - 2$ 30. $f(x) = 2x^2 - 14x + 24$

31. $f(t) = t^3 - 4t^2 + 4t$ 32. $f(x) = x^4 - x^3 - 20x^2$

33. $f(x) = \frac{1}{2}x^2 + \frac{5}{2}x - \frac{3}{2}$ 34. $f(x) = \frac{5}{3}x^2 + \frac{8}{3}x - \frac{4}{3}$

Graphical Analysis   In Exercises 35–46, (a) use a graphing utility to graph the function, (b) use the graph to approximate any zeros (accurate to three decimal places), and (c) find the zeros algebraically.

35. $f(x) = 3x^2 - 12x + 3$

36. $g(x) = 5x^2 - 10x - 5$

37. $g(t) = \frac{1}{2}t^4 - \frac{1}{2}$ 38. $y = \frac{1}{4}x^3(x^2 - 9)$

39. $f(x) = x^5 + x^3 - 6x$ 40. $g(t) = t^5 - 6t^3 + 9t$

41. $f(x) = 2x^4 - 2x^2 - 40$

42. $f(x) = 5x^4 + 15x^2 + 10$

43. $f(x) = x^3 - 4x^2 - 25x + 100$

44. $y = 4x^3 + 4x^2 - 7x + 2$

45. $y = 4x^3 - 20x^2 + 25x$

46. $y = x^5 - 5x^3 + 4x$

In Exercises 47–50, use a graphing utility to graph the function and approximate (accurate to three decimal places) any real zeros and relative extrema.

47. $f(x) = 2x^4 - 6x^2 + 1$

48. $f(x) = -\frac{3}{8}x^4 - x^3 + 2x^2 + 5$

49. $f(x) = x^5 + 3x^3 - x + 6$

50. $f(x) = -3x^3 - 4x^2 + x - 3$

In Exercises 51–60, find a polynomial function that has the given zeros. (There are many correct answers.)

51. $0, 4$ 52. $-7, 2$

53. $0, -2, -3$ 54. $0, 2, 5$

55. $4, -3, 3, 0$ 56. $-2, -1, 0, 1, 2$

57. $1 + \sqrt{3}, 1 - \sqrt{3}$ 58. $6 + \sqrt{3}, 6 - \sqrt{3}$

59. $2, 4 + \sqrt{5}, 4 - \sqrt{5}$ 60. $4, 2 + \sqrt{7}, 2 - \sqrt{7}$

In Exercises 61–72, sketch the graph of the function by (a) applying the Leading Coefficient Test, (b) finding the zeros of the polynomial, (c) plotting sufficient solution points, and (d) drawing a continuous curve through the points.

61. $f(x) = x^3 - 9x$ 62. $g(x) = x^4 - 4x^2$

63. $f(t) = \frac{1}{4}(t^2 - 2t + 15)$

64. $g(x) = -x^2 + 10x - 16$

65. $f(x) = x^3 - 3x^2$ 66. $f(x) = 3x^3 - 24x^2$

67. $f(x) = -x^3 - 5x^2$ 68. $f(x) = 3x^4 - 48x^2$

69. $f(x) = x^2(x - 4)$ 70. $h(x) = \frac{1}{3}x^3(x - 4)^2$

71. $g(t) = -\frac{1}{4}(t - 2)^2(t + 2)^2$

72. $g(x) = \frac{1}{10}(x + 1)^2(x - 3)^2$

In Exercises 73–76, (a) use the Intermediate Value Theorem and a graphing utility to find intervals of length 1 in which the polynomial function is guaranteed to have a zero, (b) use the *root* or *zero* feature of the graphing utility to approximate the zeros of the function, and (c) verify your answers in part (a) by using the *table* feature of the graphing utility.

73. $f(x) = x^3 - 3x^2 + 3$

74. $f(x) = 0.11x^3 - 2.07x^2 + 9.81x - 6.88$

75. $g(x) = 3x^4 + 4x^3 - 3$

76. $h(x) = x^4 - 10x^2 + 2$

In Exercises 77–84, use a graphing utility to graph the function. Identify any symmetry with respect to the x-axis, y-axis, or origin. Determine the number of x-intercepts of the graph.

77. $f(x) = x^2(x + 6)$ **78.** $h(x) = x^3(x - 4)^2$

79. $g(t) = -\frac{1}{2}(t - 4)^2(t + 4)^2$

80. $g(x) = \frac{1}{8}(x + 1)^2(x - 3)^3$

81. $f(x) = x^3 - 4x$ **82.** $f(x) = x^4 - 2x^2$

83. $g(x) = \frac{1}{5}(x + 1)^2(x - 3)(2x - 9)$

84. $h(x) = \frac{1}{5}(x + 2)^2(3x - 5)^2$

85. *Numerical and Graphical Analysis* An open box is to be made from a square piece of material 36 centimeters on a side by cutting equal squares with sides of length x from the corners and turning up the sides (see figure).

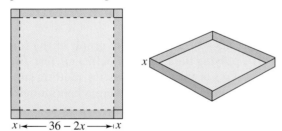

(a) Verify that the volume of the box is given by the function $V(x) = x(36 - 2x)^2$.

(b) Determine the domain of the function V.

(c) Use the *table* feature of a graphing utility to create a table that shows various box heights x and the corresponding volumes V. Use the table to estimate a range of dimensions within which the maximum volume is produced.

(d) Use a graphing utility to graph V and use the range of dimensions from part (c) to find the x-value for which $V(x)$ is maximum.

86. *Geometry* An open box with locking tabs is to be made from a square piece of material 24 inches on a side. This is done by cutting equal squares from the corners and folding along the dashed lines, as shown in the figure.

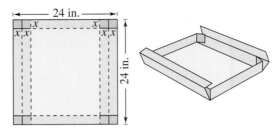

Figure for 86

(a) Verify that the volume of the box is given by the function $V(x) = 8x(6 - x)(12 - x)$.

(b) Determine the domain of the function V.

(c) Sketch the graph of the function and estimate the value of x for which $V(x)$ is maximum.

87. *Revenue* The total revenue R (in millions of dollars) for a company is related to its advertising expense by the function $R = 0.00001(-x^3 + 600x^2)$, $0 \le x \le 400$, where x is the amount spent on advertising (in tens of thousands of dollars). Use the graph of the function shown in the figure to estimate the point on the graph at which the function is increasing most rapidly. This point is called the **point of diminishing returns** because any expense above this amount will yield less return per dollar invested in advertising.

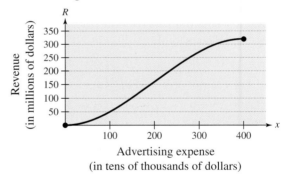

88. *Environment* The growth of a red oak tree is approximated by the function

$$G = -0.003t^3 + 0.137t^2 + 0.458t - 0.839$$

where G is the height of the tree (in feet) and t ($2 \le t \le 34$) is its age (in years). Use a graphing utility to graph the function and estimate the age of the tree when it is growing most rapidly. This point is called the **point of diminishing returns** because the increase in growth will be less with each additional year. (*Hint:* Use a viewing window in which $0 \le x \le 35$ and $0 \le y \le 60$.)

Data Analysis   In Exercises 89–92, use the table, which shows the median prices (in thousands of dollars) of new privately owned U.S. homes in the Northeast y_1 and in the South y_2 for the years 1995 through 2001. The data can be approximated by the following models.

$$y_1 = 0.1250t^3 - 1.446t^2 + 9.07t + 155.5$$

$$y_2 = -0.2000t^3 + 5.155t^2 - 37.23t + 206.8$$

In the models, t represents the year, with $t = 5$ corresponding to 1995. (Sources: U.S. Census Bureau; U.S. Department of Housing and Urban Development)

Year, t	y_1	y_2
5	180.0	124.5
6	186.0	126.2
7	190.0	129.6
8	200.0	135.8
9	210.5	145.9
10	227.4	148.0
11	246.4	155.4

89. Use a graphing utility to plot the data and graph the model for y_1 in the same viewing window. How closely does the model represent the data?

90. Use a graphing utility to plot the data and graph the model for y_2 in the same viewing window. How closely does the model represent the data?

91. Use the models to predict the median price of a new privately-owned home in both regions in 2007. Do your answers seem reasonable? Explain.

92. Use the graphs of the models in Exercises 89 and 90 to write a short paragraph about the relationship between the median prices of homes in the two regions.

Synthesis

True or False?   In Exercises 93 and 94, determine whether the statement is true or false. Justify your answer.

93. A sixth-degree polynomial can have six turning points.

94. The graph of the function

$$f(x) = 2 + x - x^2 + x^3 - x^4 + x^5 + x^6 - x^7$$

rises to the left and falls to the right.

Writing   In Exercises 95–98, match the graph of each cubic function with one of the basic shapes and write a short paragraph describing how you reached your conclusion. Is it possible for a polynomial of odd degree to have no real zeros? Explain.

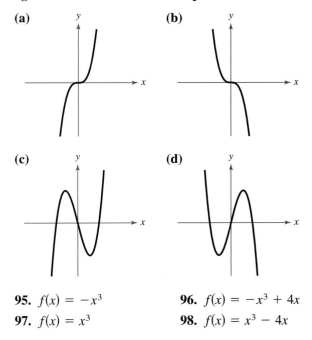

(a)   **(b)**   **(c)**   **(d)**

95. $f(x) = -x^3$

96. $f(x) = -x^3 + 4x$

97. $f(x) = x^3$

98. $f(x) = x^3 - 4x$

Review

In Exercises 99–104, let $f(x) = 14x - 3$ and $g(x) = 8x^2$. Find the indicated value.

99. $(f + g)(-4)$

100. $(g - f)(3)$

101. $(fg)\left(-\dfrac{4}{7}\right)$

102. $\left(\dfrac{f}{g}\right)(-1.5)$

103. $(f \circ g)(-1)$

104. $(g \circ f)(0)$

In Exercises 105–108, solve the inequality and sketch the solution on the real number line. Use a graphing utility to verify your solution graphically.

105. $3(x - 5) < 4x - 7$

106. $2x^2 - x \geq 1$

107. $\dfrac{5x - 2}{x - 7} \leq 4$

108. $|x + 8| - 1 \geq 15$

3.3 Real Zeros of Polynomial Functions

Long Division of Polynomials

Consider the graph of

$$f(x) = 6x^3 - 19x^2 + 16x - 4.$$

Notice in Figure 3.32 that $x = 2$ appears to be a zero of f. Because $f(2) = 0$, you know that $x = 2$ is a zero of the polynomial function f, and that $(x - 2)$ is a factor of $f(x)$. This means that there exists a second-degree polynomial $q(x)$ such that $f(x) = (x - 2) \cdot q(x)$. To find $q(x)$, you can use **long division of polynomials.**

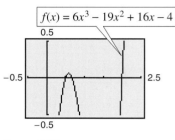

$$f(x) = 6x^3 - 19x^2 + 16x - 4$$

Figure 3.32

Example 1   Long Division of Polynomials

Divide $6x^3 - 19x^2 + 16x - 4$ by $x - 2$, and use the result to factor the polynomial completely.

Solution

Partial quotients

$$
\begin{array}{r}
6x^2 - 7x + 2 \\
x - 2 \overline{)\, 6x^3 - 19x^2 + 16x - 4} \\
\underline{6x^3 - 12x^2} \\
-7x^2 + 16x \\
\underline{-7x^2 + 14x} \\
2x - 4 \\
\underline{2x - 4} \\
0
\end{array}
$$

Multiply: $6x^2(x - 2)$.
Subtract.
Multiply: $-7x(x - 2)$.
Subtract.
Multiply: $2(x - 2)$.
Subtract.

You can see that

$$6x^3 - 19x^2 + 16x - 4 = (x - 2)(6x^2 - 7x + 2)$$

$$= (x - 2)(2x - 1)(3x - 2).$$

Note that this factorization agrees with the graph of f (see Figure 3.32) in that the three x-intercepts occur at $x = 2$, $x = \frac{1}{2}$, and $x = \frac{2}{3}$.

✓ *Checkpoint* Now try Exercise 1.

What you should learn

- Use long division to divide polynomials by other polynomials.
- Use synthetic division to divide polynomials by binomials of the form $(x - k)$.
- Use the Remainder and Factor Theorems.
- Use the Rational Zero Test to determine possible rational zeros of polynomial functions.
- Use Descartes's Rule of Signs and the Upper and Lower Bound Rules to find zeros of polynomials.

Why you should learn it

Polynomial division can help you rewrite polynomials that are used to model real-life problems. For instance, Exercise 80 on page 277 shows how polynomial division can be used to model the sales from lottery tickets in the United States from 1995 through 2001.

Reuters NewMedia, Inc./Corbis

STUDY TIP

Note that in Example 1, the division process requires $-7x^2 + 14x$ to be subtracted from $-7x^2 + 16x$. Therefore it is implied that

$$\frac{-7x^2 + 16x}{-(-7x^2 + 14x)} = \frac{-7x^2 + 16x}{7x^2 - 14x}$$

and instead is written simply as

$$\frac{-7x^2 + 16x}{-7x^2 + 14x.}$$
$$2x$$

In Example 1, $x - 2$ is a factor of the polynomial $6x^3 - 19x^2 + 16x - 4$, and the long division process produces a remainder of zero. Often, long division will produce a nonzero remainder. For instance, if you divide $x^2 + 3x + 5$ by $x + 1$, you obtain the following.

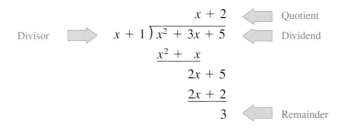

In fractional form, you can write this result as follows.

$$\underbrace{\frac{x^2 + 3x + 5}{x + 1}}_{\substack{\text{Dividend} \\ \text{Divisor}}} = \overbrace{x + 2}^{\text{Quotient}} + \underbrace{\frac{\overset{\text{Remainder}}{3}}{x + 1}}_{\text{Divisor}}$$

This implies that

$$x^2 + 3x + 5 = (x + 1)(x + 2) + 3 \qquad \text{Multiply each side by } (x + 1).$$

which illustrates the following theorem, called the **Division Algorithm.**

The Division Algorithm

If $f(x)$ and $d(x)$ are polynomials such that $d(x) \neq 0$, and the degree of $d(x)$ is less than or equal to the degree of $f(x)$, there exist unique polynomials $q(x)$ and $r(x)$ such that

$$f(x) = d(x)q(x) + r(x)$$

$\uparrow$ Dividend $\uparrow$ Quotient $\uparrow$

$\quad$ Divisor   Remainder

where $r(x) = 0$ *or* the degree of $r(x)$ is less than the degree of $d(x)$. If the remainder $r(x)$ is zero, $d(x)$ *divides evenly* into $f(x)$.

The Division Algorithm can also be written as

$$\frac{f(x)}{d(x)} = q(x) + \frac{r(x)}{d(x)}.$$

In the Division Algorithm, the rational expression $f(x)/d(x)$ is **improper** because the degree of $f(x)$ is greater than or equal to the degree of $d(x)$. On the other hand, the rational expression $r(x)/d(x)$ is **proper** because the degree of $r(x)$ is less than the degree of $d(x)$.

Before you apply the Division Algorithm, follow these steps.

1. Write the dividend and divisor in descending powers of the variable.

2. Insert placeholders with zero coefficients for missing powers of the variable.

Example 2   Long Division of Polynomials

Divide $x^3 - 1$ by $x - 1$.

Solution

Because there is no x^2-term or x-term in the dividend, you need to line up the subtraction by using zero coefficients (or leaving spaces) for the missing terms.

$$
\begin{array}{r}
x^2 + x + 1 \\
x - 1 \overline{\smash{)}\, x^3 + 0x^2 + 0x - 1} \\
\underline{x^3 - x^2} \\
x^2 + 0x \\
\underline{x^2 - x} \\
x - 1 \\
\underline{x - 1} \\
0
\end{array}
$$

So, $x - 1$ divides evenly into $x^3 - 1$, and you can write

$$\frac{x^3 - 1}{x - 1} = x^2 + x + 1, \qquad x \neq 1.$$

☑ *Checkpoint* Now try Exercise 7.

You can check the result of Example 2 by multiplying.

$$(x - 1)(x^2 + x + 1) = x^3 + x^2 + x - x^2 - x - 1$$
$$= x^3 - 1$$

Example 3   Long Division of Polynomials

Divide $2x^4 + 4x^3 - 5x^2 + 3x - 2$ by $x^2 + 2x - 3$.

Solution

$$
\begin{array}{r}
2x^2 + 1 \\
x^2 + 2x - 3 \overline{\smash{)}\, 2x^4 + 4x^3 - 5x^2 + 3x - 2} \\
\underline{2x^4 + 4x^3 - 6x^2} \\
x^2 + 3x - 2 \\
\underline{x^2 + 2x - 3} \\
x + 1
\end{array}
$$

Note that the first subtraction eliminated two terms from the dividend. When this happens, the quotient skips a term. You can write the result as

$$\frac{2x^4 + 4x^3 - 5x^2 + 3x - 2}{x^2 + 2x - 3} = 2x^2 + 1 + \frac{x + 1}{x^2 + 2x - 3}.$$

☑ *Checkpoint* Now try Exercise 9.

Synthetic Division

There is a nice shortcut for long division of polynomials when dividing by divisors of the form $x - k$. The shortcut is called **synthetic division.** The pattern for synthetic division of a cubic polynomial is summarized as follows. (The pattern for higher-degree polynomials is similar.)

Synthetic Division (of a Cubic Polynomial)

To divide $ax^3 + bx^2 + cx + d$ by $x - k$, use the following pattern.

Vertical pattern: Add terms.
Diagonal pattern: Multiply by k.

Synthetic division works *only* for divisors of the form $x - k$. [Remember that $x + k = x - (-k)$.] You cannot use synthetic division to divide a polynomial by a quadratic such as $x^2 - 3$.

Example 4   Using Synthetic Division

Use synthetic division to divide $x^4 - 10x^2 - 2x + 4$ by $x + 3$.

Solution

You should set up the array as follows. Note that a zero is included for each missing term in the dividend.

Then, use the synthetic division pattern by adding terms in columns and multiplying the results by -3.

So, you have

$$\frac{x^4 - 10x^2 - 2x + 4}{x + 3} = x^3 - 3x^2 - x + 1 + \frac{1}{x + 3}.$$

✓ *Checkpoint* Now try Exercise 19.

Exploration

Evaluate the polynomial $x^4 - 10x^2 - 2x + 4$ at $x = -3$. What do you observe?

The Remainder and Factor Theorems

The remainder obtained in the synthetic division process has an important interpretation, as described in the **Remainder Theorem.** See Appendix B for a proof of the Remainder Theorem.

> **The Remainder Theorem**
>
> If a polynomial $f(x)$ is divided by $x - k$, the remainder is
>
> $r = f(k)$.

The Remainder Theorem tells you that synthetic division can be used to evaluate a polynomial function. That is, to evaluate a polynomial function $f(x)$ when $x = k$, divide $f(x)$ by $x - k$. The remainder will be $f(k)$.

Example 5   Using the Remainder Theorem

Use the Remainder Theorem to evaluate the following function at $x = -2$.

$$f(x) = 3x^3 + 8x^2 + 5x - 7$$

Solution

Using synthetic division, you obtain the following.

$$
\begin{array}{r|rrrr}
-2 & 3 & 8 & 5 & -7 \\
   &   & -6 & -4 & -2 \\
\hline
   & 3 & 2 & 1 & -9
\end{array}
$$

Because the remainder is $r = -9$, you can conclude that

$$f(-2) = -9. \qquad {\scriptstyle r = f(k)}$$

This means that $(-2, -9)$ is a point on the graph of f. You can check this by substituting $x = -2$ in the original function.

Check

$$
\begin{aligned}
f(-2) &= 3(-2)^3 + 8(-2)^2 + 5(-2) - 7 \\
&= 3(-8) + 8(4) - 10 - 7 \\
&= -24 + 32 - 10 - 7 = -9
\end{aligned}
$$

✓ *Checkpoint* Now try Exercise 31.

Another important theorem is the **Factor Theorem.** This theorem states that you can test whether a polynomial has $(x - k)$ as a factor by evaluating the polynomial at $x = k$. If the result is 0, $(x - k)$ is a factor. See Appendix B for a proof of the Factor Theorem.

> **The Factor Theorem**
>
> A polynomial $f(x)$ has a factor $(x - k)$ if and only if $f(k) = 0$.

Example 6   Factoring a Polynomial: Repeated Division

Show that $(x - 2)$ and $(x + 3)$ are factors of

$$f(x) = 2x^4 + 7x^3 - 4x^2 - 27x - 18.$$

Then find the remaining factors of $f(x)$.

Algebraic Solution

Using synthetic division with the factor $(x - 2)$, you obtain the following.

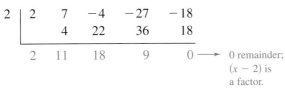

$$\begin{array}{r|rrrrr}
2 & 2 & 7 & -4 & -27 & -18 \\
& & 4 & 22 & 36 & 18 \\
\hline
& 2 & 11 & 18 & 9 & 0
\end{array}$$

0 remainder;
$(x - 2)$ is
a factor.

Take the result of this division and perform synthetic division again using the factor $(x + 3)$.

$$\begin{array}{r|rrrr}
-3 & 2 & 11 & 18 & 9 \\
& & -6 & -15 & -9 \\
\hline
& 2 & 5 & 3 & 0
\end{array}$$

$2x^2 + 5x + 3$

0 remainder;
$(x + 3)$ is
a factor.

Because the resulting quadratic factors as

$$2x^2 + 5x + 3 = (2x + 3)(x + 1)$$

the complete factorization of $f(x)$ is

$$f(x) = (x - 2)(x + 3)(2x + 3)(x + 1).$$

☑ *Checkpoint* Now try Exercise 39.

Graphical Solution

The graph of a polynomial with factors of $(x - 2)$ and $(x + 3)$ has x-intercepts at $x = 2$ and $x = -3$. Use a graphing utility to graph

$$y = 2x^4 + 7x^3 - 4x^2 - 27x - 18.$$

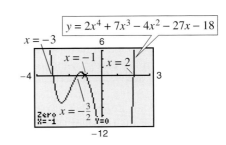

Figure 3.33

From Figure 3.33, you can see that the graph appears to cross the x-axis in two other places, near $x = -1$ and $x = -\frac{3}{2}$. Use the *zero* or *root* feature or the *zoom* and *trace* features to approximate the other two intercepts to be $x = -1$ and $x = -\frac{3}{2}$. So, the factors of f are $(x - 2)$, $(x + 3)$, $\left(x + \frac{3}{2}\right)$, and $(x + 1)$. You can rewrite the factor $\left(x + \frac{3}{2}\right)$ as $(2x + 3)$, so the complete factorization of f is $f(x) = (x - 2)(x + 3)(2x + 3)(x + 1).$

Using the Remainder in Synthetic Division

In summary, the remainder r, obtained in the synthetic division of $f(x)$ by $x - k$, provides the following information.

1. The remainder r gives the value of f at $x = k$. That is, $r = f(k)$.

2. If $r = 0$, $(x - k)$ is a factor of $f(x)$.

3. If $r = 0$, $(k, 0)$ is an x-intercept of the graph of f.

Throughout this text, the importance of developing several problem-solving strategies is emphasized. In the exercises for this section, try using more than one strategy to solve several of the exercises. For instance, if you find that $x - k$ divides evenly into $f(x)$, try sketching the graph of f. You should find that $(k, 0)$ is an x-intercept of the graph.

The Rational Zero Test

The **Rational Zero Test** relates the possible rational zeros of a polynomial (having integer coefficients) to the leading coefficient and to the constant term of the polynomial.

The Rational Zero Test

If the polynomial

$$f(x) = a_n x^n + a_{n-1} x^{n-1} + \cdots + a_2 x^2 + a_1 x + a_0$$

has integer coefficients, every rational zero of f has the form

$$\text{Rational zero} = \frac{p}{q}$$

where p and q have no common factors other than 1, p is a factor of the constant term a_0, and q is a factor of the leading coefficient a_n.

To use the Rational Zero Test, first list all rational numbers whose numerators are factors of the constant term and whose denominators are factors of the leading coefficient.

$$\text{Possible rational zeros} = \frac{\text{factors of constant term}}{\text{factors of leading coefficient}}$$

Now that you have formed this list of *possible rational zeros*, use a trial-and-error method to determine which, if any, are actual zeros of the polynomial. Note that when the leading coefficient is 1, the possible rational zeros are simply the factors of the constant term. This case is illustrated in Example 7.

Example 7 Rational Zero Test with Leading Coefficient of 1

Find the rational zeros of $f(x) = x^3 + x + 1$.

Solution

Because the leading coefficient is 1, the possible rational zeros are simply the factors of the constant term.

Possible rational zeros: ± 1

By testing these possible zeros, you can see that neither works.

$$f(1) = (1)^3 + 1 + 1 = 3$$

$$f(-1) = (-1)^3 + (-1) + 1 = -1$$

So, you can conclude that the polynomial has *no* rational zeros. Note from the graph of f in Figure 3.34 that f does have one real zero between -1 and 0. However, by the Rational Zero Test, you know that this real zero is *not* a rational number.

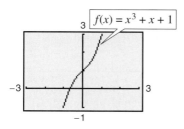

Figure 3.34

 Checkpoint Now try Exercise 45.

If the leading coefficient of a polynomial is not 1, the list of possible rational zeros can increase dramatically. In such cases the search can be shortened in several ways.

1. A programmable calculator can be used to speed up the calculations.

2. A graphing utility can give a good estimate of the locations of the zeros.

3. The Intermediate Value Theorem, along with a table generated by a graphing utility, can give approximations of zeros.

4. The Factor Theorem and synthetic division can be used to test the possible rational zeros.

Finding the first zero is often the most difficult part. After that, the search is simplified by working with the lower-degree polynomial obtained in synthetic division.

Example 8   Using the Rational Zero Test

Find the rational zeros of $f(x) = 2x^3 + 3x^2 - 8x + 3$.

Solution

The leading coefficient is 2 and the constant term is 3.

Possible rational zeros:

$$\frac{\text{Factors of 3}}{\text{Factors of 2}} = \frac{\pm 1, \pm 3}{\pm 1, \pm 2} = \pm 1, \pm 3, \pm \frac{1}{2}, \pm \frac{3}{2}$$

By synthetic division, you can determine that $x = 1$ is a rational zero.

$$
\begin{array}{r|rrrr}
1 & 2 & 3 & -8 & 3 \\
  &   & 2 & 5 & -3 \\
\hline
  & 2 & 5 & -3 & 0
\end{array}
$$

So, $f(x)$ factors as

$$f(x) = (x - 1)(2x^2 + 5x - 3) = (x - 1)(2x - 1)(x + 3)$$

and you can conclude that the rational zeros of f are $x = 1$, $x = \frac{1}{2}$, and $x = -3$, as shown in Figure 3.35.

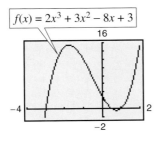

Figure 3.35

 Checkpoint Now try Exercise 47.

A graphing utility can help you determine which possible rational zeros to test, as demonstrated in Example 9.

Example 9 Finding Real Zeros of a Polynomial Function

Find all the real zeros of $f(x) = 10x^3 - 15x^2 - 16x + 12$.

Solution

Because the leading coefficient is 10 and the constant term is 12, there is a long list of possible rational zeros.

Possible rational zeros:

$$\frac{\text{Factors of } 12}{\text{Factors of } 10} = \frac{\pm 1, \pm 2, \pm 3, \pm 4, \pm 6, \pm 12}{\pm 1, \pm 2, \pm 5, \pm 10}$$

With so many possibilities (32, in fact), it is worth your time to use a graphing utility to focus on just a few. By using the *trace* feature of a graphing utility, it looks like three reasonable choices are $x = -\frac{6}{5}$, $x = \frac{1}{2}$, and $x = 2$ (see Figure 3.36). Synthetic division shows that only $x = 2$ works. (You could also use the Factor Theorem to test these choices.)

$$
\begin{array}{r|rrrr}
2 & 10 & -15 & -16 & 12 \\
  &   & 20 & 10 & -12 \\
\hline
  & 10 & 5   & -6 & 0
\end{array}
$$

So, $x = 2$ is one zero and you have

$$f(x) = (x - 2)(10x^2 + 5x - 6).$$

Using the Quadratic Formula, you find that the two additional zeros are irrational numbers.

$$x = \frac{-5 + \sqrt{265}}{20} \approx 0.5639 \quad \text{and} \quad x = \frac{-5 - \sqrt{265}}{20} \approx -1.0639$$

✓ *Checkpoint* Now try Exercise 51.

TECHNOLOGY TIP

You can use the *table* feature of a graphing utility to test the possible rational zeros of the function in Example 9, as shown below. Set the table to start at $x = -12$ and set the table step to 0.1. Look through the table to determine the values of x for which y_1 is 0.

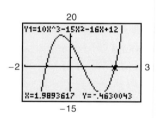

Figure 3.36

Other Tests for Zeros of Polynomials

You know that an nth-degree polynomial function can have *at most n* real zeros. Of course, many nth-degree polynomials do not have that many real zeros. For instance, $f(x) = x^2 + 1$ has no real zeros, and $f(x) = x^3 + 1$ has only one real zero. The following theorem, called **Descartes's Rule of Signs,** sheds more light on the number of real zeros of a polynomial.

> **Descartes's Rule of Signs**
>
> Let $f(x) = a_n x^n + a_{n-1} x^{n-1} + \cdots + a_2 x^2 + a_1 x + a_0$ be a polynomial with real coefficients and $a_0 \neq 0$.
>
> 1. The number of *positive real zeros* of f is either equal to the number of variations in sign of $f(x)$ or less than that number by an even integer.
> 2. The number of *negative real zeros* of f is either equal to the number of variations in sign of $f(-x)$ or less than that number by an even integer.

A **variation in sign** means that two consecutive (nonzero) coefficients have opposite signs.

When using Descartes's Rule of Signs, a zero of multiplicity k should be counted as k zeros. For instance, the polynomial $x^3 - 3x + 2$ has two variations in sign, and so has either two positive or no positive real zeros. Because

$$x^3 - 3x + 2 = (x - 1)(x - 1)(x + 2)$$

you can see that the two positive real zeros are $x = 1$ of multiplicity 2.

Example 10   Using Descartes's Rule of Signs

Describe the possible real zeros of $f(x) = 3x^3 - 5x^2 + 6x - 4$.

Solution

The original polynomial has *three* variations in sign.

$$\begin{array}{c} \overbrace{}^{+ \text{ to } -} \quad \overbrace{}^{+ \text{ to } -} \\ f(x) = 3x^3 - 5x^2 + 6x - 4 \\ \underbrace{}_{- \text{ to } +} \end{array}$$

The polynomial

$$f(-x) = 3(-x)^3 - 5(-x)^2 + 6(-x) - 4 = -3x^3 - 5x^2 - 6x - 4$$

has no variations in sign. So, from Descartes's Rule of Signs, the polynomial $f(x) = 3x^3 - 5x^2 + 6x - 4$ has either three positive real zeros or one positive real zero, and has no negative real zeros. By using the *trace* feature of a graphing utility, you can see that the function has only one real zero (it is a positive number near $x = 1$), as shown in Figure 3.37.

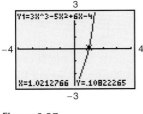

Figure 3.37

✅ *Checkpoint* Now try Exercise 57.

Another test for zeros of a polynomial function is related to the sign pattern in the last row of the synthetic division array. This test can give you an upper or lower bound of the real zeros of f, which can help you eliminate possible real zeros. A real number b is an **upper bound** for the real zeros of f if no zeros are greater than b. Similarly, b is a **lower bound** if no real zeros of f are less than b.

Upper and Lower Bound Rules

Let $f(x)$ be a polynomial with real coefficients and a positive leading coefficient. Suppose $f(x)$ is divided by $x - c$, using synthetic division.

1. If $c > 0$ and each number in the last row is either positive or zero, c is an **upper bound** for the real zeros of f.

2. If $c < 0$ and the numbers in the last row are alternately positive and negative (zero entries count as positive or negative), c is a **lower bound** for the real zeros of f.

Example 11   Finding the Zeros of a Polynomial Function

Find the real zeros of $f(x) = 6x^3 - 4x^2 + 3x - 2$.

Solution

The possible real zeros are as follows.

$$\frac{\text{Factors of 2}}{\text{Factors of 6}} = \frac{\pm 1, \pm 2}{\pm 1, \pm 2, \pm 3, \pm 6} = \pm 1, \pm\frac{1}{2}, \pm\frac{1}{3}, \pm\frac{1}{6}, \pm\frac{2}{3}, \pm 2$$

The original polynomial $f(x)$ has three variations in sign. The polynomial

$$f(-x) = 6(-x)^3 - 4(-x)^2 + 3(-x) - 2$$
$$= -6x^3 - 4x^2 - 3x - 2$$

has no variations in sign. As a result of these two findings, you can apply Descartes's Rule of Signs to conclude that there are three positive real zeros or one positive real zero, and no negative zeros. Trying $x = 1$ produces the following.

$$\begin{array}{r|rrrr} 1 & 6 & -4 & 3 & -2 \\ & & 6 & 2 & 5 \\ \hline & 6 & 2 & 5 & 3 \end{array}$$

So, $x = 1$ is not a zero, but because the last row has all positive entries, you know that $x = 1$ is an upper bound for the real zeros. Therefore, you can restrict the search to zeros between 0 and 1. By trial and error, you can determine that $x = \frac{2}{3}$ is a zero. So,

$$f(x) = \left(x - \frac{2}{3}\right)(6x^2 + 3).$$

Because $6x^2 + 3$ has no real zeros, it follows that $x = \frac{2}{3}$ is the only real zero.

✓ *Checkpoint*   Now try Exercise 67.

Before concluding this section, here are two additional hints that can help you find the real zeros of a polynomial.

1. If the terms of $f(x)$ have a common monomial factor, it should be factored out before applying the tests in this section. For instance, by writing

$$f(x) = x^4 - 5x^3 + 3x^2 + x = x(x^3 - 5x^2 + 3x + 1)$$

you can see that $x = 0$ is a zero of f and that the remaining zeros can be obtained by analyzing the cubic factor.

2. If you are able to find all but two zeros of $f(x)$, you can always use the Quadratic Formula on the remaining quadratic factor. For instance, if you succeeded in writing

$$f(x) = x^4 - 5x^3 + 3x^2 + x = x(x - 1)(x^2 - 4x - 1)$$

you can apply the Quadratic Formula to $x^2 - 4x - 1$ to conclude that the two remaining zeros are $x = 2 + \sqrt{5}$ and $x = 2 - \sqrt{5}$.

Exploration

Use a graphing utility to graph

$$y_1 = 6x^3 - 4x^2 + 3x - 2.$$

Notice that the graph intersects the x-axis at the point $\left(\frac{2}{3}, 0\right)$. How does this information relate to the real zero found in Example 11? Use a graphing utility to graph

$$y_2 = x^4 - 5x^3 + 3x^2 + x.$$

How many times does the graph intersect the x-axis? How many real zeros does y_2 have?

Exploration

Use a graphing utility to graph

$$y = x^3 + 4.9x^2 - 126x + 382.5$$

in the standard viewing window. From the graph, what do the real zeros appear to be? Discuss how the mathematical tools of this section might help you realize that the graph does not show all the important features of the polynomial function. Now use the *zoom* feature to find all the zeros of this function.

3.3   Exercises

Vocabulary Check

1. Two forms of the Division Algorithm are shown below. Identify and label each part.

$$f(x) = d(x)q(x) + r(x) \qquad \frac{f(x)}{d(x)} = q(x) + \frac{r(x)}{d(x)}$$

In Exercises 2–7, fill in the blanks.

2. The rational expression $p(x)/q(x)$ is called _____ if the degree of the numerator is greater than or equal to that of the denominator, and is called _____ if the degree of the numerator is less than that of the denominator.

3. An alternative method to long division of polynomials is called _____ , in which the divisor must be of the form $x - k$.

4. The test that gives a list of the possible rational zeros of a polynomial function is known as the _____ Test.

5. The theorem that can be used to determine the possible numbers of positive real zeros and negative real zeros of a function is called _____ of _____ .

6. The _____ states that if a polynomial $f(x)$ is divided by $x - k$, then the remainder is $r = f(k)$.

7. A real number b is an _____ for the real zeros of f if no zeros are greater than b, and is a _____ if no real zeros of f are less than b.

In Exercises 1–12, use long division to divide.

1. Divide $2x^2 + 10x + 12$ by $x + 3$.

2. Divide $5x^2 - 17x - 12$ by $x - 4$.

3. Divide $4x^3 - 7x^2 - 11x + 5$ by $4x + 5$.

4. Divide $x^4 + 5x^3 + 6x^2 - x - 2$ by $x + 2$.

5. Divide $7x + 3$ by $x + 2$.

6. Divide $8x - 5$ by $2x + 1$.

7. $(6x^3 + 10x^2 + x + 8) \div (2x^2 + 1)$

8. $(x^4 + 3x^2 + 1) \div (x^2 - 2x + 3)$

9. $(x^3 - 9) \div (x^2 + 1)$ 10. $(x^5 + 7) \div (x^3 - 1)$

11. $\dfrac{2x^3 - 4x^2 - 15x + 5}{(x - 1)^2}$ 12. $\dfrac{x^4}{(x - 1)^3}$

In Exercises 13–22, use synthetic division to divide.

13. $(3x^3 - 17x^2 + 15x - 25) \div (x - 5)$

14. $(5x^3 + 18x^2 + 7x - 6) \div (x + 3)$

15. $(6x^3 + 7x^2 - x + 26) \div (x - 3)$

16. $(2x^3 + 14x^2 - 20x + 7) \div (x + 6)$

17. $(9x^3 - 18x^2 - 16x + 32) \div (x - 2)$

18. $(5x^3 + 6x + 8) \div (x + 2)$

19. $(x^3 + 512) \div (x + 8)$

20. $(x^3 - 729) \div (x - 9)$

21. $\dfrac{4x^3 + 16x^2 - 23x - 15}{x + \frac{1}{2}}$ 22. $\dfrac{3x^3 - 4x^2 + 5}{x - \frac{3}{2}}$

Graphical Analysis **In Exercises 23 and 24, use a graphing utility to graph the two equations in the same viewing window. Use the graphs to verify that the expressions are equivalent. Verify the results algebraically.**

23. $y_1 = \dfrac{x^2}{x + 2}$, $y_2 = x - 2 + \dfrac{4}{x + 2}$

24. $y_1 = \dfrac{x^4 - 3x^2 - 1}{x^2 + 5}$, $y_2 = x^2 - 8 + \dfrac{39}{x^2 + 5}$

In Exercises 25–30, write the function in the form $f(x) = (x - k)q(x) + r(x)$ for the given value of k. Use a graphing utility to demonstrate that $f(k) = r$.

Function	*Value of k*
25. $f(x) = x^3 - x^2 - 14x + 11$	$k = 4$

	Function	Value of k
26.	$f(x) = 15x^4 + 10x^3 - 6x^2 + 14$	$k = -\frac{2}{3}$
27.	$f(x) = x^3 + 3x^2 - 2x - 14$	$k = \sqrt{2}$
28.	$f(x) = x^3 + 2x^2 - 5x - 4$	$k = -\sqrt{5}$
29.	$f(x) = 4x^3 - 6x^2 - 12x - 4$	$k = 1 - \sqrt{3}$
30.	$f(x) = -3x^3 + 8x^2 + 10x - 8$	$k = 2 + \sqrt{2}$

In Exercises 31–34, use synthetic division to find each function value. Use a graphing utility to verify your results.

31. $f(x) = 4x^3 - 13x + 10$

   (a) $f(1)$ (b) $f(-2)$ (c) $f\left(\frac{1}{2}\right)$ (d) $f(8)$

32. $g(x) = x^6 - 4x^4 + 3x^2 + 2$

   (a) $g(2)$ (b) $g(-4)$ (c) $g(3)$ (d) $g(-1)$

33. $h(x) = 3x^3 + 5x^2 - 10x + 1$

   (a) $h(3)$ (b) $h\left(\frac{1}{3}\right)$ (c) $h(-2)$ (d) $h(-5)$

34. $f(x) = 0.4x^4 - 1.6x^3 + 0.7x^2 - 2$

   (a) $f(1)$ (b) $f(-2)$ (c) $f(5)$ (d) $f(-10)$

In Exercises 35–38, use synthetic division to show that x is a solution of the third-degree polynomial equation, and use the result to factor the polynomial completely. List all the real zeros of the function.

	Polynomial Equation	Value of x
35.	$x^3 - 7x + 6 = 0$	$x = 2$
36.	$x^3 - 28x - 48 = 0$	$x = -4$
37.	$2x^3 - 15x^2 + 27x - 10 = 0$	$x = \frac{1}{2}$
38.	$48x^3 - 80x^2 + 41x - 6 = 0$	$x = \frac{2}{3}$

In Exercises 39–44, (a) verify the given factors of the function f, (b) find the remaining factors of f, (c) use your results to write the complete factorization of f, (d) list all real zeros of f, and (e) confirm your results by using a graphing utility to graph the function.

	Function	Factors
39.	$f(x) = 2x^3 + x^2 - 5x + 2$	$(x + 2), (x - 1)$
40.	$f(x) = 3x^3 + 2x^2 - 19x + 6$	$(x + 3), (x - 2)$
41.	$f(x) = x^4 - 4x^3 - 15x^2$ $+ 58x - 40$	$(x - 5), (x + 4)$
42.	$f(x) = 8x^4 - 14x^3 - 71x^2$ $- 10x + 24$	$(x + 2), (x - 4)$
43.	$f(x) = 6x^3 + 41x^2 - 9x - 14$	$(2x + 1), (3x - 2)$
44.	$f(x) = 2x^3 - x^2 - 10x + 5$	$(2x - 1), \left(x + \sqrt{5}\right)$

In Exercises 45–48, use the Rational Zero Test to list all possible rational zeros of f. Use a graphing utility to verify that the zeros of f are contained in the list.

45. $f(x) = x^3 + 3x^2 - x - 3$

46. $f(x) = x^3 - 4x^2 - 4x + 16$

47. $f(x) = 2x^4 - 17x^3 + 35x^2 + 9x - 45$

48. $f(x) = 4x^5 - 8x^4 - 5x^3 + 10x^2 + x - 2$

In Exercises 49–52, find all real solutions of the polynomial equation.

49. $z^4 - z^3 - 2z - 4 = 0$

50. $x^4 - x^3 - 29x^2 - x - 30 = 0$

51. $2y^4 + 7y^3 - 26y^2 + 23y - 6 = 0$

52. $x^5 - x^4 - 3x^3 + 5x^2 - 2x = 0$

Graphical Analysis   **In Exercises 53–56, (a) use the *zero* or *root* feature of a graphing utility to approximate (accurate to three decimal places) the zeros of the function, (b) determine one of the exact zeros and use synthetic division to verify your result, and (c) factor the polynomial completely.**

53. $h(t) = t^3 - 2t^2 - 7t + 2$

54. $f(s) = s^3 - 12s^2 + 40s - 24$

55. $h(x) = x^5 - 7x^4 + 10x^3 + 14x^2 - 24x$

56. $g(x) = 6x^4 - 11x^3 - 51x^2 + 99x - 27$

In Exercises 57–60, use Descartes's Rule of Signs to determine the possible numbers of positive and negative real zeros of the function.

57. $f(x) = 2x^4 - x^3 + 6x^2 - x + 5$

58. $f(x) = 3x^4 + 5x^3 - 6x^2 + 8x - 3$

59. $g(x) = 4x^3 - 5x + 8$

60. $g(x) = 2x^3 - 4x^2 - 5$

In Exercises 61–66, (a) use Descartes's Rule of Signs to determine the possible numbers of positive and negative real zeros of f, (b) list the possible rational zeros of f, (c) use a graphing utility to graph f so that some of the possible zeros in parts (a) and (b) can be disregarded, and (d) determine all the real zeros of f.

61. $f(x) = x^3 + x^2 - 4x - 4$

62. $f(x) = -3x^3 + 20x^2 - 36x + 16$

63. $f(x) = -2x^4 + 13x^3 - 21x^2 + 2x + 8$

64. $f(x) = 4x^4 - 17x^2 + 4$

65. $f(x) = 32x^3 - 52x^2 + 17x + 3$

66. $f(x) = 4x^3 + 7x^2 - 11x - 18$

In Exercises 67–70, use synthetic division to verify the upper and lower bounds of the real zeros of f.

67. $f(x) = x^4 - 4x^3 + 15$

Upper bound: $x = 4$;   Lower bound: $x = -1$

68. $f(x) = 2x^3 - 3x^2 - 12x + 8$

Upper bound: $x = 4$;   Lower bound: $x = -3$

69. $f(x) = x^4 - 4x^3 + 16x - 16$

Upper bound: $x = 5$;   Lower bound: $x = -3$

70. $f(x) = 2x^4 - 8x + 3$

Upper bound: $x = 3$;   Lower bound: $x = -4$

In Exercises 71–74, find the rational zeros of the polynomial function.

71. $P(x) = x^4 - \frac{25}{4}x^2 + 9 = \frac{1}{4}(4x^4 - 25x^2 + 36)$

72. $f(x) = x^3 - \frac{3}{2}x^2 - \frac{23}{2}x + 6 = \frac{1}{2}(2x^3 - 3x^2 - 23x + 12)$

73. $f(x) = x^3 - \frac{1}{4}x^2 - x + \frac{1}{4} = \frac{1}{4}(4x^3 - x^2 - 4x + 1)$

74. $f(z) = z^3 + \frac{11}{6}z^2 - \frac{1}{2}z - \frac{1}{3}$
$= \frac{1}{6}(6z^3 + 11z^2 - 3z - 2)$

In Exercises 75–78, match the cubic function with the correct number of rational and irrational zeros.

(a) **Rational zeros: 0;   Irrational zeros: 1**

(b) **Rational zeros: 3;   Irrational zeros: 0**

(c) **Rational zeros: 1;   Irrational zeros: 2**

(d) **Rational zeros: 1;   Irrational zeros: 0**

75. $f(x) = x^3 - 1$

76. $f(x) = x^3 - 2$

77. $f(x) = x^3 - x$

78. $f(x) = x^3 - 2x$

79. *Data Analysis* The average monthly rate R for basic cable television in the United States for the years 1995 through 2001 is shown in the table. The data can be approximated by the model

$R = 0.03889t^3 - 0.9064t^2 + 8.327t - 0.92$

where t represents the year, with $t = 5$ corresponding to 1995.   (Source: Kagan World Media)

Year	Rate, R
1995	23.07
1996	24.41
1997	26.48
1998	27.81
1999	28.92
2000	30.37
2001	32.87

Table for 79

(a) Use a graphing utility to plot the data and graph the model in the same viewing window. How closely does the model represent the data?

(b) Use a graphing utility and the model to create a table of estimated values for R. Compare the estimated values with the actual data.

(c) Use the Remainder Theorem to evaluate the model for the year 2008. Even though the model is relatively accurate for estimating the given data, do you think it is accurate for predicting future cable rates? Explain.

80. *Data Analysis* The table shows the sales S (in billions of dollars) from lottery tickets in the United States from 1995 to 2001. The data can be approximated by the model

$S = 0.0778t^3 - 1.931t^2 + 16.36t - 11.4$

where t represents the year, with $t = 5$ corresponding to 1995.   (Source: TLF Publications, Inc.)

Year	Sales, S
1995	31.9
1996	34.0
1997	35.5
1998	35.6
1999	36.0
2000	37.2
2001	38.4

(a) Use a graphing utility to plot the data and graph the model in the same viewing window. How closely does the model represent the data?

(b) Use a graphing utility and the model to create a table of estimated values for S. Compare the estimated values with the actual data.

(c) Use the Remainder Theorem to evaluate the model for the year 2008. Even though the model is relatively accurate for estimating the given data, would you use this model to predict the sales from lottery tickets in the future? Explain.

81. *Geometry* A rectangular package sent by a delivery service can have a maximum combined length and girth (perimeter of a cross section) of 120 inches (see figure).

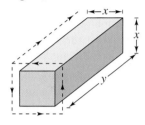

(a) Show that the volume of the package is given by the function

$$V(x) = 4x^2(30 - x).$$

(b) Use a graphing utility to graph the function and approximate the dimensions of the package that yield a maximum volume.

(c) Find values of x such that $V = 13,500$. Which of these values is a physical impossibility in the construction of the package? Explain.

82. *Automobile Emissions* The number of parts per million of nitric oxide emissions y from a car engine is approximated by the model

$$y = -5.05x^3 + 3857x - 38,411.25, \quad 13 \le x \le 18$$

where x is the air-fuel ratio.

(a) Use a graphing utility to graph the model.

(b) It is observed from the graph that two air-fuel ratios produce 2400 parts per million of nitric oxide, with one being 15. Use the graph to approximate the second air-fuel ratio.

(c) Algebraically approximate the second air-fuel ratio that produces 2400 parts per million of nitric oxide. (*Hint:* Because you know that an air-fuel ratio of 15 produces the specified nitric oxide emission, you can use synthetic division.)

Synthesis

True or False? **In Exercises 83 and 84, determine whether the statement is true or false. Justify your answer.**

83. If $(7x + 4)$ is a factor of some polynomial function f, then $\frac{4}{7}$ is a zero of f.

84. $(2x - 1)$ is a factor of the polynomial

$$6x^6 + x^5 - 92x^4 + 45x^3 + 184x^2 + 4x - 48.$$

Think About It **In Exercises 85 and 86, perform the division by assuming that n is a positive integer.**

85. $\dfrac{x^{3n} + 9x^{2n} + 27x^n + 27}{x^n + 3}$

86. $\dfrac{x^{3n} - 3x^{2n} + 5x^n - 6}{x^n - 2}$

87. *Writing* Complete each polynomial division. Write a brief description of the pattern that you obtain, and use your result to find a formula for the polynomial division $(x^n - 1)/(x - 1)$. Create a numerical example to test your formula.

(a) $\dfrac{x^2 - 1}{x - 1} = $

(b) $\dfrac{x^3 - 1}{x - 1} = $

(c) $\dfrac{x^4 - 1}{x - 1} = $

88. *Writing* Write a short paragraph explaining how you can check polynomial division. Give an example.

Review

In Exercises 89–92, use any convenient method to solve the quadratic equation.

89. $9x^2 - 25 = 0$

90. $16x^2 - 21 = 0$

91. $2x^2 + 6x + 3 = 0$

92. $8x^2 - 22x + 15 = 0$

In Exercises 93–96, find a polynomial function that has the given zeros. (There are many correct answers.)

93. $0, -12$

94. $1, -3, 8$

95. $0, -1, 2, 5$

96. $2 + \sqrt{3}, 2 - \sqrt{3}$

3.4 The Fundamental Theorem of Algebra

The Fundamental Theorem of Algebra

You know that an *n*th-degree polynomial can have at most *n* real zeros. In the complex number system, this statement can be improved. That is, in the complex number system, every *n*th-degree polynomial function has *precisely n zeros*. This important result is derived from the **Fundamental Theorem of Algebra,** first proved by the German mathematician Carl Friedrich Gauss (1777–1855).

> **The Fundamental Theorem of Algebra**
>
> If $f(x)$ is a polynomial of degree n, where $n > 0$, then f has at least one zero in the complex number system.

Using the Fundamental Theorem of Algebra and the equivalence of zeros and factors, you obtain the **Linear Factorization Theorem.** See Appendix B for a proof of the Linear Factorization Theorem.

> **Linear Factorization Theorem**
>
> If $f(x)$ is a polynomial of degree n where $n > 0$, f has precisely n linear factors
>
> $$f(x) = a_n(x - c_1)(x - c_2) \cdots (x - c_n)$$
>
> where $c_1, c_2, \ldots, c_n$ are complex numbers.

What you should learn

● Use the Fundamental Theorem of Algebra to determine the number of zeros of a polynomial function.

● Find all zeros of polynomial functions, including complex zeros.

● Find conjugate pairs of complex zeros.

● Find zeros of polynomials by factoring.

Why you should learn it

Being able to find zeros of polynomial functions is an important part of modeling real-life problems. For instance, Exercise 57 on page 285 shows how to determine whether a ball thrown with a given velocity can reach a certain height.

Jed Jacobsohn/Getty Images

Note that neither the Fundamental Theorem of Algebra nor the Linear Factorization Theorem tells you *how* to find the zeros or factors of a polynomial. Such theorems are called *existence theorems*. To find the zeros of a polynomial function, you still must rely on other techniques.

Remember that the *n* zeros of a polynomial function can be real or complex, and they may be repeated. Examples 1 and 2 illustrate several cases.

Example 1   Real Zeros of a Polynomial Function

Counting multiplicity, justify that the second-degree polynomial function

$$f(x) = x^2 - 6x + 9 = (x - 3)(x - 3)$$

has exactly *two* zeros: $x = 3$ and $x = 3$.

Solution

$$(x - 3)(x - 3) = (x - 3)^2 = 0$$

$$x - 3 = 0 \implies x = 3 \qquad \text{Repeated solution}$$

The graph in Figure 3.38 touches the *x*-axis at $x = 3$.

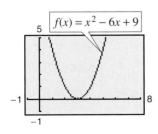

Figure 3.38

✓ *Checkpoint* Now try Exercise 1.

Example 2   Real and Complex Zeros of a Polynomial Function

Justify that the third-degree polynomial function

$$f(x) = x^3 + 4x = x(x^2 + 4)$$

has exactly three zeros: $x = 0$, $x = 2i$, and $x = -2i$.

Solution

Factor the polynomial completely as $x(x - 2i)(x + 2i)$. So, the zeros are

$$x(x - 2i)(x + 2i) = 0$$

$$x = 0$$

$$x - 2i = 0 \implies x = 2i$$

$$x + 2i = 0 \implies x = -2i.$$

In the graph in Figure 3.39, only the real zero $x = 0$ appears as an intercept.

✓ *Checkpoint* Now try Exercise 3.

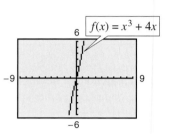

$f(x) = x^3 + 4x$

Figure 3.39

    Example 3 shows how to use the methods described in Sections 3.2 and 3.3 (the Rational Zero Test, synthetic division, and factoring) to find all the zeros of a polynomial function, including complex zeros.

Example 3   Finding the Zeros of a Polynomial Function

Write $f(x) = x^5 + x^3 + 2x^2 - 12x + 8$ as the product of linear factors, and list all the zeros of f.

Solution

The possible rational zeros are ± 1, ± 2, ± 4, and ± 8. The graph shown in Figure 3.40 indicates that 1 and -2 are likely zeros, and that 1 is possibly a repeated zero because it appears that the graph touches (but does not cross) the x-axis at this point. Using synthetic division, you can determine that -2 is a zero and 1 is a repeated zero of f. So, you have

$$f(x) = x^5 + x^3 + 2x^2 - 12x + 8 = (x - 1)(x - 1)(x + 2)(x^2 + 4).$$

By factoring $x^2 + 4$ as

$$x^2 - (-4) = \left(x - \sqrt{-4}\right)\left(x + \sqrt{-4}\right) = (x - 2i)(x + 2i)$$

you obtain

$$f(x) = (x - 1)(x - 1)(x + 2)(x - 2i)(x + 2i)$$

which gives the following five zeros of f.

$$x = 1, \ x = 1, \ x = -2, \ x = 2i, \ \text{and} \ x = -2i$$

Note from the graph of f shown in Figure 3.40 that the *real* zeros are the only ones that appear as x-intercepts.

✓ *Checkpoint* Now try Exercise 25.

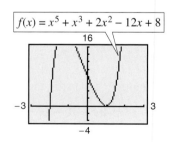

$f(x) = x^5 + x^3 + 2x^2 - 12x + 8$

Figure 3.40

Conjugate Pairs

In Example 3, note that the two complex zeros are **conjugates.** That is, they are of the forms $a + bi$ and $a - bi$.

Complex Zeros Occur in Conjugate Pairs

Let $f(x)$ be a polynomial function that has *real coefficients.* If $a + bi$, where $b \neq 0$, is a zero of the function, the conjugate $a - bi$ is also a zero of the function.

Be sure you see that this result is true only if the polynomial function has *real coefficients.* For instance, the result applies to the function $f(x) = x^2 + 1$, but not to the function $g(x) = x - i$.

Example 4   Finding a Polynomial with Given Zeros

Find a *fourth-degree* polynomial function with real coefficients that has $-1, -1,$ and $3i$ as zeros.

Solution

Because $3i$ is a zero *and* the polynomial is stated to have real coefficients, you know that the conjugate $-3i$ must also be a zero. So, from the Linear Factorization Theorem, $f(x)$ can be written as

$$f(x) = a(x + 1)(x + 1)(x - 3i)(x + 3i).$$

For simplicity, let $a = 1$ to obtain

$$f(x) = (x^2 + 2x + 1)(x^2 + 9) = x^4 + 2x^3 + 10x^2 + 18x + 9.$$

✓ *Checkpoint* Now try Exercise 37.

Factoring a Polynomial

The Linear Factorization Theorem states that you can write any nth-degree polynomial as the product of n linear factors.

$$f(x) = a(x - c_1)(x - c_2)(x - c_3) \cdots (x - c_n)$$

However, this result includes the possibility that some of the values of c_i are complex. The following theorem states that even if you do not want to get involved with "complex factors," you can still write $f(x)$ as the product of linear and/or quadratic factors. See Appendix B for a proof of this theorem.

Factors of a Polynomial

Every polynomial of degree $n > 0$ with real coefficients can be written as the product of linear and quadratic factors with real coefficients, where the quadratic factors have no real zeros.

A quadratic factor with no real zeros is said to be **prime** or **irreducible over the reals.** Be sure you see that this is not the same as being *irreducible over the rationals.* For example, the quadratic

$$x^2 + 1 = (x - i)(x + i)$$

is irreducible over the reals (and therefore over the rationals). On the other hand, the quadratic

$$x^2 - 2 = \left(x - \sqrt{2}\right)\left(x + \sqrt{2}\right)$$

is irreducible over the rationals, but *reducible* over the reals.

> **STUDY TIP**
>
> Recall that irrational and rational numbers are subsets of the set of real numbers, and the real numbers are a subset of the set of complex numbers.

Example 5   Factoring a Polynomial

Write the polynomial

$$f(x) = x^4 - x^2 - 20$$

a. as the product of factors that are irreducible over the *rationals,*

b. as the product of linear factors and quadratic factors that are irreducible over the *reals,* and

c. in completely factored form.

Solution

a. Begin by factoring the polynomial into the product of two quadratic polynomials.

$$x^4 - x^2 - 20 = (x^2 - 5)(x^2 + 4)$$

Both of these factors are irreducible over the rationals.

b. By factoring over the reals, you have

$$x^4 - x^2 - 20 = \left(x + \sqrt{5}\right)\left(x - \sqrt{5}\right)(x^2 + 4)$$

where the quadratic factor is irreducible over the reals.

c. In completely factored form, you have

$$x^4 - x^2 - 20 = \left(x + \sqrt{5}\right)\left(x - \sqrt{5}\right)(x - 2i)(x + 2i).$$

✓ *Checkpoint* Now try Exercise 41.

In Example 5, notice from the completely factored form that the fourth-degree polynomial has four zeros.

Throughout this chapter, the results and theorems have been stated in terms of zeros of polynomial functions. Be sure you see that the same results could have been stated in terms of solutions of polynomial equations. This is true because the zeros of the polynomial function

$$f(x) = a_n x^n + a_{n-1} x^{n-1} + \cdots + a_2 x^2 + a_1 x + a_0$$

are precisely the solutions of the polynomial equation

$$a_n x^n + a_{n-1} x^{n-1} + \cdots + a_2 x^2 + a_1 x + a_0 = 0.$$

Example 6   Finding the Zeros of a Polynomial Function

Find all the zeros of

$$f(x) = x^4 - 3x^3 + 6x^2 + 2x - 60$$

given that $1 + 3i$ is a zero of f.

Algebraic Solution

Because complex zeros occur in conjugate pairs, you know that $1 - 3i$ is also a zero of f. This means that both

$$x - (1 + 3i) \quad \text{and} \quad x - (1 - 3i)$$

are factors of f. Multiplying these two factors produces

$$[x - (1 + 3i)][x - (1 - 3i)] = [(x - 1) - 3i][(x - 1) + 3i]$$
$$= (x - 1)^2 - 9i^2$$
$$= x^2 - 2x + 10.$$

Using long division, you can divide $x^2 - 2x + 10$ into f to obtain the following.

$$
\begin{array}{r}
x^2 - x - 6 \\
x^2 - 2x + 10 \overline{\smash{)}\, x^4 - 3x^3 + 6x^2 + 2x - 60} \\
\underline{x^4 - 2x^3 + 10x^2} \\
-x^3 - 4x^2 + 2x \\
\underline{-x^3 + 2x^2 - 10x} \\
-6x^2 + 12x - 60 \\
\underline{-6x^2 + 12x - 60} \\
0
\end{array}
$$

So, you have

$$f(x) = (x^2 - 2x + 10)(x^2 - x - 6)$$
$$= (x^2 - 2x + 10)(x - 3)(x + 2)$$

and you can conclude that the zeros of f are $x = 1 + 3i$, $x = 1 - 3i$, $x = 3$, and $x = -2$.

☑ *Checkpoint* Now try Exercise 47.

Graphical Solution

Because complex zeros always occur in conjugate pairs, you know that $1 - 3i$ is also a zero of f. Because the polynomial is a fourth-degree polynomial, you know that there are at most two other zeros of the function. Use a graphing utility to graph

$$y = x^4 - 3x^3 + 6x^2 + 2x - 60$$

as shown in Figure 3.41.

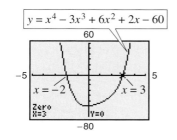

Figure 3.41

You can see that -2 and 3 appear to be x-intercepts of the graph of the function. Use the *zero* or *root* feature or the *zoom* and *trace* features of the graphing utility to confirm that $x = -2$ and $x = 3$ are x-intercepts of the graph. So, you can conclude that the zeros of f are

$$x = 1 + 3i, \ x = 1 - 3i, \ x = 3, \text{ and}$$
$$x = -2.$$

In Example 6, if you were not told that $1 + 3i$ is a zero of f, you could still find all zeros of the function by using synthetic division to find the real zeros -2 and 3. Then, you could factor the polynomial as $(x + 2)(x - 3)(x^2 - 2x + 10)$. Finally, by using the Quadratic Formula, you could determine that the zeros are $x = 1 + 3i$, $x = 1 - 3i$, $x = 3$, and $x = -2$.

3.4   Exercises

Vocabulary Check

Fill in the blanks.

1. The _____ of _____ states that if $f(x)$ is a polynomial function of degree n $(n > 0)$, then f has at least one zero in the complex number system.

2. The _____ states that if $f(x)$ is a polynomial of degree n, then f has precisely n linear factors

$$f(x) = a_n(x - c_1)(x - c_2) \cdots (x - c_n)$$

   where $c_1, c_2, \cdots, c_n$ are complex numbers.

3. A quadratic factor that cannot be factored further as a product of linear factors containing real numbers is said to be _____ over the _____ .

4. If $a + bi$ is a complex zero of a polynomial with real coefficients, then so is its _____ .

In Exercises 1–4, find all the zeros of the function.

1. $f(x) = x^2(x + 3)$ 2. $g(x) = (x - 2)(x + 4)^3$

3. $f(x) = (x + 9)(x + 2i)(x - 2i)$

4. $h(t) = (t - 3)(t - 2)(t - 3i)(t + 3i)$

Graphical and Analytical Analysis   **In Exercises 5–8, find all the zeros of the function. Is there a relationship between the number of real zeros and the number of x-intercepts of the graph? Explain.**

5. $f(x) = x^3 - 4x^2$
   $+ x - 4$

6. $f(x) = x^3 - 4x^2$
   $- 4x + 16$

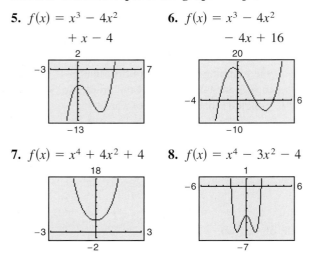

7. $f(x) = x^4 + 4x^2 + 4$ 8. $f(x) = x^4 - 3x^2 - 4$

In Exercises 9–26, find all the zeros of the function and write the polynomial as a product of linear factors. Use a graphing utility to graph the function to verify your results graphically. (If possible, use your graphing utility to verify the complex zeros.)

9. $h(x) = x^2 - 4x + 1$ 10. $g(x) = x^2 + 10x + 23$

11. $f(x) = x^2 - 12x + 26$ 12. $f(x) = x^2 + 6x - 2$

13. $f(x) = x^2 + 25$ 14. $f(x) = x^2 + 36$

15. $f(x) = x^4 - 81$ 16. $f(y) = y^4 - 625$

17. $f(z) = z^2 - z + 56$ 18. $h(x) = x^2 - 4x - 3$

19. $f(t) = t^3 - 3t^2 - 15t + 125$

20. $f(x) = x^3 + 11x^2 + 39x + 29$

21. $f(x) = 5x^3 - 9x^2 + 28x + 6$

22. $f(s) = 3s^3 - 4s^2 + 8s + 8$

23. $f(x) = x^4 + 10x^2 + 9$

24. $f(x) = x^4 + 29x^2 + 100$

25. $g(x) = x^4 - 4x^3 + 8x^2 - 16x + 16$

26. $h(x) = x^4 + 6x^3 + 10x^2 + 6x + 9$

In Exercises 27–34, (a) find all zeros of the function, (b) write the polynomial as a product of linear factors, (c) use your factorization to determine the x-intercepts of the graph of the function, and (d) use a graphing utility to verify that the real zeros are the only x-intercepts.

27. $f(x) = x^2 - 14x + 46$

28. $f(x) = x^2 - 12x + 34$

29. $f(x) = x^2 + 14x + 44$

30. $f(x) = x^2 - 16x + 62$

31. $f(x) = x^3 - 11x + 150$

32. $f(x) = x^3 + 10x^2 + 33x + 34$

33. $f(x) = x^4 + 25x^2 + 144$

34. $f(x) = x^4 - 8x^3 + 17x^2 - 8x + 16$

In Exercises 35–40, find a polynomial function with real coefficients that has the given zeros. (There are many correct answers.)

35. $3, i, -i$

36. $4, 3i, -3i$

37. $2, -4 + i, -4 - i$

38. $-1, 6 + 5i, 6 - 5i$

39. $-5, -5, 1 + \sqrt{3}i$

40. $0, 0, 4, 1 + \sqrt{2}i$

In Exercises 41–44, write the polynomial (a) as the product of factors that are irreducible over the *rationals*, (b) as the product of linear and quadratic factors that are irreducible over the *reals*, and (c) in completely factored form.

41. $f(x) = x^4 - 6x^2 - 7$

42. $f(x) = x^4 + 6x^2 - 27$

43. $f(x) = x^4 - 2x^3 - 3x^2 + 12x - 18$
   (*Hint:* One factor is $x^2 - 6$.)

44. $f(x) = x^4 - 3x^3 - x^2 - 12x - 20$
   (*Hint:* One factor is $x^2 + 4$.)

In Exercises 45–52, use the given zero to find all the zeros of the function.

Function	*Zero*
45. $f(x) = 2x^3 + 3x^2 + 50x + 75$	$5i$
46. $f(x) = x^3 + x^2 + 9x + 9$	$3i$
47. $g(x) = x^3 - 7x^2 - x + 87$	$5 + 2i$
48. $g(x) = 4x^3 + 23x^2 + 34x - 10$	$-3 + i$
49. $h(x) = 3x^3 - 4x^2 + 8x + 8$	$1 - \sqrt{3}i$
50. $f(x) = x^3 + 4x^2 + 14x + 20$	$-1 - 3i$
51. $h(x) = 8x^3 - 14x^2 + 18x - 9$	$\frac{1}{2}\left(1 - \sqrt{5}i\right)$
52. $f(x) = 25x^3 - 55x^2 - 54x - 18$	$\frac{1}{5}\left(-2 + \sqrt{2}i\right)$

Graphical Analysis   **In Exercises 53–56, (a) use the *zero* or *root* feature of a graphing utility to approximate the zeros of the function accurate to three decimal places and (b) find the exact values of the remaining zeros.**

53. $f(x) = x^4 + 3x^3 - 5x^2 - 21x + 22$

54. $f(x) = x^3 + 4x^2 + 14x + 20$

55. $h(x) = 8x^3 - 14x^2 + 18x - 9$

56. $f(x) = 25x^3 - 55x^2 - 54x - 18$

57. *Height*   A baseball is thrown upward from ground level with an initial velocity of 48 feet per second, and its height h (in feet) is given by

$$h(t) = -16t^2 + 48t, \quad 0 \le t \le 3$$

where t is the time (in seconds). You are told that the ball reaches a height of 64 feet. Is this possible? Explain.

58. *Profit*   The demand equation for a microwave is $p = 140 - 0.0001x$, where p is the unit price (in dollars) of the microwave and x is the number of units produced and sold. The cost equation for the microwave is $C = 80x + 150,000$, where C is the total cost (in dollars) and x is the number of units produced. The total profit obtained by producing and selling x units is given by

$$P = R - C = xp - C.$$

You are working in the marketing department that produces this microwave, and you are asked to determine a price p that would yield a profit of $9 million. Is this possible? Explain.

Synthesis

True or False?   **In Exercises 59 and 60, decide whether the statement is true or false. Justify your answer.**

59. It is possible for a third-degree polynomial function with integer coefficients to have no real zeros.

60. If $x = 4 + 3i$ is a zero of the function $f(x) = x^4 - 7x^3 - 13x^2 + 265x - 750$, then $x = -3i + 4$ must also be a zero of f.

61. *Exploration*   Use a graphing utility to graph the function $f(x) = x^4 - 4x^2 + k$ for different values of k. Find values of k such that the zeros of f satisfy the specified characteristics. (Some parts have many correct answers.)

   (a) Two real zeros, each of multiplicity 2

   (b) Two real zeros and two complex zeros

62. *Writing*   Compile a list of all the various techniques for factoring a polynomial that have been covered so far in the text. Give an example illustrating each technique, and write a paragraph discussing when the use of each technique is appropriate.

Review

In Exercises 63–66, sketch the graph of the quadratic function. Identify the vertex and any intercepts. Use a graphing utility to verify your results.

63. $f(x) = x^2 - 7x - 8$

64. $f(x) = -x^2 + x + 6$

65. $f(x) = 6x^2 + 5x - 6$

66. $f(x) = 4x^2 + 2x - 12$

3.5   Rational Functions and Asymptotes

Introduction to Rational Functions

A **rational function** can be written in the form

$$f(x) = \frac{N(x)}{D(x)}$$

where $N(x)$ and $D(x)$ are polynomials and $D(x)$ is not the zero polynomial.

   In general, the *domain* of a rational function of x includes all real numbers except x-values that make the denominator zero. Much of the discussion of rational functions will focus on their graphical behavior near these x-values.

Example 1   Finding the Domain of a Rational Function

Find the domain of $f(x) = 1/x$ and discuss the behavior of f near any excluded x-values.

Solution

Because the denominator is zero when $x = 0$, the domain of f is all real numbers except $x = 0$. To determine the behavior of f near this excluded value, evaluate $f(x)$ to the left and right of $x = 0$, as indicated in the following tables.

x	-1	-0.5	-0.1	-0.01	-0.001	$\rightarrow 0$
$f(x)$	-1	-2	-10	-100	-1000	$\rightarrow -\infty$

x	$0 \leftarrow$	0.001	0.01	0.1	0.5	1
$f(x)$	$\infty \leftarrow$	1000	100	10	2	1

Note that as x approaches 0 *from the left*, $f(x)$ decreases without bound. In contrast, as x approaches 0 *from the right*, $f(x)$ increases without bound. The graph of f is shown in Figure 3.42.

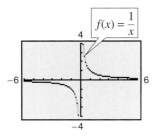

Figure 3.42

TECHNOLOGY TIP

The graphing utility graphs in this section and the next section were created using the *dot* mode. A blue curve is placed behind the graphing utility's display to indicate where the graph should appear. You will learn more about how graphing utilities graph rational functions in the next section.

Library of Functions: Rational Function

A *rational function* $f(x)$ is the quotient of two polynomials,

$$f(x) = \frac{N(x)}{D(x)}.$$

A rational function is not defined at values of x for which $D(x) = 0$. Near these values the graph of the rational function may increase or decrease without bound. The simplest type of rational function is the *reciprocal function* $f(x) = 1/x$. The basic characteristics of the reciprocal function are summarized below.

Graph of $f(x) = \dfrac{1}{x}$

Domain: $(-\infty, 0) \cup (0, \infty)$

Range: $(-\infty, 0) \cup (0, \infty)$

No intercepts

Decreasing on $(-\infty, 0)$ and $(0, \infty)$

Odd function

Origin symmetry

Vertical asymptote: y-axis

Horizontal asymptote: x-axis

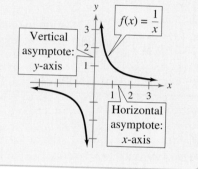

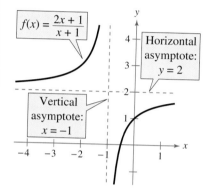

Horizontal and Vertical Asymptotes

In Example 1, the behavior of f near $x = 0$ is denoted as follows.

$$f(x) \to -\infty \text{ as } x \to 0^-$$

$f(x)$ decreases without bound as x approaches 0 from the left.

$$f(x) \to \infty \text{ as } x \to 0^+$$

$f(x)$ increases without bound as x approaches 0 from the right.

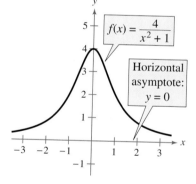

The line $x = 0$ is a **vertical asymptote** of the graph of f, as shown in the figure above. The graph of f also has a **horizontal asymptote**—the line $y = 0$. This means the values of $f(x) = 1/x$ approach zero as x increases or decreases without bound.

$$f(x) \to 0 \text{ as } x \to -\infty$$

$f(x)$ approaches 0 as x decreases without bound.

$$f(x) \to 0 \text{ as } x \to \infty$$

$f(x)$ approaches 0 as x increases without bound.

Definition of Vertical and Horizontal Asymptotes

1. The line $x = a$ is a **vertical asymptote** of the graph of f if $f(x) \to \infty$ or $f(x) \to -\infty$ as $x \to a$, either from the right or from the left.

2. The line $y = b$ is a **horizontal asymptote** of the graph of f if $f(x) \to b$ as $x \to \infty$ or $x \to -\infty$.

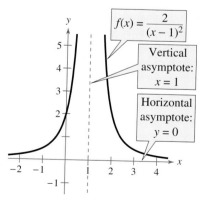

Figure 3.43 shows the horizontal and vertical asymptotes of the graphs of three rational functions.

Figure 3.43

Asymptotes of a Rational Function

Let f be the rational function

$$f(x) = \frac{N(x)}{D(x)} = \frac{a_n x^n + a_{n-1} x^{n-1} + \cdots + a_1 x + a_0}{b_m x^m + b_{m-1} x^{m-1} + \cdots + b_1 x + b_0}$$

where $N(x)$ and $D(x)$ have no common factors.

1. The graph of f has *vertical* asymptotes at the zeros of $D(x)$.

2. The graph of f has at most one *horizontal* asymptote determined by comparing the degrees of $N(x)$ and $D(x)$.

   a. If $n < m$, the graph of f has the line $y = 0$ (the x-axis) as a horizontal asymptote.

   b. If $n = m$, the graph of f has the line $y = a_n/b_m$ as a horizontal asymptote, where a_n is the leading coefficient of the numerator and b_m is the leading coefficient of the denominator.

   c. If $n > m$, the graph of f has no horizontal asymptote.

Exploration

Use a graphing utility to compare the graphs of y_1 and y_2.

$$y_1 = \frac{3x^3 - 5x^2 + 4x - 5}{2x^2 - 6x + 7}$$

$$y_2 = \frac{3x^3}{2x^2}$$

Start with a viewing window in which $-5 \le x \le 5$ and $-10 \le y \le 10$, then zoom out. Write a convincing argument that the shape of the graph of a rational function eventually behaves like the graph of $y = a_n x^n / b_m x^m$, where $a_n x^n$ is the leading term of the numerator and $b_m x^m$ is the leading term of the denominator.

Example 2   Finding Horizontal and Vertical Asymptotes

Find all horizontal and vertical asymptotes of the graph of each rational function.

a. $f(x) = \dfrac{2x}{3x^2 + 1}$ **b.** $f(x) = \dfrac{2x^2}{x^2 - 1}$

Solution

a. For this rational function, the degree of the numerator is *less than* the degree of the denominator, so the graph has the line $y = 0$ as a horizontal asymptote. To find any vertical asymptotes, set the denominator equal to zero and solve the resulting equation for x.

$$3x^2 + 1 = 0 \qquad \text{Set denominator equal to zero.}$$

Because this equation has no real solutions, you can conclude that the graph has no vertical asymptote. The graph of the function is shown in Figure 3.44.

b. For this rational function, the degree of the numerator is *equal to* the degree of the denominator. The leading coefficient of the numerator is 2 and the leading coefficient of the denominator is 1, so the graph has the line $y = 2$ as a horizontal asymptote. To find any vertical asymptotes, set the denominator equal to zero and solve the resulting equation for x.

$$x^2 - 1 = 0 \qquad \text{Set denominator equal to zero.}$$

$$(x + 1)(x - 1) = 0 \qquad \text{Factor.}$$

$$x + 1 = 0 \quad \Longrightarrow \quad x = -1 \qquad \text{Set 1st factor equal to 0.}$$

$$x - 1 = 0 \quad \Longrightarrow \quad x = 1 \qquad \text{Set 2nd factor equal to 0.}$$

This equation has two real solutions, $x = -1$ and $x = 1$, so the graph has the lines $x = -1$ and $x = 1$ as vertical asymptotes, as shown in Figure 3.45.

✓ *Checkpoint* Now try Exercise 13.

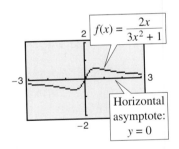

Figure 3.44

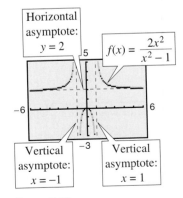

Figure 3.45

Example 3 Finding Horizontal and Vertical Asymptotes

Find all horizontal and vertical asymptotes of the graph of $f(x) = \dfrac{x^2 + x - 2}{x^2 - x - 6}$.

Solution

For this rational function the degree of the numerator is *equal to* the degree of the denominator. The leading coefficient of both the numerator and denominator is 1, so the graph has the line $y = 1$ as a horizontal asymptote. To find any vertical asymptotes, first factor the numerator and denominator as follows.

$$f(x) = \frac{x^2 + x - 2}{x^2 - x - 6} = \frac{(x - 1)\cancel{(x + 2)}}{\cancel{(x + 2)}(x - 3)} = \frac{x - 1}{x - 3}, \quad x \neq -2$$

By setting the denominator $x - 3$ (of the simplified function) equal to zero, you can determine that the graph has the line $x = 3$ as a vertical asymptote, as shown in Figure 3.46. Notice in the graph that the function appears to be defined at $x = -2$. Because the domain of the function is all real numbers except $x = -2$ and $x = 3$, you know this is not true. Graphing utilities are limited in their resolution and therefore may not show a break or hole in the graph. Using the *table* feature of a graphing utility, you can verify that the function is not defined at $x = -2$, as shown in Figure 3.47.

✓ *Checkpoint* Now try Exercise 17.

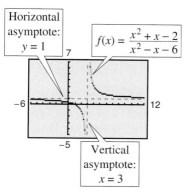

Figure 3.46

X	Y1
-3	.66667
-2	ERROR
-1	.5
0	.33333
1	0
2	-1
3	ERROR

X= -2

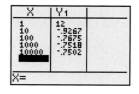

Figure 3.47

Example 4 Finding a Function's Domain and Asymptotes

For the function f, find (a) the domain of f, (b) the vertical asymptote of f, and (c) the horizontal asymptote of f.

$$f(x) = \frac{3x^3 + 7x^2 + 2}{-4x^3 + 5}$$

Algebraic Solution

a. Because the denominator is zero when $-4x^3 + 5 = 0$, solve this equation to determine that the domain of f is all real numbers except $x = \sqrt[3]{\frac{5}{4}}$.

b. Because the denominator of f has a zero at $x = \sqrt[3]{\frac{5}{4}}$, and $\sqrt[3]{\frac{5}{4}}$ is not a zero of the numerator, the graph of f has the vertical asymptote $x = \sqrt[3]{\frac{5}{4}} \approx 1.08$.

c. Because the degrees of the numerator and denominator are the same, the horizontal asymptote is given by the ratio of the leading coefficients.

$$y = \frac{\text{leading coefficient of numerator}}{\text{leading coefficient of denominator}} = -\frac{3}{4}$$

The horizontal asymptote of f is $y = -\frac{3}{4}$.

Use a graphing utility to verify the vertical and horizontal asymptotes.

✓ *Checkpoint* Now try Exercise 19.

Numerical Solution

a. See *Algebraic Solution* part (a).

b. See *Algebraic Solution* part (b).

c. You can use the table feature of a graphing utility to create tables like those shown in Figure 3.48. From the tables you can estimate that the graph of f has a horizontal asymptote at $y = -\frac{3}{4}$ because the values of $f(x)$ become closer and closer to $-\frac{3}{4}$ as x becomes increasingly large or small.

X	Y1
1	12
10	-.9267
100	-.7675
1000	-.7518
10000	-.7502

X=

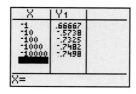

x Increases without Bound

X	Y1
-1	.66667
-10	-.5738
-100	-.7325
-1000	-.7482
-10000	-.7498

X=

x Decreases without Bound

Figure 3.48

Example 5   A Graph with Two Horizontal Asymptotes

A function that is not rational can have two horizontal asymptotes—one to the left and one to the right. For instance, the graph of

$$f(x) = \frac{x + 10}{|x| + 2}$$

is shown in Figure 3.49. It has the line $y = -1$ as a horizontal asymptote to the left and the line $y = 1$ as a horizontal asymptote to the right. You can confirm this by rewriting the function as follows.

$$f(x) = \begin{cases} \dfrac{x + 10}{-x + 2}, & x < 0 \qquad |x| = -x \text{ for } x < 0 \\[2ex] \dfrac{x + 10}{x + 2}, & x \geq 0 \qquad |x| = x \text{ for } x \geq 0 \end{cases}$$

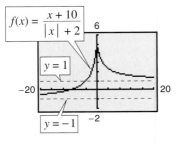

Figure 3.49

 Checkpoint Now try Exercise 21.

Applications

There are many examples of asymptotic behavior in real life. For instance, Example 6 shows how a vertical asymptote can be used to analyze the cost of removing pollutants from smokestack emissions.

Example 6   Cost-Benefit Model

A utility company burns coal to generate electricity. The cost C (in dollars) of removing $p\%$ of the smokestack pollutants is given by $C = 80{,}000p/(100 - p)$ for $0 \leq p < 100$. Use a graphing utility to graph this function. You are a member of a state legislature that is considering a law that would require utility companies to remove 90% of the pollutants from their smokestack emissions. The current law requires 85% removal. How much additional cost would there be to the utility company because of the new law?

Solution

The graph of this function is shown in Figure 3.50. Note that the graph has a vertical asymptote at $p = 100$. Because the current law requires 85% removal, the current cost to the utility company is

$$C = \frac{80{,}000(85)}{100 - 85} \approx \$453{,}333. \qquad \text{Evaluate } C \text{ when } p = 85.$$

If the new law increases the percent removal to 90%, the cost will be

$$C = \frac{80{,}000(90)}{100 - 90} = \$720{,}000. \qquad \text{Evaluate } C \text{ when } p = 90.$$

So, the new law would require the utility company to spend an additional

$$720{,}000 - 453{,}333 = \$266{,}667. \qquad \begin{array}{l}\text{Subtract 85\% removal cost from}\\\text{90\% removal cost.}\end{array}$$

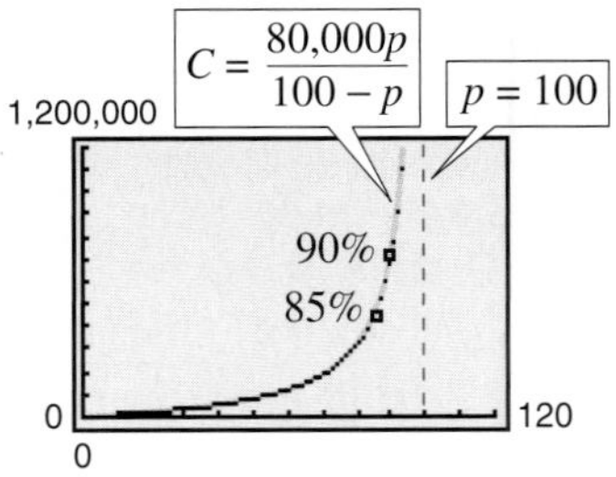

 Checkpoint Now try Exercise 35.

Figure 3.50

Example 7   Ultraviolet Radiation

For a person with sensitive skin, the amount of time T (in hours) the person can be exposed to the sun with a minimal burning can be modeled by

$$T = \frac{0.37s + 23.8}{s}, \quad 0 < s \leq 120$$

where s is the Sunsor Scale reading. The Sunsor Scale is based on the level of intensity of UVB rays.   (Source: Sunsor, Inc.)

a. Find the amount of time a person with sensitive skin can be exposed to the sun with minimal burning when $s = 10$, $s = 25$, and $s = 100$.

b. If the model were valid for all $s > 0$, what would be the horizontal asymptote of this function, and what would it represent?

TECHNOLOGY SUPPORT

For instructions on how to use the *value* feature, see Appendix A; for specific keystrokes, go to the text website at *college.hmco.com*.

Algebraic Solution

a. When $s = 10$, $T = \dfrac{0.37(10) + 23.8}{10}$

$= 2.75$ hours.

When $s = 25$, $T = \dfrac{0.37(25) + 23.8}{25}$

≈ 1.32 hours.

When $s = 100$, $T = \dfrac{0.37(100) + 23.8}{100}$

≈ 0.61 hour.

b. Because the degree of the numerator and denominator are the same for

$$T = \frac{0.37s + 23.8}{s}$$

the horizontal asymptote is given by the ratio of the leading coefficients of the numerator and denominator. So, the graph has the line $T = 0.37$ as a horizontal asymptote. This line represents the shortest possible exposure time with minimal burning.

Graphical Solution

a. Use a graphing utility to graph the function

$$y_1 = \frac{0.37x + 23.8}{x}$$

using a viewing window similar to that shown in Figure 3.51. Then use the *trace* or *value* feature to approximate the value of y_1 when $x = 10$, $x = 25$, and $x = 100$. You should obtain the following values.

When $x = 10$, $y_1 = 2.75$ hours.

When $x = 25$, $y_1 \approx 1.32$ hours.

When $x = 100$, $y_1 \approx 0.61$ hour.

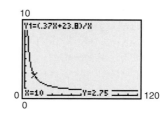

Figure 3.51

b. Continue to use the *trace* or *value* feature to approximate values of $f(x)$ for larger and larger values of x (see Figure 3.52). From this, you can estimate the horizontal asymptote to be $y = 0.37$. This line represents the shortest possible exposure time with minimal burning.

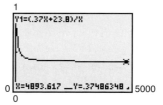

Figure 3.52

✓ *Checkpoint* Now try Exercise 39.

3.5   Exercises

Vocabulary Check

Fill in the blanks.

1. Functions of the form $f(x) = N(x)/D(x)$, where $N(x)$ and $D(x)$ are polynomials and $D(x)$ is not the zero polynomial, are called _____ .
2. If $f(x) \to \pm\infty$ as $x \to a$ from the left (or right), then $x = a$ is a _____ of the graph of f.
3. If $f(x) \to b$ as $x \to \pm\infty$, then $y = b$ is a _____ of the graph of f.

In Exercises 1–6, (a) complete each table, (b) determine the vertical and horizontal asymptotes of the function, and (c) find the domain of the function.

x	$f(x)$
0.5	
0.9	
0.99	
0.999	

x	$f(x)$
1.5	
1.1	
1.01	
1.001	

x	$f(x)$
5	
10	
100	
1000	

x	$f(x)$
-5	
-10	
-100	
-1000	

1. $f(x) = \dfrac{1}{x-1}$

2. $f(x) = \dfrac{5x}{x-1}$

3. $f(x) = \dfrac{3x}{|x-1|}$

4. $f(x) = \dfrac{3}{|x-1|}$

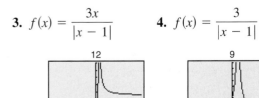

5. $f(x) = \dfrac{3x^2}{x^2-1}$

6. $f(x) = \dfrac{4x}{x^2-1}$

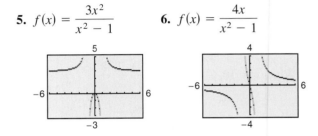

In Exercises 7–12, match the function with its graph. [The graphs are labeled (a), (b), (c), (d), (e), and (f).]

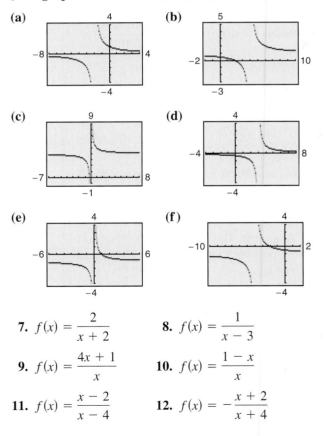

7. $f(x) = \dfrac{2}{x+2}$

8. $f(x) = \dfrac{1}{x-3}$

9. $f(x) = \dfrac{4x+1}{x}$

10. $f(x) = \dfrac{1-x}{x}$

11. $f(x) = \dfrac{x-2}{x-4}$

12. $f(x) = -\dfrac{x+2}{x+4}$

In Exercises 13–22, (a) find the domain of the function, (b) identify any horizontal and vertical asymptotes, and (c) verify your answer to part (a) both graphically by using a graphing utility and numerically by creating a table of values.

13. $f(x) = \dfrac{1}{x^2}$

14. $f(x) = \dfrac{3}{(x-2)^3}$

15. $f(x) = \dfrac{2+x}{2-x}$

16. $f(x) = \dfrac{1-5x}{1+2x}$

17. $f(x) = \dfrac{x^2+2x}{2x^2-x}$

18. $f(x) = \dfrac{x^2-25}{x^2+5x}$

19. $f(x) = \dfrac{3x^2+x-5}{x^2+1}$

20. $f(x) = \dfrac{3x^2+1}{x^2+x+9}$

21. $f(x) = \dfrac{x-3}{|x|}$

22. $f(x) = \dfrac{x+1}{|x|+1}$

Analytical and Numerical Explanation   In Exercises 23–26, (a) determine the domains of f and g, (b) simplify f and find any vertical asymptotes of f, (c) complete the table, and (d) explain how the two functions differ.

23. $f(x) = \dfrac{x^2-4}{x+2}$, $g(x) = x-2$

x	-4	-3	-2.5	-2	-1.5	-1	0
$f(x)$							
$g(x)$							

24. $f(x) = \dfrac{x^2(x-3)}{x^2-3x}$, $g(x) = x$

x	-1	0	1	2	3	3.5	4
$f(x)$							
$g(x)$							

25. $f(x) = \dfrac{x-3}{x^2-3x}$, $g(x) = \dfrac{1}{x}$

x	-1	-0.5	0	0.5	2	3	4
$f(x)$							
$g(x)$							

26. $f(x) = \dfrac{2x-8}{x^2-9x+20}$, $g(x) = \dfrac{2}{x-5}$

x	0	1	2	3	4	5	6
$f(x)$							
$g(x)$							

Exploration   In Exercises 27–30, determine the value that the function f approaches as the magnitude of x increases. Is $f(x)$ greater than or less than this functional value when x is positive and large in magnitude? What about when x is negative and large in magnitude?

27. $f(x) = 4 - \dfrac{1}{x}$

28. $f(x) = 2 + \dfrac{1}{x-3}$

29. $f(x) = \dfrac{2x-1}{x-3}$

30. $f(x) = \dfrac{2x-1}{x^2+1}$

In Exercises 31–34, find the zeros (if any) of the rational function. Use a graphing utility to verify your answer.

31. $g(x) = \dfrac{x^2-4}{x+3}$

32. $g(x) = \dfrac{x^3-8}{x^2+4}$

33. $f(x) = 1 - \dfrac{2}{x-5}$

34. $h(x) = 5 + \dfrac{3}{x^2+1}$

35. *Environment*   The cost C (in millions of dollars) of removing $p\%$ of the industrial and municipal pollutants discharged into a river is given by

$$C = \dfrac{255p}{100-p}, \quad 0 \le p < 100.$$

(a) Find the cost of removing 10% of the pollutants.

(b) Find the cost of removing 40% of the pollutants.

(c) Find the cost of removing 75% of the pollutants.

(d) Use a graphing utility to graph the cost function. Be sure to choose an appropriate viewing window. Explain why you chose the values that you used in your viewing window.

(e) According to this model, would it be possible to remove 100% of the pollutants? Explain.

36. *Environment* In a pilot project, a rural township is given recycling bins for separating and storing recyclable products. The cost C (in dollars) for supplying bins to $p\%$ of the population is given by

$$C = \frac{25,000p}{100 - p}, \quad 0 \le p < 100.$$

(a) Find the cost of supplying bins to 15% of the population.

(b) Find the cost of supplying bins to 50% of the population.

(c) Find the cost of supplying bins to 90% of the population.

(d) Use a graphing utility to graph the cost function. Be sure to choose an appropriate viewing window. Explain why you chose the values that you used in your viewing window.

(e) According to this model, would it be possible to supply bins to 100% of the residents? Explain.

37. *Data Analysis* The endpoints of the interval over which distinct vision is possible are called the *near point* and *far point* of the eye (see figure). With increasing age these points normally change. The table shows the approximate near points y (in inches) for various ages x (in years).

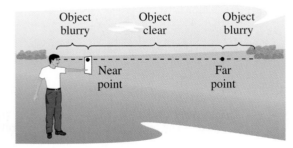

Age, x	Near point, y
16	3.0
32	4.7
44	9.8
50	19.7
60	39.4

(a) Find a rational model for the data. Take the reciprocals of the near points to generate the points $(x, 1/y)$. Use the *regression* feature of a graphing utility to find a linear model for the data. The resulting line has the form

$$\frac{1}{y} = ax + b.$$

Solve for y.

(b) Use the *table* feature of a graphing utility to create a table showing the predicted near point based on the model for each of the ages in the original table.

(c) Do you think the model can be used to predict the near point for a person who is 70 years old? Explain.

38. *Data Analysis* Consider a physics laboratory experiment designed to determine an unknown mass. A flexible metal meter stick is clamped to a table with 50 centimeters overhanging the edge (see figure). Known masses M ranging from 200 grams to 2000 grams are attached to the end of the meter stick. For each mass, the meter stick is displaced vertically and then allowed to oscillate. The average time t (in seconds) of one oscillation for each mass is recorded in the table.

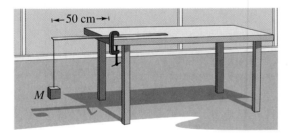

Mass, M	Time, t
200	0.450
400	0.597
600	0.721
800	0.831
1000	0.906
1200	1.003
1400	1.088
1600	1.168
1800	1.218
2000	1.338

A model for the data is given by

$$t = \frac{38M + 16{,}965}{10(M + 5000)}.$$

(a) Use the *table* feature of a graphing utility to create a table showing the estimated time based on the model for each of the masses shown in the table. What can you conclude?

(b) Use the model to approximate the mass of an object when the average time for one oscillation is 1.056 seconds.

39. *Wildlife* The game commission introduces 100 deer into newly acquired state game lands. The population N of the herd is given by

$$N = \frac{20(5 + 3t)}{1 + 0.04t}, \quad t \geq 0$$

where t is the time in years.

(a) Use a graphing utility to graph the model.

(b) Find the population when $t = 5$, $t = 10$, and $t = 25$.

(c) What is the limiting size of the herd as time increases? Explain.

40. *Wildlife* The table shows the number N of threatened and endangered species in the United States from 1993 to 2002. The data can be approximated by the model

$$N = \frac{42.58t^2 + 690}{0.03t^2 + 1}$$

where t represents the year, with $t = 3$ corresponding to 1993. (Source: U.S. Fish and Wildlife Service)

Year	Number, N
1993	813
1994	941
1995	962
1996	1053
1997	1132
1998	1194
1999	1205
2000	1244
2001	1254
2002	1262

(a) Use a graphing utility to plot the data and graph the model in the same viewing window. How closely does the model represent the data?

(b) Use the model to estimate the number of threatened and endangered species in 2006.

(c) Would this model be useful for estimating the number of threatened and endangered species in future years? Explain.

Synthesis

True or False? In Exercises 41 and 42, determine whether the statement is true or false. Justify your answer.

41. A rational function can have infinitely many vertical asymptotes.

42. $f(x) = x^3 - 2x^2 - 5x + 6$ is a rational function.

Think About It In Exercises 43–46, write a rational function f having the specified characteristics. (There are many correct answers.)

43. Vertical asymptotes: $x = -2$, $x = 1$

44. Vertical asymptote: None
    Horizontal asymptote: $y = 0$

45. Vertical asymptote: None
    Horizontal asymptote: $y = 2$

46. Vertical asymptotes: $x = 0$, $x = \frac{5}{2}$
    Horizontal asymptote: $y = -3$

Review

In Exercises 47–50, write the general form of the equation of the line that passes through the points.

47. $(3, 2), (0, -1)$ **48.** $(-6, 1), (4, -5)$

49. $(2, 7), (3, 10)$ **50.** $(0, 0), (-9, 4)$

In Exercises 52–54, divide using long division.

51. $(x^2 + 5x + 6) \div (x - 4)$

52. $(x^2 - 10x + 15) \div (x - 3)$

53. $(2x^2 + x - 11) \div (x + 5)$

54. $(4x^2 + 3x - 10) \div (x + 6)$

The Graph of a Rational Function

To sketch the graph of a rational function, use the following guidelines.

> **Guidelines for Graphing Rational Functions**
>
> Let $f(x) = N(x)/D(x)$, where $N(x)$ and $D(x)$ are polynomials.
>
> **1.** Simplify f, if possible.
>
> **2.** Find and plot the y-intercept (if any) by evaluating $f(0)$.
>
> **3.** Find the zeros of the numerator (if any) by solving the equation $N(x) = 0$. Then plot the corresponding x-intercepts.
>
> **4.** Find the zeros of the denominator (if any) by solving the equation $D(x) = 0$. Then sketch the corresponding vertical asymptotes using dashed vertical lines.
>
> **5.** Find and sketch the horizontal asymptote (if any) of the graph using a dashed horizontal line.
>
> **6.** Plot at least one point *between* and one point *beyond* each x-intercept and vertical asymptote.
>
> **7.** Use smooth curves to complete the graph between and beyond the vertical asymptotes.

What you should learn

- Analyze and sketch graphs of rational functions.
- Sketch graphs of rational functions that have slant asymptotes.
- Use rational functions to model and solve real-life problems.

Why you should learn it

The graph of a rational function provides a good indication of the future behavior of a mathematical model. Exercise 72 on page 304 models the average room rate for hotels in the U.S. and enables you to estimate the average room rate in the coming years.

Michael Keller/Corbis

TECHNOLOGY TIP Some graphing utilities have difficulty graphing rational functions that have vertical asymptotes. Often, the utility will connect parts of the graph that are not supposed to be connected. For instance, notice that the graph in Figure 3.53(a) should consist of two *unconnected* portions—one to the left of $x = 2$ and the other to the right of $x = 2$. To eliminate this problem, you can try changing the *mode* of the graphing utility to *dot mode* [see Figure 3.53(b)]. The problem with this mode is that the graph is then represented as a collection of dots rather than as a smooth curve, as shown in Figure 3.53(c). In this text, a blue curve is placed behind the graphing utility's display to indicate where the graph should appear. [See Figure 3.53(c).]

TECHNOLOGY SUPPORT

For instructions on how to use the *connected* mode and the *dot* mode, see Appendix A; for specific keystrokes, go to the text website at *college.hmco.com*.

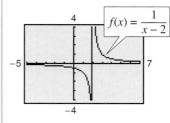

(a) Connected mode

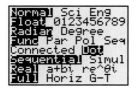

(b) Mode screen

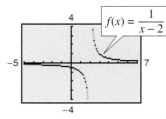

(c) Dot mode

Figure 3.53

Example 1   Sketching the Graph of a Rational Function

Sketch the graph of $g(x) = \dfrac{3}{x-2}$ by hand.

Solution

y-Intercept:	$\left(0, -\frac{3}{2}\right)$, because $g(0) = -\frac{3}{2}$
x-Intercept:	None, because $3 \neq 0$
Vertical Asymptote:	$x = 2$, zero of denominator
Horizontal Asymptote:	$y = 0$, because degree of $N(x) <$ degree of $D(x)$
Additional Points:	

x	-4	1	2	3	5
$g(x)$	-0.5	-3	Undefined	3	1

By plotting the intercept, asymptotes, and a few additional points, you can obtain the graph shown in Figure 3.54. Confirm this with a graphing utility.

 Checkpoint Now try Exercise 9.

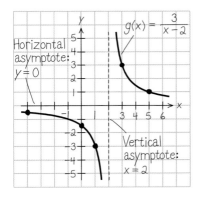

Figure 3.54

Note that the graph of g in Example 1 is a vertical stretch and a right shift of the graph of

$$f(x) = \frac{1}{x}$$

because

$$g(x) = \frac{3}{x-2} = 3\left(\frac{1}{x-2}\right) = 3f(x-2).$$

Example 2   Sketching the Graph of a Rational Function

Sketch the graph of $f(x) = \dfrac{2x-1}{x}$ by hand.

Solution

y-Intercept:	None, because $x = 0$ is not in the domain
x-Intercept:	$\left(\frac{1}{2}, 0\right)$, because $2x - 1 = 0$
Vertical Asymptote:	$x = 0$, zero of denominator
Horizontal Asymptote:	$y = 2$, because degree of $N(x) =$ degree of $D(x)$
Additional Points:	

x	-4	-1	0	$\frac{1}{4}$	4
$f(x)$	2.25	3	Undefined	-2	1.75

By plotting the intercept, asymptotes, and a few additional points, you can obtain the graph shown in Figure 3.55. Confirm this with a graphing utility.

 Checkpoint Now try Exercise 13.

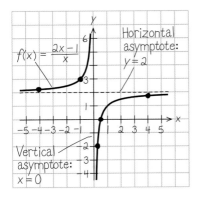

Figure 3.55

Example 3   Sketching the Graph of a Rational Function

Sketch the graph of $f(x) = \dfrac{x}{x^2 - x - 2}$.

Solution

Factor the denominator to determine more easily the zeros of the denominator.

$$f(x) = \frac{x}{x^2 - x - 2} = \frac{x}{(x + 1)(x - 2)}.$$

y-Intercept: $(0, 0)$, because $f(0) = 0$

x-Intercept: $(0, 0)$

Vertical Asymptotes: $x = -1$, $x = 2$, zeros of denominator

Horizontal Asymptote: $y = 0$, because degree of $N(x) <$ degree of $D(x)$

Additional Points:

x	-3	-1	-0.5	1	2	3
$f(x)$	-0.3	Undefined	0.4	-0.5	Undefined	0.75

The graph is shown in Figure 3.56.

☑ *Checkpoint* Now try Exercise 21.

Example 4   Sketching the Graph of a Rational Function

Sketch the graph of $f(x) = \dfrac{x^2 - 9}{x^2 - 2x - 3}$.

Solution

By factoring the numerator and denominator, you have

$$f(x) = \frac{x^2 - 9}{x^2 - 2x - 3} = \frac{(x - 3)(x + 3)}{(x - 3)(x + 1)} = \frac{x + 3}{x + 1}, \quad x \neq 3.$$

y-Intercept: $(0, 3)$, because $f(0) = 3$

x-Intercept: $(-3, 0)$

Vertical Asymptote: $x = -1$, zero of (simplified) denominator

Horizontal Asymptote: $y = 1$, because degree of $N(x) =$ degree of $D(x)$

Additional Points:

x	-5	-2	-1	1	-0.5	3	4
$f(x)$	0.5	-1	Undefined	2	5	Undefined	1.4

The graph is shown in Figure 3.57. Notice there is a hole in the graph at $x = 3$. This is because the function is not defined when $x = 3$.

☑ *Checkpoint* Now try Exercise 23.

Exploration

Use a graphing utility to graph

$$f(x) = 1 + \frac{1}{x - \dfrac{1}{x}}.$$

Set the graphing utility to *dot* mode and use a decimal viewing window. Use the *trace* feature to find three "holes" or "breaks" in the graph. Do all three holes represent zeros of the denominator

$$x - \frac{1}{x}?$$

Explain.

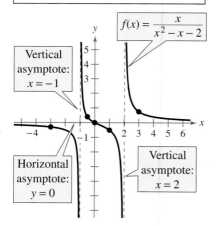

Figure 3.56

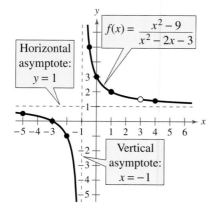

Figure 3.57   Hole at $x = 3$

Slant Asymptotes

Consider a rational function whose denominator is of degree 1 or greater. If the degree of the numerator is exactly *one more* than the degree of the denominator, the graph of the function has a **slant** (or **oblique**) **asymptote.** For example, the graph of

$$f(x) = \frac{x^2 - x}{x + 1}$$

has a slant asymptote, as shown in Figure 3.58. To find the equation of a slant asymptote, use long division. For instance, by dividing $x + 1$ into $x^2 - x$, you have

$$f(x) = \frac{x^2 - x}{x + 1} = \underbrace{x - 2}_{\substack{\text{Slant asymptote} \\ (y = x - 2)}} + \frac{2}{x + 1}.$$

As x increases or decreases without bound, the remainder term $2/(x + 1)$ approaches 0, so the graph of f approaches the line $y = x - 2$, as shown in Figure 3.58.

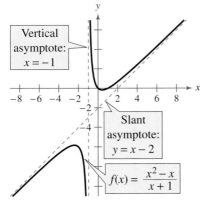

Figure 3.58

Example 5   A Rational Function with a Slant Asymptote

Sketch the graph of $f(x) = \dfrac{x^2 - x - 2}{x - 1}$.

Solution
First write $f(x)$ in two different ways. Factoring the numerator

$$f(x) = \frac{x^2 - x - 2}{x - 1} = \frac{(x - 2)(x + 1)}{x - 1}$$

enables you to recognize the x-intercepts. Long division

$$f(x) = \frac{x^2 - x - 2}{x - 1} = x - \frac{2}{x - 1}$$

enables you to recognize that the line $y = x$ is a slant asymptote of the graph.

y-Intercept: (0, 2), because $f(0) = 2$
x-Intercepts: $(-1, 0)$ and $(2, 0)$
Vertical Asymptote: $x = 1$, zero of denominator
Horizontal Asymptote: None, because degree of $N(x)$ > degree of $D(x)$
Slant Asymptote: $y = x$
Additional Points:

x	-2	0.5	1	1.5	3
$f(x)$	-1.33	4.5	Undefined	-2.5	2

The graph is shown in Figure 3.59.

✓ *Checkpoint* Now try Exercise 45.

Exploration

Do you think it is possible for the graph of a rational function to cross its horizontal asymptote or its slant asymptote? Use the graphs of the following functions to investigate this question. Write a summary of your conclusion. Explain your reasoning.

$$f(x) = \frac{x}{x^2 + 1}$$

$$g(x) = \frac{2x}{3x^2 - 2x + 1}$$

$$h(x) = \frac{x^3}{x^2 + 1}$$

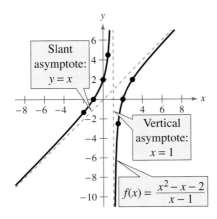

Figure 3.59

Application

Example 6   Finding a Minimum Area

A rectangular page is designed to contain 48 square inches of print. The margins on each side of the page are $1\frac{1}{2}$ inches wide. The margins at the top and bottom are each 1 inch deep. What should the dimensions of the page be so that the minimum amount of paper is used?

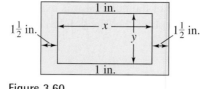

Figure 3.60

Graphical Solution

Let A be the area to be minimized. From Figure 3.60, you can write

$$A = (x + 3)(y + 2).$$

The printed area inside the margins is modeled by $48 = xy$ or $y = 48/x$. To find the minimum area, rewrite the equation for A in terms of just one variable by substituting $48/x$ for y.

$$A = (x + 3)\left(\frac{48}{x} + 2\right) = \frac{(x + 3)(48 + 2x)}{x}, \quad x > 0$$

The graph of this rational function is shown in Figure 3.61. Because x represents the width of the printed area, you need consider only the portion of the graph for which x is positive. Using the *minimum* feature or the *zoom* and *trace* features of a graphing utility, you can approximate the minimum value of A to occur when $x \approx 8.5$ inches. The corresponding value of y is $48/8.5 \approx 5.6$ inches. So, the dimensions should be

$$x + 3 \approx 11.5 \text{ inches by } y + 2 \approx 7.6 \text{ inches.}$$

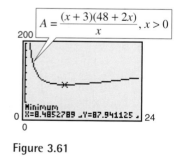

Figure 3.61

✓ *Checkpoint* Now try Exercise 65.

Numerical Solution

Let A be the area to be minimized. From Figure 3.60, you can write

$$A = (x + 3)(y + 2).$$

The printed area inside the margins is modeled by $48 = xy$ or $y = 48/x$. To find the minimum area, rewrite the equation for A in terms of just one variable by substituting $48/x$ for y.

$$A = (x + 3)\left(\frac{48}{x} + 2\right) = \frac{(x + 3)(48 + 2x)}{x}, \quad x > 0$$

Use the *table* feature of a graphing utility to create a table of values for the function

$$y_1 = \frac{(x + 3)(48 + 2x)}{x}$$

beginning at $x = 1$. From the table, you can see that the minimum value of y_1 occurs when x is somewhere between 8 and 9, as shown in Figure 3.62. To approximate the minimum value of y_1 to one decimal place, change the table to begin at $x = 8$ and set the table step to 0.1. The minimum value of y_1 occurs when $x \approx 8.5$, as shown in Figure 3.63. The corresponding value of y is $48/8.5 \approx 5.6$ inches. So, the dimensions should be $x + 3 \approx 11.5$ inches by $y + 2 \approx 7.6$ inches.

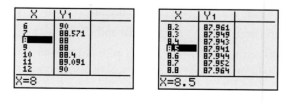

Figure 3.62   Figure 3.63

If you go on to take a course in calculus, you will learn an analytic technique for finding the exact value of x that produces a minimum area in Example 6. In this case, that value is $x = 6\sqrt{2} \approx 8.485$.

3.6 Exercises

Vocabulary Check

Fill in the blanks.

1. For the rational function $f(x) = N(x)/D(x)$, if the degree of $N(x)$ is exactly one more than the degree of $D(x)$, then the graph of f has a _____ (or oblique) _____ .

2. The graph of $f(x) = 1/x$ has a _____ asymptote at $x = 0$.

In Exercises 1–4, use a graphing utility to graph $f(x) = 2/x$ and the function g in the same viewing window. Describe the relationship between the two graphs.

1. $g(x) = f(x) + 1$
2. $g(x) = f(x - 1)$
3. $g(x) = -f(x)$
4. $g(x) = \frac{1}{2}f(x + 2)$

In Exercises 5–8, use a graphing utility to graph $f(x) = 2/x^2$ and the function g in the same viewing window. Describe the relationship between the two graphs.

5. $g(x) = f(x) - 2$
6. $g(x) = -f(x)$
7. $g(x) = f(x - 2)$
8. $g(x) = \frac{1}{4}f(x)$

In Exercises 9–26, sketch the graph of the rational function by hand. As sketching aids, check for intercepts, vertical asymptotes, and horizontal asymptotes. Use a graphing utility to verify your graph.

9. $f(x) = \dfrac{1}{x + 2}$
10. $f(x) = \dfrac{1}{x - 6}$

11. $C(x) = \dfrac{5 + 2x}{1 + x}$
12. $P(x) = \dfrac{1 - 3x}{1 - x}$

13. $f(t) = \dfrac{1 - 2t}{t}$
14. $g(x) = \dfrac{1}{x + 2} + 2$

15. $f(x) = \dfrac{x^2}{x^2 - 4}$
16. $g(x) = \dfrac{x}{x^2 - 9}$

17. $f(x) = \dfrac{x}{x^2 - 1}$
18. $f(x) = -\dfrac{1}{(x - 2)^2}$

19. $g(x) = \dfrac{4(x + 1)}{x(x - 4)}$
20. $h(x) = \dfrac{2}{x^2(x - 3)}$

21. $f(x) = \dfrac{3x}{x^2 - x - 2}$
22. $f(x) = \dfrac{2x}{x^2 + x - 2}$

23. $f(x) = \dfrac{x^2 + 3x}{x^2 + x - 6}$
24. $g(x) = \dfrac{5(x + 4)}{x^2 + x - 12}$

25. $f(x) = \dfrac{x^2 - 1}{x + 1}$
26. $f(x) = \dfrac{x^2 - 16}{x - 4}$

In Exercises 27–36, use a graphing utility to graph the function. Determine its domain and identify any vertical or horizontal asymptotes.

27. $f(x) = \dfrac{2 + x}{1 - x}$
28. $f(x) = \dfrac{3 - x}{2 - x}$

29. $f(t) = \dfrac{3t + 1}{t}$
30. $h(x) = \dfrac{x - 2}{x - 3}$

31. $h(t) = \dfrac{4}{t^2 + 1}$
32. $g(x) = -\dfrac{x}{(x - 2)^2}$

33. $f(x) = \dfrac{x + 1}{x^2 - x - 6}$
34. $f(x) = \dfrac{x + 4}{x^2 + x - 6}$

35. $f(x) = \dfrac{20x}{x^2 + 1} - \dfrac{1}{x}$

36. $f(x) = 5\left(\dfrac{1}{x - 4} - \dfrac{1}{x + 2}\right)$

Exploration **In Exercises 37–42, use a graphing utility to graph the function. What do you observe about its asymptotes?**

37. $h(x) = \dfrac{6x}{\sqrt{x^2 + 1}}$
38. $f(x) = -\dfrac{x}{\sqrt{9 + x^2}}$

39. $g(x) = \dfrac{4|x - 2|}{x + 1}$
40. $f(x) = -\dfrac{8|3 + x|}{x - 2}$

41. $f(x) = \dfrac{4(x - 1)^2}{x^2 - 4x + 5}$
42. $g(x) = \dfrac{3x^4 - 5x + 3}{x^4 + 1}$

In Exercises 43–50, sketch the graph of the rational function by hand. As sketching aids, check for intercepts, vertical asymptotes, and slant asymptotes.

43. $f(x) = \dfrac{2x^2 + 1}{x}$
44. $g(x) = \dfrac{1 - x^2}{x}$

45. $h(x) = \dfrac{x^2}{x-1}$

46. $f(x) = \dfrac{x^3}{x^2-1}$

47. $g(x) = \dfrac{x^3}{2x^2-8}$

48. $f(x) = \dfrac{x^2-1}{x^2+4}$

49. $f(x) = \dfrac{x^3+2x^2+4}{2x^2+1}$

50. $f(x) = \dfrac{2x^2-5x+5}{x-2}$

Graphical Reasoning In Exercises 51–54, (a) use the graph to estimate any x-intercepts of the rational function and (b) set y = 0 and solve the resulting equation to confirm your result in part (a).

51. $y = \dfrac{x+1}{x-3}$

52. $y = \dfrac{2x}{x-3}$

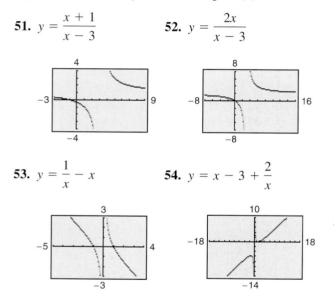

53. $y = \dfrac{1}{x} - x$

54. $y = x - 3 + \dfrac{2}{x}$

In Exercises 55–58, use a graphing utility to graph the rational function. Determine the domain of the function and identify any asymptotes.

55. $y = \dfrac{2x^2+x}{x+1}$

56. $y = \dfrac{x^2+5x+8}{x+3}$

57. $y = \dfrac{1+3x^2-x^3}{x^2}$

58. $y = \dfrac{12-2x-x^2}{2(4+x)}$

Graphical Reasoning In Exercises 59–62, (a) use a graphing utility to graph the function and determine any x-intercepts, and (b) set y = 0 and solve the resulting equation to confirm your result in part (a).

59. $y = \dfrac{1}{x+5} + \dfrac{4}{x}$

60. $y = 20\left(\dfrac{2}{x+1} - \dfrac{3}{x}\right)$

61. $y = x - \dfrac{6}{x-1}$

62. $y = x - \dfrac{9}{x}$

63. *Concentration of a Mixture*   A 1000-liter tank contains 50 liters of a 25% brine solution. You add x liters of a 75% brine solution to the tank.

(a) Show that the concentration C, the proportion of brine to the total solution, of the final mixture is given by

$$C = \dfrac{3x+50}{4(x+50)}.$$

(b) Determine the domain of the function based on the physical constraints of the problem.

(c) Use a graphing utility to graph the function. As the tank is filled, what happens to the rate at which the concentration of brine increases? What percent does the concentration of brine appear to approach?

64. *Geometry*   A rectangular region of length x and width y has an area of 500 square meters.

(a) Write the width y as a function of x.

(b) Determine the domain of the function based on the physical constraints of the problem.

(c) Sketch a graph of the function and determine the width of the rectangle when x = 30 meters.

65. *Page Design*   A page that is x inches wide and y inches high contains 30 square inches of print. The margins at the top and bottom are 2 inches deep and the margins on each side are 1 inch wide (see figure).

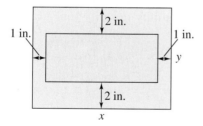

(a) Show that the total area A of the page is given by

$$A = \dfrac{2x(2x+11)}{x-2}.$$

(b) Determine the domain of the function based on the physical constraints of the problem.

(c) Use a graphing utility to graph the area function and approximate the page size such that the minimum amount of paper will be used. Verify your answer numerically using the *table* feature of a graphing utility.

66. Geometry A right triangle is formed in the first quadrant by the *x*-axis, the *y*-axis, and a line segment through the point $(3, 2)$ (see figure).

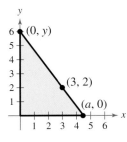

(a) Show that an equation of the line segment is given by

$$y = \frac{2(a - x)}{a - 3}, \quad 0 \le x \le a.$$

(b) Show that the area of the triangle is given by

$$A = \frac{a^2}{a - 3}.$$

(c) Use a graphing utility to graph the area function and estimate the value of *a* that yields a minimum area. Estimate the minimum area. Verify your answer numerically using the *table* feature of a graphing utility.

67. Cost The ordering and transportation cost *C* (in thousands of dollars) for the components used in manufacturing a product is given by

$$C = 100\left(\frac{200}{x^2} + \frac{x}{x + 30}\right), \quad x \ge 1$$

where *x* is the order size (in hundreds). Use a graphing utility to graph the cost function. From the graph, estimate the order size that minimizes cost.

68. Average Cost The cost *C* of producing *x* units of a product is given by $C = 0.2x^2 + 10x + 5$, and the average cost per unit is given by

$$\overline{C} = \frac{C}{x} = \frac{0.2x^2 + 10x + 5}{x}, \quad x > 0.$$

Sketch the graph of the average cost function, and estimate the number of units that should be produced to minimize the average cost per unit.

69. Medicine The concentration *C* of a chemical in the bloodstream *t* hours after injection into muscle tissue is given by

$$C = \frac{3t^2 + t}{t^3 + 50}, \quad t \ge 0.$$

(a) Determine the horizontal asymptote of the function and interpret its meaning in the context of the problem.

(b) Use a graphing utility to graph the function and approximate the time when the bloodstream concentration is greatest.

(c) Use a graphing utility to determine when the concentration is less than 0.345.

70. Numerical and Graphical Analysis A driver averaged 50 miles per hour on the round trip between Baltimore, Maryland and Philadelphia, Pennsylvania, 100 miles away. The average speeds for going and returning were *x* and miles per hour, respectively.

(a) Show that $y = (25x)/(x - 25)$.

(b) Determine the vertical and horizontal asymptotes of the function.

(c) Use a graphing utility to complete the table. What do you observe?

x	30	35	40	45	50	55	60
y							

(d) Use a graphing utility to graph the function.

(e) Is it possible to average 20 miles per hour in one direction and still average 50 miles per hour on the round trip? Explain.

71. Comparing Models The attendance *A* (in millions) at women's Division I college basketball games from 1995 to 2002 is shown in the table. (Source: NCAA)

Year	Attendance, A
1995	4.0
1996	4.2
1997	4.9
1998	5.4
1999	5.8
2000	6.4
2001	6.5
2002	6.9

For each of the following, let t represent the year, with $t = 5$ corresponding to 1995.

(a) Use the *regression* feature of a graphing utility to find a linear model for the data. Use a graphing utility to plot the data and graph the model in the same viewing window.

(b) Find a rational model for the data. Take the reciprocal of A to generate the points $(t, 1/A)$. Use the *regression* feature of a graphing utility to find a linear model for this data. The resulting line has the form

$$\frac{1}{A} = at + b.$$

Solve for A. Use a graphing utility to plot the data and graph the rational model in the same viewing window.

(c) Use the *table* feature of a graphing utility to create a table showing the predicted attendance based on each model for each of the years in the original table. Which model do you prefer? Why?

72. **Comparing Models**   The table shows the average room rate R (in dollars) for hotels in the United States from 1995 to 2001. The data can be approximated by the model

$$R = \frac{6.245t + 44.05}{0.025t + 1.00}, \quad 5 \le t \le 11$$

where t represents the year, with $t = 5$ corresponding to 1995. (Source: American Hotel & Lodging Association)

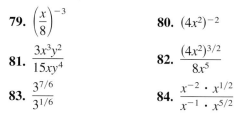

Year	Rate, R
1995	66.65
1996	70.93
1997	75.31
1998	78.62
1999	81.33
2000	85.89
2001	88.27

(a) Use a graphing utility to plot the data and graph the model in the same viewing window.

(b) Use the *regression* feature of a graphing utility to find a linear model for the data. Then use a graphing utility to plot the data and graph the linear model in the same viewing window.

(c) Which of the two models would you recommend as a predictor of the average room rate for a hotel for the years following 2001? Explain your reasoning.

Synthesis

True or False?   **In Exercises 73 and 74, determine whether the statement is true or false. Justify your answer.**

73. If the graph of a rational function f has a vertical asymptote at $x = 5$, it is possible to sketch the graph without lifting your pencil from the paper.

74. The graph of a rational function can never cross one of its asymptotes.

Think About It   **In Exercises 75 and 76, use a graphing utility to graph the function. Explain why there is no vertical asymptote when a superficial examination of the function might indicate that there should be one.**

75. $h(x) = \dfrac{6 - 2x}{3 - x}$ 76. $g(x) = \dfrac{x^2 + x - 2}{x - 1}$

Think About It   **In Exercises 77 and 78, write a rational function satisfying the following criteria.**

77. Vertical asymptote: $x = 2$
    Slant asymptote: $y = x + 1$
    Zero of the function: $x = -2$

78. Vertical asymptote: $x = -4$
    Slant asymptote: $y = x - 2$
    Zero of the function: $x = 3$

Review

In Exercises 79–84, simplify the expression.

79. $\left(\dfrac{x}{8}\right)^{-3}$ 80. $(4x^2)^{-2}$

81. $\dfrac{3x^3y^2}{15xy^4}$ 82. $\dfrac{(4x^2)^{3/2}}{8x^5}$

83. $\dfrac{3^{7/6}}{3^{1/6}}$ 84. $\dfrac{x^{-2} \cdot x^{1/2}}{x^{-1} \cdot x^{5/2}}$

In Exercises 85–88, use a graphing utility to graph the function and find its domain and range.

85. $f(x) = \sqrt{6 + x^2}$ 86. $f(x) = \sqrt{121 - x^2}$
87. $f(x) = -|x + 9|$ 88. $f(x) = -x^2 + 9$

3.7 Exploring Data: Quadratic Models

Classifying Scatter Plots

In real life, many relationships between two variables are parabolic, as in Section 3.1, Example 5. A scatter plot can be used to give you an idea of which type of model will best fit a set of data.

Example 1   Classifying Scatter Plots

Decide whether each set of data could best be modeled by a linear model, $y = ax + b$, or a quadratic model, $y = ax^2 + bx + c$.

a. $(0.9, 1.4)$, $(1.3, 1.5)$, $(1.3, 1.9)$, $(1.4, 2.1)$, $(1.6, 2.8)$, $(1.8, 2.9)$, $(2.1, 3.4)$, $(2.1, 3.4)$, $(2.5, 3.6)$, $(2.9, 3.7)$, $(3.2, 4.2)$, $(3.3, 4.3)$, $(3.6, 4.4)$, $(4.0, 4.5)$, $(4.2, 4.8)$, $(4.3, 5.0)$

b. $(0.9, 2.5)$, $(1.3, 4.03)$, $(1.3, 4.1)$, $(1.4, 4.4)$, $(1.6, 5.1)$, $(1.8, 6.05)$, $(2.1, 7.48)$, $(2.1, 7.6)$, $(2.5, 9.8)$, $(2.9, 12.4)$, $(3.2, 14.3)$, $(3.3, 15.2)$, $(3.6, 18.1)$, $(4.0, 19.9)$, $(4.2, 23.0)$, $(4.3, 23.9)$

Solution

Begin by entering the data into a graphing utility as shown in Figure 3.64.

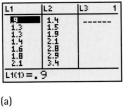
(a)

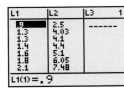
(b)

Figure 3.64

Then display the scatter plots as shown in Figure 3.65.

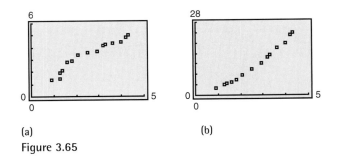

(a)   (b)

Figure 3.65

From the scatter plots, it appears that the data in part (a) follows a linear pattern. So, it can be modeled by a linear function. The data in part (b) follows a parabolic pattern. So, it can be modeled by a quadratic function.

✓ *Checkpoint* Now try Exercise 3.

Fitting a Quadratic Model to Data

In Section 2.6, you created scatter plots of data and used a graphing utility to find the least squares regression lines for the data. You can use a similar procedure to find a model for nonlinear data. Once you have used a scatter plot to determine the type of model that would best fit a set of data, there are several ways that you can actually find the model. Each method is best used with a computer or calculator, rather than with hand calculations.

Example 2   Fitting a Quadratic Model to Data

A study was done to compare the speed x (in miles per hour) with the mileage y (in miles per gallon) of an automobile. The results are shown in the table. (Source: Federal Highway Administration)

a. Use a graphing utility to create a scatter plot of the data.

b. Use the *regression* feature of the graphing utility to find a model that best fits the data.

c. Approximate the speed at which the mileage is the greatest.

Speed, x	Mileage, y
15	22.3
20	25.5
25	27.5
30	29.0
35	28.8
40	30.0
45	29.9
50	30.2
55	30.4
60	28.8
65	27.4
70	25.3
75	23.3

Solution

a. Begin by entering the data into a graphing utility and displaying the scatter plot, as shown in Figure 3.66. From the scatter plot, you can see that the data has a parabolic trend.

b. Using the *regression* feature of a graphing utility, you can find the quadratic model, as shown in Figure 3.67. So, the quadratic equation that best fits the data is given by

$$y = -0.0082x^2 + 0.746x + 13.47.$$ Quadratic model

c. Graph the data and the model in the same viewing window, as shown in Figure 3.68. Use the *maximum* feature or *zoom* and *trace* features of the graphing utility to approximate the speed at which the mileage is greatest. You should obtain a maximum of approximately (47, 30), as shown in Figure 3.68. So, the speed at which the mileage is greatest is about 47 miles per hour.

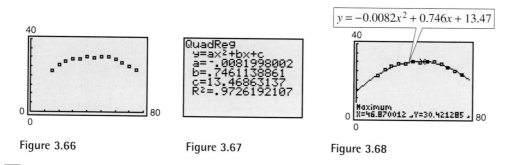

Figure 3.66   Figure 3.67   Figure 3.68

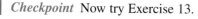 *Checkpoint* Now try Exercise 13.

TECHNOLOGY SUPPORT   For instructions on how to use the *regression* feature, see Appendix A; for specific keystrokes, go to the text website at *college.hmco.com*.

Example 3   Fitting a Quadratic Model to Data

A basketball is dropped from a height of about 5.25 feet. The height of the basketball is recorded 23 times at intervals of about 0.02 second.* The results are shown in the table. Use a graphing utility to find a model that best fits the data. Then use the model to predict the time when the basketball will hit the ground.

Time, x	Height, y
0.0	5.23594
0.02	5.20353
0.04	5.16031
0.06	5.09910
0.08	5.02707
0.099996	4.95146
0.119996	4.85062
0.139992	4.74979
0.159988	4.63096
0.179988	4.50132
0.199984	4.35728
0.219984	4.19523
0.23998	4.02958
0.25993	3.84593
0.27998	3.65507
0.299976	3.44981
0.319972	3.23375
0.339961	3.01048
0.359961	2.76921
0.379951	2.52074
0.399941	2.25786
0.419941	1.98058
0.439941	1.63488

Solution

Begin by entering the data into a graphing utility and displaying the scatter plot, as shown in Figure 3.69. From the scatter plot, you can see that the data has a parabolic trend. So, using the *regression* feature of the graphing utility, you can find the quadratic model, as shown in Figure 3.70. The quadratic model that best fits the data is given by

$$y = -15.449x^2 - 1.30x + 5.2.$$ Quadratic model

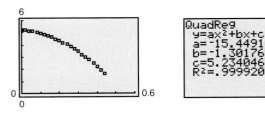

Figure 3.69   Figure 3.70

Using this model, you can predict the time when the basketball will hit the ground by substituting 0 for y and solving the resulting equation for x.

$y = -15.449x^2 - 1.30x + 5.2$ Write original model.

$0 = -15.449x^2 - 1.30x + 5.2$ Substitute 0 for y.

$x = \dfrac{-b \pm \sqrt{b^2 - 4ac}}{2a}$ Quadratic Formula

$\quad = \dfrac{-(-1.30) \pm \sqrt{(-1.30)^2 - 4(-15.449)(5.2)}}{2(-15.449)}$ Substitute for a, b, and c.

$\quad \approx 0.54$ Choose positive solution.

So, the solution is about 0.54 second. In other words, the basketball will continue to fall for about $0.54 - 0.44 = 0.1$ second more before hitting the ground.

✓ *Checkpoint* Now try Exercise 15.

Choosing a Model

Sometimes it is not easy to distinguish from a scatter plot which type of model a set of data can best be modeled by. You should first find several models for the data and then choose the model that best fits the data by comparing the y-values of each model with the actual y-values.

*Data was collected with a Texas Instruments CBL (Calculator-Based Laboratory) System.

Example 4 Choosing a Model

The table shows the amount y (in billions of dollars) spent on books and maps in the United States for the years 1990 to 2000. Use the *regression* feature of a graphing utility to find a linear model and a quadratic model for the data. Determine which model best fits the data. (Source: U.S. Bureau of Economic Analysis)

Year	Amount, y
1990	16.5
1991	16.9
1992	17.7
1993	18.8
1994	20.8
1995	23.1
1996	24.9
1997	26.3
1998	28.2
1999	30.7
2000	33.9

Solution

Let x represent the year, with $x = 0$ corresponding to 1990. Begin by entering the data into the graphing utility. Then use the *regression* feature to find a linear model (see Figure 3.71) and a quadratic model (see Figure 3.72) for the data.

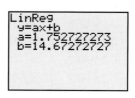

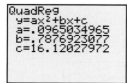

Figure 3.71 Linear model

Figure 3.72 Quadratic model

So, a linear model for the data is given by

$$y = 1.75x + 14.7 \qquad \text{Linear model}$$

and a quadratic model for the data is given by

$$y = 0.097x^2 + 0.79x + 16.1. \qquad \text{Quadratic model}$$

Plot the data and the linear model in the same viewing window, as shown in Figure 3.73. Then plot the data and the quadratic model in the same viewing window, as shown in Figure 3.74. To determine which model best fits the data, compare the y-values given by each model with the actual y-values. The model whose y-values are closest to the actual values is the one that fits best. In this case, the best-fitting model is the quadratic model.

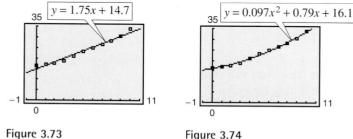

Figure 3.73

Figure 3.74

✓ *Checkpoint* Now try Exercise 21.

TECHNOLOGY TIP Recall from Section 2.6 that when you use the *regression* feature of a graphing utility, the program may output a correlation coefficient. The correlation coefficient for the linear model in Example 4 is $r^2 \approx 0.972$ and the correlation coefficient for the quadratic model is $r^2 \approx 0.995$. Because the correlation coefficient for the quadratic model is closer to 1, the quadratic model better fits the data.

3.7 Exercises

Vocabulary Check

Fill in the blanks.

1. A scatter plot with either a positive or a negative correlation could be modeled by a _____ equation.

2. A scatter plot that appears parabolic could be modeled by a _____ equation.

In Exercises 1–6, determine whether the scatter plot could best be modeled by a linear model, a quadratic model, or neither.

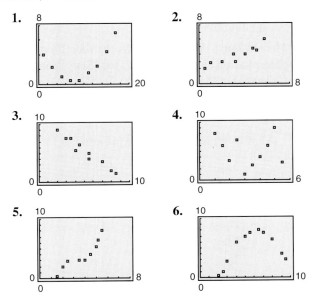

In Exercises 7–12, (a) use a graphing utility to create a scatter plot of the data, (b) determine whether the data could best be modeled by a linear model or a quadratic model, (c) use the *regression* feature of a graphing utility to find a model for the data, (d) use a graphing utility to graph the model with the scatter plot from part (a), and (e) create a table to compare the original data with the data given by the model.

7. $(0, 2.1)$, $(1, 2.4)$, $(2, 2.5)$, $(3, 2.8)$, $(4, 2.9)$, $(5, 3.0)$, $(6, 3.0)$, $(7, 3.2)$, $(8, 3.4)$, $(9, 3.5)$, $(10, 3.6)$

8. $(0, 10.0)$, $(1, 9.7)$, $(2, 9.4)$, $(3, 9.3)$, $(4, 9.1)$, $(5, 8.9)$, $(6, 8.6)$, $(7, 8.4)$, $(8, 8.4)$, $(9, 8.2)$, $(10, 8.0)$

9. $(0, 3480)$, $(5, 2235)$, $(10, 1250)$, $(15, 565)$, $(20, 150)$, $(25, 12)$, $(30, 145)$, $(35, 575)$, $(40, 1275)$, $(45, 2225)$, $(50, 3500)$, $(55, 5010)$

10. $(0, 6140)$, $(2, 6815)$, $(4, 7335)$, $(6, 7710)$, $(8, 7915)$, $(10, 7590)$, $(12, 7975)$, $(14, 7700)$, $(16, 7325)$, $(18, 6820)$, $(20, 6125)$, $(22, 5325)$

11. $(-5, 3.8)$, $(-4, 4.7)$, $(-3, 5.5)$, $(-2, 6.2)$, $(-1, 7.1)$, $(0, 7.9)$, $(1, 8.1)$, $(2, 7.7)$, $(3, 6.9)$, $(4, 6.0)$, $(5, 5.6)$, $(6, 4.4)$, $(7, 3.2)$

12. $(-9, -8.7)$, $(-8, -6.5)$, $(-7, -4.5)$, $(-6, -2.4)$, $(-5, -1.2)$, $(-4, 0.3)$, $(-3, 1.5)$, $(-2, 2.5)$, $(-1, 3.3)$, $(0, 3.9)$, $(1, 4.5)$, $(2, 4.6)$

13. *Education* The table shows the percent P of public schools in the United States with access to the Internet from 1994 to 2000. (Source: U.S. National Center for Education Statistics)

Year	Percent, P
1994	35
1995	50
1996	65
1997	78
1998	89
1999	95
2000	98

(a) Use a graphing utility to create a scatter plot of the data. Let t represent the year, with $t = 4$ corresponding to 1994.

(b) Use the *regression* feature of a graphing utility to find a quadratic model for the data.

(c) Use a graphing utility to graph the model with the scatter plot from part (a). Is the quadratic model a good fit for the data?

(d) Use the model to determine when 100% of public schools will have access to the Internet.

(e) Can the model be used to predict the percent of public schools with Internet access in the future? Explain.

14. *Entertainment* The table on the next page shows the number H of hours spent per person playing video games in the United States from 1995 to 2000. (Source: Veronis Suhler Stevenson)

Year	Hours, *H*
1995	24
1996	25
1997	36
1998	43
1999	61
2000	70

Table for 14

(a) Use a graphing utility to create a scatter plot of the data. Let t represent the year, with $t = 5$ corresponding to 1995.

(b) Use the *regression* feature of a graphing utility to find a quadratic model for the data.

(c) Use a graphing utility to graph the model with the scatter plot from part (a). Is the quadratic model a good fit for the data?

(d) The projected number H^* of hours spent per person playing video games for the years 2001 to 2005 is shown in the table. Use the model obtained in part (b) to predict the number of hours for the same years.

Year	2001	2002	2003	2004	2005
H^*	79	90	97	103	115

(e) Compare your predictions from part (d) with those given in the table. Explain why the values may differ.

15. *Medicine* The table shows the number H (in thousands) of hospitals in the United States for selected years from 1960 to 2000.   (Source: Health Forum)

Year	Hospitals, *H*
1960	6876
1965	7123
1970	7123
1975	7156
1980	6965
1985	6872
1990	6649
1995	6291
2000	5810

(a) Use a graphing utility to create a scatter plot of the data. Let t represent the year, with $t = 0$ corresponding to 1960.

(b) Use the *regression* feature of a graphing utility to find a quadratic model for the data.

(c) Use a graphing utility to graph the model with the scatter plot from part (a). Is the quadratic model a good fit for the data?

(d) Use the graph from part (c) to determine in which year the number of hospitals reached a maximum.

(e) Do you think the model can be used to predict the number of hospitals in the United States in the future? Explain.

16. *Meteorology* The table shows the monthly normal precipitation P (in inches) for San Francisco, California.   (Source: U.S. National Oceanic and Atmospheric Administration)

Month	Precipitation, *P*
January	4.45
February	4.01
March	3.26
April	1.17
May	0.38
June	0.11
July	0.03
August	0.07
September	0.20
October	1.04
November	2.49
December	2.89

(a) Use a graphing utility to create a scatter plot of the data. Let t represent the month, with $t = 1$ corresponding to January.

(b) Use the *regression* feature of a graphing utility to find a quadratic model for the data.

(c) Use a graphing utility to graph the model with the scatter plot from part (a).

(d) Use the graph from part (c) to determine in which month the normal precipitation in San Francisco is the least.

In Exercises 17–20, (a) use the *regression* **feature of a graphing utility to find a linear model and a quadratic model for the data, (b) determine the correlation coefficient for each model, and (c) use the correlation coefficient to determine which model best fits the data.**

17. (1, 4.0), (2, 6.5), (3, 8.8), (4, 10.6), (5, 13.9), (6, 15.0), (7, 17.5), (8, 20.1), (9, 24.0), (10, 27.1)

18. (1, 1.1), (2, 3.0), (3, 5.1), (4, 7.3), (5, 9.3), (6, 11.5), (7, 13.6), (8, 15.5), (9, 17.8), (10, 20.0)

19. (−8, 7.4), (−6, 5.7), (−4, 3.7), (−2, 2.1), (0, 0.2), (2, −1.6), (4, −3.4), (6, −5.1), (8, −6.9), (10, −8.6)

20. (−20, 805), (−15, 744), (−10, 704), (−5, 653), (0, 587), (5, 551), (10, 512), (15, 478), (20, 436), (25, 430)

21. *Sales*   The table shows the sales S (in millions of dollars) for Guitar Center, Inc. from 1996 to 2002. (Source: Guitar Center, Inc.)

Year	Sales, S
1996	213.3
1997	296.7
1998	391.7
1999	620.1
2000	785.7
2001	938.2
2002	1101.1

(a) Use a graphing utility to create a scatter plot of the data. Let t represent the year, with $t = 6$ corresponding to 1996.

(b) Use the *regression* feature of a graphing utility to find a linear model for the data.

(c) Use a graphing utility to graph the model with the scatter plot from part (a).

(d) Use the *regression* feature of a graphing utility to find a quadratic model for the data.

(e) Use a graphing utility to graph the quadratic model with the scatter plot from part (a).

(f) Determine which model best fits the data and use the model you chose to predict the sales for Guitar Center, Inc. in 2007.

22. *Writing*   Explain why the parabola shown in the figure is not a good fit for the data.

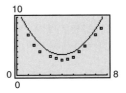

Synthesis

True or False?   **In Exercises 23 and 24, determine whether the statement is true or false. Justify your answer.**

23. The graph of a quadratic model with a negative leading coefficient will have a maximum value at its vertex.

24. The graph of a quadratic model with a positive leading coefficient will have a minimum value at its vertex.

Review

In Exercises 25–28, find (a) $f \circ g$ and (b) $g \circ f$.

25. $f(x) = 2x - 1,$ $g(x) = x^2 + 3$

26. $f(x) = 5x + 8,$ $g(x) = 2x^2 - 1$

27. $f(x) = x^3 - 1,$ $g(x) = \sqrt[3]{x + 1}$

28. $f(x) = \sqrt[3]{x + 5},$ $g(x) = x^3 - 5$

In Exercises 29–32, determine algebraically whether the function is one-to-one. If it is, find its inverse function. Verify your answer graphically.

29. $f(x) = 2x + 5$

30. $f(x) = \dfrac{x - 4}{5}$

31. $f(x) = x^2 + 5, x \geq 0$

32. $f(x) = 2x^2 - 3, x \geq 0$

In Exercises 33–36, plot the complex number in the complex plane.

33. $1 - 3i$

34. $-2 + 4i$

35. $-5i$

36. $8i$

3 Chapter Summary

What did you learn?

Section 3.1	Review Exercises
☐ Analyze graphs of quadratic functions.	1, 2
☐ Write quadratic functions in standard form and use the results to sketch graphs of functions.	3–10
☐ Find minimum and maximum values of functions in real-life applications.	11, 12

Section 3.2
☐ Use transformations to sketch graphs of polynomial functions.   13–16
☐ Use the Leading Coefficient Test to determine the end behavior of graphs of polynomial functions.   17–22
☐ Find and use zeros of polynomial functions as sketching aids.   23–28
☐ Use the Intermediate Value Theorem to help locate zeros of polynomial functions.   29–32

Section 3.3
☐ Use long division to divide polynomials by other polynomials.   35–42
☐ Use synthetic division to divide polynomials by binomials of the form $(x - k)$.   43–48
☐ Use the Remainder and Factor Theorems.   49–54
☐ Use the Rational Zero Test to determine possible rational zeros of polynomial functions.   55–60
☐ Use Descartes's Rule of Signs and the Upper and Lower Bound Rules to find zeros of polynomials.   61–64

Section 3.4
☐ Use the Fundamental Theorem of Algebra to determine the number of zeros of a polynomial function.   65–68
☐ Find all zeros of polynomial functions, including complex zeros.   69–78
☐ Find conjugate pairs of complex zeros.   79–82
☐ Find zeros of polynomials by factoring.   83–86

Section 3.5
☐ Find the domains of rational functions.   87–98
☐ Find horizontal and vertical asymptotes of graphs of rational functions.   87–98
☐ Use rational functions to model and solve real-life problems.   99, 100

Section 3.6
☐ Analyze and sketch graphs of rational functions.   101–110
☐ Sketch graphs of rational functions that have slant asymptotes.   111–114
☐ Use rational functions to model and solve real-life problems.   115, 116

Section 3.7
☐ Classify scatter plots.   117–120
☐ Use scatter plots and a graphing utility to find quadratic models for data.   121
☐ Choose a model that best fits a set of data.   122

3 Review Exercises

3.1 In Exercises 1 and 2, use a graphing utility to graph each function in the same viewing window. Describe how the graph of each function is related to the graph of $y = x^2$.

1. (a) $y = 2x^2$ (b) $y = -2x^2$
   (c) $y = x^2 + 2$ (d) $y = (x + 5)^2$

2. (a) $y = x^2 - 4$ (b) $y = 4 - x^2$
   (c) $y = (x - 1)^2$ (d) $y = \frac{1}{2}x^2 - 1$

In Exercises 3–6, sketch the graph of the quadratic function. Identify the vertex and the intercept(s).

3. $f(x) = \left(x + \frac{3}{2}\right)^2 + 1$

4. $f(x) = (x - 4)^2 - 4$

5. $f(x) = \frac{1}{3}(x^2 + 5x - 4)$

6. $f(x) = 3x^2 - 12x + 11$

In Exercises 7–10, write the standard form of the quadratic function that has the indicated vertex and whose graph passes through the given point. Verify your result with a graphing utility.

7. Vertex: $(1, -4)$;   Point: $(2, -3)$

8. Vertex: $(2, 3)$;   Point: $(0, 2)$

9. Vertex: $(-2, -2)$;   Point: $(-1, 0)$

10. Vertex: $\left(-\frac{1}{4}, \frac{3}{2}\right)$;   Point: $(-2, 0)$

11. *Numerical, Graphical, and Analytical Analysis*   A rectangle is inscribed in the region bounded by the x-axis, the y-axis, and the graph of $x + 2y - 8 = 0$, as shown in the figure.

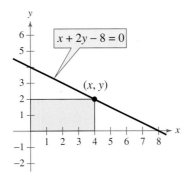

(a) Write the area A as a function of x. Determine the domain of the function in the context of the problem.

(b) Use the *table* feature of a graphing utility to create a table showing possible values of x and the corresponding areas of the rectangle. Use the table to estimate the dimensions that will produce a maximum area.

(c) Use a graphing utility to graph the area function. Use the graph to approximate the dimensions that will produce a maximum area.

(d) Write the area function in standard form to find algebraically the dimensions that will produce a maximum area.

(e) Compare your results from parts (b), (c), and (d).

12. *Cost*   A textile manufacturer has daily production costs of

$$C = 10,000 - 110x + 0.45x^2$$

where C is the total cost (in dollars) and x is the number of units produced. Use the *table* feature of a graphing utility to determine how many units should be produced each day to yield a minimum cost.

3.2 In Exercises 13–16, sketch the graph of $y = x^n$ and each specified transformation.

13. $y = x^4$
   (a) $f(x) = (x + 5)^4$ (b) $f(x) = x^4 - 4$
   (c) $f(x) = 3 + x^4$ (d) $f(x) = \frac{1}{4}(x - 2)^4$

14. $y = x^5$
   (a) $f(x) = (x + 4)^5$ (b) $f(x) = 6 + x^5$
   (c) $f(x) = 3 - \frac{1}{2}x^5$ (d) $f(x) = 2(x + 3)^5$

15. $y = x^6$
   (a) $f(x) = x^6 - 2$ (b) $f(x) = -\frac{1}{4}x^6$
   (c) $f(x) = -\frac{1}{2}x^6 - 5$ (d) $f(x) = -(x + 7)^6 - 5$

16. $y = x^3$
   (a) $f(x) = -x^3 + 4$ (b) $f(x) = (x + 2)^3 - 1$
   (c) $f(x) = -\frac{1}{3}x^3 + 1$ (d) $f(x) = -(x + 8)^3$

Graphical Analysis   In Exercises 17 and 18, use a graphing utility to graph the functions *f* and *g* in the same viewing window. Zoom out far enough so that the right-hand and left-hand behaviors of *f* and *g* appear identical.

17. $f(x) = \frac{1}{2}x^3 - 2x + 1,$ $g(x) = \frac{1}{2}x^3$

18. $f(x) = -x^4 + 2x^3,$ $g(x) = -x^4$

In Exercises 19–22, use the Leading Coefficient Test to determine the right-hand and left-hand behavior of the graph of the polynomial function.

19. $f(x) = -x^2 + 6x + 9$

20. $f(x) = \frac{1}{3}x^3 + 2x$

21. $g(x) = \frac{3}{4}(x^4 + 3x^2 + 2)$

22. $h(x) = -x^5 - 7x^2 + 10x$

In Exercises 23–28, (a) use a graphing utility to graph the function, (b) use the graph to approximate any zeros, and (c) find the zeros algebraically.

23. $g(x) = x^4 - x^3 - 2x^2$

24. $h(x) = -2x^3 - x^2 + x$

25. $f(t) = t^3 - 3t$

26. $f(x) = -(x + 6)^3 - 8$

27. $f(x) = x(x + 3)^2$

28. $f(t) = t^4 - 4t^2$

In Exercises 29–32, (a) use the Intermediate Value Theorem and a graphing utility to find intervals of length 1 in which the polynomial function is guaranteed to have a zero, (b) use the *zero* or *root* feature of a graphing utility to approximate the zeros of the function, and (c) verify your results in part (a) by using the *table* feature of a graphing utility.

29. $f(x) = x^3 + 2x^2 - x - 1$

30. $f(x) = 0.24x^3 - 2.6x - 1.4$

31. $f(x) = x^4 - 6x^2 - 4$

32. $f(x) = 2x^4 + \frac{7}{2}x^3 - 2$

3.3 *Graphical Analysis*   In Exercises 33 and 34, use a graphing utility to graph the two equations in the same viewing window. Use the graphs to verify that the expressions are equivalent. Verify the results algebraically.

33. $y_1 = \dfrac{x^2}{x - 2},$ $y_2 = x + 2 + \dfrac{4}{x - 2}$

34. $y_1 = \dfrac{x^4 + 1}{x^2 + 2},$ $y_2 = x^2 - 2 + \dfrac{5}{x^2 + 2}$

In Exercises 35–42, use long division to divide.

35. $\dfrac{24x^2 - x - 8}{3x - 2}$

36. $\dfrac{4x^2 + 7}{3x - 2}$

37. $\dfrac{x^4 - 3x^2 + 2}{x^2 - 1}$

38. $\dfrac{3x^4}{x^2 - 1}$

39. $(5x^3 - 13x^2 - x + 2) \div (x^2 - 3x + 1)$

40. $(x^4 + x^3 - x^2 + 2x) \div (x^2 + 2x)$

41. $\dfrac{6x^4 + 10x^3 + 13x^2 - 5x + 2}{2x^2 - 1}$

42. $\dfrac{x^4 - 3x^3 + 4x^2 - 6x + 3}{x^2 + 2}$

In Exercises 43–48, use synthetic division to divide.

43. $(0.25x^4 - 4x^3) \div (x + 2)$

44. $(0.1x^3 + 0.3x^2 - 0.5) \div (x - 5)$

45. $(6x^4 - 4x^3 - 27x^2 + 18x) \div \left(x - \frac{2}{3}\right)$

46. $(2x^3 + 2x^2 - x + 2) \div \left(x - \frac{1}{2}\right)$

47. $(3x^3 - 10x^2 + 12x - 22) \div (x - 4)$

48. $(2x^3 + 6x^2 - 14x + 9) \div (x - 1)$

In Exercises 49 and 50, use synthetic division to find each function value. Use a graphing utility to verify your results.

49. $f(x) = x^4 + 10x^3 - 24x^2 + 20x + 44$

    (a) $f(-3)$ (b) $f(-1)$

50. $g(t) = 2t^5 - 5t^4 - 8t + 20$

    (a) $g(-4)$ (b) $g(\sqrt{2})$

In Exercises 51–54, (a) verify the given factor(s) of the function *f*, (b) find the remaining factors of *f*, (c) use your results to write the complete factorization of *f*, (d) list all real zeros of *f*, and (e) confirm your results by using a graphing utility to graph the function.

	Function	*Factor(s)*
51.	$f(x) = x^3 + 4x^2 - 25x - 28$	$(x - 4)$
52.	$f(x) = 2x^3 + 11x^2 - 21x - 90$	$(x + 6)$
53.	$f(x) = x^4 - 4x^3 - 7x^2 + 22x + 24$	$(x + 2),$
		$(x - 3)$
54.	$f(x) = x^4 - 11x^3 + 41x^2 - 61x + 30$	$(x - 2),$
		$(x - 5)$

In Exercises 55 and 56, use the Rational Zero Test to list all possible rational zeros of f. Use a graphing utility to verify that the zeros of f are contained in the list.

55. $f(x) = 4x^3 - 11x^2 + 10x - 3$

56. $f(x) = 10x^3 + 21x^2 - x - 6$

In Exercises 57–60, find all the zeros of the function.

57. $f(x) = 6x^3 - 5x^2 + 24x - 20$

58. $f(x) = x^3 - 1.3x^2 - 1.7x + 0.6$

59. $f(x) = 6x^4 - 25x^3 + 14x^2 + 27x - 18$

60. $f(x) = 5x^4 + 126x^2 + 25$

In Exercises 61 and 62, use Descartes's Rule of Signs to determine the possible numbers of positive and negative real zeros of the function.

61. $g(x) = 5x^3 + 3x^2 - 6x + 9$

62. $h(x) = -2x^5 + 4x^3 - 2x^2 + 5$

In Exercises 63 and 64, use synthetic division to verify the upper and lower bounds of the real zeros of f.

63. $f(x) = 4x^3 - 3x^2 + 4x - 3$

Upper bound: $x = 1$; Lower bound: $x = -\frac{1}{4}$

64. $f(x) = 2x^3 - 5x^2 - 14x + 8$

Upper bound: $x = 8$; Lower bound: $x = -4$

3.4 In Exercises 65–68, find all the zeros of the function.

65. $f(x) = 3x(x - 2)^2$

66. $f(x) = (x - 4)(x + 9)^2$

67. $f(x) = (x + 4)(x - 6)(x - 2i)(x + 2i)$

68. $g(t) = (t - 8)(t - 5)^2(t - 3 + i)(t - 3 - i)$

In Exercises 69–72, find all the zeros of the function and write the polynomial as a product of linear factors. Use a graphing utility to graph the function to verify your results graphically.

69. $f(x) = 2x^4 - 5x^3 + 10x - 12$

70. $g(x) = 3x^4 - 4x^3 + 7x^2 + 10x - 4$

71. $h(x) = x^3 - 7x^2 + 18x - 24$

72. $f(x) = 2x^3 - 5x^2 - 9x + 40$

In Exercises 73–78, (a) find all the zeros of the function, (b) write the polynomial as a product of linear factors, (c) use your factorization to determine the x-intercepts of the graph of the function, and (d) use a graphing utility to verify that the real zeros are the only x-intercepts.

73. $f(x) = x^3 - 4x^2 + 6x - 4$

74. $f(x) = x^3 - 5x^2 - 7x + 51$

75. $f(x) = x^3 + 6x^2 + 11x + 12$

76. $f(x) = 2x^3 - 9x^2 + 22x - 30$

77. $f(x) = x^4 + 34x^2 + 225$

78. $f(x) = x^4 + 10x^3 + 26x^2 + 10x + 25$

In Exercises 79–82, find a polynomial function with real coefficients that has the given zeros. (There are many correct answers.)

79. $-2, -2, -5i$

80. $4, 4, 2i$

81. $1, -4, -3 + 5i$

82. $-4, -4, 1 + \sqrt{3}i$

In Exercises 83–86, write the polynomial (a) as the product of factors that are irreducible over the *rationals*, (b) as the product of linear and quadratic factors that are irreducible over the *reals*, and (c) in completely factored form.

83. $f(x) = x^4 + 2x^2 - 8$

84. $f(x) = x^4 - x^3 - x^2 + 5x - 20$

(*Hint:* One factor is $x^2 - 5$.)

85. $f(x) = x^4 - 2x^3 + 8x^2 - 18x - 9$

(*Hint:* One factor is $x^2 + 9$.)

86. $f(x) = x^4 - 4x^3 + 3x^2 + 8x - 16$

(*Hint:* One factor is $x^2 - x - 4$.)

3.5 In Exercises 87–98, (a) find the domain of the function and (b) identify any horizontal and vertical asymptotes.

87. $f(x) = \dfrac{x - 8}{1 - x}$

88. $f(x) = \dfrac{5x}{x + 12}$

89. $f(x) = \dfrac{2}{x^2 - 3x - 18}$

90. $f(x) = \dfrac{2x^2 + 3}{x^2 + x + 3}$

91. $f(x) = \dfrac{7 + x}{7 - x}$

92. $f(x) = \dfrac{6x}{x^2 - 1}$

93. $f(x) = \dfrac{4x^2}{2x^2 - 3}$

94. $f(x) = \dfrac{3x^2 - 11x - 4}{x^2 + 2}$

95. $f(x) = \dfrac{2x - 10}{x^2 - 2x - 15}$

96. $f(x) = \dfrac{x^3 - 4x^2}{x^2 + 3x + 2}$

97. $f(x) = \dfrac{x - 2}{|x| + 2}$

98. $f(x) = \dfrac{2x}{|2x - 1|}$

99. *Seizure of Illegal Drugs* The cost C in millions of dollars for the U.S. government to seize $p\%$ of an illegal drug as it enters the country is given by

$$C = \frac{528p}{100 - p}, \quad 0 \le p < 100.$$

(a) Find the cost of seizing 25%, 50%, and 75% of the illegal drug.

(b) Use a graphing utility to graph the function. Be sure to choose an appropriate viewing window. Explain why you chose the values you used in your viewing window.

(c) According to this model, would it be possible to seize 100% of the drug? Explain.

100. *Wildlife* A biology class performs an experiment comparing the quantity of food consumed by a certain kind of moth with the quantity supplied. The model for the experimental data is given by

$$y = \frac{1.568x - 0.001}{6.360x + 1}, \quad x > 0$$

where x is the quantity (in milligrams) of food supplied and y is the quantity (in milligrams) eaten (see figure). At what level of consumption will the moth become satiated?

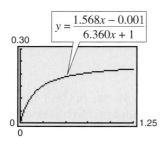

3.6 **In Exercises 101–110, sketch the graph of the rational function by hand. As sketching aids, check for intercepts, vertical asymptotes, and horizontal asymptotes. Use a graphing utility to verify your graph.**

101. $f(x) = \dfrac{2x - 1}{x - 5}$

102. $f(x) = \dfrac{x - 3}{x - 2}$

103. $f(x) = \dfrac{2x}{x^2 + 4}$

104. $f(x) = \dfrac{2x^2}{x^2 - 4}$

105. $f(x) = \dfrac{x^2}{x^2 + 1}$

106. $f(x) = \dfrac{5x}{x^2 + 1}$

107. $f(x) = \dfrac{2}{(x + 1)^2}$

108. $f(x) = \dfrac{4}{(x - 1)^2}$

109. $f(x) = \dfrac{2(x^2 - 16)}{x^2 + 2x - 8}$

110. $f(x) = \dfrac{3x^2 - 6x}{x^2 - 4}$

In Exercises 111–114, sketch the graph of the rational function by hand. As sketching aids, check for intercepts, vertical asymptotes, horizontal asymptotes, and slant asymptotes.

111. $f(x) = \dfrac{2x^3}{x^2 + 1}$

112. $f(x) = \dfrac{x^3}{3x^2 - 6}$

113. $f(x) = \dfrac{x^2 - x + 1}{x - 3}$

114. $f(x) = \dfrac{2x^2 + 7x + 3}{x + 1}$

115. *Wildlife* The Parks and Wildlife Commission introduces 80,000 fish into a large human-made lake. The population N of the fish in thousands is given by

$$N = \frac{20(4 + 3t)}{1 + 0.05t}, \quad t \ge 0$$

where t is time in years.

(a) Use a graphing utility to graph the function.

(b) Use the graph from part (a) to find the populations when $t = 5$, $t = 10$, and $t = 25$.

(c) What is the maximum number of fish in the lake as time increases? Explain your reasoning.

116. *Page Design* A page that is x inches wide and y inches high contains 30 square inches of print. The top and bottom margins are 2 inches deep and the margins on each side are 2 inches wide.

(a) Draw a diagram that illustrates the problem.

(b) Show that the total area A of the page is given by

$$A = \frac{2x(2x + 7)}{x - 4}.$$

(c) Determine the domain of the function based on the physical constraints of the problem.

(d) Use a graphing utility to graph the area function and approximate the page size such that the minimum amount of paper will be used. Verify your answer numerically using the *table* feature of a graphing utility.

3.7 In Exercises 117–120, determine whether the scatter plot could best be modeled by a linear model, a quadratic model, or neither.

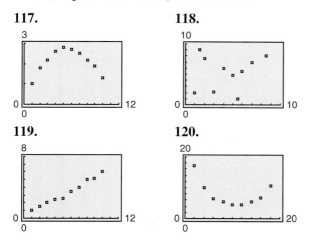

117.

118.

119.

120.

121. *Revenue* The table shows the revenue R (in millions of dollars) for OfficeMax, Inc. from 1994 to 2001. (Source: OfficeMax, Inc.)

Year	Revenue, R
1994	1841.2
1995	2542.5
1996	3179.3
1997	3765.4
1998	4337.8
1999	4842.7
2000	5156.4
2001	4636.0

(a) Use a graphing utility to create a scatter plot of the data. Let t represent the year, with $t = 4$ corresponding to 1994.

(b) Use the *regression* feature of a graphing utility to find a quadratic model for the data.

(c) Use a graphing utility to graph the model with the scatter plot from part (a). Is the quadratic model a good fit for the data?

(d) Use the graph from part (c) to determine in which year the revenue for OfficeMax, Inc. was the greatest.

(e) Do you think the model can be used to predict the revenue for OfficeMax, Inc. in the future? Explain.

122. *Consumer Awareness* The table shows the average price P (in dollars) for a personal computer from 1997 to 2002. (Source: Consumer Electronics Association)

Year	Average price, P
1997	1450
1998	1300
1999	1100
2000	1000
2001	900
2002	855

(a) Use a graphing utility to create a scatter plot of the data. Let t represent the year, with $t = 7$ corresponding to 1997.

(b) Use the *regression* feature of a graphing utility to find a linear model for the data.

(c) Use a graphing utility to graph the linear model with the scatter plot from part (a).

(d) Use the *regression* feature of a graphing utility to find a quadratic model for the data.

(e) Use a graphing utility to graph the quadratic model with the scatter plot from part (a).

(f) Determine which model best fits the data and use the model you chose to predict the average price for a personal computer in 2008. Does your answer seem reasonable? Explain.

Synthesis

True or False? **In Exercises 123 and 124, determine whether the statement is true or false. Justify your answer.**

123. The graph of $f(x) = \dfrac{2x^3}{x + 1}$ has a slant asymptote.

124. A fourth-degree polynomial with real coefficients can have -5, $-8i$, $4i$, and 5 as its zeros.

125. *Think About It* What does it mean for a divisor to divide evenly into a dividend?

126. *Writing* Write a paragraph discussing whether every rational function has a vertical asymptote.

3 | Chapter Test

Take this test as you would take a test in class. After you are finished, check your work against the answers given in the back of the book.

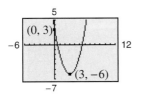

Figure for 3

1. Describe how the graph of g differs from the graph of $f(x) = x^2$.
   (a) $g(x) = 6 - x^2$ (b) $g(x) = \left(x - \frac{3}{2}\right)^2$

2. Identify the vertex and intercepts of the graph of $y = x^2 + 4x + 3$.

3. Write an equation of the parabola shown at the right.

4. The path of a ball is given by $y = -\frac{1}{20}x^2 + 3x + 5$, where y is the height (in feet) and x is the horizontal distance (in feet).
   (a) Find the maximum height of the ball.
   (b) Which term determines the height at which the ball was thrown? Does changing this term change the maximum height of the ball? Explain.

5. Divide using long division: $(3x^3 + 4x - 1) \div (x^2 + 1)$.

6. Divide using synthetic division: $(2x^4 - 5x^2 - 3) \div (x - 2)$.

In Exercises 7 and 8, list all the possible rational zeros of the function. Use a graphing utility to graph the function and find all the rational zeros.

7. $g(t) = 2t^4 - 3t^3 + 16t - 24$ 8. $h(x) = 3x^5 + 2x^4 - 3x - 2$

In Exercises 9 and 10, use the *zero* or *root* feature of a graphing utility to approximate (accurate to three decimal places) the real zeros of the function.

9. $f(x) = x^4 - x^3 - 1$ 10. $f(x) = 3x^5 + 2x^4 - 12x - 8$

In Exercises 11–13, find a polynomial function with real coefficients that has the given zeros. (There are many correct answers.)

11. $0, 3, 3 + i, 3 - i$ 12. $1 - \sqrt{3}i, 2, 2$ 13. $0, -5, 1 + i$

In Exercises 14–16, sketch the graph of the rational function. As sketching aids, check for intercepts, vertical asymptotes, horizontal asymptotes, and slant asymptotes.

14. $h(x) = \dfrac{4}{x^2} - 1$ 15. $g(x) = \dfrac{x^2 + 2}{x - 1}$ 16. $f(x) = \dfrac{2x^2 + 9}{5x^2 + 2}$

17. The table shows the number C of U.S. Supreme Court cases waiting to be tried for the years 1995 to 2000. (Source: Office of the Clerk, Supreme Court of the United States)
    (a) Use a graphing utility to create a scatter plot of the data. Let t represent the year, with $t = 5$ corresponding to 1995.
    (b) Use the *regression* feature of a graphing utility to find a quadratic model for the data.
    (c) Use a graphing utility to graph the model with the scatter plot from part (a). Is the quadratic model a good fit for the data?
    (d) Use the model to predict the year in which there will be 15,000 U.S. Supreme Court cases waiting to be tried.

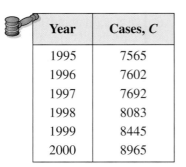

Year	Cases, C
1995	7565
1996	7602
1997	7692
1998	8083
1999	8445
2000	8965

Exponential models are widely used in the financial world. The growth pattern of a savings account and the calculation of mortgage rates both require exponential functions.

Ryan McVay/Photodisc/Getty Images

Exponential and Logarithmic Functions

What You Should Learn

In this chapter, you will learn how to:

■ Recognize, evaluate, and graph exponential and logarithmic functions.

■ Rewrite logarithmic functions with different bases.

■ Use properties of logarithms to evaluate, rewrite, expand, or condense logarithmic expressions.

■ Solve exponential and logarithmic equations.

■ Use exponential growth models, exponential decay models, Gaussian models, logistic models, and logarithmic models to solve real-life problems.

■ Fit exponential, logarithmic, power, and logistic models to sets of data.

4.1 Exponential Functions and Their Graphs

Exponential Functions

So far, this text has dealt mainly with **algebraic functions,** which include polynomial functions and rational functions. In this chapter you will study two types of nonalgebraic functions—*exponential functions* and *logarithmic functions*. These functions are examples of **transcendental functions.**

Definition of Exponential Function

The **exponential function** f **with base** a is denoted by

$$f(x) = a^x$$

where $a > 0$, $a \neq 1$, and x is any real number.

Note that in the definition of an exponential function, the base $a = 1$ is excluded because it yields $f(x) = 1^x = 1$. This is a constant function, not an exponential function.

You have already evaluated a^x for integer and rational values of x. For example, you know that $4^3 = 64$ and $4^{1/2} = 2$. However, to evaluate 4^x for any real number x, you need to interpret forms with *irrational* exponents. For the purposes of this text, it is sufficient to think of

$$a^{\sqrt{2}} \left(\text{where } \sqrt{2} \approx 1.41421356\right)$$

as the number that has the successively closer approximations

$$a^{1.4}, a^{1.41}, a^{1.414}, a^{1.4142}, a^{1.41421}, \ldots$$

Example 1 shows how to use a calculator to evaluate exponential functions.

OSF/Animals Animals

Example 1　Evaluating Exponential Functions

Use a calculator to evaluate each function at the indicated value of x.

Function	Value
a. $f(x) = 2^x$	$x = -3.1$
b. $f(x) = 2^{-x}$	$x = \pi$
c. $f(x) = 0.6^x$	$x = \frac{3}{2}$

Solution

Function Value	Graphing Calculator Keystrokes	Display
a. $f(-3.1) = 2^{-3.1}$	2 [^] [(−)] 3.1 [ENTER]	0.1166291
b. $f(\pi) = 2^{-\pi}$	2 [^] [(−)] π [ENTER]	0.1133147
c. $f\left(\frac{3}{2}\right) = (0.6)^{3/2}$	.6 [^] [(] 3 [÷] 2 [)] [ENTER]	0.4647580

✓ *Checkpoint*　Now try Exercise 3.

Graphs of Exponential Functions

The graphs of all exponential functions have similar characteristics, as shown in Examples 2, 3, and 4.

Example 2   Graphs of $y = a^x$

In the same coordinate plane, sketch the graph of each function by hand.

a. $f(x) = 2^x$ **b.** $g(x) = 4^x$

Solution

The table below lists some values for each function. By plotting these points and connecting them with a smooth curve, you obtain the graphs shown in Figure 4.1. Note that both graphs are increasing. Moreover, the graph of $g(x) = 4^x$ is increasing more rapidly than the graph of $f(x) = 2^x$.

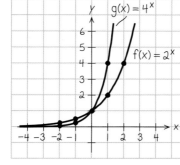

Figure 4.1

x	-2	-1	0	1	2	3
2^x	$\frac{1}{4}$	$\frac{1}{2}$	1	2	4	8
4^x	$\frac{1}{16}$	$\frac{1}{4}$	1	4	16	64

✓ *Checkpoint* Now try Exercise 7.

Example 3   Graphs of $y = a^{-x}$

In the same coordinate plane, sketch the graph of each function by hand.

a. $F(x) = 2^{-x}$ **b.** $G(x) = 4^{-x}$

Solution

The table below lists some values for each function. By plotting these points and connecting them with a smooth curve, you obtain the graphs shown in Figure 4.2. Note that both graphs are decreasing. Moreover, the graph of $G(x) = 4^{-x}$ is decreasing more rapidly than the graph of $F(x) = 2^{-x}$.

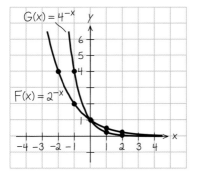

Figure 4.2

x	-3	-2	-1	0	1	2
2^{-x}	8	4	2	1	$\frac{1}{2}$	$\frac{1}{4}$
4^{-x}	64	16	4	1	$\frac{1}{4}$	$\frac{1}{16}$

✓ *Checkpoint* Now try Exercise 9.

The properties of exponents presented in Section P.2 can also be applied to real-number exponents. For review, these properties are listed below.

1. $a^x a^y = a^{x+y}$ **2.** $\dfrac{a^x}{a^y} = a^{x-y}$ **3.** $a^{-x} = \dfrac{1}{a^x} = \left(\dfrac{1}{a}\right)^x$ **4.** $a^0 = 1$

5. $(ab)^x = a^x b^x$ **6.** $(a^x)^y = a^{xy}$ **7.** $\left(\dfrac{a}{b}\right)^x = \dfrac{a^x}{b^x}$ **8.** $|a^2| = |a|^2 = a^2$

STUDY TIP

In Example 3, note that the functions $F(x) = 2^{-x}$ and $G(x) = 4^{-x}$ can be rewritten with positive exponents.

$$F(x) = 2^{-x} = \left(\frac{1}{2}\right)^x \quad \text{and}$$

$$G(x) = 4^{-x} = \left(\frac{1}{4}\right)^x$$

Comparing the functions in Examples 2 and 3, observe that

$$F(x) = 2^{-x} = f(-x) \qquad \text{and} \qquad G(x) = 4^{-x} = g(-x).$$

Consequently, the graph of F is a reflection (in the y-axis) of the graph of f, as shown in Figure 4.3. The graphs of G and g have the same relationship, as shown in Figure 4.4.

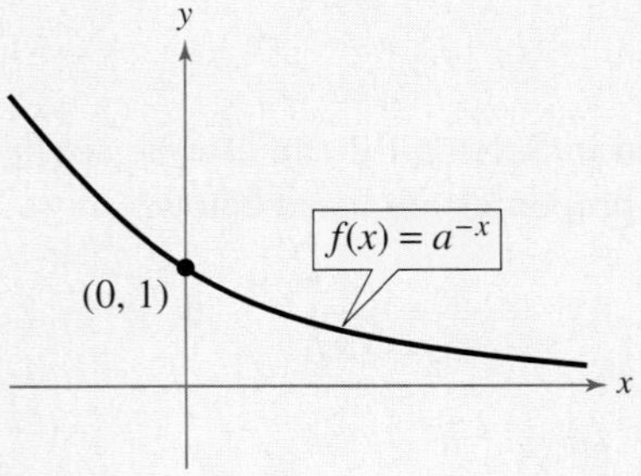

STUDY TIP

Notice that the range of an exponential function is $(0, \infty)$, which means that $a^x > 0$ for all values of x.

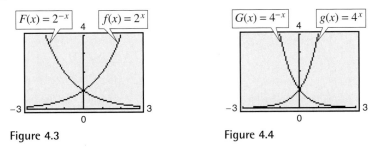

Figure 4.3

Figure 4.4

The graphs in Figures 4.1 and 4.2 are typical of the graphs of the exponential functions $f(x) = a^x$ and $f(x) = a^{-x}$. They have one y-intercept and one horizontal asymptote (the x-axis), and they are continuous.

Library of Functions: Exponential Function

The *exponential function*

$$f(x) = a^x, \quad a > 0, \quad a \neq 1$$

is different from all the functions you have studied so far because the variable x is an *exponent*. A distinguishing characteristic of an exponential function is its rapid increase as x increases (for $a > 1$). Many real-life phenomena with patterns of rapid growth (or decline) can be modeled by exponential functions. The basic characteristics of the exponential function are summarized below.

Graph of $f(x) = a^x$, $a > 1$

Domain: $(-\infty, \infty)$

Range: $(0, \infty)$

Intercept: $(0, 1)$

Increasing on $(-\infty, \infty)$

x-axis is a horizontal asymptote

$(a^x \to 0$ as $x \to -\infty)$

Continuous

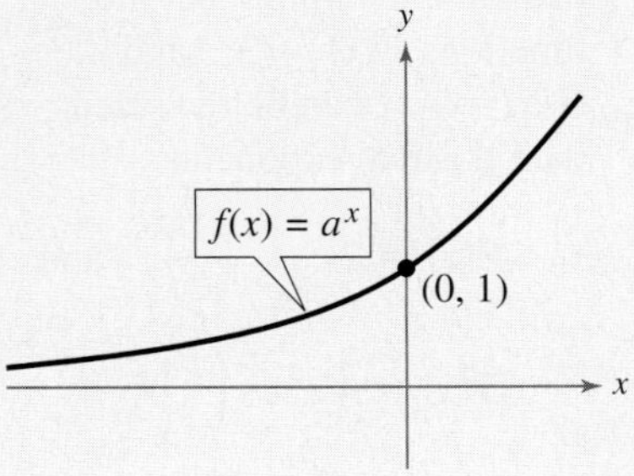

Graph of $f(x) = a^{-x}$, $a > 1$

Domain: $(-\infty, \infty)$

Range: $(0, \infty)$

Intercept: $(0, 1)$

Decreasing on $(-\infty, \infty)$

x-axis is a horizontal asymptote

$(a^{-x} \to 0$ as $x \to \infty)$

Continuous

Exploration

Use a graphing utility to graph $y = a^x$ for $a = 3, 5,$ and 7 in the same viewing window. (Use a viewing window in which $-2 \le x \le 1$ and $0 \le y \le 2$.) How do the graphs compare with each other? Which graph is on the top in the interval $(-\infty, 0)$? Which is on the bottom? Which graph is on the top in the interval $(0, \infty)$? Which is on the bottom? Repeat this experiment with the graphs of $y = b^x$ for $b = \frac{1}{3}, \frac{1}{5},$ and $\frac{1}{7}$. (Use a viewing window in which $-1 \le x \le 2$ and $0 \le y \le 2$.) What can you conclude about the shape of the graph of $y = b^x$ and the value of b?

In the following example, notice how the graph of $y = a^x$ can be used to sketch the graphs of functions of the form $f(x) = b \pm a^{x+c}$.

Example 4   Transformations of Graphs of Exponential Functions

Each of the following graphs is a transformation of the graph of $f(x) = 3^x$.

a. Because $g(x) = 3^{x+1} = f(x + 1)$, the graph of g can be obtained by shifting the graph of f one unit to the *left*, as shown in Figure 4.5.

b. Because $h(x) = 3^x - 2 = f(x) - 2$, the graph of h can be obtained by shifting the graph of f *downward* two units, as shown in Figure 4.6.

c. Because $k(x) = -3^x = -f(x)$, the graph of k can be obtained by *reflecting* the graph of f in the x-axis, as shown in Figure 4.7.

d. Because $j(x) = 3^{-x} = f(-x)$, the graph of j can be obtained by *reflecting* the graph of f in the y-axis, as shown in Figure 4.8.

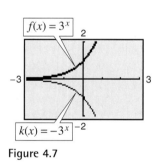

Figure 4.5

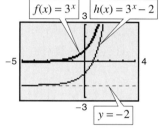

Figure 4.6

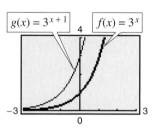

Figure 4.7

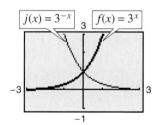

Figure 4.8

Exploration

The following table shows some points of the graphs in Figure 4.5. The functions $f(x)$ and $g(x)$ are represented by Y_1 and Y_2, respectively. Explain how you can use the table to describe the transformation.

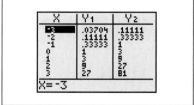

✓ *Checkpoint*   Now try Exercise 19.

Notice that the transformations in Figures 4.5, 4.7, and 4.8 keep the x-axis ($y = 0$) as a horizontal asymptote, but the transformation in Figure 4.6 yields a new horizontal asymptote of $y = -2$. Also, be sure to note how the y-intercept is affected by each transformation.

The Natural Base e

For many applications, the convenient choice for a base is the irrational number

$$e \approx 2.718281828.$$

This number is called the **natural base.** The function $f(x) = e^x$ is called the **natural exponential function** and its graph is shown in Figure 4.9. The graph of the exponential function has the same basic characteristics as the graph of the function $f(x) = a^x$ (see page 322). Be sure you see that for the exponential function $f(x) = e^x, e$ is the constant $2.718281828 \ldots$, whereas x is the variable.

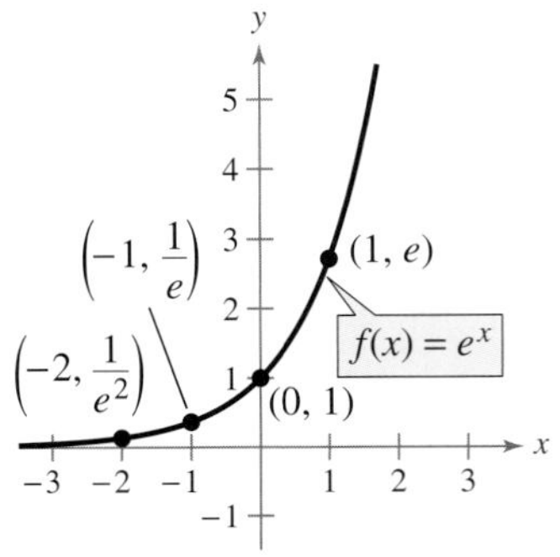

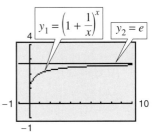

Figure 4.9   The Natural Exponential Function

Exploration

Use your graphing utility to graph the functions

$$y_1 = 2^x$$

$$y_2 = e^x$$

$$y_3 = 3^x$$

in the same viewing window. From the relative positions of these graphs, make a guess as to the value of the real number e. Then try to find a number a such that the graphs of $y_2 = e^x$ and $y_4 = a^x$ are as close as possible.

In Example 5, you will see that the number e can be approximated by the expression

$$\left(1 + \frac{1}{x}\right)^x \text{ for large values of } x.$$

TECHNOLOGY SUPPORT

For instructions on how to use the *trace* feature and the *table* feature, see Appendix A; for specific keystrokes, go to the text website at *college.hmco.com*.

Example 5   Approximation of the Number e

Evaluate the expression $[1 + (1/x)]^x$ for several large values of x to see that the values approach $e \approx 2.718281828$ as x increases without bound.

Graphical Solution

Use a graphing utility to graph

$$y_1 = [1 + (1/x)]^x \qquad \text{and} \qquad y_2 = e$$

in the same viewing window, as shown in Figure 4.10. Use the *trace* feature of the graphing utility to verify that as x increases, the graph of y_1 gets closer and closer to the line $y_2 = e$.

Figure 4.10

Numerical Solution

Use the *table* feature (in *ask* mode) of a graphing utility to create a table of values for the function $y = [1 + (1/x)]^x$, beginning at $x = 10$ and increasing the x-values as shown in Figure 4.11.

X	Y1
10	2.5937
100	2.7048
1000	2.7169
10000	2.7181
100000	2.7183
1E6	2.7183

X=

Figure 4.11

From the table, it seems reasonable to conclude that

$$\left(1 + \frac{1}{x}\right)^x \to e \text{ as } x \to \infty.$$

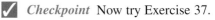
Checkpoint Now try Exercise 37.

Example 6   Evaluating the Natural Exponential Function

Use a calculator to evaluate the function $f(x) = e^x$ at each indicated value of x.

a. $x = -2$ **b.** $x = 0.25$ **c.** $x = -0.4$

Solution

Function Value	*Graphing Calculator Keystrokes*	*Display*
a. $f(-2) = e^{-2}$	$\boxed{e^x}$ $\boxed{(-)}$ 2 $\boxed{\text{ENTER}}$	0.1353353
b. $f(0.25) = e^{0.25}$	$\boxed{e^x}$ .25 $\boxed{\text{ENTER}}$	1.2840254
c. $f(-0.4) = e^{-0.4}$	$\boxed{e^x}$ $\boxed{(-)}$ .4 $\boxed{\text{ENTER}}$	0.6703200

✓ *Checkpoint* Now try Exercise 23.

Exploration

Use a graphing utility to graph $y = (1 + x)^{1/x}$. Describe the behavior of the graph near $x = 0$. Is there a y-intercept? How does the behavior of the graph near $x = 0$ relate to the result of Example 5? Use the *table* feature of a graphing utility to create a table that shows values of y for values of x near $x = 0$, to help you describe the behavior of the graph near this point.

Example 7   Graphing Natural Exponential Functions

Sketch the graph of each natural exponential function.

a. $f(x) = 2e^{0.24x}$ **b.** $g(x) = \frac{1}{2}e^{-0.58x}$

Solution

To sketch these two graphs, you can use a calculator to construct a table of values, as shown below.

x	-3	-2	-1	0	1	2	3
$f(x)$	0.974	1.238	1.573	2.000	2.542	3.232	4.109
$g(x)$	2.849	1.595	0.893	0.500	0.280	0.157	0.088

After constructing the table, plot the points and connect them with smooth curves. Note that the graph in Figure 4.12 is increasing, whereas the graph in Figure 4.13 is decreasing. Use a graphing calculator to verify these graphs.

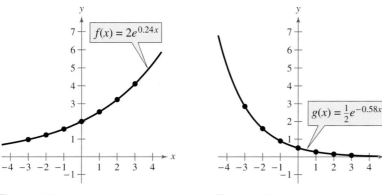

Figure 4.12   Figure 4.13

✓ *Checkpoint* Now try Exercise 35.

Applications

One of the most familiar examples of exponential growth is that of an investment earning *continuously compounded interest.* Suppose a principal P is invested at an annual interest rate r, compounded once a year. If the interest is added to the principal at the end of the year, the new balance P_1 is $P_1 = P + Pr = P(1 + r)$. This pattern of multiplying the previous principal by $1 + r$ is then repeated each successive year, as shown in the table.

Time in Years	Balance After Each Compounding
0	$P = P$
1	$P_1 = P(1 + r)$
2	$P_2 = P_1(1 + r) = P(1 + r)(1 + r) = P(1 + r)^2$
$\vdots$	$\vdots$
t	$P_t = P(1 + r)^t$

> ### Exploration
>
> Use the formula
> $$A = P\left(1 + \frac{r}{n}\right)^{nt}$$
> to calculate the amount in an account when $P = \$3000$, $r = 6\%$, $t = 10$ years, and the number of compoundings is (a) by the day, (b) by the hour, (c) by the minute, and (d) by the second. Does increasing the number of compoundings per year result in unlimited growth of the amount in the account? Explain.

To accommodate more frequent (quarterly, monthly, or daily) compounding of interest, let n be the number of compoundings per year and let t be the number of years. (The product nt represents the total number of times the interest will be compounded.) Then the interest rate per compounding period is r/n, and the account balance after t years is

$$A = P\left(1 + \frac{r}{n}\right)^{nt}. \qquad \text{Amount (balance) with } n \text{ compoundings per year}$$

If you let the number of compoundings n increase without bound, the process approaches what is called **continuous compounding.** In the formula for n compoundings per year, let $m = n/r$. This produces

$$A = P\left(1 + \frac{r}{n}\right)^{nt} = P\left(1 + \frac{1}{m}\right)^{mrt} = P\left[\left(1 + \frac{1}{m}\right)^{m}\right]^{rt}.$$

As m increases without bound, you know from Example 5 that $[1 + (1/m)]^m$ approaches e. So, for continuous compounding, it follows that

$$P\left[\left(1 + \frac{1}{m}\right)^{m}\right]^{rt} \rightarrow P[e]^{rt}$$

and you can write $A = Pe^{rt}$. This result is part of the reason that e is the "natural" choice for a base of an exponential function.

> ### STUDY TIP
>
> The interest rate r in the formula for compound interest should be written as a decimal. For example, an interest rate of 7% would be written as $r = 0.07$.

> #### Formulas for Compound Interest
>
> After t years, the balance A in an account with principal P and annual interest rate r (in decimal form) is given by the following formulas.
>
> **1.** For n compoundings per year: $A = P\left(1 + \dfrac{r}{n}\right)^{nt}$
>
> **2.** For continuous compounding: $A = Pe^{rt}$

Example 8   Finding the Balance for Compound Interest

A total of $9000 is invested at an annual interest rate of 2.5%, compounded annually. Find the balance in the account after 5 years.

Algebraic Solution

In this case,

$$P = 9000, \; r = 2.5\% = 0.025, \; n = 1, \; t = 5.$$

Using the formula for compound interest with n compoundings per year, you have

$$A = P\left(1 + \frac{r}{n}\right)^{nt} \qquad \text{Formula for compound interest}$$

$$= 9000\left(1 + \frac{0.025}{1}\right)^{1(5)} \qquad \begin{array}{l}\text{Substitute for } P, r,\\ n, \text{ and } t.\end{array}$$

$$= 9000(1.025)^5 \qquad \text{Simplify.}$$

$$\approx \$10{,}182.67. \qquad \text{Use a calculator.}$$

So, the balance in the account after 5 years will be about $10,182.67.

Graphical Solution

Substitute the values for P, r, and n into the formula for compound interest with n compoundings per year as follows.

$$A = P\left(1 + \frac{r}{n}\right)^{nt} \qquad \text{Formula for compound interest}$$

$$= 9000\left(1 + \frac{0.025}{1}\right)^{(1)t} \qquad \text{Substitute for } P, r, \text{ and } n.$$

$$= 9000(1.025)^t \qquad \text{Simplify.}$$

Use a graphing utility to graph $y = 9000(1.025)^x$. Using the *value* feature or *zoom* and *trace* features, you can approximate the value of y when $x = 5$ to be about 10,182.67, as shown in Figure 4.14. So, the balance in the account after 5 years will be about $10,182.67.

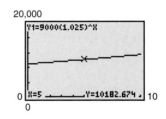

Figure 4.14

 Checkpoint   Now try Exercise 55.

Example 9   Finding Compound Interest

A total of $12,000 is invested at an annual interest rate of 3%. Find the balance after 4 years if the interest is compounded (a) quarterly and (b) continuously.

Solution

a. For quarterly compoundings, $n = 4$. So, after 4 years at 3%, the balance is

$$A = P\left(1 + \frac{r}{n}\right)^{nt} = 12{,}000\left(1 + \frac{0.03}{4}\right)^{4(4)}$$

$$\approx \$13{,}523.91.$$

b. For continuous compounding, the balance is

$$A = Pe^{rt} = 12{,}000e^{0.03(4)}$$

$$\approx \$13{,}529.96.$$

Note that a continuous-compounding account yields more than a quarterly-compounding account.

 Checkpoint   Now try Exercise 57.

Example 10   Radioactive Decay

Let y represent a mass of radioactive strontium (^{90}Sr), in grams, whose half-life is 28 years. The quantity of strontium present after t years is $y = 10\left(\frac{1}{2}\right)^{t/28}$.

a. What is the initial mass (when $t = 0$)?

b. How much of the initial mass is present after 80 years?

Algebraic Solution

a. $y = 10\left(\frac{1}{2}\right)^{t/28}$ Write original equation.

$= 10\left(\frac{1}{2}\right)^{0/28}$ Substitute 0 for t.

$= 10$ Simplify.

So, the initial mass is 10 grams.

b. $y = 10\left(\frac{1}{2}\right)^{t/28}$ Write original equation.

$= 10\left(\frac{1}{2}\right)^{80/28}$ Substitute 80 for t.

$\approx 10\left(\frac{1}{2}\right)^{2.857}$ Simplify.

≈ 1.380 Use a calculator.

So, about 1.380 grams is present after 80 years.

 Checkpoint Now try Exercise 65.

Graphical Solution

Use a graphing utility to graph $y = 10\left(\frac{1}{2}\right)^{x/28}$.

a. Use the *value* feature or the *zoom* and *trace* features of the graphing utility to determine that the value of y when $x = 0$ is 10, as shown in Figure 4.15. So, the initial mass is 10 grams.

b. Use the *value* feature or the *zoom* and *trace* features of the graphing utility to determine that the value of y when $x = 80$ is about 1.380, as shown in Figure 4.16. So, about 1.380 grams is present after 80 years.

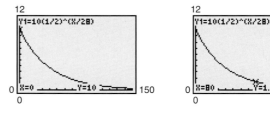

Figure 4.15   Figure 4.16

Example 11   Population Growth

The approximate number of fruit flies in an experimental population after t hours is given by $Q(t) = 20e^{0.03t}$, where $t \geq 0$.

a. Find the initial number of fruit flies in the population.

b. How large is the population of fruit flies after 72 hours?

c. Graph Q.

Solution

a. To find the initial population, evaluate $Q(t)$ at $t = 0$.

$Q(0) = 20e^{0.03(0)} = 20e^0 = 20(1) = 20$ flies

b. After 72 hours, the population size is

$Q(72) = 20e^{0.03(72)} = 20e^{2.16} \approx 173$ flies.

c. The graph of Q is shown in Figure 4.17.

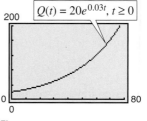

Figure 4.17

 Checkpoint Now try Exercise 67.

4.1   Exercises

Vocabulary Check

Fill in the blanks.

1. Polynomial and rational functions are examples of _____ functions.

2. Exponential and logarithmic functions are examples of nonalgebraic functions, also called _____ functions.

3. The exponential function $f(x) = e^x$ is called the _____ function, and the base e is called the _____ base.

4. To find the amount A in an account after t years with principal P and annual interest rate r compounded n times per year, you can use the formula _____ .

5. To find the amount A in an account after t years with principal P and annual interest rate r compounded continuously, you can use the formula _____ .

In Exercises 1–6, use a calculator to evaluate the function at the indicated value of x. Round your result to three decimal places.

Function	*Value*
1. $f(x) = 3.4^x$	$x = 6.8$
2. $f(x) = 1.2^x$	$x = \frac{1}{3}$
3. $g(x) = 5^x$	$x = -\pi$
4. $g(x) = 5000(2^x)$	$x = -1.5$
5. $h(x) = 17^{2x}$	$x = \sqrt{3}$
6. $h(x) = 8.6^{-3x}$	$x = -\sqrt{2}$

In Exercises 7–14, graph the exponential function by hand. Identify any asymptotes and intercepts and determine whether the graph of the function is increasing or decreasing.

7. $g(x) = 5^x$

8. $f(x) = \left(\frac{3}{2}\right)^x$

9. $f(x) = \left(\frac{1}{5}\right)^x = 5^{-x}$

10. $h(x) = \left(\frac{3}{2}\right)^{-x}$

11. $h(x) = 5^{x-2}$

12. $g(x) = \left(\frac{3}{2}\right)^{x+2}$

13. $g(x) = 5^{-x} - 3$

14. $f(x) = \left(\frac{3}{2}\right)^{-x} + 2$

In Exercises 15–18, use the graph of $y = 2^x$ to match the function with its graph. [The graphs are labeled (a), (b), (c), and (d).]

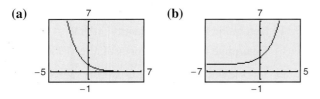

(a)

(b)

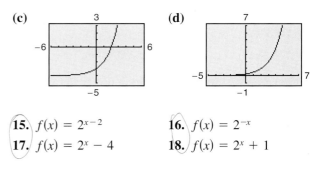

(c)

(d)

15. $f(x) = 2^{x-2}$

16. $f(x) = 2^{-x}$

17. $f(x) = 2^x - 4$

18. $f(x) = 2^x + 1$

In Exercises 19–22, use the graph of f to describe the transformation that yields the graph of g.

19. $f(x) = 3^x$, $g(x) = 3^{x-5}$

20. $f(x) = -2^x$, $g(x) = 5 - 2^x$

21. $f(x) = \left(\frac{3}{5}\right)^x$, $g(x) = -\left(\frac{3}{5}\right)^{x+4}$

22. $f(x) = 0.3^x$, $g(x) = -0.3^x + 5$

In Exercises 23–28, use a calculator to evaluate the function at the indicated value of x. Round your result to three decimal places.

Function	*Value*
23. $f(x) = e^x$	$x = 9.2$
24. $f(x) = e^{-x}$	$x = -\frac{3}{4}$
25. $g(x) = 50e^{4x}$	$x = 0.02$
26. $g(x) = 100e^{0.01x}$	$x = 12$
27. $h(x) = 2.5e^x$	$x = -\frac{1}{2}$
28. $h(x) = -5.5e^{-x}$	$x = 200$

In Exercises 29–38, use a graphing utility to construct a table of values for the function. Then sketch the graph of the function.

29. $f(x) = \left(\frac{5}{2}\right)^x$ **30.** $f(x) = \left(\frac{5}{2}\right)^{-x}$

31. $f(x) = 6^x$ **32.** $f(x) = 2^{x-1}$

33. $f(x) = 3^{x+2}$ **34.** $f(x) = e^{-x}$

35. $f(x) = 3e^{x+4}$ **36.** $f(x) = 2e^{-0.5x}$

37. $f(x) = 2 + e^{x-5}$ **38.** $f(x) = 4^{x-3} + 3$

In Exercises 39–46, use a graphing utility to graph the exponential function. Identify any asymptotes of the graph.

39. $y = 2^{-x^2}$ **40.** $y = 3^{-|x|}$

41. $y = 3^{x-2} + 1$ **42.** $y = 4^{x+1} - 2$

43. $g(x) = 2 - e^{-x}$ **44.** $s(t) = 3e^{-0.2t}$

45. $s(t) = 2e^{0.12t}$ **46.** $g(x) = 1 + e^{-x}$

In Exercises 47–50, use a graphing utility to (a) graph the function and (b) find any asymptotes numerically by creating a table of values for the function.

47. $f(x) = \dfrac{8}{1 + e^{-0.5x}}$ **48.** $g(x) = \dfrac{8}{1 + e^{-0.5/x}}$

49. $f(x) = -\dfrac{6}{2 - e^{0.2x}}$ **50.** $f(x) = \dfrac{6}{2 - e^{0.2/x}}$

In Exercises 51–54, (a) use a graphing utility to graph the function, (b) use the graph to find the open intervals on which the function is increasing and decreasing, and (c) approximate any relative maximum or minimum values.

51. $f(x) = x^2 e^{-x}$ **52.** $f(x) = 2x^2 e^{x+1}$

53. $f(x) = x(2^{3-x})$ **54.** $f(x) = -\left(\frac{1}{2}x\right)3^{x+4}$

Compound Interest In Exercises 55–58, complete the table to determine the balance A for P dollars invested at rate r for t years and compounded n times per year.

n	1	2	4	12	365	Continuous
A						

55. $P = \$2500$, $r = 2.5\%$, $t = 10$ years

56. $P = \$1000$, $r = 6\%$, $t = 10$ years

57. $P = \$2500$, $r = 4\%$, $t = 20$ years

58. $P = \$1000$, $r = 3\%$, $t = 40$ years

Compound Interest In Exercises 59–62, complete the table to determine the balance A for \$12,000 invested at a rate r for t years, compounded continuously.

t	1	10	20	30	40	50
A						

59. $r = 4\%$ **60.** $r = 6\%$

61. $r = 3.5\%$ **62.** $r = 2.5\%$

63. *Demand* The demand function for a product is given by

$$p = 5000\left(1 - \frac{4}{4 + e^{-0.002x}}\right)$$

where p is the price and x is the number of units.

(a) Use a graphing utility to graph the demand function for $x > 0$ and $p > 0$.

(b) Find the price p for a demand of $x = 500$ units.

(c) Use the graph in part (a) to approximate the highest price that will still yield a demand of at least 600 units.

(d) Verify your answers to parts (b) and (c) numerically by creating a table of values for the function.

64. *Compound Interest* There are three options for investing \$500. The first earns 7% compounded annually, the second earns 7% compounded quarterly, and the third earns 7% compounded continuously.

(a) Find equations that model each investment growth and use a graphing utility to graph each model in the same viewing window over a 20-year period.

(b) Use the graph from part (a) to determine which investment yields the highest return after 20 years. What is the difference in earnings between each investment?

65. *Radioactive Decay* Let Q represent a mass of radioactive radium (^{226}Ra), in grams, whose half-life is 1620 years. The quantity of radium present after t years is given by $Q = 25\left(\frac{1}{2}\right)^{t/1620}$.

(a) Determine the initial quantity (when $t = 0$).

(b) Determine the quantity present after 1000 years.

(c) Use a graphing utility to graph the function over the interval $t = 0$ to $t = 5000$.

(d) When will the quantity of radium be 0 grams? Explain.

66. Radioactive Decay Let Q represent a mass of carbon 14 (^{14}C), in grams, whose half-life is 5730 years. The quantity present after t years is given by $Q = 10(\frac{1}{2})^{t/5730}$.

(a) Determine the initial quantity (when $t = 0$).

(b) Determine the quantity present after 2000 years.

(c) Sketch the graph of the function over the interval $t = 0$ to $t = 10,000$.

67. Bacteria Growth A certain type of bacteria increases according to the model $P(t) = 100e^{0.2197t}$, where t is the time in hours.

(a) Use a graphing utility to graph the model.

(b) Use a graphing utility to approximate $P(0)$, $P(5)$, and $P(10)$.

(c) Verify your answers in part (b) algebraically.

68. Population Growth The population of a town increases according to the model $P(t) = 2500e^{0.0293t}$, where t is the time in years, with $t = 0$ corresponding to 2000.

(a) Use a graphing utility to graph the function for the years 2000 through 2025.

(b) Use a graphing utility to approximate the population in 2015 and 2025.

(c) Verify your answers in part (b) algebraically.

69. Inflation If the annual rate of inflation averages 4% over the next 10 years, the approximate cost C of goods or services during any year in that decade will be modeled by $C(t) = P(1.04)^t$, where t is the time (in years) and P is the present cost. The price of an oil change for your car is presently $23.95.

(a) Use a graphing utility to graph the function.

(b) Use the graph in part (a) to approximate the price of an oil change 10 years from now.

(c) Verify your answer in part (b) algebraically.

70. Depreciation After t years, the value of a car that costs $20,000 is modeled by $V(t) = 20,000(\frac{3}{4})^t$.

(a) Use a graphing utility to graph the function.

(b) Use a graphing utility to create a table of values that shows the value V for $t = 1$ to $t = 10$ years.

Synthesis

True or False? **In Exercises 71 and 72, determine whether the statement is true or false. Justify your answer.**

71. $f(x) = 1^x$ is not an exponential function.

72. $e = \dfrac{271,801}{99,990}$

73. Exploration Use a graphing utility to graph $y_1 = e^x$ and each of the functions $y_2 = x^2$, $y_3 = x^3$, $y_4 = \sqrt{x}$, and $y_5 = |x|$.

(a) Which function increases at the fastest rate for "large" values of x?

(b) Use the result of part (a) to make a conjecture about the rates of growth of $y_1 = e^x$ and $y = x^n$, where n is a natural number and x is "large."

(c) Use the results of parts (a) and (b) to describe what is implied when it is stated that a quantity is growing exponentially.

74. Exploration Consider the functions $f(x) = 3^x$ and $g(x) = 4^x$.

(a) Use a graphing utility to complete the table, and use the table to estimate the solution of the inequality $4^x < 3^x$.

x	-1	-0.5	0	0.5	1
$f(x)$					
$g(x)$					

(b) Use a graphing utility to graph f and g in the same viewing window. Use the graphs to solve the inequalities (i) $4^x < 3^x$ and (ii) $4^x > 3^x$.

75. Graphical Analysis Use a graphing utility to graph $f(x) = (1 + 0.5/x)^x$ and $g(x) = e^{0.5}$ in the same viewing window. What is the relationship between f and g as x increases without bound?

76. Think About It Which functions are exponential? Explain.

(a) $3x$ (b) $3x^2$ (c) 3^x (d) 2^{-x}

Review

In Exercises 77–80, determine whether the function has an inverse function. If it does, find f^{-1}.

77. $f(x) = 5x - 7$

78. $f(x) = -\frac{2}{3}x + \frac{5}{2}$

79. $f(x) = \sqrt[3]{x + 8}$

80. $f(x) = \sqrt{x^2 + 6}$

In Exercises 81 and 82, sketch the graph of the rational function.

81. $f(x) = \dfrac{2x}{x - 7}$

82. $f(x) = \dfrac{x^2 + 3}{x + 1}$

4.2 Logarithmic Functions and Their Graphs

Logarithmic Functions

In Section 1.7, you studied the concept of an inverse function. There, you learned that if a function is one-to-one—that is, if the function has the property such that no horizontal line intersects its graph more than once—the function must have an inverse function. By looking back at the graphs of the exponential functions introduced in Section 4.1, you will see that every function of the form

$$f(x) = a^x, \qquad a > 0, \, a \neq 1$$

passes the Horizontal Line Test and therefore must have an inverse function. This inverse function is called the **logarithmic function with base a.**

Definition of Logarithmic Function

For $x > 0$, $a > 0$, and $a \neq 1$,

$$y = \log_a x \qquad \text{if and only if} \qquad x = a^y.$$

The function given by

$$f(x) = \log_a x \qquad \text{Read as “log base } a \text{ of } x.\text{”}$$

is called the **logarithmic function with base a.**

The equations

$$y = \log_a x \quad \text{and} \quad x = a^y$$

are equivalent. The first equation is in logarithmic form and the second is in exponential form.

When evaluating logarithms, remember that *a logarithm is an exponent.* This means that $\log_a x$ is the exponent to which a must be raised to obtain x. For instance, $\log_2 8 = 3$ because 2 must be raised to the third power to get 8.

Example 1 Evaluating Logarithms

Use the definition of logarithmic function to evaluate each logarithm at the indicated value of x.

a. $f(x) = \log_2 x, \, x = 32$ **b.** $f(x) = \log_3 x, \, x = 1$
c. $f(x) = \log_4 x, \, x = 2$ **d.** $f(x) = \log_{10} x, \, x = \frac{1}{100}$

Solution

a. $f(32) = \log_2 32 = 5$ because $2^5 = 32.$
b. $f(1) = \log_3 1 = 0$ because $3^0 = 1.$
c. $f(2) = \log_4 2 = \frac{1}{2}$ because $4^{1/2} = \sqrt{4} = 2.$
d. $f\left(\frac{1}{100}\right) = \log_{10} \frac{1}{100} = -2$ because $10^{-2} = \frac{1}{10^2} = \frac{1}{100}.$

✓ *Checkpoint* Now try Exercise 17.

What you should learn

- Recognize and evaluate logarithmic functions with base a.
- Graph logarithmic functions.
- Recognize, evaluate, and graph natural logarithmic functions.
- Use logarithmic functions to model and solve real-life problems.

Why you should learn it

Logarithmic functions are useful in modeling data that represents quantities that increase or decrease slowly. For instance, Exercise 76 on page 341 shows how to use a logarithmic function to model the minimum required ventilation rates in public school classrooms.

Mark Richards/PhotoEdit

The logarithmic function with base 10 is called the **common logarithmic function.** On most calculators, this function is denoted by ⌊LOG⌋. Example 2 shows how to use a calculator to evaluate common logarithmic functions. You will learn how to use a calculator to calculate logarithms to any base in the next section.

Example 2   Evaluating Common Logarithms on a Calculator

Use a calculator to evaluate the function $f(x) = \log_{10} x$ at each value of x.

a. $x = 10$ **b.** $x = 2.5$ **c.** $x = -2$ **d.** $x = \frac{1}{4}$

Solution

Function Value	Graphing Calculator Keystrokes	Display
a. $f(10) = \log_{10} 10$	⌊LOG⌋ 10 ⌊ENTER⌋	1
b. $f(2.5) = \log_{10} 2.5$	⌊LOG⌋ 2.5 ⌊ENTER⌋	0.3979400
c. $f(-2) = \log_{10}(-2)$	⌊LOG⌋ ⌊(-)⌋ 2 ⌊ENTER⌋	ERROR
d. $f\left(\frac{1}{4}\right) = \log_{10} \frac{1}{4}$	⌊LOG⌋ ⌊(⌋ 1 ⌊÷⌋ 4 ⌊)⌋ ⌊ENTER⌋	−0.6020600

Note that the calculator displays an error message when you try to evaluate $\log_{10}(-2)$. The reason for this is that the domain of every logarithmic function is the set of *positive* real numbers. In this case, there is no *real* power to which 10 can be raised to obtain -2.

 Checkpoint Now try Exercise 21.

The following properties follow directly from the definition of the logarithmic function with base a.

Properties of Logarithms

1. $\log_a 1 = 0$ because $a^0 = 1$.

2. $\log_a a = 1$ because $a^1 = a$.

3. $\log_a a^x = x$ and $a^{\log_a x} = x$.   Inverse Properties

4. If $\log_a x = \log_a y$, then $x = y$.   One-to-One Property

---

Example 3   Using Properties of Logarithms

a. Solve for x: $\log_2 x = \log_2 3$ **b.** Solve for x: $\log_4 4 = x$

c. Simplify: $\log_5 5^x$ **d.** Simplify: $7^{\log_7 14}$

Solution

a. Using the One-to-One Property (Property 4), you can conclude that $x = 3$.

b. Using Property 2, you can conclude that $x = 1$.

c. Using the Inverse Property (Property 3), it follows that $\log_5 5^x = x$.

d. Using the Inverse Property (Property 3), it follows that $7^{\log_7 14} = 14$.

 Checkpoint Now try Exercise 25.

Graphs of Logarithmic Functions

To sketch the graph of $y = \log_a x$, you can use the fact that the graphs of inverse functions are reflections of each other in the line $y = x$.

Example 4 Graphs of Exponential and Logarithmic Functions

In the same coordinate plane, sketch the graph of each function by hand.

a. $f(x) = 2^x$ **b.** $g(x) = \log_2 x$

Solution

a. For $f(x) = 2^x$, construct a table of values. By plotting these points and connecting them with a smooth curve, you obtain the graph of f shown in Figure 4.18.

x	-2	-1	0	1	2	3
$f(x) = 2^x$	$\frac{1}{4}$	$\frac{1}{2}$	1	2	4	8

b. Because $g(x) = \log_2 x$ is the inverse function of $f(x) = 2^x$, the graph of g is obtained by plotting the points $(f(x), x)$ and connecting them with a smooth curve. The graph of g is a reflection of the graph of f in the line $y = x$, as shown in Figure 4.18.

 Checkpoint Now try Exercise 35.

Before you can confirm the result of Example 4 using a graphing utility, you need to know how to enter $\log_2 x$. You will learn how to do this using the *change-of-base formula* discussed in Section 4.3.

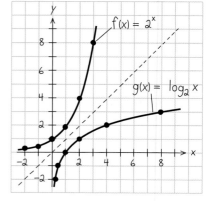

Figure 4.18

Example 5 Sketching the Graph of a Logarithmic Function

Sketch the graph of the common logarithmic function $f(x) = \log_{10} x$ by hand.

Solution

Begin by constructing a table of values. Note that some of the values can be obtained without a calculator by using the Inverse Property of Logarithms. Others require a calculator. Next, plot the points and connect them with a smooth curve, as shown in Figure 4.19.

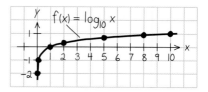

Figure 4.19

	Without Calculator				With Calculator		
x	$\frac{1}{100}$	$\frac{1}{10}$	1	10	2	5	8
$f(x) = \log_{10} x$	-2	-1	0	1	0.301	0.699	0.903

 Checkpoint Now try Exercise 41.

The nature of the graph in Figure 4.19 is typical of functions of the form $f(x) = \log_a x$, $a > 1$. They have one x-intercept and one vertical asymptote. Notice how slowly the graph rises for $x > 1$.

STUDY TIP

In Example 5, you can also sketch the graph of $f(x) = \log_{10} x$ by evaluating the inverse function of f, $g(x) = 10^x$, for several values of x. Plot the points, sketch the graph of g, and then reflect the graph in the line $y = x$ to obtain the graph of f.

Library of Functions: Logarithmic Function

The *logarithmic function*

$$f(x) = \log_a x, \quad a > 0, \ a \neq 1$$

is the inverse function of the exponential function. Its domain is the set of positive real numbers and its range is the set of all real numbers. This is the opposite of the exponential function. Moreover, the logarithmic function has the *y*-axis as a vertical asymptote, whereas the exponential function has the *x*-axis as a horizontal asymptote. Many real-life phenomena with a slow rate of growth can be modeled by logarithmic functions. The basic characteristics of the logarithmic function are summarized below.

Graph of $f(x) = \log_a x, \ a > 1$

Domain: $(0, \infty)$

Range: $(-\infty, \infty)$

Intercept: $(1, 0)$

Increasing on $(0, \infty)$

y-axis is a vertical asymptote
$(\log_a x \to -\infty$ as $x \to 0^+)$

Continuous

Reflection of graph of $f(x) = a^x$
in the line $y = x$

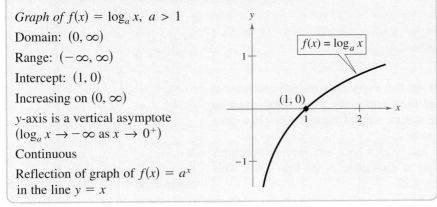

Example 6   Transformations of Graphs of Logarithmic Functions

Each of the following functions is a transformation of the graph of $f(x) = \log_{10} x$.

a. Because $g(x) = \log_{10}(x - 1) = f(x - 1)$, the graph of g can be obtained by shifting the graph of f one unit to the *right*, as shown in Figure 4.20.

b. Because $h(x) = 2 + \log_{10} x = 2 + f(x)$, the graph of h can be obtained by shifting the graph of f two units *upward*, as shown in Figure 4.21.

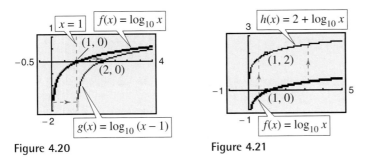

Figure 4.20   Figure 4.21

Notice that the transformation in Figure 4.21 keeps the *y*-axis as a vertical asymptote, but the transformation in Figure 4.20 yields the new vertical asymptote $x = 1$.

✓ *Checkpoint* Now try Exercise 49.

TECHNOLOGY TIP

When a graphing utility graphs a logarithmic function, it may appear that the graph has an endpoint. Recall from Section 1.1 that this is because some graphing utilities have a limited resolution. So, in this text a blue or light red curve is placed behind the graphing utility's display to indicate where the graph should appear.

The Natural Logarithmic Function

By looking back at the graph of the natural exponential function introduced in Section 4.1, you will see that $f(x) = e^x$ is one-to-one and so has an inverse function. This inverse function is called the **natural logarithmic function** and is denoted by the special symbol ln x, read as "the natural log of x" or "el en of x."

The Natural Logarithmic Function

For $x > 0$,

$$y = \ln x \text{ if and only if } x = e^y.$$

The function given by

$$f(x) = \log_e x = \ln x$$

is called the **natural logarithmic function.**

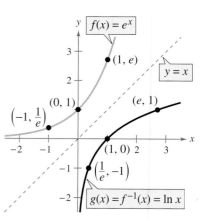

Reflection of graph of $f(x) = e^x$ in the line $y = x$

Figure 4.22

From the above definition, you can see that every logarithmic equation can be written in an equivalent exponential form and every exponential equation can be written in logarithmic form. Note that the natural logarithm ln x is written without a base. The base is understood to be e.

Because the functions $f(x) = e^x$ and $g(x) = \ln x$ are inverse functions of each other, their graphs are reflections of each other in the line $y = x$. This reflective property is illustrated in Figure 4.22.

Example 7 Evaluating the Natural Logarithmic Function

Use a calculator to evaluate the function $f(x) = \ln x$ at each indicated value of x.

a. $x = 2$ **b.** $x = 0.3$ **c.** $x = -1$

Solution

Function Value	Graphing Calculator Keystrokes	Display
a. $f(2) = \ln 2$	[LN] 2 [ENTER]	0.6931472
b. $f(0.3) = \ln 0.3$	[LN] .3 [ENTER]	-1.2039728
c. $f(-1) = \ln(-1)$	[LN] [(–)] 1 [ENTER]	ERROR

✓ *Checkpoint* Now try Exercise 53.

The four properties of logarithms listed on page 333 are also valid for natural logarithms.

Properties of Natural Logarithms

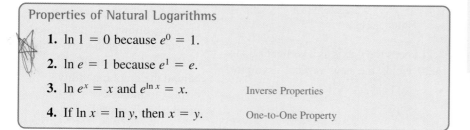

1. $\ln 1 = 0$ because $e^0 = 1$.

2. $\ln e = 1$ because $e^1 = e$.

3. $\ln e^x = x$ and $e^{\ln x} = x$.   Inverse Properties

4. If $\ln x = \ln y$, then $x = y$.   One-to-One Property

TECHNOLOGY TIP

On most calculators, the natural logarithm is denoted by [LN], as illustrated in Example 7.

STUDY TIP

In Example 7(c), be sure you see that $\ln(-1)$ gives an error message on most calculators. This occurs because the domain of ln x is the set of *positive* real numbers (see Figure 4.22). So, $\ln(-1)$ is undefined.

Example 8   Using Properties of Natural Logarithms

Use the properties of natural logarithms to rewrite each expression.

a. $\ln \dfrac{1}{e}$ **b.** $e^{\ln 5}$ **c.** $\ln e^0$ **d.** $2 \ln e$

Solution

a. $\ln \dfrac{1}{e} = \ln e^{-1} = -1$ Inverse Property   **b.** $e^{\ln 5} = 5$ Inverse Property

c. $\ln e^0 = \ln 1 = 0$ Property 1   **d.** $2 \ln e = 2(1) = 2$ Property 2

✓ *Checkpoint* Now try Exercise 57.

Example 9   Finding the Domains of Logarithmic Functions

Find the domain of each function.

a. $f(x) = \ln(x - 2)$ **b.** $g(x) = \ln(2 - x)$ **c.** $h(x) = \ln x^2$

Algebraic Solution

a. Because $\ln(x - 2)$ is defined only if

$$x - 2 > 0,$$

it follows that the domain of f is $(2, \infty)$.

b. Because $\ln(2 - x)$ is defined only if

$$2 - x > 0,$$

it follows that the domain of g is $(-\infty, 2)$.

c. Because $\ln x^2$ is defined only if

$$x^2 > 0,$$

it follows that the domain of h is all real numbers except $x = 0$.

✓ *Checkpoint* Now try Exercise 61.

Graphical Solution

Use a graphing utility to graph each function using an appropriate viewing window. Then use the *trace* feature to determine the domain of each function.

a. From Figure 4.23, you can see that the x-coordinates of the points on the graph appear to extend from the right of 2 to $+\infty$. So, you can estimate the domain to be $(2, \infty)$.

b. From Figure 4.24, you can see that the x-coordinates of the points on the graph appear to extend from $-\infty$ to the left of 2. So, you can estimate the domain to be $(-\infty, 2)$.

c. From Figure 4.25, you can see that the x-coordinates of the points on the graph appear to include all real numbers except $x = 0$. So, you can estimate the domain to be all real numbers except $x = 0$.

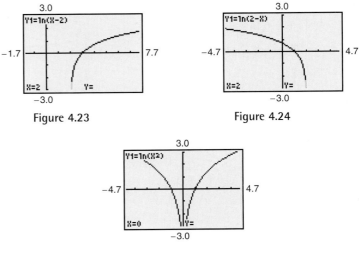

Figure 4.23

Figure 4.24

Figure 4.25

In Example 9, suppose you had been asked to analyze the function $h(x) = \ln|x - 2|$. How would the domain of this function compare with the domains of the functions given in parts (a) and (b) of the example?

Application

Logarithmic functions are used to model many situations in real life, as shown in the next example.

Example 10   Human Memory Model

TECHNOLOGY SUPPORT

For instructions on how to use the *value* feature and the *zoom* and *trace* features, see Appendix A; for specific keystrokes, go to the text website at *college.hmco.com*.

Students participating in a psychology experiment attended several lectures on a subject and were given an exam. Every month for a year after the exam, the students were retested to see how much of the material they remembered. The average scores for the group are given by the *human memory model*

$$f(t) = 75 - 6 \ln(t + 1), \qquad 0 \le t \le 12$$

where t is the time in months.

a. What was the average score on the original ($t = 0$) exam?

b. What was the average score at the end of $t = 2$ months?

c. What was the average score at the end of $t = 6$ months?

Algebraic Solution

a. The original average score was

$$f(0) = 75 - 6 \ln(0 + 1)$$

$$= 75 - 6 \ln 1$$

$$= 75 - 6(0)$$

$$= 75.$$

b. After 2 months, the average score was

$$f(2) = 75 - 6 \ln(2 + 1)$$

$$= 75 - 6 \ln 3$$

$$\approx 75 - 6(1.0986)$$

$$\approx 68.4.$$

c. After 6 months, the average score was

$$f(6) = 75 - 6 \ln(6 + 1)$$

$$= 75 - 6 \ln 7$$

$$\approx 75 - 6(1.9459)$$

$$\approx 63.3.$$

 Checkpoint Now try Exercise 69.

Graphical Solution

Use a graphing utility to graph the model $y = 75 - 6 \ln(x + 1)$. Then use the *value* or *trace* feature to approximate the following.

a. When $x = 0$, $y = 75$ (see Figure 4.26). So, the original average score was 75.

b. When $x = 2$, $y \approx 68.4$ (see Figure 4.27). So, the average score after 2 months was about 68.4.

c. When $x = 6$, $y \approx 63.3$ (see Figure 4.28). So, the average score after 6 months was about 63.3.

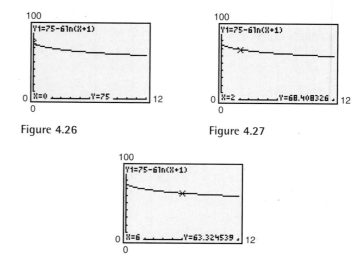

Figure 4.26

Figure 4.27

Figure 4.28

4.2   Exercises

Vocabulary Check

Fill in the blanks.

1. The inverse function of the exponential function $f(x) = a^x$ is called the _____ with base a.
2. The common logarithmic function has base _____ .
3. The logarithmic function $f(x) = \ln x$ is called the _____ function.
4. The inverse property of logarithms states that $\log_a a^x = x$ and _____ .
5. The one-to-one property of natural logarithms states that if $\ln x = \ln y$, then _____ .

In Exercises 1–8, write the logarithmic equation in exponential form. For example, the exponential form of $\log_5 25 = 2$ is $5^2 = 25$.

1. $\log_4 64 = 3$
2. $\log_3 81 = 4$
3. $\log_7 \frac{1}{49} = -2$
4. $\log_{10} \frac{1}{1000} = -3$
5. $\log_{32} 4 = \frac{2}{5}$
6. $\log_{16} 8 = \frac{3}{4}$
7. $\ln 1 = 0$
8. $\ln 4 = 1.386 \ldots$

In Exercises 9–16, write the exponential equation in logarithmic form. For example, the logarithmic form of $2^3 = 8$ is $\log_2 8 = 3$.

9. $5^3 = 125$
10. $8^2 = 64$
11. $81^{1/4} = 3$
12. $9^{3/2} = 27$
13. $6^{-2} = \frac{1}{36}$
14. $10^{-3} = 0.001$
15. $e^3 = 20.0855 \ldots$
16. $e^x = 4$

In Exercises 17–20, evaluate the function at the indicated value of x without using a calculator.

Function	Value
17. $f(x) = \log_2 x$	$x = 16$
18. $f(x) = \log_{16} x$	$x = \frac{1}{4}$
19. $g(x) = \log_{10} x$	$x = 0.01$
20. $g(x) = \log_{10} x$	$x = 10$

In Exercises 21–24, use a calculator to evaluate the function at the indicated value of x. Round your result to three decimal places.

Function	Value
21. $f(x) = \log_{10} x$	$x = 345$
22. $f(x) = \log_{10} x$	$x = \frac{4}{5}$
23. $h(x) = 6 \log_{10} x$	$x = 14.8$
24. $h(x) = 1.9 \log_{10} x$	$x = 4.3$

In Exercises 25–30, solve the equation for x.

25. $\log_7 x = \log_7 9$
26. $\log_5 5 = x$
27. $\log_6 6^2 = x$
28. $\log_2 2^{-1} = x$
29. $\log_8 x = \log_8 10^{-1}$
30. $\log_3 4^3 = x$

In Exercises 31–34, describe the relationship between the graphs of f and g.

31. $f(x) = 3^x$
    $g(x) = \log_3 x$
32. $f(x) = 5^x$
    $g(x) = \log_5 x$
33. $f(x) = e^x$
    $g(x) = \ln x$
34. $f(x) = 10^x$
    $g(x) = \log_{10} x$

In Exercises 35–44, find the domain, vertical asymptote, and x-intercept of the logarithmic function, and sketch its graph by hand. Verify using a graphing utility.

35. $f(x) = \log_4 x$
36. $g(x) = \log_6 x$
37. $f(x) = \log_{10}\left(\frac{x}{5}\right)$
38. $g(x) = \log_2(-x)$
39. $h(x) = \log_4(x - 3)$
40. $f(x) = -\log_6(x + 2)$
41. $y = -\log_{10} x + 2$
42. $y = \log_{10}(x - 1) + 4$
43. $f(x) = 6 + \log_6(x - 3)$
44. $f(x) = -\log_3(x + 2) - 4$

In Exercises 45–48, use the graph of $y = \log_3 x$ **to match the function with its graph. [The graphs are labeled (a), (b), (c), and (d).]**

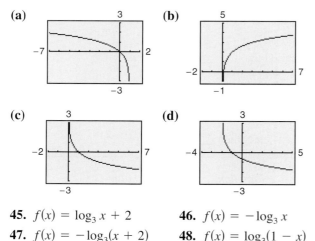

(a)

(b)

(c)

(d)

45. $f(x) = \log_3 x + 2$

46. $f(x) = -\log_3 x$

47. $f(x) = -\log_3(x + 2)$

48. $f(x) = \log_3(1 - x)$

In Exercises 49–52, use the graph of f to describe the transformation that yields the graph of g.

49. $f(x) = \log_{10} x, \quad g(x) = -\log_{10} x$

50. $f(x) = \log_{10} x, \quad g(x) = \log_{10}(x + 7)$

51. $f(x) = \log_2 x, \quad g(x) = 4 - \log_2 x$

52. $f(x) = \log_2 x, \quad g(x) = 3 + \log_2 x$

In Exercises 53–56, use a calculator to evaluate the function at the indicated value of x. Round your result to three decimal places.

Function	*Value*
53. $f(x) = \ln x$	$x = \sqrt{42}$
54. $f(x) = \ln x$	$x = 18.31$
55. $f(x) = -\ln x$	$x = \frac{1}{2}$
56. $f(x) = 3 \ln x$	$x = 0.75$

In Exercises 57–60, use the properties of natural logarithms to rewrite the expression.

57. $\ln e^2$

58. $-\ln e$

59. $e^{\ln 1.8}$

60. $7 \ln e^0$

In Exercises 61–64, use a graphing utility to graph the logarithmic function. Determine the domain and identify any vertical asymptote and x-intercept. ·

61. $f(x) = \ln(x - 1)$

62. $h(x) = \ln(x + 1)$

63. $g(x) = \ln(-x)$

64. $f(x) = \ln(3 - x)$

In Exercises 65–68, (a) use a graphing utility to graph the function, (b) find the domain, (c) use the graph to find the open intervals on which the function is increasing and decreasing, and (d) approximate any relative maximum or minimum values of the function.

65. $f(x) = \dfrac{x}{2} - \ln \dfrac{x}{4}$

66. $g(x) = \dfrac{12 \ln x}{x}$

67. $h(x) = 4x \ln x$

68. $f(x) = \dfrac{x}{\ln x}$

69. ***Human Memory Model***   Students in a mathematics class were given an exam and then tested monthly with an equivalent exam. The average scores for the class are given by the human memory model

$$f(t) = 80 - 17 \log_{10}(t + 1), \quad 0 \le t \le 12$$

where t is the time in months.

(a) What was the average score on the original exam $(t = 0)$?

(b) What was the average score after 4 months?

(c) What was the average score after 10 months?

(d) Verify your answers in parts (a), (b), and (c) using a graphing utility.

70. ***Data Analysis***   The table shows the temperatures T (in °F) at which water boils at selected pressures p (in pounds per square inch).   (Source: Standard Handbook of Mechanical Engineers)

Pressure, *p*	Temperature, *T*
5	162.24°
10	193.21°
14.696 (1 atm)	212.00°
20	227.96°
30	250.33°
40	267.25°
60	292.71°
80	312.03°
100	327.81°

A model that approximates this data is given by

$$T = 87.97 + 34.96 \ln p + 7.91 \sqrt{p}.$$

(a) Use a graphing utility to plot the data and graph the model in the same viewing window. How well does the model fit the data?

(b) Use the graph to estimate the pressure required for the boiling point of water to exceed 300°F.

(c) Calculate T when the pressure is 74 pounds per square inch. Verify your answer graphically.

71. *Compound Interest* A principal P, invested at $5\frac{1}{2}\%$ and compounded continuously, increases to an amount K times the original principal after t years, where $t = (\ln K)/0.055$.

(a) Complete the table and interpret your results.

K	1	2	4	6	8	10	12
t							

(b) Use a graphing utility to graph the function.

72. *Population* The time t in years for the world population to double if it is increasing at a continuous rate of r is given by

$$t = \frac{\ln 2}{r}.$$

(a) Complete the table and interpret your results.

r	0.005	0.010	0.015	0.020	0.025	0.030
t						

(b) Use a graphing utility to graph the function.

73. *Sound Intensity* The relationship between the number of decibels β and the intensity of a sound I in watts per square meter is given by

$$\beta = 10 \log_{10}\left(\frac{I}{10^{-12}}\right).$$

(a) Determine the number of decibels of a sound with an intensity of 1 watt per square meter.

(b) Determine the number of decibels of a sound with an intensity of 10^{-2} watt per square meter.

(c) The intensity of the sound in part (a) is 100 times as great as that in part (b). Is the number of decibels 100 times as great? Explain.

74. *Home Mortgage* The model

$$t = 16.625 \ln\left(\frac{x}{x - 750}\right), \quad x > 750$$

approximates the length of a home mortgage of $150,000 at 6% in terms of the monthly payment. In the model, t is the length of the mortgage in years and x is the monthly payment in dollars.

(a) Use the model to approximate the length of a $150,000 mortgage at 6% when the monthly payment is $897.72 and when the monthly payment is $1659.24.

(b) Approximate the total amount paid over the term of the mortgage with a monthly payment of $897.72 and with a monthly payment of $1659.24. What amount of the total is interest costs for each payment?

Ventilation Rates In Exercises 75 and 76, use the model

$$y = 80.4 - 11 \ln x, \quad 100 \le x \le 1500$$

which approximates the minimum required ventilation rate in terms of the air space per child in a public school classroom. In the model, x is the air space per child (in cubic feet) and y is the ventilation rate per child (in cubic feet per minute).

75. Use a graphing utility to graph the function and approximate the required ventilation rate when there is 300 cubic feet of air space per child.

76. A classroom is designed for 30 students. The air-conditioning system in the room has the capacity to move 450 cubic feet of air per minute.

(a) Determine the ventilation rate per child, assuming that the room is filled to capacity.

(b) Use the graph in Exercise 75 to estimate the air space required per child.

(c) Determine the minimum number of square feet of floor space required for the room if the ceiling height is 30 feet.

Synthesis

True or False? **In Exercises 77 and 78, determine whether the statement is true or false. Justify your answer.**

77. You can determine the graph of $f(x) = \log_6 x$ by graphing $g(x) = 6^x$ and reflecting it about the x-axis.

78. The graph of $f(x) = \log_3 x$ contains the point $(27, 3)$.

79. *Writing* Explain why $\log_a x$ is defined only for $0 < a < 1$ and $a > 1$.

80. *Graphical Analysis* Use a graphing utility to graph f and g in the same viewing window and determine which is increasing at the greater rate as x approaches $+\infty$. What can you conclude about the rate of growth of the natural logarithmic function?

(a) $f(x) = \ln x, \quad g(x) = \sqrt{x}$

(b) $f(x) = \ln x, \quad g(x) = \sqrt[4]{x}$

81. *Exploration* The following table of values was obtained by evaluating a function. Determine which of the statements may be true and which must be false.

x	1	2	8
y	0	1	3

(a) y is an exponential function of x.

(b) y is a logarithmic function of x.

(c) x is an exponential function of y.

(d) y is a linear function of x.

82. *Pattern Recognition*

(a) Use a graphing utility to compare the graph of the function $y = \ln x$ with the graph of each function.

$$y_1 = x - 1, \ y_2 = (x - 1) - \tfrac{1}{2}(x - 1)^2,$$

$$y_3 = (x - 1) - \tfrac{1}{2}(x - 1)^2 + \tfrac{1}{3}(x - 1)^3$$

(b) Identify the pattern of successive polynomials given in part (a). Extend the pattern one more term and compare the graph of the resulting polynomial function with the graph of $y = \ln x$. What do you think the pattern implies?

83. *Numerical and Graphical Analysis*

(a) Use a graphing utility to complete the table for the function

$$f(x) = \frac{\ln x}{x}.$$

x	1	5	10	10^2	10^4	10^6
$f(x)$						

(b) Use the table in part (a) to determine what value $f(x)$ approaches as x increases without bound.

(c) Use a graphing utility to confirm the result of part (b).

84. *Writing* Use a graphing utility to determine how many months it would take for the average score in Example 10 to decrease to 60. Explain your method of solving the problem. Describe another way that you can use a graphing utility to determine the answer. Also, make a statement about the general shape of the model. Would a student forget more quickly soon after the test or as time passes? Explain your reasoning.

Review

In Exercises 85–92, factor the polynomial.

85. $x^2 + 2x - 3$ **86.** $2x^2 + 3x - 5$

87. $12x^2 + 5x - 3$ **88.** $16x^2 + 16x + 7$

89. $16x^2 - 25$ **90.** $36x^2 - 49$

91. $2x^3 + x^2 - 45x$ **92.** $3x^2 - 5x^2 - 12x$

In Exercises 93–96, evaluate the function for $f(x) = 3x + 2$ and $g(x) = x^3 - 1$.

93. $(f + g)(2)$ **94.** $(f - g)(-1)$

95. $(fg)(6)$ **96.** $\left(\dfrac{f}{g}\right)(0)$

In Exercises 97–100, solve the equation graphically.

97. $5x - 7 = x + 4$ **98.** $-2x + 3 = 8x$

99. $\sqrt{3x - 2} = 9$ **100.** $\sqrt{x - 11} = x + 2$

In Exercises 101–106, find the vertical and horizontal asymptotes of the rational function.

101. $f(x) = \dfrac{4}{-8 - x}$ **102.** $f(x) = \dfrac{2x^3 - 3}{x^2}$

103. $f(x) = \dfrac{x + 5}{2x^2 + x - 15}$ **104.** $f(x) = \dfrac{2x^2(x - 5)}{x - 7}$

105. $g(x) = \dfrac{x^2 - 4}{x^2 - 4x - 12}$

106. $g(x) = \dfrac{x^2 + 3x}{2x^2 + 3x - 2}$

4.3   Properties of Logarithms

Change of Base

Most calculators have only two types of log keys, one for common logarithms (base 10) and one for natural logarithms (base e). Although common logs and natural logs are the most frequently used, you may occasionally need to evaluate logarithms to other bases. To do this, you can use the following **change-of-base formula.**

What you should learn

- Rewrite logarithms with different bases.
- Use properties of logarithms to evaluate or rewrite logarithmic expressions.
- Use properties of logarithms to expand or condense logarithmic expressions.
- Use logarithmic functions to model and solve real-life problems.

Why you should learn it

Logarithmic functions can be used to model and solve real-life problems, such as the human memory model in Exercise 82 on page 348.

> **Change-of-Base Formula**
>
> Let a, b, and x be positive real numbers such that $a \neq 1$ and $b \neq 1$. Then $\log_a x$ can be converted to a different base using any of the following formulas.
>
> | Base b | Base 10 | Base e |
> |---|---|---|
> | $\log_a x = \dfrac{\log_b x}{\log_b a}$ | $\log_a x = \dfrac{\log_{10} x}{\log_{10} a}$ | $\log_a x = \dfrac{\ln x}{\ln a}$ |

One way to look at the change-of-base formula is that logarithms to base a are simply *constant multiples* of logarithms to base b. The constant multiplier is $1/(\log_b a)$.

Gary Conner/PhotoEdit

Example 1   Changing Bases Using Common Logarithms

a. $\log_4 25 = \dfrac{\log_{10} 25}{\log_{10} 4}$ $\log_a x = \dfrac{\log_{10} x}{\log_{10} a}$

$\approx \dfrac{1.39794}{0.60206} \approx 2.3219$ Use a calculator.

b. $\log_2 12 = \dfrac{\log_{10} 12}{\log_{10} 2} \approx \dfrac{1.07918}{0.30103} \approx 3.5850$

✓ *Checkpoint* Now try Exercise 9.

Example 2   Changing Bases Using Natural Logarithms

a. $\log_4 25 = \dfrac{\ln 25}{\ln 4}$ $\log_a x = \dfrac{\ln x}{\ln a}$

$\approx \dfrac{3.21888}{1.38629} \approx 2.3219$ Use a calculator.

b. $\log_2 12 = \dfrac{\ln 12}{\ln 2} \approx \dfrac{2.48491}{0.69315} \approx 3.5850$

✓ *Checkpoint* Now try Exercise 11.

STUDY TIP

Notice in Examples 1 and 2 that the result is the same whether common logarithms or natural logarithms are used in the change-of-base formula.

Properties of Logarithms

You know from the previous section that the logarithmic function with base a is the *inverse function* of the exponential function with base a. So, it makes sense that the properties of exponents (see Section 4.1) should have corresponding properties involving logarithms. For instance, the exponential property $a^0 = 1$ has the corresponding logarithmic property $\log_a 1 = 0$.

Properties of Logarithms

Let a be a positive number such that $a \neq 1$, and let n be a real number. If u and v are positive real numbers, the following properties are true.

Logarithm with Base a	*Natural Logarithm*
1. $\log_a(uv) = \log_a u + \log_a v$	**1.** $\ln(uv) = \ln u + \ln v$
2. $\log_a \dfrac{u}{v} = \log_a u - \log_a v$	**2.** $\ln \dfrac{u}{v} = \ln u - \ln v$
3. $\log_a u^n = n \log_a u$	**3.** $\ln u^n = n \ln u$

See Appendix B for a proof of Property 1.

STUDY TIP

There is no general property that can be used to rewrite $\log_a(u \pm v)$. Specifically, $\log_a(x + y)$ is *not* equal to $\log_a x + \log_a y$.

Example 3   Using Properties of Logarithms

Write each logarithm in terms of $\ln 2$ and $\ln 3$.

a. $\ln 6$ **b.** $\ln \dfrac{2}{27}$

Solution

a. $\ln 6 = \ln(2 \cdot 3)$ Rewrite 6 as $2 \cdot 3$.

$\quad\quad = \ln 2 + \ln 3$ Property 1

b. $\ln \dfrac{2}{27} = \ln 2 - \ln 27$ Property 2

$\quad\quad\quad = \ln 2 - \ln 3^3$ Rewrite 27 as 3^3.

$\quad\quad\quad = \ln 2 - 3 \ln 3$ Property 3

✓ *Checkpoint* Now try Exercise 19.

Example 4   Using Properties of Logarithms

Use the properties of logarithms to verify that $-\log_{10} \frac{1}{100} = \log_{10} 100$.

Solution

$\quad -\log_{10} \frac{1}{100} = -\log_{10} (100^{-1})$ Rewrite $\frac{1}{100}$ as 100^{-1}.

$\quad\quad\quad\quad\quad = -(-1) \log_{10} 100 = \log_{10} 100$ Property 3 and simplify.

✓ *Checkpoint* Now try Exercise 21.

Rewriting Logarithmic Expressions

The properties of logarithms are useful for rewriting logarithmic expressions in forms that simplify the operations of algebra. This is true because they convert complicated products, quotients, and exponential forms into simpler sums, differences, and products, respectively.

Example 5   Expanding Logarithmic Expressions

Use the properties of logarithms to expand each expression.

a. $\log_4 5x^3y$ **b.** $\ln \dfrac{\sqrt{3x-5}}{7}$

Solution

a. $\log_4 5x^3y = \log_4 5 + \log_4 x^3 + \log_4 y$ Property 1

$\qquad\qquad\quad = \log_4 5 + 3\log_4 x + \log_4 y$ Property 3

b. $\ln \dfrac{\sqrt{3x-5}}{7} = \ln\left[\dfrac{(3x-5)^{1/2}}{7}\right]$ Rewrite rational exponent.

$\qquad\qquad\quad = \ln(3x-5)^{1/2} - \ln 7$ Property 2

$\qquad\qquad\quad = \tfrac{1}{2}\ln(3x-5) - \ln 7$ Property 3

✓ *Checkpoint* Now try Exercise 39.

In Example 5, the properties of logarithms were used to *expand* logarithmic expressions. In Example 6, this procedure is reversed and the properties of logarithms are used to *condense* logarithmic expressions.

Example 6   Condensing Logarithmic Expressions

Use the properties of logarithms to condense each logarithmic expression.

a. $\tfrac{1}{2}\log_{10} x + 3\log_{10}(x+1)$ **b.** $2\ln(x+2) - \ln x$

c. $\tfrac{1}{3}[\log_2 x + \log_2(x-4)]$

Solution

a. $\tfrac{1}{2}\log_{10} x + 3\log_{10}(x+1) = \log_{10} x^{1/2} + \log_{10}(x+1)^3$ Property 3

$\qquad\qquad\qquad\qquad = \log_{10}\left[\sqrt{x}(x+1)^3\right]$ Property 1

b. $2\ln(x+2) - \ln x = \ln(x+2)^2 - \ln x$ Property 3

$\qquad\qquad\qquad = \ln\dfrac{(x+2)^2}{x}$ Property 2

c. $\tfrac{1}{3}[\log_2 x + \log_2(x-4)] = \tfrac{1}{3}\{\log_2[x(x-4)]\}$ Property 1

$\qquad\qquad\qquad = \log_2[x(x-4)]^{1/3} = \log_2 \sqrt[3]{x(x-4)}$ Property 3

✓ *Checkpoint* Now try Exercise 57.

Exploration

Use a graphing utility to graph the functions

$$y = \ln x - \ln(x-3)$$

and

$$y = \ln \dfrac{x}{x-3}$$

in the same viewing window. Does the graphing utility show the functions with the same domain? If so, should it? Explain your reasoning.

Example 7 Finding a Mathematical Model

The table shows the mean distance from the sun x and the period (the time it takes a planet to orbit the sun) y for each of the six planets that are closest to the sun. In the table, the mean distance is given in astronomical units (where the Earth's mean distance is defined as 1.0), and the period is given in years. Find an equation that relates y and x.

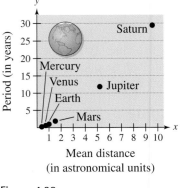

Figure 4.29

Planet	Mercury	Venus	Earth	Mars	Jupiter	Saturn
Mean distance, x	0.387	0.723	1.000	1.524	5.203	9.555
Period, y	0.241	0.615	1.000	1.881	11.860	29.420

Algebraic Solution

The points in the table are plotted in Figure 4.29. From this figure it is not clear how to find an equation that relates y and x. To solve this problem, take the natural log of each of the x- and y-values in the table. This produces the following results.

Planet	Mercury	Venus	Earth
$\ln x = X$	-0.949	-0.324	0.000
$\ln y = Y$	-1.423	-0.486	0.000

Planet	Mars	Jupiter	Saturn
$\ln x = X$	0.421	1.649	2.257
$\ln y = Y$	0.632	2.473	3.382

Now, by plotting the points in the table, you can see that all six of the points appear to lie in a line. Choose any two points to determine the slope of the line. Using the two points $(0.421, 0.632)$ and $(0, 0)$, you can determine that the slope of the line is

$$m = \frac{0.632 - 0}{0.421 - 0} \approx 1.5 = \frac{3}{2}.$$

By the point-slope form, the equation of the line is $Y = \frac{3}{2}X$, where $Y = \ln y$ and $X = \ln x$. You can therefore conclude that $\ln y = \frac{3}{2} \ln x$.

✓ *Checkpoint* Now try Exercise 83.

Graphical Solution

The points in the table are plotted in Figure 4.29. From this figure it is not clear how to find an equation that relates y and x. To solve this problem, take the natural log of each of the x- and y-values in the table. This produces the following results.

Planet	Mercury	Venus	Earth	Mars	Jupiter	Saturn
$\ln x = X$	-0.949	-0.324	0.000	0.421	1.649	2.257
$\ln y = Y$	-1.423	-0.486	0.000	0.632	2.473	3.382

Now, by plotting the points in the table, you can see that all six of the points appear to lie in a line, as shown in Figure 4.30. Using the *linear regression* feature of a graphing utility, you can find a linear model for the data, as shown in Figure 4.31. You can approximate this model to be $Y = 1.5X = \frac{3}{2}X$, where $Y = \ln y$ and $X = \ln x$. From the model, you can see that the slope of the line is $\frac{3}{2}$. So, you can conclude that $\ln y = \frac{3}{2} \ln x$.

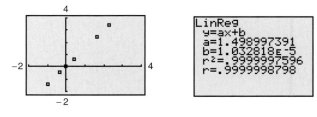

Figure 4.30   Figure 4.31

In Example 7, try to convert the final equation to $y = f(x)$ form. You will get a function of the form $y = ax^b$, which is called a *power model*.

4.3   Exercises

Vocabulary Check

Fill in the blanks.

1. To evaluate logarithms to any base, you can use the _____ formula.
2. The change-of-base formula for base e is given by $\log_a x = $ _____ .
3. _____ $= n \log_a u$
4. $\ln(uv) = $ _____

In Exercises 1–8, rewrite the logarithm as a ratio of (a) a common logarithm and (b) a natural logarithm.

1. $\log_5 x$
2. $\log_3 x$
3. $\log_{1/5} x$
4. $\log_{1/3} x$
5. $\log_a \frac{3}{10}$
6. $\log_a \frac{3}{4}$
7. $\log_{2.6} x$
8. $\log_{7.1} x$

In Exercises 9–16, evaluate the logarithm using the change-of-base formula. Round your result to three decimal places.

9. $\log_3 7$
10. $\log_7 4$
11. $\log_{1/2} 4$
12. $\log_{1/8} 64$
13. $\log_9(0.8)$
14. $\log_3(0.015)$
15. $\log_{15} 1460$
16. $\log_{20} 135$

In Exercises 17–20, use the properties of logarithms to rewrite and simplify the logarithmic expression.

17. $\log_4 8$
18. $\log_2 (4^2 \cdot 3^4)$
19. $\ln(5e^6)$
20. $\ln \dfrac{6}{e^2}$

In Exercises 21 and 22, use the properties of logarithms to verify the equation.

21. $\log_5 \frac{1}{250} = -3 - \log_5 2$
22. $-\ln 24 = -(3 \ln 2 + \ln 3)$

In Exercises 23–42, use the properties of logarithms to expand the expression as a sum, difference, and/or constant multiple of logarithms. (Assume all variables are positive.)

23. $\log_{10} 5x$
24. $\log_{10} 10z$
25. $\log_{10} \dfrac{5}{x}$
26. $\log_{10} \dfrac{y}{2}$
27. $\log_8 x^4$
28. $\log_6 z^{-3}$
29. $\ln \sqrt{z}$
30. $\ln \sqrt[3]{t}$
31. $\ln xyz$
32. $\ln \dfrac{xy}{z}$
33. $\ln(a^2 \sqrt{a - 1}), \quad a > 1$
34. $\ln[z(z - 1)^2], \quad z > 1$
35. $\ln \sqrt[3]{\dfrac{x}{y}}$
36. $\ln \sqrt{\dfrac{x^2}{y^3}}$
37. $\ln\left(\dfrac{x^2 - 1}{x^3}\right), \quad x > 1$
38. $\ln \dfrac{x}{\sqrt{x^2 + 1}}$
39. $\ln \dfrac{x^4 \sqrt{y}}{z^5}$
40. $\ln \sqrt{x^2(x + 2)}$
41. $\log_b \dfrac{x^2}{y^2 z^3}$
42. $\log_b \dfrac{\sqrt{x} y^4}{z^4}$

Graphical Analysis   **In Exercises 43 and 44, (a) use a graphing utility to graph the two equations in the same viewing window and (b) use the *table* feature of the graphing utility to create a table of values for each equation. (c) What do the graphs and tables suggest? Explain your reasoning.**

43. $y_1 = \ln[x^3(x + 4)], \qquad y_2 = 3 \ln x + \ln(x + 4)$
44. $y_1 = \ln\left(\dfrac{\sqrt{x}}{x - 2}\right), \qquad y_2 = \frac{1}{2} \ln x - \ln(x - 2)$

In Exercises 45–62, condense the expression to the logarithm of a single quantity.

45. $\ln x + \ln 4$
46. $\ln y + \ln z$
47. $\log_4 z - \log_4 y$
48. $\log_5 8 - \log_5 t$
49. $2 \log_2(x + 3)$
50. $\frac{5}{2} \log_7(z - 4)$
51. $\frac{1}{3} \log_3 7x$
52. $-6 \log_6 2x$
53. $\ln x - 3 \ln(x + 1)$
54. $2 \ln 8 + 5 \ln z$

55. $\ln(x - 2) - \ln(x + 2)$

56. $3 \ln x + 2 \ln y - 4 \ln z$

57. $\ln x - 2[\ln(x + 2) + \ln(x - 2)]$

58. $4[\ln z + \ln(z + 5)] - 2 \ln(z - 5)$

59. $\frac{1}{3}[2 \ln(x + 3) + \ln x - \ln(x^2 - 1)]$

60. $2[\ln x - \ln(x + 1) - \ln(x - 1)]$

61. $\frac{1}{3}[\ln y + 2 \ln(y + 4)] - \ln(y - 1)$

62. $\frac{1}{2}[\ln(x + 1) + 2 \ln(x - 1)] + 3 \ln x$

Graphical Analysis In Exercises 63 and 64, (a) use a graphing utility to graph the two equations in the same viewing window and (b) use the *table* feature of the graphing utility to create a table of values for each equation. (c) What do the graphs and tables suggest? Verify your conclusion algebraically.

63. $y_1 = 2[\ln 8 - \ln(x^2 + 1)]$, $\quad y_2 = \ln\left[\dfrac{64}{(x^2 + 1)^2}\right]$

64. $y_1 = \ln x + \frac{1}{2} \ln(x + 1)$, $\quad y_2 = \ln(x\sqrt{x + 1})$

Think About It In Exercises 65 and 66, (a) use a graphing utility to graph the two equations in the same viewing window and (b) use the *table* feature of the graphing utility to create a table of values for each equation. (c) Are the expressions equivalent? Explain.

65. $y_1 = \ln x^2$, $\quad y_2 = 2 \ln x$

66. $y_1 = \frac{1}{4} \ln[x^4(x^2 + 1)]$, $\quad y_2 = \ln x + \frac{1}{4} \ln(x^2 + 1)$

In Exercises 67–80, find the exact value of the logarithm without using a calculator. If this is not possible, state the reason.

67. $\log_3 9$

68. $\log_6 \sqrt[3]{6}$

69. $\log_4 16^{3.4}$

70. $\log_5\left(\frac{1}{125}\right)$

71. $\log_2(-4)$

72. $\log_4(-16)$

73. $\log_5 75 - \log_5 3$

74. $\log_4 2 + \log_4 32$

75. $\ln e^3 - \ln e^7$

76. $\ln e^6 - 2 \ln e^5$

77. $2 \ln e^4$

78. $\ln e^{4.5}$

79. $\ln \dfrac{1}{\sqrt{e}}$

80. $\ln \sqrt[5]{e^3}$

81. *Sound Intensity* The relationship between the number of decibels β and the intensity of a sound I in watts per square meter is given by

$$\beta = 10 \log_{10}\left(\frac{I}{10^{-12}}\right).$$

(a) Use the properties of logarithms to write the formula in a simpler form.

(b) Use a graphing utility to complete the table.

I	10^{-4}	10^{-6}	10^{-8}	10^{-10}	10^{-12}	10^{-14}
β						

(c) Verify your answers in part (b) algebraically.

82. *Human Memory Model* Students participating in a psychology experiment attended several lectures and were given an exam. Every month for the next year, the students were retested to see how much of the material they remembered. The average scores for the group are given by the human memory model

$$f(t) = 90 - 15 \log_{10}(t + 1), \quad 0 \le t \le 12$$

where t is the time (in months).

(a) Use a graphing utility to graph the function over the specified domain.

(b) What was the average score on the original exam $(t = 0)$?

(c) What was the average score after 6 months?

(d) What was the average score after 12 months?

(e) When will the average score decrease to 75?

83. *Comparing Models* A cup of water at an initial temperature of 78°C is placed in a room at a constant temperature of 21°C. The temperature of the water is measured every 5 minutes during a half-hour period. The results are recorded as ordered pairs of the form (t, T), where t is the time (in minutes) and T is the temperature (in degrees Celsius).

$(0, 78.0°)$, $(5, 66.0°)$, $(10, 57.5°)$, $(15, 51.2°)$, $(20, 46.3°)$, $(25, 42.5°)$, $(30, 39.6°)$

(a) The graph of the model for the data should be asymptotic with the graph of the temperature of the room. Subtract the room temperature from each of the temperatures in the ordered pairs. Use a graphing utility to plot the data points (t, T) and $(t, T - 21)$.

(b) An exponential model for the data $(t, T - 21)$ is given by

$$T - 21 = 54.4(0.964)^t.$$

Solve for T and graph the model. Compare the result with the plot of the original data.

(c) Take the natural logarithms of the revised temperatures. Use a graphing utility to plot the points $(t, \ln(T - 21))$ and observe that the points appear linear. Use the *regression* feature of a graphing utility to fit a line to this data. The resulting line has the form

$$\ln(T - 21) = at + b.$$

Use the properties of logarithms to solve for T. Verify that the result is equivalent to the model in part (b).

(d) Fit a rational model to the data. Take the reciprocals of the y-coordinates of the revised data points to generate the points

$$\left(t, \frac{1}{T - 21} \right).$$

Use a graphing utility to plot these points and observe that they appear linear. Use the *regression* feature of a graphing utility to fit a line to this data. The resulting line has the form

$$\frac{1}{T - 21} = at + b.$$

Solve for T, and use a graphing utility to graph the rational function and the original data points.

84. Writing Write a short paragraph explaining why the transformations of the data in Exercise 83 were necessary to obtain the models. Why did taking the logarithms of the temperatures lead to a linear scatter plot? Why did taking the reciprocals of the temperatures lead to a linear scatter plot?

Synthesis

True or False? **In Exercises 85–91, determine whether the statement is true or false given that $f(x) = \ln x$. Justify your answer.**

85. $f(0) = 0$

86. $f(1) = 1$

87. $f(ax) = f(a) + f(x), \quad a > 0, x > 0$

88. $f(x - 2) = f(x) - f(2), \quad x > 2$

89. $\sqrt{f(x)} = \frac{1}{2}f(x)$

90. If $f(u) = 2f(v)$, then $v = u^2$.

91. If $f(x) < 0$, then $0 < x < 1$.

92. *Proof* Prove that $\log_b \dfrac{u}{v} = \log_b u - \log_b v$.

93. *Proof* Prove that $\log_b u^n = n \log_b u$.

94. *Proof* Prove that $\dfrac{\log_a x}{\log_{a/b} x} = 1 + \log_a \dfrac{1}{b}$.

In Exercises 95–100, use the change-of-base formula to rewrite the logarithm as a ratio of logarithms. Then use a graphing utility to graph the ratio.

95. $f(x) = \log_2 x$ **96.** $f(x) = \log_4 x$

97. $f(x) = \log_3 \sqrt{x}$ **98.** $f(x) = \log_2 \sqrt[3]{x}$

99. $f(x) = \log_5 \dfrac{x}{3}$ **100.** $f(x) = \log_3 \dfrac{x}{5}$

101. *Think About It* Use a graphing utility to graph

$$f(x) = \ln \frac{x}{2}, \quad g(x) = \frac{\ln x}{\ln 2}, \quad h(x) = \ln x - \ln 2$$

in the same viewing window. Which two functions have identical graphs? Explain why.

102. *Exploration* For how many integers between 1 and 20 can the natural logarithms be approximated given that $\ln 2 \approx 0.6931$, $\ln 3 \approx 1.0986$, and $\ln 5 \approx 1.6094$? Approximate these logarithms. (Do not use a calculator.)

Review

In Exercises 103–106, simplify the expression.

103. $\dfrac{24xy^{-2}}{16x^{-3}y}$ **104.** $\left(\dfrac{2x^2}{3y} \right)^{-3}$

105. $(18x^3y^4)^{-3}(18x^3y^4)^3$ **106.** $xy(x^{-1} + y^{-1})^{-1}$

In Exercises 107–112, find all solutions of the equation. Be sure to check all your solutions.

107. $x^2 - 6x + 2 = 0$

108. $2x^3 + 20x^2 + 50x = 0$

109. $x^4 - 19x^2 + 48 = 0$

110. $9x^4 - 37x^2 + 4 = 0$

111. $x^3 - 6x^2 - 4x + 24 = 0$

112. $9x^4 - 226x^2 + 25 = 0$

4.4 Solving Exponential and Logarithmic Equations

Introduction

So far in this chapter, you have studied the definitions, graphs, and properties of exponential and logarithmic functions. In this section, you will study procedures for *solving equations* involving exponential and logarithmic functions.

There are two basic strategies for solving exponential or logarithmic equations. The first is based on the One-to-One Properties and the second is based on the Inverse Properties. For $a > 0$ and $a \neq 1$, the following properties are true for all x and y for which $\log_a x$ and $\log_a y$ are defined.

One-to-One Properties

$a^x = a^y$ if and only if $x = y$.

$\log_a x = \log_a y$ if and only if $x = y$.

Inverse Properties

$a^{\log_a x} = x$

$\log_a a^x = x$

What you should learn

● Solve simple exponential and logarithmic equations.

● Solve more complicated exponential equations.

● Solve more complicated logarithmic equations.

● Use exponential and logarithmic equations to model and solve real-life problems.

Why you should learn it

Exponential and logarithmic equations can be used to model and solve real-life problems. For instance, Exercise 115 on page 359 shows how to use an exponential function to model the average heights of men and women.

Charles Gupton/Corbis

Example 1   Solving Simple Exponential and Logarithmic Equations

Original Equation	Rewritten Equation	Solution	Property
a. $2^x = 32$	$2^x = 2^5$	$x = 5$	One-to-One
b. $\ln x - \ln 3 = 0$	$\ln x = \ln 3$	$x = 3$	One-to-One
c. $\left(\frac{1}{3}\right)^x = 9$	$3^{-x} = 3^2$	$x = -2$	One-to-One
d. $e^x = 7$	$\ln e^x = \ln 7$	$x = \ln 7$	Inverse
e. $\ln x = -3$	$e^{\ln x} = e^{-3}$	$x = e^{-3}$	Inverse
f. $\log_{10} x = -1$	$10^{\log_{10} x} = 10^{-1}$	$x = 10^{-1} = \frac{1}{10}$	Inverse

✓ *Checkpoint*   Now try Exercise 21.

STUDY TIP

In Example 1(d), remember that $\ln x$ has a base of e. That is, $\ln e^x = \ln_e e^x$.

The strategies used in Example 1 are summarized as follows.

Strategies for Solving Exponential and Logarithmic Equations

1. Rewrite the original equation in a form that allows the use of the One-to-One Properties of exponential or logarithmic functions.

2. Rewrite an *exponential* equation in logarithmic form and apply the Inverse Property of logarithmic functions.

3. Rewrite a *logarithmic* equation in exponential form and apply the Inverse Property of exponential functions.

Solving Exponential Equations

Example 2   Solving Exponential Equations

Solve each equation.   **a.** $e^x = 72$ **b.** $3(2^x) = 42$

Algebraic Solution

a.

$e^x = 72$	Write original equation.
$\ln e^x = \ln 72$	Take natural log of each side.
$x = \ln 72 \approx 4.277$	Inverse Property

The solution is $x = \ln 72 \approx 4.277$. Check this in the original equation.

b.

$3(2^x) = 42$	Write original equation.
$2^x = 14$	Divide each side by 3.
$\log_2 2^x = \log_2 14$	Take log (base 2) of each side.
$x = \log_2 14$	Inverse Property
$x = \dfrac{\ln 14}{\ln 2} \approx 3.807$	Change-of-base formula

The solution is $x = \log_2 14 \approx 3.807$. Check this in the original equation.

✓ *Checkpoint* Now try Exercise 45.

Graphical Solution

a. Use a graphing utility to graph the left- and right-hand sides of the equation as $y_1 = e^x$ and $y_2 = 72$ in the same viewing window. Use the *intersect* feature or the *zoom* and *trace* features of the graphing utility to approximate the intersection point, as shown in Figure 4.32. So, the approximate solution is $x \approx 4.277$.

b. Use a graphing utility to graph $y_1 = 3(2^x)$ and $y_2 = 42$ in the same viewing window. Use the *intersect* feature or the *zoom* and *trace* features to approximate the intersection point, as shown in Figure 4.33. So, the approximate solution is $x \approx 3.807$.

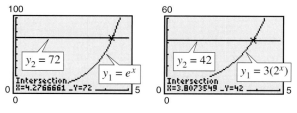

Figure 4.32   **Figure 4.33**

Example 3   Solving an Exponential Equation

Solve $4e^{2x} - 3 = 2$.

Algebraic Solution

$4e^{2x} - 3 = 2$	Write original equation.
$4e^{2x} = 5$	Add 3 to each side.
$e^{2x} = \dfrac{5}{4}$	Divide each side by 4.
$\ln e^{2x} = \ln \dfrac{5}{4}$	Take logarithm of each side.
$2x = \ln \dfrac{5}{4}$	Inverse Property
$x = \dfrac{1}{2} \ln \dfrac{5}{4} \approx 0.112$	Divide each side by 2.

The solution is $x = \frac{1}{2} \ln \frac{5}{4} \approx 0.112$. Check this in the original equation.

✓ *Checkpoint* Now try Exercise 49.

Graphical Solution

Rather than using the procedure in Example 2, another way to graphically solve the equation is to first rewrite the equation as $4e^{2x} - 5 = 0$, then use a graphing utility to graph $y = 4e^{2x} - 5$. Use the *zero* or *root* feature or the *zoom* and *trace* features of the graphing utility to approximate the value of x for which $y = 0$. From Figure 4.34, you can see that the zero occurs at $x \approx 0.112$. So, the solution is $x \approx 0.112$.

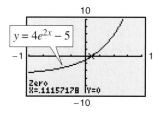

Figure 4.34

Example 4   Solving an Exponential Equation

Solve $2(3^{2t-5}) - 4 = 11$.

Solution

$2(3^{2t-5}) - 4 = 11$	Write original equation.
$2(3^{2t-5}) = 15$	Add 4 to each side.
$3^{2t-5} = \frac{15}{2}$	Divide each side by 2.
$\log_3 3^{2t-5} = \log_3 \frac{15}{2}$	Take log (base 3) of each side.
$2t - 5 = \log_3 \frac{15}{2}$	Inverse Property
$2t = 5 + \log_3 7.5$	Add 5 to each side.
$t = \frac{5}{2} + \frac{1}{2}\log_3 7.5$	Divide each side by 2.
$t \approx 3.417$	Use a calculator.

The solution is $t = \frac{5}{2} + \frac{1}{2}\log_3 7.5 \approx 3.417$. Check this in the original equation.

✓ *Checkpoint* Now try Exercise 53.

> **STUDY TIP**
>
> Remember that to evaluate a logarithm such as $\log_3 7.5$, you need to use the change-of-base formula.
>
> $$\log_3 7.5 = \frac{\ln 7.5}{\ln 3} \approx 1.834$$

When an equation involves two or more exponential expressions, you can still use a procedure similar to that demonstrated in the previous three examples. However, the algebra is a bit more complicated.

Example 5   Solving an Exponential Equation in Quadratic Form

Solve $e^{2x} - 3e^x + 2 = 0$.

Algebraic Solution

$e^{2x} - 3e^x + 2 = 0$	Write original equation.
$(e^x)^2 - 3e^x + 2 = 0$	Write in quadratic form.
$(e^x - 2)(e^x - 1) = 0$	Factor.
$e^x - 2 = 0$	Set 1st factor equal to 0.
$e^x = 2$	Add 2 to each side.
$x = \ln 2$	Solution
$e^x - 1 = 0$	Set 2nd factor equal to 0.
$e^x = 1$	Add 1 to each side.
$x = \ln 1$	Inverse Property
$x = 0$	Solution

The solutions are $x = \ln 2 \approx 0.693$ and $x = 0$. Check these in the original equation.

✓ *Checkpoint* Now try Exercise 55.

Graphical Solution

Use a graphing utility to graph $y = e^{2x} - 3e^x + 2$. Use the *zero* or *root* feature or the *zoom* and *trace* features of the graphing utility to approximate the values of x for which $y = 0$. In Figure 4.35, you can see that the zeros occur at $x = 0$ and at $x \approx 0.693$. So, the solutions are $x = 0$ and $x \approx 0.693$.

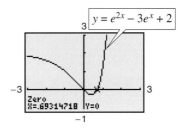

$y = e^{2x} - 3e^x + 2$

Zero
X=.69314718 Y=0

Figure 4.35

Solving Logarithmic Equations

To solve a logarithmic equation, you can write it in exponential form.

$\ln x = 3$ Logarithmic form

$e^{\ln x} = e^3$ Exponentiate each side.

$x = e^3$ Exponential form

This procedure is called *exponentiating* each side of an equation. It is applied after the logarithmic expression has been isolated.

Example 6   Solving Logarithmic Equations

Solve each logarithmic equation.

a. $\ln x = 2$ **b.** $\log_3(5x - 1) = \log_3(x + 7)$

Solution

a. $\ln x = 2$ Write original equation.

    $e^{\ln x} = e^2$ Exponentiate each side.

      $x = e^2 \approx 7.389$ Inverse Property

The solution is $x = e^2 \approx 7.389$. Check this in the original equation.

b. $\log_3(5x - 1) = \log_3(x + 7)$ Write original equation.

       $5x - 1 = x + 7$ One-to-One Property

          $4x = 8$ Add $-x$ and 1 to each side.

           $x = 2$ Divide each side by 4.

The solution is $x = 2$. Check this in the original equation.

✓ *Checkpoint* Now try Exercise 75.

Example 7   Solving a Logarithmic Equation

Solve $5 + 2 \ln x = 4$.

Algebraic Solution

$5 + 2 \ln x = 4$ Write original equation.

   $2 \ln x = -1$ Subtract 5 from each side.

     $\ln x = -\frac{1}{2}$ Divide each side by 2.

    $e^{\ln x} = e^{-1/2}$ Exponentiate each side.

       $x = e^{-1/2}$ Inverse Property

       $x \approx 0.607$ Use a calculator.

The solution is $x = e^{-1/2} \approx 0.607$. Check this in the original equation.

✓ *Checkpoint* Now try Exercise 77.

Graphical Solution

Use a graphing utility to graph $y_1 = 5 + 2 \ln x$ and $y_2 = 4$ in the same viewing window. Use the *intersect* feature or the *zoom* and *trace* features to approximate the intersection point, as shown in Figure 4.36. So, the solution is $x \approx 0.607$.

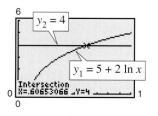

Figure 4.36

Example 8 Solving a Logarithmic Equation

Solve $2 \log_5 3x = 4$.

Solution

$2 \log_5 3x = 4$ Write original equation.

$\log_5 3x = 2$ Divide each side by 2.

$5^{\log_5 3x} = 5^2$ Exponentiate each side (base 5).

$3x = 25$ Inverse Property

$x = \frac{25}{3}$ Divide each side by 3.

The solution is $x = \frac{25}{3}$. Check this in the original equation. Or, perform a graphical check by graphing

$$y_1 = 2 \log_5 3x = 2\left(\frac{\log_{10} 3x}{\log_{10} 5}\right) \quad \text{and} \quad y_2 = 4$$

in the same viewing window. The two graphs should intersect at $x = \frac{25}{3} \approx 8.333$ and $y = 4$, as shown in Figure 4.37.

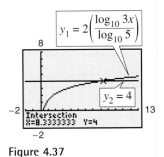

Figure 4.37

 Checkpoint Now try Exercise 81.

Because the domain of a logarithmic function generally does not include all real numbers, you should be sure to check for extraneous solutions of logarithmic equations, as shown in the next example.

Example 9 Checking for Extraneous Solutions

Solve $\ln(x - 2) + \ln(2x - 3) = 2 \ln x$.

Algebraic Solution

$\ln(x - 2) + \ln(2x - 3) = 2 \ln x$ Write original equation.

$\ln[(x - 2)(2x - 3)] = \ln x^2$ Use properties of logarithms.

$\ln(2x^2 - 7x + 6) = \ln x^2$ Multiply binomials.

$2x^2 - 7x + 6 = x^2$ One-to-One Property

$x^2 - 7x + 6 = 0$ Write in general form.

$(x - 6)(x - 1) = 0$ Factor.

$x - 6 = 0 \implies x = 6$ Set 1st factor equal to 0.

$x - 1 = 0 \implies x = 1$ Set 2nd factor equal to 0.

Finally, by checking these two "solutions" in the original equation, you can conclude that $x = 1$ is not valid. This is because when $x = 1$, $\ln(x - 2) + \ln(2x - 3) = \ln(-1) + \ln(-1)$, which is invalid because -1 is not in the domain of the natural logarithmic function. So, the only solution is $x = 6$.

 Checkpoint Now try Exercise 89.

Graphical Solution

First rewrite the original equation as $\ln(x - 2) + \ln(2x - 3) - 2 \ln x = 0$. Then use a graphing utility to graph $y = \ln(x - 2) + \ln(2x - 3) - 2 \ln x$. Use the *zero* or *root* feature or the *zoom* and *trace* features of the graphing utility to determine that $x = 6$ is an approximate solution, as shown in Figure 4.38. Verify that 6 is an exact solution algebraically.

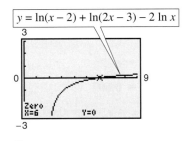

Figure 4.38

Example 10   The Change-of-Base Formula

Prove the change-of-base formula: $\log_a x = \dfrac{\log_b x}{\log_b a}$.

Solution

Begin by letting $y = \log_a x$ and writing the equivalent exponential form $a^y = x$. Now, taking the logarithms *with base b* of each side produces the following.

$$\log_b a^y = \log_b x$$

$$y \log_b a = \log_b x \qquad \text{Property of logarithms}$$

$$y = \frac{\log_b x}{\log_b a} \qquad \text{Divide each side by } \log_b a.$$

$$\log_a x = \frac{\log_b x}{\log_a a} \qquad \text{Replace } y \text{ with } \log_a x.$$

Equations that involve combinations of algebraic functions, exponential functions, and/or logarithmic functions can be very difficult to solve by algebraic procedures. Here again, you can take advantage of a graphing utility.

Example 11   Approximating the Solution of an Equation

Approximate (to three decimal places) the solution of $\ln x = x^2 - 2$.

Solution

To begin, write the equation so that all terms on one side are equal to 0.

$$\ln x - x^2 + 2 = 0$$

Then use a graphing utility to graph

$$y = -x^2 + 2 + \ln x$$

as shown in Figure 4.39. From this graph, you can see that the equation has two solutions. Next, using the *zero* or *root* feature or the *zoom* and *trace* features, you can approximate the two solutions to be $x \approx 0.138$ and $x \approx 1.564$.

Check

$$\ln x = x^2 - 2 \qquad \text{Write original equation.}$$

$$\ln(0.138) \overset{?}{\approx} (0.138)^2 - 2 \qquad \text{Substitute 0.138 for } x.$$

$$-1.9805 \approx -1.9810 \qquad \text{Solution checks. } \checkmark$$

$$\ln(1.564) \overset{?}{\approx} (1.564)^2 - 2 \qquad \text{Substitute 1.564 for } x.$$

$$0.4472 \approx 0.4461 \qquad \text{Solution checks. } \checkmark$$

So, the two solutions $x \approx 0.138$ and $x \approx 1.564$ seem reasonable.

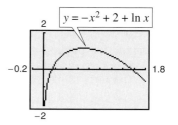

Figure 4.39

 Checkpoint Now try Exercise 97.

Applications

Example 12   Doubling an Investment

You have deposited $500 in an account that pays 6.75% interest, compounded continuously. How long will it take your money to double?

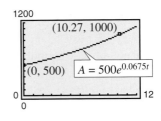

Figure 4.40

Solution

Using the formula for continuous compounding, you can find that the balance in the account is

$$A = Pe^{rt} = 500e^{0.0675t}.$$

To find the time required for the balance to double, let $A = 1000$, and solve the resulting equation for t.

$500e^{0.0675t} = 1000$	Substitute 1000 for A.
$e^{0.0675t} = 2$	Divide each side by 500.
$\ln e^{0.0675t} = \ln 2$	Take natural log of each side.
$0.0675t = \ln 2$	Inverse Property
$t = \dfrac{\ln 2}{0.0675} \approx 10.27$	Divide each side by 0.0675.

The balance in the account will double after approximately 10.27 years. This result is demonstrated graphically in Figure 4.40.

 Checkpoint Now try Exercise 109.

Example 13   Average Salary for Public School Teachers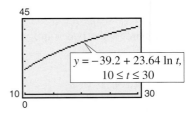

For selected years from 1980 to 2000, the average salary y (in thousands of dollars) for public school teachers for the year t can be modeled by the equation

$$y = -39.2 + 23.64 \ln t, \quad 10 \leq t \leq 30$$

where $t = 10$ represents 1980 (see Figure 4.41). During which year did the average salary for public school teachers reach $40.0 thousand?   (Source: National Education Association)

Solution

$-39.2 + 23.64 \ln t = y$	Write original equation.
$-39.2 + 23.64 \ln t = 40.0$	Substitute 40.0 for y.
$23.64 \ln t = 79.2$	Add 39.2 to each side.
$\ln t \approx 3.350$	Divide each side by 23.64.
$e^{\ln t} = e^{3.350}$	Exponentiate each side.
$t \approx 28.5$	Inverse Property

The solution is $t \approx 28.5$ years. Because $t = 10$ represents 1980, it follows that the average salary for public school teachers reached $40.0 thousand in 1998.

Figure 4.41

 Checkpoint Now try Exercise 118.

4.4 Exercises

Vocabulary Check

Fill in the blanks.

1. To _____ an equation in x means to find all values of x for which the equation is true.
2. To solve exponential and logarithmic equations, you can use the following one-to-one and inverse properties.
   (a) $a^x = a^y$ if and only if _____ .   (b) $\log_a x = \log_a y$ if and only if _____ .
   (c) $a^{\log_a x} =$ _____   (d) $\log_a a^x =$ _____
3. An _____ solution does not satisfy the original equation.

In Exercises 1–8, determine whether each x-value is a solution of the equation.

1. $4^{2x-7} = 64$
   (a) $x = 5$
   (b) $x = 2$

2. $2^{3x+1} = 32$
   (a) $x = -1$
   (b) $x = 2$

3. $3e^{x+2} = 75$
   (a) $x = -2 + e^{25}$
   (b) $x = -2 + \ln 25$
   (c) $x \approx 1.2189$

4. $4e^{x-1} = 60$
   (a) $x = 1 + \ln 15$
   (b) $x \approx 3.7081$
   (c) $x = \ln 16$

5. $\log_4(3x) = 3$
   (a) $x \approx 21.3560$
   (b) $x = -4$
   (c) $x = \frac{64}{3}$

6. $\log_6\left(\frac{5}{3}x\right) = 2$
   (a) $x \approx 20.2882$
   (b) $x = \frac{108}{5}$
   (c) $x = 7.2$

7. $\ln(x - 1) = 3.8$
   (a) $x = 1 + e^{3.8}$
   (b) $x \approx 45.7012$
   (c) $x = 1 + \ln 3.8$

8. $\ln(2 + x) = 2.5$
   (a) $x = e^{2.5} - 2$
   (b) $x \approx \frac{4073}{400}$
   (c) $x = \frac{1}{2}$

In Exercises 9–16, use a graphing utility to graph f and g in the same viewing window. Approximate the point of intersection of the graphs of f and g. Then solve the equation $f(x) = g(x)$ algebraically.

9. $f(x) = 2^x$
   $g(x) = 8$

10. $f(x) = 27^x$
    $g(x) = 9$

11. $f(x) = 5^{x-2} - 15$
    $g(x) = 10$

12. $f(x) = 2^{-x+1} - 3$
    $g(x) = 13$

13. $f(x) = 4 \log_3 x$
    $g(x) = 20$

14. $f(x) = 3 \log_5 x$
    $g(x) = 6$

15. $f(x) = \ln e^{x+1}$
    $g(x) = 2x + 5$

16. $f(x) = \ln e^{x-2}$
    $g(x) = 3x + 2$

In Exercises 17–36, solve for x.

17. $4^x = 16$
18. $3^x = 243$
19. $5^x = \frac{1}{625}$
20. $7^x = \frac{1}{49}$
21. $\left(\frac{1}{8}\right)^x = 64$
22. $\left(\frac{1}{2}\right)^x = 32$
23. $\left(\frac{2}{3}\right)^x = \frac{81}{16}$
24. $\left(\frac{3}{4}\right)^x = \frac{27}{64}$
25. $e^x = 4$
26. $e^x = 0$
27. $\ln x - \ln 5 = 0$
28. $\ln x - \ln 2 = 0$
29. $\ln x = -7$
30. $\ln x = -1$
31. $\log_x 625 = 4$
32. $\log_x 25 = 2$
33. $\log_{10} x = -1$
34. $\log_{10} x = -\frac{1}{2}$
35. $\ln(2x - 1) = 5$
36. $\ln(3x + 5) = 8$

In Exercises 37–42, simplify the expression.

37. $\ln e^{x^2}$
38. $\ln e^{2x-1}$
39. $e^{\ln(5x+2)}$
40. $-1 + \ln e^{2x}$
41. $e^{\ln x^2}$
42. $-8 + e^{\ln x^3}$

In Exercises 43–60, solve the exponential equation algebraically. Round your result to three decimal places. Use a graphing utility to verify your answer.

43. $8^{3x} = 360$
44. $6^{5x} = 3000$
45. $2e^{5x} = 18$
46. $4e^{2x} = 40$
47. $500e^{-x} = 300$
48. $1000e^{-4x} = 75$
49. $7 - 2e^x = 5$
50. $-14 + 3e^x = 11$
51. $5^{-t/2} = 0.20$
52. $4^{-3t} = 0.10$
53. $2^{3-x} = 565$
54. $8^{-2-x} = 431$
55. $e^{2x} - 4e^x - 5 = 0$
56. $e^{2x} - 5e^x + 6 = 0$
57. $\dfrac{400}{1 + e^{-x}} = 350$
58. $\dfrac{525}{1 + e^{-x}} = 275$

59. $\left(1 + \dfrac{0.10}{12}\right)^{12t} = 2$ **60.** $\left(16 + \dfrac{0.878}{26}\right)^{3t} = 30$

71. $g(t) = e^{0.09t} - 3$ **72.** $h(t) = e^{0.125t} - 8$

In Exercises 61–64, complete the table to find an interval containing the solution of the equation. Then use a graphing utility to graph both sides of the equation to estimate the solution. Round your result to three decimal places.

61. $e^{3x} = 12$

x	0.6	0.7	0.8	0.9	1.0
e^{3x}					

62. $e^{2x} = 50$

x	1.6	1.7	1.8	1.9	2.0
e^{2x}					

63. $20(100 - e^{x/2}) = 500$

x	5	6	7	8	9
$20(100 - e^{x/2})$					

64. $\dfrac{400}{1 + e^{-x}} = 350$

x	0	1	2	3	4
$\dfrac{400}{1 + e^{-x}}$					

In Exercises 65–68, use the *zero* or *root* feature or the *zoom* and *trace* features of a graphing utility to approximate the solution of the exponential equation accurate to three decimal places.

65. $\left(1 + \dfrac{0.065}{365}\right)^{365t} = 4$ **66.** $\left(4 - \dfrac{2.471}{40}\right)^{9t} = 21$

67. $\dfrac{3000}{2 + e^{2x}} = 2$ **68.** $\dfrac{119}{e^{6x} - 14} = 7$

In Exercises 69–72, use a graphing utility to graph the function and approximate its zero accurate to three decimal places.

69. $g(x) = 6e^{1-x} - 25$ **70.** $f(x) = 3e^{3x/2} - 962$

In Exercises 73–92, solve the logarithmic equation algebraically. Round the result to three decimal places. Verify your answer using a graphing utility.

73. $\ln x = -3$ **74.** $\ln x = -2$

75. $\ln 4x = 2.1$ **76.** $\ln 2x = 1.5$

77. $-2 + 2 \ln 3x = 17$ **78.** $3 + 2 \ln x = 10$

79. $\log_{10}(z - 3) = 2$ **80.** $\log_{10} x^2 = 6$

81. $7 \log_4(0.6x) = 12$ **82.** $4 \log_{10}(x - 6) = 11$

83. $\ln \sqrt{x + 2} = 1$ **84.** $\ln \sqrt{x - 8} = 5$

85. $\ln(x + 1)^2 = 2$ **86.** $\ln(x^2 + 1) = 8$

87. $\log_4 x - \log_4(x - 1) = \frac{1}{2}$

88. $\log_3 x + \log_3(x - 8) = 2$

89. $\ln(x + 5) = \ln(x - 1) - \ln(x + 1)$

90. $\ln(x + 1) - \ln(x - 2) = \ln x$

91. $\log_{10} 8x - \log_{10}\left(1 + \sqrt{x}\right) = 2$

92. $\log_{10} 4x - \log_{10}\left(12 + \sqrt{x}\right) = 2$

In Exercises 93–96, complete the table to find an interval containing the solution of the equation. Then use a graphing utility to graph both sides of the equation to estimate the solution. Round your result to three decimal places.

93. $\ln 2x = 2.4$

x	2	3	4	5	6
$\ln 2x$					

94. $3 \ln 5x = 10$

x	4	5	6	7	8
$3 \ln 5x$					

95. $6 \log_3(0.5x) = 11$

x	12	13	14	15	16
$6 \log_3(0.5x)$					

96. $5 \log_{10}(x - 2) = 11$

x	150	155	160	165	170
$5 \log_{10}(x - 2)$					

In Exercises 97–102, use the *zero* or *root* feature or the *zoom* and *trace* features of a graphing utility to approximate the solution of the logarithmic equation accurate to three decimal places.

97. $\log_{10} x = x^3 - 3$

98. $\log_{10} x^2 = 4$

99. $\log_3 x + \log_3(x - 3) = 1$

100. $\log_2 x + \log_2(x + 5) = 4$

101. $\ln(x - 3) + \ln(x + 3) = 1$

102. $\ln x + \ln(x^2 + 4) = 10$

In Exercises 103–108, use a graphing utility to approximate the point of intersection of the graphs. Round your result to three decimal places.

103. $y_1 = 7$
    $y_2 = 2^{x-1} - 5$

104. $y_1 = 4$
    $y_2 = 3^{x+1} - 2$

105. $y_1 = 80$
    $y_2 = 4e^{-0.2x}$

106. $y_1 = 500$
    $y_2 = 1500e^{-x/2}$

107. $y_1 = 3.25$
    $y_2 = \frac{1}{2}\ln(x + 2)$

108. $y_1 = 1.05$
    $y_2 = \ln\sqrt{x - 2}$

Compound Interest **In Exercises 109 and 110, find the time required for a $1000 investment to (a) double at interest rate *r*, compounded continuously, and (b) triple at interest rate *r*, compounded continuously.**

109. $r = 0.085$

110. $r = 0.12$

111. *Demand* The demand equation for a camera is given by

$$p = 500 - 0.5(e^{0.004x}).$$

Find the demand x for a price of (a) $p = \$350$ and (b) $p = \$300$.

112. *Demand* The demand equation for a hand-held electronic organizer is given by

$$p = 5000\left(1 - \frac{4}{4 + e^{-0.002x}}\right).$$

Find the demand x for a price of (a) $p = \$600$ and (b) $p = \$400$.

113. *Forestry* The number of trees per acre N of a certain species is approximated by the model

$$N = 68(10^{-0.04x}), \quad 5 \le x \le 40$$

where x is the average diameter of the trees (in inches) three feet above the ground. Use the model to approximate the average diameter of the trees in a test plot for which $N = 21$.

114. *Forestry* The yield V (in millions of cubic feet per acre) for a forest at age t years is given by

$$V = 6.7e^{-48.1/t}.$$

(a) Use a graphing utility to graph the function.

(b) Determine the horizontal asymptote of the function. Interpret its meaning in the context of the problem.

(c) Find the time necessary to obtain a yield of 1.3 million cubic feet.

115. *Average Heights* The percent m of American males between the ages of 18 and 24 who are no more than x inches tall is modeled by

$$m(x) = \frac{100}{1 + e^{-0.6114(x - 69.71)}}$$

and the percent f of American females between the ages of 18 and 24 who are no more than x inches tall is modeled by

$$f(x) = \frac{100}{1 + e^{-0.66607(x - 64.51)}}.$$

(Source: U.S. National Center for Health Statistics)

(a) Use a graphing utility to graph the two functions in the same viewing window.

(b) Use the graphs in part (a) to determine the horizontal asymptotes of the functions. Interpret their meaning in the context of the problem.

(c) What is the average height for each sex?

116. *Human Memory Model* In a group project in learning theory, a mathematical model for the proportion P of correct responses after n trials was found to be

$$P = \frac{0.83}{1 + e^{-0.2n}}.$$

(a) Use a graphing utility to graph the function.

(b) Use the graph in part (a) to determine any horizontal asymptotes of the function. Interpret the meaning of the upper asymptote in the context of the problem.

(c) After how many trials will 60% of the responses be correct?

117. *Data Analysis* An object at a temperature of 160°C was removed from a furnace and placed in a room at 20°C. The temperature T of the object was measured after each hour h and recorded in the table. A model for this data is given by $T = 20\left[1 + 7(2^{-h})\right]$.

Hour, h	Temperature, T
0	160°
1	90°
2	56°
3	38°
4	29°
5	24°

(a) Use a graphing utility to plot the data and graph the model in the same viewing window.

(b) Identify the horizontal asymptote of the graph of the model and interpret the asymptote in the context of the problem.

(c) Approximate the time when the temperature of the object is 100°C.

118. *Finance* The table shows the number N (in thousands) of banks in the United States from 1995 to 2001. The data can be modeled by the logarithmic function $N = 17.02 - 3.096 \ln t$, where t represents the year, with $t = 5$ corresponding to 1995. (Source: Federal Deposit Insurance Corp.)

Year	Number, N
1995	11.97
1996	11.67
1997	10.92
1998	10.46
1999	10.22
2000	9.91
2001	9.63

(a) Use the model to determine during which year the number of banks reached 10,000.

(b) Use a graphing utility to graph the model.

(c) Use the graph from part (b) to verify your answer in part (a).

Synthesis

True or False? In Exercises 119 and 120, determine whether the statement is true or false. Justify your answer.

119. You can approximate the solution of the equation $\frac{2}{3}e^x = 42$ by graphing $y = \frac{2}{3}e^x - 42$ and finding its x-intercept.

120. A logarithmic equation can have at most one extraneous solution.

121. *Writing* Write two or three sentences stating the general guidelines that you follow when (a) solving exponential equations and (b) solving logarithmic equations.

122. *Graphical Analysis* Let $f(x) = \log_a x$ and $g(x) = a^x$, where $a > 1$.

(a) Let $a = 1.2$ and use a graphing utility to graph the two functions in the same viewing window. What do you observe? Approximate any points of intersection of the two graphs.

(b) Determine the value(s) of a for which the two graphs have one point of intersection.

(c) Determine the value(s) of a for which the two graphs have two points of intersection.

123. *Think About It* Is the time required for an investment to quadruple twice as long as the time required for it to double? Give a reason for your answer and verify your answer algebraically.

124. *Writing* Write a paragraph explaining whether or not the time required for an investment to double is dependent on the size of the investment.

Review

In Exercises 125–130, sketch the graph of the function.

125. $f(x) = 3x^3 - 4$

126. $f(x) = -(x + 1)^3 + 2$

127. $f(x) = |x| + 9$

128. $f(x) = |x + 2| - 8$

129. $f(x) = \begin{cases} 2x, & x < 0 \\ -x^2 + 4, & x \geq 0 \end{cases}$

130. $f(x) = \begin{cases} x - 9, & x \leq -1 \\ x^2 + 1, & x > -1 \end{cases}$

4.5 Exponential and Logarithmic Models

Introduction

The five most common types of mathematical models involving exponential functions and logarithmic functions are as follows.

1. **Exponential growth model:** $y = ae^{bx}, \qquad b > 0$
2. **Exponential decay model:** $y = ae^{-bx}, \qquad b > 0$
3. **Gaussian model:** $y = ae^{-(x-b)^2/c}$
4. **Logistic growth model:** $y = \dfrac{a}{1 + be^{-rx}}$
5. **Logarithmic models:** $y = a + b \ln x, \qquad y = a + b \log_{10} x$

The basic shapes of these graphs are shown in Figure 4.42.

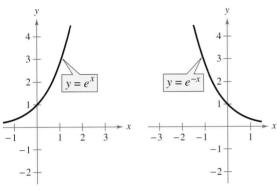

Exponential Growth Model

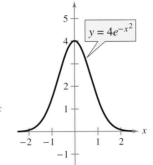

Exponential Decay Model

Gaussian Model

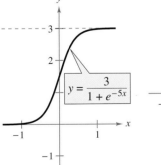

Logistic Growth Model

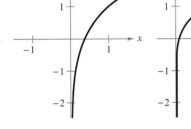

Natural Logarithmic Model   Common Logarithmic Model

Figure 4.42

What you should learn

- Recognize the five most common types of models involving exponential or logarithmic functions.
- Use exponential growth and decay functions to model and solve real-life problems.
- Use Gaussian functions to model and solve real-life problems.
- Use logistic growth functions to model and solve real-life problems.
- Use logarithmic functions to model and solve real-life problems

Why you should learn it

Exponential and logarithmic functions can be used to model and solve a variety of business applications. In Exercise 34 on page 370, you will compare an exponential decay model and a linear model for the depreciation of a computer over 3 years.

Spencer Grant/PhotoEdit

You can often gain quite a bit of insight into a situation modeled by an exponential or logarithmic function by identifying and interpreting the function's asymptotes. Use the graphs in Figure 4.42 to identify the asymptotes of each function.

Exponential Growth and Decay

Example 1   Population Growth

Estimates of the world population (in millions) from 1995 through 2004 are shown in the table. A scatter plot of the data is shown in Figure 4.43.   (Source: U.S. Bureau of the Census)

Year	Population, P
1995	5685
1996	5764
1997	5844
1998	5923
1999	6002

Year	Population, P
2000	6079
2001	6154
2002	6228
2003	6302
2004	6376

An exponential growth model that approximates this data is given by

$$P = 5344e^{0.012744t}, \qquad 5 \le t \le 14$$

where P is the population (in millions) and $t = 5$ represents 1995. Compare the values given by the model with the estimates shown in the table. According to this model, when will the world population reach 6.8 billion?

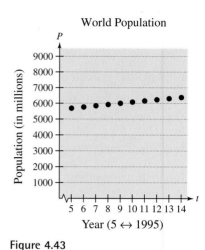

Figure 4.43

Algebraic Solution

The following table compares the two sets of population figures.

Year	1995	1996	1997	1998	1999	2000	2001	2002	2003	2004
Population	5685	5764	5844	5923	6002	6079	6154	6228	6302	6376
Model	5696	5769	5843	5918	5993	6070	6148	6227	6307	6388

To find when the world population will reach 6.8 billion, let $P = 6800$ in the model and solve for t.

$5344e^{0.012744t} = P$	Write original model.
$5344e^{0.012744t} = 6800$	Substitute 6800 for P.
$e^{0.012744t} \approx 1.27246$	Divide each side by 5344.
$\ln e^{0.012744t} \approx \ln 1.27246$	Take natural log of each side.
$0.012744t \approx 0.24095$	Inverse Property
$t \approx 18.9$	Divide each side by 0.012744.

According to the model, the world population will reach 6.8 billion in 2008.

✓ *Checkpoint* Now try Exercise 27.

Graphical Solution

Use a graphing utility to graph the model $y = 5344e^{0.012744x}$ and the data in the same viewing window. You can see in Figure 4.44 that the model appears to closely fit the data.

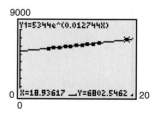

Figure 4.44

Use the *zoom* and *trace* features of the graphing utility to find that the approximate value of x for $y = 6800$ is $x \approx 18.9$. So, according to the model, the world population will reach 6.8 billion in 2008.

An exponential model increases (or decreases) by the same percent each year. What is the annual percent increase for the model in Example 1?

In Example 1, you were given the exponential growth model. Sometimes you must find such a model. One technique for doing this is shown in Example 2.

Example 2   Modeling Population Growth

In a research experiment, a population of fruit flies is increasing according to the law of exponential growth. After 2 days there are 100 flies, and after 4 days there are 300 flies. How many flies will there be after 5 days?

Solution

Let y be the number of flies at time t (in days). From the given information, you know that $y = 100$ when $t = 2$ and $y = 300$ when $t = 4$. Substituting this information into the model $y = ae^{bt}$ produces

$$100 = ae^{2b} \quad \text{and} \quad 300 = ae^{4b}.$$

To solve for b, solve for a in the first equation.

$$100 = ae^{2b} \quad \Longrightarrow \quad a = \frac{100}{e^{2b}} \qquad \text{Solve for } a \text{ in the first equation.}$$

Then substitute the result into the second equation.

$$300 = ae^{4b} \qquad \qquad \text{Write second equation.}$$

$$300 = \left(\frac{100}{e^{2b}}\right)e^{4b} \qquad \text{Substitute } \tfrac{100}{e^{2b}} \text{ for } a.$$

$$\frac{300}{100} = e^{2b} \qquad \qquad \text{Divide each side by 100.}$$

$$\ln \frac{300}{100} = \ln e^{2b} \qquad \qquad \text{Take natural log of each side.}$$

$$\ln 3 = 2b \qquad \qquad \text{Inverse Property}$$

$$\frac{1}{2} \ln 3 = b \qquad \qquad \text{Solve for } b.$$

Using $b = \tfrac{1}{2} \ln 3$ and the equation you found for a, you can determine that

$$a = \frac{100}{e^{2[(1/2)\ln 3]}} \qquad \text{Substitute } \tfrac{1}{2} \ln 3 \text{ for } b.$$

$$= \frac{100}{e^{\ln 3}} \qquad \qquad \text{Simplify.}$$

$$= \frac{100}{3} \approx 33.33. \qquad \text{Inverse Property}$$

So, with $a \approx 33.33$ and $b = \tfrac{1}{2} \ln 3 \approx 0.5493$, the exponential growth model is

$$y = 33.33e^{0.5493t},$$

as shown in Figure 4.45. This implies that after 5 days, the population will be

$$y = 33.33e^{0.5493(5)} \approx 520 \text{ flies.}$$

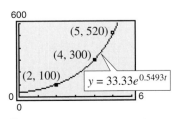

Figure 4.45

 Checkpoint Now try Exercise 29.

In living organic material, the ratio of the content of radioactive carbon isotopes (carbon 14) to the content of nonradioactive carbon isotopes (carbon 12) is about 1 to 10^{12}. When organic material dies, its carbon 12 content remains fixed, whereas its radioactive carbon 14 begins to decay with a half-life of 5730 years. To estimate the age of dead organic material, scientists use the following formula, which denotes the ratio of carbon 14 to carbon 12 present at any time t (in years).

$$R = \frac{1}{10^{12}} e^{-t/8267}$$ Carbon dating model

The graph of R is shown in Figure 4.46. Note that R decreases as t increases.

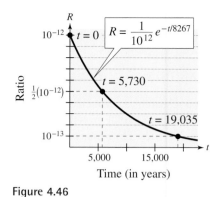

Figure 4.46

Example 3   Carbon Dating

The ratio of carbon 14 to carbon 12 in a newly discovered fossil is

$$R = \frac{1}{10^{13}}.$$

Estimate the age of the fossil.

Algebraic Solution

In the carbon dating model, substitute the given value of R to obtain the following.

$\dfrac{1}{10^{12}} e^{-t/8267} = R$	Write original model.
$\dfrac{e^{-t/8267}}{10^{12}} = \dfrac{1}{10^{13}}$	Substitute $\frac{1}{10^{13}}$ for R.
$e^{-t/8267} = \dfrac{1}{10}$	Multiply each side by 10^{12}.
$\ln e^{-t/8267} = \ln \dfrac{1}{10}$	Take natural log of each side.
$-\dfrac{t}{8267} \approx -2.3026$	Inverse Property
$t \approx 19{,}036$	Multiply each side by -8267.

So, to the nearest thousand years, you can estimate the age of the fossil to be 19,000 years.

 Checkpoint Now try Exercise 32.

Graphical Solution

Use a graphing utility to graph the formula for the ratio of carbon 14 to carbon 12 at any time t as

$$y_1 = \frac{1}{10^{12}} e^{-x/8267}.$$

In the same viewing window, graph $y_2 = 1/(10^{13})$. Use the *intersect* feature or the *zoom* and *trace* features of the graphing utility to estimate that $x \approx 19{,}035$ when $y = 1/(10^{13})$, as shown in Figure 4.47.

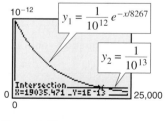

Figure 4.47

So, to the nearest thousand years, you can estimate the age of the fossil to be 19,000 years.

The carbon dating model in Example 3 assumed that the carbon 14 to carbon 12 ratio was one part in 10,000,000,000,000. Suppose an error in measurement occurred and the actual ratio was only one part in 8,000,000,000,000. The fossil age corresponding to the actual ratio would then be approximately 17,000 years. Try checking this result.

Gaussian Models

As mentioned at the beginning of this section, Gaussian models are of the form

$$y = ae^{-(x-b)^2/c}.$$

This type of model is commonly used in probability and statistics to represent populations that are **normally distributed.** For *standard* normal distributions, the model takes the form

$$y = \frac{1}{\sqrt{2\pi}}e^{-x^2/2}.$$

The graph of a Gaussian model is called a **bell-shaped curve.** Try graphing the normal distribution curve with a graphing utility. Can you see why it is called a bell-shaped curve?

The average value for a population can be found from the bell-shaped curve by observing where the maximum y-value of the function occurs. The x-value corresponding to the maximum y-value of the function represents the average value of the independent variable—in this case, x.

Example 4   SAT Scores

In 2002, the Scholastic Aptitude Test (SAT) mathematics scores for college-bound seniors roughly followed the normal distribution

$$y = 0.0035e^{-(x-516)^2/25,992}, \qquad 200 \le x \le 800$$

where x is the SAT score for mathematics. Use a graphing utility to graph this function and estimate the average SAT score.   (Source: College Board)

Solution

The graph of the function is shown in Figure 4.48. On this bell-shaped curve, the maximum value of the curve represents the average score. Using the *maximum* feature or the *zoom* and *trace* features of the graphing utility, you can see that the average mathematics score for college-bound seniors in 2002 was 516.

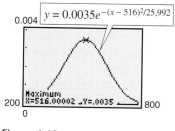

Figure 4.48

TECHNOLOGY SUPPORT

For instructions on how to use the *maximum* feature, see Appendix A; for specific keystrokes, go to the text website at *college.hmco.com.*

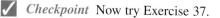 *Checkpoint* Now try Exercise 37.

In Example 4, note that 50% of the seniors who took the test received a score lower than 516.

Logistic Growth Models

Some populations initially have rapid growth, followed by a declining rate of growth, as indicated by the graph in Figure 4.49. One model for describing this type of growth pattern is the **logistic curve** given by the function

$$y = \frac{a}{1 + be^{-rx}}$$

where y is the population size and x is the time. An example is a bacteria culture that is initially allowed to grow under ideal conditions, and then under less favorable conditions that inhibit growth. A logistic growth curve is also called a **sigmoidal curve**.

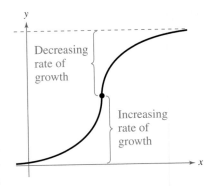

Figure 4.49   Logistic Curve

Example 5   Spread of a Virus

On a college campus of 5000 students, one student returns from vacation with a contagious flu virus. The spread of the virus is modeled by

$$y = \frac{5000}{1 + 4999e^{-0.8t}}, \qquad t \geq 0$$

where y is the total number infected after t days. The college will cancel classes when 40% or more of the students are infected. (a) How many students are infected after 5 days? (b) After how many days will the college cancel classes?

Algebraic Solution

a. After 5 days, the number of students infected is

$$y = \frac{5000}{1 + 4999e^{-0.8(5)}} = \frac{5000}{1 + 4999e^{-4}} \approx 54.$$

b. Classes are cancelled when the number of infected students is $(0.40)(5000) = 2000$.

$$2000 = \frac{5000}{1 + 4999e^{-0.8t}}$$

$$1 + 4999e^{-0.8t} = 2.5$$

$$e^{-0.8t} \approx \frac{1.5}{4999}$$

$$\ln e^{-0.8t} \approx \ln \frac{1.5}{4999}$$

$$-0.8t \approx \ln \frac{1.5}{4999}$$

$$t = -\frac{1}{0.8} \ln \frac{1.5}{4999} \approx 10.14$$

So, after about 10 days, at least 40% of the students will be infected, and classes will be canceled.

✓ *Checkpoint* Now try Exercise 39.

Graphical Solution

a. Use a graphing utility to graph $y = \dfrac{5000}{1 + 4999e^{-0.8x}}$.

Use the *value* feature or the *zoom* and *trace* features of the graphing utility to estimate that $y = 54$ when $x = 5$. So, after 5 days, about 54 students will be infected.

b. Classes are cancelled when the number of infected students is $(0.40)(5000) = 2000$. Use a graphing utility to graph

$$y_1 = \frac{5000}{1 + 4999e^{-0.8x}} \qquad \text{and} \qquad y_2 = 2000$$

in the same viewing window. Use the *intersect* feature or the *zoom* and *trace* features of the graphing utility to find the point of intersection of the graphs. In Figure 4.50, you can see that the point of intersection occurs near $x \approx 10.14$. So, after about 10 days, at least 40% of the students will be infected, and classes will be canceled.

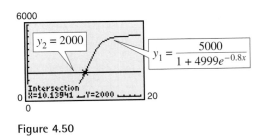

Figure 4.50

Logarithmic Models

On the Richter scale, the magnitude R of an earthquake of intensity I is given by

$$R = \log_{10} \frac{I}{I_0}$$

where $I_0 = 1$ is the minimum intensity used for comparison. Intensity is a measure of the wave energy of an earthquake.

On January 22, 2003, an earthquake of magnitude 7.6 in Colima, Mexico killed at least 29 people and left 10,000 people homeless.

Example 6   Magnitudes of Earthquakes

In 2001, the coast of Peru experienced an earthquake that measured 8.4 on the Richter scale. In 2003, Colima, Mexico experienced an earthquake that measured 7.6 on the Richter scale. Find the intensity of each earthquake and compare the two intensities.

Solution

Because $I_0 = 1$ and $R = 8.4$, you have

$$8.4 = \log_{10} \frac{I}{1} \qquad \text{Substitute 1 for } I_0 \text{ and 8.4 for } R.$$

$$10^{8.4} = 10^{\log_{10} I} \qquad \text{Exponentiate each side.}$$

$$10^{8.4} = I \qquad \text{Inverse Property}$$

$$251,189,000 \approx I. \qquad \text{Use a calculator.}$$

For $R = 7.6$, you have

$$7.6 = \log_{10} \frac{I}{1} \qquad \text{Substitute 1 for } I_0 \text{ and 7.6 for } R.$$

$$10^{7.6} = 10^{\log_{10} I} \qquad \text{Exponentiate each side.}$$

$$10^{7.6} = I \qquad \text{Inverse Property}$$

$$39,811,000 \approx I. \qquad \text{Use a calculator.}$$

Note that an increase of 0.8 unit on the Richter scale (from 7.6 to 8.4) represents an increase in intensity by a factor of

$$\frac{251,189,000}{39,811,000} \approx 6.$$

In other words, the 2001 earthquake had an intensity about 6 times greater than that of the 2003 earthquake.

 Checkpoint Now try Exercise 41.

4.5 Exercises

Vocabulary Check

Fill in the blanks.

1. An exponential growth model has the form _____ .

2. A logarithmic model has the form _____ or _____ .

3. A _____ model has the form $y = \dfrac{a}{1 + be^{-rx}}$.

4. The graph of a Gaussian model is called a _____ .

5. A logistic curve is also called a _____ curve.

In Exercises 1–6, match the function with its graph. [The graphs are labeled (a), (b), (c), (d), (e), and (f).]

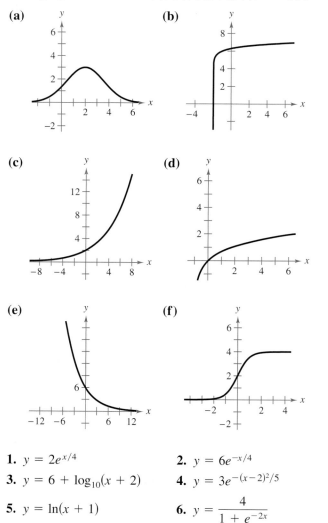

(a)

(b)

(c)

(d)

(e)

(f)

1. $y = 2e^{x/4}$

2. $y = 6e^{-x/4}$

3. $y = 6 + \log_{10}(x + 2)$

4. $y = 3e^{-(x-2)^2/5}$

5. $y = \ln(x + 1)$

6. $y = \dfrac{4}{1 + e^{-2x}}$

Compound Interest **In Exercises 7–14, complete the table for a savings account in which interest is compounded continuously.**

	Initial Investment	Annual % Rate	Time to Double	Amount After 10 Years
7.	$1000	3.5%		
8.	$20,000	$10\frac{1}{2}\%$		
9.	$750		$7\frac{3}{4}$ yr	
10.	$10,000		12 yr	
11.	$500			$1292.85
12.	$600			$1505.00
13.		4.5%		$10,000.00
14.		2%		$2000.00

15. *Compound Interest* Complete the table for the time t necessary for P dollars to triple if interest is compounded continuously at rate r. Create a scatter plot of the data.

r	2%	4%	6%	8%	10%	12%
t						

16. *Compound Interest* Complete the table for the time t necessary for P dollars to triple if interest is compounded annually at rate r. Create a scatter plot of the data.

r	2%	4%	6%	8%	10%	12%
t						

17. ***Comparing Investments***   If $1 is invested in an account over a 10-year period, the amount in the account, where t represents the time in years, is given by

$$A = 1 + 0.075[\![t]\!] \quad \text{or} \quad A = e^{0.07t}$$

depending on whether the account pays simple interest at $7\frac{1}{2}\%$ or continuous compound interest at 7%. Use a graphing utility to graph each function in the same viewing window. Which grows at a faster rate? (Remember that $[\![t]\!]$ is the greatest integer function discussed in Section 1.4.)

18. ***Comparing Investments***   If $1 is invested in an account over a 10-year period, the amount in the account, where t represents the time in years, is given by

$$A = 1 + 0.06[\![t]\!] \quad \text{or} \quad A = \left(1 + \frac{0.055}{365}\right)^{[\![365t]\!]}$$

depending on whether the account pays simple interest at 6% or compound interest at $5\frac{1}{2}\%$ compounded daily. Use a graphing utility to graph each function in the same viewing window. Which grows at a faster rate?

Radioactive Decay   **In Exercises 19–22, complete the table for the radioactive isotope.**

Isotope	Half-Life (years)	Initial Quantity	Amount After 1000 Years
19. ^{226}Ra	1600	10 g	
20. ^{226}Ra	1600		1.5 g
21. ^{14}C	5730	3 g	
22. ^{239}Pu	24,110		0.4 g

In Exercises 23–26, find the exponential model $y = ae^{bx}$ that fits the points given in the graph or table.

23.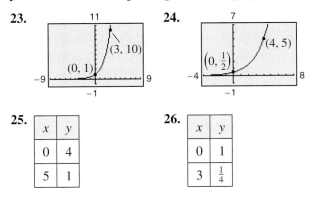

24.

25.

x	y
0	4
5	1

26.

x	y
0	1
3	$\frac{1}{4}$

27. ***Population***   The table shows the populations (in millions) of five countries in 2000 and the projected populations (in millions) for the year 2010. (Source: U.S. Census Bureau)

Country	2000	2010
Australia	19.2	20.9
Canada	31.3	34.3
Philippines	81.2	97.9
South Africa	43.4	41.1
Turkey	65.7	73.3

(a) Find the exponential growth or decay model, $y = ae^{bt}$ or $y = ae^{-bt}$, for the population of each country by letting $t = 0$ correspond to 2000. Use the model to predict the population of each country in 2030.

(b) You can see that the populations of Australia and Turkey are growing at different rates. What constant in the equation $y = ae^{bt}$ is determined by these different growth rates? Discuss the relationship between the different growth rates and the magnitude of the constant.

(c) You can see that the population of Canada is increasing while the population of South Africa is decreasing. What constant in the equation $y = ae^{bt}$ reflects this difference? Explain.

28. ***Population***   The population P (in thousands) of Bellevue, Washington is given by

$$P = 110e^{kt}$$

where $t = 0$ represents the year 2000. In 1980, the population was 74,000. Find the value of k and use this result to predict the population in the year 2020. (Source: U.S. Census Bureau)

29. ***Bacteria Growth***   The number N of bacteria in a culture is given by the model

$$N = 100e^{kt}$$

where t is the time (in hours). If $N = 300$ when $t = 5$, estimate the time required for the population to double in size. Verify your estimate graphically.

30. ***Bacteria Growth***   The number N of bacteria in a culture is given by the model $N = 250e^{kt}$, where t is the time (in hours). If $N = 280$ when $t = 10$, estimate the time required for the population to double in size. Verify your estimate graphically.

31. *Radioactive Decay* The half-life of radioactive radium (^{226}Ra) is 1620 years. What percent of a present amount of radioactive radium will remain after 100 years?

32. *Carbon Dating* Carbon 14 (^{14}C) dating assumes that the carbon dioxide on Earth today has the same radioactive content as it did centuries ago. If this is true, the amount of ^{14}C absorbed by a tree that grew several centuries ago should be the same as the amount of ^{14}C absorbed by a tree growing today. A piece of ancient charcoal contains only 15% as much radioactive carbon as a piece of modern charcoal. How long ago was the tree burned to make the ancient charcoal if the half-life of ^{14}C is 5730 years?

33. *Depreciation* A sport utility vehicle (SUV) that cost $32,000 new has a book value of $18,000 after 2 years.

  (a) Find the linear model $V = mt + b$.

  (b) Find the exponential model $V = ae^{kt}$.

  (c) Use a graphing utility to graph the two models in the same viewing window. Which model depreciates faster in the first year?

  (d) Use each model to find the book values of the SUV after 1 year and after 3 years.

  (e) Interpret the slope of the linear model.

34. *Depreciation* A computer that cost $2000 new has a book value of $500 after 2 years.

  (a) Find the linear model $V = mt + b$.

  (b) Find the exponential model $V = ae^{kt}$.

  (c) Use a graphing utility to graph the two models in the same viewing window. Which model depreciates faster in the first year?

  (d) Use each model to find the book values of the computer after 1 year and after 3 years.

  (e) Interpret the slope of the linear model.

35. *Sales* The sales S (in thousands of units) of a new CD burner after it has been on the market t years are given by $S = 100(1 - e^{kt})$. Fifteen thousand units of the new product were sold the first year.

  (a) Complete the model by solving for k.

  (b) Use a graphing utility to graph the model.

  (c) Use the graph in part (b) to estimate the number of units sold after 5 years.

36. *Sales* The sales S (in thousands of units) of a cleaning solution after x hundred dollars is spent on advertising are given by

$$S = 10(1 - e^{kx}).$$

When $500 is spent on advertising, 2500 units are sold.

  (a) Complete the model by solving for k.

  (b) Estimate the number of units that will be sold if advertising expenditures are raised to $700.

37. *IQ Scores* The IQ scores for adults roughly follow the normal distribution

$$y = 0.0266e^{-(x-100)^2/450}, \quad 70 \le x \le 115$$

where x is the IQ score.

  (a) Use a graphing utility to graph the function.

  (b) From the graph in part (a), estimate the average IQ score.

38. *Education* The time (in hours per week) a student uses a math lab roughly follows the normal distribution

$$y = 0.7979e^{-(x-5.4)^2/0.5}, \quad 4 \le x \le 7$$

where x is the time spent in the lab.

  (a) Use a graphing utility to graph the function.

  (b) From the graph in part (a), estimate the average time a student spends per week in the math lab.

39. *Wildlife* A conservation organization releases 100 animals of an endangered species into a game preserve. The organization believes that the preserve has a carrying capacity of 1000 animals and that the growth of the herd will follow the logistic curve

$$p(t) = \frac{1000}{1 + 9e^{-0.1656t}}$$

where t is measured in months.

  (a) Use a graphing utility to graph the function. Use the graph to determine the values of p at which the horizontal asymptotes occur. Interpret the meaning of the larger asymptote in the context of the problem.

  (b) Estimate the population after 5 months.

  (c) When will the population reach 500?

40. *Yeast Growth*   The amount Y of yeast in a culture is given by the model

$$Y = \frac{663}{1 + 72e^{-0.547t}}, \quad 0 \le t \le 18$$

where t represents the time (in hours).

(a) Use a graphing utility to graph the model.

(b) Use the model to predict the population for the 19th hour and the 30th hour.

(c) According to this model, what is the limiting value of the population?

(d) Why do you think the population of yeast follows a logistic growth model instead of an exponential growth model?

Geology   **In Exercises 41 and 42, use the Richter scale (see page 367) for measuring the magnitudes of earthquakes.**

41. Find the intensities I of the following earthquakes measuring R on the Richter scale (let $I_0 = 1$).

(a) Figi Islands in 2003, $R = 6.5$

(b) Central Alaska in 2002, $R = 7.9$

(c) Northern California in 2000, $R = 5.2$

42. Find the magnitudes R of the following earthquakes of intensity I (let $I_0 = 1$).

(a) $I = 39,811,000$ (b) $I = 12,589,000$

(c) $I = 251,200$

Sound Intensity   **In Exercises 43–46, use the following information for determining sound intensity. The level of sound β (in decibels) with an intensity I is $\beta = 10 \log_{10}(I/I_0)$, where I_0 is an intensity of 10^{-12} watt per square meter, corresponding roughly to the faintest sound that can be heard by the human ear. In Exercises 43 and 44, find the level of sound β.**

43. (a) $I = 10^{-10}$ watt per m^2 (quiet room)

(b) $I = 10^{-5}$ watt per m^2 (busy street corner)

(c) $I \approx 10^0$ watt per m^2 (threshold of pain)

44. (a) $I = 10^{-4}$ watt per m^2 (door slamming)

(b) $I = 10^{-3}$ watt per m^2 (loud car horn)

(c) $I = 10^{-2}$ watt per m^2 (siren at 30 meters)

45. As a result of the installation of a muffler, the noise level of an engine was reduced from 88 to 72 decibels. Find the percent decrease in the intensity level of the noise due to the installation of the muffler.

46. As a result of the installation of noise suppression materials, the noise level in an auditorium was reduced from 93 to 80 decibels. Find the percent decrease in the intensity level of the noise due to the installation of these materials.

pH Levels   **In Exercises 47–50, use the acidity model given by pH $= -\log_{10}[\text{H}^+]$, where acidity (pH) is a measure of the hydrogen ion concentration $[\text{H}^+]$ (measured in moles of hydrogen per liter) of a solution.**

47. Find the pH if $[\text{H}^+] = 2.3 \times 10^{-5}$.

48. Compute $[\text{H}^+]$ for a solution for which pH $= 5.8$.

49. A grape has a pH of 3.5, and milk of magnesia has a pH of 10.5. The hydrogen ion concentration of the grape is how many times that of the milk of magnesia?

50. The pH of a solution is decreased by one unit. The hydrogen ion concentration is increased by what factor?

51. *Home Mortgage*   A \$120,000 home mortgage for 30 years at $7\frac{1}{2}\%$ has a monthly payment of \$839.06. Part of the monthly payment goes toward the interest charge on the unpaid balance, and the remainder of the payment is used to reduce the principal. The amount that goes toward the interest is given by

$$u = M - \left(M - \frac{Pr}{12}\right)\left(1 + \frac{r}{12}\right)^{12t}$$

and the amount that goes toward reduction of the principal is given by

$$v = \left(M - \frac{Pr}{12}\right)\left(1 + \frac{r}{12}\right)^{12t}.$$

In these formulas, P is the size of the mortgage, r is the interest rate, M is the monthly payment, and t is the time (in years).

(a) Use a graphing utility to graph each function in the same viewing window. (The viewing window should show all 30 years of mortgage payments.)

(b) In the early years of the mortgage, the larger part of the monthly payment goes for what purpose? Approximate the time when the monthly payment is evenly divided between interest and principal reduction.

(c) Repeat parts (a) and (b) for a repayment period of 20 years ($M = \$966.71$). What can you conclude?

52. Home Mortgage The total interest u paid on a home mortgage of P dollars at interest rate r for t years is given by

$$u = P\left[\frac{rt}{1 - \left(\dfrac{1}{1 + r/12}\right)^{12t}} - 1\right].$$

Consider a \$120,000 home mortgage at $7\frac{1}{2}\%$.

(a) Use a graphing utility to graph the total interest function.

(b) Approximate the length of the mortgage when the total interest paid is the same as the size of the mortgage. Is it possible that a person could pay twice as much in interest charges as the size of his or her mortgage?

53. Newton's Law of Cooling At 8:30 A.M., a coroner was called to the home of a person who had died during the night. In order to estimate the time of death, the coroner took the person's temperature twice. At 9:00 A.M. the temperature was $85.7°$F, and at 11:00 A.M. the temperature was $82.8°$F. From these two temperatures the coroner was able to determine that the time elapsed since death and the body temperature were related by the formula

$$t = -10 \ln \frac{T - 70}{98.6 - 70}$$

where t is the time (in hours elapsed since the person died) and T is the temperature (in degrees Fahrenheit) of the person's body. Assume that the person had a normal body temperature of $98.6°$F at death and that the room temperature was a constant $70°$F. Use the formula to estimate the time of death of the person. (This formula is derived from a general cooling principle called Newton's Law of Cooling.)

54. Newton's Law of Cooling You take a five-pound package of steaks out of a freezer at 11 A.M. and place it in the refrigerator. Will the steaks be thawed in time to be grilled at 6 P.M.? Assume that the refrigerator temperature is $40°$F and the freezer temperature is $0°$F. Use the formula for Newton's Law of Cooling

$$t = -5.05 \ln \frac{T - 40}{0 - 40}$$

where t is the time in hours (with $t = 0$ corresponding to 11 A.M.) and T is the temperature of the package of steaks (in degrees Fahrenheit).

Synthesis

True or False? In Exercises 55–58, determine whether the statement is true or false. Justify your answer.

55. The domain of a logistic growth function cannot be the set of real numbers.

56. The graph of a logistic growth function will always have an x-intercept.

57. The graph of a Gaussian model will never have an x-intercept.

58. The graph of a Gaussian model will always have a maximum point.

Review

In Exercises 59–62, match the equation with its graph, and identify any intercepts. [The graphs are labeled (a), (b), (c), and (d).]

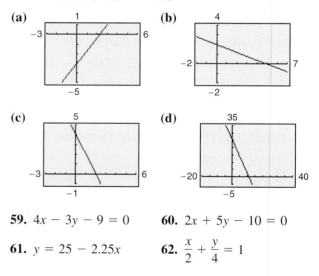

(a)

(b)

(c)

(d)

59. $4x - 3y - 9 = 0$

60. $2x + 5y - 10 = 0$

61. $y = 25 - 2.25x$

62. $\dfrac{x}{2} + \dfrac{y}{4} = 1$

In Exercises 63–66, use the Leading Coefficient Test to determine the right-hand and left-hand behavior of the graph of the polynomial function.

63. $f(x) = 2x^3 - 3x^2 + x - 1$

64. $f(x) = 5 - x^2 - 4x^4$

65. $g(x) = -1.6x^5 + 4x^2 - 2$

66. $g(x) = 7x^6 + 9.1x^5 - 3.2x^4 + 25x^3$

In Exercises 67 and 68, divide using synthetic division.

67. $(2x^3 - 8x^2 + 3x - 9) \div (x - 4)$

68. $(x^4 - 3x + 1) \div (x + 5)$

4.6 Exploring Data: Nonlinear Models

Classifying Scatter Plots

In Section 2.6, you saw how to fit linear models to data and in Section 3.7, you saw how to fit quadratic models to data. In real life, many relationships between two variables are represented by different types of growth patterns. A scatter plot can be used to give you an idea of which type of model will best fit a set of data.

Example 1   Classifying Scatter Plots

Decide whether each set of data could best be modeled by an exponential model $y = ab^x$ or a logarithmic model $y = a + b \ln x$.

a. $(2, 1)$, $(2.5, 1.2)$, $(3, 1.3)$, $(3.5, 1.5)$, $(4, 1.8)$, $(4.5, 2)$, $(5, 2.4)$, $(5.5, 2.5)$, $(6, 3.1)$, $(6.5, 3.8)$, $(7, 4.5)$, $(7.5, 5)$, $(8, 6.5)$, $(8.5, 7.8)$, $(9, 9)$, $(9.5, 10)$

b. $(2, 2)$, $(2.5, 3.1)$, $(3, 3.8)$, $(3.5, 4.3)$, $(4, 4.6)$, $(4.5, 5.3)$, $(5, 5.6)$, $(5.5, 5.9)$, $(6, 6.2)$, $(6.5, 6.4)$, $(7, 6.9)$, $(7.5, 7.2)$, $(8, 7.6)$, $(8.5, 7.9)$, $(9, 8)$, $(9.5, 8.2)$

Solution

Begin by entering the data into a graphing utility. You should obtain the scatter plots shown in Figure 4.51.

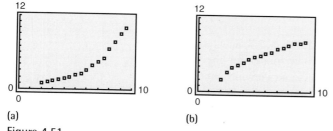

(a)   (b)

Figure 4.51

From the scatter plots, it appears that the data in part (a) can be modeled by an exponential function and the data in part (b) can be modeled by a logarithmic function.

 Checkpoint Now try Exercise 9.

You can change an exponential model of the form $y = ab^x$ to one of the form $y = ae^{cx}$ by rewriting b in the form

$$b = e^{\ln b}.$$

For instance, $y = 3(2^x)$ can be written as

$$y = 3(2^x) = 3e^{(\ln 2)x} \approx 3e^{0.693x}.$$

Fitting Nonlinear Models to Data

Once you have used a scatter plot to determine the type of model that would best fit a set of data, there are several ways that you can actually find the model. Each method is best used with a computer or calculator, rather than with hand calculations.

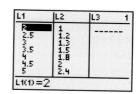

From Example 1(a), you already know that the data can be modeled by an exponential function. In the next example you will determine whether an exponential model best fits the data.

Example 2   Fitting a Model to Data

Fit the following data from Example 1(a) to a quadratic model, an exponential model, and a power model. Determine which model best fits the data.

(2, 1), (2.5, 1.2), (3, 1.3), (3.5, 1.5), (4, 1.8), (4.5, 2), (5, 2.4), (5.5, 2.5), (6, 3.1), (6.5, 3.8), (7, 4.5), (7.5, 5), (8, 6.5), (8.5, 7.8), (9, 9), (9.5, 10)

Solution

Begin by entering the data into a graphing utility. Then use the *regression* feature of the graphing utility to find quadratic, exponential, and power models for the data, as shown in Figure 4.52.

TECHNOLOGY SUPPORT

For instructions on how to use the *regression* feature, see Appendix A; for specific keystrokes, go to the text website at *college.hmco.com*.

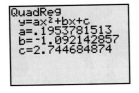

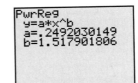

Quadratic Model   Exponential Model   Power Model

Figure 4.52

So, a quadratic model for the data is $y = 0.195x^2 - 1.09x + 2.7$; an exponential model for the data is $y = 0.507(1.368)^x$; and a power model for the data is $y = 0.249x^{1.518}$. Plot the data and each model in the same viewing window, as shown in Figure 4.53. To determine which model best fits the data, compare the y-values given by each model with the actual y-values. The model whose y-values are closest to the actual values is the one that fits best. In this case, the best-fitting model is the exponential model.

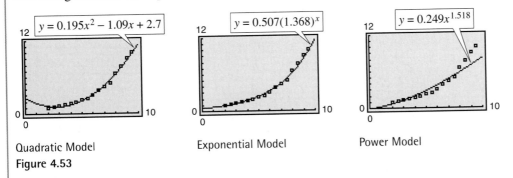

Quadratic Model   Exponential Model   Power Model

Figure 4.53

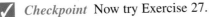 *Checkpoint* Now try Exercise 27.

Deciding which model best fits a set of data is a question that is studied in detail in statistics. Recall from Section 2.6 that the model that best fits a set of data is the one whose *sum of squared differences* is the least. In Example 2, the sums of squared differences are 0.89 for the quadratic model, 0.85 for the exponential model, and 14.39 for the power model.

Example 3   Fitting a Model to Data

The table shows the yield y (in milligrams) of a chemical reaction after x minutes. Use a graphing utility to find a logarithmic model and a linear model for the data. Determine which model best fits the data.

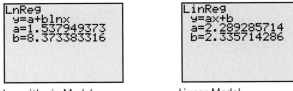

Minutes, x	Yield, y
1	1.5
2	7.4
3	10.2
4	13.4
5	15.8
6	16.3
7	18.2
8	18.3

Solution

Begin by entering the data into a graphing utility. Then use the *regression* feature of the graphing utility to find logarithmic and linear models for the data, as shown in Figure 4.54.

```
LnReg
y=a+blnx
a=1.537949373
b=8.373383316
```

```
LinReg
y=ax+b
a=2.289285714
b=2.335714286
```

Logarithmic Model   Linear Model

Figure 4.54

So, a logarithmic model for the data is $y = 1.538 + 8.373 \ln x$ and a linear model for the data is $y = 2.29x + 2.3$. Plot the data and each model in the same viewing window, as shown in Figure 4.55. To determine which model best fits the data, compare the y-values given by each model with the actual y-values. The model whose y-values are closest to the actual values is the one that fits best. In this case, the best-fitting model is the logarithmic model.

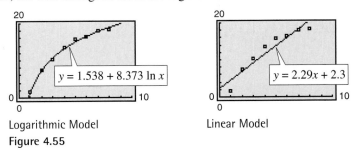

Logarithmic Model   Linear Model

Figure 4.55

In Example 3, the sum of the squared differences for the logarithmic model is 1.55 and the sum of the squared differences for the linear model is 23.86.

Exploration

Use a graphing utility to find a quadratic model for the data in Example 3. Do you think this model fits the data better than the logarithmic model from Example 3? Explain your reasoning.

Modeling With Exponential and Logistic Functions

Example 4   Fitting an Exponential Model to Data

The table at the right shows the revenue R (in billions of dollars) collected by the Internal Revenue Service (IRS) for selected years from 1960 to 2000. Use a graphing utility to find a model for the data. Then use the model to estimate the revenue collected in 2008.   (Source: Internal Revenue Service)

Year	Revenue, R
1960	91.8
1965	114.4
1970	195.7
1975	293.8
1980	519.4
1985	742.9
1990	1056.4
1995	1375.7
2000	2096.9

Solution

Let x represent the year, with $x = 0$ corresponding to 1960. Begin by entering the data into a graphing utility and displaying the scatter plot, as shown in Figure 4.56.

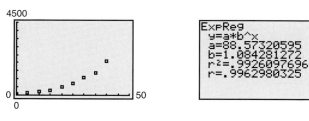

Figure 4.56   Figure 4.57

From the scatter plot, it appears that an exponential model is a good fit. Use the *regression* feature of the graphing utility to find the exponential model, as shown in Figure 4.57. Change the model to a natural exponential model, as follows.

$$R = 88.57(1.084)^x \qquad \text{Write original model.}$$

$$= 88.57e^{(\ln 1.084)x} \qquad b = e^{\ln b}$$

$$\approx 88.57e^{0.0807x} \qquad \text{Simplify.}$$

Graph the data and the model in the same viewing window, as shown in Figure 4.58. From the model, you can see that the revenue collected by the IRS from 1960 to 2000 had an average annual increase of 8%. From this model, you can estimate the 2008 revenue to be

$$R = 88.57e^{0.0807x} \qquad \text{Write original model.}$$

$$= 88.57e^{0.0807(48)} \approx \$4261.6 \text{ billion} \qquad \text{Substitute 48 for } x.$$

which is more than twice the amount collected in 2000. You can also use the *value* feature or the *zoom* and *trace* features of a graphing utility to approximate the revenue in 2008 to be $4261.6 billion, as shown in Figure 4.58.

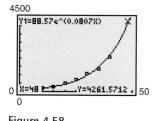

Figure 4.58

 Checkpoint Now try Exercise 33.

The next example demonstrates how to use a graphing utility to fit a logistic model to data.

Example 5   Fitting a Logistic Model to Data

To estimate the amount of defoliation caused by the gypsy moth during a given year, a forester counts the number x of egg masses on $\frac{1}{40}$ of an acre (circle of radius 18.6 feet) in the fall. The percent of defoliation y the next spring is shown in the table.   (Source: USDA, Forest Service)

Egg masses, x	Percent of defoliation, y
0	12
25	44
50	81
75	96
100	99

a. Use the *regression* feature of a graphing utility to find a logistic model for the data.

b. How closely does the model represent the data?

Graphical Solution

a. Enter the data into the graphing utility. Using the *regression* feature of the graphing utility, you can find the logistic model, as shown in Figure 4.59. You can approximate this model to be

$$y = \frac{100}{1 + 7e^{-0.069x}}.$$

b. You can use a graphing utility to graph the actual data and the model in the same viewing window. From Figure 4.60, it appears that the model is a good fit for the actual data.

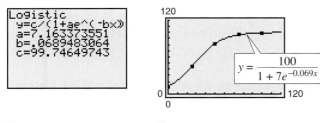

Figure 4.59   Figure 4.60

Numerical Solution

a. Enter the data into the graphing utility. Using the *regression* feature of the graphing utility, you can approximate the logistic model to be

$$y = \frac{100}{1 + 7e^{-0.069x}}.$$

b. You can see how well the model fits the data by comparing the actual values of y with the values of y given by the model, which are labeled y^* in the table below.

x	0	25	50	75	100
y	12	44	81	96	99
y^*	12.5	44.5	81.8	96.2	99.3

From the table, you can see that the model appears to be a good fit for the actual data.

 Checkpoint Now try Exercise 34.

4.6　Exercises

Vocabulary Check

Fill in the blanks.

1. A linear model has the form _____ .
2. A _____ model has the form $y = ax^2 + bx + c$.
3. A power model has the form _____ .
4. One way of determining which model best fits a set of data is to compare the _____ of _____ .
5. An exponential model has the form _____ or _____ .

In Exercises 1–8, determine whether the scatter plot could best be modeled by a linear model, a quadratic model, an exponential model, a logarithmic model, or a logistic model.

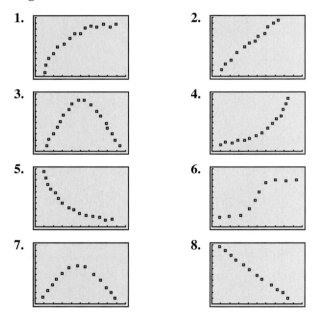

1.

2.

3.

4.

5.

6.

7.

8.

In Exercises 9–14, use a graphing utility to create a scatter plot of the data. Decide whether the data could best be modeled by a linear model, an exponential model, or a logarithmic model.

9. $(1, 2.0)$, $(1.5, 3.5)$, $(2, 4.0)$, $(4, 5.8)$, $(6, 7.0)$, $(8, 7.8)$
10. $(1, 5.8)$, $(1.5, 6.0)$, $(2, 6.5)$, $(4, 7.6)$, $(6, 8.9)$, $(8, 10.0)$
11. $(1, 4.4)$, $(1.5, 4.7)$, $(2, 5.5)$, $(4, 9.9)$, $(6, 18.1)$, $(8, 33.0)$
12. $(1, 11.0)$, $(1.5, 9.6)$, $(2, 8.2)$, $(4, 4.5)$, $(6, 2.5)$, $(8, 1.4)$
13. $(1, 7.5)$, $(1.5, 7.0)$, $(2, 6.8)$, $(4, 5.0)$, $(6, 3.5)$, $(8, 2.0)$

14. $(1, 5.0)$, $(1.5, 6.0)$, $(2, 6.4)$, $(4, 7.8)$, $(6, 8.6)$, $(8, 9.0)$

In Exercises 15–18, use the *regression* feature of a graphing utility to find an exponential model $y = ab^x$ for the data. Use the graphing utility to plot the data and graph the model in the same viewing window.

15. $(0, 4)$, $(1, 5)$, $(2, 6)$, $(3, 8)$, $(4, 12)$
16. $(0, 6.0)$, $(2, 8.9)$, $(4, 20.0)$, $(6, 34.3)$, $(8, 61.1)$, $(10, 120.5)$
17. $(0, 10.0)$, $(1, 6.1)$, $(2, 4.2)$, $(3, 3.8)$, $(4, 3.6)$
18. $(-3, 120.2)$, $(0, 80.5)$, $(3, 64.8)$, $(6, 58.2)$, $(10, 55.0)$

In Exercises 19–22, use the *regression* feature of a graphing utility to find a logarithmic model $y = a + b \ln x$ for the data. Use the graphing utility to plot the data and graph the model in the same viewing window.

19. $(1, 2.0)$, $(2, 3.0)$, $(3, 3.5)$, $(4, 4.0)$, $(5, 4.1)$, $(6, 4.2)$, $(7, 4.5)$
20. $(1, 8.5)$, $(2, 11.4)$, $(4, 12.8)$, $(6, 13.6)$, $(8, 14.2)$, $(10, 14.6)$
21. $(1, 10)$, $(2, 6)$, $(3, 6)$, $(4, 5)$, $(5, 3)$, $(6, 2)$
22. $(3, 14.6)$, $(6, 11.0)$, $(9, 9.0)$, $(12, 7.6)$, $(15, 6.5)$

In Exercises 23–26, use the *regression* feature of a graphing utility to find a power model $y = ax^b$ for the data. Use the graphing utility to plot the data and graph the model in the same viewing window.

23. $(1, 2.0)$, $(2, 3.4)$, $(5, 6.7)$, $(6, 7.3)$, $(10, 12.0)$
24. $(0.5, 1.0)$, $(2, 12.5)$, $(4, 33.2)$, $(6, 65.7)$, $(8, 98.5)$, $(10, 150.0)$

25. $(1, 10.0), (2, 4.0), (3, 0.7), (4, 0.1)$

26. $(2, 450), (4, 385), (6, 345), (8, 332), (10, 312)$

27. *Elections*   The table shows the number R (in millions) of registered voters in the United States for presidential election years from 1972 to 2000. (Source: Federal Election Commission)

Year	Number of voters, R
1972	97.3
1976	105.0
1980	113.0
1984	124.2
1988	126.4
1992	133.8
1996	146.2
2000	156.4

(a) Use the *regression* feature of a graphing utility to find a quadratic model, an exponential model, and a power model for the data. Let x represent the year, with $x = 2$ corresponding to 1972.

(b) Use a graphing utility to graph each model with the original data.

(c) Determine which model best fits the data.

(d) Use the model you chose in part (c) to predict the number of registered voters in 2004.

28. *Consumer Awareness*   The table shows the retail price P (in dollars) of a half-gallon package of ice cream for each year from 1995 to 2001.   (Source: U.S. Bureau of Labor Statistics)

Year	Retail price, P
1995	2.68
1996	2.94
1997	3.02
1998	3.30
1999	3.40
2000	3.66
2001	3.84

(a) Use the *regression* feature of a graphing utility to find a quadratic model, an exponential model, and a power model for the data. Let x represent the year, with $x = 5$ corresponding to 1995.

(b) Use a graphing utility to graph each model with the original data.

(c) Determine which model best fits the data.

(d) Use the model you chose in part (c) to predict the price of a half-gallon package of ice cream in 2007.

29. *Population*   The population y (in millions) of the United States for the years 1992 through 2001 is shown in the table, where x represents the year, with $x = 2$ corresponding to 1992. (Source: U.S. Census Bureau)

Year, x	Population, y
2	257
3	260
4	263
5	267
6	270
7	273
8	276
9	279
10	282
11	285

(a) Use the *regression* feature of a graphing utility to find a linear model for the data.

(b) Use the *regression* feature of a graphing utility to find an exponential model for the data.

(c) Population growth is often exponential. For the 10 years of data given, is the exponential model a better fit than the linear model? Explain.

(d) Use each model to predict the population in the year 2008.

30. *Atmospheric Pressure*   The atmospheric pressure decreases with increasing altitude. At sea level, the average air pressure is approximately 1.03323 kilograms per square centimeter, and this pressure is called one atmosphere. Variations in weather conditions cause changes in the atmospheric pressure of up to ± 5 percent. The table shows the pressures p (in atmospheres) for different altitudes h (in kilometers).

Altitude, h	Pressure, p
0	1
5	0.55
10	0.25
15	0.12
20	0.06
25	0.02

Table for 30

(a) Use the *regression* feature of a graphing utility to attempt to find the logarithmic model $p = a + b \ln h$ for the data. Explain why the result is an error message.

(b) Use the *regression* feature of a graphing utility to find the logarithmic model $h = a + b \ln p$ for the data.

(c) Use a graphing utility to plot the data and graph the logarithmic model in the same viewing window.

(d) Use the model to estimate the altitude at which the pressure is 0.75 atmosphere.

(e) Use the graph in part (c) to estimate the pressure at an altitude of 13 kilometers.

31. **Data Analysis** A cup of water at an initial temperature of 78°C is placed in a room at a constant temperature of 21°C. The temperature of the water is measured every 5 minutes for a period of $\frac{1}{2}$ hour. The results are recorded in the table, where t is the time (in minutes) and T is the temperature (in degrees Celsius).

Time, t	Temperature, T
0	78.0°
5	66.0°
10	57.5°
15	51.2°
20	46.3°
25	42.5°
30	39.6°

(a) Use the *regression* feature of a graphing utility to find a linear model for the data. Use the graphing utility to plot the data and graph the model in the same viewing window. Does the data appear linear? Explain.

(b) Use the *regression* feature of a graphing utility to find a quadratic model for the data. Use the graphing utility to plot the data and graph the model in the same viewing window. Does the data appear quadratic? Even though the quadratic model appears to be a good fit, explain why it might not be a good model for predicting the temperature of the water when $t = 60$.

(c) The graph of the model should be asymptotic with the graph of the temperature of the room. Subtract the room temperature from each of the temperatures in the table. Use the *regression* feature of a graphing utility to find an exponential model for the revised data. Add the room temperature to this model. Use a graphing utility to plot the original data and graph the model in the same viewing window.

(d) Explain why the procedure in part (c) was necessary for finding the exponential model.

32. **Sales** The table shows the sales S (in billions of dollars) for Home Depot, Inc. from 1996 to 2001. (Source: The Home Depot, Inc.)

Year	Sales, S
1996	19.5
1997	24.2
1998	30.2
1999	38.4
2000	45.7
2001	53.6

(a) Use the *regression* feature of a graphing utility to find an exponential model for the data. Let x represent the year, with $x = 6$ corresponding to 1996.

(b) Use the graphing utility to graph the model with the original data.

(c) How closely does the model represent the data?

(d) Use the model to estimate the sales for Home Depot, Inc. in 2007.

33. **Sales** The table on the next page shows the sales S (in millions of dollars) for Carnival Corporation from 1996 to 2001. (Source: Carnival Corporation)

Year	Sales, S
1996	2212.6
1997	2447.5
1998	3009.3
1999	3497.5
2000	3778.5
2001	4535.8

Table for 33

(a) Use the *regression* feature of a graphing utility to find an exponential model for the data. Let x represent the year, with $x = 6$ corresponding to 1996.

(b) Use the graphing utility to graph the model with the original data.

(c) How closely does the model represent the data?

(d) Use the model to estimate the sales for Carnival Corporation in 2007.

34. *Vital Statistics*   The table shows the percent P of men who have never been married for different age groups (in years).   (Source: U.S. Census Bureau)

Age Group	Percent, P
18–19	98.3
20–24	83.7
25–29	51.7
30–34	30.0
35–39	20.3
40–44	15.7
45–54	9.5
55–64	5.5
65–74	4.3
75 and over	4.1

(a) Use the *regression* feature of a graphing utility to find a logistic model for the data. Let x represent the age group, with $x = 1$ corresponding to the 18–19 age group.

(b) Use the graphing utility to graph the model with the original data.

(c) How closely does the model represent the data?

35. *Comparing Models*   The amounts y (in billions of dollars) donated to charity (by individuals, foundations, corporations, and charitable bequests) in the United States from 1996 to 2001 are shown in the table, where x represents the year, with $x = 6$ corresponding to 1996.   (Source: AAFRC Trust for Philanthropy)

Year, x	Amount, y
6	138.6
7	157.1
8	174.8
9	199.0
10	210.9
11	212.0

Table for 35

(a) Use the *regression* feature of a graphing utility to find a linear model, a logarithmic model, a quadratic model, an exponential model, and a power model for the data.

(b) Use the graphing utility to graph each model with the original data. Use the graphs to choose the model that you think best fits the data.

(c) For each model, find the sum of the squared differences. Use the results to choose the model that best fits the data.

(d) For each model, find the r^2-value determined by the graphing utility. Use the results to choose the model that best fits the data.

(e) Compare your results from parts (b), (c), and (d).

Synthesis

36. *Writing*   In your own words, explain how to fit a model to a set of data using a graphing utility.

True or False?   **In Exercises 37 and 38, determine whether the statement is true or false. Justify your answer.**

37. The exponential model $y = ae^{bx}$ represents a growth model if $b > 0$.

38. To change an exponential model of the form $y = ab^x$ to one of the form $y = ae^{cx}$, rewrite b as $b = \ln e^b$.

Review

In Exercises 39–42, find the slope and y-intercept of the equation of the line. Then sketch the line by hand.

39. $2x + 5y = 10$

40. $3x - 2y = 9$

41. $1.2x + 3.5y = 10.5$

42. $0.4x - 2.5y = 12.0$

4 Chapter Summary

What did you learn?

Section 4.1	Review Exercises
☐ Recognize and evaluate exponential functions with base a.	1–4
☐ Graph exponential functions.	5–12
☐ Recognize, evaluate, and graph exponential functions with base e.	13–28
☐ Use exponential functions to model and solve real-life problems.	29–32

Section 4.2	
☐ Recognize and evaluate logarithmic functions with base a.	33–40
☐ Graph logarithmic functions.	41–44
☐ Recognize, evaluate, and graph natural logarithmic functions.	45–54
☐ Use logarithmic functions to model and solve real-life problems.	55, 56

Section 4.3	
☐ Rewrite logarithms with different bases.	57–60
☐ Use properties of logarithms to evaluate or rewrite logarithmic expressions.	61–64
☐ Use properties of logarithms to expand or condense logarithmic expressions.	65–76
☐ Use logarithmic functions to model and solve real-life problems.	77, 78

Section 4.4	
☐ Solve simple exponential and logarithmic equations.	79–86
☐ Solve more complicated exponential equations.	87–96
☐ Solve more complicated logarithmic equations.	97–108
☐ Use exponential and logarithmic equations to model and solve real-life problems.	109, 110

Section 4.5	
☐ Recognize the five most common types of models involving exponential or logarithmic functions.	111–116
☐ Use exponential growth and decay functions to model and solve real-life problems.	117–123
☐ Use Gaussian functions to model and solve real-life problems.	124
☐ Use logistic growth functions to model and solve real-life problems.	125
☐ Use logarithmic functions to model and solve real-life problems.	126

Section 4.6	
☐ Classify scatter plots.	127–130
☐ Use scatter plots and a graphing utility to find models for data and choose a model that best fits a set of data.	131
☐ Use a graphing utility to find exponential and logistic models for data.	132, 133

4 Review Exercises

4.1 In Exercises 1–4, use a calculator to evaluate the function at the indicated value of x. Round your result to four decimal places.

Function	Value
1. $f(x) = 1.45^x$	$x = 2\pi$
2. $f(x) = 7^x$	$x = -\sqrt{11}$
3. $g(x) = 60^{2x}$	$x = -1.1$
4. $g(x) = 25^{-3x}$	$x = \frac{3}{2}$

In Exercises 5–8, match the function with its graph. [The graphs are labeled (a), (b), (c), and (d).]

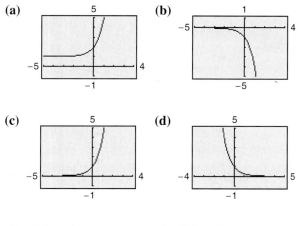

(a)

(b)

(c)

(d)

5. $f(x) = 4^x$

6. $f(x) = 4^{-x}$

7. $f(x) = -4^x$

8. $f(x) = 4^x + 1$

In Exercises 9–12, graph the exponential function by hand. Identify any asymptotes and intercepts and determine whether the graph of the function is increasing or decreasing.

9. $f(x) = 6^x$

10. $f(x) = 0.3^{x+1}$

11. $g(x) = 1 + 6^{-x}$

12. $g(x) = 0.3^{-x}$

In Exercises 13–16, use a calculator to evaluate the function $f(x) = e^x$ for the indicated value of x. Round your result to three decimal places.

13. $x = 8$

14. $x = \sqrt{5}$

15. $x = -2.1$

16. $x = -\frac{3}{5}$

In Exercises 17–22, use a graphing utility to construct a table of values for the function. Then sketch the graph of the function.

17. $h(x) = e^{x-1}$

18. $f(x) = e^{x+2}$

19. $h(x) = -e^x$

20. $f(x) = 3 - e^{-x}$

21. $f(x) = 4e^{-0.5x}$

22. $f(x) = 2 + e^{x+3}$

In Exercises 23–28, use a graphing utility to graph the exponential function. Identify any asymptotes of the graph.

23. $g(t) = 8 - 0.5e^{-t/4}$

24. $h(x) = 12(1 + e^{-x/2})$

25. $g(x) = 200e^{4/x}$

26. $f(x) = -8e^{-4/x}$

27. $f(x) = \dfrac{10}{1 + 2^{-0.05x}}$

28. $f(x) = -\dfrac{12}{1 + 4^{-x}}$

Compound Interest In Exercises 29 and 30, complete the table to determine the balance A for \$10,000 invested at rate r for t years, compounded continuously.

t	1	10	20	30	40	50
A						

29. $r = 8\%$

30. $r = 3\%$

31. ***Depreciation*** After t years, the value of a car that costs \$26,000 is modeled by

$$V(t) = 26,000\left(\frac{3}{4}\right)^t.$$

(a) Use a graphing utility to graph the function.

(b) Find the value of the car 2 years after it was purchased.

(c) According to the model, when does the car depreciate most rapidly? Is this realistic? Explain.

32. ***Radioactive Decay*** Let Q represent a mass of plutonium 241 (^{241}Pu), in grams whose half-life is 14 years. The quantity of plutonium present after t years is given by

$$Q = 100\left(\frac{1}{2}\right)^{t/14}.$$

(a) Determine the initial quantity (when $t = 0$).

(b) Determine the quantity present after 10 years.

(c) Use a graphing utility to graph the function over the interval $t = 0$ to $t = 100$.

4.2 In Exercises 33–36, write the exponential equation in logarithmic form.

33. $4^3 = 64$

34. $3^5 = 243$

35. $25^{3/2} = 125$

36. $12^{-1} = \frac{1}{12}$

In Exercises 37–40, evaluate the function at the indicated value of x without using a calculator.

Function	Value
37. $f(x) = \log_6 x$	$x = 216$
38. $f(x) = \log_7 x$	$x = 1$
39. $f(x) = \log_4 x$	$x = \frac{1}{4}$
40. $f(x) = \log_{10} x$	$x = 0.001$

In Exercises 41–44, find the domain, vertical asymptote, and x-intercept of the logarithmic function, and sketch its graph by hand. Verify using a graphing utility.

41. $g(x) = -\log_2 x + 5$

42. $g(x) = \log_5(x - 3)$

43. $f(x) = \log_2(x - 1) + 6$

44. $f(x) = \log_5(x + 2) - 3$

In Exercises 45–50, use a calculator to evaluate the function $f(x) = \ln x$ at the indicated value of x. Round your result to three decimal places, if necessary.

45. $x = 21.5$

46. $x = 0.98$

47. $x = e^7$

48. $x = e^{-1/2}$

49. $x = \sqrt{6}$

50. $x = \frac{2}{5}$

In Exercises 51–54, use a graphing utility to graph the logarithmic function. Determine the domain and identify any vertical asymptote and x-intercept.

51. $f(x) = \ln x + 3$

52. $f(x) = \ln(x - 3)$

53. $h(x) = \frac{1}{2}\ln x$

54. $f(x) = \frac{1}{4}\ln x$

55. *Climb Rate*   The time t (in minutes) for a small plane to climb to an altitude of h feet is given by

$$t = 50 \log_{10} \frac{18{,}000}{18{,}000 - h}$$

where 18,000 feet is the plane's absolute ceiling.

(a) Determine the domain of the function appropriate for the context of the problem.

(b) Use a graphing utility to graph the function and identify any asymptotes.

(c) As the plane approaches its absolute ceiling, what can be said about the time required to further increase its altitude?

(d) Find the amount of time it will take for the plane to climb to an altitude of 4000 feet.

56. *Home Mortgage*   The model

$$t = 12.542 \ln\left(\frac{x}{x - 1000}\right), \quad x > 1000$$

approximates the length of a home mortgage of $150,000 at 8% in terms of the monthly payment. In the model, t is the length of the mortgage in years and x is the monthly payment in dollars.

(a) Use the model to approximate the length of a $150,000 mortgage at 8% when the monthly payment is $1254.68.

(b) Approximate the total amount paid over the term of the mortgage with a monthly payment of $1254.68. What amount of the total is interest costs?

4.3 In Exercises 57–60, evaluate the logarithm using the change-of-base formula. Do each problem twice, once with common logarithms and once with natural logarithms. Round your results to three decimal places.

57. $\log_4 9$

58. $\log_{1/2} 5$

59. $\log_{12} 200$

60. $\log_3 0.28$

In Exercises 61–64, use the properties of logarithms to rewrite and simplify the logarithmic expression.

61. $\ln 20$

62. $\ln(3e^{-4})$

63. $\log_5\left(\frac{1}{15}\right)$

64. $\log_{10} \frac{9}{300}$

In Exercises 65–70, use the properties of logarithms to expand the expression as a sum, difference, and/or constant multiple of logarithms. (Assume all variables are positive.)

65. $\log_5 5x^2$

66. $\log_4 3xy^2$

67. $\log_{10} \frac{5\sqrt{y}}{x^2}$

68. $\ln \frac{\sqrt{x}}{4}$

69. $\ln\left(\frac{x + 3}{xy}\right)$

70. $\ln \frac{xy^5}{\sqrt{z}}$

In Exercises 71–76, condense the expression to the logarithm of a single quantity.

71. $\log_2 5 + \log_2 x$ **72.** $\log_6 y - 2\log_6 z$

73. $\frac{1}{2}\ln(2x - 1) - 2\ln(x + 1)$

74. $5\ln(x - 2) - \ln(x + 2) - 3\ln(x)$

75. $\ln 3 + \frac{1}{3}\ln(4 - x^2) - \ln x$

76. $3[\ln x - 2\ln(x^2 + 1)] + 2\ln 5$

77. Snow Removal   The number of miles s of roads cleared of snow is approximated by the model

$$s = 25 - \frac{13\ln(h/12)}{\ln 3}, \quad 2 \le h \le 15$$

where h is the depth of the snow (in inches).

(a) Use a graphing utility to graph the function.

(b) Complete the table.

h	4	6	8	10	12	14
s						

(c) Using the graph of the function and the table, what conclusion can you make about the miles of roads cleared as the depth of the snow increases?

78. Human Memory Model   Students in a sociology class were given an exam and then retested monthly with an equivalent exam. The average scores for the class are given by the human memory model $f(t) = 85 - 14\log_{10}(t + 1)$, where t is the time in months and $0 \le t \le 10$. When will the average score decrease to 71?

4.4 In Exercises 79–86, solve for x.

79. $8^x = 512$ **80.** $3^x = 729$

81. $6^x = \frac{1}{216}$ **82.** $6^{x-2} = 1296$

83. $\log_7 x = 4$ **84.** $\log_x 243 = 5$

85. $\ln x = 4$ **86.** $\ln x = -3$

In Exercises 87–96, solve the exponential equation algebraically. Round your result to three decimal places.

87. $e^x = 12$ **88.** $e^{3x} = 25$

89. $3e^{-5x} = 132$ **90.** $14e^{3x+2} = 560$

91. $2^x + 13 = 35$ **92.** $6^x - 28 = -8$

93. $-4(5^x) = -68$ **94.** $2(12^x) = 190$

95. $e^{2x} - 7e^x + 10 = 0$ **96.** $e^{2x} - 6e^x + 8 = 0$

In Exercises 97–108, solve the logarithmic equation algebraically. Round your result to three decimal places.

97. $\ln 3x = 8.2$ **98.** $\ln 5x = 7.2$

99. $2\ln 4x = 15$ **100.** $4\ln 3x = 15$

101. $\ln x - \ln 3 = 2$ **102.** $\ln\sqrt{x + 8} = 3$

103. $\ln\sqrt{x + 1} = 2$ **104.** $\ln x - \ln 5 = 4$

105. $\log_{10}(x - 1) = \log_{10}(x - 2) - \log_{10}(x + 2)$

106. $\log_{10}(x + 2) - \log_{10} x = \log_{10}(x + 5)$

107. $\log_{10}(1 - x) = -1$ **108.** $\log_{10}(-x - 4) = 2$

109. Compound Interest   You deposit $7550 into an account that pays 7.25% interest, compounded continuously. How long will it take for the money to triple?

110. Demand   The demand equation for a 32-inch television is modeled by $p = 500 - 0.5e^{0.004x}$. Find the demand x for a price of (a) $p = \$450$ and (b) $p = \$400$.

4.5 In Exercises 111–116, match the function with its graph. [The graphs are labeled (a), (b), (c), (d), (e), and (f).]

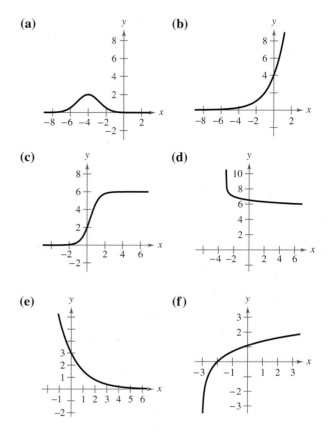

(a)

(b)

(c)

(d)

(e)

(f)

111. $y = 3e^{-2x/3}$

112. $y = 4e^{2x/3}$

113. $y = \ln(x + 3)$

114. $y = 7 - \log_{10}(x + 3)$

115. $y = 2e^{-(x+4)^2/3}$

116. $y = \dfrac{6}{1 + 2e^{-2x}}$

In Exercises 117–120, find the exponential model $y = ae^{bx}$ that fits the two points.

117. $(0, 2), (4, 3)$

118. $(0, 2), (5, 1)$

119. $\left(0, \frac{1}{2}\right), (5, 5)$

120. $(0, 4), \left(5, \frac{1}{2}\right)$

121. *Population* The population P (in thousands) of Colorado Springs, Colorado is given by

$$P = 361e^{kt}$$

where $t = 0$ represents the year 2000. In 1980, the population was 215,000. Find the value of k and use this result to predict the population in the year 2020. (Source: U.S. Census Bureau)

122. *Radioactive Decay* The half-life of radioactive uranium II (^{234}U) is 245,500 years. What percent of the present amount of radioactive uranium II will remain after 5000 years?

123. *Compound Interest* A deposit of $10,000 is made in a savings account for which the interest is compounded continuously. The balance will double in 12 years.

(a) What is the annual interest rate for this account?

(b) Find the balance after 1 year.

124. *Test Scores* The test scores for a biology test follow a normal distribution modeled by

$$y = 0.0499e^{-(x-71)^2/128}$$

where x is the test score.

(a) Use a graphing utility to graph the function.

(b) From the graph in part (a), estimate the average test score.

125. *Typing Speed* In a typing class, the average number of words per minute N typed after t weeks of lessons was found to be modeled by

$$N = \frac{157}{1 + 5.4e^{-0.12t}}.$$

Find the number of weeks necessary to type (a) 50 words per minute and (b) 75 words per minute.

126. *Geology* On the Richter scale, the magnitude R of an earthquake of intensity I is modeled by

$$R = \log_{10}\frac{I}{I_0}$$

where $I_0 = 1$ is the minimum intensity used for comparison. Find the intensities I of the following earthquakes measuring R on the Richter scale.

(a) $R = 8.4$ (b) $R = 6.85$ (c) $R = 9.1$

4.6 **In Exercises 127–130, determine whether the scatter plot could best be modeled by a linear model, a quadratic model, an exponential model, a logarithmic model, or a logistic model.**

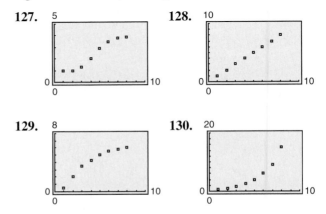

127.

128.

129.

130.

131. *Entertainment* The table shows the number M (in thousands) of movie theater screens in the United States for selected years from 1975 to 2000. (Source: Motion Picture Association of America)

Year	Number of screens, M
1975	11
1980	14
1985	18
1990	23
1995	27
2000	37

(a) Use the *regression* feature of a graphing utility to find a quadratic model, an exponential model, and a power model for the data. Let x represent the year, with $x = 5$ corresponding to 1975.

(b) Use a graphing utility to graph each model with the original data.

(c) Determine which model best fits the data.

(d) Use the model you chose in part (c) to predict the number of movie theater screens in 2007.

132. *Sports*   The table shows the number G of municipal golf facilities in the United States for selected years from 1975 to 2000.   (Source: National Golf Foundation)

Year	Number of facilities, G
1975	1586
1980	1794
1985	1912
1990	2012
1995	2259
2000	2438

(a) Use the *regression* feature of a graphing utility to find an exponential model for the data. Let x represent the year, with $x = 5$ corresponding to 1975.

(b) Use a graphing utility to graph the model with the original data.

(c) How closely does the model represent the data?

(d) Use the model to estimate the number of municipal golf facilities in 2010.

133. *Wildlife*   A lake is stocked with 500 fish, and the fish population P increases every month. The local fish commission records this increase as shown in the table.

Month, x	Population, P
0	500
6	1488
12	3672
18	6583
24	8650
30	9550
36	9860

(a) Use the *regression* feature of a graphing utility to find a logistic model for the data. Let x represent the month.

(b) Use a graphing utility to graph the model with the original data.

(c) How closely does the model represent the data?

(d) What is the limiting size of the population?

Synthesis

134. *Think About It*   Without using a calculator, explain why you know that $2^{\sqrt{2}}$ is greater than 2, but less than 4.

True or False?   In Exercises 135–140, determine whether the equation or statement is true or false. Justify your answer.

135. $\log_b b^{2x} = 2x$

136. $e^{x-1} = \dfrac{e^x}{e}$

137. $\ln(x + y) = \ln x + \ln y$

138. $\ln(x + y) = \ln(xy)$

139. The domain of the function $f(x) = \ln x$ is the set of all real numbers.

140. The logarithm of the quotient of two numbers is equal to the difference of the logarithms of the numbers.

141. *Pattern Recognition*

(a) Use a graphing utility to compare the graph of the function $y = e^x$ with the graph of each function. [$n!$ (read as "n factorial") is defined as $n! = 1 \cdot 2 \cdot 3 \cdot \, \cdots \, (n-1) \cdot n$.]

$$y_1 = 1 + \frac{x}{1!}, \quad y_2 = 1 + \frac{x}{1!} + \frac{x^2}{2!},$$

$$y_3 = 1 + \frac{x}{1!} + \frac{x^2}{2!} + \frac{x^3}{3!}$$

(b) Identify the pattern of successive polynomials given in part (a). Extend the pattern one more term and compare the graph of the resulting polynomial function with the graph of $y = e^x$. What do you think this pattern implies?

4 Chapter Test

Take this test as you would take a test in class. After you are finished, check your work against the answers given in the back of the book.

In Exercises 1–4, evaluate the expression. Round your result to three decimal places.

1. $12.4^{2.79}$ **2.** $4^{3\pi/2}$ **3.** $e^{-7/10}$ **4.** $e^{3.1}$

In Exercises 5–7, use a graphing utility to construct a table of values for the function. Then sketch a graph of the function.

5. $f(x) = 10^{-x}$ **6.** $f(x) = -6^{x-2}$ **7.** $f(x) = 1 - e^{2x}$

8. Evaluate (a) $\log_7 7^{-0.89}$ and (b) $4.6 \ln e^2$.

In Exercises 9–11, use a graphing utility to graph the function. Determine the domain and identify any vertical asymptote and x-intercept.

9. $f(x) = -\log_{10} x - 6$ **10.** $f(x) = \ln(x - 4)$ **11.** $f(x) = 1 + \ln(x + 6)$

In Exercises 12–14, evaluate the logarithm using the change-of-base formula. Round your result to three decimal places.

12. $\log_7 44$ **13.** $\log_{2/5} 0.9$ **14.** $\log_{24} 68$

In Exercises 15 and 16, use the properties of logarithms to expand the expression as a sum, difference, and/or multiple of logarithms.

15. $\log_2 3a^4$ **16.** $\ln \dfrac{5\sqrt{x}}{6}$

In Exercises 17 and 18, condense the expression to the logarithm of a single quantity.

17. $\log_3 13 + \log_3 y$ **18.** $4 \ln x - 4 \ln y$

In Exercises 19 and 20, solve the equation algebraically. Round your result to three decimal places.

19. $\dfrac{1025}{8 + e^{4x}} = 5$ **20.** $\log_{10} x - \log_{10}(8 - 5x) = 2$

21. The half-life of radioactive actinium (^{227}Ac) is 22 years. What percent of a present amount of radioactive actinium will remain after 19 years?

22. The table at the right shows the mail revenues R (in billions of dollars) for the U.S. Postal Service from 1995 to 2001.   (Source: U.S. Postal Service)

(a) Use the *regression* feature of a graphing utility to find a quadratic model, an exponential model, and a power model for the data. Let x represent the year, with $x = 5$ corresponding to 1995.

(b) Use a graphing utility to graph each model with the original data.

(c) Determine which model best fits the data.

(d) Use the model you chose in part (c) to predict the mail revenues in 2007.

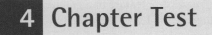

Year	Revenues, R
1995	52.5
1996	54.5
1997	56.3
1998	58.0
1999	60.4
2000	62.3
2001	63.4

3-4 Cumulative Test

Take this test to review the material from earlier chapters. After you are finished, check your work against the answers given in the back of the book.

In Exercises 1–3, sketch the graph of the function. Use a graphing utility to verify the graph.

1. $f(x) = -\frac{1}{2}(x^2 + 4x)$

2. $f(x) = \frac{1}{4}x(x - 2)^2$

3. $f(x) = x^3 + 2x^2 - 9x - 18$

4. Find all the zeros of $f(x) = x^3 + 2x^2 + 4x + 8$.

5. Use a graphing utility to approximate any real zeros of $g(x) = x^3 + 4x^2 - 11$ accurate to three decimal places.

6. Divide $(4x^2 + 14x - 9)$ by $(x + 3)$ using long division.

7. Divide $(2x^3 - 5x^2 + 6x - 20)$ by $(x - 6)$ using synthetic division.

8. Find a polynomial function with real coefficients that has the zeros $0, -3$, and $1 + \sqrt{5}i$.

In Exercises 9–11, sketch the graph of the rational function. Identify any asymptotes. Use a graphing utility to verify the graph.

9. $f(x) = \dfrac{2x}{x - 3}$

10. $f(x) = \dfrac{5x}{x^2 + x - 6}$

11. $f(x) = \dfrac{x^2 - 3x + 8}{x - 2}$

12. Write a rational function whose graph has no vertical asymptotes and has a horizontal asymptote at $y = 4$.

In Exercises 13–16, evaluate the expression without using a calculator.

13. $\log_2 64$

14. $\log_2\left(\dfrac{1}{16}\right)$

15. $\ln e^{10}$

16. $\ln \dfrac{1}{e^3}$

In Exercises 17–20, use a calculator to evaluate the expression. Round your answer to three decimal places.

17. $(1.85)^{3.1}$

18. $58^{\sqrt{5}}$

19. $e^{-20/11}$

20. $4e^{2.56}$

In Exercises 21–24, sketch the graph of the function by hand. Use a graphing utility to verify the graph.

21. $f(x) = -3^{x+4} - 5$

22. $f(x) = -\left(\frac{1}{2}\right)^{-x} - 3$

23. $f(x) = 4 + \log_{10}(x - 3)$

24. $f(x) = \ln(4 - x)$

In Exercises 25–28, evaluate the logarithm using the change-of-base formula. Round your result to three decimal places.

25. $\log_5 21$

26. $\log_9 6.8$

27. $\log_{3/4}(8.61)$

28. $\log_{7/8}\left(\frac{3}{2}\right)$

29. Write $2 \ln x - \frac{1}{2}\ln(x + 5)$ as a logarithm of a single quantity.

30. Use the properties of logarithms to expand $\ln\left(x\sqrt[3]{x - 5}\right)$.

In Exercises 31–36, solve the equation algebraically. Round your result to three decimal places and verify your result graphically.

31. $6e^{2x} = 72$ **32.** $4e^{x-3} + 21 = 30$ **33.** $e^{2x} - 11e^x + 24 = 0$

34. $-3 + \ln 4x = 0$ **35.** $\ln \sqrt{x + 2} = 3$ **36.** $\log_2 x + \log_2 5 = 6$

37. The profit P (in thousands of dollars) of an office supply company is given by

$$P = 230 + 20x - \tfrac{1}{2}x^2$$

where x is the amount (in hundreds of dollars) the company spends on advertising. What amount will yield a maximum profit?

38. The average cost $\overline{C}$ (in dollars) of recycling a waste product x (in pounds) is given by

$$\overline{C}(x) = \frac{450,000 + 5x}{x}, \quad x > 0.$$

Find the average cost $\overline{C}$ of recycling $x = 10,000$ pounds, $x = 100,000$ pounds, and $x = 1,000,000$ pounds. According to this model, what is the limiting average cost as the number of pounds increases?

39. You deposit $2500 in an account earning 7.5% interest, compounded continuously. Find the balance after 25 years.

40. If the inflation rate averages 4.5% over the next 10 years, the approximate cost C of goods or services t years from now is given by $C(t) = P(1.045)^t$, where P is the present cost. If the price of a tire is presently $69.95, estimate the price 10 years from now.

41. The population P (in thousands) of Baton Rouge, Louisiana is given by $P = 228e^{kt}$, where t represents the year, with $t = 0$ corresponding to 2000. In 1970, the population was 166,000. Find the value of k, and use this result to predict the population in the year 2010.   (Source: U.S. Census Bureau)

42. The population p of a species t years after it is introduced into a new habitat is given by $p(t) = 1200/(1 + 3e^{-t/5})$.

  (a) Determine the population size that was introduced into the habitat.

  (b) Determine the population after 5 years.

  (c) After how many years will the population be 800?

43. The table at the right shows the numbers y (in thousands) of pilots and copilots in the U.S. scheduled airline industry from 1994 to 2000.   (Source: Air Transport Association of America)

  (a) Use the *regression* feature of a graphing utility to find a quadratic model, an exponential model, and a power model for the data. Let x represent the year, with $x = 4$ corresponding to 1994.

  (b) Use a graphing utility to graph each model with the original data.

  (c) Determine which model best fits the data.

  (d) Use the model you chose in part (c) to predict the number of pilots and copilots in 2006.

Year	Number of pilots and copilots, y
1994	52.9
1995	55.4
1996	57.6
1997	60.4
1998	64.1
1999	67.2
2000	72.6

Trigonometric functions are used in medicine to calculate the pressure against the walls of blood vessels. One cycle of the trigonometric model corresponds to one heartbeat.

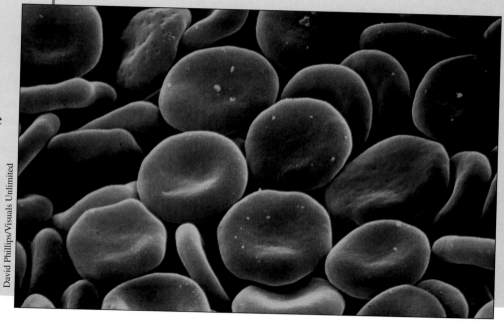

David Phillips/Visuals Unlimited

5 Trigonometric Functions

What You Should Learn

In this chapter, you will learn how to:

- Describe an angle and convert between degree and radian measures.

- Evaluate trigonometric functions of any angle.

- Use fundamental trigonometric identities.

- Sketch graphs of trigonometric functions.

- Evaluate inverse trigonometric functions.

- Evaluate the composition of trigonometric functions.

- Use trigonometric functions to model and solve real-life problems.

5.1   Angles and Their Measure

Angles

As derived from the Greek language, the word **trigonometry** means "measurement of triangles." Initially, trigonometry dealt with relationships among the sides and angles of triangles and was used in the development of astronomy, navigation, and surveying. With the development of calculus and the physical sciences in the 17th century, a different perspective arose—one that viewed the classic trigonometric relationships as *functions* having the set of real numbers as their domains. Consequently, the applications of trigonometry expanded to include a vast number of physical phenomena involving rotations and vibrations, including sound waves, light rays, planetary orbits, vibrating strings, pendulums, and orbits of atomic particles.

This text incorporates *both* perspectives, starting with angles and their measure.

What you should learn

- Describe angles.
- Use degree measure.
- Use radian measure.
- Convert between degree and radian measures.
- Use angles to model and solve real-life problems.

Why you should learn it

Radian measures of angles are involved in numerous aspects of our daily lives. For instance, in Exercise 89 on page 402, you are asked to determine measures of angles of figure-skating jumps.

Reuters NewMedia, Inc./Corbis

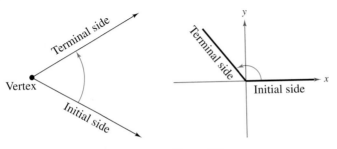

Figure 5.1   Figure 5.2

An **angle** is determined by rotating a ray (half-line) about its endpoint. The starting position of the ray is the **initial side** of the angle, and the position after rotation is the **terminal side,** as shown in Figure 5.1. The endpoint of the ray is the **vertex** of the angle. This perception of an angle fits a coordinate system in which the origin is the vertex and the initial side coincides with the positive *x*-axis. Such an angle is in **standard position,** as shown in Figure 5.2. **Positive angles** are generated by counterclockwise rotation, and **negative angles** by clockwise rotation, as shown in Figure 5.3. Angles are labeled with Greek letters such as α (alpha), β (beta), and θ (theta), as well as uppercase letters such as A, B, and C. In Figure 5.4, note that angles α and β have the same initial and terminal sides. Such angles are **coterminal.**

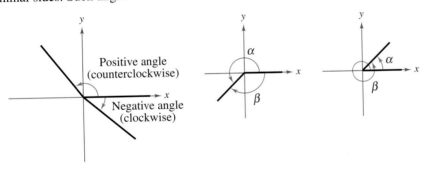

Figure 5.3   Figure 5.4

Degree Measure

The measure of an angle is determined by the amount of rotation from the initial side to the terminal side. The most common unit of angle measure is the **degree,** denoted by the symbol °. A measure of one degree (1°) is equivalent to a rotation of $\frac{1}{360}$ of a complete revolution about the vertex. To measure angles, it is convenient to mark degrees on the circumference of a circle, as shown in Figure 5.5. So, a full revolution (counterclockwise) corresponds to 360°, a half revolution to 180°, a quarter revolution to 90°, and so on.

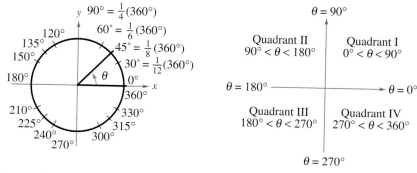

Figure 5.5   **Figure 5.6**

Recall that the four quadrants in a coordinate system are numbered I, II, III, and IV. Figure 5.6 shows which angles between 0° and 360° lie in each of the four quadrants. Figure 5.7 shows several common angles with their degree measures. Note that angles between 0° and 90° are **acute** and angles between 90° and 180° are **obtuse.**

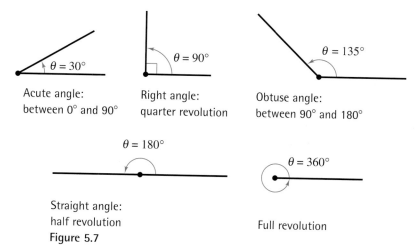

Figure 5.7

Two angles are coterminal if they have the same initial and terminal sides. For instance, the angles 0° and 360° are coterminal, as are the angles 30° and 390°. You can find an angle that is coterminal to a given angle θ by adding or subtracting 360° (one revolution), as demonstrated in Example 1. A given angle θ has infinitely many coterminal angles. For instance, $\theta = 30°$ is coterminal with

$$30° + n(360°)$$

where n is an integer.

Example 1   Sketching and Finding Coterminal Angles

Find two coterminal angles (one positive and one negative) for (a) $\theta = 390°$ and (b) $\theta = -120°$.

Solution

a. For the positive angle $\theta = 390°$, subtract $360°$ to obtain a positive coterminal angle.

$$390° - 360° = 30° \qquad \text{See Figure 5.8.}$$

Subtract $2(360°) = 720°$ to obtain a negative coterminal angle.

$$390° - 720° = -330°$$

b. For the negative angle $\theta = -120°$, add $360°$ to obtain a positive coterminal angle.

$$-120° + 360° = 240° \qquad \text{See Figure 5.9.}$$

Subtract $360°$ to obtain a negative coterminal angle.

$$-120° - 360° = -480°$$

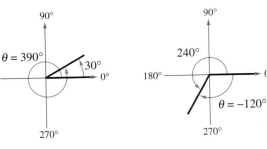

Figure 5.8   **Figure 5.9**

✓ *Checkpoint* Now try Exercise 15.

Two positive angles α and β are **complementary** (complements of each other) if their sum is $90°$. Two positive angles are **supplementary** (supplements of each other) if their sum is $180°$. See Figure 5.10.

Example 2   Complementary and Supplementary Angles

If possible, find the complement and the supplement of (a) $72°$ and (b) $148°$.

Solution

a. The complement of $72°$ is $90° - 72° = 18°$. The supplement of $72°$ is $180° - 72° = 108°$.

b. Because $148°$ is greater than $90°$, $148°$ has no complement. (Remember that complements are *positive* angles.) The supplement is

$$180° - 148° = 32°.$$

✓ *Checkpoint* Now try Exercise 29.

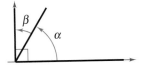

Complementary angles

Supplementary angles
Figure 5.10

Radian Measure

A second way to measure angles is in radians. This type of measure is especially useful in calculus. To define a radian, you can use a **central angle** of a circle, one whose vertex is the center of the circle, as shown in Figure 5.11.

> **Definition of a Radian**
>
> One **radian** is the measure of a central angle θ that intercepts an arc s equal in length to the radius r of the circle. See Figure 5.11.

Because the circumference of a circle is $2\pi r$ units, it follows that a central angle of one full revolution (counterclockwise) corresponds to an arc length of $s = 2\pi r$. Moreover, because $2\pi \approx 6.28$, there are just over six radius lengths in a full circle, as shown in Figure 5.12. In general, the radian measure of a central angle θ is obtained by dividing the arc length s by r. That is, $s/r = \theta$, where θ is *measured in radians*. Because the units of measure for s and r are the same, this ratio has no units—it is simply a real number.

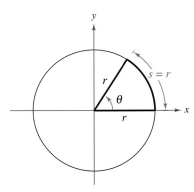

Arc length = radius when θ = 1 radian
Figure 5.11

Example 3   Finding Angles

Find each angle.

a. The complement of $\theta = \pi/12$ **b.** The supplement of $\theta = 5\pi/6$
c. A coterminal angle to $\theta = 17\pi/6$

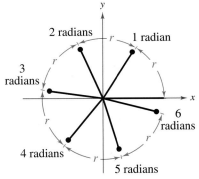

Figure 5.12

Solution

a. In radian measure, the complement of an angle is found by subtracting the angle from $\pi/2$ ($\pi/2 = 90°$). So, the complement of $\theta = \pi/12$ is $\pi/2 - \theta$, which is

$$\pi/2 - \pi/12 = 6\pi/12 - \pi/12 = 5\pi/12. \qquad \text{See Figure 5.13.}$$

b. In radian measure, the supplement of an angle is found by subtracting the angle from π ($\pi = 180°$). So, the supplement of $\theta = 5\pi/6$ is $\pi - \theta$, which is

$$\pi - 5\pi/6 = 6\pi/6 - 5\pi/6 = \pi/6. \qquad \text{See Figure 5.14.}$$

c. In radian measure, a coterminal angle is found by adding or subtracting 2π. For $\theta = 17\pi/6$, subtract 2π to obtain a coterminal angle.

$$17\pi/6 - 2\pi = 17\pi/6 - 12\pi/6 = 5\pi/6 \qquad \text{See Figure 5.15.}$$

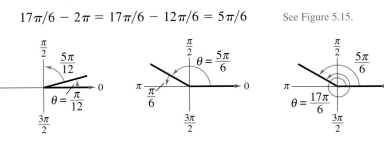

Figure 5.13   **Figure 5.14**   **Figure 5.15**

 Checkpoint Now try Exercise 49.

Conversion of Angle Measure

Because 2π radians corresponds to one complete revolution, degrees and radians are related by the equations

$$360° = 2\pi \text{ rad} \quad \text{and} \quad 180° = \pi \text{ rad}.$$

From the second equation, you obtain

$$1° = \frac{\pi}{180} \text{ rad} \quad \text{and} \quad 1 \text{ rad} = \left(\frac{180}{\pi}\right)°$$

which lead to the following conversion rules.

$\frac{\pi}{6}$
30°

$\frac{\pi}{4}$
45°

$\frac{\pi}{3}$
60°

$\frac{\pi}{2}$
90°

π
180°

2π
360°

Figure 5.16

Conversions Between Degrees and Radians

1. To convert degrees to radians, multiply degrees by $\dfrac{\pi \text{ rad}}{180°}$.

2. To convert radians to degrees, multiply radians by $\dfrac{180°}{\pi \text{ rad}}$.

To apply these two conversion rules, use the basic relationship $\pi \text{ rad} = 180°$. (See Figure 5.16.)

Example 4   Converting from Degrees to Radians

a. $135° = (135 \text{ deg})\left(\dfrac{\pi \text{ rad}}{180 \text{ deg}}\right) = \dfrac{3\pi}{4}$ radians   Multiply by $\dfrac{\pi}{180}$.

b. $-270° = (-270 \text{ deg})\left(\dfrac{\pi \text{ rad}}{180 \text{ deg}}\right) = -\dfrac{3\pi}{2}$ radians   Multiply by $\dfrac{\pi}{180}$.

✓ *Checkpoint* Now try Exercise 51.

Example 5   Converting from Radians to Degrees

a. $-\dfrac{\pi}{2} \text{ rad} = \left(-\dfrac{\pi}{2} \text{ rad}\right)\left(\dfrac{180 \text{ deg}}{\pi \text{ rad}}\right) = -90°$ Multiply by $\dfrac{180}{\pi}$.

b. $2 \text{ rad} = (2 \text{ rad})\left(\dfrac{180 \text{ deg}}{\pi \text{ rad}}\right) = \dfrac{360}{\pi} \approx 114.59°$ Multiply by $\dfrac{180}{\pi}$.

✓ *Checkpoint* Now try Exercise 55.

STUDY TIP

Note that when no units of angle measure are specified, *radian measure is implied*. For instance, if you write $\theta = \pi$ or $\theta = 2$, you imply that $\theta = \pi$ radians or $\theta = 2$ radians.

Linear and Angular Speed

The *radian measure* formula $\theta = s/r$ can be used to measure arc length along a circle. Specifically, for a circle of radius r, a central angle θ (θ is measured in radians) intercepts an arc of length s given by

$$s = r\theta. \qquad \text{Length of circular arc}$$

Example 6   Finding Arc Length

A circle has a radius of 4 inches. Find the length of the arc intercepted by a central angle of 240°, as shown in Figure 5.17.

Solution

To use the formula $s = r\theta$, first convert 240° to radian measure.

$$240° = (240 \text{ deg})\left(\frac{\pi \text{ rad}}{180 \text{ deg}}\right) \qquad \text{Convert from degrees to radians.}$$

$$= \frac{4\pi}{3} \text{ radians} \qquad \text{Simplify.}$$

Then, using a radius of $r = 4$ inches, you can find the arc length to be

$$s = r\theta \qquad \text{Length of circular arc}$$

$$= 4\left(\frac{4\pi}{3}\right) \qquad \text{Substitute for } r \text{ and } \theta.$$

$$= \frac{16\pi}{3} \qquad \text{Simplify.}$$

$$\approx 16.76 \text{ inches} \qquad \text{Use a calculator.}$$

Note that the units for $r\theta$ are determined by the units for r because θ is given in radian measure and therefore has no units.

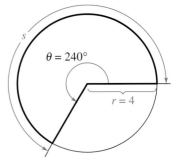

Figure 5.17

✓ *Checkpoint* Now try Exercise 79.

The formula for the length of a circular arc can be used to analyze the motion of a particle moving at a constant speed along a circular path.

Linear and Angular Speed

Consider a particle moving at a constant speed along a circular arc of radius r. If s is the length of the arc traveled in time t, then the **linear speed** of the particle is

$$\text{Linear speed} = \frac{\text{arc length}}{\text{time}} = \frac{s}{t}.$$

Moreover, if θ is the angle (in radian measure) corresponding to the arc length s, then the **angular speed** of the particle is

$$\text{Angular speed} = \frac{\text{central angle}}{\text{time}} = \frac{\theta}{t}.$$

Linear speed measures how fast the particle moves, and angular speed measures how fast the angle changes.

Example 7   Finding Linear Speed

The second hand of a clock is 10.2 centimeters long, as shown in Figure 5.18.
Find the linear speed of the tip of this second hand.

Solution

In one revolution, the arc length traveled is

$$s = 2\pi r$$

$$= 2\pi(10.2) \qquad \text{Substitute for } r.$$

$$= 20.4\pi \text{ centimeters.}$$

The time required for the second hand to travel this distance is

$$t = 1 \text{ minute} = 60 \text{ seconds.}$$

So, the linear speed of the tip of the second hand is

$$\text{Linear speed} = \frac{s}{t}$$

$$= \frac{20.4\pi \text{ centimeters}}{60 \text{ seconds}} \approx 1.068 \text{ centimeters per second.}$$

Figure 5.18

✓ *Checkpoint* Now try Exercise 91.

Example 8   Finding Angular and Linear Speed

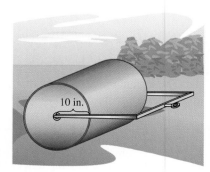

A lawn roller with a 10-inch radius makes 1.2 revolutions per second (see Figure 5.19).

a. Find the angular speed of the roller in radians per second.

b. Find the speed of the tractor that is pulling the roller.

Solution

a. Because each revolution generates 2π radians, it follows that the roller turns $(1.2)(2\pi) = 2.4\pi$ radians per second. In other words, the angular speed is

$$\text{Angular speed} = \frac{\theta}{t}$$

$$= \frac{2.4\pi \text{ radians}}{1 \text{ second}} = 2.4\pi \text{ radians per second.}$$

Figure 5.19

b. The linear speed is

$$\text{Linear speed} = \frac{s}{t} = \frac{r\theta}{t}$$

$$= \frac{10(2.4\pi) \text{ inches}}{1 \text{ second}} \approx 75.4 \text{ inches per second.}$$

✓ *Checkpoint* Now try Exercise 92.

5.1   Exercises

Vocabulary Check

Fill in the blanks.

1. _____ means "measurement of triangles."

2. An _____ is determined by rotating a ray about its endpoint.

3. An angle whose initial side coincides with the positive *x*-axis and that has the origin as its vertex is said to be in _____.

4. Two angles that have the same initial and terminal sides are _____ .

5. The angle measure that is equivalent to $\frac{1}{360}$ of a complete revolution about an angle's vertex is one _____ .

6. Two positive angles that have a sum of 90° are _____ angles.

7. Two positive angles that have a sum of 180° are _____ angles.

8. One _____ is the measure of a central angle that intercepts an arc equal to the radius of the circle.

9. The _____ speed of a particle is the ratio of the arc length traveled to the time traveled.

10. The _____ speed of a particle is the ratio of the change in the central angle to time.

In Exercises 1–4, estimate the number of degrees in the angle.

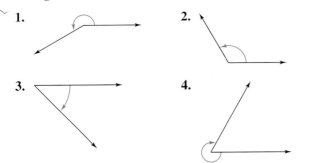

1.

2.

3.

4.

In Exercises 5–8, determine the quadrant in which each angle lies.

5. (a) 150°   (b) 282°

6. (a) 7.9°   (b) 257.5°

7. (a) −132° 50′   (b) −336° 30′

8. (a) −260.25°   (b) −2.4°

In Exercises 9–12, sketch each angle in standard position.

9. (a) 30°   (b) 150°

10. (a) −270°   (b) −120°

11. (a) 405°   (b) 780°

12. (a) −450°   (b) −600°

In Exercises 13–16, determine two coterminal angles in degree measure (one positive and one negative) for each angle.

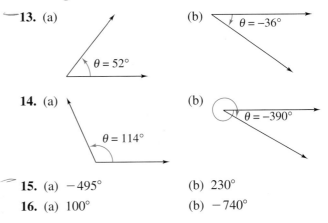

13. (a)   (b) $\theta = -36°$

$\theta = 52°$

14. (a)   (b) $\theta = -390°$

$\theta = 114°$

15. (a) −495°   (b) 230°

16. (a) 100°   (b) −740°

In Exercises 17–22, use the angle conversion capabilities of a graphing utility to convert the angle measure to decimal degree form. Round your answer to three decimal places, if necessary.

17. 64° 45′   18. −124° 30′

19. 85° 18′30″   20. −408° 16′25″

21. −125° 36″   22. 330° 25″

In Exercises 23–28, use the angle conversion capabilities of a graphing utility to convert the angle measure to D° M′S″ form.

23. 280.6°   24. −115.8°

25. $-345.12°$ **26.** $310.75°$

27. -0.355 **28.** 0.7865

In Exercises 29 and 30, find (if possible) the complement and supplement of each angle.

29. (a) $24°$ (b) $126°$ **30.** (a) $87°$ (b) $167°$

In Exercises 31–34, estimate the angle to the nearest one-half radian.

31.   **32.**

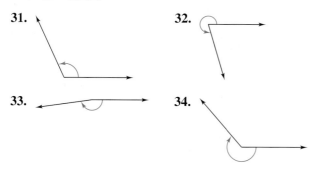

33.   **34.**

In Exercises 35–40, determine the quadrant in which each angle lies. (The angle is given in radians.)

35. (a) $\dfrac{\pi}{5}$ (b) $\dfrac{7\pi}{5}$ **36.** (a) $\dfrac{7\pi}{4}$ (b) $\dfrac{11\pi}{4}$

37. (a) $-\dfrac{\pi}{12}$ (b) $-\dfrac{11\pi}{9}$ **38.** (a) -1 (b) -2

39. (a) 3.5 (b) 2.25 **40.** (a) 5.63 (b) -2.25

In Exercises 41–44, sketch each angle in standard position.

41. (a) $\dfrac{3\pi}{4}$ (b) $\dfrac{5\pi}{6}$ **42.** (a) $-\dfrac{7\pi}{4}$ (b) $-\dfrac{7\pi}{2}$

43. (a) $\dfrac{11\pi}{6}$ (b) -3 **44.** (a) 4 (b) 5π

In Exercises 45–48, determine two coterminal angles in radian measure (one positive and one negative) for each angle.

45. (a)   (b)

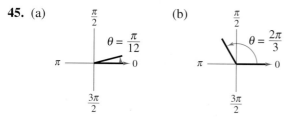

46. (a)   (b)

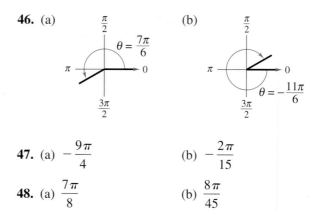

47. (a) $-\dfrac{9\pi}{4}$ (b) $-\dfrac{2\pi}{15}$

48. (a) $\dfrac{7\pi}{8}$ (b) $\dfrac{8\pi}{45}$

In Exercises 49 and 50, find (if possible) the complement and supplement of each angle.

49. (a) $\dfrac{\pi}{3}$ (b) $\dfrac{3\pi}{4}$

50. (a) 1 (b) 2

In Exercises 51–54, rewrite each angle in radian measure as a multiple of π. (Do not use a calculator.)

51. (a) $30°$ (b) $150°$

52. (a) $315°$ (b) $120°$

53. (a) $-20°$ (b) $-240°$

54. (a) $-270°$ (b) $144°$

In Exercises 55–58, rewrite each angle in degree measure. (Do not use a calculator.)

55. (a) $\dfrac{3\pi}{2}$ (b) $-\dfrac{7\pi}{6}$

56. (a) $-\dfrac{7\pi}{12}$ (b) $\dfrac{\pi}{9}$

57. (a) $\dfrac{7\pi}{3}$ (b) $-\dfrac{13\pi}{60}$

58. (a) $\dfrac{15\pi}{6}$ (b) $\dfrac{28\pi}{15}$

In Exercises 59–64, convert the angle measure from degrees to radians. Round your answer to three decimal places.

59. $126°$ **60.** $83.7°$

61. $-216.35°$ **62.** $-46.52°$

63. $-0.78°$ **64.** $383°$

In Exercises 65–70, convert the angle measure from radians to degrees. Round your answer to three decimal places.

65. $\dfrac{\pi}{7}$

66. $\dfrac{2\pi}{11}$

67. $\dfrac{13\pi}{8}$

68. 6.5π

69. -2

70. -0.39

In Exercises 71–74, find the angle in radians.

71.

72.

73.

74.

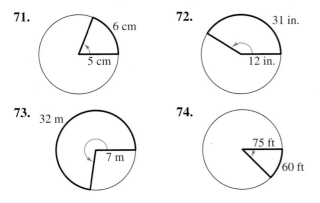

In Exercises 75–78, find the radian measure of the central angle of a circle of radius r that intercepts an arc of length s.

	Radius r	Arc Length s
75.	29 inches	8 inches
76.	14 feet	8 feet
77.	14.5 centimeters	35 centimeters
78.	80 kilometers	160 kilometers

In Exercises 79–82, find the length of the arc on a circle of radius r intercepted by a central angle θ.

	Radius r	Central Angle θ
79.	15 inches	$180°$
80.	9 feet	$60°$
81.	2 meters	1 radian
82.	40 centimeters	$\dfrac{3\pi}{4}$ radians

Distance   **In Exercises 83 and 84, find the distance between the cities. Assume that Earth is a sphere of radius 4000 miles and that the cities are on the same longitude (one city is due north of the other).**

	City	Latitude
83.	Miami	$25°\,46'\,26''$N
	Erie	$42°\,7'\,45''$N
84.	Johannesburg, South Africa	$26°\,11'$S
	Jerusalem, Israel	$31°\,47'$N

85. ***Difference in Latitudes***   Assuming that Earth is a sphere of radius 6378 kilometers, what is the difference in latitudes of Syracuse, New York and Annapolis, Maryland, where Syracuse is 450 kilometers due north of Annapolis?

86. ***Difference in Latitudes***   Assuming that Earth is a sphere of radius 6378 kilometers, what is the difference in latitudes of Lynchburg, Virginia and Myrtle Beach, South Carolina, where Lynchburg is 400 kilometers due north of Myrtle Beach?

87. ***Instrumentation***   A voltmeter's pointer is 6 centimeters in length (see figure). Find the angle through which it rotates when it moves 2.5 centimeters on the scale.

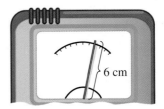

88. ***Electric Hoist***   An electric hoist is used to lift a beam 2 feet (see figure). The diameter of the drum on the hoist is 10 inches. Find the number of degrees through which the drum must rotate.

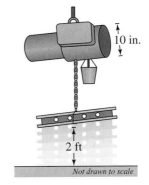

89. *Sports*   The number of revolutions made by a figure skater for each type of axel jump is given. Determine the measure of the angle generated as the skater performs each jump. Give the answer in both degrees and radians.

(a) Single axel: $1\frac{1}{2}$ (b) Double axel: $2\frac{1}{2}$

(c) Triple axel: $3\frac{1}{2}$

90. *Angular Speed*   A car is moving at a rate of 65 miles per hour, and the diameter of its wheels is 2.5 feet.

(a) Find the number of revolutions per minute the wheels are rotating.

(b) Find the angular speed of the wheels in radians per minute.

91. *Floppy Disk*   The radius of the magnetic disk in a 3.5-inch diskette is 1.68 inches. Find the linear speed of a point on the circumference of the disk if it is rotating at a speed of 360 revolutions per minute.

92. *Construction*   The circular blade on a saw has a diameter of 7.5 inches and rotates at 2400 revolutions per minute (see figure).

(a) Find the angular speed in radians per second.

(b) Find the linear speed of the saw teeth (in feet per second) as they contact the wood being cut.

├──── 7.5 in.────┤

Synthesis

True or False?   **In Exercises 93 and 94, determine whether the statement is true or false. Justify your answer.**

93. A degree is a larger unit of measure than a radian.

94. An angle that measures $-1260°$ lies in Quadrant III.

95. *Writing*   In your own words, explain the meanings of (a) an angle in standard position, (b) a negative angle, (c) coterminal angles, and (d) an obtuse angle.

96. *Geometry*   Show that the area of a circular sector of radius r with central angle θ is $A = \frac{1}{2}r^2\theta$, where θ is measured in radians.

Geometry   **In Exercises 97 and 98, use the result of Exercise 96 to find the area of the sector.**

97.   **98.**

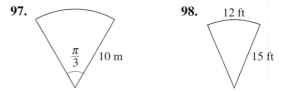

99. *Graphical Reasoning*   The formulas for the area of a circular sector and arc length are $A = \frac{1}{2}r^2\theta$ and $s = r\theta$, respectively. (r is the radius and θ is the angle measured in radians.)

(a) If $\theta = 0.8$, write the area and arc length as functions of r. What is the domain of each function? Use a graphing utility to graph the functions. Use the graphs to determine which function changes more rapidly as r increases. Explain.

(b) If $r = 10$ centimeters, write the area and arc length as functions of θ. What is the domain of each function? Use a graphing utility to graph and identify the functions.

100. *Writing*   A fan motor turns at a given angular speed. How does the speed of the tips of the blades change if a fan of greater diameter is installed on the motor? Explain.

Review

In Exercises 101 and 102, use the *regression* feature of a graphing utility to find a linear model that approximates the set of data.

101.

x	2	3	4	5	6	7
y	25	31	33	40	45	47

102.

x	-3	-2	-1	0	1	2
y	58.3	49.9	45.0	38.1	33.2	23.6

In Exercises 103–106, find all the real zeros of the polynomial function.

103. $f(x) = x^2 + 11x + 28$

104. $f(x) = 54x^2 - 6x^4$

105. $f(x) = x^3 + 3x^2 - 10x$

106. $f(x) = 4x^4 + 44x^3 + 96x^2$

5.2 | Right Triangle Trigonometry

The Six Trigonometric Functions

Our first look at the trigonometric functions is from a *right triangle* perspective. Consider a right triangle with one acute angle labeled θ, as shown in Figure 5.20. Relative to the angle θ, the three sides of the triangle are the **hypotenuse,** the **opposite side** (the side opposite the angle θ), and the **adjacent side** (the side adjacent to the angle θ).

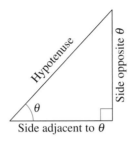

Figure 5.20

Using the lengths of these three sides, you can form six ratios that define the six trigonometric functions of the acute angle θ.

sine   cosecant   cosine   secant   tangent   cotangent

These six functions are normally abbreviated as sin, csc, cos, sec, tan, and cot, respectively. In the following definitions it is important to see that $0° < \theta < 90°$ (θ lies in the first quadrant) and that for such angles the value of each trigonometric function is *positive*.

> **Right Triangle Definitions of Trigonometric Functions**
>
> Let θ be an *acute* angle of a right triangle. Then the six trigonometric functions *of the angle* θ are defined as follows. (Note that the functions in the second row are the *reciprocals* of the corresponding functions in the first row.)
>
> $$\sin \theta = \frac{\text{opp}}{\text{hyp}} \qquad \cos \theta = \frac{\text{adj}}{\text{hyp}} \qquad \tan \theta = \frac{\text{opp}}{\text{adj}}$$
>
> $$\csc \theta = \frac{\text{hyp}}{\text{opp}} \qquad \sec \theta = \frac{\text{hyp}}{\text{adj}} \qquad \cot \theta = \frac{\text{adj}}{\text{opp}}$$
>
> The abbreviations "opp," "adj," and "hyp" represent the lengths of the three sides of a right triangle.
>
> opp = the length of the side *opposite* θ
>
> adj = the length of the side *adjacent* to θ
>
> hyp = the length of the *hypotenuse*

What you should learn

- Evaluate trigonometric functions of acute angles.
- Use the fundamental trigonometric identities.
- Use a calculator to evaluate trigonometric functions.
- Use trigonometric functions to model and solve real-life problems.

Why you should learn it

You can use trigonometry to analyze all aspects of a geometric figure. For instance, Exercise 62 on page 412 shows you how trigonometric functions can be used to approximate the height of the Jin Mao Building in China.

Chen Yixin/China Stock

Example 1 Evaluating Trigonometric Functions

Use the triangle in Figure 5.21 to find the exact values of the six trigonometric functions of θ.

Solution

By the Pythagorean Theorem, $(\text{hyp})^2 = (\text{opp})^2 + (\text{adj})^2$, it follows that

$$\text{hyp} = \sqrt{4^2 + 3^2} = \sqrt{25} = 5.$$

So, the six trigonometric functions of θ are

$$\sin \theta = \frac{\text{opp}}{\text{hyp}} = \frac{4}{5} \qquad \csc \theta = \frac{\text{hyp}}{\text{opp}} = \frac{5}{4}$$

$$\cos \theta = \frac{\text{adj}}{\text{hyp}} = \frac{3}{5} \qquad \sec \theta = \frac{\text{hyp}}{\text{adj}} = \frac{5}{3}$$

$$\tan \theta = \frac{\text{opp}}{\text{adj}} = \frac{4}{3} \qquad \cot \theta = \frac{\text{adj}}{\text{opp}} = \frac{3}{4}.$$

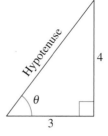

Figure 5.21

✓ *Checkpoint* Now try Exercise 3.

In Example 1, you were given the lengths of two sides of the right triangle, but not the angle θ. Often you will be asked to find the trigonometric functions for a *given* acute angle θ. To do this, construct a right triangle having θ as one of its angles.

Example 2 Evaluating Trigonometric Functions of 45°

Find the values of $\sin 45°$, $\cos 45°$, and $\tan 45°$.

Solution

Construct a right triangle having $45°$ as one of its acute angles, as shown in Figure 5.22. Choose 1 as the length of the adjacent side. From geometry, you know that the other acute angle is also $45°$. So, the triangle is isosceles, and the length of the opposite side is also 1. Using the Pythagorean Theorem, you find the length of the hypotenuse to be $\sqrt{2}$.

$$\sin 45° = \frac{\text{opp}}{\text{hyp}} = \frac{1}{\sqrt{2}} = \frac{\sqrt{2}}{2}$$

$$\cos 45° = \frac{\text{adj}}{\text{hyp}} = \frac{1}{\sqrt{2}} = \frac{\sqrt{2}}{2}$$

$$\tan 45° = \frac{\text{opp}}{\text{adj}} = \frac{1}{1} = 1$$

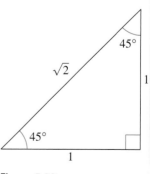

Figure 5.22

✓ *Checkpoint* Now try Exercise 17.

TECHNOLOGY TIP You can use a calculator to convert the answers in Example 2 to decimals. However, the radical form is the exact value and in most cases, the exact value is preferred.

Example 3   Evaluating Trigonometric Functions of 30° and 60°

Use the equilateral triangle shown in Figure 5.23 to find the values of sin 60°, cos 60°, sin 30°, and cos 30°.

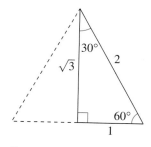

Figure 5.23

Solution

Use the Pythagorean Theorem and the equilateral triangle to verify the lengths of the sides given in Figure 5.23. For $\theta = 60°$, you have adj $= 1$, opp $= \sqrt{3}$, and hyp $= 2$. So,

$$\sin 60° = \frac{\text{opp}}{\text{hyp}} = \frac{\sqrt{3}}{2} \qquad \text{and} \qquad \cos 60° = \frac{\text{adj}}{\text{hyp}} = \frac{1}{2}.$$

For $\theta = 30°$, adj $= \sqrt{3}$, opp $= 1$, and hyp $= 2$. So,

$$\sin 30° = \frac{\text{opp}}{\text{hyp}} = \frac{1}{2} \qquad \text{and} \qquad \cos 30° = \frac{\text{adj}}{\text{hyp}} = \frac{\sqrt{3}}{2}.$$

 Checkpoint Now try Exercise 19.

Sines, Cosines, and Tangents of Special Angles

$$\sin 30° = \sin \frac{\pi}{6} = \frac{1}{2} \qquad \cos 30° = \cos \frac{\pi}{6} = \frac{\sqrt{3}}{2} \qquad \tan 30° = \tan \frac{\pi}{6} = \frac{\sqrt{3}}{3}$$

$$\sin 45° = \sin \frac{\pi}{4} = \frac{\sqrt{2}}{2} \qquad \cos 45° = \cos \frac{\pi}{4} = \frac{\sqrt{2}}{2} \qquad \tan 45° = \tan \frac{\pi}{4} = 1$$

$$\sin 60° = \sin \frac{\pi}{3} = \frac{\sqrt{3}}{2} \qquad \cos 60° = \cos \frac{\pi}{3} = \frac{1}{2} \qquad \tan 60° = \tan \frac{\pi}{3} = \sqrt{3}$$

In the box, note that $\sin 30° = \frac{1}{2} = \cos 60°$. This occurs because 30° and 60° are complementary angles, and, in general, it can be shown from the right triangle definitions that *cofunctions of complementary angles are equal*. That is, if θ is an acute angle, the following relationships are true.

$$\sin(90° - \theta) = \cos \theta \qquad \cos(90° - \theta) = \sin \theta$$

$$\tan(90° - \theta) = \cot \theta \qquad \cot(90° - \theta) = \tan \theta$$

$$\sec(90° - \theta) = \csc \theta \qquad \csc(90° - \theta) = \sec \theta$$

Trigonometric Identities

In trigonometry, a great deal of time is spent studying relationships between trigonometric functions (identities).

Fundamental Trigonometric Identities

Reciprocal Identities

$$\sin \theta = \frac{1}{\csc \theta} \qquad \cos \theta = \frac{1}{\sec \theta} \qquad \tan \theta = \frac{1}{\cot \theta}$$

$$\csc \theta = \frac{1}{\sin \theta} \qquad \sec \theta = \frac{1}{\cos \theta} \qquad \cot \theta = \frac{1}{\tan \theta}$$

Quotient Identities

$$\tan \theta = \frac{\sin \theta}{\cos \theta} \qquad \cot \theta = \frac{\cos \theta}{\sin \theta}$$

Pythagorean Identities

$$\sin^2 \theta + \cos^2 \theta = 1$$

$$1 + \tan^2 \theta = \sec^2 \theta$$

$$1 + \cot^2 \theta = \csc^2 \theta$$

Note that $\sin^2 \theta$ represents $(\sin \theta)^2$, $\cos^2 \theta$ represents $(\cos \theta)^2$, and so on.

Exploration

Select a number t and use your graphing utility to calculate $(\sin t)^2 + (\cos t)^2$. Repeat this experiment for other values of t and explain why the answer is always the same. Is the result true in both *radian* and *degree* modes?

Example 4   Applying Trigonometric Identities

Let θ be an acute angle such that $\sin \theta = 0.6$. Find the values of (a) $\cos \theta$ and (b) $\tan \theta$ using trigonometric identities.

Solution

a. To find the value of $\cos \theta$, use the Pythagorean identity

$$\sin^2 \theta + \cos^2 \theta = 1.$$

So, you have

$$(0.6)^2 + \cos^2 \theta = 1 \qquad \text{Substitute 0.6 for } \sin \theta.$$

$$\cos^2 \theta = 1 - (0.6)^2 = 0.64 \qquad \text{Subtract } (0.6)^2 \text{ from each side.}$$

$$\cos \theta = \sqrt{0.64} = 0.8. \qquad \text{Extract positive square root.}$$

b. Now, knowing the sine and cosine of θ, you can find the tangent of θ to be

$$\tan \theta = \frac{\sin \theta}{\cos \theta} = \frac{0.6}{0.8} = 0.75.$$

Use the definitions of $\cos \theta$ and $\tan \theta$ and the triangle shown in Figure 5.24 to check these results.

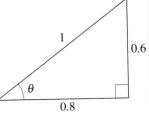

Figure 5.24

 Checkpoint Now try Exercise 31.

Example 5   Using Trigonometric Identities

Use trigonometric identities to transform one side of the equation into the other $(0 < \theta < \pi/2)$.

a. $\cos \theta \sec \theta = 1$ **b.** $(\sec \theta + \tan \theta)(\sec \theta - \tan \theta) = 1$

Solution

Simplify the expression on the left-hand side of the equation until you obtain the right-hand side.

a. $\cos \theta \sec \theta = \left(\dfrac{1}{\sec \theta} \right) \sec \theta$ Reciprocal identity

$\qquad\qquad\qquad = 1$ Divide out common factor.

b. $(\sec \theta + \tan \theta)(\sec \theta - \tan \theta)$

$\qquad = \sec^2 \theta - \sec \theta \tan \theta + \sec \theta \tan \theta - \tan^2 \theta$ Distributive Property

$\qquad = \sec^2 \theta - \tan^2 \theta$ Simplify.

$\qquad = 1$ Pythagorean identity

✓ *Checkpoint* Now try Exercise 33.

Evaluating Trigonometric Functions with a Calculator

When evaluating a trigonometric function with a calculator, you need to set the calculator to the desired *mode* of measurement (degrees or radians).

Most calculators do not have keys for the cosecant, secant, and cotangent functions. To evaluate these functions, you can use the $\boxed{x^{-1}}$ key with their respective reciprocal functions sine, cosine, and tangent. For example, to evaluate $\csc(\pi/8)$, use the fact that

$$\csc \frac{\pi}{8} = \frac{1}{\sin(\pi/8)}$$

and enter the following keystroke sequence in *radian* mode.

$\boxed{(}$ $\boxed{\text{SIN}}$ π $\boxed{\div}$ 8 $\boxed{)}$ $\boxed{)}$ $\boxed{x^{-1}}$ $\boxed{\text{ENTER}}$ Display 2.6131259

TECHNOLOGY TIP

The reciprocal identities for sine, cosine, and tangent can be used to evaluate the cosecant, secant, and cotangent functions with a calculator. For instance, you could use the following alternative keystroke sequence to evaluate the function in Example 6(c).

1 $\boxed{\div}$ $\boxed{\text{TAN}}$ 1.5 $\boxed{)}$ $\boxed{\text{ENTER}}$

You should obtain 0.0709148.

Example 6   Using a Calculator

Function	Mode	Graphing Calculator Keystrokes	Display
a. $\sin 76.4°$	Degree	$\boxed{\text{SIN}}$ 76.4 $\boxed{\text{ENTER}}$	0.9719610
b. $\cos 89°$	Degree	$\boxed{\text{COS}}$ 89 $\boxed{\text{ENTER}}$	0.0174524
c. $\cot 1.5$	Radian	$\boxed{(}$ $\boxed{\text{TAN}}$ 1.5 $\boxed{)}$ $\boxed{x^{-1}}$ $\boxed{\text{ENTER}}$	0.0709148

✓ *Checkpoint* Now try Exercise 41.

TECHNOLOGY TIP   When evaluating trigonometric functions with a calculator, remember to enclose all fractional angle measures in parentheses. For instance, if you want to evaluate $\sin \theta$ for $\theta = \pi/6$, you should enter

SIN π ÷ 6) ENTER.

These keystrokes yield the correct value of 0.5.

Applications Involving Right Triangles

Many applications of trigonometry involve a process called **solving right triangles.** In this type of application, you are usually given one side of a right triangle and one of the acute angles and asked to find one of the other sides, *or* you are given two sides and asked to find one of the acute angles. In Example 7, the angle you are given is the **angle of elevation,** which represents the angle from the horizontal upward to the object. In other applications you may be given the **angle of depression,** which represents the angle from the horizontal downward to the object.

Example 7   Using Trigonometry to Solve a Right Triangle

A surveyor is standing 50 feet from the base of a large tree, as shown in Figure 5.25. The surveyor measures the angle of elevation to the top of the tree as 71.5°. How tall is the tree?

Solution

From Figure 5.25, you can see that

$$\tan 71.5° = \frac{\text{opp}}{\text{adj}} = \frac{y}{x}$$

where $x = 50$ and y is the height of the tree. So, the height of the tree is

$$y = x \tan 71.5° \approx 50(2.9887) \approx 149.4 \text{ feet.}$$

✓ *Checkpoint* Now try Exercise 59.

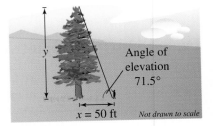

Angle of elevation 71.5°

$x = 50$ ft   *Not drawn to scale*

Figure 5.25

Example 8   Using Trigonometry to Solve a Right Triangle

You are 200 yards from a river. Rather than walking directly to the river, you walk 400 yards along a straight path to the river's edge. Find the acute angle θ between this path and the river's edge, as illustrated in Figure 5.26.

Solution

From Figure 5.26, you can see that the sine of the angle θ is

$$\sin \theta = \frac{\text{opp}}{\text{hyp}} = \frac{200}{400} = \frac{1}{2}.$$

Now, you should recognize that $\theta = 30°$.

✓ *Checkpoint* Now try Exercise 61.

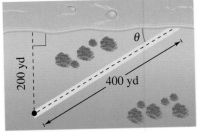

200 yd θ 400 yd

Figure 5.26

In Example 8, you were able to recognize that $\theta = 30°$ is the acute angle that satisfies the equation $\sin \theta = \frac{1}{2}$. Suppose, however, that you were given the equation $\sin \theta = 0.6$ and were asked to find the acute angle θ. Because

$$\sin 30° = \frac{1}{2} = 0.5000$$

and

$$\sin 45° = \frac{1}{\sqrt{2}} \approx 0.7071$$

you might guess that θ lies somewhere between $30°$ and $45°$. In a later section, you will study a method by which a more precise value of θ can be determined.

TECHNOLOGY TIP

Calculators and graphing utilities have both *degree* and *radian* modes. As you progress through this chapter, be sure you use the correct mode.

Example 9   Solving a Right Triangle

Find the length c of the skateboard ramp shown in Figure 5.27.

Figure 5.27

Solution

From Figure 5.27, you can see that

$$\sin 18.4° = \frac{\text{opp}}{\text{hyp}}$$

$$= \frac{4}{c}.$$

So, the length of the ramp is

$$c = \frac{4}{\sin 18.4°}$$

$$\approx \frac{4}{0.3156}$$

$$\approx 12.7 \text{ feet.}$$

✓ *Checkpoint* Now try Exercise 63.

5.2   Exercises

Vocabulary Check

1. Match the trigonometric function with its right triangle definition.

   (a) Sine   (b) Cosine   (c) Tangent

   (d) Cosecant   (e) Secant   (f) Cotangent

   (i) $\dfrac{\text{hyp}}{\text{adj}}$ (ii) $\dfrac{\text{opp}}{\text{adj}}$ (iii) $\dfrac{\text{opp}}{\text{hyp}}$

   (iv) $\dfrac{\text{adj}}{\text{opp}}$ (v) $\dfrac{\text{hyp}}{\text{opp}}$ (vi) $\dfrac{\text{adj}}{\text{hyp}}$

In Exercises 2 and 3, fill in the blanks.

2. Relative to the angle θ, the three sides of a right triangle are the _____ , the _____ side, and the _____ side.

3. An angle that measures from the horizontal upward to an object is called the angle of _____ , whereas an angle that measures from the horizontal downward to an object is called the angle of _____ .

In Exercises 1–4, find the exact values of the six trigonometric functions of the angle θ shown in the figure. (Use the Pythagorean Theorem to find the third side of the triangle.)

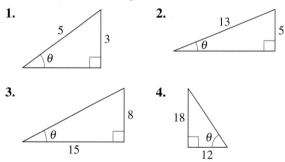

1.

2.

3.

4.

In Exercises 5–8, find the exact values of the six trigonometric functions of the angle θ for each of the triangles. Explain why the function values are the same.

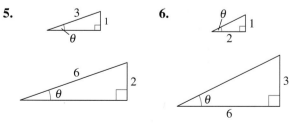

5.

6.

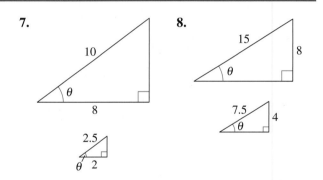

7.

8.

In Exercises 9–16, sketch a right triangle corresponding to the trigonometric function of the acute angle θ. Use the Pythagorean Theorem to determine the third side and then find the other five trigonometric functions of θ.

9. $\sin \theta = \dfrac{2}{3}$ 10. $\cot \theta = \dfrac{3}{7}$

11. $\sec \theta = 4$ 12. $\cos \theta = \dfrac{1}{5}$

13. $\tan \theta = 3$ 14. $\csc \theta = 2$

15. $\cot \theta = \dfrac{9}{4}$ 16. $\sin \theta = \dfrac{3}{8}$

In Exercises 17–26, construct an appropriate triangle to complete the table. ($0 \le \theta \le 90°, 0 \le \theta \le \pi/2$)

	Function	θ (deg)	θ (rad)	Function Value
17.	sin	30°		
18.	cos	45°		

Function	θ (deg)	θ (rad)	Function Value
19. tan		$\dfrac{\pi}{3}$	
20. sec		$\dfrac{\pi}{4}$	
21. cot			$\dfrac{\sqrt{3}}{3}$
22. csc			$\sqrt{2}$
23. cos		$\dfrac{\pi}{6}$	
24. sin		$\dfrac{\pi}{4}$	
25. cot			1
26. tan			$\dfrac{1}{\sqrt{3}}$

In Exercises 27–32, use the given function value(s) and trigonometric identities to find the indicated trigonometric functions.

27. $\sin 60° = \dfrac{\sqrt{3}}{2}, \quad \cos 60° = \dfrac{1}{2}$

(a) $\tan 60°$ (b) $\sin 30°$

(c) $\cos 30°$ (d) $\cot 60°$

28. $\sin 30° = \dfrac{1}{2}, \quad \tan 30° = \dfrac{\sqrt{3}}{3}$

(a) $\csc 30°$ (b) $\cot 60°$

(c) $\cos 30°$ (d) $\cot 30°$

29. $\csc \theta = 3, \quad \sec \theta = \dfrac{3\sqrt{2}}{4}$

(a) $\sin \theta$ (b) $\cos \theta$

(c) $\tan \theta$ (d) $\sec(90° - \theta)$

30. $\sec \theta = 5, \quad \tan \theta = 2\sqrt{6}$

(a) $\cos \theta$ (b) $\cot \theta$

(c) $\cot(90° - \theta)$ (d) $\sin \theta$

31. $\cos \alpha = \dfrac{1}{4}$

(a) $\sec \alpha$ (b) $\sin \alpha$

(c) $\cot \alpha$ (d) $\sin(90° - \alpha)$

32. $\tan \beta = 5$

(a) $\cot \beta$ (b) $\cos \beta$

(c) $\tan(90° - \beta)$ (d) $\csc \beta$

In Exercises 33–40, use trigonometric identities to transform one side of the equation into the other $(0 < \theta < \pi/2)$.

33. $\tan \theta \cot \theta = 1$

34. $\csc \theta \tan \theta = \sec \theta$

35. $\tan \theta \cos \theta = \sin \theta$

36. $\cot \theta \sin \theta = \cos \theta$

37. $(1 + \cos \theta)(1 - \cos \theta) = \sin^2 \theta$

38. $(1 + \sin \theta)(1 - \sin \theta) = \cos^2 \theta$

39. $\dfrac{\sin \theta}{\cos \theta} + \dfrac{\cos \theta}{\sin \theta} = \csc \theta \sec \theta$

40. $\dfrac{\tan \theta + \cot \theta}{\tan \theta} = \csc^2 \theta$

In Exercises 41–46, use a calculator to evaluate each function. Round your answers to four decimal places. (Be sure the calculator is in the correct angle mode.)

41. (a) $\sin 12°$ (b) $\cos 72°$

42. (a) $\tan 18.5°$ (b) $\cot 71.5°$

43. (a) $\sec 42° \, 12'$ (b) $\csc 48° \, 7'$

44. (a) $\cos 8° \, 50' \, 25''$ (b) $\sec 8° \, 50' \, 25''$

45. (a) $\cot \dfrac{\pi}{16}$ (b) $\tan \dfrac{\pi}{16}$

46. (a) $\sec 0.75$ (b) $\cos 0.75$

In Exercises 47–52, find each value of θ in degrees $(0° < \theta < 90°)$ and radians $(0 < \theta < \pi/2)$ without using a calculator.

47. (a) $\sin \theta = \tfrac{1}{2}$ (b) $\csc \theta = 2$

48. (a) $\cos \theta = \dfrac{\sqrt{2}}{2}$ (b) $\tan \theta = 1$

49. (a) $\sec \theta = 2$ (b) $\cot \theta = 1$

50. (a) $\tan \theta = \sqrt{3}$ (b) $\cos \theta = \tfrac{1}{2}$

51. (a) $\csc \theta = \dfrac{2\sqrt{3}}{3}$ (b) $\sin \theta = \dfrac{\sqrt{2}}{2}$

52. (a) $\cot \theta = \dfrac{\sqrt{3}}{3}$ (b) $\sec \theta = \sqrt{2}$

In Exercises 53–56, solve for *x*, *y*, or *r*, as indicated.

53. Solve for *y*.

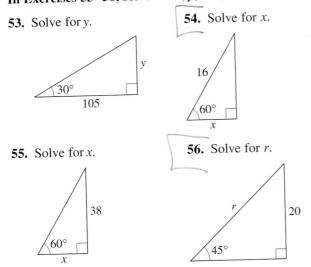

54. Solve for *x*.

55. Solve for *x*.

56. Solve for *r*.

57. *Height*   A six-foot person walks from the base of a streetlight directly toward the tip of the shadow cast by the streetlight. When the person is 16 feet from the streetlight and 5 feet from the tip of the street-light's shadow, the person's shadow starts to appear beyond the streetlight's shadow.

(a) Draw a right triangle that gives a visual representation of the problem. Show the known quantities of the triangle and use a variable to indicate the height of the streetlight.

(b) Use a trigonometric function to write an equation involving the unknown quantity.

(c) What is the height of the streetlight?

58. *Height*   A 30-meter line is used to tether a helium-filled balloon. Because of a breeze, the line makes an angle of approximately 75° with the ground.

(a) Draw a right triangle that gives a visual representation of the problem. Show the known quantities of the triangle and use a variable to indicate the height of the balloon.

(b) Use a trigonometric function to write an equation involving the unknown quantity.

(c) What is the height of the balloon?

59. *Width*   A biologist wants to know the width *w* of a river in order to properly set instruments for studying the pollutants in the water. From point A, the biologist walks downstream 100 feet and sights to point *C*. From this sighting, it is determined that $\theta = 58°$. How wide is the river? Verify your result numerically.

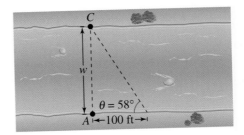

Figure for 59

60. *Height*   In traveling across flat land, you notice a mountain directly in front of you. Its angle of elevation (to the peak) is 3.5°. After you drive 13 miles closer to the mountain, the angle of elevation is 9°. Approximate the height of the mountain.

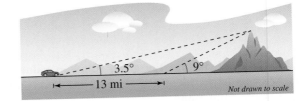

61. *Angle of Elevation*   A ramp 20 feet in length rises to a loading platform that is $3\frac{1}{3}$ feet off the ground.

(a) Draw a right triangle that gives a visual representation of the problem. Show the known quantities of the triangle and use a variable to indicate the angle of elevation of the ramp.

(b) Use a trigonometric function to write an equation involving the unknown quantity.

(c) Use a graphing utility to approximate the angle of elevation numerically.

62. *Jin Mao Building*   You are standing 65 meters from the base of the Jin Mao Building in Shanghai, China. You estimate that the angle of elevation to the top of the 88th floor (sightseeing level) is 80°. What is the approximate height of the building? One of your friends is on the sightseeing level. What is the distance between you and your friend?

63. *Length*   A guywire is stretched from the top of a 200-foot broadcasting tower to an anchor making an angle of 58° with the ground.

(a) How long is the wire?

(b) How far is the anchor from the base of the tower?

64. Machine Shop Calculations A steel plate has the form of one-fourth of a circle with a radius of 60 centimeters. Two two-centimeter holes are to be drilled in the plate, positioned as shown in the figure. Find the coordinates of the center of each hole.

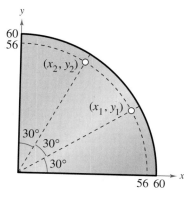

65. Machine Shop Calculations A tapered shaft has a diameter of 5 centimeters at the small end and is 15 centimeters long (see figure). The taper is 3°. Find the diameter d of the large end of the shaft.

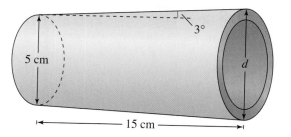

66. Geometry Use a compass to sketch a quarter of a circle of radius 10 centimeters. Using a protractor, construct an angle of 20° in standard position (see figure). Drop a perpendicular from the point of intersection of the terminal side of the angle and the arc of the circle. By actual measurement, calculate the coordinates (x, y) of the point of intersection and use these measurements to approximate the six trigonometric functions of a 20° angle.

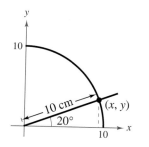

Synthesis

True or False? **In Exercises 67–70, determine whether the statement is true or false. Justify your answer.**

67. $\sin 60° \csc 60° = 1$ **68.** $\sec 30° = \csc 60°$

69. $\sin 45° + \cos 45° = 1$

70. $\cot^2 10° - \csc^2 10° = -1$

71. Exploration

(a) Use a graphing utility to complete the table. Round your results to four decimal places.

θ	0°	20°	40°	60°	80°
$\sin \theta$					
$\cos \theta$					
$\tan \theta$					

(b) Classify each of the three trigonometric functions as increasing or decreasing for the table values.

(c) From the values in the table, verify that the tangent function is the quotient of the sine and cosine functions.

72. Exploration Use a graphing utility to complete the table and make a conjecture about the relationship between $\cos \theta$ and $\sin(90° - \theta)$. What are the angles θ and $90° - \theta$ called?

θ	0°	20°	40°	60°	80°
$\cos \theta$					
$\sin (90° - \theta)$					

Review

In Exercises 73–76, sketch the graph of the equation and identify all x- and y-intercepts.

73. $y = -x - 9$ **74.** $2x + y = 10$

75. $-3x + 8y = 16$ **76.** $12x - 7y = 22$

In Exercises 77–80, use a calculator to evaluate the expression. Round your result to three decimal places.

77. $2.16^{3.8}$ **78.** $4^{2\pi}$

79. $\sqrt[3]{5286}$ **80.** $\sqrt[5]{10,321}$

5.3   Trigonometric Functions of Any Angle

Introduction

In Section 5.2, the definitions of trigonometric functions were restricted to acute angles. In this section, the definitions are extended to cover *any* angle. If θ is an *acute* angle, the definitions here coincide with those given in the preceding section.

Definitions of Trigonometric Functions of Any Angle

Let θ be an angle in standard position with (x, y) a point on the terminal side of θ and $r = \sqrt{x^2 + y^2} \neq 0$.

$$\sin \theta = \frac{y}{r} \qquad\qquad \cos \theta = \frac{x}{r}$$

$$\tan \theta = \frac{y}{x}, \quad x \neq 0 \qquad \cot \theta = \frac{x}{y}, \quad y \neq 0$$

$$\sec \theta = \frac{r}{x}, \quad x \neq 0 \qquad \csc \theta = \frac{r}{y}, \quad y \neq 0$$

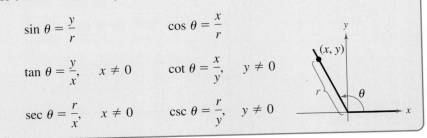

Because $r = \sqrt{x^2 + y^2}$ *cannot* be zero, it follows that the sine and cosine functions are defined for any real value of θ. However, if $x = 0$, the tangent and secant of θ are undefined. For example, the tangent of $90°$ is undefined. Similarly, if $y = 0$, the cotangent and cosecant of θ are undefined.

Example 1   Evaluating Trigonometric Functions

Let $(-3, 4)$ be a point on the terminal side of θ. Find the sine, cosine, and tangent of θ.

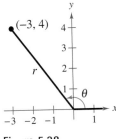

Figure 5.28

Solution

Referring to Figure 5.28, you can see that $x = -3$, $y = 4$, and

$$r = \sqrt{x^2 + y^2} = \sqrt{(-3)^2 + 4^2} = \sqrt{25} = 5.$$

So, you have $\sin \theta = \dfrac{y}{r} = \dfrac{4}{5}$, $\cos \theta = \dfrac{x}{r} = -\dfrac{3}{5}$, and $\tan \theta = \dfrac{y}{x} = -\dfrac{4}{3}$.

✓ *Checkpoint* Now try Exercise 1.

The *signs* of the trigonometric functions in the four quadrants can be determined easily from the definitions of the functions. For instance, because $\cos \theta = x/r$, it follows that $\cos \theta$ is positive wherever $x > 0$, which is in Quadrants I and IV. (Remember, r is always positive.) In a similar manner, you can verify the results shown in Figure 5.29.

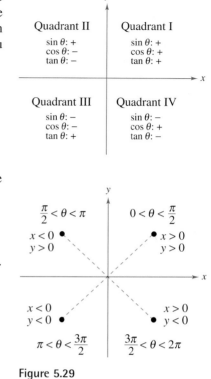

Quadrant II	Quadrant I
$\sin \theta$: +	$\sin \theta$: +
$\cos \theta$: −	$\cos \theta$: +
$\tan \theta$: −	$\tan \theta$: +
Quadrant III	Quadrant IV
$\sin \theta$: −	$\sin \theta$: −
$\cos \theta$: −	$\cos \theta$: +
$\tan \theta$: +	$\tan \theta$: −

Example 2   Evaluating Trigonometric Functions

Given $\tan \theta = -\frac{5}{4}$ and $\cos \theta > 0$, find $\sin \theta$ and $\sec \theta$.

Solution

Note that θ lies in Quadrant IV because that is the only quadrant in which the tangent is negative and the cosine is positive. Moreover, using

$$\tan \theta = \frac{y}{x} = -\frac{5}{4}$$

and the fact that y is negative in Quadrant IV, you can let $y = -5$ and $x = 4$. So, $r = \sqrt{16 + 25} = \sqrt{41}$, and you have the following.

$$\sin \theta = \frac{y}{r} = \frac{-5}{\sqrt{41}} \qquad \text{Exact value}$$

$$\approx -0.7809 \qquad \text{Approximate value}$$

$$\sec \theta = \frac{r}{x} = \frac{\sqrt{41}}{4} \qquad \text{Exact value}$$

$$\approx 1.6008 \qquad \text{Approximate value}$$

Figure 5.29

✓ *Checkpoint*   Now try Exercise 19.

Example 3   Trigonometric Functions of Quadrant Angles

Evaluate the sine and cosine functions at the angles 0, $\dfrac{\pi}{2}$, π, and $\dfrac{3\pi}{2}$.

Solution

To begin, choose a point on the terminal side of each angle, as shown in Figure 5.30. For each of the four given points, $r = 1$, and you have the following.

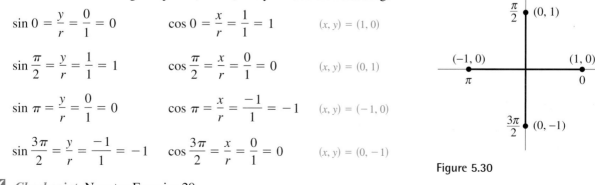

$$\sin 0 = \frac{y}{r} = \frac{0}{1} = 0 \qquad \cos 0 = \frac{x}{r} = \frac{1}{1} = 1 \qquad (x, y) = (1, 0)$$

$$\sin \frac{\pi}{2} = \frac{y}{r} = \frac{1}{1} = 1 \qquad \cos \frac{\pi}{2} = \frac{x}{r} = \frac{0}{1} = 0 \qquad (x, y) = (0, 1)$$

$$\sin \pi = \frac{y}{r} = \frac{0}{1} = 0 \qquad \cos \pi = \frac{x}{r} = \frac{-1}{1} = -1 \qquad (x, y) = (-1, 0)$$

$$\sin \frac{3\pi}{2} = \frac{y}{r} = \frac{-1}{1} = -1 \qquad \cos \frac{3\pi}{2} = \frac{x}{r} = \frac{0}{1} = 0 \qquad (x, y) = (0, -1)$$

Figure 5.30

✓ *Checkpoint*   Now try Exercise 29.

Reference Angles

The values of the trigonometric functions of angles greater than $90°$ (or less than $0°$) can be determined from their values at corresponding acute angles called **reference angles.**

> **Definition of Reference Angle**
>
> Let θ be an angle in standard position. Its **reference angle** is the acute angle θ' formed by the terminal side of θ and the horizontal axis.

Figure 5.31 shows the reference angles for θ in Quadrants II, III, and IV.

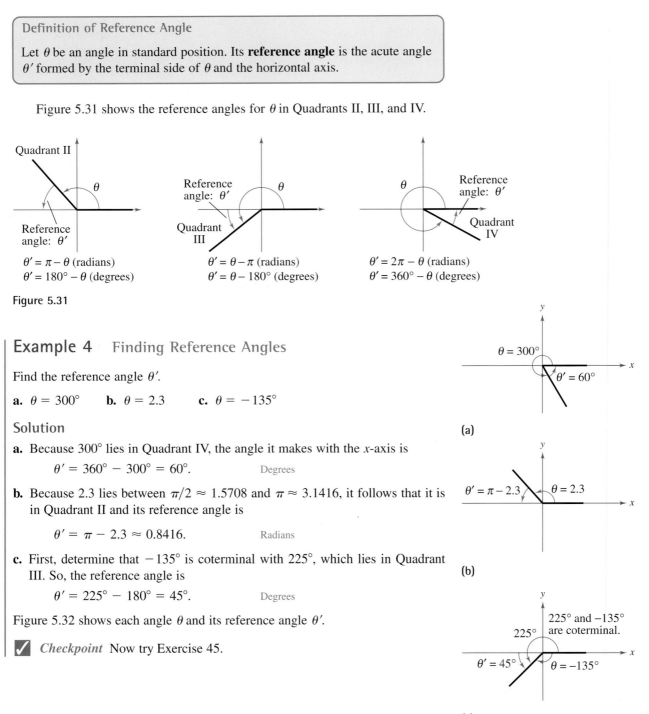

$\theta' = \pi - \theta$ (radians)
$\theta' = 180° - \theta$ (degrees)

$\theta' = \theta - \pi$ (radians)
$\theta' = \theta - 180°$ (degrees)

$\theta' = 2\pi - \theta$ (radians)
$\theta' = 360° - \theta$ (degrees)

Figure 5.31

Example 4   Finding Reference Angles

Find the reference angle θ'.

a. $\theta = 300°$ **b.** $\theta = 2.3$ **c.** $\theta = -135°$

Solution

a. Because $300°$ lies in Quadrant IV, the angle it makes with the x-axis is

$\qquad \theta' = 360° - 300° = 60°.$ Degrees

b. Because 2.3 lies between $\pi/2 \approx 1.5708$ and $\pi \approx 3.1416$, it follows that it is in Quadrant II and its reference angle is

$\qquad \theta' = \pi - 2.3 \approx 0.8416.$ Radians

c. First, determine that $-135°$ is coterminal with $225°$, which lies in Quadrant III. So, the reference angle is

$\qquad \theta' = 225° - 180° = 45°.$ Degrees

Figure 5.32 shows each angle θ and its reference angle θ'.

✓ *Checkpoint* Now try Exercise 45.

(a)

(b)

(c)
Figure 5.32

To see how a reference angle is used to evaluate a trigonometric function, consider the point (x, y) on the terminal side of θ, as shown in Figure 5.33. By definition, you know that

$$\sin \theta = \frac{y}{r} \quad \text{and} \quad \tan \theta = \frac{y}{x}.$$

For the right triangle with acute angle θ' and sides of lengths $|x|$ and $|y|$, you have

$$\sin \theta' = \frac{\text{opp}}{\text{hyp}} = \frac{|y|}{r}$$

and

$$\tan \theta' = \frac{\text{opp}}{\text{adj}} = \frac{|y|}{|x|}.$$

So, it follows that $\sin \theta$ and $\sin \theta'$ are equal, *except possibly in sign*. The same is true for $\tan \theta$ and $\tan \theta'$, and for the other four trigonometric functions. In all cases, the sign of the function value can be determined by the quadrant in which θ lies.

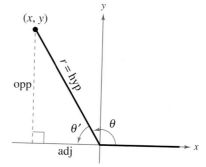

Figure 5.33

Evaluating Trigonometric Functions of Any Angle

To find the value of a trigonometric function of any angle θ:

1. Determine the function value for the associated reference angle θ'.

2. Depending on the quadrant in which θ lies, affix the appropriate sign to the function value.

By using reference angles and the special angles discussed in the preceding section, you can greatly extend the scope of *exact* trigonometric values. For instance, knowing the function values of 30° means that you know the function values of all angles for which 30° is a reference angle. For convenience, the following table shows the exact values of the trigonometric functions of special angles and quadrant angles.

Trigonometric Values of Common Angles

θ (degrees)	0°	30°	45°	60°	90°	180°	270°
θ (radians)	0	$\dfrac{\pi}{6}$	$\dfrac{\pi}{4}$	$\dfrac{\pi}{3}$	$\dfrac{\pi}{2}$	π	$\dfrac{3\pi}{2}$
$\sin \theta$	0	$\dfrac{1}{2}$	$\dfrac{\sqrt{2}}{2}$	$\dfrac{\sqrt{3}}{2}$	1	0	-1
$\cos \theta$	1	$\dfrac{\sqrt{3}}{2}$	$\dfrac{\sqrt{2}}{2}$	$\dfrac{1}{2}$	0	-1	0
$\tan \theta$	0	$\dfrac{\sqrt{3}}{3}$	1	$\sqrt{3}$	Undef.	0	Undef.

STUDY TIP

Learning the table of values at the left is worth the effort because doing so will increase both your efficiency and your confidence. Here is a pattern for the sine function that may help you remember the values.

θ	0°	30°	45°	60°	90°
$\sin \theta$	$\dfrac{\sqrt{0}}{2}$	$\dfrac{\sqrt{1}}{2}$	$\dfrac{\sqrt{2}}{2}$	$\dfrac{\sqrt{3}}{2}$	$\dfrac{\sqrt{4}}{2}$

Reverse the order to get the cosine values of the same angles.

Example 5   Trigonometric Functions of Nonacute Angles

Evaluate each trigonometric function.

a. $\cos \dfrac{4\pi}{3}$ **b.** $\tan(-210°)$ **c.** $\csc \dfrac{11\pi}{4}$

Solution

a. Because $\theta = 4\pi/3$ lies in Quadrant III, the reference angle, θ', is $(4\pi/3) - \pi = \pi/3$, as shown in Figure 5.34. Moreover, the cosine is negative in Quadrant III, so

$$\cos \frac{4\pi}{3} = (-)\cos \frac{\pi}{3} = -\frac{1}{2}.$$

b. Because $-210° + 360° = 150°$, it follows that $-210°$ is coterminal with the second-quadrant angle $150°$. Therefore, the reference angle, θ', is $180° - 150° = 30°$, as shown in Figure 5.35. Finally, because the tangent is negative in Quadrant II, you have

$$\tan(-210°) = (-)\tan 30° = -\frac{\sqrt{3}}{3}.$$

c. Because $(11\pi/4) - 2\pi = 3\pi/4$, it follows that $11\pi/4$ is coterminal with the second-quadrant angle $3\pi/4$. Therefore, the reference angle, θ', is $\pi - (3\pi/4) = \pi/4$, as shown in Figure 5.36. Because the cosecant is positive in Quadrant II, you have

$$\csc \frac{11\pi}{4} = (+)\csc \frac{\pi}{4}$$

$$= \frac{1}{\sin(\pi/4)}$$

$$= \sqrt{2}.$$

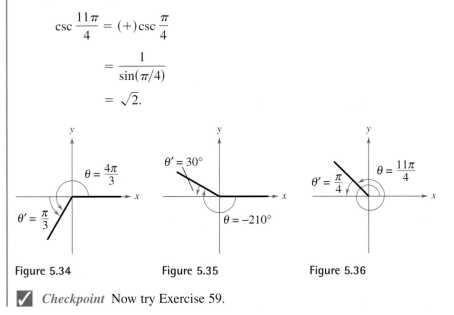

Figure 5.34   Figure 5.35   Figure 5.36

✓ *Checkpoint* Now try Exercise 59.

The fundamental trigonometric identities listed in the preceding section (for an acute angle θ) are also valid when θ is any angle in the domain of the function.

Example 6   Using Trigonometric Identities

Let θ be an angle in Quadrant II such that $\sin \theta = \frac{1}{3}$. Find (a) $\cos \theta$ and (b) $\tan \theta$ by using trigonometric identities.

Solution

a. Using the Pythagorean identity $\sin^2 \theta + \cos^2 \theta = 1$, you obtain

$$\left(\tfrac{1}{3}\right)^2 + \cos^2 \theta = 1$$

$$\cos^2 \theta = 1 - \frac{1}{9}$$

$$\cos^2 \theta = \frac{8}{9}$$

Because $\cos \theta < 0$ in Quadrant II, you can use the negative root to obtain

$$\cos \theta = -\frac{\sqrt{8}}{\sqrt{9}} = -\frac{2\sqrt{2}}{3}.$$

b. Using the trigonometric identity $\tan \theta = \sin \theta / \cos \theta$, you obtain

$$\tan \theta = \frac{1/3}{-2\sqrt{2}/3}$$

$$= -\frac{1}{2\sqrt{2}}$$

$$= -\frac{\sqrt{2}}{4}.$$

 Checkpoint Now try Exercise 67.

> **Exploration**
>
> Set your graphing utility to *degree* mode and enter tan 90. What happens? Why? Now set your graphing utility to *radian* mode and enter $\tan(\pi/2)$. Explain the graphing utility's answer.

You can use a calculator to evaluate trigonometric functions, as shown in the next example.

Example 7   Using a Calculator

Use a calculator to evaluate each trigonometric function.

a. $\cot 410°$ **b.** $\sin(-7)$ **c.** $\sec \dfrac{\pi}{9}$

Solution

Function	Mode	Graphing Calculator Keystrokes	Display
a. $\cot 410°$	Degree	(TAN 410) x^{-1} ENTER	0.8390996
b. $\sin(-7)$	Radian	SIN (−) 7) ENTER	−0.6569866
c. $\sec \dfrac{\pi}{9}$	Radian	(COS π ÷ 9) x^{-1} ENTER	1.0641777

 Checkpoint Now try Exercise 75.

Trigonometric Functions of Real Numbers

To define a trigonometric function of a real number (rather than an angle), let t represent any real number. Then imagine that the real number line is wrapped around a *unit circle*, as shown in Figure 5.37. Note that positive numbers correspond to a counterclockwise wrapping, and negative numbers correspond to a clockwise wrapping.

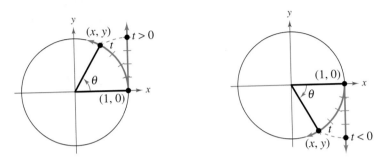

Figure 5.37

As the real number line is wrapped around the unit circle, each real number t will correspond with a central angle θ. Moreover, because the circle has a radius of 1, the arc intercepted by the angle θ will have a length of t. The point is that if θ *is measured in radians*, then $t = \theta$. So, you can define $\sin t$ as $\sin t = \sin(t \text{ radians})$. Similarly, $\cos t = \cos(t \text{ radians})$, $\tan t = \tan(t \text{ radians})$, and so on.

Example 8   Evaluating Trigonometric Functions

Evaluate $f(t) = \sin t$ for (a) $t = 1$ and (b) $t = 3\pi/2$.

Algebraic Solution

a. $f(1) = \sin 1$ Radian mode

≈ 0.841471

b. $f\left(\dfrac{3\pi}{2}\right) = \sin\dfrac{3\pi}{2}$ Common angle

$= -1$

 Checkpoint Now try Exercise 99.

Graphical Solution

Use a graphing utility set in *parametric* and *radian* modes to graph $X_{1T} = \cos T$ and $Y_{1T} = \sin T$ using the following settings.

$$\text{Tmin} = 0, \quad \text{Tmax} = 6.3, \quad \text{Tstep} = 0.1$$
$$\text{Xmin} = -1.5, \quad \text{Xmax} = 1.5, \quad \text{Xscl} = 1$$
$$\text{Ymin} = -1, \quad \text{Ymax} = 1, \quad \text{Yscl} = 1$$

a. Use the *value* or *trace* feature to estimate that $\sin 1 \approx 0.841471$, as shown in Figure 5.38.

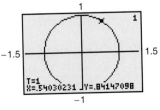

Figure 5.38

b. Use the *value* or *trace* feature to estimate that $\sin\dfrac{3\pi}{2} \approx$ $\sin 4.7 \approx -1$.

The *domain* of the sine and cosine functions is the set of all real numbers. To determine the *range* of these two functions, consider the unit circle shown in Figure 5.39. Because $r = 1$, it follows that $\sin t = y$ and $\cos t = x$. Moreover, because (x, y) is on the unit circle, you know that $-1 \le y \le 1$ and $-1 \le x \le 1$. So, the values of sine and cosine also range between -1 and 1.

$$-1 \le y \quad \le 1 \qquad -1 \le x \quad \le 1$$
$$\text{and}$$
$$-1 \le \sin t \le 1 \qquad -1 \le \cos t \le 1$$

You can add 2π to each value of t in the interval $[0, 2\pi]$, completing a second revolution around the unit circle, as shown in Figure 5.40. The values of $\sin(t + 2\pi)$ and $\cos(t + 2\pi)$ correspond to those of $\sin t$ and $\cos t$. Similar results can be obtained for repeated revolutions (positive or negative) on the unit circle. This leads to the general result

$$\sin(t + 2\pi m) = \sin t \qquad \text{and} \qquad \cos(t + 2\pi n) = \cos t$$

for any integer n and real number t. Functions that behave in such a repetitive (or cyclic) manner are called **periodic.**

Exploration

With your graphing utility in *radian* mode, select a number t and calculate $\cos t$ and $\cos(t + 2\pi)$. Repeat this experiment for other values of t and explain why the answers are always the same. Perform a similar experiment with $\tan t$ and $\tan(t + \pi)$.

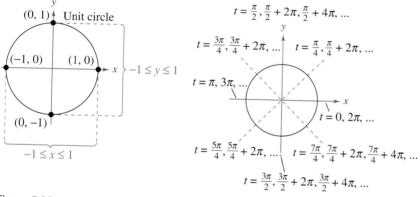

Figure 5.39   Figure 5.40

Definition of Periodic Function

A function f is **periodic** if there exists a positive real number c such that

$$f(t + c) = f(t)$$

for all t in the domain of f. The least number c for which f is periodic is called the **period** of f.

From this definition it follows that the sine and cosine functions are periodic and have a period of 2π. The other four trigonometric functions are also periodic, and more will be said about this in Section 5.5.

Recall from Section 1.4 that a function f is *even* if $f(-t) = f(t)$ and is *odd* if $f(-t) = -f(t)$.

Even and Odd Trigonometric Functions

The cosine and secant functions are *even*.

$$\cos(-t) = \cos t \qquad \sec(-t) = \sec t$$

The sine, cosecant, tangent, and cotangent functions are *odd*.

$$\sin(-t) = -\sin t \qquad \csc(-t) = -\csc t$$

$$\tan(-t) = -\tan t \qquad \cot(-t) = -\cot t$$

Exploration

Select a number t and use your graphing utility to calculate $\cos t$ and $\cos(-t)$. Repeat this experiment for other values of t and explain why the answer is always the same. Perform a similar experiment to determine the relationship between $\sin t$ and $\sin(-t)$.

Library of Functions: Trigonometric Functions

Trigonometric functions are transcendental functions. The six trigonometric functions, sine, cosine, tangent, cosecant, secant, and cotangent, have important uses in construction, surveying, and navigation. Their periodic behavior makes them useful for modeling phenomena such as business cycles, planetary orbits, pendulums, wave motion, and light rays.

The six trigonometric functions can be defined in three different ways.

1. As the ratio of two sides of a right triangle [see Figure 5.41(a)].

2. As coordinates of a point (x, y) in the plane and its distance r from the origin [see Figure 5.41(b)].

3. As functions of any real number, such as time t.

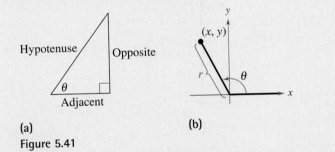

(a)

(b)

Figure 5.41

To be efficient in the use of trigonometric functions, you should learn the trigonometric function values of common angles, such as those listed on page 417. Because pairs of trigonometric functions are related to each other by a variety of identities, it is useful to know the fundamental identities presented in Section 5.2. Finally, trigonometric functions and their identity relationships play a prominent role in calculus.

STUDY TIP

For your convenience, a summary of basic trigonometry is included on the inside back cover of this text.

At this point, you have completed your introduction to basic trigonometry. You have measured angles in both degrees and radians. You have defined the six trigonometric functions from a right triangle perspective and as functions of real numbers. In your remaining work with trigonometry, you should continue to rely on both perspectives. For instance, in the next two sections on graphing techniques, it helps to think of the trigonometric functions as functions of real numbers. Later, in Section 5.7, you will look at applications involving angles and triangles.

5.3 Exercises

Vocabulary Check

Fill in the blanks.

1. The acute positive angle that is formed by the terminal side of the angle θ and the horizontal axis is called the _____ angle of θ and is denoted by θ'.

2. A function f is _____ if there exists a positive real number c such that $f(t + c) = f(t)$ for all t in the domain of f.

3. A function f is _____ if $f(-t) = -f(t)$.

4. A function f is _____ if $f(-t) = f(t)$.

In Exercises 1–4, determine the exact values of the six trigonometric functions of the angle θ.

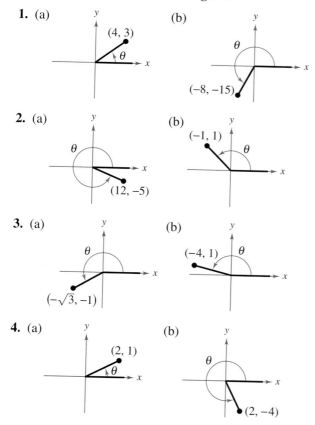

1. (a) $(4, 3)$ (b) $(-8, -15)$

2. (a) $(12, -5)$ (b) $(-1, 1)$

3. (a) $(-\sqrt{3}, -1)$ (b) $(-4, 1)$

4. (a) $(2, 1)$ (b) $(2, -4)$

In Exercises 5–12, the point is on the terminal side of an angle in standard position. Determine the exact values of the six trigonometric functions of the angle.

5. $(7, 24)$

6. $(8, 15)$

7. $(5, -12)$

8. $(-24, 10)$

9. $(-4, 10)$

10. $(-5, -2)$

11. $(-2, 9)$

12. $(6, -14)$

In Exercises 13–16, state the quadrant in which θ lies.

13. $\sin \theta < 0$ and $\cos \theta < 0$

14. $\sin \theta > 0$ and $\cos \theta > 0$

15. $\sec \theta > 0$ and $\cot \theta < 0$

16. $\tan \theta > 0$ and $\csc \theta < 0$

In Exercises 17–24, find the values of the six trigonometric functions of θ.

	Function Value	Constraint
17.	$\sin \theta = \frac{3}{5}$	θ lies in Quadrant II.
18.	$\cos \theta = -\frac{4}{5}$	θ lies in Quadrant III.
19.	$\tan \theta = -\frac{15}{8}$	$\sin \theta < 0$
20.	$\csc \theta = 4$	$\cot \theta < 0$
21.	$\sec \theta = -2$	$0 \le \theta \le \pi$
22.	$\sin \theta = 0$	$\sec \theta = -1$
23.	$\cot \theta$ is undefined.	$\dfrac{\pi}{2} \le \theta \le \dfrac{3\pi}{2}$
24.	$\tan \theta$ is undefined.	$\pi \le \theta \le 2\pi$

In Exercises 25–28, the terminal side of θ lies on the given line in the specified quadrant. Find the values of the six trigonometric functions of θ by finding a point on the line.

	Line	Quadrant
25.	$y = -x$	II
26.	$y = \frac{1}{3}x$	III
27.	$2x - y = 0$	III
28.	$4x + 3y = 0$	IV

In Exercises 29–36, evaluate the trigonometric function of the quadrant angle.

29. $\sec \pi$

30. $\tan \dfrac{\pi}{2}$

31. $\cot \dfrac{3\pi}{2}$

32. $\csc \pi$

33. $\sec 0$

34. $\csc \dfrac{3\pi}{2}$

35. $\cot \pi$

36. $\csc \dfrac{\pi}{2}$

In Exercises 37–44, find the reference angle θ' for the special angle θ. Then sketch θ and θ' in standard position.

37. $\theta = 120°$

38. $\theta = 225°$

39. $\theta = -135°$

40. $\theta = -330°$

41. $\theta = \dfrac{5\pi}{3}$

42. $\theta = \dfrac{3\pi}{4}$

43. $\theta = -\dfrac{5\pi}{6}$

44. $\theta = -\dfrac{2\pi}{3}$

In Exercises 45–52, find the reference angle θ' and sketch θ and θ' in standard position.

45. $\theta = 208°$

46. $\theta = 322°$

47. $\theta = -292°$

48. $\theta = -95°$

49. $\theta = \dfrac{11\pi}{5}$

50. $\theta = \dfrac{17\pi}{7}$

51. $\theta = 3.5$

52. $\theta = -1.72$

In Exercises 53–66, evaluate the sine, cosine, and tangent of the angle without using a calculator.

53. $225°$

54. $300°$

55. $-750°$

56. $-495°$

57. $-240°$

58. $-330°$

59. $\dfrac{5\pi}{3}$

60. $\dfrac{3\pi}{4}$

61. $-\dfrac{7\pi}{6}$

62. $-\dfrac{4\pi}{3}$

63. $\dfrac{11\pi}{4}$

64. $\dfrac{10\pi}{3}$

65. $-\dfrac{17\pi}{6}$

66. $-\dfrac{20\pi}{3}$

In Exercises 67–72, find the indicated trigonometric value in the specified quadrant.

Function	Quadrant	Trigonometric Value
67. $\sin \theta = -\frac{3}{5}$	IV	$\cos \theta$
68. $\cot \theta = -3$	II	$\sin \theta$
69. $\tan \theta = \frac{3}{2}$	III	$\sec \theta$
70. $\csc \theta = -2$	IV	$\cot \theta$
71. $\cos \theta = \frac{5}{8}$	I	$\sec \theta$
72. $\sec \theta = -\frac{9}{4}$	III	$\tan \theta$

In Exercises 73–88, use a calculator to evaluate the trigonometric function. Round your answer to four decimal places. (Be sure the calculator is set to the correct angle mode.)

73. $\sin 10°$

74. $\sec 235°$

75. $\tan 245°$

76. $\csc 320°$

77. $\cos(-110°)$

78. $\cot(-220°)$

79. $\sec(-280°)$

80. $\sin(-195°)$

81. $\sin 0.65$

82. $\sin(-0.65)$

83. $\cos(-1.81)$

84. $\sec 0.33$

85. $\tan \dfrac{2\pi}{9}$

86. $\tan \dfrac{11\pi}{9}$

87. $\csc\left(-\dfrac{8\pi}{9}\right)$

88. $\cos\left(-\dfrac{15\pi}{14}\right)$

In Exercises 89–94, find two solutions of the equation. Give your answers in degrees ($0° \le \theta < 360°$) and radians ($0 \le \theta < 2\pi$). Do not use a calculator.

89. (a) $\sin \theta = \frac{1}{2}$ 　　(b) $\sin \theta = -\frac{1}{2}$

90. (a) $\cos \theta = \dfrac{\sqrt{2}}{2}$ 　　(b) $\cos \theta = -\dfrac{\sqrt{2}}{2}$

91. (a) $\csc \theta = \dfrac{2\sqrt{3}}{3}$ 　　(b) $\cot \theta = -1$

92. (a) $\csc \theta = -\sqrt{2}$ 　　(b) $\csc \theta = 2$

93. (a) $\sec \theta = -\dfrac{2\sqrt{3}}{3}$ 　　(b) $\cos \theta = \dfrac{1}{2}$

94. (a) $\cot \theta = -\sqrt{3}$ 　　(b) $\sec \theta = \sqrt{2}$

In Exercises 95–102, find the point (x, y) on the unit circle that corresponds to the real number t. Use the result to evaluate $\sin t$, $\cos t$, and $\tan t$.

95. $t = \dfrac{\pi}{4}$

96. $t = \dfrac{\pi}{3}$

97. $t = \dfrac{5\pi}{6}$

98. $t = \dfrac{5\pi}{4}$

99. $t = \dfrac{4\pi}{3}$

100. $t = \dfrac{11\pi}{6}$

101. $t = \dfrac{3\pi}{2}$

102. $t = \pi$

Estimation In Exercises 103 and 104, use the figure and a straightedge to approximate the value of each trigonometric function. Check your approximation using a graphing utility. To print an enlarged copy of the graph, go to the website *www.mathgraphs.com*.

103. (a) $\sin 5$ (b) $\cos 2$

104. (a) $\sin 0.75$ (b) $\cos 2.5$

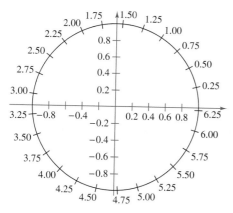

Estimation In Exercises 105 and 106, use the figure in Exercises 103 and 104 to approximate the solution of each equation, where $0 \le t < 2\pi$. Check your approximation using a graphing utility. To print an enlarged copy of the graph, go to the website *www.mathgraphs.com*.

105. (a) $\sin t = 0.25$ (b) $\cos t = -0.25$

106. (a) $\sin t = -0.75$ (b) $\cos t = 0.75$

107. *Meteorology* The monthly normal temperature T (in degrees Fahrenheit) for Santa Fe, New Mexico is given by

$$T = 49.5 + 20.5 \cos\left(\frac{\pi t}{6} - \frac{7\pi}{6}\right)$$

where t is the time in months, with $t = 1$ corresponding to January. Find the monthly normal temperature for each month. (Source: National Climatic Data Center)

(a) January   (b) July   (c) December

108. *Sales* A company that produces water skis, which are seasonal products, forecasts monthly sales over a two-year period to be

$$S = 23.1 + 0.442t + 4.3 \sin\frac{\pi t}{6}$$

where S is measured in thousands of units and t is the time (in months), with $t = 1$ representing January 2004. Estimate sales for each month.

(a) January 2004   (b) February 2005

(c) May 2004   (d) June 2004

109. *Harmonic Motion* The displacement from equilibrium of an oscillating weight suspended by a spring is given by

$$y(t) = 2e^{-t} \cos 6t$$

where y is the displacement (in centimeters) and t is the time (in seconds).

(a) What is the initial displacement $(t = 0)$?

(b) Use a graphing utility to complete the table.

t	0.50	1.02	1.54	2.07	2.59
y					

(c) The approximate times when the weight is at its maximum distance from equilibrium are shown in the table in part (b). Explain why the magnitude of the maximum displacement is decreasing. What causes this decrease in maximum displacement in the physical system? What factor in the model measures this decrease?

(d) Find the first two times that the weight is at the equilibrium point $(y = 0)$.

110. *Harmonic Motion* The displacement from equilibrium of an oscillating weight suspended by a spring is given by

$$y(t) = 2 \cos 6t$$

where y is the displacement (in centimeters) and t is the time (in seconds). Find the displacement when (a) $t = 0$, (b) $t = \frac{1}{4}$, and (c) $t = \frac{1}{2}$.

111. *Electric Circuits* The initial current and charge in an electric circuit are zero. The current when 100 volts is applied to the circuit is given by

$$I = 5e^{-2t} \sin t$$

where the resistance, inductance, and capacitance are 80 ohms, 20 henrys, and 0.01 farad, respectively. Approximate the current (in amperes) $t = 0.7$ second after the voltage is applied.

112. *Distance* An airplane, flying at an altitude of 6 miles, is on a flight path that passes directly over an observer (see figure). If θ is the angle of elevation from the observer to the plane, find the distance from the observer to the plane when (a) $\theta = 30°$, (b) $\theta = 90°$, and (c) $\theta = 120°$.

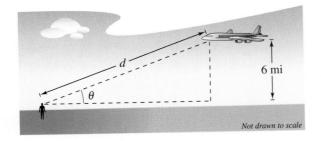

Not drawn to scale

Synthesis

True or False? **In Exercises 113–116, determine whether the statement is true or false. Justify your answer.**

113. $\sin 151° = \sin 29°$

114. $\tan 24° = \tan 156°$

115. $\csc\left(-\dfrac{7\pi}{6}\right) = \csc\left(-\dfrac{11\pi}{6}\right)$

116. $-\cot\left(\dfrac{3\pi}{4}\right) = \cot\left(-\dfrac{\pi}{4}\right)$

117. *Conjecture*

(a) Use a graphing utility to complete the table.

θ	0°	20°	40°	60°	80°
$\sin \theta$					
$\sin(180° - \theta)$					

(b) Make a conjecture about the relationship between $\sin \theta$ and $\sin(180° - \theta)$.

118. *Conjecture*

(a) Use a graphing utility to complete the table.

θ	0	0.3	0.6	0.9	1.2	1.5
$\cos\left(\dfrac{3\pi}{2} - \theta\right)$						
$-\sin \theta$						

(b) Make a conjecture about the relationship between $\cos\left(\dfrac{3\pi}{2} - \theta\right)$ and $-\sin \theta$.

119. *Writing* Create a table of the six trigonometric functions comparing their domains, ranges, evenness, oddness, periods, and zeros. Then identify and write a short paragraph describing any inherent patterns in the trigonometric functions. What can you conclude?

120. *Think About It* Because $f(t) = \sin t$ is an odd function and $g(t) = \cos t$ is an even function, what can be said about the function $h(t) = f(t)g(t)$?

Review

In Exercises 121–124, sketch the graph of the function. Identify any intercepts and asymptotes.

121. $y = 2^{x-1}$ **122.** $y = 3^{x+2}$

123. $y = \ln(x - 1)$ **124.** $y = \ln(x + 1)$

In Exercises 125–128, solve the equation. Round your answer to three decimal places.

125. $4^{3-x} = 726$ **126.** $\dfrac{4500}{4 + e^{2x}} = 50$

127. $\ln x = -6$ **128.** $\ln \sqrt{x + 10} = 1$

5.4 Graphs of Sine and Cosine Functions

Basic Sine and Cosine Curves

In this section you will study techniques for sketching the graphs of the sine and cosine functions. The graph of the sine function is a **sine curve.** In Figure 5.42, the black portion of the graph represents one period of the function and is called **one cycle** of the sine curve. The gray portion of the graph indicates that the basic sine wave repeats indefinitely to the right and left. The graph of the cosine function is shown in Figure 5.43. To produce these graphs with a graphing utility, make sure you have set the graphing utility to *radian* mode.

Recall from Section 5.3 that the domain of the sine and cosine functions is the set of all real numbers. Moreover, the range of each function is the interval $[-1, 1]$, and each function has a period of 2π. Do you see how this information is consistent with the basic graphs shown in Figures 5.42 and 5.43?

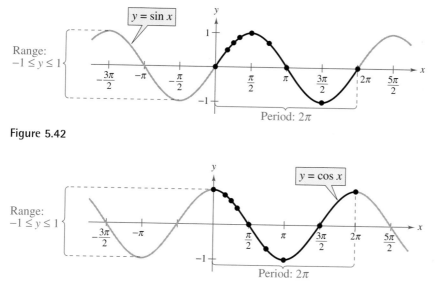

Figure 5.42

Figure 5.43

Jerry Lodriguss/Photo Researchers, Inc.

> ### What you should learn
> - Sketch the graphs of basic sine and cosine functions.
> - Use amplitude and period to help sketch the graphs of sine and cosine functions.
> - Sketch translations of graphs of sine and cosine functions.
> - Use sine and cosine functions to model real-life data.
>
> ### Why you should learn it
> Sine and cosine functions are often used in scientific calculations. For instance, in Exercise 79 on page 436, you can use a trigonometric function to model the percent of the moon's face that is illuminated for any given day in 2006.

The table below lists key points on the graphs of $y = \sin x$ and $y = \cos x$.

x	0	$\dfrac{\pi}{6}$	$\dfrac{\pi}{4}$	$\dfrac{\pi}{3}$	$\dfrac{\pi}{2}$	$\dfrac{3\pi}{4}$	π	$\dfrac{3\pi}{2}$	2π
$\sin x$	0	$\dfrac{1}{2}$	$\dfrac{\sqrt{2}}{2}$	$\dfrac{\sqrt{3}}{2}$	1	$\dfrac{\sqrt{2}}{2}$	0	-1	0
$\cos x$	1	$\dfrac{\sqrt{3}}{2}$	$\dfrac{\sqrt{2}}{2}$	$\dfrac{1}{2}$	0	$-\dfrac{\sqrt{2}}{2}$	-1	0	1

Note from Figures 5.42 and 5.43 that the sine graph is symmetric with respect to the *origin*, whereas the cosine graph is symmetric with respect to the *y-axis*. These properties of symmetry follow from the fact that the sine function is odd whereas the cosine function is even.

To sketch the graphs of the basic sine and cosine functions by hand, it helps to note five *key points* in one period of each graph: the *intercepts*, the *maximum points*, and the *minimum points*. See Figure 5.44.

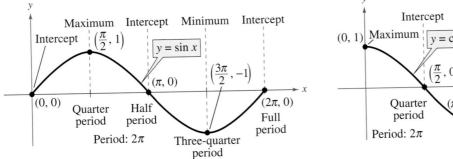

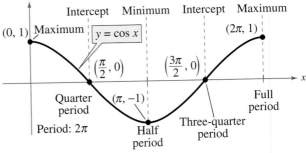

Figure 5.44

Example 1   Using Key Points to Sketch a Sine Curve

Sketch the graph of $y = 2 \sin x$ by hand on the interval $[-\pi, 4\pi]$.

Solution

Note that

$$y = 2 \sin x = 2(\sin x)$$

indicates that the y-values of the key points will have twice the magnitude of those on the graph of $y = \sin x$. Divide the period 2π into four equal parts to get the key points

Intercept	Maximum	Intercept	Minimum		Intercept
$(0, 0),$	$\left(\dfrac{\pi}{2}, 2\right),$	$(\pi, 0),$	$\left(\dfrac{3\pi}{2}, -2\right),$	and	$(2\pi, 0).$

By connecting these key points with a smooth curve and extending the curve in both directions over the interval $[-\pi, 4\pi]$, you obtain the graph shown in Figure 5.45. Use a graphing utility to confirm this graph. Be sure to set the graphing utility to *radian* mode.

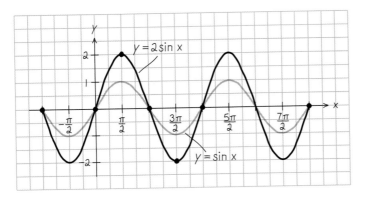

Figure 5.45

✓ *Checkpoint* Now try Exercise 39.

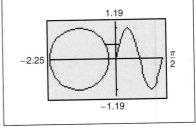

Amplitude and Period of Sine and Cosine Curves

In the rest of this section you will study the graphic effect of each of the constants a, b, c, and d in equations of the forms

$$y = d + a \sin(bx - c) \quad \text{and} \quad y = d + a \cos(bx - c).$$

A quick review of the transformations studied in Section 1.5 should help in this investigation.

The constant factor a in $y = a \sin x$ acts as a *scaling factor*—a *vertical stretch* or *vertical shrink* of the basic sine curve. If $|a| > 1$, the basic sine curve is stretched, and if $|a| < 1$, the basic sine curve is shrunk. The result is that the graph of $y = a \sin x$ ranges between $-a$ and a instead of between -1 and 1. The absolute value of a is the **amplitude** of the function $y = a \sin x$. The range of the function $y = a \sin x$ for $a > 0$ is $-a \le y \le a$.

Definition of Amplitude of Sine and Cosine Curves

The **amplitude** of $y = a \sin x$ and $y = a \cos x$ represents half the distance between the maximum and minimum values of the function and is given by

$$\text{Amplitude} = |a|.$$

Example 2   Scaling: Vertical Shrinking and Stretching

On the same set of coordinate axes, sketch the graph of each function by hand.

a. $y = \frac{1}{2} \cos x$ **b.** $y = 3 \cos x$

Solution

a. Because the amplitude of $y = \frac{1}{2} \cos x$ is $\frac{1}{2}$, the maximum value is $\frac{1}{2}$ and the minimum value is $-\frac{1}{2}$. Divide one cycle, $0 \le x \le 2\pi$, into four equal parts to get the key points

Maximum	*Intercept*	*Minimum*	*Intercept*	*Maximum*
$\left(0, \frac{1}{2}\right),$	$\left(\frac{\pi}{2}, 0\right),$	$\left(\pi, -\frac{1}{2}\right),$	$\left(\frac{3\pi}{2}, 0\right),$ and	$\left(2\pi, \frac{1}{2}\right).$

b. A similar analysis shows that the amplitude of $y = 3 \cos x$ is 3, and the key points are

Maximum	*Intercept*	*Minimum*	*Intercept*	*Maximum*
$(0, 3),$	$\left(\frac{\pi}{2}, 0\right),$	$(\pi, -3),$	$\left(\frac{3\pi}{2}, 0\right),$ and	$(2\pi, 3).$

The graphs of these two functions are shown in Figure 5.46. Notice that the graph of $y = \frac{1}{2} \cos x$ is a vertical shrink of the graph of $y = \cos x$ and the graph of $y = 3 \cos x$ is a vertical stretch of the graph of $y = \cos x$. Use a graphing utility to confirm these graphs.

✓ *Checkpoint* Now try Exercise 40.

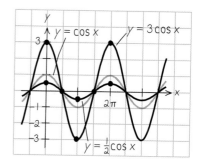

Figure 5.46

You know from Section 1.5 that the graph of $y = -f(x)$ is a *reflection* in the x-axis of the graph of $y = f(x)$. For instance, the graph of $y = -3 \cos x$ is a reflection of the graph of $y = 3 \cos x$, as shown in Figure 5.47.

Because $y = a \sin x$ completes one cycle from $x = 0$ to $x = 2\pi$, it follows that $y = a \sin bx$ completes one cycle from $x = 0$ to $x = 2\pi/b$.

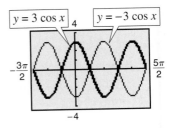

Figure 5.47

> ### Period of Sine and Cosine Functions
>
> Let b be a positive real number. The **period** of $y = a \sin bx$ and $y = a \cos bx$ is given by
>
> $$\text{Period} = \frac{2\pi}{b}.$$

Note that if $0 < b < 1$, the period of $y = a \sin bx$ is greater than 2π and represents a *horizontal stretching* of the graph of $y = a \sin x$. Similarly, if $b < 1$, the period of $y = a \sin bx$ is less than 2π and represents a *horizontal shrinking* of the graph of $y = a \sin x$. If b is negative, the identities $\sin(-x) = -\sin x$ and $\cos(-x) = \cos x$ are used to rewrite the function.

Example 3   Scaling: Horizontal Stretching

Sketch the graph of $y = \sin \dfrac{x}{2}$ by hand.

Solution

The amplitude is 1. Moreover, because $b = \frac{1}{2}$, the period is

$$\frac{2\pi}{b} = \frac{2\pi}{\frac{1}{2}} = 4\pi. \qquad \text{Substitute for } b.$$

Now, divide the period-interval $[0, 4\pi]$ into four equal parts with the values π, 2π, and 3π to obtain the key points on the graph

Intercept	*Maximum*	*Intercept*	*Minimum*		*Intercept*
$(0, 0),$	$(\pi, 1),$	$(2\pi, 0),$	$(3\pi, -1),$	and	$(4\pi, 0).$

The graph is shown in Figure 5.48. Use a graphing utility to confirm this graph.

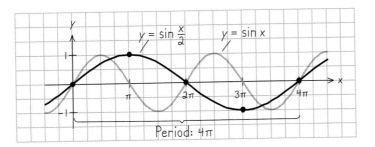

Figure 5.48

✓ *Checkpoint* Now try Exercise 41.

> ### STUDY TIP
>
> In general, to divide a period-interval into four equal parts, successively add "period/4," starting with the left endpoint of the interval. For instance, for the period-interval $\left[-\pi/6, \pi/2\right]$ of length $2\pi/3$, you would successively add
>
> $$\frac{2\pi/3}{4} = \frac{\pi}{6}$$
>
> to get $-\pi/6, 0, \pi/6, \pi/3,$ and $\pi/2$ as the key points on the graph.

Translations of Sine and Cosine Curves

The constant c in the general equations

$$y = a \sin(bx - c) \qquad \text{and} \qquad y = a \cos(bx - c)$$

creates *horizontal translations* (shifts) of the basic sine and cosine curves. Comparing $y = a \sin bx$ with $y = a \sin(bx - c)$, you find that the graph of $y = a \sin(bx - c)$ completes one cycle from $bx - c = 0$ to $bx - c = 2\pi$. By solving for x, you can find the interval for one cycle to be

Left endpoint   Right endpoint

$$\frac{c}{b} \le x \le \frac{c}{b} + \frac{2\pi}{b}.$$

Period

This implies that the period of $y = a \sin(bx - c)$ is $2\pi/b$, and the graph of $y = a \sin bx$ is shifted by an amount c/b. The number c/b is the **phase shift.**

Graphs of Sine and Cosine Functions

The graphs of $y = a \sin(bx - c)$ and $y = a \cos(bx - c)$ have the following characteristics. (Assume $b > 0$.)

$$\text{Amplitude} = |a| \qquad \text{Period} = 2\pi/b$$

The left and right endpoints of a one-cycle interval can be determined by solving the equations $bx - c = 0$ and $bx - c = 2\pi$.

TECHNOLOGY SUPPORT

For instructions on how to use the *minimum* feature, the *maximum* feature, and the *zero* or *root* feature, see Appendix A; for specific keystrokes, go to the text website at *college.hmco.com.*

Example 4   Horizontal Translation

Analyze the graph of $y = \frac{1}{2} \sin(x - \pi/3)$.

Algebraic Solution

The amplitude is $\frac{1}{2}$ and the period is 2π. By solving the equations

$$x - \frac{\pi}{3} = 0 \qquad \text{and} \qquad x - \frac{\pi}{3} = 2\pi$$

$$x = \frac{\pi}{3} \qquad\qquad\qquad x = \frac{7\pi}{3}$$

you see that the interval $[\pi/3, 7\pi/3]$ corresponds to one cycle of the graph. Dividing this interval into four equal parts produces the following key points.

Intercept	Maximum	Intercept	Minimum	Intercept
$\left(\dfrac{\pi}{3}, 0\right)$,	$\left(\dfrac{5\pi}{6}, \dfrac{1}{2}\right)$,	$\left(\dfrac{4\pi}{3}, 0\right)$,	$\left(\dfrac{11\pi}{6}, -\dfrac{1}{2}\right)$,	$\left(\dfrac{7\pi}{3}, 0\right)$

✓ *Checkpoint* Now try Exercise 45.

Graphical Solution

Use a graphing utility set in *radian* mode to graph $y = (1/2) \sin(x - \pi/3)$, as shown in Figure 5.49. Use the *minimum, maximum,* and *zero* or *root* features of the graphing utility to approximate the key points $(1.047, 0)$, $(2.618, 0.5)$, $(4.189, 0)$, $(5.760, -0.5)$, and $(7.330, 0)$.

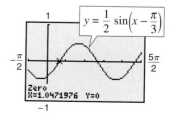

Figure 5.49

Example 5   Horizontal Translation

Use a graphing utility to analyze the graph of $y = -3 \cos(2\pi x + 4\pi)$.

Solution

The amplitude is 3 and the period is $2\pi/2\pi = 1$. By solving the equations

$$2\pi x + 4\pi = 0 \qquad \text{and} \qquad 2\pi x + 4\pi = 2\pi$$

$$2\pi x = -4\pi \qquad\qquad\qquad 2\pi x = -2\pi$$

$$x = -2 \qquad\qquad\qquad\qquad x = -1$$

you see that the interval $[-2, -1]$ corresponds to one cycle of the graph. Dividing this interval into four equal parts produces the key points

Minimum	Intercept	Maximum	Intercept	Minimum
$(-2, -3)$,	$(-7/4, 0)$,	$(-3/2, 3)$,	$(-5/4, 0)$, and	$(-1, -3)$.

The graph is shown in Figure 5.50.

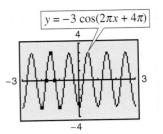
$y = -3 \cos(2\pi x + 4\pi)$

Figure 5.50

✓ *Checkpoint*   Now try Exercise 47.

The final type of transformation is the *vertical translation* caused by the constant d in the equations

$$y = d + a \sin(bx - c) \qquad \text{and} \qquad y = d + a \cos(bx - c).$$

The shift is d units upward for $d > 0$ and d units downward for $d < 0$. In other words, the graph oscillates about the horizontal line $y = d$ instead of about the x-axis.

Example 6   Vertical Translation

Use a graphing utility to analyze the graph of $y = 2 + 3 \cos 2x$.

Solution

The amplitude is 3 and the period is π. The key points over the interval $[0, \pi]$ are

$$(0, 5), \qquad (\pi/4, 2), \qquad (\pi/2, -1), \qquad (3\pi/4, 2), \qquad \text{and} \qquad (\pi, 5).$$

The graph is shown in Figure 5.51. Compared with the graph of $f(x) = 3 \cos 2x$, the graph of $y = 2 + 3 \cos 2x$ is shifted upward two units.

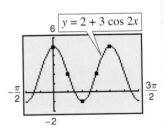
$y = 2 + 3 \cos 2x$

Figure 5.51

✓ *Checkpoint*   Now try Exercise 49.

Example 7   Finding an Equation for a Graph

Find the amplitude, period, and phase shift for the sine function whose graph is shown in Figure 5.52. Write an equation for this graph.

Solution

The amplitude of this sine curve is 2. The period is 2π, and there is a right phase shift of $\pi/2$. So, you can write $y = 2 \sin(x - \pi/2)$.

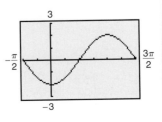

Figure 5.52

✓ *Checkpoint*   Now try Exercise 67.

Mathematical Modeling

Sine and cosine functions can be used to model many real-life situations, including electric currents, musical tones, radio waves, tides, and weather patterns.

Example 8   Finding a Trigonometric Model

Throughout the day, the depth of the water at the end of a dock in Bangor, Washington varies with the tides. The table shows the depths (in feet) at various times during the morning.   (Source: Nautical Software, Inc.)

Time	Depth, y
Midnight	3.1
2 A.M.	7.8
4 A.M.	11.3
6 A.M.	10.9
8 A.M.	6.6
10 A.M.	1.7
Noon	0.9

a. Use a trigonometric function to model this data.

b. A boat needs at least 10 feet of water to moor at the dock. During what times in the evening can it safely dock?

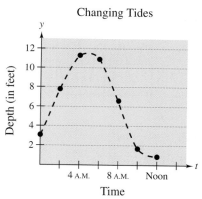

Changing Tides

Figure 5.53

Solution

a. Begin by graphing the data, as shown in Figure 5.53. You can use either a sine or cosine model. Suppose you use a cosine model of the form

$$y = a \cos(bt - c) + d.$$

The difference between the maximum height and minimum height of the graph is twice the amplitude of the function. So, the amplitude is

$$a = \tfrac{1}{2}[(\text{maximum depth}) - (\text{minimum depth})] = \tfrac{1}{2}(11.3 - 0.9) = 5.2.$$

The cosine function completes one half of a cycle between the times at which the maximum and minimum depths occur. So, the period is

$$p = 2[(\text{time of min. depth}) - (\text{time of max. depth})] = 2(12 - 4) = 16$$

which implies that $b = 2\pi/p \approx 0.393$. Because high tide occurs 4 hours after midnight, consider the left endpoint to be $c/b = 4$, so $c \approx 1.571$. Moreover, because the average depth is $\tfrac{1}{2}(11.3 + 0.9) = 6.1$, it follows that $d = 6.1$. So, you can model the depth with the function

$$y = 5.2 \cos(0.393t - 1.571) + 6.1.$$

b. Using a graphing utility, graph the model with the line $y = 10$. Using the *intersect* feature, you can determine that the depth is at least 10 feet between 6:06 P.M. ($t \approx 18.1$) and 9:48 P.M. ($t \approx 21.8$), as shown in Figure 5.54.

✓ *Checkpoint*   Now try Exercise 79.

TECHNOLOGY SUPPORT

For instructions on how to use the *intersect* feature, see Appendix A; for specific keystrokes, go to the text website at *college.hmco.com*.

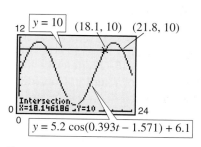

Figure 5.54

5.4 Exercises

Vocabulary Check

Fill in the blanks.

1. The _____ of a sine or cosine curve represents half the distance between the maximum and minimum values of the function.

2. One period of a sine function is called _____ of the sine curve.

3. The period of a sine or cosine function is given by _____ .

4. For the equation $y = a \sin(bx - c)$, $\dfrac{c}{b}$ is the _____ of the graph of the equation.

In Exercises 1–14, find the period and amplitude.

1. $y = 3 \sin 2x$

2. $y = 2 \cos 3x$

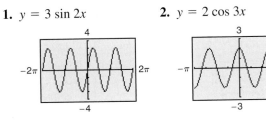

3. $y = \dfrac{5}{2} \cos \dfrac{x}{2}$

4. $y = -3 \sin \dfrac{x}{3}$

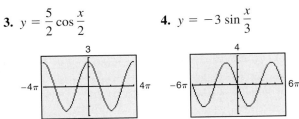

5. $y = \dfrac{2}{3} \sin \pi x$

6. $y = \dfrac{3}{2} \cos \dfrac{\pi x}{2}$

7. $y = -2 \sin x$

8. $y = -\cos \dfrac{2x}{5}$

9. $y = 3 \sin 10x$

10. $y = \dfrac{1}{3} \sin 10x$

11. $y = \dfrac{1}{4} \cos \dfrac{2x}{3}$

12. $y = \dfrac{5}{2} \cos \dfrac{x}{4}$

13. $y = \dfrac{1}{3} \sin 4\pi x$

14. $y = \dfrac{3}{4} \cos \dfrac{\pi x}{12}$

In Exercises 15–22, describe the relationship between the graphs of *f* and *g*. Consider amplitudes, periods, and shifts.

15. $f(x) = \sin x$
 $g(x) = \sin(x - \pi)$

16. $f(x) = \cos x$
 $g(x) = \cos(x + \pi)$

17. $f(x) = \cos 2x$
 $g(x) = -\cos 2x$

18. $f(x) = \sin 3x$
 $g(x) = \sin(-3x)$

19. $f(x) = \cos x$
 $g(x) = -5 \cos x$

20. $f(x) = \sin x$
 $g(x) = -\dfrac{1}{2} \sin x$

21. $f(x) = \sin 2x$
 $g(x) = 5 + \sin 2x$

22. $f(x) = \cos 4x$
 $g(x) = -6 + \cos 4x$

In Exercises 23–26, describe the relationship between the graphs of *f* and *g*. Consider amplitudes, periods, and shifts.

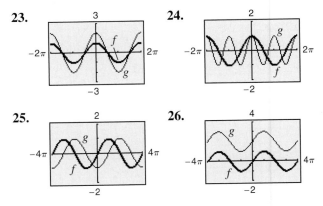

In Exercises 27–34, sketch the graphs of *f* and *g* in the same coordinate plane. (Include two full periods.)

27. $f(x) = \sin x$
 $g(x) = -4 \sin x$

28. $f(x) = \sin x$
 $g(x) = \sin \dfrac{x}{3}$

29. $f(x) = \cos x$
 $g(x) = 4 + \cos x$

30. $f(x) = 2 \cos 2x$
 $g(x) = -\cos 4x$

31. $f(x) = -\dfrac{1}{2} \sin \dfrac{x}{2}$

$g(x) = 3 - \dfrac{1}{2} \sin \dfrac{x}{2}$

32. $f(x) = 4 \sin \pi x$

$g(x) = 4 \sin \pi x - 2$

33. $f(x) = 2 \cos x$

$g(x) = 2 \cos(x + \pi)$

34. $f(x) = -\cos x$

$g(x) = -\cos\left(x - \dfrac{\pi}{2}\right)$

57. $y = \cos\left(2\pi x - \dfrac{\pi}{2}\right) + 1$

58. $y = 3 \cos\left(\dfrac{\pi x}{2} + \dfrac{\pi}{2}\right) - 3$

59. $y = 5 \sin(\pi - 2x) + 10$

60. $y = 5 \cos(\pi - 2x) + 6$

61. $y = \dfrac{1}{100} \sin 120\pi t$ **62.** $y = -\dfrac{1}{100} \cos 50\pi t$

Conjecture In Exercises 35–38, use a graphing utility to graph f and g in the same viewing window. (Include two full periods.) Make a conjecture about the functions.

35. $f(x) = \sin x$

$g(x) = \cos\left(x - \dfrac{\pi}{2}\right)$

36. $f(x) = \sin x$

$g(x) = -\cos\left(x + \dfrac{\pi}{2}\right)$

37. $f(x) = \cos x$

$g(x) = -\sin\left(x - \dfrac{\pi}{2}\right)$

38. $f(x) = \cos x$

$g(x) = -\cos(x - \pi)$

In Exercises 39–54, sketch the graph of the function by hand. Use a graphing utility to verify your sketch. (Include two full periods.)

39. $y = 3 \sin x$

40. $y = \dfrac{1}{4} \cos x$

41. $y = \cos \dfrac{x}{2}$

42. $y = \sin 4x$

43. $y = -2 \sin \dfrac{2\pi x}{3}$

44. $y = -10 \cos \dfrac{\pi x}{6}$

45. $y = \sin\left(x - \dfrac{\pi}{4}\right)$

46. $y = \sin(x - \pi)$

47. $y = -8 \cos(x + \pi)$

48. $y = 6 \cos\left(x + \dfrac{\pi}{3}\right)$

49. $y = 1 + \dfrac{1}{2} \cos 4\pi t$

50. $y = -4 + 5 \cos \dfrac{\pi t}{12}$

51. $y = 2 - 2 \sin \dfrac{2\pi x}{3}$

52. $y = 2 \cos x - 3$

53. $y = \dfrac{2}{3} \cos\left(\dfrac{x}{2} - \dfrac{\pi}{4}\right)$

54. $y = -3 \cos(6x + \pi)$

In Exercises 55–62, use a graphing utility to graph the function. (Include two full periods.) Identify the amplitude and period of the graph.

55. $y = -2 \sin(4x + \pi)$ **56.** $y = -4 \sin\left(\dfrac{2}{3}x - \dfrac{\pi}{3}\right)$

Graphical Reasoning In Exercises 63–66, find a and d for the function $f(x) = a \cos x + d$ such that the graph of f matches the figure.

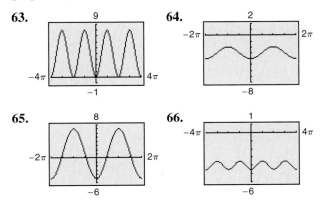

63.   **64.**

65.   **66.**

Graphical Reasoning In Exercises 67–70, find $a, b,$ and c for the function $f(x) = a \sin(bx - c)$ such that the graph of f matches the graph shown.

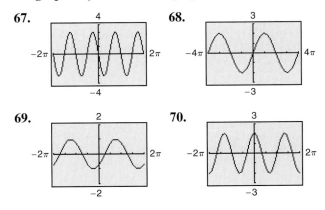

67.   **68.**

69.   **70.**

In Exercises 71 and 72, use a graphing utility to graph y_1 and y_2 for all real numbers x in the interval $[-2\pi, 2\pi]$. Use the graphs to find the real numbers x such that $y_1 = y_2$.

71. $y_1 = \sin x$

$y_2 = -\dfrac{1}{2}$

72. $y_1 = \cos x$

$y_2 = -1$

73. *Health* For a person at rest, the velocity v (in liters per second) of air flow during a respiratory cycle (the time from the beginning of one breath to the beginning of the next) is given by

$$v = 0.85 \sin \frac{\pi t}{3}$$

where t is the time (in seconds). (Inhalation occurs when $v > 0$, and exhalation occurs when $v < 0$.)

(a) Use a graphing utility to graph v.

(b) Find the time for one full respiratory cycle.

(c) Find the number of cycles per minute.

(d) The model is for a person at rest. How might the model change for a person who is exercising? Explain.

74. *Sales* A company that produces snowboards, which are seasonal products, forecasts monthly sales for 1 year to be

$$S = 74.50 + 43.75 \cos \frac{\pi t}{6}$$

where S is the sales in thousands of units and t is the time in months, with $t = 1$ corresponding to January.

(a) Use a graphing utility to graph the sales function over the one-year period.

(b) Use the graph in part (a) to determine the months of maximum and minimum sales.

75. *Recreation* You are riding a Ferris wheel. Your height h (in feet) above the ground at any time t (in seconds) can be modeled by

$$h = 25 \sin \frac{\pi}{15}(t - 75) + 30.$$

The Ferris wheel turns for 135 seconds before it stops to let the first passengers off.

(a) Graph the model.

(b) What are the minimum and maximum heights above the ground?

76. *Health* The pressure P (in millimeters of mercury) against the walls of the blood vessels of a person is modeled by

$$P = 100 - 20 \cos \frac{8\pi}{3} t$$

where t is the time (in seconds). Graph the model. One cycle is equivalent to one heartbeat. What is the person's pulse rate in heartbeats per minute?

77. *Fuel Consumption* The daily consumption C (in gallons) of diesel fuel on a farm is modeled by

$$C = 30.3 + 21.6 \sin\left(\frac{2\pi t}{365} + 10.9\right)$$

where t is the time in days, with $t = 1$ corresponding to January 1.

(a) What is the period of the model? Is it what you expected? Explain.

(b) What is the average daily fuel consumption? Which term of the model did you use? Explain.

(c) Use a graphing utility to graph the model. Use the graph to approximate the time of the year when consumption exceeds 40 gallons per day.

78. *Data Analysis* The motion of an oscillating weight suspended from a spring was measured by a motion detector. The data was collected, and the approximate maximum displacements from equilibrium ($y = 2$) are labeled in the figure. The distance y from the motion detector is measured in centimeters and the time t is measured in seconds.

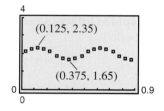

(a) Is y a function of t? Explain.

(b) Approximate the amplitude and period.

(c) Find a model for the data.

(d) Use a graphing utility to graph the model in part (c). Compare the result with the data in the figure.

79. *Data Analysis* The percent y of the moon's face that is illuminated on day x of the year 2006, where $x = 1$ represents January 1, is shown in the table. (Source: U.S. Naval Observatory)

Day, x	Percent, y
29	0.0
36	0.5
44	1.0
52	0.5
58	0.0
65	0.5

(a) Create a scatter plot of the data.

(b) Find a trigonometric model for the data.

(c) Add the graph of your model in part (b) to the scatter plot. How well does the model fit the data?

(d) What is the period of the model?

(e) Estimate the percent illumination of the moon on June 22, 2007.

80. **Data Analysis**   The table shows the average daily high temperatures for Nantucket, Massachusetts N and Athens, Georgia A (in degrees Fahrenheit) for month t, with $t = 1$ corresponding to January. (Source: U.S. Weather Bureau and the National Weather Service)

Month, t	Nantucket, N	Athens, A
1	40	52
2	41	56
3	42	65
4	53	73
5	62	81
6	71	87
7	78	90
8	76	88
9	70	83
10	59	74
11	48	64
12	40	55

(a) A model for the temperature in Nantucket is given by

$$N(t) = 58 + 19 \sin\left(\frac{2\pi t}{11} - \frac{21\pi}{25}\right).$$

Find a trigonometric model for Athens.

(b) Use a graphing utility to graph the data and the model for the temperatures in Nantucket in the same viewing window. How well does the model fit the data?

(c) Use a graphing utility to graph the data and the model for the temperatures in Athens in the same viewing window. How well does the model fit the data?

(d) Use the models to estimate the average daily high temperature in each city. Which term of the models did you use? Explain.

(e) What is the period of each model? Are the periods what you expected? Explain.

(f) Which city has the greater variability in temperature throughout the year? Which factor of the models determines this variability? Explain.

Synthesis

True or False?   **In Exercises 81–83, determine whether the statement is true or false. Justify your answer.**

81. The graph of $y = 6 - \frac{3}{4} \sin \frac{3x}{10}$ has a period of $\frac{20\pi}{3}$.

82. The function $y = \frac{1}{2} \cos 2x$ has an amplitude that is twice that of the function $y = \cos x$.

83. The graph of $y = -\cos x$ is a reflection of the graph of $y = \sin(x + \pi/2)$ in the x-axis.

84. **Writing**   Use a graphing utility to graph the function

$$y = d + a \sin(bx - c)$$

for different values of a, b, c, and d. Write a paragraph describing the changes in the graph corresponding to changes in each variable.

85. **Exploration**   In Section 5.3 it was shown that $f(x) = \cos x$ is an even function and $g(x) = \sin x$ is an odd function. Use a graphing utility to graph h and use the graph to determine whether h is even, odd, or neither.

(a) $h(x) = \cos^2 x$ (b) $h(x) = \sin^2 x$

(c) $h(x) = \sin x \cos x$

86. **Conjecture**   If f is an even function and g is an odd function, use the results of Exercise 85 to make a conjecture about each of the following.

(a) $h(x) = [f(x)]^2$ (b) $h(x) = [g(x)]^2$

(c) $h(x) = f(x)g(x)$

Review

In Exercises 87 and 88, plot the points and find the slope of the line passing through the points.

87. $(0, 1), (2, 7)$ 88. $(-1, 4), (3, -2)$

In Exercises 89 and 90, convert the angle measure from radians to degrees. Round your answer to three decimal places.

89. 8.5 90. -0.48

5.5 Graphs of Other Trigonometric Functions

Graph of the Tangent Function

Recall that the tangent function is odd. That is, $\tan(-x) = -\tan x$. Consequently, the graph of $y = \tan x$ is symmetric with respect to the origin. You also know from the identity $\tan x = \sin x/\cos x$ that the tangent function is undefined at values at which $\cos x = 0$. Two such values are $x = \pm\pi/2 \approx \pm 1.5708$.

x	$-\dfrac{\pi}{2}$	-1.57	-1.5	$-\dfrac{\pi}{4}$	0	$\dfrac{\pi}{4}$	1.5	1.57	$\dfrac{\pi}{2}$
$\tan x$	Undef.	-1255.8	-14.1	-1	0	1	14.1	1255.8	Undef.

$\Longleftarrow$ tan x approaches $-\infty$ as x approaches $-\pi/2$ from the right.

tan x approaches ∞ as x approaches $\pi/2$ from the left. $\Longrightarrow$

As indicated in the table, $\tan x$ increases without bound as x approaches $\pi/2$ from the left, and it decreases without bound as x approaches $-\pi/2$ from the right. So, the graph of $y = \tan x$ has *vertical asymptotes* at $x = \pi/2$ and $-\pi/2$, as shown in Figure 5.55. Moreover, because the period of the tangent function is π, vertical asymptotes also occur at $x = \pi/2 + n\pi$, where n is an integer. The domain of the tangent function is the set of all real numbers other than $x = \pi/2 + n\pi$, and the range is the set of all real numbers.

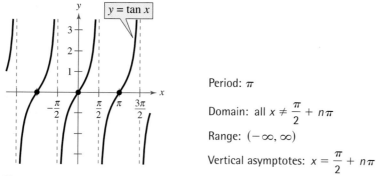

$y = \tan x$

Period: π

Domain: all $x \neq \dfrac{\pi}{2} + n\pi$

Range: $(-\infty, \infty)$

Vertical asymptotes: $x = \dfrac{\pi}{2} + n\pi$

Figure 5.55

Sketching the graph of $y = a\tan(bx - c)$ is similar to sketching the graph of $y = a\sin(bx - c)$ in that you locate key points that identify the intercepts and asymptotes. Two consecutive asymptotes can be found by solving the equations $bx - c = -\pi/2$ and $bx - c = \pi/2$. The midpoint between two consecutive asymptotes is an x-intercept of the graph. The period of the function $y = a\tan(bx - c)$ is the distance between two consecutive asymptotes. The amplitude of a tangent function is not defined. After plotting the asymptotes and the x-intercept, plot a few additional points between the two asymptotes and sketch one cycle. Finally, sketch one or two additional cycles to the left and right.

Example 1   Sketching the Graph of a Tangent Function

Sketch the graph of $y = \tan \dfrac{x}{2}$ by hand.

Solution

By solving the equations $x/2 = -\pi/2$ and $x/2 = \pi/2$, you can see that two consecutive asymptotes occur at $x = -\pi$ and $x = \pi$. Between these two asymptotes, plot a few points, including the x-intercept, as shown in the table. Three cycles of the graph are shown in Figure 5.56. Use a graphing utility to confirm this graph.

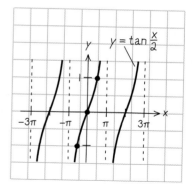

Figure 5.56

x	$-\pi$	$-\dfrac{\pi}{2}$	0	$\dfrac{\pi}{2}$	π
$\tan \dfrac{x}{2}$	Undef.	-1	0	1	Undef.

 Checkpoint Now try Exercise 7.

Example 2   Sketching the Graph of a Tangent Function

Sketch the graph of $y = -3 \tan 2x$.

Solution

By solving the equations $2x = -\pi/2$ and $2x = \pi/2$, you can see that two consecutive asymptotes occur at $x = -\pi/4$ and $x = \pi/4$. Between these two asymptotes, plot a few points, including the x-intercept, as shown in the table. Three complete cycles of the graph are shown in Figure 5.57.

x	$-\dfrac{\pi}{4}$	$-\dfrac{\pi}{8}$	0	$\dfrac{\pi}{8}$	$\dfrac{\pi}{4}$
$-3 \tan 2x$	Undef.	3	0	-3	Undef.

> **TECHNOLOGY TIP**
>
> Your graphing utility may connect parts of the graphs of tangent, cotangent, secant, and cosecant functions that are not supposed to be connected. So, in this text, these functions are graphed on a graphing utility using the *dot* mode. A blue curve is placed behind the graphing utility's display to indicate where the graph should appear. For instructions on how to use the *dot* mode, see Appendix A; for specific keystrokes, go to the text website at *college.hmco.com*.

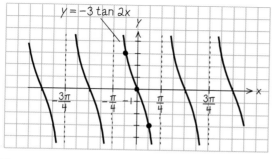

Figure 5.57

 Checkpoint Now try Exercise 9.

TECHNOLOGY TIP Graphing utilities are helpful in verifying sketches of trigonometric functions. You can use a graphing utility set in *radian* and *dot* modes to graph the function $y = -3 \tan 2x$ from Example 2, as shown in Figure 5.58. You can use the *zero* or *root* feature or the *zoom* and *trace* features to approximate the key points of the graph.

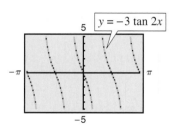

Figure 5.58

By comparing the graphs in Examples 1 and 2, you can see that the graph of $y = a \tan(bx - c)$ increases between consecutive vertical asymptotes when $a > 0$ and decreases between consecutive vertical asymptotes when $a < 0$. In other words, the graph for $a < 0$ is a reflection in the *x*-axis of the graph for $a > 0$.

Graph of the Cotangent Function

The graph of the cotangent function is similar to the graph of the tangent function. It also has a period of π. However, from the identity

$$y = \cot x = \frac{\cos x}{\sin x}$$

you can see that the cotangent function has vertical asymptotes when $\sin x$ is zero, which occurs at $x = n\pi$, where n is an integer. The graph of the cotangent function is shown in Figure 5.59.

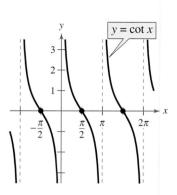

Period: π
Domain: all $x \neq n\pi$
Range: $(-\infty, \infty)$
Vertical asymptotes: $x = n\pi$
Figure 5.59

Example 3 Sketching the Graph of a Cotangent Function

Sketch the graph of $y = 2 \cot \dfrac{x}{3}$ by hand.

Solution

To locate two consecutive vertical asymptotes of the graph, solve the equations $x/3 = 0$ and $x/3 = \pi$ to see that two consecutive asymptotes occur at $x = 0$ and $x = 3\pi$. Then, between these two asymptotes, plot a few points, including the *x*-intercept, as shown in the table. Three cycles of the graph are shown in Figure 5.60. Use a graphing utility to confirm this graph. [Enter the function as $y = 2/\tan(x/3)$.] Note that the period is 3π, the distance between consecutive asymptotes.

x	0	$\dfrac{3\pi}{4}$	$\dfrac{3\pi}{2}$	$\dfrac{9\pi}{4}$	3π
$2 \cot \dfrac{x}{3}$	Undef.	2	0	-2	Undef.

✓ *Checkpoint* Now try Exercise 19.

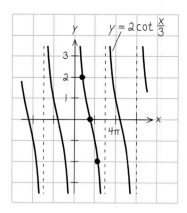

Figure 5.60

E x p l o r a t i o n

Use a graphing utility to graph the functions $y_1 = \cos x$ and $y_2 = \sec x = 1/\cos x$ in the same viewing window. How are the graphs related? What happens to the graph of the secant function as x approaches the zeros of the cosine function?

Graphs of the Reciprocal Functions

The graphs of the two remaining trigonometric functions can be obtained from the graphs of the sine and cosine functions using the reciprocal identities

$$\csc x = \frac{1}{\sin x} \quad \text{and} \quad \sec x = \frac{1}{\cos x}.$$

For instance, at a given value of x, the y-coordinate for $\sec x$ is the reciprocal of the y-coordinate for $\cos x$. Of course, when $\cos x = 0$, the reciprocal does not exist. Near such values of x, the behavior of the secant function is similar to that of the tangent function. In other words, the graphs of

$$\tan x = \frac{\sin x}{\cos x} \quad \text{and} \quad \sec x = \frac{1}{\cos x}$$

have vertical asymptotes at $x = \pi/2 + n\pi$, where n is an integer (i.e., the values at which the cosine is zero). Similarly,

$$\cot x = \frac{\cos x}{\sin x} \quad \text{and} \quad \csc x = \frac{1}{\sin x}$$

have vertical asymptotes where $\sin x = 0$, that is, at $x = n\pi$.

To sketch the graph of a secant or cosecant function, you should first make a sketch of its reciprocal function. For instance, to sketch the graph of $y = \csc x$, first sketch the graph of $y = \sin x$. Then take the reciprocals of the y-coordinates to obtain points on the graph of $y = \csc x$. You can use this procedure to obtain the graphs shown in Figure 5.61.

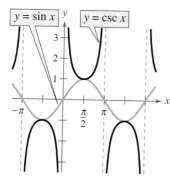

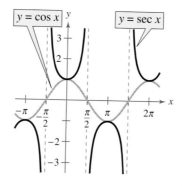

Period: 2π

Domain: all $x \neq n\pi$

Range: $(-\infty, -1] \cup [1, \infty)$

Vertical asymptotes: $x = n\pi$

Symmetry: origin

Period: 2π

Domain: all $x \neq \dfrac{\pi}{2} + n\pi$

Range: $(-\infty, -1] \cup [1, \infty)$

Vertical asymptotes: $x = \dfrac{\pi}{2} + n\pi$

Symmetry: y-axis

Figure 5.61

In comparing the graphs of the secant and cosecant functions with those of the sine and cosine functions, note that the "hills" and "valleys" are interchanged. For example, a hill (or maximum point) on the sine curve corresponds to a valley (a local minimum) on the cosecant curve, and a valley (or minimum point) on the sine curve corresponds to a hill (a local maximum) on the cosecant curve, as shown in Figure 5.62. Additionally, x-intercepts of the sine and cosine functions become vertical asymptotes of the cosecant and secant functions, respectively (see Figure 5.62).

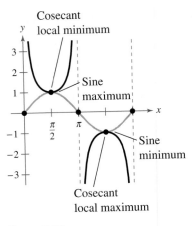

Cosecant
local minimum

Sine
maximum

Sine
minimum

Cosecant
local maximum

Figure 5.62

Example 4   Comparing Trigonometric Graphs

Use a graphing utility to compare the graphs of

$$y = 2 \sin\left(x + \frac{\pi}{4}\right) \quad \text{and} \quad y = 2 \csc\left(x + \frac{\pi}{4}\right).$$

Solution

The two graphs are shown in Figure 5.63. Note how the hills and valleys of the graphs are related. For the function $y = 2 \sin[x + (\pi/4)]$, the amplitude is 2 and the period is 2π. By solving the equations

$$x + \frac{\pi}{4} = 0 \quad \text{and} \quad x + \frac{\pi}{4} = 2\pi$$

you can see that one cycle of the sine function corresponds to the interval from $x = -\pi/4$ to $x = 7\pi/4$. The graph of this sine function is represented by the thick curve in Figure 5.63. Because the sine function is zero at the endpoints of this interval, the corresponding cosecant function

$$y = 2 \csc\left(x + \frac{\pi}{4}\right) = 2\left(\frac{1}{\sin[x + (\pi/4)]}\right)$$

has vertical asymptotes at $x = -\pi/4, 3\pi/4, 7\pi/4$, and so on.

✓ *Checkpoint* Now try Exercise 31.

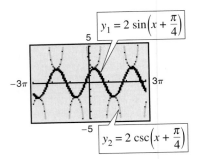

$y_1 = 2 \sin\left(x + \frac{\pi}{4}\right)$

$y_2 = 2 \csc\left(x + \frac{\pi}{4}\right)$

Figure 5.63

Example 5   Comparing Trigonometric Graphs

Use a graphing utility to compare the graphs of $y = \cos 2x$ and $y = \sec 2x$.

Solution

Begin by graphing $y_1 = \cos 2x$ and $y_2 = \sec 2x = 1/\cos 2x$ in the same viewing window, as shown in Figure 5.64. Note that the x-intercepts of $y = \cos 2x$

$$\left(-\frac{\pi}{4}, 0\right), \quad \left(\frac{\pi}{4}, 0\right), \quad \left(\frac{3\pi}{4}, 0\right), \ldots$$

correspond to the vertical asymptotes

$$x = -\frac{\pi}{4}, \quad x = \frac{\pi}{4}, \quad x = \frac{3\pi}{4}, \ldots$$

of the graph of $y = \sec 2x$. Moreover, notice that the period of $y = \cos 2x$ and $y = \sec 2x$ is π.

✓ *Checkpoint* Now try Exercise 33.

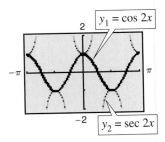

$y_1 = \cos 2x$

$y_2 = \sec 2x$

Figure 5.64

Damped Trigonometric Graphs

A *product* of two functions can be graphed using properties of the individual functions. For instance, consider the function

$$f(x) = x \sin x$$

as the product of the functions $y = x$ and $y = \sin x$. Using properties of absolute value and the fact that $|\sin x| \leq 1$, you have $0 \leq |x| \, |\sin x| \leq |x|$. Consequently,

$$-|x| \leq x \sin x \leq |x|$$

which means that the graph of $f(x) = x \sin x$ lies between the lines $y = -x$ and $y = x$. Furthermore, because

$$f(x) = x \sin x = \pm x \qquad \text{at} \qquad x = \frac{\pi}{2} + n\pi$$

and

$$f(x) = x \sin x = 0 \qquad \text{at} \qquad x = n\pi$$

the graph of f touches the line $y = -x$ or the line $y = x$ at $x = \pi/2 + n\pi$ and has x-intercepts at $x = n\pi$. A sketch of f is shown in Figure 5.65. In the function $f(x) = x \sin x$, the factor x is called the **damping factor.**

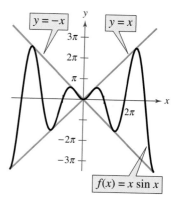

Figure 5.65

Example 6   Analyzing a Damped Sine Curve

Analyze the graph of

$$f(x) = e^{-x} \sin 3x.$$

Solution

Consider $f(x)$ as the product of the two functions

$$y = e^{-x} \qquad \text{and} \qquad y = \sin 3x$$

each of which has the set of real numbers as its domain. For any real number x, you know that $e^{-x} \geq 0$ and $|\sin 3x| \leq 1$. So, $|e^{-x}| \, |\sin 3x| \leq e^{-x}$, which means that

$$-e^{-x} \leq e^{-x} \sin 3x \leq e^{-x}.$$

Furthermore, because

$$f(x) = e^{-x} \sin 3x = \pm e^{-x} \qquad \text{at} \qquad x = \frac{\pi}{6} + \frac{n\pi}{3}$$

and

$$f(x) = e^{-x} \sin 3x = 0 \qquad \text{at} \qquad x = \frac{n\pi}{3}$$

the graph of f touches the curves $y = -e^{-x}$ and $y = e^{-x}$ at $x = \pi/6 + n\pi/3$ and has intercepts at $x = n\pi/3$. The graph is shown in Figure 5.66.

✓ *Checkpoint* Now try Exercise 57.

> ## STUDY TIP
>
> Do you see why the graph of $f(x) = x \sin x$ touches the lines $y = \pm x$ at $x = \pi/2 + n\pi$ and why the graph has x-intercepts at $x = n\pi$? Recall that the sine function is equal to ± 1 at $\pi/2, 3\pi/2, 5\pi/2, \ldots$ (odd multiples of $\pi/2$) and is equal to 0 at $\pi, 2\pi, 3\pi, \ldots$ (multiples of π).

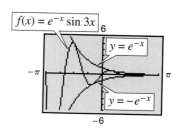

Figure 5.66

Figure 5.67 summarizes the six basic trigonometric functions.

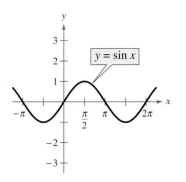

Domain: all reals
Range: $[-1, 1]$
Period: 2π

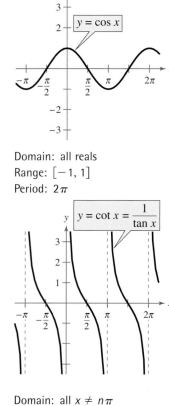

Domain: all reals
Range: $[-1, 1]$
Period: 2π

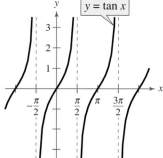

Domain: all $x \neq \dfrac{\pi}{2} + n\pi$

Range: $(-\infty, \infty)$
Period: π

Domain: all $x \neq n\pi$
Range: $(-\infty, \infty)$
Period: π

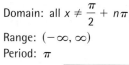

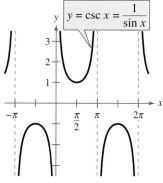

Domain: all $x \neq n\pi$
Range: $(-\infty, -1] \cup [1, \infty)$
Period: 2π

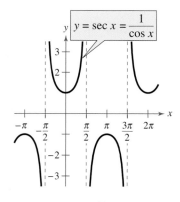

Domain: all $x \neq \dfrac{\pi}{2} + n\pi$

Range: $(-\infty, -1] \cup [1, \infty)$
Period: 2π

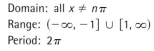

Figure 5.67

5.5 Exercises

Vocabulary Check

Fill in the blanks.

1. The graphs of the tangent, cotangent, secant, and cosecant functions have _____ asymptotes.

2. To sketch the graph of a secant or cosecant function, first make a sketch of its _____ function.

3. For the function $f(x) = g(x) \sin x$, $g(x)$ is called the _____ factor of the function.

In Exercises 1–6, match the function with its graph. State the period of the function. [The graphs are labeled (a), (b), (c), (d), (e), and (f).]

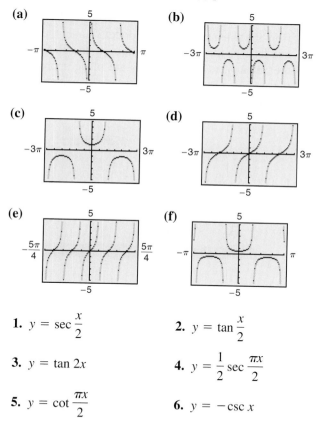

(a)

(b)

(c)

(d)

(e)

(f)

1. $y = \sec \dfrac{x}{2}$

2. $y = \tan \dfrac{x}{2}$

3. $y = \tan 2x$

4. $y = \dfrac{1}{2} \sec \dfrac{\pi x}{2}$

5. $y = \cot \dfrac{\pi x}{2}$

6. $y = -\csc x$

In Exercises 7–30, sketch the graph of the function. (Include two full periods.) Use a graphing utility to verify your result.

7. $y = \dfrac{1}{2} \tan x$

8. $y = \dfrac{1}{4} \tan x$

9. $y = -2 \tan 2x$

10. $y = -3 \tan 4x$

11. $y = -\dfrac{1}{2} \sec x$

12. $y = \dfrac{1}{4} \sec x$

13. $y = -\sec \pi x$

14. $y = 2 \sec \pi x$

15. $y = \sec \pi x - 3$

16. $y = -2 \sec 4x + 2$

17. $y = 3 \csc \dfrac{x}{2}$

18. $y = -\csc \dfrac{x}{3}$

19. $y = \dfrac{1}{2} \cot \dfrac{x}{2}$

20. $y = 3 \cot \pi x$

21. $y = 2 \tan \dfrac{\pi x}{4}$

22. $y = -\dfrac{1}{2} \tan \pi x$

23. $y = \dfrac{1}{2} \sec 2x$

24. $y = \sec(x + \pi)$

25. $y = \csc(\pi - x)$

26. $y = \csc(2x - \pi)$

27. $y = 2 \cot\left(x - \dfrac{\pi}{2}\right)$

28. $y = \dfrac{1}{4} \cot(x + \pi)$

29. $y = \tan\left(x - \dfrac{\pi}{4}\right)$

30. $y = \dfrac{1}{2} \tan\left(\dfrac{\pi x}{4} + \dfrac{\pi}{4}\right)$

In Exercises 31–36, use a graphing utility to graph the function. (Include two full periods.) Compare the graph of the function with the graph of the corresponding reciprocal function. Describe your viewing window.

31. $y = 2 \csc 3x$

32. $y = -\csc(4x - \pi)$

33. $y = -2 \sec 4x$

34. $y = \dfrac{1}{4} \sec \pi x$

35. $y = \dfrac{1}{3} \sec\left(\dfrac{\pi x}{2} + \dfrac{\pi}{2}\right)$

36. $y = \dfrac{1}{2} \csc(2x - \pi)$

In Exercises 37–40, use a graph to solve the equation on the interval $[-2\pi, 2\pi]$.

37. $\tan x = 1$

38. $\cot x = -\sqrt{3}$

39. $\sec x = -2$

40. $\csc x = \sqrt{2}$

In Exercises 41–44, use the graph of the function to determine whether the function is even, odd, or neither.

41. $f(x) = \sec x$

42. $f(x) = \tan x$

43. $f(x) = \csc 2x$

44. $f(x) = \cot 2x$

In Exercises 45–48, use a graphing utility to graph the two equations in the same viewing window. Use the graphs to determine whether the expressions are equivalent. Verify the results algebraically.

45. $y_1 = \sin x \csc x$, $y_2 = 1$

46. $y_1 = \sin x \sec x$, $y_2 = \tan x$

47. $y_1 = \dfrac{\cos x}{\sin x}$, $y_2 = \cot x$

48. $y_1 = \sec^2 x - 1$, $y_2 = \tan^2 x$

In Exercises 49–52, match the function with its graph. Describe the behavior of the function as x approaches zero. [The graphs are labeled (a), (b), (c), and (d).]

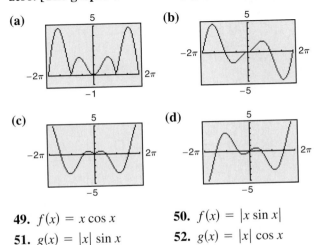

(a)

(b)

(c)

(d)

49. $f(x) = x \cos x$

50. $f(x) = |x \sin x|$

51. $g(x) = |x| \sin x$

52. $g(x) = |x| \cos x$

Conjecture In Exercises 53–56, use a graphing utility to graph the functions f and g. Use the graphs to make a conjecture about the relationship between the functions.

53. $f(x) = \sin x + \cos\left(x + \dfrac{\pi}{2}\right)$, $g(x) = 0$

54. $f(x) = \sin x - \cos\left(x + \dfrac{\pi}{2}\right)$, $g(x) = 2 \sin x$

55. $f(x) = \sin^2 x$, $g(x) = \tfrac{1}{2}(1 - \cos 2x)$

56. $f(x) = \cos^2 \dfrac{\pi x}{2}$, $g(x) = \dfrac{1}{2}(1 + \cos \pi x)$

In Exercises 57–60, use a graphing utility to graph the function and the damping factor of the function in the same viewing window. Describe the behavior of the function as x increases without bound.

57. $f(x) = e^{-x} \cos x$

58. $f(x) = 2^{-x/4} \cos \pi x$

59. $g(x) = e^{-x^2/2} \sin x$

60. $h(x) = 2^{-x^2/4} \sin x$

Exploration In Exercises 61–66, use a graphing utility to graph the function. Describe the behavior of the function as x approaches zero.

61. $f(x) = \dfrac{6}{x} + \cos x$

62. $f(x) = \sin x - \dfrac{4}{x}$

63. $f(x) = \dfrac{\sin x}{x}$

64. $f(x) = \dfrac{1 - \cos x}{x}$

65. $f(x) = \dfrac{\tan x}{x}$

66. $f(x) = \dfrac{x}{\cot x}$

67. *Distance* A plane flying at an altitude of 5 miles over level ground will pass directly over a radar antenna (see figure). Let d be the ground distance from the antenna to the point directly under the plane and let x be the angle of elevation to the plane from the antenna. (d is positive as the plane approaches the antenna.) Write d as a function of x and graph the function over the interval $0 < x < \pi$.

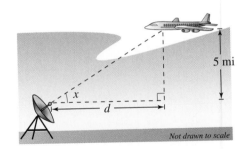

5 mi

Not drawn to scale

68. *Television Coverage* A television camera is on a reviewing platform 36 meters from the street on which a parade will be passing from left to right (see figure). Write the distance d from the camera to a particular unit in the parade as a function of the angle x, and graph the function over the interval $-\pi/2 < x < \pi/2$. (Consider x as negative when a unit in the parade approaches from the left.)

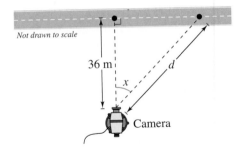

Not drawn to scale

36 m

d

x

Camera

69. Predator-Prey Model   The population of coyotes (a predator) at time t (in months) in a region is estimated to be $P = 10,000 + 3000 \sin(\pi t/12)$ and the population of rabbits (its prey) is estimated to be $p = 15,000 + 5000 \cos(\pi t/12)$. Use the graph of the models to explain the oscillations in the size of each population.

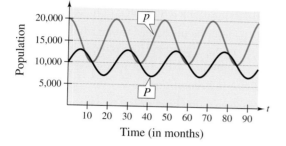

70. Meteorology   The normal monthly high temperatures in degrees Fahrenheit for Erie, Pennsylvania, are approximated by

$$H(t) = 54.33 - 20.38 \cos \frac{\pi t}{6} - 15.69 \sin \frac{\pi t}{6}$$

and the normal monthly low temperatures are approximated by

$$L(t) = 39.36 - 15.70 \cos \frac{\pi t}{6} - 14.16 \sin \frac{\pi t}{6}$$

where t is the time (in months), with $t = 1$ corresponding to January.   (Source: National Oceanic and Atmospheric Association)

(a) Use a graphing utility to graph each function. What is the period of each function?

(b) During what part of the year is the difference between the normal high and normal low temperatures greatest? When is it smallest?

(c) The sun is the farthest north in the sky around June 21, but the graph shows the warmest temperatures at a later date. Approximate the lag time of the temperatures relative to the position of the sun.

71. Harmonic Motion   An object weighing W pounds is suspended from a ceiling by a steel spring (see figure). The weight is pulled downward (positive direction) from its equilibrium position and released. The resulting motion of the weight is described by the function $y = \frac{1}{2}e^{-t/4} \cos 4t$, where y is the distance in feet and t is the time in seconds ($t > 0$).

(a) Use a graphing utility to graph the function.

(b) Describe the behavior of the displacement function for increasing values of time t.

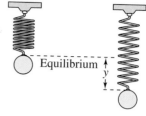

Figure for 71

72. Numerical and Graphical Reasoning   A crossed belt connects a 10-centimeter pulley on an electric motor with a 20-centimeter pulley on a saw arbor (see figure). The electric motor runs at 1700 revolutions per minute.

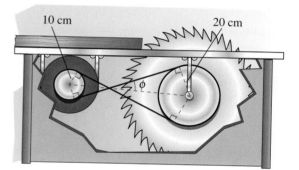

(a) Determine the number of revolutions per minute of the saw.

(b) How does crossing the belt affect the saw in relation to the motor?

(c) Let L be the total length of the belt. Write L as a function of ϕ, where ϕ is measured in radians. What is the domain of the function? (*Hint:* Add the lengths of the straight sections of the belt and the length of belt around each pulley.)

(d) Use a graphing utility to complete the table.

ϕ	0.3	0.6	0.9	1.2	1.5
L					

(e) As ϕ increases, do the lengths of the straight sections of the belt change faster or slower than the lengths of the belts around each pulley?

(f) Use a graphing utility to graph the function over the appropriate domain.

73. *Data Analysis* The motion of an oscillating weight suspended by a spring was measured by a motion detector. The data was collected, and the approximate maximum (positive and negative) displacements from equilibrium are shown in the graph. The displacement y is measured in centimeters and the time t is measured in seconds.

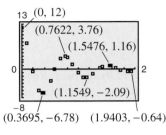

(a) Is y a function of t? Explain.

(b) Approximate the frequency of the oscillations.

(c) Fit a model of the form $y = ab^t \cos ct$ to the data. Use the result of part (b) to approximate c. Use the *regression* feature of a graphing utility to fit an exponential model to the positive maximum displacements of the weight.

(d) Rewrite the model in the form $y = ae^{kt} \cos ct$.

(e) Use a graphing utility to graph the model. Compare the result with the data in the graph above.

74. *Writing* Write a short paragraph describing the specified change in the physical system of Exercise 73.

(a) A spring of less stiffness is used, and so the length of time for each oscillation is greater.

(b) The effect of friction is decreased.

Synthesis

True or False? **In Exercises 75 and 76, determine whether the statement is true or false. Justify your answer.**

75. The graph of $y = -\dfrac{1}{8} \tan\left(\dfrac{x}{2} + \pi\right)$ has an asymptote at $x = -3\pi$.

76. For the graph of $y = 2^x \sin x$, as x approaches $-\infty$, y approaches 0.

77. *Writing* Describe the behavior of $f(x) = \tan x$ as x approaches $\pi/2$ from the left and from the right.

78. *Writing* Describe the behavior of $f(x) = \csc x$ as x approaches π from the left and from the right.

79. *Graphical Reasoning* Consider the functions $f(x) = 2 \sin x$ and $g(x) = \dfrac{1}{2} \csc x$ on the interval $(0, \pi)$.

(a) Use a graphing utility to graph f and g in the same viewing window.

(b) Approximate the interval in which $f > g$.

(c) Describe the behavior of each of the functions as x approaches π. How is the behavior of g related to the behavior of f as x approaches π?

80. *Pattern Recognition*

(a) Use a graphing utility to graph each function.

$$y_1 = \frac{4}{\pi}\left(\sin \pi x + \frac{1}{3} \sin 3\pi x\right)$$

$$y_2 = \frac{4}{\pi}\left(\sin \pi x + \frac{1}{3} \sin 3\pi x + \frac{1}{5} \sin 5\pi x\right)$$

(b) Identify the pattern in part (a) and find a function y_3 that continues the pattern one more term. Use a graphing utility to graph y_3.

(c) The graphs in parts (a) and (b) approximate the periodic function in the figure. Find a function y_4 that is a better approximation.

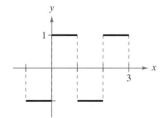

Review

In Exercises 81–84, identify the rule of algebra illustrated by the statement.

81. $5(a - 9) = 5a - 45$ **82.** $7\left(\frac{1}{7}\right) = 1$

83. $(3 + x) + 0 = 3 + x$

84. $(a + b) + 10 = a + (b + 10)$

In Exercises 85–88, determine whether the function is one-to-one. If it is, find its inverse function.

85. $f(x) = -10$ **86.** $f(x) = (x - 7)^2 + 3$

87. $f(x) = \sqrt{3x - 14}$ **88.** $f(x) = \sqrt[3]{x - 5}$

5.6 | Inverse Trigonometric Functions

Inverse Sine Function

Recall from Section 1.7 that for a function to have an inverse function, it must be one-to-one—that is, it must pass the Horizontal Line Test. From Figure 5.68 it is obvious that $y = \sin x$ does not pass the test because different values of x yield the same y-value.

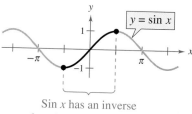

Sin x has an inverse
function on this interval.

Figure 5.68

However, if you restrict the domain to the interval $-\pi/2 \le x \le \pi/2$ (corresponding to the black portion of the graph in Figure 5.68), the following properties hold.

1. On the interval $[-\pi/2, \pi/2]$, the function $y = \sin x$ is increasing.

2. On the interval $[-\pi/2, \pi/2]$, $y = \sin x$ takes on its full range of values, $-1 \le \sin x \le 1$.

3. On the interval $[-\pi/2, \pi/2]$, $y = \sin x$ is one-to-one.

So, on the restricted domain $-\pi/2 \le x \le \pi/2$, $y = \sin x$ has a unique inverse function called the **inverse sine function.** It is denoted by

$$y = \arcsin x \qquad \text{or} \qquad y = \sin^{-1} x.$$

The notation $\sin^{-1} x$ is consistent with the inverse function notation $f^{-1}(x)$. The arcsin x notation (read as "the arcsine of x") comes from the association of a central angle with its intercepted *arc length* on a unit circle. So, arcsin x means the angle (or arc) whose sine is x. Both notations, arcsin x and $\sin^{-1} x$, are commonly used in mathematics, so remember that $\sin^{-1} x$ denotes the *inverse* sine function rather than $1/\sin x$. The values of arcsin x lie in the interval $-\pi/2 \le \arcsin x \le \pi/2$. The graph of $y = \arcsin x$ is shown in Example 2.

Definition of Inverse Sine Function

The **inverse sine function** is defined by

$$y = \arcsin x \qquad \text{if and only if} \qquad \sin y = x$$

where $-1 \le x \le 1$ and $-\pi/2 \le y \le \pi/2$. The domain of $y = \arcsin x$ is $[-1, 1]$ and the range is $[-\pi/2, \pi/2]$.

When evaluating the inverse sine function, it helps to remember the phrase "the arcsine of x is the angle (or number) whose sine is x."

Example 1   Evaluating the Inverse Sine Function

If possible, find the exact value.

a. $\arcsin\left(-\dfrac{1}{2}\right)$ **b.** $\sin^{-1}\dfrac{\sqrt{3}}{2}$ **c.** $\sin^{-1}2$

Solution

a. Because $\sin\left(-\dfrac{\pi}{6}\right) = -\dfrac{1}{2}$, and $-\dfrac{\pi}{6}$ lies in $\left[-\dfrac{\pi}{2},\dfrac{\pi}{2}\right]$, it follows that

$$\arcsin\left(-\dfrac{1}{2}\right) = -\dfrac{\pi}{6}.$$ Angle whose sine is $-\frac{1}{2}$

b. Because $\sin\left(\dfrac{\pi}{3}\right) = \dfrac{\sqrt{3}}{2}$, and $\dfrac{\pi}{3}$ lies in $\left[-\dfrac{\pi}{2},\dfrac{\pi}{2}\right]$, it follows that

$$\sin^{-1}\dfrac{\sqrt{3}}{2} = \dfrac{\pi}{3}.$$ Angle whose sine is $\sqrt{3}/2$

c. It is not possible to evaluate $y = \sin^{-1}x$ at $x = 2$ because there is no angle whose sine is 2. Remember that the domain of the inverse sine function is $[-1, 1]$.

✓ *Checkpoint* Now try Exercise 1.

Example 2   Graphing the Arcsine Function

Sketch a graph of $y = \arcsin x$ by hand.

Solution

By definition, the equations

$$y = \arcsin x \qquad \text{and} \qquad \sin y = x$$

are equivalent for $-\pi/2 \le y \le \pi/2$. So, their graphs are the same. For the interval $[-\pi/2, \pi/2]$, you can assign values to y in the second equation to make a table of values.

y	$-\dfrac{\pi}{2}$	$-\dfrac{\pi}{4}$	$-\dfrac{\pi}{6}$	0	$\dfrac{\pi}{6}$	$\dfrac{\pi}{4}$	$\dfrac{\pi}{2}$
$x = \sin y$	-1	$-\dfrac{\sqrt{2}}{2}$	$-\dfrac{1}{2}$	0	$\dfrac{1}{2}$	$\dfrac{\sqrt{2}}{2}$	1

Then plot the points and connect them with a smooth curve. The resulting graph of $y = \arcsin x$ is shown in Figure 5.69. Note that it is the reflection (in the line $y = x$) of the black portion of the graph in Figure 5.68. Use a graphing utility to confirm this graph. Be sure you see that Figure 5.69 shows the *entire* graph of the inverse sine function. Remember that the domain of $y = \arcsin x$ is the closed interval $[-1, 1]$ and the range is the closed interval $[-\pi/2, \pi/2]$.

✓ *Checkpoint* Now try Exercise 8.

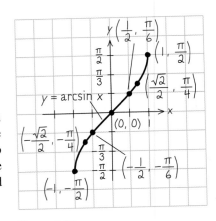

Figure 5.69

Other Inverse Trigonometric Functions

The cosine function is decreasing and one-to-one on the interval $0 \leq x \leq \pi$, as shown in Figure 5.70.

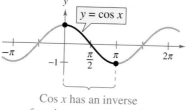

Figure 5.70

Consequently, on this interval the cosine function has an inverse function—the **inverse cosine function**—denoted by

$$y = \arccos x \qquad \text{or} \qquad y = \cos^{-1} x.$$

Because $y = \arccos x$ and $x = \cos y$ are equivalent for $0 \leq y \leq \pi$, their graphs are the same, and can be confirmed by the following table of values.

y	0	$\dfrac{\pi}{6}$	$\dfrac{\pi}{3}$	$\dfrac{\pi}{2}$	$\dfrac{2\pi}{3}$	$\dfrac{5\pi}{6}$	π
$x = \cos y$	1	$\dfrac{\sqrt{3}}{2}$	$\dfrac{1}{2}$	0	$-\dfrac{1}{2}$	$-\dfrac{\sqrt{3}}{2}$	-1

Similarly, you can define an **inverse tangent function** by restricting the domain of $y = \tan x$ to the interval $(-\pi/2, \pi/2)$. The following list summarizes the definitions of the three most common inverse trigonometric functions. The remaining three are defined in Exercises 79–81.

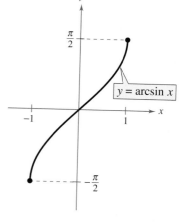

Domain: $[-1, 1]$; Range: $\left[-\dfrac{\pi}{2}, \dfrac{\pi}{2}\right]$

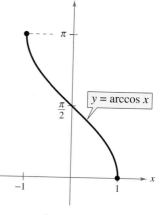

Domain: $[-1, 1]$; Range: $[0, \pi]$

Definition of the Inverse Trigonometric Functions

Function	Domain	Range
$y = \arcsin x$ if and only if $\sin y = x$	$-1 \leq x \leq 1$	$-\dfrac{\pi}{2} \leq y \leq \dfrac{\pi}{2}$
$y = \arccos x$ if and only if $\cos y = x$	$-1 \leq x \leq 1$	$0 \leq y \leq \pi$
$y = \arctan x$ if and only if $\tan y = x$	$-\infty < x < \infty$	$-\dfrac{\pi}{2} < y < \dfrac{\pi}{2}$

The graphs of these three inverse trigonometric functions are shown in Figure 5.71.

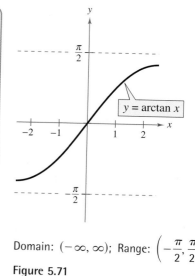

Domain: $(-\infty, \infty)$; Range: $\left(-\dfrac{\pi}{2}, \dfrac{\pi}{2}\right)$

Figure 5.71

Example 3 Evaluating Inverse Trigonometric Functions

Find the exact value.

a. $\arccos \dfrac{\sqrt{2}}{2}$ **b.** $\cos^{-1}(-1)$ **c.** $\arctan 0$ **d.** $\tan^{-1}(-1)$

Solution

a. Because $\cos(\pi/4) = \sqrt{2}/2$, and $\pi/4$ lies in $[0, \pi]$, it follows that

$$\arccos \frac{\sqrt{2}}{2} = \frac{\pi}{4}. \qquad \text{Angle whose cosine is } \frac{\sqrt{2}}{2}$$

b. Because $\cos \pi = -1$, and π lies in $[0, \pi]$, it follows that

$$\cos^{-1}(-1) = \pi. \qquad \text{Angle whose cosine is } -1$$

c. Because $\tan 0 = 0$, and 0 lies in $(-\pi/2, \pi/2)$, it follows that

$$\arctan 0 = 0. \qquad \text{Angle whose tangent is } 0$$

d. Because $\tan(-\pi/4) = -1$ and $-\pi/4$ lies in $(-\pi/2, \pi/2)$, it follows that

$$\tan^{-1}(-1) = -\frac{\pi}{4}. \qquad \text{Angle whose tangent is } -1$$

✓ *Checkpoint* Now try Exercise 3.

Example 4 Calculators and Inverse Trigonometric Functions

Use a calculator to approximate the value (if possible).

a. $\arctan(-8.45)$ **b.** $\sin^{-1} 0.2447$ **c.** $\arccos 2$

Solution

Function	*Mode*	*Graphing Calculator Keystrokes*
a. $\arctan(-8.45)$	Radian	[TAN⁻¹] [(-)] 8.45 [ )] [ENTER]

From the display, it follows that $\arctan(-8.45) \approx -1.4530010$.

| **b.** $\sin^{-1} 0.2447$ | Radian | [SIN⁻¹] 0.2447 [ENTER] |

From the display, it follows that $\sin^{-1} 0.2447 \approx 0.2472103$.

| **c.** $\arccos 2$ | Radian | [COS⁻¹] 2 [ENTER] |

In *real number* mode, the calculator should display an *error message* because the domain of the inverse cosine function is $[-1, 1]$.

✓ *Checkpoint* Now try Exercise 13.

TECHNOLOGY TIP In Example 4, if you had set the calculator to *degree* mode, the display would have been in degrees rather than in radians. This convention is peculiar to calculators. By definition, the values of inverse trigonometric functions are always in *radians*.

TECHNOLOGY TIP

You can use the [SIN⁻¹], [COS⁻¹], and [TAN⁻¹] keys on your calculator to approximate values of inverse trigonometric functions. To evaluate the inverse cosecant function, the inverse secant function, or the inverse cotangent function, you can use the inverse sine, inverse cosine, and inverse tangent functions, respectively. For instance, to evaluate $\sec^{-1} 3.4$, enter the expression as shown below.

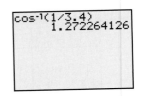

```
cos⁻¹(1/3.4)
        1.272264126
```

Compositions of Functions

Recall from Section 1.7 that for all x in the domains of f and f^{-1}, inverse functions have the properties

$$f(f^{-1}(x)) = x \quad \text{and} \quad f^{-1}(f(x)) = x.$$

Inverse Properties

If $-1 \le x \le 1$ and $-\pi/2 \le y \le \pi/2$, then

$$\sin(\arcsin x) = x \quad \text{and} \quad \arcsin(\sin y) = y.$$

If $-1 \le x \le 1$ and $0 \le y \le \pi$, then

$$\cos(\arccos x) = x \quad \text{and} \quad \arccos(\cos y) = y.$$

If x is a real number and $-\pi/2 < y < \pi/2$, then

$$\tan(\arctan x) = x \quad \text{and} \quad \arctan(\tan y) = y.$$

Exploration

Use a graphing utility to graph $y = \arcsin(\sin x)$. What are the domain and range of this function? Explain why $\arcsin(\sin 4)$ does not equal 4.

Now graph $y = \sin(\arcsin x)$ and determine the domain and range. Explain why $\sin(\arcsin 4)$ is not defined.

Keep in mind that these inverse properties do not apply for arbitrary values of x and y. For instance,

$$\arcsin\left(\sin \frac{3\pi}{2} \right) = \arcsin(-1) = -\frac{\pi}{2} \ne \frac{3\pi}{2}.$$

In other words, the property $\arcsin(\sin y) = y$ is not valid for values of y outside the interval $[-\pi/2, \pi/2]$.

Example 5   Using Inverse Properties

If possible, find the exact value.

a. $\tan[\arctan(-5)]$ **b.** $\arcsin\left(\sin \dfrac{5\pi}{3} \right)$ **c.** $\cos(\cos^{-1} \pi)$

Solution

a. Because -5 lies in the domain of the arctan function, the inverse property applies, and you have $\tan[\arctan(-5)] = -5$.

b. In this case, $5\pi/3$ does not lie within the range of the arcsine function, $-\pi/2 \le y \le \pi/2$. However, $5\pi/3$ is coterminal with

$$\frac{5\pi}{3} - 2\pi = -\frac{\pi}{3}$$

which does lie in the range of the arcsine function, and you have

$$\arcsin\left(\sin \frac{5\pi}{3} \right) = \arcsin\left[\sin\left(-\frac{\pi}{3} \right) \right] = -\frac{\pi}{3}.$$

c. The expression $\cos(\cos^{-1} \pi)$ is not defined because $\cos^{-1} \pi$ is not defined. Remember that the domain of the inverse cosine function is $[-1, 1]$.

✓ *Checkpoint* Now try Exercise 33.

Example 6 shows how to use right triangles to find exact values of compositions of inverse functions.

Example 6   Evaluating Compositions of Functions

Find the exact value.

a. $\tan\left(\arccos\dfrac{2}{3}\right)$ **b.** $\cos\left[\arcsin\left(-\dfrac{3}{5}\right)\right]$

Algebraic Solution

a. If you let $u = \arccos\frac{2}{3}$, then $\cos u = \frac{2}{3}$. Because $\cos u$ is positive, u is a first-quadrant angle. You can sketch and label angle u as shown in Figure 5.72.

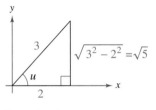

Figure 5.72

Consequently,

$$\tan\left(\arccos\dfrac{2}{3}\right) = \tan u = \dfrac{\text{opp}}{\text{adj}} = \dfrac{\sqrt{5}}{2}.$$

b. If you let $u = \arcsin\left(-\frac{3}{5}\right)$, then $\sin u = -\frac{3}{5}$. Because $\sin u$ is negative, u is a fourth-quadrant angle. You can sketch and label angle u as shown in Figure 5.73.

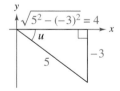

Figure 5.73

Consequently,

$$\cos\left[\arcsin\left(-\dfrac{3}{5}\right)\right] = \cos u = \dfrac{\text{adj}}{\text{hyp}} = \dfrac{4}{5}.$$

Graphical Solution

a. Use a graphing utility set in *radian* mode to graph $y = \tan(\arccos x)$, as shown in Figure 5.74. Use the *value* feature or the *zoom* and *trace* features of the graphing utility to find that the value of the composition of functions when $x = \frac{2}{3} \approx 0.66667$ is

$$y = 1.118 \approx \dfrac{\sqrt{5}}{2}.$$

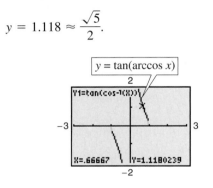

Figure 5.74

b. Use a graphing utility set in *radian* mode to graph $y = \cos(\arcsin x)$, as shown in Figure 5.75. Use the *value* feature or the *zoom* and *trace* features of the graphing utility to find that the value of the composition of functions when $x = -\frac{3}{5} = -0.6$ is

$$y = 0.8 = \dfrac{4}{5}.$$

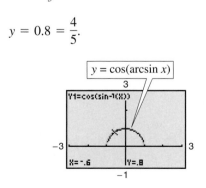

Figure 5.75

✓ *Checkpoint* Now try Exercise 41.

Library of Functions: Inverse Trigonometric Functions

The inverse trigonometric functions are obtained from the trigonometric functions in much the same way that the logarithmic function was developed from the exponential function. However, unlike the exponential function, the trigonometric functions are not one-to-one, and so it is necessary to restrict their domains to intervals on which they pass the Horizontal Line Test. Consequently, the inverse trigonometric functions have restricted domains and ranges, and they are not periodic.

One prominent role played by inverse trigonometric functions is in solving trigonometric equations in which the argument (angle) of the trigonometric function is the unknown quantity in the equation. You will learn how to solve such equations in the next chapter.

Inverse trigonometric functions play a unique role in calculus. There are two basic operations of calculus. One operation (called *differentiation*) transforms an inverse trigonometric function (a transcendental function) into an algebraic function. The other operation (called *integration*) produces the opposite transformation—from algebraic to transcendental.

Example 7   Some Problems from Calculus

Write each of the following as an algebraic expression in x.

a. $\sin(\arccos 3x)$, $0 \le x \le \frac{1}{3}$ **b.** $\cot(\arccos 3x)$, $0 \le x < \frac{1}{3}$

Solution

If you let $u = \arccos 3x$, then $\cos u = 3x$, where $-1 \le 3x \le 1$. Because

$$\cos u = \frac{\text{adj}}{\text{hyp}} = \frac{3x}{1}$$

you can sketch a right triangle with acute angle u, as shown in Figure 5.76. From this triangle, you can easily convert each expression to algebraic form.

a. $\sin(\arccos 3x) = \sin u = \dfrac{\text{opp}}{\text{hyp}} = \sqrt{1 - 9x^2}$, $0 \le x \le \dfrac{1}{3}$

b. $\cot(\arccos 3x) = \cot u = \dfrac{\text{adj}}{\text{opp}} = \dfrac{3x}{\sqrt{1 - 9x^2}}$, $0 \le x < \dfrac{1}{3}$

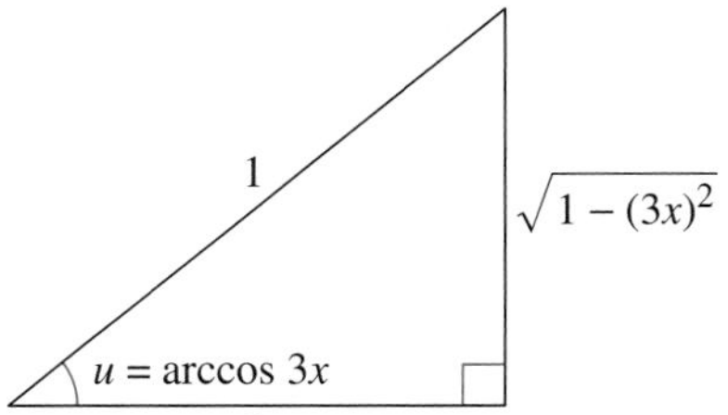

(right triangle with hypotenuse 1, vertical side $\sqrt{1 - (3x)^2}$, horizontal side $3x$, and angle $u = \arccos 3x$)

Figure 5.76

A similar argument can be made here for x-values lying in the interval $\left[-\frac{1}{3}, 0\right]$.

✓ *Checkpoint* Now try Exercise 47.

5.6 Exercises

Vocabulary Check

Fill in the blanks.

Function	Alternative Notation	Domain	Range
1. $y = \arcsin x$	_____	_____	$-\dfrac{\pi}{2} \le y \le \dfrac{\pi}{2}$
2. _____	$y = \cos^{-1} x$	$-1 \le x \le 1$	_____
3. $y = \arctan x$	_____	_____	_____

In Exercises 1–7, find the exact value of each expression without using a calculator.

1. (a) $\arcsin \frac{1}{2}$ (b) $\arcsin 0$

2. (a) $\arccos \frac{1}{2}$ (b) $\arccos 0$

3. (a) $\arctan \dfrac{\sqrt{3}}{3}$ (b) $\arctan(-1)$

4. (a) $\cos^{-1}\!\left(-\dfrac{\sqrt{2}}{2}\right)$ (b) $\sin^{-1}\!\left(-\dfrac{\sqrt{2}}{2}\right)$

5. (a) $\arctan(-\sqrt{3})$ (b) $\arctan \sqrt{3}$

6. (a) $\arccos\!\left(-\dfrac{1}{2}\right)$ (b) $\arcsin \dfrac{\sqrt{2}}{2}$

7. (a) $\sin^{-1} \dfrac{\sqrt{3}}{2}$ (b) $\tan^{-1}\!\left(-\dfrac{\sqrt{3}}{3}\right)$

8. *Numerical and Graphical Analysis* Consider the function $y = \arcsin x$.

(a) Use a graphing utility to complete the table.

x	-1	-0.8	-0.6	-0.4	-0.2
y					

x	0	0.2	0.4	0.6	0.8	1
y						

(b) Plot the points from the table in part (a) and graph the function. (Do not use a graphing utility.)

(c) Use a graphing utility to graph the inverse sine function and compare the result with your hand-drawn graph in part (b).

(d) Determine any intercepts and symmetry of the graph.

9. *Numerical and Graphical Analysis* Consider the function $y = \arccos x$.

(a) Use a graphing utility to complete the table.

x	-1	-0.8	-0.6	-0.4	-0.2
y					

x	0	0.2	0.4	0.6	0.8	1
y						

(b) Plot the points from the table in part (a) and graph the function. (Do not use a graphing utility.)

(c) Use a graphing utility to graph the inverse cosine function and compare the result with your hand-drawn graph in part (b).

(d) Determine any intercepts and symmetry of the graph.

10. *Numerical and Graphical Analysis* Consider the function $y = \arctan x$.

(a) Use a graphing utility to complete the table.

x	-10	-8	-6	-4	-2
y					

x	0	2	4	6	8	10
y						

(b) Plot the points from the table in part (a) and graph the function. (Do not use a graphing utility.)

(c) Use a graphing utility to graph the inverse tangent function and compare the result with your hand-drawn graph in part (b).

(d) Determine the horizontal asymptotes of the graph.

In Exercises 11 and 12, determine the missing coordinates of the points on the graph of the function.

11.   **12.**

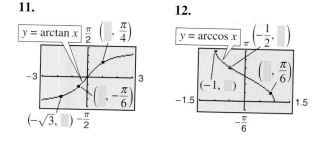

In Exercises 13–24, use a calculator to approximate the value of the expression. Round your answer to two decimal places.

13. $\cos^{-1} 0.75$

14. $\sin^{-1} 0.56$

15. $\arcsin(-0.75)$

16. $\arccos(-0.7)$

17. $\arctan(-6)$

18. $\arctan(-18)$

19. $\sin^{-1} 0.19$

20. $\cos^{-1} 0.21$

21. $\arccos(-0.51)$

22. $\arcsin(-0.125)$

23. $\tan^{-1} 1.32$

24. $\tan^{-1} 5.9$

In Exercises 25 and 26, use a graphing utility to graph $f, g,$ and $y = x$ in the same viewing window to verify geometrically that g is the inverse function of f. (Be sure to properly restrict the domain of f.)

25. $f(x) = \tan x, \quad g(x) = \arctan x$

26. $f(x) = \sin x, \quad g(x) = \arcsin x$

In Exercises 27–30, use an inverse trigonometric function to write θ as a function of x.

27.   **28.**

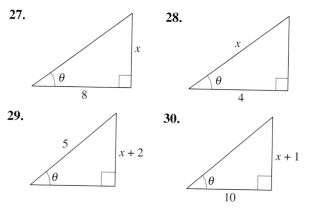

29.   **30.**

In Exercises 31–38, use the properties of inverse functions to find the exact value of the expression.

31. $\sin(\arcsin 0.7)$

32. $\tan(\arctan 35)$

33. $\cos[\arccos(-0.3)]$

34. $\sin[\arcsin(-0.1)]$

35. $\arcsin(\sin 3\pi)$

36. $\arccos\left(\cos \dfrac{7\pi}{2}\right)$

37. $\tan^{-1}\left(\tan \dfrac{11\pi}{6}\right)$

38. $\sin^{-1}\left(\sin \dfrac{7\pi}{4}\right)$

In Exercises 39–46, find the exact value of the expression. Use a graphing utility to verify your result. (*Hint:* Make a sketch of a right triangle.)

39. $\sin\left(\arctan \frac{4}{3}\right)$

40. $\sec\left(\arcsin \frac{3}{5}\right)$

41. $\cos\left(\arcsin \frac{24}{25}\right)$

42. $\csc\left[\arctan\left(-\frac{12}{5}\right)\right]$

43. $\sec\left[\arctan\left(-\frac{3}{5}\right)\right]$

44. $\tan\left[\arcsin\left(-\frac{3}{4}\right)\right]$

45. $\sin\left[\arccos\left(-\frac{2}{3}\right)\right]$

46. $\cot\left(\arctan \frac{5}{8}\right)$

In Exercises 47–54, write an algebraic expression that is equivalent to the expression. (*Hint:* Sketch a right triangle, as demonstrated in Example 7.)

47. $\cot(\arctan x)$

48. $\sin(\arctan x)$

49. $\sin[\arccos(x + 2)]$

50. $\sec[\arcsin(x - 1)]$

51. $\tan\left(\arccos \dfrac{x}{5}\right)$

52. $\cot\left(\arctan \dfrac{4}{x}\right)$

53. $\csc\left(\arctan \dfrac{x}{\sqrt{7}}\right)$

54. $\cos\left(\arcsin \dfrac{x - h}{r}\right)$

In Exercises 55 and 56, use a graphing utility to graph f and g in the same viewing window to verify that the two functions are equal. Explain why they are equal. Identify any asymptotes of the graphs.

55. $f(x) = \sin(\arctan 2x), \quad g(x) = \dfrac{2x}{\sqrt{1 + 4x^2}}$

56. $f(x) = \tan\left(\arccos \dfrac{x}{2}\right), \quad g(x) = \dfrac{\sqrt{4 - x^2}}{x}$

In Exercises 57–60, complete the equation.

57. $\arctan \dfrac{14}{x} = \arcsin(\quad), \quad x > 0$

58. $\arcsin \dfrac{\sqrt{36 - x^2}}{6} = \arccos(\quad), \quad 0 \le x \le 6$

59. $\arccos \dfrac{3}{\sqrt{x^2 - 2x + 10}} = \arcsin(\quad)$

60. $\arccos \dfrac{x - 2}{2} = \arctan(\quad), \quad 2 < x < 4$

In Exercises 61–68, use a graphing utility to graph the function.

61. $y = 2 \arccos x$

62. $y = \arcsin \dfrac{x}{2}$

63. $f(x) = \arcsin(x - 2)$

64. $g(t) = \arccos(t + 2)$

65. $f(x) = \arctan 2x$

66. $f(x) = \pi + \arctan x$

67. $h(v) = \tan(\arccos v)$

68. $f(x) = \arccos \dfrac{x}{4}$

In Exercises 69 and 70, write the function in terms of the sine function by using the identity

$$A \cos \omega t + B \sin \omega t = \sqrt{A^2 + B^2} \sin\left(\omega t + \arctan \dfrac{A}{B}\right).$$

Use a graphing utility to graph both forms of the function. What does the graph imply?

69. $f(t) = 3 \cos 2t + 3 \sin 2t$

70. $f(t) = 4 \cos \pi t + 3 \sin \pi t$

71. Docking a Boat A boat is pulled in by means of a winch located on a dock 10 feet above the deck of the boat (see figure). Let θ be the angle of elevation from the boat to the winch and let s be the length of the rope from the winch to the boat.

(a) Write θ as a function of s.

(b) Find θ when $s = 52$ feet and when $s = 26$ feet.

72. Granular Angle of Repose Different types of granular substances naturally settle at different angles when stored in cone-shaped piles. This angle θ is called the *angle of repose*. When rock salt is stored in a cone-shaped pile 11 feet high, the diameter of the pile's base is about 34 feet. (Source: Bulk-Store Structures, Inc.)

(a) Draw a diagram that gives a visual representation of the problem. Label all known and unknown quantities.

(b) Find the angle of repose for rock salt.

(c) How tall is a pile of rock salt that has a base diameter of 40 feet?

73. Photography A television camera at ground level is filming the lift-off of a space shuttle at a point 750 meters from the launch pad (see figure). Let θ be the angle of elevation to the shuttle and let s be the height of the shuttle.

(a) Write θ as a function of s.

(b) Find θ when $s = 400$ meters and when $s = 1600$ meters.

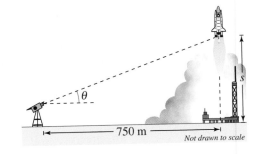

Not drawn to scale

74. Photography A photographer is taking a picture of a three-foot painting hung in an art gallery. The camera lens is 1 foot below the lower edge of the painting (see figure). The angle β subtended by the camera lens x feet from the painting is

$$\beta = \arctan \dfrac{3x}{x^2 + 4}, \quad x > 0.$$

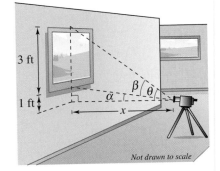

Not drawn to scale

(a) Use a graphing utility to graph β as a function of x.

(b) Move the cursor along the graph to approximate the distance from the picture when β is maximum.

(c) Identify the asymptote of the graph and discuss its meaning in the context of the problem.

75. *Angle of Elevation* An airplane flies at an altitude of 6 miles toward a point directly over an observer. Consider θ and x as shown in the figure.

(a) Write θ as a function of x.

(b) Find θ when $x = 10$ miles and $x = 3$ miles.

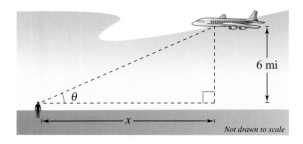

76. *Security Patrol* A security car with its spotlight on is parked 20 meters from a long warehouse. Consider θ and x as shown in the figure.

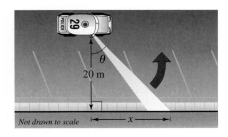

(a) Write θ as a function of x.

(b) Find θ when $x = 5$ meters and when $x = 12$ meters.

Synthesis

True or False? **In Exercises 77 and 78, determine whether the statement is true or false. Justify your answer.**

77. $\sin \dfrac{5\pi}{6} = \dfrac{1}{2}$ $\implies$ $\arcsin \dfrac{1}{2} = \dfrac{5\pi}{6}$

78. $\arctan x = \dfrac{\arcsin x}{\arccos x}$

79. Define the inverse cotangent function by restricting the domain of the cotangent function to the interval $(0, \pi)$, and sketch the inverse function's graph.

80. Define the inverse secant function by restricting the domain of the secant function to the intervals $[0, \pi/2)$ and $(\pi/2, \pi]$, and sketch the inverse function's graph.

81. Define the inverse cosecant function by restricting the domain of the cosecant function to the intervals $[-\pi/2, 0)$ and $(0, \pi/2]$, and sketch the inverse function's graph.

82. Use the results of Exercises 79–81 to evaluate the following without using a calculator.

(a) $\operatorname{arcsec} \sqrt{2}$ (b) $\operatorname{arcsec} 1$

(c) $\operatorname{arccot}\left(-\sqrt{3}\right)$ (d) $\operatorname{arccsc} 2$

Proof **In Exercises 83–85, prove the identity.**

83. $\arcsin(-x) = -\arcsin x$

84. $\arctan(-x) = -\arctan x$

85. $\arcsin x + \arccos x = \dfrac{\pi}{2}$

86. *Area* In calculus, it is shown that the area of the region bounded by the graphs of $y = 0$, $y = 1/(x^2 + 1)$, $x = a$, and $x = b$ is given by

$$\text{Area} = \arctan b - \arctan a$$

(see figure). Find the areas for each value of a and b.

(a) $a = 0, b = 1$ (b) $a = -1, b = 1$

(c) $a = 0, b = 3$ (d) $a = -1, b = 3$

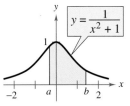

Review

In Exercises 87–90, simplify the radical expression.

87. $\dfrac{4}{4\sqrt{2}}$ **88.** $\dfrac{2}{\sqrt{3}}$

89. $\dfrac{2\sqrt{3}}{6}$ **90.** $\dfrac{5\sqrt{5}}{2\sqrt{10}}$

In Exercises 91–94, sketch a right triangle corresponding to the trigonometric function of the acute angle θ. Use the Pythagorean Theorem to determine the third side and then find the other five trigonometric functions of θ.

91. $\sin \theta = \dfrac{5}{6}$ **92.** $\tan \theta = 2$

93. $\sin \theta = \dfrac{3}{4}$ **94.** $\sec \theta = 3$

5.7 Applications and Models

Applications Involving Right Triangles

In this section, the three angles of a right triangle are denoted by the letters A, B, and C (where C is the right angle), and the lengths of the sides opposite these angles by the letters a, b, and c (where c is the hypotenuse).

Example 1 Solving a Right Triangle

Solve the right triangle shown in Figure 5.77 for all unknown sides and angles.

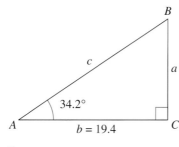

Figure 5.77

Solution

Because $C = 90°$, it follows that $A + B = 90°$ and $B = 90° - 34.2° = 55.8°$. To solve for a, use the fact that

$$\tan A = \frac{\text{opp}}{\text{adj}} = \frac{a}{b} \qquad \Longrightarrow \qquad a = b \tan A.$$

So, $a = 19.4 \tan 34.2° \approx 13.18$. Similarly, to solve for c, use the fact that

$$\cos A = \frac{\text{adj}}{\text{hyp}} = \frac{b}{c} \qquad \Longrightarrow \qquad c = \frac{b}{\cos A}.$$

So, $c = \dfrac{19.4}{\cos 34.2°} \approx 23.46$.

✓ *Checkpoint* Now try Exercise 1.

Recall from Section 5.2 that the term *angle of elevation* denotes the angle from the horizontal upward to an object and that the term *angle of depression* denotes the angle from the horizontal downward to an object. An angle of elevation and an angle of depression are shown in Figure 5.78.

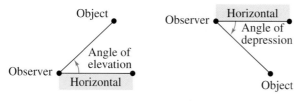

Figure 5.78

What you should learn

● Solve real-life problems involving right triangles.
● Solve real-life problems involving directional bearings.
● Solve real-life problems involving harmonic motion.

Why you should learn it

You can use trigonometric functions to model and solve real-life problems. For instance, Exercise 60 on page 470 shows you how a trigonometric function can be used to model the harmonic motion of a buoy.

Mary Kate Denny/PhotoEdit

Example 2   Finding a Side of a Right Triangle

A safety regulation states that the maximum angle of elevation for a rescue ladder is 72°. A fire department's longest ladder is 110 feet. What is the maximum safe rescue height?

Solution

A sketch is shown in Figure 5.79. From the equation $\sin A = a/c$, it follows that

$$a = c \sin A = 110 \sin 72° \approx 104.6.$$

So, the maximum safe rescue height is about 104.6 feet above the height of the fire truck.

✓ *Checkpoint* Now try Exercise 17.

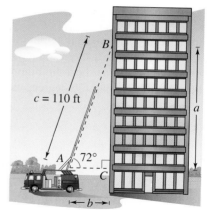

Figure 5.79

Example 3   Finding a Side of a Right Triangle

At a point 200 feet from the base of a building, the angle of elevation to the *bottom* of a smokestack is 35°, and the angle of elevation to the *top* is 53°, as shown in Figure 5.80. Find the height s of the smokestack alone.

Solution

This problem involves two right triangles. For the smaller right triangle, use the fact that $\tan 35° = a/200$ to conclude that the height of the building is

$$a = 200 \tan 35°.$$

Now, for the larger right triangle, use the equation

$$\tan 53° = \frac{a + s}{200}$$

to conclude that $s = 200 \tan 53° - a$. So, the height of the smokestack is

$$s = 200 \tan 53° - a = 200 \tan 53° - 200 \tan 35° \approx 125.4 \text{ feet.}$$

✓ *Checkpoint* Now try Exercise 19.

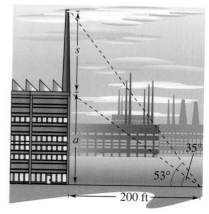

Figure 5.80

Example 4   Finding an Angle of Depression

A swimming pool is 20 meters long and 12 meters wide. The bottom of the pool is slanted so that the water depth is 1.3 meters at the shallow end and 4 meters at the deep end, as shown in Figure 5.81. Find the angle of depression of the bottom of the pool.

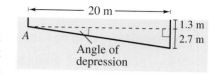

Figure 5.81

Solution

Using the tangent function, you see that

$$\tan A = \frac{\text{opp}}{\text{adj}} = \frac{2.7}{20} = 0.135.$$

So, the angle of depression is $A = \arctan 0.135 \approx 0.13419$ radian $\approx 7.69°$.

✓ *Checkpoint* Now try Exercise 25.

Trigonometry and Bearings

In surveying and navigation, directions are generally given in terms of **bearings.** A bearing measures the acute angle a path or line of sight makes with a fixed north–south line, as shown in Figure 5.82. For instance, the bearing of S 35° E in Figure 5.82(a) means 35 degrees east of south.

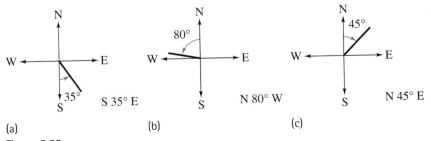

(a)   (b)   (c)

Figure 5.82

Example 5   Finding Directions in Terms of Bearings

A ship leaves port at noon and heads due west at 20 knots, or 20 nautical miles (nm) per hour. At 2 P.M. the ship changes course to N 54° W, as shown in Figure 5.83. Find the ship's bearing and distance from the port of departure at 3 P.M.

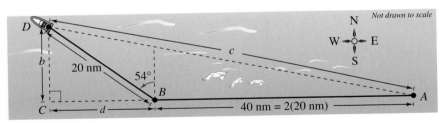

Figure 5.83

Solution

For triangle BCD, you have $B = 90° - 54° = 36°$. The two sides of this triangle can be determined to be

$$b = 20 \sin 36° \quad \text{and} \quad d = 20 \cos 36°.$$

In triangle ACD, you can find angle A as follows.

$$\tan A = \frac{b}{d + 40} = \frac{20 \sin 36°}{20 \cos 36° + 40} \approx 0.2092494$$

$$A \approx \arctan 0.2092494 \approx 0.2062732 \text{ radian} \approx 11.82°$$

The angle with the north-south line is $90° - 11.82° = 78.18°$. So, the bearing of the ship is N 78.18° W. Finally, from triangle ACD, you have $\sin A = b/c$, which yields

$$c = \frac{b}{\sin A} = \frac{20 \sin 36°}{\sin 11.82°} \approx 57.4 \text{ nautical miles.} \qquad \text{Distance from port}$$

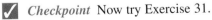 **Checkpoint** Now try Exercise 31.

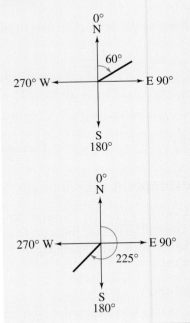

Harmonic Motion

The periodic nature of the trigonometric functions is useful for describing the motion of a point on an object that vibrates, oscillates, rotates, or is moved by wave motion.

For example, consider a ball that is bobbing up and down on the end of a spring, as shown in Figure 5.84. Suppose that 10 centimeters is the maximum distance the ball moves vertically upward or downward from its equilibrium (at-rest) position. Suppose further that the time it takes for the ball to move from its maximum displacement above zero to its maximum displacement below zero and back again is $t = 4$ seconds. Assuming the ideal conditions of perfect elasticity and no friction or air resistance, the ball would continue to move up and down in a uniform and regular manner.

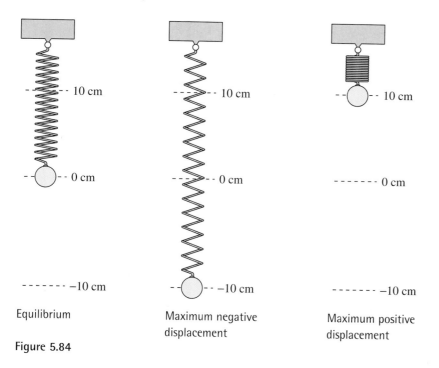

Equilibrium   Maximum negative   Maximum positive
                   displacement   displacement

Figure 5.84

From this spring you can conclude that the period (time for one complete cycle) of the motion is

Period = 4 seconds

its amplitude (maximum displacement from equilibrium) is

Amplitude = 10 centimeters

and its **frequency** (number of cycles per second) is

Frequency = $\frac{1}{4}$ cycle per second.

Motion of this nature can be described by a sine or cosine function, and is called **simple harmonic motion.**

Definition of Simple Harmonic Motion

A point that moves on a coordinate line is said to be in **simple harmonic motion** if its distance d from the origin at time t is given by either

$$d = a \sin \omega t \quad \text{or} \quad d = a \cos \omega t$$

where a and ω are real numbers such that $\omega > 0$. The motion has amplitude $|a|$, period $2\pi/\omega$, and frequency $\omega/(2\pi)$.

Example 6   Simple Harmonic Motion

Write the equation for the simple harmonic motion of the ball illustrated in Figure 5.84, where the period is 4 seconds. What is the frequency of this motion?

Solution

Because the spring is at equilibrium $(d = 0)$ when $t = 0$, you use the equation

$$d = a \sin \omega t.$$

Moreover, because the maximum displacement from zero is 10 and the period is 4, you have the following.

Amplitude $= |a| = 10$

$$\text{Period} = \frac{2\pi}{\omega} = 4 \quad \Longrightarrow \quad \omega = \frac{\pi}{2}$$

Consequently, the equation of motion is

$$d = 10 \sin \frac{\pi}{2} t.$$

Note that the choice of $a = 10$ or $a = -10$ depends on whether the ball initially moves up or down. The frequency is

$$\begin{aligned}
\text{Frequency} &= \frac{\omega}{2\pi} \\
&= \frac{\pi/2}{2\pi} \\
&= \frac{1}{4} \text{ cycle per second.}
\end{aligned}$$

 Checkpoint Now try Exercise 51.

Figure 5.85

One illustration of the relationship between sine waves and harmonic motion is the wave motion that results when a stone is dropped into a calm pool of water. The waves move outward in roughly the shape of sine (or cosine) waves, as shown in Figure 5.85. As an example, suppose you are fishing and your fishing bob is attached so that it does not move horizontally. As the waves move outward from the dropped stone, your fishing bob will move up and down in simple harmonic motion, as shown in Figure 5.86.

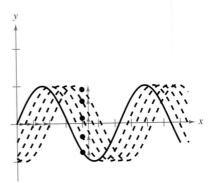

Figure 5.86

Example 7   Simple Harmonic Motion

Given the equation for simple harmonic motion

$$d = 6 \cos \frac{3\pi}{4}t$$

find (a) the maximum displacement, (b) the frequency, (c) the value of d when $t = 4$, and (d) the least positive value of t for which $d = 0$.

Algebraic Solution

The given equation has the form $d = a \cos \omega t$, with $a = 6$ and $\omega = 3\pi/4$.

a. The maximum displacement (from the point of equilibrium) is given by the amplitude. So, the maximum displacement is 6.

b. Frequency $= \dfrac{\omega}{2\pi}$

$$= \frac{3\pi/4}{2\pi}$$

$$= \frac{3}{8} \text{ cycle per unit of time}$$

c. $d = 6 \cos\left[\dfrac{3\pi}{4}(4)\right]$

$$= 6 \cos 3\pi$$

$$= 6(-1)$$

$$= -6$$

d. To find the least positive value of t for which $d = 0$, solve the equation

$$d = 6 \cos \frac{3\pi}{4}t = 0.$$

First divide each side by 6 to obtain

$$\cos \frac{3\pi}{4}t = 0.$$

You know that $\cos t = 0$ when

$$t = \frac{\pi}{2}, \frac{3\pi}{2}, \frac{5\pi}{2}, \cdots .$$

Multiply these values by $4/(3\pi)$ to obtain

$$t = \frac{2}{3}, 2, \frac{10}{3}, \cdots .$$

So, the least positive value of t is $t = \frac{2}{3}$.

✓ *Checkpoint* Now try Exercise 55.

Graphical Solution

Use a graphing utility set in *radian* mode to graph

$$y = 6 \cos \frac{3\pi}{4}x.$$

a. Use the *maximum* feature of the graphing utility to estimate that the maximum displacement from the point of equilibrium $y = 0$ is 6, as shown in Figure 5.87.

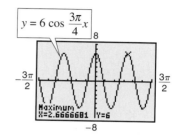

Figure 5.87

b. The period is the time for the graph to complete one cycle, which is $x \approx 2.667$. You can estimate the frequency as follows.

$$\text{Frequency} \approx \frac{1}{2.667} \approx 0.375 \text{ cycle per unit of time}$$

c. Use the *value* or *trace* feature to estimate that the value of y when $x = 4$ is $y = -6$, as shown in Figure 5.88.

d. Use the *zero* or *root* feature to estimate that the least positive value of x for which $y = 0$ is $x \approx 0.6667$, as shown in Figure 5.89.

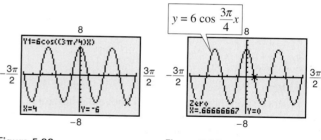

Figure 5.88   Figure 5.89

5.7 Exercises

Vocabulary Check

Fill in the blanks.

1. An angle that measures from the horizontal upward to an object is called the angle of _____ , whereas an angle that measures from the horizontal downward to an object is called the angle of _____ .

2. A _____ measures the acute angle a path or line of sight makes with a fixed north-south line.

3. A point that moves on a coordinate line is said to be in simple _____ if its distance from the origin at time t is given by either $d = a \sin \omega t$ or $d = a \cos \omega t$.

In Exercises 1–10, solve the right triangle shown in the figure. (Round your answers to two decimal places.)

1. $A = 20°$, $b = 10$

2. $B = 54°$, $c = 15$

3. $B = 71°$, $b = 24$

4. $A = 7.4°$, $a = 40.5$

5. $a = 6$, $b = 16$

6. $a = 25$, $c = 35$

7. $b = 16$, $c = 48$

8. $b = 1.32$, $c = 9.45$

9. $A = 12° 15'$, $c = 430.5$

10. $B = 65° 12'$, $a = 14.2$

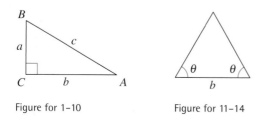

Figure for 1–10   Figure for 11–14

In Exercises 11–14, find the altitude of the isosceles triangle shown in the figure. Round your answer to two decimal places.

11. $\theta = 52°$, $b = 4$ inches

12. $\theta = 18°$, $b = 10$ meters

13. $\theta = 41.6°$, $b = 14.2$ feet

14. $\theta = 72.94°$, $b = 3.36$ centimeters

15. **Length** A shadow of length L is created by a 60-foot silo when the sun is $\theta°$ above the horizon.

   (a) Draw a right triangle that gives a visual representation of the problem. Label the known and unknown quantities.

   (b) Write L as a function of θ.

   (c) Use a graphing utility to complete the table.

θ	10°	20°	30°	40°	50°
L					

   (d) The angle measure increases in equal increments in the table. Does the length of the shadow change in equal increments? Explain.

16. **Length** A shadow of length L is created by an 850-foot building when the sun is $\theta°$ above the horizon.

   (a) Draw a right triangle that gives a visual representation of the problem. Label the known and unknown quantities.

   (b) Write L as a function of θ.

   (c) Use a graphing utility to complete the table.

θ	10°	20°	30°	40°	50°
L					

   (d) The angle measure increases in equal increments in the table. Does the length of the shadow change in equal increments? Explain.

17. **Height** A ladder 20 feet long leans against the side of a house. The angle of elevation of the ladder is 80°. Find the height from the top of the ladder to the ground.

18. **Height** The angle of elevation from the base to the top of a waterslide is 13°. The slide extends horizontally 58.2 meters. Approximate the height of the waterslide.

19. *Height*   From a point 50 feet in front of a church, the angles of elevation to the base of the steeple and the top of the steeple are 35° and 47° 40′, respectively.

(a) Draw right triangles that give a visual representation of the problem. Label the known and unknown quantities.

(b) Use a trigonometric function to write an equation involving the unknown quantity.

(c) Find the height of the steeple.

20. *Height*   From a point 100 feet in front of a public library, the angles of elevation to the base of the flagpole and the top of the flagpole are 28° and 39° 45′, respectively. The flagpole is mounted on the front of the library's roof. Find the height of the flagpole.

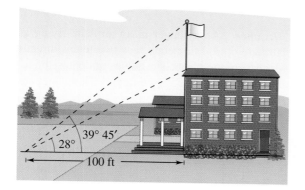

21. *Depth*   The sonar of a navy cruiser detects a submarine that is 4000 feet from the cruiser. The angle between the water level and the submarine is 31.5°. How deep is the submarine?

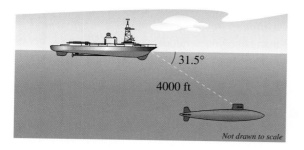

22. *Height*   A 100-foot line is attached to a kite. When the kite has pulled the line taut, the angle of elevation to the kite is approximately 50°. Approximate the height of the kite.

23. *Angle of Elevation*   An engineer erects a 75-foot vertical cellular-phone tower. Find the angle of elevation to the top of the tower from a point on level ground 95 feet from its base.

24. *Angle of Elevation*   The height of an outdoor basketball backboard is $12\frac{1}{2}$ feet, and the backboard casts a shadow $17\frac{1}{3}$ feet long.

(a) Draw a right triangle that gives a visual representation of the problem. Label the known and unknown quantities.

(b) Use a trigonometric function to write an equation involving the unknown quantity.

(c) Find the angle of elevation of the sun.

25. *Angle of Depression*   A Global Positioning System satellite orbits 12,500 miles above Earth's surface. Find the angle of depression from the satellite to the horizon. Assume the radius of Earth is 4000 miles.

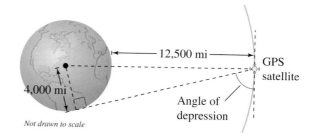

26. *Angle of Depression*   Find the angle of depression from the top of a lighthouse 250 feet above water level to the water line of a ship $2\frac{1}{2}$ miles offshore.

27. *Airplane Ascent*   When an airplane leaves the runway, its angle of climb is 18° and its speed is 275 feet per second. Find the plane's altitude after 1 minute.

28. *Airplane Ascent*   How long will it take the plane in Exercise 27 to climb to an altitude of 10,000 feet? 16,000 feet?

29. *Mountain Descent*   A sign on the roadway at the top of a mountain indicates that for the next 4 miles the grade is 9.5° (see figure). Find the change in elevation for a car descending the mountain.

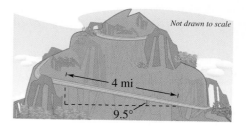

30. Ski Slope A ski slope on a mountain has an angle of elevation of 25.2°. The vertical height of the slope is 1808 feet. How long is the slope?

31. Navigation A ship leaves port at noon and has a bearing of S 29° W. The ship sails at 20 knots. How many nautical miles south and how many nautical miles west will the ship have traveled by 6:00 P.M.?

32. Navigation An airplane flying at 600 miles per hour has a bearing of 52°. After flying 1.5 hours, how far north and how far east has the plane traveled from its point of departure?

33. Surveying A surveyor wants to find the distance across a swamp. The bearing from A to B is N 32° W. The surveyor walks 50 meters from A, and at the point C the bearing to B is N 68° W. Find (a) the bearing from A to C and (b) the distance from A to B.

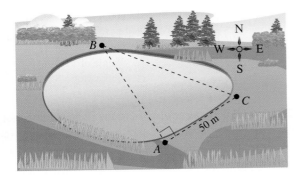

34. Location of a Fire Two fire towers are 30 kilometers apart, where tower A is due west of tower B. A fire is spotted from the towers, and the bearings from A and B are E 14° N and W 34° N, respectively. Find the distance d of the fire from the line segment AB.

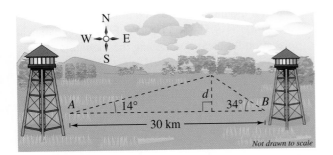

Not drawn to scale

35. Navigation A ship is 45 miles east and 30 miles south of port. The captain wants to sail directly to port. What bearing should be taken?

36. Navigation A plane is 160 miles north and 85 miles east of an airport. The pilot wants to fly directly to the airport. What bearing should be taken?

37. Distance An observer in a lighthouse 350 feet above sea level observes two ships directly offshore. The angles of depression to the ships are 4° and 6.5° (see figure). How far apart are the ships?

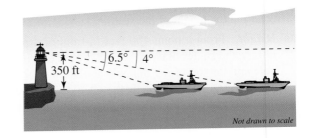

350 ft

Not drawn to scale

38. Distance A passenger in an airplane flying at an altitude of 10 kilometers sees two towns directly to the east of the plane. The angles of depression to the towns are 28° and 55° (see figure). How far apart are the towns?

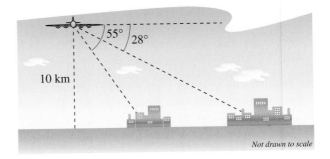

55° 28°

10 km

Not drawn to scale

39. Altitude A plane is observed approaching your home and you assume its speed is 550 miles per hour. The angle of elevation to the plane is 16° at one time and 57° one minute later. Approximate the altitude of the plane.

40. Height While traveling across flat land, you notice a mountain directly in front of you. The angle of elevation to the peak is 2.5°. After you drive 18 miles closer to the mountain, the angle of elevation is 10°. Approximate the height of the mountain.

Geometry In Exercises 41 and 42, find the angle α between two nonvertical lines L_1 and L_2. The angle α satisfies the equation

$$\tan \alpha = \left| \frac{m_2 - m_1}{1 + m_2 m_1} \right|$$

where m_1 and m_2 are the slopes of L_1 and L_2, respectively. (Assume $m_1 m_2 \neq -1$.)

41. L_1: $3x - 2y = 5$
L_2: $x + y = 1$

42. L_1: $2x + y = 8$
L_2: $x - 5y = -4$

43. Geometry Determine the angle between the diagonal of a cube and the diagonal of its base, as shown in the figure.

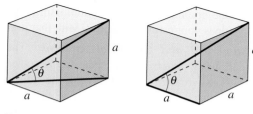

Figure for 43　　　　Figure for 44

44. Geometry Determine the angle between the diagonal of a cube and its edge, as shown in the figure.

45. Hardware Write the distance y across the flat sides of a hexagonal nut as a function of r, as shown in the figure.

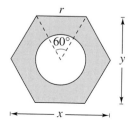

46. Hardware The figure shows a circular piece of sheet metal of diameter 40 centimeters. The sheet contains 12 equally spaced bolt holes. Determine the straight-line distance between the centers of two consecutive bolt holes.

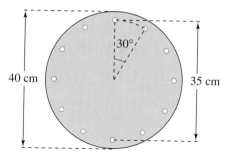

47. Geometry A regular pentagon (a pentagon with congruent sides and angles) is inscribed in a circle of radius 25 inches. Find the length of the sides of the pentagon.

48. Geometry A regular hexagon (a hexagon with congruent sides and angles) is inscribed in a circle of radius 25 inches. Find the length of the sides of the hexagon.

Trusses In Exercises 49 and 50, find the lengths of all the unknown members of the truss.

49.

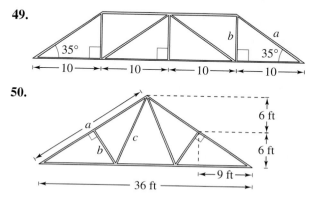

50.

Harmonic Motion In Exercises 51–54, find a model for simple harmonic motion satisfying the specified conditions.

Displacement ($t = 0$)	Amplitude	Period
51. 0	8 centimeters	2 seconds
52. 0	3 meters	6 seconds
53. 3 inches	3 inches	1.5 seconds
54. 2 feet	2 feet	10 seconds

Harmonic Motion In Exercises 55–58, for the simple harmonic motion described by the trigonometric function, find (a) the maximum displacement, (b) the frequency, (c) the value of d when $t = 5$, and (d) the least positive value of t for which $d = 0$. Use a graphing utility to verify your results.

55. $d = 4 \cos 8\pi t$

56. $d = \frac{1}{2} \cos 20\pi t$

57. $d = \frac{1}{16} \sin 140\pi t$

58. $d = \frac{1}{64} \sin 792\pi t$

59. Tuning Fork A point on the end of a tuning fork moves in the simple harmonic motion described by $d = a \sin \omega t$. Find ω given that the tuning fork for middle C has a frequency of 264 vibrations per second.

60. *Wave Motion* A buoy oscillates in simple harmonic motion as waves go past. At a given time it is noted that the buoy moves a total of 3.5 feet from its low point to its high point (see figure), and that it returns to its high point every 10 seconds. Write an equation that describes the motion of the buoy if it is at its high point at time $t = 0$.

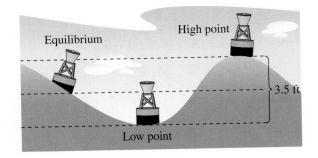

61. *Springs* A ball that is bobbing up and down on the end of a spring has a maximum displacement of 3 inches. Its motion (in ideal conditions) is modeled by

$$y = \tfrac{1}{4} \cos 16t, \quad t > 0$$

where y is measured in feet and t is the time in seconds.

(a) Use a graphing utility to graph the function.

(b) What is the period of the oscillations?

(c) Determine the first time the ball passes the point of equilibrium ($y = 0$).

62. *Numerical and Graphical Analysis* A two-meter-high fence is 3 meters from the side of a grain storage bin. A grain elevator must reach from ground level outside the fence to the storage bin (see figure). The objective is to determine the shortest elevator that meets the constraints.

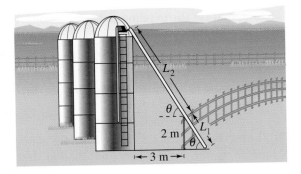

(a) Complete four rows of the table.

θ	L_1	L_2	$L_1 + L_2$
0.1	$\dfrac{2}{\sin 0.1}$	$\dfrac{3}{\cos 0.1}$	23.0
0.2	$\dfrac{2}{\sin 0.2}$	$\dfrac{3}{\cos 0.2}$	13.1

(b) Use the *table* feature of a graphing utility to generate additional rows of the table. Use the table to estimate the minimum length of the elevator.

(c) Write the length $L_1 + L_2$ as a function of θ.

(d) Use a graphing utility to graph the function. Use the graph to estimate the minimum length. How does your estimate compare with that in part (b)?

63. *Numerical and Graphical Analysis* The cross sections of an irrigation canal are isosceles trapezoids, where the length of three of the sides is 8 feet (see figure). The objective is to find the angle θ that maximizes the area of the cross sections. [*Hint:* The area of a trapezoid is given by $(h/2)(b_1 + b_2)$.]

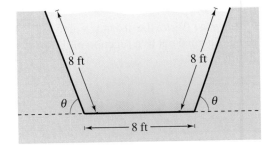

(a) Complete seven rows of the table.

Base 1	Base 2	Altitude	Area
8	$8 + 16 \cos 10°$	$8 \sin 10°$	22.1
8	$8 + 16 \cos 20°$	$8 \sin 20°$	42.5

(b) Use the *table* feature of a graphing utility to generate additional rows of the table. Use the table to estimate the maximum cross-sectional area.

(c) Write the area A as a function of θ.

(d) Use a graphing utility to graph the function. Use the graph to estimate the maximum cross-sectional area. How does your estimate compare with that in part (b)?

64. *Data Analysis*   The times S of sunset (Greenwich Mean Time) at 40° north latitude on the 15th of each month are: 1(16:59), 2(17:35), 3(18:06), 4(18:38), 5(19:08), 6(19:30), 7(19:28), 8(18:57), 9(18:09), 10(17:21), 11(16:44), 12(16:36). The month is represented by t, with $t = 1$ corresponding to January. A model (in which minutes have been converted to the decimal parts of an hour) for this data is given by

$$S(t) = 18.09 + 1.41 \sin\left(\frac{\pi t}{6} + 4.60\right).$$

(a) Use a graphing utility to graph the data points and the model in the same viewing window.

(b) What is the period of the model? Is it what you expected? Explain.

(c) What is the amplitude of the model? What does it represent in the context of the problem? Explain.

65. *Data Analysis*   The table shows the average sales S (in millions of dollars) of an outerwear manufacturer for each month t, where $t = 1$ represents January.

Month, t	Sales, S
1	13.46
2	11.15
3	8.00
4	4.85
5	2.54
6	1.70
7	2.54
8	4.85
9	8.00
10	11.15
11	13.46
12	14.30

(a) Create a scatter plot of the data.

(b) Find a trigonometric model that fits the data. Graph the model on your scatter plot. How well does the model fit?

(c) What is the period of the model? Do you think it is reasonable given the context? Explain your reasoning.

(d) Interpret the meaning of the model's amplitude in the context of the problem.

66. *Writing*   Is it true that N 24° E means 24 degrees north of east? Explain.

Synthesis

True or False?   **In Exercises 67 and 68, determine whether the statement is true or false. Justify your answer.**

67. In the right triangle shown below, $a = \dfrac{22.56}{\tan 41.9°}$.

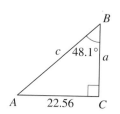

68. For the harmonic motion of a ball bobbing up and down on the end of a spring, one period can be described as the length of one coil of the spring.

Review

In Exercises 69–72, write the standard form of the equation of the line that has the specified characteristics.

69. $m = 4$, passes through $(-1, 2)$

70. $m = -\frac{1}{2}$, passes through $\left(\frac{1}{3}, 0\right)$

71. Passes through $(-2, 6)$ and $(3, 2)$

72. Passes through $\left(\frac{1}{4}, -\frac{2}{3}\right)$ and $\left(-\frac{1}{2}, \frac{1}{3}\right)$

In Exercises 73–80, find the domain of the function.

73. $f(x) = 3x + 8$

74. $f(x) = -x^2 - 1$

75. $g(x) = \sqrt[3]{x + 2}$

76. $g(x) = \sqrt{7 - x}$

77. $h(x) = \dfrac{2}{x^2 - 2x}$

78. $h(x) = \dfrac{x}{3x + 5}$

79. $f(x) = 4e^{-x}$

80. $f(x) = \ln(x - 2)$

In Exercises 81–84, solve the equation. Round your answer to three decimal places.

81. $e^{2x} = 54$

82. $\dfrac{300}{1 + e^{-x}} = 100$

83. $\ln(x^2 + 1) = 3.2$

84. $\log_8 x + \log_8(x - 1) = \frac{1}{3}$

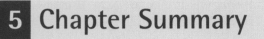

5 Chapter Summary

What did you learn?

5 Review Exercises

5.1 In Exercises 1 and 2, estimate the number of degrees in the angle.

1.

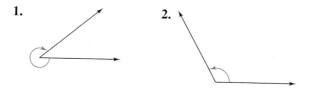

2.

In Exercises 3–6, (a) sketch the angle in standard position, (b) determine the quadrant in which the angle lies, and (c) list one positive and one negative coterminal angle.

3. $40°$

4. $190°$

5. $-110°$

6. $-405°$

In Exercises 7–10, find (if possible) the complement and supplement of the angle.

7. $5°$

8. $94°$

9. $171°$

10. $36°$

In Exercises 11–14, use the angle-conversion capabilities of a graphing utility to convert the angle measure to decimal degree form. Round your answer to three decimal places.

11. $135° \, 16' \, 45''$

12. $-234° \, 40''$

13. $5° \, 22' \, 53''$

14. $280° \, 8' \, 50''$

In Exercises 15–18, use the angle-conversion capabilities of a graphing utility to convert the angle measure to D°M′S″ form.

15. $135.29°$

16. $25.8°$

17. $-85.36°$

18. $-327.93°$

In Exercises 19–22, (a) sketch the angle in standard position, (b) determine the quadrant in which the angle lies, and (c) list one positive and one negative coterminal angle.

19. $\dfrac{\pi}{16}$

20. $\dfrac{40\pi}{47}$

21. $-\dfrac{9\pi}{15}$

22. $-\dfrac{11\pi}{3}$

In Exercises 23–26, find (if possible) the complement and supplement of the angle.

23. $\dfrac{\pi}{8}$

24. $\dfrac{\pi}{12}$

25. $\dfrac{3\pi}{10}$

26. $\dfrac{2\pi}{21}$

In Exercises 27–30, convert the angle measure from degrees to radians. Round your answer to three decimal places.

27. $480°$

28. $-16.5°$

29. $-33°$

30. $84°$

In Exercises 31–34, convert the angle measure from radians to degrees. Round your answer to three decimal places.

31. $\dfrac{5\pi}{7}$

32. $-\dfrac{3\pi}{5}$

33. -3.5

34. 1.55

35. Find the radian measure of the central angle of a circle with a radius of 12 feet that intercepts an arc of length 25 feet.

36. Find the radian measure of the central angle of a circle with a radius of 60 inches that intercepts an arc of length 245 inches.

37. Find the length of the arc on a circle with a radius of 20 meters intercepted by a central angle of $138°$.

38. Find the length of the arc on a circle with a radius of 15 centimeters intercepted by a central angle of $60°$.

39. **Music**   The radius of a compact disc is 6 centimeters. Find the linear speed of a point on the circumference of the disc if it is rotating at a speed of 500 revolutions per minute.

40. **Angular Speed**   A car is moving at a rate of 45 miles per hour, and the diameter of its wheels is about $2\frac{1}{3}$ feet.

    (a) Find the number of revolutions per minute the wheels are rotating.

    (b) Find the angular speed of the wheels in radians per minute.

5.2 In Exercises 41–44, find the exact values of the six trigonometric functions of the angle θ shown in the figure.

41.   **42.**

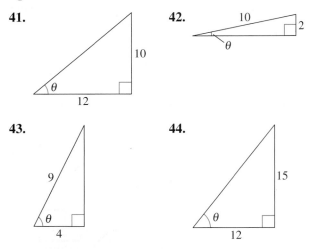

43.   **44.**

In Exercises 45 and 46, use trigonometric identities to transform one side of the equation into the other.

45. $(\csc \theta + \cot \theta)(\csc \theta - \cot \theta) = 1$

46. $\dfrac{\cot \theta + \tan \theta}{\cot \theta} = \sec^2 \theta$

In Exercises 47–50, use a calculator to evaluate each function. Round your answers to four decimal places.

47. (a) $\cos 84°$ 　　　(b) $\sin 6°$

48. (a) $\csc 52° \, 12'$ 　　(b) $\sec 54° \, 7'$

49. (a) $\cos \dfrac{\pi}{4}$ 　　　(b) $\sec \dfrac{\pi}{4}$

50. (a) $\tan \dfrac{3\pi}{20}$ 　　(b) $\cot \dfrac{3\pi}{20}$

51. *Width* An engineer is trying to determine the width of a river. From point P, the engineer walks downstream 125 feet and sights to point Q. From this sighting, it is determined that $\theta = 62°$. How wide is the river?

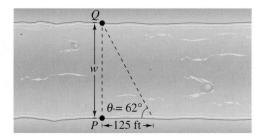

52. *Height* An escalator 152 feet in length rises to a platform and makes a 30° angle with the ground.

(a) Draw a right triangle that gives a visual representation of the problem. Show the known quantities of the triangle and use a variable to indicate the height of the platform above the ground.

(b) Use a trigonometric function to write an equation involving the unknown quantity.

(c) Find the height of the platform above the ground.

5.3 In Exercises 53–58, the point is on the terminal side of an angle in standard position. Determine the exact values of the six trigonometric functions of the angle.

53. $(12, 16)$ 　　　　**54.** $(-4, -6)$

55. $(-7, 2)$ 　　　　**56.** $(3, -4)$

57. $\left(\dfrac{2}{3}, \dfrac{5}{2}\right)$ 　　　**58.** $\left(-\dfrac{10}{3}, -\dfrac{2}{3}\right)$

In Exercises 59–62, find the values of the five trigonometric functions of θ satisfying the given condition.

59. $\sec \theta = \dfrac{6}{5}, \quad \tan \theta < 0$

60. $\tan \theta = -\dfrac{12}{5}, \quad \sin \theta > 0$

61. $\sin \theta = \dfrac{3}{8}, \quad \cos \theta < 0$

62. $\cos \theta = -\dfrac{2}{5}, \quad \sin \theta > 0$

In Exercises 63–66, find the reference angle θ' and sketch θ and θ' in standard position.

63. $\theta = 264°$ 　　　**64.** $\theta = 635°$

65. $\theta = -\dfrac{6\pi}{5}$ 　　**66.** $\theta = \dfrac{17\pi}{3}$

In Exercises 67–74, evaluate the sine, cosine, and tangent of the angle without using a calculator.

67. $240°$ 　　　　**68.** $315°$

69. $-210°$ 　　　**70.** $-315°$

71. $-\dfrac{9\pi}{4}$ 　　　**72.** $-\dfrac{11\pi}{6}$

73. $\dfrac{\pi}{2}$ 　　　　**74.** $\dfrac{\pi}{3}$

In Exercises 75–78, find the point (x, y) on the unit circle that corresponds to the given real number t. Use the result to evaluate $\sin t$, $\cos t$, and $\tan t$.

75. $t = \dfrac{2\pi}{3}$

76. $t = \dfrac{7\pi}{4}$

77. $t = \dfrac{7\pi}{6}$

78. $t = \dfrac{3\pi}{4}$

5.4 In Exercises 79–82, sketch the graph of the function.

79. $f(x) = 3 \sin x$

80. $f(x) = 2 \cos x$

81. $f(x) = \frac{1}{4} \cos x$

82. $f(x) = \frac{7}{2} \sin x$

In Exercises 83–86, find the period and amplitude.

83.

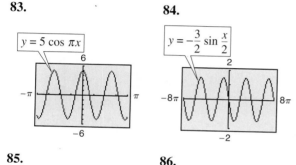

$y = 5 \cos \pi x$

84.

$y = -\dfrac{3}{2} \sin \dfrac{x}{2}$

85.

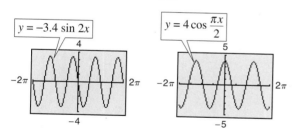

$y = -3.4 \sin 2x$

86.

$y = 4 \cos \dfrac{\pi x}{2}$

In Exercises 87–98, sketch the graph of the function by hand. (Include two full periods.)

87. $f(x) = 3 \cos 2\pi x$

88. $f(x) = -2 \sin \pi x$

89. $f(x) = 5 \sin \dfrac{2x}{5}$

90. $f(x) = 8 \cos\left(-\dfrac{x}{4}\right)$

91. $f(x) = -\dfrac{5}{2} \cos \dfrac{x}{4}$

92. $f(x) = -\dfrac{1}{2} \sin \dfrac{\pi x}{4}$

93. $f(x) = \frac{5}{2} \sin(x - \pi)$

94. $f(x) = 3 \cos(x + \pi)$

95. $f(x) = 2 - \cos \dfrac{\pi x}{2}$

96. $f(x) = \frac{1}{2} \sin(\pi x) - 3$

97. $f(x) = -3 \sin\left(\dfrac{x}{2} - \dfrac{\pi}{4}\right)$

98. $f(x) = 4 - 2 \cos(4x + \pi)$

Graphical Reasoning In Exercises 99–102, find a, b, and c for the function $f(x) = a \cos(bx - c)$ such that the graph of f matches the graph shown.

99.

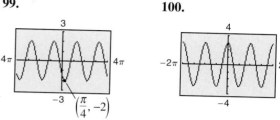

$\left(\dfrac{\pi}{4}, -2\right)$

100.

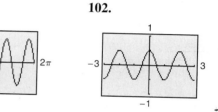

101.

102.

Sales In Exercises 103 and 104, use a graphing utility to graph the sales function over 1 year, where S is the sales (in thousands of units) and t is the time (in months), with $t = 1$ corresponding to January. Determine the months of maximum and minimum sales.

103. $S = 48.4 - 6.1 \cos \dfrac{\pi t}{6}$

104. $S = 56.25 + 9.50 \sin \dfrac{\pi t}{6}$

5.5 In Exercises 105–124, sketch the graph of the function. (Include two full periods.)

105. $f(x) = -\tan \dfrac{\pi x}{4}$

106. $f(x) = 4 \tan \pi x$

107. $f(x) = -\dfrac{1}{4} \tan \dfrac{\pi x}{2}$

108. $f(x) = \tan\left(x + \dfrac{\pi}{4}\right)$

109. $f(x) = \dfrac{1}{4} \tan\left(x - \dfrac{\pi}{2}\right)$

110. $f(x) = 2 + 2 \tan \dfrac{x}{3}$

111. $f(x) = 3 \cot \dfrac{x}{2}$

112. $f(x) = \dfrac{1}{2} \cot \dfrac{\pi x}{2}$

113. $f(x) = \dfrac{1}{2} \cot\left(x - \dfrac{\pi}{2}\right)$

114. $f(x) = 4 \cot\left(x + \dfrac{\pi}{4}\right)$

115. $f(x) = \frac{1}{4} \sec x$

116. $f(x) = \frac{1}{2} \csc x$

117. $f(x) = \frac{1}{4} \csc 2x$

118. $f(x) = \frac{1}{2} \sec 2\pi x$

119. $f(x) = \sec\left(x - \frac{\pi}{4}\right)$

120. $f(x) = \frac{1}{2} \csc(2x + \pi)$

121. $f(x) = 2 \sec(x - \pi)$

122. $f(x) = -2 \csc(x - \pi)$

123. $f(x) = \csc\left(3x - \frac{\pi}{2}\right)$

124. $f(x) = 3 \csc\left(2x + \frac{\pi}{4}\right)$

In Exercises 125–128, use a graphing utility to graph the function and the damping factor of the function in the same viewing window. Describe the behavior of the function as x increases without bound.

125. $f(x) = e^x \sin 2x$

126. $f(x) = 2x \cos x$

127. $f(x) = e^x \cos x$

128. $f(x) = x \sin \pi x$

5.6 In Exercises 129–132, find the value of each expression without using a calculator.

129. (a) arcsin 1

(b) arcsin 4

130. (a) $\arcsin \frac{\sqrt{2}}{2}$

(b) $\arcsin\left(-\frac{\sqrt{3}}{2}\right)$

131. (a) $\cos^{-1} \frac{\sqrt{2}}{2}$

(b) $\cos^{-1}\left(-\frac{\sqrt{3}}{2}\right)$

132. (a) $\tan^{-1}\left(-\sqrt{3}\right)$

(b) $\tan^{-1} 1$

In Exercises 133–140, use a calculator to approximate the value of the expression. Round your answer to two decimal places.

133. arccos 0.42

134. arcsin 0.63

135. $\sin^{-1}(-0.94)$

136. $\cos^{-1}(-0.12)$

137. $\arctan(-12)$

138. arctan 21

139. $\tan^{-1} 0.81$

140. $\tan^{-1} 6.4$

In Exercises 141 and 142, use an inverse trigonometric function to write θ as a function of x.

141.

142.

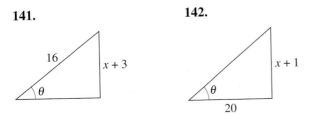

In Exercises 143–146, write an algebraic expression that is equivalent to the expression.

143. $\sec[\arcsin(x - 1)]$

144. $\tan\left(\arccos \frac{x}{2}\right)$

145. $\sin\left(\arccos \frac{x^2}{4 - x^2}\right)$

146. $\csc(\arcsin 10x)$

5.7

147. *Angle of Elevation*   The height of a radio transmission tower is 70 meters, and it casts a shadow of length 45 meters (see figure). Find the angle of elevation of the sun.

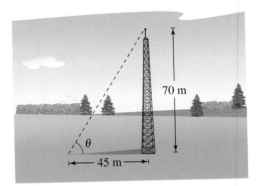

148. *Height*   An observer 2.5 miles from the launch pad of a space shuttle launch measures the angle of elevation to the base of the shuttle to be $25°$ soon after lift-off (see figure). How high is the shuttle at that instant? (Assume the shuttle is still moving vertically.)

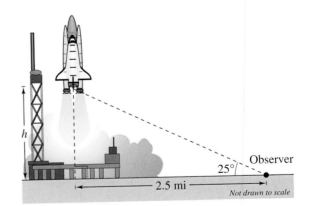

149. *Mountain Descent*   A road sign at the top of a mountain indicates that for the next 4 miles the grade is 12%. Find the angle of the grade and the change in elevation for a car descending the mountain.

150. *Railroad Grade* A train travels 3.5 kilometers on a straight track with a grade of 1° 10′. What is the vertical rise of the train in that distance?

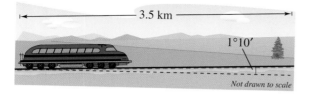

Not drawn to scale

151. *Distance* A passenger in an airplane flying at an altitude of 37,000 feet sees two towns directly to the west of the airplane. The angles of depression to the towns are 32° and 76° (see figure). How far apart are the towns?

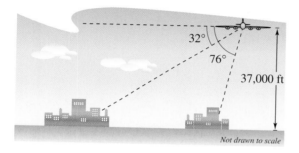

Not drawn to scale

152. *Distance* From city A to city B, a plane flies 650 miles at a bearing of 48°. From city B to city C, the plane flies 810 miles at a bearing of 115°. Find the distance from A to C and the bearing from A to C.

153. *Wave Motion* A buoy oscillates in simple harmonic motion as waves go past. At a given time it is noted that the buoy moves a total of 6 feet from its low point to its high point (see figure), and that it returns to its high point every 15 seconds. Write an equation that describes the motion of the buoy if it is at its high point at $t = 0$.

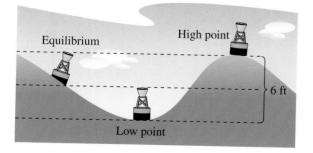

154. *Wave Motion* Your fishing bobber oscillates in simple harmonic motion from the waves in the lake where you fish. Your bobber moves a total of 1.5 inches from its high point to its low point and returns to its high point every 3 seconds. Write an equation modeling the motion of your bobber if it is at its high point at $t = 0$.

Synthesis

True or False? **In Exercises 155 and 156, determine whether the statement is true or false. Justify your answer.**

155. $y = \sin \theta$ is not a function because $\sin 30° = \sin 150°$.

156. The tangent function is often useful for modeling simple harmonic motion.

157. *Writing* Consider an angle in standard position with $r = 12$ centimeters, as shown in the figure. Write a short paragraph describing the changes in the values of x, y, $\sin \theta$, $\cos \theta$, and $\tan \theta$ as θ increases continually from 0° to 90°.

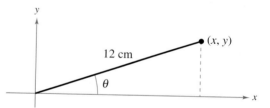

158. *Writing* Describe the behavior of $f(x) = \sec x$ at the zeros of $g(x) = \cos x$. Explain your reasoning.

159. *Approximation* In calculus it can be shown that the arctangent function can be approximated by the polynomial

$$\arctan x \approx x - \frac{x^3}{3} + \frac{x^5}{5} - \frac{x^7}{7}$$

where x is in radians.

(a) Use a graphing utility to graph the arctangent function and its polynomial approximation in the same viewing window. How do the graphs compare?

(b) Study the pattern in the polynomial approximation of the arctangent function and guess the next term. Then repeat part (a). How does the accuracy of the approximation change when additional terms are added?

5 Chapter Test

Take this test as you would take a test in class. After you are finished, check your work against the answers given in the back of the book.

1. Consider an angle that measures $\dfrac{5\pi}{4}$ radians.

   (a) Sketch the angle in standard position.

   (b) Determine two coterminal angles (one positive and one negative).

   (c) Convert the angle to degree measure.

2. A truck is moving at a rate of 90 kilometers per hour, and the diameter of its wheels is 1 meter. Find the angular speed of the wheels in radians per minute.

3. Find the exact values of the six trigonometric functions of the angle θ shown in the figure.

4. Given that $\tan \theta = \frac{6}{5}$, find the other five trigonometric functions of θ.

5. Determine the reference angle θ' of the angle $\theta = 255°$ and sketch θ and θ' in standard position.

6. Determine the quadrant in which θ lies if $\sec \theta < 0$ and $\tan \theta > 0$.

7. Find two exact values of θ in degrees $(0 \le \theta < 360°)$ if $\cos \theta = -\sqrt{3}/2$.

8. Use a calculator to approximate two values of θ in radians $(0 \le \theta < 2\pi)$ if $\csc \theta = 1.030$. Round your answer to two decimal places.

9. Find the five remaining trigonometric functions of θ, given that $\cos \theta = -\frac{3}{5}$ and $\sin \theta > 0$.

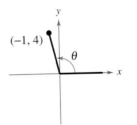

(−1, 4)

θ

Figure for 3

In Exercises 10–15, sketch the graph of the function. (Include two full periods.)

10. $g(x) = -2 \sin\left(x - \dfrac{\pi}{4}\right)$

11. $f(\alpha) = \dfrac{1}{2} \tan 2\alpha$

12. $f(x) = \frac{1}{2} \sec(x - \pi)$

13. $f(x) = 2 \cos(\pi - 2x) + 3$

14. $f(x) = 2 \csc\left(x + \dfrac{\pi}{2}\right)$

15. $f(x) = \dfrac{1}{4} \cot\left(x - \dfrac{\pi}{2}\right)$

In Exercises 16 and 17, use a graphing utility to graph the function. If the function is periodic, find its period.

16. $y = \sin 2\pi x + 2 \cos \pi x$

17. $y = 6e^{-0.12t} \cos(0.25t), \quad 0 \le t \le 32$

18. Find a, b, and c for the function $f(x) = a \sin(bx + c)$ such that the graph of f matches the graph at the right.

19. Find the exact value of $\tan\left(\arccos \frac{2}{3}\right)$ without using a calculator.

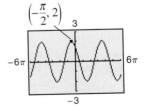

$\left(-\dfrac{\pi}{2}, 2\right)$

Figure for 18

In Exercises 20–22, use a graphing utility to graph the function.

20. $f(x) = 2 \arcsin\left(\dfrac{1}{2}x\right)$

21. $f(x) = 2 \arccos x$

22. $f(x) = \arctan \dfrac{x}{2}$

23. A plane is 160 miles north and 110 miles east of an airport. What bearing should be taken to fly directly to the airport?

Trigonometry can be used to model the flight of a baseball. Given the angle at which the ball leaves the bat and the initial velocity, you can determine the distance the ball will travel.

Al Bello/Getty Images

6 Analytic Trigonometry

What You Should Learn

In this chapter, you will learn how to:

■ Use fundamental trigonometric identities to evaluate trigonometric functions and simplify trigonometric expressions.

■ Verify trigonometric identities.

■ Use standard algebraic techniques and inverse trigonometric functions to solve trigonometric equations.

■ Use sum and difference formulas, multiple-angle formulas, power-reducing formulas, half-angle formulas, and product-to-sum formulas to rewrite and evaluate trigonometric functions.

6.1 Using Fundamental Identities

Introduction

In Chapter 5, you studied the basic definitions, properties, graphs, and applications of the individual trigonometric functions. In this chapter, you will learn how to use the fundamental identities to do the following.

1. Evaluate trigonometric functions.
2. Simplify trigonometric expressions.
3. Develop additional trigonometric identities.
4. Solve trigonometric equations.

Fundamental Trigonometric Identities

Reciprocal Identities

$$\sin u = \frac{1}{\csc u} \qquad \cos u = \frac{1}{\sec u} \qquad \tan u = \frac{1}{\cot u}$$

$$\csc u = \frac{1}{\sin u} \qquad \sec u = \frac{1}{\cos u} \qquad \cot u = \frac{1}{\tan u}$$

Quotient Identities

$$\tan u = \frac{\sin u}{\cos u} \qquad \cot u = \frac{\cos u}{\sin u}$$

Pythagorean Identities

$$\sin^2 u + \cos^2 u = 1 \qquad 1 + \tan^2 u = \sec^2 u \qquad 1 + \cot^2 u = \csc^2 u$$

Cofunction Identities

$$\sin\left(\frac{\pi}{2} - u\right) = \cos u \qquad \cos\left(\frac{\pi}{2} - u\right) = \sin u$$

$$\tan\left(\frac{\pi}{2} - u\right) = \cot u \qquad \cot\left(\frac{\pi}{2} - u\right) = \tan u$$

$$\sec\left(\frac{\pi}{2} - u\right) = \csc u \qquad \csc\left(\frac{\pi}{2} - u\right) = \sec u$$

Even/Odd Identities

$$\sin(-u) = -\sin u \qquad \cos(-u) = \cos u \qquad \tan(-u) = -\tan u$$

$$\csc(-u) = -\csc u \qquad \sec(-u) = \sec u \qquad \cot(-u) = -\cot u$$

Using the Fundamental Identities

One common use of trigonometric identities is to use given values of trigonometric functions to evaluate other trigonometric functions.

What you should learn

- Recognize and write the fundamental trigonometric identities.
- Use the fundamental trigonometric identities to evaluate trigonometric functions, simplify trigonometric expressions, and rewrite trigonometric expressions.

Why you should learn it

The fundamental trigonometric identities can be used to simplify trigonometric expressions. For instance, Exercise 97 on page 487 shows you how trigonometric identities can be used to simplify an expression for the rate of change of a function, a concept used in calculus.

STUDY TIP

Pythagorean identities are sometimes used in radical form such as

$$\sin u = \pm\sqrt{1 - \cos^2 u}$$

or

$$\tan u = \pm\sqrt{\sec^2 u - 1}$$

where the sign depends on the choice of u.

Example 1   Using Identities to Evaluate a Function

Use the values $\sec u = -\frac{3}{2}$ and $\tan u > 0$ to find the values of all six trigonometric functions.

Solution

Using a reciprocal identity, you have

$$\cos u = \frac{1}{\sec u} = \frac{1}{-3/2} = -\frac{2}{3}.$$

Using a Pythagorean identity, you have

$$\sin^2 u = 1 - \cos^2 u \qquad \text{Pythagorean identity}$$

$$= 1 - \left(-\frac{2}{3}\right)^2 \qquad \text{Substitute } -\frac{2}{3} \text{ for } \cos u.$$

$$= 1 - \frac{4}{9} = \frac{5}{9}. \qquad \text{Simplify.}$$

Because $\sec u < 0$ and $\tan u > 0$, it follows that u lies in Quadrant III. Moreover, because $\sin u$ is negative when u is in Quadrant III, you can choose the negative root and obtain $\sin u = -\sqrt{5}/3$. Now, knowing the values of the sine and cosine, you can find the values of all six trigonometric functions.

$$\sin u = -\frac{\sqrt{5}}{3} \qquad\qquad \csc u = \frac{1}{\sin u} = -\frac{3}{\sqrt{5}} = -\frac{3\sqrt{5}}{5}$$

$$\cos u = -\frac{2}{3} \qquad\qquad \sec u = \frac{1}{\cos u} = -\frac{3}{2}$$

$$\tan u = \frac{\sin u}{\cos u} = \frac{-\sqrt{5}/3}{-2/3} = \frac{\sqrt{5}}{2} \qquad \cot u = \frac{1}{\tan u} = \frac{2}{\sqrt{5}} = \frac{2\sqrt{5}}{5}$$

✓ *Checkpoint* Now try Exercise 11.

Example 2   Simplifying a Trigonometric Expression

Simplify $\sin x \cos^2 x - \sin x$.

Solution

First factor out a common monomial factor and then use a fundamental identity.

$$\sin x \cos^2 x - \sin x = \sin x(\cos^2 x - 1) \qquad \text{Factor out monomial factor.}$$

$$= -\sin x(1 - \cos^2 x) \qquad \text{Distributive Property}$$

$$= -\sin x(\sin^2 x) \qquad \text{Pythagorean identity}$$

$$= -\sin^3 x \qquad \text{Multiply.}$$

✓ *Checkpoint* Now try Exercise 29.

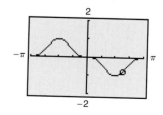

Example 3   Verifying a Trigonometric Identity

Determine whether the equation appears to be an identity.

$$\cos 3x \overset{?}{=} 4\cos^3 x - 3\cos x$$

Numerical Solution

Use the *table* feature of a graphing utility set in *radian* mode to create a table that shows the values of $y_1 = \cos 3x$ and $y_2 = 4\cos^3 x - 3\cos x$ for different values of x, as shown in Figure 6.1. The values of y_1 and y_2 appear to be identical, so $\cos 3x = 4\cos^3 x - 3\cos x$ appears to be an identity.

X	Y1	Y2
-.5	.07074	.07074
-.25	.73169	.73169
0	1	1
.25	.73169	.73169
.5	.07074	.07074
.75	-.6282	-.6282
1	-.99	-.99

X=1

Figure 6.1

Note that if the values of y_1 and y_2 were not identical, then the equation would not be an identity.

✓ *Checkpoint* Now try Exercise 39.

Graphical Solution

Use a graphing utility set in *radian* mode to graph $y_1 = \cos 3x$ and $y_2 = 4\cos^3 x - 3\cos x$ in the same viewing window, as shown in Figure 6.2. (Select the *line* style for y_1 and the *path* style for y_2.) Because the graphs appear to coincide, $\cos 3x = 4\cos^3 x - 3\cos x$ appears to be an identity.

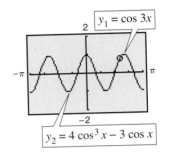

$y_1 = \cos 3x$

$y_2 = 4\cos^3 x - 3\cos x$

Figure 6.2

Note that if the graphs of y_1 and y_2 did not coincide, then the equation would not be an identity.

Example 4   Verifying a Trigonometric Identity

Verify the identity $\dfrac{\sin \theta}{1 + \cos \theta} + \dfrac{\cos \theta}{\sin \theta} = \csc \theta.$

Algebraic Solution

$$\frac{\sin \theta}{1 + \cos \theta} + \frac{\cos \theta}{\sin \theta} = \frac{(\sin \theta)(\sin \theta) + (\cos \theta)(1 + \cos \theta)}{(1 + \cos \theta)(\sin \theta)}$$

$$= \frac{\sin^2 \theta + \cos^2 \theta + \cos \theta}{(1 + \cos \theta)(\sin \theta)} \quad \text{Multiply.}$$

$$= \frac{1 + \cos \theta}{(1 + \cos \theta)(\sin \theta)} \quad \text{Pythagorean identity}$$

$$= \csc \theta \quad \begin{array}{l}\text{Divide out common factor}\\\text{and use reciprocal identity.}\end{array}$$

Notice how the identity is verified. You start with the left side of the equation (the more complicated side) and use the fundamental trigonometric identities to simplify it until you obtain the right side.

✓ *Checkpoint* Now try Exercise 45.

Graphical Solution

Use a graphing utility set in *radian* and *dot* modes to graph y_1 and y_2 in the same viewing window, as shown in Figure 6.3. Because the graphs appear to coincide, this equation appears to be an identity.

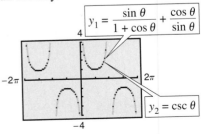

$y_1 = \dfrac{\sin \theta}{1 + \cos \theta} + \dfrac{\cos \theta}{\sin \theta}$

$y_2 = \csc \theta$

Figure 6.3

When factoring trigonometric expressions, it is helpful to find a polynomial form that fits the expression, as shown in Example 5.

Example 5   Factoring Trigonometric Expressions

Factor (a) $\sec^2 \theta - 1$ and (b) $4 \tan^2 \theta + \tan \theta - 3$.

Solution

a. Here the expression is a difference of two squares, which factors as

$$\sec^2 \theta - 1 = (\sec \theta - 1)(\sec \theta + 1).$$

b. This expression has the polynomial form $ax^2 + bx + c$ and it factors as

$$4 \tan^2 \theta + \tan \theta - 3 = (4 \tan \theta - 3)(\tan \theta + 1).$$

☑ *Checkpoint* Now try Exercise 51.

On occasion, factoring or simplifying can best be done by first rewriting the expression in terms of just *one* trigonometric function or in terms of *sine or cosine alone*. These strategies are illustrated in Examples 6 and 7.

Example 6   Factoring a Trigonometric Expression

Factor $\csc^2 x - \cot x - 3$.

Solution

Use the identity $\csc^2 x = 1 + \cot^2 x$ to rewrite the expression in terms of the cotangent.

$$
\begin{aligned}
\csc^2 x - \cot x - 3 &= (1 + \cot^2 x) - \cot x - 3 \qquad &\text{Pythagorean identity}\\
&= \cot^2 x - \cot x - 2 \qquad &\text{Combine like terms.}\\
&= (\cot x - 2)(\cot x + 1) \qquad &\text{Factor.}
\end{aligned}
$$

☑ *Checkpoint* Now try Exercise 57.

Example 7   Simplifying a Trigonometric Expression

Simplify $\sin t + \cot t \cos t$.

Solution

Begin by rewriting $\cot t$ in terms of sine and cosine.

$$
\begin{aligned}
\sin t + \cot t \cos t &= \sin t + \left(\frac{\cos t}{\sin t}\right)\cos t \qquad &\text{Quotient identity}\\
&= \frac{\sin^2 t + \cos^2 t}{\sin t} \qquad &\text{Add fractions.}\\
&= \frac{1}{\sin t} = \csc t \qquad &\text{Pythagorean identity and}\\
& &\text{reciprocal identity}
\end{aligned}
$$

☑ *Checkpoint* Now try Exercise 67.

The last two examples in this section involve techniques for rewriting expressions into forms that are used in calculus.

Example 8   Rewriting a Trigonometric Expression

Rewrite $\dfrac{1}{1 + \sin x}$ so that it is *not* in fractional form.

Solution

From the Pythagorean identity $\cos^2 x = 1 - \sin^2 x = (1 - \sin x)(1 + \sin x)$, you can see that multiplying both the numerator and the denominator by $(1 - \sin x)$ will produce a monomial denominator.

$$\frac{1}{1 + \sin x} = \frac{1}{1 + \sin x} \cdot \frac{1 - \sin x}{1 - \sin x} \qquad \text{Multiply numerator and denominator by } (1 - \sin x).$$

$$= \frac{1 - \sin x}{1 - \sin^2 x} \qquad \text{Multiply.}$$

$$= \frac{1 - \sin x}{\cos^2 x} \qquad \text{Pythagorean identity}$$

$$= \frac{1}{\cos^2 x} - \frac{\sin x}{\cos^2 x} \qquad \text{Write as separate fractions.}$$

$$= \frac{1}{\cos^2 x} - \frac{\sin x}{\cos x} \cdot \frac{1}{\cos x} \qquad \text{Write as separate fractions.}$$

$$= \sec^2 x - \tan x \sec x \qquad \text{Reciprocal and quotient identities}$$

✓ *Checkpoint* Now try Exercise 69.

Example 9   Trigonometric Substitution

Use the substitution $x = 2 \tan \theta$, $0 < \theta < \pi/2$, to write $\sqrt{4 + x^2}$ as a trigonometric function of θ.

Solution

Begin by letting $x = 2 \tan \theta$. Then you can obtain

$$\sqrt{4 + x^2} = \sqrt{4 + (2 \tan \theta)^2} \qquad \text{Substitute } 2 \tan \theta \text{ for } x.$$

$$= \sqrt{4(1 + \tan^2 \theta)} \qquad \text{Distributive Property}$$

$$= \sqrt{4 \sec^2 \theta} \qquad \text{Pythagorean identity}$$

$$= 2 \sec \theta. \qquad \sec \theta > 0 \text{ for } 0 < \theta < \frac{\pi}{2}$$

✓ *Checkpoint* Now try Exercise 81.

Figure 6.4 shows the right triangle illustration of the substitution in Example 9. For $0 < \theta < \pi/2$, you have

$$\text{opp} = x, \text{ adj} = 2, \text{ and hyp} = \sqrt{4 + x^2}.$$

Try using these expressions to obtain the result shown in Example 9.

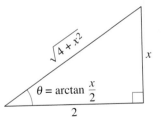

Figure 6.4

6.1 Exercises

Vocabulary Check

Fill in the blank to complete the trigonometric identity.

1. $\dfrac{1}{\cos u} = $ _____

2. $\dfrac{1}{\cot u} = $ _____

3. $\dfrac{\cos u}{\sin u} = $ _____

4. $\dfrac{1}{\sin u} = $ _____

5. $1 + $ _____ $= \csc^2 u$

6. $1 + \tan^2 u = $ _____

7. $\cos\left(\dfrac{\pi}{2} - u\right) = $ _____

8. $\sec\left(\dfrac{\pi}{2} - u\right) = $ _____

9. $\sin(-u) = $ _____

10. $\cos(-u) = $ _____

In Exercises 1–14, use the given values to evaluate (if possible) all six trigonometric functions.

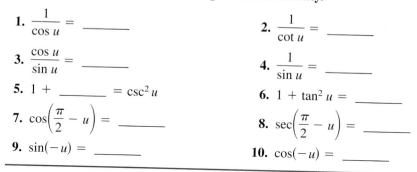

1. $\sin x = \dfrac{\sqrt{3}}{2}, \quad \cos x = \dfrac{1}{2}$

2. $\csc \theta = 2, \quad \tan \theta = \dfrac{\sqrt{3}}{3}$

3. $\sec \theta = \sqrt{2}, \quad \sin \theta = -\dfrac{\sqrt{2}}{2}$

4. $\tan x = \dfrac{\sqrt{3}}{3}, \quad \cos x = -\dfrac{\sqrt{3}}{2}$

5. $\tan x = \dfrac{7}{24}, \quad \sec x = -\dfrac{25}{24}$

6. $\cot \phi = -5, \quad \sin \phi = \dfrac{\sqrt{26}}{26}$

7. $\sec \phi = -\dfrac{13}{12}, \quad \sin \phi = \dfrac{5}{13}$

8. $\cos\left(\dfrac{\pi}{2} - x\right) = \dfrac{3}{5}, \quad \cos x = \dfrac{4}{5}$

9. $\sin(-x) = -\dfrac{2}{3}, \quad \tan x = -\dfrac{2\sqrt{5}}{5}$

10. $\csc(-x) = -5, \quad \cos x = \dfrac{\sqrt{24}}{5}$

11. $\tan \theta = 2, \quad \sin \theta < 0$

12. $\sec \theta = -5, \quad \tan \theta < 0$

13. $\csc \theta$ is undefined, $\quad \cos \theta < 0$

14. $\tan \theta$ is undefined, $\quad \sin \theta > 0$

In Exercises 15–20, match the trigonometric expression with one of the following.

(a) $\sec x$ (b) -1 (c) $\cot x$

(d) 1 (e) $-\tan x$ (f) $\sin x$

15. $\sec x \cos x$

16. $\tan x \csc x$

17. $\cot^2 x - \csc^2 x$

18. $(1 - \cos^2 x)(\csc x)$

19. $\dfrac{\sin(-x)}{\cos(-x)}$

20. $\dfrac{\sin[(\pi/2) - x]}{\cos[(\pi/2) - x]}$

In Exercises 21–26, match the trigonometric expression with one of the following.

(a) $\csc x$ (b) $\tan x$ (c) $\sin^2 x$

(d) $\sin x \tan x$ (e) $\sec^2 x$ (f) $\sec^2 x + \tan^2 x$

21. $\sin x \sec x$

22. $\cos^2 x(\sec^2 x - 1)$

23. $\sec^4 x - \tan^4 x$

24. $\cot x \sec x$

25. $\dfrac{\sec^2 x - 1}{\sin^2 x}$

26. $\dfrac{\cos^2[(\pi/2) - x]}{\cos x}$

In Exercises 27–38, use the fundamental identities to simplify the expression. Use the *table* feature of a graphing utility to check your result numerically.

27. $\cot x \sin x$

28. $\cos \beta \tan \beta$

29. $\sin \phi(\csc \phi - \sin \phi)$

30. $\sec^2 x(1 - \sin^2 x)$

31. $\dfrac{\cot x}{\csc x}$

32. $\dfrac{\sec \theta}{\csc \theta}$

33. $\sec \alpha \cdot \dfrac{\sin \alpha}{\tan \alpha}$

34. $\dfrac{\tan^2 \theta}{\sec^2 \theta}$

35. $\sin\left(\dfrac{\pi}{2} - x\right)\csc x$

36. $\cot\left(\dfrac{\pi}{2} - x\right)\cos x$

37. $\dfrac{\cos^2 y}{1 - \sin y}$

38. $\dfrac{1}{\tan^2 x + 1}$

In Exercises 39–44, verify the identity algebraically. Use the *table* feature of a graphing utility to check your result numerically.

39. $\sin \theta + \cos \theta \cot \theta = \csc \theta$

40. $(\sec \theta - \tan \theta)(\csc \theta + 1) = \cot \theta$

41. $\dfrac{\cos \theta}{1 - \sin \theta} = \sec \theta + \tan \theta$

42. $\dfrac{1 + \csc \theta}{\cot \theta + \cos \theta} = \sec \theta$

43. $\dfrac{1 + \cos \theta}{\sin \theta} + \dfrac{\sin \theta}{1 + \cos \theta} = 2 \csc \theta$

44. $\dfrac{\sin \theta + \cos \theta}{\sin \theta} - \dfrac{\cos \theta - \sin \theta}{\cos \theta} = \sec \theta \csc \theta$

In Exercises 45–50, verify the identity algebraically. Use a graphing utility to check your result graphically.

45. $\csc \theta \tan \theta = \sec \theta$

46. $\cos \theta \sec \theta - \cos^2 \theta = \sin^2 \theta$

47. $1 - \dfrac{\sin^2 \theta}{1 - \cos \theta} = -\cos \theta$

48. $\dfrac{\tan \theta}{1 + \sec \theta} + \dfrac{1 + \sec \theta}{\tan \theta} = 2 \csc \theta$

49. $\dfrac{\cot(-\theta)}{\csc \theta} = -\cos \theta$

50. $\dfrac{\csc\left(\dfrac{\pi}{2} - \theta\right)}{\tan(-\theta)} = \csc \theta$

In Exercises 51–60, factor the expression and use the fundamental identities to simplify. Use a graphing utility to check your result graphically.

51. $\cot^2 x - \cot^2 x \cos^2 x$

52. $\sec^2 x \tan^2 x + \sec^2 x$

53. $\dfrac{\cos^2 x - 4}{\cos x - 2}$

54. $\dfrac{\csc^2 x - 1}{\csc x - 1}$

55. $\tan^4 x + 2 \tan^2 x + 1$

56. $1 - 2 \sin^2 x + \sin^4 x$

57. $\sin^4 x - \cos^4 x$

58. $\sec^4 x - \tan^4 x$

59. $\csc^3 x - \csc^2 x - \csc x + 1$

60. $\sec^3 x - \sec^2 x - \sec x + 1$

In Exercises 61–64, perform the multiplication and use the fundamental identities to simplify.

61. $(\sin x + \cos x)^2$

62. $(\cot x + \csc x)(\cot x - \csc x)$

63. $(\sec x + 1)(\sec x - 1)$

64. $(3 - 3 \sin x)(3 + 3 \sin x)$

In Exercises 65–68, perform the addition or subtraction and use the fundamental identities to simplify.

65. $\dfrac{1}{1 + \cos x} + \dfrac{1}{1 - \cos x}$

66. $\dfrac{1}{\sec x + 1} - \dfrac{1}{\sec x - 1}$

67. $\tan x - \dfrac{\sec^2 x}{\tan x}$

68. $\dfrac{\cos x}{1 + \sin x} + \dfrac{1 + \sin x}{\cos x}$

In Exercises 69–72, rewrite the expression so that it is *not* in fractional form.

69. $\dfrac{\sin^2 y}{1 - \cos y}$

70. $\dfrac{5}{\tan x + \sec x}$

71. $\dfrac{3}{\sec x - \tan x}$

72. $\dfrac{\tan^2 x}{\csc x + 1}$

Numerical and Graphical Analysis **In Exercises 73–76, use a graphing utility to complete the table and graph the functions in the same viewing window. Make a conjecture about y_1 and y_2.**

x	0.2	0.4	0.6	0.8	1.0	1.2	1.4
y_1							
y_2							

73. $y_1 = \cos\left(\dfrac{\pi}{2} - x\right)$, $y_2 = \sin x$

74. $y_1 = \cos x + \sin x \tan x$, $y_2 = \sec x$

75. $y_1 = \dfrac{\cos x}{1 - \sin x}$, $y_2 = \dfrac{1 + \sin x}{\cos x}$

76. $y_1 = \sec^4 x - \sec^2 x$, $y_2 = \tan^2 x + \tan^4 x$

In Exercises 77–80, use a graphing utility to determine which of the six trigonometric functions is equal to the expression.

77. $\cos x \cot x + \sin x$

78. $\sin x(\cot x + \tan x)$

79. $\sec x - \dfrac{\cos x}{1 + \sin x}$

80. $\dfrac{1}{2}\left(\dfrac{1 + \sin \theta}{\cos \theta} + \dfrac{\cos \theta}{1 + \sin \theta}\right)$

In Exercises 81–84, use the trigonometric substitution to write the algebraic expression as a trigonometric function of θ, where $0 < \theta < \pi/2$.

81. $\sqrt{25 - x^2}$, $x = 5 \sin \theta$

82. $\sqrt{64 - 16x^2}$, $x = 2 \cos \theta$

83. $\sqrt{x^2 - 9}$, $x = 3 \sec \theta$

84. $\sqrt{x^2 + 100}$, $x = 10 \tan \theta$

In Exercises 85–88, use a graphing utility to solve the equation for θ, where $0 \le \theta < 2\pi$.

85. $\sin \theta = \sqrt{1 - \cos^2 \theta}$

86. $\cos \theta = -\sqrt{1 - \sin^2 \theta}$

87. $\sec \theta = \sqrt{1 + \tan^2 \theta}$

88. $\tan \theta = \sqrt{\sec^2 \theta - 1}$

In Exercises 89–92, rewrite the expression as a single logarithm and simplify the result.

89. $\ln|\cos \theta| - \ln|\sin \theta|$

90. $\ln|\csc \theta| + \ln|\tan \theta|$

91. $\ln(1 + \sin x) - \ln|\sec x|$

92. $\ln|\cot t| + \ln(1 + \tan^2 t)$

In Exercises 93–96, use the *table* feature of a graphing utility to demonstrate the identity for each value of θ.

93. $\csc^2 \theta - \cot^2 \theta = 1$, (a) $\theta = 132°$ (b) $\theta = \dfrac{2\pi}{7}$

94. $\tan^2 \theta + 1 = \sec^2 \theta$, (a) $\theta = 346°$ (b) $\theta = 3.1$

95. $\cos\left(\dfrac{\pi}{2} - \theta\right) = \sin \theta$, (a) $\theta = 80°$ (b) $\theta = 0.8$

96. $\sin(-\theta) = -\sin \theta$, (a) $\theta = 250°$ (b) $\theta = \frac{1}{2}$

97. *Rate of Change*   The rate of change of the function $f(x) = -\csc x - \sin x$ is given by the expression $\csc x \cot x - \cos x$. Show that this expression can also be written as $\cos x \cot^2 x$.

98. *Rate of Change*   The rate of change of the function $f(x) = \sec x + \cos x$ is given by the expression $\sec x \tan x - \sin x$. Show that this expression can also be written as $\sin x \tan^2 x$.

Synthesis

True or False?   In Exercises 99 and 100, determine whether the statement is true or false. Justify your answer.

99. $\sin \theta \csc \theta = 1$

100. $\cos \theta \sec \phi = 1$

In Exercises 101–104, fill in the blanks. (*Note:* $x \to c^+$ indicates that x approaches c from the right, and $x \to c^-$ indicates that x approaches c from the left.)

101. As $x \to \dfrac{\pi^-}{2}$, $\sin x \to$ ▨ and $\csc x \to$ ▨.

102. As $x \to 0^+$, $\cos x \to$ ▨ and $\sec x \to$ ▨.

103. As $x \to \dfrac{\pi^-}{2}$, $\tan x \to$ ▨ and $\cot x \to$ ▨.

104. As $x \to \pi^+$, $\sin x \to$ ▨ and $\csc x \to$ ▨.

105. Write each of the other trigonometric functions of θ in terms of $\sin \theta$.

106. Write each of the other trigonometric functions of θ in terms of $\cos \theta$.

107. Use the definitions of sine and cosine to derive the Pythagorean identity $\sin^2 \theta + \cos^2 \theta = 1$.

108. *Writing*   Use the Pythagorean identity $\sin^2 \theta + \cos^2 \theta = 1$ to derive the other Pythagorean identities $1 + \tan^2 \theta = \sec^2 \theta$ and $1 + \cot^2 \theta = \csc^2 \theta$. Discuss how to remember these identities and other fundamental identities.

Review

In Exercises 109–112, sketch the graph of the function. (Include two full periods.)

109. $f(x) = \dfrac{1}{2} \sin \pi x$

110. $f(x) = -2 \tan \dfrac{\pi x}{2}$

111. $f(x) = \dfrac{1}{2} \cot\left(x + \dfrac{\pi}{4}\right)$

112. $f(x) = \dfrac{3}{2} \cos(x - \pi) + 3$

6.2 Verifying Trigonometric Identities

What you should learn

● Verify trigonometric identities.

Why you should learn it

You can use trigonometric identities to rewrite trigonometric expressions. For instance, Exercise 67 on page 495 shows you how trigonometric identities can be used to solve a problem about the coefficient of friction for an object on an inclined plane.

Nancy Dudley/Stock Boston

Introduction

In this section, you will study techniques for verifying trigonometric identities. In the next section, you will study techniques for solving trigonometric equations. The key to both verifying identities *and* solving equations is your ability to use the fundamental identities and the rules of algebra to rewrite trigonometric expressions.

Remember that a *conditional equation* is an equation that is true for only some of the values in its domain. For example, the conditional equation

$$\sin x = 0 \qquad \text{Conditional equation}$$

is true only for $x = n\pi$, where n is an integer. When you find these values, you are *solving* the equation.

On the other hand, an equation that is true for all real values in the domain of the variable is an *identity*. For example, the familiar equation

$$\sin^2 x = 1 - \cos^2 x \qquad \text{Identity}$$

is true for all real numbers x. So, it is an identity.

Verifying Trigonometric Identities

Verifying that a trigonometric equation is an identity is quite different from solving an equation. There is no well-defined set of rules to follow in verifying trigonometric identities, and the process is best learned by practice.

Guidelines for Verifying Trigonometric Identities

1. Work with one side of the equation at a time. It is often better to work with the more complicated side first.

2. Look for opportunities to factor an expression, add fractions, square a binomial, or create a monomial denominator.

3. Look for opportunities to use the fundamental identities. Note which functions are in the final expression you want. Sines and cosines pair up well, as do secants and tangents, and cosecants and cotangents.

4. If the preceding guidelines do not help, try converting all terms to sines and cosines.

5. Always try *something*. Even making an attempt that leads to a dead end provides insight.

Verifying trigonometric identities is a useful process if you need to convert a trigonometric expression into a form that is more useful algebraically. When you verify an identity, you cannot assume that the two sides of the equation are equal because you are trying to verify that they are equal. As a result, when verifying identities, you cannot use operations such as adding the same quantity to each side of the equation or cross multiplication.

Example 1 Verifying a Trigonometric Identity

Verify the identity $\dfrac{\sec^2 \theta - 1}{\sec^2 \theta} = \sin^2 \theta$.

Solution

Because the left side is more complicated, start with it.

$$\begin{aligned}
\frac{\sec^2 \theta - 1}{\sec^2 \theta} &= \frac{(\tan^2 \theta + 1) - 1}{\sec^2 \theta} && \text{Pythagorean identity} \\
&= \frac{\tan^2 \theta}{\sec^2 \theta} && \text{Simplify.} \\
&= \tan^2 \theta(\cos^2 \theta) && \text{Reciprocal identity} \\
&= \frac{\sin^2 \theta}{\cancel{\cos^2 \theta}}(\cancel{\cos^2 \theta}) && \text{Quotient identity} \\
&= \sin^2 \theta && \text{Simplify.}
\end{aligned}$$

✓ *Checkpoint* Now try Exercise 5.

STUDY TIP

Remember that an identity is only true for all real values in the domain of the variable. For instance, in Example 1 the identity is not true when $\theta = \pi/2$ because $\sec^2 \theta$ is not defined when $\theta = \pi/2$.

There can be more than one way to verify an identity. Here is another way to verify the identity in Example 1.

$$\begin{aligned}
\frac{\sec^2 \theta - 1}{\sec^2 \theta} &= \frac{\sec^2 \theta}{\sec^2 \theta} - \frac{1}{\sec^2 \theta} && \text{Rewrite as the difference of fractions.} \\
&= 1 - \cos^2 \theta && \text{Reciprocal identity} \\
&= \sin^2 \theta && \text{Pythagorean identity}
\end{aligned}$$

Example 2 Combining Fractions Before Using Identities

Verify the identity $\dfrac{1}{1 - \sin \alpha} + \dfrac{1}{1 + \sin \alpha} = 2 \sec^2 \alpha$.

Algebraic Solution

$$\begin{aligned}
\frac{1}{1 - \sin \alpha} + \frac{1}{1 + \sin \alpha} &= \frac{1 + \sin \alpha + 1 - \sin \alpha}{(1 - \sin \alpha)(1 + \sin \alpha)} && \text{Add fractions.} \\
&= \frac{2}{1 - \sin^2 \alpha} && \text{Simplify.} \\
&= \frac{2}{\cos^2 \alpha} && \text{Pythagorean identity} \\
&= 2 \sec^2 \alpha && \text{Reciprocal identity}
\end{aligned}$$

✓ *Checkpoint* Now try Exercise 31.

Numerical Solution

Use the *table* feature of a graphing utility set in *radian* mode to create a table that shows the values of $y_1 = 1/(1 - \sin x) + 1/(1 + \sin x)$ and $y_2 = 2/\cos^2 x$ for different values of x, as shown in Figure 6.5. From the table, you can see that the values appear to be identical, so $1/(1 - \sin x) + 1/(1 + \sin x) = 2 \sec^2 x$ appears to be an identity.

X	Y1	Y2
-.5	2.5969	2.5969
-.25	2.1304	2.1304
0	2	2
.25	2.1304	2.1304
.5	2.5969	2.5969
.75	3.7357	3.7357
1	6.851	6.851

X=-.5

Figure 6.5

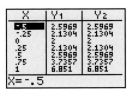

Example 3   Verifying a Trigonometric Identity

Verify the identity $(\tan^2 x + 1)(\cos^2 x - 1) = -\tan^2 x$.

Algebraic Solution

By applying identities before multiplying, you obtain the following.

$$(\tan^2 x + 1)(\cos^2 x - 1) = (\sec^2 x)(-\sin^2 x) \qquad \text{Pythagorean identities}$$

$$= -\frac{\sin^2 x}{\cos^2 x} \qquad \text{Reciprocal identity}$$

$$= -\left(\frac{\sin x}{\cos x}\right)^2 \qquad \text{Rule of exponents}$$

$$= -\tan^2 x \qquad \text{Quotient identity}$$

✓ *Checkpoint* Now try Exercise 39.

Graphical Solution

Use a graphing utility set in *radian* mode to graph the left side of the identity $y_1 = (\tan^2 x + 1)(\cos^2 x - 1)$ and the right side of the identity $y_2 = -\tan^2 x$ in the same viewing window, as shown in Figure 6.6. (Select the *line* style for y_1 and the *path* style for y_2.) Because the graphs appear to coincide, $(\tan^2 x + 1) \cdot (\cos^2 x - 1) = -\tan^2 x$ appears to be an identity.

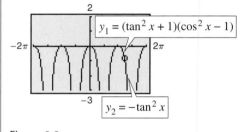

Figure 6.6

Example 4   Converting to Sines and Cosines

Verify the identity $\tan x + \cot x = \sec x \csc x$.

Solution

In this case there appear to be no fractions to add, no products to find, and no opportunities to use the Pythagorean identities. So, try converting the left side into sines and cosines.

$$\tan x + \cot x = \frac{\sin x}{\cos x} + \frac{\cos x}{\sin x} \qquad \text{Quotient identities}$$

$$= \frac{\sin^2 x + \cos^2 x}{\cos x \sin x} \qquad \text{Add fractions.}$$

$$= \frac{1}{\cos x \sin x} \qquad \text{Pythagorean identity}$$

$$= \frac{1}{\cos x} \cdot \frac{1}{\sin x} \qquad \text{Product of fractions}$$

$$= \sec x \csc x \qquad \text{Reciprocal identities}$$

✓ *Checkpoint* Now try Exercise 41.

TECHNOLOGY TIP

Although a graphing utility can be useful in helping to verify an identity, you must use algebraic techniques to produce a valid proof. For example, graph the two functions

$$y_1 = \sin 50x$$

$$y_2 = \sin 2x$$

in a trigonometric viewing window. Although their graphs seem identical, $\sin 50x \neq \sin 2x$.

Recall from algebra that *rationalizing the denominator* using conjugates is, on occasion, a powerful simplification technique. A related form of this technique works for simplifying trigonometric expressions as well. For instance, to simplify $1/(1 - \cos x)$, multiply the numerator and the denominator by $1 + \cos x$.

$$\frac{1}{1 - \cos x} = \frac{1}{1 - \cos x}\left(\frac{1 + \cos x}{1 + \cos x}\right)$$

$$= \frac{1 + \cos x}{1 - \cos^2 x}$$

$$= \frac{1 + \cos x}{\sin^2 x}$$

$$= \csc^2 x(1 + \cos x)$$

As shown above, $\csc^2 x(1 + \cos x)$ is considered a simplified form of $1/(1 - \cos x)$ because the expression does not contain any fractions.

TECHNOLOGY SUPPORT

For instructions on how to use the *radian* and *dot* modes, see Appendix A; for specific keystrokes, go to the text website at *college.hmco.com.*

Example 5   Verifying a Trigonometric Identity

Verify the identity

$$\sec x + \tan x = \frac{\cos x}{1 - \sin x}.$$

Algebraic Solution

Work with the *right* side because you can create a monomial denominator by multiplying the numerator and denominator by $(1 + \sin x)$.

$$\frac{\cos x}{1 - \sin x} = \frac{\cos x}{1 - \sin x}\left(\frac{1 + \sin x}{1 + \sin x}\right) \quad \text{Multiply numerator and denominator by } (1 + \sin x).$$

$$= \frac{\cos x + \cos x \sin x}{1 - \sin^2 x} \quad \text{Multiply.}$$

$$= \frac{\cos x + \cos x \sin x}{\cos^2 x} \quad \text{Pythagorean identity}$$

$$= \frac{\cos x}{\cos^2 x} + \frac{\cos x \sin x}{\cos^2 x} \quad \text{Separate fractions.}$$

$$= \frac{1}{\cos x} + \frac{\sin x}{\cos x} \quad \text{Simplify.}$$

$$= \sec x + \tan x \quad \text{Identities}$$

Graphical Solution

Use a graphing utility set in *radian* and *dot* modes to graph $y_1 = \sec x + \tan x = 1/\cos x + \tan x$ and $y_2 = \cos x/(1 - \sin x)$ in the same viewing window, as shown in Figure 6.7. Because the graphs appear to coincide, $\sec x + \tan x = \cos x/(1 - \sin x)$ appears to be an identity.

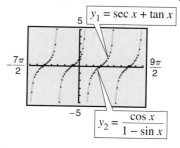

Figure 6.7

✓ *Checkpoint* Now try Exercise 47.

In Examples 1 through 5, you have been verifying trigonometric identities by working with one side of the equation and converting it to the form given on the other side. On occasion it is practical to work with each side *separately* to obtain one common form equivalent to both sides. This is illustrated in Example 6.

Example 6 Working with Each Side Separately

Verify the identity $\dfrac{\cot^2 \theta}{1 + \csc \theta} = \dfrac{1 - \sin \theta}{\sin \theta}$.

Solution

Working with the left side, you have

$$\frac{\cot^2 \theta}{1 + \csc \theta} = \frac{\csc^2 \theta - 1}{1 + \csc \theta} \qquad \text{Pythagorean identity}$$

$$= \frac{(\csc \theta - 1)(\cancel{\csc \theta + 1})}{\cancel{1 + \csc \theta}} \qquad \text{Factor.}$$

$$= \csc \theta - 1. \qquad \text{Simplify.}$$

Now, simplifying the right side, you have

$$\frac{1 - \sin \theta}{\sin \theta} = \frac{1}{\sin \theta} - \frac{\sin \theta}{\sin \theta} \qquad \text{Separate fractions.}$$

$$= \csc \theta - 1. \qquad \text{Reciprocal identity}$$

The identity is verified because both sides are equal to $\csc \theta - 1$.

✓ *Checkpoint* Now try Exercise 49.

In Example 7, powers of trigonometric functions are rewritten as more complicated sums of products of trigonometric functions. This is a common procedure used in calculus.

Example 7 Examples from Calculus

Verify each identity.

a. $\tan^4 x = \tan^2 x \sec^2 x - \tan^2 x$ **b.** $\sin^3 x \cos^4 x = (\cos^4 x - \cos^6 x)\sin x$

Solution

a. $\tan^4 x = (\tan^2 x)(\tan^2 x)$ Write as separate factors.

$\qquad\quad = \tan^2 x(\sec^2 x - 1)$ Pythagorean identity

$\qquad\quad = \tan^2 x \sec^2 x - \tan^2 x$ Multiply.

b. $\sin^3 x \cos^4 x = \sin^2 x \cos^4 x \sin x$ Write as separate factors.

$\qquad\qquad\quad = (1 - \cos^2 x)\cos^4 x \sin x$ Pythagorean identity

$\qquad\qquad\quad = (\cos^4 x - \cos^6 x)\sin x$ Multiply.

✓ *Checkpoint* Now try Exercise 63.

TECHNOLOGY TIP Remember that you can use a graphing utility to assist in verifying an identity by creating a table or by graphing.

6.2 Exercises

Vocabulary Check

In Exercises 1 and 2, fill in the blanks.

1. An equation that is true for only some values in its domain is called a _____ equation.

2. An equation that is true for all real values in its domain is called an _____ .

In Exercises 3–10, fill in the blank to complete the trigonometric identity.

3. $\dfrac{1}{\tan u} =$ _____

4. $\dfrac{1}{\csc u} =$ _____

5. $\dfrac{\sin u}{\cos u} =$ _____

6. $\dfrac{1}{\sec u} =$ _____

7. $\sin^2 u +$ _____ $= 1$

8. $\sin\left(\dfrac{\pi}{2} - u\right) =$ _____

9. $\tan(-u) =$ _____

10. $\sec(-u) =$ _____

In Exercises 1–10, verify the identity.

1. $\sin t \csc t = 1$

2. $\sec y \cos y = 1$

3. $\dfrac{\csc^2 x}{\cot x} = \csc x \sec x$

4. $\dfrac{\sin^2 t}{\tan^2 t} = \cos^2 t$

5. $\cos^2 \beta - \sin^2 \beta = 1 - 2\sin^2 \beta$

6. $\cos^2 \beta - \sin^2 \beta = 2\cos^2 \beta - 1$

7. $\tan^2 \theta + 5 = \sec^2 \theta + 4$

8. $2 - \csc^2 z = 1 - \cot^2 z$

9. $(1 + \sin x)(1 - \sin x) = \cos^2 x$

10. $\cot^2 y(\sec^2 y - 1) = 1$

Numerical, Graphical, and Algebraic Analysis In Exercises 11–18, use a graphing utility to complete the table and graph the functions in the same viewing window. Use both the table and the graph as evidence that $y_1 = y_2$. Then verify the identity algebraically.

x	0.2	0.4	0.6	0.8	1.0	1.2	1.4
y_1							
y_2							

11. $y_1 = \dfrac{1}{\sec x \tan x}$, $y_2 = \csc x - \sin x$

12. $y_1 = \dfrac{\csc x - 1}{1 - \sin x}$, $y_2 = \csc x$

13. $y_1 = \csc x - \sin x$, $y_2 = \cos x \cot x$

14. $y_1 = \sec x - \cos x$, $y_2 = \sin x \tan x$

15. $y_1 = \sin x + \cos x \cot x$, $y_2 = \csc x$

16. $y_1 = \cos x + \sin x \tan x$, $y_2 = \sec x$

17. $y_1 = \dfrac{1}{\tan x} + \dfrac{1}{\cot x}$, $y_2 = \tan x + \cot x$

18. $y_1 = \dfrac{1}{\sin x} - \dfrac{1}{\csc x}$, $y_2 = \csc x - \sin x$

Error Analysis In Exercises 19 and 20, describe the error.

19.

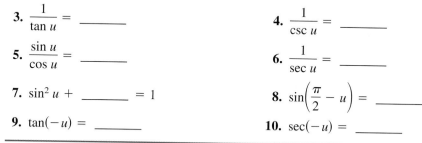

$(1 + \tan x)[1 + \cot(-x)]$
$= (1 + \tan x)(1 + \cot x)$
$= 1 + \cot x + \tan x + \tan x \cot x$
$= 1 + \cot x + \tan x + 1$
$= 2 + \cot x + \tan x$

20.

$\dfrac{1 + \sec(-\theta)}{\sin(-\theta) + \tan(-\theta)} = \dfrac{1 - \sec \theta}{\sin \theta - \tan \theta}$

$= \dfrac{1 - \sec \theta}{(\sin \theta)\left[1 - \left(\dfrac{1}{\cos \theta}\right)\right]}$

$= \dfrac{1 - \sec \theta}{\sin \theta(1 - \sec \theta)}$

$= \dfrac{1}{\sin \theta} = \csc \theta$

In Exercises 21–30, verify the identity.

21. $\sin^{1/2} x \cos x - \sin^{5/2} x \cos x = \cos^3 x \sqrt{\sin x}$

22. $\sec^6 x(\sec x \tan x) - \sec^4 x(\sec x \tan x) = \sec^5 x \tan^3 x$

23. $\tan\left(\dfrac{\pi}{2} - x\right) \sec x = \csc x$

24. $\dfrac{\cos[(\pi/2) - x]}{\sin[(\pi/2) - x]} = \tan x$

25. $\dfrac{\csc(-x)}{\sec(-x)} = -\cot x$

26. $(1 + \sin y)[1 + \sin(-y)] = \cos^2 y$

27. $\dfrac{\cos(-\theta)}{1 + \sin(-\theta)} = \sec \theta + \tan \theta$

28. $\dfrac{1 + \csc(-\theta)}{\cos(-\theta) + \cot(-\theta)} = \sec \theta$

29. $\dfrac{\sin x \cos y + \cos x \sin y}{\cos x \cos y - \sin x \sin y} = \dfrac{\tan x + \tan y}{1 - \tan x \tan y}$

30. $\dfrac{\tan x + \tan y}{1 - \tan x \tan y} = \dfrac{\cot x + \cot y}{\cot x \cot y - 1}$

In Exercises 31–38, verify the identity algebraically. Use the *table* feature of a graphing utility to check your result numerically.

31. $\dfrac{\cos x - \cos y}{\sin x + \sin y} + \dfrac{\sin x - \sin y}{\cos x + \cos y} = 0$

32. $\dfrac{\tan x + \cot y}{\tan x \cot y} = \tan y + \cot x$

33. $\sqrt{\dfrac{1 + \sin \theta}{1 - \sin \theta}} = \dfrac{1 + \sin \theta}{|\cos \theta|}$

34. $\sqrt{\dfrac{1 - \cos \theta}{1 + \cos \theta}} = \dfrac{1 - \cos \theta}{|\sin \theta|}$

35. $\cos^2 x + \cos^2\left(\dfrac{\pi}{2} - x\right) = 1$

36. $\sec^2 y - \cot^2\left(\dfrac{\pi}{2} - y\right) = 1$

37. $\sin x \csc\left(\dfrac{\pi}{2} - x\right) = \tan x$

38. $\csc^2\left(\dfrac{\pi}{2} - x\right) - 1 = \tan^2 x$

In Exercises 39–50, verify the identity algebraically. Use a graphing utility to check your result graphically.

39. $2 \sec^2 x - 2 \sec^2 x \sin^2 x - \sin^2 x - \cos^2 x = 1$

40. $\csc x(\csc x - \sin x) + \dfrac{\sin x - \cos x}{\sin x} + \cot x = \csc^2 x$

41. $\dfrac{\tan x \cot x}{\cos x} = \sec x$

42. $\dfrac{1 + \csc \theta}{\sec \theta} - \cot \theta = \cos \theta$

43. $\csc^4 x - 2 \csc^2 x + 1 = \cot^4 x$

44. $\sin x(1 - 2 \cos^2 x + \cos^4 x) = \sin^5 x$

45. $\sec^4 \theta - \tan^4 \theta = 1 + 2 \tan^2 \theta$

46. $\csc^4 \theta - \cot^4 \theta = 2 \csc^2 \theta - 1$

47. $\dfrac{\sin \beta}{1 - \cos \beta} = \dfrac{1 + \cos \beta}{\sin \beta}$

48. $\dfrac{\cot \alpha}{\csc \alpha - 1} = \dfrac{\csc \alpha + 1}{\cot \alpha}$

49. $\dfrac{\tan^3 \alpha - 1}{\tan \alpha - 1} = \tan^2 \alpha + \tan \alpha + 1$

50. $\dfrac{\sin^3 \beta + \cos^3 \beta}{\sin \beta + \cos \beta} = 1 - \sin \beta \cos \beta$

Conjecture **In Exercises 51–54, use a graphing utility to graph the trigonometric function. Use the graph to make a conjecture about a simplification of the expression. Verify the resulting identity algebraically.**

51. $y = \dfrac{1}{\cot x + 1} + \dfrac{1}{\tan x + 1}$

52. $y = \dfrac{\cos x}{1 - \tan x} + \dfrac{\sin x \cos x}{\sin x - \cos x}$

53. $y = \dfrac{1}{\sin x} - \dfrac{\cos^2 x}{\sin x}$

54. $y = \sin t + \dfrac{\cot^2 t}{\csc t}$

In Exercises 55–58, use the properties of logarithms and trigonometric identities to verify the identity.

55. $\ln|\cot \theta| = \ln|\cos \theta| - \ln|\sin \theta|$

56. $\ln|\sec \theta| = -\ln|\cos \theta|$

57. $-\ln(1 + \cos \theta) = \ln(1 - \cos \theta) - 2 \ln|\sin \theta|$

58. $-\ln|\csc \theta + \cot \theta| = \ln|\csc \theta - \cot \theta|$

In Exercises 59–62, use the cofunction identities to evaluate the expression without using a calculator.

59. $\sin^2 25° + \sin^2 65°$

60. $\cos^2 14° + \cos^2 76°$

61. $\cos^2 20° + \cos^2 52° + \cos^2 38° + \cos^2 70°$

62. $\sin^2 12° + \sin^2 40° + \sin^2 50° + \sin^2 78°$

∫ **In Exercises 63–66, powers of trigonometric functions are rewritten to be useful in calculus. Verify the identity.**

63. $\tan^5 x = \tan^3 x \sec^2 x - \tan^3 x$

64. $\sec^4 x \tan^2 x = (\tan^2 x + \tan^4 x)\sec^2 x$

65. $\cos^3 x \sin^2 x = (\sin^2 x - \sin^4 x)\cos x$

66. $\sin^4 x + \cos^4 x = 1 - 2 \cos^2 x + 2 \cos^4 x$

67. *Friction*   The forces acting on an object weighing W units on an inclined plane positioned at an angle of θ with the horizontal (see figure) are modeled by

$$\mu W \cos \theta = W \sin \theta$$

where μ is the coefficient of friction. Solve the equation for μ and simplify the result.

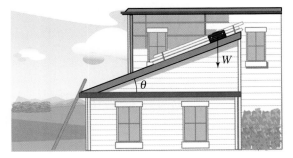

68. *Shadow Length*   The length s of the shadow cast by a vertical *gnomon* (a device used to tell time) of height h when the angle of the sun above the horizon is θ can be modeled by the equation

$$s = \frac{h \sin(90° - \theta)}{\sin \theta}.$$

Show that the equation is equivalent to $s = h \cot \theta$.

Synthesis

True or False?   **In Exercises 69–72, determine whether the statement is true or false. Justify your answer.**

69. There can be more than one way to verify a trigonometric identity.

70. Of the six trigonometric functions, two are even.

71. The equation $\sin^2 \theta + \cos^2 \theta = 1 + \tan^2 \theta$ is an identity, because $\sin^2(0) + \cos^2(0) = 1$ and $1 + \tan^2(0) = 1$.

72. $\sin(x^2) = \sin^2(x)$

Think About It   **In Exercises 73 and 74, explain why the equation is not an identity and find one value of the variable for which the equation is not true.**

73. $\sqrt{\tan^2 x} = \tan x$ **74.** $\sin \theta = \sqrt{1 - \cos^2 \theta}$

75. Verify that for all integers n,

$$\cos\left[\frac{(2n + 1)\pi}{2}\right] = 0.$$

76. Verify that for all integers n,

$$\sin\left[\frac{(12n + 1)\pi}{6}\right] = \frac{1}{2}.$$

Review

In Exercises 77–80, find a polynomial function with real coefficients that has the given zeros. (There are many correct answers.)

77. $1, 8i, -8i$ **78.** $i, -i, 4i, -4i$

79. $4, 6 + i, 6 - i$ **80.** $0, 0, 2, 1 - i$

In Exercises 81–84, sketch the graph of the function by hand.

81. $f(x) = 2^x + 3$ **82.** $f(x) = -2^{x-3}$

83. $f(x) = 2^{-x} + 1$ **84.** $f(x) = 2^{x-1} + 3$

In Exercises 85–88, state the quadrant in which θ lies.

85. $\csc \theta > 0$ and $\tan \theta < 0$

86. $\cot \theta > 0$ and $\cos \theta < 0$

87. $\sec \theta > 0$ and $\sin \theta < 0$

88. $\cot \theta > 0$ and $\sec \theta < 0$

In Exercises 89–92, solve the right triangle shown in the figure. Round your answer to two decimal places.

89. $B = 80°, a = 16$ **90.** $A = 28°, c = 20$

91. $a = 14, b = 8$ **92.** $b = 6.2, c = 12.54$

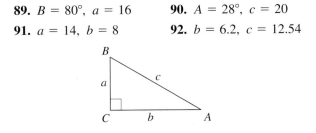

6.3 Solving Trigonometric Equations

Introduction

To solve a trigonometric equation, use standard algebraic techniques such as collecting like terms and factoring. Your preliminary goal is to isolate the trigonometric function involved in the equation.

Example 1   Solving a Trigonometric Equation

$2 \sin x - 1 = 0$ Original equation

$\quad 2 \sin x = 1$ Add 1 to each side.

$\quad\quad \sin x = \frac{1}{2}$ Divide each side by 2.

To solve for x, note in Figure 6.8 that the equation $\sin x = \frac{1}{2}$ has solutions $x = \pi/6$ and $x = 5\pi/6$ in the interval $[0, 2\pi)$. Moreover, because $\sin x$ has a period of 2π, there are infinitely many other solutions, which can be written as

$$x = \frac{\pi}{6} + 2n\pi \quad \text{and} \quad x = \frac{5\pi}{6} + 2n\pi \quad \text{General solution}$$

where n is an integer, as shown in Figure 6.8.

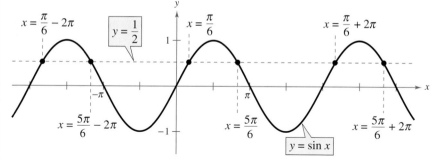

Figure 6.8

✓ *Checkpoint* Now try Exercise 7.

Figure 6.9 verifies that the equation $\sin x = \frac{1}{2}$ has infinitely many solutions. Any angles that are coterminal with $\pi/6$ or $5\pi/6$ are also solutions of the equation.

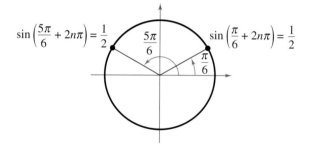

Figure 6.9

Example 2   Collecting Like Terms

Find all solutions of $\sin x + \sqrt{2} = -\sin x$ in the interval $[0, 2\pi)$.

Algebraic Solution

Rewrite the equation so that $\sin x$ is isolated on one side of the equation.

$$\sin x + \sqrt{2} = -\sin x \qquad \text{Write original equation.}$$

$$\sin x + \sin x = -\sqrt{2} \qquad \begin{array}{l}\text{Add } \sin x \text{ to and subtract}\\ \sqrt{2} \text{ from each side.}\end{array}$$

$$2 \sin x = -\sqrt{2} \qquad \text{Combine like terms.}$$

$$\sin x = -\frac{\sqrt{2}}{2} \qquad \text{Divide each side by 2.}$$

The solutions in the interval $[0, 2\pi)$ are

$$x = \frac{5\pi}{4} \qquad \text{and} \qquad x = \frac{7\pi}{4}.$$

Numerical Solution

Use the *table* feature of a graphing utility set in *radian* mode to create a table that shows the values of $y_1 = \sin x + \sqrt{2}$ and $y_2 = -\sin x$ for different values of x. Your table should go from $x = 0$ to $x = 2\pi$ using increments of $\pi/8$, as shown in Figure 6.10. From the table, you can see that the values of y_1 and y_2 appear to be identical when $x \approx 3.927 \approx 5\pi/4$ and $x \approx 5.4978 \approx 7\pi/4$. These values are the approximate solutions of $\sin x + \sqrt{2} = -\sin x$.

Figure 6.10

 Checkpoint Now try Exercise 17.

Example 3   Extracting Square Roots

Solve $3 \tan^2 x - 1 = 0$.

Solution

Rewrite the equation so that $\tan x$ is isolated on one side of the equation.

$$3 \tan^2 x = 1 \qquad \text{Add 1 to each side.}$$

$$\tan^2 x = \frac{1}{3} \qquad \text{Divide each side by 3.}$$

$$\tan x = \pm\frac{1}{\sqrt{3}} \qquad \text{Extract square roots.}$$

Because $\tan x$ has a period of π, first find all solutions in the interval $[0, \pi)$. These are $x = \pi/6$ and $x = 5\pi/6$. Finally, add multiples of π to each of these solutions to get the general form

$$x = \frac{\pi}{6} + n\pi \qquad \text{and} \qquad x = \frac{5\pi}{6} + n\pi \qquad \text{General solution}$$

where n is an integer. The graph of $y = 3 \tan^2 x - 1$, shown in Figure 6.11, confirms this result.

TECHNOLOGY SUPPORT

For instructions on how to use the *table* feature, see Appendix A; for specific keystrokes, go to the text website at *college.hmco.com*.

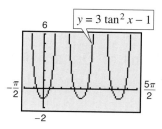

Figure 6.11

 Checkpoint Now try Exercise 19.

Recall that the solutions of an equation correspond to the x-intercepts of the graph of the equation. For instance, the graph in Figure 6.11 has x-intercepts at $\pi/6$, $5\pi/6$, $7\pi/6$, and so on.

The equations in Examples 1, 2, and 3 involved only one trigonometric function. When two or more functions occur in the same equation, collect all terms on one side and try to separate the functions by factoring or by using appropriate identities. This may produce factors that yield no solutions, as illustrated in Example 4.

Example 4   Factoring

Solve $\cot x \cos^2 x = 2 \cot x$.

Solution

Begin by rewriting the equation so that all terms are collected on one side of the equation.

$$\cot x \cos^2 x = 2 \cot x \qquad \text{Write original equation.}$$

$$\cot x \cos^2 x - 2 \cot x = 0 \qquad \text{Subtract 2 cot } x \text{ from each side.}$$

$$\cot x(\cos^2 x - 2) = 0 \qquad \text{Factor.}$$

By setting each of these factors equal to zero, you obtain the following.

$$\cot x = 0 \qquad \text{and} \qquad \cos^2 x - 2 = 0$$

$$\cos^2 x = 2$$

$$\cos x = \pm\sqrt{2}$$

The equation $\cot x = 0$ has the solution $x = \pi/2$ [in the interval $(0, \pi)$]. No solution is obtained for $\cos x = \pm\sqrt{2}$ because $\pm\sqrt{2}$ are outside the range of the cosine function. Because $\cot x$ has a period of π, the general form of the solution is obtained by adding multiples of π to $x = \pi/2$, to get

$$x = \frac{\pi}{2} + n\pi \qquad \text{General solution}$$

where n is an integer. The graph of $y = \cot x \cos^2 x - 2 \cot x$ (in *dot* mode), shown in Figure 6.12, confirms this result.

✓ *Checkpoint* Now try Exercise 21.

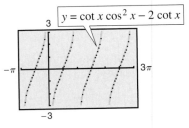

$y = \cot x \cos^2 x - 2 \cot x$

Figure 6.12

Equations of Quadratic Type

Many trigonometric equations are of quadratic type. Here are a few examples.

Quadratic in sin x	*Quadratic in sec x*
$2\sin^2 x - \sin x - 1 = 0$	$\sec^2 x - 3\sec x - 2 = 0$
$2(\sin x)^2 - \sin x - 1 = 0$	$(\sec x)^2 - 3\sec x - 2 = 0$

To solve equations of this type, factor the quadratic or, if factoring is not possible, use the Quadratic Formula.

Example 5   Factoring an Equation of Quadratic Type

Find all solutions of $2 \sin^2 x - \sin x - 1 = 0$ in the interval $[0, 2\pi)$.

Algebraic Solution

Treating the equation as quadratic in $\sin x$ and factoring produces the following.

$$2 \sin^2 x - \sin x - 1 = 0 \quad \text{Write original equation.}$$

$$(2 \sin x + 1)(\sin x - 1) = 0 \quad \text{Factor.}$$

Setting each factor equal to zero, you obtain the following solutions in the interval $[0, 2\pi)$.

$$2 \sin x + 1 = 0 \quad \text{and} \quad \sin x - 1 = 0$$

$$\sin x = -\frac{1}{2} \qquad\qquad \sin x = 1$$

$$x = \frac{7\pi}{6}, \frac{11\pi}{6} \qquad\qquad x = \frac{\pi}{2}$$

✓ *Checkpoint* Now try Exercise 31.

Graphical Solution

Use a graphing utility set in *radian* mode to graph $y = 2 \sin^2 x - \sin x - 1$ for $0 \le x < 2\pi$, as shown in Figure 6.13. Use the *zero* or *root* feature or the *zoom* and *trace* features to approximate the x-intercepts to be

$$x \approx 1.571 \approx \frac{\pi}{2}, \quad x \approx 3.665 \approx \frac{7\pi}{6}, \quad x \approx 5.760 \approx \frac{11\pi}{6}.$$

These values are the approximate solutions of $2 \sin^2 x - \sin x - 1 = 0$.

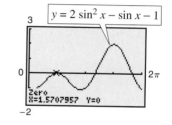

$y = 2 \sin^2 x - \sin x - 1$

Figure 6.13

When working with an equation of quadratic type, be sure that the equation involves a *single* trigonometric function, as shown in the next example.

Example 6   Rewriting with a Single Trigonometric Function

Solve $2 \sin^2 x + 3 \cos x - 3 = 0$.

Solution

Begin by rewriting the equation so that it has only cosine functions.

$$2 \sin^2 x + 3 \cos x - 3 = 0 \qquad \text{Write original equation.}$$

$$2(1 - \cos^2 x) + 3 \cos x - 3 = 0 \qquad \text{Pythagorean identity}$$

$$2 \cos^2 x - 3 \cos x + 1 = 0 \qquad \begin{array}{l}\text{Combine like terms and} \\ \text{multiply each side by } -1.\end{array}$$

$$(2 \cos x - 1)(\cos x - 1) = 0 \qquad \text{Factor.}$$

By setting each factor equal to zero, you can find the solutions in the interval $[0, 2\pi)$ to be $x = 0$, $x = \pi/3$, and $x = 5\pi/3$. Because $\cos x$ has a period of 2π, the general solution is

$$x = 2n\pi, \qquad x = \frac{\pi}{3} + 2n\pi, \qquad x = \frac{5\pi}{3} + 2n\pi \qquad \text{General solution}$$

where n is an integer. The graph of $y = 2 \sin^2 x + 3 \cos x - 3$, shown in Figure 6.14, confirms this result.

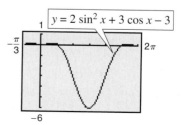

$y = 2 \sin^2 x + 3 \cos x - 3$

Figure 6.14

✓ *Checkpoint* Now try Exercise 33.

Sometimes you must square each side of an equation to obtain a quadratic. Because this procedure can introduce extraneous solutions, you should check any solutions in the original equation to see whether they are valid or extraneous.

Example 7   Squaring and Converting to Quadratic Type

Find all solutions of $\cos x + 1 = \sin x$ in the interval $[0, 2\pi)$.

Solution

It is not clear how to rewrite this equation in terms of a single trigonometric function. See what happens when you square each side of the equation.

$\cos x + 1 = \sin x$	Write original equation.
$\cos^2 x + 2 \cos x + 1 = \sin^2 x$	Square each side.
$\cos^2 x + 2 \cos x + 1 = 1 - \cos^2 x$	Pythagorean identity
$2 \cos^2 x + 2 \cos x = 0$	Combine like terms.
$2 \cos x(\cos x + 1) = 0$	Factor.

Setting each factor equal to zero produces the following.

$$2 \cos x = 0 \quad \text{and} \quad \cos x + 1 = 0$$

$$\cos x = 0 \qquad\qquad \cos x = -1$$

$$x = \frac{\pi}{2}, \frac{3\pi}{2} \qquad\qquad x = \pi$$

Because you squared the original equation, check for extraneous solutions.

Check

$$\cos \frac{\pi}{2} + 1 \overset{?}{=} \sin \frac{\pi}{2} \qquad\qquad \text{Substitute } \pi/2 \text{ for } x.$$

$$0 + 1 = 1 \qquad\qquad \text{Solution checks.} \checkmark$$

$$\cos \frac{3\pi}{2} + 1 \overset{?}{=} \sin \frac{3\pi}{2} \qquad\qquad \text{Substitute } 3\pi/2 \text{ for } x.$$

$$0 + 1 \neq -1 \qquad\qquad \text{Solution does not check.}$$

$$\cos \pi + 1 \overset{?}{=} \sin \pi \qquad\qquad \text{Substitute } \pi \text{ for } x.$$

$$-1 + 1 = 0 \qquad\qquad \text{Solution checks.} \checkmark$$

Of the three possible solutions, $x = 3\pi/2$ is extraneous. So, in the interval $[0, 2\pi)$, the only solutions are $x = \pi/2$ and $x = \pi$. The graph of $y = \cos x + 1 - \sin x$, shown in Figure 6.15, confirms this result because the graph has two x-intercepts (at $x = \pi/2$ and $x = \pi$) in the interval $[0, 2\pi)$.

 **Checkpoint** Now try Exercise 35.

> **Exploration**
>
> Use a graphing utility to confirm the solutions found in Example 7 in two different ways. Do both methods produce the same x-values? Which method do you prefer? Why?
>
> 1. Graph both sides of the equation and find the x-coordinates of the points at which the graphs intersect.
>
>   *Left side:* $y = \cos x + 1$
>
>   *Right side:* $y = \sin x$
>
> 2. Graph the equation $y = \cos x + 1 - \sin x$ and find the x-intercepts of the graph.

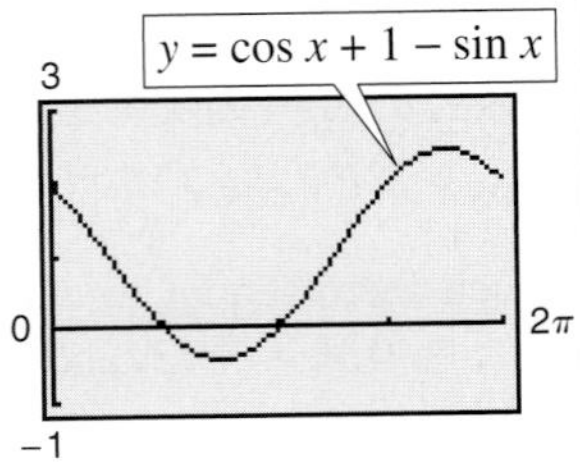

$y = \cos x + 1 - \sin x$

Figure 6.15

Functions Involving Multiple Angles

Example 8   Functions of Multiple Angles

Solve $2 \cos 3t - 1 = 0$.

Solution

$$2 \cos 3t - 1 = 0 \qquad \text{Write original equation.}$$

$$2 \cos 3t = 1 \qquad \text{Add 1 to each side.}$$

$$\cos 3t = \frac{1}{2} \qquad \text{Divide each side by 2.}$$

In the interval $[0, 2\pi)$, you know that $3t = \pi/3$ and $3t = 5\pi/3$ are the only solutions. So in general, you have $3t = \pi/3 + 2n\pi$ and $3t = 5\pi/3 + 2n\pi$. Dividing this result by 3, you obtain the general solution

$$t = \frac{\pi}{9} + \frac{2n\pi}{3} \qquad \text{and} \qquad t = \frac{5\pi}{9} + \frac{2n\pi}{3} \qquad \text{General solution}$$

where n is an integer. This solution is confirmed graphically in Figure 6.16.

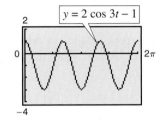

Figure 6.16

✓ *Checkpoint* Now try Exercise 39.

Example 9   Functions of Multiple Angles

Solve $3 \tan \dfrac{x}{2} + 3 = 0$.

Solution

$$3 \tan \frac{x}{2} + 3 = 0 \qquad \text{Write original equation.}$$

$$3 \tan \frac{x}{2} = -3 \qquad \text{Subtract 3 from each side.}$$

$$\tan \frac{x}{2} = -1 \qquad \text{Divide each side by 3.}$$

In the interval $[0, \pi)$, you know that $x/2 = 3\pi/4$ is the only solution. So in general, you have $x/2 = 3\pi/4 + n\pi$. Multiplying this result by 2, you obtain the general solution

$$x = \frac{3\pi}{2} + 2n\pi \qquad \text{General solution}$$

where n is an integer. This solution is confirmed graphically in Figure 6.17.

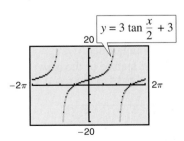

Figure 6.17

✓ *Checkpoint* Now try Exercise 45.

Using Inverse Functions

Example 10   Using Inverse Functions

Find all solutions of $\sec^2 x - 2 \tan x = 4$.

Solution

$$\sec^2 x - 2 \tan x = 4 \qquad \text{Write original equation.}$$

$$1 + \tan^2 x - 2 \tan x - 4 = 0 \qquad \text{Pythagorean identity}$$

$$\tan^2 x - 2 \tan x - 3 = 0 \qquad \text{Combine like terms.}$$

$$(\tan x - 3)(\tan x + 1) = 0 \qquad \text{Factor.}$$

Setting each factor equal to zero, you obtain two solutions in the interval $(-\pi/2, \pi/2)$. [Recall that the range of the inverse tangent function is $(-\pi/2, \pi/2)$.]

$$\tan x = 3 \qquad \text{and} \qquad \tan x = -1$$

$$x = \arctan 3 \qquad x = \arctan(-1) = -\frac{\pi}{4}$$

Finally, because $\tan x$ has a period of π, add multiples of π to obtain

$$x = \arctan 3 + n\pi \qquad \text{and} \qquad x = -\frac{\pi}{4} + n\pi \qquad \text{General solution}$$

where n is an integer. This solution is confirmed graphically in Figure 6.18.

✓ *Checkpoint* Now try Exercise 29.

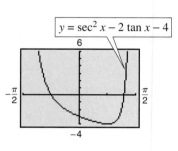

Figure 6.18

With some trigonometric equations, there is no reasonable way to find the solutions algebraically. In such cases, you can still use a graphing utility to approximate the solutions.

Example 11   Approximating Solutions

Approximate the solutions of $x = 2 \sin x$ in the interval $[-\pi, \pi]$.

Solution

Use a graphing utility to graph $y = x - 2 \sin x$ in the interval $[-\pi, \pi]$. Using the *zero* or *root* feature or the *zoom* and *trace* features, you can see that the solutions are $x \approx -1.8955$, $x = 0$, and $x \approx 1.8955$. (See Figure 6.19.)

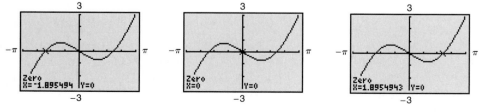

Figure 6.19 $y = x - 2 \sin x$

✓ *Checkpoint* Now try Exercise 59.

Example 12   Surface Area of a Honeycomb

The surface area of a honeycomb is given by the equation

$$S = 6hs + \frac{3}{2}s^2\left(\frac{\sqrt{3} - \cos\theta}{\sin\theta}\right), \qquad 0 < \theta \le 90°$$

where $h = 2.4$ inches, $s = 0.75$ inch, and θ is the angle indicated in Figure 6.20.

a. What value of θ gives a surface area of 12 square inches?
b. What value of θ gives the minimum surface area?

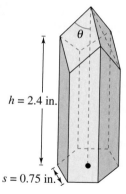

$h = 2.4$ in.

$s = 0.75$ in.

Figure 6.20

Solution

a. Let $h = 2.4$, $s = 0.75$, and $S = 12$.

$$S = 6hs + \frac{3}{2}s^2\left(\frac{\sqrt{3} - \cos\theta}{\sin\theta}\right)$$

$$12 = 6(2.4)(0.75) + \frac{3}{2}(0.75)^2\left(\frac{\sqrt{3} - \cos\theta}{\sin\theta}\right)$$

$$12 = 10.8 + 0.84375\left(\frac{\sqrt{3} - \cos\theta}{\sin\theta}\right)$$

$$0 = 0.84375\left(\frac{\sqrt{3} - \cos\theta}{\sin\theta}\right) - 1.2$$

Using a graphing utility set to *degree* mode, you can graph the function

$$y = 0.84375\left(\frac{\sqrt{3} - \cos x}{\sin x}\right) - 1.2.$$

Using the *zero* or *root* feature or the *zoom* and *trace* features, you can determine that $\theta \approx 49.9°$ and $\theta \approx 59.9°$. (See Figure 6.21.)

TECHNOLOGY SUPPORT

For instructions on how to use the *degree* mode and the *minimum* feature, see Appendix A; for specific keystrokes, go to the text website at *college.hmco.com*.

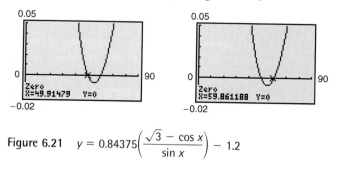

Figure 6.21 $y = 0.84375\left(\dfrac{\sqrt{3} - \cos x}{\sin x}\right) - 1.2$

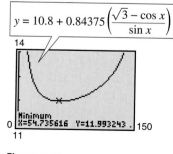

Figure 6.22

b. From part (a), let $h = 2.4$ and $s = 0.75$ to obtain

$$S = 10.8 + 0.84375\left(\frac{\sqrt{3} - \cos\theta}{\sin\theta}\right).$$

Graph this function using a graphing utility set to *degree* mode. Use the *minimum* feature or the *zoom* and *trace* features to approximate the minimum point on the graph, which occurs at $\theta \approx 54.7°$, as shown in Figure 6.22.

✓ *Checkpoint* Now try Exercise 63.

STUDY TIP

By using calculus, it can be shown that the exact minimum value is

$$\theta = \arccos\left(\frac{1}{\sqrt{3}}\right) \approx 54.7356°.$$

6.3 Exercises

Vocabulary Check

Fill in the blanks.

1. The equation $2 \cos x - 1 = 0$ has the solutions $x = \dfrac{\pi}{3} + 2n\pi$ and $x = \dfrac{5\pi}{3} + 2n\pi$, which are called _____ solutions.

2. The equation $\tan^2 x - 5 \tan x + 6 = 0$ is an equation of _____ type.

3. A solution to an equation that does not satisfy the original equation is called an _____ solution.

In Exercises 1–6, verify that each x-value is a solution of the equation.

1. $2 \cos x - 1 = 0$

    (a) $x = \dfrac{\pi}{3}$ (b) $x = \dfrac{5\pi}{3}$

2. $\sec x - 2 = 0$

    (a) $x = \dfrac{\pi}{3}$ (b) $x = \dfrac{5\pi}{3}$

3. $3 \tan^2 2x - 1 = 0$

    (a) $x = \dfrac{\pi}{12}$ (b) $x = \dfrac{5\pi}{12}$

4. $4 \cos^2 2x - 2 = 0$

    (a) $x = \dfrac{\pi}{8}$ (b) $x = \dfrac{7\pi}{8}$

5. $2 \sin^2 x - \sin x - 1 = 0$

    (a) $x = \dfrac{\pi}{2}$ (b) $x = \dfrac{7\pi}{6}$

6. $\sec^4 x - 3 \sec^2 x - 4 = 0$

    (a) $x = \dfrac{2\pi}{3}$ (b) $x = \dfrac{5\pi}{3}$

In Exercises 7–16, solve the equation.

7. $2 \cos x + 1 = 0$ 8. $\sqrt{2} \sin x + 1 = 0$

9. $\sqrt{3} \sec x - 2 = 0$

10. $\cot x + 1 = 0$

11. $3 \csc^2 x - 4 = 0$

12. $3 \cot^2 x - 1 = 0$

13. $4 \cos^2 x - 1 = 0$

14. $\cos x (\cos x - 1) = 0$

15. $\sin^2 x = 3 \cos^2 x$

16. $(3 \tan^2 x - 1)(\tan^2 x - 3) = 0$

In Exercises 17–30, find all solutions of the equation in the interval $[0, 2\pi)$ algebraically. Use the *table* feature of a graphing utility to check your answers numerically.

17. $\tan x + \sqrt{3} = 0$ 18. $2 \sin x + 1 = 0$

19. $\csc^2 x - 2 = 0$ 20. $\tan^2 x - 1 = 0$

21. $3 \tan^3 x = \tan x$

22. $2 \sin^2 x = 2 + \cos x$

23. $\sec^2 x - \sec x = 2$

24. $\sec x \csc x = 2 \csc x$

25. $2 \sin x + \csc x = 0$

26. $\sec x + \tan x = 1$

27. $\cos x + \sin x \tan x = 2$

28. $\sin^2 x + \cos x + 1 = 0$

29. $\sec^2 x + \tan x = 3$

30. $2 \cos 2x + \cos x - 1 = 0$

In Exercises 31–38, use a graphing utility to approximate the solutions of the equation in the interval $[0, 2\pi)$ by setting the equation equal to 0, graphing the new equation, and using the *zero* or *root* feature to approximate the x-intercepts of the graph.

31. $2 \sin^2 x + 3 \sin x + 1 = 0$

32. $2 \sec^2 x + \tan^2 x - 3 = 0$

33. $4 \sin^2 x = 2 \cos x + 1$

34. $\csc^2 x = 3 \csc x + 4$

35. $\csc x + \cot x = 1$

36. $4 \sin x = \cos x - 2$

37. $\dfrac{\cos x \cot x}{1 - \sin x} = 3$

38. $\dfrac{1 + \sin x}{\cos x} + \dfrac{\cos x}{1 + \sin x} = 4$

In Exercises 39–46, solve the multiple angle equation.

39. $\sin 2x = -\dfrac{\sqrt{3}}{2}$

40. $\sec 4x = 2$

41. $2 \sin^2 2x = 1$

42. $\tan^2 3x = 3$

43. $\tan 3x(\tan x - 1) = 0$

44. $\cos 2x(2 \cos x + 1) = 0$

45. $\cos \dfrac{x}{2} = \dfrac{\sqrt{2}}{2}$

46. $\tan \dfrac{x}{3} = 1$

In Exercises 47–50, approximate the x-intercepts of the graph. Use a graphing utility to check your solutions.

47. $y = \sin \dfrac{\pi x}{2} + 1$

48. $y = \sin \pi x + \cos \pi x$

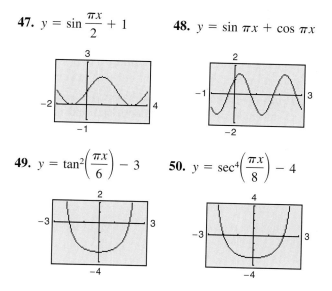

49. $y = \tan^2 \left(\dfrac{\pi x}{6} \right) - 3$

50. $y = \sec^4 \left(\dfrac{\pi x}{8} \right) - 4$

In Exercises 51–58, use a graphing utility to approximate the solutions of the equation in the interval $[0, 2\pi)$.

51. $2 \cos x - \sin x = 0$

52. $2 \sin x + \cos x = 0$

53. $x \tan x - 1 = 0$

54. $2x \sin x - 2 = 0$

55. $\sec^2 x + 0.5 \tan x = 1$

56. $\csc^2 x + 0.5 \cot x = 5$

57. $12 \sin^2 x - 13 \sin x + 3 = 0$

58. $3 \tan^2 x + 4 \tan x - 4 = 0$

In Exercises 59–62, use a graphing utility to approximate the solutions (to three decimal places) of the equation in the given interval.

59. $3 \tan^2 x + 5 \tan x - 4 = 0$, $\left[-\dfrac{\pi}{2}, \dfrac{\pi}{2} \right]$

60. $\cos^2 x - 2 \cos x - 1 = 0$, $[0, \pi]$

61. $4 \cos^2 x - 2 \sin x + 1 = 0$, $\left[-\dfrac{\pi}{2}, \dfrac{\pi}{2} \right]$

62. $2 \sec^2 x + \tan x - 6 = 0$, $\left[-\dfrac{\pi}{2}, \dfrac{\pi}{2} \right]$

In Exercises 63 and 64, (a) use a graphing utility to graph the function and approximate the maximum and minimum points of the graph in the interval $[0, 2\pi]$, and (b) solve the trigonometric equation and verify that the x-coordinates of the maximum and minimum points of f are among its solutions (the trigonometric equation is found using calculus).

	Function	*Trigonometric Equation*
63.	$f(x) = \sin x + \cos x$	$\cos x - \sin x = 0$
64.	$f(x) = 2 \sin x + \cos 2x$	$2 \cos x - 4 \sin x \cos x = 0$

Fixed Point   In Exercises 65 and 66, find the smallest positive fixed point of the function f. [A *fixed point* of a function f is a real number c such that $f(c) = c$.]

65. $f(x) = \tan \dfrac{\pi x}{4}$

66. $f(x) = \cos x$

67. *Graphical Reasoning*   Consider the function

$$f(x) = \cos \dfrac{1}{x}$$

and its graph shown in the figure.

(a) What is the domain of the function?

(b) Identify any symmetry or asymptotes of the graph.

(c) Describe the behavior of the function as $x \to 0$.

(d) How many solutions does the equation

$$\cos \dfrac{1}{x} = 0$$

have in the interval $[-1, 1]$? Find the solutions.

(e) Does the equation $\cos(1/x) = 0$ have a greatest solution? If so, approximate the solution. If not, explain.

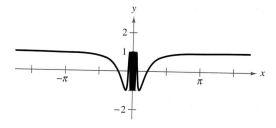

68. *Graphical Reasoning*   Consider the function

$$f(x) = \frac{\sin x}{x}$$

and its graph shown in the figure.

(a) What is the domain of the function?

(b) Identify any symmetry or asymptotes of the graph.

(c) Describe the behavior of the function as $x \to 0$.

(d) How many solutions does the equation

$$\frac{\sin x}{x} = 0$$

have in the interval $[-8, 8]$? Find the solutions.

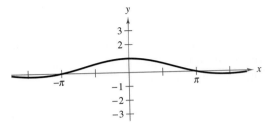

69. *Harmonic Motion*   A weight is oscillating on the end of a spring (see figure). The position of the weight relative to the point of equilibrium is given by

$$y = \frac{1}{12}(\cos 8t - 3 \sin 8t)$$

where y is the displacement (in meters) and t is the time (in seconds). Find the times when the weight is at the point of equilibrium $(y = 0)$ for $0 \le t \le 1$.

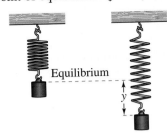

Equilibrium

70. *Damped Harmonic Motion*   The displacement from equilibrium of a weight oscillating on the end of a spring is given by $y = 1.56e^{-0.22t} \cos 4.9t$, where y is the displacement (in feet) and t is the time (in seconds). Use a graphing utility to graph the displacement function for $0 \le t \le 10$. Find the time beyond which the displacement does not exceed 1 foot from equilibrium.

71. *Sales*   The monthly sales S (in thousands of units) of lawn mowers are approximated by

$$S = 74.50 - 43.75 \cos \frac{\pi t}{6}$$

where t is the time (in months), with $t = 1$ corresponding to January. Determine the months during which sales exceed 100,000 units.

72. *Position of the Sun*   Cheyenne, Wyoming has a latitude of $41°$ N. At this latitude, the position of the sun at sunrise can be modeled by

$$D = 31 \sin\left(\frac{2\pi}{365} t - 1.4\right)$$

where t is the time (in days) and $t = 1$ represents January 1. In this model, D represents the number of degrees north or south of due east that the sun rises. Use a graphing utility to determine the days that the sun is more than $20°$ north of due east at sunrise.

73. *Projectile Motion*   A batted baseball leaves the bat at an angle of θ with the horizontal and an initial velocity of $v_0 = 100$ feet per second. The ball is caught by an outfielder 300 feet from home plate (see figure). Find θ if the range r of a projectile is given by

$$r = \frac{1}{32}v_0^2 \sin 2\theta.$$

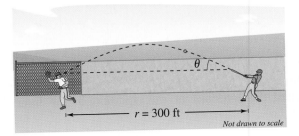

$r = 300$ ft

Not drawn to scale

74. *Area*   The area of a rectangle inscribed in one arch of the graph of $y = \cos x$ (see figure) is given by

$$A = 2x \cos x, \quad 0 \le x \le \frac{\pi}{2}.$$

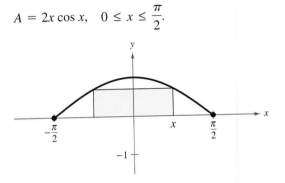

(a) Use a graphing utility to graph the area function, and approximate the area of the largest inscribed rectangle.

(b) Determine the values of x for which $A \geq 1$.

75. *Data Analysis* The table shows the unemployment rate r for the years 1990 through 2001 in the United States. The time t is measured in years, with $t = 0$ corresponding to 1990. (Source: U.S. Bureau of Labor Statistics)

Time, t	Rate, r
0	5.6
1	6.8
2	7.5
3	6.9
4	6.1
5	5.6
6	5.4
7	4.9
8	4.5
9	4.2
10	4.0
11	4.8

(a) Use a graphing utility to create a scatter plot of the data.

(b) A model for the data is given by $r = 1.39 \sin (0.48t + 0.42) + 5.51$. Graph the model with the scatter plot in part (a). Is the model a good fit for the original data? Explain.

(c) What term in the model gives the average unemployment rate? What is the rate?

(d) Economists study the lengths of business cycles, such as cycles in the unemployment rate. Based on this short span of time, use the model to give the length of this cycle.

(e) Use the model to estimate the next time the unemployment rate will be 6.5% or less.

76. *Quadratic Approximation* Consider the function

$f(x) = 3 \sin(0.6x - 2)$.

(a) Approximate the zero of the function in the interval $[0, 6]$.

(b) A quadratic approximation agreeing with f at $x = 5$ is

$g(x) = -0.45x^2 + 5.52x - 13.70.$

Use a graphing utility to graph f and g in the same viewing window. Describe the result.

(c) Use the Quadratic Formula to find the zeros of g. Compare the zero in the interval $[0, 6]$ with the result of part (a).

Synthesis

True or False? In Exercises 77–79, determine whether the statement is true or false. Justify your answer.

77. All trigonometric equations have either an infinite number of solutions or no solution.

78. The solutions of any trigonometric equation can always be found from its solutions in the interval $[0, 2\pi)$.

79. If you correctly solve a trigonometric equation down to the statement $\sin x = 3.4$, then you can finish solving the equation by using an inverse trigonometric function.

80. *Writing* Describe the difference between verifying an identity and solving an equation.

Review

In Exercises 81–84, convert the angle measure from degrees to radians. Round your answer to three decimal places.

81. $124°$

82. $486°$

83. $-0.41°$

84. $-210.55°$

In Exercises 85 and 86, solve for x.

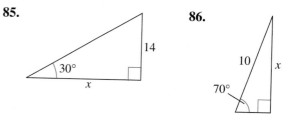

85.

14 / 30° / x

86.

10 / x / 70°

87. *Distance* From the 100-foot roof of a condominium on the coast, a tourist sights a cruise ship. The angle of depression is 2.5°. How far is the ship from the shoreline?

6.4   Sum and Difference Formulas

Using Sum and Difference Formulas

In this section and the following section, you will study the uses of several trigonometric identities and formulas. See Appendix B for proofs of these formulas.

Sum and Difference Formulas

$$\sin(u + v) = \sin u \cos v + \cos u \sin v \qquad \tan(u + v) = \frac{\tan u + \tan v}{1 - \tan u \tan v}$$

$$\sin(u - v) = \sin u \cos v - \cos u \sin v$$

$$\cos(u + v) = \cos u \cos v - \sin u \sin v \qquad \tan(u - v) = \frac{\tan u - \tan v}{1 + \tan u \tan v}$$

$$\cos(u - v) = \cos u \cos v + \sin u \sin v$$

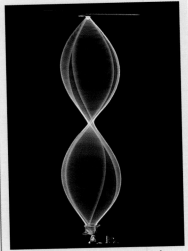

Richard Megna/Fundamental Photographs

Exploration

Use a graphing utility to graph $y_1 = \cos(x + 2)$ and $y_2 = \cos x + \cos 2$ in
the same viewing window. What can you conclude about the graphs? Is it
true that $\cos(x + 2) = \cos x + \cos 2$?

Use a graphing utility to graph $y_1 = \sin(x + 4)$ and $y_2 = \sin x + \sin 4$ in
the same viewing window. What can you conclude about the graphs? Is it
true that $\sin(x + 4) = \sin x + \sin 4$?

Examples 1 and 2 show how **sum and difference formulas** can be used to
find exact values of trigonometric functions involving sums or differences of
special angles.

Example 1   Evaluating a Trigonometric Function

Find the exact value of $\cos 75°$.

Solution

To find the exact value of $\cos 75°$, use the fact that $75° = 30° + 45°$.
Consequently, the formula for $\cos(u + v)$ yields

$$\cos 75° = \cos(30° + 45°)$$

$$= \cos 30° \cos 45° - \sin 30° \sin 45°$$

$$= \frac{\sqrt{3}}{2}\left(\frac{\sqrt{2}}{2}\right) - \frac{1}{2}\left(\frac{\sqrt{2}}{2}\right)$$

$$= \frac{\sqrt{6} - \sqrt{2}}{4}.$$

Try checking this result on your calculator. You will find that $\cos 75° \approx 0.259$.

✓ *Checkpoint* Now try Exercise 1.

Example 2 Evaluating a Trigonometric Function

Find the exact value of $\sin \dfrac{\pi}{12}$.

Solution

Using the fact that $\pi/12 = \pi/3 - \pi/4$ together with the formula for $\sin(u - v)$, you obtain

$$\sin \frac{\pi}{12} = \sin\left(\frac{\pi}{3} - \frac{\pi}{4}\right) = \sin \frac{\pi}{3} \cos \frac{\pi}{4} - \cos \frac{\pi}{3} \sin \frac{\pi}{4}$$

$$= \frac{\sqrt{3}}{2}\left(\frac{\sqrt{2}}{2}\right) - \frac{1}{2}\left(\frac{\sqrt{2}}{2}\right) = \frac{\sqrt{6} - \sqrt{2}}{4}.$$

✓ *Checkpoint* Now try Exercise 3.

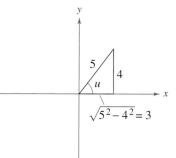

Figure 6.23

Example 3 Evaluating a Trigonometric Expression

Find the exact value of $\sin(u + v)$ given

$$\sin u = \frac{4}{5}, \text{ where } 0 < u < \frac{\pi}{2} \quad \text{and} \quad \cos v = -\frac{12}{13}, \text{ where } \frac{\pi}{2} < v < \pi.$$

Solution

Because $\sin u = 4/5$ and u is in Quadrant I, $\cos u = 3/5$, as shown in Figure 6.23. Because $\cos v = -12/13$ and v is in Quadrant II, $\sin v = 5/13$, as shown in Figure 6.24. You can find $\sin(u + v)$ as follows.

$$\sin(u + v) = \sin u \cos v + \cos u \sin v$$

$$= \left(\frac{4}{5}\right)\left(-\frac{12}{13}\right) + \left(\frac{3}{5}\right)\left(\frac{5}{13}\right) = -\frac{48}{65} + \frac{15}{65} = -\frac{33}{65}$$

✓ *Checkpoint* Now try Exercise 35.

Figure 6.24

Example 4 An Application of a Sum Formula

Write $\cos(\arctan 1 + \arccos x)$ as an algebraic expression.

Solution

This expression fits the formula for $\cos(u + v)$. Angles $u = \arctan 1$ and $v = \arccos x$ are shown in Figure 6.25.

$$\cos(u + v) = \cos(\arctan 1)\cos(\arccos x) - \sin(\arctan 1)\sin(\arccos x)$$

$$= \frac{1}{\sqrt{2}} \cdot x - \frac{1}{\sqrt{2}} \cdot \sqrt{1 - x^2} = \frac{x - \sqrt{1 - x^2}}{\sqrt{2}}.$$

✓ *Checkpoint* Now try Exercise 43.

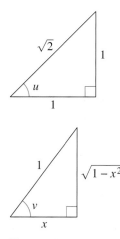

Figure 6.25

Example 5 Proving a Cofunction Identity

Prove the cofunction identity $\cos\left(\dfrac{\pi}{2} - x\right) = \sin x$.

Solution
Using the formula for $\cos(u - v)$, you have

$$\cos\left(\frac{\pi}{2} - x\right) = \cos\frac{\pi}{2}\cos x + \sin\frac{\pi}{2}\sin x$$

$$= (0)(\cos x) + (1)(\sin x)$$

$$= \sin x.$$

✓ *Checkpoint* Now try Exercise 47.

Sum and difference formulas can be used to derive **reduction formulas** involving expressions such as

$$\sin\left(\theta + \frac{n\pi}{2}\right) \quad \text{and} \quad \cos\left(\theta + \frac{n\pi}{2}\right), \text{ where } n \text{ is an integer.}$$

Example 6 Deriving Reduction Formulas

Simplify each expression.

a. $\cos\left(\theta - \dfrac{3\pi}{2}\right)$ **b.** $\tan(\theta + 3\pi)$

Solution
a. Using the formula for $\cos(u - v)$, you have

$$\cos\left(\theta - \frac{3\pi}{2}\right) = \cos\theta\cos\frac{3\pi}{2} + \sin\theta\sin\frac{3\pi}{2}$$

$$= (\cos\theta)(0) + (\sin\theta)(-1)$$

$$= -\sin\theta.$$

b. Using the formula for $\tan(u + v)$, you have

$$\tan(\theta + 3\pi) = \frac{\tan\theta + \tan 3\pi}{1 - \tan\theta\tan 3\pi}$$

$$= \frac{\tan\theta + 0}{1 - (\tan\theta)(0)}$$

$$= \tan\theta.$$

Note that the period of $\tan\theta$ is π, so the period of $\tan(\theta + 3\pi)$ is the same as the period of $\tan\theta$.

✓ *Checkpoint* Now try Exercise 51.

Example 7   Solving a Trigonometric Equation

Find all solutions of $\sin\left(x + \frac{\pi}{4}\right) + \sin\left(x - \frac{\pi}{4}\right) = -1$ in the interval $[0, 2\pi)$.

Algebraic Solution

Using sum and difference formulas, rewrite the equation as

$$\sin x \cos \frac{\pi}{4} + \cos x \sin \frac{\pi}{4} + \sin x \cos \frac{\pi}{4} - \cos x \sin \frac{\pi}{4} = -1$$

$$2 \sin x \cos \frac{\pi}{4} = -1$$

$$2(\sin x)\left(\frac{\sqrt{2}}{2}\right) = -1$$

$$\sin x = -\frac{1}{\sqrt{2}}$$

$$\sin x = -\frac{\sqrt{2}}{2}.$$

So, the only solutions in the interval $[0, 2\pi)$ are

$$x = \frac{5\pi}{4} \qquad \text{and} \qquad x = \frac{7\pi}{4}.$$

✓ *Checkpoint* Now try Exercise 55.

Graphical Solution

Use a graphing utility set in *radian* mode to graph $y = \sin\left(x + \frac{\pi}{4}\right) + \sin\left(x - \frac{\pi}{4}\right) + 1$, as shown in Figure 6.26. Use the *zero* or *root* feature or the *zoom* and *trace* features to approximate the *x*-intercepts in the interval $[0, 2\pi)$ to be

$$x \approx 3.927 \approx \frac{5\pi}{4} \text{ and } x \approx 5.498 \approx \frac{7\pi}{4}.$$

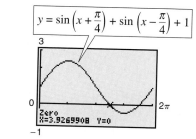

Figure 6.26

The next example was taken from calculus. It is used to derive the formula for the derivative of the sine function.

Example 8   An Application from Calculus   ∫ algebra of calculus

Verify $\dfrac{\sin(x + h) - \sin x}{h} = (\cos x)\left(\dfrac{\sin h}{h}\right) - (\sin x)\left(\dfrac{1 - \cos h}{h}\right), h \neq 0.$

Solution

Using the formula for $\sin(u + v)$, you have

$$\frac{\sin(x + h) - \sin x}{h} = \frac{\sin x \cos h + \cos x \sin h - \sin x}{h}$$

$$= \frac{\cos x \sin h - \sin x(1 - \cos h)}{h}$$

$$= (\cos x)\left(\frac{\sin h}{h}\right) - (\sin x)\left(\frac{1 - \cos h}{h}\right).$$

✓ *Checkpoint* Now try Exercise 77.

TECHNOLOGY SUPPORT

For instructions on how to use the *zero* or *root* feature and the *zoom* and *trace* features, see Appendix A; for specific keystrokes, go to the text website at *college.hmco.com*.

6.4 Exercises

Vocabulary Check

Fill in the blank to complete the trigonometric formula.

1. $\sin(u - v) = $ _____

2. $\cos(u + v) = $ _____

3. $\tan(u + v) = $ _____

4. $\sin(u + v) = $ _____

5. $\cos(u - v) = $ _____

6. $\tan(u - v) = $ _____

In Exercises 1–6, find the exact value of each expression.

1. (a) $\cos(240° - 0°)$ (b) $\cos 240° - \cos 0°$

2. (a) $\sin(390° + 120°)$ (b) $\sin 390° + \sin 120°$

3. (a) $\cos\left(\dfrac{\pi}{6} + \dfrac{\pi}{3}\right)$ (b) $\cos\dfrac{\pi}{6} + \cos\dfrac{\pi}{3}$

4. (a) $\sin\left(\dfrac{2\pi}{3} + \dfrac{3\pi}{4}\right)$ (b) $\sin\dfrac{2\pi}{3} + \sin\dfrac{3\pi}{4}$

5. (a) $\sin(315° - 60°)$ (b) $\sin 315° - \sin 60°$

6. (a) $\sin\left(\dfrac{7\pi}{6} - \dfrac{\pi}{3}\right)$ (b) $\sin\dfrac{7\pi}{6} - \sin\dfrac{\pi}{3}$

In Exercises 7–22, find the exact values of the sine, cosine, and tangent of the angle.

7. $105° = 60° + 45°$

8. $165° = 135° + 30°$

9. $195° = 225° - 30°$

10. $285° = 330° - 45°$

11. $\dfrac{11\pi}{12} = \dfrac{3\pi}{4} + \dfrac{\pi}{6}$

12. $\dfrac{17\pi}{12} = \dfrac{7\pi}{6} + \dfrac{\pi}{4}$

13. $-\dfrac{\pi}{12} = \dfrac{\pi}{6} - \dfrac{\pi}{4}$

14. $-\dfrac{19\pi}{12} = \dfrac{2\pi}{3} - \dfrac{9\pi}{4}$

15. $75°$

16. $15°$

17. $-225°$

18. $-105°$

19. $\dfrac{13\pi}{12}$

20. $\dfrac{5\pi}{12}$

21. $-\dfrac{7\pi}{12}$

22. $-\dfrac{13\pi}{12}$

In Exercises 23–30, write the expression as the sine, cosine, or tangent of an angle.

23. $\cos 60° \cos 10° - \sin 60° \sin 10°$

24. $\sin 110° \cos 80° + \cos 110° \sin 80°$

25. $\dfrac{\tan 325° - \tan 86°}{1 + \tan 325° \tan 86°}$

26. $\dfrac{\tan 152° - \tan 47°}{1 + \tan 152° \tan 47°}$

27. $\sin 3.5 \cos 1.2 - \cos 3.5 \sin 1.2$

28. $\cos 0.88 \cos 0.34 + \sin 0.88 \sin 0.34$

29. $\cos\dfrac{\pi}{7}\cos\dfrac{\pi}{5} - \sin\dfrac{\pi}{7}\sin\dfrac{\pi}{5}$

30. $\sin\dfrac{4\pi}{9}\cos\dfrac{\pi}{8} + \cos\dfrac{4\pi}{9}\sin\dfrac{\pi}{8}$

Numerical, Graphical, and Algebraic Analysis **In Exercises 31–34, use a graphing utility to complete the table and graph the two functions in the same viewing window. Use both the table and the graph as evidence that $y_1 = y_2$. Then verify the identity algebraically.**

x	0.2	0.4	0.6	0.8	1.0	1.2	1.4
y_1							
y_2							

31. $y_1 = \sin\left(\dfrac{\pi}{6} + x\right)$, $y_2 = \dfrac{1}{2}\left(\cos x + \sqrt{3}\sin x\right)$

32. $y_1 = \cos\left(\dfrac{5\pi}{4} - x\right)$, $y_2 = -\dfrac{\sqrt{2}}{2}\left(\cos x + \sin x\right)$

33. $y_1 = \cos(x + \pi)\cos(x - \pi)$, $y_2 = \cos^2 x$

34. $y_1 = \sin(x + \pi)\sin(x - \pi)$, $y_2 = \sin^2 x$

In Exercises 35–38, find the exact value of the trigonometric function given that $\sin u = \dfrac{5}{13}$ and $\cos v = -\dfrac{3}{5}$. (Both u and v are in Quadrant II.)

35. $\sin(u + v)$

36. $\cos(v - u)$

37. $\tan(u + v)$

38. $\sin(u - v)$

In Exercises 39–42, find the exact value of the trigonometric function given that $\sin u = -\dfrac{7}{25}$ and $\cos v = -\dfrac{4}{5}$. (Both u and v are in Quadrant III.)

39. $\cos(u + v)$

40. $\tan(u + v)$

41. $\sin(v - u)$

42. $\cos(u - v)$

In Exercises 43–46, write the trigonometric expression as an algebraic expression.

43. $\sin(\arcsin x + \arccos x)$

44. $\cos(\arccos x - \arcsin x)$

45. $\sin(\arctan 2x - \arccos x)$

46. $\cos(\arcsin x - \arctan 2x)$

In Exercises 47–54, verify the identity.

47. $\sin\left(\dfrac{\pi}{2} + x\right) = \cos x$ **48.** $\sin(3\pi - x) = \sin x$

49. $\tan(x + \pi) - \tan(\pi - x) = 2 \tan x$

50. $\tan\left(\dfrac{\pi}{4} - \theta\right) = \dfrac{1 - \tan \theta}{1 + \tan \theta}$

51. $\sin(x + y) + \sin(x - y) = 2 \sin x \cos y$

52. $\cos(x + y) + \cos(x - y) = 2 \cos x \cos y$

53. $\cos(x + y) \cos(x - y) = \cos^2 x - \sin^2 y$

54. $\sin(x + y) \sin(x - y) = \sin^2 x - \sin^2 y$

In Exercises 55–58, find the solutions of the equation in the interval $[0, 2\pi)$. Use a graphing utility to verify your results.

55. $\sin\left(x + \dfrac{\pi}{3}\right) + \sin\left(x - \dfrac{\pi}{3}\right) = 1$

56. $\cos\left(x + \dfrac{\pi}{4}\right) - \cos\left(x - \dfrac{\pi}{4}\right) = 1$

57. $\tan(x + \pi) + 2 \sin(x + \pi) = 0$

58. $2 \sin\left(x + \dfrac{\pi}{2}\right) + 3 \tan(\pi - x) = 0$

In Exercises 59–62, use a graphing utility to approximate the solutions of the equation in the interval $[0, 2\pi)$.

59. $\cos\left(x + \dfrac{\pi}{4}\right) + \cos\left(x - \dfrac{\pi}{4}\right) = 1$

60. $\sin\left(x + \dfrac{\pi}{2}\right) - \cos\left(x + \dfrac{3\pi}{2}\right) = 0$

61. $\tan(x + \pi) - \cos\left(x + \dfrac{\pi}{2}\right) = 0$

62. $\tan(\pi - x) + 2 \cos\left(x + \dfrac{3\pi}{2}\right) = 0$

63. *Standing Waves*   The equation of a standing wave is obtained by adding the displacements of two waves traveling in opposite directions (see figure).

Assume that each of the waves has amplitude A, period T, and wavelength λ. If the models for these waves are

$$y_1 = A \cos 2\pi\left(\dfrac{t}{T} - \dfrac{x}{\lambda}\right) \text{ and } y_2 = A \cos 2\pi\left(\dfrac{t}{T} + \dfrac{x}{\lambda}\right)$$

show that

$$y_1 + y_2 = 2A \cos \dfrac{2\pi t}{T} \cos \dfrac{2\pi x}{\lambda}.$$

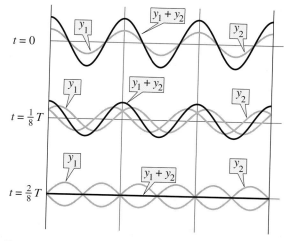

Figure for 63

64. *Harmonic Motion*   A weight is attached to a spring suspended vertically from a ceiling. When a driving force is applied to the system, the weight moves vertically from its equilibrium position, and this motion is modeled by

$$y = \tfrac{1}{3} \sin 2t + \tfrac{1}{4} \cos 2t$$

where y is the distance from equilibrium (in feet) and t is the time (in seconds).

(a) Use a graphing utility to graph the model.

(b) Use the identity

$$a \sin B\theta + b \cos B\theta = \sqrt{a^2 + b^2} \sin(B\theta + C)$$

where $C = \arctan(b/a)$, $a > 0$, to write the model in the form

$$y = \sqrt{a^2 + b^2} \sin(Bt + C).$$

Use a graphing utility to verify your result.

(c) Find the amplitude of the oscillations of the weight.

(d) Find the frequency of the oscillations of the weight.

Synthesis

True or False? In Exercises 65 and 66, determine whether the statement is true or false. Justify your answer.

65. $\cos(u \pm v) = \cos u \pm \cos v$

66. $\sin\left(x - \dfrac{11\pi}{2}\right) = \cos x$

In Exercises 67–70, verify the identity.

67. $\cos(n\pi + \theta) = (-1)^n \cos \theta$, n is an integer.

68. $\sin(n\pi + \theta) = (-1)^n \sin \theta$, n is an integer.

69. $a \sin B\theta + b \cos B\theta = \sqrt{a^2 + b^2}\, \sin(B\theta + C)$, where $C = \arctan(b/a)$ and $a > 0$.

70. $a \sin B\theta + b \cos B\theta = \sqrt{a^2 + b^2}\, \cos(B\theta - C)$, where $C = \arctan(a/b)$ and $b > 0$.

In Exercises 71–74, use the formulas given in Exercises 69 and 70 to write the expression in the following forms. Use a graphing utility to verify your results.

(a) $\sqrt{a^2 + b^2}\, \sin(B\theta + C)$

(b) $\sqrt{a^2 + b^2}\, \cos(B\theta - C)$

71. $\sin \theta + \cos \theta$

72. $3 \sin 2\theta + 4 \cos 2\theta$

73. $12 \sin 3\theta + 5 \cos 3\theta$

74. $\sin 2\theta - \cos 2\theta$

In Exercises 75 and 76, use the formulas given in Exercises 69 and 70 to write the trigonometric expression in the form $a \sin B\theta + b \cos B\theta$.

75. $2 \sin\left(\theta + \dfrac{\pi}{2}\right)$

76. $5 \cos\left(\theta + \dfrac{\pi}{4}\right)$

77. Verify the following identity used in calculus.

$$\frac{\cos(x + h) - \cos x}{h} = \frac{\cos x(\cos h - 1)}{h} - \frac{\sin x \sin h}{h}$$

78. **Exploration** Let $x = \pi/6$ in the identity in Exercise 77 and define the functions f and g as follows.

$$f(h) = \frac{\cos(\pi/6 + h) - \cos(\pi/6)}{h}$$

$$g(h) = \cos\frac{\pi}{6}\left(\frac{\cos h - 1}{h}\right) - \sin\frac{\pi}{6}\left(\frac{\sin h}{h}\right)$$

(a) What are the domains of the functions f and g?

(b) Use a graphing utility to complete the table.

h	0.01	0.02	0.05	0.1	0.2	0.5
$f(h)$						
$g(h)$						

(c) Use a graphing utility to graph the functions f and g.

(d) Use the table and graph to make a conjecture about the values of the functions f and g as $h \to 0$.

79. **Conjecture** Three squares of side s are placed side by side (see figure). Make a conjecture about the relationship between the sum $u + v$ and w. Prove your conjecture by using the identity for the tangent of the sum of two angles.

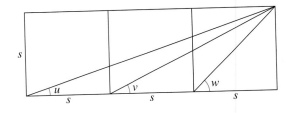

80. (a) Write a sum formula for $\sin(u + v + w)$.

(b) Write a sum formula for $\tan(u + v + w)$.

Review

In Exercises 81–84, find the x- and y-intercepts of the graph of the equation. Use a graphing utility to verify your results.

81. $y = -\frac{1}{2}(x - 10) + 14$

82. $y = x^2 - 3x - 40$

83. $y = |2x - 9| - 5$

84. $y = 2x\sqrt{x + 7}$

In Exercises 85–88, evaluate the expression without using a calculator.

85. $\arccos\left(\dfrac{\sqrt{3}}{2}\right)$

86. $\arctan\left(-\sqrt{3}\right)$

87. $\sin^{-1} 1$

88. $\tan^{-1} 0$

6.5   Multiple-Angle and Product-to-Sum Formulas

Multiple-Angle Formulas

What you should learn

• Use multiple-angle formulas to rewrite and evaluate trigonometric functions.
• Use power-reducing formulas to rewrite and evaluate trigonometric functions.
• Use half-angle formulas to rewrite and evaluate trigonometric functions.
• Use product-to-sum and sum-to-product formulas to rewrite and evaluate trigonometric functions.

In this section you will study four additional categories of trigonometric identities.

1. The first category involves *functions of multiple angles* such as $\sin ku$ and $\cos ku$.

2. The second category involves *squares of trigonometric functions* such as $\sin^2 u$.

3. The third category involves *functions of half-angles* such as $\sin(u/2)$.

4. The fourth category involves *products of trigonometric functions* such as $\sin u \cos v$.

You should learn the **double-angle formulas** below because they are used most often. See Appendix B for proofs of the double-angle formulas.

Why you should learn it

You can use a variety of trigonometric formulas to rewrite trigonometric functions in more convenient forms. For instance, Exercise 115 on page 524 shows you how to use a half-angle formula to determine the apex angle of a sound wave cone caused by the speed of an airplane.

Double-Angle Formulas

$$\sin 2u = 2 \sin u \cos u \qquad \cos 2u = \cos^2 u - \sin^2 u$$

$$\tan 2u = \frac{2 \tan u}{1 - \tan^2 u} \qquad\quad = 2 \cos^2 u - 1$$

$$= 1 - 2 \sin^2 u$$

NASA-Liaison Agency

Example 1   Solving a Multiple-Angle Equation

Solve $2 \cos x + \sin 2x = 0$.

Solution

Begin by rewriting the equation so that it involves functions of x (rather than $2x$). Then factor and solve as usual.

$2 \cos x + \sin 2x = 0$	Write original equation.
$2 \cos x + 2 \sin x \cos x = 0$	Double-angle formula
$2 \cos x(1 + \sin x) = 0$	Factor.
$\cos x = 0 \qquad\qquad 1 + \sin x = 0$	Set factors equal to zero.
$x = \dfrac{\pi}{2}, \dfrac{3\pi}{2} \qquad\qquad x = \dfrac{3\pi}{2}$	Solutions in $[0, 2\pi)$

So, the general solution is

$$x = \frac{\pi}{2} + 2n\pi \qquad \text{and} \qquad x = \frac{3\pi}{2} + 2n\pi \qquad \text{General solution}$$

where n is an integer. Try verifying this solution graphically.

✓ *Checkpoint*   Now try Exercise 9.

Example 2   Using Double-Angle Formulas to Analyze Graphs

Analyze the graph of $y = 4 \cos^2 x - 2$ in the interval $[0, 2\pi]$.

Solution

Using a double-angle formula, you can rewrite the original function as

$$y = 4 \cos^2 x - 2$$

$$= 2(2 \cos^2 x - 1)$$

$$= 2 \cos 2x.$$

Using the techniques discussed in Section 5.4, you can recognize that the graph of this function has an amplitude of 2 and a period of π. The key points in the interval $[0, \pi]$ are as follows.

Maximum	Intercept	Minimum	Intercept	Maximum
$(0, 2)$	$\left(\dfrac{\pi}{4}, 0\right)$	$\left(\dfrac{\pi}{2}, -2\right)$	$\left(\dfrac{3\pi}{4}, 0\right)$	$(\pi, 2)$

Two cycles of the graph are shown in Figure 6.27.

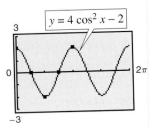

Figure 6.27

✓ *Checkpoint* Now try Exercise 13.

Example 3   Evaluating Functions Involving Double Angles

Use the following to find $\sin 2\theta$, $\cos 2\theta$, and $\tan 2\theta$.

$$\cos \theta = \frac{5}{13}, \quad \frac{3\pi}{2} < \theta < 2\pi$$

Solution

From Figure 6.28, you can see that $\sin \theta = y/r = -12/13$. Consequently, using each of the double-angle formulas, you can write

$$\sin 2\theta = 2 \sin \theta \cos \theta = 2\left(-\frac{12}{13}\right)\left(\frac{5}{13}\right) = -\frac{120}{169}$$

$$\cos 2\theta = 2 \cos^2 \theta - 1 = 2\left(\frac{25}{169}\right) - 1 = -\frac{119}{169}$$

$$\tan 2\theta = \frac{\sin 2\theta}{\cos 2\theta} = \frac{120}{119}.$$

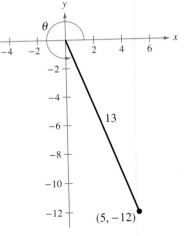

Figure 6.28

✓ *Checkpoint* Now try Exercise 17.

The double-angle formulas are not restricted to the angles 2θ and θ. Other *double* combinations, such as 4θ and 2θ or 6θ and 3θ, are also valid. Here are two examples.

$$\sin 4\theta = 2 \sin 2\theta \cos 2\theta \quad \text{and} \quad \cos 6\theta = \cos^2 3\theta - \sin^2 3\theta$$

By using double-angle formulas together with the sum formulas derived in the preceding section, you can form other multiple-angle formulas.

Example 4   Deriving a Triple-Angle Formula

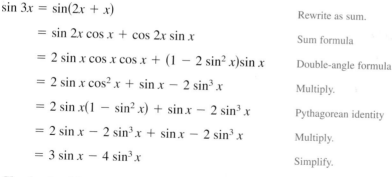

$\sin 3x = \sin(2x + x)$ Rewrite as sum.

$\quad = \sin 2x \cos x + \cos 2x \sin x$ Sum formula

$\quad = 2 \sin x \cos x \cos x + (1 - 2 \sin^2 x)\sin x$ Double-angle formula

$\quad = 2 \sin x \cos^2 x + \sin x - 2 \sin^3 x$ Multiply.

$\quad = 2 \sin x(1 - \sin^2 x) + \sin x - 2 \sin^3 x$ Pythagorean identity

$\quad = 2 \sin x - 2 \sin^3 x + \sin x - 2 \sin^3 x$ Multiply.

$\quad = 3 \sin x - 4 \sin^3 x$ Simplify.

✓ *Checkpoint* Now try Exercise 23.

Power–Reducing Formulas

The double-angle formulas can be used to obtain the following **power-reducing formulas**. See Appendix B for proofs of the power-reducing formulas.

> **Power-Reducing Formulas**
>
> $$\sin^2 u = \frac{1 - \cos 2u}{2} \qquad \cos^2 u = \frac{1 + \cos 2u}{2} \qquad \tan^2 u = \frac{1 - \cos 2u}{1 + \cos 2u}$$

Example 5   Reducing a Power

Rewrite $\sin^4 x$ as a sum of first powers of the cosines of multiple angles.

Solution

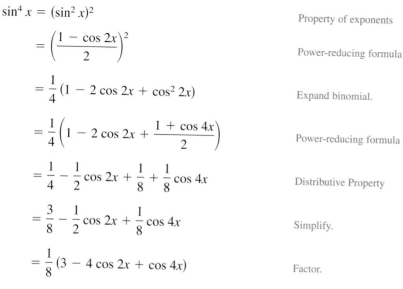

$\sin^4 x = (\sin^2 x)^2$ Property of exponents

$\quad = \left(\dfrac{1 - \cos 2x}{2}\right)^2$ Power-reducing formula

$\quad = \dfrac{1}{4}(1 - 2\cos 2x + \cos^2 2x)$ Expand binomial.

$\quad = \dfrac{1}{4}\left(1 - 2\cos 2x + \dfrac{1 + \cos 4x}{2}\right)$ Power-reducing formula

$\quad = \dfrac{1}{4} - \dfrac{1}{2}\cos 2x + \dfrac{1}{8} + \dfrac{1}{8}\cos 4x$ Distributive Property

$\quad = \dfrac{3}{8} - \dfrac{1}{2}\cos 2x + \dfrac{1}{8}\cos 4x$ Simplify.

$\quad = \dfrac{1}{8}(3 - 4\cos 2x + \cos 4x)$ Factor.

✓ *Checkpoint* Now try Exercise 27.

STUDY TIP

Power-reducing formulas are often used in calculus. Example 5 shows a typical power reduction that is used in calculus. Note the repeated use of power-reducing formulas.

Half-Angle Formulas

You can derive some useful alternative forms of the power-reducing formulas by replacing u with $u/2$. The results are called **half-angle formulas.**

Half-Angle Formulas

$$\sin \frac{u}{2} = \pm \sqrt{\frac{1 - \cos u}{2}} \qquad \cos \frac{u}{2} = \pm \sqrt{\frac{1 + \cos u}{2}}$$

$$\tan \frac{u}{2} = \frac{1 - \cos u}{\sin u} = \frac{\sin u}{1 + \cos u}$$

The signs of $\sin \dfrac{u}{2}$ and $\cos \dfrac{u}{2}$ depend on the quadrant in which $\dfrac{u}{2}$ lies.

Example 6 Using a Half-Angle Formula

Find the exact value of $\sin 105°$.

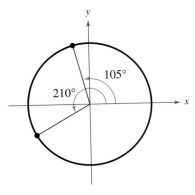

Figure 6.29

STUDY TIP

To find the exact value of a trigonometric function with an angle in D°M′S″ form using a half-angle formula, first convert the angle measure to decimal degree form. Then multiply the angle measure by 2.

Solution

Begin by noting that $105°$ is half of $210°$. Then, using the half-angle formula for $\sin(u/2)$ and the fact that $105°$ lies in Quadrant II (see Figure 6.29), you have

$$\sin 105° = \sqrt{\frac{1 - \cos 210°}{2}}$$

$$= \sqrt{\frac{1 - (-\cos 30°)}{2}}$$

$$= \sqrt{\frac{1 + \left(\sqrt{3}/2\right)}{2}} = \frac{\sqrt{2 + \sqrt{3}}}{2}.$$

The positive square root is chosen because $\sin \theta$ is positive in Quadrant II.

 Checkpoint Now try Exercise 41.

TECHNOLOGY TIP

Use your calculator to verify the result obtained in Example 6. That is, evaluate $\sin 105°$ and $\left(\sqrt{2 + \sqrt{3}}\right)/2$. You will notice that both expressions yield the same result.

Example 7   Solving a Trigonometric Equation

Find all solutions of $2 - \sin^2 x = 2\cos^2 \dfrac{x}{2}$ in the interval $[0, 2\pi)$.

Algebraic Solution

$$2 - \sin^2 x = 2\cos^2 \frac{x}{2} \qquad \text{Write original equation.}$$

$$2 - \sin^2 x = 2\left(\pm\sqrt{\frac{1 + \cos x}{2}}\right)^2 \qquad \text{Half-angle formula}$$

$$2 - \sin^2 x = 2\left(\frac{1 + \cos x}{2}\right) \qquad \text{Simplify.}$$

$$2 - \sin^2 x = 1 + \cos x \qquad \text{Simplify.}$$

$$2 - (1 - \cos^2 x) = 1 + \cos x \qquad \text{Pythagorean identity}$$

$$\cos^2 x - \cos x = 0 \qquad \text{Simplify.}$$

$$\cos x(\cos x - 1) = 0 \qquad \text{Factor.}$$

By setting the factors $\cos x$ and $(\cos x - 1)$ equal to zero, you find that the solutions in the interval $[0, 2\pi)$ are

$$x = \frac{\pi}{2}, \quad x = \frac{3\pi}{2}, \quad \text{and} \quad x = 0.$$

✓ *Checkpoint* Now try Exercise 59.

Graphical Solution

Use a graphing utility set in *radian* mode to graph $y = 2 - \sin^2 x - 2\cos^2 (x/2)$, as shown in Figure 6.30. Use the *zero* or *root* feature or the *zoom* and *trace* features to approximate the x-intercepts in the interval $[0, 2\pi)$ to be

$$x = 0, \quad x \approx 1.5708 \approx \frac{\pi}{2}, \quad \text{and} \quad x \approx 4.7124 \approx \frac{3\pi}{2}.$$

These values are the approximate solutions of $2 - \sin^2 x = 2\cos^2 \dfrac{x}{2}$ in the interval $[0, 2\pi)$.

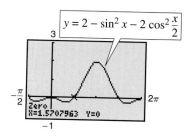

Figure 6.30

Product-to-Sum Formulas

Each of the following **product-to-sum formulas** is easily verified using the sum and difference formulas discussed in the preceding section.

Product-to-Sum Formulas

$$\sin u \sin v = \frac{1}{2}[\cos(u - v) - \cos(u + v)]$$

$$\cos u \cos v = \frac{1}{2}[\cos(u - v) + \cos(u + v)]$$

$$\sin u \cos v = \frac{1}{2}[\sin(u + v) + \sin(u - v)]$$

$$\cos u \sin v = \frac{1}{2}[\sin(u + v) - \sin(u - v)]$$

Product-to-sum formulas are used in calculus to evaluate integrals involving the products of sines and cosines of two different angles.

Example 8   Writing Products as Sums

Rewrite the product as a sum or difference.

$$\cos 5x \sin 4x$$

Solution

$$\cos 5x \sin 4x = \frac{1}{2}[\sin(5x + 4x) - \sin(5x - 4x)]$$

$$= \frac{1}{2}\sin 9x - \frac{1}{2}\sin x$$

✓ *Checkpoint* Now try Exercise 65.

Occasionally, it is useful to reverse the procedure and write a sum of trigonometric functions as a product. This can be accomplished with the following **sum-to-product formulas.** See Appendix B for a proof of the first formula.

Sum-to-Product Formulas

$$\sin u + \sin v = 2 \sin\left(\frac{u + v}{2}\right)\cos\left(\frac{u - v}{2}\right)$$

$$\sin u - \sin v = 2 \cos\left(\frac{u + v}{2}\right)\sin\left(\frac{u - v}{2}\right)$$

$$\cos u + \cos v = 2 \cos\left(\frac{u + v}{2}\right)\cos\left(\frac{u - v}{2}\right)$$

$$\cos u - \cos v = -2 \sin\left(\frac{u + v}{2}\right)\sin\left(\frac{u - v}{2}\right)$$

Example 9   Using a Sum-to-Product Formula

Find the exact value of $\cos 195° + \cos 105°$.

Solution

Using the appropriate sum-to-product formula, you obtain

$$\cos 195° + \cos 105° = 2 \cos\left(\frac{195° + 105°}{2}\right)\cos\left(\frac{195° - 105°}{2}\right)$$

$$= 2 \cos 150° \cos 45°$$

$$= 2\left(-\frac{\sqrt{3}}{2}\right)\left(\frac{\sqrt{2}}{2}\right)$$

$$= -\frac{\sqrt{6}}{2}.$$

✓ *Checkpoint* Now try Exercise 75.

Example 10   Solving a Trigonometric Equation

Find all solutions of $\sin 5x + \sin 3x = 0$ in the interval $[0, 2\pi)$.

Algebraic Solution

$$\sin 5x + \sin 3x = 0 \qquad \text{Write original equation.}$$

$$2\sin\left(\frac{5x + 3x}{2}\right)\cos\left(\frac{5x - 3x}{2}\right) = 0 \qquad \text{Sum-to-product formula}$$

$$2\sin 4x \cos x = 0 \qquad \text{Simplify.}$$

By setting the factor $\sin 4x$ equal to zero, you can find that the solutions in the interval $[0, 2\pi)$ are

$$x = 0, \frac{\pi}{4}, \frac{\pi}{2}, \frac{3\pi}{4}, \pi, \frac{5\pi}{4}, \frac{3\pi}{2}, \frac{7\pi}{4}.$$

Moreover, the equation $\cos x = 0$ yields no additional solutions.

✓ *Checkpoint*   Now try Exercise 79.

Graphical Solution

Use a graphing utility set in *radian* mode to graph $y = \sin 5x + \sin 3x$, as shown in Figure 6.31. Use the *zero* or *root* feature or the *zoom* and *trace* features to approximate the x-intercepts in the interval $[0, 2\pi)$ to be

$$x \approx 0, \ x \approx 0.7854 \approx \frac{\pi}{4}, x \approx 1.5708 \approx \frac{\pi}{2},$$

$$x \approx 2.3562 \approx \frac{3\pi}{4}, \ x \approx 3.1416 \approx \pi, x \approx 3.9270 \approx \frac{5\pi}{4},$$

$$x \approx 4.7124 \approx \frac{3\pi}{2}, \ x \approx 5.4978 \approx \frac{7\pi}{4}.$$

These values are the approximate solutions of $\sin 5x + \sin 3x = 0$ in the interval $[0, 2\pi)$.

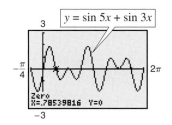

Figure 6.31

Note in Example 10 that the general solution is $x = \dfrac{n\pi}{4}$, where n is an integer.

Example 11   Verifying a Trigonometric Identity

Verify the identity $\dfrac{\sin t + \sin 3t}{\cos t + \cos 3t} = \tan 2t$.

Solution

Using appropriate sum-to-product formulas, you have

$$\frac{\sin t + \sin 3t}{\cos t + \cos 3t} = \frac{2\sin 2t \cos(-t)}{2\cos 2t \cos(-t)}$$

$$= \frac{\sin 2t}{\cos 2t}$$

$$= \tan 2t.$$

✓ *Checkpoint*   Now try Exercise 97.

6.5 Exercises

Vocabulary Check

Fill in the blank to complete the trigonometric formula.

1. $\sin 2u = $ _____

2. $\cos^2 u = $ _____

3. _____ $= 1 - 2 \sin^2 u$

4. _____ $= \dfrac{\sin u}{1 + \cos u}$

5. $\tan 2u = $ _____

6. $\cos u \cos v = $ _____

7. _____ $= \dfrac{1 - \cos 2u}{2}$

8. _____ $= \pm \sqrt{\dfrac{1 + \cos u}{2}}$

9. $\sin u \cos v = $ _____

10. $\sin u + \sin v = $ _____

In Exercises 1–8, use the figure to find the exact value of the trigonometric function.

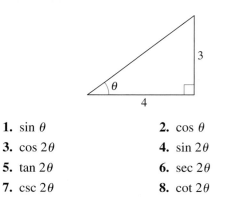

1. $\sin \theta$

2. $\cos \theta$

3. $\cos 2\theta$

4. $\sin 2\theta$

5. $\tan 2\theta$

6. $\sec 2\theta$

7. $\csc 2\theta$

8. $\cot 2\theta$

In Exercises 9–16, use a graphing utility to approximate the solutions of the equation in the interval $[0, 2\pi)$. If possible, find the exact solutions algebraically.

9. $\sin 2x - \sin x = 0$

10. $\sin 2x + \cos x = 0$

11. $4 \sin x \cos x = 1$

12. $\sin 2x \sin x = \cos x$

13. $\cos 2x - \cos x = 0$

14. $\tan 2x - \cot x = 0$

15. $\sin 4x = -2 \sin 2x$

16. $(\sin 2x + \cos 2x)^2 = 1$

In Exercises 17–22, find the exact values of $\sin 2u$, $\cos 2u$, and $\tan 2u$ using the double-angle formulas.

17. $\sin u = \frac{3}{5}, \quad 0 < u < \pi/2$

18. $\cos u = -\frac{2}{7}, \quad \pi/2 < u < \pi$

19. $\tan u = \frac{1}{2}, \quad \pi < u < 3\pi/2$

20. $\cot u = -6, \quad 3\pi/2 < u < 2\pi$

21. $\sec u = -\frac{5}{2}, \quad \pi/2 < u < \pi$

22. $\csc u = 3, \quad \pi/2 < u < \pi$

In Exercises 23–26, use a double-angle formula to rewrite the expression. Use a graphing utility to graph both expressions to verify that both forms are the same.

23. $8 \sin x \cos x$

24. $4 \sin x \cos x + 1$

25. $6 - 12 \sin^2 x$

26. $(\cos x + \sin x)(\cos x - \sin x)$

In Exercises 27–32, rewrite the expression in terms of the first power of the cosine. Use a graphing utility to graph both expressions to verify that both forms are the same.

27. $\cos^4 x$

28. $\sin^4 x$

29. $\sin^2 x \cos^2 x$

30. $\cos^6 x$

31. $\sin^2 x \cos^4 x$

32. $\sin^4 x \cos^2 x$

In Exercises 33–40, use the figure to find the exact value of the trigonometric function.

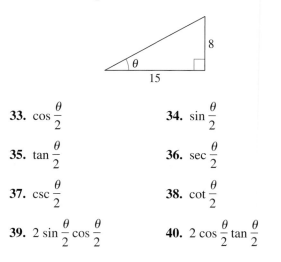

33. $\cos \dfrac{\theta}{2}$

34. $\sin \dfrac{\theta}{2}$

35. $\tan \dfrac{\theta}{2}$

36. $\sec \dfrac{\theta}{2}$

37. $\csc \dfrac{\theta}{2}$

38. $\cot \dfrac{\theta}{2}$

39. $2 \sin \dfrac{\theta}{2} \cos \dfrac{\theta}{2}$

40. $2 \cos \dfrac{\theta}{2} \tan \dfrac{\theta}{2}$

In Exercises 41–48, use the half-angle formulas to determine the exact values of the sine, cosine, and tangent of the angle.

41. $15°$

42. $165°$

43. $112° \, 30'$

44. $157° \, 30'$

45. $\dfrac{\pi}{8}$

46. $\dfrac{\pi}{12}$

47. $\dfrac{3\pi}{8}$

48. $\dfrac{7\pi}{12}$

In Exercises 49–54, find the exact values of $\sin(u/2)$, $\cos(u/2)$, and $\tan(u/2)$ using the half-angle formulas.

49. $\sin u = \frac{5}{13}, \quad \pi/2 < u < \pi$

50. $\cos u = \frac{7}{25}, \quad 0 < u < \pi/2$

51. $\tan u = -\frac{8}{5}, \quad 3\pi/2 < u < 2\pi$

52. $\cot u = 7, \quad \pi < u < 3\pi/2$

53. $\csc u = -\frac{5}{3}, \quad \pi < u < 3\pi/2$

54. $\sec u = -\frac{7}{2}, \quad \pi/2 < u < \pi$

In Exercises 55–58, use the half-angle formulas to simplify the expression.

55. $\sqrt{\dfrac{1 - \cos 6x}{2}}$

56. $\sqrt{\dfrac{1 + \cos 4x}{2}}$

57. $-\sqrt{\dfrac{1 - \cos 8x}{1 + \cos 8x}}$

58. $-\sqrt{\dfrac{1 - \cos(x - 1)}{2}}$

In Exercises 59–62, find the solutions of the equation in the interval $[0, 2\pi)$. Use a graphing utility to verify your answers.

59. $\sin \dfrac{x}{2} - \cos x = 0$

60. $\sin \dfrac{x}{2} + \cos x - 1 = 0$

61. $\cos \dfrac{x}{2} - \sin x = 0$

62. $\tan \dfrac{x}{2} - \sin x = 0$

In Exercises 63–68, use the product-to-sum formulas to write the product as a sum or difference.

63. $6 \sin \dfrac{\pi}{3} \cos \dfrac{\pi}{3}$

64. $4 \sin \dfrac{\pi}{3} \cos \dfrac{5\pi}{6}$

65. $\sin 5\theta \cos 3\theta$

66. $5 \sin 3\alpha \sin 4\alpha$

67. $5 \cos(-5\beta) \cos 3\beta$

68. $\cos 2\theta \cos 4\theta$

In Exercises 69–74, use the sum-to-product formulas to write the sum or difference as a product.

69. $\sin 5\theta - \sin \theta$

70. $\sin x + \sin 7x$

71. $\sin(\alpha + \beta) - \sin(\alpha - \beta)$

72. $\cos(\phi + 2\pi) + \cos \phi$

73. $\cos\left(\theta + \dfrac{\pi}{2}\right) - \cos\left(\theta - \dfrac{\pi}{2}\right)$

74. $\sin\left(x + \dfrac{\pi}{2}\right) + \sin\left(x - \dfrac{\pi}{2}\right)$

In Exercises 75–78, use the sum-to-product formulas to find the exact value of the expression.

75. $\sin 195° + \sin 105°$

76. $\cos 165° - \cos 75°$

77. $\cos \dfrac{5\pi}{12} + \cos \dfrac{\pi}{12}$

78. $\sin \dfrac{11\pi}{12} - \sin \dfrac{7\pi}{12}$

In Exercises 79–82, find the solutions of the equation in the interval $[0, 2\pi)$. Use a graphing utility to verify your answers.

79. $\sin 6x + \sin 2x = 0$

80. $\cos 2x - \cos 6x = 0$

81. $\dfrac{\cos 2x}{\sin 3x - \sin x} - 1 = 0$

82. $\sin^2 3x - \sin^2 x = 0$

In Exercises 83–86, use the figure and trigonometric identities to find the exact value of the trigonometric function in two ways.

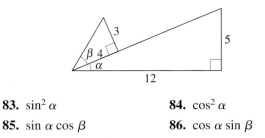

83. $\sin^2 \alpha$

84. $\cos^2 \alpha$

85. $\sin \alpha \cos \beta$

86. $\cos \alpha \sin \beta$

In Exercises 87–98, verify the identity algebraically. Use a graphing utility to check your result graphically.

87. $\csc 2\theta = \dfrac{\csc \theta}{2 \cos \theta}$

88. $\sec 2\theta = \dfrac{\sec^2 \theta}{2 - \sec^2 \theta}$

89. $\cos^2 2\alpha - \sin^2 2\alpha = \cos 4\alpha$

90. $\cos^4 x - \sin^4 x = \cos 2x$

91. $(\sin x + \cos x)^2 = 1 + \sin 2x$

92. $1 + \cos 10y = 2 \cos^2 5y$

93. $\sec \dfrac{u}{2} = \pm \sqrt{\dfrac{2 \tan u}{\tan u + \sin u}}$

94. $\tan \dfrac{u}{2} = \csc u - \cot u$

95. $\cos 3\beta = \cos^3 \beta - 3 \sin^2 \beta \cos \beta$

96. $\sin 4\beta = 4 \sin \beta \cos \beta(1 - 2 \sin^2 \beta)$

97. $\dfrac{\cos 4x - \cos 2x}{2 \sin 3x} = -\sin x$

98. $\dfrac{\cos 3x - \cos x}{\sin 3x - \sin x} = -\tan 2x$

In Exercises 99–102, graph the function by using the power-reducing formulas.

99. $f(x) = \sin^2 x$

100. $f(x) = \cos^2 x$

101. $f(x) = \cos^4 x$

102. $f(x) = \sin^3 x$

ƒ **In Exercises 103–106, (a) use a graphing utility to graph the function and approximate the maximum and minimum points of the graph in the interval $[0, 2\pi]$, and (b) solve the trigonometric equation and verify that the x-coordinates of the maximum and minimum points of f are among its solutions (calculus is required to find the trigonometric equation).**

Function	*Trigonometric Equation*

103. $f(x) = 4 \sin \dfrac{x}{2} + \cos x \quad 2 \cos \dfrac{x}{2} - \sin x = 0$

104. $f(x) = \cos 2x - 2 \sin x \quad -2 \cos x(2 \sin x + 1) = 0$

105. $f(x) = 2 \cos \dfrac{x}{2} + \sin 2x \quad 2 \cos 2x - \sin \dfrac{x}{2} = 0$

106. $f(x) = 2 \sin \dfrac{x}{2} -$

$\qquad 5 \cos\left(2x - \dfrac{\pi}{4}\right) \qquad 10 \sin\left(2x - \dfrac{\pi}{4}\right) +$

$\qquad\qquad\qquad\qquad\qquad \cos \dfrac{x}{2} = 0$

In Exercises 107–112, write the trigonometric expression as an algebraic expression.

107. $\sin(2 \arcsin x)$

108. $\cos(2 \arccos x)$

109. $\cos(2 \arcsin x)$

110. $\sin(2 \arccos x)$

111. $\cos(2 \arctan x)$

112. $\sin(2 \arctan x)$

113. *Projectile Motion* The range of a projectile fired at an angle θ with the horizontal and with an initial velocity of v_0 feet per second is given by

$$r = \tfrac{1}{32} v_0^2 \sin 2\theta$$

where r is measured in feet. Rewrite the expression for the range in terms of θ.

114. *Geometry* The length of each of the two equal sides of an isosceles triangle is 10 meters (see figure). The angle between the two sides is θ.

(a) Write the area of the triangle as a function of $\theta/2$.

(b) Write the area of the triangle as a function of θ and determine the value of θ such that the area is a maximum.

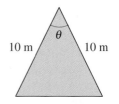

115. *Mach Number* The mach number M of an airplane is the ratio of its speed to the speed of sound. When an airplane travels faster than the speed of sound, the sound waves form a cone behind the airplane (see figure). The mach number is related to the apex angle θ of the cone by

$$\sin \dfrac{\theta}{2} = \dfrac{1}{M}.$$

(a) Find the angle θ that corresponds to a mach number of 1.

(b) Find the angle θ that corresponds to a mach number of 4.5.

(c) The speed of sound is about 760 miles per hour. Determine the speed of an object having the mach numbers in parts (a) and (b).

(d) Rewrite the equation as a trigonometric function of θ.

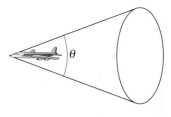

116. *Railroad Track*   When two railroad tracks merge, the overlapping portions of the tracks are in the shape of a circular arc (see figure). The radius of each arc r (in feet) and the angle θ are related by

$$\frac{x}{2} = 2r \sin^2 \frac{\theta}{2}.$$

Write a formula for x in terms of $\cos \theta$.

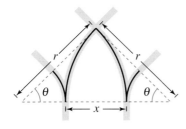

Synthesis

True or False?   **In Exercises 117 and 118, determine whether the statement is true or false. Justify your answer.**

117. $\sin \dfrac{x}{2} = -\sqrt{\dfrac{1 - \cos x}{2}}, \qquad \pi \le x \le 2\pi$

118. The graph of $y = 4 - 8 \sin^2 x$ has a maximum at $(\pi, 4)$.

119. *Conjecture*   Consider the function

$$f(x) = 2 \sin x \left(2 \cos^2 \frac{x}{2} - 1 \right).$$

(a) Use a graphing utility to graph the function.

(b) Make a conjecture about the function that is an identity with f.

(c) Verify your conjecture algebraically.

120. *Exploration*   Consider the function

$$f(x) = \sin^4 x + \cos^4 x.$$

(a) Use the power-reducing formulas to write the function in terms of cosine to the first power.

(b) Determine another way of rewriting the function. Use a graphing utility to rule out incorrectly rewritten functions.

(c) Add a trigonometric term to the function so that it becomes a perfect square trinomial. Rewrite the function as a perfect square trinomial minus the term that you added. Use a graphing utility to rule out incorrectly rewritten functions.

(d) Rewrite the result of part (c) in terms of the sine of a double angle. Use a graphing utility to rule out incorrectly rewritten functions.

(e) When you rewrite a trigonometric expression, the result may not be the same as a friend's. Does this mean that one of you is wrong? Explain.

121. *Writing*   Describe how you can use a double-angle formula or a half-angle formula to derive a formula for the area of an isosceles triangle. Use a labeled sketch to illustrate your derivation. Then write two examples that show how your formula can be used.

122. (a) Write a formula for $\cos 3\theta$.

(b) Write a formula for $\cos 4\theta$.

Review

In Exercises 123–126, (a) plot the points, (b) find the distance between the points, and (c) find the midpoint of the line segment connecting the points.

123. $(5, 2), (-1, 4)$ **124.** $(-4, -3), (6, 10)$

125. $\left(0, \frac{1}{2}\right), \left(\frac{4}{3}, \frac{5}{2}\right)$ **126.** $\left(\frac{1}{3}, \frac{2}{3}\right), \left(-1, -\frac{3}{2}\right)$

In Exercises 127–130, find (if possible) the complement and supplement of each angle.

127. (a) $55°$ (b) $162°$

128. (a) $109°$ (b) $78°$

129. (a) $\dfrac{\pi}{18}$ (b) $\dfrac{9\pi}{20}$

130. (a) 0.95 (b) 2.76

131. Find the radian measure of the central angle of a circle with a radius of 15 inches that intercepts an arc of length 7 inches.

132. Find the length of the arc on a circle of radius 21 centimeters intercepted by a central angle of $35°$.

In Exercises 133–136, sketch a graph of the function. (Include two full periods.) Use a graphing utility to verify your graph.

133. $f(x) = \dfrac{3}{2} \cos 2x$ **134.** $f(x) = \dfrac{5}{2} \sin \dfrac{1}{2} x$

135. $f(x) = \dfrac{1}{2} \tan 2\pi x$ **136.** $f(x) = \dfrac{1}{4} \sec \dfrac{\pi x}{2}$

6 Chapter Summary

What did you learn?

	Review Exercises

Section 6.1

☐ Recognize and write the fundamental trigonometric identities.   1–10

☐ Use the fundamental trigonometric identities to evaluate trigonometric functions,
simplify trigonometric expressions, and rewrite trigonometric expressions.   11–24

Section 6.2

☐ Verify trigonometric identities.   25–36

Section 6.3

☐ Use standard algebraic techniques to solve trigonometric equations.   37–48

☐ Solve trigonometric equations of quadratic type.   49–52

☐ Solve trigonometric equations involving multiple angles.   53–58

☐ Use inverse trigonometric functions to solve trigonometric equations.   59–62

Section 6.4

☐ Use sum and difference formulas to evaluate trigonometric functions,
verify identities, and solve trigonometric equations.   63–84

Section 6.5

☐ Use multiple-angle formulas to rewrite and evaluate trigonometric functions.   85–94

☐ Use power-reducing formulas to rewrite and evaluate trigonometric functions.   95–98

☐ Use half-angle formulas to rewrite and evaluate trigonometric functions.   99–110

☐ Use product-to-sum and sum-to-product formulas to rewrite and evaluate
trigonometric functions.   111–122

6 Review Exercises

6.1 **In Exercises 1–10, name the trigonometric function equivalent to the expression.**

1. $\dfrac{1}{\cos x}$

2. $\dfrac{1}{\sin x}$

3. $\dfrac{1}{\sec x}$

4. $\dfrac{1}{\tan x}$

5. $\sqrt{1 - \cos^2 x}$

6. $\sqrt{1 + \tan^2 x}$

7. $\csc\left(\dfrac{\pi}{2} - x\right)$

8. $\cot\left(\dfrac{\pi}{2} - x\right)$

9. $\sec(-x)$

10. $\tan(-x)$

In Exercises 11–14, use the given values to evaluate (if possible) all six trigonometric functions of the angle.

11. $\sin x = \dfrac{4}{5}, \qquad \cos x = \dfrac{3}{5}$

12. $\tan \theta = \dfrac{2}{3}, \qquad \sec \theta = \dfrac{\sqrt{13}}{3}$

13. $\sin\left(\dfrac{\pi}{2} - x\right) = \dfrac{1}{\sqrt{2}}, \qquad \sin x = -\dfrac{1}{\sqrt{2}}$

14. $\csc\left(\dfrac{\pi}{2} - \theta\right) = 3, \qquad \sin \theta = \dfrac{2\sqrt{2}}{3}$

In Exercises 15–22, use the fundamental identities to simplify the expression. Use the *table* feature of a graphing utility to check your result numerically.

15. $\dfrac{1}{\cot^2 x + 1}$

16. $\dfrac{\sec^2 x - 1}{\sec x - 1}$

17. $\dfrac{\sin^2 \alpha - \cos^2 \alpha}{\sin^2 \alpha - \sin \alpha \cos \alpha}$

18. $\dfrac{\sin^3 \beta + \cos^3 \beta}{\sin \beta + \cos \beta}$

19. $\tan^2 \theta (\csc^2 \theta - 1)$

20. $\csc^2 x(1 - \cos^2 x)$

21. $\tan\left(\dfrac{\pi}{2} - x\right) \sec x$

22. $\dfrac{\sin(-x) \cot x}{\sin\left(\dfrac{\pi}{2} - x\right)}$

23. *Rate of Change* The rate of change of the function $f(x) = 2\sqrt{\sin x}$ is given by the expression $\sin^{-1/2} x \cos x$. Show that this expression can also be written as $\cot x \sqrt{\sin x}$.

24. *Rate of Change* The rate of change of the function $f(x) = \csc x - \cot x$ is the expression $\csc^2 x - \csc x \cot x$. Show that this expression can also be written as $(1 - \cos x)/\sin^2 x$.

6.2 **In Exercises 25–36, verify the identity.**

25. $\cos x(\tan^2 x + 1) = \sec x$

26. $\sec^2 x \cot x - \cot x = \tan x$

27. $\sin^3 \theta + \sin \theta \cos^2 \theta = \sin \theta$

28. $\cot^2 x - \cos^2 x = \cot^2 x \cos^2 x$

29. $\sin^5 x \cos^2 x = (\cos^2 x - 2\cos^4 x + \cos^6 x)\sin x$

30. $\cos^3 x \sin^2 x = (\sin^2 x - \sin^4 x)\cos x$

31. $\sqrt{\dfrac{1 - \sin \theta}{1 + \sin \theta}} = \dfrac{1 - \sin \theta}{|\cos \theta|}$

32. $\sqrt{1 - \cos x} = \dfrac{|\sin x|}{\sqrt{1 + \cos x}}$

33. $\dfrac{\csc(-x)}{\sec(-x)} = -\cot x$

34. $\dfrac{1 + \sec(-x)}{\sin(-x) + \tan(-x)} = -\csc x$

35. $\sin^2 x + \sin^2\left(\dfrac{\pi}{2} - x\right) = 1$

36. $\csc x \sin\left(\dfrac{\pi}{2} - x\right) = \cot x$

6.3 **In Exercises 37–48, solve the equation.**

37. $2 \sin x - 1 = 0$

38. $\tan x + 1 = 0$

39. $\sin x = \sqrt{3} - \sin x$

40. $4 \cos x = 1 + 2 \cos x$

41. $3\sqrt{3} \tan x = 3$

42. $\frac{1}{2} \sec x - 1 = 0$

43. $3 \csc^2 x = 4$

44. $4 \tan^2 x - 1 = \tan^2 x$

45. $4 \cos^2 x - 3 = 0$

46. $\sin x(\sin x + 1) = 0$

47. $\sin x - \tan x = 0$

48. $\csc x - 2 \cot x = 0$

In Exercises 49–58, find all solutions of the equation in the interval $[0, 2\pi)$. Use a graphing utility to check your answers.

49. $2 \cos^2 x - \cos x = 1$

50. $2 \sin^2 x - 3 \sin x = -1$

51. $\cos^2 x + \sin x = 1$

52. $\sin^2 x + 2 \cos x = 2$

53. $2 \sin 2x - \sqrt{2} = 0$

54. $\sqrt{3} \tan 3x = 0$

55. $\cos 4x(\cos x - 1) = 0$

56. $3 \csc^2 5x = -4$

57. $\cos 4x - 7 \cos 2x = 8$ **58.** $\sin 4x - \sin 2x = 0$

In Exercises 59–62, use the inverse functions where necessary to find all solutions of the equation in the interval $[0, 2\pi)$.

59. $\sin^2 x - 2 \sin x = 0$

60. $2 \cos^2 x + 3 \cos x = 0$

61. $\tan^2 \theta + \tan \theta - 12 = 0$

62. $\sec^2 x + 6 \tan x + 4 = 0$

6.4 In Exercises 63–66, find the exact values of the sine, cosine, and tangent of the angle.

63. $285° = 315° - 30°$ **64.** $345° = 300° + 45°$

65. $\dfrac{25\pi}{12} = \dfrac{11\pi}{6} + \dfrac{\pi}{4}$ **66.** $\dfrac{19\pi}{12} = \dfrac{11\pi}{6} - \dfrac{\pi}{4}$

In Exercises 67–70, write the expression as the sine, cosine, or tangent of an angle.

67. $\sin 140° \cos 50° + \cos 140° \sin 50°$

68. $\cos 45° \cos 120° - \sin 45° \sin 120°$

69. $\dfrac{\tan 25° + \tan 10°}{1 - \tan 25° \tan 10°}$ **70.** $\dfrac{\tan 68° - \tan 115°}{1 + \tan 68° \tan 115°}$

In Exercises 71–76, find the exact value of the trigonometric function given that $\sin u = \frac{3}{4}$ and $\cos v = -\frac{5}{13}$. (Both u and v are in Quadrant II.)

71. $\sin(u + v)$ **72.** $\tan(u + v)$

73. $\tan(u - v)$ **74.** $\sin(u - v)$

75. $\cos(u + v)$ **76.** $\cos(u - v)$

In Exercises 77–82, verify the identity.

77. $\cos\left(x + \dfrac{\pi}{2}\right) = -\sin x$ **78.** $\sin\left(x - \dfrac{3\pi}{2}\right) = \cos x$

79. $\cot\left(\dfrac{\pi}{2} - x\right) = \tan x$ **80.** $\sin(\pi - x) = \sin x$

81. $\cos 3x = 4 \cos^3 x - 3 \cos x$

82. $\dfrac{\sin(\alpha + \beta)}{\cos \alpha \cos \beta} = \tan \alpha + \tan \beta$

In Exercises 83 and 84, find the solutions of the equation in the interval $[0, 2\pi)$.

83. $\sin\left(x + \dfrac{\pi}{2}\right) - \sin\left(x - \dfrac{\pi}{2}\right) = \sqrt{3}$

84. $\cos\left(x + \dfrac{3\pi}{4}\right) - \cos\left(x - \dfrac{3\pi}{4}\right) = 0$

6.5 In Exercises 85–88, find the exact values of $\sin 2u$, $\cos 2u$, and $\tan 2u$ using the double-angle formulas.

85. $\sin u = -\dfrac{5}{7}$, $\quad \pi < u < \dfrac{3\pi}{2}$

86. $\cos u = \dfrac{4}{5}$, $\quad \dfrac{3\pi}{2} < u < 2\pi$

87. $\tan u = -\dfrac{2}{9}$, $\quad \dfrac{\pi}{2} < u < \pi$

88. $\cos u = -\dfrac{2}{\sqrt{5}}$, $\quad \dfrac{\pi}{2} < u < \pi$

In Exercises 89–92, use double-angle formulas to verify the identity algebraically. Use a graphing utility to check your result graphically.

89. $6 \sin x \cos x = 3 \sin 2x$

90. $4 \sin x \cos x + 2 = 2 \sin 2x + 2$

91. $1 - 4 \sin^2 x \cos^2 x = \cos^2 2x$

92. $\sin 4x = 8 \cos^3 x \sin x - 4 \cos x \sin x$

93. *Projectile Motion* A baseball leaves the hand of the first baseman at an angle of θ with the horizontal and with an initial velocity of $v_0 = 80$ feet per second. The ball is caught by the second baseman 100 feet away. Find θ if the range r of a projectile is given by $r = \frac{1}{32} v_0^2 \sin 2\theta$.

94. *Projectile Motion* Use the equation in Exercise 93 to find θ when a golf ball is hit with an initial velocity of $v_0 = 50$ feet per second and lands 77 feet away.

In Exercises 95–98, use the power-reducing formulas to rewrite the expression in terms of the first power of the cosine.

95. $\sin^6 x$ **96.** $\cos^4 x \sin^4 x$

97. $\cos^4 2x$ **98.** $\sin^4 2x$

In Exercises 99–102, use the half-angle formulas to determine the exact values of the sine, cosine, and tangent of the angle.

99. $105°$ **100.** $67° \, 30'$

101. $\dfrac{7\pi}{8}$ **102.** $\dfrac{11\pi}{12}$

In Exercises 103–106, find the exact values of $\sin(u/2)$, $\cos(u/2)$, and $\tan(u/2)$ using the half-angle formulas.

103. $\sin u = \dfrac{3}{5}, \qquad 0 < u < \dfrac{\pi}{2}$

104. $\tan u = \dfrac{5}{8}, \qquad \pi < u < \dfrac{3\pi}{2}$

105. $\cos u = -\dfrac{2}{7}, \qquad \dfrac{\pi}{2} < u < \pi$

106. $\sec u = -6, \qquad \dfrac{\pi}{2} < u < \pi$

In Exercises 107 and 108, use the half-angle formulas to simplify the expression.

107. $-\sqrt{\dfrac{1 + \cos 10x}{2}}$

108. $\dfrac{\sin 6x}{1 + \cos 6x}$

Geometry In Exercises 109 and 110, a trough for feeding cattle is 4 meters long and its cross sections are isosceles triangles with two equal sides of $\frac{1}{2}$ meter (see figure). The angle between the equal sides is θ.

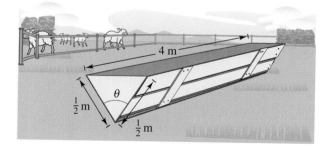

109. Write the trough's volume as a function of $\theta/2$.

110. Write the volume of the trough as a function of θ and determine the value of θ such that the volume is maximum.

In Exercises 111–114, use the product-to-sum formulas to write the product as a sum or difference.

111. $6 \sin \dfrac{\pi}{4} \cos \dfrac{\pi}{4}$

112. $4 \sin 15° \sin 45°$

113. $\sin 3\alpha \sin 2\alpha$

114. $\cos 4\theta \sin 6\theta$

In Exercises 115–118, use the sum-to-product formulas to write the sum or difference as a product.

115. $\cos 3\theta + \cos 2\theta$

116. $\sin 5\theta + \sin 3\theta$

117. $\sin\left(x + \dfrac{\pi}{4}\right) - \sin\left(x - \dfrac{\pi}{4}\right)$

118. $\cos\left(x + \dfrac{\pi}{6}\right) - \cos\left(x - \dfrac{\pi}{6}\right)$

Harmonic Motion In Exercises 119–122, a weight is attached to a spring suspended vertically from a ceiling. When a driving force is applied to the system, the weight moves vertically from its equilibrium position. This motion is described by the model

$$y = 1.5 \sin 8t - 0.5 \cos 8t$$

where y is the distance from equilibrium in feet and t is the time in seconds.

119. Write the model in the form
$$y = \sqrt{a^2 + b^2} \sin(Bt + C).$$

120. Use a graphing utility to graph the model.

121. Find the amplitude of the oscillations of the weight.

122. Find the frequency of the oscillations of the weight.

Synthesis

True or False? In Exercises 123–126, determine whether the statement is true or false. Justify your answer.

123. If $\dfrac{\pi}{2} < \theta < \pi$, then $\cos \dfrac{\theta}{2} < 0$.

124. $\sin(x + y) = \sin x + \sin y$

125. $4 \sin(-x) \cos(-x) = -2 \sin 2x$

126. $4 \sin 45° \cos 15° = 1 + \sqrt{3}$

127. List the reciprocal identities, quotient identities, and Pythagorean identities from memory.

128. Is $\cos \theta = \sqrt{1 - \sin^2 \theta}$ an identity? Explain.

In Exercises 129 and 130, use the graphs of y_1 and y_2 to determine how to change y_2 to a new function y_3 such that $y_1 = y_3$.

129. $y_1 = \sec^2\left(\dfrac{\pi}{2} - x\right)$

$y_2 = \cot^2 x$

130. $y_1 = \dfrac{\cos 3x}{\cos x}$

$y_2 = (2 \sin x)^2$

6 | Chapter Test

Take this test as you would take a test in class. After you are finished, check your work against the answers given in the back of the book.

1. If $\tan \theta = \frac{6}{5}$ and $\cos \theta < 0$, use the fundamental identities to evaluate the other five trigonometric functions of θ.

2. Use the fundamental identities to simplify $\csc^2 \beta(1 - \cos^2 \beta)$.

3. Factor and simplify $\dfrac{\sec^4 x - \tan^4 x}{\sec^2 x + \tan^2 x}$.

4. Add and simplify $\dfrac{\cos \theta}{\sin \theta} + \dfrac{\sin \theta}{\cos \theta}$.

5. Determine the values of θ, $0 \le \theta < 2\pi$, for which $\tan \theta = -\sqrt{\sec^2 \theta - 1}$ is true.

6. Use a graphing utility to graph the functions $y_1 = \cos x + \sin x \tan x$ and $y_2 = \sec x$. Make a conjecture about y_1 and y_2. Verify your result algebraically.

In Exercises 7–12, verify the identity.

7. $\sin \theta \sec \theta = \tan \theta$

8. $\sec^2 x \tan^2 x + \sec^2 x = \sec^4 x$

9. $\dfrac{\csc \alpha + \sec \alpha}{\sin \alpha + \cos \alpha} = \cot \alpha + \tan \alpha$

10. $\cos\left(x + \dfrac{\pi}{2}\right) = -\sin x$

11. $\cos(\pi - \theta) + \sin\left(\dfrac{\pi}{2} + \theta\right) = 0$

12. $(\sin x + \cos x)^2 = 1 + \sin 2x$

13. Find the exact value of $\tan 105°$.

14. Rewrite $\sin^4 x \tan^2 x$ in terms of the first power of the cosine.

15. Use a half-angle formula to simplify the expression $\dfrac{\sin 4\theta}{1 + \cos 4\theta}$.

16. Write $6 \sin 4\theta \sin 6\theta$ as a sum or difference.

17. Write $\cos 5\theta + \cos 3\theta$ as a product.

In Exercises 18–21, find all solutions of the equation in the interval $[0, 2\pi)$.

18. $\tan^2 x + \tan x = 0$

19. $\sin 2\alpha - \cos \alpha = 0$

20. $4 \cos^2 x - 3 = 0$

21. $\csc^2 x - \csc x - 2 = 0$

22. Use a graphing utility to approximate the solutions of the equation $5 \cos x - x = 0$ in the interval $[0, 2\pi)$ accurate to three decimal places.

23. Use the figure at the right to find the exact values of $\sin 2u$, $\cos 2u$, and $\tan 2u$.

24. The *index of refraction n* of a transparent material is the ratio of the speed of light in a vacuum to the speed of light in the material. For the glass triangular prism in the figure at the right, $n = 1.5$ and $\alpha = 60°$. Find the angle θ for the glass prism if

$$n = \frac{\sin(\theta/2 + \alpha/2)}{\sin(\theta/2)}.$$

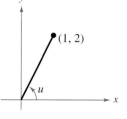

Figure for 23

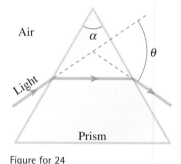

Figure for 24

The Law of Sines can be used to determine the angles and sides of triangles that occur in real-life applications. For example, you can model an airplane's glide path at landing and determine the altitude of the plane at the moment it begins its descent.

Zefa Visual Media-Germany/Index Stock

7 Additional Topics in Trigonometry

What You Should Learn

In this chapter, you will learn how to:

- Use the Law of Sines and the Law of Cosines to solve oblique triangles.

- Find areas of oblique triangles.

- Represent vectors as directed line segments and perform mathematical operations on vectors.

- Find direction angles of vectors.

- Find the dot product of two vectors and use properties of the dot product.

- Multiply and divide complex numbers written in trigonometric form.

- Find powers and nth roots of complex numbers.

Introduction

In Chapter 5 you looked at techniques for solving right triangles. In this section and the next, you will solve **oblique triangles**—triangles that have no right angles. As standard notation, the angles of a triangle are labeled A, B, and C, and their opposite sides are labeled a, b, and c, as shown in Figure 7.1.

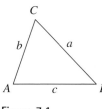

Figure 7.1

To solve an oblique triangle, you need to know the measure of at least one side and the measures of any two other parts of the triangle—two sides, two angles, or one angle and one side. This breaks down into the following four cases.

1. Two angles and any side (AAS or ASA)

2. Two sides and an angle opposite one of them (SSA)

3. Three sides (SSS)

4. Two sides and their included angle (SAS)

The first two cases can be solved using the **Law of Sines,** whereas the last two cases require the Law of Cosines (Section 7.2). See Appendix B for a proof of the Law of Sines.

Law of Sines

If ABC is a triangle with sides a, b, and c, then

$$\frac{a}{\sin A} = \frac{b}{\sin B} = \frac{c}{\sin C}.$$

Oblique Triangles

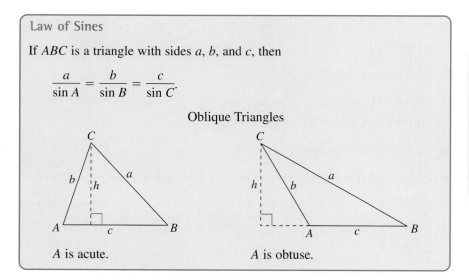

A is acute. A is obtuse.

The Law of Sines can also be written in the reciprocal form

$$\frac{\sin A}{a} = \frac{\sin B}{b} = \frac{\sin C}{c}.$$

Example 1   Given Two Angles and One Side—AAS

For the triangle in Figure 7.2, $C = 102.3°$, $B = 28.7°$, and $b = 27.4$ feet. Find the remaining angle and sides.

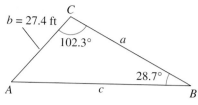

Figure 7.2

Solution

The third angle of the triangle is

$$A = 180° - B - C$$

$$= 180° - 28.7° - 102.3°$$

$$= 49.0°.$$

By the Law of Sines, you have

$$\frac{a}{\sin A} = \frac{b}{\sin B} = \frac{c}{\sin C}.$$

Using $b = 27.4$ produces

$$a = \frac{b}{\sin B}(\sin A) = \frac{27.4}{\sin 28.7°}(\sin 49.0°) \approx 43.06 \text{ feet}$$

and

$$c = \frac{b}{\sin B}(\sin C) = \frac{27.4}{\sin 28.7°}(\sin 102.3°) \approx 55.75 \text{ feet}.$$

 Checkpoint Now try Exercise 3.

Example 2   Given Two Angles and One Side—ASA

A pole tilts *toward* the sun at an 8° angle from the vertical, and it casts a 22-foot shadow. The angle of elevation from the tip of the shadow to the top of the pole is 43°. How tall is the pole?

Solution

In Figure 7.3, $A = 43°$ and $B = 90° + 8° = 98°$. So, the third angle is

$$C = 180° - A - B = 180° - 43° - 98° = 39°.$$

By the Law of Sines, you have

$$\frac{a}{\sin A} = \frac{c}{\sin C}.$$

Because $c = 22$ feet, the length of the pole is

$$a = \frac{c}{\sin C}(\sin A) = \frac{22}{\sin 39°}(\sin 43°) \approx 23.84 \text{ feet}.$$

 Checkpoint Now try Exercise 27.

For practice, try reworking Example 2 for a pole that tilts *away from* the sun under the same conditions.

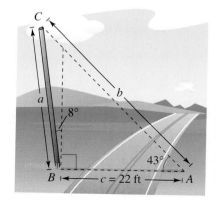

Figure 7.3

> **STUDY TIP**
>
> When you are solving triangles, a careful sketch is useful as a quick test for the feasibility of an answer. Remember that the longest side lies opposite the largest angle, and the shortest side lies opposite the smallest angle.

The Ambiguous Case (SSA)

In Examples 1 and 2 you saw that two angles and one side determine a unique triangle. However, if two sides and one opposite angle are given, three possible situations can occur: (1) no such triangle exists, (2) one such triangle exists, or (3) two distinct triangles satisfy the conditions.

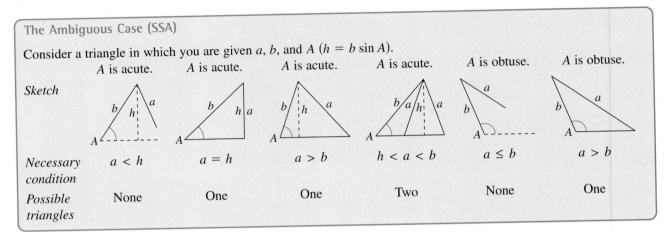

The Ambiguous Case (SSA)

Consider a triangle in which you are given a, b, and A ($h = b \sin A$).

	A is acute.	A is acute.	A is acute.	A is acute.	A is obtuse.	A is obtuse.
Necessary condition	$a < h$	$a = h$	$a > b$	$h < a < b$	$a \leq b$	$a > b$
Possible triangles	None	One	One	Two	None	One

Example 3   Single-Solution Case–SSA

For the triangle in Figure 7.4, $a = 22$ inches, $b = 12$ inches, and $A = 42°$. Find the remaining side and angles.

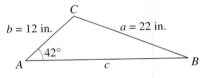

Figure 7.4   One solution: $a > b$

Solution

By the Law of Sines, you have

$$\frac{\sin B}{b} = \frac{\sin A}{a} \qquad \text{Reciprocal form}$$

$$\sin B = b\left(\frac{\sin A}{a}\right) \qquad \text{Multiply each side by } b.$$

$$\sin B = 12\left(\frac{\sin 42°}{22}\right) \qquad \text{Substitute for } A, a, \text{ and } b.$$

$$B \approx 21.41°. \qquad B \text{ is acute.}$$

Now you can determine that

$$C \approx 180° - 42° - 21.41° = 116.59°.$$

Then the remaining side is given by

$$\frac{c}{\sin C} = \frac{a}{\sin A}$$

$$c = \frac{a}{\sin A}(\sin C) = \frac{22}{\sin 42°}(\sin 116.59°) \approx 29.40 \text{ inches.}$$

✓ *Checkpoint* Now try Exercise 15.

Example 4   No-Solution Case—SSA

Show that there is no triangle for which $a = 15$, $b = 25$, and $A = 85°$.

Solution

Begin by making the sketch shown in Figure 7.5. From this figure it appears that no triangle is formed. You can verify this by using the Law of Sines.

$$\frac{\sin B}{b} = \frac{\sin A}{a} \qquad \text{Reciprocal form}$$

$$\sin B = b\left(\frac{\sin A}{a}\right) \qquad \text{Multiply each side by } b.$$

$$\sin B = 25\left(\frac{\sin 85°}{15}\right) \approx 1.660 > 1$$

This contradicts the fact that $|\sin B| \leq 1$. So, no triangle can be formed having sides $a = 15$ and $b = 25$ and an angle of $A = 85°$.

☑ *Checkpoint* Now try Exercise 17.

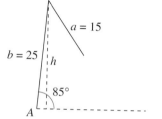

Figure 7.5   No solution: $a < h$

Example 5   Two-Solution Case—SSA

Find two triangles for which $a = 12$ meters, $b = 31$ meters, and $A = 20.5°$.

Solution

Because $h = b \sin A = 31(\sin 20.5°) \approx 10.86$ meters, you can conclude that there are two possible triangles (because $h < a < b$). By the Law of Sines, you have

$$\frac{\sin B}{b} = \frac{\sin A}{a} \qquad \text{Reciprocal form}$$

$$\sin B = b\left(\frac{\sin A}{a}\right) = 31\left(\frac{\sin 20.5°}{12}\right) \approx 0.9047.$$

There are two angles $B_1 \approx 64.8°$ and $B_2 \approx 180° - 64.8° = 115.2°$ between $0°$ and $180°$ whose sine is 0.9047. For $B_1 \approx 64.8°$, you obtain

$$C \approx 180° - 20.5° - 64.8° = 94.7°$$

$$c = \frac{a}{\sin A}(\sin C) = \frac{12}{\sin 20.5°}(\sin 94.7°) \approx 34.15 \text{ meters.}$$

For $B_2 \approx 115.2°$, you obtain

$$C \approx 180° - 20.5° - 115.2° = 44.3°$$

$$c = \frac{a}{\sin A}(\sin C) = \frac{12}{\sin 20.5°}(\sin 44.3°) \approx 23.93 \text{ meters.}$$

The resulting triangles are shown in Figure 7.6.

☑ *Checkpoint* Now try Exercise 19.

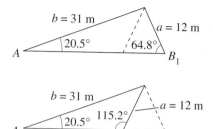

Figure 7.6   Two solutions: $h < a < b$

Area of an Oblique Triangle

The procedure used to prove the Law of Sines leads to a simple formula for the area of an oblique triangle. Referring to Figure 7.7, note that each triangle has a height of $h = b \sin A$. To see this when A is obtuse, substitute the reference angle $180° - A$ for A. Now the height of the triangle is given by

$$h = b \sin(180° - A).$$

Using the difference formula for sine, the height is given by

$$h = b(\sin 180° \cos A - \cos 180° \sin A) \qquad \text{{\small $\sin(u - v) = \sin u \cos v - \cos u \sin v$}}$$

$$= b[0 \cdot \cos A - (-1) \cdot \sin A]$$

$$= b \sin A.$$

Consequently, the area of each triangle is given by

$$\text{Area} = \frac{1}{2}(\text{base})(\text{height}) = \frac{1}{2}(c)(b \sin A) = \frac{1}{2}bc \sin A.$$

By similar arguments, you can develop the formulas

$$\text{Area} = \frac{1}{2}ab \sin C = \frac{1}{2}ac \sin B.$$

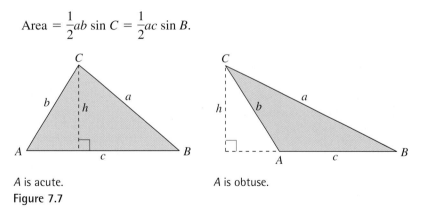

A is acute.   *A* is obtuse.

Figure 7.7

Area of an Oblique Triangle

The area of any triangle is one-half the product of the lengths of two sides times the sine of their included angle. That is,

$$\text{Area} = \frac{1}{2}bc \sin A = \frac{1}{2}ab \sin C = \frac{1}{2}ac \sin B.$$

Note that if angle A is $90°$, the formula gives the area of a right triangle as

$$\text{Area} = \frac{1}{2}bc = \frac{1}{2}(\text{base})(\text{height}).$$

Similar results are obtained for angles C and B equal to $90°$.

Example 6   Finding the Area of an Oblique Triangle

Find the area of a triangular lot having two sides of lengths 90 meters and 52 meters and an included angle of 102°.

Solution

Consider $a = 90$ meters, $b = 52$ meters, and $C = 102°$, as shown in Figure 7.8. Then the area of the triangle is

$$\text{Area} = \frac{1}{2}ab \sin C = \frac{1}{2}(90)(52)(\sin 102°) \approx 2289 \text{ square meters.}$$

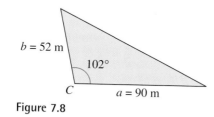

Figure 7.8

 Checkpoint Now try Exercise 21.

Example 7   An Application of the Law of Sines

The course for a boat race starts at point A and proceeds in the direction S 52° W to point B, then in the direction S 40° E to point C, and finally back to A, as shown in Figure 7.9. Point C lies 8 kilometers directly south of point A. Approximate the total distance of the race course.

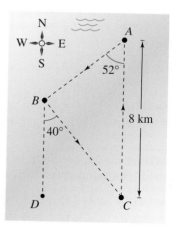

Figure 7.9

Solution

Because lines BD and AC are parallel, it follows that $\angle BCA \cong \angle DBC$. Consequently, triangle ABC has the measures shown in Figure 7.10. For angle B, you have $B = 180° - 52° - 40° = 88°$. Using the Law of Sines

$$\frac{a}{\sin 52°} = \frac{b}{\sin 88°} = \frac{c}{\sin 40°}$$

you can let $b = 8$ and obtain

$$a = \frac{8}{\sin 88°}(\sin 52°) \approx 6.308$$

and

$$c = \frac{8}{\sin 88°}(\sin 40°) \approx 5.145.$$

The total length of the course is approximately

$$\text{Length} \approx 8 + 6.308 + 5.145 = 19.453 \text{ kilometers.}$$

 Checkpoint Now try Exercise 29.

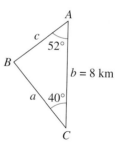

Figure 7.10

7.1   Exercises

Vocabulary Check

Fill in the blanks.

1. An _____ triangle is one that has no right angles.

2. Law of Sines: $\dfrac{a}{\sin A} =$ _____ $= \dfrac{c}{\sin C}$

3. To find the area of any triangle, use one of the following three formulas: Area = _____ , _____ , or _____ .

In Exercises 1–14, use the Law of Sines to solve the triangle.

1. $A = 30°$, $a = 12$, $B = 45°$
2. $C = 105°$, $c = 20$, $B = 40°$
3. $A = 10°$, $a = 4.5$, $B = 60°$
4. $C = 135°$, $c = 45$, $B = 10°$
5. $A = 36°$, $a = 8$, $b = 5$
6. $A = 60°$, $a = 9$, $c = 10$
7. $A = 102.4°$, $C = 16.7°$, $a = 21.6$
8. $A = 24.3°$, $C = 54.6°$, $c = 2.68$
9. $A = 83° \, 20'$, $C = 54.6°$, $c = 18.1$
10. $A = 5° \, 40'$, $B = 8° \, 15'$, $b = 4.8$
11. $B = 15° \, 30'$, $a = 4.5$, $b = 6.8$
12. $C = 85° \, 20'$, $a = 35$, $c = 50$
13. $A = 110° \, 15'$, $a = 48$, $b = 16$
14. $B = 2° \, 45'$, $b = 6.2$, $c = 5.8$

In Exercises 15–20, use the Law of Sines to solve the triangle. If two solutions exist, find both.

15. $A = 110°$, $a = 125$, $b = 100$
16. $A = 110°$, $a = 125$, $b = 200$
17. $A = 76°$, $a = 18$, $b = 20$
18. $A = 76°$, $a = 34$, $b = 21$
19. $A = 58°$, $a = 11.4$, $b = 12.8$
20. $A = 58°$, $a = 4.5$, $b = 12.8$

In Exercises 21–26, find the area of the triangle having the indicated angle and sides.

21. $C = 110°$, $a = 6$, $b = 10$

22. $B = 130°$, $a = 92$, $c = 30$
23. $A = 38° \, 45'$, $b = 67$, $c = 85$
24. $A = 5° \, 15'$, $b = 4.5$, $c = 22$
25. $B = 74° \, 30'$, $a = 103$, $c = 58$
26. $C = 84° \, 30'$, $a = 16$, $b = 20$

27. *Height*   A flagpole at a right angle to the horizontal is located on a slope that makes an angle of 14° with the horizontal. The flagpole casts a 16-meter shadow up the slope when the angle of elevation from the tip of the shadow to the sun is 20°.

    (a) Draw a triangle that represents the problem. Show the known quantities on the triangle and use a variable to indicate the height of the flagpole.

    (b) Write an equation involving the unknown quantity.

    (c) Find the height of the flagpole.

28. *Height*   You are standing 40 meters from the base of a tree that is leaning 8° from vertical away from you. The angle of elevation from your feet to the top of the tree is 20° 50′.

    (a) Draw a triangle that represents the problem. Show the known quantities on the triangle and use a variable to indicate the height of the tree.

    (b) Write an equation involving the unknown height of the tree.

    (c) Find the height of the tree.

29. *Flight Path*   A plane flies 500 kilometers with a bearing of 316° (clockwise from north) from Naples to Elgin (see figure on next page). The plane then flies 720 kilometers from Elgin to Canton. Find the bearing of the flight from Elgin to Canton.

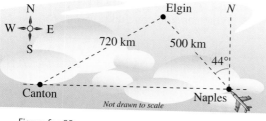

Figure for 29

30. Bridge Design   A bridge is to be built across a small lake from a gazebo to a dock (see figure). The bearing from the gazebo to the dock is S 41° W. From a tree 100 meters from the gazebo, the bearings to the gazebo and the dock are S 74° E and S 28° E, respectively. Find the distance from the gazebo to the dock.

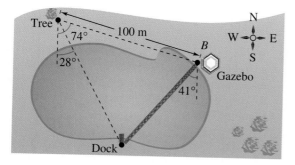

31. Railroad Track Design   The circular arc of a railroad curve has a chord of length 3000 feet and a central angle of 40°.

(a) Draw a diagram that visually represents the problem. Show the known quantities on the diagram and use the variables r and s to represent the radius of the arc and the length of the arc, respectively.

(b) Find the radius r of the circular arc.

(c) Find the length s of the circular arc.

32. Glide Path   A pilot has just started on the glide path for landing at an airport with a runway of length 9000 feet. The angles of depression from the plane to the ends of the runway are 17.5° and 18.8°.

(a) Draw a diagram that visually represents the problem.

(b) Find the air distance the plane must travel until touching down on the near end of the runway.

(c) Find the ground distance the plane must travel until touching down.

(d) Find the altitude of the plane when the pilot begins the descent.

33. Locating a Fire   The bearing from the Pine Knob fire tower to the Colt Station fire tower is N 65° E, and the two towers are 30 kilometers apart. A fire spotted by rangers in each tower has a bearing of N 80° E from Pine Knob and S 70° E from Colt Station. Find the distance of the fire from each tower.

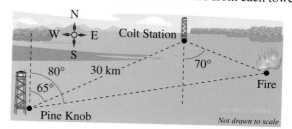

34. Distance   A boat is sailing due east parallel to the shoreline at a speed of 10 miles per hour. At a given time the bearing to the lighthouse is S 70° E, and 15 minutes later the bearing is S 63° E (see figure). The lighthouse is located at the shoreline. Find the distance from the boat to the shoreline.

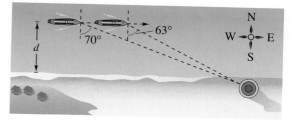

35. Angle of Elevation   A 10-meter telephone pole casts a 17-meter shadow directly down a slope when the angle of elevation of the sun is 42° (see figure). Find θ, the angle of elevation of the ground.

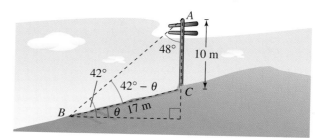

36. Distance   The angles of elevation θ and ϕ to an airplane are being continuously monitored at two observation points A and B, respectively, which are 2 miles apart, and the airplane is east of both points in the same vertical plane.

(a) Draw a diagram that illustrates the problem.

(b) Write an equation giving the distance d between the plane and point B in terms of θ and ϕ.

37. Shadow Length The Leaning Tower of Pisa in Italy leans because it was built on unstable soil—a mixture of clay, sand, and water. The tower is approximately 58.36 meters tall from its foundation (see figure). The top of the tower leans about 5.45 meters off center.

(a) Find the angle of lean α of the tower.

(b) Write β as a function of d and θ, where θ is the angle of elevation to the sun.

(c) Use the Law of Sines to write an equation for the length d of the shadow cast by the tower in terms of θ.

(d) Use a graphing utility to complete the table.

θ	10°	20°	30°	40°	50°	60°
d						

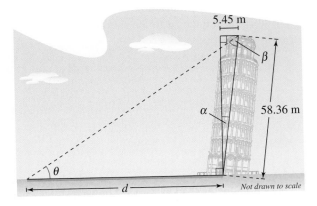

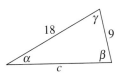

38. Graphical and Numerical Analysis In the figure, α and β are positive angles.

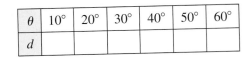

(a) Write α as a function of β.

(b) Use a graphing utility to graph the function. Determine its domain and range.

(c) Use the result of part (b) to write c as a function of β.

(d) Use a graphing utility to graph the function in part (c). Determine its domain and range.

(e) Use a graphing utility to complete the table. What can you conclude?

β	0.4	0.8	1.2	1.6	2.0	2.4	2.8
α							
c							

Synthesis

True or False? **In Exercises 39 and 40, determine whether the statement is true or false. Justify your answer.**

39. If any three sides or angles of an oblique triangle are known, then the triangle can be solved.

40. If a triangle contains an obtuse angle, then it must be oblique.

41. Writing Can the Law of Sines be used to solve a right triangle? If so, write a short paragraph explaining how to use the Law of Sines to solve the following triangle. Is there an easier way to solve the triangle? Explain.

$$B = 50°, \ C = 90°, \ a = 10$$

42. Think About It Given $A = 36°$ and $a = 5$, find a value for b such that the triangle has (a) one solution, (b) two solutions, and (c) no solution.

Review

In Exercises 43–46, use the given values to find (if possible) the values of the six trigonometric functions of θ.

43. $\cos \theta = \frac{5}{13}, \ \sin \theta = -\frac{12}{13}$

44. $\tan \theta = \frac{2}{9}, \ \csc \theta = -\frac{\sqrt{85}}{2}$

45. $\sec \theta = -\frac{\sqrt{122}}{11}, \ \cot \theta = -11$

46. $\cot \theta$ is undefined, $\cos \theta < 0$

In Exercises 47–50, write the product as a sum or difference.

47. $6 \sin 8\theta \cos 3\theta$

48. $2 \cos 2\theta \cos 5\theta$

49. $3 \cos \frac{\pi}{6} \sin \frac{5\pi}{3}$

50. $\frac{5}{2} \sin \frac{3\pi}{4} \sin \frac{5\pi}{6}$

7.2 Law of Cosines

Introduction

Two cases remain in the list of conditions needed to solve an oblique triangle—SSS and SAS. To use the Law of Sines, you must know at least one side and its opposite angle. If you are given three sides (SSS), or two sides and their included angle (SAS), none of the ratios in the Law of Sines would be complete. In such cases you can use the **Law of Cosines.** See Appendix B for a proof of the Law of Cosines.

Law of Cosines

Standard Form	*Alternative Form*
$a^2 = b^2 + c^2 - 2bc \cos A$	$\cos A = \dfrac{b^2 + c^2 - a^2}{2bc}$
$b^2 = a^2 + c^2 - 2ac \cos B$	$\cos B = \dfrac{a^2 + c^2 - b^2}{2ac}$
$c^2 = a^2 + b^2 - 2ab \cos C$	$\cos C = \dfrac{a^2 + b^2 - c^2}{2ab}$

Jed Jacobsohn/Getty Images

Example 1   Three Sides of a Triangle—SSS

Find the three angles of the triangle shown in Figure 7.11.

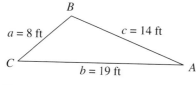

Figure 7.11

Solution

It is a good idea first to find the angle opposite the longest side—side b in this case. Using the Law of Cosines, you find that

$$\cos B = \frac{a^2 + c^2 - b^2}{2ac} = \frac{8^2 + 14^2 - 19^2}{2(8)(14)} \approx -0.45089.$$

Because $\cos B$ is negative, you know that B is an *obtuse* angle given by $B \approx 116.80°$. At this point it is simpler to use the Law of Sines to determine A.

$$\sin A = a\left(\frac{\sin B}{b}\right) \approx 8\left(\frac{\sin 116.80°}{19}\right) \approx 0.37583$$

Because B is obtuse, A must be acute, because a triangle can have at most one obtuse angle. So, $A \approx 22.08°$ and $C \approx 180° - 22.08° - 116.80° = 41.12°$.

 Checkpoint Now try Exercise 1.

Do you see why it was wise to find the largest angle *first* in Example 1? Knowing the cosine of an angle, you can determine whether the angle is acute or obtuse. That is,

$\cos \theta > 0$ for $0° < \theta < 90°$ Acute

$\cos \theta < 0$ for $90° < \theta < 180°$.   Obtuse

So, in Example 1, once you found that angle B was obtuse, you knew that angles A and C were both acute. Furthermore, if the largest angle is acute, the remaining two angles are also acute.

Exploration

What familiar formula do you obtain when you use the third form of the Law of Cosines

$$c^2 = a^2 + b^2 - 2ab \cos C$$

and you let $C = 90°$? What is the relationship between the Law of Cosines and this formula?

Example 2   Two Sides and the Included Angle—SAS

Find the remaining angles and the side of the triangle shown in Figure 7.12.

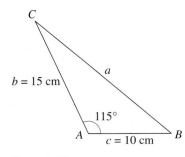

Figure 7.12

Solution

Use the Law of Cosines to find the unknown side a in the figure.

$a^2 = b^2 + c^2 - 2bc \cos A$

$a^2 = 15^2 + 10^2 - 2(15)(10) \cos 115°$

$a^2 \approx 451.79$

$a \approx 21.26$

Because $a \approx 21.26$ centimeters, you now know the ratio $\sin A / a$ and you can use the reciprocal form of the Law of Sines

$$\frac{\sin B}{b} = \frac{\sin A}{a}$$

to solve for B.

$\sin B = b\left(\dfrac{\sin A}{a}\right)$

$\quad = 15\left(\dfrac{\sin 115°}{21.26}\right)$

$\quad \approx 0.63945$

So, $B \approx \arcsin 0.63945 \approx 39.75°$, and $C \approx 180° - 115° - 39.75° = 25.25°$.

 Checkpoint Now try Exercise 3.

Applications

Example 3   An Application of the Law of Cosines

The pitcher's mound on a women's softball field is 43 feet from home plate and the distance between the bases is 60 feet, as shown in Figure 7.13. (The pitcher's mound is *not* halfway between home plate and second base.) How far is the pitcher's mound from first base?

Solution

In triangle *HPF*, $H = 45°$ (line *HP* bisects the right angle at *H*), $f = 43$, and $p = 60$. Using the Law of Cosines for this SAS case, you have

$$h^2 = f^2 + p^2 - 2fp \cos H$$

$$= 43^2 + 60^2 - 2(43)(60) \cos 45°$$

$$\approx 1800.3.$$

So, the approximate distance from the pitcher's mound to first base is

$$h \approx \sqrt{1800.3} \approx 42.43 \text{ feet.}$$

✓ *Checkpoint* Now try Exercise 25.

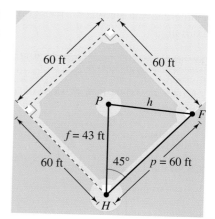

60 ft   60 ft

P   *h*

f = 43 ft   *F*

60 ft   45°   *p* = 60 ft

H

Figure 7.13

Example 4   An Application of the Law of Cosines

A ship travels 60 miles due east, then adjusts its course northward, as shown in Figure 7.14. After traveling 80 miles in the new direction, the ship is 139 miles from its point of departure. Describe the bearing from point *B* to point *C*.

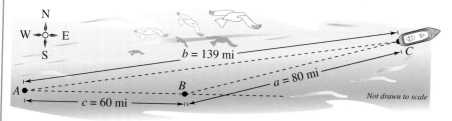

N
W →◇→ E
S

b = 139 mi   *C*

B

a = 80 mi

A

c = 60 mi

Not drawn to scale

Figure 7.14

Solution

You have $a = 80$, $b = 139$, and $c = 60$; so, using the alternative form of the Law of Cosines, you have

$$\cos B = \frac{a^2 + c^2 - b^2}{2ac} = \frac{80^2 + 60^2 - 139^2}{2(80)(60)} \approx -0.97094.$$

So, $B \approx \arccos(-0.97094) \approx 166.15°$. Therefore, the bearing measured from due north from point *B* to point *C* is $166.15° - 90° = 76.15°$, or N 76.15° E.

✓ *Checkpoint* Now try Exercise 27.

Heron's Area Formula

The Law of Cosines can be used to establish the following formula for the area of a triangle. This formula is called **Heron's Area Formula** after the Greek mathematician Heron (ca. 100 B.C.). See Appendix B for a proof of Heron's Area Formula.

Heron's Area Formula

Given any triangle with sides of lengths a, b, and c, the area of the triangle is given by

$$\text{Area} = \sqrt{s(s - a)(s - b)(s - c)}$$

where $s = \dfrac{a + b + c}{2}$.

Example 5   Using Heron's Area Formula

Find the area of a triangle having sides of lengths $a = 43$ meters, $b = 53$ meters, and $c = 72$ meters.

Solution

Because $s = (a + b + c)/2 = 168/2 = 84$, Heron's Area Formula yields

$$\text{Area} = \sqrt{s(s - a)(s - b)(s - c)}$$
$$= \sqrt{84(84 - 43)(84 - 53)(84 - 72)}$$
$$= \sqrt{84(41)(31)(12)}$$
$$\approx 1131.89 \text{ square meters.}$$

✓ *Checkpoint* Now try Exercise 35.

You have now studied three different formulas for the area of a triangle.

Formulas for Area of a Triangle

1. **Standard Formula:** $\text{Area} = \frac{1}{2}bh$

2. **Oblique Triangle:** $\text{Area} = \frac{1}{2}bc \sin A = \frac{1}{2}ab \sin C = \frac{1}{2}ac \sin B$

3. **Heron's Area Formula:** $\text{Area} = \sqrt{s(s - a)(s - b)(s - c)}$

Exploration

Can the above formulas be used to find the area of any type of triangle? Explain the advantages and disadvantages of using one formula over another.

7.2 Exercises

Vocabulary Check

Fill in the blanks.

1. The standard form of the Law of Cosines for $\cos C = \dfrac{a^2 + b^2 - c^2}{2ab}$ is _____ .

2. _____ Formula is established by using the Law of Cosines.

3. Three different formulas for the area of a triangle are given by Area = _____ , Area = $\frac{1}{2}bc \sin A$ = $\frac{1}{2}ab \sin C = \frac{1}{2}ac \sin B$, and Area = _____ .

In Exercises 1–12, use the Law of Cosines to solve the triangle.

1. $a = 6$, $b = 8$, $c = 12$
2. $a = 9$, $b = 3$, $c = 11$
3. $A = 50°$, $b = 15$, $c = 30$
4. $C = 108°$, $a = 10$, $b = 7$
5. $a = 9$, $b = 12$, $c = 15$
6. $a = 45$, $b = 30$, $c = 72$
7. $a = 75.4$, $b = 48$, $c = 48$
8. $a = 1.42$, $b = 0.75$, $c = 1.25$
9. $B = 8° 15'$, $a = 26$, $c = 18$
10. $B = 10° 35'$, $a = 40$, $c = 30$
11. $B = 75° 20'$, $a = 6.2$, $c = 9.5$
12. $B = 15° 15'$, $a = 6.25$, $b = 2.15$

In Exercises 13–18, complete the table by solving the parallelogram shown in the figure. (The lengths of the diagonals are given by c and d.)

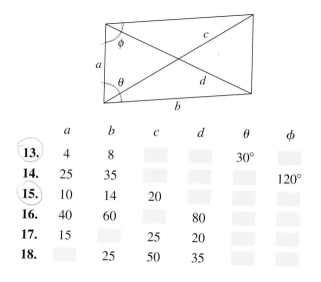

	a	b	c	d	θ	ϕ
13.	4	8			30°	
14.	25	35				120°
15.	10	14	20			
16.	40	60		80		
17.	15		25	20		
18.		25	50	35		

In Exercises 19–24, use Heron's Area Formula to find the area of the triangle.

19. $a = 5$, $b = 8$, $c = 10$
20. $a = 14$, $b = 17$, $c = 7$
21. $a = 3.5$, $b = 10.2$, $c = 9$
22. $a = 75.4$, $b = 52$, $c = 52$
23. $a = 10.59$, $b = 6.65$, $c = 12.31$
24. $a = 4.45$, $b = 1.85$, $c = 3.00$

25. **Navigation** A plane flies 810 miles from Franklin to Centerville with a bearing of 75° (clockwise from north). Then it flies 648 miles from Centerville to Rosemont with a bearing of 32°. Draw a diagram that visually represents the problem, and find the straight-line distance and bearing from Rosemont to Franklin.

26. **Navigation** A boat race runs along a triangular course marked by buoys A, B, and C. The race starts with the boats headed west for 3600 meters. The other two sides of the course lie to the north of the first side, and their lengths are 1500 meters and 2800 meters. Draw a diagram that visually represents the problem, and find the bearings for the last two legs of the race.

27. **Surveying** To approximate the length of a marsh, a surveyor walks 380 meters from point A to point B. Then the surveyor turns 80° and walks 240 meters to point C (see figure). Approximate the length AC of the marsh.

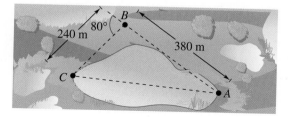

28. *Streetlight Design*   Determine the angle θ in the design of the streetlight shown in the figure.

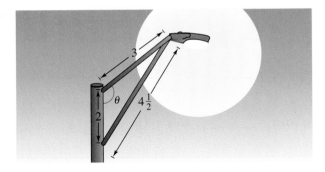

29. *Distance*   Two ships leave a port at 9 A.M. One travels at a bearing of N 53° W at 12 miles per hour, and the other travels at a bearing of S 67° W at 16 miles per hour. Approximate how far apart the ships are at noon that day.

30. *Length*   A 100-foot vertical tower is to be erected on the side of a hill that makes a 6° angle with the horizontal (see figure). Find the length of each of the two guy wires that will be anchored 75 feet uphill and downhill from the base of the tower.

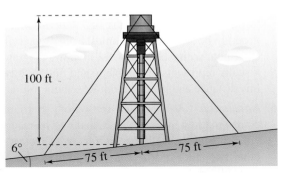

31. *Navigation*   On a map, Orlando is 178 millimeters due south of Niagara Falls, Denver is 273 millimeters from Orlando, and Denver is 235 millimeters from Niagara Falls (see figure).

(a) Find the bearing of Denver from Orlando.

(b) Find the bearing of Denver from Niagara Falls.

32. *Sports*   The baseball player in center field is playing approximately 330 feet from the television camera that is behind home plate. A batter hits a fly ball that goes to the wall 420 feet from the camera (see figure). The camera turns 6° to follow the play. Approximate the distance the center fielder has to run to make the catch.

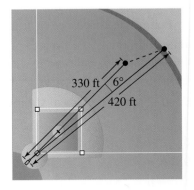

33. *Trusses* Q is the midpoint of the line segment $\overline{PR}$ in the truss rafter shown in the figure. What are the lengths of the line segments $\overline{PQ}$, $\overline{QS}$, and $\overline{RS}$?

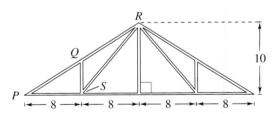

34. *Awning Design*   A retractable awning above a patio lowers at an angle of 50° from the exterior wall at a height of 10 feet above the ground (see figure). No direct sunlight is to enter the door when the angle of elevation of the sun is greater than 70°. What is the length x of the awning?

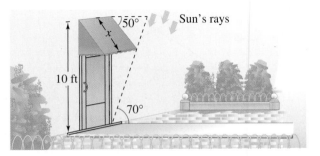

35. *Landau Building* The Landau Building in Cambridge, Massachusetts has a triangular-shaped base. The lengths of the sides of the triangular base are 145 feet, 257 feet, and 290 feet. Find the area of the base of the building.

36. *Geometry* A parking lot has the shape of a parallelogram (see figure). The lengths of two adjacent sides are 70 meters and 100 meters. The angle between the two sides is 70°. What is the area of the parking lot?

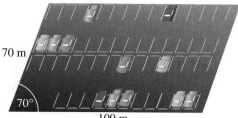

70 m

70°

100 m

37. *Engine Design* An engine has a seven-inch connecting rod fastened to a crank (see figure).

(a) Use the Law of Cosines to write an equation giving the relationship between x and θ.

(b) Write x as a function of θ. (Select the sign that yields positive values of x.)

(c) Use a graphing utility to graph the function in part (b).

(d) Use the graph in part (c) to determine the total distance the piston moves in one cycle.

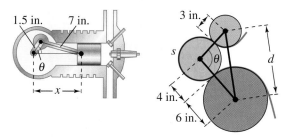

1.5 in.   7 in.

θ

x

3 in.

s

θ

d

4 in.

6 in.

Figure for 37   Figure for 38

38. *Manufacturing* In a process with continuous paper, the paper passes across three rollers of radii 3 inches, 4 inches, and 6 inches (see figure). The centers of the three-inch and six-inch rollers are d inches apart, and the length of the arc in contact with the paper on the four-inch roller is s inches.

(a) Use the Law of Cosines to write an equation giving the relationship between d and θ.

(b) Write θ as a function of d.

(c) Write s as a function of θ.

(d) Complete the table.

d (inches)	9	10	12	13	14	15	16
θ (degrees)							
s (inches)							

Synthesis

True or False? **In Exercises 39–41, determine whether the statement is true or false. Justify your answer.**

39. A triangle with side lengths of 10 feet, 16 feet, and 5 feet can be solved using the Law of Cosines.

40. Two sides and their included angle determine a unique triangle.

41. In Heron's Area Formula, s is the average of the lengths of the three sides of the triangle.

42. *Proof* Use the Law of Cosines to prove each of the following.

(a) $\dfrac{1}{2} bc (1 + \cos A) = \left(\dfrac{a+b+c}{2}\right)\left(\dfrac{-a+b+c}{2}\right)$

(b) $\dfrac{1}{2} bc (1 - \cos A) = \left(\dfrac{a-b+c}{2}\right)\left(\dfrac{a+b-c}{2}\right)$

Review

In Exercises 43–46, evaluate the expression without using a calculator.

43. $\arcsin(-1)$

44. $\cos^{-1} 0$

45. $\tan^{-1} \sqrt{3}$

46. $\arcsin\left(-\dfrac{\sqrt{3}}{2}\right)$

In Exercises 47 and 48, write the sum or difference as a product.

47. $\cos \dfrac{5\pi}{6} - \cos \dfrac{\pi}{3}$

48. $\sin\left(x - \dfrac{\pi}{2}\right) - \sin\left(x + \dfrac{\pi}{2}\right)$

7.3 Vectors in the Plane

Introduction

Many quantities in geometry and physics, such as area, time, and temperature, can be represented by a single real number. Other quantities, such as force and velocity, involve both *magnitude* and *direction* and cannot be completely characterized by a single real number. To represent such a quantity, you can use a **directed line segment,** as shown in Figure 7.15. The directed line segment $\overrightarrow{PQ}$ has **initial point** P and **terminal point** Q. Its **magnitude,** or **length,** is denoted by $\|\overrightarrow{PQ}\|$ and can be found by using the Distance Formula.

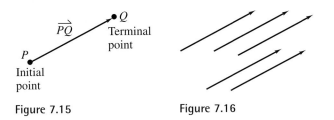

Figure 7.15   **Figure 7.16**

Two directed line segments that have the same magnitude and direction are *equivalent.* For example, the directed line segments in Figure 7.16 are all equivalent. The set of all directed line segments that are equivalent to a given directed line segment $\overrightarrow{PQ}$ is a **vector v in the plane,** written $\mathbf{v} = \overrightarrow{PQ}$. Vectors are denoted by lowercase, boldface letters such as $\mathbf{u}$, $\mathbf{v}$, and $\mathbf{w}$.

Example 1   Equivalent Directed Line Segments

Let $\mathbf{u}$ be represented by the directed line segment from $P = (0, 0)$ to $Q = (3, 2)$, and let $\mathbf{v}$ be represented by the directed line segment from $R = (1, 2)$ to $S = (4, 4)$, as shown in Figure 7.17. Show that $\mathbf{u} = \mathbf{v}$.

Solution

From the Distance Formula, it follows that $\overrightarrow{PQ}$ and $\overrightarrow{RS}$ have the *same magnitude.*

$$\|\overrightarrow{PQ}\| = \sqrt{(3 - 0)^2 + (2 - 0)^2} = \sqrt{13}$$

$$\|\overrightarrow{RS}\| = \sqrt{(4 - 1)^2 + (4 - 2)^2} = \sqrt{13}$$

Moreover, both line segments have the *same direction,* because they are both directed toward the upper right on lines having the same slope.

$$\text{Slope of } \overrightarrow{PQ} = \frac{2 - 0}{3 - 0} = \frac{2}{3}$$

$$\text{Slope of } \overrightarrow{RS} = \frac{4 - 2}{4 - 1} = \frac{2}{3}$$

So, $\overrightarrow{PQ}$ and $\overrightarrow{RS}$ have the same magnitude and direction, and it follows that $\mathbf{u} = \mathbf{v}$.

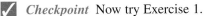 *Checkpoint* Now try Exercise 1.

What you should learn

- Represent vectors as directed line segments.
- Write the component form of vectors.
- Perform basic vector operations and represent vectors graphically.
- Write vectors as linear combinations of unit vectors.
- Find the direction angles of vectors.
- Use vectors to model and solve real-life problems.

Why you should learn it

Vectors are used to analyze numerous aspects of everyday life. Exercise 78 on page 559 shows you how vectors can be used to determine the tension in the cables of two cranes lifting an object.

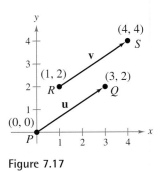

Figure 7.17

Component Form of a Vector

The directed line segment whose initial point is the origin is often the most convenient representative of a set of equivalent directed line segments. This representative of the vector **v** is in **standard position.**

A vector whose initial point is at the origin $(0, 0)$ can be uniquely represented by the coordinates of its terminal point (v_1, v_2). This is the **component form of a vector v,** written as

$$\mathbf{v} = \langle v_1, v_2 \rangle.$$

The coordinates v_1 and v_2 are the *components* of **v.** If both the initial point and the terminal point lie at the origin, **v** is the **zero vector** and is denoted by $\mathbf{0} = \langle 0, 0 \rangle$.

Component Form of a Vector

The component form of the vector with initial point $P = (p_1, p_2)$ and terminal point $Q = (q_1, q_2)$ is given by

$$\overrightarrow{PQ} = \langle q_1 - p_1, q_2 - p_2 \rangle = \langle v_1, v_2 \rangle = \mathbf{v}.$$

The **magnitude** (or length) of **v** is given by

$$\|\mathbf{v}\| = \sqrt{(q_1 - p_1)^2 + (q_2 - p_2)^2} = \sqrt{v_1^2 + v_2^2}.$$

If $\|\mathbf{v}\| = 1$, **v** is a **unit vector.** Moreover, $\|\mathbf{v}\| = 0$ if and only if **v** is the zero vector **0**.

Two vectors $\mathbf{u} = \langle u_1, u_2 \rangle$ and $\mathbf{v} = \langle v_1, v_2 \rangle$ are *equal* if and only if $u_1 = v_1$ and $u_2 = v_2$. For instance, in Example 1, the vector **u** from $P = (0, 0)$ to $Q = (3, 2)$ is

$$\mathbf{u} = \overrightarrow{PQ} = \langle 3 - 0, 2 - 0 \rangle = \langle 3, 2 \rangle$$

and the vector **v** from $R = (1, 2)$ to $S = (4, 4)$ is

$$\mathbf{v} = \overrightarrow{RS} = \langle 4 - 1, 4 - 2 \rangle = \langle 3, 2 \rangle.$$

Example 2   Finding the Component Form of a Vector

Find the component form and magnitude of the vector **v** that has initial point $(4, -7)$ and terminal point $(-1, 5)$.

Solution

Let $P = (4, -7) = (p_1, p_2)$ and $Q = (-1, 5) = (q_1, q_2)$, as shown in Figure 7.18. Then, the components of $\mathbf{v} = \langle v_1, v_2 \rangle$ are

$$v_1 = q_1 - p_1 = -1 - 4 = -5$$

$$v_2 = q_2 - p_2 = 5 - (-7) = 12.$$

So, $\mathbf{v} = \langle -5, 12 \rangle$ and the magnitude of **v** is

$$\|\mathbf{v}\| = \sqrt{(-5)^2 + 12^2} = \sqrt{169} = 13.$$

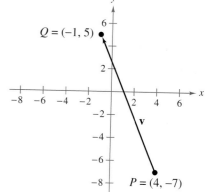

Figure 7.18

 Checkpoint Now try Exercise 11.

Vector Operations

The two basic vector operations are **scalar multiplication** and **vector addition.** Geometrically, the product of a vector **v** and a scalar k is the vector that is $|k|$ times as long as **v**. If k is positive, $k\mathbf{v}$ has the same direction as **v**, and if k is negative, $k\mathbf{v}$ has the opposite direction of **v**, as shown in Figure 7.19.

To add two vectors geometrically, position them (without changing their lengths or directions) so that the initial point of one coincides with the terminal point of the other. The sum **u** + **v** is formed by joining the initial point of the second vector **v** with the terminal point of the first vector **u**, as shown in Figure 7.20. This technique is called the **parallelogram law** for vector addition because the vector **u** + **v**, often called the **resultant** of vector addition, is the diagonal of a parallelogram having **u** and **v** as its adjacent sides.

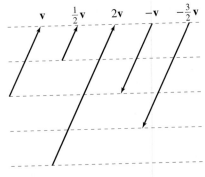

Figure 7.19

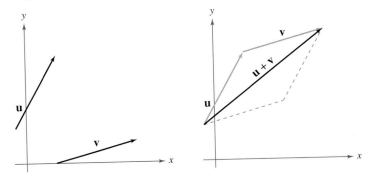

Figure 7.20

> **Definition of Vector Addition and Scalar Multiplication**
>
> Let $\mathbf{u} = \langle u_1, u_2 \rangle$ and $\mathbf{v} = \langle v_1, v_2 \rangle$ be vectors and let k be a scalar (a real number). Then the **sum** of **u** and **v** is the vector
>
> $$\mathbf{u} + \mathbf{v} = \langle u_1 + v_1, u_2 + v_2 \rangle \qquad \text{Sum}$$
>
> and the **scalar multiple** of k times **u** is the vector
>
> $$k\mathbf{u} = k\langle u_1, u_2 \rangle = \langle ku_1, ku_2 \rangle. \qquad \text{Scalar multiple}$$

The **negative** of $\mathbf{v} = \langle v_1, v_2 \rangle$ is

$$-\mathbf{v} = (-1)\mathbf{v}$$

$$= \langle -v_1, -v_2 \rangle \qquad \text{Negative}$$

and the **difference** of **u** and **v** is

$$\mathbf{u} - \mathbf{v} = \mathbf{u} + (-\mathbf{v}) \qquad \text{Add } (-\mathbf{v}). \text{ See Figure 7.21.}$$

$$= \langle u_1 - v_1, u_2 - v_2 \rangle. \qquad \text{Difference}$$

Figure 7.21

To represent **u** − **v** geometrically, you can use directed line segments with the *same* initial point. The difference **u** − **v** is the vector from the terminal point of **v** to the terminal point of **u**, which is equal to **u** + (−**v**), as shown in Figure 7.21.

The component definitions of vector addition and scalar multiplication are illustrated in Example 3. In this example, notice that each of the vector operations can be interpreted geometrically.

Example 3   Vector Operations

Let $\mathbf{v} = \langle -2, 5 \rangle$ and $\mathbf{w} = \langle 3, 4 \rangle$, and find each of the following vectors.

a. $2\mathbf{v}$ **b.** $\mathbf{w} - \mathbf{v}$ **c.** $\mathbf{v} + 2\mathbf{w}$

Solution

a. Because $\mathbf{v} = \langle -2, 5 \rangle$, you have

$$2\mathbf{v} = 2\langle -2, 5 \rangle$$
$$= \langle 2(-2), 2(5) \rangle$$
$$= \langle -4, 10 \rangle.$$

A sketch of $2\mathbf{v}$ is shown in Figure 7.22.

b. The difference of $\mathbf{w}$ and $\mathbf{v}$ is

$$\mathbf{w} - \mathbf{v} = \langle 3 - (-2), 4 - 5 \rangle$$
$$= \langle 5, -1 \rangle.$$

A sketch of $\mathbf{w} - \mathbf{v}$ is shown in Figure 7.23.

c. The sum of $\mathbf{v}$ and $2\mathbf{w}$ is

$$\mathbf{v} + 2\mathbf{w} = \langle -2, 5 \rangle + 2\langle 3, 4 \rangle$$
$$= \langle -2, 5 \rangle + \langle 2(3), 2(4) \rangle$$
$$= \langle -2, 5 \rangle + \langle 6, 8 \rangle$$
$$= \langle -2 + 6, 5 + 8 \rangle$$
$$= \langle 4, 13 \rangle.$$

A sketch of $\mathbf{v} + 2\mathbf{w}$ is shown in Figure 7.24.

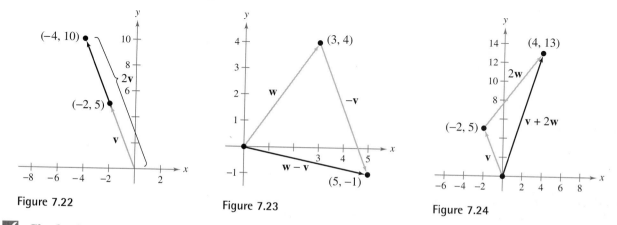

Figure 7.22

Figure 7.23

Figure 7.24

✓ *Checkpoint* Now try Exercise 19.

Note that Figure 7.23 shows the vector difference $\mathbf{w} - \mathbf{v}$ as the sum $\mathbf{w} + (-\mathbf{v})$.

Vector addition and scalar multiplication share many of the properties of ordinary arithmetic.

Properties of Vector Addition and Scalar Multiplication

Let **u**, **v**, and **w** be vectors and let c and d be scalars. Then the following properties are true.

1. $\mathbf{u} + \mathbf{v} = \mathbf{v} + \mathbf{u}$ 2. $(\mathbf{u} + \mathbf{v}) + \mathbf{w} = \mathbf{u} + (\mathbf{v} + \mathbf{w})$

3. $\mathbf{u} + \mathbf{0} = \mathbf{u}$ 4. $\mathbf{u} + (-\mathbf{u}) = \mathbf{0}$

5. $c(d\mathbf{u}) = (cd)\mathbf{u}$ 6. $(c + d)\mathbf{u} = c\mathbf{u} + d\mathbf{u}$

7. $c(\mathbf{u} + \mathbf{v}) = c\mathbf{u} + c\mathbf{v}$ 8. $1(\mathbf{u}) = \mathbf{u},\ 0(\mathbf{u}) = \mathbf{0}$

9. $\|c\mathbf{v}\| = |c|\,\|\mathbf{v}\|$

> **STUDY TIP**
>
> Property 9 can be stated as follows: The magnitude of the vector $c\mathbf{v}$ is the absolute value of c times the magnitude of **v**.

Unit Vectors

In many applications of vectors, it is useful to find a unit vector that has the same direction as a given nonzero vector **v**. To do this, you can divide **v** by its length to obtain

$$\mathbf{u} = \text{unit vector} = \frac{\mathbf{v}}{\|\mathbf{v}\|} = \left(\frac{1}{\|\mathbf{v}\|}\right)\mathbf{v}. \qquad \text{Unit vector in direction of } \mathbf{v}$$

Note that **u** is a scalar multiple of **v**. The vector **u** has a magnitude of 1 and the same direction as **v**. The vector **u** is called a **unit vector in the direction of v.**

Example 4 Finding a Unit Vector

Find a unit vector in the direction of $\mathbf{v} = \langle -2, 5 \rangle$ and verify that the result has a magnitude of 1.

Solution
The unit vector in the direction of **v** is

$$\frac{\mathbf{v}}{\|\mathbf{v}\|} = \frac{\langle -2, 5 \rangle}{\sqrt{(-2)^2 + (5)^2}}$$

$$= \frac{1}{\sqrt{29}}\langle -2, 5 \rangle$$

$$= \left\langle \frac{-2}{\sqrt{29}}, \frac{5}{\sqrt{29}} \right\rangle.$$

This vector has a magnitude of 1 because

$$\sqrt{\left(\frac{-2}{\sqrt{29}}\right)^2 + \left(\frac{5}{\sqrt{29}}\right)^2} = \sqrt{\frac{4}{29} + \frac{25}{29}} = \sqrt{\frac{29}{29}} = 1.$$

✓ *Checkpoint* Now try Exercise 27.

The unit vectors $\langle 1, 0 \rangle$ and $\langle 0, 1 \rangle$ are called the **standard unit vectors** and are denoted by

$$\mathbf{i} = \langle 1, 0 \rangle \qquad \text{and} \qquad \mathbf{j} = \langle 0, 1 \rangle$$

as shown in Figure 7.25. (Note that the lowercase letter $\mathbf{i}$ is written in boldface to distinguish it from the imaginary number $i = \sqrt{-1}$.) These vectors can be used to represent any vector $\mathbf{v} = \langle v_1, v_2 \rangle$ as follows.

$$\mathbf{v} = \langle v_1, v_2 \rangle$$

$$= v_1 \langle 1, 0 \rangle + v_2 \langle 0, 1 \rangle$$

$$= v_1 \mathbf{i} + v_2 \mathbf{j}$$

The scalars v_1 and v_2 are called the **horizontal and vertical components of v,** respectively. The vector sum

$$v_1 \mathbf{i} + v_2 \mathbf{j}$$

is called a **linear combination** of the vectors $\mathbf{i}$ and $\mathbf{j}$. Any vector in the plane can be written as a linear combination of the standard unit vectors $\mathbf{i}$ and $\mathbf{j}$.

Figure 7.25

Example 5   Writing a Linear Combination of Unit Vectors

Let $\mathbf{u}$ be the vector with initial point $(2, -5)$ and terminal point $(-1, 3)$. Write $\mathbf{u}$ as a linear combination of the standard unit vectors $\mathbf{i}$ and $\mathbf{j}$.

Solution

Begin by writing the component form of the vector $\mathbf{u}$.

$$\mathbf{u} = \langle -1 - 2, 3 - (-5) \rangle$$

$$= \langle -3, 8 \rangle$$

$$= -3\mathbf{i} + 8\mathbf{j}$$

This result is shown graphically in Figure 7.26.

✓ *Checkpoint* Now try Exercise 41.

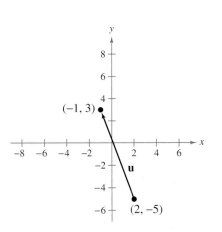

Figure 7.26

Example 6   Vector Operations

Let $\mathbf{u} = -3\mathbf{i} + 8\mathbf{j}$ and $\mathbf{v} = 2\mathbf{i} - \mathbf{j}$. Find $2\mathbf{u} - 3\mathbf{v}$.

Solution

You could solve this problem by converting $\mathbf{u}$ and $\mathbf{v}$ to component form. This, however, is not necessary. It is just as easy to perform the operations in unit vector form.

$$2\mathbf{u} - 3\mathbf{v} = 2(-3\mathbf{i} + 8\mathbf{j}) - 3(2\mathbf{i} - \mathbf{j})$$

$$= -6\mathbf{i} + 16\mathbf{j} - 6\mathbf{i} + 3\mathbf{j}$$

$$= -12\mathbf{i} + 19\mathbf{j}$$

✓ *Checkpoint* Now try Exercise 47.

Direction Angles

If **u** is a *unit vector* such that θ is the angle (measured counterclockwise) from the positive x-axis to **u**, the terminal point of **u** lies on the unit circle and you have

$$\mathbf{u} = \langle x, y \rangle = \langle \cos\theta, \sin\theta \rangle = (\cos\theta)\mathbf{i} + (\sin\theta)\mathbf{j}$$

as shown in Figure 7.27. The angle θ is the **direction angle** of the vector **u**.

Suppose that **u** is a unit vector with direction angle θ. If $\mathbf{v} = a\mathbf{i} + b\mathbf{j}$ is any vector that makes an angle θ with the positive x-axis, then it has the same direction as **u** and you can write

$$\mathbf{v} = \|\mathbf{v}\| \langle \cos\theta, \sin\theta \rangle$$

$$= \|\mathbf{v}\| (\cos\theta)\mathbf{i} + \|\mathbf{v}\| (\sin\theta)\mathbf{j}.$$

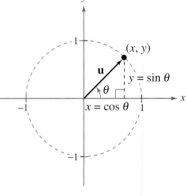

Figure 7.27

Because $\mathbf{v} = a\mathbf{i} + b\mathbf{j} = \|\mathbf{v}\| (\cos\theta)\mathbf{i} + \|\mathbf{v}\| (\sin\theta)\mathbf{j}$, it follows that the direction angle θ for **v** is determined from

$$\tan\theta = \frac{\sin\theta}{\cos\theta} \qquad \text{Quotient identity}$$

$$= \frac{\|\mathbf{v}\| \sin\theta}{\|\mathbf{v}\| \cos\theta} \qquad \text{Multiply numerator and denominator by } \|\mathbf{v}\|.$$

$$= \frac{b}{a}. \qquad \text{Simplify.}$$

Example 7   Finding Direction Angles of Vectors

Find the direction angle of each vector.

a. $\mathbf{u} = 3\mathbf{i} + 3\mathbf{j}$ **b.** $\mathbf{v} = 3\mathbf{i} - 4\mathbf{j}$

Solution

a. The direction angle is

$$\tan\theta = \frac{b}{a} = \frac{3}{3} = 1.$$

So, $\theta = 45°$, as shown in Figure 7.28.

b. The direction angle is

$$\tan\theta = \frac{b}{a} = \frac{-4}{3}.$$

Moreover, because $\mathbf{v} = 3\mathbf{i} - 4\mathbf{j}$ lies in Quadrant IV, θ lies in Quadrant IV and its reference angle is

$$\theta' = \left| \arctan\left(-\frac{4}{3}\right) \right| \approx |-53.13°| = 53.13°.$$

So, it follows that $\theta \approx 360° - 53.13° = 306.87°$, as shown in Figure 7.29.

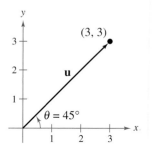

Figure 7.28

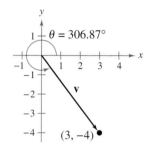

Figure 7.29

✓ *Checkpoint* Now try Exercise 55.

Applications of Vectors

Example 8   Finding the Component Form of a Vector

Find the component form of the vector that represents the velocity of an airplane descending at a speed of 100 miles per hour at an angle of 30° below the horizontal, as shown in Figure 7.30.

Solution

The velocity vector $\mathbf{v}$ has a magnitude of 100 and a direction angle of $\theta = 210°$.

$$\mathbf{v} = \|\mathbf{v}\| (\cos \theta)\mathbf{i} + \|\mathbf{v}\| (\sin \theta)\mathbf{j}$$

$$= 100(\cos 210°)\mathbf{i} + 100(\sin 210°)\mathbf{j}$$

$$= 100\left(-\frac{\sqrt{3}}{2}\right)\mathbf{i} + 100\left(-\frac{1}{2}\right)\mathbf{j}$$

$$= -50\sqrt{3}\,\mathbf{i} - 50\mathbf{j} = \langle -50\sqrt{3}, -50\rangle$$

You can check that $\mathbf{v}$ has a magnitude of 100 as follows.

$$\|\mathbf{v}\| = \sqrt{\left(-50\sqrt{3}\right)^2 + (-50)^2}$$

$$= \sqrt{7500 + 2500} = \sqrt{10{,}000} = 100 \qquad \text{Solution checks.} \checkmark$$

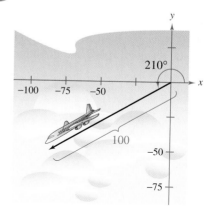

Figure 7.30

✓ *Checkpoint* Now try Exercise 75.

Example 9   Using Vectors to Determine Weight

A force of 600 pounds is required to pull a boat and trailer up a ramp inclined at 15° from the horizontal. Find the combined weight of the boat and trailer.

Solution

Based on Figure 7.31, you can make the following observations.

$\|\overrightarrow{BA}\|$ = force of gravity = combined weight of boat and trailer

$\|\overrightarrow{BC}\|$ = force against ramp

$\|\overrightarrow{AC}\|$ = force required to move boat up ramp = 600 pounds

By construction, triangles BWD and ABC are similar. So, angle ABC is 15°. In triangle ABC you have

$$\sin 15° = \frac{\|\overrightarrow{AC}\|}{\|\overrightarrow{BA}\|} = \frac{600}{\|\overrightarrow{BA}\|}$$

$$\|\overrightarrow{BA}\| = \frac{600}{\sin 15°} \approx 2318.$$

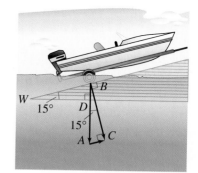

Figure 7.31

So, the combined weight is approximately 2318 pounds. (In Figure 7.31, note that $\overrightarrow{AC}$ is parallel to the ramp.)

✓ *Checkpoint* Now try Exercise 77.

Example 10   Using Vectors to Find Speed and Direction

An airplane is traveling at a speed of 500 miles per hour with a bearing of 330° at a fixed altitude with a negligible wind velocity, as shown in Figure 7.32(a). As the airplane reaches a certain point, it encounters a wind blowing with a velocity of 70 miles per hour in the direction N 45° E, as shown in Figure 7.32(b). What are the resultant speed and direction of the airplane?

STUDY TIP

Recall from Section 5.7 that in air navigation, bearings are measured in degrees clockwise from north.

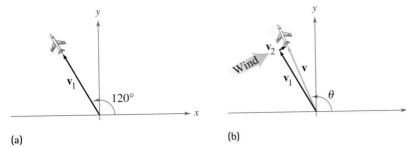

(a)   (b)

Figure 7.32

Solution

Using Figure 7.32, the velocity of the airplane (alone) is

$$\mathbf{v}_1 = 500\langle \cos 120°, \sin 120° \rangle$$
$$= \langle -250, 250\sqrt{3} \rangle$$

and the velocity of the wind is

$$\mathbf{v}_2 = 70\langle \cos 45°, \sin 45° \rangle$$
$$= \langle 35\sqrt{2}, 35\sqrt{2} \rangle.$$

So, the velocity of the airplane (in the wind) is

$$\mathbf{v} = \mathbf{v}_1 + \mathbf{v}_2$$
$$= \langle -250 + 35\sqrt{2}, 250\sqrt{3} + 35\sqrt{2} \rangle$$
$$\approx \langle -200.5, 482.5 \rangle$$

and the resultant speed of the airplane is

$$\|\mathbf{v}\| = \sqrt{(-200.5)^2 + (482.5)^2}$$
$$\approx 522.5 \text{ miles per hour.}$$

Finally, if θ is the direction angle of the flight path, you have

$$\tan \theta = \frac{482.5}{-200.5} \approx -2.4065$$

which implies that

$$\theta \approx 180° + \arctan(-2.4065) \approx 180° - 67.4° = 112.6°.$$

So, the true direction of the airplane is 337.4°.

✓ *Checkpoint* Now try Exercise 81.

7.3 Exercises

Vocabulary Check

Fill in the blanks.

1. A _____ can be used to represent a quantity that involves both magnitude and direction.

2. The directed line segment $\overrightarrow{PQ}$ has _____ point P and _____ point Q.

3. The _____ of the directed line segment $\overrightarrow{PQ}$ is denoted by $\|\overrightarrow{PQ}\|$.

4. The set of all directed line segments that are equivalent to a given directed line segment $\overrightarrow{PQ}$ is a _____ **v** in the plane.

5. The directed line segment whose initial point is the origin is said to be in _____ .

6. A vector that has a magnitude of 1 is called a _____ .

7. The two basic vector operations are scalar _____ and vector _____ .

8. The vector $\mathbf{u} + \mathbf{v}$ is called the _____ of vector addition.

9. The vector sum $v_1\mathbf{i} + v_2\mathbf{j}$ is called a _____ of the vectors $\mathbf{i}$ and $\mathbf{j}$, and the scalars v_1 and v_2 are called the _____ and _____ components of **v**, respectively.

In Exercises 1 and 2, show that u = v.

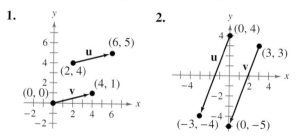

1.
2.

7.
8.

In Exercises 3–12, find the component form and the magnitude of the vector v.

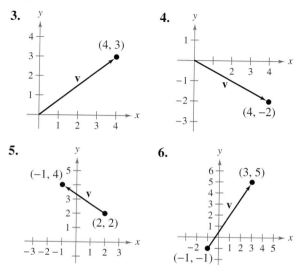

3.
4.

5.
6.

Initial Point	Terminal Point
9. $\left(\frac{5}{2}, 1\right)$	$\left(-2, -\frac{3}{2}\right)$
10. $(3.4, 0)$	$(0, 5.8)$
11. $(-3, -5)$	$(5, 1)$
12. $(-3, 11)$	$(9, 40)$

In Exercises 13–18, use the figure to sketch a graph of the specified vector. To print an enlarged copy of the graph, go to the website *www.mathgraphs.com*.

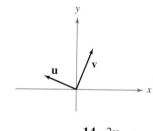

13. $-\mathbf{v}$

14. $3\mathbf{u}$

15. $\mathbf{u} + \mathbf{v}$

16. $\mathbf{u} - \mathbf{v}$

17. $\mathbf{u} + 2\mathbf{v}$

18. $\mathbf{v} - \frac{1}{2}\mathbf{u}$

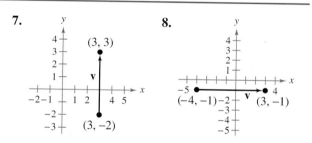

In Exercises 19–24, find (a) u + v, (b) u − v, (c) 2u − 3v, and (d) v + 4u.

19. $\mathbf{u} = \langle 4, 2 \rangle, \quad \mathbf{v} = \langle 7, 1 \rangle$

20. $\mathbf{u} = \langle 5, 3 \rangle, \quad \mathbf{v} = \langle -4, 0 \rangle$

21. $\mathbf{u} = \langle -6, -8 \rangle, \quad \mathbf{v} = \langle 2, 4 \rangle$

22. $\mathbf{u} = \langle 0, -5 \rangle, \quad \mathbf{v} = \langle -3, 9 \rangle$

23. $\mathbf{u} = \mathbf{i} + \mathbf{j}, \quad \mathbf{v} = 2\mathbf{i} - 3\mathbf{j}$

24. $\mathbf{u} = 2\mathbf{i} - \mathbf{j}, \quad \mathbf{v} = -\mathbf{i} + \mathbf{j}$

In Exercises 25–34, find a unit vector in the direction of the given vector.

25. $\mathbf{u} = \langle 6, 0 \rangle$

26. $\mathbf{u} = \langle 0, -2 \rangle$

27. $\mathbf{v} = \langle -1, 1 \rangle$

28. $\mathbf{v} = \langle 3, -4 \rangle$

29. $\mathbf{v} = \langle -24, -7 \rangle$

30. $\mathbf{v} = \langle 8, -20 \rangle$

31. $\mathbf{v} = 4\mathbf{i} - 3\mathbf{j}$

32. $\mathbf{w} = \mathbf{i} - 2\mathbf{j}$

33. $\mathbf{w} = 2\mathbf{j}$

34. $\mathbf{w} = -3\mathbf{i}$

In Exercises 35–40, find the vector v with the given magnitude and the same direction as u.

Magnitude	Direction
35. $\|\mathbf{v}\| = 8$	$\mathbf{u} = \langle 5, 6 \rangle$
36. $\|\mathbf{v}\| = 3$	$\mathbf{u} = \langle 4, -4 \rangle$
37. $\|\mathbf{v}\| = 7$	$\mathbf{u} = 3\mathbf{i} + 4\mathbf{j}$
38. $\|\mathbf{v}\| = 10$	$\mathbf{u} = 2\mathbf{i} - 3\mathbf{j}$
39. $\|\mathbf{v}\| = 8$	$\mathbf{u} = -2\mathbf{i}$
40. $\|\mathbf{v}\| = 4$	$\mathbf{u} = 5\mathbf{j}$

In Exercises 41–44, the initial and terminal points of a vector are given. Write a linear combination of the standard unit vectors i and j.

Initial Point	Terminal Point
41. $(-3, 1)$	$(4, 5)$
42. $(0, -2)$	$(3, 6)$
43. $(-1, -5)$	$(2, 3)$
44. $(-6, 4)$	$(0, 1)$

In Exercises 45–50, find the component form of v and sketch the specified vector operations geometrically, where u = 2i − j and w = i + 2j.

45. $\mathbf{v} = \frac{3}{2}\mathbf{u}$

46. $\mathbf{v} = \frac{2}{3}\mathbf{w}$

47. $\mathbf{v} = \mathbf{u} + 2\mathbf{w}$

48. $\mathbf{v} = -\mathbf{u} + \mathbf{w}$

49. $\mathbf{v} = \frac{1}{2}(3\mathbf{u} + \mathbf{w})$

50. $\mathbf{v} = 2\mathbf{u} - 2\mathbf{w}$

In Exercises 51–56, find the magnitude and direction angle of the vector v.

51. $\mathbf{v} = 5(\cos 30°\mathbf{i} + \sin 30°\mathbf{j})$

52. $\mathbf{v} = 8(\cos 135°\mathbf{i} + \sin 135°\mathbf{j})$

53. $\mathbf{v} = 6\mathbf{i} - 6\mathbf{j}$

54. $\mathbf{v} = -4\mathbf{i} - 7\mathbf{j}$

55. $\mathbf{v} = -2\mathbf{i} + 5\mathbf{j}$

56. $\mathbf{v} = 12\mathbf{i} + 15\mathbf{j}$

In Exercises 57–62, find the component form of v given its magnitude and the angle it makes with the positive x-axis. Sketch v.

Magnitude	Angle
57. $\|\mathbf{v}\| = 3$	$\theta = 0°$
58. $\|\mathbf{v}\| = 1$	$\theta = 45°$
59. $\|\mathbf{v}\| = 3\sqrt{2}$	$\theta = 150°$
60. $\|\mathbf{v}\| = 4\sqrt{3}$	$\theta = 90°$
61. $\|\mathbf{v}\| = 2$	v in the direction $\mathbf{i} + 3\mathbf{j}$
62. $\|\mathbf{v}\| = 3$	v in the direction $3\mathbf{i} + 4\mathbf{j}$

In Exercises 63–66, find the component form of the sum of u and v with direction angles θ_u and θ_v.

Magnitude	Angle
63. $\|\mathbf{u}\| = 5$	$\theta_u = 60°$
$\|\mathbf{v}\| = 5$	$\theta_v = 90°$
64. $\|\mathbf{u}\| = 2$	$\theta_u = 30°$
$\|\mathbf{v}\| = 2$	$\theta_v = 90°$
65. $\|\mathbf{u}\| = 20$	$\theta_u = 45°$
$\|\mathbf{v}\| = 50$	$\theta_v = 150°$
66. $\|\mathbf{u}\| = 35$	$\theta_u = 25°$
$\|\mathbf{v}\| = 50$	$\theta_v = 120°$

In Exercises 67 and 68, use the Law of Cosines to find the angle α between the vectors. (Assume $0° \le \alpha \le 180°$.)

67. $\mathbf{v} = \mathbf{i} + \mathbf{j}, \quad \mathbf{w} = 2(\mathbf{i} - \mathbf{j})$

68. $\mathbf{v} = 3\mathbf{i} + \mathbf{j}, \quad \mathbf{w} = 2\mathbf{i} - \mathbf{j}$

In Exercises 69 and 70, graph the vectors and the resultant of the vectors. Find the magnitude and direction of the resultant.

69.   **70.**

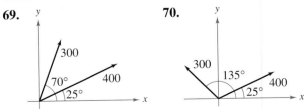

Resultant Force   In Exercises 71 and 72, find the angle between the forces given the magnitude of their resultant. (*Hint:* Write force 1 as a vector in the direction of the positive x-axis and force 2 as a vector at an angle θ with the positive x-axis.)

	Force 1	Force 2	Resultant Force
71.	45 pounds	60 pounds	90 pounds
72.	3000 pounds	1000 pounds	3750 pounds

73. ***Resultant Force***   Forces with magnitudes of 2000 newtons and 900 newtons act on a machine part at angles of 30° and −45°, respectively, with the positive x-axis (see figure). Find the direction and magnitude of the resultant of these forces.

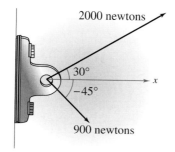

74. ***Resultant Force***

(a) Three forces with magnitudes of 70 pounds, 40 pounds, and 60 pounds act on an object at angles of −30°, 45°, and 135°, respectively, with the positive x-axis. Find the direction and magnitude of the resultant of these forces.

(b) Three forces with magnitudes of 75 pounds, 100 pounds, and 125 pounds act on an object at angles of 30°, 45°, and 120°, respectively, with the positive x-axis. Find the direction and magnitude of the resultant of these forces.

75. ***Velocity***   A ball is thrown with an initial velocity of 70 feet per second, at an angle of 40° with the horizontal (see figure). Find the vertical and horizontal components of the velocity.

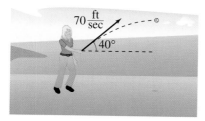

76. ***Velocity***   A gun with a muzzle velocity of 1200 feet per second is fired at an angle of 4° with the horizontal. Find the vertical and horizontal components of the velocity.

77. ***Tension***   Use the figure to determine the tension in each cable supporting the load.

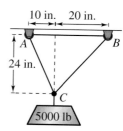

78. ***Tension***   The cranes shown in the figure are lifting an object that weighs 20,240 pounds. Find the tension in the cable of each crane.

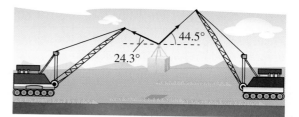

79. ***Numerical and Graphical Analysis***   A loaded barge is being towed by two tugboats, and the magnitude of the resultant is 6000 pounds directed along the axis of the barge (see figure). Each tow line makes an angle of θ degrees with the axis of the barge.

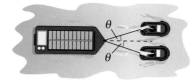

(a) Write the resultant tension T of each line as a function of θ. Determine the domain of the function.

(b) Use a graphing utility to complete the table.

θ	10°	20°	30°	40°	50°	60°
T						

(c) Use a graphing utility to graph the tension function.

(d) Explain why the tension increases as θ increases.

80. *Numerical and Graphical Analysis* To carry a 100-pound cylindrical weight, two people lift on the ends of short ropes that are tied to an eyelet on the top center of the cylinder. Each rope makes an angle of θ degrees with the vertical (see figure).

100 lb

(a) Write the tension T of each rope as a function of θ. Determine the domain of the function.

(b) Use a graphing utility to complete the table.

θ	10°	20°	30°	40°	50°	60°
T						

(c) Use a graphing utility to graph the tension function.

(d) Explain why the tension increases as θ increases.

81. *Navigation* An airplane is flying in the direction 148° with an airspeed of 860 kilometers per hour. Because of the wind, its groundspeed and direction are, respectively, 800 kilometers per hour and 140°. Find the direction and speed of the wind.

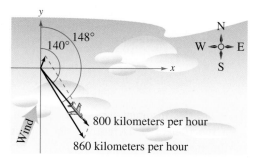

82. *Navigation* A commercial jet is flying from Miami to Seattle. The jet's velocity with respect to the air is 580 miles per hour, and its bearing is 332°. The wind, at the altitude of the plane, is blowing from the southwest with a velocity of 60 miles per hour.

(a) Draw a figure that gives a visual representation of the problem.

(b) Write the velocity of the wind as a vector in component form.

(c) Write the velocity of the jet relative to the air as a vector in component form.

(d) What is the speed of the jet with respect to the ground?

(e) What is the true direction of the jet?

83. *Numerical and Graphical Analysis* Forces with magnitudes of 150 newtons and 220 newtons act on a hook (see figure).

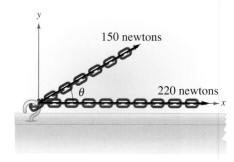

(a) Find the direction and magnitude of the resultant of the forces when $\theta = 30°$.

(b) Write the magnitude M of the resultant and the direction α of the resultant as functions of θ, where $0° \le \theta \le 180°$.

(c) Use a graphing utility to complete the table.

θ	0°	30°	60°	90°	120°	150°	180°
M							
α							

(d) Use a graphing utility to graph the two functions.

(e) Explain why one function decreases for increasing θ, whereas the other doesn't.

84. *Numerical and Graphical Analysis* A tetherball weighing 1 pound is pulled outward from the pole by a horizontal force **u** until the rope makes an angle of θ degrees with the pole (see figure).

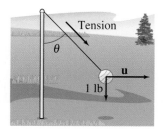

(a) Write the tension T in the rope and the magnitude of **u** as functions of θ. Determine the domains of the functions.

(b) Use a graphing utility to complete the table.

θ	0°	10°	20°	30°	40°	50°	60°
T							
$\|\mathbf{u}\|$							

(c) Use a graphing utility to graph the two functions for $0° \le \theta \le 60°$.

(d) Compare T and $\|\mathbf{u}\|$ as θ increases.

Synthesis

True or False?   In Exercises 85–88, determine whether the statement is true or false. Justify your answer.

85. If **u** and **v** have the same magnitude and direction, then $\mathbf{u} = \mathbf{v}$.

86. If **u** is a unit vector in the direction of **v**, then $\mathbf{v} = \|\mathbf{v}\|\,\mathbf{u}$.

87. If $\mathbf{v} = a\mathbf{i} + b\mathbf{j} = \mathbf{0}$, then $a = -b$.

88. If $\mathbf{u} = a\mathbf{i} + b\mathbf{j}$ is a unit vector, then $a^2 + b^2 = 1$.

89. **Think About It**   Consider two forces of equal magnitude acting on a point.

(a) If the magnitude of the resultant is the sum of the magnitudes of the two forces, make a conjecture about the angle between the forces.

(b) If the resultant of the forces is **0**, make a conjecture about the angle between the forces.

(c) Can the magnitude of the resultant be greater than the sum of the magnitudes of the two forces? Explain.

90. **Graphical Reasoning**   Consider two forces

$\mathbf{F}_1 = \langle 10, 0 \rangle$ and $\mathbf{F}_2 = 5\langle \cos \theta, \sin \theta \rangle$.

(a) Find $\|\mathbf{F}_1 + \mathbf{F}_2\|$ as a function of θ.

(b) Use a graphing utility to graph the function for $0 \le \theta < 2\pi$.

(c) Use the graph in part (b) to determine the range of the function. What is its maximum, and for what value of θ does it occur? What is its minimum, and for what value of θ does it occur?

(d) Explain why the magnitude of the resultant is never 0.

91. **Proof**   Prove that $(\cos \theta)\mathbf{i} + (\sin \theta)\mathbf{j}$ is a unit vector for any value of θ.

92. **Technology**   Write a program for your graphing utility that graphs two vectors and their difference given the vectors in component form.

In Exercises 93 and 94, use the program in Exercise 92 to find the difference of the vectors shown in the graph.

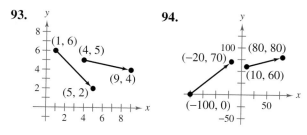

93.

94.

Review

In Exercises 95–100, simplify the expression.

95. $\left(\dfrac{6x^4}{7y^{-2}}\right)(14x^{-1}y^5)$

96. $(5s^5t^{-5})\left(\dfrac{3s^{-2}}{50t^{-1}}\right)$

97. $(18x)^0(4xy)^2(3x^{-1})$

98. $(5ab^2)(a^{-3}b^0)(2a^0b)^{-2}$

99. $(2.1 \times 10^9)(3.4 \times 10^{-4})$

100. $(6.5 \times 10^6)(3.8 \times 10^4)$

In Exercises 101–104, use the trigonometric substitution to write the algebraic expression as a trigonometric function of θ, where $0 < \theta < \pi/2$.

101. $\sqrt{49 - x^2}, \quad x = 7 \sin \theta$

102. $\sqrt{x^2 - 49}, \quad x = 7 \sec \theta$

103. $\sqrt{x^2 + 100}, \quad x = 10 \cot \theta$

104. $\sqrt{x^2 - 4}, \quad x = 2 \csc \theta$

In Exercises 105–108, solve the equation.

105. $\cos x(\cos x + 1) = 0$

106. $\sin x(2 \sin x + \sqrt{2}) = 0$

107. $3 \sec x + 4 = 10$

108. $\cos x \cot x - \cos x = 0$

7.4   Vectors and Dot Products

The Dot Product of Two Vectors

So far you have studied two vector operations—vector addition and multiplication by a scalar—each of which yields another vector. In this section you will study a third vector operation, the **dot product.** This product yields a scalar, rather than a vector.

Definition of Dot Product

The **dot product** of $\mathbf{u} = \langle u_1, u_2 \rangle$ and $\mathbf{v} = \langle v_1, v_2 \rangle$ is given by

$$\mathbf{u} \cdot \mathbf{v} = u_1 v_1 + u_2 v_2.$$

Properties of the Dot Product

Let $\mathbf{u}$, $\mathbf{v}$, and $\mathbf{w}$ be vectors in the plane or in space and let c be a scalar.

1. $\mathbf{u} \cdot \mathbf{v} = \mathbf{v} \cdot \mathbf{u}$

2. $\mathbf{0} \cdot \mathbf{v} = 0$

3. $\mathbf{u} \cdot (\mathbf{v} + \mathbf{w}) = \mathbf{u} \cdot \mathbf{v} + \mathbf{u} \cdot \mathbf{w}$

4. $\mathbf{v} \cdot \mathbf{v} = \|\mathbf{v}\|^2$

5. $c(\mathbf{u} \cdot \mathbf{v}) = c\mathbf{u} \cdot \mathbf{v} = \mathbf{u} \cdot c\mathbf{v}$

See Appendix B for proofs of the properties of the dot product.

What you should learn

- Find the dot product of two vectors and use properties of the dot product.
- Find angles between vectors and determine whether two vectors are orthogonal.
- Write vectors as sums of two vector components.
- Use vectors to find the work done by a force.

Why you should learn it

You can use the dot product of two vectors to solve real-life problems involving two vector quantities. For instance, Exercise 55 on page 570 shows you how the dot product can be used to find the force necessary to keep a truck from rolling down a hill.

Alan Thornton/Getty Images

Example 1   Finding Dot Products

Find each dot product.

a. $\langle 4, 5 \rangle \cdot \langle 2, 3 \rangle$

b. $\langle 2, -1 \rangle \cdot \langle 1, 2 \rangle$

c. $\langle 0, 3 \rangle \cdot \langle 4, -2 \rangle$

Solution

a. $\langle 4, 5 \rangle \cdot \langle 2, 3 \rangle = 4(2) + 5(3) = 8 + 15 = 23$

b. $\langle 2, -1 \rangle \cdot \langle 1, 2 \rangle = 2(1) + (-1)(2) = 2 - 2 = 0$

c. $\langle 0, 3 \rangle \cdot \langle 4, -2 \rangle = 0(4) + 3(-2) = 0 - 6 = -6$

✓ *Checkpoint* Now try Exercise 1.

In Example 1, be sure you see that the dot product of two vectors is a scalar (a real number), not a vector. Moreover, notice that the dot product can be positive, zero, or negative.

Example 2   Using Properties of Dot Products

Let $\mathbf{u} = \langle -1, 3 \rangle$, $\mathbf{v} = \langle 2, -4 \rangle$, and $\mathbf{w} = \langle 1, -2 \rangle$. Find each dot product.

a. $(\mathbf{u} \cdot \mathbf{v})\mathbf{w}$ **b.** $\mathbf{u} \cdot 2\mathbf{v}$

Solution

Begin by finding the dot product of $\mathbf{u}$ and $\mathbf{v}$.

$$
\begin{aligned}
\mathbf{u} \cdot \mathbf{v} &= \langle -1, 3 \rangle \cdot \langle 2, -4 \rangle \\
&= (-1)(2) + 3(-4) \\
&= -14
\end{aligned}
$$

a.
$$
\begin{aligned}
(\mathbf{u} \cdot \mathbf{v})\mathbf{w} &= -14 \langle 1, -2 \rangle \\
&= \langle -14, 28 \rangle
\end{aligned}
$$

b.
$$
\begin{aligned}
\mathbf{u} \cdot 2\mathbf{v} &= 2(\mathbf{u} \cdot \mathbf{v}) \\
&= 2(-14) \\
&= -28
\end{aligned}
$$

Notice that the product in part (a) is a vector, whereas the product in part (b) is a scalar. Can you see why?

✓ *Checkpoint* Now try Exercise 7.

Example 3   Dot Product and Magnitude

The dot product of $\mathbf{u}$ with itself is 5. What is the magnitude of $\mathbf{u}$?

Solution

Because $\|\mathbf{u}\|^2 = \mathbf{u} \cdot \mathbf{u} = 5$, it follows that

$$
\begin{aligned}
\|\mathbf{u}\| &= \sqrt{\mathbf{u} \cdot \mathbf{u}} \\
&= \sqrt{5}.
\end{aligned}
$$

✓ *Checkpoint* Now try Exercise 11.

The Angle Between Two Vectors

The **angle between two nonzero vectors** is the angle θ, $0 \le \theta \le \pi$, between their respective standard position vectors, as shown in Figure 7.33. This angle can be found using the dot product. (Note that the angle between the zero vector and another vector is not defined.) See Appendix B for a proof.

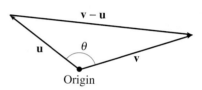

Figure 7.33

Angle Between Two Vectors

If θ is the angle between two nonzero vectors $\mathbf{u}$ and $\mathbf{v}$, then

$$
\cos \theta = \frac{\mathbf{u} \cdot \mathbf{v}}{\|\mathbf{u}\| \, \|\mathbf{v}\|}.
$$

Example 4   Finding the Angle Between Two Vectors

Find the angle between $\mathbf{u} = \langle 4, 3 \rangle$ and $\mathbf{v} = \langle 3, 5 \rangle$.

Solution

$$\cos\theta = \frac{\mathbf{u} \cdot \mathbf{v}}{\|\mathbf{u}\|\,\|\mathbf{v}\|}$$

$$= \frac{\langle 4, 3 \rangle \cdot \langle 3, 5 \rangle}{\|\langle 4, 3 \rangle\|\,\|\langle 3, 5 \rangle\|}$$

$$= \frac{27}{5\sqrt{34}}$$

This implies that the angle between the two vectors is

$$\theta = \arccos\frac{27}{5\sqrt{34}} \approx 22.2°,$$

as shown in Figure 7.34.

✓ *Checkpoint* Now try Exercise 17.

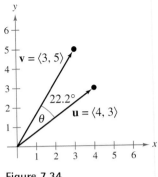

Figure 7.34

TECHNOLOGY TIP

The graphing utility program Finding the Angle Between Two Vectors, found on our website *college.hmco.com*, graphs two vectors $\mathbf{u} = \langle a, b \rangle$ and $\mathbf{v} = \langle c, d \rangle$ in standard position and finds the measure of the angle between them. Use the program to verify Example 4.

Rewriting the expression for the angle between two vectors in the form

$$\mathbf{u} \cdot \mathbf{v} = \|\mathbf{u}\|\,\|\mathbf{v}\|\cos\theta \qquad \text{Alternative form of dot product}$$

produces an alternative way to calculate the dot product. From this form, you can see that because $\|\mathbf{u}\|$ and $\|\mathbf{v}\|$ are always positive, $\mathbf{u} \cdot \mathbf{v}$ and $\cos\theta$ will always have the same sign. Figure 7.35 shows the five possible orientations of two vectors.

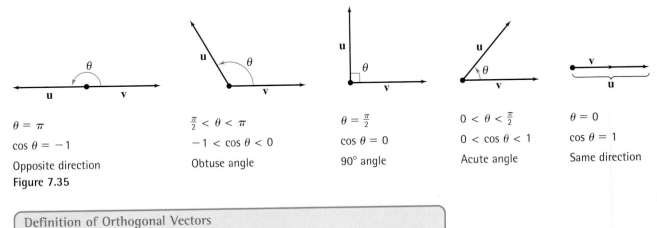

$\theta = \pi$

$\cos\theta = -1$

Opposite direction

Figure 7.35

$\frac{\pi}{2} < \theta < \pi$

$-1 < \cos\theta < 0$

Obtuse angle

$\theta = \frac{\pi}{2}$

$\cos\theta = 0$

90° angle

$0 < \theta < \frac{\pi}{2}$

$0 < \cos\theta < 1$

Acute angle

$\theta = 0$

$\cos\theta = 1$

Same direction

Definition of Orthogonal Vectors

The vectors $\mathbf{u}$ and $\mathbf{v}$ are **orthogonal** if $\mathbf{u} \cdot \mathbf{v} = 0$.

The terms *orthogonal* and *perpendicular* mean essentially the same thing—meeting at right angles. Even though the angle between the zero vector and another vector is not defined, it is convenient to extend the definition of orthogonality to include the zero vector. In other words, the zero vector is orthogonal to every vector $\mathbf{u}$ because $\mathbf{0} \cdot \mathbf{u} = 0$.

Example 5   Determining Orthogonal Vectors

Are the vectors $\mathbf{u} = \langle 2, -3 \rangle$ and $\mathbf{v} = \langle 6, 4 \rangle$ orthogonal?

Solution

Begin by finding the dot product of the two vectors.

$$\mathbf{u} \cdot \mathbf{v} = \langle 2, -3 \rangle \cdot \langle 6, 4 \rangle = 2(6) + (-3)(4) = 0$$

Because the dot product is 0, the two vectors are orthogonal, as shown in Figure 7.36.

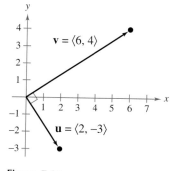

Figure 7.36

✔ *Checkpoint* Now try Exercise 35.

Finding Vector Components

You have already seen applications in which two vectors are added to produce a resultant vector. Many applications in physics and engineering pose the reverse problem—decomposing a given vector into the sum of two **vector components.**

Consider a boat on an inclined ramp, as shown in Figure 7.37. The force **F** due to gravity pulls the boat *down* the ramp and *against* the ramp. These two orthogonal forces, $\mathbf{w}_1$ and $\mathbf{w}_2$, are vector components of **F**. That is,

$$\mathbf{F} = \mathbf{w}_1 + \mathbf{w}_2. \qquad \text{Vector components of } \mathbf{F}$$

The negative of component $\mathbf{w}_1$ represents the force needed to keep the boat from rolling down the ramp, and $\mathbf{w}_2$ represents the force that the tires must withstand against the ramp. A procedure for finding $\mathbf{w}_1$ and $\mathbf{w}_2$ is shown on the next page.

Figure 7.37

Definition of Vector Components

Let $\mathbf{u}$ and $\mathbf{v}$ be nonzero vectors such that

$$\mathbf{u} = \mathbf{w}_1 + \mathbf{w}_2$$

where $\mathbf{w}_1$ and $\mathbf{w}_2$ are orthogonal and $\mathbf{w}_1$ is parallel to (or a scalar multiple of) $\mathbf{v}$, as shown in Figure 7.38. The vectors $\mathbf{w}_1$ and $\mathbf{w}_2$ are called **vector components** of $\mathbf{u}$. The vector $\mathbf{w}_1$ is the **projection** of $\mathbf{u}$ onto $\mathbf{v}$ and is denoted by

$$\mathbf{w}_1 = \text{proj}_\mathbf{v}\mathbf{u}.$$

The vector $\mathbf{w}_2$ is given by $\mathbf{w}_2 = \mathbf{u} - \mathbf{w}_1$.

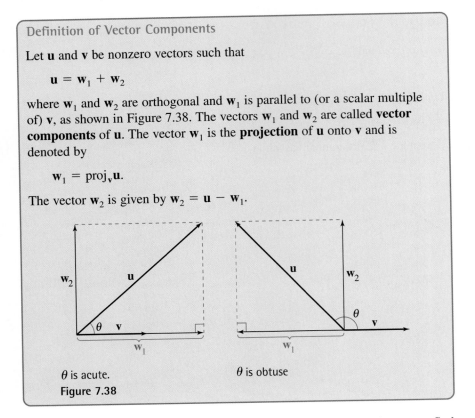

θ is acute.

θ is obtuse

Figure 7.38

From the definition of vector components, you can see that it is easy to find the component $\mathbf{w}_2$ once you have found the projection of $\mathbf{u}$ onto $\mathbf{v}$. To find the projection, you can use the dot product, as follows.

$$\mathbf{u} = \mathbf{w}_1 + \mathbf{w}_2 = c\mathbf{v} + \mathbf{w}_2 \qquad \text{$\mathbf{w}_1$ is a scalar multiple of $\mathbf{v}$.}$$

$$\mathbf{u} \cdot \mathbf{v} = (c\mathbf{v} + \mathbf{w}_2) \cdot \mathbf{v} \qquad \text{Take dot product of each side with $\mathbf{v}$.}$$

$$= c\mathbf{v} \cdot \mathbf{v} + \mathbf{w}_2 \cdot \mathbf{v}$$

$$= c\|\mathbf{v}\|^2 + 0 \qquad \text{$\mathbf{w}_2$ and $\mathbf{v}$ are orthogonal.}$$

So,

$$c = \frac{\mathbf{u} \cdot \mathbf{v}}{\|\mathbf{v}\|^2}$$

and

$$\mathbf{w}_1 = \text{proj}_\mathbf{v}\mathbf{u} = c\mathbf{v} = \frac{\mathbf{u} \cdot \mathbf{v}}{\|\mathbf{v}\|^2}\mathbf{v}.$$

Projection of $\mathbf{u}$ onto $\mathbf{v}$

Let $\mathbf{u}$ and $\mathbf{v}$ be nonzero vectors. The projection of $\mathbf{u}$ onto $\mathbf{v}$ is given by

$$\text{proj}_\mathbf{v}\mathbf{u} = \left(\frac{\mathbf{u} \cdot \mathbf{v}}{\|\mathbf{v}\|^2}\right)\mathbf{v}.$$

Example 6   Decomposing a Vector into Components

Find the projection of $\mathbf{u} = \langle 3, -5 \rangle$ onto $\mathbf{v} = \langle 6, 2 \rangle$. Then write $\mathbf{u}$ as the sum of two orthogonal vectors, one of which is $\text{proj}_\mathbf{v}\mathbf{u}$.

Solution

The projection of $\mathbf{u}$ onto $\mathbf{v}$ is

$$\mathbf{w}_1 = \text{proj}_\mathbf{v}\mathbf{u} = \left(\frac{\mathbf{u} \cdot \mathbf{v}}{\|\mathbf{v}\|^2}\right)\mathbf{v} = \left(\frac{8}{40}\right)\langle 6, 2 \rangle = \left\langle \frac{6}{5}, \frac{2}{5} \right\rangle$$

as shown in Figure 7.39. The other component, $\mathbf{w}_2$, is

$$\mathbf{w}_2 = \mathbf{u} - \mathbf{w}_1 = \langle 3, -5 \rangle - \left\langle \frac{6}{5}, \frac{2}{5} \right\rangle = \left\langle \frac{9}{5}, -\frac{27}{5} \right\rangle.$$

So, $\mathbf{u} = \mathbf{w}_1 + \mathbf{w}_2 = \left\langle \frac{6}{5}, \frac{2}{5} \right\rangle + \left\langle \frac{9}{5}, -\frac{27}{5} \right\rangle = \langle 3, -5 \rangle.$

✓ *Checkpoint*   Now try Exercise 39.

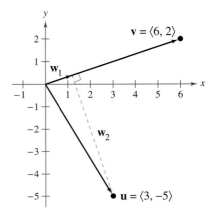

Figure 7.39

Example 7   Finding a Force

A 200-pound cart sits on a ramp inclined at 30°, as shown in Figure 7.40. What force is required to keep the cart from rolling down the ramp?

Solution

Because the force due to gravity is vertical and downward, you can represent the gravitational force by the vector

$$\mathbf{F} = -200\mathbf{j}. \qquad \text{Force due to gravity}$$

To find the force required to keep the cart from rolling down the ramp, project $\mathbf{F}$ onto a unit vector $\mathbf{v}$ in the direction of the ramp, as follows.

$$\mathbf{v} = (\cos 30°)\mathbf{i} + (\sin 30°)\mathbf{j} = \frac{\sqrt{3}}{2}\mathbf{i} + \frac{1}{2}\mathbf{j} \qquad \text{Unit vector along ramp}$$

Therefore, the projection of $\mathbf{F}$ onto $\mathbf{v}$ is

$$\mathbf{w}_1 = \text{proj}_\mathbf{v}\mathbf{F} = \left(\frac{\mathbf{F} \cdot \mathbf{v}}{\|\mathbf{v}\|^2}\right)\mathbf{v}$$

$$= (\mathbf{F} \cdot \mathbf{v})\mathbf{v}$$

$$= (-200)\left(\frac{1}{2}\right)\mathbf{v}$$

$$= -100\left(\frac{\sqrt{3}}{2}\mathbf{i} + \frac{1}{2}\mathbf{j}\right).$$

The magnitude of this force is 100, and therefore a force of 100 pounds is required to keep the cart from rolling down the ramp.

✓ *Checkpoint*   Now try Exercise 55.

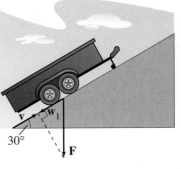

Figure 7.40

Work

The work W done by a constant force $\mathbf{F}$ acting along the line of motion of an object is given by

$$W = (\text{magnitude of force})(\text{distance}) = \|\mathbf{F}\| \, \|\overrightarrow{PQ}\|$$

as shown in Figure 7.41. If the constant force $\mathbf{F}$ is not directed along the line of motion (see Figure 7.42), the work W done by the force is given by

$$W = \|\text{proj}_{\overrightarrow{PQ}} \mathbf{F}\| \, \|\overrightarrow{PQ}\|$$ Projection form for work

$$= (\cos \theta) \|\mathbf{F}\| \, \|\overrightarrow{PQ}\|$$ $\|\text{proj}_{\overrightarrow{PQ}} \mathbf{F}\| = (\cos \theta)\|\mathbf{F}\|$

$$= \mathbf{F} \cdot \overrightarrow{PQ}.$$ Dot product form for work

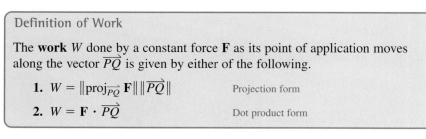

Force acts along the line of motion.

Figure 7.41

Force acts at angle θ with the line of motion.

Figure 7.42

This notion of work is summarized in the following definition.

Definition of Work

The **work** W done by a constant force $\mathbf{F}$ as its point of application moves along the vector $\overrightarrow{PQ}$ is given by either of the following.

  1. $W = \|\text{proj}_{\overrightarrow{PQ}} \mathbf{F}\| \, \|\overrightarrow{PQ}\|$ Projection form

  2. $W = \mathbf{F} \cdot \overrightarrow{PQ}$ Dot product form

---

Example 8 Finding Work

To close a barn's sliding door, a person pulls on a rope with a constant force of 50 pounds at a constant angle of $60°$, as shown in Figure 7.43. Find the work done in moving the door 12 feet to its closed position.

Solution

Using a projection, you can calculate the work as follows.

$$W = \|\text{proj}_{\overrightarrow{PQ}} \mathbf{F}\| \, \|\overrightarrow{PQ}\|$$ Projection form for work

$$= (\cos 60°)\|\mathbf{F}\| \, \|\overrightarrow{PQ}\|$$

$$= \frac{1}{2}(50)(12) = 300 \text{ foot-pounds}$$

So, the work done is 300 foot-pounds. You can verify this result by finding the vectors $\mathbf{F}$ and $\overrightarrow{PQ}$ and calculating their dot product.

Figure 7.43

✅ *Checkpoint* Now try Exercise 57.

7.4   Exercises

Vocabulary Check

Fill in the blanks.

1. The _____ of two vectors yields a scalar, rather than a vector.
2. If θ is the angle between two nonzero vectors **u** and **v**, then $\cos \theta =$ _____ .
3. The vectors **u** and **v** are _____ if $\mathbf{u} \cdot \mathbf{v} = 0$.
4. The projection of **u** onto **v** is given by $\text{proj}_\mathbf{v}\mathbf{u} =$ _____ .
5. The work W done by a constant force **F** as its point of application moves along the vector $\overrightarrow{PQ}$ is given by either $W =$ _____ or $W =$ _____ .

In Exercises 1–4, find the dot product of u and v.

1. $\mathbf{u} = \langle 6, 3 \rangle$
   $\mathbf{v} = \langle 2, -4 \rangle$

2. $\mathbf{u} = \langle 1, 8 \rangle$
   $\mathbf{v} = \langle -3, 2 \rangle$

3. $\mathbf{u} = 5\mathbf{i} + \mathbf{j}$
   $\mathbf{v} = 3\mathbf{i} - \mathbf{j}$

4. $\mathbf{u} = 8\mathbf{i} - 2\mathbf{j}$
   $\mathbf{v} = 4\mathbf{i} - 3\mathbf{j}$

In Exercises 5–10, use the vectors $\mathbf{u} = \langle 2, 2 \rangle$, $\mathbf{v} = \langle -3, 4 \rangle$, and $\mathbf{w} = \langle 1, -4 \rangle$ to find the indicated quantity. State whether the result is a vector or a scalar.

5. $\mathbf{u} \cdot \mathbf{u}$

6. $\|\mathbf{u}\| - 2$

7. $(\mathbf{u} \cdot \mathbf{v})\mathbf{w}$

8. $(\mathbf{w} \cdot \mathbf{u})\mathbf{v}$

9. $\mathbf{u} \cdot 2\mathbf{v}$

10. $4\mathbf{u} \cdot \mathbf{v}$

In Exercises 11–16, use the dot product to find the magnitude of u.

11. $\mathbf{u} = \langle -5, 12 \rangle$

12. $\mathbf{u} = \langle 2, -4 \rangle$

13. $\mathbf{u} = 20\mathbf{i} + 25\mathbf{j}$

14. $\mathbf{u} = 6\mathbf{i} - 10\mathbf{j}$

15. $\mathbf{u} = -4\mathbf{j}$

16. $\mathbf{u} = 9\mathbf{i}$

In Exercises 17–24, find the angle θ between the vectors.

17. $\mathbf{u} = \langle -1, 0 \rangle$
   $\mathbf{v} = \langle 0, 2 \rangle$

18. $\mathbf{u} = \langle 4, 4 \rangle$
   $\mathbf{v} = \langle -2, 0 \rangle$

19. $\mathbf{u} = 3\mathbf{i} + 4\mathbf{j}$
   $\mathbf{v} = -2\mathbf{i} + 3\mathbf{j}$

20. $\mathbf{u} = 2\mathbf{i} - 3\mathbf{j}$
   $\mathbf{v} = \mathbf{i} - 2\mathbf{j}$

21. $\mathbf{u} = 2\mathbf{i}$
   $\mathbf{v} = -3\mathbf{j}$

22. $\mathbf{u} = 4\mathbf{j}$
   $\mathbf{v} = -3\mathbf{i}$

23. $\mathbf{u} = \cos\left(\dfrac{\pi}{3}\right)\mathbf{i} + \sin\left(\dfrac{\pi}{3}\right)\mathbf{j}$
   $\mathbf{v} = \cos\left(\dfrac{3\pi}{4}\right)\mathbf{i} + \sin\left(\dfrac{3\pi}{4}\right)\mathbf{j}$

24. $\mathbf{u} = \cos\left(\dfrac{\pi}{4}\right)\mathbf{i} + \sin\left(\dfrac{\pi}{4}\right)\mathbf{j}$
   $\mathbf{v} = \cos\left(\dfrac{2\pi}{3}\right)\mathbf{i} + \sin\left(\dfrac{2\pi}{3}\right)\mathbf{j}$

In Exercises 25–28, graph the vectors and find the degree measure of the angle between the vectors.

25. $\mathbf{u} = 3\mathbf{i} + 4\mathbf{j}$
   $\mathbf{v} = -7\mathbf{i} + 5\mathbf{j}$

26. $\mathbf{u} = -6\mathbf{i} - 3\mathbf{j}$
   $\mathbf{v} = -8\mathbf{i} + 4\mathbf{j}$

27. $\mathbf{u} = 5\mathbf{i} + 5\mathbf{j}$
   $\mathbf{v} = -8\mathbf{i} + 8\mathbf{j}$

28. $\mathbf{u} = 2\mathbf{i} - 3\mathbf{j}$
   $\mathbf{v} = 4\mathbf{i} + 3\mathbf{j}$

In Exercises 29 and 30, use vectors to find the interior angles of the triangle with the given vertices.

29. $(1, 2), (3, 4), (2, 5)$

30. $(-3, 0), (2, 2), (0, 6)$

In Exercises 31 and 32, find $\mathbf{u} \cdot \mathbf{v}$, where θ is the angle between u and v.

31. $\|\mathbf{u}\| = 4, \|\mathbf{v}\| = 10, \theta = \dfrac{2\pi}{3}$

32. $\|\mathbf{u}\| = 100, \|\mathbf{v}\| = 250, \theta = \dfrac{\pi}{6}$

In Exercises 33–38, determine whether u and v are orthogonal, parallel, or neither.

33. $\mathbf{u} = \langle -12, 30 \rangle$
$\quad \mathbf{v} = \langle \frac{1}{2}, -\frac{5}{4} \rangle$

34. $\mathbf{u} = \langle 15, 45 \rangle$
$\quad \mathbf{v} = \langle -5, 12 \rangle$

35. $\mathbf{u} = \frac{1}{4}(3\mathbf{i} - \mathbf{j})$
$\quad \mathbf{v} = 5\mathbf{i} + 6\mathbf{j}$

36. $\mathbf{u} = \mathbf{j}$
$\quad \mathbf{v} = \mathbf{i} - 2\mathbf{j}$

37. $\mathbf{u} = 2\mathbf{i} - 2\mathbf{j}$
$\quad \mathbf{v} = -\mathbf{i} - \mathbf{j}$

38. $\mathbf{u} = 8\mathbf{i} + 4\mathbf{j}$
$\quad \mathbf{v} = -2\mathbf{i} - \mathbf{j}$

In Exercises 39–42, find the projection of u onto v. Then write u as the sum of two orthogonal vectors, one of which is $\text{proj}_\mathbf{v}\, \mathbf{u}$.

39. $\mathbf{u} = \langle 3, 4 \rangle$
$\quad \mathbf{v} = \langle 8, 2 \rangle$

40. $\mathbf{u} = \langle 4, 2 \rangle$
$\quad \mathbf{v} = \langle 1, -2 \rangle$

41. $\mathbf{u} = \langle 0, 3 \rangle$
$\quad \mathbf{v} = \langle 2, 15 \rangle$

42. $\mathbf{u} = \langle -5, -1 \rangle$
$\quad \mathbf{v} = \langle -1, 1 \rangle$

In Exercises 43–46, use the graph to mentally determine the projection of u onto v. (The coordinates of the terminal points of the vectors in standard position are given.) Use the formula for the projection of u onto v to verify your result.

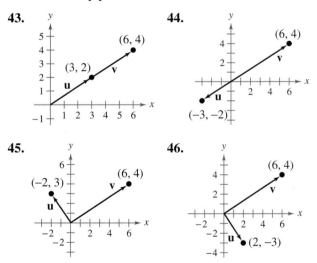

43.
44.
45.
46.

In Exercises 47–50, find two vectors in opposite directions that are orthogonal to the vector u. (There are many correct answers.)

47. $\mathbf{u} = \langle 2, 6 \rangle$

48. $\mathbf{u} = \langle -7, 5 \rangle$

49. $\mathbf{u} = \frac{1}{2}\mathbf{i} - \frac{3}{4}\mathbf{j}$

50. $\mathbf{u} = -\frac{5}{2}\mathbf{i} - 3\mathbf{j}$

Work **In Exercises 51 and 52, find the work done in moving a particle from P to Q if the magnitude and direction of the force are given by v.**

51. $P = (0, 0), \quad Q = (4, 7), \quad \mathbf{v} = \langle 1, 4 \rangle$

52. $P = (1, 3), \quad Q = (-3, 5), \quad \mathbf{v} = -2\mathbf{i} + 3\mathbf{j}$

53. *Revenue* The vector $\mathbf{u} = \langle 1245, 2600 \rangle$ gives the numbers of units of two types of picture frames produced by a company. The vector $\mathbf{v} = \langle 12.20, 8.50 \rangle$ gives the price (in dollars) of each frame, respectively. Find the dot product $\mathbf{u} \cdot \mathbf{v}$ and explain what information it gives.

54. *Revenue* Repeat Exercise 53 after increasing the prices by 5%. Identify the vector operation used to increase the prices by 5%.

55. *Braking Load* A truck with a gross weight of 30,000 pounds is parked on a slope of $d°$ (see figure). Assume that the only force to overcome is the force of gravity.

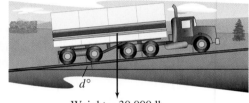

Weight = 30,000 lb

(a) Find the force required to keep the truck from rolling down the hill in terms of the slope d.

(b) Use a graphing utility to complete the table.

d	0°	1°	2°	3°	4°	5°
Force						

d	6°	7°	8°	9°	10°
Force					

(c) Find the force perpendicular to the hill when $d = 5°$.

56. *Braking Load* A sport utility vehicle with a gross weight of 5400 pounds is parked on a slope of 10°. Assume that the only force to overcome is the force of gravity. Find the force required to keep the vehicle from rolling down the hill. Find the force perpendicular to the hill.

57. **Work**   A tractor pulls a log d meters and the tension in the cable connecting the tractor and log is approximately 1600 kilograms (15,691 newtons). The direction of the force is 30° above the horizontal (see figure).

(a) Find the work done in terms of the distance d.

(b) Use a graphing utility to complete the table.

d	0	200	400	800
Work				

58. **Work**   A force of 45 pounds in the direction of 30° above the horizontal is required to slide a table across a floor. Find the work done if the table is dragged 20 feet.

59. **Work**   A toy wagon is pulled by exerting a force of 20 pounds on a handle that makes a 25° angle with the horizontal. Find the work done in pulling the wagon 40 feet.

60. **Work**   A mover exerts a horizontal force of 25 pounds on a crate as it is pushed up a ramp that is 12 feet long and inclined at an angle of 20° above the horizontal. Find the work done on the crate.

Synthesis

True or False?   **In Exercises 61 and 62, determine whether the statement is true or false. Justify your answer.**

61. The vectors $\mathbf{u} = \langle 0, 0 \rangle$ and $\mathbf{v} = \langle -12, 6 \rangle$ are orthogonal.

62. The work W done by a constant force $\mathbf{F}$ acting along the line of motion of an object is represented by a vector.

63. If $\mathbf{u} = \langle \cos \theta, \sin \theta \rangle$ and $\mathbf{v} = \langle \sin \theta, -\cos \theta \rangle$, are $\mathbf{u}$ and $\mathbf{v}$ orthogonal, parallel, or neither? Explain.

64. **Think About It**   What is known about θ, the angle between two nonzero vectors $\mathbf{u}$ and $\mathbf{v}$, if each of the following is true?

(a) $\mathbf{u} \cdot \mathbf{v} = 0$ (b) $\mathbf{u} \cdot \mathbf{v} > 0$ (c) $\mathbf{u} \cdot \mathbf{v} < 0$

65. **Think About It**   What can be said about the vectors $\mathbf{u}$ and $\mathbf{v}$ under each condition?

(a) The projection of $\mathbf{u}$ onto $\mathbf{v}$ equals $\mathbf{u}$.

(b) The projection of $\mathbf{u}$ onto $\mathbf{v}$ equals $\mathbf{0}$.

66. **Proof**   Use vectors to prove that the diagonals of a rhombus are perpendicular.

67. **Proof**   Prove the following.

$$\|\mathbf{u} - \mathbf{v}\|^2 = \|\mathbf{u}\|^2 + \|\mathbf{v}\|^2 - 2\mathbf{u} \cdot \mathbf{v}$$

68. **Proof**   Prove that if $\mathbf{u}$ is orthogonal to $\mathbf{v}$ and $\mathbf{w}$, then $\mathbf{u}$ is orthogonal to $c\mathbf{v} + d\mathbf{w}$ for any scalars c and d.

Review

In Exercises 69–72, describe how the graph of g is related to the graph of f.

69. $g(x) = f(x - 4)$ **70.** $g(x) = -f(x)$

71. $g(x) = f(x) + 6$ **72.** $g(x) = f(2x)$

In Exercises 73–80, perform the operation and write the result in standard form.

73. $\sqrt{-4} - 1$ **74.** $\sqrt{-8} + 5$

75. $3i(4 - 5i)$ **76.** $-2i(1 + 6i)$

77. $(1 + 3i)(1 - 3i)$ **78.** $(7 - 4i)(7 + 4i)$

79. $\dfrac{3}{1 + i} + \dfrac{2}{2 - 3i}$ **80.** $\dfrac{6}{4 - i} - \dfrac{3}{1 + i}$

In Exercises 81–84, plot the complex number in the complex plane.

81. $-2i$ **82.** $3i$

83. $1 + 8i$ **84.** $9 - 7i$

85. **Partnership Costs**   A group of people agree to share equally in the cost of a \$250,000 endowment to a college. If they could find two more people to join the group, each person's share of the cost would decrease by \$6250. How many people are presently in the group?

86. **Current Speed**   A boat travels at a speed of 18 miles per hour in still water. It travels 35 miles upstream and then returns to the starting point in a total of 4 hours. Find the speed of the current.

7.5 Trigonometric Form of a Complex Number

The Complex Plane

Recall from Section 2.3 that you can represent a complex number $z = a + bi$ as the point (a, b) in a coordinate plane (the complex plane). The horizontal axis is called the real axis and the vertical axis is called the imaginary axis, as shown in Figure 7.44.

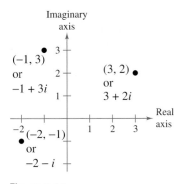

Figure 7.44

The **absolute value of a complex number** $a + bi$ is defined as the distance between the origin $(0, 0)$ and the point (a, b).

> **Definition of the Absolute Value of a Complex Number**
>
> The **absolute value** of the complex number $z = a + bi$ is given by
> $$|a + bi| = \sqrt{a^2 + b^2}.$$

If the complex number $a + bi$ is a real number (that is, if $b = 0$), then this definition agrees with that given for the absolute value of a real number

$$|a + 0i| = \sqrt{a^2 + 0^2} = |a|.$$

Example 1 Finding the Absolute Value of a Complex Number

Plot $z = -2 + 5i$ and find its absolute value.

Solution

The number is plotted in Figure 7.45. It has an absolute value of

$$|z| = \sqrt{(-2)^2 + 5^2}$$
$$= \sqrt{29}.$$

✓ *Checkpoint* Now try Exercise 5.

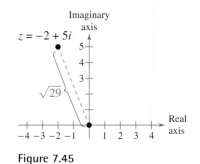

Figure 7.45

What you should learn

- Find absolute values of complex numbers.
- Write trigonometric forms of complex numbers.
- Multiply and divide complex numbers written in trigonometric form.
- Use DeMoivre's Theorem to find powers of complex numbers.
- Find *n*th roots of complex numbers.

Why you should learn it

You can use the trigonometric form of a complex number to perform operations with complex numbers. For instance, in Exercises 109–116 on page 582, you can use the trigonometric form of a complex number to help you solve polynomial equations.

Trigonometric Form of a Complex Number

In Section 2.3 you learned how to add, subtract, multiply, and divide complex numbers. To work effectively with *powers* and *roots* of complex numbers, it is helpful to write complex numbers in trigonometric form. In Figure 7.46, consider the nonzero complex number $a + bi$. By letting θ be the angle from the positive real axis (measured counterclockwise) to the line segment connecting the origin and the point (a, b), you can write

$$a = r \cos \theta \quad \text{and} \quad b = r \sin \theta$$

where $r = \sqrt{a^2 + b^2}$. Consequently, you have

$$a + bi = (r \cos \theta) + (r \sin \theta)i$$

from which you can obtain the **trigonometric form of a complex number.**

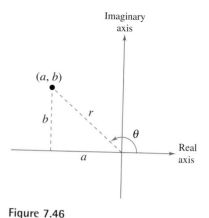

Figure 7.46

> ### Trigonometric Form of a Complex Number
>
> The **trigonometric form** of the complex number $z = a + bi$ is given by
>
> $$z = r(\cos \theta + i \sin \theta)$$
>
> where $a = r \cos \theta$, $b = r \sin \theta$, $r = \sqrt{a^2 + b^2}$, and $\tan \theta = b/a$.
> The number r is the **modulus** of z, and θ is called an **argument** of z.

The trigonometric form of a complex number is also called the *polar form.* Because there are infinitely many choices for θ, the trigonometric form of a complex number is not unique. Normally, θ is restricted to the interval $0 \le \theta < 2\pi$, although on occasion it is convenient to use $\theta < 0$.

Example 2   Writing a Complex Number in Trigonometric Form

Write the complex number $z = -2 - 2\sqrt{3}i$ in trigonometric form.

Solution

The absolute value of z is

$$r = \left| -2 - 2\sqrt{3}i \right| = \sqrt{(-2)^2 + \left(-2\sqrt{3}\right)^2} = \sqrt{16} = 4$$

and the angle θ is given by

$$\tan \theta = \frac{b}{a} = \frac{-2\sqrt{3}}{-2} = \sqrt{3}.$$

Because $\tan(\pi/3) = \sqrt{3}$ and $z = -2 - 2\sqrt{3}i$ lies in Quadrant III, choose θ to be $\theta = \pi + \pi/3 = 4\pi/3$. So, the trigonometric form is

$$z = r(\cos \theta + i \sin \theta) = 4\left(\cos \frac{4\pi}{3} + i \sin \frac{4\pi}{3} \right).$$

See Figure 7.47.

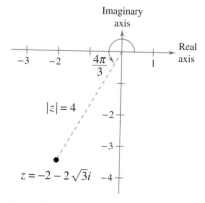

Figure 7.47

✓ *Checkpoint* Now try Exercise 15.

Example 3   Writing a Complex Number in Standard Form

Write the complex number in standard form $a + bi$.

$$z = \sqrt{8}\left[\cos\left(-\frac{\pi}{3}\right) + i\sin\left(-\frac{\pi}{3}\right)\right]$$

Solution

Because $\cos(-\pi/3) = 1/2$ and $\sin(-\pi/3) = -\sqrt{3}/2$, you can write

$$z = \sqrt{8}\left[\cos\left(-\frac{\pi}{3}\right) + i\sin\left(-\frac{\pi}{3}\right)\right]$$

$$= \sqrt{8}\left[\frac{1}{2} - \frac{\sqrt{3}}{2}i\right]$$

$$= 2\sqrt{2}\left[\frac{1}{2} - \frac{\sqrt{3}}{2}i\right] = \sqrt{2} - \sqrt{6}i.$$

 Checkpoint Now try Exercise 33.

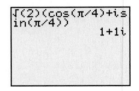
Multiplication and Division of Complex Numbers

The trigonometric form adapts nicely to multiplication and division of complex numbers. Suppose you are given two complex numbers

$$z_1 = r_1(\cos \theta_1 + i \sin \theta_1) \quad \text{and} \quad z_2 = r_2(\cos \theta_2 + i \sin \theta_2).$$

The product of z_1 and z_2 is

$$z_1 z_2 = r_1 r_2 (\cos \theta_1 + i \sin \theta_1)(\cos \theta_2 + i \sin \theta_2)$$

$$= r_1 r_2 [(\cos \theta_1 \cos \theta_2 - \sin \theta_1 \sin \theta_2) + i(\sin \theta_1 \cos \theta_2 + \cos \theta_1 \sin \theta_2)].$$

Using the sum and difference formulas for cosine and sine, you can rewrite this equation as

$$z_1 z_2 = r_1 r_2 [\cos(\theta_1 + \theta_2) + i \sin(\theta_1 + \theta_2)].$$

This establishes the first part of the following rule. The second part is left to you (see Exercise 120).

Product and Quotient of Two Complex Numbers

Let $z_1 = r_1(\cos \theta_1 + i \sin \theta_1)$ and $z_2 = r_2(\cos \theta_2 + i \sin \theta_2)$ be complex numbers.

$$z_1 z_2 = r_1 r_2 [\cos(\theta_1 + \theta_2) + i \sin(\theta_1 + \theta_2)] \qquad \text{Product}$$

$$\frac{z_1}{z_2} = \frac{r_1}{r_2}[\cos(\theta_1 - \theta_2) + i \sin(\theta_1 - \theta_2)], \qquad z_2 \neq 0 \qquad \text{Quotient}$$

Note that this rule says that to *multiply* two complex numbers you multiply moduli and add arguments, whereas to *divide* two complex numbers you divide moduli and subtract arguments.

Example 4   Multiplying Complex Numbers in Trigonometric Form

Find the product z_1z_2 of the complex numbers.

$$z_1 = 2\left(\cos\frac{2\pi}{3} + i\sin\frac{2\pi}{3}\right) \qquad z_2 = 8\left(\cos\frac{11\pi}{6} + i\sin\frac{11\pi}{6}\right)$$

Solution

$$z_1z_2 = 2\left(\cos\frac{2\pi}{3} + i\sin\frac{2\pi}{3}\right) \cdot 8\left(\cos\frac{11\pi}{6} + i\sin\frac{11\pi}{6}\right)$$

$$= 16\left[\cos\left(\frac{2\pi}{3} + \frac{11\pi}{6}\right) + i\sin\left(\frac{2\pi}{3} + \frac{11\pi}{6}\right)\right]$$

$$= 16\left(\cos\frac{5\pi}{2} + i\sin\frac{5\pi}{2}\right)$$

$$= 16\left(\cos\frac{\pi}{2} + i\sin\frac{\pi}{2}\right)$$

$$= 16[0 + i(1)] = 16i$$

You can check this result by first converting to the standard forms $z_1 = -1 + \sqrt{3}i$ and $z_2 = 4\sqrt{3} - 4i$ and then multiplying algebraically, as in Section 2.3.

$$z_1z_2 = \left(-1 + \sqrt{3}i\right)\left(4\sqrt{3} - 4i\right)$$

$$= -4\sqrt{3} + 4i + 12i + 4\sqrt{3} = 16i$$

✓ *Checkpoint* Now try Exercise 51.

> **TECHNOLOGY TIP**
>
> Some graphing utilities can multiply and divide complex numbers in trigonometric form. If you have access to such a graphing utility, use it to find z_1z_2 and z_1/z_2 in Examples 4 and 5.

Example 5   Dividing Complex Numbers in Trigonometric Form

Find the quotient z_1/z_2 of the complex numbers.

$$z_1 = 24(\cos 300° + i\sin 300°) \qquad z_2 = 8(\cos 75° + i\sin 75°)$$

Solution

$$\frac{z_1}{z_2} = \frac{24(\cos 300° + i\sin 300°)}{8(\cos 75° + i\sin 75°)}$$

$$= \frac{24}{8}[\cos(300° - 75°) + i\sin(300° - 75°)]$$

$$= 3(\cos 225° + i\sin 225°)$$

$$= 3\left[\left(-\frac{\sqrt{2}}{2}\right) + i\left(-\frac{\sqrt{2}}{2}\right)\right] = -\frac{3\sqrt{2}}{2} - \frac{3\sqrt{2}}{2}i$$

✓ *Checkpoint* Now try Exercise 57.

Powers of Complex Numbers

The trigonometric form of a complex number is used to raise a complex number to a power. To accomplish this, consider repeated use of the multiplication rule.

$z = r(\cos \theta + i \sin \theta)$

$z^2 = r(\cos \theta + i \sin \theta)r(\cos \theta + i \sin \theta) = r^2(\cos 2\theta + i \sin 2\theta)$

$z^3 = r^2(\cos 2\theta + i \sin 2\theta)r(\cos \theta + i \sin \theta) = r^3(\cos 3\theta + i \sin 3\theta)$

$z^4 = r^4(\cos 4\theta + i \sin 4\theta)$

$z^5 = r^5(\cos 5\theta + i \sin 5\theta)$

$\vdots$

This pattern leads to **DeMoivre's Theorem**, which is named after the French mathematician Abraham DeMoivre (1667–1754).

DeMoivre's Theorem

If $z = r(\cos \theta + i \sin \theta)$ is a complex number and n is a positive integer, then

$$z^n = [r(\cos \theta + i \sin \theta)]^n$$

$$= r^n(\cos n\theta + i \sin n\theta).$$

Exploration

Plot the numbers i, i^2, i^3, i^4, and i^5 in the complex plane. Write each number in trigonometric form and describe what happens to the angle θ as you form higher powers of i^n.

Example 6 Finding Powers of a Complex Number

Use DeMoivre's Theorem to find $\left(-1 + \sqrt{3}i\right)^{12}$.

Solution

First convert the complex number to trigonometric form using

$$r = \sqrt{(-1)^2 + \left(\sqrt{3}\right)^2} = 2 \quad \text{and} \quad \theta = \arctan \frac{\sqrt{3}}{-1} = \frac{2\pi}{3}.$$

So, the trigonometric form is

$$-1 + \sqrt{3}i = 2\left(\cos \frac{2\pi}{3} + i \sin \frac{2\pi}{3}\right).$$

Then, by DeMoivre's Theorem, you have

$$\left(-1 + \sqrt{3}i\right)^{12} = \left[2\left(\cos \frac{2\pi}{3} + i \sin \frac{2\pi}{3}\right)\right]^{12}$$

$$= 2^{12}\left[\cos\left(12 \cdot \frac{2\pi}{3}\right) + i \sin\left(12 \cdot \frac{2\pi}{3}\right)\right]$$

$$= 4096(\cos 8\pi + i \sin 8\pi)$$

$$= 4096(1 + 0) = 4096.$$

✓ *Checkpoint* Now try Exercise 73.

Roots of Complex Numbers

Recall that a consequence of the Fundamental Theorem of Algebra is that a polynomial equation of degree n has n solutions in the complex number system. So, an equation such as $x^6 = 1$ has six solutions, and in this particular case you can find the six solutions by factoring and using the Quadratic Formula.

$$x^6 - 1 = (x^3 - 1)(x^3 + 1)$$
$$= (x - 1)(x^2 + x + 1)(x + 1)(x^2 - x + 1)$$
$$= 0$$

Consequently, the solutions are

$$x = \pm 1, \qquad x = \frac{-1 \pm \sqrt{3}i}{2}, \qquad \text{and} \qquad x = \frac{1 \pm \sqrt{3}i}{2}.$$

Each of these numbers is a sixth root of 1. In general, the **nth root of a complex number** is defined as follows.

Definition of nth Root of a Complex Number

The complex number $u = a + bi$ is an **nth root** of the complex number z if

$$z = u^n = (a + bi)^n.$$

To find a formula for an nth root of a complex number, let u be an nth root of z, where

$$u = s(\cos \beta + i \sin \beta) \qquad \text{and} \qquad z = r(\cos \theta + i \sin \theta).$$

By DeMoivre's Theorem and the fact that $u^n = z$, you have

$$s^n(\cos n\beta + i \sin n\beta) = r(\cos \theta + i \sin \theta).$$

Taking the absolute value of each side of this equation, it follows that $s^n = r$. Substituting back into the previous equation and dividing by r, you get

$$\cos n\beta + i \sin n\beta = \cos \theta + i \sin \theta.$$

So, it follows that

$$\cos n\beta = \cos \theta \qquad \text{and} \qquad \sin n\beta = \sin \theta.$$

Because both sine and cosine have a period of 2π, these last two equations have solutions if and only if the angles differ by a multiple of 2π. Consequently, there must exist an integer k such that

$$n\beta = \theta + 2\pi k$$
$$\beta = \frac{\theta + 2\pi k}{n}.$$

By substituting this value of β into the trigonometric form of u, you get the result stated in the theorem on the following page.

Exploration

The nth roots of a complex number are useful for solving some polynomial equations. For instance, explain how you can use DeMoivre's Theorem to solve the polynomial equation

$$x^4 + 16 = 0.$$

[*Hint:* Write -16 as

$$16(\cos \pi + i \sin \pi).]$$

> **nth Roots of a Complex Number**
>
> For a positive integer n, the complex number $z = r(\cos \theta + i \sin \theta)$ has exactly n distinct nth roots given by
>
> $$\sqrt[n]{r}\left(\cos \frac{\theta + 2\pi k}{n} + i \sin \frac{\theta + 2\pi k}{n}\right)$$
>
> where $k = 0, 1, 2, \ldots, n - 1$.

When $k > n - 1$ the roots begin to repeat. For instance, if $k = n$, the angle

$$\frac{\theta + 2\pi n}{n} = \frac{\theta}{n} + 2\pi$$

is coterminal with θ/n, which is also obtained when $k = 0$.

The formula for the nth roots of a complex number z has a nice geometrical interpretation, as shown in Figure 7.48. Note that because the nth roots of z all have the same magnitude $\sqrt[n]{r}$, they all lie on a circle of radius $\sqrt[n]{r}$ with center at the origin. Furthermore, because successive nth roots have arguments that differ by $2\pi/n$, the n roots are equally spaced around the circle.

You have already found the sixth roots of 1 by factoring and by using the Quadratic Formula. Example 7 shows how you can solve the same problem with the formula for nth roots.

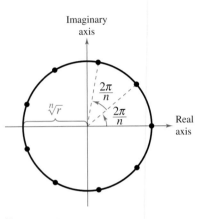

Figure 7.48

Example 7 Finding the nth Roots of a Real Number

Find all the sixth roots of 1.

Solution

First write 1 in the trigonometric form $1 = 1(\cos 0 + i \sin 0)$. Then, by the nth root formula with $n = 6$ and $r = 1$, the roots have the form

$$\sqrt[6]{1}\left(\cos \frac{0 + 2\pi k}{6} + i \sin \frac{0 + 2\pi k}{6}\right) = \cos \frac{\pi k}{3} + i \sin \frac{\pi k}{3}.$$

So, for $k = 0, 1, 2, 3, 4,$ and 5, the sixth roots are as follows. (See Figure 7.49.)

$$\cos 0 + i \sin 0 = 1$$

$$\cos \frac{\pi}{3} + i \sin \frac{\pi}{3} = \frac{1}{2} + \frac{\sqrt{3}}{2}i \qquad \text{Incremented by } \frac{2\pi}{n} = \frac{2\pi}{6} = \frac{\pi}{3}$$

$$\cos \frac{2\pi}{3} + i \sin \frac{2\pi}{3} = -\frac{1}{2} + \frac{\sqrt{3}}{2}i$$

$$\cos \pi + i \sin \pi = -1$$

$$\cos \frac{4\pi}{3} + i \sin \frac{4\pi}{3} = -\frac{1}{2} - \frac{\sqrt{3}}{2}i$$

$$\cos \frac{5\pi}{3} + i \sin \frac{5\pi}{3} = \frac{1}{2} - \frac{\sqrt{3}}{2}i$$

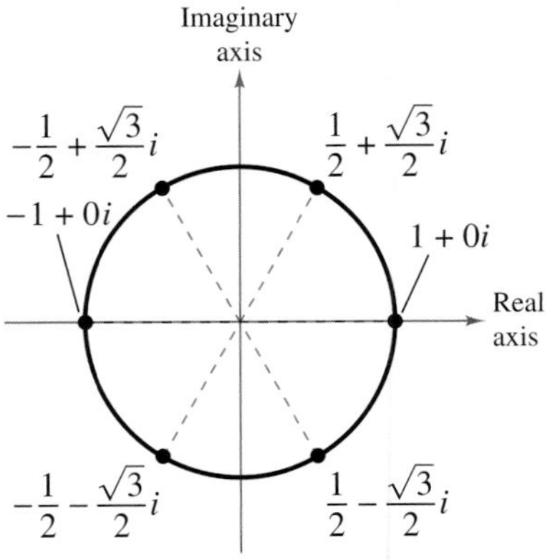

 Checkpoint Now try Exercise 103.

Figure 7.49

In Figure 7.49, notice that the roots obtained in Example 7 all have a magnitude of 1 and are equally spaced around the unit circle. Also notice that the complex roots occur in conjugate pairs, as discussed in Section 3.4. The n distinct nth roots of 1 are called the **nth roots of unity.**

Example 8   Finding the nth Roots of a Complex Number

Find the three cube roots of $z = -2 + 2i$.

Solution

The absolute value of z is

$$r = |-2 + 2i| = \sqrt{(-2)^2 + 2^2} = \sqrt{8}$$

and the angle θ is given by

$$\tan \theta = \frac{b}{a} = \frac{2}{-2} = -1.$$

Because z lies in Quadrant II, the trigonometric form of z is

$$z = -2 + 2i$$
$$= \sqrt{8}(\cos 135° + i \sin 135°).$$

By the formula for nth roots, the cube roots have the form

$$\sqrt[6]{8}\left(\cos \frac{135° + 360°k}{3} + i \sin \frac{135° + 360°k}{3} \right).$$

Finally, for $k = 0$, 1, and 2, you obtain the roots

$$\sqrt[6]{8}\left(\cos \frac{135° + 360°(0)}{3} + i \sin \frac{135° + 360°(0)}{3} \right)$$
$$= \sqrt{2}(\cos 45° + i \sin 45°)$$
$$= 1 + i$$

$$\sqrt[6]{8}\left(\cos \frac{135° + 360°(1)}{3} + i \sin \frac{135° + 360°(1)}{3} \right)$$
$$= \sqrt{2}(\cos 165° + i \sin 165°)$$
$$\approx -1.3660 + 0.3660i$$

$$\sqrt[6]{8}\left(\cos \frac{135° + 360°(2)}{3} + i \sin \frac{135° + 360°(2)}{3} \right)$$
$$= \sqrt{2}(\cos 285° + i \sin 285°)$$
$$\approx 0.3660 - 1.3660i.$$

See Figure 7.50.

✓ *Checkpoint*   Now try Exercise 93.

Exploration

Use a graphing utility set in *parametric* and *radian* modes to display the graphs of

$$\text{X1T} = \cos T$$

and

$$\text{Y1T} = \sin T.$$

Set the viewing window so that $-1.5 \le X \le 1.5$ and $-1 \le Y \le 1$. Then, using $0 \le T \le 2\pi$, set the "Tstep" to $2\pi/n$ for various values of n. Explain how the graphing utility can be used to obtain the nth roots of unity.

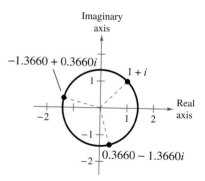

Figure 7.50

7.5 Exercises

Vocabulary Check

Fill in the blanks.

1. The _____ of a complex number $a + bi$ is the distance between the origin $(0, 0)$ and the point (a, b).

2. The _____ of a complex number $z = a + bi$ is given by $z = r(\cos \theta + i \sin \theta)$, where r is the _____ of z and θ is the _____ of z.

3. _____ Theorem states that if $z = r(\cos \theta + i \sin \theta)$ is a complex number and n is a positive integer, then $z^n = r^n(\cos n\theta + i \sin n\theta)$.

4. The complex number $u = a + bi$ is an _____ of the complex number z if $z = u^n = (a + bi)^n$.

In Exercises 1–8, plot the complex number and find its absolute value.

1. $6i$
2. $-2i$
3. -4
4. 7
5. $-4 + 4i$
6. $-5 - 12i$
7. $3 + 6i$
8. $10 - 3i$

In Exercises 9–12, write the complex number in trigonometric form.

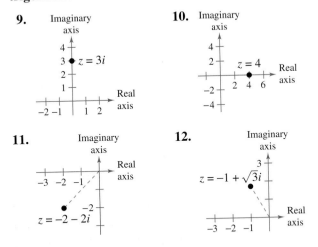

9.
10.
11.
12.

In Exercises 13–32, represent the complex number graphically, and find the trigonometric form of the number.

13. $5 - 5i$
14. $2 + 2i$
15. $\sqrt{3} + i$
16. $-1 - \sqrt{3}i$
17. $-2(1 + \sqrt{3}i)$
18. $\frac{5}{2}(\sqrt{3} - i)$
19. $-8i$
20. $4i$
21. $-7 + 4i$
22. $5 - i$
23. 3
24. 6
25. $3 + \sqrt{3}i$
26. $2\sqrt{2} - i$
27. $-1 - 2i$
28. $1 + 3i$
29. $5 + 2i$
30. $-3 + i$
31. $3\sqrt{2} - 7i$
32. $-8 - 5\sqrt{3}i$

In Exercises 33–44, represent the complex number graphically, and find the standard form of the number.

33. $2(\cos 120° + i \sin 120°)$
34. $5(\cos 135° + i \sin 135°)$
35. $\frac{3}{2}(\cos 330° + i \sin 330°)$
36. $\frac{3}{4}(\cos 315° + i \sin 315°)$
37. $3.75\left(\cos \dfrac{3\pi}{4} + i \sin \dfrac{3\pi}{4}\right)$
38. $1.5\left(\cos \dfrac{\pi}{2} + i \sin \dfrac{\pi}{2}\right)$
39. $6\left(\cos \dfrac{\pi}{3} + i \sin \dfrac{\pi}{3}\right)$
40. $8\left(\cos \dfrac{5\pi}{6} + i \sin \dfrac{5\pi}{6}\right)$
41. $4\left(\cos \dfrac{3\pi}{2} + i \sin \dfrac{3\pi}{2}\right)$
42. $9(\cos 0 + i \sin 0)$
43. $3[\cos(18° \ 45') + i \sin(18° \ 45')]$
44. $6[\cos(230° \ 30') + i \sin(230° \ 30')]$

In Exercises 45–48, use a graphing utility to represent the complex number in standard form.

45. $5\left(\cos \dfrac{\pi}{9} + i \sin \dfrac{\pi}{9}\right)$

46. $12\left(\cos \dfrac{3\pi}{5} + i \sin \dfrac{3\pi}{5}\right)$

47. $9(\cos 58° + i \sin 58°)$

48. $4(\cos 216.5° + i \sin 216.5°)$

In Exercises 49 and 50, represent the powers z, z^2, z^3, and z^4 graphically. Describe the pattern.

49. $z = \dfrac{\sqrt{2}}{2}(1 + i)$

50. $z = \dfrac{1}{2}\left(1 + \sqrt{3}i\right)$

In Exercises 51–62, perform the operation and leave the result in trigonometric form.

51. $\left[3\left(\cos \dfrac{\pi}{3} + i \sin \dfrac{\pi}{3}\right)\right]\left[4\left(\cos \dfrac{\pi}{6} + i \sin \dfrac{\pi}{6}\right)\right]$

52. $\left[\dfrac{3}{2}\left(\cos \dfrac{\pi}{6} + i \sin \dfrac{\pi}{6}\right)\right]\left[6\left(\cos \dfrac{\pi}{4} + i \sin \dfrac{\pi}{4}\right)\right]$

53. $\left[\dfrac{5}{3}(\cos 140° + i \sin 140°)\right]\left[\dfrac{2}{3}(\cos 60° + i \sin 60°)\right]$

54. $\left[\dfrac{1}{2}(\cos 115° + i \sin 115°)\right]\left[\dfrac{4}{5}(\cos 300° + i \sin 300°)\right]$

55. $\left[\dfrac{11}{20}(\cos 290° + i \sin 290°)\right]\left[\dfrac{2}{5}(\cos 200° + i \sin 200°)\right]$

56. $(\cos 5° + i \sin 5°)(\cos 20° + i \sin 20°)$

57. $\dfrac{\cos 50° + i \sin 50°}{\cos 20° + i \sin 20°}$

58. $\dfrac{5(\cos 4.3 + i \sin 4.3)}{4(\cos 2.1 + i \sin 2.1)}$

59. $\dfrac{2(\cos 120° + i \sin 120°)}{4(\cos 40° + i \sin 40°)}$

60. $\dfrac{\cos\left(\dfrac{7\pi}{4}\right) + i \sin\left(\dfrac{7\pi}{4}\right)}{\cos \pi + i \sin \pi}$

61. $\dfrac{18(\cos 54° + i \sin 54°)}{3(\cos 102° + i \sin 102°)}$

62. $\dfrac{9(\cos 20° + i \sin 20°)}{5(\cos 75° + i \sin 75°)}$

In Exercises 63–68, (a) write the trigonometric forms of the complex numbers, (b) perform the indicated operation using the trigonometric forms, and (c) perform the indicated operation using the standard forms and check your result with that of part (b).

63. $(2 + 2i)(1 - i)$

64. $\left(\sqrt{3} + i\right)(1 + i)$

65. $-2i(1 + i)$

66. $\dfrac{3 + 4i}{1 - \sqrt{3}i}$

67. $\dfrac{5}{2 + 3i}$

68. $\dfrac{4i}{-4 + 2i}$

In Exercises 69–72, sketch the graph of all complex numbers z satisfying the given condition.

69. $|z| = 2$

70. $|z| = 5$

71. $\theta = \dfrac{\pi}{6}$

72. $\theta = \dfrac{\pi}{4}$

In Exercises 73–88, use DeMoivre's Theorem to find the indicated power of the complex number. Write the result in standard form.

73. $(1 + i)^3$

74. $(2 + 2i)^6$

75. $(-1 + i)^{10}$

76. $(1 - i)^8$

77. $2\left(\sqrt{3} + i\right)^5$

78. $4\left(1 - \sqrt{3}i\right)^3$

79. $[5(\cos 20° + i \sin 20°)]^3$

80. $[3(\cos 150° + i \sin 150°)]^4$

81. $\left(\cos \dfrac{5\pi}{4} + i \sin \dfrac{5\pi}{4}\right)^{10}$

82. $\left[2\left(\cos \dfrac{\pi}{2} + i \sin \dfrac{\pi}{2}\right)\right]^{12}$

83. $[4(\cos 2.8 + i \sin 2.8)]^5$

84. $(\cos 0 + i \sin 0)^{20}$

85. $(3 - 2i)^5$

86. $\left(\sqrt{5} - 4i\right)^4$

87. $[3(\cos 15° + i \sin 15°)]^4$

88. $\left[2\left(\cos \dfrac{\pi}{10} + i \sin \dfrac{\pi}{10}\right)\right]^5$

89. Show that $-\dfrac{1}{2}\left(1 + \sqrt{3}i\right)$ is a sixth root of 1.

90. Show that $2^{-1/4}(1 - i)$ is a fourth root of -2.

Graphical Reasoning   In Exercises 91 and 92, use the graph of the roots of a complex number. (a) Write each of the roots in trigonometric form. (b) Identify the complex number whose roots are given. (c) Use a graphing utility to verify the results of part (b).

91.

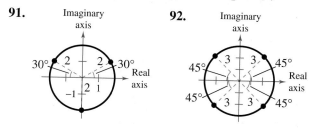

92.

In Exercises 93–108, (a) use the theorem on page 578 to find the indicated roots of the complex number, (b) represent each of the roots graphically, and (c) write each of the roots in standard form.

93. Square roots of $5(\cos 120° + i \sin 120°)$

94. Square roots of $16(\cos 60° + i \sin 60°)$

95. Fourth roots of $16\left(\cos \dfrac{4\pi}{3} + i \sin \dfrac{4\pi}{3}\right)$

96. Fifth roots of $32\left(\cos \dfrac{5\pi}{6} + i \sin \dfrac{5\pi}{6}\right)$

97. Cube roots of $-27i$

98. Fourth roots of $625i$

99. Cube roots of $-\dfrac{125}{2}(1 + \sqrt{3}i)$

100. Cube roots of $-4\sqrt{2}(1 - i)$

101. Cube roots of $64i$

102. Fourth roots of i

103. Fifth roots of 1

104. Cube roots of 1000

105. Cube roots of -125

106. Fourth roots of -4

107. Fifth roots of $128(-1 + i)$

108. Sixth roots of $729i$

In Exercises 109–116, use the theorem on page 578 to find all the solutions of the equation, and represent the solutions graphically.

109. $x^4 - i = 0$

110. $x^3 + 27 = 0$

111. $x^5 + 243 = 0$

112. $x^4 - 81 = 0$

113. $x^4 + 16i = 0$

114. $x^6 - 64i = 0$

115. $x^3 - (1 - i) = 0$

116. $x^4 + (1 + i) = 0$

Synthesis

True or False? **In Exercises 117–119, determine whether the statement is true or false. Justify your answer.**

117. $\frac{1}{2}(1 - \sqrt{3}i)$ is a ninth root of -1.

118. $\sqrt{3} + i$ is a solution of the equation $x^2 - 8i = 0$.

119. Geometrically, the nth roots of any complex number z are all equally spaced around the unit circle centered at the origin.

120. Given two complex numbers $z_1 = r_1(\cos \theta_1 + i \sin \theta_1)$ and $z_2 = r_2(\cos \theta_2 + i \sin \theta_2)$, $z_2 \neq 0$, show that

$$\frac{z_1}{z_2} = \frac{r_1}{r_2}[\cos(\theta_1 - \theta_2) + i \sin(\theta_1 - \theta_2)].$$

121. Show that $\bar{z} = r[\cos(-\theta) + i \sin(-\theta)]$ is the complex conjugate of $z = r(\cos \theta + i \sin \theta)$.

122. Use the trigonometric forms of z and $\bar{z}$ in Exercise 121 to find (a) $z\bar{z}$ and (b) $z/\bar{z}$, $\bar{z} \neq 0$.

123. Show that the negative of $z = r(\cos \theta + i \sin \theta)$ is $-z = r[\cos(\theta + \pi) + i \sin(\theta + \pi)]$.

124. *Writing* The famous formula

$$e^{a+bi} = e^a(\cos b + i \sin b)$$

is called Euler's Formula, after the Swiss mathematician Leonhard Euler (1707–1783). This formula gives rise to the equation

$$e^{\pi i} + 1 = 0.$$

This equation relates the five most famous numbers in mathematics—0, 1, π, e, and i—in a single equation. Show how Euler's Formula can be used to derive this equation. Write a short paragraph summarizing your work.

Review

Harmonic Motion **In Exercises 125–128, for the simple harmonic motion described by the trigonometric function, find the maximum displacement from equilibrium and the lowest possible positive value of t for which $d = 0$.**

125. $d = 16 \cos \dfrac{\pi}{4}t$

126. $d = \dfrac{1}{16} \sin \dfrac{5\pi}{4}t$

127. $d = \frac{1}{8} \cos 12\pi t$

128. $d = \dfrac{1}{12} \sin 60\pi t$

In Exercises 129–132, find all solutions of the equation in the interval $[0, 2\pi)$. Use a graphing utility to verify your answers.

129. $2 \cos(x + \pi) + 2 \cos(x - \pi) = 0$

130. $\sin\left(x + \dfrac{3\pi}{2}\right) - \sin\left(x - \dfrac{3\pi}{2}\right) = 0$

131. $\sin\left(x - \dfrac{\pi}{3}\right) - \sin\left(x + \dfrac{\pi}{3}\right) = \dfrac{3}{2}$

132. $\tan(x + \pi) - \cos\left(x + \dfrac{5\pi}{2}\right) = 0$

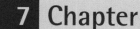 **Chapter Summary**

What did you learn?

7 Review Exercises

7.1 In Exercises 1–10, use the Law of Sines to solve the triangle. If two solutions exist, find both.

1. $A = 21°$, $B = 42°$, $a = 6$
2. $B = 110°$, $C = 30°$, $c = 11$
3. $A = 75°$, $a = 2.5$, $b = 16.5$
4. $A = 130°$, $a = 60$, $b = 48$
5. $B = 115°$, $a = 9$, $b = 14.5$
6. $C = 50°$, $a = 25$, $c = 22$
7. $A = 15°$, $a = 5$, $b = 10$
8. $B = 150°$, $a = 64$, $b = 10$
9. $B = 25°$, $a = 6.2$, $b = 4$
10. $A = 74°$, $b = 12.8$, $a = 12.5$

In Exercises 11–14, find the area of the triangle having the indicated angle and sides.

11. $A = 27°$, $b = 5$, $c = 8$
12. $B = 80°$, $a = 4$, $c = 8$
13. $C = 122°$, $b = 18$, $a = 29$
14. $C = 100°$, $a = 120$, $b = 74$

15. **Height** From a distance of 50 meters, the angle of elevation to the top of a building is 17°. Approximate the height of the building.

16. **Distance** A family is traveling due west on a road that passes a famous landmark. At a given time the bearing to the landmark is N 62° W, and after the family travels 5 miles farther, the bearing is N 38° W. What is the closest the family will come to the landmark while on the road?

17. **Height** A tree stands on a hillside of slope 28° from the horizontal. From a point 75 feet down the hill, the angle of elevation to the top of the tree is 45° (see figure). Find the height of the tree.

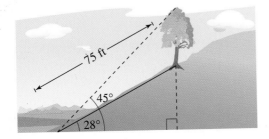

18. **Width** A surveyor finds that a tree on the opposite bank of a river has a bearing of N 22° 30′ E from a certain point and a bearing of N 15° W from a point 400 feet downstream. Find the width of the river.

7.2 In Exercises 19–28, use the Law of Cosines to solve the triangle.

19. $a = 9$, $b = 12$, $c = 20$
20. $a = 7$, $b = 15$, $c = 19$
21. $C = 45°$, $a = 6$, $b = 9$
22. $B = 90°$, $a = 5$, $c = 12$
23. $B = 110°$, $a = 4$, $c = 4$
24. $B = 12°$, $a = 32$, $c = 36$
25. $B = 150°$, $a = 10$, $c = 20$
26. $a = 42$, $b = 25$, $c = 58$
27. $a = 8.9$, $b = 6.1$, $c = 10.5$
28. $a = 7.5$, $b = 9.8$, $c = 4.5$

29. **Geometry** The lengths of the diagonals of a parallelogram are 10 feet and 16 feet. Find the lengths of the sides of the parallelogram if the diagonals intersect at an angle of 28°.

30. **Geometry** The lengths of the diagonals of a parallelogram are 30 meters and 40 meters. Find the lengths of the sides of the parallelogram if the diagonals intersect at an angle of 34°.

31. **Navigation** Two planes leave Washington, D.C.'s Dulles International Airport at approximately the same time. One is flying at 425 miles per hour at a bearing of 355°, and the other is flying at 530 miles per hour at a bearing of 67° (see figure). Determine the distance between the planes after they have flown for 2 hours.

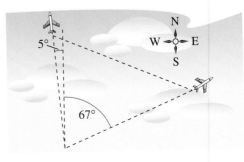

32. *Surveying*   To approximate the length of a marsh, a surveyor walks 425 meters from point A to point B. The surveyor then turns $65°$ and walks 300 meters to point C. Approximate the length AC of the marsh.

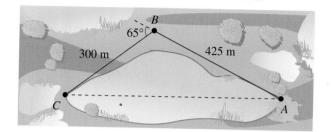

In Exercises 33–36, use Heron's Area Formula to find the area of the triangle with the given side lengths.

33. $a = 4, \quad b = 5, \quad c = 7$

34. $a = 15, \quad b = 8, \quad c = 10$

35. $a = 64.8, \quad b = 49.2, \quad c = 24.1$

36. $a = 8.55, \quad b = 5.14, \quad c = 12.73$

7.3 In Exercises 37 and 38, show that $u = v$.

37.

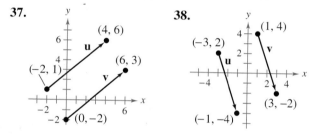

38.

In Exercises 39–44, find the component form of the vector v satisfying the given conditions.

39.

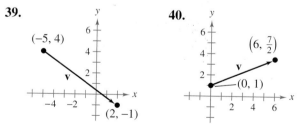

40.

41. Initial point: $(0, 10)$; terminal point: $(7, 3)$

42. Initial point: $(1, 5)$; terminal point: $(15, 9)$

43. $\|v\| = 8, \quad \theta = 120°$ **44.** $\|v\| = \frac{1}{2}, \quad \theta = 225°$

In Exercises 45–52, find (a) $u + v$, (b) $u - v$, (c) $3u$, and (d) $2v + 5u$.

45. $u = \langle -1, -3 \rangle, v = \langle -3, 6 \rangle$

46. $u = \langle 4, 5 \rangle, v = \langle 0, -1 \rangle$

47. $u = \langle -5, 2 \rangle, v = \langle 4, 4 \rangle$

48. $u = \langle 1, -8 \rangle, v = \langle 3, -2 \rangle$

49. $u = 2i - j, v = 5i + 3j$

50. $u = -6j, v = i + j$

51. $u = 4i, \quad v = -i + 6j$

52. $u = -7i - 3j, \quad v = 4i - j$

In Exercises 53–56, find the component form of w and sketch the specified vector operations geometrically, where $u = 6i - 5j$ and $v = 10i + 3j$.

53. $w = 3v$ **54.** $w = \frac{1}{2}v$

55. $w = 4u + 5v$ **56.** $w = 3v - 2u$

In Exercises 57–60, find a unit vector in the direction of the given vector.

57. $u = \langle 0, -6 \rangle$ **58.** $v = \langle -12, -5 \rangle$

59. $v = 5i - 2j$ **60.** $w = -7i$

In Exercises 61 and 62, write a linear combination of the standard unit vectors i and j for the given initial and terminal points.

61. Initial point: $(-8, 3)$
Terminal point: $(1, -5)$

62. Initial point: $(2, -3.2)$
Terminal point: $(-6.4, 10.8)$

In Exercises 63 and 64, write the vector v in the form $\|v\|[(\cos \theta)i + (\sin \theta)j]$.

63. $v = -10i + 10j$ **64.** $v = 4i - j$

In Exercises 65 and 66, graph the vectors and the resultant of the vectors. Find the magnitude and direction of the resultant.

65.

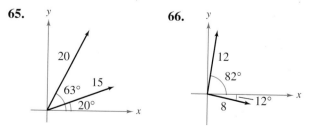

66.

67. Resultant Force Three forces with magnitudes of 250 pounds, 100 pounds, and 200 pounds act on an object at angles of 60°, 150°, and −90°, respectively, with the positive x-axis. Find the direction and magnitude of the resultant of these forces.

68. Resultant Force Forces with magnitudes of 85 pounds and 50 pounds act on a single point. The angle between the forces is 15°. Describe the resultant force.

69. Tension A 180-pound weight is supported by two ropes, as shown in the figure. Find the tension in each rope.

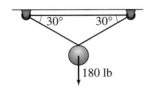

70. Cable Tension In a manufacturing process, an electric hoist lifts 200-pound ingots. Find the tension in the supporting cables (see figure).

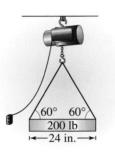

71. Navigation An airplane has an airspeed of 430 miles per hour at a bearing of 135°. The wind velocity is 35 miles per hour in the direction N 30° E. Find the resultant speed and direction of the plane.

72. Navigation An airplane has an airspeed of 724 kilometers per hour at a bearing of 30°. The wind velocity is from the west at 32 kilometers per hour. Find the resultant speed and direction of the plane.

7.4 In Exercises 73–76, find the dot product of u and v.

73. $u = \langle 0, -2 \rangle$
$v = \langle 1, 10 \rangle$

74. $u = \langle -4, 5 \rangle$
$v = \langle 3, -1 \rangle$

75. $u = 6i - j$
$v = 2i + 5j$

76. $u = 8i - 7j$
$v = 3i - 4j$

In Exercises 77–80, use the vectors $u = \langle -3, -4 \rangle$ and $v = \langle 2, 1 \rangle$ to find the indicated quantity.

77. $u \cdot u$

78. $\|v\| - 3$

79. $4u \cdot v$

80. $(u \cdot v)u$

In Exercises 81–84, find the angle θ between the vectors.

81. $u = \langle 2\sqrt{2}, -4 \rangle$, $v = \langle -\sqrt{2}, 1 \rangle$

82. $u = \langle 3, 1 \rangle$, $v = \langle 4, 5 \rangle$

83. $u = \cos \dfrac{7\pi}{4} i + \sin \dfrac{7\pi}{4} j$, $v = \cos \dfrac{5\pi}{6} i + \sin \dfrac{5\pi}{6} j$

84. $u = \cos 45° i + \sin 45° j$
$v = \cos 300° i + \sin 300° j$

In Exercises 85–88, graph the vectors and find the degree measure of the angle between the vectors.

85. $u = 4i + j$
$v = i - 4j$

86. $u = 6i + 2j$
$v = -3i - j$

87. $u = 7i - 5j$
$v = 10i + 3j$

88. $u = -5.3i + 2.8j$
$v = -8.1i - 4j$

In Exercises 89–92, determine whether u and v are orthogonal, parallel, or neither.

89. $u = \langle 39, -12 \rangle$
$v = \langle -26, 8 \rangle$

90. $u = \langle 8, -4 \rangle$
$v = \langle 5, 10 \rangle$

91. $u = \langle 8, 5 \rangle$
$v = \langle -2, 4 \rangle$

92. $u = \langle -15, 51 \rangle$
$v = \langle 20, -68 \rangle$

In Exercises 93–96, find the projection of u onto v. Then write u as the sum of two orthogonal vectors, one of which is proj$_v$ u.

93. $u = \langle -4, 3 \rangle$, $v = \langle -8, -2 \rangle$

94. $u = \langle 5, 6 \rangle$, $v = \langle 10, 0 \rangle$

95. $u = \langle 2, 7 \rangle$, $v = \langle 1, -1 \rangle$

96. $u = \langle -3, 5 \rangle$, $v = \langle -5, 2 \rangle$

97. Work Determine the work done by a crane lifting an 18,000-pound truck 48 inches.

98. Braking Force A 500-pound motorcycle is headed up a hill inclined at 12°. What force is required to keep the motorcycle from rolling back down the hill when stopped at a red light?

7.5 In Exercises 99–102, plot the complex number and find its absolute value.

99. $-i$

100. $5i$

101. $7 - 5i$

102. $-3 + 9i$

In Exercises 103–108, write the complex number in trigonometric form.

103.

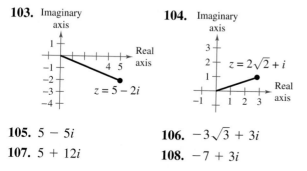

104.

$z = 2\sqrt{2} + i$

105. $5 - 5i$

106. $-3\sqrt{3} + 3i$

107. $5 + 12i$

108. $-7 + 3i$

In Exercises 109–112, perform the operation and leave the result in trigonometric form.

109. $\left[\dfrac{5}{2}\left(\cos\dfrac{\pi}{2} + i\sin\dfrac{\pi}{2}\right)\right]\left[4\left(\cos\dfrac{\pi}{4} + i\sin\dfrac{\pi}{4}\right)\right]$

110. $\left[2\left(\cos\dfrac{2\pi}{3} + i\sin\dfrac{2\pi}{3}\right)\right]\left[3\left(\cos\dfrac{\pi}{6} + i\sin\dfrac{\pi}{6}\right)\right]$

111. $\dfrac{20(\cos 320° + i\sin 320°)}{5(\cos 80° + i\sin 80°)}$

112. $\dfrac{3(\cos 230° + i\sin 230°)}{9(\cos 95° + i\sin 95°)}$

In Exercises 113–116, use DeMoivre's Theorem to find the indicated power of the complex number. Write the result in standard form.

113. $\left[5\left(\cos\dfrac{\pi}{12} + i\sin\dfrac{\pi}{12}\right)\right]^4$

114. $\left[2\left(\cos\dfrac{4\pi}{15} + i\sin\dfrac{4\pi}{15}\right)\right]^5$

115. $(2 + 3i)^6$

116. $(1 - i)^8$

In Exercises 117–120, (a) use the theorem on page 578 to find the indicated roots of the complex number, (b) represent each of the roots graphically, and (c) write each of the roots in standard form.

117. Sixth roots of $-729i$ **118.** Fourth roots of $256i$

119. Cube roots of 8 **120.** Fifth roots of -1024

In Exercises 121–124, use the theorem on page 578 to find all solutions of the equation, and represent the solutions graphically.

121. $x^4 + 256 = 0$

122. $x^5 - 32i = 0$

123. $x^3 + 8i = 0$

124. $x^4 + 81 = 0$

Synthesis

In Exercises 125 and 126, determine whether the statement is true or false. Justify your answer.

125. The Law of Sines is true if one of the angles in the triangle is a right angle.

126. When the Law of Sines is used, the solution is always unique.

127. What characterizes a vector in the plane?

128. Which vectors in the figure appear to be equivalent?

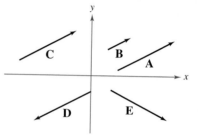

129. The figure shows z_1 and z_2. Describe $z_1 z_2$ and z_1/z_2.

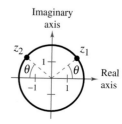

130. One of the fourth roots of a complex number z is shown in the graph.

(a) How many roots are not shown?

(b) Describe the other roots.

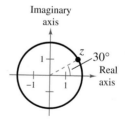

7 | Chapter Test

Take this test as you would take a test in class. After you are finished, check your work against the answers given in the back of the book.

In Exercises 1–6, use the given information to solve the triangle. If two solutions exist, find both.

1. $A = 36°$, $B = 98°$, $c = 18$

2. $a = 4$, $b = 7$, $c = 9$

3. $A = 35°$, $b = 8$, $c = 11$

4. $A = 25°$, $b = 28$, $a = 15$

5. $B = 130°$, $c = 10.1$, $b = 5.2$

6. $A = 150°$, $b = 4.8$, $a = 9.4$

7. Find the length of the pond shown at the right.

8. A triangular parcel of land has borders of lengths 55 meters, 85 meters, and 100 meters. Find the area of the parcel of land.

9. Find the component form and magnitude of the vector **w** that has initial point $(-8, -12)$ and terminal point $(4, 1)$.

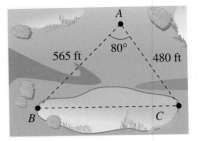

Figure for 7

In Exercises 10–13, find (a) 2v + u, (b) u − 3v, and (c) 5u − v.

10. $\mathbf{u} = \langle 0, -4 \rangle$, $\mathbf{v} = \langle -2, 4 \rangle$

11. $\mathbf{u} = \langle -2, -3 \rangle$, $\mathbf{v} = \langle 1, 5 \rangle$

12. $\mathbf{u} = \mathbf{i} - \mathbf{j}$, $\mathbf{v} = 6\mathbf{i} + 9\mathbf{j}$

13. $\mathbf{u} = 2\mathbf{i} + 3\mathbf{j}$, $\mathbf{v} = -\mathbf{i} - 2\mathbf{j}$

14. Find a unit vector in the direction of $\mathbf{v} = 7\mathbf{i} + 4\mathbf{j}$.

15. Find the component form of the vector **v** with $\|\mathbf{v}\| = 12$, in the same direction as $\mathbf{u} = \langle 3, -5 \rangle$.

16. Forces with magnitudes of 250 pounds and 130 pounds act on an object at angles of $45°$ and $-60°$, respectively, with the positive x-axis. Find the direction and magnitude of the resultant of these forces.

17. Find the dot product of $\mathbf{u} = \langle -9, 4 \rangle$ and $\mathbf{v} = \langle 1, 3 \rangle$.

18. Find the angle between the vectors $\mathbf{u} = 7\mathbf{i} + 2\mathbf{j}$ and $\mathbf{v} = -4\mathbf{j}$.

19. Are the vectors $\mathbf{u} = \langle 6, -4 \rangle$ and $\mathbf{v} = \langle 2, -3 \rangle$ orthogonal? Explain.

20. Find the projection of $\mathbf{u} = \langle 6, 7 \rangle$ onto $\mathbf{v} = \langle -5, -1 \rangle$. Then write **u** as the sum of two orthogonal vectors.

21. Write the complex number $z = -2 + 2i$ in trigonometric form.

22. Write the complex number $100(\cos 240° + i \sin 240°)$ in standard form.

In Exercises 23 and 24, use DeMoivre's Theorem to find the indicated power of the complex number. Write the result in standard form.

23. $\left[3\left(\cos \dfrac{5\pi}{6} + i \sin \dfrac{5\pi}{6} \right) \right]^8$

24. $(3 - 3i)^6$

25. Find the fourth roots of $128(1 + \sqrt{3}i)$.

26. Find all solutions of the equation $x^4 - 625i = 0$ and represent the solutions graphically.

5–7 Cumulative Test

Take this test to review the material from earlier chapters. After you are finished, check your work against the answers given in the back of the book.

1. Consider the angle $\theta = -120°$.

   (a) Sketch the angle in standard position.

   (b) Determine a coterminal angle in the interval $[0°, 360°)$.

   (c) Convert the angle to radian measure.

   (d) Find the reference angle θ'.

   (e) Find the exact values of the six trigonometric functions of θ.

2. Convert the angle $\theta = 2.35$ radians to degrees. Round your answer to one decimal place.

3. Find $\cos \theta$ if $\tan \theta = -\frac{4}{3}$ and $\sin \theta > 0$.

In Exercises 4–6, sketch the graph of the function by hand. (Include two full periods.) Use a graphing utility to verify your graph.

4. $f(x) = 3 - 2 \sin \pi x$ 5. $f(x) = \tan 3x$ 6. $f(x) = \frac{1}{2} \sec(x + \pi)$

7. Find a, b, and c such that the graph of the function $h(x) = a \cos(bx + c)$ matches the graph in the figure at the right.

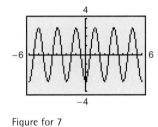

Figure for 7

In Exercises 8 and 9, find the exact value of the expression without using a calculator.

8. $\tan(\arctan 6.7)$

9. $\tan\left(\arcsin \frac{3}{5}\right)$

10. Write an algebraic expression equivalent to $\sin(\arccos 2x)$.

11. Subtract and simplify: $\dfrac{\sin \theta - 1}{\cos \theta} - \dfrac{\cos \theta}{\sin \theta - 1}$.

In Exercises 12–14, verify the identity.

12. $\cot^2 \alpha(\sec^2 \alpha - 1) = 1$ 13. $\sin(x + y) \sin(x - y) = \sin^2 x - \sin^2 y$

14. $\sin^2 x \cos^2 x = \frac{1}{8}(1 - \cos 4x)$

In Exercises 15 and 16, solve the equation.

15. $\sin^2 x + 2 \sin x + 1 = 0$ 16. $3 \tan \theta - \cot \theta = 0$

17. Approximate the solutions to the equation $\cos^2 x - 5 \cos x - 1 = 0$ in the interval $[0, 2\pi)$.

In Exercises 18 and 19, use a graphing utility to graph the function and approximate its zeros in the interval $[0, 2\pi)$. If possible, find the exact values of the zeros algebraically.

18. $y = \dfrac{1 + \sin x}{\cos x} + \dfrac{\cos x}{1 + \sin x} - 4$

19. $y = \tan^3 x - \tan^2 x + 3 \tan x - 3$

20. Given that $\sin u = \frac{12}{13}$, $\cos v = \frac{3}{5}$, and angles u and v are both in Quadrant I, find $\tan(u - v)$.

21. If $\tan \theta = \dfrac{1}{2}$, find the exact value of $\tan 2\theta$, $0 < \theta < \dfrac{\pi}{2}$.

22. If $\tan \theta = \dfrac{4}{3}$, find the exact value of $\sin \dfrac{\theta}{2}$, $\pi < \theta < \dfrac{3\pi}{2}$.

23. Write $\cos 8x + \cos 4x$ as a product.

In Exercises 24–27, verify the identity.

24. $\tan x(1 - \sin^2 x) = \frac{1}{2} \sin 2x$ **25.** $\sin 3\theta \sin \theta = \frac{1}{2}(\cos 2\theta - \cos 4\theta)$

26. $\sin 3x \cos 2x = \frac{1}{2}(\sin 5x + \sin x)$ **27.** $\dfrac{2 \cos 3x}{\sin 4x - \sin 2x} = \csc x$

In Exercises 28–31, use the information to solve the triangle shown at the right.

28. $A = 46°$, $a = 14$, $b = 5$ **29.** $A = 30°$, $b = 8$, $c = 10$

30. $A = 24°$, $C = 101°$, $a = 10$ **31.** $a = 24$, $b = 30$, $c = 47$

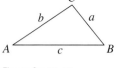

Figure for 28–31

32. Two sides of a triangle have lengths 14 inches and 19 inches. Their included angle measures 82°. Find the area of the triangle.

33. Find the area of a triangle with sides of lengths 11 inches, 16 inches, and 17 inches.

34. Write the vector $\mathbf{u} = \langle 3, 5 \rangle$ as a linear combination of the standard unit vectors $\mathbf{i}$ and $\mathbf{j}$.

35. Find a unit vector in the direction of $\mathbf{v} = \mathbf{i} + \mathbf{j}$.

36. Find $\mathbf{u} \cdot \mathbf{v}$ for $\mathbf{u} = 3\mathbf{i} + 4\mathbf{j}$ and $\mathbf{v} = \mathbf{i} - 2\mathbf{j}$.

37. Find the projection of $\mathbf{u} = \langle 8, -2 \rangle$ onto $\mathbf{v} = \langle 1, 5 \rangle$. Then write $\mathbf{u}$ as the sum of two orthogonal vectors.

38. Find the trigonometric form of the complex number shown at the right.

39. Write the complex number $8\left(\cos \dfrac{5\pi}{6} + i \sin \dfrac{5\pi}{6} \right)$ in standard form.

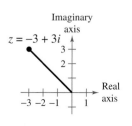

Figure for 38

40. Find the product $[4(\cos 30° + i \sin 30°)][6(\cos 120° + i \sin 120°)]$. Write the answer in standard form.

41. Find the three cube roots of 1.

42. Write all the solutions of the equation $x^5 + 243 = 0$.

43. From a point 200 feet from a flagpole, the angles of elevation to the bottom and top of the flag are 16° 45′ and 18°, respectively. Approximate the height of the flag to the nearest foot.

44. Write a model for a particle in simple harmonic motion with a displacement of 4 inches and a period of 8 seconds.

45. An airplane's velocity with respect to the air is 500 kilometers per hour, with a bearing of 30°. The wind at the altitude of the plane has a velocity of 50 kilometers per hour with a bearing of N 60° E. What is the true direction of the plane, and what is its speed relative to the ground?

46. Forces of 60 pounds and 100 pounds have a resultant force of 125 pounds. Find the angle between the two forces.

Systems of equations can be used to model the change in sales of consumer products. The growth in sales of DVD players is closely tied to the decline in sales of VCR players.

Bonn Sequenz/Imapress/The Image Works

8 Linear Systems and Matrices

What You Should Learn

In this chapter, you will learn how to:

■ Solve systems of equations by substitution, by elimination, by Gaussian elimination, by Gauss-Jordan elimination, by using inverse matrices, by Cramer's Rule, and graphically.

■ Recognize a linear system in row-echelon form and use back-substitution to solve the system.

■ Solve nonsquare systems of equations.

■ Use systems of equations to model and solve real-life problems.

■ Write matrices, identify their order, and perform elementary row operations.

■ Perform operations with matrices.

■ Find inverses of matrices.

■ Find the determinants of square matrices.

8.1 Solving Systems of Equations

The Method of Substitution

Up to this point in the text, most problems have involved either a function of one variable or a single equation in two variables. However, many problems in science, business, and engineering involve two or more equations in two or more variables. To solve such problems, you need to find solutions of **systems of equations.** Here is an example of a system of two equations in two unknowns, x and y.

$$\begin{cases} 2x + \; y = 5 & \text{Equation 1} \\ 3x - 2y = 4 & \text{Equation 2} \end{cases}$$

A **solution** of this system is an ordered pair that satisfies each equation in the system. Finding the set of all such solutions is called **solving the system of equations.** For instance, the ordered pair $(2, 1)$ is a solution of this system. To check this, you can substitute 2 for x and 1 for y in *each* equation.

In this chapter you will study six ways to solve systems of equations, beginning with the **method of substitution.**

Method	Section	Type of System
1. Substitution	8.1	Linear or nonlinear, two variables
2. Graphical	8.1	Linear or nonlinear, two variables
3. Elimination	8.2	Linear, two variables
4. Gaussian Elimination	8.3	Linear, three or more variables
5. Matrices	8.4	Linear, two or more variables
6. Cramer's Rule	8.8	Linear, two or more variables

The Method of Substitution

1. Solve one of the equations for one variable in terms of the other.

2. Substitute the expression found in Step 1 into the other equation to obtain an equation in one variable.

3. Solve the equation obtained in Step 2.

4. Back-substitute the value obtained in Step 3 into the expression obtained in Step 1 to find the value of the other variable.

5. Check that the solution satisfies *each* of the original equations.

In the algebraic solution of Example 1, you use the method of substitution to solve the system of equations. In the graphical solution, note that the solution of the system corresponds to the **point of intersection** of the graphs.

What you should learn

- Use the method of substitution and the graphical method to solve systems of equations in two variables.
- Use systems of equations to model and solve real-life problems.

Why you should learn it

You can use systems of equations in situations in which the variables must satisfy two or more conditions. For instance, Exercise 72 on page 601 shows how to use a system of equations to compare two models for estimating the number of board feet in a 16-foot log.

Bruce Hands/Getty Images

Example 1   Solving a System of Equations

Solve the system of equations.

$$\begin{cases} x + y = 4 & \text{Equation 1} \\ x - y = 2 & \text{Equation 2} \end{cases}$$

Algebraic Solution

Begin by solving for y in Equation 1.

$y = 4 - x$ Solve for y in Equation 1.

Next, substitute this expression for y into Equation 2 and solve the resulting single-variable equation for x.

$x - y = 2$	Write Equation 2.
$x - (4 - x) = 2$	Substitute $4 - x$ for y.
$x - 4 + x = 2$	Distributive Property
$2x = 6$	Combine like terms.
$x = 3$	Divide each side by 2.

Finally, you can solve for y by *back-substituting* $x = 3$ into the equation $y = 4 - x$ to obtain

$y = 4 - x$	Write revised Equation 1.
$y = 4 - 3$	Substitute 3 for x.
$y = 1.$	Solve for y.

The solution is the ordered pair $(3, 1)$. Check this as follows.

Check $(3, 1)$ in Equation 1:

$x + y = 4$	Write Equation 1.
$3 + 1 \overset{?}{=} 4$	Substitute for x and y.
$4 = 4$	Solution checks in Equation 1. ✓

Check $(3, 1)$ in Equation 2:

$x - y = 2$	Write Equation 2.
$3 - 1 \overset{?}{=} 2$	Substitute for x and y.
$2 = 2$	Solution checks in Equation 2. ✓

✓ **Checkpoint** Now try Exercise 5.

Graphical Solution

Begin by solving both equations for y. Then use a graphing utility to graph the equations $y_1 = 4 - x$ and $y_2 = x - 2$ in the same viewing window. Use the *intersect* feature (see Figure 8.1) or the *zoom* and *trace* features of the graphing utility to approximate the point of intersection of the graphs.

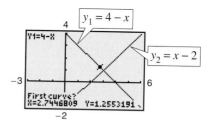

Figure 8.1

The point of intersection is $(3, 1)$, as shown in Figure 8.2.

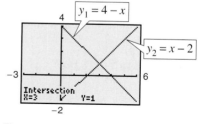

Figure 8.2

Check that $(3, 1)$ is the exact solution as follows.

Check $(3, 1)$ in Equation 1:

$3 + 1 \overset{?}{=} 4$	Substitute for x and y in Equation 1.
$4 = 4$	Solution checks in Equation 1. ✓

Check $(3, 1)$ in Equation 2:

$3 - 1 \overset{?}{=} 2$	Substitute for x and y in Equation 2.
$2 = 2$	Solution checks in Equation 2. ✓

In the algebraic solution of Example 1, note that the term *back-substitution* implies that you work *backwards*. First you solve for one of the variables, and then you substitute that value *back* into one of the equations in the system to find the value of the other variable.

Example 2   Solving a System by Substitution

A total of $12,000 is invested in two funds paying 9% and 11% simple interest. The yearly interest is $1180. How much is invested at each rate?

Solution

Verbal Model:

$$\boxed{\begin{matrix}9\% \\ \text{fund}\end{matrix}} + \boxed{\begin{matrix}11\% \\ \text{fund}\end{matrix}} = \boxed{\begin{matrix}\text{Total} \\ \text{investment}\end{matrix}}$$

$$\boxed{\begin{matrix}9\% \\ \text{interest}\end{matrix}} + \boxed{\begin{matrix}11\% \\ \text{interest}\end{matrix}} = \boxed{\begin{matrix}\text{Total} \\ \text{interest}\end{matrix}}$$

Labels: Amount in 9% fund = x Amount in 11% fund = y (dollars)

Interest for 9% fund = $0.09x$ Interest for 11% fund = $0.11y$ (dollars)

Total investment = $12,000$ Total interest = 1180 (dollars)

System:
$$\begin{cases} x + y = 12,000 & \text{Equation 1} \\ 0.09x + 0.11y = 1,180 & \text{Equation 2} \end{cases}$$

To begin, it is convenient to multiply each side of Equation 2 by 100. This eliminates the need to work with decimals.

$$9x + 11y = 118,000 \qquad \text{Revised Equation 2}$$

To solve this system, you can solve for x in Equation 1.

$$x = 12,000 - y \qquad \text{Revised Equation 1}$$

Next, substitute this expression for x into revised Equation 2 and solve the resulting equation for y.

$$9x + 11y = 118,000 \qquad \text{Write revised Equation 2.}$$
$$9(12,000 - y) + 11y = 118,000 \qquad \text{Substitute } 12,000 - y \text{ for } x.$$
$$108,000 - 9y + 11y = 118,000 \qquad \text{Distributive Property}$$
$$2y = 10,000 \qquad \text{Combine like terms.}$$
$$y = 5000 \qquad \text{Divide each side by 2.}$$

Finally, back-substitute the value $y = 5000$ to solve for x.

$$x = 12,000 - y \qquad \text{Write revised Equation 1.}$$
$$x = 12,000 - 5000 \qquad \text{Substitute 5000 for } y.$$
$$x = 7000 \qquad \text{Simplify.}$$

The solution is (7000, 5000). So, $7000 is invested at 9% and $5000 is invested at 11% to yield yearly interest of $1180. Check this in the original system.

✓ *Checkpoint* Now try Exercise 71.

The equations in Examples 1 and 2 are linear. Substitution can also be used to solve systems in which one or both of the equations are nonlinear.

Example 3   Substitution: No-Solution Case

Solve the system of equations.

$$\begin{cases} -x + y = 4 \\ x^2 + y = 3 \end{cases}$$ Equation 1

Equation 2

Solution

Begin by solving for y in Equation 1 to obtain $y = x + 4$. Next, substitute this expression for y into Equation 2 and solve for x.

$$x^2 + y = 3$$ Write Equation 2.

$$x^2 + (x + 4) = 3$$ Substitute $x + 4$ for y.

$$x^2 + x + 1 = 0$$ Simplify.

$$x = \frac{-1 \pm \sqrt{3}i}{2}$$ Quadratic Formula

Because this yields two complex values, the equation $x^2 + x + 1 = 0$ has no *real* solution. So, the original system of equations has no *real* solution.

☑ *Checkpoint* Now try Exercise 23.

> **STUDY TIP**
>
> When using substitution, solve for the variable that is not raised to a power in either equation. For instance, in Example 3 it would not be practical to solve for x in Equation 2. Can you see why?

> **Exploration**
>
> Graph the system of equations in Example 3. Do the graphs of the equations intersect? Why or why not?

Example 4   Substitution: Two-Solution Case

Solve the system of equations: $\begin{cases} x^2 + 4x - y = 7 \\ 2x - y = -1 \end{cases}$ Equation 1

Equation 2

Algebraic Solution

Begin by solving for y in Equation 2 to obtain $y = 2x + 1$. Next, substitute this expression for y into Equation 1 and solve for x.

$$x^2 + 4x - y = 7$$ Write Equation 1.

$$x^2 + 4x - (2x + 1) = 7$$ Substitute $2x + 1$ for y.

$$x^2 + 4x - 2x - 1 = 7$$ Distributive Property

$$x^2 + 2x - 8 = 0$$ Write in general form.

$$(x + 4)(x - 2) = 0$$ Factor.

$$x + 4 = 0 \implies x = -4$$ Set 1st factor equal to 0.

$$x - 2 = 0 \implies x = 2$$ Set 2nd factor equal to 0.

Back-substituting these values of x into Equation 2 produces

$$y = 2(-4) + 1 = -7 \quad \text{and} \quad y = 2(2) + 1 = 5.$$

So, the solutions are $(-4, -7)$ and $(2, 5)$. Check these in the original system.

☑ *Checkpoint* Now try Exercise 27.

Graphical Solution

To graph each equation, first solve both equations for y. Then use a graphing utility to graph the equations in the same viewing window. Use the *intersect* feature or the *zoom* and *trace* features to approximate the points of intersection of the graphs. The points of intersection are $(-4, -7)$ and $(2, 5)$, as shown in Figure 8.3. Check that $(-4, -7)$ and $(2, 5)$ are the exact solutions by substituting *both* ordered pairs into *both* equations.

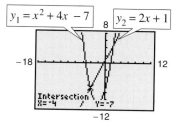

Figure 8.3

From Examples 2, 3, and 4, you can see that a system of two equations in two unknowns can have exactly one solution, more than one solution, or no solution. For instance, in Figure 8.4, the two equations graph as two lines with a *single point* of intersection. The two equations in Example 4 graph as a parabola and a line with *two points* of intersection, as shown in Figure 8.5. The two equations in Example 3 graph as a line and a parabola that have *no points* of intersection, as shown in Figure 8.6.

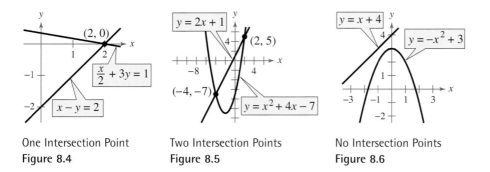

One Intersection Point
Figure 8.4

Two Intersection Points
Figure 8.5

No Intersection Points
Figure 8.6

Example 5 shows the value of a graphical approach to solving systems of equations in two variables. Notice what would happen if you tried only the substitution method in Example 5. You would obtain the equation $x + \ln x = 1$. It would be difficult to solve this equation for x using standard algebraic techniques. In such cases, a graphical approach to solving a system of equations is more convenient.

Example 5 Solving a System of Equations Graphically

Solve the system of equations.

$$\begin{cases} y = \ln x & \text{Equation 1} \\ x + y = 1 & \text{Equation 2} \end{cases}$$

Solution

From the graphs of these equations, it is clear that there is only one point of intersection. Use the *intersect* feature or the *zoom* and *trace* features of a graphing utility to approximate the solution point as $(1, 0)$, as shown in Figure 8.7. You can confirm this by substituting $(1, 0)$ into *both* equations.

Check $(1, 0)$ *in Equation 1:*

$y = \ln x$ Write Equation 1.

$0 = \ln 1$ Equation 1 checks. ✓

Check $(1, 0)$ *in Equation 2:*

$x + y = 1$ Write Equation 2.

$1 + 0 = 1$ Equation 2 checks. ✓

✓ *Checkpoint* Now try Exercise 45.

TECHNOLOGY SUPPORT

For instructions on how to use the *intersect* feature and the *zoom* and *trace* features, see Appendix A; for specific keystrokes, go to the text website at *college.hmco.com*.

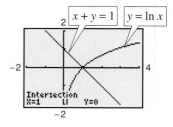

Figure 8.7

Points of Intersection and Applications

The total cost C of producing x units of a product typically has two components: the initial cost and the cost per unit. When enough units have been sold that the total revenue R equals the total cost C, the sales are said to have reached the **break-even point.** You will find that the break-even point corresponds to the point of intersection of the cost and revenue curves.

Example 6   Break-Even Analysis

A small business invests \$10,000 in equipment to produce a new soft drink. Each bottle of the soft drink costs \$0.65 to produce and is sold for \$1.20. How many items must be sold before the business breaks even?

Solution

The total cost of producing x bottles is

| Total cost | = | Cost per bottle | · | Number of bottles | + | Initial cost |

$$C = 0.65x + 10,000. \qquad \text{Equation 1}$$

The revenue obtained by selling x bottles is

| Total revenue | = | Price per bottle | · | Number of bottles |

$$R = 1.20x. \qquad \text{Equation 2}$$

Because the break-even point occurs when $R = C$, you have $C = 1.20x$, and the system of equations to solve is

$$\begin{cases} C = 0.65x + 10,000 \\ C = 1.20x \end{cases}.$$

Now you can solve by substitution.

$$1.20x = 0.65x + 10,000 \qquad \text{Substitute } 1.20x \text{ for } C \text{ in Equation 1.}$$

$$0.55x = 10,000 \qquad \text{Subtract } 0.65x \text{ from each side.}$$

$$x = \frac{10,000}{0.55} \approx 18,182 \text{ bottles.} \qquad \text{Divide each side by } 0.55.$$

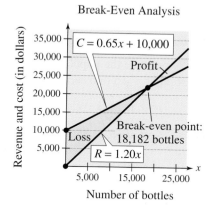

Break-Even Analysis

Figure 8.8

Note in Figure 8.8 that revenue less than the break-even point corresponds to an overall loss, whereas revenue greater than the break-even point corresponds to a profit. Verify the break-even point using the *intersect* feature or the *zoom* and *trace* features of a graphing utility.

 Checkpoint Now try Exercise 67.

Another way to view the solution in Example 6 is to consider the profit function $P = R - C$. The break-even point occurs when the profit is 0, which is the same as saying that $R = C$.

Example 7 State Populations

From 1991 to 2001, the population of Idaho was increasing at a faster rate than the population of New Hampshire. Two models that approximate the populations P (in thousands) are

$$\begin{cases} P = 1019 + 28.5t & \text{Idaho} \\ P = 1080 + 15.7t & \text{New Hampshire} \end{cases}$$

where t represents the year, with $t = 1$ corresponding to 1991. (Source: U.S. Census Bureau)

a. According to these two models, when would you expect the population of Idaho to have exceeded the population of New Hampshire?

b. Use the two models to estimate the population of both states in 2006.

TECHNOLOGY SUPPORT

For instructions on how to use the *value* feature, see Appendix A; for specific keystrokes, go to the text website at *college.hmco.com*.

Algebraic Solution

a. Because the first equation has already been solved for P in terms of t, you can substitute this value into the second equation and solve for t, as follows.

$$1019 + 28.5t = 1080 + 15.7t$$

$$28.5t - 15.7t = 1080 - 1019$$

$$12.8t = 61$$

$$t \approx 4.8$$

So, from the given models, you would expect that the population of Idaho exceeded the population of New Hampshire after $t \approx 4.8$ years, which was sometime during 1994.

b. To estimate the population of both states in 2006, substitute $t = 16$ into each model and evaluate, as follows.

$$P = 1019 + 28.5t \qquad \text{Model for Idaho}$$

$$= 1019 + 28.5(16) \qquad \text{Substitute 16 for } t.$$

$$= 1475 \qquad \text{Simplify.}$$

$$P = 1080 + 15.7t \qquad \text{Model for New Hampshire}$$

$$= 1080 + 15.7(16) \qquad \text{Substitute 16 for } t.$$

$$= 1331.2 \qquad \text{Simplify.}$$

So, according to the models, Idaho's population in 2006 will be 1475 thousand and New Hampshire's population in 2006 will be 1331.2 thousand.

✓ *Checkpoint* Now try Exercise 73.

Graphical Solution

a. Use a graphing utility to graph $y_1 = 1019 + 28.5x$ and $y_2 = 1080 + 15.7x$ in the same viewing window. Use the *intersect* feature or the *zoom* and *trace* features of the graphing utility to approximate the point of intersection of the graphs. The point of intersection occurs at $x \approx 4.8$, as shown in Figure 8.9. So, it appears that the population of Idaho exceeded the population of New Hampshire sometime during 1994.

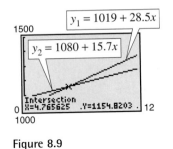

Figure 8.9

b. To estimate the population of both states in 2006, use the *value* feature or *zoom* and *trace* features of the graphing utility to find the value of y when $x = 16$. (Be sure to adjust your viewing window.) So, from Figure 8.10, you can see that Idaho's population in 2006 will be 1475 thousand and New Hampshire's population in 2006 will be 1331.2 thousand.

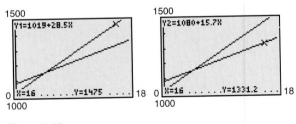

Figure 8.10

8.1   Exercises

Vocabulary Check

Fill in the blanks.

1. A set of two or more equations in two or more unknowns is called a _____ of _____ .
2. A _____ of a system of equations is an ordered pair that satisfies each equation in the system.
3. The first step in solving a system of equations by the _____ of _____ is to solve one of the equations for one variable in terms of the other variable.
4. Graphically, the solution to a system of equations is called the _____ of _____ .
5. In business applications, the _____ occurs when revenue equals cost.

In Exercises 1–4, determine whether each ordered pair is a solution of the system of equations.

1. $\begin{cases} 4x - y = 1 \\ 6x + y = -6 \end{cases}$
   (a) $(0, -3)$ (b) $(-1, -5)$
   (c) $\left(-\frac{3}{2}, 3\right)$ (d) $\left(-\frac{1}{2}, -3\right)$

2. $\begin{cases} 4x^2 + y = 3 \\ -x - y = 11 \end{cases}$
   (a) $(2, -13)$ (b) $(-2, -9)$
   (c) $\left(-\frac{3}{2}, 6\right)$ (d) $\left(-\frac{7}{4}, -\frac{37}{4}\right)$

3. $\begin{cases} y = -2e^x \\ 3x - y = 2 \end{cases}$
   (a) $(-2, 0)$ (b) $(0, -2)$
   (c) $(0, -3)$ (d) $(-1, -5)$

4. $\begin{cases} -\log_{10} x + 3 = y \\ \frac{1}{9}x + y = \frac{28}{9} \end{cases}$
   (a) $(100, 1)$ (b) $(10, 2)$
   (c) $(1, 3)$ (d) $(1, 1)$

In Exercises 5–14, solve the system by the method of substitution. Check your solution graphically.

5. $\begin{cases} 2x + y = 6 \\ -x + y = 0 \end{cases}$

6. $\begin{cases} x - y = -4 \\ x + 2y = 5 \end{cases}$

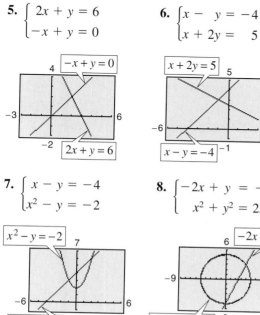

7. $\begin{cases} x - y = -4 \\ x^2 - y = -2 \end{cases}$

8. $\begin{cases} -2x + y = -5 \\ x^2 + y^2 = 25 \end{cases}$

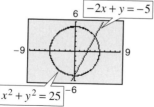
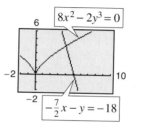

9. $\begin{cases} 3x + y = 2 \\ x^3 - 2 + y = 0 \end{cases}$

10. $\begin{cases} x + y = 0 \\ x^3 - 5x - y = 0 \end{cases}$

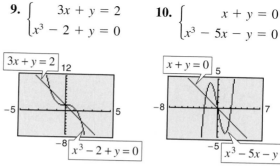
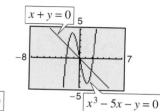

11. $\begin{cases} x^2 + y = 0 \\ x^2 - 4x - y = 0 \end{cases}$

12. $\begin{cases} y = -2x^2 + 2 \\ y = 2(x^4 - 2x^2 + 1) \end{cases}$

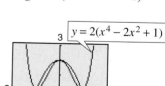

13. $\begin{cases} -\frac{7}{2}x - y = -18 \\ 8x^2 - 2y^3 = 0 \end{cases}$

14. $\begin{cases} y = x^3 - 3x^2 + 4 \\ y = -2x + 4 \end{cases}$

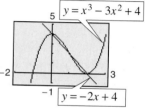

In Exercises 15–28, solve the system by the method of substitution. Use a graphing utility to verify your results.

15. $\begin{cases} x - y = 0 \\ 5x - 3y = 10 \end{cases}$ 16. $\begin{cases} x + 2y = 1 \\ 5x - 4y = -23 \end{cases}$

17. $\begin{cases} 2x - y + 2 = 0 \\ 4x + y - 5 = 0 \end{cases}$ 18. $\begin{cases} 6x - 3y - 4 = 0 \\ x + 2y - 4 = 0 \end{cases}$

19. $\begin{cases} 1.5x + 0.8y = 2.3 \\ 0.3x - 0.2y = 0.1 \end{cases}$ 20. $\begin{cases} 0.5x + 3.2y = 9.0 \\ 0.2x - 1.6y = -3.6 \end{cases}$

21. $\begin{cases} \frac{1}{5}x + \frac{1}{2}y = 8 \\ x + y = 20 \end{cases}$ 22. $\begin{cases} \frac{1}{2}x + \frac{3}{4}y = 10 \\ \frac{3}{4}x - y = 4 \end{cases}$

23. $\begin{cases} 6x + 5y = -3 \\ -x - \frac{5}{6}y = -7 \end{cases}$ 24. $\begin{cases} 2x - y = 4 \\ -4x + 2y = -12 \end{cases}$

25. $\begin{cases} -\frac{5}{3}x + y = 5 \\ -5x + 3y = 6 \end{cases}$ 26. $\begin{cases} -\frac{2}{3}x + y = 2 \\ 2x - 3y = 6 \end{cases}$

27. $\begin{cases} x^3 - y = 0 \\ x - y = 0 \end{cases}$ 28. $\begin{cases} y = -x \\ y = x^3 + 3x^2 + 2x \end{cases}$

In Exercises 29–36, solve the system graphically. Verify your solutions algebraically.

29. $\begin{cases} -x + 2y = 2 \\ 3x + y = 15 \end{cases}$ 30. $\begin{cases} x + y = 0 \\ 3x - 2y = 10 \end{cases}$

31. $\begin{cases} x - 3y = -2 \\ 5x + 3y = 17 \end{cases}$ 32. $\begin{cases} -x + 2y = 1 \\ x - y = 2 \end{cases}$

33. $\begin{cases} x + y = 4 \\ x^2 + y^2 - 4x = 0 \end{cases}$

34. $\begin{cases} -x + y = 3 \\ x^2 + y^2 - 6x - 27 = 0 \end{cases}$

35. $\begin{cases} x - y + 3 = 0 \\ x^2 - 4x + 7 = y \end{cases}$

36. $\begin{cases} y^2 - 4x + 11 = 0 \\ -\frac{1}{2}x + y = -\frac{1}{2} \end{cases}$

In Exercises 37–50, use a graphing utility to approximate all points of intersection of the graph of the system of equations. Verify your solutions by checking them in the original system.

37. $\begin{cases} 7x + 8y = 24 \\ x - 8y = 8 \end{cases}$ 38. $\begin{cases} x - y = 0 \\ 5x - 2y = 6 \end{cases}$

39. $\begin{cases} 2x - y + 3 = 0 \\ x^2 + y^2 - 4x = 0 \end{cases}$ 40. $\begin{cases} 3x - 2y = 0 \\ x^2 + y^2 = 4 \end{cases}$

41. $\begin{cases} x^2 + y^2 = 8 \\ y = x^2 \end{cases}$ 42. $\begin{cases} x^2 + y^2 = 25 \\ (x - 8)^2 + y^2 = 41 \end{cases}$

43. $\begin{cases} y = e^x \\ x - y + 1 = 0 \end{cases}$ 44. $\begin{cases} y = -4e^{-x} \\ y + 3x + 8 = 0 \end{cases}$

45. $\begin{cases} x + 2y = 8 \\ y = 2 + \ln x \end{cases}$

46. $\begin{cases} y = -2 + \ln(x - 1) \\ 3y + 2x = 9 \end{cases}$

47. $\begin{cases} y = \sqrt{x} + 4 \\ y = 2x + 1 \end{cases}$ 48. $\begin{cases} x - y = 3 \\ \sqrt{x} - y = 1 \end{cases}$

49. $\begin{cases} x^2 + y^2 = 169 \\ x^2 - 8y = 104 \end{cases}$ 50. $\begin{cases} x^2 + y^2 = 4 \\ 2x^2 - y = 2 \end{cases}$

In Exercises 51–62, solve the system graphically or algebraically. Explain your choice of method.

51. $\begin{cases} 2x - y = 0 \\ x^2 - y = -1 \end{cases}$ 52. $\begin{cases} x + y = 4 \\ x^2 + y = 2 \end{cases}$

53. $\begin{cases} 3x - 7y = -6 \\ x^2 - y^2 = 4 \end{cases}$ 54. $\begin{cases} x^2 + y^2 = 25 \\ 2x + y = 10 \end{cases}$

55. $\begin{cases} y = 2x + 1 \\ y = \sqrt{x + 2} \end{cases}$ 56. $\begin{cases} y = 2x - 1 \\ y = \sqrt{x + 1} \end{cases}$

57. $\begin{cases} y - e^{-x} = 1 \\ y - \ln x = 3 \end{cases}$ 58. $\begin{cases} 2 \ln x + y = 4 \\ e^x - y = 0 \end{cases}$

59. $\begin{cases} y = x^3 - 2x^2 + 1 \\ y = 1 - x^2 \end{cases}$ 60. $\begin{cases} y = x^3 - 2x^2 + x - 1 \\ y = -x^2 + 3x - 1 \end{cases}$

61. $\begin{cases} xy - 1 = 0 \\ 2x - 4y + 7 = 0 \end{cases}$ 62. $\begin{cases} xy - 2 = 0 \\ 3x - 2y + 4 = 0 \end{cases}$

Break-Even Analysis In Exercises 63–66, use a graphing utility to graph the cost and revenue functions in the same viewing window. Find the sales x necessary to break even $(R = C)$ and the corresponding revenue R obtained by selling x units. (Round to the nearest whole unit.)

	Cost	*Revenue*
63.	$C = 8650x + 250{,}000$	$R = 9950x$
64.	$C = 2.65x + 350{,}000$	$R = 4.15x$
65.	$C = 5.5\sqrt{x} + 10{,}000$	$R = 3.29x$
66.	$C = 7.8\sqrt{x} + 18{,}500$	$R = 12.84x$

67. Break-Even Analysis   A small software company invests $16,000 to produce a software package that will sell for $55.95. Each unit can be produced for $35.45.

(a) Write the cost and revenue functions for x units produced and sold.

(b) Use a graphing utility to graph the cost and revenue functions in the same viewing window. Use the graph to approximate the number of units that must be sold to break even.

(c) Verify the result of part (b) algebraically.

68. Break-Even Analysis   A small fast-food restaurant invests $5000 to produce a new food item that will sell for $3.49. Each item can be produced for $2.16.

(a) Write the cost and revenue functions for x items produced and sold.

(b) Use a graphing utility to graph the cost and revenue functions in the same viewing window. Use the graph to approximate the number of items that must be sold to break even.

(c) Verify the result of part (b) algebraically.

69. Choice of Two Jobs   You are offered two different jobs selling dental supplies. One company offers a straight commission of 6% of sales. The other company offers a salary of $350 per week plus 3% of sales. How much would you have to sell in a week in order to make the straight commission offer the better offer?

70. Choice of Two Jobs   You are offered two jobs selling college textbooks. One company offers an annual salary of $25,000 plus a year-end bonus of 1% of your total sales. The other company offers an annual salary of $20,000 plus a year-end bonus of 2% of your total sales. How much would you have to sell in a year to make the second offer the better offer?

71. Investment   A total of $20,000 is invested in two funds paying 6.5% and 8.5% simple interest. The 6.5% investment has a lower risk. The investor wants a yearly interest check of $1600 from the investments.

(a) Write a system of equations in which one equation represents the total amount invested and the other equation represents the $1600 required in interest. Let x and y represent the amounts invested at 6.5% and 8.5%, respectively.

(b) Use a graphing utility to graph the two equations in the same viewing window. As the amount invested at 6.5% increases, how does the amount invested at 8.5% change? How does the amount of interest change? Explain.

(c) What amount should be invested at 6.5% to meet the requirement of $1600 per year in interest?

72. Log Volume   You are offered two different rules for estimating the number of board feet in a 16-foot log. (A board foot is a unit of measure for lumber equal to a board 1 foot square and 1 inch thick.) One rule is the *Doyle Log Rule* and is modeled by

$$V = (D - 4)^2, \quad 5 \le D \le 40$$

and the other rule is the *Scribner Log Rule* and is modeled by

$$V = 0.79D^2 - 2D - 4, \quad 5 \le D \le 40$$

where D is the diameter (in inches) of the log and V is its volume in board feet.

(a) Use a graphing utility to graph the two log rules in the same viewing window.

(b) For what diameter do the two rules agree?

(c) You are selling large logs by the board foot. Which rule would you use? Explain your reasoning.

73. Sales   The table shows the factory sales F (in millions of dollars) of VCRs and DVD players from 1997 to 2001. (Source: Consumer Electronics Association)

Year	VCR sales, F	DVD player sales, F
1997	2618	171
1998	2409	421
1999	2333	1099
2000	1869	1717
2001	1099	2145

(a) Use the *regression* feature of a graphing utility to find quadratic models for the data. Let x represent the year, with $x = 7$ corresponding to 1997.

(b) Use a graphing utility to graph the models with the original data in the same viewing window.

(c) Use the graph in part (b) to determine the year in which DVD player sales exceeded VCR sales.

(d) Algebraically determine the year in which DVD player sales exceeded VCR sales.

(e) Compare your results from parts (c) and (d).

74. Sales The table shows the sales S (in billions of dollars) for grocery stores and general merchandise stores from 1995 to 2001. (Source: U.S. Census Bureau)

Year	Grocery store sales, S_1	General merchandise store sales, S_2
1995	356.9	300.6
1996	366.1	315.4
1997	373.1	331.5
1998	382.4	351.8
1999	401.8	381.4
2000	415.3	405.9
2001	425.4	430.5

(a) Use the *regression* feature of a graphing utility to find quadratic models for the data. Let x represent the year, with $x = 5$ corresponding to 1995.

(b) Use a graphing utility to graph the models with the original data in the same viewing window.

(c) Use the graph in part (b) to determine the year in which general merchandise store sales exceeded grocery store sales.

(d) Algebraically determine the year in which general merchandise store sales exceeded grocery store sales.

(e) Compare your results from parts (c) and (d).

Geometry In Exercises 75 and 76, find the dimensions of the rectangle meeting the specified conditions.

75. The perimeter is 30 meters and the length is 3 meters greater than the width.

76. The perimeter is 280 centimeters and the width is 20 centimeters less than the length.

77. Geometry What are the dimensions of a rectangular tract of land if its perimeter is 40 miles and its area is 96 square miles?

78. Geometry What are the dimensions of an isosceles right triangle with a two-inch hypotenuse and an area of 1 square inch?

Synthesis

True or False? In Exercises 79 and 80, determine whether the statement is true or false. Justify your answer.

79. In order to solve a system of equations by substitution, you must always solve for y in one of the two equations and then back-substitute.

80. If a system consists of a parabola and a circle, then it can have at most two solutions.

81. Think About It When solving a system of equations by substitution, how do you recognize that the system has no solution?

82. Writing Write a brief paragraph describing any advantages of substitution over the graphical method of solving a system of equations.

83. Exploration Find an equation of a line whose graph intersects the graph of the parabola $y = x^2$ at (a) two points, (b) one point, and (c) no points. (There are many correct answers.)

84. Conjecture Consider the system of equations
$$\begin{cases} y = b^x \\ y = x^b \end{cases}$$

(a) Use a graphing utility to graph the system of equations for $b = 2$ and $b = 4$.

(b) For a fixed value of $b > 1$, make a conjecture about the number of points of intersection of the graphs in part (a).

Review

In Exercises 85–90, find the general form of the equation of the line passing through the two points.

85. $(-2, 7), (5, 5)$ **86.** $(3.5, 4), (10, 6)$

87. $(6, 3), (10, 3)$ **88.** $(4, -2), (4, 5)$

89. $\left(\frac{3}{5}, 0\right), (4, 6)$ **90.** $\left(-\frac{7}{3}, 8\right), \left(\frac{5}{2}, \frac{1}{2}\right)$

In Exercises 91–94, find the domain of the function and identify any horizontal or vertical asymptotes.

91. $f(x) = \dfrac{5}{x - 6}$ **92.** $f(x) = \dfrac{2x - 7}{3x + 2}$

93. $f(x) = \dfrac{x^2 + 2}{x^2 - 16}$ **94.** $f(x) = 3 - \dfrac{2}{x^2}$

8.2 Systems of Linear Equations in Two Variables

The Method of Elimination

In Section 8.1, you studied two methods for solving a system of equations: substitution and graphing. Now you will study the **method of elimination.** The key step in this method is to obtain, for one of the variables, coefficients that differ only in sign so that *adding* the equations eliminates the variable.

$$
\begin{array}{ll}
3x + 5y = 7 & \text{Equation 1} \\
\underline{-3x - 2y = -1} & \text{Equation 2} \\
3y = 6 & \text{Add equations.}
\end{array}
$$

Note that by adding the two equations, you eliminate the x-terms and obtain a single equation in y. Solving this equation for y produces $y = 2$, which you can then back-substitute into one of the original equations to solve for x.

Example 1   Solving a System by Elimination

Solving the system of linear equations.

$$
\begin{cases}
3x + 2y = 4 & \text{Equation 1} \\
5x - 2y = 8 & \text{Equation 2}
\end{cases}
$$

Solution

You can eliminate the y-terms by adding the two equations.

$$
\begin{array}{ll}
3x + 2y = 4 & \text{Write Equation 1.} \\
\underline{5x - 2y = 8} & \text{Write Equation 2.} \\
8x = 12 & \text{Add equations.}
\end{array}
$$

So, $x = \frac{3}{2}$. By back-substituting into Equation 1, you can solve for y.

$$
\begin{array}{ll}
3x + 2y = 4 & \text{Write Equation 1.} \\
3\left(\frac{3}{2}\right) + 2y = 4 & \text{Substitute } \frac{3}{2} \text{ for } x. \\
y = -\frac{1}{4} & \text{Solve for } y.
\end{array}
$$

The solution is $\left(\frac{3}{2}, -\frac{1}{4}\right)$. You can check the solution *algebraically* by substituting into the original system, or graphically as shown in Section 8.1.

Check

$$
\begin{array}{ll}
3\left(\frac{3}{2}\right) + 2\left(-\frac{1}{4}\right) \overset{?}{=} 4 & \text{Substitute into Equation 1.} \\
\frac{9}{2} - \frac{1}{2} = 4 & \text{Equation 1 checks. } \checkmark \\
5\left(\frac{3}{2}\right) - 2\left(-\frac{1}{4}\right) \overset{?}{=} 8 & \text{Substitute into Equation 2.} \\
\frac{15}{2} + \frac{1}{2} = 8 & \text{Equation 2 checks. } \checkmark
\end{array}
$$

✓ *Checkpoint* Now try Exercise 7.

Frank Siteman/PhotoEdit

Exploration

Use the method of substitution to solve the system given in Example 1. Which method is easier?

The Method of Elimination

To use the **method of elimination** to solve a system of two linear equations in x and y, perform the following steps.

1. Obtain coefficients for x (or y) that differ only in sign by multiplying all terms of one or both equations by suitably chosen constants.

2. Add the equations to eliminate one variable; solve the resulting equation.

3. Back-substitute the value obtained in Step 2 into either of the original equations and solve for the other variable.

4. Check your solution in both of the original equations.

Example 2 Solving a System by Elimination

Solve the system of linear equations.

$$\begin{cases} 5x + 3y = 9 & \text{Equation 1} \\ 2x - 4y = 14 & \text{Equation 2} \end{cases}$$

Algebraic Solution

You can obtain coefficients that differ only in sign by multiplying Equation 1 by 4 and multiplying Equation 2 by 3.

$$
\begin{array}{llll}
5x + 3y = 9 & \Longrightarrow & 20x + 12y = 36 & \text{Multiply Equation 1 by 4.} \\
2x - 4y = 14 & \Longrightarrow & \underline{6x - 12y = 42} & \text{Multiply Equation 2 by 3.} \\
& & \ \ 26x \qquad = 78 & \text{Add equations.}
\end{array}
$$

From this equation, you can see that $x = 3$. By back-substituting this value of x into Equation 2, you can solve for y.

$$
\begin{array}{ll}
2x - 4y = 14 & \text{Write Equation 2.} \\
2(3) - 4y = 14 & \text{Substitute 3 for } x. \\
-4y = 8 & \text{Combine like terms.} \\
y = -2 & \text{Solve for } y.
\end{array}
$$

The solution is $(3, -2)$. You can check the solution algebraically by substituting into the original system.

 Checkpoint Now try Exercise 11.

Graphical Solution

Solve each equation for y. Then use a graphing utility to graph $y_1 = 3 - \frac{5}{3}x$ and $y_2 = -\frac{7}{2} + \frac{1}{2}x$ in the same viewing window. Use the *intersect* feature or the *zoom* and *trace* features to approximate the point of intersection of the graphs. The point of intersection is $(3, -2)$, as shown in Figure 8.11. You can determine that this is the exact solution by checking $(3, -2)$ in both equations.

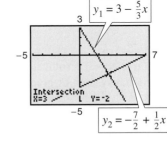

Figure 8.11

In Example 2, the original system and the system obtained by multiplying by constants are called **equivalent systems** because they have precisely the same solution set. The operations that can be performed on a system of linear equations to produce an equivalent system are (1) interchanging any two equations, (2) multiplying an equation by a nonzero constant, and (3) adding a multiple of one equation to any other equation in the system.

Graphical Interpretation of Two-Variable Systems

It is possible for a *general* system of equations to have exactly one solution, two or more solutions, or no solution. If a system of *linear* equations has two different solutions, it must have an *infinite* number of solutions. To see why this is true, consider the following graphical interpretations of a system of two linear equations in two variables.

Graphical Interpretation of Solutions

For a system of two linear equations in two variables, the number of solutions is one of the following.

Number of Solutions	*Graphical Interpretation*
1. Exactly one solution	The two lines intersect at one point.
2. Infinitely many solutions	The two lines are coincident (identical).
3. No solution	The two lines are parallel.

A system of linear equations is **consistent** if it has at least one solution. It is **inconsistent** if it has no solution.

Example 3   Recognizing Graphs of Linear Systems

Match each system of linear equations (a, b, c) with its graph (i, ii, iii) in Figure 8.12. Describe the number of solutions. Then state whether the system is consistent or inconsistent.

a. $\begin{cases} 2x - 3y = 3 \\ -4x + 6y = 6 \end{cases}$ **b.** $\begin{cases} 2x - 3y = 3 \\ x + 2y = 5 \end{cases}$ **c.** $\begin{cases} 2x - 3y = 3 \\ -4x + 6y = -6 \end{cases}$

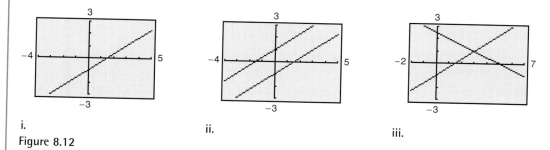

i.   ii.   iii.

Figure 8.12

Solution

a. The graph is a pair of parallel lines (ii). The lines have no point of intersection, so the system has no solution. The system is inconsistent.

b. The graph is a pair of intersecting lines (iii). The lines have one point of intersection, so the system has exactly one solution. The system is consistent.

c. The graph is a pair of lines that coincide (i). The lines have infinitely many points of intersection, so the system has infinitely many solutions. The system is consistent.

✓ *Checkpoint* Now try Exercises 17–20.

In Examples 4 and 5, note how you can use the method of elimination to determine that a system of linear equations has no solution or infinitely many solutions.

Example 4   The Method of Elimination: No Solution Case

Solve the system of linear equations.

$$\begin{cases} x - 2y = 3 & \text{Equation 1} \\ -2x + 4y = 1 & \text{Equation 2} \end{cases}$$

Algebraic Solution

To obtain coefficients that differ only in sign, multiply Equation 1 by 2.

$$\begin{array}{rcl} x - 2y = 3 & \Rightarrow & 2x - 4y = 6 \\ -2x + 4y = 1 & \Rightarrow & \underline{-2x + 4y = 1} \\ & & 0 = 7 \end{array}$$

By adding the equations, you obtain $0 = 7$. Because there are no values of x and y for which $0 = 7$, this is a false statement. So, you can conclude that the system is inconsistent and has no solution.

Graphical Solution

Solving each equation for y yields $y_1 = -\frac{3}{2} + \frac{1}{2}x$ and $y_2 = \frac{1}{4} + \frac{1}{2}x$. Notice that the lines have the same slope and different y-intercepts, so they are parallel. You can use a graphing utility to verify this by graphing both equations in the same viewing window, as shown in Figure 8.13. Then try using the *intersect* feature to find a point of intersection. Because the graphing utility cannot find a point of intersection, you will get an error message. Therefore, the system has no solution.

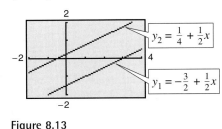

Figure 8.13

 Checkpoint Now try Exercise 23.

Example 5   The Method of Elimination: Infinitely Many Solutions Case

Solve the system of linear equations: $\begin{cases} 2x - y = 1 & \text{Equation 1} \\ 4x - 2y = 2 & \text{Equation 2} \end{cases}$.

Solution

To obtain coefficients that differ only in sign, multiply Equation 2 by $-\frac{1}{2}$.

$$\begin{array}{rcll} 2x - y = 1 & \Rightarrow & 2x - y = 1 & \text{Write Equation 1.} \\ 4x - 2y = 2 & \Rightarrow & \underline{-2x + y = -1} & \text{Multiply Equation 2 by } -\frac{1}{2}. \\ & & 0 = 0 & \text{Add equations.} \end{array}$$

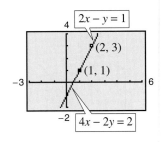

Figure 8.14

Because $0 = 0$ for all values of x and y, the two equations turn out to be equivalent (have the same solution set). You can conclude that the system has infinitely many solutions. The solution set consists of all points (x, y) lying on the line $2x - y = 1$, as shown in Figure 8.14.

 Checkpoint Now try Exercise 25.

In Example 4, note that the occurrence of a false statement, such as $0 = 7$, indicates that the system has no solution. In Example 5, note that the occurrence of a statement that is true for all values of the variables—in this case, $0 = 0$—indicates that the system has infinitely many solutions.

Example 6 illustrates a strategy for solving a system of linear equations that has decimal coefficients.

Example 6   A Linear System Having Decimal Coefficients

Solve the system of linear equations.

$$\begin{cases} 0.02x - 0.05y = -0.38 & \text{Equation 1} \\ 0.03x + 0.04y = 1.04 & \text{Equation 2} \end{cases}$$

Solution

Because the coefficients in this system have two decimal places, you can begin by multiplying each equation by 100 to produce a system with integer coefficients.

$$\begin{cases} 2x - 5y = -38 & \text{Revised Equation 1} \\ 3x + 4y = 104 & \text{Revised Equation 2} \end{cases}$$

Now, to obtain coefficients that differ only in sign, multiply revised Equation 1 by 3 and multiply revised Equation 2 by -2.

$$\begin{array}{ll} 2x - 5y = -38 & \qquad 6x - 15y = -114 \qquad \text{Multiply revised Equation 1 by 3.} \\ \underline{3x + 4y = 104} & \qquad \underline{-6x - 8y = -208} \qquad \text{Multiply revised Equation 2 by } -2. \\ & \qquad \qquad -23y = -322 \qquad \text{Add equations.} \end{array}$$

So, you can conclude that

$$y = \frac{-322}{-23}$$

$$= 14.$$

Back-substituting this value into revised Equation 2 produces the following.

$$\begin{array}{ll} 3x + 4y = 104 & \text{Write revised Equation 2.} \\ 3x + 4(14) = 104 & \text{Substitute 14 for } y. \\ 3x = 48 & \text{Combine like terms.} \\ x = 16 & \text{Solve for } x. \end{array}$$

The solution is $(16, 14)$. Check this in the original system.

✓ *Checkpoint* Now try Exercise 31.

STUDY TIP

The general solution of the linear system

$$\begin{cases} ax + by = c \\ dx + ey = f \end{cases}$$

is $x = (ce - bf)/(ae - bd)$ and $y = (af - cd)/(ae - bd)$. If $ae - bd = 0$, the system does not have a unique solution. A program (called Systems of Linear Equations) for solving such a system is available on our website, *college.hmco.com*. Try using this program to check the solution of the system in Example 6.

Application

At this point, you may be asking the question "How can I tell which application problems can be solved using a system of linear equations?" The answer comes from the following considerations.

1. Does the problem involve more than one unknown quantity?

2. Are there two (or more) equations or conditions to be satisfied?

If one or both of these conditions are met, the appropriate mathematical model for the problem may be a system of linear equations.

Example 7 An Application of a Linear System

An airplane flying into a headwind travels the 2000-mile flying distance between Fresno, California and Cleveland, Ohio in 4 hours and 24 minutes. On the return flight, the same distance is traveled in 4 hours. Find the airspeed of the plane and the speed of the wind, assuming that both remain constant.

Solution

The two unknown quantities are the speeds of the wind and the plane. If r_1 is the speed of the plane and r_2 is the speed of the wind, then

$r_1 - r_2 =$ speed of the plane *against* the wind

$r_1 + r_2 =$ speed of the plane *with* the wind

as shown in Figure 8.15. Using the formula distance $=$ (rate)(time) for these two speeds, you obtain the following equations.

$$2000 = (r_1 - r_2)\left(4 + \frac{24}{60}\right)$$

$$2000 = (r_1 + r_2)(4)$$

Original flight

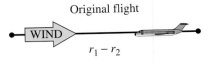

These two equations simplify as follows.

$$\begin{cases} 5000 = 11r_1 - 11r_2 & \text{Equation 1} \\ 500 = r_1 + r_2 & \text{Equation 2} \end{cases}$$

Return flight

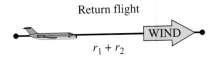

Figure 8.15

To solve this system by elimination, multiply Equation 2 by 11.

$$5000 = 11r_1 - 11r_2 \quad\Longrightarrow\quad 5000 = 11r_1 - 11r_2 \qquad \text{Write Equation 1.}$$

$$500 = r_1 + r_2 \quad\Longrightarrow\quad \underline{5500 = 11r_1 + 11r_2} \qquad \begin{matrix}\text{Multiply Equation 2}\\ \text{by 11.}\end{matrix}$$

$$10{,}500 = 22r_1 \qquad \text{Add equations.}$$

So,

$$r_1 = \frac{10{,}500}{22} = \frac{5250}{11} \approx 477.27 \text{ miles per hour} \qquad \text{Speed of plane}$$

$$r_2 = 500 - \frac{5250}{11} = \frac{250}{11} \approx 22.73 \text{ miles per hour.} \qquad \text{Speed of wind}$$

Check this solution in the original statement of the problem.

 Checkpoint Now try Exercise 61.

8.2   Exercises

Vocabulary Check

Fill in the blanks.

1. The first step in solving a system of equations by the _____ of _____ is to obtain coefficients for x (or y) that differ only in sign.

2. Two systems of equations that have the same solution set are called _____ systems.

3. A system of linear equations that has at least one solution is called _____ , whereas a system of linear equations that has no solution is called _____ .

In Exercises 1–6, solve the system by the method of elimination. Label each line with its equation.

1. $\begin{cases} 2x + y = 5 \\ x - y = 1 \end{cases}$ 2. $\begin{cases} x + 3y = 1 \\ -x + 2y = 4 \end{cases}$

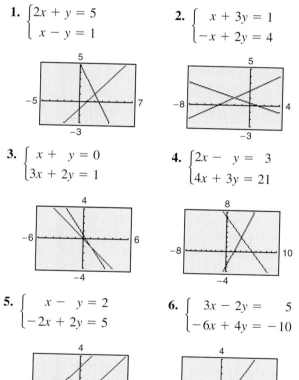

3. $\begin{cases} x + y = 0 \\ 3x + 2y = 1 \end{cases}$ 4. $\begin{cases} 2x - y = 3 \\ 4x + 3y = 21 \end{cases}$

5. $\begin{cases} x - y = 2 \\ -2x + 2y = 5 \end{cases}$ 6. $\begin{cases} 3x - 2y = 5 \\ -6x + 4y = -10 \end{cases}$

In Exercises 7–16, solve the system by the method of elimination and check any solutions algebraically.

7. $\begin{cases} x + 2y = 4 \\ x - 2y = 1 \end{cases}$ 8. $\begin{cases} 3x - 2y = 5 \\ x + 2y = 7 \end{cases}$

9. $\begin{cases} 2x + 3y = 18 \\ 5x - y = 11 \end{cases}$ 10. $\begin{cases} x + 7y = 12 \\ 3x - 5y = 10 \end{cases}$

11. $\begin{cases} 3r + 2s = 10 \\ 2r + 5s = 3 \end{cases}$ 12. $\begin{cases} 2r + 4s = 5 \\ 16r + 50s = 55 \end{cases}$

13. $\begin{cases} 5u + 6v = 24 \\ 3u + 5v = 18 \end{cases}$ 14. $\begin{cases} 3u + 11v = 4 \\ -2u - 5v = 9 \end{cases}$

15. $\begin{cases} 1.8x + 1.2y = 4 \\ 9x + 6y = 3 \end{cases}$ 16. $\begin{cases} 3.1x - 2.9y = -10.2 \\ 31x - 12y = 34 \end{cases}$

In Exercises 17–20, match the system of linear equations with its graph. [The graphs are labeled (a), (b), (c), and (d).]

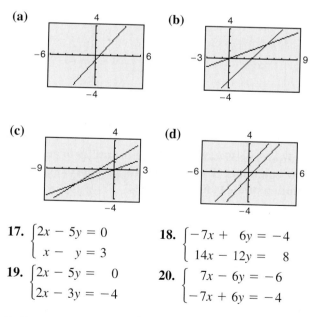

17. $\begin{cases} 2x - 5y = 0 \\ x - y = 3 \end{cases}$ 18. $\begin{cases} -7x + 6y = -4 \\ 14x - 12y = 8 \end{cases}$

19. $\begin{cases} 2x - 5y = 0 \\ 2x - 3y = -4 \end{cases}$ 20. $\begin{cases} 7x - 6y = -6 \\ -7x + 6y = -4 \end{cases}$

In Exercises 21–32, solve the system by the method of elimination and check any solutions using a graphing utility.

21. $\begin{cases} 4x + 3y = 3 \\ 3x + 11y = 13 \end{cases}$ 22. $\begin{cases} 2x + 5y = 8 \\ 5x + 8y = 10 \end{cases}$

23. $\begin{cases} \dfrac{x}{4} + \dfrac{y}{6} = 1 \\ -3x - 2y = 0 \end{cases}$ **24.** $\begin{cases} \dfrac{2}{3}x + \dfrac{1}{6}y = \dfrac{2}{3} \\ 4x + y = 4 \end{cases}$

25. $\begin{cases} \frac{3}{4}x + y = \frac{1}{8} \\ \frac{9}{4}x + 3y = \frac{3}{8} \end{cases}$ **26.** $\begin{cases} \frac{2}{5}x - \frac{3}{2}y = 4 \\ \frac{1}{5}x - \frac{3}{4}y = -2 \end{cases}$

27. $\begin{cases} \dfrac{x+3}{4} + \dfrac{y-1}{3} = 1 \\ 2x - y = 12 \end{cases}$ **28.** $\begin{cases} \dfrac{x-1}{2} + \dfrac{y+2}{3} = 4 \\ x - 2y = 5 \end{cases}$

29. $\begin{cases} 2.5x - 3y = 1.5 \\ 2x - 2.4y = 1.2 \end{cases}$ **30.** $\begin{cases} 6.3x + 7.2y = 5.4 \\ 5.6x + 6.4y = 4.8 \end{cases}$

31. $\begin{cases} 0.2x - 0.5y = -27.8 \\ 0.3x + 0.4y = 68.7 \end{cases}$

32. $\begin{cases} 0.05x - 0.03y = 0.21 \\ 0.07x + 0.02y = 0.16 \end{cases}$

In Exercises 33–38, use a graphing utility to graph the lines in the system. Use the graphs to determine whether the system is consistent or inconsistent. If the system is consistent, determine the solution. Verify your results algebraically.

33. $\begin{cases} 2x - 5y = 0 \\ x - y = 3 \end{cases}$ **34.** $\begin{cases} 2x + y = 5 \\ x - 2y = -1 \end{cases}$

35. $\begin{cases} \frac{3}{5}x - y = 3 \\ -3x + 5y = 9 \end{cases}$ **36.** $\begin{cases} 4x - 6y = 9 \\ \frac{16}{3}x - 8y = 12 \end{cases}$

37. $\begin{cases} 8x - 14y = 5 \\ 2x - 3.5y = 1.25 \end{cases}$ **38.** $\begin{cases} -x + 7y = 3 \\ -\frac{1}{7}x + y = 5 \end{cases}$

In Exercises 39–44, use a graphing utility to graph the two equations. Use the graphs to approximate the solution of the system.

39. $\begin{cases} 6y = 42 \\ 6x - y = 16 \end{cases}$ **40.** $\begin{cases} 4y = -8 \\ 7x - 2y = 25 \end{cases}$

41. $\begin{cases} \frac{3}{2}x - \frac{1}{5}y = 8 \\ -2x + 3y = 3 \end{cases}$ **42.** $\begin{cases} \frac{3}{4}x - \frac{5}{2}y = -9 \\ -x + 6y = 28 \end{cases}$

43. $\begin{cases} 0.5x + 2.2y = 9 \\ 6x + 0.4y = -22 \end{cases}$ **44.** $\begin{cases} 2.4x + 3.8y = -17.6 \\ 4x - 0.2y = -3.2 \end{cases}$

In Exercises 45–52, use any method to solve the system.

45. $\begin{cases} 3x - 5y = 7 \\ 2x + y = 9 \end{cases}$ **46.** $\begin{cases} -x + 3y = 17 \\ 4x + 3y = 7 \end{cases}$

47. $\begin{cases} y = 4x + 3 \\ y = -5x - 12 \end{cases}$ **48.** $\begin{cases} 7x + 3y = 16 \\ y = x + 1 \end{cases}$

49. $\begin{cases} x - 5y = 21 \\ 6x + 5y = 21 \end{cases}$ **50.** $\begin{cases} y = -3x - 8 \\ y = 15 - 2x \end{cases}$

51. $\begin{cases} -2x + 8y = 19 \\ y = x - 3 \end{cases}$ **52.** $\begin{cases} 4x - 3y = 6 \\ -5x + 7y = -1 \end{cases}$

Exploration   In Exercises 53–56, find a system of linear equations that has the given solution. (There are many correct answers.)

53. $(0, 8)$ **54.** $(3, -4)$

55. $\left(3, \frac{5}{2}\right)$ **56.** $\left(-\frac{2}{3}, -10\right)$

Supply and Demand   In Exercises 57–60, find the point of equilibrium of the demand and supply equations. The point of equilibrium is the price p and the number of units x that satisfy both the demand and supply equations.

	Demand	Supply
57.	$p = 50 - 0.5x$	$p = 0.125x$
58.	$p = 100 - 0.05x$	$p = 25 + 0.1x$
59.	$p = 140 - 0.00002x$	$p = 80 + 0.00001x$
60.	$p = 400 - 0.0002x$	$p = 225 + 0.0005x$

61. *Airplane Speed*   An airplane flying into a headwind travels the 1800-mile flying distance between Albuquerque, New Mexico and New York City in 3 hours and 36 minutes. On the return flight, the same distance is traveled in 3 hours. Find the airspeed of the plane and the speed of the wind, assuming that both remain constant.

62. *Airplane Speed*   Two planes start from Boston's Logan International Airport and fly in opposite directions. The second plane starts $\frac{1}{2}$ hour after the first plane, but its speed is 80 kilometers per hour faster. Find the airspeed of each plane if 2 hours after the first plane departs, the planes are 3200 kilometers apart.

63. *Acid Mixture*   Twenty liters of a 50% acid solution is obtained by mixing a 40% and a 65% solution.

(a) Write a system of equations in which one equation represents the amount of final mixture required and the other represents the amount of acid in the final mixture. Let x and y represent the amounts of 40% and 65% solutions, respectively.

(b) Use a graphing utility to graph the two equations in part (a) in the same viewing window. As the amount of the 40% solution increases, how does the amount of the 65% solution change?

(c) How much of each solution is required to obtain the specified concentration of the final mixture?

64. **Fuel Mixture**   Five hundred gallons of 89 octane gasoline is obtained by mixing 87 octane gasoline with 92 octane gasoline.

(a) Write a system of equations in which one equation represents the amount of final mixture required and the other represents the amounts of 87 octane and 92 octane gasoline in the final mixture. Let x and y represent the gallons of 87 octane and 92 octane gasoline, respectively.

(b) Use a graphing utility to graph the two equations in part (a) in the same viewing window. As the amount of 87 octane gasoline increases, how does the amount of 92 octane gasoline change?

(c) How much of each type of gasoline is required to obtain the 500 gallons of 89 octane gasoline?

65. **Investment Portfolio**   A total of $15,000 is invested in two corporate bonds that pay 7.5% and 6% simple interest. The investor wants an annual interest income of $990 from the investments. What is the most that can be invested in the 6% bond?

66. **Investment Portfolio**   A total of $39,000 is invested in two municipal bonds that pay 5.75% and 6.25% simple interest. The investor wants an annual interest income of $2400 from the investments. What is the most that can be invested in the 5.75% bond?

67. **Ticket Sales**   Five hundred tickets were sold for one performance of a play. The tickets for adults and children sold for $7.50 and $4.00, respectively, and the receipts for the performance totaled $3312.50. How many of each type of ticket were sold?

68. **Sales**   On Saturday night, the manager of a shoe store evaluates the receipts of the previous week's sales. Two hundred and fifty pairs of two different styles of running shoes were sold. One style sold for $75.50 and the other sold for $89.95. The receipts totaled $20,031. The cash register that was supposed to record the number of each type of shoe sold malfunctioned. Can you recover the information? If so, how many shoes of each type were sold?

Fitting a Line to Data   In Exercises 69 and 70, find the least squares regression line $y = ax + b$ for the points $(x_1, y_1), (x_2, y_2), \dots , (x_n, y_n)$ by solving the system for a and b. Then use the *regression* feature of a graphing utility to confirm your result. (For an explanation of how the coefficients of a and b in the system are obtained, see Appendix C.)

69. $\begin{cases} 5b + 10a = 20.2 \\ 10b + 30a = 50.1 \end{cases}$ 70. $\begin{cases} 5b + 10a = 11.7 \\ 10b + 30a = 25.6 \end{cases}$

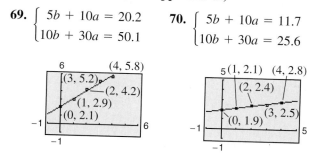

71. **Data Analysis**   A farmer used four test plots to determine the relationship between wheat yield (in bushels per acre) and the amount of fertilizer applied (in hundreds of pounds per acre). The results are shown in the table.

Fertilizer, x	Yield, y
1.0	32
1.5	41
2.0	48
2.5	53

(a) Find the least squares regression line $y = ax + b$ for the data by solving the system for a and b.

$\begin{cases} 4b + 7.0a = 174 \\ 7b + 13.5a = 322 \end{cases}$

(b) Use the *regression* feature of a graphing utility to confirm the result in part (a).

(c) Use a graphing utility to plot the data and graph the linear model in the same viewing window.

(d) Use the linear model to predict the yield for a fertilizer application of 160 pounds per acre.

72. *Data Analysis* A candy store manager wants to know the demand for a candy bar as a function of the price. The daily sales for different prices of the product are shown in the table.

Price, x	Demand, y
$1.00	45
$1.20	37
$1.50	23

(a) Find the least squares regression line $y = ax + b$ for the data by solving the system for a and b.

$$\begin{cases} 3.00b + 3.70a = 105.00 \\ 3.70b + 4.69a = 123.90 \end{cases}$$

(b) Use the *regression* feature of a graphing utility to confirm the result in part (a).

(c) Use a graphing utility to plot the data and graph the linear model in part (a) in the same viewing window.

(d) Use the linear model in part (a) to predict the demand when the price is $1.75.

Synthesis

True or False? **In Exercises 73 and 74, determine whether the statement is true or false. Justify your answer.**

73. If a system of linear equations has two distinct solutions, then it has an infinite number of solutions.

74. If a system of linear equations has no solution, then the lines must be parallel.

Think About It **In Exercises 75 and 76, the graphs of the two equations appear to be parallel. Yet, when the system is solved algebraically, it is found that the system does have a solution. Find the solution and explain why it does not appear on the portion of the graph that is shown.**

75. $\begin{cases} 100y - x = 200 \\ 99y - x = -198 \end{cases}$

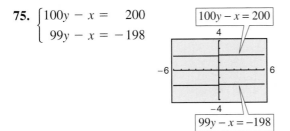

76. $\begin{cases} 21x - 20y = 0 \\ 13x - 12y = 120 \end{cases}$

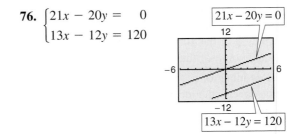

77. *Writing* Briefly explain whether or not it is possible for a consistent system of linear equations to have exactly two solutions.

78. *Think About It* Give examples of (a) a system of linear equations that has no solution and (b) a system that has an infinite number of solutions.

In Exercises 79 and 80, find the value of k such that the system of equations is inconsistent.

79. $\begin{cases} 4x - 8y = -3 \\ 2x + ky = 16 \end{cases}$

80. $\begin{cases} 15x + 3y = 6 \\ -10x + ky = 9 \end{cases}$

Advanced Applications **In Exercises 81 and 82, solve the system of equations for u and v. While solving for these variables, consider the transcendental functions as constants. (Systems of this type are found in a course in differential equations.)**

81. $\begin{cases} u \sin x + v \cos x = 0 \\ u \cos x - v \sin x = \sec x \end{cases}$

82. $\begin{cases} u \cos 2x + v \sin 2x = 0 \\ u(-2 \sin 2x) + v(2 \cos 2x) = \csc 2x \end{cases}$

Review

In Exercises 83–88, solve the inequality and graph the solution on a real number line.

83. $-11 - 6x \geq 33$

84. $-6 \leq 3x - 10 < 6$

85. $|x - 8| < 10$

86. $|x + 10| \geq -3$

87. $2x^2 + 3x - 35 < 0$

88. $3x^2 + 12x > 0$

In Exercises 89–92, write the expression as the logarithm of a single quantity.

89. $\ln x + \ln 6$

90. $\ln x - 5 \ln(x + 3)$

91. $\log_9 12 - \log_9 x$

92. $\frac{1}{4} \log_6 3 + \frac{1}{4} \log_6 x$

8.3 Multivariable Linear Systems

Row-Echelon Form and Back-Substitution

The method of elimination can be applied to a system of linear equations in more than two variables. When elimination is used to solve a system of linear equations, the goal is to rewrite the system in a form to which back-substitution can be applied. To see how this works, consider the following two systems of linear equations.

System of Three Linear Equations in Three Variables (See Example 2):

$$\begin{cases} x - 2y + 3z = 9 \\ -x + 3y \quad\;\; = -4 \\ 2x - 5y + 5z = 17 \end{cases}$$

Equivalent System in Row-Echelon Form (See Example 1):

$$\begin{cases} x - 2y + 3z = 9 \\ y + 3z = 5 \\ z = 2 \end{cases}$$

The second system is said to be in **row-echelon form,** which means that it has a "stair-step" pattern with leading coefficients of 1. After comparing the two systems, it should be clear that it is easier to solve the system in row-echelon form, using back-substitution.

Example 1 Using Back-Substitution in Row-Echelon Form

Solve the system of linear equations.

$$\begin{cases} x - 2y + 3z = 9 & \text{Equation 1} \\ y + 3z = 5 & \text{Equation 2} \\ z = 2 & \text{Equation 3} \end{cases}$$

Solution

From Equation 3, you know the value of z. To solve for y, substitute $z = 2$ into Equation 2 to obtain

$$y + 3(2) = 5 \qquad \text{Substitute 2 for } z.$$
$$y = -1. \qquad \text{Solve for } y.$$

Finally, substitute $y = -1$ and $z = 2$ into Equation 1 to obtain

$$x - 2(-1) + 3(2) = 9 \qquad \text{Substitute } -1 \text{ for } y \text{ and 2 for } z.$$
$$x = 1. \qquad \text{Solve for } x.$$

The solution is $x = 1$, $y = -1$, and $z = 2$, which can be written as the **ordered triple** $(1, -1, 2)$. Check this in the original system of equations.

✓ *Checkpoint* Now try Exercise 5.

What you should learn

- Use back-substitution to solve linear systems in row-echelon form.
- Use Gaussian elimination to solve systems of linear equations.
- Solve nonsquare systems of linear equations.
- Graphically interpret three-variable linear systems.
- Use systems of linear equations to write partial fraction decompositions of rational expressions.
- Use systems of linear equations in three or more variables to model and solve real-life problems.

Why you should learn it

Systems of linear equations in three or more variables can be used to model and solve real-life problems. For instance, Exercise 99 on page 627 shows how to use a system of linear equations to analyze an automobile's braking system.

Andy Sacks/Getty Images

Gaussian Elimination

Two systems of equations are *equivalent* if they have the same solution set. To solve a system that is not in row-echelon form, first convert it to an *equivalent* system that *is* in row-echelon form by using one or more of the elementary row operations shown below. This process is called **Gaussian elimination,** after the German mathematician Carl Friedrich Gauss (1777–1855).

> **Elementary Row Operations**
>
> **1.** Interchange two equations.
>
> **2.** Multiply one of the equations by a nonzero constant.
>
> **3.** Add a multiple of one equation to another equation.

Example 2 Using Gaussian Elimination to Solve a System

Solve the system of linear equations.

$$\begin{cases} x - 2y + 3z = 9 & \text{Equation 1} \\ -x + 3y = -4 & \text{Equation 2} \\ 2x - 5y + 5z = 17 & \text{Equation 3} \end{cases}$$

Solution

Because the leading coefficient of the first equation is 1, you can begin by saving the x at the upper left and eliminating the other x-terms from the first column.

$$\begin{cases} x - 2y + 3z = 9 \\ y + 3z = 5 \\ 2x - 5y + 5z = 17 \end{cases}$$

Adding the first equation to the second equation produces a new second equation.

$$\begin{cases} x - 2y + 3z = 9 \\ y + 3z = 5 \\ -y - z = -1 \end{cases}$$

Adding -2 times the first equation to the third equation produces a new third equation.

Now that all but the first x have been eliminated from the first column, go to work on the second column. (You need to eliminate y from the third equation.)

$$\begin{cases} x - 2y + 3z = 9 \\ y + 3z = 5 \\ 2z = 4 \end{cases}$$

Adding the second equation to the third equation produces a new third equation.

Finally, you need a coefficient of 1 for z in the third equation.

$$\begin{cases} x - 2y + 3z = 9 \\ y + 3z = 5 \\ z = 2 \end{cases}$$

Multiplying the third equation by $\frac{1}{2}$ produces a new third equation.

This is the same system that was solved in Example 1. As in that example, you can conclude that the solution is $x = 1$, $y = -1$, and $z = 2$, written as $(1, -1, 2)$.

✓ *Checkpoint* Now try Exercise 13.

The goal of Gaussian elimination is to use elementary row operations on a system in order to isolate one variable. You can then solve for the value of the variable and use back-substitution to find the values of the remaining variables.

The next example involves an *inconsistent* system—one that has no solution. The key to recognizing an inconsistent system is that at some stage in the elimination process, you obtain a false statement such as $0 = -2$.

Example 3   An Inconsistent System

Solve the system of linear equations.

$$\begin{cases} x - 3y + z = 1 \\ 2x - y - 2z = 2 \\ x + 2y - 3z = -1 \end{cases}$$
Equation 1
Equation 2
Equation 3

Solution

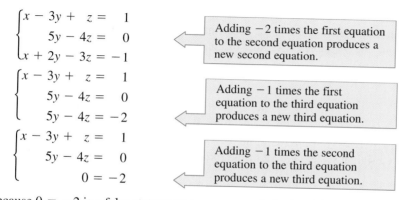

$$\begin{cases} x - 3y + z = 1 \\ \quad\quad 5y - 4z = 0 \\ x + 2y - 3z = -1 \end{cases}$$

Adding -2 times the first equation to the second equation produces a new second equation.

$$\begin{cases} x - 3y + z = 1 \\ \quad\quad 5y - 4z = 0 \\ \quad\quad 5y - 4z = -2 \end{cases}$$

Adding -1 times the first equation to the third equation produces a new third equation.

$$\begin{cases} x - 3y + z = 1 \\ \quad\quad 5y - 4z = 0 \\ \quad\quad\quad\quad 0 = -2 \end{cases}$$

Adding -1 times the second equation to the third equation produces a new third equation.

Because $0 = -2$ is a false statement, you can conclude that this system is inconsistent and so has no solution. Moreover, because this system is equivalent to the original system, you can conclude that the original system also has no solution.

✓ *Checkpoint* Now try Exercise 19.

As with a system of linear equations in two variables, the number of solutions of a system of linear equations in more than two variables must fall into one of three categories.

The Number of Solutions of a Linear System

For a system of linear equations, exactly one of the following is true.

1. There is exactly one solution.

2. There are infinitely many solutions.

3. There is no solution.

A system of linear equations is called *consistent* if it has at least one solution. A consistent system with exactly one solution is **independent.** A consistent system with infinitely many solutions is **dependent.** A system of linear equations is called *inconsistent* if it has no solution.

Example 4 A System with Infinitely Many Solutions

Solve the system of linear equations.

$$\begin{cases} x + y - 3z = -1 & \text{Equation 1} \\ y - z = 0 & \text{Equation 2} \\ -x + 2y = 1 & \text{Equation 3} \end{cases}$$

Solution

$$\begin{cases} x + y - 3z = -1 \\ y - z = 0 \\ 3y - 3z = 0 \end{cases}$$

Adding the first equation to the third equation produces a new third equation.

$$\begin{cases} x + y - 3z = -1 \\ y - z = 0 \\ 0 = 0 \end{cases}$$

Adding -3 times the second equation to the third equation produces a new third equation.

This result means that Equation 3 depends on Equations 1 and 2 in the sense that it gives us no additional information about the variables. So, the original system is equivalent to the system

$$\begin{cases} x + y - 3z = -1 \\ y - z = 0 \end{cases}.$$

In the last equation, solve for y in terms of z to obtain $y = z$. Back-substituting for y into the previous equation produces $x = 2z - 1$. Finally, letting $z = a$, where a is a real number, the solutions to the original system are all of the form

$$x = 2a - 1, \qquad y = a, \qquad \text{and} \qquad z = a.$$

So, every ordered triple of the form

$$(2a - 1, a, a), \qquad a \text{ is a real number}$$

is a solution of the system.

✓ *Checkpoint* Now try Exercise 23.

In Example 4, there are other ways to write the same infinite set of solutions. For instance, the solutions could have been written as

$$\left(b, \tfrac{1}{2}(b + 1), \tfrac{1}{2}(b + 1)\right), \qquad b \text{ is a real number.}$$

This description produces the same set of solutions, as shown below.

Substitution	*Solution*	
$a = 0$	$(2(0) - 1, 0, 0) = (-1, 0, 0)$	Same solution
$b = -1$	$\left(-1, \tfrac{1}{2}(-1 + 1), \tfrac{1}{2}(-1 + 1)\right) = (-1, 0, 0)$	
$a = 1$	$(2(1) - 1, 1, 1) = (1, 1, 1)$	Same solution
$b = 1$	$\left(1, \tfrac{1}{2}(1 + 1), \tfrac{1}{2}(1 + 1)\right) = (1, 1, 1)$	
$a = 2$	$(2(2) - 1, 2, 2) = (3, 2, 2)$	Same solution
$b = 3$	$\left(3, \tfrac{1}{2}(3 + 1), \tfrac{1}{2}(3 + 1)\right) = (3, 2, 2)$	

STUDY TIP

There are an infinite number of solutions to Example 4, but they are all of a specific form. By selecting, for example, a-values of 0, 1, and 3, you can verify that $(-1, 0, 0)$, $(1, 1, 1)$, and $(5, 3, 3)$ are specific solutions. It is incorrect to say simply that the solution to Example 4 is "infinite." You must also specify the form of the solutions.

Nonsquare Systems

So far, each system of linear equations you have looked at has been *square*, which means that the number of equations is equal to the number of variables. In a **nonsquare system of equations,** the number of equations differs from the number of variables. A system of linear equations cannot have a unique solution unless there are at least as many equations as there are variables in the system.

Example 5   A System with Fewer Equations Than Variables

Solve the system of linear equations.

$$\begin{cases} x - 2y + z = 2 & \text{Equation 1} \\ 2x - y - z = 1 & \text{Equation 2} \end{cases}$$

Solution

Begin by rewriting the system in row-echelon form.

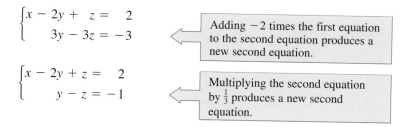

$$\begin{cases} x - 2y + z = 2 \\ 3y - 3z = -3 \end{cases}$$

Adding -2 times the first equation to the second equation produces a new second equation.

$$\begin{cases} x - 2y + z = 2 \\ y - z = -1 \end{cases}$$

Multiplying the second equation by $\frac{1}{3}$ produces a new second equation.

Solve for y in terms of z to obtain $y = z - 1$. By back-substituting into Equation 1, you can solve for x as follows.

$$x - 2(z - 1) + z = 2 \qquad \text{Substitute for } y \text{ in Equation 1.}$$

$$x - 2z + 2 + z = 2 \qquad \text{Distributive Property}$$

$$x = z \qquad \text{Solve for } x.$$

Finally, by letting $z = a$ where a is a real number, you have the solution $x = a$, $y = a - 1$, and $z = a$. So, every ordered triple of the form

$$(a, a - 1, a), \qquad a \text{ is a real number}$$

is a solution of the system.

✓ *Checkpoint* Now try Exercise 31.

In Example 5, try choosing some values of a to obtain different solutions of the system, such as $(1, 0, 1)$, $(2, 1, 2)$, and $(3, 2, 3)$. Then check each of the solutions in the original system as follows.

Check: $(1, 0, 1)$

$$1 - 2(0) + 1 \overset{?}{=} 2$$
$$2 = 2 \checkmark$$

$$2(1) - 0 - 1 \overset{?}{=} 1$$
$$1 = 1 \checkmark$$

Check: $(2, 1, 2)$

$$2 - 2(1) + 2 \overset{?}{=} 2$$
$$2 = 2 \checkmark$$

$$2(2) - 1 - 2 \overset{?}{=} 1$$
$$1 = 1 \checkmark$$

Check: $(3, 2, 3)$

$$3 - 2(2) + 3 \overset{?}{=} 2$$
$$2 = 2 \checkmark$$

$$2(3) - 2 - 3 \overset{?}{=} 1$$
$$1 = 1 \checkmark$$

Graphical Interpretation of Three-Variable Systems

Solutions of equations in three variables can be pictured using a **three-dimensional coordinate system.** To construct such a system, begin with the xy-coordinate plane in a horizontal position. Then draw the z-axis as a vertical line through the origin.

Every ordered triple (x, y, z) corresponds to a point on the three-dimensional coordinate system. For instance, the points corresponding to

$$(-2, 5, 4), \quad (2, -5, 3), \quad \text{and} \quad (3, 3, -2)$$

are shown in Figure 8.16.

The **graph of an equation in three variables** consists of all points (x, y, z) that are solutions of the equation. The graph of a linear equation in three variables is a *plane*. Sketching graphs on a three-dimensional coordinate system is difficult because the sketch itself is only two-dimensional.

One technique for sketching a plane is to find the three points at which the plane intersects the axes. For instance, the plane

$$3x + 2y + 4z = 12$$

intersects the x-axis at the point $(4, 0, 0)$, the y-axis at the point $(0, 6, 0)$, and the z-axis at the point $(0, 0, 3)$. By plotting these three points, connecting them with line segments, and shading the resulting triangular region, you can sketch a portion of the graph, as shown in Figure 8.17.

The graph of a system of three linear equations in three variables consists of *three* planes. When these planes intersect in a single point, the system has exactly one solution (see Figure 8.18). When the three planes have no point in common, the system has no solution (see Figures 8.19 and 8.20). When the three planes intersect in a line or a plane, the system has infinitely many solutions (see Figures 8.21 and 8.22).

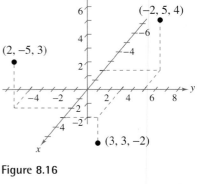

Figure 8.16

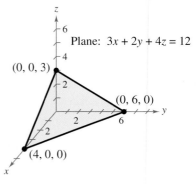

Figure 8.17

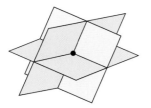

Solution: One point
Figure 8.18

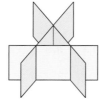

Solution: None
Figure 8.19

Solution: None
Figure 8.20

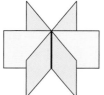

Solution: One line
Figure 8.21

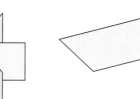

Solution: One plane
Figure 8.22

TECHNOLOGY TIP

Three-dimensional graphing utilities and computer algebra systems, such as *Derive* and *Mathematica*, are very efficient in producing three-dimensional graphs. They are good tools to use while studying calculus. If you have access to such a utility, try reproducing the plane shown in Figure 8.17.

Partial Fraction Decomposition and Other Applications

A rational expression can often be written as the sum of two or more simpler rational expressions. For example, the rational expression

$$\frac{x + 7}{x^2 - x - 6}$$

can be written as the sum of two fractions with linear denominators. That is,

$$\frac{x + 7}{x^2 - x - 6} = \underbrace{\frac{2}{x - 3}}_{\substack{\text{Partial} \\ \text{fraction}}} + \underbrace{\frac{-1}{x + 2}}_{\substack{\text{Partial} \\ \text{fraction}}}.$$

Each fraction on the right side of the equation is a **partial fraction,** and together they make up the **partial fraction decomposition** of the left side.

Decomposition of $N(x)/D(x)$ into Partial Fractions

1. *Divide if improper:* If $N(x)/D(x)$ is an improper fraction [degree of $N(x) \geq$ degree of $D(x)$], divide the denominator into the numerator to obtain

   $$\frac{N(x)}{D(x)} = (\text{polynomial}) + \frac{N_1(x)}{D(x)}$$

   and apply Steps 2, 3, and 4 (below) to the proper rational expression $N_1(x)/D(x)$.

2. *Factor denominator:* Completely factor the denominator into factors of the form

   $$(px + q)^m \quad \text{and} \quad (ax^2 + bx + c)^n$$

   where $(ax^2 + bx + c)$ is irreducible.

3. *Linear factors:* For *each* factor of the form $(px + q)^m$, the partial fraction decomposition must include the following sum of m fractions.

   $$\frac{A_1}{(px + q)} + \frac{A_2}{(px + q)^2} + \cdot \cdot \cdot + \frac{A_m}{(px + q)^m}$$

4. *Quadratic factors:* For *each* factor of the form $(ax^2 + bx + c)^n$, the partial fraction decomposition must include the following sum of n fractions.

   $$\frac{B_1 x + C_1}{ax^2 + bx + c} + \frac{B_2 x + C_2}{(ax^2 + bx + c)^2} + \cdot \cdot \cdot + \frac{B_n x + C_n}{(ax^2 + bx + c)^n}$$

One of the most important applications of partial fractions is in calculus. If you go on to take a course in calculus, you will learn how partial fractions can be used in a calculus operation called antidifferentiation.

Example 6   Partial Fraction Decomposition: Distinct Linear Factors

Write the partial fraction decomposition of

$$\frac{x + 7}{x^2 - x - 6}.$$

Solution

Because $x^2 - x - 6 = (x - 3)(x + 2)$, you should include one partial fraction with a constant numerator for each linear factor of the denominator and write

$$\frac{x + 7}{x^2 - x - 6} = \frac{A}{x - 3} + \frac{B}{x + 2}.$$

Multiplying each side of this equation by the least common denominator, $(x - 3)(x + 2)$, leads to the **basic equation**

$$
\begin{aligned}
x + 7 &= A(x + 2) + B(x - 3) &&\text{Basic equation}\\
&= Ax + 2A + Bx - 3B &&\text{Distributive Property}\\
&= (A + B)x + 2A - 3B. &&\text{Write in polynomial form.}
\end{aligned}
$$

By equating coefficients of like terms on opposite sides of the equation, you obtain the following system of linear equations.

$$
\begin{cases}
A + B = 1 &\qquad \text{Equation 1}\\
2A - 3B = 7 &\qquad \text{Equation 2}
\end{cases}
$$

You can solve the system of linear equations as follows.

$$
\begin{array}{ll}
A + B = 1 \;\Longrightarrow\; & 3A + 3B = 3 \qquad \text{Multiply Equation 1 by 3.}\\
2A - 3B = 7 \;\Longrightarrow\; & \underline{2A - 3B = 7} \qquad \text{Write Equation 2.}\\
& 5A \quad\;\; = 10 \qquad \text{Add equations.}
\end{array}
$$

From this equation, you can see that $A = 2$. By back-substituting this value of A into Equation 1, you can determine that $B = -1$. So, the partial fraction decomposition is

$$\frac{x + 7}{x^2 - x - 6} = \frac{2}{x - 3} - \frac{1}{x + 2}.$$

Check this result by combining the two partial fractions on the right side of the equation.

✓ *Checkpoint*   Now try Exercise 59.

TECHNOLOGY TIP   You can graphically check the decomposition found in Example 6. To do this, use a graphing utility to graph

$$y_1 = \frac{x + 7}{x^2 - x - 6} \quad \text{and} \quad y_2 = \frac{2}{x - 3} - \frac{1}{x + 2}$$

in the same viewing window. The graphs should be identical, as shown in Figure 8.23.

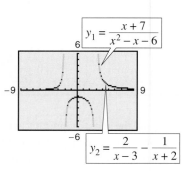

$$y_1 = \frac{x + 7}{x^2 - x - 6}$$

$$y_2 = \frac{2}{x - 3} - \frac{1}{x + 2}$$

Figure 8.23

The next example shows how to find the partial fraction decomposition for a rational function whose denominator has a repeated linear factor.

Example 7   Partial Fraction Decomposition: Repeated Linear Factors

Write the partial fraction decomposition of $\dfrac{5x^2 + 20x + 6}{x^3 + 2x^2 + x}$.

Solution

Because the denominator factors as

$$x^3 + 2x^2 + x = x(x^2 + 2x + 1)$$

$$= x(x + 1)^2$$

you should include one partial fraction with a constant numerator for each power of x and $(x + 1)$ and write

$$\frac{5x^2 + 20x + 6}{x^3 + 2x^2 + x} = \frac{A}{x} + \frac{B}{x + 1} + \frac{C}{(x + 1)^2}.$$

Multiplying by the LCD, $x(x + 1)^2$, leads to the basic equation

$$5x^2 + 20x + 6 = A(x + 1)^2 + Bx(x + 1) + Cx \qquad \text{Basic equation}$$

$$= Ax^2 + 2Ax + A + Bx^2 + Bx + Cx \qquad \text{Expand.}$$

$$= (A + B)x^2 + (2A + B + C)x + A. \qquad \text{Polynomial form}$$

By equating coefficients of like terms on opposite sides of the equation, you obtain the following system of linear equations.

$$\begin{cases} A + B & = 5 \\ 2A + B + C = 20 \\ A & = 6 \end{cases}$$

Substituting 6 for A in the first equation produces

$$6 + B = 5$$

$$B = -1.$$

Substituting 6 for A and -1 for B in the second equation produces

$$2(6) + (-1) + C = 20$$

$$C = 9.$$

So, the partial fraction decomposition is

$$\frac{5x^2 + 20x + 6}{x^3 + 2x^2 + x} = \frac{6}{x} - \frac{1}{x + 1} + \frac{9}{(x + 1)^2}.$$

Check this result by combining the three partial fractions on the right side of the equation.

✓ *Checkpoint*   Now try Exercise 63.

Exploration

Partial fraction decomposition is practical only for rational functions whose denominators factor "nicely." For example, the factorization of the expression $x^2 - x - 5$ is

$$\left(x - \frac{1 - \sqrt{21}}{2}\right)\left(x - \frac{1 + \sqrt{21}}{2}\right).$$

Write the basic equation and try to complete the decomposition for

$$\frac{x + 7}{x^2 - x - 5}.$$

What problems do you encounter?

Example 8 Vertical Motion

The height at time t of an object that is moving in a (vertical) line with constant acceleration a is given by the *position equation* $s = \frac{1}{2}at^2 + v_0t + s_0$. The height s is measured in feet, t is measured in seconds, v_0 is the initial velocity (in feet per second) at $t = 0$, and s_0 is the initial height. Find the values of a, v_0, and s_0 if $s = 52$ at $t = 1$, $s = 52$ at $t = 2$, and $s = 20$ at $t = 3$, as shown in Figure 8.24.

Solution

You can obtain three linear equations in a, v_0, and s_0 as follows.

When $t = 1$: $\quad \frac{1}{2}a(1)^2 + v_0(1) + s_0 = 52 \quad \Longrightarrow \quad a + 2v_0 + 2s_0 = 104$

When $t = 2$: $\quad \frac{1}{2}a(2)^2 + v_0(2) + s_0 = 52 \quad \Longrightarrow \quad 2a + 2v_0 + s_0 = 52$

When $t = 3$: $\quad \frac{1}{2}a(3)^2 + v_0(3) + s_0 = 20 \quad \Longrightarrow \quad 9a + 6v_0 + 2s_0 = 40$

Solving this system yields $a = -32$, $v_0 = 48$, and $s_0 = 20$.

✓ *Checkpoint* Now try Exercise 73.

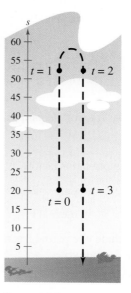

Figure 8.24

Example 9 Data Analysis: Curve-Fitting

Find a quadratic equation $y = ax^2 + bx + c$ whose graph passes through the points $(-1, 3)$, $(1, 1)$, and $(2, 6)$.

Solution

Because the graph of $y = ax^2 + bx + c$ passes through the points $(-1, 3)$, $(1, 1)$, and $(2, 6)$, you can write the following.

When $x = -1$, $y = 3$: $\quad a(-1)^2 + b(-1) + c = 3$

When $x = 1$, $\quad y = 1$: $\quad a(1)^2 + b(1) + c = 1$

When $x = 2$, $\quad y = 6$: $\quad a(2)^2 + b(2) + c = 6$

This produces the following system of linear equations.

$$\begin{cases} a - b + c = 3 & \text{Equation 1} \\ a + b + c = 1 & \text{Equation 2} \\ 4a + 2b + c = 6 & \text{Equation 3} \end{cases}$$

The solution of this system is $a = 2$, $b = -1$, and $c = 0$. So, the equation of the parabola is $y = 2x^2 - x$, and its graph is shown in Figure 8.25.

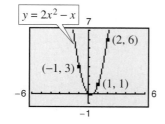

Figure 8.25

✓ *Checkpoint* Now try Exercise 77.

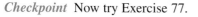

STUDY TIP

When you use a system of linear equations to solve an application problem, it is wise to interpret your solution in the context of the problem to see if it makes sense. For instance, in Example 8 the solution results in the position equation

$$s = -16t^2 + 48t + 20$$

which implies that the object was thrown upward at a velocity of 48 feet per second from a height of 20 feet. The object undergoes a constant downward acceleration of 32 feet per second squared. (In physics, this is the value of the acceleration due to gravity.)

8.3 Exercises

Vocabulary Check

Fill in the blanks.

1. A system of equations that is in _____ form has a "stair-step" pattern with leading coefficients of 1.

2. A solution to a system of three linear equations in three unknowns can be written as an _____ , which has the form (x, y, z).

3. The process used to write a system of equations in row-echelon form is called _____ elimination.

4. A system of linear equations that has exactly one solution is called _____ , whereas a system of linear equations that has infinitely many solutions is called _____ .

5. A system of equations is called _____ if the number of equations differs from the number of variables in the system.

6. Solutions of equations in three variables can be pictured using a _____ coordinate system.

7. The process of writing a rational expression as the sum of two or more simpler rational expressions is called _____ .

In Exercises 1–4, determine whether each ordered triple is a solution of the system of equations.

1. $\begin{cases} 3x - y + z = 1 \\ 2x - 3z = -14 \\ 5y + 2z = 8 \end{cases}$

   (a) $(2, 5, 0)$ (b) $(-2, 0, 4)$
   (c) $(0, -1, 3)$ (d) $(-1, 0, 4)$

2. $\begin{cases} 3x + 4y - z = 17 \\ 5x - y + 2z = -2 \\ 2x - 3y + 7z = -21 \end{cases}$

   (a) $(3, 2, 0)$ (b) $(1, 3, -2)$
   (c) $(4, 1, -3)$ (d) $(1, 5, -1)$

3. $\begin{cases} 4x + y - z = 0 \\ -8x - 6y + z = -\frac{7}{4} \\ 3x - y = -\frac{9}{4} \end{cases}$

   (a) $(0, 1, 1)$ (b) $\left(-\frac{3}{2}, \frac{5}{4}, -\frac{5}{4}\right)$
   (c) $\left(-\frac{1}{2}, \frac{3}{4}, -\frac{5}{4}\right)$ (d) $\left(-\frac{1}{2}, 2, 0\right)$

4. $\begin{cases} -4x - y - 8z = -6 \\ y + z = 0 \\ 4x - 7y = 6 \end{cases}$

   (a) $(-2, -2, 2)$ (b) $\left(-\frac{33}{2}, -10, 10\right)$
   (c) $\left(\frac{1}{8}, -\frac{1}{2}, \frac{1}{2}\right)$ (d) $\left(-\frac{11}{2}, -4, 4\right)$

In Exercises 5–10, use back-substitution to solve the system of linear equations.

5. $\begin{cases} 2x - y + 5z = 24 \\ y + 2z = 4 \\ z = 6 \end{cases}$

6. $\begin{cases} 4x - 3y - 2z = 21 \\ 6y - 5z = -8 \\ z = -2 \end{cases}$

7. $\begin{cases} 2x + y - 3z = 10 \\ y + z = 12 \\ z = 2 \end{cases}$

8. $\begin{cases} x - y + 2z = 22 \\ 3y - 8z = -9 \\ z = -3 \end{cases}$

9. $\begin{cases} 4x - 2y + z = 8 \\ -y + z = 4 \\ z = 2 \end{cases}$

10. $\begin{cases} 5x - 8z = 22 \\ 3y - 5z = 10 \\ z = -4 \end{cases}$

In Exercises 11 and 12, perform the row operation and write the equivalent system.

11. Add Equation 1 to Equation 2.

   $\begin{cases} x - 2y + 3z = 5 & \text{Equation 1} \\ -x + 3y - 5z = 4 & \text{Equation 2} \\ 2x - 3z = 0 & \text{Equation 3} \end{cases}$

   What did this operation accomplish?

12. Add -2 times Equation 1 to Equation 3.

   $\begin{cases} x - 2y + 3z = 5 & \text{Equation 1} \\ -x + 3y - 5z = 4 & \text{Equation 2} \\ 2x - 3z = 0 & \text{Equation 3} \end{cases}$

   What did this operation accomplish?

In Exercises 13–38, solve the system of linear equations and check any solution algebraically.

13. $\begin{cases} x + y + z = 6 \\ 2x - y + z = 3 \\ 3x \quad\;\; - z = 0 \end{cases}$
14. $\begin{cases} x + y + z = 2 \\ -x + 3y + 2z = 8 \\ 4x + y \quad\;\; = 4 \end{cases}$

15. $\begin{cases} 2x \quad\;\; + 2z = 2 \\ 5x + 3y \quad\;\; = 4 \\ \quad\;\; 3y - 4z = 4 \end{cases}$
16. $\begin{cases} 4x + y - 3z = 11 \\ 2x - 3y + 2z = 9 \\ x + y + z = -3 \end{cases}$

17. $\begin{cases} \quad\;\; 6y + 4z = -18 \\ 3x + 3y \quad\;\; = 9 \\ 2x \quad\;\; - 3z = 12 \end{cases}$

18. $\begin{cases} 2x + 4y + z = -4 \\ 2x - 4y + 6z = 13 \\ 4x - 2y + z = 6 \end{cases}$

19. $\begin{cases} 3x - 2y + 4z = 1 \\ x + y - 2z = 3 \\ 2x - 3y + 6z = 8 \end{cases}$
20. $\begin{cases} 5x - 3y + 2z = 3 \\ 2x + 4y - z = 7 \\ x - 11y + 4z = 3 \end{cases}$

21. $\begin{cases} 3x + 3y + 5z = 1 \\ 3x + 5y + 9z = 0 \\ 5x + 9y + 17z = 0 \end{cases}$
22. $\begin{cases} 2x + y + 3z = 1 \\ 2x + 6y + 8z = 3 \\ 6x + 8y + 18z = 5 \end{cases}$

23. $\begin{cases} x + 2y - 7z = -4 \\ 2x + y + z = 13 \\ 3x + 9y - 36z = -33 \end{cases}$

24. $\begin{cases} 2x + y - 3z = 4 \\ 4x \quad\;\; + 2z = 10 \\ -2x + 3y - 13z = -8 \end{cases}$

25. $\begin{cases} 3x - 3y + 6z = 6 \\ x + 2y - z = 5 \\ 5x - 8y + 13z = 7 \end{cases}$

26. $\begin{cases} x \quad\;\; + 4z = 13 \\ 4x - 2y + z = 7 \\ 2x - 2y - 7z = -19 \end{cases}$

27. $\begin{cases} x - 2y + 3z = 4 \\ 3x - y + 2z = 0 \\ x + 3y - 4z = -2 \end{cases}$

28. $\begin{cases} -x + 3y + z = 4 \\ 4x - 2y - 5z = -7 \\ 2x + 4y - 3z = 12 \end{cases}$

29. $\begin{cases} x \quad\;\; + 4z = 1 \\ x + y + 10z = 10 \\ 2x - y + 2z = -5 \end{cases}$

30. $\begin{cases} 3x - 2y - 6z = -4 \\ -3x + 2y + 6z = 1 \\ x - y - 5z = -3 \end{cases}$

31. $\begin{cases} x - 2y + 5z = 2 \\ 4x \quad\;\; - z = 0 \end{cases}$
32. $\begin{cases} 12x + 5y + z = 0 \\ 23x + 4y - z = 0 \end{cases}$

33. $\begin{cases} 2x - 3y + z = -2 \\ -4x + 9y = 7 \end{cases}$
34. $\begin{cases} 10x - 3y + 2z = 0 \\ 19x - 5y - z = 0 \end{cases}$

35. $\begin{cases} x - 3y + 2z = 18 \\ 5x - 13y + 12z = 80 \end{cases}$

36. $\begin{cases} 2x + 3y + 3z = 7 \\ 4x + 18y + 15z = 44 \end{cases}$

37. $\begin{cases} x \quad\;\;\quad\;\; + 3w = 4 \\ 2y - z - w = 0 \\ 3y \quad\;\; - 2w = 1 \\ 2x - y + 4z \quad\;\; = 5 \end{cases}$

38. $\begin{cases} x + y + z + w = 6 \\ 2x + 3y \quad\;\; - w = 0 \\ -3x + 4y + z + 2w = 4 \\ x + 2y - z + w = 0 \end{cases}$

Exploration **In Exercises 39–42, find a system of linear equations that has the given solution. (There are many correct answers.)**

39. $(4, -1, 2)$
40. $(-5, -2, 1)$
41. $\left(3, -\frac{1}{2}, \frac{7}{4}\right)$
42. $\left(-\frac{3}{2}, 4, -7\right)$

Three-Dimensional Graphics **In Exercises 43–46, sketch the plane represented by the linear equation. Then list four points that lie in the plane.**

43. $2x + 3y + 4z = 12$
44. $x + y + z = 6$
45. $2x + y + z = 4$
46. $x + 2y + 2z = 6$

In Exercises 47–52, write the form of the partial fraction decomposition of the rational expression. Do not solve for the constants.

47. $\dfrac{7}{x^2 - 14x}$
48. $\dfrac{x - 2}{x^2 + 4x + 3}$

49. $\dfrac{12}{x^3 - 10x^2}$
50. $\dfrac{x^2 - 3x + 2}{4x^3 + 11x^2}$

51. $\dfrac{4x^2 + 3}{(x - 5)^3}$
52. $\dfrac{6x + 5}{(x + 2)^4}$

In Exercises 53–70, **(a)** write the partial fraction decomposition for the rational expression, **(b)** check your result algebraically by combining the fractions, and **(c)** check your result graphically by using a graphing utility to graph the rational expression and the partial fractions in the same viewing window.

53. $\dfrac{1}{x^2 - 1}$

54. $\dfrac{1}{4x^2 - 9}$

55. $\dfrac{1}{x^2 + x}$

56. $\dfrac{3}{x^2 - 3x}$

57. $\dfrac{1}{2x^2 + x}$

58. $\dfrac{5}{x^2 + x - 6}$

59. $\dfrac{5 - x}{2x^2 + x - 1}$

60. $\dfrac{x - 2}{x^2 + 4x + 3}$

61. $\dfrac{x^2 + 12x + 12}{x^3 - 4x}$

62. $\dfrac{x^2 + 12x - 9}{x^3 - 9x}$

63. $\dfrac{4x^2 + 2x - 1}{x^2(x + 1)}$

64. $\dfrac{2x - 3}{(x - 1)^2}$

65. $\dfrac{27 - 7x}{x(x - 3)^2}$

66. $\dfrac{x^2 - x + 2}{x(x - 1)^2}$

67. $\dfrac{2x^3 - x^2 + x + 5}{x^2 + 3x + 2}$

68. $\dfrac{x^3 + 2x^2 - x + 1}{x^2 + 3x - 4}$

69. $\dfrac{x^4}{(x - 1)^3}$

70. $\dfrac{4x^4}{(2x - 1)^3}$

Graphical Analysis   In Exercises 71 and 72, write the partial fraction decomposition for the rational function. Identify the graph of the rational function and the graphs of each term of its decomposition. State any relationship between the vertical asymptotes of the rational function and the vertical asymptotes of the terms of the decomposition.

71. $y = \dfrac{x - 12}{x(x - 4)}$

72. $y = \dfrac{2(4x - 3)}{x^2 - 9}$

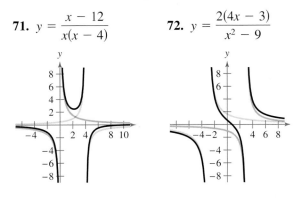

Vertical Motion   In Exercises 73–76, an object moving vertically is at the given heights at the specified times. Find the position equation $s = \frac{1}{2}at^2 + v_0 t + s_0$ for the object.

73. At $t = 1$ second, $s = 128$ feet.
   At $t = 2$ seconds, $s = 80$ feet.
   At $t = 3$ seconds, $s = 0$ feet.

74. At $t = 1$ second, $s = 48$ feet.
   At $t = 2$ seconds, $s = 64$ feet.
   At $t = 3$ seconds, $s = 48$ feet.

75. At $t = 1$ second, $s = 452$ feet.
   At $t = 2$ seconds, $s = 372$ feet.
   At $t = 3$ seconds, $s = 260$ feet.

76. At $t = 1$ second, $s = 132$ feet.
   At $t = 2$ seconds, $s = 100$ feet.
   At $t = 3$ seconds, $s = 36$ feet.

In Exercises 77–80, find the equation of the parabola

$$y = ax^2 + bx + c$$

that passes through the points. To verify your result, use a graphing utility to plot the points and graph the parabola.

77. $(0, 0), (2, -2), (4, 0)$ **78.** $(0, 3), (1, 4), (2, 3)$

79. $(2, 0), (3, -1), (4, 0)$ **80.** $(1, 3), (2, 2), (3, -3)$

In Exercises 81–84, find the equation of the circle

$$x^2 + y^2 + Dx + Ey + F = 0$$

that passes through the points. To verify your result, use a graphing utility to plot the points and graph the circle.

81. $(0, 0), (2, 2), (4, 0)$ **82.** $(0, 0), (0, 6), (3, 3)$

83. $(-3, -1), (2, 4), (-6, 8)$

84. $(-6, -1), (-4, 3), (2, -5)$

85. *Borrowing*   A small corporation borrowed $775,000 to expand its software line. Some of the money was borrowed at 8%, some at 9%, and some at 10%. How much was borrowed at each rate if the annual interest was $67,000 and the amount borrowed at 8% was four times the amount borrowed at 10%?

86. *Borrowing* A small corporation borrowed $1,000,000 to expand its line of toys. Some of the money was borrowed at 8%, some at 10%, and some at 12%. How much was borrowed at each rate if the annual interest was $97,200 and the amount borrowed at 8% was two times the amount borrowed at 10%?

Investment Portfolio **In Exercises 87 and 88, consider an investor with a portfolio totaling $500,000 that is invested in certificates of deposit, municipal bonds, blue-chip stocks, and growth or speculative stocks. How much is invested in each type of investment?**

87. The certificates of deposit pay 8% annually, and the municipal bonds pay 9% annually. Over a five-year period, the investor expects the blue-chip stocks to return 12% annually and the growth stocks to return 15% annually. The investor wants a combined annual return of 10% and also wants to have only one-fourth of the portfolio invested in municipal bonds.

88. The certificates of deposit pay 9% annually, and the municipal bonds pay 5% annually. Over a five-year period, the investor expects the blue-chip stocks to return 12% annually and the growth stocks to return 14% annually. The investor wants a combined annual return of 10% and also wants to have only one-fourth of the portfolio invested in stocks.

89. *Agriculture* A mixture of 12 liters of chemical A, 16 liters of chemical B, and 26 liters of chemical C is required to kill a destructive crop insect. Commercial spray X contains 1, 2, and 2 parts, respectively, of these chemicals. Commercial spray Y contains only chemical C. Commercial spray Z contains only chemicals A and B in equal amounts. How much of each type of commercial spray is needed to obtain the desired mixture?

90. *Acid Mixture* A chemist needs 10 liters of a 25% acid solution. The solution is to be mixed from three solutions whose concentrations are 10%, 20%, and 50%. How many liters of each solution should the chemist use so that as little as possible of the 50% solution is used?

91. *Truck Scheduling* A small company that manufactures two models of exercise machines has an order for 15 units of the standard model and 16 units of the deluxe model. The company has trucks of three different sizes that can haul the products, as shown in the table. How many trucks of each size are needed to deliver the order? Give two possible solutions.

	Truck	Standard	Deluxe
	Large	6	3
	Medium	4	4
	Small	0	3

Table for 91

92. *Sports* The University of Georgia and Florida State University scored a total of 39 points during the 2003 Sugar Bowl. The points came from a total of 11 different scoring plays, which were a combination of touchdowns, extra-point kicks, and field goals, worth 6, 1, and 3 points, respectively. The same numbers of touchdowns and field goals were scored. How many touchdowns, extra-point kicks, and field goals were scored during the game? (Source: espn.com)

93. *Electrical Networks* When Kirchhoff's Laws are applied to the electrical network in the figure, the currents I_1, I_2, and I_3 are the solution of the system

$$\begin{cases} I_1 - I_2 + I_3 = 0 \\ 3I_1 + 2I_2 \quad\quad = 7. \\ \quad\quad 2I_2 + 4I_3 = 8 \end{cases}$$

Find the currents.

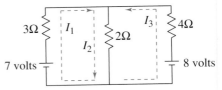

94. *Pulley System* A system of pulleys is loaded with 128-pound and 32-pound weights (see figure). The tensions t_1 and t_2 in the ropes and the acceleration a of the 32-pound weight are modeled by the system

$$\begin{cases} t_1 - 2t_2 \quad\quad = \quad 0 \\ t_1 \quad\quad - 2a = 128 \\ \quad\quad t_2 + a = \quad 32 \end{cases}$$

where t_1 and t_2 are measured in pounds and a is in feet per second squared. Solve the system.

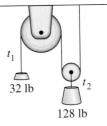

Fitting a Parabola   **In Exercises 95–98, find the least squares regression parabola** $y = ax^2 + bx + c$ **for the points** $(x_1, y_1), (x_2, y_2), \ldots, (x_n, y_n)$ **by solving the following system of linear equations for** a, b, **and** c. **Then use the** *regression* **feature of a graphing utility to confirm your result. (For an explanation of how the coefficients of** a, b, **and** c **in the system are obtained, see Appendix C.)**

95.
$$\begin{cases} 4c & + 40a = 19 \\ 40b & = -12 \\ 40c & + 544a = 160 \end{cases}$$

96.
$$\begin{cases} 5c & + 10a = 8 \\ 10b & = 12 \\ 10c & + 34a = 22 \end{cases}$$

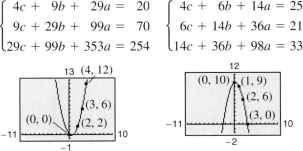

97.
$$\begin{cases} 4c + 9b + 29a = 20 \\ 9c + 29b + 99a = 70 \\ 29c + 99b + 353a = 254 \end{cases}$$

98.
$$\begin{cases} 4c + 6b + 14a = 25 \\ 6c + 14b + 36a = 21 \\ 14c + 36b + 98a = 33 \end{cases}$$

99. *Data Analysis*   During the testing of a new automobile braking system, the speeds x (in miles per hour) and the stopping distances y (in feet) were recorded in the table.

Speed, x	Stopping distance, y
30	55
40	105
50	188

(a) Use the data to create a system of linear equations. Then find the least squares regression parabola for the data by solving the system.

(b) Use a graphing utility to graph the parabola and the data in the same viewing window.

(c) Use the model to estimate the stopping distance for a speed of 70 miles per hour.

100. *Data Analysis*   A wildlife management team studied the reproduction rates of deer in three five-acre tracts of a wildlife preserve. In each tract, the number of females x and the percent of females y that had offspring the following year were recorded. The results are shown in the table.

Number, x	Percent, y
120	68
140	55
160	30

(a) Use the data to create a system of linear equations. Then find the least squares regression parabola for the data by solving the system.

(b) Use a graphing utility to graph the parabola and the data in the same viewing window.

(c) Use the model to predict the percent of females that had offspring when there were 170 females.

101. *Thermodynamics*   The magnitude of the range R of exhaust temperatures (in degrees Fahrenheit) in an experimental diesel engine is approximated by the model

$$R = \frac{2000(4 - 3x)}{(11 - 7x)(7 - 4x)}, \quad 0 \le x \le 1$$

where x is the relative load (in foot-pounds).

(a) Write the partial fraction decomposition for the rational function.

(b) The decomposition in part (a) is the difference of two fractions. The absolute values of the terms give the expected maximum and minimum temperatures of the exhaust gases. Use a graphing utility to graph each term.

102. *Environment*   The predicted cost C (in thousands of dollars) for a company to remove $p\%$ of a chemical from its waste water is given by the model

$$C = \frac{120p}{10,000 - p^2}, \quad 0 \le p < 100.$$

(a) Write the partial fraction decomposition for the rational function.

(b) Verify your result by using the *table* feature of a graphing utility to create a table comparing the original function to the partial fractions.

Synthesis

True or False?   In Exercises 103–105, determine whether the statement is true or false. Justify your answer.

103. The system $\begin{cases} x + 4y - 5z = 8 \\ \quad\;\; 2y + \;\; z = 5 \\ \qquad\qquad z = 1 \end{cases}$ is in row-echelon form.

104. If a system of three linear equations is inconsistent, then its graph has no points common to all three equations.

105. For the rational expression

$$\frac{x}{(x + 10)(x - 10)^2}$$

the partial fraction decomposition is of the form

$$\frac{A}{x + 10} + \frac{B}{(x - 10)^2}.$$

106. *Error Analysis*   You are tutoring a student in algebra. In trying to find a partial fraction decomposition, your student writes the following.

$$\frac{x^2 + 1}{x(x - 1)} = \frac{A}{x} + \frac{B}{x - 1}$$

$$x^2 + 1 = A(x - 1) + Bx \qquad \text{Basic equation}$$

$$x^2 + 1 = (A + B)x - A$$

Your student then forms the following system of linear equations.

$$\begin{cases} \;\; A + B = 0 \\ -A \qquad\;\; = 1 \end{cases}$$

Solve the system and check the partial fraction decomposition it yields. Has your student worked the problem correctly? If not, what went wrong?

In Exercises 107–110, write the partial fraction decomposition for the rational expression. Check your result algebraically. Then assign a value to the constant *a* and check the result graphically.

107. $\dfrac{1}{a^2 - x^2}$

108. $\dfrac{1}{(x + 1)(a - x)}$

109. $\dfrac{1}{y(a - y)}$

110. $\dfrac{1}{x(x + a)}$

111. *Think About It*   Are the two systems of equations equivalent? Give reasons for your answer.

$$\begin{cases} x + 3y - \;\; z = 6 \\ 2x - \;\; y + 2z = 1 \\ 3x + 2y - \;\; z = 2 \end{cases} \qquad \begin{cases} x + 3y - \;\; z = \quad 6 \\ \quad\; -7y + 4z = \quad 1 \\ \quad\; -7y - 4z = -16 \end{cases}$$

112. *Writing*   When using Gaussian elimination to solve a system of linear equations, explain how you can recognize that the system has no solution. Give an example that illustrates your answer.

Advanced Applications   **In Exercises 113–116, find values of *x*, *y*, and λ that satisfy the system. These systems arise in certain optimization problems in calculus. (λ is called a *Lagrange multiplier*.)**

113. $\begin{cases} \qquad\quad y + \lambda = 0 \\ \qquad x \quad + \lambda = 0 \\ x + y - 10 \quad\;\; = 0 \end{cases}$

114. $\begin{cases} 2x \qquad + \lambda = 0 \\ \quad\; 2y + \lambda = 0 \\ x + y - 4 \quad\;\; = 0 \end{cases}$

115. $\begin{cases} 2x - 2x\lambda = 0 \\ -2y + \lambda = 0 \\ \quad y - x^2 = 0 \end{cases}$

116. $\begin{cases} 2 + 2x + 2\lambda = 0 \\ 2x + 1 + \lambda = 0 \\ 2x + y - 100 = 0 \end{cases}$

Review

In Exercises 117–122, sketch the graph of the function.

117. $f(x) = -3x + 7$

118. $f(x) = 6 - x$

119. $f(x) = -2x^2$

120. $f(x) = \frac{1}{4}x^2 + 1$

121. $f(x) = -x^2(x - 3)$

122. $f(x) = \frac{1}{2}x^3 - 1$

In Exercises 123–126, (a) determine the real zeros of *f* and (b) sketch the graph of *f*.

123. $f(x) = x^3 + x^2 - 12x$

124. $f(x) = -8x^4 + 32x^2$

125. $f(x) = 2x^3 + 5x^2 - 21x - 36$

126. $f(x) = 6x^3 - 29x^2 - 6x + 5$

In Exercises 127–130, use a graphing utility to create a table of values for the function. Then sketch the graph of the function by hand.

127. $y = 4^{-x-4} - 5$

128. $y = \left(\frac{5}{2}\right)^{x-1} - 4$

129. $y = 2.9^{0.8x} - 3$

130. $y = -3.5^{x+2} - 6$

8.4   Matrices and Systems of Equations

Matrices

What you should learn

● Write matrices and identify their orders.
● Perform elementary row operations on matrices.
● Use matrices and Gaussian elimination to solve systems of linear equations.
● Use matrices and Gauss-Jordan elimination to solve systems of linear equations.

Why you should learn it

Matrices can be used to solve systems of linear equations in two or more variables. For instance, Exercise 76 on page 642 shows how a matrix can be used to help find a model for the parabolic path of a baseball.

In this section you will study a streamlined technique for solving systems of linear equations. This technique involves the use of a rectangular array of real numbers called a **matrix.** The plural of matrix is *matrices*.

Definition of Matrix

If m and n are positive integers, an $m \times n$ (read "m by n") matrix is a rectangular array

$$
\begin{array}{ccccc}
 & \text{Column 1} & \text{Column 2} & \text{Column 3} & \cdots & \text{Column } n \\
\text{Row 1} & a_{11} & a_{12} & a_{13} & \cdots & a_{1n} \\
\text{Row 2} & a_{21} & a_{22} & a_{23} & \cdots & a_{2n} \\
\text{Row 3} & a_{31} & a_{32} & a_{33} & \cdots & a_{3n} \\
\vdots & \vdots & \vdots & \vdots & & \vdots \\
\text{Row } m & a_{m1} & a_{m2} & a_{m3} & \cdots & a_{mn}
\end{array}
$$

in which each **entry** a_{ij} of the matrix is a real number. An $m \times n$ matrix has m rows and n columns.

Michael Steele/Getty Images

The entry in the ith row and jth column is denoted by the *double subscript* notation a_{ij}. For instance, the entry a_{23} is the entry in the second row and third column. A matrix having m rows and n columns is said to be of **order** $m \times n$. If $m = n$, the matrix is **square** of order n. For a square matrix, the entries $a_{11}, a_{22}, a_{33}, \ldots$ are the **main diagonal** entries.

Example 1   Order of Matrices

Determine the order of each matrix.

a. $[2]$ **b.** $\begin{bmatrix} 1 & -3 & 0 & \frac{1}{2} \end{bmatrix}$ **c.** $\begin{bmatrix} 0 & 0 \\ 0 & 0 \end{bmatrix}$ **d.** $\begin{bmatrix} 5 & 0 \\ 2 & -2 \\ -7 & 4 \end{bmatrix}$

Solution

a. This matrix has *one* row and *one* column. The order of the matrix is 1×1.

b. This matrix has *one* row and *four* columns. The order of the matrix is 1×4.

c. This matrix has *two* rows and *two* columns. The order of the matrix is 2×2.

d. This matrix has *three* rows and *two* columns. The order of the matrix is 3×2.

✓ *Checkpoint* Now try Exercise 3.

A matrix that has only one row [such as the matrix in Example 1(b)] is called a **row matrix,** and a matrix that has only one column is called a **column matrix.**

A matrix derived from a system of linear equations (each written in standard form with the constant term on the right) is the **augmented matrix** of the system. Moreover, the matrix derived from the coefficients of the system (but not including the constant terms) is the **coefficient matrix** of the system.

System
$$\begin{cases} x - 4y + 3z = 5 \\ -x + 3y - z = -3 \\ 2x \quad\quad - 4z = 6 \end{cases}$$

Augmented Matrix
$$\begin{bmatrix} 1 & -4 & 3 & \vdots & 5 \\ -1 & 3 & -1 & \vdots & -3 \\ 2 & 0 & -4 & \vdots & 6 \end{bmatrix}$$

Coefficient Matrix
$$\begin{bmatrix} 1 & -4 & 3 \\ -1 & 3 & -1 \\ 2 & 0 & -4 \end{bmatrix}$$

Note the use of 0 for the missing coefficient of the y-variable in the third equation, and also note the fourth column (of constant terms) in the augmented matrix. The optional dotted line in the augmented matrix helps to separate the coefficients of the linear system from the constant terms.

When forming either the coefficient matrix or the augmented matrix of a system, you should begin by vertically aligning the variables in the equations and using 0's for any missing coefficients of variables.

Example 2   Writing an Augmented Matrix

Write the augmented matrix for the system of linear equations.

$$\begin{cases} x + 3y = 9 \\ -y + 4z = -2 \\ x - 5z = 0 \end{cases}$$

Solution

Begin by writing the linear system and aligning the variables.

$$\begin{cases} x + 3y \quad\quad = 9 \\ -y + 4z = -2 \\ x \quad\quad - 5z = 0 \end{cases}$$

Next, use the coefficients and constant terms as the matrix entries. Include zeros for each missing coefficient.

$$\begin{matrix} R_1 \\ R_2 \\ R_3 \end{matrix} \begin{bmatrix} 1 & 3 & 0 & \vdots & 9 \\ 0 & -1 & 4 & \vdots & -2 \\ 1 & 0 & -5 & \vdots & 0 \end{bmatrix}$$

The notation R_n is used to designate each row in the matrix. For example, Row 1 is represented by R_1.

✓ *Checkpoint* Now try Exercise 7.

Elementary Row Operations

In Section 8.3, you studied three operations that can be used on a system of linear equations to produce an equivalent system. These operations are: interchange two equations, multiply an equation by a nonzero constant, and add a multiple of an equation to another equation. In matrix terminology these three operations correspond to **elementary row operations.** An elementary row operation on an augmented matrix of a given system of linear equations produces a new augmented matrix corresponding to a new (but equivalent) system of linear equations. Two matrices are **row-equivalent** if one can be obtained from the other by a sequence of elementary row operations.

Elementary Row Operations

1. Interchange two rows.

2. Multiply a row by a nonzero constant.

3. Add a multiple of a row to another row.

Although elementary row operations are simple to perform, they involve a lot of arithmetic. Because it is easy to make a mistake, you should get in the habit of noting the elementary row operations performed in each step so that you can go back and check your work.

Example 3 demonstrates the elementary row operations described above.

Example 3   Elementary Row Operations

a. Interchange the first and second rows of the original matrix.

Original Matrix

$$\begin{bmatrix} 0 & 1 & 3 & 4 \\ -1 & 2 & 0 & 3 \\ 2 & -3 & 4 & 1 \end{bmatrix}$$

New Row-Equivalent Matrix

$$\begin{matrix} R_2 \\ R_1 \end{matrix} \begin{bmatrix} -1 & 2 & 0 & 3 \\ 0 & 1 & 3 & 4 \\ 2 & -3 & 4 & 1 \end{bmatrix}$$

b. Multiply the first row of the original matrix by $\frac{1}{2}$.

Original Matrix

$$\begin{bmatrix} 2 & -4 & 6 & -2 \\ 1 & 3 & -3 & 0 \\ 5 & -2 & 1 & 2 \end{bmatrix}$$

New Row-Equivalent Matrix

$$\frac{1}{2}R_1 \rightarrow \begin{bmatrix} 1 & -2 & 3 & -1 \\ 1 & 3 & -3 & 0 \\ 5 & -2 & 1 & 2 \end{bmatrix}$$

c. Add -2 times the first row of the original matrix to the third row.

Original Matrix

$$\begin{bmatrix} 1 & 2 & -4 & 3 \\ 0 & 3 & -2 & -1 \\ 2 & 1 & 5 & -2 \end{bmatrix}$$

New Row-Equivalent Matrix

$$\begin{bmatrix} 1 & 2 & -4 & 3 \\ 0 & 3 & -2 & -1 \\ -2R_1 + R_3 \rightarrow \ 0 & -3 & 13 & -8 \end{bmatrix}$$

Note that the elementary row operation is written beside the row that is *changed*.

✓ *Checkpoint* Now try Exercise 21.

TECHNOLOGY TIP

Most graphing utilities can perform elementary row operations on matrices. The top screen below shows how one graphing utility displays the original matrix in Example 3(a). The bottom screen below shows the new row-equivalent matrix in Example 3(a). The new row-equivalent matrix is obtained by using the *row swap* feature of the graphing utility. For instructions on how to use the *matrix* feature and the *row swap* feature (and other elementary row operations features) of a graphing utlity, see Appendix A; for specific keystrokes, go to the text website at *college.hmco.com*.

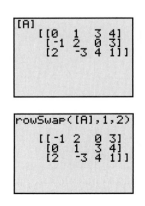

Gaussian Elimination with Back Substitution

In Example 2 of Section 8.3, you used Gaussian elimination with back-substitution to solve a system of linear equations. The next example demonstrates the matrix version of Gaussian elimination. The two methods are essentially the same. The basic difference is that with matrices you do not need to keep writing the variables.

Example 4   Comparing Linear Systems and Matrix Operations

Linear System   *Associated Augmented Matrix*

$$\begin{cases} x - 2y + 3z = 9 \\ -x + 3y \quad\quad = -4 \\ 2x - 5y + 5z = 17 \end{cases}$$

$$\begin{bmatrix} 1 & -2 & 3 & \vdots & 9 \\ -1 & 3 & 0 & \vdots & -4 \\ 2 & -5 & 5 & \vdots & 17 \end{bmatrix}$$

Add the first equation to the second equation.

Add the first row to the second row $(R_1 + R_2)$.

$$\begin{cases} x - 2y + 3z = 9 \\ y + 3z = 5 \\ 2x - 5y + 5z = 17 \end{cases}$$

$$R_1 + R_2 \rightarrow \begin{bmatrix} 1 & -2 & 3 & \vdots & 9 \\ 0 & 1 & 3 & \vdots & 5 \\ 2 & -5 & 5 & \vdots & 17 \end{bmatrix}$$

Add -2 times the first equation to the third equation.

Add -2 times the first row to the third row $(-2R_1 + R_3)$.

$$\begin{cases} x - 2y + 3z = 9 \\ y + 3z = 5 \\ -y - z = -1 \end{cases}$$

$$-2R_1 + R_3 \rightarrow \begin{bmatrix} 1 & -2 & 3 & \vdots & 9 \\ 0 & 1 & 3 & \vdots & 5 \\ 0 & -1 & -1 & \vdots & -1 \end{bmatrix}$$

Add the second equation to the third equation.

Add the second row to the third row $(R_2 + R_3)$.

$$\begin{cases} x - 2y + 3z = 9 \\ y + 3z = 5 \\ 2z = 4 \end{cases}$$

$$R_2 + R_3 \rightarrow \begin{bmatrix} 1 & -2 & 3 & \vdots & 9 \\ 0 & 1 & 3 & \vdots & 5 \\ 0 & 0 & 2 & \vdots & 4 \end{bmatrix}$$

Multiply the third equation by $\frac{1}{2}$.

Multiply the third row by $\frac{1}{2}$.

$$\begin{cases} x - 2y + 3z = 9 \\ y + 3z = 5 \\ z = 2 \end{cases}$$

$$\tfrac{1}{2}R_3 \rightarrow \begin{bmatrix} 1 & -2 & 3 & \vdots & 9 \\ 0 & 1 & 3 & \vdots & 5 \\ 0 & 0 & 1 & \vdots & 2 \end{bmatrix}$$

At this point, you can use back-substitution to find that the solution is $x = 1$, $y = -1$, and $z = 2$, as was done in Example 2 of Section 8.3.

Remember that you should check a solution by substituting the values of x, y, and z into each equation in the original system.

The last matrix in Example 4 is in **row-echelon form.** The term *echelon* refers to the stair-step pattern formed by the nonzero elements of the matrix. To be in this form, a matrix must have the properties listed on the next page.

> **Row-Echelon Form and Reduced Row-Echelon Form**
>
> A matrix in **row-echelon form** has the following properties.
>
> **1.** Any rows consisting entirely of zeros occur at the bottom of the matrix.
>
> **2.** For each row that does not consist entirely of zeros, the first nonzero entry is 1 (called a **leading 1**).
>
> **3.** For two successive (nonzero) rows, the leading 1 in the higher row is farther to the left than the leading 1 in the lower row.
>
> A matrix in *row-echelon form* is in **reduced row-echelon form** if every column that has a leading 1 has zeros in every position above and below its leading 1.

It is worth mentioning that the row-echelon form of a matrix is not unique. That is, two different sequences of elementary row operations may yield different row-echelon forms.

Example 5   Row–Echelon Form

Determine whether each matrix is in row-echelon form. If it is, determine whether the matrix is in reduced row-echelon form.

a. $\begin{bmatrix} 1 & 2 & -1 & 4 \\ 0 & 1 & 0 & 3 \\ 0 & 0 & 1 & -2 \end{bmatrix}$ **b.** $\begin{bmatrix} 1 & 2 & -1 & 2 \\ 0 & 0 & 0 & 0 \\ 0 & 1 & 2 & -4 \end{bmatrix}$

c. $\begin{bmatrix} 1 & -5 & 2 & -1 & 3 \\ 0 & 0 & 1 & 3 & -2 \\ 0 & 0 & 0 & 1 & 4 \\ 0 & 0 & 0 & 0 & 1 \end{bmatrix}$ **d.** $\begin{bmatrix} 1 & 0 & 0 & -1 \\ 0 & 1 & 0 & 2 \\ 0 & 0 & 1 & 3 \\ 0 & 0 & 0 & 0 \end{bmatrix}$

e. $\begin{bmatrix} 1 & 2 & -3 & 4 \\ 0 & 2 & 1 & -1 \\ 0 & 0 & 1 & -3 \end{bmatrix}$ **f.** $\begin{bmatrix} 0 & 1 & 0 & 5 \\ 0 & 0 & 1 & 3 \\ 0 & 0 & 0 & 0 \end{bmatrix}$

Solution

The matrices in (a), (c), (d), and (f) are in row-echelon form. The matrices in (d) and (f) are in *reduced* row-echelon form because every column that has a leading 1 has zeros in every position above and below its leading 1. The matrix in (b) is not in row-echelon form because a row of all zeros does not occur at the bottom of the matrix. The matrix in (e) is not in row-echelon form because the first nonzero entry in row 2 is not a leading 1.

 Checkpoint   Now try Exercise 23.

TECHNOLOGY TIP

Some graphing utilities can auto-matically transform a matrix to row-echelon form and reduced row-echelon form. The screen below shows how one graphing utility displays the row-echelon form of the matrix

$$\begin{bmatrix} 1 & 2 & -6 \\ 1 & 3 & 1 \\ 2 & 0 & -4 \end{bmatrix}.$$

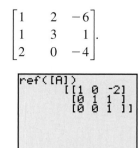

For instructions on how to use the *row-echelon form* feature and the *reduced row-echelon* feature of a graphing utility, see Appendix A; for specific keystrokes, go to the text website at *college.hmco.com*.

STUDY TIP

You have seen that the row-echelon form of a given matrix is *not* unique; however, the *reduced* row-echelon form of a given matrix *is* unique.

Every matrix is row-equivalent to a matrix in row-echelon form. For instance, in Example 5, you can change the matrix in part (e) to row-echelon form by multiplying its second row by $\frac{1}{2}$. What elementary row operation could you perform on the matrix in part (b) so that it would be in row-echelon form?

Gaussian elimination with back-substitution works well for solving systems of linear equations by hand or with a computer. For this algorithm, the order in which the elementary row operations are performed is important. You should operate *from left to right by columns*, using elementary row operations to obtain zeros in all entries directly below the leading 1's.

Example 6   Gaussian Elimination with Back-Substitution

Solve the system $\begin{cases} \quad\ y + \ z - 2w = \ -3 \\ x + 2y - \ z \qquad\quad = \quad 2 \\ 2x + 4y + \ z - 3w = \ -2 \\ x - 4y - 7z - \ w = -19 \end{cases}$.

Solution

$$\begin{bmatrix} 0 & 1 & 1 & -2 & \vdots & -3 \\ 1 & 2 & -1 & 0 & \vdots & 2 \\ 2 & 4 & 1 & -3 & \vdots & -2 \\ 1 & -4 & -7 & -1 & \vdots & -19 \end{bmatrix}$$

Write augmented matrix.

$$\begin{matrix} R_2 \\ R_1 \end{matrix} \begin{bmatrix} 1 & 2 & -1 & 0 & \vdots & 2 \\ 0 & 1 & 1 & -2 & \vdots & -3 \\ 2 & 4 & 1 & -3 & \vdots & -2 \\ 1 & -4 & -7 & -1 & \vdots & -19 \end{bmatrix}$$

Interchange R_1 and R_2 so first column has leading 1 in upper left corner.

$$\begin{matrix} \\ \\ -2R_1 + R_3 \rightarrow \\ -R_1 + R_4 \rightarrow \end{matrix} \begin{bmatrix} 1 & 2 & -1 & 0 & \vdots & 2 \\ 0 & 1 & 1 & -2 & \vdots & -3 \\ 0 & 0 & 3 & -3 & \vdots & -6 \\ 0 & -6 & -6 & -1 & \vdots & -21 \end{bmatrix}$$

Perform operations on R_3 and R_4 so first column has zeros below its leading 1.

$$\begin{matrix} \\ \\ \\ 6R_2 + R_4 \rightarrow \end{matrix} \begin{bmatrix} 1 & 2 & -1 & 0 & \vdots & 2 \\ 0 & 1 & 1 & -2 & \vdots & -3 \\ 0 & 0 & 3 & -3 & \vdots & -6 \\ 0 & 0 & 0 & -13 & \vdots & -39 \end{bmatrix}$$

Perform operations on R_4 so second column has zeros below its leading 1.

$$\begin{matrix} \\ \\ \tfrac{1}{3}R_3 \rightarrow \\ -\tfrac{1}{13}R_4 \rightarrow \end{matrix} \begin{bmatrix} 1 & 2 & -1 & 0 & \vdots & 2 \\ 0 & 1 & 1 & -2 & \vdots & -3 \\ 0 & 0 & 1 & -1 & \vdots & -2 \\ 0 & 0 & 0 & 1 & \vdots & 3 \end{bmatrix}$$

Perform operations on R_3 and R_4 so third and fourth columns have leading 1's.

The matrix is now in row-echelon form, and the corresponding system is

$$\begin{cases} x + 2y - z \qquad\quad = \quad 2 \\ \quad\ y + z - 2w = -3 \\ \qquad\quad z - \ w = -2 \\ \qquad\qquad\ \ w = \quad 3 \end{cases}.$$

Using back-substitution, you can determine that the solution is $x = -1$, $y = 2$, $z = 1$, and $w = 3$. Check this in the original system of equations.

✓ *Checkpoint* Now try Exercise 47.

The following steps summarize the procedure used in Example 6.

> **Gaussian Elimination with Back-Substitution**
>
> **1.** Write the augmented matrix of the system of linear equations.
>
> **2.** Use elementary row operations to rewrite the augmented matrix in row-echelon form.
>
> **3.** Write the system of linear equations corresponding to the matrix in row-echelon form and use back-substitution to find the solution.

Remember that it is possible for a system to have no solution. If, in the elimination process, you obtain a row with zeros except for the last entry, you can conclude that the system is inconsistent.

Example 7   A System with No Solution

Solve the system $\begin{cases} x - y + 2z = 4 \\ x \quad + z = 6 \\ 2x - 3y + 5z = 4 \\ 3x + 2y - z = 1 \end{cases}$.

Solution

$$\begin{bmatrix} 1 & -1 & 2 & \vdots & 4 \\ 1 & 0 & 1 & \vdots & 6 \\ 2 & -3 & 5 & \vdots & 4 \\ 3 & 2 & -1 & \vdots & 1 \end{bmatrix}$$ Write augmented matrix.

$$\begin{matrix} \\ -R_1 + R_2 \rightarrow \\ -2R_1 + R_3 \rightarrow \\ -3R_1 + R_4 \rightarrow \end{matrix} \begin{bmatrix} 1 & -1 & 2 & \vdots & 4 \\ 0 & 1 & -1 & \vdots & 2 \\ 0 & -1 & 1 & \vdots & -4 \\ 0 & 5 & -7 & \vdots & -11 \end{bmatrix}$$ Perform row operations.

$$\begin{matrix} \\ \\ R_2 + R_3 \rightarrow \\ \\ \end{matrix} \begin{bmatrix} 1 & -1 & 2 & \vdots & 4 \\ 0 & 1 & -1 & \vdots & 2 \\ 0 & 0 & 0 & \vdots & -2 \\ 0 & 5 & -7 & \vdots & -11 \end{bmatrix}$$ Perform row operations.

Note that the third row of this matrix consists of zeros except for the last entry. This means that the original system of linear equations is *inconsistent*. You can see why this is true by converting back to a system of linear equations. Because the third equation is not possible, the system has no solution.

$$\begin{cases} x - y + 2z = 4 \\ y - z = 2 \\ 0 = -2 \\ 5y - 7z = -11 \end{cases}$$

✓ *Checkpoint* Now try Exercise 49.

Gauss–Jordan Elimination

With Gaussian elimination, elementary row operations are applied to a matrix to obtain a (row-equivalent) row-echelon form of the matrix. A second method of elimination, called **Gauss-Jordan elimination** after Carl Friedrich Gauss (1777–1855) and Wilhelm Jordan (1842–1899), continues the reduction process until a *reduced* row-echelon form is obtained. This procedure is demonstrated in Example 8.

Example 8   Gauss–Jordan Elimination

Use Gauss-Jordan elimination to solve the system.

$$\begin{cases} x - 2y + 3z = 9 \\ -x + 3y \quad\;\;\; = -4 \\ 2x - 5y + 5z = 17 \end{cases}$$

Solution

In Example 4, Gaussian elimination was used to obtain the row-echelon form

$$\begin{bmatrix} 1 & -2 & 3 & \vdots & 9 \\ 0 & 1 & 3 & \vdots & 5 \\ 0 & 0 & 1 & \vdots & 2 \end{bmatrix}.$$

Now, rather than using back-substitution, apply additional elementary row operations until you obtain a matrix in *reduced* row-echelon form. To do this, you must produce zeros above each of the leading 1's, as follows.

$$2R_2 + R_1 \rightarrow \begin{bmatrix} 1 & 0 & 9 & \vdots & 19 \\ 0 & 1 & 3 & \vdots & 5 \\ 0 & 0 & 1 & \vdots & 2 \end{bmatrix}$$

Perform operations on R_1 so second column has a zero above its leading 1.

$$\begin{matrix} -9R_3 + R_1 \rightarrow \\ -3R_3 + R_2 \rightarrow \end{matrix} \begin{bmatrix} 1 & 0 & 0 & \vdots & 1 \\ 0 & 1 & 0 & \vdots & -1 \\ 0 & 0 & 1 & \vdots & 2 \end{bmatrix}$$

Perform operations on R_1 and R_2 so third column has zeros above its leading 1.

The matrix is now in reduced row-echelon form. Converting back to a system of linear equations, you have

$$\begin{cases} x = 1 \\ y = -1 \\ z = 2 \end{cases}$$

which is the same solution that was obtained using Gaussian elimination.

✓ *Checkpoint* Now try Exercise 55.

The beauty of Gauss-Jordan elimination is that, from the reduced row-echelon form, you can simply read the solution. Which technique do you prefer: Gaussian elimination or Gauss-Jordan elimination?

TECHNOLOGY TIP

For a demonstration of a graphical approach to Gauss-Jordan elimination on a 2×3 matrix, see the Visualizing Row Operations Program, available for several models of graphing calculators at our website *college.hmco.com.*

The elimination procedures described in this section employ an algorithmic approach that is easily adapted to computer programs. However, the procedure makes no effort to avoid fractional coefficients. For instance, in the elimination procedure for the system

$$\begin{cases} 2x - 5y + 5z = 17 \\ 3x - 2y + 3z = 11 \\ -3x + 3y = -6 \end{cases}$$

you may be inclined to multiply the first row by $\frac{1}{2}$ to produce a leading 1, which will result in working with fractional coefficients. For hand computations, you can sometimes avoid fractions by judiciously choosing the order in which you apply elementary row operations.

Example 9   A System with an Infinite Number of Solutions

Solve the system $\begin{cases} 2x + 4y - 2z = 0 \\ 3x + 5y = 1 \end{cases}$.

Solution

$$\begin{bmatrix} 2 & 4 & -2 & \vdots & 0 \\ 3 & 5 & 0 & \vdots & 1 \end{bmatrix} \qquad \frac{1}{2}R_1 \rightarrow \begin{bmatrix} 1 & 2 & -1 & \vdots & 0 \\ 3 & 5 & 0 & \vdots & 1 \end{bmatrix}$$

$$-3R_1 + R_2 \rightarrow \begin{bmatrix} 1 & 2 & -1 & \vdots & 0 \\ 0 & -1 & 3 & \vdots & 1 \end{bmatrix}$$

$$-R_2 \rightarrow \begin{bmatrix} 1 & 2 & -1 & \vdots & 0 \\ 0 & 1 & -3 & \vdots & -1 \end{bmatrix}$$

$$-2R_2 + R_1 \rightarrow \begin{bmatrix} 1 & 0 & 5 & \vdots & 2 \\ 0 & 1 & -3 & \vdots & -1 \end{bmatrix}$$

The corresponding system of equations is

$$\begin{cases} x + 5z = 2 \\ y - 3z = -1 \end{cases}.$$

Solving for x and y in terms of z, you have $x = -5z + 2$ and $y = 3z - 1$. To write a solution of the system that does not use any of the three variables of the system, let a represent any real number and let $z = a$. Now substitute a for z in the equations for x and y.

$$x = -5z + 2 = -5a + 2$$

$$y = 3z - 1 = 3a - 1$$

So, the solution set has the form

$$(-5a + 2, 3a - 1, a).$$

Recall from Section 8.3 that a solution set of this form represents an infinite number of solutions. Try substituting values for a to obtain a few solutions. Then check each solution in the original system of equations.

✓ *Checkpoint* Now try Exercise 57.

Example 10   Analysis of a Network

Set up a system of linear equations representing the network shown in Figure 8.26. In a network, it is assumed that the total flow into a junction (blue circle) is equal to the total flow out of the junction.

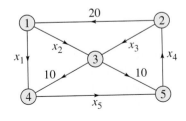

Figure 8.26

Solution

Because Junction 1 in Figure 8.26 has 20 units flowing into it, there must be 20 units flowing out of it. This is represented by the linear equation $x_1 + x_2 = 20$. Because Junction 2 has 20 units flowing out of it, there must be 20 units flowing into it. This is represented by $x_4 - x_3 = 20$ or $-x_3 + x_4 = 20$. A linear equation can be written for each of the network's five junctions, so the network is modeled by the following system.

$$\begin{cases} x_1 + x_2 & = 20 & \text{Junction 1} \\ -x_3 + x_4 & = 20 & \text{Junction 2} \\ x_2 + x_3 & = 20 & \text{Junction 3} \\ -x_1 & + x_5 = 10 & \text{Junction 4} \\ x_4 - x_5 = 10 & & \text{Junction 5} \end{cases}$$

Using Gauss-Jordan elimination on the augmented matrix produces the matrix in reduced row-echelon form.

Augmented Matrix

$$\begin{bmatrix} 1 & 1 & 0 & 0 & 0 & \vdots & 20 \\ 0 & 0 & -1 & 1 & 0 & \vdots & 20 \\ 0 & 1 & 1 & 0 & 0 & \vdots & 20 \\ -1 & 0 & 0 & 0 & 1 & \vdots & 10 \\ 0 & 0 & 0 & 1 & -1 & \vdots & 10 \end{bmatrix}$$

Matrix in Reduced Row-Echelon Form

$$\begin{bmatrix} 1 & 0 & 0 & 0 & -1 & \vdots & -10 \\ 0 & 1 & 0 & 0 & 1 & \vdots & 30 \\ 0 & 0 & 1 & 0 & -1 & \vdots & -10 \\ 0 & 0 & 0 & 1 & -1 & \vdots & 10 \\ 0 & 0 & 0 & 0 & 0 & \vdots & 0 \end{bmatrix}$$

Letting $x_5 = t$, where t is a real number, you have $x_1 = t - 10$, $x_2 = -t + 30$, $x_3 = t - 10$, and $x_4 = t + 10$. So, this system has an infinite number of solutions.

✓ *Checkpoint* Now try Exercise 79.

8.4 Exercises

Vocabulary Check

Fill in the blanks.

1. A rectangular array of real numbers that can be used to solve a system of linear equations is called a _____ .

2. A matrix is _____ if the number of rows equals the number of columns.

3. A matrix with only one row is called a _____ and a matrix with only one column is called a _____ .

4. The matrix derived from a system of linear equations is called the _____ of the system.

5. The matrix derived from the coefficients of a system of linear equations is called the _____ of the system.

6. Two matrices are called _____ if one of the matrices can be obtained from the other by a sequence of elementary row operations.

7. A matrix in row-echelon form is in _____ if every column that has a leading 1 has zeros in every position above and below its leading one.

8. The process of using row operations to write a matrix in reduced row-echelon form is called _____ .

In Exercises 1–6, determine the order of the matrix.

1. $\begin{bmatrix} 7 & 0 \end{bmatrix}$

2. $\begin{bmatrix} 6 & -3 & 10 & 8 \end{bmatrix}$

3. $\begin{bmatrix} 4 \\ 32 \\ 3 \end{bmatrix}$

4. $\begin{bmatrix} -3 & 7 & 15 & 0 \\ 0 & 0 & 3 & 3 \\ 1 & 1 & 6 & 7 \end{bmatrix}$

5. $\begin{bmatrix} 33 & 45 \\ -9 & 20 \end{bmatrix}$

6. $\begin{bmatrix} 3 & -1 & 4 \\ 6 & 0 & -5 \end{bmatrix}$

In Exercises 7–10, write the augmented matrix for the system of linear equations.

7. $\begin{cases} 4x - 5y = 33 \\ -x + 5y = -27 \end{cases}$

8. $\begin{cases} 7x + 4y = 22 \\ 5x - 9y = 15 \end{cases}$

9. $\begin{cases} x + 10y - 2z = 2 \\ 5x - 3y + 4z = 0 \\ 2x + y = 6 \end{cases}$

10. $\begin{cases} x - 3y + z = 1 \\ 4y = 0 \\ 7z = -5 \end{cases}$

In Exercises 11–14, write the system of linear equations represented by the augmented matrix. (Use the variables x, y, z, and w if applicable.)

11. $\begin{bmatrix} 1 & 2 & \vdots & 7 \\ 2 & -3 & \vdots & 4 \end{bmatrix}$

12. $\begin{bmatrix} 7 & -5 & \vdots & 0 \\ 8 & 3 & \vdots & -2 \end{bmatrix}$

13. $\begin{bmatrix} 9 & 12 & 3 & \vdots & 0 \\ -2 & 18 & 5 & \vdots & 10 \\ 1 & 7 & -8 & \vdots & -4 \end{bmatrix}$

14. $\begin{bmatrix} 6 & 2 & -1 & -5 & \vdots & -25 \\ -1 & 0 & 7 & 3 & \vdots & 7 \\ 4 & -1 & -10 & 6 & \vdots & 23 \\ 0 & 8 & 1 & -11 & \vdots & -21 \end{bmatrix}$

In Exercises 15–18, fill in the blanks using elementary row operations to form a row-equivalent matrix.

15. $\begin{bmatrix} 1 & 4 & 3 \\ 2 & 10 & 5 \end{bmatrix}$

$\begin{bmatrix} 1 & 4 & 3 \\ 0 & & -1 \end{bmatrix}$

16. $\begin{bmatrix} 3 & 6 & 8 \\ 4 & -3 & 6 \end{bmatrix}$

$\begin{bmatrix} 1 & & \frac{8}{3} \\ 4 & -3 & 6 \end{bmatrix}$

17. $\begin{bmatrix} 1 & 1 & 4 & -1 \\ 3 & 8 & 10 & 3 \\ -2 & 1 & 12 & 6 \end{bmatrix}$

$\begin{bmatrix} 1 & 1 & 4 & -1 \\ 0 & 5 & & \\ 0 & 3 & & \end{bmatrix}$

$\begin{bmatrix} 1 & 1 & 4 & -1 \\ 0 & 1 & -\frac{2}{5} & \frac{6}{5} \\ 0 & 3 & & \end{bmatrix}$

18. $\begin{bmatrix} 2 & 4 & 8 & 3 \\ 1 & -1 & -3 & 2 \\ 2 & 6 & 4 & 9 \end{bmatrix}$

$\begin{bmatrix} 1 & & & \\ 1 & -1 & -3 & 2 \\ 2 & 6 & 4 & 9 \end{bmatrix}$

$\begin{bmatrix} 1 & 2 & 4 & \frac{3}{2} \\ 0 & & -7 & \frac{1}{2} \\ 0 & 2 & & \end{bmatrix}$

In Exercises 19–22, identify the elementary row operation performed to obtain the new row-equivalent matrix.

19.
Original Matrix

$\begin{bmatrix} -2 & 5 & 1 \\ 3 & -1 & -8 \end{bmatrix}$

New Row-Equivalent Matrix

$\begin{bmatrix} 13 & 0 & -39 \\ 3 & -1 & -8 \end{bmatrix}$

20.
Original Matrix

$\begin{bmatrix} 3 & -1 & -4 \\ -4 & 3 & 7 \end{bmatrix}$

New Row-Equivalent Matrix

$\begin{bmatrix} 3 & -1 & -4 \\ 5 & 0 & -5 \end{bmatrix}$

21.
Original Matrix

$\begin{bmatrix} 0 & -1 & -5 & 5 \\ -1 & 3 & -7 & 6 \\ 4 & -5 & 1 & 3 \end{bmatrix}$

New Row-Equivalent Matrix

$\begin{bmatrix} -1 & 3 & -7 & 6 \\ 0 & -1 & -5 & 5 \\ 4 & -5 & 1 & 3 \end{bmatrix}$

22.
Original Matrix

$\begin{bmatrix} -1 & -2 & 3 & -2 \\ 2 & -5 & 1 & -7 \\ 5 & 4 & -7 & 6 \end{bmatrix}$

New Row-Equivalent Matrix

$\begin{bmatrix} -1 & -2 & 3 & -2 \\ 2 & -5 & 1 & -7 \\ 0 & -6 & 8 & -4 \end{bmatrix}$

In Exercises 23–26, determine whether the matrix is in row-echelon form. If it is, determine if it is also in reduced row-echelon form.

23. $\begin{bmatrix} 1 & 0 & 0 & 0 \\ 0 & 1 & 1 & 5 \\ 0 & 0 & 0 & 0 \end{bmatrix}$ 24. $\begin{bmatrix} 1 & 3 & 0 & 0 \\ 0 & 0 & 1 & 8 \\ 0 & 0 & 0 & 0 \end{bmatrix}$

25. $\begin{bmatrix} 2 & 0 & 4 & 0 \\ 0 & -1 & 3 & 6 \\ 0 & 0 & 1 & 5 \end{bmatrix}$ 26. $\begin{bmatrix} 1 & 0 & 2 & 1 \\ 0 & 1 & -3 & 10 \\ 0 & 0 & 1 & 0 \end{bmatrix}$

27. Perform the sequence of row operations on the matrix. What did the operations accomplish?

$\begin{bmatrix} 1 & 2 & 3 \\ 2 & -1 & -4 \\ 3 & 1 & -1 \end{bmatrix}$

(a) Add -2 times R_1 to R_2.

(b) Add -3 times R_1 to R_3.

(c) Add -1 times R_2 to R_3.

(d) Multiply R_2 by $-\frac{1}{5}$.

(e) Add -2 times R_2 to R_1.

28. Perform the sequence of row operations on the matrix. What did the operations accomplish?

$\begin{bmatrix} 7 & 1 \\ 0 & 2 \\ -3 & 4 \\ 4 & 1 \end{bmatrix}$

(a) Add R_3 to R_4.

(b) Interchange R_1 and R_4.

(c) Add 3 times R_1 to R_3.

(d) Add -7 times R_1 to R_4.

(e) Multiply R_2 by $\frac{1}{2}$.

(f) Add the appropriate multiples of R_2 to R_1, R_3, and R_4.

29. Repeat steps (a) through (e) in Exercise 27 using a graphing utility.

30. Repeat steps (a) through (f) in Exercise 28 using a graphing utility.

In Exercises 31–34, write the matrix in row-echelon form. Remember that the row-echelon form of a matrix is not unique.

31. $\begin{bmatrix} 1 & 1 & 0 & 5 \\ -2 & -1 & 2 & -10 \\ 3 & 6 & 7 & 14 \end{bmatrix}$

32. $\begin{bmatrix} 1 & 2 & -1 & 3 \\ 3 & 7 & -5 & 14 \\ -2 & -1 & -3 & 8 \end{bmatrix}$

33. $\begin{bmatrix} 1 & -1 & -1 & 1 \\ 5 & -4 & 1 & 8 \\ -6 & 8 & 18 & 0 \end{bmatrix}$

34. $\begin{bmatrix} 1 & -3 & 0 & -7 \\ -3 & 10 & 1 & 23 \\ 4 & -10 & 2 & -24 \end{bmatrix}$

In Exercises 35–38, use the matrix capabilities of a graphing utility to write the matrix in reduced row-echelon form.

35. $\begin{bmatrix} 3 & 3 & 3 \\ -1 & 0 & -4 \\ 2 & 4 & -2 \end{bmatrix}$ 36. $\begin{bmatrix} 1 & 3 & 2 \\ 5 & 15 & 9 \\ 2 & 6 & 10 \end{bmatrix}$

37. $\begin{bmatrix} -3 & 5 & 1 & 12 \\ 1 & -1 & 1 & 4 \end{bmatrix}$

38. $\begin{bmatrix} 5 & 1 & 2 & 4 \\ -1 & 5 & 10 & -32 \end{bmatrix}$

In Exercises 39–42, write the system of linear equations represented by the augmented matrix. Then use back-substitution to find the solution. (Use the variables x, y, and z, if applicable.)

39. $\left[\begin{array}{cc:c} 1 & -2 & 4 \\ 0 & 1 & -3 \end{array}\right]$ 40. $\left[\begin{array}{cc:c} 1 & 5 & 0 \\ 0 & 1 & -1 \end{array}\right]$

41. $\begin{bmatrix} 1 & -1 & 2 & \vdots & 4 \\ 0 & 1 & -1 & \vdots & 2 \\ 0 & 0 & 1 & \vdots & -2 \end{bmatrix}$

42. $\begin{bmatrix} 1 & 2 & -2 & \vdots & -1 \\ 0 & 1 & 1 & \vdots & 9 \\ 0 & 0 & 1 & \vdots & -3 \end{bmatrix}$

In Exercises 43–46, an augmented matrix that represents a system of linear equations (in the variables x and y or x, y, and z) has been reduced using Gauss-Jordan elimination. Write the solution represented by the augmented matrix.

43. $\begin{bmatrix} 1 & 0 & \vdots & 7 \\ 0 & 1 & \vdots & -5 \end{bmatrix}$ **44.** $\begin{bmatrix} 1 & 0 & \vdots & -2 \\ 0 & 1 & \vdots & 4 \end{bmatrix}$

45. $\begin{bmatrix} 1 & 0 & 0 & \vdots & -4 \\ 0 & 1 & 0 & \vdots & -8 \\ 0 & 0 & 1 & \vdots & 2 \end{bmatrix}$

46. $\begin{bmatrix} 1 & 0 & 0 & \vdots & 3 \\ 0 & 1 & 0 & \vdots & -1 \\ 0 & 0 & 1 & \vdots & 0 \end{bmatrix}$

In Exercises 47–60, use matrices to solve the system of equations if possible. Use Gaussian elimination with back-substitution or Gauss-Jordan elimination.

47. $\begin{cases} x + 2y = 7 \\ 2x + y = 8 \end{cases}$

48. $\begin{cases} 2x + 6y = 16 \\ 2x + 3y = 7 \end{cases}$

49. $\begin{cases} -x + y = -22 \\ 3x + 4y = 4 \\ 4x - 8y = 32 \end{cases}$

50. $\begin{cases} x + 2y = 0 \\ x + y = 6 \\ 3x - 2y = 8 \end{cases}$

51. $\begin{cases} 8x - 4y = 13 \\ 5x + 2y = 7 \end{cases}$

52. $\begin{cases} x - 3y = 5 \\ -2x + 6y = -10 \end{cases}$

53. $\begin{cases} -x + 2y = 1.5 \\ 2x - 4y = 3 \end{cases}$

54. $\begin{cases} 2x - y = -0.1 \\ 3x + 2y = 1.6 \end{cases}$

55. $\begin{cases} x - 3z = -2 \\ 3x + y - 2z = 5 \\ 2x + 2y + z = 4 \end{cases}$

56. $\begin{cases} 2x - y + 3z = 24 \\ 2y - z = 14 \\ 7x - 5y = 6 \end{cases}$

57. $\begin{cases} x + y - 5z = 3 \\ x - 2z = 1 \\ 2x - y - z = 0 \end{cases}$

58. $\begin{cases} 2x + 3z = 3 \\ 4x - 3y + 7z = 5 \\ 8x - 9y + 15z = 9 \end{cases}$

59. $\begin{cases} -x + y - z = -14 \\ 2x - y + z = 21 \\ 3x + 2y + z = 19 \end{cases}$

60. $\begin{cases} 2x + 2y - z = 2 \\ x - 3y + z = 28 \\ -x + y = 14 \end{cases}$

In Exercises 61–66, use the matrix capabilities of a graphing utility to reduce the augmented matrix corresponding to the system of equations, and solve the system.

61. $\begin{cases} 3x + 3y + 12z = 6 \\ x + y + 4z = 2 \\ 2x + 5y + 20z = 10 \\ -x + 2y + 8z = 4 \end{cases}$

62. $\begin{cases} 2x + 10y + 2z = 6 \\ x + 5y + 2z = 6 \\ x + 5y + z = 3 \\ -3x - 15y - 3z = -9 \end{cases}$

63. $\begin{cases} 2x + y - z + 2w = -6 \\ 3x + 4y + w = 1 \\ x + 5y + 2z + 6w = -3 \\ 5x + 2y - z - w = 3 \end{cases}$

64. $\begin{cases} x + 2y + 2z + 4w = 11 \\ 3x + 6y + 5z + 12w = 30 \end{cases}$

65. $\begin{cases} x + y + z = 0 \\ 2x + 3y + z = 0 \\ 3x + 5y + z = 0 \end{cases}$ **66.** $\begin{cases} x + 2y + z + 3w = 0 \\ x - y + w = 0 \\ y - z + 2w = 0 \end{cases}$

In Exercises 67–70, determine whether the two systems of linear equations yield the same solution. If so, find the solution.

67. (a) $\begin{cases} x - 2y + z = -6 \\ y - 5z = 16 \\ z = -3 \end{cases}$ (b) $\begin{cases} x + y - 2z = 6 \\ y + 3z = -8 \\ z = -3 \end{cases}$

68. (a) $\begin{cases} x - 3y + 4z = -11 \\ y - z = -4 \\ z = 2 \end{cases}$ (b) $\begin{cases} x + 4y = -11 \\ y + 3z = 4 \\ z = 2 \end{cases}$

69. (a) $\begin{cases} x - 4y + 5z = 27 \\ y - 7z = -54 \\ z = 8 \end{cases}$ (b) $\begin{cases} x - 6y + z = 15 \\ y + 5z = 42 \\ z = 8 \end{cases}$

70. (a) $\begin{cases} x + 3y - z = 19 \\ y + 6z = -18 \\ z = -4 \end{cases}$ (b) $\begin{cases} x - y + 3z = -15 \\ y - 2z = 14 \\ z = -4 \end{cases}$

In Exercises 71 and 72, use a system of equations to find the equation of the parabola $y = ax^2 + bx + c$ that passes through the points. Solve the system using matrices. Use a graphing utility to verify your result.

71.

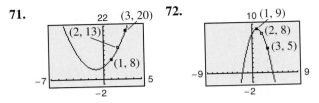

72.

(a) Use a system of equations to find the equation of the parabola $y = ax^2 + bx + c$ that passes through the points. Solve the system using matrices.

(b) Use a graphing utility to graph the parabola.

(c) Graphically approximate the maximum height of the ball and the point at which the ball strikes the ground.

(d) Algebraically approximate the maximum height of the ball and the point at which the ball strikes the ground.

73. **Borrowing Money** A small corporation borrowed $1,500,000 to expand its line of shoes. Some of the money was borrowed at 7%, some at 8%, and some at 10%. Use a system of equations to determine how much was borrowed at each rate if the annual interest was $130,500 and the amount borrowed at 10% was four times the amount borrowed at 7%. Solve the system using matrices.

74. **Borrowing Money** A small corporation borrowed $500,000 to build a new office building. Some of the money was borrowed at 9%, some at 10%, and some at 12%. Use a system of equations to determine how much was borrowed at each rate if the annual interest was $52,000 and the amount borrowed at 10% was $2\frac{1}{2}$ times the amount borrowed at 9%. Solve the system using matrices.

75. **Electrical Network** The currents in an electrical network are given by the solution of the system

$$\begin{cases} I_1 - I_2 + I_3 = 0 \\ 2I_1 + 2I_2 \quad\quad = 7 \\ \quad\quad 2I_2 + 4I_3 = 8 \end{cases}$$

where I_1, I_2, and I_3 are measured in amperes. Solve the system of equations using matrices.

76. **Mathematical Modeling** A videotape of the path of a ball thrown by a baseball player was analyzed with a grid covering the TV screen. The tape was paused three times, and the position of the ball was measured each time. The coordinates obtained are shown in the table (x and y are measured in feet).

Horizontal distance, x	Height, y
0	5.0
15	9.6
30	12.4

77. **Data Analysis** The table shows the average price y (in dollars) of shares traded on the New York Stock Exchange from 1999 to 2001. (Source: New York Stock Exchange)

Year	Average price, y
1999	43.90
2000	42.10
2001	34.10

(a) Use a system of equations to find the equation of the parabola $y = at^2 + bt + c$ that passes through the points. Let $t = 9$ represent 1999. Solve the system using matrices.

(b) Use a graphing utility to graph the parabola.

(c) Use the equation in part (a) to estimate the average price of shares traded in 2002.

(d) Use the equation in part (a) to estimate the average price of shares traded in 2005. Is the estimate reasonable? Explain.

78. **Data Analysis** The table shows the average monthly bill y (in dollars) for cellular telephone subscribers from 1999 to 2001. (Source: Cellular Telecommunications & Internet Association)

Year	Average monthly bill, y
1999	41.24
2000	45.27
2001	47.37

(a) Use a system of equations to find the equation of the parabola $y = at^2 + bt + c$ that passes through the points. Let $t = 9$ represent 1999. Solve the system using matrices.

(b) Use a graphing utility to graph the parabola.

(c) Use the equation in part (a) to estimate the average monthly bill in 2002.

(d) Use the equation in part (a) to estimate the average monthly bill in 2005. Is the estimate reasonable? Explain.

Network Analysis   **In Exercises 79 and 80, answer the questions about the specified network.**

79. Water flowing through a network of pipes (in thousands of cubic meters per hour) is shown below.

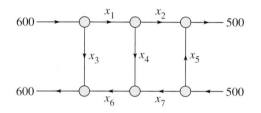

(a) Use matrices to solve this system for the water flow represented by x_i, $i = 1, 2, 3, 4, 5, 6,$ and 7.

(b) Find the network flow pattern when $x_6 = 0$ and $x_7 = 0$.

(c) Find the network flow pattern when $x_5 = 1000$ and $x_6 = 0$.

80. The flow of traffic (in vehicles per hour) through a network of streets is shown below.

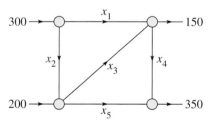

(a) Use matrices to solve this system for the traffic flow represented by x_i, $i = 1, 2, 3, 4,$ and 5.

(b) Find the traffic flow when $x_2 = 200$ and $x_3 = 50$.

(c) Find the traffic flow when $x_2 = 150$ and $x_3 = 0$.

Synthesis

True or False?   **In Exercises 81 and 82, determine whether the statement is true or false. Justify your answer.**

81. $\begin{bmatrix} 6 & 0 & -3 & 10 \\ -2 & 5 & -6 & 2 \end{bmatrix}$ is a 4×2 matrix.

82. Gaussian elimination reduces a matrix until a reduced row-echelon form is obtained.

83. ***Think About It***   The augmented matrix represents a system of linear equations (in the variables x, y, and z) that has been reduced using Gauss-Jordan elimination. Write a system of equations with *nonzero* coefficients that is represented by the reduced matrix. (There are many correct answers.)

$$\begin{bmatrix} 1 & 0 & 3 & \vdots & -2 \\ 0 & 1 & 4 & \vdots & 1 \\ 0 & 0 & 0 & \vdots & 0 \end{bmatrix}$$

84. ***Think About It***

(a) Describe the row-echelon form of an augmented matrix that corresponds to a system of linear equations that is inconsistent.

(b) Describe the row-echelon form of an augmented matrix that corresponds to a system of linear equations that has an infinite number of solutions.

85. ***Error Analysis***   One of your classmates has submitted the following steps for the solution of a system by Gauss-Jordan elimination. Find the error(s) in the solution. Write a short paragraph explaining the error(s) to your classmate.

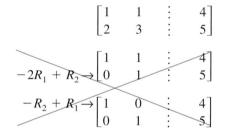

86. ***Writing***   In your own words, describe the difference between a matrix in row-echelon form and a matrix in reduced row-echelon form.

Review

In Exercises 87–92, sketch the graph of the function. Identify any asymptotes.

87. $f(x) = \dfrac{7}{-x - 1}$

88. $f(x) = \dfrac{4x}{5x^2 + 2}$

89. $f(x) = \dfrac{x^2 - 2x - 3}{x - 4}$

90. $f(x) = \dfrac{x^2 - 36}{x + 1}$

91. $f(x) = \dfrac{2x^2 - 4x}{3x - x^2}$

92. $f(x) = \dfrac{x^2 - 2x + 1}{x^2 - 1}$

8.5   Operations with Matrices

What you should learn

- Decide whether two matrices are equal.
- Add and subtract matrices and multiply matrices by a scalar.
- Multiply two matrices.
- Use matrix operations to model and solve real-life problems.

Why you should learn it

Matrix algebra provides a systematic way of performing mathematical operations on large arrays of numbers. In Exercise 70 on page 657, you will use matrix multiplication to help analyze the labor and wage requirements for a boat manufacturer.

Michael St. Maur Sheil/Corbis

Equality of Matrices

In Section 8.4, you used matrices to solve systems of linear equations. There is a rich mathematical theory of matrices, and its applications are numerous. This section and the next two introduce some fundamentals of matrix theory. It is standard mathematical convention to represent matrices in any of the following three ways.

Representation of Matrices

1. A matrix can be denoted by an uppercase letter such as A, B, or C.

2. A matrix can be denoted by a representative element enclosed in brackets, such as $[a_{ij}]$, $[b_{ij}]$, or $[c_{ij}]$.

3. A matrix can be denoted by a rectangular array of numbers such as

$$A = [a_{ij}] = \begin{bmatrix} a_{11} & a_{12} & a_{13} & \cdots & a_{1n} \\ a_{21} & a_{22} & a_{23} & \cdots & a_{2n} \\ a_{31} & a_{32} & a_{33} & \cdots & a_{3n} \\ \vdots & \vdots & \vdots & & \vdots \\ a_{m1} & a_{m2} & a_{m3} & \cdots & a_{mn} \end{bmatrix}.$$

Two matrices $A = [a_{ij}]$ and $B = [b_{ij}]$ are **equal** if they have the same order $(m \times n)$ and all of their corresponding entries are equal. For instance, using the matrix equation

$$\begin{bmatrix} a_{11} & a_{12} \\ a_{21} & a_{22} \end{bmatrix} = \begin{bmatrix} 2 & -1 \\ -3 & 0 \end{bmatrix}$$

you can conclude that $a_{11} = 2$, $a_{12} = -1$, and $a_{22} = 0$.

Matrix Addition and Scalar Multiplication

You can add two matrices (of the same order) by adding their corresponding entries.

Definition of Matrix Addition

If $A = [a_{ij}]$ and $B = [b_{ij}]$ are matrices of order $m \times n$, their sum is the $m \times n$ matrix given by

$$A + B = [a_{ij} + b_{ij}].$$

The sum of two matrices of different orders is undefined.

Example 1 Addition of Matrices

a. $\begin{bmatrix} -1 & 2 \\ 0 & 1 \end{bmatrix} + \begin{bmatrix} 1 & 3 \\ -1 & 2 \end{bmatrix} = \begin{bmatrix} -1+1 & 2+3 \\ 0-1 & 1+2 \end{bmatrix} = \begin{bmatrix} 0 & 5 \\ -1 & 3 \end{bmatrix}$

b. $\begin{bmatrix} 1 \\ -3 \\ -2 \end{bmatrix} + \begin{bmatrix} -1 \\ 3 \\ 2 \end{bmatrix} = \begin{bmatrix} 0 \\ 0 \\ 0 \end{bmatrix}$

c. The sum of

$$A = \begin{bmatrix} 2 & 1 & 0 \\ 4 & 0 & -1 \end{bmatrix} \quad \text{and} \quad B = \begin{bmatrix} 0 & 1 \\ -1 & 3 \end{bmatrix}$$

is undefined because A is of order 2×3 and B is of order 2×2.

✓ *Checkpoint* Now try Exercise 7(a).

TECHNOLOGY TIP Most graphing utilities can perform matrix operations. Example 2 shows how a graphing utility can be used to add two matrices.

Example 2 Addition of Matrices

Use a graphing utility to find the sum of

$$A = \begin{bmatrix} 0 & 1 & -2 \\ 1 & 2 & 3 \end{bmatrix} \quad \text{and} \quad B = \begin{bmatrix} 0 & 0 & 0 \\ 0 & 0 & 0 \end{bmatrix}.$$

Solution

Use the *matrix editor* to enter A and B in the graphing utility (see Figure 8.27). Then, find the sum as shown in Figure 8.28.

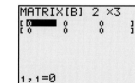
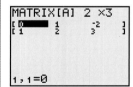
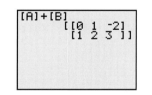

Matrix *A*   Matrix *B*
Figure 8.27   **Figure 8.28**

✓ *Checkpoint* Now try Exercise 17.

In operations with matrices, numbers are usually referred to as **scalars.** In this text, scalars will always be real numbers. You can multiply a matrix A by a scalar c by multiplying each entry in A by c.

Definition of Scalar Multiplication

If $A = [a_{ij}]$ is an $m \times n$ matrix and c is a scalar, the **scalar multiple** of A by c is the $m \times n$ matrix given by

$$cA = [ca_{ij}].$$

The symbol $-A$ represents the negation of A, which is the scalar product $(-1)A$. Moreover, if A and B are of the same order, then $A - B$ represents the sum of A and $(-1)B$. That is,

$$A - B = A + (-1)B.$$ Subtraction of matrices

Example 3   Scalar Multiplication and Matrix Subtraction

For the following matrices, find (a) $3A$, (b) $-B$, and (c) $3A - B$.

$$A = \begin{bmatrix} 2 & 2 & 4 \\ -3 & 0 & -1 \\ 2 & 1 & 2 \end{bmatrix} \quad \text{and} \quad B = \begin{bmatrix} 2 & 0 & 0 \\ 1 & -4 & 3 \\ -1 & 3 & 2 \end{bmatrix}$$

Solution

a. $3A = 3\begin{bmatrix} 2 & 2 & 4 \\ -3 & 0 & -1 \\ 2 & 1 & 2 \end{bmatrix}$ Scalar multiplication

$$= \begin{bmatrix} 3(2) & 3(2) & 3(4) \\ 3(-3) & 3(0) & 3(-1) \\ 3(2) & 3(1) & 3(2) \end{bmatrix}$$ Multiply each entry by 3.

$$= \begin{bmatrix} 6 & 6 & 12 \\ -9 & 0 & -3 \\ 6 & 3 & 6 \end{bmatrix}$$ Simplify.

b. $-B = (-1)\begin{bmatrix} 2 & 0 & 0 \\ 1 & -4 & 3 \\ -1 & 3 & 2 \end{bmatrix}$ Definition of negation

$$= \begin{bmatrix} -2 & 0 & 0 \\ -1 & 4 & -3 \\ 1 & -3 & -2 \end{bmatrix}$$ Multiply each entry by -1.

c. $3A - B = \begin{bmatrix} 6 & 6 & 12 \\ -9 & 0 & -3 \\ 6 & 3 & 6 \end{bmatrix} - \begin{bmatrix} 2 & 0 & 0 \\ 1 & -4 & 3 \\ -1 & 3 & 2 \end{bmatrix}$ Matrix subtraction

$$= \begin{bmatrix} 4 & 6 & 12 \\ -10 & 4 & -6 \\ 7 & 0 & 4 \end{bmatrix}$$ Subtract corresponding entries.

✓ *Checkpoint* Now try Exercises 7(b), (c), and (d).

It is often convenient to rewrite the scalar multiple cA by factoring c out of every entry in the matrix. For instance, in the following example, the scalar $\frac{1}{2}$ has been factored out of the matrix.

$$\begin{bmatrix} \frac{1}{2} & -\frac{3}{2} \\ \frac{5}{2} & \frac{1}{2} \end{bmatrix} = \begin{bmatrix} \frac{1}{2}(1) & \frac{1}{2}(-3) \\ \frac{1}{2}(5) & \frac{1}{2}(1) \end{bmatrix} = \frac{1}{2}\begin{bmatrix} 1 & -3 \\ 5 & 1 \end{bmatrix}$$

STUDY TIP

The order of operations for matrix expressions is similar to that for real numbers. In particular, you perform scalar multiplication before matrix addition and subtraction, as shown in Example 3(c).

Exploration

What do you observe about the relationship between the corresponding entries of A and B below? Use a graphing utility to find $A + B$. What conclusion can you make about the entries of A and B and the sum of $A + B$?

$$A = \begin{bmatrix} -1 & 5 \\ 2 & -6 \end{bmatrix}$$

$$B = \begin{bmatrix} 1 & -5 \\ -2 & 6 \end{bmatrix}$$

Example 4   Scalar Multiplication and Matrix Subtraction

For the following matrices, use a graphing utility to find $2A - 4B$.

$$A = \begin{bmatrix} -1 & 8 \\ 6 & 2 \end{bmatrix} \quad \text{and} \quad B = \begin{bmatrix} 0 & 4 \\ 5 & -3 \end{bmatrix}$$

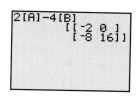

Figure 8.29

Solution

Use the matrix editor to enter A and B into the graphing utility. Then, find $2A - 4B$ as shown in Figure 8.29

✓ *Checkpoint*   Now try Exercise 19.

The properties of matrix addition and scalar multiplication are similar to those of addition and multiplication of real numbers.

Properties of Matrix Addition and Scalar Multiplication

Let A, B, and C be $m \times n$ matrices and let c and d be scalars.

1. $A + B = B + A$ Commutative Property of Matrix Addition
2. $A + (B + C) = (A + B) + C$ Associative Property of Matrix Addition
3. $(cd)A = c(dA)$ Associative Property of Scalar Multiplication
4. $1A = A$ Scalar Identity
5. $c(A + B) = cA + cB$ Distributive Property
6. $(c + d)A = cA + dA$ Distributive Property

STUDY TIP

Note that the Associative Property of Matrix Addition allows you to write expressions such as $A + B + C$ without ambiguity because the same sum occurs no matter how the matrices are grouped. This same reasoning applies to sums of four or more matrices.

Example 5   Addition of More Than Two Matrices

By adding corresponding entries, you obtain the following sum of four matrices.

$$\begin{bmatrix} 1 \\ 2 \\ -3 \end{bmatrix} + \begin{bmatrix} -1 \\ -1 \\ 2 \end{bmatrix} + \begin{bmatrix} 0 \\ 1 \\ 4 \end{bmatrix} + \begin{bmatrix} 2 \\ -3 \\ -2 \end{bmatrix} = \begin{bmatrix} 2 \\ -1 \\ 1 \end{bmatrix}$$

✓ *Checkpoint*   Now try Exercise 13.

Example 6   Using the Distributive Property

$$3\left(\begin{bmatrix} -2 & 0 \\ 4 & 1 \end{bmatrix} + \begin{bmatrix} 4 & -2 \\ 3 & 7 \end{bmatrix} \right) = 3\begin{bmatrix} -2 & 0 \\ 4 & 1 \end{bmatrix} + 3\begin{bmatrix} 4 & -2 \\ 3 & 7 \end{bmatrix}$$

$$= \begin{bmatrix} -6 & 0 \\ 12 & 3 \end{bmatrix} + \begin{bmatrix} 12 & -6 \\ 9 & 21 \end{bmatrix} = \begin{bmatrix} 6 & -6 \\ 21 & 24 \end{bmatrix}$$

STUDY TIP

In Example 6, you could add the two matrices first and then multiply the resulting matrix by 3. The result would be the same.

✓ *Checkpoint*   Now try Exercise 15.

One important property of addition of real numbers is that the number 0 is the additive identity. That is, $c + 0 = c$ for any real number c. For matrices, a similar property holds. That is, if A is an $m \times n$ matrix and O is the $m \times n$ **zero matrix** consisting entirely of zeros, then $A + O = A$.

In other words, O is the **additive identity** for the set of all $m \times n$ matrices. For example, the following matrices are the additive identities for the sets of all 2×3 and 2×2 matrices.

$$O = \begin{bmatrix} 0 & 0 & 0 \\ 0 & 0 & 0 \end{bmatrix} \quad \text{and} \quad O = \begin{bmatrix} 0 & 0 \\ 0 & 0 \end{bmatrix}$$

2×3 zero matrix $\qquad\qquad$ 2×2 zero matrix

The algebra of real numbers and the algebra of matrices have many similarities. For example, compare the following solutions.

Real Numbers (Solve for x.)	$m \times n$ Matrices (Solve for X.)
$x + a = b$	$X + A = B$
$x + a + (-a) = b + (-a)$	$X + A + (-A) = B + (-A)$
$x + 0 = b - a$	$X + O = B - A$
$x = b - a$	$X = B - A$

The algebra of real numbers and the algebra of matrices also have important differences, which will be discussed later.

Example 7 Solving a Matrix Equation

Solve for X in the equation $3X + A = B$, where

$$A = \begin{bmatrix} 1 & -2 \\ 0 & 3 \end{bmatrix} \quad \text{and} \quad B = \begin{bmatrix} -3 & 4 \\ 2 & 1 \end{bmatrix}.$$

Solution

Begin by solving the equation for X to obtain

$$3X = B - A$$

$$X = \frac{1}{3}(B - A).$$

Now, using the matrices A and B, you have

$$X = \frac{1}{3}\left(\begin{bmatrix} -3 & 4 \\ 2 & 1 \end{bmatrix} - \begin{bmatrix} 1 & -2 \\ 0 & 3 \end{bmatrix} \right) \qquad \text{Substitute the matrices.}$$

$$= \frac{1}{3}\begin{bmatrix} -4 & 6 \\ 2 & -2 \end{bmatrix} \qquad \text{Subtract matrix } A \text{ from matrix } B.$$

$$= \begin{bmatrix} -\frac{4}{3} & 2 \\ \frac{2}{3} & -\frac{2}{3} \end{bmatrix}. \qquad \text{Multiply the resulting matrix by } \frac{1}{3}.$$

✓ *Checkpoint* Now try Exercise 25.

Matrix Multiplication

The third basic matrix operation is **matrix multiplication.** At first glance, the following definition may seem unusual. You will see later, however, that this definition of the product of two matrices has many practical applications.

> **Definition of Matrix Multiplication**
>
> If $A = [a_{ij}]$ is an $m \times n$ matrix and $B = [b_{ij}]$ is an $n \times p$ matrix, the product AB is an $m \times p$ matrix given by
>
> $$AB = [c_{ij}]$$
>
> where $c_{ij} = a_{i1}b_{1j} + a_{i2}b_{2j} + a_{i3}b_{3j} + \cdots + a_{in}b_{nj}$.

The definition of matrix multiplication indicates a *row-by-column* multiplication, where the entry in the ith row and jth column of the product AB is obtained by multiplying the entries in the ith row of A by the corresponding entries in the jth column of B and then adding the results. The general pattern for matrix multiplication is as follows.

$$\begin{bmatrix} a_{11} & a_{12} & a_{13} & \cdots & a_{1n} \\ a_{21} & a_{22} & a_{23} & \cdots & a_{2n} \\ a_{31} & a_{32} & a_{33} & \cdots & a_{3n} \\ \vdots & \vdots & \vdots & & \vdots \\ a_{i1} & a_{i2} & a_{i3} & \cdots & a_{in} \\ \vdots & \vdots & \vdots & & \vdots \\ a_{m1} & a_{m2} & a_{m3} & \cdots & a_{mn} \end{bmatrix} \begin{bmatrix} b_{11} & b_{12} & \cdots & b_{1j} & \cdots & b_{1p} \\ b_{21} & b_{22} & \cdots & b_{2j} & \cdots & b_{2p} \\ b_{31} & b_{32} & \cdots & b_{3j} & \cdots & b_{3p} \\ \vdots & \vdots & & \vdots & & \vdots \\ b_{n1} & b_{n2} & \cdots & b_{nj} & \cdots & b_{np} \end{bmatrix} = \begin{bmatrix} c_{11} & c_{12} & \cdots & c_{1j} & \cdots & c_{1p} \\ c_{21} & c_{22} & \cdots & c_{2j} & \cdots & c_{2p} \\ \vdots & \vdots & & \vdots & & \vdots \\ c_{i1} & c_{i2} & \cdots & c_{ij} & \cdots & c_{ip} \\ \vdots & \vdots & & \vdots & & \vdots \\ c_{m1} & c_{m2} & \cdots & c_{mj} & \cdots & c_{mp} \end{bmatrix}$$

$$a_{i1}b_{1j} + a_{i2}b_{2j} + a_{i3}b_{3j} + \cdots + a_{in}b_{nj} = c_{ij}$$

Example 8 Finding the Product of Two Matrices

Find the product AB using $A = \begin{bmatrix} -1 & 3 \\ 4 & -2 \\ 5 & 0 \end{bmatrix}$ and $B = \begin{bmatrix} -3 & 2 \\ -4 & 1 \end{bmatrix}$.

Solution

First, note that the product AB is defined because the number of columns of A is equal to the number of rows of B. Moreover, the product AB has order 3×2. To find the entries of the product, multiply each row of A by each column of B.

$$AB = \begin{bmatrix} -1 & 3 \\ 4 & -2 \\ 5 & 0 \end{bmatrix} \begin{bmatrix} -3 & 2 \\ -4 & 1 \end{bmatrix}$$

$$= \begin{bmatrix} (-1)(-3) + (3)(-4) & (-1)(2) + (3)(1) \\ (4)(-3) + (-2)(-4) & (4)(2) + (-2)(1) \\ (5)(-3) + (0)(-4) & (5)(2) + (0)(1) \end{bmatrix} = \begin{bmatrix} -9 & 1 \\ -4 & 6 \\ -15 & 10 \end{bmatrix}$$

✓ *Checkpoint* Now try Exercise 29.

Be sure you understand that for the product of two matrices to be defined, the number of *columns* of the first matrix must equal the number of *rows* of the second matrix. That is, the middle two indices must be the same. The outside two indices give the order of the product, as shown in the following diagram.

$$\begin{array}{ccccc} A & \times & B & = & AB \\ m \times n & & n \times p & & m \times p \end{array}$$

Equal
Order of AB

Example 9   Matrix Multiplication

a. $\begin{bmatrix} 1 & 0 & 3 \\ 2 & -1 & -2 \end{bmatrix} \begin{bmatrix} -2 & 4 & 2 \\ 1 & 0 & 0 \\ -1 & 1 & -1 \end{bmatrix} = \begin{bmatrix} -5 & 7 & -1 \\ -3 & 6 & 6 \end{bmatrix}$

$\qquad 2 \times 3 \qquad\qquad 3 \times 3 \qquad\qquad 2 \times 3$

b. $\begin{bmatrix} 3 & 4 \\ -2 & 5 \end{bmatrix} \begin{bmatrix} 1 & 0 \\ 0 & 1 \end{bmatrix} = \begin{bmatrix} 3 & 4 \\ -2 & 5 \end{bmatrix}$

$\quad 2 \times 2 \qquad 2 \times 2 \qquad 2 \times 2$

c. $\begin{bmatrix} 1 & 2 \\ 1 & 1 \end{bmatrix} \begin{bmatrix} -1 & 2 \\ 1 & -1 \end{bmatrix} = \begin{bmatrix} 1 & 0 \\ 0 & 1 \end{bmatrix}$

$\quad 2 \times 2 \qquad 2 \times 2 \qquad 2 \times 2$

d. $\begin{bmatrix} 1 & -2 & -3 \end{bmatrix} \begin{bmatrix} 2 \\ -1 \\ 1 \end{bmatrix} = \begin{bmatrix} 1 \end{bmatrix}$

$\quad 1 \times 3 \qquad 3 \times 1 \quad 1 \times 1$

e. $\begin{bmatrix} 2 \\ -1 \\ 1 \end{bmatrix} \begin{bmatrix} 1 & -2 & -3 \end{bmatrix} = \begin{bmatrix} 2 & -4 & -6 \\ -1 & 2 & 3 \\ 1 & -2 & -3 \end{bmatrix}$

$\quad 3 \times 1 \qquad 1 \times 3 \qquad\qquad 3 \times 3$

f. The product *AB* for the following matrices is not defined.

$$A = \begin{bmatrix} -2 & 1 \\ 1 & -3 \\ 1 & 4 \end{bmatrix} \quad \text{and} \quad B = \begin{bmatrix} -2 & 3 & 1 & 4 \\ 0 & 1 & -1 & 2 \\ 2 & -1 & 0 & 1 \end{bmatrix}$$

$\qquad\qquad 3 \times 2 \qquad\qquad\qquad 3 \times 4$

✓ *Checkpoint* Now try Exercise 31.

In parts (d) and (e) of Example 9, note that the two products are different. Matrix multiplication is not, in general, commutative. That is, for most matrices, $AB \neq BA$. This is one way in which the algebra of real numbers and the algebra of matrices differ.

Exploration

Use the following matrices to find *AB*, *BA*, (*AB*)*C*, and *A*(*BC*). What do your results tell you about matrix multiplication and commutativity and associativity?

$$A = \begin{bmatrix} 1 & 2 \\ 3 & 4 \end{bmatrix},$$

$$B = \begin{bmatrix} 0 & 1 \\ 2 & 3 \end{bmatrix},$$

$$C = \begin{bmatrix} 3 & 0 \\ 0 & 1 \end{bmatrix}$$

Example 10   Matrix Multiplication

Use a graphing utility to find the product AB using

$$A = \begin{bmatrix} 1 & 2 & 3 \\ 2 & -5 & 1 \end{bmatrix} \quad \text{and} \quad B = \begin{bmatrix} -3 & 2 & 1 \\ 4 & -2 & 0 \\ 1 & 2 & 3 \end{bmatrix}.$$

Solution

Note that the order of A is 2×3 and the order of B is 3×3. So, the product will have order 2×3. Use the matrix editor to enter A and B into the graphing utility. Then, find the product as shown in Figure 8.30.

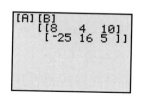

Figure 8.30

✓ *Checkpoint* Now try Exercise 41.

Properties of Matrix Multiplication

Let A, B, and C be matrices and let c be a scalar.

1. $A(BC) = (AB)C$ Associative Property of Matrix Multiplication

2. $A(B + C) = AB + AC$ Left Distributive Property

3. $(A + B)C = AC + BC$ Right Distributive Property

4. $c(AB) = (cA)B = A(cB)$ Associative Property of Scalar Multiplication

---

Definition of Identity Matrix

The $n \times n$ matrix that consists of 1's on its main diagonal and 0's elsewhere is called the **identity matrix of order n** and is denoted by

$$I_n = \begin{bmatrix} 1 & 0 & 0 & \dots & 0 \\ 0 & 1 & 0 & \dots & 0 \\ 0 & 0 & 1 & \dots & 0 \\ \vdots & \vdots & \vdots & & \vdots \\ 0 & 0 & 0 & \dots & 1 \end{bmatrix}. \quad \text{Identity matrix}$$

Note that an identity matrix must be *square*. When the order is understood to be n, you can denote I_n simply by I.

If A is an $n \times n$ matrix, the identity matrix has the property that $AI_n = A$ and $I_nA = A$. For example,

$$\begin{bmatrix} 3 & -2 & 5 \\ 1 & 0 & 4 \\ -1 & 2 & -3 \end{bmatrix}\begin{bmatrix} 1 & 0 & 0 \\ 0 & 1 & 0 \\ 0 & 0 & 1 \end{bmatrix} = \begin{bmatrix} 3 & -2 & 5 \\ 1 & 0 & 4 \\ -1 & 2 & -3 \end{bmatrix} \quad AI = A$$

and

$$\begin{bmatrix} 1 & 0 & 0 \\ 0 & 1 & 0 \\ 0 & 0 & 1 \end{bmatrix}\begin{bmatrix} 3 & -2 & 5 \\ 1 & 0 & 4 \\ -1 & 2 & -3 \end{bmatrix} = \begin{bmatrix} 3 & -2 & 5 \\ 1 & 0 & 4 \\ -1 & 2 & -3 \end{bmatrix}. \quad IA = A$$

Applications

Matrix multiplication can be used to represent a system of linear equations. Note how the system

$$\begin{cases} a_{11}x_1 + a_{12}x_2 + a_{13}x_3 = b_1 \\ a_{21}x_1 + a_{22}x_2 + a_{23}x_3 = b_2 \\ a_{31}x_1 + a_{32}x_2 + a_{33}x_3 = b_3 \end{cases}$$

can be written as the matrix equation $AX = B$, where A is the *coefficient matrix* of the system and X and B are column matrices.

$$\underset{A}{\begin{bmatrix} a_{11} & a_{12} & a_{13} \\ a_{21} & a_{22} & a_{23} \\ a_{31} & a_{32} & a_{33} \end{bmatrix}} \underset{X}{\begin{bmatrix} x_1 \\ x_2 \\ x_3 \end{bmatrix}} = \underset{B}{\begin{bmatrix} b_1 \\ b_2 \\ b_3 \end{bmatrix}}$$

$$A \quad \times \quad X \quad = \quad B$$

Example 11 Solving a System of Linear Equations

Consider the system of linear equations $\begin{cases} x_1 - 2x_2 + x_3 = -4 \\ x_2 + 2x_3 = 4. \\ 2x_1 + 3x_2 - 2x_3 = 2 \end{cases}$

a. Write this system as a matrix equation $AX = B$.

b. Use Gauss-Jordan elimination on $[A \vdots B]$ to solve for the matrix X.

Solution

a. In matrix form $AX = B$, the system can be written as follows.

$$\begin{bmatrix} 1 & -2 & 1 \\ 0 & 1 & 2 \\ 2 & 3 & -2 \end{bmatrix} \begin{bmatrix} x_1 \\ x_2 \\ x_3 \end{bmatrix} = \begin{bmatrix} -4 \\ 4 \\ 2 \end{bmatrix}$$

b. The augmented matrix is

$$[A \vdots B] = \begin{bmatrix} 1 & -2 & 1 & \vdots & -4 \\ 0 & 1 & 2 & \vdots & 4 \\ 2 & 3 & -2 & \vdots & 2 \end{bmatrix}.$$

Using Gauss-Jordan elimination, you can rewrite this equation as

$$[I \vdots X] = \begin{bmatrix} 1 & 0 & 0 & \vdots & -1 \\ 0 & 1 & 0 & \vdots & 2 \\ 0 & 0 & 1 & \vdots & 1 \end{bmatrix}.$$

So, the solution of the system of linear equations is $x_1 = -1$, $x_2 = 2$, and $x_3 = 1$. The solution of the matrix equation is

$$X = \begin{bmatrix} x_1 \\ x_2 \\ x_3 \end{bmatrix} = \begin{bmatrix} -1 \\ 2 \\ 1 \end{bmatrix}.$$

✓ *Checkpoint* Now try Exercise 59.

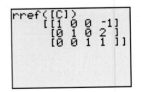

Example 12   Health Care

A company offers three types of health care plans with two levels of coverage to its employees. The current annual costs for these plans are represented by the matrix A. If the annual costs are expected to increase by 4% next year, what will be the annual costs for each plan next year?

$$A = \begin{bmatrix} 694 & 451 & 489 \\ 1725 & 1187 & 1248 \end{bmatrix} \begin{matrix} \text{Single} \\ \text{Family} \end{matrix}$$

Plan: Premium, HMO, HMO Plus; Coverage level: Single, Family

Solution

Because an increase of 4% corresponds to 100% + 4%, multiply A by 104% or 1.04. So, the annual costs for each health care plan next year are as follows.

$$1.04A = 1.04 \begin{bmatrix} 694 & 451 & 489 \\ 1725 & 1187 & 1248 \end{bmatrix} = \begin{bmatrix} 722 & 469 & 509 \\ 1794 & 1234 & 1298 \end{bmatrix} \begin{matrix} \text{Single} \\ \text{Family} \end{matrix}$$

Plan: Premium, HMO, HMO Plus; Coverage level: Single, Family

✓ *Checkpoint* Now try Exercise 65.

Example 13   Softball Team Expenses

Two softball teams submit equipment lists to their sponsors, as shown in the table at the right. Each bat costs $80, each ball costs $6, and each glove costs $60. Use matrices to find the total cost of equipment for each team.

Equipment	Women's Team	Men's Team
Bats	12	15
Balls	45	38
Gloves	15	17

Solution

The equipment lists E and the costs per item C can be written in matrix form as

$$E = \begin{bmatrix} 12 & 15 \\ 45 & 38 \\ 15 & 17 \end{bmatrix} \quad \text{and} \quad C = \begin{bmatrix} 80 & 6 & 60 \end{bmatrix}.$$

You can find the total cost of the equipment for each team using the product CE because the number of columns of C (3 columns) equals the number of rows of E (3 rows). Therefore, the total cost of equipment for each team is given by

$$CE = \begin{bmatrix} 80 & 6 & 60 \end{bmatrix} \begin{bmatrix} 12 & 15 \\ 45 & 38 \\ 15 & 17 \end{bmatrix}$$

$$= \begin{bmatrix} 80(12) + 6(45) + 60(15) & 80(15) + 6(38) + 60(17) \end{bmatrix}$$

$$= \begin{bmatrix} 2130 & 2448 \end{bmatrix}.$$

So, the total cost of equipment for the women's team is $2130, and the total cost of equipment for the men's team is $2448.

✓ *Checkpoint* Now try Exercise 67.

STUDY TIP

Notice in Example 13 that you cannot find the total cost using the product EC because EC is not defined. That is, the number of columns of E (2 columns) does not equal the number of rows of C (1 row).

8.5   Exercises

Vocabulary Check

In Exercises 1–4, fill in the blanks.

1. Two matrices are _____ if all of their corresponding entries are equal.

2. When working with matrices, real numbers are often referred to as _____ .

3. A matrix consisting entirely of zeros is called a _____ matrix and is denoted by _____ .

4. The $n \times n$ matrix consisting of 1's on its main diagonal and 0's elsewhere is called the _____ matrix of order n.

In Exercises 5 and 6, match the matrix property with the correct form. A, B, and C are matrices, and c and d are scalars.

5. (a) $(cd)A = c(dA)$

    (b) $A + B = B + A$

    (c) $1A = A$

    (d) $c(A + B) = cA + cB$

    (e) $A + (B + C) = (A + B) + C$

    (i) Commutative Property of Matrix Addition

    (ii) Associative Property of Matrix Addition

    (iii) Associative Property of Scalar Multiplication

    (iv) Scalar Identity

    (v) Distributive Property

6. (a) $A(B + C) = AB + AC$

    (b) $c(AB) = (cA)B = A(cB)$

    (c) $A(BC) = (AB)C$

    (d) $(A + B)C = AC + BC$

    (i) Associative Property of Matrix Multiplication

    (ii) Left Distributive Property

    (iii) Right Distributive Property

    (iv) Associative Property of Scalar Multiplication

In Exercises 1–4, find x, y, and z.

1. $\begin{bmatrix} x & -2 \\ 7 & y \end{bmatrix} = \begin{bmatrix} -4 & -2 \\ 7 & 22 \end{bmatrix}$

2. $\begin{bmatrix} -5 & x \\ y & 8 \end{bmatrix} = \begin{bmatrix} -5 & 13 \\ 12 & 8 \end{bmatrix}$

3. $\begin{bmatrix} 16 & 4 & 5 & 4 \\ -3 & 13 & 15 & 12 \\ 0 & 2 & 4 & 0 \end{bmatrix} = \begin{bmatrix} 16 & 4 & 2x + 7 & 4 \\ -3 & 13 & & 15 & 3y \\ 0 & 2 & 3z - 14 & 0 \end{bmatrix}$

4. $\begin{bmatrix} x + 4 & 8 & -3 \\ 1 & 22 & 2y \\ 7 & -2 & z + 2 \end{bmatrix} = \begin{bmatrix} 2x + 9 & 8 & -3 \\ 1 & 22 & -8 \\ 7 & -2 & 11 \end{bmatrix}$

In Exercises 5–12, find, if possible, (a) $A + B$, (b) $A - B$, (c) $3A$, and (d) $3A - 2B$. Use the matrix capabilities of a graphing utility to verify your results.

5. $A = \begin{bmatrix} 1 & -1 \\ 2 & -1 \end{bmatrix}$, $B = \begin{bmatrix} 2 & -1 \\ -1 & 8 \end{bmatrix}$

6. $A = \begin{bmatrix} 1 & 2 \\ 2 & 1 \end{bmatrix}$, $B = \begin{bmatrix} -3 & -2 \\ 4 & 2 \end{bmatrix}$

7. $A = \begin{bmatrix} 8 & -1 \\ 2 & 3 \\ -4 & 5 \end{bmatrix}$, $B = \begin{bmatrix} 1 & 6 \\ -1 & -5 \\ 1 & 10 \end{bmatrix}$

8. $A = \begin{bmatrix} 2 & 3 & 1 \\ 0 & -1 & 4 \end{bmatrix}$, $B = \begin{bmatrix} 5 & 0 & 7 \\ -4 & 1 & -2 \end{bmatrix}$

9. $A = \begin{bmatrix} 4 & 5 & -1 & 3 & 4 \\ 1 & 2 & -2 & -1 & 0 \end{bmatrix}$,

    $B = \begin{bmatrix} 1 & 0 & -1 & 1 & 0 \\ -6 & 8 & 2 & -3 & -7 \end{bmatrix}$

10. $A = \begin{bmatrix} -1 & 4 & 0 \\ 3 & -2 & 2 \\ 5 & 4 & -1 \\ 0 & 8 & -6 \\ -4 & -1 & 0 \end{bmatrix}$, $B = \begin{bmatrix} -3 & 5 & 1 \\ 2 & -4 & -7 \\ 10 & -9 & -1 \\ 3 & 2 & -4 \\ 0 & 1 & -2 \end{bmatrix}$

11. $A = \begin{bmatrix} 6 & 0 & 3 \\ -1 & -4 & 0 \end{bmatrix}$, $B = \begin{bmatrix} 8 & -1 \\ 4 & -3 \end{bmatrix}$

12. $A = \begin{bmatrix} 3 \\ 2 \\ -1 \end{bmatrix}$, $B = \begin{bmatrix} -4 & 6 & 2 \end{bmatrix}$

In Exercises 13–16, evaluate the expression.

13. $\begin{bmatrix} -5 & 0 \\ 3 & -6 \end{bmatrix} + \begin{bmatrix} 7 & 1 \\ -2 & -1 \end{bmatrix} + \begin{bmatrix} -10 & -8 \\ 14 & 6 \end{bmatrix}$

14. $\begin{bmatrix} 6 & 9 \\ -1 & 0 \\ 7 & 1 \end{bmatrix} + \begin{bmatrix} 0 & 5 \\ -2 & -1 \\ 3 & -6 \end{bmatrix} + \begin{bmatrix} -13 & -7 \\ 4 & -1 \\ -6 & 0 \end{bmatrix}$

15. $4\left(\begin{bmatrix} -4 & 0 & 1 \\ 0 & 2 & 3 \end{bmatrix} - \begin{bmatrix} 2 & 1 & -2 \\ 3 & -6 & 0 \end{bmatrix} \right)$

16. $\frac{1}{2}([5 \quad -2 \quad 4 \quad 0] + [14 \quad 6 \quad -18 \quad 9])$

In Exercises 17–22, use the matrix capabilities of a graphing utility to evaluate the expression. Round your results to three decimal places, if necessary.

17. $\begin{bmatrix} 2 & 5 \\ -1 & -4 \end{bmatrix} + \begin{bmatrix} -3 & 0 \\ 2 & 2 \end{bmatrix}$

18. $\begin{bmatrix} 14 & -11 \\ -22 & 19 \end{bmatrix} + \begin{bmatrix} -22 & 20 \\ 13 & 6 \end{bmatrix}$

19. $-\frac{1}{2} \begin{bmatrix} 3.211 & 6.829 \\ -1.004 & 4.914 \\ 0.055 & -3.889 \end{bmatrix} - 8 \begin{bmatrix} 1.630 & -3.090 \\ 5.256 & 8.335 \\ -9.768 & 4.251 \end{bmatrix}$

20. $-12\left(\begin{bmatrix} 6 & 20 \\ 1 & -9 \\ -2 & 5 \end{bmatrix} + \begin{bmatrix} 14 & -15 \\ -8 & -6 \\ 7 & 0 \end{bmatrix} + \begin{bmatrix} -31 & -19 \\ 16 & 10 \\ 24 & 10 \end{bmatrix} \right)$

21. $-3\left(\begin{bmatrix} 0 & -3 \\ 7 & 2 \end{bmatrix} + \begin{bmatrix} -6 & 3 \\ 8 & 1 \end{bmatrix} \right) - \begin{bmatrix} 4 & -4 \\ 7 & -9 \end{bmatrix}$

22. $-1\begin{bmatrix} 4 & 11 \\ -2 & -1 \\ 9 & 3 \end{bmatrix} + \frac{1}{6}\left(\begin{bmatrix} -5 & -1 \\ 3 & 4 \\ 0 & 13 \end{bmatrix} + \begin{bmatrix} 7 & 5 \\ -9 & -1 \\ 6 & -1 \end{bmatrix} \right)$

In Exercises 23–26, solve for X when

$$A = \begin{bmatrix} -2 & -1 \\ 1 & 0 \\ 3 & -4 \end{bmatrix} \quad \text{and} \quad B = \begin{bmatrix} 0 & 3 \\ 2 & 0 \\ -4 & -1 \end{bmatrix}.$$

23. $X = 3A - 2B$

24. $2X = 2A - B$

25. $2X + 3A = B$

26. $2A + 4B = -2X$

In Exercises 27–34, find AB, if possible.

27. $A = \begin{bmatrix} 2 & 1 \\ -3 & 4 \\ -1 & 6 \end{bmatrix}$, $B = \begin{bmatrix} 0 & -3 & 0 \\ 4 & 0 & 2 \\ 8 & -2 & 7 \end{bmatrix}$

28. $A = \begin{bmatrix} 0 & -1 & 2 \\ 6 & 0 & 3 \\ 7 & -1 & 8 \end{bmatrix}$, $B = \begin{bmatrix} 2 & -1 \\ 4 & -5 \\ 1 & 6 \end{bmatrix}$

29. $A = \begin{bmatrix} -1 & 6 \\ -4 & 5 \\ 0 & 3 \end{bmatrix}$, $B = \begin{bmatrix} 2 & 3 \\ 0 & 9 \end{bmatrix}$

30. $A = \begin{bmatrix} 1 & 0 & 0 \\ 0 & 4 & 0 \\ 0 & 0 & -2 \end{bmatrix}$, $B = \begin{bmatrix} 3 & 0 & 0 \\ 0 & -1 & 0 \\ 0 & 0 & 5 \end{bmatrix}$

31. $A = \begin{bmatrix} 5 & 0 & 0 \\ 0 & -8 & 0 \\ 0 & 0 & 7 \end{bmatrix}$, $B = \begin{bmatrix} \frac{1}{5} & 0 & 0 \\ 0 & -\frac{1}{8} & 0 \\ 0 & 0 & \frac{1}{2} \end{bmatrix}$

32. $A = \begin{bmatrix} 0 & 0 & 5 \\ 0 & 0 & -3 \\ 0 & 0 & 4 \end{bmatrix}$, $B = \begin{bmatrix} 6 & -11 & 4 \\ 8 & 16 & 4 \\ 0 & 0 & 0 \end{bmatrix}$

33. $A = \begin{bmatrix} 5 \\ 6 \end{bmatrix}$, $B = [-3 \quad -1 \quad -5 \quad -9]$

34. $A = \begin{bmatrix} 1 & 0 & 3 & -2 \\ 6 & 13 & 8 & -17 \end{bmatrix}$, $B = \begin{bmatrix} 1 & 6 \\ 4 & 2 \end{bmatrix}$

In Exercises 35–40, find, if possible, (a) AB, (b) BA, and (c) A^2. (Note: $A^2 = AA$.) Use the matrix capabilities of a graphing utility to verify your results.

35. $A = \begin{bmatrix} 1 & 2 \\ 5 & 2 \end{bmatrix}$, $B = \begin{bmatrix} 2 & -1 \\ -1 & 8 \end{bmatrix}$

36. $A = \begin{bmatrix} 6 & 3 \\ -2 & -4 \end{bmatrix}$, $B = \begin{bmatrix} -2 & 0 \\ 2 & 4 \end{bmatrix}$

37. $A = \begin{bmatrix} 3 & -1 \\ 1 & 3 \end{bmatrix}$, $B = \begin{bmatrix} 1 & -3 \\ 3 & 1 \end{bmatrix}$

38. $A = \begin{bmatrix} 1 & -1 \\ 1 & 1 \end{bmatrix}$, $B = \begin{bmatrix} 1 & 3 \\ -3 & 1 \end{bmatrix}$

39. $A = \begin{bmatrix} 7 \\ 8 \\ -1 \end{bmatrix}$, $B = [1 \quad 1 \quad 2]$

40. $A = [3 \quad 2 \quad 1]$, $B = \begin{bmatrix} 2 \\ 3 \\ 0 \end{bmatrix}$

In Exercises 41–46, use the matrix capabilities of a graphing utility to find AB, if possible.

41. $A = \begin{bmatrix} 7 & 5 & -4 \\ -2 & 5 & 1 \\ 10 & -4 & -7 \end{bmatrix}$, $B = \begin{bmatrix} 2 & -2 & 3 \\ 8 & 1 & 4 \\ -4 & 2 & -8 \end{bmatrix}$

42. $A = \begin{bmatrix} 1 & -12 & 4 \\ 14 & 10 & 12 \\ 6 & -15 & 3 \end{bmatrix}$, $B = \begin{bmatrix} 12 & 10 \\ -6 & 12 \\ 10 & 16 \end{bmatrix}$

43. $A = \begin{bmatrix} -3 & 8 & -6 & 8 \\ -12 & 15 & 9 & 6 \\ 5 & -1 & 1 & 5 \end{bmatrix}$, $B = \begin{bmatrix} 3 & 1 & 6 \\ 24 & 15 & 14 \\ 16 & 10 & 21 \\ 8 & -4 & 10 \end{bmatrix}$

44. $A = \begin{bmatrix} -2 & 6 & 12 \\ 21 & -5 & 6 \\ 13 & -2 & 9 \end{bmatrix}$, $B = \begin{bmatrix} 3 & 0 \\ -7 & 18 \\ 34 & 14 \\ 0.5 & 1.4 \end{bmatrix}$

45. $A = \begin{bmatrix} 9 & 10 & -38 & 18 \\ 100 & -50 & 250 & 75 \end{bmatrix}$,

$B = \begin{bmatrix} 52 & -85 & 27 & 45 \\ 40 & -35 & 60 & 82 \end{bmatrix}$

46. $A = \begin{bmatrix} 16 & -18 \\ -4 & 13 \\ -9 & 21 \end{bmatrix}$, $B = \begin{bmatrix} -7 & 20 & -1 \\ 7 & 15 & 26 \end{bmatrix}$

In Exercises 47–50, use the matrix capabilities of a graphing utility to evaluate the expression.

47. $\begin{bmatrix} 3 & 1 \\ 0 & -2 \end{bmatrix} \begin{bmatrix} 1 & 0 \\ -2 & 2 \end{bmatrix} \begin{bmatrix} 1 & 0 \\ 2 & 4 \end{bmatrix}$

48. $-3 \left(\begin{bmatrix} 6 & 5 & -1 \\ 1 & -2 & 0 \end{bmatrix} \begin{bmatrix} 0 & 3 \\ -1 & -3 \\ 4 & 1 \end{bmatrix} \right)$

49. $\begin{bmatrix} 0 & 2 & -2 \\ 4 & 1 & 2 \end{bmatrix} \left(\begin{bmatrix} 4 & 0 \\ 0 & -1 \\ -1 & 2 \end{bmatrix} + \begin{bmatrix} -2 & 3 \\ -3 & 5 \\ 0 & -3 \end{bmatrix} \right)$

50. $\begin{bmatrix} 3 \\ -1 \\ 5 \\ 7 \end{bmatrix} ([5 \quad -6] + [7 \quad -1] + [-8 \quad 9])$

In Exercises 51–54, use matrix multiplication to determine whether each matrix is a solution of the system of equations. Use a graphing utility to verify your results.

51. $\begin{cases} x + 2y = 4 \\ 3x + 2y = 0 \end{cases}$

(a) $\begin{bmatrix} 2 \\ 1 \end{bmatrix}$ (b) $\begin{bmatrix} -2 \\ 3 \end{bmatrix}$

(c) $\begin{bmatrix} -4 \\ 4 \end{bmatrix}$ (d) $\begin{bmatrix} 2 \\ -3 \end{bmatrix}$

52. $\begin{cases} 6x + 2y = 0 \\ -x + 5y = 16 \end{cases}$

(a) $\begin{bmatrix} -1 \\ 3 \end{bmatrix}$ (b) $\begin{bmatrix} 2 \\ -6 \end{bmatrix}$

(c) $\begin{bmatrix} 3 \\ -9 \end{bmatrix}$ (d) $\begin{bmatrix} -3 \\ 9 \end{bmatrix}$

53. $\begin{cases} -2x - 3y = -6 \\ 4x + 2y = 20 \end{cases}$

(a) $\begin{bmatrix} 3 \\ 0 \end{bmatrix}$ (b) $\begin{bmatrix} 6 \\ -2 \end{bmatrix}$

(c) $\begin{bmatrix} -6 \\ 6 \end{bmatrix}$ (d) $\begin{bmatrix} 4 \\ 2 \end{bmatrix}$

54. $\begin{cases} 5x - 7y = -15 \\ 3x + y = 17 \end{cases}$

(a) $\begin{bmatrix} 4 \\ 5 \end{bmatrix}$ (b) $\begin{bmatrix} 5 \\ 2 \end{bmatrix}$

(c) $\begin{bmatrix} -4 \\ -5 \end{bmatrix}$ (d) $\begin{bmatrix} 2 \\ 11 \end{bmatrix}$

In Exercises 55–62, (a) write each system of equations as a matrix equation $AX = B$ and (b) use Gauss-Jordan elimination on the augmented matrix $[A \vdots B]$ to solve for the matrix X. Use a graphing utility to check your solution.

55. $\begin{cases} -x_1 + x_2 = 4 \\ -2x_1 + x_2 = 0 \end{cases}$

56. $\begin{cases} 2x_1 + 3x_2 = 5 \\ x_1 + 4x_2 = 10 \end{cases}$

57. $\begin{cases} -2x_1 - 3x_2 = -4 \\ 6x_1 + x_2 = -36 \end{cases}$

58. $\begin{cases} -4x_1 + 9x_2 = -13 \\ x_1 - 3x_2 = 12 \end{cases}$

59. $\begin{cases} x_1 - 2x_2 + 3x_3 = 9 \\ -x_1 + 3x_2 - x_3 = -6 \\ 2x_1 - 5x_2 + 5x_3 = 17 \end{cases}$

60. $\begin{cases} x_1 + x_2 - 3x_3 = -1 \\ -x_1 + 2x_2 = 1 \\ x_1 - x_2 + x_3 = 2 \end{cases}$

61. $\begin{cases} x_1 - 5x_2 + 2x_3 = -20 \\ -3x_1 + x_2 - x_3 = 8 \\ -2x_2 + 5x_3 = -16 \end{cases}$

62. $\begin{cases} x_1 - x_2 + 4x_3 = 17 \\ x_1 + 3x_2 = -11 \\ -6x_2 + 5x_3 = 40 \end{cases}$

In Exercises 63 and 64, use the matrix capabilities of a graphing utility to find $f(A) = a_0 I_n + a_1 A + a_2 A^2$.

63. $f(x) = x^2 - 5x + 2$, $A = \begin{bmatrix} 2 & 0 \\ 4 & 5 \end{bmatrix}$

64. $f(x) = x^2 - 7x + 6$, $A = \begin{bmatrix} 5 & 4 \\ 1 & 2 \end{bmatrix}$

65. Manufacturing   A corporation has three factories, each of which manufactures acoustic guitars and electric guitars. The number of units of guitars produced at factory j in one day is represented by a_{ij} in the matrix

$$A = \begin{bmatrix} 70 & 50 & 25 \\ 35 & 100 & 70 \end{bmatrix}.$$

Find the production levels if production is increased by 20%.

66. Manufacturing   A corporation has four factories, each of which manufactures sport utility vehicles and pickup trucks. The number of units of vehicle i produced at factory j in one day is represented by a_{ij} in the matrix

$$A = \begin{bmatrix} 100 & 90 & 70 & 30 \\ 40 & 20 & 60 & 60 \end{bmatrix}.$$

Find the production levels if production is increased by 10%.

67. Agriculture   A fruit grower raises two crops, apples and peaches. Each of these crops is shipped to three different outlets. The number of units of crop i that are shipped to outlet j is represented by a_{ij} in the matrix

$$A = \begin{bmatrix} 125 & 100 & 75 \\ 100 & 175 & 125 \end{bmatrix}.$$

The profit per unit is represented by the matrix

$$B = [\$3.50 \quad \$6.00].$$

Find the product BA and state what each entry of the product represents.

68. Revenue   A manufacturer produces three models of portable CD players, which are shipped to two warehouses. The number of units of model i that are shipped to warehouse j is represented by a_{ij} in the matrix

$$A = \begin{bmatrix} 5{,}000 & 4{,}000 \\ 6{,}000 & 10{,}000 \\ 8{,}000 & 5{,}000 \end{bmatrix}.$$

The price per unit is represented by the matrix

$$B = [\$39.50 \quad \$44.50 \quad \$56.50].$$

Compute BA and state what each entry of the product represents.

69. Inventory   A company sells five models of computers through three retail outlets. The inventories are given by S. The wholesale and retail prices are given by T. Compute ST and interpret the result.

Model

$$S = \begin{bmatrix} 3 & 2 & 2 & 3 & 0 \\ 0 & 2 & 3 & 4 & 3 \\ 4 & 2 & 1 & 3 & 2 \end{bmatrix} \begin{matrix} 1 \\ 2 \\ 3 \end{matrix} \text{ Outlet}$$

(column labels: A B C D E)

Price

$$T = \begin{bmatrix} \$840 & \$1100 \\ \$1200 & \$1350 \\ \$1450 & \$1650 \\ \$2650 & \$3000 \\ \$3050 & \$3200 \end{bmatrix} \begin{matrix} A \\ B \\ C \\ D \\ E \end{matrix} \text{ Model}$$

(column labels: Wholesale Retail)

70. Labor/Wage Requirements   A company that manufactures boats has the following labor-hour and wage requirements. Compute ST and interpret the result.

Labor per Boat

Department

$$S = \begin{bmatrix} 1.0 \text{ hr} & 0.5 \text{ hr} & 0.2 \text{ hr} \\ 1.6 \text{ hr} & 1.0 \text{ hr} & 0.2 \text{ hr} \\ 2.5 \text{ hr} & 2.0 \text{ hr} & 0.4 \text{ hr} \end{bmatrix} \begin{matrix} \text{Small} \\ \text{Medium} \\ \text{Large} \end{matrix} \begin{matrix} \text{Boat} \\ \text{Size} \end{matrix}$$

(column labels: Cutting Assembly Packaging)

Wages per Hour

Plant

$$T = \begin{bmatrix} \$12 & \$10 \\ \$9 & \$8 \\ \$6 & \$5 \end{bmatrix} \begin{matrix} \text{Cutting} \\ \text{Assembly} \\ \text{Packaging} \end{matrix} \text{ Department}$$

(column labels: A B)

71. Voting Preference   The matrix

From

$$P = \begin{bmatrix} 0.6 & 0.1 & 0.1 \\ 0.2 & 0.7 & 0.1 \\ 0.2 & 0.2 & 0.8 \end{bmatrix} \begin{matrix} R \\ D \\ I \end{matrix} \text{ To}$$

(column labels: R D I)

is called a *stochastic matrix*. Each entry $p_{ij} (i \neq j)$ represents the proportion of the voting population that changes from party i to party j, and p_{ii} represents the proportion that remains loyal to the party from one election to the next. Compute and interpret P^2.

72. *Voting Preference* Use a graphing utility to find P^3, P^4, P^5, P^6, P^7, and P^8 for the matrix given in Exercise 71. Can you detect a pattern as P is raised to higher powers?

Synthesis

True or False? **In Exercises 73 and 74, determine whether the statement is true or false. Justify your answer.**

73. Two matrices can be added only if they have the same order.

74. Matrix multiplication is commutative.

Think About It **In Exercises 75–82, let matrices A, B, C, and D be of orders 2×3, 2×3, 3×2, and 2×2, respectively. Determine whether the matrices are of proper order to perform the operation(s). If so, give the order of the answer.**

75. $A + 2C$

76. $B - 3C$

77. AB

78. BC

79. $BC - D$

80. $CB - D$

81. $D(A - 3B)$

82. $(BC - D)A$

83. *Think About It* If a, b, and c are real numbers such that $c \neq 0$ and $ac = bc$, then $a = b$. However, if A, B, and C are nonzero matrices such that $AC = BC$, then A is *not necessarily* equal to B. Illustrate this using the following matrices.

$$A = \begin{bmatrix} 0 & 1 \\ 0 & 1 \end{bmatrix}, \quad B = \begin{bmatrix} 1 & 0 \\ 1 & 0 \end{bmatrix}, \quad C = \begin{bmatrix} 2 & 3 \\ 2 & 3 \end{bmatrix}$$

84. *Think About It* If a and b are real numbers such that $ab = 0$, then $a = 0$ or $b = 0$. However, if A and B are matrices such that $AB = O$, it is *not necessarily* true that $A = O$ or $B = O$. Illustrate this using the following matrices.

$$A = \begin{bmatrix} 3 & 3 \\ 4 & 4 \end{bmatrix}, \quad B = \begin{bmatrix} 1 & -1 \\ -1 & 1 \end{bmatrix}$$

85. *Exploration* Let $i = \sqrt{-1}$ and let

$$A = \begin{bmatrix} i & 0 \\ 0 & i \end{bmatrix} \quad \text{and} \quad B = \begin{bmatrix} 0 & -i \\ i & 0 \end{bmatrix}.$$

(a) Find A^2, A^3, and A^4. Identify any similarities with i^2, i^3, and i^4.

(b) Find and identify B^2.

86. *Conjecture* Let A and B be unequal diagonal matrices of the same order. (A *diagonal matrix* is a square matrix in which each entry not on the main diagonal is zero.) Determine the products AB for several pairs of such matrices. Make a conjecture about a quick rule for such products.

87. *Exploration* Consider matrices of the form

$$A = \begin{bmatrix} 0 & a_{12} & a_{13} & a_{14} & \cdots & a_{1n} \\ 0 & 0 & a_{23} & a_{24} & \cdots & a_{2n} \\ 0 & 0 & 0 & a_{34} & \cdots & a_{3n} \\ \vdots & \vdots & \vdots & \vdots & \cdots & \vdots \\ 0 & 0 & 0 & 0 & \cdots & a_{(n-1)n} \\ 0 & 0 & 0 & 0 & \cdots & 0 \end{bmatrix}.$$

(a) Write a 2×2 matrix and a 3×3 matrix in the form of A.

(b) Use a graphing utility to raise each of the matrices to higher powers. Describe the result.

(c) Use the result of part (b) to make a conjecture about powers of A if A is a 4×4 matrix. Use a graphing utility to test your conjecture.

(d) Use the results of parts (b) and (c) to make a conjecture about powers of an $n \times n$ matrix A.

88. *Writing* Two competing companies offer cable television to a city with 100,000 households. Gold Cable Company has 25,000 subscribers and Galaxy Cable Company has 30,000 subscribers. (The other 45,000 households do not subscribe.) The percent changes in cable subscriptions each year are shown below. Write a short paragraph explaining how matrix multiplication can be used to find the number of subscribers each company will have in 1 year.

| | | Percent Changes | |
		From Gold	From Galaxy	From Nonsubscriber
Percent Changes	To Gold	0.70	0.15	0.15
	To Galaxy	0.20	0.80	0.15
	To Nonsubscriber	0.10	0.05	0.70

Review

In Exercises 89–92, condense the expression to the logarithm of a single quantity.

89. $3 \ln 4 - \frac{1}{3} \ln(x^2 + 3)$

90. $\ln x - 3[\ln(x + 6) + \ln(x - 6)]$

91. $\frac{1}{2}[2 \ln(x + 5) + \ln x - \ln(x - 8)]$

92. $\frac{3}{2} \ln 7t^4 - \frac{3}{5} \ln t^5$

8.6 The Inverse of a Square Matrix

The Inverse of a Matrix

This section further develops the algebra of matrices. To begin, consider the real number equation $ax = b$. To solve this equation for x, multiply each side of the equation by a^{-1} (provided that $a \neq 0$).

$$ax = b$$
$$(a^{-1}a)x = a^{-1}b$$
$$(1)x = a^{-1}b$$
$$x = a^{-1}b$$

The number a^{-1} is called the *multiplicative inverse of a* because $a^{-1}a = 1$. The definition of the multiplicative **inverse of a matrix** is similar.

Definition of the Inverse of a Square Matrix

Let A be an $n \times n$ matrix and let I_n be the $n \times n$ identity matrix. If there exists a matrix A^{-1} such that

$$AA^{-1} = I_n = A^{-1}A$$

then A^{-1} is called the **inverse** of A. The symbol A^{-1} is read "A inverse."

Example 1   The Inverse of a Matrix

Show that B is the inverse of A, where $A = \begin{bmatrix} -1 & 2 \\ -1 & 1 \end{bmatrix}$ and $B = \begin{bmatrix} 1 & -2 \\ 1 & -1 \end{bmatrix}$.

Solution

To show that B is the inverse of A, show that $AB = I = BA$, as follows.

$$AB = \begin{bmatrix} -1 & 2 \\ -1 & 1 \end{bmatrix}\begin{bmatrix} 1 & -2 \\ 1 & -1 \end{bmatrix} = \begin{bmatrix} -1 + 2 & 2 - 2 \\ -1 + 1 & 2 - 1 \end{bmatrix} = \begin{bmatrix} 1 & 0 \\ 0 & 1 \end{bmatrix}$$

$$BA = \begin{bmatrix} 1 & -2 \\ 1 & -1 \end{bmatrix}\begin{bmatrix} -1 & 2 \\ -1 & 1 \end{bmatrix} = \begin{bmatrix} -1 + 2 & 2 - 2 \\ -1 + 1 & 2 - 1 \end{bmatrix} = \begin{bmatrix} 1 & 0 \\ 0 & 1 \end{bmatrix}$$

As you can see, $AB = I = BA$. This is an example of a square matrix that has an inverse. Note that not all square matrices have an inverse.

✓ *Checkpoint* Now try Exercise 3.

Recall that it is not always true that $AB = BA$, even if both products are defined. However, if A and B are both square matrices and $AB = I_n$, it can be shown that $BA = I_n$. So, in Example 1, you need only check that $AB = I_2$.

What you should learn

- Verify that two matrices are inverses of each other.
- Use Gauss-Jordan elimination to find inverses of matrices.
- Use a formula to find inverses of 2×2 matrices.
- Use inverse matrices to solve systems of linear equations.

Why you should learn it

A system of equations can be solved using the inverse of the coefficient matrix. This method is particularly useful when the coefficients are the same for several systems, but the constants are different. Exercise 61 on page 667 shows how to use an inverse matrix to find unknown currents in electrical circuits.

Firefly Productions/Corbis

If a matrix A has an inverse, A is called **invertible** (or **nonsingular**); otherwise, A is called **singular.** A nonsquare matrix cannot have an inverse. To see this, note that if A is of order $m \times n$ and B is of order $n \times m$ (where $m \neq n$), the products AB and BA are of different orders and so cannot be equal to each other. Not all square matrices have inverses, as you will see at the bottom of page 662. If, however, a matrix does have an inverse, that inverse is unique. Example 2 shows how to use systems of equations to find the inverse of a matrix.

Example 2   Finding the Inverse of a Matrix

Find the inverse of

$$A = \begin{bmatrix} 1 & 4 \\ -1 & -3 \end{bmatrix}.$$

Solution

To find the inverse of A, try to solve the matrix equation $AX = I$ for X.

$$\underset{A}{\begin{bmatrix} 1 & 4 \\ -1 & -3 \end{bmatrix}} \underset{X}{\begin{bmatrix} x_{11} & x_{12} \\ x_{21} & x_{22} \end{bmatrix}} = \underset{I}{\begin{bmatrix} 1 & 0 \\ 0 & 1 \end{bmatrix}}$$

$$\begin{bmatrix} x_{11} + 4x_{21} & x_{12} + 4x_{22} \\ -x_{11} - 3x_{21} & -x_{12} - 3x_{22} \end{bmatrix} = \begin{bmatrix} 1 & 0 \\ 0 & 1 \end{bmatrix}$$

Equating corresponding entries, you obtain the following two systems of linear equations.

$$\begin{cases} x_{11} + 4x_{21} = 1 \\ -x_{11} - 3x_{21} = 0 \end{cases} \qquad \begin{cases} x_{12} + 4x_{22} = 0 \\ -x_{12} - 3x_{22} = 1 \end{cases}$$

Solve the first system using elementary row operations to determine that $x_{11} = -3$ and $x_{21} = 1$. From the second system you can determine that $x_{12} = -4$ and $x_{22} = 1$. Therefore, the inverse of A is

$$X = A^{-1}$$

$$= \begin{bmatrix} -3 & -4 \\ 1 & 1 \end{bmatrix}.$$

You can use matrix multiplication to check this result.

Check

$$AA^{-1} = \begin{bmatrix} 1 & 4 \\ -1 & -3 \end{bmatrix} \begin{bmatrix} -3 & -4 \\ 1 & 1 \end{bmatrix} = \begin{bmatrix} 1 & 0 \\ 0 & 1 \end{bmatrix} \checkmark$$

$$A^{-1}A = \begin{bmatrix} -3 & -4 \\ 1 & 1 \end{bmatrix} \begin{bmatrix} 1 & 4 \\ -1 & -3 \end{bmatrix} = \begin{bmatrix} 1 & 0 \\ 0 & 1 \end{bmatrix} \checkmark$$

✓ *Checkpoint* Now try Exercise 13.

Exploration

Most graphing utilities are capable of finding the inverse of a square matrix. Try using a graphing utility to find the inverse of the matrix

$$A = \begin{bmatrix} 2 & -3 & 1 \\ -1 & 2 & -1 \\ -2 & 0 & 1 \end{bmatrix}.$$

After you find A^{-1}, store it as $[B]$ and use the graphing utility to find $[A] \times [B]$ and $[B] \times [A]$. What can you conclude?

Finding Inverse Matrices

In Example 2, note that the two systems of linear equations have the *same coefficient matrix A*. Rather than solve the two systems represented by

$$\begin{bmatrix} 1 & 4 & \vdots & 1 \\ -1 & -3 & \vdots & 0 \end{bmatrix}$$

and

$$\begin{bmatrix} 1 & 4 & \vdots & 0 \\ -1 & -3 & \vdots & 1 \end{bmatrix}$$

separately, you can solve them *simultaneously* by *adjoining* the identity matrix to the coefficient matrix to obtain

$$\overset{A}{} \qquad \overset{I}{}$$

$$\begin{bmatrix} 1 & 4 & \vdots & 1 & 0 \\ -1 & -3 & \vdots & 0 & 1 \end{bmatrix}.$$

> ### Exploration
>
> Select two 2×2 matrices A and B that have inverses. Enter them into your graphing utility and calculate $(AB)^{-1}$. Then calculate $B^{-1}A^{-1}$ and $A^{-1}B^{-1}$. Make a conjecture about the inverse of the product of two invertible matrices.

This "doubly augmented" matrix can be represented as $[A \vdots I]$. By applying Gauss-Jordan elimination to this matrix, you can solve *both* systems with a single elimination process.

$$\begin{bmatrix} 1 & 4 & \vdots & 1 & 0 \\ -1 & -3 & \vdots & 0 & 1 \end{bmatrix}$$

$$R_1 + R_2 \rightarrow \begin{bmatrix} 1 & 4 & \vdots & 1 & 0 \\ 0 & 1 & \vdots & 1 & 1 \end{bmatrix}$$

$$-4R_2 + R_1 \rightarrow \begin{bmatrix} 1 & 0 & \vdots & -3 & -4 \\ 0 & 1 & \vdots & 1 & 1 \end{bmatrix}$$

So, from the "doubly augmented" matrix $[A \vdots I]$, you obtained the matrix $[I \vdots A^{-1}]$.

$$\overset{A}{} \qquad \overset{I}{} \qquad\qquad \overset{I}{} \qquad \overset{A^{-1}}{}$$

$$\begin{bmatrix} 1 & 4 & \vdots & 1 & 0 \\ -1 & -3 & \vdots & 0 & 1 \end{bmatrix} \implies \begin{bmatrix} 1 & 0 & \vdots & -3 & -4 \\ 0 & 1 & \vdots & 1 & 1 \end{bmatrix}$$

This procedure (or algorithm) works for any square matrix that has an inverse.

Finding an Inverse Matrix

Let A be a square matrix of order n.

1. Write the $n \times 2n$ matrix that consists of the given matrix A on the left and the $n \times n$ identity matrix I on the right to obtain $[A \vdots I]$.

2. If possible, row reduce A to I using elementary row operations on the *entire* matrix $[A \vdots I]$. The result will be the matrix $[I \vdots A^{-1}]$. If this is not possible, A is not invertible.

3. Check your work by multiplying to see that $AA^{-1} = I = A^{-1}A$.

Example 3   Finding the Inverse of a Matrix

Find the inverse of $A = \begin{bmatrix} 1 & -1 & 0 \\ 1 & 0 & -1 \\ 6 & -2 & -3 \end{bmatrix}$.

Solution

Begin by adjoining the identity matrix to A to form the matrix

$$[A \; \vdots \; I] = \begin{bmatrix} 1 & -1 & 0 & \vdots & 1 & 0 & 0 \\ 1 & 0 & -1 & \vdots & 0 & 1 & 0 \\ 6 & -2 & -3 & \vdots & 0 & 0 & 1 \end{bmatrix}.$$

Use elementary row operations to obtain the form $[I \; \vdots \; A^{-1}]$, as follows.

$$\begin{bmatrix} 1 & 0 & 0 & \vdots & -2 & -3 & 1 \\ 0 & 1 & 0 & \vdots & -3 & -3 & 1 \\ 0 & 0 & 1 & \vdots & -2 & -4 & 1 \end{bmatrix}$$

Therefore, the matrix A is invertible and its inverse is

$$A^{-1} = \begin{bmatrix} -2 & -3 & 1 \\ -3 & -3 & 1 \\ -2 & -4 & 1 \end{bmatrix}.$$

Try using a graphing utility to confirm this result by multiplying A by A^{-1} to obtain I.

 Checkpoint   Now try Exercise 21.

TECHNOLOGY TIP

Most graphing utilities can find the inverse of a matrix. A graphing utility can be used to check matrix operations. This saves valuable time otherwise spent doing minor arithmetic calculations.

The algorithm shown in Example 3 applies to any $n \times n$ matrix A. When using this algorithm, if the matrix A does not reduce to the identity matrix, then A does not have an inverse. For instance, the following matrix has no inverse.

$$A = \begin{bmatrix} 1 & 2 & 0 \\ 3 & -1 & 2 \\ -2 & 3 & -2 \end{bmatrix}$$

To confirm that matrix A above has no inverse, begin by adjoining the identity matrix to A to form

$$[A \; \vdots \; I] = \begin{bmatrix} 1 & 2 & 0 & \vdots & 1 & 0 & 0 \\ 3 & -1 & 2 & \vdots & 0 & 1 & 0 \\ -2 & 3 & -2 & \vdots & 0 & 0 & 1 \end{bmatrix}.$$

Then use elementary row operations to obtain

$$\begin{bmatrix} 1 & 2 & 0 & \vdots & 1 & 0 & 0 \\ 0 & -7 & 2 & \vdots & -3 & 1 & 0 \\ 0 & 0 & 0 & \vdots & -1 & 1 & 1 \end{bmatrix}.$$

At this point in the elimination process you can see that it is impossible to obtain the identity matrix I on the left. Therefore, A is not invertible.

Example 4   Finding the Inverse of a Matrix

Use a graphing utility to find the inverse of $A = \begin{bmatrix} 1 & 2 & -2 \\ 1 & -1 & 0 \\ 0 & -1 & 4 \end{bmatrix}$.

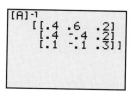

Solution

Use the *matrix editor* to enter A into the graphing utility. Use the inverse key $\boxed{x^{-1}}$ to find the inverse of the matrix, as shown in Figure 8.31. Check this result algebraically by multiplying A by A^{-1} to obtain I.

Figure 8.31

 Checkpoint Now try Exercise 27.

The Inverse of a 2 × 2 Matrix

Using Gauss-Jordan elimination to find the inverse of a matrix works well (even as a computer technique) for matrices of order 3×3 or greater. For 2×2 matrices, however, many people prefer to use a formula for the inverse rather than Gauss-Jordan elimination. This simple formula, which works *only* for 2×2 matrices, is explained as follows. If A is the 2×2 matrix given by

$$A = \begin{bmatrix} a & b \\ c & d \end{bmatrix}$$

then A is invertible if and only if $ad - bc \neq 0$. If $ad - bc \neq 0$, the inverse is given by

$$A^{-1} = \frac{1}{ad - bc} \begin{bmatrix} d & -b \\ -c & a \end{bmatrix}. \qquad \text{Formula for inverse of matrix } A$$

The denominator $ad - bc$ is called the *determinant* of the 2×2 matrix A. You will study determinants in the next section.

Example 5   Finding the Inverse of a 2 × 2 Matrix

If possible, find the inverse of $A = \begin{bmatrix} 3 & -1 \\ -2 & 2 \end{bmatrix}$.

Solution

Apply the formula for the inverse of a 2×2 matrix to obtain

$$ad - bc = (3)(2) - (-1)(-2) = 4.$$

Because this quantity is not zero, the inverse is formed by interchanging the entries on the main diagonal, changing the signs of the other two entries, and multiplying by the scalar $\frac{1}{4}$, as follows.

$$A^{-1} = \frac{1}{4} \begin{bmatrix} 2 & 1 \\ 2 & 3 \end{bmatrix} = \begin{bmatrix} \frac{1}{2} & \frac{1}{4} \\ \frac{1}{2} & \frac{3}{4} \end{bmatrix}$$

 Checkpoint Now try Exercise 33.

Exploration

Use a graphing utility to find the inverse of the matrix

$$A = \begin{bmatrix} 1 & -3 \\ -2 & 6 \end{bmatrix}.$$

What message appears on the screen? Why does the graphing utility display this message?

Systems of Linear Equations

You know that a system of linear equations can have exactly one solution, infinitely many solutions, or no solution. If the coefficient matrix A of a *square* system (a system that has the same number of equations as variables) is invertible, the system has a unique solution, which is defined as follows.

A System of Equations with a Unique Solution

If A is an invertible matrix, the system of linear equations represented by $AX = B$ has a unique solution given by

$$X = A^{-1}B.$$

The formula $X = A^{-1}B$ is used on most graphing utilities to solve linear systems that have invertible coefficient matrices. That is, you enter the $n \times n$ coefficient matrix $[A]$ and the $n \times 1$ column matrix $[B]$. The solution X is given by $[A]^{-1}[B]$.

Example 6   Solving a System of Equations Using an Inverse

Use an inverse matrix to solve the system.

$$\begin{cases} 2x + 3y + z = -1 \\ 3x + 3y + z = 1 \\ 2x + 4y + z = -2 \end{cases}$$

Solution

Begin by writing the system as $AX = B$.

$$\begin{bmatrix} 2 & 3 & 1 \\ 3 & 3 & 1 \\ 2 & 4 & 1 \end{bmatrix} \begin{bmatrix} x \\ y \\ z \end{bmatrix} = \begin{bmatrix} -1 \\ 1 \\ -2 \end{bmatrix}$$

Then, use Gauss-Jordan elimination to find A^{-1}.

$$A^{-1} = \begin{bmatrix} -1 & 1 & 0 \\ -1 & 0 & 1 \\ 6 & -2 & -3 \end{bmatrix}$$

Finally, multiply B by A^{-1} on the left to obtain the solution.

$$\begin{aligned} X &= A^{-1}B \\ &= \begin{bmatrix} -1 & 1 & 0 \\ -1 & 0 & 1 \\ 6 & -2 & -3 \end{bmatrix} \begin{bmatrix} -1 \\ 1 \\ -2 \end{bmatrix} = \begin{bmatrix} 2 \\ -1 \\ -2 \end{bmatrix} \end{aligned}$$

So, the solution is $x = 2$, $y = -1$, and $z = -2$. Use a graphing utility to verify A^{-1} for the system of equations.

✓ *Checkpoint* Now try Exercise 51.

> **STUDY TIP**
>
> Remember that matrix multiplication is not commutative. So, you must multiply matrices in the correct order. For instance, in Example 6, you must multiply B by A^{-1} on the left.

8.6   Exercises

Vocabulary Check

Fill in the blanks.

1. In a _____ matrix, the number of rows equals the number of columns.
2. If there exists an $n \times n$ matrix A^{-1} such that $AA^{-1} = I_n = A^{-1}A$, then A^{-1} is called the _____ of A.
3. If a matrix A has an inverse, it is called invertible or _____ ; if it does not have an inverse, it is called _____ .

In Exercises 1–8, show that B is the inverse of A.

1. $A = \begin{bmatrix} 2 & 1 \\ 5 & 3 \end{bmatrix}$, $B = \begin{bmatrix} 3 & -1 \\ -5 & 2 \end{bmatrix}$

2. $A = \begin{bmatrix} 1 & -1 \\ -1 & 2 \end{bmatrix}$, $B = \begin{bmatrix} 2 & 1 \\ 1 & 1 \end{bmatrix}$

3. $A = \begin{bmatrix} 1 & 2 \\ 3 & 4 \end{bmatrix}$, $B = \begin{bmatrix} -2 & 1 \\ \frac{3}{2} & -\frac{1}{2} \end{bmatrix}$

4. $A = \begin{bmatrix} 1 & -1 \\ 2 & 3 \end{bmatrix}$, $B = \begin{bmatrix} \frac{3}{5} & \frac{1}{5} \\ -\frac{2}{5} & \frac{1}{5} \end{bmatrix}$

5. $A = \begin{bmatrix} 2 & -17 & 11 \\ -1 & 11 & -7 \\ 0 & 3 & -2 \end{bmatrix}$, $B = \begin{bmatrix} 1 & 1 & 2 \\ 2 & 4 & -3 \\ 3 & 6 & -5 \end{bmatrix}$

6. $A = \begin{bmatrix} -4 & 1 & 5 \\ -1 & 2 & 4 \\ 0 & -1 & -1 \end{bmatrix}$, $B = \begin{bmatrix} -\frac{1}{2} & 1 & \frac{3}{2} \\ \frac{1}{4} & -1 & -\frac{11}{4} \\ -\frac{1}{4} & 1 & \frac{7}{4} \end{bmatrix}$

7. $A = \begin{bmatrix} -2 & 2 & 3 \\ 1 & -1 & 0 \\ 0 & 1 & 4 \end{bmatrix}$, $B = \frac{1}{3}\begin{bmatrix} -4 & -5 & 3 \\ -4 & -8 & 3 \\ 1 & 2 & 0 \end{bmatrix}$

8. $A = \begin{bmatrix} 1 & 0 & -1 \\ -1 & 1 & 0 \\ 1 & 2 & 0 \end{bmatrix}$, $B = \frac{1}{3}\begin{bmatrix} 0 & -2 & 1 \\ 0 & 1 & 1 \\ -3 & -2 & 1 \end{bmatrix}$

In Exercises 9–12, use the matrix capabilities of a graphing utility to show that B is the inverse of A.

9. $A = \begin{bmatrix} -1 & -4 \\ 1 & 2 \end{bmatrix}$, $B = \begin{bmatrix} 1 & 2 \\ -\frac{1}{2} & -\frac{1}{2} \end{bmatrix}$

10. $A = \begin{bmatrix} 11 & -12 \\ 2 & -2 \end{bmatrix}$, $B = \begin{bmatrix} -1 & 6 \\ -1 & \frac{11}{2} \end{bmatrix}$

11. $A = \begin{bmatrix} 1.6 & 2 \\ -3.5 & -4.5 \end{bmatrix}$, $B = \begin{bmatrix} 22.5 & 10 \\ -17.5 & -8 \end{bmatrix}$

12. $A = \begin{bmatrix} 4 & 0 & -2 \\ 1 & 2 & -4 \\ 0 & 3 & 1 \end{bmatrix}$, $B = \begin{bmatrix} 0.28 & -0.12 & 0.08 \\ -0.02 & 0.08 & 0.28 \\ 0.06 & -0.24 & 0.16 \end{bmatrix}$

In Exercises 13–24, find the inverse of the matrix (if it exists).

13. $\begin{bmatrix} 2 & 0 \\ 0 & 3 \end{bmatrix}$

14. $\begin{bmatrix} 1 & 2 \\ 3 & 7 \end{bmatrix}$

15. $\begin{bmatrix} 1 & -2 \\ 2 & -3 \end{bmatrix}$

16. $\begin{bmatrix} -7 & 33 \\ 4 & -19 \end{bmatrix}$

17. $\begin{bmatrix} -1 & 1 \\ -2 & 1 \end{bmatrix}$

18. $\begin{bmatrix} 2 & -4 \\ 4 & 8 \end{bmatrix}$

19. $\begin{bmatrix} 2 & 7 & 1 \\ -3 & -9 & 2 \end{bmatrix}$

20. $\begin{bmatrix} -2 & 5 \\ 6 & -15 \\ 0 & 1 \end{bmatrix}$

21. $\begin{bmatrix} 1 & 1 & 1 \\ 3 & 5 & 4 \\ 3 & 6 & 5 \end{bmatrix}$

22. $\begin{bmatrix} 1 & 2 & 2 \\ 3 & 7 & 9 \\ -1 & -4 & -7 \end{bmatrix}$

23. $\begin{bmatrix} -5 & 0 & 0 \\ 2 & 0 & 0 \\ -1 & 5 & 7 \end{bmatrix}$

24. $\begin{bmatrix} 1 & 0 & 0 \\ 3 & 5 & 0 \\ 2 & 5 & 0 \end{bmatrix}$

In Exercises 25–32, use the matrix capabilities of a graphing utility to find the inverse of the matrix (if it exists).

25. $\begin{bmatrix} 1 & 0 & -2 \\ 4 & 0 & 0 \\ -2 & 0 & 3 \end{bmatrix}$

26. $\begin{bmatrix} 1 & 2 & -1 \\ 3 & 7 & -10 \\ -5 & -7 & -15 \end{bmatrix}$

27. $\begin{bmatrix} -\frac{1}{2} & \frac{3}{4} & \frac{1}{4} \\ 1 & 0 & -\frac{3}{2} \\ 0 & -1 & \frac{1}{2} \end{bmatrix}$

28. $\begin{bmatrix} -\frac{5}{6} & \frac{1}{3} & \frac{11}{6} \\ 0 & \frac{2}{3} & 2 \\ 1 & -\frac{1}{2} & -\frac{5}{2} \end{bmatrix}$

29. $\begin{bmatrix} 0.1 & 0.2 & 0.3 \\ -0.3 & 0.2 & 0.2 \\ 0.5 & 0.4 & 0.4 \end{bmatrix}$

30. $\begin{bmatrix} 0.6 & 0 & -0.3 \\ 0.7 & -1 & 0.2 \\ 1 & 0 & -0.9 \end{bmatrix}$

31. $\begin{bmatrix} -1 & 0 & 1 & 0 \\ 0 & 2 & 0 & -1 \\ 2 & 0 & -1 & 0 \\ 0 & -1 & 0 & 1 \end{bmatrix}$

32. $\begin{bmatrix} 1 & -2 & -1 & -2 \\ 3 & -5 & -2 & -3 \\ 2 & -5 & -2 & -5 \\ -1 & 4 & 4 & 11 \end{bmatrix}$

In Exercises 33–36, use the formula on page 663 to find the inverse of the 2×2 matrix.

33. $\begin{bmatrix} 5 & 1 \\ -2 & -2 \end{bmatrix}$

34. $\begin{bmatrix} -8 & 0 \\ 11 & -10 \end{bmatrix}$

35. $\begin{bmatrix} \frac{7}{2} & -\frac{3}{4} \\ \frac{1}{5} & \frac{4}{5} \end{bmatrix}$

36. $\begin{bmatrix} -\frac{1}{4} & -\frac{2}{3} \\ \frac{1}{3} & \frac{8}{9} \end{bmatrix}$

In Exercises 37–40, use the inverse matrix found in Exercise 15 to solve the system of linear equations.

37. $\begin{cases} x - 2y = 5 \\ 2x - 3y = 10 \end{cases}$

38. $\begin{cases} x - 2y = 0 \\ 2x - 3y = 3 \end{cases}$

39. $\begin{cases} x - 2y = 4 \\ 2x - 3y = 2 \end{cases}$

40. $\begin{cases} x - 2y = 1 \\ 2x - 3y = -2 \end{cases}$

In Exercises 41 and 42, use the inverse matrix found in Exercise 21 to solve the system of linear equations.

41. $\begin{cases} x + y + z = 0 \\ 3x + 5y + 4z = 5 \\ 3x + 6y + 5z = 2 \end{cases}$

42. $\begin{cases} x + y + z = -1 \\ 3x + 5y + 4z = 2 \\ 3x + 6y + 5z = 0 \end{cases}$

In Exercises 43 and 44, use the inverse matrix found in Exercise 32 and the matrix capabilities of a graphing utility to solve the system of linear equations.

43. $\begin{cases} x_1 - 2x_2 - x_3 - 2x_4 = 0 \\ 3x_1 - 5x_2 - 2x_3 - 3x_4 = 1 \\ 2x_1 - 5x_2 - 2x_3 - 5x_4 = -1 \\ -x_1 + 4x_2 + 4x_3 + 11x_4 = 2 \end{cases}$

44. $\begin{cases} x_1 - 2x_2 - x_3 - 2x_4 = 1 \\ 3x_1 - 5x_2 - 2x_3 - 3x_4 = -2 \\ 2x_1 - 5x_2 - 2x_3 - 5x_4 = 0 \\ -x_1 + 4x_2 + 4x_3 + 11x_4 = -3 \end{cases}$

In Exercises 45–52, use an inverse matrix to solve (if possible) the system of linear equations.

45. $\begin{cases} 3x + 4y = -2 \\ 5x + 3y = 4 \end{cases}$

46. $\begin{cases} 18x + 12y = 13 \\ 30x + 24y = 23 \end{cases}$

47. $\begin{cases} -0.4x + 0.8y = 1.6 \\ 2x - 4y = 5 \end{cases}$

48. $\begin{cases} 0.2x - 0.6y = 2.4 \\ -x + 1.4y = -8.8 \end{cases}$

49. $\begin{cases} -\frac{1}{4}x + \frac{3}{8}y = -2 \\ \frac{3}{2}x + \frac{3}{4}y = -12 \end{cases}$

50. $\begin{cases} \frac{5}{6}x - y = -20 \\ \frac{4}{3}x - \frac{7}{2}y = -51 \end{cases}$

51. $\begin{cases} 4x - y + z = -5 \\ 2x + 2y + 3z = 10 \\ 5x - 2y + 6z = 1 \end{cases}$

52. $\begin{cases} 4x - 2y + 3z = -2 \\ 2x + 2y + 5z = 16 \\ 8x - 5y - 2z = 4 \end{cases}$

In Exercises 53–56, use the matrix capabilities of a graphing utility to solve (if possible) the system of linear equations.

53. $\begin{cases} 5x - 3y + 2z = 2 \\ 2x + 2y - 3z = 3 \\ -x + 7y - 8z = 4 \end{cases}$

54. $\begin{cases} 2x + 3y + 5z = 4 \\ 3x + 5y - 9z = 7 \\ 5x + 9y + 17z = 13 \end{cases}$

55. $\begin{cases} 7x - 3y + 2w = 41 \\ -2x + y - w = -13 \\ 4x + z - 2w = 12 \\ -x + y - w = -8 \end{cases}$

56. $\begin{cases} 2x + 5y + w = 11 \\ x + 4y + 2z - 2w = -7 \\ 2x - 2y + 5z + w = 3 \\ x - 3w = -1 \end{cases}$

Investment Portfolio In Exercises 57–60, consider a person who invests in AAA-rated bonds, A-rated bonds, and B-rated bonds. The average yields are 6.5% on AAA bonds, 7% on A bonds, and 9% on B bonds. The person invests twice as much in B bonds as in A bonds. Let x, y, and z represent the amounts invested in AAA, A, and B bonds, respectively.

$$\begin{cases} x + y + z = \text{(total investment)} \\ 0.065x + 0.07y + 0.09z = \text{(annual return)} \\ 2y - z = 0 \end{cases}$$

Use the inverse of the coefficient matrix of this system to find the amount invested in each type of bond.

	Total Investment	Annual Return
57.	$25,000	$1900
58.	$10,000	$760

Total Investment	Annual Return
59. $65,000	$5050
60. $500,000	$38,000

61. *Circuit Analysis* Consider the circuit in the figure. The currents I_1, I_2, and I_3, in amperes, are given by the solution of the system of linear equations

$$\begin{cases} 2I_1 \quad\quad + 4I_3 = E_1 \\ \quad\quad I_2 + 4I_3 = E_2 \\ I_1 + I_2 - \; I_3 = 0 \end{cases}$$

where E_1 and E_2 are voltages. Use the inverse of the coefficient matrix of this system to find the unknown currents for the voltages.

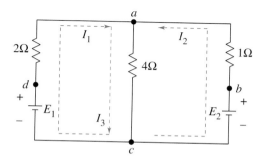

(a) $E_1 = 14$ volts, $E_2 = 28$ volts

(b) $E_1 = 10$ volts, $E_2 = 10$ volts

62. *Data Analysis* The table shows the numbers y (in millions) of motor vehicle registrations in the United States from 1999 to 2001. (Source: U.S. Federal Highway Administration)

Year	Registrations, y
1999	216.3
2000	221.5
2001	230.4

(a) This data can be approximated by a parabola. Create a system of linear equations for the data. Let t represent the year, with $t = 9$ corresponding to 1999.

(b) Use the matrix capabilities of a graphing utility to find an inverse matrix to solve the system in part (a) and find the least squares regression parabola $y = at^2 + bt + c$.

(c) Use the result of part (b) to determine the year in which the number of vehicle registrations will reach 300 million.

Synthesis

True or False? In Exercises 63 and 64, determine whether the statement is true or false. Justify your answer.

63. Multiplication of an invertible matrix and its inverse is commutative.

64. No nonsquare matrices have inverses.

65. If A is a 2×2 matrix given by $A = \begin{bmatrix} a & b \\ c & d \end{bmatrix}$, then A is invertible if and only if $ad - bc \neq 0$. If $ad - bc \neq 0$, verify that the inverse is

$$A^{-1} = \frac{1}{ad - bc} \begin{bmatrix} d & -b \\ -c & a \end{bmatrix}.$$

66. *Exploration* Consider the matrices of the form

$$A = \begin{bmatrix} a_{11} & 0 & 0 & 0 & \cdots & 0 \\ 0 & a_{22} & 0 & 0 & \cdots & 0 \\ 0 & 0 & a_{33} & 0 & \cdots & 0 \\ \vdots & \vdots & \vdots & \vdots & \cdots & \vdots \\ 0 & 0 & 0 & 0 & \cdots & a_{nn} \end{bmatrix}.$$

(a) Write a 2×2 matrix and a 3×3 matrix in the form of A. Find the inverse of each.

(b) Use the result of part (a) to make a conjecture about the inverse of a matrix in the form of A.

Review

In Exercises 67–70, simplify the complex fraction.

67. $\dfrac{\left(\dfrac{9}{x}\right)}{\left(\dfrac{6}{x} + 2\right)}$

68. $\dfrac{\left(1 + \dfrac{2}{x}\right)}{\left(1 - \dfrac{4}{x}\right)}$

69. $\dfrac{\left(\dfrac{4}{x^2 - 9} + \dfrac{2}{x - 2}\right)}{\left(\dfrac{1}{x + 3} + \dfrac{1}{x - 3}\right)}$

70. $\dfrac{\left(\dfrac{1}{x + 1} + \dfrac{1}{2}\right)}{\left(\dfrac{3}{2x^2 + 4x + 2}\right)}$

In Exercises 71–74, solve the equation algebraically. Round your result to three decimal places.

71. $e^{2x} + 2e^x - 15 = 0$

72. $e^{2x} - 10e^x + 24 = 0$

73. $7 \ln 3x = 12$

74. $\ln(x + 9) = 2$

8.7 The Determinant of a Square Matrix

The Determinant of a 2 × 2 Matrix

Every *square* matrix can be associated with a real number called its **determinant.** Determinants have many uses, and several will be discussed in this and the next section. Historically, the use of determinants arose from special number patterns that occur when systems of linear equations are solved. For instance, the system

$$\begin{cases} a_1 x + b_1 y = c_1 \\ a_2 x + b_2 y = c_2 \end{cases}$$

has a solution

$$x = \frac{c_1 b_2 - c_2 b_1}{a_1 b_2 - a_2 b_1}$$

and

$$y = \frac{a_1 c_2 - a_2 c_1}{a_1 b_2 - a_2 b_1}$$

provided that $a_1 b_2 - a_2 b_1 \neq 0$. Note that the denominator of each fraction is the same. This denominator is called the *determinant* of the coefficient matrix of the system.

Coefficient Matrix	*Determinant*

$$A = \begin{bmatrix} a_1 & b_1 \\ a_2 & b_2 \end{bmatrix} \qquad \det(A) = a_1 b_2 - a_2 b_1$$

The determinant of the matrix A can also be denoted by vertical bars on both sides of the matrix, as indicated in the following definition.

Definition of the Determinant of a 2 × 2 Matrix

The **determinant** of the matrix

$$A = \begin{bmatrix} a_1 & b_1 \\ a_2 & b_2 \end{bmatrix}$$

is given by

$$\det(A) = |A| = \begin{vmatrix} a_1 & b_1 \\ a_2 & b_2 \end{vmatrix}$$

$$= a_1 b_2 - a_2 b_1.$$

In this text, $\det(A)$ and $|A|$ are used interchangeably to represent the determinant of A. Although vertical bars are also used to denote the absolute value of a real number, the context will show which use is intended.

What you should learn

- Find the determinants of 2 × 2 matrices.
- Find minors and cofactors of square matrices.
- Find the determinants of square matrices.
- Find the determinants of triangular matrices.

Why you should learn it

Determinants are often used in other branches of mathematics. For instance, Exercises 53–58 on page 675 show some types of determinants that are useful in calculus.

A convenient method for remembering the formula for the determinant of a 2×2 matrix is shown in the following diagram.

$$\det(A) = \begin{vmatrix} a_1 & b_1 \\ a_2 & b_2 \end{vmatrix} = a_1 b_2 - a_2 b_1$$

Note that the determinant is the difference of the products of the two diagonals of the matrix.

Example 1   The Determinant of a 2×2 Matrix

Find the determinant of each matrix.

a. $A = \begin{bmatrix} 2 & -3 \\ 1 & 2 \end{bmatrix}$ **b.** $B = \begin{bmatrix} 2 & 1 \\ 4 & 2 \end{bmatrix}$ **c.** $C = \begin{bmatrix} 0 & \frac{3}{2} \\ 2 & 4 \end{bmatrix}$

Solution

a. $\det(A) = \begin{vmatrix} 2 & -3 \\ 1 & 2 \end{vmatrix} = 2(2) - 1(-3)$

$$= 4 + 3 = 7$$

b. $\det(B) = \begin{vmatrix} 2 & 1 \\ 4 & 2 \end{vmatrix} = 2(2) - 4(1)$

$$= 4 - 4 = 0$$

c. $\det(C) = \begin{vmatrix} 0 & \frac{3}{2} \\ 2 & 4 \end{vmatrix} = 0(4) - 2\left(\frac{3}{2}\right)$

$$= 0 - 3 = -3$$

✓ *Checkpoint* Now try Exercise 5.

Notice in Example 1 that the determinant of a matrix can be positive, zero, or negative.

The determinant of a matrix of order 1×1 is defined simply as the entry of the matrix. For instance, if $A = [-2]$, then $\det(A) = -2$.

TECHNOLOGY TIP   Most graphing utilities can evaluate the determinant of a matrix. For instance, you can evaluate the determinant of the matrix A in Example 1(a) by entering the matrix as $[A]$ (see Figure 8.32) and then choosing the *determinant* feature. The result should be 7, as in Example 1(a) (see Figure 8.33).

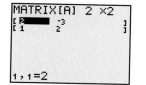

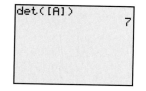

Figure 8.32   Figure 8.33

Minors and Cofactors

To define the determinant of a square matrix of order 3×3 or higher, it is helpful to introduce the concepts of **minors** and **cofactors.**

> **Minors and Cofactors of a Square Matrix**
>
> If A is a square matrix, the **minor** M_{ij} of the entry a_{ij} is the determinant of the matrix obtained by deleting the ith row and jth column of A. The **cofactor** C_{ij} of the entry a_{ij} is given by
>
> $$C_{ij} = (-1)^{i+j} M_{ij}.$$

Example 2   Finding the Minors and Cofactors of a Matrix

Find all the minors and cofactors of

$$A = \begin{bmatrix} 0 & 2 & 1 \\ 3 & -1 & 2 \\ 4 & 0 & 1 \end{bmatrix}.$$

Solution

To find the minor M_{11}, delete the first row and first column of A and evaluate the determinant of the resulting matrix.

$$\begin{bmatrix} 0 & 2 & 1 \\ 3 & -1 & 2 \\ 4 & 0 & 1 \end{bmatrix}, \quad M_{11} = \begin{vmatrix} -1 & 2 \\ 0 & 1 \end{vmatrix} = -1(1) - 0(2) = -1$$

Similarly, to find M_{12}, delete the first row and second column.

$$\begin{bmatrix} 0 & 2 & 1 \\ 3 & -1 & 2 \\ 4 & 0 & 1 \end{bmatrix}, \quad M_{12} = \begin{vmatrix} 3 & 2 \\ 4 & 1 \end{vmatrix} = 3(1) - 4(2) = -5$$

Continuing this pattern, you obtain the minors.

$$M_{11} = -1 \quad M_{12} = -5 \quad M_{13} = 4$$
$$M_{21} = 2 \quad M_{22} = -4 \quad M_{23} = -8$$
$$M_{31} = 5 \quad M_{32} = -3 \quad M_{33} = -6$$

Now, to find the cofactors, combine these minors with the checkerboard pattern of signs for a 3×3 matrix shown at the upper right.

$$C_{11} = -1 \quad C_{12} = 5 \quad C_{13} = 4$$
$$C_{21} = -2 \quad C_{22} = -4 \quad C_{23} = 8$$
$$C_{31} = 5 \quad C_{32} = 3 \quad C_{33} = -6$$

✓ *Checkpoint* Now try Exercise 17.

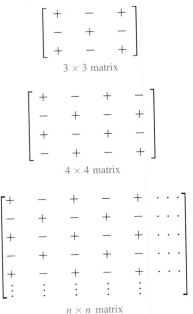

Sign Pattern for Cofactors

$$\begin{bmatrix} + & - & + \\ - & + & - \\ + & - & + \end{bmatrix}$$
3×3 matrix

$$\begin{bmatrix} + & - & + & - \\ - & + & - & + \\ + & - & + & - \\ - & + & - & + \end{bmatrix}$$
4×4 matrix

$$\begin{bmatrix} + & - & + & - & + & \cdots \\ - & + & - & + & - & \cdots \\ + & - & + & - & + & \cdots \\ - & + & - & + & - & \cdots \\ + & - & + & - & + & \cdots \\ \vdots & \vdots & \vdots & \vdots & \vdots & \end{bmatrix}$$
$n \times n$ matrix

STUDY TIP

In the sign pattern for cofactors above, notice that *odd* positions (where $i + j$ is odd) have negative signs and *even* positions (where $i + j$ is even) have positive signs.

The Determinant of a Square Matrix

The following definition is called *inductive* because it uses determinants of matrices of order $n - 1$ to define determinants of matrices of order n.

Determinant of a Square Matrix

If A is a square matrix (of order 2×2 or greater), the determinant of A is the sum of the entries in any row (or column) of A multiplied by their respective cofactors. For instance, expanding along the first row yields

$$|A| = a_{11}C_{11} + a_{12}C_{12} + \cdots + a_{1n}C_{1n}.$$

Applying this definition to find a determinant is called **expanding by cofactors.**

Try checking that for a 2×2 matrix

$$A = \begin{bmatrix} a_1 & b_1 \\ a_2 & b_2 \end{bmatrix}$$

the definition of the determinant above yields

$$|A| = a_1 b_2 - a_2 b_1$$

as previously defined.

Example 3   The Determinant of a Matrix of Order 3×3

Find the determinant of $A = \begin{bmatrix} 0 & 2 & 1 \\ 3 & -1 & 2 \\ 4 & 0 & 1 \end{bmatrix}$.

Solution

Note that this is the same matrix that was in Example 2. There you found the cofactors of the entries in the first row to be

$$C_{11} = -1, \qquad C_{12} = 5, \qquad \text{and} \qquad C_{13} = 4.$$

So, by the definition of the determinant of a square matrix, you have

$$\begin{aligned} |A| &= a_{11}C_{11} + a_{12}C_{12} + a_{13}C_{13} \qquad \text{First-row expansion} \\ &= 0(-1) + 2(5) + 1(4) \\ &= 14. \end{aligned}$$

✓ *Checkpoint* Now try Exercise 23.

In Example 3, the determinant was found by expanding by the cofactors in the first row. You could have used any row or column. For instance, you could have expanded along the second row to obtain

$$\begin{aligned} |A| &= a_{21}C_{21} + a_{22}C_{22} + a_{23}C_{23} \qquad \text{Second-row expansion} \\ &= 3(-2) + (-1)(-4) + 2(8) \\ &= 14. \end{aligned}$$

When expanding by cofactors, you do not need to find cofactors of zero entries, because zero times its cofactor is zero.

$$a_{ij}C_{ij} = (0)C_{ij} = 0$$

So, the row (or column) containing the most zeros is usually the best choice for expansion by cofactors.

Triangular Matrices

Evaluating determinants of matrices of order 4 or higher can be tedious. There is, however, an important exception: the determinant of a **triangular** matrix. A triangular matrix is a square matrix with all zero entries either below or above its main diagonal. A square matrix is **upper triangular** if it has all zero entries below its main diagonal and **lower triangular** if it has all zero entries above its main diagonal. A matrix that is both upper and lower triangular is called **diagonal.** That is, a diagonal matrix is a square matrix in which all entries above and below the main diagonal are zero.

Upper Triangular Matrix

$$\begin{bmatrix} a_{11} & a_{12} & a_{13} & \cdots & a_{1n} \\ 0 & a_{22} & a_{23} & \cdots & a_{2n} \\ 0 & 0 & a_{33} & \cdots & a_{3n} \\ \vdots & \vdots & \vdots & & \vdots \\ 0 & 0 & 0 & \cdots & a_{nn} \end{bmatrix}$$

Lower Triangular Matrix

$$\begin{bmatrix} a_{11} & 0 & 0 & \cdots & 0 \\ a_{21} & a_{22} & 0 & \cdots & 0 \\ a_{31} & a_{32} & a_{33} & \cdots & 0 \\ \vdots & \vdots & \vdots & & \vdots \\ a_{n1} & a_{n2} & a_{n3} & \cdots & a_{nn} \end{bmatrix}$$

Diagonal Matrix

$$\begin{bmatrix} a_{11} & 0 & 0 & \cdots & 0 \\ 0 & a_{22} & 0 & \cdots & 0 \\ 0 & 0 & a_{33} & \cdots & 0 \\ \vdots & \vdots & \vdots & & \vdots \\ 0 & 0 & 0 & \cdots & a_{nn} \end{bmatrix}$$

To find the determinant of a triangular matrix of any order, simply form the product of the entries on the main diagonal.

Example 4 The Determinant of a Triangular Matrix

a.
$$\begin{vmatrix} 2 & 0 & 0 & 0 \\ 4 & -2 & 0 & 0 \\ -5 & 6 & 1 & 0 \\ 1 & 5 & 3 & 3 \end{vmatrix} = (2)(-2)(1)(3) = -12$$

b.
$$\begin{vmatrix} -1 & 0 & 0 & 0 & 0 \\ 0 & 3 & 0 & 0 & 0 \\ 0 & 0 & 2 & 0 & 0 \\ 0 & 0 & 0 & 4 & 0 \\ 0 & 0 & 0 & 0 & -2 \end{vmatrix} = (-1)(3)(2)(4)(-2) = 48$$

✓ *Checkpoint* Now try Exercise 31.

8.7   Exercises

Vocabulary Check

Fill in the blanks.

1. Both det(A) and $|A|$ represent the _____ of the matrix A.

2. The _____ M_{ij} of the entry a_{ij} is the determinant of the matrix obtained by deleting the ith row and jth column of the square matrix A.

3. The _____ C_{ij} of the entry a_{ij} is given by $(-1)^{i+j}M_{ij}$.

4. One way of finding the determinant of a matrix of order 2×2 or greater is _____ .

5. A square matrix with all zero entries either above or below its main diagonal is called a _____ matrix.

6. A matrix that is both upper and lower triangular is called a _____ matrix.

In Exercises 1–12, find the determinant of the matrix.

1. $[4]$

2. $[-10]$

3. $\begin{bmatrix} 8 & 4 \\ 2 & 3 \end{bmatrix}$

4. $\begin{bmatrix} -9 & 0 \\ 6 & 2 \end{bmatrix}$

5. $\begin{bmatrix} 6 & 2 \\ -5 & 3 \end{bmatrix}$

6. $\begin{bmatrix} 3 & -3 \\ 4 & -8 \end{bmatrix}$

7. $\begin{bmatrix} -7 & 6 \\ \frac{1}{2} & 3 \end{bmatrix}$

8. $\begin{bmatrix} 4 & -3 \\ 0 & 0 \end{bmatrix}$

9. $\begin{bmatrix} 2 & -1 & 0 \\ 4 & 2 & 1 \\ 4 & 2 & 1 \end{bmatrix}$

10. $\begin{bmatrix} -2 & 2 & 3 \\ 1 & -1 & 0 \\ 0 & 1 & 4 \end{bmatrix}$

11. $\begin{bmatrix} -1 & 2 & -5 \\ 0 & 3 & 4 \\ 0 & 0 & 3 \end{bmatrix}$

12. $\begin{bmatrix} 1 & 0 & 0 \\ -4 & -1 & 0 \\ 5 & 1 & 5 \end{bmatrix}$

In Exercises 13 and 14, use the matrix capabilities of a graphing utility to find the determinant of the matrix.

13. $\begin{bmatrix} 0.3 & 0.2 & 0.2 \\ 0.2 & 0.2 & 0.2 \\ -0.4 & 0.4 & 0.3 \end{bmatrix}$

14. $\begin{bmatrix} 0.1 & 0.2 & 0.3 \\ -0.3 & 0.2 & 0.2 \\ 0.5 & 0.4 & 0.4 \end{bmatrix}$

In Exercises 15–18, find all (a) minors and (b) cofactors of the matrix.

15. $\begin{bmatrix} 3 & 4 \\ 2 & -5 \end{bmatrix}$

16. $\begin{bmatrix} 11 & 0 \\ -3 & 2 \end{bmatrix}$

17. $\begin{bmatrix} 3 & -2 & 8 \\ 3 & 2 & -6 \\ -1 & 3 & 6 \end{bmatrix}$

18. $\begin{bmatrix} -2 & 9 & 4 \\ 7 & -6 & 0 \\ 6 & 7 & -6 \end{bmatrix}$

In Exercises 19–22, find the determinant of the matrix by the method of expansion by cofactors. Expand using the indicated row or column.

19. $\begin{bmatrix} -3 & 2 & 1 \\ 4 & 5 & 6 \\ 2 & -3 & 1 \end{bmatrix}$

   (a) Row 1

   (b) Column 2

20. $\begin{bmatrix} -3 & 4 & 2 \\ 6 & 3 & 1 \\ 4 & -7 & -8 \end{bmatrix}$

   (a) Row 2

   (b) Column 3

21. $\begin{bmatrix} 6 & 0 & -3 & 5 \\ 4 & 13 & 6 & -8 \\ -1 & 0 & 7 & 4 \\ 8 & 6 & 0 & 2 \end{bmatrix}$

   (a) Row 2

   (b) Column 2

22. $\begin{bmatrix} 10 & 8 & 3 & -7 \\ 4 & 0 & 5 & -6 \\ 0 & 3 & 2 & 7 \\ 1 & 0 & -3 & 2 \end{bmatrix}$

   (a) Row 3

   (b) Column 1

In Exercises 23–30, find the determinant of the matrix. Expand by cofactors on the row or column that appears to make the computations easiest.

23. $\begin{bmatrix} 1 & 4 & -2 \\ 3 & 2 & 0 \\ -1 & 4 & 3 \end{bmatrix}$

24. $\begin{bmatrix} 2 & -1 & 3 \\ 1 & 4 & 4 \\ 1 & 0 & 2 \end{bmatrix}$

25. $\begin{bmatrix} 2 & 4 & 6 \\ 0 & 3 & 1 \\ 0 & 0 & -5 \end{bmatrix}$

26. $\begin{bmatrix} -3 & 0 & 0 \\ 7 & 11 & 0 \\ 1 & 2 & 2 \end{bmatrix}$

27. $\begin{bmatrix} 2 & 6 & 6 & 2 \\ 2 & 7 & 3 & 6 \\ 1 & 5 & 0 & 1 \\ 3 & 7 & 0 & 7 \end{bmatrix}$

28. $\begin{bmatrix} 3 & 6 & -5 & 4 \\ -2 & 0 & 6 & 0 \\ 1 & 1 & 2 & 2 \\ 0 & 3 & -1 & -1 \end{bmatrix}$

29. $\begin{bmatrix} 3 & 2 & 4 & -1 & 5 \\ -2 & 0 & 1 & 3 & 2 \\ 1 & 0 & 0 & 4 & 0 \\ 6 & 0 & 2 & -1 & 0 \\ 3 & 0 & 5 & 1 & 0 \end{bmatrix}$

30. $\begin{bmatrix} 5 & 2 & 0 & 0 & -2 \\ 0 & 1 & 4 & 3 & 2 \\ 0 & 0 & 2 & 6 & 3 \\ 0 & 0 & 3 & 4 & 1 \\ 0 & 0 & 0 & 0 & 2 \end{bmatrix}$

In Exercises 31–34, evaluate the determinant. Do not use a graphing utility.

31. $\begin{vmatrix} 4 & 0 & 0 & 0 \\ 6 & -5 & 0 & 0 \\ 1 & 3 & 1 & 0 \\ 1 & -2 & 7 & 3 \end{vmatrix}$

32. $\begin{vmatrix} 5 & -10 & 1 & 1 \\ 0 & 6 & 3 & 4 \\ 0 & 0 & -2 & -1 \\ 0 & 0 & 0 & -1 \end{vmatrix}$

33. $\begin{vmatrix} -6 & 7 & 2 & 0 & 5 \\ 0 & -1 & 3 & 4 & -3 \\ 0 & 0 & -7 & 0 & 4 \\ 0 & 0 & 0 & -2 & 1 \\ 0 & 0 & 0 & 0 & -2 \end{vmatrix}$

34. $\begin{vmatrix} -2 & 0 & 0 & 0 & 0 \\ -1 & 4 & 0 & 0 & 0 \\ 3 & 5 & 1 & 0 & 0 \\ 6 & -11 & 8 & 10 & 0 \\ 0 & 13 & -9 & 0 & -3 \end{vmatrix}$

In Exercises 35–38, use the matrix capabilities of a graphing utility to evaluate the determinant.

35. $\begin{vmatrix} 1 & -1 & 8 & 4 \\ 2 & 6 & 0 & -4 \\ 2 & 0 & 2 & 6 \\ 0 & 2 & 8 & 0 \end{vmatrix}$

36. $\begin{vmatrix} 0 & -3 & 8 & 2 \\ 8 & 1 & -1 & 6 \\ -4 & 6 & 0 & 9 \\ -7 & 0 & 0 & 14 \end{vmatrix}$

37. $\begin{vmatrix} 3 & -2 & 4 & 3 & 1 \\ -1 & 0 & 2 & 1 & 0 \\ 5 & -1 & 0 & 3 & 2 \\ 4 & 7 & -8 & 0 & 0 \\ 1 & 2 & 3 & 0 & 2 \end{vmatrix}$

38. $\begin{vmatrix} -2 & 0 & 0 & 0 & 0 \\ 0 & 3 & 0 & 0 & 0 \\ 0 & 0 & -1 & 0 & 0 \\ 0 & 0 & 0 & 2 & 0 \\ 0 & 0 & 0 & 0 & -4 \end{vmatrix}$

In Exercises 39–42, find (a) $|A|$, (b) $|B|$, (c) AB, and (d) $|AB|$.

39. $A = \begin{bmatrix} -1 & 0 \\ 0 & 3 \end{bmatrix}$, $B = \begin{bmatrix} 2 & 0 \\ 0 & -1 \end{bmatrix}$

40. $A = \begin{bmatrix} 4 & 0 \\ 3 & -2 \end{bmatrix}$, $B = \begin{bmatrix} -1 & 1 \\ -2 & 2 \end{bmatrix}$

41. $A = \begin{bmatrix} -1 & 2 & 1 \\ 1 & 0 & 1 \\ 0 & 1 & 0 \end{bmatrix}$, $B = \begin{bmatrix} -1 & 0 & 0 \\ 0 & 2 & 0 \\ 0 & 0 & 3 \end{bmatrix}$

42. $A = \begin{bmatrix} 2 & 0 & 1 \\ 1 & -1 & 2 \\ 3 & 1 & 0 \end{bmatrix}$, $B = \begin{bmatrix} 2 & -1 & 4 \\ 0 & 1 & 3 \\ 3 & -2 & 1 \end{bmatrix}$

In Exercises 43 and 44, use the matrix capabilities of a graphing utility to find (a) $|A|$, (b) $|B|$, (c) AB, and (d) $|AB|$.

43. $A = \begin{bmatrix} 6 & 4 & 0 & 1 \\ 2 & -3 & -2 & -4 \\ 0 & 1 & 5 & 0 \\ -1 & 0 & -1 & 1 \end{bmatrix}$,

$B = \begin{bmatrix} 0 & -5 & 0 & -2 \\ -2 & 4 & -1 & -4 \\ 3 & 0 & 1 & 0 \\ 1 & -2 & 3 & 0 \end{bmatrix}$

44. $A = \begin{bmatrix} -1 & 5 & 2 & 0 \\ 0 & 0 & 1 & 1 \\ 3 & -3 & -1 & 0 \\ 4 & 2 & 4 & -1 \end{bmatrix}$,

$B = \begin{bmatrix} 1 & 5 & 0 & 0 \\ 10 & -1 & 2 & 4 \\ 2 & 0 & 0 & 1 \\ -3 & 2 & 5 & 0 \end{bmatrix}$

In Exercises 45–50, evaluate the determinants to verify the equation.

45. $\begin{vmatrix} w & x \\ y & z \end{vmatrix} = - \begin{vmatrix} y & z \\ w & x \end{vmatrix}$

46. $\begin{vmatrix} w & cx \\ y & cz \end{vmatrix} = c \begin{vmatrix} w & x \\ y & z \end{vmatrix}$

47. $\begin{vmatrix} w & x \\ y & z \end{vmatrix} = \begin{vmatrix} w & x + cw \\ y & z + cy \end{vmatrix}$

48. $\begin{vmatrix} w & x \\ cw & cx \end{vmatrix} = 0$

49. $\begin{vmatrix} 1 & x & x^2 \\ 1 & y & y^2 \\ 1 & z & z^2 \end{vmatrix} = (y - x)(z - x)(z - y)$

50. $\begin{vmatrix} a + b & a & a \\ a & a + b & a \\ a & a & a + b \end{vmatrix} = b^2(3a + b)$

In Exercises 51 and 52, solve for x.

51. $\begin{vmatrix} x + 3 & 2 \\ 1 & x + 2 \end{vmatrix} = 0$ **52.** $\begin{vmatrix} x - 2 & -1 \\ -3 & x \end{vmatrix} = 0$

In Exercises 53–58, evaluate the determinant, in which the entries are functions. Determinants of this type occur when changes of variables are made in calculus.

53. $\begin{vmatrix} 4u & -1 \\ -1 & 2v \end{vmatrix}$

54. $\begin{vmatrix} 3x^2 & -3y^2 \\ 1 & 1 \end{vmatrix}$

55. $\begin{vmatrix} e^{2x} & e^{3x} \\ 2e^{2x} & 3e^{3x} \end{vmatrix}$

56. $\begin{vmatrix} e^{-x} & xe^{-x} \\ -e^{-x} & (1 - x)e^{-x} \end{vmatrix}$

57. $\begin{vmatrix} x & \ln x \\ 1 & 1/x \end{vmatrix}$

58. $\begin{vmatrix} x & x \ln x \\ 1 & 1 + \ln x \end{vmatrix}$

Synthesis

True or False? In Exercises 59 and 60, determine whether the statement is true or false. Justify your answer.

59. If a square matrix has an entire row of zeros, the determinant will always be zero.

60. If two columns of a square matrix are the same, the determinant of the matrix will be zero.

61. Exploration   Find square matrices A and B to demonstrate that $|A + B| \neq |A| + |B|$.

62. Conjecture   Consider square matrices in which the entries are consecutive integers. An example of such a matrix is

$$\begin{bmatrix} 4 & 5 & 6 \\ 7 & 8 & 9 \\ 10 & 11 & 12 \end{bmatrix}.$$

Use a graphing utility to evaluate four determinants of this type. Make a conjecture based on the results. Then verify your conjecture.

In Exercises 63–65, a property of determinants is given (A and B are square matrices). State how the property has been applied to the given determinants and use a graphing utility to verify the results.

63. If B is obtained from A by interchanging two rows of A or by interchanging two columns of A, then $|B| = -|A|$.

(a) $\begin{vmatrix} 1 & 3 & 4 \\ -7 & 2 & -5 \\ 6 & 1 & 2 \end{vmatrix} = - \begin{vmatrix} 1 & 4 & 3 \\ -7 & -5 & 2 \\ 6 & 2 & 1 \end{vmatrix}$

(b) $\begin{vmatrix} 1 & 3 & 4 \\ -2 & 2 & 0 \\ 1 & 6 & 2 \end{vmatrix} = - \begin{vmatrix} 1 & 6 & 2 \\ -2 & 2 & 0 \\ 1 & 3 & 4 \end{vmatrix}$

64. If B is obtained from A by adding a multiple of a row of A to another row of A or by adding a multiple of a column of A to another column of A, then $|B| = |A|$.

(a) $\begin{vmatrix} 1 & -3 \\ 5 & 2 \end{vmatrix} = \begin{vmatrix} 1 & -3 \\ 0 & 17 \end{vmatrix}$

(b) $\begin{vmatrix} 5 & 4 & 2 \\ 2 & -3 & 4 \\ 7 & 6 & 3 \end{vmatrix} = \begin{vmatrix} 1 & 10 & -6 \\ 2 & -3 & 4 \\ 7 & 6 & 3 \end{vmatrix}$

65. If B is obtained from A by multiplying a row of A by a nonzero constant c or by multiplying a column of A by a nonzero constant c, then $|B| = c|A|$.

(a) $\begin{vmatrix} 5 & 10 \\ 2 & -3 \end{vmatrix} = 5 \begin{vmatrix} 1 & 2 \\ 2 & -3 \end{vmatrix}$

(b) $\begin{vmatrix} 1 & 8 & -3 \\ 3 & -12 & 6 \\ 7 & 4 & 9 \end{vmatrix} = 12 \begin{vmatrix} 1 & 2 & -1 \\ 3 & -3 & 2 \\ 7 & 1 & 3 \end{vmatrix}$

66. Writing   Write an argument that explains why the determinant of a 3×3 triangular matrix is the product of its main diagonal entries.

Review

In Exercises 67–70, factor the expression.

67. $x^2 - 3x + 2$

68. $x^2 + 5x + 6$

69. $4y^2 - 12y + 9$

70. $4y^2 - 28y + 49$

In Exercises 71 and 72, solve the system of equations using the method of substitution or the method of elimination.

71. $\begin{cases} 3x - 10y = 46 \\ x + y = -2 \end{cases}$

72. $\begin{cases} 5x + 7y = 23 \\ -4x - 2y = -4 \end{cases}$

8.8 Applications of Matrices and Determinants

Area of a Triangle

In this section, you will study some additional applications of matrices and determinants. The first involves a formula for finding the area of a triangle whose vertices are given by three points on a rectangular coordinate system.

> **Area of a Triangle**
>
> The area of a triangle with vertices (x_1, y_1), (x_2, y_2), and (x_3, y_3) is
>
> $$\text{Area} = \pm\frac{1}{2} \begin{vmatrix} x_1 & y_1 & 1 \\ x_2 & y_2 & 1 \\ x_3 & y_3 & 1 \end{vmatrix}$$
>
> where the symbol $(\pm)$ indicates that the appropriate sign should be chosen to yield a positive area.

Example 1   Finding the Area of a Triangle

Find the area of the triangle whose vertices are $(1, 0)$, $(2, 2)$, and $(4, 3)$, as shown in Figure 8.34.

Solution

Let $(x_1, y_1) = (1, 0)$, $(x_2, y_2) = (2, 2)$, and $(x_3, y_3) = (4, 3)$. Then, to find the area of the triangle, evaluate the determinant

$$\begin{vmatrix} x_1 & y_1 & 1 \\ x_2 & y_2 & 1 \\ x_3 & y_3 & 1 \end{vmatrix} = \begin{vmatrix} 1 & 0 & 1 \\ 2 & 2 & 1 \\ 4 & 3 & 1 \end{vmatrix}$$

$$= 1(-1)^2 \begin{vmatrix} 2 & 1 \\ 3 & 1 \end{vmatrix} + 0(-1)^3 \begin{vmatrix} 2 & 1 \\ 4 & 1 \end{vmatrix} + 1(-1)^4 \begin{vmatrix} 2 & 2 \\ 4 & 3 \end{vmatrix}$$

$$= 1(-1) + 0 + 1(-2)$$

$$= -3.$$

Using this value, you can conclude that the area of the triangle is

$$\text{Area} = -\frac{1}{2} \begin{vmatrix} 1 & 0 & 1 \\ 2 & 2 & 1 \\ 4 & 3 & 1 \end{vmatrix}$$

$$= -\frac{1}{2}(-3)$$

$$= \frac{3}{2} \text{ square units.}$$

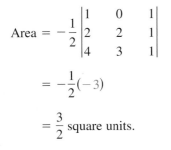 *Checkpoint* Now try Exercise 1.

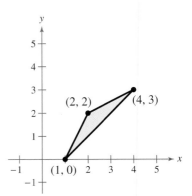

Layne Kennedy/Corbis

Figure 8.34

Collinear Points

What if the three points in Example 1 had been on the same line? What would have happened had the area formula been applied to three such points? The answer is that the determinant would have been zero. Consider, for instance, the three collinear points $(0, 1)$, $(2, 2)$, and $(4, 3)$, as shown in Figure 8.35. The area of the "triangle" that has these three points as vertices is

$$\frac{1}{2}\begin{vmatrix} 0 & 1 & 1 \\ 2 & 2 & 1 \\ 4 & 3 & 1 \end{vmatrix} = \frac{1}{2}\left[0(-1)^2\begin{vmatrix} 2 & 1 \\ 3 & 1 \end{vmatrix} + 1(-1)^3\begin{vmatrix} 2 & 1 \\ 4 & 1 \end{vmatrix} + 1(-1)^4\begin{vmatrix} 2 & 2 \\ 4 & 3 \end{vmatrix} \right]$$

$$= \frac{1}{2}[0 - 1(-2) + 1(-2)]$$

$$= 0$$

This result is generalized as follows.

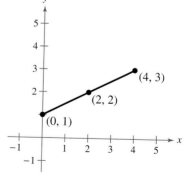

Figure 8.35

> **Test for Collinear Points**
>
> Three points (x_1, y_1), (x_2, y_2), and (x_3, y_3) are **collinear** (lie on the same line) if and only if
>
> $$\begin{vmatrix} x_1 & y_1 & 1 \\ x_2 & y_2 & 1 \\ x_3 & y_3 & 1 \end{vmatrix} = 0.$$

Example 2   Testing for Collinear Points

Determine whether the points $(-2, -2)$, $(1, 1)$, and $(7, 5)$ lie on the same line. (See Figure 8.36.)

Solution

Letting $(x_1, y_1) = (-2, -2)$, $(x_2, y_2) = (1, 1)$, and $(x_3, y_3) = (7, 5)$, you have

$$\begin{vmatrix} x_1 & y_1 & 1 \\ x_2 & y_2 & 1 \\ x_3 & y_3 & 1 \end{vmatrix} = \begin{vmatrix} -2 & -2 & 1 \\ 1 & 1 & 1 \\ 7 & 5 & 1 \end{vmatrix}$$

$$= -2(-1)^2\begin{vmatrix} 1 & 1 \\ 5 & 1 \end{vmatrix} + (-2)(-1)^3\begin{vmatrix} 1 & 1 \\ 7 & 1 \end{vmatrix} + 1(-1)^4\begin{vmatrix} 1 & 1 \\ 7 & 5 \end{vmatrix}$$

$$= -2(-4) + 2(-6) + 1(-2)$$

$$= -6.$$

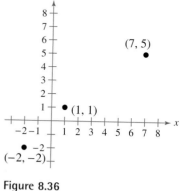

Figure 8.36

Because the value of this determinant is *not* zero, you can conclude that the three points do not lie on the same line.

 Checkpoint Now try Exercise 9.

Cramer's Rule

So far, you have studied three methods for solving a system of linear equations: substitution, elimination with equations, and elimination with matrices. You will now study one more method, **Cramer's Rule,** named after Gabriel Cramer (1704–1752). This rule uses determinants to write the solution of a system of linear equations. To see how Cramer's Rule works, take another look at the solution described at the beginning of Section 8.7. There, it was pointed out that the system

$$\begin{cases} a_1x + b_1y = c_1 \\ a_2x + b_2y = c_2 \end{cases}$$

has a solution

$$x = \frac{c_1b_2 - c_2b_1}{a_1b_2 - a_2b_1} \quad \text{and} \quad y = \frac{a_1c_2 - a_2c_1}{a_1b_2 - a_2b_1}$$

provided that $a_1b_2 - a_2b_1 \neq 0$. Each numerator and denominator in this solution can be expressed as a determinant, as follows.

$$x = \frac{c_1b_2 - c_2b_1}{a_1b_2 - a_2b_1} = \frac{\begin{vmatrix} c_1 & b_1 \\ c_2 & b_2 \end{vmatrix}}{\begin{vmatrix} a_1 & b_1 \\ a_2 & b_2 \end{vmatrix}}$$

$$y = \frac{a_1c_2 - a_2c_1}{a_1b_2 - a_2b_1} = \frac{\begin{vmatrix} a_1 & c_1 \\ a_2 & c_2 \end{vmatrix}}{\begin{vmatrix} a_1 & b_1 \\ a_2 & b_2 \end{vmatrix}}$$

Relative to the original system, the denominators of x and y are simply the determinant of the *coefficient* matrix of the system. This determinant is denoted by D. The numerators of x and y are denoted by D_x and D_y, respectively. They are formed by using the column of constants as replacements for the coefficients of x and y, as follows.

Coefficient Matrix	D	D_x	D_y
$\begin{bmatrix} a_1 & b_1 \\ a_2 & b_2 \end{bmatrix}$	$\begin{vmatrix} a_1 & b_1 \\ a_2 & b_2 \end{vmatrix}$	$\begin{vmatrix} c_1 & b_1 \\ c_2 & b_2 \end{vmatrix}$	$\begin{vmatrix} a_1 & c_1 \\ a_2 & c_2 \end{vmatrix}$

For example, given the system

$$\begin{cases} 2x - 5y = 3 \\ -4x + 3y = 8 \end{cases}$$

the coefficient matrix, D, D_x, and D_y are as follows.

Coefficient Matrix	D	D_x	D_y
$\begin{bmatrix} 2 & -5 \\ -4 & 3 \end{bmatrix}$	$\begin{vmatrix} 2 & -5 \\ -4 & 3 \end{vmatrix}$	$\begin{vmatrix} 3 & -5 \\ 8 & 3 \end{vmatrix}$	$\begin{vmatrix} 2 & 3 \\ -4 & 8 \end{vmatrix}$

Cramer's Rule generalizes easily to systems of n equations in n variables. The value of each variable is given as the quotient of two determinants. The denominator is the determinant of the coefficient matrix, and the numerator is the determinant of the matrix formed by replacing the column corresponding to the variable being solved for with the column representing the constants. For instance, the solution for x_3 in the following system is shown.

$$\begin{cases} a_{11}x_1 + a_{12}x_2 + a_{13}x_3 = b_1 \\ a_{21}x_1 + a_{22}x_2 + a_{23}x_3 = b_2 \\ a_{31}x_1 + a_{32}x_2 + a_{33}x_3 = b_3 \end{cases} \qquad x_3 = \frac{|A_3|}{|A|} = \frac{\begin{vmatrix} a_{11} & a_{12} & b_1 \\ a_{21} & a_{22} & b_2 \\ a_{31} & a_{32} & b_3 \end{vmatrix}}{\begin{vmatrix} a_{11} & a_{12} & a_{13} \\ a_{21} & a_{22} & a_{23} \\ a_{31} & a_{32} & a_{33} \end{vmatrix}}$$

Cramer's Rule

If a system of n linear equations in n variables has a coefficient matrix A with a *nonzero* determinant $|A|$, the solution of the system is

$$x_1 = \frac{|A_1|}{|A|}, \quad x_2 = \frac{|A_2|}{|A|}, \quad \ldots, \quad x_n = \frac{|A_n|}{|A|}$$

where the ith column of A_i is the column of constants in the system of equations. If the determinant of the coefficient matrix is zero, the system has either no solution or infinitely many solutions.

STUDY TIP

Cramer's Rule does not apply when the determinant of the coefficient matrix is zero. This would create division by zero, which is undefined.

Example 3   Using Cramer's Rule for a 2 × 2 System

Use Cramer's Rule to solve the system $\begin{cases} 4x - 2y = 10 \\ 3x - 5y = 11 \end{cases}$.

Solution

To begin, find the determinant of the coefficient matrix.

$$D = \begin{vmatrix} 4 & -2 \\ 3 & -5 \end{vmatrix} = -20 - (-6) = -14$$

Because this determinant is not zero, apply Cramer's Rule.

$$x = \frac{D_x}{D} = \frac{\begin{vmatrix} 10 & -2 \\ 11 & -5 \end{vmatrix}}{-14} = \frac{(-50) - (-22)}{-14} = \frac{-28}{-14} = 2$$

$$y = \frac{D_y}{D} = \frac{\begin{vmatrix} 4 & 10 \\ 3 & 11 \end{vmatrix}}{-14} = \frac{44 - 30}{-14} = \frac{14}{-14} = -1$$

So, the solution is $x = 2$ and $y = -1$. Check this in the original system.

✓ *Checkpoint* Now try Exercise 15.

Example 4   Using Cramer's Rule for a 3×3 System

Use Cramer's Rule and a graphing utility, if possible, to solve the system of linear equations.

$$\begin{cases} -x \quad\quad + \ z = \quad 4 \\ 2x - y + \ z = -3 \\ \quad\quad y - 3z = \quad 1 \end{cases}$$

Solution

Using a graphing utility to evaluate the determinant of the coefficient matrix A, you find that Cramer's Rule cannot be applied because $|A| = 0$.

 Checkpoint Now try Exercise 17.

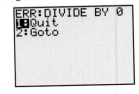
Example 5   Using Cramer's Rule for a 3×3 System

Use Cramer's Rule, if possible, to solve the system of linear equations.

$$\begin{cases} -x + 2y - 3z = 1 \\ 2x \quad\quad + \ z = 0 \\ 3x - 4y + 4z = 2 \end{cases}$$

Coefficient Matrix

$$\begin{bmatrix} -1 & 2 & -3 \\ 2 & 0 & 1 \\ 3 & -4 & 4 \end{bmatrix}$$

Solution

The coefficient matrix above can be expanded along the second row, as follows.

$$D = 2(-1)^3 \begin{vmatrix} 2 & -3 \\ -4 & 4 \end{vmatrix} + 0(-1)^4 \begin{vmatrix} -1 & -3 \\ 3 & 4 \end{vmatrix} + 1(-1)^5 \begin{vmatrix} -1 & 2 \\ 3 & -4 \end{vmatrix}$$

$$= -2(-4) + 0 - 1(-2) = 10$$

Because this determinant is not zero, you can apply Cramer's Rule.

$$x = \frac{D_x}{D} = \frac{\begin{vmatrix} 1 & 2 & -3 \\ 0 & 0 & 1 \\ 2 & -4 & 4 \end{vmatrix}}{10} = \frac{8}{10} = \frac{4}{5}$$

$$y = \frac{D_y}{D} = \frac{\begin{vmatrix} -1 & 1 & -3 \\ 2 & 0 & 1 \\ 3 & 2 & 4 \end{vmatrix}}{10} = \frac{-15}{10} = -\frac{3}{2}$$

$$z = \frac{D_z}{D} = \frac{\begin{vmatrix} -1 & 2 & 1 \\ 2 & 0 & 0 \\ 3 & -4 & 2 \end{vmatrix}}{10} = \frac{-16}{10} = -\frac{8}{5}$$

The solution is $\left(\frac{4}{5}, -\frac{3}{2}, -\frac{8}{5}\right)$. Check this in the original system.

 Checkpoint Now try Exercise 21.

Cryptography

A **cryptogram** is a message written according to a secret code. (The Greek word *kryptos* means "hidden.") Matrix multiplication can be used to encode and decode messages. To begin, you need to assign a number to each letter in the alphabet (with 0 assigned to a blank space), as follows.

0 = _	9 = I	18 = R
1 = A	10 = J	19 = S
2 = B	11 = K	20 = T
3 = C	12 = L	21 = U
4 = D	13 = M	22 = V
5 = E	14 = N	23 = W
6 = F	15 = O	24 = X
7 = G	16 = P	25 = Y
8 = H	17 = Q	26 = Z

Then the message is converted to numbers and partitioned into **uncoded row matrices,** each having n entries, as demonstrated in Example 6.

Example 6   Forming Uncoded Row Matrices

Write the uncoded row matrices of order 1×3 for the message

MEET ME MONDAY.

Solution

Partitioning the message (including blank spaces, but ignoring punctuation) into groups of three produces the following uncoded row matrices.

$$[13 \quad 5 \quad 5] \quad [20 \quad 0 \quad 13] \quad [5 \quad 0 \quad 13] \quad [15 \quad 14 \quad 4] \quad [1 \quad 25 \quad 0]$$

   M E E   T   M   E   M   O N D A Y

Note that a blank space is used to fill out the last uncoded row matrix.

✓ *Checkpoint* Now try Exercise 27.

To encode a message, choose an $n \times n$ invertible matrix A by using the techniques demonstrated in Section 8.6 and multiply the uncoded row matrices by A (on the right) to obtain **coded row matrices.** Here is an example.

Uncoded Matrix   *Encoding Matrix A*   *Coded Matrix*

$$[13 \quad 5 \quad 5] \begin{bmatrix} 1 & -2 & 2 \\ -1 & 1 & 3 \\ 1 & -1 & -4 \end{bmatrix} = [13 \quad -26 \quad 21]$$

This technique is further illustrated in Example 7.

Example 7   Encoding a Message

Use the following matrix to encode the message MEET ME MONDAY.

$$A = \begin{bmatrix} 1 & -2 & 2 \\ -1 & 1 & 3 \\ 1 & -1 & -4 \end{bmatrix}$$

Solution

The coded row matrices are obtained by multiplying each of the uncoded row matrices found in Example 6 by the matrix A, as follows.

Uncoded Matrix   *Encoding Matrix A*   *Coded Matrix*

$$\begin{bmatrix} 13 & 5 & 5 \end{bmatrix} \begin{bmatrix} 1 & -2 & 2 \\ -1 & 1 & 3 \\ 1 & -1 & -4 \end{bmatrix} = \begin{bmatrix} 13 & -26 & 21 \end{bmatrix}$$

$$\begin{bmatrix} 20 & 0 & 13 \end{bmatrix} \begin{bmatrix} 1 & -2 & 2 \\ -1 & 1 & 3 \\ 1 & -1 & -4 \end{bmatrix} = \begin{bmatrix} 33 & -53 & -12 \end{bmatrix}$$

$$\begin{bmatrix} 5 & 0 & 13 \end{bmatrix} \begin{bmatrix} 1 & -2 & 2 \\ -1 & 1 & 3 \\ 1 & -1 & -4 \end{bmatrix} = \begin{bmatrix} 18 & -23 & -42 \end{bmatrix}$$

$$\begin{bmatrix} 15 & 14 & 4 \end{bmatrix} \begin{bmatrix} 1 & -2 & 2 \\ -1 & 1 & 3 \\ 1 & -1 & -4 \end{bmatrix} = \begin{bmatrix} 5 & -20 & 56 \end{bmatrix}$$

$$\begin{bmatrix} 1 & 25 & 0 \end{bmatrix} \begin{bmatrix} 1 & -2 & 2 \\ -1 & 1 & 3 \\ 1 & -1 & -4 \end{bmatrix} = \begin{bmatrix} -24 & 23 & 77 \end{bmatrix}$$

So, the sequence of coded row matrices is

$$\begin{bmatrix} 13 & -26 & 21 \end{bmatrix}\begin{bmatrix} 33 & -53 & -12 \end{bmatrix}\begin{bmatrix} 18 & -23 & -42 \end{bmatrix}\begin{bmatrix} 5 & -20 & 56 \end{bmatrix}\begin{bmatrix} -24 & 23 & 77 \end{bmatrix}.$$

Finally, removing the matrix notation produces the following cryptogram.

$$13 \ -26 \ 21 \ 33 \ -53 \ -12 \ 18 \ -23 \ -42 \ 5 \ -20 \ 56 \ -24 \ 23 \ 77$$

✓ *Checkpoint* Now try Exercise 29.

For those who do not know the encoding matrix A, decoding the cryptogram found in Example 7 is difficult. But for an authorized receiver who knows the encoding matrix A, decoding is simple. The receiver need only multiply the coded row matrices by A^{-1} (on the right) to retrieve the uncoded row matrices. Here is an example.

$$\underbrace{\begin{bmatrix} 13 & -26 & 21 \end{bmatrix}}_{\text{Coded}} \underbrace{\begin{bmatrix} -1 & -10 & -8 \\ -1 & -6 & -5 \\ 0 & -1 & -1 \end{bmatrix}}_{A^{-1}} = \underbrace{\begin{bmatrix} 13 & 5 & 5 \end{bmatrix}}_{\text{Uncoded}}$$

TECHNOLOGY TIP

An efficient method for encoding the message at the left with your graphing utility is to enter A as a 3×3 matrix. Let B be the 5×3 matrix whose rows are the uncoded row matrices

$$B = \begin{bmatrix} 13 & 5 & 5 \\ 20 & 0 & 13 \\ 5 & 0 & 13 \\ 15 & 14 & 4 \\ 1 & 25 & 0 \end{bmatrix}.$$

The product BA gives the coded row matrices.

Example 8 Decoding a Message

Use the inverse of the matrix

$$A = \begin{bmatrix} 1 & -2 & 2 \\ -1 & 1 & 3 \\ 1 & -1 & -4 \end{bmatrix}$$

to decode the cryptogram

13 −26 21 33 −53 −12 18 −23 −42 5 −20 56 −24 23 77

Solution

First find A^{-1} by using the techniques demonstrated in Section 8.6. A^{-1} is the decoding matrix. Next partition the message into groups of three to form the coded row matrices. Then multiply each coded row matrix by A^{-1} (on the right).

Coded Matrix   *Decoding Matrix A^{-1}*   *Decoded Matrix*

$$\begin{bmatrix} 13 & -26 & 21 \end{bmatrix} \begin{bmatrix} -1 & -10 & -8 \\ -1 & -6 & -5 \\ 0 & -1 & -1 \end{bmatrix} = \begin{bmatrix} 13 & 5 & 5 \end{bmatrix}$$

$$\begin{bmatrix} 33 & -53 & -12 \end{bmatrix} \begin{bmatrix} -1 & -10 & -8 \\ -1 & -6 & -5 \\ 0 & -1 & -1 \end{bmatrix} = \begin{bmatrix} 20 & 0 & 13 \end{bmatrix}$$

$$\begin{bmatrix} 18 & -23 & -42 \end{bmatrix} \begin{bmatrix} -1 & -10 & -8 \\ -1 & -6 & -5 \\ 0 & -1 & -1 \end{bmatrix} = \begin{bmatrix} 5 & 0 & 13 \end{bmatrix}$$

$$\begin{bmatrix} 5 & -20 & 56 \end{bmatrix} \begin{bmatrix} -1 & -10 & -8 \\ -1 & -6 & -5 \\ 0 & -1 & -1 \end{bmatrix} = \begin{bmatrix} 15 & 14 & 4 \end{bmatrix}$$

$$\begin{bmatrix} -24 & 23 & 77 \end{bmatrix} \begin{bmatrix} -1 & -10 & -8 \\ -1 & -6 & -5 \\ 0 & -1 & -1 \end{bmatrix} = \begin{bmatrix} 1 & 25 & 0 \end{bmatrix}$$

So, the message is as follows.

$\begin{bmatrix} 13 & 5 & 5 \end{bmatrix}$ $\begin{bmatrix} 20 & 0 & 13 \end{bmatrix}$ $\begin{bmatrix} 5 & 0 & 13 \end{bmatrix}$ $\begin{bmatrix} 15 & 14 & 4 \end{bmatrix}$ $\begin{bmatrix} 1 & 25 & 0 \end{bmatrix}$

M E E   T   M E   M   O N D   A Y

✓ *Checkpoint* Now try Exercise 33.

TECHNOLOGY TIP   An efficient method for decoding the cryptogram in Example 8 with your graphing utility is to enter A as a 3×3 matrix and then find A^{-1}. Let B be the 5×3 matrix whose rows are the coded row matrices, as shown at the right. The product BA^{-1} gives the decoded row matrices.

$$B = \begin{bmatrix} 13 & -26 & 21 \\ 33 & -53 & -12 \\ 18 & -23 & -42 \\ 5 & -20 & 56 \\ -24 & 23 & 77 \end{bmatrix}$$

8.8   Exercises

Vocabulary Check

Fill in the blanks.

1. Three points are _____ if they lie on the same line.
2. The method of using determinants to solve a system of linear equations is called _____ .
3. A message written according to a secret code is called a _____ .
4. To encode a message, choose an invertible matrix A and multiply the _____ row matrices by A (on the right) to obtain _____ row matrices.

In Exercises 1–6, use a determinant to find the area of the triangle with the given vertices.

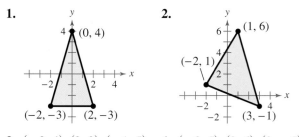

1.   2.

3. $(-2, 4), (2, 3), (-1, 5)$ 4. $(-3, 5), (2, 6), (3, -5)$
5. $\left(0, \frac{1}{2}\right), \left(\frac{5}{2}, 0\right), (4, 3)$ 6. $\left(\frac{9}{2}, 0\right), (2, 6), \left(0, -\frac{3}{2}\right)$

In Exercises 7 and 8, find x such that the triangle has an area of 4 square units.

7. $(-5, 1), (0, 2), (-2, x)$
8. $(-4, 2), (-3, 5), (-1, x)$

In Exercises 9–12, use a determinant to determine whether the points are collinear.

9. $(3, -1), (0, -3), (12, 5)$
10. $(3, -5), (6, 1), (4, 2)$
11. $\left(2, -\frac{1}{2}\right), (-4, 4), (6, -3)$
12. $\left(0, \frac{1}{2}\right), (2, -1), \left(-4, \frac{7}{2}\right)$

In Exercises 13 and 14, find x such that the points are collinear.

13. $(2, -5), (4, x), (5, -2)$
14. $(-6, 2), (-5, x), (-3, 5)$

In Exercises 15–22, use Cramer's Rule to solve (if possible) the system of equations.

15. $\begin{cases} -7x + 11y = -1 \\ 3x - 9y = 9 \end{cases}$ 16. $\begin{cases} 4x - 3y = -10 \\ 6x + 9y = 12 \end{cases}$

17. $\begin{cases} 3x + 2y = -2 \\ 6x + 4y = 4 \end{cases}$ 18. $\begin{cases} 6x - 5y = 17 \\ -13x + 3y = -76 \end{cases}$

19. $\begin{cases} -0.4x + 0.8y = 1.6 \\ 0.2x + 0.3y = 2.2 \end{cases}$ 20. $\begin{cases} 2.4x - 0.8y = 10.8 \\ 4.6x + 1.2y = 24.8 \end{cases}$

21. $\begin{cases} 4x - y + z = -5 \\ 2x + 2y + 3z = 10 \\ 5x - 2y + 6z = 1 \end{cases}$ 22. $\begin{cases} 4x - 2y + 3z = -2 \\ 2x + 2y + 5z = 16 \\ 8x - 5y - 2z = 4 \end{cases}$

In Exercises 23 and 24, use a graphing utility and Cramer's Rule to solve (if possible) the system of equations.

23. $\begin{cases} 3x + 3y + 5z = 1 \\ 3x + 5y + 9z = 2 \\ 5x + 9y + 17z = 4 \end{cases}$ 24. $\begin{cases} 2x + 3y + 5z = 4 \\ 3x + 5y + 9z = 7 \\ 5x + 9y + 17z = 13 \end{cases}$

25. *Area of a Region*   A large region of forest has been infected with gypsy moths. The region is roughly triangular, as shown in the figure. From the northernmost vertex A of the region, the distances to the other vertices are 25 miles south and 10 miles east (for vertex B), and 20 miles south and 28 miles east (for vertex C). Use a graphing utility to approximate the number of square miles in this region.

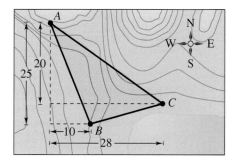

26. *Data Analysis* The table shows the numbers y (in millions) of families in the United States from 1998 to 2000. (Source: U.S. Census Bureau)

Year	Families, y
1998	70.9
1999	71.5
2000	72.0

(a) This data can be approximated by a parabola. Create a system of linear equations for the data. Let t represent the years with $t = 8$ corresponding to 1998.

(b) Use Cramer's Rule to solve the system in part (a) and find the least squares regression parabola $y = at^2 + bt + c$.

(c) Use the result of part (b) to estimate the number of families in the United States in 2006. Is the estimate reasonable? Explain.

In Exercises 27 and 28, find the uncoded 1×3 row matrices for the message. Then encode the message using the encoding matrix.

Message   *Encoding Matrix*

27. CALL ME TOMORROW
$$\begin{bmatrix} 1 & -1 & 0 \\ 1 & 0 & -1 \\ -6 & 2 & 3 \end{bmatrix}$$

28. PLEASE SEND MONEY
$$\begin{bmatrix} 4 & 2 & 1 \\ -3 & -3 & -1 \\ 3 & 2 & 1 \end{bmatrix}$$

In Exercises 29 and 30, write a cryptogram for the message using the matrix A.

$$A = \begin{bmatrix} 1 & 2 & 2 \\ 3 & 7 & 9 \\ -1 & -4 & -7 \end{bmatrix}$$

29. GONE FISHING   **30.** HAPPY BIRTHDAY

In Exercises 31 and 32, use A^{-1} to decode the cryptogram.

31. $A = \begin{bmatrix} 1 & 2 \\ 3 & 5 \end{bmatrix}$ 11 21 64 112 25 50 29 53
23 46 40 75 55 92

32. $A = \begin{bmatrix} 2 & 3 \\ 3 & 4 \end{bmatrix}$ 85 120 6 8 10 15 84 117
42 56 90 125 60 80 30
45 19 26

33. Decode the cryptogram by using the inverse of the matrix A.

$$A = \begin{bmatrix} 1 & 2 & 2 \\ 3 & 7 & 9 \\ -1 & -4 & -7 \end{bmatrix}$$

16 -1 -48 5 -20 -65 8 4 -14
41 83 89 76 177 227

34. The following cryptogram was encoded with a 2×2 matrix.

8 21 -15 -10 -13 -13 5 10 5 25 5 19
-1 6 20 40 -18 -18 1 16

The last word of the message is _RON. What is the message?

Synthesis

True or False? **In Exercises 35 and 36, determine whether the statement is true or false. Justify your answer.**

35. Cramer's Rule cannot be used to solve a system of linear equations if the determinant of the coefficient matrix is zero.

36. In a system of linear equations, if the determinant of the coefficient matrix is zero, the system has no solution.

37. *Writing* At this point in the book, you have learned several methods for solving a system of linear equations. Briefly describe which method(s) you find easiest to use and which method(s) you find most difficult to use.

38. *Writing* Use your school's library, the Internet, or some other reference source to research a few current real-life uses of cryptography. Write a short summary of these uses. Include a description of how messages are encoded and decoded in each case.

Review

In Exercises 39–42, find the general form of the equation of the line that passes through the two points.

39. $(-1, 5), (7, 3)$ **40.** $(0, -6), (-2, 10)$

41. $(3, -3), (10, -1)$ **42.** $(-4, 12), (4, 2)$

In Exercises 43 and 44, sketch the graph of the rational function. Identify any asymptotes.

43. $f(x) = \dfrac{2x^2}{x^2 + 4}$ **44.** $f(x) = \dfrac{2x}{x^2 + 3x - 18}$

8 Chapter Summary

What did you learn?

	Review Exercises
Section 8.1	
☐ Use the method of substitution and the graphical method to solve systems of equations in two variables.	1–12
☐ Use systems of equations to model and solve real-life problems.	13–16
Section 8.2	
☐ Use the method of elimination to solve systems of linear equations in two variables.	17–24
☐ Graphically interpret the number of solutions of a system of linear equations in two variables.	25–30
☐ Use systems of linear equations in two variables to model and solve real-life problems.	31–34
Section 8.3	
☐ Use back-substitution to solve linear systems in row-echelon form.	35, 36
☐ Use Gaussian elimination to solve systems of linear equations.	37–40
☐ Solve nonsquare systems of linear equations.	41, 42
☐ Graphically interpret three-variable linear systems.	43, 44
☐ Use systems of linear equations to write partial fraction decompositions of rational expressions.	45–48
☐ Use systems of linear equations in three or more variables to model and solve real-life problems.	49–52
Section 8.4	
☐ Write matrices and identify their orders.	53–62
☐ Perform elementary row operations on matrices.	63–68
☐ Use matrices and Gaussian elimination to solve systems of linear equations.	69–76
☐ Use matrices and Gauss-Jordan elimination to solve systems of linear equations.	77–84
Section 8.5	
☐ Decide whether two matrices are equal.	85–88
☐ Add and subtract matrices and multiply matrices by a scalar.	89–102
☐ Multiply two matrices.	103–110
☐ Use matrix operations to model and solve real-life problems.	111, 112
Section 8.6	
☐ Verify that two matrices are inverses of each other.	113, 114
☐ Use Gauss-Jordan elimination to find inverses of matrices.	115–122
☐ Use a formula to find inverses of 2×2 matrices.	123–126
☐ Use inverse matrices to solve systems of linear equations.	127–136

Section 8.7

Section 8.8

8 Review Exercises

8.1 In Exercises 1–6, solve the system by the method of substitution.

1. $\begin{cases} x + y = 2 \\ x - y = 0 \end{cases}$

2. $\begin{cases} 2x - 3y = 3 \\ x - y = 0 \end{cases}$

3. $\begin{cases} x^2 - y^2 = 9 \\ x - y = 1 \end{cases}$

4. $\begin{cases} x^2 + y^2 = 169 \\ 3x + 2y = 39 \end{cases}$

5. $\begin{cases} y = 2x^2 \\ y = x^4 - 2x^2 \end{cases}$

6. $\begin{cases} x = y + 3 \\ x = y^2 + 1 \end{cases}$

In Exercises 7–12, use a graphing utility to approximate all points of intersection of the graph of the system of equations. Verify your solutions by checking them in the original system.

7. $\begin{cases} 5x + 6y = 7 \\ -x - 4y = 0 \end{cases}$

8. $\begin{cases} 8x - 3y = -3 \\ 2x + 5y = 28 \end{cases}$

9. $\begin{cases} y^2 - 2y + x = 0 \\ x + y = 0 \end{cases}$

10. $\begin{cases} y = 2x^2 - 4x + 1 \\ y = x^2 - 4x + 3 \end{cases}$

11. $\begin{cases} y = 2(6 - x) \\ y = 2^{x-2} \end{cases}$

12. $\begin{cases} y = \ln(x - 1) - 3 \\ y = 4 - \frac{1}{2}x \end{cases}$

13. *Break-Even Analysis* You set up a business and make an initial investment of $10,000. The unit cost of the product is $2.85 and the selling price is $4.95. How many units must you sell to break even?

14. *Choice of Two Jobs* You are offered two sales jobs. One company offers an annual salary of $22,500 plus a year-end bonus of 1.5% of your total sales. The other company offers a salary of $20,000 plus a year-end bonus of 2% of your total sales. How much would you have to sell in a year to make the second offer the better offer?

15. *Geometry* The perimeter of a rectangle is 480 meters and its length is 1.5 times its width. Find the dimensions of the rectangle.

16. *Geometry* The perimeter of a rectangle is 68 feet and its width is $\frac{8}{9}$ times its length. Find the dimensions of the rectangle.

8.2 In Exercises 17–24, solve the system by the method of elimination.

17. $\begin{cases} 2x - y = 2 \\ 6x + 8y = 39 \end{cases}$

18. $\begin{cases} 40x + 30y = 24 \\ 20x - 50y = -14 \end{cases}$

19. $\begin{cases} \frac{1}{5}x + \frac{3}{10}y = \frac{7}{50} \\ \frac{2}{5}x + \frac{1}{2}y = \frac{1}{5} \end{cases}$

20. $\begin{cases} \frac{5}{12}x - \frac{3}{4}y = \frac{25}{4} \\ -x + \frac{7}{8}y = -\frac{38}{5} \end{cases}$

21. $\begin{cases} 3x - 2y = 0 \\ 3x + 2(y + 5) = 10 \end{cases}$

22. $\begin{cases} 7x + 12y = 63 \\ 2x + 3y = 15 \end{cases}$

23. $\begin{cases} 1.25x - 2y = 3.5 \\ 5x - 8y = 14 \end{cases}$

24. $\begin{cases} 1.5x + 2.5y = 8.5 \\ 6x + 10y = 24 \end{cases}$

In Exercises 25–30, use a graphing utility to graph the lines in the system. Use the graphs to determine whether the system is consistent or inconsistent. If the system is consistent, determine the solution. Verify your results algebraically.

25. $\begin{cases} 3x + 2y = 0 \\ x - y = 4 \end{cases}$

26. $\begin{cases} x + y = 6 \\ -2x - 2y = -12 \end{cases}$

27. $\begin{cases} \frac{1}{4}x - \frac{1}{5}y = 2 \\ -5x + 4y = 8 \end{cases}$

28. $\begin{cases} \frac{7}{2}x - 7y = -1 \\ -x + 2y = 4 \end{cases}$

29. $\begin{cases} 2x - 2y = 8 \\ 4x - 1.5y = -5.5 \end{cases}$

30. $\begin{cases} -x + 3.2y = 10.4 \\ -2x - 9.6y = 6.4 \end{cases}$

Supply and Demand In Exercises 31 and 32, find the point of equilibrium of the demand and supply equations.

Demand Function	Supply Function
31. $p = 37 - 0.0002x$	$p = 22 + 0.00001x$
32. $p = 120 - 0.0001x$	$p = 45 + 0.0002x$

33. *Airplane Speed* Two planes leave Pittsburgh and Philadelphia at the same time, each going to the other city. One plane flies 25 miles per hour faster than the other. Find the airspeed of each plane if the cities are 275 miles apart and the planes pass each other after 40 minutes of flying time.

34. *Investment Portfolio* A total of $46,000 is invested in two corporate bonds that pay 6.75% and 7.25% simple interest. The investor wants an annual interest income of $3245 from the investments. What is the most that can be invested in the 6.75% bond?

8.3 **In Exercises 35 and 36, use back-substitution to solve the system of linear equations.**

35. $\begin{cases} x - 4y + 3z = 3 \\ -y + z = -1 \\ z = -5 \end{cases}$ **36.** $\begin{cases} x - 7y + 8z = 85 \\ y - 9z = -35 \\ z = 3 \end{cases}$

In Exercises 37–42, solve the system of linear equations and check any solution algebraically.

37. $\begin{cases} x + 3y - z = 13 \\ 2x - 5z = 23 \\ 4x - y - 2z = 14 \end{cases}$ **38.** $\begin{cases} x + 2y + 6z = 4 \\ 3x - 2y + z = 4 \\ 4x + 2z = 0 \end{cases}$

39. $\begin{cases} x - 2y + z = -6 \\ 2x - 3y = -7 \\ -x + 3y - 3z = 11 \end{cases}$

40. $\begin{cases} 2x + 6z = -9 \\ 3x - 2y + 11z = -16 \\ 3x - y + 7z = -11 \end{cases}$

41. $\begin{cases} 5x - 12y + 7z = 16 \\ 3x - 7y + 4z = 9 \end{cases}$

42. $\begin{cases} 2x + 5y - 19z = 34 \\ 3x + 8y - 31z = 54 \end{cases}$

In Exercises 43 and 44, sketch the plane represented by the linear equation. Then list four points that lie in the plane.

43. $2x - 4y + z = 8$ **44.** $3x + 3y - z = 9$

In Exercises 45–48, (a) write the partial fraction decomposition for the rational expression, (b) check your result algebraically by combining the fractions, and (c) check your result graphically by using a graphing utility to graph the rational expression and the partial fractions in the same viewing window.

45. $\dfrac{4 - x}{x^2 + 6x + 8}$ **46.** $\dfrac{-x}{x^2 + 3x + 2}$

47. $\dfrac{x^2 + 2x}{x^3 - x^2 + x - 1}$ **48.** $\dfrac{3x^3 + 4x}{(x^2 + 1)^2}$

In Exercises 49 and 50, find the equation of the parabola $y = ax^2 + bx + c$ that passes through the points. To verify your result, use a graphing utility to plot the points and graph the parabola.

49. $(0, -5), (1, -2), (2, 5)$ **50.** $(-5, 6), (1, 0), (2, 20)$

51. *Agriculture* A mixture of 6 gallons of chemical A, 8 gallons of chemical B, and 13 gallons of chemical C is required to kill a destructive crop insect. Commercial spray X contains 1, 2, and 2 parts, respectively, of these chemicals. Commercial spray Y contains only chemical C. Commercial spray Z contains chemicals A, B, and C in equal amounts. How much of each type of commercial spray is needed to obtain the desired mixture?

52. *Investment Portfolio* An inheritance of $20,000 is divided among three investments yielding $1780 in interest per year. The interest rates for the three investments are 7%, 9%, and 11%. Find the amount of each investment if the second and third were $3000 and $1000 less than the first, respectively.

8.4 **In Exercises 53–56, determine the order of the matrix.**

53. $\begin{bmatrix} -3 \\ 1 \\ 10 \end{bmatrix}$ **54.** $\begin{bmatrix} 3 & -1 & 0 & 6 \\ -2 & 7 & 1 & 4 \end{bmatrix}$

55. $[14]$ **56.** $[6 \quad 7 \quad -5 \quad 0 \quad -8]$

In Exercises 57–60, write the augmented matrix for the system of linear equations.

57. $\begin{cases} 3x - 10y = 15 \\ 5x + 4y = 22 \end{cases}$ **58.** $\begin{cases} -x + y = 12 \\ 10x - 4y = -90 \end{cases}$

59. $\begin{cases} 8x - 7y + 4z = 12 \\ 3x - 5y + 2z = 20 \\ 5x + 3y - 3z = 26 \end{cases}$

60. $\begin{cases} 3x - 5y + z = 25 \\ -4x - 2z = -14 \\ 6x + y = 15 \end{cases}$

In Exercises 61 and 62, write the system of linear equations represented by the augmented matrix. (Use the variables x, y, z, and w, if applicable.)

61. $\begin{bmatrix} 5 & 1 & 7 & \vdots & -9 \\ 4 & 2 & 0 & \vdots & 10 \\ 9 & 4 & 2 & \vdots & 3 \end{bmatrix}$

62. $\begin{bmatrix} 13 & 16 & 7 & 3 & \vdots & 2 \\ 1 & 21 & 8 & 5 & \vdots & 12 \\ 4 & 10 & -4 & 3 & \vdots & -1 \end{bmatrix}$

In Exercises 63 and 64, write the matrix in row-echelon form. Remember that the row-echelon form of a matrix is not unique.

63. $\begin{bmatrix} 0 & 1 & 1 \\ 1 & 2 & 3 \\ 2 & 2 & 2 \end{bmatrix}$ 64. $\begin{bmatrix} 3 & 5 & 2 \\ 1 & -2 & 4 \\ -2 & 0 & 5 \end{bmatrix}$

In Exercises 65–68, use the matrix capabilities of a graphing utility to write the matrix in reduced row-echelon form.

65. $\begin{bmatrix} 3 & -2 & 1 & 0 \\ 4 & -3 & 0 & 1 \end{bmatrix}$

66. $\begin{bmatrix} 1 & 1 & 2 & 1 & 0 & 0 \\ -1 & 0 & 3 & 0 & 1 & 0 \\ 1 & 2 & 8 & 0 & 0 & 1 \end{bmatrix}$

67. $\begin{bmatrix} 1 & 3 & 4 \\ 0 & 1 & 1 \\ 2 & 4 & 6 \end{bmatrix}$ 68. $\begin{bmatrix} 4 & 8 & 16 \\ 3 & -1 & 2 \\ -2 & 10 & 12 \end{bmatrix}$

In Exercises 69–76, use matrices to solve the system of equations if possible. Use Gaussian elimination with back-substitution.

69. $\begin{cases} 5x + 4y = 2 \\ -x + y = -22 \end{cases}$ 70. $\begin{cases} 2x - 5y = 2 \\ 3x - 7y = 1 \end{cases}$

71. $\begin{cases} 2x + y = 0.3 \\ 3x - y = -1.3 \end{cases}$ 72. $\begin{cases} 0.2x - 0.1y = 0.07 \\ 0.4x - 0.5y = -0.01 \end{cases}$

73. $\begin{cases} 2x + 3y + 3z = 3 \\ 6x + 6y + 12z = 13 \\ 12x + 9y - z = 2 \end{cases}$

74. $\begin{cases} x + 2y + 6z = 1 \\ 2x + 5y + 15z = 4 \\ 3x + y + 3z = -6 \end{cases}$

75. $\begin{cases} 3x + 21y - 29z = -1 \\ 2x + 15y - 21z = 0 \end{cases}$

76. $\begin{cases} x + 2y + w = 3 \\ -3y + 3z = 0 \\ 4x + 4y + z + 2w = 0 \\ 2x - y + z = 3 \end{cases}$

In Exercises 77–80, use matrices to solve the system of equations if possible. Use Gauss-Jordan elimination.

77. $\begin{cases} -x + y + 2z = 1 \\ 2x + 3y + z = -2 \\ 5x + 4y + 2z = 4 \end{cases}$

78. $\begin{cases} 4x + 4y + 4z = 5 \\ 4x - 2y - 8z = 1 \\ 5x + 3y + 8z = 6 \end{cases}$

79. $\begin{cases} 2x - y + 9z = -8 \\ -x - 3y + 4z = -15 \\ 5x + 2y - z = 17 \end{cases}$

80. $\begin{cases} -3x + y + 7z = -20 \\ 5x - 2y - z = 34 \\ -x + y + 4z = -8 \end{cases}$

In Exercises 81–84, use the matrix capabilities of a graphing utility to reduce the augmented matrix corresponding to the system of equations and solve the system.

81. $\begin{cases} x + 2y - z = 7 \\ -y - z = 4 \\ 4x - z = 16 \end{cases}$ 82. $\begin{cases} 3x + 6z = 0 \\ -2x + y = 5 \\ y + 2z = 3 \end{cases}$

83. $\begin{cases} 3x - y + 5z - 2w = -44 \\ x + 6y + 4z - w = 1 \\ 5x - y + z + 3w = -15 \\ 4y - z - 8w = 58 \end{cases}$

84. $\begin{cases} 4x + 12y + 2z = 20 \\ x + 6y + 4z = 12 \\ x + 6y + z = 8 \\ -2x - 10y - 2z = -10 \end{cases}$

8.5 In Exercises 85–88, find x and y.

85. $\begin{bmatrix} -1 & x \\ y & 9 \end{bmatrix} = \begin{bmatrix} -1 & 12 \\ -7 & 9 \end{bmatrix}$

86. $\begin{bmatrix} -1 & 0 \\ x & 5 \\ -4 & y \end{bmatrix} = \begin{bmatrix} -1 & 0 \\ 8 & 5 \\ -4 & 0 \end{bmatrix}$

87. $\begin{bmatrix} x + 3 & 4 & -4y \\ 0 & -3 & 2 \\ -2 & y + 5 & 6x \end{bmatrix} = \begin{bmatrix} 5x - 1 & 4 & -44 \\ 0 & -3 & 2 \\ -2 & 16 & 6 \end{bmatrix}$

88. $\begin{bmatrix} -9 & 4 & 2 & -5 \\ 0 & -3 & 7 & -4 \\ 6 & -1 & 1 & 0 \end{bmatrix} = \begin{bmatrix} -9 & 4 & x - 10 & -5 \\ 0 & -3 & 7 & 2y \\ \frac{1}{2}x & -1 & 1 & 0 \end{bmatrix}$

In Exercises 89–92, find, if possible, (a) $A + B$, (b) $A - B$, (c) $4A$, and (d) $A + 3B$.

89. $A = \begin{bmatrix} 7 & 3 \\ -1 & 5 \end{bmatrix}$, $B = \begin{bmatrix} 10 & -20 \\ 14 & -3 \end{bmatrix}$

90. $A = \begin{bmatrix} -11 & 16 & 19 \\ -7 & -2 & 1 \end{bmatrix}$, $B = \begin{bmatrix} 6 & 0 \\ 8 & -4 \\ -2 & 10 \end{bmatrix}$

91. $A = \begin{bmatrix} 6 & 0 & 7 \\ 5 & -1 & 2 \\ 3 & 2 & 3 \end{bmatrix}$, $B = \begin{bmatrix} 0 & 5 & 1 \\ -4 & 8 & 6 \\ 2 & -1 & 1 \end{bmatrix}$

92. $A = \begin{bmatrix} 2 & -3 & 6 \\ 0 & 4 & 1 \end{bmatrix}$, $B = \begin{bmatrix} -3 & 5 & 5 \\ 1 & 1 & 1 \end{bmatrix}$

In Exercises 93–96, evaluate the expression. If it is not possible, explain why.

93. $\begin{bmatrix} 2 & 1 & 0 \\ 0 & 5 & -4 \end{bmatrix} - 3\begin{bmatrix} 5 & 3 & -6 \\ 0 & -2 & 5 \end{bmatrix}$

94. $-2\begin{bmatrix} 1 & 2 \\ 5 & -4 \\ 6 & 0 \end{bmatrix} + 8\begin{bmatrix} 7 & 1 \\ 1 & 2 \\ 1 & 4 \end{bmatrix}$

95. $-\begin{bmatrix} 8 & -1 & 8 \\ -2 & 4 & 12 \\ 0 & -6 & 0 \end{bmatrix} - 5\begin{bmatrix} -2 & 0 & -4 \\ 3 & -1 & 1 \\ 6 & 12 & -8 \end{bmatrix}$

96. $6\begin{bmatrix} -4 & -1 & -3 & 4 \\ 2 & -5 & 7 & -10 \end{bmatrix} + 2\begin{bmatrix} -1 & 1 & 13 & -7 \\ 14 & -3 & 8 & -1 \end{bmatrix}$

In Exercises 97 and 98, use the matrix capabilities of a graphing utility to evaluate the expression.

97. $3\begin{bmatrix} 8 & -2 & 5 \\ 1 & 3 & -1 \end{bmatrix} + 6\begin{bmatrix} 4 & -2 & -3 \\ 2 & 7 & 6 \end{bmatrix}$

98. $-5\begin{bmatrix} 2 & 0 \\ 7 & -2 \\ 8 & 2 \end{bmatrix} + 4\begin{bmatrix} 4 & -2 \\ 6 & 11 \\ -1 & 3 \end{bmatrix}$

In Exercises 99–102, solve for X when

$$A = \begin{bmatrix} -4 & 0 \\ 1 & -5 \\ -3 & 2 \end{bmatrix} \quad \text{and} \quad B = \begin{bmatrix} 1 & 2 \\ -2 & 1 \\ 4 & 4 \end{bmatrix}.$$

99. $X = 3A - 2B$ **100.** $6X = 4A + 3B$

101. $3X + 2A = B$ **102.** $2A - 5B = 3X$

In Exercises 103–106, find AB if possible.

103. $A = \begin{bmatrix} 1 & 2 \\ 5 & -4 \\ 6 & 0 \end{bmatrix}$, $B = \begin{bmatrix} 6 & -2 & 8 \\ 4 & 0 & 0 \end{bmatrix}$

104. $A = \begin{bmatrix} 1 & 5 & 6 \\ 2 & -4 & 0 \end{bmatrix}$, $B = \begin{bmatrix} 7 & 5 & 2 \\ 0 & 1 & 0 \end{bmatrix}$

105. $A = \begin{bmatrix} 3 & -2 & 0 \\ 1 & 4 & 9 \end{bmatrix}$, $B = \begin{bmatrix} 7 & 0 \\ 5 & 3 \\ -1 & 3 \end{bmatrix}$

106. $A = \begin{bmatrix} 1 & 3 & 2 \\ 0 & 2 & -4 \\ 1 & -1 & 3 \end{bmatrix}$, $B = \begin{bmatrix} 4 & -3 & 2 \\ 0 & 3 & -1 \\ 0 & 6 & 2 \end{bmatrix}$

In Exercises 107–110, use the matrix capabilities of a graphing utility to evaluate the expression.

107. $\begin{bmatrix} 4 & 1 \\ 11 & -7 \\ 12 & 3 \end{bmatrix}\begin{bmatrix} 3 & -5 & 6 \\ 2 & -2 & -2 \end{bmatrix}$

108. $\begin{bmatrix} -2 & 3 & 10 \\ 4 & -2 & 2 \end{bmatrix}\begin{bmatrix} 1 & 1 \\ -5 & 2 \\ 3 & 2 \end{bmatrix}$

109. $\begin{bmatrix} 2 & 1 \\ 6 & 0 \end{bmatrix}\left(\begin{bmatrix} 4 & 2 \\ -3 & 1 \end{bmatrix} + \begin{bmatrix} -2 & 4 \\ 0 & 4 \end{bmatrix}\right)$

110. $\begin{bmatrix} 1 & -1 \\ 4 & 2 \end{bmatrix}\left(\begin{bmatrix} 0 & 3 \\ 1 & 2 \end{bmatrix}\begin{bmatrix} 1 & 0 \\ 5 & -3 \end{bmatrix}\right)$

111. *Manufacturing* A corporation has four factories, each of which manufactures three types of cordless power tools. The number of units of cordless power tools produced at factory j in one day is represented by a_{ij} in the matrix

$$A = \begin{bmatrix} 80 & 70 & 90 & 40 \\ 50 & 30 & 80 & 20 \\ 90 & 60 & 100 & 50 \end{bmatrix}.$$

Find the production levels if production is increased by 20%.

112. *Manufacturing* A manufacturing company produces three kinds of computer games that are shipped to two warehouses. The number of units of game i that are shipped to warehouse j is represented by a_{ij} in the matrix

$$A = \begin{bmatrix} 8200 & 7400 \\ 6500 & 9800 \\ 5400 & 4800 \end{bmatrix}.$$

The price per unit is represented by the matrix

$$B = [\$10.25 \quad \$14.50 \quad \$17.75].$$

Compute BA and state what each entry of the product represents.

8.6 In Exercises 113 and 114, show that B is the inverse of A.

113. $A = \begin{bmatrix} -4 & -1 \\ 7 & 2 \end{bmatrix}, \quad B = \begin{bmatrix} -2 & -1 \\ 7 & 4 \end{bmatrix}$

114. $A = \begin{bmatrix} 1 & 1 & 0 \\ 1 & 0 & 1 \\ 6 & 2 & 3 \end{bmatrix}, \quad B = \begin{bmatrix} -2 & -3 & 1 \\ 3 & 3 & -1 \\ 2 & 4 & -1 \end{bmatrix}$

In Exercises 115–118, find the inverse of the matrix (if it exists).

115. $\begin{bmatrix} -6 & 5 \\ -5 & 4 \end{bmatrix}$ **116.** $\begin{bmatrix} -3 & -5 \\ 2 & 3 \end{bmatrix}$

117. $\begin{bmatrix} -1 & -2 & -2 \\ 3 & 7 & 9 \\ 1 & 4 & 7 \end{bmatrix}$ **118.** $\begin{bmatrix} 0 & -2 & 1 \\ -5 & -2 & -3 \\ 7 & 3 & 4 \end{bmatrix}$

In Exercises 119–122, use the matrix capabilities of a graphing utility to find the inverse of the matrix (if it exists).

119. $\begin{bmatrix} 2 & 6 \\ 3 & -6 \end{bmatrix}$ **120.** $\begin{bmatrix} 3 & -10 \\ 4 & 2 \end{bmatrix}$

121. $\begin{bmatrix} 2 & 0 & 3 \\ -1 & 1 & 1 \\ 2 & -2 & 1 \end{bmatrix}$ **122.** $\begin{bmatrix} 1 & 4 & 6 \\ 2 & -3 & 1 \\ -1 & 18 & 16 \end{bmatrix}$

In Exercises 123–126, use the formula on page 663 to find the inverse of the 2 × 2 matrix.

123. $\begin{bmatrix} -7 & 2 \\ -8 & 2 \end{bmatrix}$ **124.** $\begin{bmatrix} 10 & 4 \\ 7 & 3 \end{bmatrix}$

125. $\begin{bmatrix} -1 & 20 \\ \frac{3}{10} & -6 \end{bmatrix}$ **126.** $\begin{bmatrix} -6 & -5 \\ 3 & 3 \end{bmatrix}$

In Exercises 127–132, use an inverse matrix to solve (if possible) the system of linear equations.

127. $\begin{cases} -x + 4y = 8 \\ 2x - 7y = -5 \end{cases}$ **128.** $\begin{cases} 5x - y = 13 \\ -9x + 2y = -24 \end{cases}$

129. $\begin{cases} 3x + 2y - z = 6 \\ x - y + 2z = -1 \\ 5x + y + z = 7 \end{cases}$

130. $\begin{cases} -x + 4y - 2z = 12 \\ 2x - 9y + 5z = -25 \\ -x + 5y - 4z = 10 \end{cases}$

131. $\begin{cases} -2x + y + 2z = -13 \\ -x - 4y + z = -11 \\ -y - z = 0 \end{cases}$

132. $\begin{cases} 3x - y + 5z = -14 \\ -x + y + 6z = 8 \\ -8x + 4y - z = 44 \end{cases}$

In Exercises 133–136, use the matrix capabilities of a graphing utility to solve (if possible) the system of linear equations.

133. $\begin{cases} x + 2y = -1 \\ 3x + 4y = -5 \end{cases}$ **134.** $\begin{cases} x + 3y = 23 \\ -6x + 2y = -18 \end{cases}$

135. $\begin{cases} -3x - 3y - 4z = 2 \\ y + z = -1 \\ 4x + 3y + 4z = -1 \end{cases}$

136. $\begin{cases} 2x + 3y - 4z = 1 \\ x - y + 2z = -4 \\ 3x + 7y - 10z = 0 \end{cases}$

8.7 In Exercises 137–140, find the determinant of the matrix.

137. $\begin{bmatrix} 8 & 5 \\ 2 & -4 \end{bmatrix}$ **138.** $\begin{bmatrix} -9 & 11 \\ 7 & -4 \end{bmatrix}$

139. $\begin{bmatrix} 50 & -30 \\ 10 & 5 \end{bmatrix}$ **140.** $\begin{bmatrix} 14 & -24 \\ 12 & -15 \end{bmatrix}$

In Exercises 141–144, find all (a) minors and (b) cofactors of the matrix.

141. $\begin{bmatrix} 2 & -1 \\ 7 & 4 \end{bmatrix}$ **142.** $\begin{bmatrix} 3 & 6 \\ 5 & -4 \end{bmatrix}$

143. $\begin{bmatrix} 3 & 2 & -1 \\ -2 & 5 & 0 \\ 1 & 8 & 6 \end{bmatrix}$ **144.** $\begin{bmatrix} 8 & 3 & 4 \\ 6 & 5 & -9 \\ -4 & 1 & 2 \end{bmatrix}$

In Exercises 145–150, find the determinant of the matrix. Expand by cofactors on the row or column that appears to make the computations easiest.

145. $\begin{bmatrix} -2 & 4 & 1 \\ -6 & 0 & 2 \\ 5 & 3 & 4 \end{bmatrix}$ **146.** $\begin{bmatrix} 4 & 7 & -1 \\ 2 & -3 & 4 \\ -5 & 1 & -1 \end{bmatrix}$

147. $\begin{bmatrix} 1 & 0 & -2 \\ 0 & 1 & 0 \\ -2 & 0 & 1 \end{bmatrix}$ **148.** $\begin{bmatrix} 0 & 3 & 1 \\ 5 & -2 & 1 \\ 1 & 6 & 1 \end{bmatrix}$

149. $\begin{bmatrix} 3 & 0 & -4 & 0 \\ 0 & 8 & 1 & 2 \\ 6 & 1 & 8 & 2 \\ 0 & 3 & -4 & 1 \end{bmatrix}$

150. $\begin{bmatrix} -5 & 6 & 0 & 0 \\ 0 & 1 & -1 & 2 \\ -3 & 4 & -5 & 1 \\ 1 & 6 & 0 & 3 \end{bmatrix}$

In Exercises 151 and 152, evaluate the determinant. Do not use a graphing utility.

151. $\begin{vmatrix} 8 & 6 & 0 & 2 \\ 0 & -1 & 1 & -4 \\ 0 & 0 & 4 & 5 \\ 0 & 0 & 0 & 3 \end{vmatrix}$

152. $\begin{vmatrix} -5 & 0 & 0 & 0 \\ 7 & -2 & 0 & 0 \\ 11 & 21 & 2 & 0 \\ -6 & 9 & 12 & 14 \end{vmatrix}$

8.8 **In Exercises 153–156, use a determinant to find the area of the triangle with the given vertices.**

153. $(1, 0), (5, 0), (5, 8)$ **154.** $(-4, 0), (4, 0), (0, 6)$

155. $\left(\frac{1}{2}, 1\right), \left(2, -\frac{5}{2}\right), \left(\frac{3}{2}, 1\right)$ **156.** $\left(\frac{3}{2}, 1\right), \left(4, -\frac{1}{2}\right), (4, 2)$

In Exercises 157 and 158, use a determinant to determine whether the points are collinear.

157. $(-1, 7), (2, 5), (4, 1)$ **158.** $(0, -5), (2, 1), (4, 7)$

In Exercises 159–164, use Cramer's Rule to solve (if possible) the system of equations.

159. $\begin{cases} x + 2y = 5 \\ -x + y = 1 \end{cases}$ **160.** $\begin{cases} 2x - y = -10 \\ 3x + 2y = -1 \end{cases}$

161. $\begin{cases} 5x - 2y = 6 \\ -11x + 3y = -23 \end{cases}$ **162.** $\begin{cases} 3x + 8y = -7 \\ 9x - 5y = 37 \end{cases}$

163. $\begin{cases} -2x + 3y - 5z = -11 \\ 4x - y + z = -3 \\ -x - 4y + 6z = 15 \end{cases}$

164. $\begin{cases} 5x - 2y + z = 15 \\ 3x - 3y - z = -7 \\ 2x - y - 7z = -3 \end{cases}$

In Exercises 165 and 166, use a graphing utility and Cramer's Rule to solve (if possible) the system of equations.

165. $\begin{cases} x - 3y + 2z = 2 \\ 2x + 2y - 3z = 3 \\ x - 7y + 8z = -4 \end{cases}$

166. $\begin{cases} 14x - 21y - 7z = 10 \\ -4x + 2y - 2z = 4 \\ 56x - 21y + 7z = 5 \end{cases}$

In Exercises 167 and 168, find the uncoded 1×3 row matrices for the message. Then encode the message using the encoding matrix.

Message	*Encoding Matrix*

167. LOOK OUT BELOW $\begin{bmatrix} 2 & -2 & 0 \\ 3 & 0 & -3 \\ -6 & 2 & 3 \end{bmatrix}$

168. CONGRATULATIONS $\begin{bmatrix} 2 & 1 & 0 \\ -6 & -6 & -2 \\ 3 & 2 & 1 \end{bmatrix}$

In Exercises 169 and 170, decode the cryptogram by using the inverse of the matrix

$$A = \begin{bmatrix} -5 & 4 & -3 \\ 10 & -7 & 6 \\ 8 & -6 & 5 \end{bmatrix}.$$

169. -5 11 -2 370 -265 225 -57 48 -33 32
-15 20 245 -171 147

170. 67 -43 43 84 -62 53 -17 14 -10 -30
26 -17 17 -9 12 -60 48 -36

Synthesis

True or False? **In Exercises 171 and 172, determine whether the statement is true or false. Justify your answer.**

171. Solving a system of equations graphically will always give an exact solution.

172. $\begin{vmatrix} a_{11} & a_{12} & a_{13} \\ a_{21} & a_{22} & a_{23} \\ a_{31} + c_1 & a_{32} + c_2 & a_{33} + c_3 \end{vmatrix} =$

$\begin{vmatrix} a_{11} & a_{12} & a_{13} \\ a_{21} & a_{22} & a_{23} \\ a_{31} & a_{32} & a_{33} \end{vmatrix} + \begin{vmatrix} a_{11} & a_{12} & a_{13} \\ a_{21} & a_{22} & a_{23} \\ c_1 & c_2 & c_3 \end{vmatrix}$

173. What is the relationship between the three elementary row operations performed on an augmented matrix and the operations that lead to equivalent systems of equations?

174. Under what conditions does a matrix have an inverse?

8 Chapter Test

Take this test as you would take a test in class. After you are finished, check your work against the answers given in the back of the book.

In Exercises 1–3, solve the system by the method of substitution. Check your solution graphically.

1. $\begin{cases} x - y = 6 \\ 3x + 5y = 2 \end{cases}$ **2.** $\begin{cases} y = x - 1 \\ y = (x - 1)^3 \end{cases}$ **3.** $\begin{cases} 4x - y^2 = 7 \\ x - y = 3 \end{cases}$

In Exercises 4–6, solve the system by the method of elimination.

4. $\begin{cases} 2x + 5y = -11 \\ 5x - y = 19 \end{cases}$ **5.** $\begin{cases} x - 2y + 3z = -5 \\ 2x - z = -4 \\ 3y + z = 17 \end{cases}$ **6.** $\begin{cases} 5x + 5y - z = 0 \\ 10x + 5y + 2z = 0 \\ 5x + 15y - 9z = 0 \end{cases}$

7. Find the equation of the parabola $y = ax^2 + bx + c$ that passes through the points $(0, 6)$, $(-2, 2)$, and $(3, \frac{9}{2})$.

8. Write the partial fraction decomposition for the rational expression $\dfrac{5x - 2}{(x - 1)^2}$.

In Exercises 9 and 10, use matrices to solve the system of equations if possible.

9. $\begin{cases} 2x + y + 2z = 4 \\ 2x + 2y = 5 \\ 2x - y + 6z = 2 \end{cases}$ **10.** $\begin{cases} 2x + 3y + z = 10 \\ 2x - 3y - 3z = 22 \\ 4x - 2y + 3z = -2 \end{cases}$

$\begin{cases} -6x + 4y = 10 \\ 10x - 5y = 20 \end{cases}$

System for 12

11. If possible, find (a) $A - B$, (b) $3A$, (c) $3A - 2B$, and (d) AB.

$$A = \begin{bmatrix} 5 & 4 & 4 \\ -4 & -4 & 0 \\ 1 & 2 & 0 \end{bmatrix}, \quad B = \begin{bmatrix} 4 & 4 & 0 \\ 3 & 2 & 1 \\ 1 & -2 & 0 \end{bmatrix}$$

12. Find A^{-1} for $A = \begin{bmatrix} -6 & 4 \\ 10 & -5 \end{bmatrix}$ and use A^{-1} to solve the system at the right.

In Exercises 13 and 14, find the determinant of the matrix.

13. $\begin{bmatrix} -25 & 18 \\ 6 & -7 \end{bmatrix}$ **14.** $\begin{bmatrix} 4 & 0 & 3 \\ 1 & -8 & 2 \\ 3 & 2 & 2 \end{bmatrix}$

15. Use a determinant to find the area of the triangle shown at the right.

16. Use Cramer's Rule to solve (if possible) $\begin{cases} 20x + 8y = 11 \\ 12x - 24y = 21 \end{cases}$.

17. The flow of traffic (in vehicles per hour) through a network of streets is shown at the right. Solve the system for the traffic flow represented by x_i, $i = 1, 2, 3, 4,$ and 5.

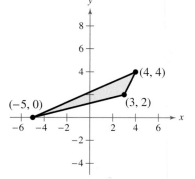

Figure for 15

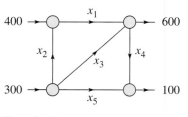

Figure for 17

Many states have established lotteries to increase revenues. You can use the probability theory developed in this chapter to calculate the odds of winning a state lottery.

Michael Simpson/Getty Images

9 Sequences, Series, and Probability

What You Should Learn

In this chapter, you will learn how to:

- Use sequence, factorial, and summation notation to write the terms and sums of sequences.

- Recognize, write, and use arithmetic sequences and geometric sequences.

- Use mathematical induction to prove statements involving a positive integer n.

- Use the Binomial Theorem and Pascal's Triangle to calculate binomial coefficients and write binomial expansions.

- Solve counting problems using the Fundamental Counting Principle, permutations, and combinations.

- Find the probabilities of events and their complements.

695

9.1 Sequences and Series

Sequences

In mathematics, the word *sequence* is used in much the same way as in ordinary English. Saying that a collection is listed *in sequence* means that it is ordered so that it has a first member, a second member, a third member, and so on.

Mathematically, you can think of a sequence as a *function* whose domain is the set of positive integers. Instead of using function notation, sequences are usually written using subscript notation, as shown in the following definition.

Definition of Sequence

An **infinite sequence** is a function whose domain is the set of positive integers. The function values

$$a_1, a_2, a_3, a_4, \ldots, a_n, \ldots$$

are the **terms** of the sequence. If the domain of a function consists of the first n positive integers only, the sequence is a **finite sequence.**

On occasion, it is convenient to begin subscripting a sequence with 0 instead of 1 so that the terms of the sequence become $a_0, a_1, a_2, a_3, \ldots$.

Example 1   Writing the Terms of a Sequence

Write the first four terms of the sequences given by

a. $a_n = 3n - 2$ **b.** $a_n = 3 + (-1)^n$.

Solution

a. The first four terms of the sequence given by $a_n = 3n - 2$ are

$$a_1 = 3(1) - 2 = 1 \qquad \text{1st term}$$
$$a_2 = 3(2) - 2 = 4 \qquad \text{2nd term}$$
$$a_3 = 3(3) - 2 = 7 \qquad \text{3rd term}$$
$$a_4 = 3(4) - 2 = 10. \qquad \text{4th term}$$

b. The first four terms of the sequence given by $a_n = 3 + (-1)^n$ are

$$a_1 = 3 + (-1)^1 = 3 - 1 = 2 \qquad \text{1st term}$$
$$a_2 = 3 + (-1)^2 = 3 + 1 = 4 \qquad \text{2nd term}$$
$$a_3 = 3 + (-1)^3 = 3 - 1 = 2 \qquad \text{3rd term}$$
$$a_4 = 3 + (-1)^4 = 3 + 1 = 4. \qquad \text{4th term}$$

✓ *Checkpoint* Now try Exercise 1.

What you should learn

- Use sequence notation to write the terms of sequences.
- Use factorial notation.
- Use summation notation to write sums.
- Find sums of infinite series.
- Use sequences and series to model and solve real-life problems.

Why you should learn it

Sequences and series are useful in modeling sets of values in order to identify a pattern. For instance, Exercise 111 on page 705 shows how a sequence can be used to model the number of children enrolled in Head Start programs from 1993 to 2001.

Cathy Melloan Resources/PhotoEdit

TECHNOLOGY TIP

To graph a sequence using a graphing utility, set the mode to *dot* and *sequence* and enter the sequence. Try graphing the sequences in Example 1 and using the *value* or *trace* feature to identify the terms. For instructions on how to use the *dot* mode, *sequence* mode, *value* feature, and *trace* feature, see Appendix A; for specific keystrokes, go to the text website at *college.hmco.com.*

Example 2   Writing the Terms of a Sequence

Write the first five terms of the sequence given by $a_n = \dfrac{(-1)^n}{2n-1}$.

Algebraic Solution

The first five terms of the sequence are as follows.

$a_1 = \dfrac{(-1)^1}{2(1)-1} = \dfrac{-1}{2-1} = -1$ 1st term

$a_2 = \dfrac{(-1)^2}{2(2)-1} = \dfrac{1}{4-1} = \dfrac{1}{3}$ 2nd term

$a_3 = \dfrac{(-1)^3}{2(3)-1} = \dfrac{-1}{6-1} = -\dfrac{1}{5}$ 3rd term

$a_4 = \dfrac{(-1)^4}{2(4)-1} = \dfrac{1}{8-1} = \dfrac{1}{7}$ 4th term

$a_5 = \dfrac{(-1)^5}{2(5)-1} = \dfrac{-1}{10-1} = -\dfrac{1}{9}$ 5th term

✓ *Checkpoint* Now try Exercise 11.

Numerical Solution

Set your graphing utility to *sequence* mode. Enter the sequence into your graphing utility as shown in Figure 9.1. Use the *table* feature (in *ask* mode) to create a table showing the terms of the sequence u_n for $n = 1, 2, 3, 4,$ and 5. From Figure 9.2, you can estimate the first five terms of the sequence as follows.

$u_1 = -1,$ $u_2 = 0.33333 \approx \frac{1}{3},$ $u_3 = -0.2 = -\frac{1}{5},$

$u_4 = 0.14286 \approx \frac{1}{7},$ and $u_5 = -0.1111 \approx -\frac{1}{9}$

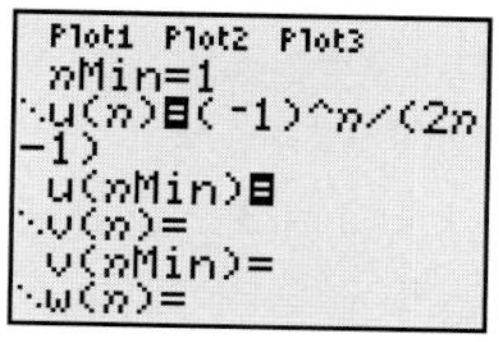

Figure 9.1

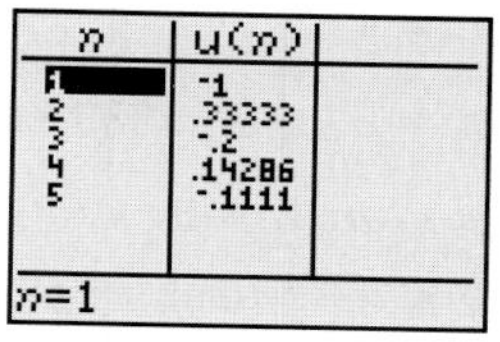

Figure 9.2

Simply listing the first few terms is not sufficient to define a unique sequence—the *n*th term *must be given*. To see this, consider the following sequences, both of which have the same first three terms.

$$\frac{1}{2}, \frac{1}{4}, \frac{1}{8}, \frac{1}{16}, \ldots, \frac{1}{2^n}, \ldots$$

$$\frac{1}{2}, \frac{1}{4}, \frac{1}{8}, \frac{1}{15}, \ldots, \frac{6}{(n+1)(n^2-n+6)}, \ldots$$

TECHNOLOGY SUPPORT

For instructions on how to use the *table* feature, see Appendix A; for specific keystrokes, go to the text website at *college.hmco.com*.

Example 3   Finding the *n*th Term of a Sequence

Write an expression for the apparent *n*th term (a_n) of each sequence.

a. 1, 3, 5, 7, . . .   **b.** 2, 5, 10, 17, . . .

Solution

a.   *n:* 1 2 3 4 . . . *n*

Terms: 1 3 5 7 . . . a_n

Apparent Pattern: Each term is 1 less than twice *n*. So, the apparent *n*th term is $a_n = 2n - 1$.

b.   *n:* 1 2 3 4 . . . *n*

Terms: 2 5 10 17 . . . a_n

Apparent Pattern: Each term is 1 more than the square of *n*. So, the apparent *n*th term is $a_n = n^2 + 1$.

✓ *Checkpoint* Now try Exercise 39.

Some sequences are defined **recursively.** To define a sequence recursively, you need to be given one or more of the first few terms. All other terms of the sequence are then defined using previous terms. A well-known example is the Fibonacci sequence, shown in Example 4.

Example 4 The Fibonacci Sequence: A Recursive Sequence

The Fibonacci sequence is defined recursively as follows.

$$a_0 = 1, \ a_1 = 1, \ a_k = a_{k-2} + a_{k-1}, \qquad \text{where } k \geq 2$$

Write the first six terms of this sequence.

Solution

$a_0 = 1$	0th term is given.
$a_1 = 1$	1st term is given.
$a_2 = a_{2-2} + a_{2-1} = a_0 + a_1 = 1 + 1 = 2$	Use recursion formula.
$a_3 = a_{3-2} + a_{3-1} = a_1 + a_2 = 1 + 2 = 3$	Use recursion formula.
$a_4 = a_{4-2} + a_{4-1} = a_2 + a_3 = 2 + 3 = 5$	Use recursion formula.
$a_5 = a_{5-2} + a_{5-1} = a_3 + a_4 = 3 + 5 = 8$	Use recursion formula.

✓ *Checkpoint* Now try Exercise 53.

Factorial Notation

Some very important sequences in mathematics involve terms that are defined with special types of products called **factorials.**

> **Definition of Factorial**
>
> If n is a positive integer, n **factorial** is defined as
>
> $$n! = 1 \cdot 2 \cdot 3 \cdot 4 \cdots (n-1) \cdot n.$$
>
> As a special case, zero factorial is defined as $0! = 1$.

Here are some values of $n!$ for the first few nonnegative integers. Notice that $0! = 1$ by definition.

$$0! = 1$$
$$1! = 1$$
$$2! = 1 \cdot 2 = 2$$
$$3! = 1 \cdot 2 \cdot 3 = 6$$
$$4! = 1 \cdot 2 \cdot 3 \cdot 4 = 24$$
$$5! = 1 \cdot 2 \cdot 3 \cdot 4 \cdot 5 = 120$$

The value of n does not have to be very large before the value of $n!$ becomes huge. For instance, $10! = 3,628,800$.

> **Exploration**
>
> Most graphing utilities have the capability to compute $n!$. Use your graphing utility to compare $3 \cdot 5!$ and $(3 \cdot 5)!$. How do they differ? How large a value of $n!$ will your graphing utility allow you to compute?

Factorials follow the same conventions for order of operations as do exponents. For instance,

$$2n! = 2(n!) = 2(1 \cdot 2 \cdot 3 \cdot 4 \cdots n)$$

whereas $(2n)! = 1 \cdot 2 \cdot 3 \cdot 4 \cdots 2n$.

Example 5   Writing the Terms of a Sequence Involving Factorials

Write the first five terms of the sequence given by $a_n = \dfrac{2^n}{n!}$. Begin with $n = 0$.

Algebraic Solution

$$a_0 = \frac{2^0}{0!} = \frac{1}{1} = 1 \qquad \text{0th term}$$

$$a_1 = \frac{2^1}{1!} = \frac{2}{1} = 2 \qquad \text{1st term}$$

$$a_2 = \frac{2^2}{2!} = \frac{4}{2} = 2 \qquad \text{2nd term}$$

$$a_3 = \frac{2^3}{3!} = \frac{8}{6} = \frac{4}{3} \qquad \text{3rd term}$$

$$a_4 = \frac{2^4}{4!} = \frac{16}{24} = \frac{2}{3} \qquad \text{4th term}$$

✓ *Checkpoint* Now try Exercise 61.

Graphical Solution

Using a graphing utility set to *dot* and *sequence* modes, enter the sequence $u_n = 2^n/n!$, as shown in Figure 9.3. Set the viewing window to $0 \le n \le 4$, $0 \le x \le 6$, and $0 \le y \le 4$. Then graph the sequence as shown in Figure 9.4. Use the *value* or *trace* feature to approximate the first five terms as follows.

$$u_0 = 1, \quad u_1 = 2, \quad u_2 = 2, \quad u_3 \approx 1.333 \approx \tfrac{4}{3}, \quad u_4 \approx 0.666 \approx \tfrac{2}{3}$$

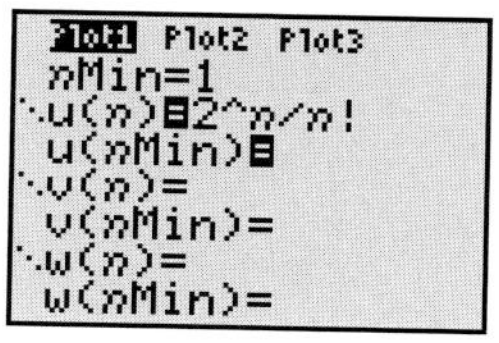

Figure 9.3

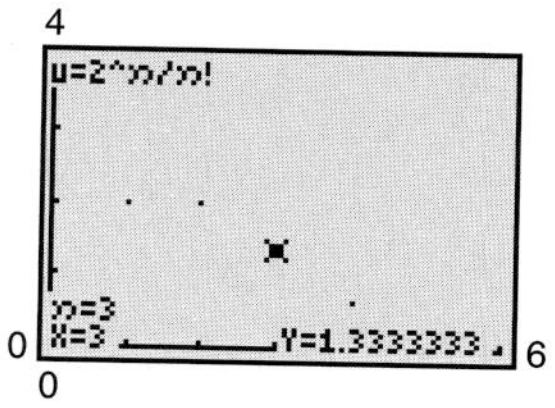

Figure 9.4

When working with fractions involving factorials, you will often find that the fractions can be reduced to simplify the computations.

Example 6   Evaluating Factorial Expressions

Evaluate each factorial expression.

a. $\dfrac{8!}{2! \cdot 6!}$
b. $\dfrac{2! \cdot 6!}{3! \cdot 5!}$
c. $\dfrac{n!}{(n-1)!}$

Solution

a. $\dfrac{8!}{2! \cdot 6!} = \dfrac{1 \cdot 2 \cdot 3 \cdot 4 \cdot 5 \cdot 6 \cdot 7 \cdot 8}{1 \cdot 2 \cdot 1 \cdot 2 \cdot 3 \cdot 4 \cdot 5 \cdot 6} = \dfrac{7 \cdot 8}{2} = 28$

b. $\dfrac{2! \cdot 6!}{3! \cdot 5!} = \dfrac{1 \cdot 2 \cdot 1 \cdot 2 \cdot 3 \cdot 4 \cdot 5 \cdot 6}{1 \cdot 2 \cdot 3 \cdot 1 \cdot 2 \cdot 3 \cdot 4 \cdot 5} = \dfrac{6}{3} = 2$

c. $\dfrac{n!}{(n-1)!} = \dfrac{1 \cdot 2 \cdot 3 \cdots (n-1) \cdot n}{1 \cdot 2 \cdot 3 \cdots (n-1)} = n$

✓ *Checkpoint* Now try Exercise 71.

STUDY TIP

Note in Example 6(a) that you can simplify the computation as follows.

$$\frac{8!}{2! \cdot 6!} = \frac{8 \cdot 7 \cdot 6!}{2! \cdot 6!}$$

$$= \frac{8 \cdot 7}{2 \cdot 1} = 28$$

Summation Notation

There is a convenient notation for the sum of the terms of a finite sequence. It is called **summation notation** or **sigma notation** because it involves the use of the uppercase Greek letter sigma, written as Σ.

Definition of Summation Notation

The sum of the first n terms of a sequence is represented by

$$\sum_{i=1}^{n} a_i = a_1 + a_2 + a_3 + a_4 + \cdots + a_n$$

where i is called the **index of summation,** n is the **upper limit of summation,** and 1 is the **lower limit of summation.**

STUDY TIP

Summation notation is an instruction to add the terms of a sequence. From the definition at the left, the upper limit of summation tells you where to end the sum. Summation notation helps you generate the appropriate terms of the sequence prior to finding the actual sum, which may be unclear.

Example 7 Sigma Notation for Sums

a. $\displaystyle\sum_{i=1}^{5} 3i = 3(1) + 3(2) + 3(3) + 3(4) + 3(5)$

$= 3(1 + 2 + 3 + 4 + 5)$

$= 3(15)$

$= 45$

b. $\displaystyle\sum_{k=3}^{6} (1 + k^2) = (1 + 3^2) + (1 + 4^2) + (1 + 5^2) + (1 + 6^2)$

$= 10 + 17 + 26 + 37 = 90$

c. $\displaystyle\sum_{n=0}^{8} \frac{1}{n!} = \frac{1}{0!} + \frac{1}{1!} + \frac{1}{2!} + \frac{1}{3!} + \frac{1}{4!} + \frac{1}{5!} + \frac{1}{6!} + \frac{1}{7!} + \frac{1}{8!}$

$= 1 + 1 + \dfrac{1}{2} + \dfrac{1}{6} + \dfrac{1}{24} + \dfrac{1}{120} + \dfrac{1}{720} + \dfrac{1}{5040} + \dfrac{1}{40{,}320}$

≈ 2.71828

For the summation in part (c), note that the sum is very close to the irrational number $e \approx 2.718281828$. It can be shown that as more terms of the sequence whose nth term is $1/n!$ are added, the sum becomes closer and closer to e.

✓ *Checkpoint* Now try Exercise 75.

In Example 7, note that the lower limit of a summation does not have to be 1. Also note that the index of summation does not have to be the letter i. For instance, in part (b), the letter k is the index of summation.

TECHNOLOGY TIP Most graphing utilities are able to sum the first n terms of a sequence. Figure 9.5 shows an example of how one graphing utility displays the sum of the terms of the sequence below using the *sum sequence* feature.

$$a_n = \frac{1}{n!} \quad \text{from} \quad n = 0 \quad \text{to} \quad n = 8$$

TECHNOLOGY SUPPORT

For instructions on how to use the *sum sequence* feature, see Appendix A; for specific keystrokes, go to the text website at *college.hmco.com*.

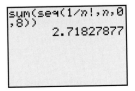

Figure 9.5

Properties of Sums

1. $\displaystyle\sum_{i=1}^{n} c = cn,$ c is a constant.   2. $\displaystyle\sum_{i=1}^{n} ca_i = c\sum_{i=1}^{n} a_i,$ c is a constant.

3. $\displaystyle\sum_{i=1}^{n} (a_i + b_i) = \sum_{i=1}^{n} a_i + \sum_{i=1}^{n} b_i$ 4. $\displaystyle\sum_{i=1}^{n} (a_i - b_i) = \sum_{i=1}^{n} a_i - \sum_{i=1}^{n} b_i$

See Appendix B for a proof of Property 2.

Series

Many applications involve the sum of the terms of a finite or an infinite sequence. Such a sum is called a **series.**

Definition of a Series

Consider the infinite sequence $a_1, a_2, a_3, \ldots , a_i, \ldots$.

1. The sum of the first n terms of the sequence is called a **finite series** or the **partial sum** of the sequence and is denoted by

$$a_1 + a_2 + a_3 + \cdots + a_n = \sum_{i=1}^{n} a_i.$$

2. The sum of all the terms of the infinite sequence is called an **infinite series** and is denoted by

$$a_1 + a_2 + a_3 + \cdots + a_i + \cdots = \sum_{i=1}^{\infty} a_i.$$

Example 8   Finding the Sum of a Series

For the series $\displaystyle\sum_{i=1}^{\infty} \frac{3}{10^i}$, find (a) the third partial sum and (b) the sum.

Solution

a. The third partial sum is

$$\sum_{i=1}^{3} \frac{3}{10^i} = \frac{3}{10^1} + \frac{3}{10^2} + \frac{3}{10^3} = 0.3 + 0.03 + 0.003 = 0.333.$$

b. The sum of the series is

$$\sum_{i=1}^{\infty} \frac{3}{10^i} = \frac{3}{10^1} + \frac{3}{10^2} + \frac{3}{10^3} + \frac{3}{10^4} + \frac{3}{10^5} + \cdots$$

$$= 0.3 + 0.03 + 0.003 + 0.0003 + 0.00003 + \cdots$$

$$= 0.33333\ldots = \frac{1}{3}.$$

✓ *Checkpoint* Now try Exercise 101.

Notice in Example 8(b) that the sum of an infinite series can be a finite number.

STUDY TIP

Variations in the upper and lower limits of summation can produce quite different-looking summation notations for *the same sum.* For example, the following two sums have identical terms.

$$\sum_{i=1}^{3} 3(2^i) = 3(2^1 + 2^2 + 2^3)$$

$$\sum_{i=0}^{2} 3(2^{i+1}) = 3(2^1 + 2^2 + 2^3)$$

Application

Sequences have many applications in situations that involve a recognizable pattern. One such model is illustrated in Example 9.

Example 9 Population of the United States

From 1970 to 2001, the resident population of the United States can be approximated by the model

$$a_n = 205.7 + 1.78n + 0.025n^2, \qquad n = 0, 1, \ldots, 31$$

where a_n is the population in millions and n represents the year, with $n = 0$ corresponding to 1970. Find the last five terms of this finite sequence. (Source: U.S. Census Bureau)

Algebraic Solution

The last five terms of this finite sequence are as follows.

$a_{27} = 205.7 + 1.78(27) + 0.025(27)^2$

≈ 272.0 1997 population

$a_{28} = 205.7 + 1.78(28) + 0.025(28)^2$

≈ 275.1 1998 population

$a_{29} = 205.7 + 1.78(29) + 0.025(29)^2$

≈ 278.3 1999 population

$a_{30} = 205.7 + 1.78(30) + 0.025(30)^2$

$= 281.6$ 2000 population

$a_{31} = 205.7 + 1.78(31) + 0.025(31)^2$

≈ 284.9 2001 population

✓ *Checkpoint* Now try Exercise 111.

Graphical Solution

Using a graphing utility set to *dot* and *sequence* modes, enter the sequence

$$u_n = 205.7 + 1.78n + 0.025n^2.$$

Set the viewing window to $0 \le n \le 32, 0 \le x \le 32$, and $200 \le y \le 300$. Then graph the sequence. Use the *value* or *trace* feature to approximate the last five terms, as shown in Figure 9.6.

$a_{27} \approx 272.0,$

$a_{28} \approx 275.1,$

$a_{29} \approx 278.3,$

$a_{30} = 281.6,$

$a_{31} \approx 284.9$

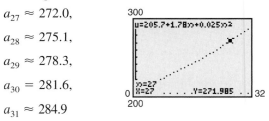

Figure 9.6

Exploration

A $3 \times 3 \times 3$ cube is created using 27 unit cubes (a unit cube has a length, width, and height of 1 unit) and only the faces of each cube that are visible are painted blue (see Figure 9.7). Complete the table below to determine how many unit cubes of the $3 \times 3 \times 3$ cube have 0 blue faces, 1 blue face, 2 blue faces, and 3 blue faces. Do the same for a $4 \times 4 \times 4$ cube, a $5 \times 5 \times 5$ cube, and a $6 \times 6 \times 6$ cube and add your results to the table below. What type of pattern do you observe in the table? Write a formula you could use to determine the column values for an $n \times n \times n$ cube.

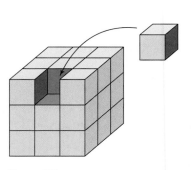

Figure 9.7

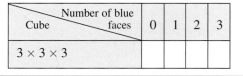

Cube \ Number of blue faces	0	1	2	3
$3 \times 3 \times 3$				

9.1　Exercises

Vocabulary Check

Fill in the blanks.

1. An _____ is a function whose domain is the set of positive integers.
2. The function values $a_1, a_2, a_3, a_4, \ldots, a_n, \ldots$ are called the _____ of a sequence.
3. A sequence is a _____ sequence if the domain of the function consists of the first n positive integers.
4. If you are given one or more of the first few terms of a sequence, and all other terms of the sequence are defined using previous terms, then the sequence is defined _____ .
5. If n is a positive integer, n _____ is defined as $n! = 1 \cdot 2 \cdot 3 \cdot 4 \cdots (n-1) \cdot n$.
6. The notation used to represent the sum of the terms of a finite sequence is _____ or sigma notation.
7. For the sum $\displaystyle\sum_{i=1}^{n} a_i$, i is called the _____ of summation, n is the _____ of summation, and 1 is the _____ of summation.
8. The sum of the terms of a finite or an infinite sequence is called a _____ .
9. The _____ of a sequence is the sum of the first n terms of the sequence.

In Exercises 1–20, write the first five terms of the sequence. (Assume n begins with 1.) Use the *table* feature of a graphing utility to verify your results.

1. $a_n = 2n + 5$
2. $a_n = 4n - 7$
3. $a_n = 2^n$
4. $a_n = \left(\frac{1}{2}\right)^n$
5. $a_n = \left(-\frac{1}{2}\right)^n$
6. $a_n = (-2)^n$
7. $a_n = \dfrac{n+1}{n}$
8. $a_n = \dfrac{n}{n+1}$
9. $a_n = \dfrac{n}{n^2 + 1}$
10. $a_n = \dfrac{2n}{n+1}$
11. $a_n = \dfrac{1 + (-1)^n}{n}$
12. $a_n = \dfrac{1 + (-1)^n}{2n}$
13. $a_n = 1 - \dfrac{1}{2^n}$
14. $a_n = \dfrac{3^n}{4^n}$
15. $a_n = \dfrac{1}{n^{3/2}}$
16. $a_n = \dfrac{1}{\sqrt{n}}$
17. $a_n = \dfrac{(-1)^n}{n^2}$
18. $a_n = (-1)^n \left(\dfrac{n}{n+1}\right)$
19. $a_n = (2n - 1)(2n + 1)$　20. $a_n = n(n-1)(n-2)$

In Exercises 21–24, find the indicated term of the sequence.

21. $a_n = (-1)^n(3n - 2)$

　$a_{25} = $

22. $a_n = (-1)^{n-1}[n(n-1)]$

　$a_{16} = $

23. $a_n = \dfrac{n^2}{n^2 + 1}$

　$a_{10} = $

24. $a_n = \dfrac{n^2}{2n + 1}$

　$a_5 = $

In Exercises 25–30, use a graphing utility to graph the first 10 terms of the sequence. (Assume n begins with 1.)

25. $a_n = \dfrac{2}{3}n$
26. $a_n = 2 - \dfrac{4}{n}$
27. $a_n = 16(-0.5)^{n-1}$
28. $a_n = 8(0.75)^{n-1}$
29. $a_n = \dfrac{2n}{n+1}$
30. $a_n = \dfrac{3n^2}{n^2 + 1}$

In Exercises 31–34, use the *table* feature of a graphing utility to find the first 10 terms of the sequence. (Assume n begins with 1.)

31. $a_n = 2(3n - 1) + 5$
32. $a_n = 2n(n + 1)(n + 2)$
33. $a_n = 1 + \dfrac{n+1}{n}$
34. $a_n = \dfrac{4n^2}{n + 2}$

In Exercises 35–38, match the sequence with its graph. [The graphs are labeled (a), (b), (c), and (d).]

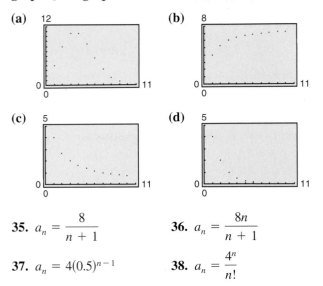

(a) 12 ... 11, 0

(b) 8 ... 11, 0

(c) 5 ... 11, 0

(d) 5 ... 11, 0

35. $a_n = \dfrac{8}{n+1}$ **36.** $a_n = \dfrac{8n}{n+1}$

37. $a_n = 4(0.5)^{n-1}$ **38.** $a_n = \dfrac{4^n}{n!}$

In Exercises 39–52, write an expression for the *apparent* nth term of the sequence. (Assume n begins with 1.)

39. $1, 4, 7, 10, 13, \ldots$ **40.** $3, 7, 11, 15, 19, \ldots$

41. $0, 3, 8, 15, 24, \ldots$ **42.** $1, \frac{1}{4}, \frac{1}{9}, \frac{1}{16}, \frac{1}{25}, \ldots$

43. $\frac{2}{3}, \frac{3}{4}, \frac{4}{5}, \frac{5}{6}, \frac{6}{7}, \ldots$

44. $\frac{2}{1}, \frac{3}{3}, \frac{4}{5}, \frac{5}{7}, \frac{6}{9}, \ldots$

45. $\frac{1}{2}, \frac{-1}{4}, \frac{1}{8}, \frac{-1}{16}, \ldots$

46. $\frac{1}{3}, -\frac{2}{9}, \frac{4}{27}, -\frac{8}{81}, \ldots$

47. $1 + \frac{1}{1}, 1 + \frac{1}{2}, 1 + \frac{1}{3}, 1 + \frac{1}{4}, 1 + \frac{1}{5}, \ldots$

48. $1 + \frac{1}{2}, 1 + \frac{3}{4}, 1 + \frac{7}{8}, 1 + \frac{15}{16}, 1 + \frac{31}{32}, \ldots$

49. $1, \frac{1}{2}, \frac{1}{6}, \frac{1}{24}, \frac{1}{120}, \ldots$

50. $1, 2, \dfrac{2^2}{2}, \dfrac{2^3}{6}, \dfrac{2^4}{24}, \dfrac{2^5}{120}, \ldots$

51. $1, 3, 1, 3, 1, \ldots$

52. $1, -1, 1, -1, 1, \ldots$

In Exercises 53–56, write the first five terms of the sequence defined recursively.

53. $a_1 = 28, \quad a_{k+1} = a_k - 4$

54. $a_1 = 15, \quad a_{k+1} = a_k + 3$

55. $a_1 = 3, \quad a_{k+1} = 2(a_k - 1)$

56. $a_1 = 32, \quad a_{k+1} = \frac{1}{2}a_k$

In Exercises 57–60, write the first five terms of the sequence defined recursively. Use the pattern to write the nth term of the sequence as a function of n. (Assume n begins with 1.)

57. $a_1 = 6, \quad a_{k+1} = a_k + 2$

58. $a_1 = 25, \quad a_{k+1} = a_k - 5$

59. $a_1 = 81, \quad a_{k+1} = \frac{1}{3}a_k$

60. $a_1 = 14, \quad a_{k+1} = -2a_k$

In Exercises 61–66, write the first five terms of the sequence. (Assume n begins with 0.) Use the *table* feature of a graphing utility to verify your results.

61. $a_n = \dfrac{1}{n!}$ **62.** $a_n = \dfrac{1}{(n+1)!}$

63. $a_n = \dfrac{n!}{2n+1}$ **64.** $a_n = \dfrac{n^2}{(n+1)!}$

65. $a_n = \dfrac{(-1)^{2n}}{(2n)!}$ **66.** $a_n = \dfrac{(-1)^{2n+1}}{(2n+1)!}$

In Exercises 67–74, simplify the factorial expression.

67. $\dfrac{2!}{4!}$ **68.** $\dfrac{5!}{7!}$

69. $\dfrac{12!}{4! \cdot 8!}$ **70.** $\dfrac{10! \cdot 3!}{4! \cdot 6!}$

71. $\dfrac{(n+1)!}{n!}$ **72.** $\dfrac{(n+2)!}{n!}$

73. $\dfrac{(2n-1)!}{(2n+1)!}$ **74.** $\dfrac{(2n+2)!}{(2n)!}$

In Exercises 75–86, find the sum.

75. $\displaystyle\sum_{i=1}^{5}(2i+1)$ **76.** $\displaystyle\sum_{i=1}^{6}(3i-1)$

77. $\displaystyle\sum_{k=1}^{4}10$ **78.** $\displaystyle\sum_{k=1}^{5}6$

79. $\displaystyle\sum_{i=0}^{4}i^2$ **80.** $\displaystyle\sum_{i=0}^{5}3i^2$

81. $\displaystyle\sum_{k=0}^{3}\dfrac{1}{k^2+1}$ **82.** $\displaystyle\sum_{j=3}^{5}\dfrac{1}{j}$

83. $\displaystyle\sum_{i=1}^{4}[(i-1)^2 + (i+1)^3]$

84. $\displaystyle\sum_{k=2}^{5}(k+1)(k-3)$

85. $\displaystyle\sum_{i=1}^{4} 2^i$

86. $\displaystyle\sum_{j=0}^{4} (-2)^j$

In Exercises 87–90, use a graphing utility to find the sum.

87. $\displaystyle\sum_{j=1}^{6} (24 - 3j)$

88. $\displaystyle\sum_{j=1}^{10} \frac{3}{j+1}$

89. $\displaystyle\sum_{k=0}^{4} \frac{(-1)^k}{k+1}$

90. $\displaystyle\sum_{k=0}^{4} \frac{(-1)^k}{k!}$

In Exercises 91–100, use sigma notation to write the sum. Then use a graphing utility to find the sum.

91. $\dfrac{1}{3(1)} + \dfrac{1}{3(2)} + \dfrac{1}{3(3)} + \cdots + \dfrac{1}{3(9)}$

92. $\dfrac{5}{1+1} + \dfrac{5}{1+2} + \dfrac{5}{1+3} + \cdots + \dfrac{5}{1+15}$

93. $\left[2\left(\frac{1}{8}\right) + 3\right] + \left[2\left(\frac{2}{8}\right) + 3\right] + \cdots + \left[2\left(\frac{8}{8}\right) + 3\right]$

94. $\left[1 - \left(\frac{1}{6}\right)^2\right] + \left[1 - \left(\frac{2}{6}\right)^2\right] + \cdots + \left[1 - \left(\frac{6}{6}\right)^2\right]$

95. $3 - 9 + 27 - 81 + 243 - 729$

96. $1 - \frac{1}{2} + \frac{1}{4} - \frac{1}{8} + \cdots - \frac{1}{128}$

97. $\dfrac{1}{1^2} - \dfrac{1}{2^2} + \dfrac{1}{3^2} - \dfrac{1}{4^2} + \cdots - \dfrac{1}{20^2}$

98. $\dfrac{1}{1 \cdot 3} + \dfrac{1}{2 \cdot 4} + \dfrac{1}{3 \cdot 5} + \cdots + \dfrac{1}{10 \cdot 12}$

99. $\frac{1}{4} + \frac{3}{8} + \frac{7}{16} + \frac{15}{32} + \frac{31}{64}$

100. $\frac{1}{2} + \frac{2}{4} + \frac{6}{8} + \frac{24}{16} + \frac{120}{32} + \frac{720}{64}$

In Exercises 101–104, find the indicated partial sum of the series.

101. $\displaystyle\sum_{i=1}^{\infty} 5\left(\frac{1}{2}\right)^i$

Fourth partial sum

102. $\displaystyle\sum_{i=1}^{\infty} 2\left(\frac{1}{3}\right)^i$

Fifth partial sum

103. $\displaystyle\sum_{n=1}^{\infty} 4\left(-\frac{1}{2}\right)^n$

Third partial sum

104. $\displaystyle\sum_{n=1}^{\infty} 8\left(-\frac{1}{4}\right)^n$

Fourth partial sum

In Exercises 105–108, find the sum of the infinite series.

105. $\displaystyle\sum_{i=1}^{\infty} 6\left(\frac{1}{10}\right)^i$

106. $\displaystyle\sum_{k=1}^{\infty} 4\left(\frac{1}{10}\right)^k$

107. $\displaystyle\sum_{k=1}^{\infty} \left(\frac{1}{10}\right)^k$

108. $\displaystyle\sum_{i=1}^{\infty} 2\left(\frac{1}{10}\right)^i$

109. *Compound Interest*   A deposit of $5000 is made in an account that earns 3% interest compounded quarterly. The balance in the account after n quarters is given by

$$A_n = 5000\left(1 + \frac{0.03}{4}\right)^n, \quad n = 1, 2, 3, \dots$$

(a) Compute the first eight terms of this sequence.

(b) Find the balance in this account after 10 years by computing the 40th term of the sequence.

110. *Compound Interest*   A deposit of $100 is made *each month* in an account that earns 12% interest compounded monthly. The balance in the account after n months is given by

$$A_n = 100(101)[(1.01)^n - 1], \quad n = 1, 2, 3, \dots$$

(a) Compute the first six terms of this sequence.

(b) Find the balance in this account after 5 years by computing the 60th term of the sequence.

(c) Find the balance in this account after 20 years by computing the 240th term of the sequence.

111. *Education*   The number a_n (in thousands) of children enrolled in Head Start programs from 1993 to 2001 can be approximated by the model

$$a_n = 1.37n^2 + 3.1n + 698, \quad n = 3, 4, \dots, 11$$

where n is the year, with $n = 3$ corresponding to 1993.   (Source: U.S. Administration for Children and Families)

(a) Find the terms of this finite sequence and use a graphing utility to graph the sequence.

(b) What does the graph in part (a) say about the future enrollment in Head Start programs?

112. *Federal Debt*   From 1990 to 2002, the federal debt rose from just over $3 trillion to over $6 trillion. The federal debt a_n (in trillions of dollars) from 1990 to 2002 is approximated by the model

$$a_n = -0.0140n^2 + 0.394n + 3.25,$$
$$n = 0, 1, \dots, 12$$

where n is the year, with $n = 0$ corresponding to 1990.   (Source: Office of Management and Budget)

(a) Find the terms of this finite sequence and construct a bar graph that represents the sequence.

(b) What does the pattern in the bar graph in part (a) say about the future of the federal debt?

113. *Net Profit*   The net profits a_n (in millions of dollars) of Avon Products, Inc. for the years 1994 through 2002 are shown in the bar graph. These profits can be approximated by the model

$$a_n = 2.151n^2 + 235.9, \qquad n = 4, 5, \ldots, 12$$

where n is the year, with $n = 4$ corresponding to 1994. Use this model to approximate the total net profits from 1994 through 2002. Compare this sum with the result of adding the profits shown in the bar graph.   (Source: Avon Products, Inc.)

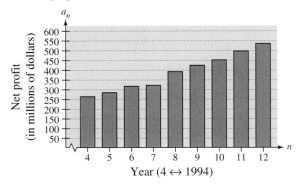

Year (4 ↔ 1994)

114. *Sales*   The sales a_n (in millions of dollars) of Abercrombie & Fitch Company for the years 1996 through 2002 are shown in the bar graph. These sales can be approximated by the model

$$a_n = -2985.8 + 1829.9 \ln n, \qquad n = 6, 7, \ldots, 12$$

where $n = 6$ represents 1996. Use this model to approximate the total sales from 1996 through 2002. Compare this sum with the result of adding the sales shown in the bar graph.   (Source: Abercrombie & Fitch Company)

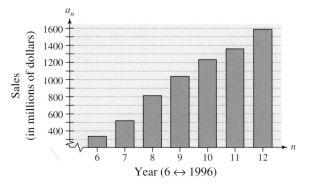

Year (6 ↔ 1996)

Synthesis

True or False?   **In Exercises 115 and 116, determine whether the statement is true or false. Justify your answer.**

115. $\displaystyle\sum_{i=1}^{4} (i^2 + 2i) = \sum_{i=1}^{4} i^2 + 2\sum_{i=1}^{4} i$

116. $\displaystyle\sum_{j=1}^{4} 2^j = \sum_{j=3}^{6} 2^{j-2}$

Fibonacci Sequence   **In Exercises 117 and 118, use the Fibonacci sequence. (See Example 4.)**

117. Write the first 12 terms of the Fibonacci sequence a_n and the first 10 terms of the sequence given by

$$b_n = \frac{a_{n+1}}{a_n}, \quad n > 0.$$

118. Using the definition of b_n given in Exercise 117, show that b_n can be defined recursively by

$$b_n = 1 + \frac{1}{b_{n-1}}.$$

In Exercises 119–122, write the first five terms of the sequence.

119. $a_n = \dfrac{x^n}{n!}$

120. $a_n = \dfrac{(-1)^n x^{2n+1}}{2n+1}$

121. $a_n = \dfrac{(-1)^n x^{2n}}{(2n)!}$

122. $a_n = \dfrac{(-1)^n x^{2n+1}}{(2n+1)!}$

Review

In Exercises 123–126, find, if possible, (a) $A - B$, (b) $2B - 3A$, (c) AB, and (d) BA.

123. $A = \begin{bmatrix} 6 & 5 \\ 3 & 4 \end{bmatrix}$, $B = \begin{bmatrix} -2 & 4 \\ 6 & -3 \end{bmatrix}$

124. $A = \begin{bmatrix} 10 & 7 \\ -4 & 6 \end{bmatrix}$, $B = \begin{bmatrix} 0 & -12 \\ 8 & 11 \end{bmatrix}$

125. $A = \begin{bmatrix} -2 & -3 & 6 \\ 4 & 5 & 7 \\ 1 & 7 & 4 \end{bmatrix}$, $B = \begin{bmatrix} 1 & 4 & 2 \\ 0 & 1 & 6 \\ 0 & 3 & 1 \end{bmatrix}$

126. $A = \begin{bmatrix} -1 & 4 & 0 \\ 5 & 1 & 2 \\ 0 & -1 & 3 \end{bmatrix}$, $B = \begin{bmatrix} 0 & 4 & 0 \\ 3 & 1 & -2 \\ -1 & 0 & 2 \end{bmatrix}$

9.2   Arithmetic Sequences and Partial Sums

Arithmetic Sequences

A sequence whose consecutive terms have a common difference is called an **arithmetic sequence.**

What you should learn

- Recognize, write, and find the nth terms of arithmetic sequences.
- Find nth partial sums of arithmetic sequences.
- Use arithmetic sequences to model and solve real-life problems.

Why you should learn it

Arithmetic sequences can reduce the amount of time it takes to find the sum of a sequence of numbers with a common difference. In Exercise 77 on page 714, you will use an arithmetic sequence to find the number of bricks needed to lay a brick patio.

> **Definition of Arithmetic Sequence**
>
> A sequence is **arithmetic** if the differences between consecutive terms are the same. So, the sequence
>
> $$a_1, a_2, a_3, a_4, \ldots, a_n, \ldots$$
>
> is arithmetic if there is a number d such that
>
> $$a_2 - a_1 = a_3 - a_2 = a_4 - a_3 = \cdots = d.$$
>
> The number d is the **common difference** of the arithmetic sequence.

Index Stock

Example 1   Examples of Arithmetic Sequences

a. The sequence whose nth term is $4n + 3$ is arithmetic. For this sequence, the common difference between consecutive terms is 4.

$$7, 11, 15, 19, \ldots, 4n + 3, \ldots \qquad \text{Begin with } n = 1.$$

$$11 - 7 = 4$$

b. The sequence whose nth term is $7 - 5n$ is arithmetic. For this sequence, the common difference between consecutive terms is -5.

$$2, -3, -8, -13, \ldots, 7 - 5n, \ldots \qquad \text{Begin with } n = 1.$$

$$-3 - 2 = -5$$

c. The sequence whose nth term is $\frac{1}{4}(n + 3)$ is arithmetic. For this sequence, the common difference between consecutive terms is $\frac{1}{4}$.

$$1, \frac{5}{4}, \frac{3}{2}, \frac{7}{4}, \ldots, \frac{n + 3}{4}, \ldots \qquad \text{Begin with } n = 1.$$

$$\frac{5}{4} - 1 = \frac{1}{4}$$

 Checkpoint Now try Exercise 9.

The sequence $1, 4, 9, 16, \ldots$, whose nth term is n^2, is *not* arithmetic. The difference between the first two terms is

$$a_2 - a_1 = 4 - 1 = 3$$

but the difference between the second and third terms is

$$a_3 - a_2 = 9 - 4 = 5.$$

In Example 1, notice that each of the arithmetic sequences has an nth term that is of the form $dn + c$, where the common difference of the sequence is d.

> **The nth Term of an Arithmetic Sequence**
>
> The nth term of an arithmetic sequence has the form
>
> $$a_n = dn + c$$
>
> where d is the common difference between consecutive terms of the sequence and $c = a_1 - d$.

An arithmetic sequence $a_n = dn + c$ can be thought of as "counting by d's" after a shift of c units from d. For instance, the sequence

$$2, 6, 10, 14, 18, \ldots$$

has a common difference of 4, so you are counting by 4's after a shift of two units below 4 (beginning with $a_1 = 2$). So, the nth term is $4n - 2$. Similarly, the nth term of the sequence

$$6, 11, 16, 21, \ldots$$

is $5n + 1$ because you are counting by 5's after a shift of one unit above 5 (beginning with $a_1 = 6$).

Example 2   Finding the nth Term of an Arithmetic Sequence

Find a formula for the nth term of the arithmetic sequence whose common difference is 3 and whose first term is 2.

Solution

Because the sequence is arithmetic, you know that the formula for the nth term is of the form $a_n = dn + c$. Moreover, because the common difference is $d = 3$, the formula must have the form $a_n = 3n + c$. Because $a_1 = 2$, it follows that

$$c = a_1 - d = 2 - 3 = -1.$$

So, the formula for the nth term is $a_n = 3n - 1$. The sequence therefore has the following form.

$$2, 5, 8, 11, 14, \ldots, 3n - 1, \ldots$$

A graph of the first 15 terms of the sequence is shown in Figure 9.8. Notice that the points lie on a line. This makes sense because a_n is a linear function of n. In other words, the terms "arithmetic" and "linear" are closely connected.

 Checkpoint Now try Exercise 17.

Another way to find a formula for the nth term of the sequence in Example 2 is to begin by writing the terms of the sequence.

a_1	a_2	a_3	a_4	a_5	a_6	a_7	
2	$2 + 3$	$5 + 3$	$8 + 3$	$11 + 3$	$14 + 3$	$17 + 3$	$\ldots$
2	5	8	11	14	17	20	$\ldots$

From these terms, you can reason that the nth term is of the form

$$a_n = dn + c = 3n - 1.$$

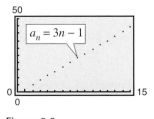

$a_n = 3n - 1$

Figure 9.8

Example 3   Writing the Terms of an Arithmetic Sequence

The fourth term of an arithmetic sequence is 20, and the 13th term is 65. Write the first several terms of this sequence.

Solution

The fourth and 13th terms of the sequence are related by

$$a_{13} = a_4 + 9d.$$

Using $a_4 = 20$ and $a_{13} = 65$, you have $65 = 20 + 9d$. So, you can conclude that $d = 5$, which implies that the sequence is as follows.

a_1	a_2	a_3	a_4	a_5	a_6	a_7	a_8	a_9	a_{10}	a_{11}	a_{12}	a_{13}	
5,	10,	15,	20,	25,	30,	35,	40,	45,	50,	55,	60,	65,	. . .

✓ *Checkpoint* Now try Exercise 31.

If you know the nth term of an arithmetic sequence *and* you know the common difference of the sequence, you can find the $(n + 1)$th term by using the *recursion formula*

$$a_{n+1} = a_n + d. \qquad \text{Recursion formula}$$

With this formula, you can find any term of an arithmetic sequence, *provided* that you know the preceding term. For instance, if you know the first term, you can find the second term. Then, knowing the second term, you can find the third term, and so on.

If you substitute $a_1 - d$ for c in the formula $a_n = dn + c$, the nth term of an arithmetic sequence has the alternative recursion formula

$$a_n = a_1 + (n - 1)d. \qquad \text{Alternative recursion formula}$$

Use this formula to solve Example 4. You should obtain the same answer.

Example 4   Using a Recursion Formula

Find the seventh term of the arithmetic sequence whose first two terms are 2 and 9.

Solution

For this sequence, the common difference is $d = 9 - 2 = 7$. Next find a formula for the nth term. Because the first term is 2, it follows that

$$c = a_1 - d = 2 - 7 = -5.$$

Therefore, a formula for the nth term is

$$a_n = dn + c = 7n - 5.$$

which implies that the seventh term is

$$a_7 = 7(7) - 5 = 44.$$

✓ *Checkpoint* Now try Exercise 39.

STUDY TIP

Another way to find the seventh term in Example 4 is to determine the common difference, $d = 7$, and then simply write out the first seven terms (by repeatedly adding 7).

2, 9, 16, 23, 30, 37, 44

As you can see, the seventh term is 44.

The Sum of a Finite Arithmetic Sequence

There is a simple formula for the *sum* of a finite arithmetic sequence. See Appendix B for a proof of this formula.

The Sum of a Finite Arithmetic Sequence

The sum of a finite arithmetic sequence with n terms is given by

$$S_n = \frac{n}{2}(a_1 + a_n).$$

Be sure you see that this formula works only for *arithmetic* sequences. Using this formula reduces the amount of time it takes to find the sum of an arithmetic sequence, as you will see in the following example.

Example 5   Finding the Sum of a Finite Arithmetic Sequence

Find each sum.

a. $1 + 3 + 5 + 7 + 9 + 11 + 13 + 15 + 17 + 19$

b. Sum of the integers from 1 to 100

Solution

a. To begin, notice that the sequence is arithmetic (with a common difference of 2). Moreover, the sequence has 10 terms. So, the sum of the sequence is

$$S_n = 1 + 3 + 5 + 7 + 9 + 11 + 13 + 15 + 17 + 19$$

$$= \frac{n}{2}(a_1 + a_n) \qquad \text{Formula for sum of an arithmetic sequence}$$

$$= \frac{10}{2}(1 + 19) \qquad \text{Substitute 10 for } n, 1 \text{ for } a_1, \text{ and 19 for } a_{10}.$$

$$= 5(20) = 100. \qquad \text{Simplify.}$$

b. The integers from 1 to 100 form an arithmetic sequence that has 100 terms. So, you can use the formula for the sum of an arithmetic sequence, as follows.

$$S_n = 1 + 2 + 3 + 4 + 5 + 6 + \cdots + 99 + 100$$

$$= \frac{n}{2}(a_1 + a_n) \qquad \text{Formula for sum of an arithmetic sequence}$$

$$= \frac{100}{2}(1 + 100) \qquad \text{Substitute 100 for } n, 1 \text{ for } a_1, \text{ and 100 for } a_{100}.$$

$$= 50(101) = 5050 \qquad \text{Simplify.}$$

✓ *Checkpoint* Now Try Exercise 53.

The sum of the first n terms of an infinite sequence is called the ***n*th partial sum.** The nth partial sum can be found by using the formula for the sum of a finite arithmetic sequence.

Example 6   Finding a Partial Sum of an Arithmetic Sequence

Find the 150th partial sum of the arithmetic sequence 5, 16, 27, 38, 49, ...

Solution

For this arithmetic sequence, you have $a_1 = 5$ and $d = 16 - 5 = 11$. So,

$$c = a_1 - d = 5 - 11 = -6$$

and the nth term is $a_n = 11n - 6$. Therefore, $a_{150} = 11(150) - 6 = 1644$, and the sum of the first 150 terms is

$$S_{150} = \frac{n}{2}(a_1 + a_{150}) \qquad \text{\textit{n}th partial sum formula}$$

$$= \frac{150}{2}(5 + 1644) \qquad \text{Substitute 150 for } n, 5 \text{ for } a_1, \text{ and } 1644 \text{ for } a_{150}.$$

$$= 75(1649) = 123{,}675. \qquad \text{Simplify.}$$

✓ *Checkpoint* Now try Exercise 63.

Applications

Example 7   Seating Capacity

An auditorium has 20 rows of seats. There are 20 seats in the first row, 21 seats in the second row, 22 seats in the third row, and so on (see Figure 9.9). How many seats are there in all 20 rows?

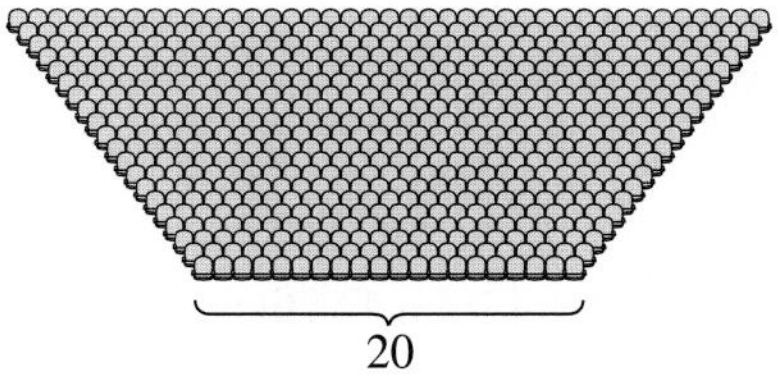

Solution

The numbers of seats in the 20 rows form an arithmetic sequence for which the common difference is $d = 1$. Because

$$c = a_1 - d = 20 - 1 = 19$$

you can determine that the formula for the nth term of the sequence is $a_n = n + 19$. So, the 20th term of the sequence is $a_{20} = 20 + 19 = 39$, and the total number of seats is

$$S_{20} = 20 + 21 + 22 + \cdots + 39$$

$$= \frac{20}{2}(20 + 39) \qquad \text{Substitute 20 for } n, 20 \text{ for } a_1, \text{ and 39 for } a_{20}.$$

$$= 10(59) = 590. \qquad \text{Simplify.}$$

✓ *Checkpoint* Now try Exercise 79.

20

Figure 9.9

Example 8 Total Sales

A small business sells $10,000 worth of sports memorabilia during its first year. The owner of the business has set a goal of increasing annual sales by $7500 each year for 19 years. Assuming that this goal is met, find the total sales during the first 20 years this business is in operation.

Algebraic Solution

The annual sales form an arithmetic sequence in which $a_1 = 10,000$ and $d = 7500$. So,

$$c = a_1 - d$$

$$= 10,000 - 7500$$

$$= 2500$$

and the nth term of the sequence is

$$a_n = 7500n + 2500.$$

This implies that the 20th term of the sequence is

$$a_{20} = 7500(20) + 2500$$

$$= 152,500.$$

The sum of the first 20 terms of the sequence is

$$S_{20} = \frac{n}{2}(a_1 + a_{20}) \qquad \text{\textit{n}th partial sum formula}$$

$$= \frac{20}{2}(10,000 + 152,500) \qquad \text{Substitute 20 for } n, 10,000 \text{ for } a_1, \text{ and } 152,500 \text{ for } a_{20}.$$

$$= 10(162,500) \qquad \text{Simplify.}$$

$$= 1,625,000. \qquad \text{Simplify.}$$

So, the total sales for the first 20 years are $1,625,000.

Numerical Solution

The annual sales form an arithmetic sequence in which $a_1 = 10,000$ and $d = 7500$. So,

$$c = a_1 - d$$

$$= 10,000 - 7500$$

$$= 2500.$$

So, the nth term of the sequence is given by

$$u_n = 7500n + 2500.$$

You can use the *list editor* of a graphing utility to create a table that shows the sales for each of the 20 years. First, enter the numbers 1 through 20 in L_1. Then enter $7500*L_1 + 2500$ for L_2. You should obtain a table like the one shown in Figure 9.10. Finally, use the *sum* feature of the graphing utility to find the sum of the data in L_2, as shown in Figure 9.11. So, the total sales for the first 20 years are $1,625,000.

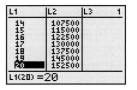

Figure 9.10

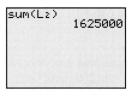

Figure 9.11

✓ *Checkpoint* Now try Exercise 81.

If you go on to take a course in calculus, you will study sequences and series in detail. You will learn that sequences and series play a major role in the study of calculus.

TECHNOLOGY SUPPORT

For instructions on how to use the *list editor* and *sum* features, see Appendix A; for specific keystrokes, go to the text website at *college.hmco.com*.

9.2   Exercises

Vocabulary Check

Fill in the blanks.

1. A sequence is called an _____ sequence if the differences between consecutive terms are the same. This difference is called the _____ difference.

2. The nth term of an arithmetic sequence has the form _____ .

3. The formula $S_n = \dfrac{n}{2}(a_1 + a_n)$ can be used to find the sum of the first n terms of an arithmetic sequence, called the _____ .

In Exercises 1–8, determine whether or not the sequence is arithmetic. If it is, find the common difference.

1. $10, 8, 6, 4, 2, \ldots$

2. $4, 9, 14, 19, 24, \ldots$

3. $3, \frac{5}{2}, 2, \frac{3}{2}, 1, \ldots$

4. $\frac{1}{3}, \frac{2}{3}, \frac{4}{3}, \frac{8}{3}, \frac{16}{3}, \ldots$

5. $-24, -16, -8, 0, 8, \ldots$

6. $\ln 1, \ln 2, \ln 3, \ln 4, \ln 5, \ldots$

7. $3.7, 4.3, 4.9, 5.5, 6.1, \ldots$

8. $1^2, 2^2, 3^2, 4^2, 5^2, \ldots$

In Exercises 9–16, write the first five terms of the sequence. Determine whether or not the sequence is arithmetic. If it is, find the common difference. (Assume n begins with 1.)

9. $a_n = 8 + 13n$

10. $a_n = (2^n)n$

11. $a_n = \dfrac{1}{n+1}$

12. $a_n = 1 + (n-1)4$

13. $a_n = 150 - 7n$

14. $a_n = 2^{n-1}$

15. $a_n = 3 + \dfrac{(-1)^n 2}{n}$

16. $a_n = 3 - 4(n + 6)$

In Exercises 17–26, find a formula for a_n for the arithmetic sequence.

17. $a_1 = 1, d = 3$

18. $a_1 = 15, d = 4$

19. $a_1 = 100, d = -8$

20. $a_1 = 0, d = -\frac{2}{3}$

21. $4, \frac{3}{2}, -1, -\frac{7}{2}, \ldots$

22. $10, 5, 0, -5, -10, \ldots$

23. $a_1 = 5, a_4 = 15$

24. $a_1 = -4, a_5 = 16$

25. $a_3 = 94, a_6 = 85$

26. $a_5 = 190, a_{10} = 115$

In Exercises 27–34, write the first five terms of the arithmetic sequence. Use the *table* feature of a graphing utility to verify your results.

27. $a_1 = 5, d = 6$

28. $a_1 = 5, d = -\frac{3}{4}$

29. $a_1 = -10, d = -12$

30. $a_4 = 16, a_{10} = 46$

31. $a_8 = 26, a_{12} = 42$

32. $a_6 = -38, a_{11} = -73$

33. $a_3 = 19, a_{15} = -1.7$

34. $a_5 = 16, a_{14} = 38.5$

In Exercises 35–38, write the first five terms of the arithmetic sequence. Find the common difference and write the nth term of the sequence as a function of n.

35. $a_1 = 15, \quad a_{k+1} = a_k + 4$

36. $a_1 = 200, \quad a_{k+1} = a_k - 10$

37. $a_1 = \frac{7}{2}, \quad a_{k+1} = a_k - \frac{1}{4}$

38. $a_1 = 0.375, \quad a_{k+1} = a_k + 0.25$

In Exercises 39–42, the first two terms of the arithmetic sequence are given. Find the missing term. Use the *table* feature of a graphing utility to verify your results.

39. $a_1 = 5, \quad a_2 = 11, \quad a_{10} = $ ▪

40. $a_1 = 3, \quad a_2 = 13, \quad a_9 = $ ▪

41. $a_1 = 4.2, \quad a_2 = 6.6, \quad a_7 = $ ▪

42. $a_1 = -0.7, \quad a_2 = -13.8, \quad a_8 = $ ▪

In Exercises 43–46, use a graphing utility to graph the first 10 terms of the sequence. (Assume n begins with 1.)

43. $a_n = 15 - \frac{3}{2}n$

44. $a_n = -5 + 2n$

45. $a_n = 0.5n + 4$

46. $a_n = -0.9n + 2$

In Exercises 47–52, use the *table* feature of a graphing utility to find the first 10 terms of the sequence. (Assume n begins with 1.)

47. $a_n = 4n - 5$

48. $a_n = 17 + 3n$

49. $a_n = 20 - \frac{3}{4}n$

50. $a_n = \frac{4}{5}n + 12$

51. $a_n = 1.5 + 0.005n$

52. $a_n = -12.4n + 9$

In Exercises 53–58, find the indicated nth partial sum of the arithmetic sequence.

53. $8, 20, 32, 44, \ldots,\quad n = 10$

54. $-6, -2, 2, 6, \ldots,\quad n = 50$

55. $0.5, 1.3, 2.1, 2.9, \ldots,\quad n = 10$

56. $4.2, 3.7, 3.2, 2.7, \ldots,\quad n = 12$

57. $a_1 = 100,\ a_{25} = 220,\quad n = 25$

58. $a_1 = 15,\ a_{100} = 307,\quad n = 100$

59. Find the sum of the first 100 positive odd integers.

60. Find the sum of the integers from -10 to 50.

In Exercises 61–68, find the partial sum without using a graphing utility.

61. $\displaystyle\sum_{n=1}^{50} n$

62. $\displaystyle\sum_{n=1}^{100} 2n$

63. $\displaystyle\sum_{n=1}^{100} 5n$

64. $\displaystyle\sum_{n=51}^{100} 7n$

65. $\displaystyle\sum_{n=11}^{30} n - \sum_{n=1}^{10} n$

66. $\displaystyle\sum_{n=51}^{100} n - \sum_{n=1}^{50} n$

67. $\displaystyle\sum_{n=1}^{500} (n + 8)$

68. $\displaystyle\sum_{n=1}^{250} (1000 - n)$

In Exercises 69–74, use a graphing utility to find the partial sum.

69. $\displaystyle\sum_{n=1}^{20} (2n + 5)$

70. $\displaystyle\sum_{n=0}^{50} (100 - 5n)$

71. $\displaystyle\sum_{n=1}^{100} \frac{n + 4}{2}$

72. $\displaystyle\sum_{n=0}^{100} \frac{8 - 3n}{16}$

73. $\displaystyle\sum_{i=1}^{60} \left(250 - \frac{8}{3}i\right)$

74. $\displaystyle\sum_{j=1}^{200} (4.5 + 0.025j)$

Job Offer In Exercises 75 and 76, consider a job offer with the given starting salary and guaranteed salary increase for the first 5 years of employment.

(a) Determine the salary during the sixth year of employment.

(b) Determine the total compensation from the company through 6 full years of employment.

(c) Verify your results in parts (a) and (b) numerically.

	Starting Salary	Annual Raise
75.	$32,500	$1500
76.	$36,800	$1750

77. *Brick Pattern* A brick patio has the approximate shape of a trapezoid, as shown in the figure. The patio has 18 rows of bricks. The first row has 14 bricks and the 18th row has 31 bricks. How many bricks are in the patio?

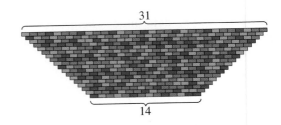

78. *Number of Logs* Logs are stacked in a pile, as shown in the figure. The top row has 15 logs and the bottom row has 24 logs. How many logs are in the stack?

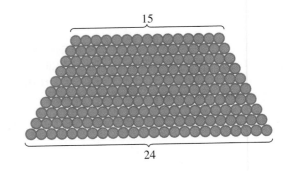

79. *Seating Capacity* Each row in a small auditorium has two more seats than the preceding row, as shown in the figure. Find the seating capacity of the auditorium if the front row seats 25 people and there are 15 rows of seats.

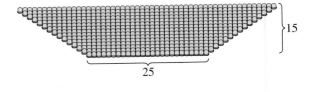

80. *Baling Hay* In the first two trips around a field baling hay, a farmer makes 93 bales and 89 bales, respectively, as shown in the figure. Because each trip is shorter than the preceding trip, the farmer estimates that the same pattern will continue. Estimate the total number of bales made if there are another six trips around the field.

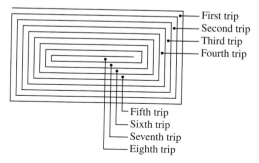

81. *Sales* A small hardware store makes a profit of $20,000 during its first year. The store owner sets a goal of increasing profits by $5000 each year for 4 years. Assuming that this goal is met, find the total profit during the first 5 years of business.

82. *Falling Object* An object with negligible air resistance is dropped from an airplane. During the first second of fall, the object falls 4.9 meters; during the second second, it falls 14.7 meters; during the third second, it falls 24.5 meters; and during the fourth second, it falls 34.3 meters. If this arithmetic pattern continues, how many meters will the object fall in 10 seconds?

Synthesis

True or False? In Exercises 83 and 84, determine whether the statement is true or false. Justify your answer.

83. Given an arithmetic sequence for which only the first and second terms are known, it is possible to find the nth term.

84. If the only known information about a finite arithmetic sequence is its first term and its last term, then it is possible to find the sum of the sequence.

In Exercises 85 and 86, find the first 10 terms of the sequence.

85. $a_1 = x, d = 2x$

86. $a_1 = -y, d = 5y$

87. *Think About It* The sum of the first 20 terms of an arithmetic sequence with a common difference of 3 is 650. Find the first term.

88. *Think About It* The sum of the first n terms of an arithmetic sequence with first term a_1 and common difference d is S_n. Determine the sum if each term is increased by 5. Explain.

89. *Pattern Recognition*

(a) Compute the following sums of positive odd integers.

$$1 + 3 = \boxed{}$$

$$1 + 3 + 5 = \boxed{}$$

$$1 + 3 + 5 + 7 = \boxed{}$$

$$1 + 3 + 5 + 7 + 9 = \boxed{}$$

$$1 + 3 + 5 + 7 + 9 + 11 = \boxed{}$$

(b) Use the sums in part (a) to make a conjecture about the sums of positive odd integers. Check your conjecture for the sum

$$1 + 3 + 5 + 7 + 9 + 11 + 13 = \boxed{}.$$

(c) Verify your conjecture algebraically.

90. *Think About It* Decide whether it is possible to fill in the blanks in each of the sequences such that the resulting sequence is arithmetic. If so, find a recursion formula for the sequence. Write a short paragraph explaining how you made your decisions.

(a) $-7, \boxed{}, \boxed{}, \boxed{}, \boxed{}, \boxed{}, 11$

(b) $17, \boxed{}, \boxed{}, \boxed{}, \boxed{}, \boxed{}, \boxed{}, 59$

(c) $2, 6, \boxed{}, \boxed{}, 162$

(d) $4, 7.5, \boxed{}, \boxed{}, \boxed{}, \boxed{}, 28.5$

(e) $8, 12, \boxed{}, \boxed{}, \boxed{}, 60.75$

Review

In Exercises 91 and 92, use Gauss-Jordan elimination to solve the system of equations.

91. $\begin{cases} 2x - y + 7z = -10 \\ 3x + 2y - 4z = 17 \\ 6x - 5y + z = -20 \end{cases}$

92. $\begin{cases} -x + 4y + 10z = 4 \\ 5x - 3y + z = 31 \\ 8x + 2y - 3z = -5 \end{cases}$

In Exercises 93 and 94, use a determinant to find the area of the triangle with the given vertices.

93. $(0, 0), (4, -3), (2, 6)$ **94.** $(-1, 2), (5, 1), (3, 8)$

9.3　Geometric Sequences and Series

Geometric Sequences

In Section 9.2, you learned that a sequence whose consecutive terms have a common *difference* is an arithmetic sequence. In this section, you will study another important type of sequence called a **geometric sequence.** Consecutive terms of a geometric sequence have a common *ratio*.

Definition of Geometric Sequence

A sequence is **geometric** if the ratios of consecutive terms are the same. So, the sequence $a_1, a_2, a_3, a_4, \ldots, a_n, \ldots$ is geometric if there is a number r such that

$$\frac{a_2}{a_1} = \frac{a_3}{a_2} = \frac{a_4}{a_3} = \cdots = r, \qquad r \neq 0.$$

The number r is the **common ratio** of the sequence.

What you should learn

- Recognize, write, and find the nth terms of geometric sequences.
- Find nth partial sums of geometric sequences.
- Find sums of infinite geometric series.
- Use geometric sequences to model and solve real-life problems.

Why you should learn it

Geometric sequences can reduce the amount of time it takes to find the sum of a sequence of numbers with a common ratio. For instance, Exercise 89 on page 724 shows how to use a geometric sequence to estimate the population growth of New Zealand.

Michael S. Yamashita/Corbis

Example 1　Examples of Geometric Sequences

a. The sequence whose nth term is 2^n is geometric. For this sequence, the common ratio between consecutive terms is 2.

$$2, 4, 8, 16, \ldots, 2^n, \ldots \qquad \text{Begin with } n = 1.$$

$$\frac{4}{2} = 2$$

b. The sequence whose nth term is $4(3^n)$ is geometric. For this sequence, the common ratio between consecutive terms is 3.

$$12, 36, 108, 324, \ldots, 4(3^n), \ldots \qquad \text{Begin with } n = 1.$$

$$\frac{36}{12} = 3$$

c. The sequence whose nth term is $\left(-\frac{1}{3}\right)^n$ is geometric. For this sequence, the common ratio between consecutive terms is $-\frac{1}{3}$.

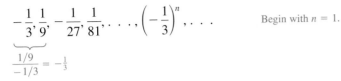

$$-\frac{1}{3}, \frac{1}{9}, -\frac{1}{27}, \frac{1}{81}, \ldots, \left(-\frac{1}{3}\right)^n, \ldots \qquad \text{Begin with } n = 1.$$

$$\frac{1/9}{-1/3} = -\frac{1}{3}$$

 Checkpoint Now try Exercise 1.

The sequence $1, 4, 9, 16, \ldots$, whose nth term is n^2, is *not* geometric. The ratio of the second term to first term is

$$\frac{a_2}{a_1} = \frac{4}{1} = 4$$

but the ratio of the third term to the second term is $\dfrac{a_3}{a_2} = \dfrac{9}{4}.$

STUDY TIP

In Example 1, notice that each of the geometric sequences has an nth term of the form ar^n, where r is the common ratio of the sequence.

> **The *n*th Term of a Geometric Sequence**
>
> The *n*th term of a geometric sequence has the form
>
> $$a_n = a_1 r^{n-1}$$
>
> where r is the common ratio of consecutive terms of the sequence. So, every geometric sequence can be written in the following form.
>
> $$a_1, \quad a_2, \quad a_3, \quad a_4, \quad a_5, \quad \ldots, \quad a_n, \quad \ldots$$
>
> $$a_1, a_1 r, a_1 r^2, a_1 r^3, a_1 r^4, \ldots, a_1 r^{n-1}, \ldots$$

If you know the *n*th term of a geometric sequence, you can find the $(n + 1)$th term by multiplying by r. That is, $a_{n+1} = ra_n$.

Example 2   Finding the Terms of a Geometric Sequence

Write the first five terms of the geometric sequence whose first term is $a_1 = 3$ and whose common ratio is $r = 2$.

Solution

Starting with 3, repeatedly multiply by 2 to obtain the following.

$a_1 = 3$	1st term	$a_4 = 3(2^3) = 24$	4th term
$a_2 = 3(2^1) = 6$	2nd term	$a_5 = 3(2^4) = 48$	5th term
$a_3 = 3(2^2) = 12$	3rd term		

✓ *Checkpoint* Now try Exercise 11.

TECHNOLOGY TIP

You can use a graphing utility to generate the geometric sequence in Example 2 by using the following steps.

3 [ENTER]

2 [×] [ANS]

Now press the enter key repeatedly to generate the terms of the sequence.

Most graphing utilities have a built-in function that will display the terms of a geometric sequence.

Example 3   Finding a Term of a Geometric Sequence

Find the 15th term of the geometric sequence whose first term is 20 and whose common ratio is 1.05.

Algebraic Solution

$$a_n = a_1 r^{n-1} \qquad \text{Formula for a geometric sequence}$$

$$a_{15} = 20(1.05)^{15-1} \qquad \text{Substitute 20 for } a_1, \text{ 1.05 for } r, \text{ and 15 for } n.$$

$$\approx 39.599 \qquad \text{Use a calculator.}$$

Numerical Solution

For this sequence, $r = 1.05$ and $a_1 = 20$. So, $a_n = 20(1.05)^{n-1}$. Use the *table* feature of a graphing utility to create a table that shows the values of $u_n = 20(1.05)^{n-1}$ for $n = 1$ through $n = 15$. From Figure 9.12, the number in the 15th row is approximately 39.599, so the 15th term of the geometric sequence is about 39.599.

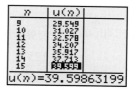

n	$u(n)$
9	29.549
10	31.027
11	32.578
12	34.207
13	35.917
14	37.713
15	39.599

$u(n)=39.59863199$

Figure 9.12

✓ *Checkpoint* Now try Exercise 25.

Example 4   Finding a Term of a Geometric Sequence

Find a formula for the nth term of the following geometric sequence. What is the ninth term of the sequence?

5, 15, 45, . . .

Solution

The common ratio of this sequence is

$$r = \frac{15}{5} = 3.$$

Because the first term is $a_1 = 5$, the formula must have the form

$$a_n = a_1 r^{n-1} = 5(3)^{n-1}.$$

You can determine the ninth term $(n = 9)$ to be

$$a_9 = 5(3)^{9-1} \qquad \text{Substitute 9 for } n.$$

$$= 5(6561) = 32{,}805. \qquad \text{Use a calculator.}$$

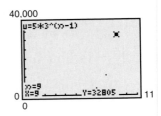

Figure 9.13

A graph of the first nine terms of the sequence is shown in Figure 9.13. Notice that the points lie on an exponential curve. This makes sense because a_n is an exponential function of n.

✓ *Checkpoint* Now try Exercise 33.

If you know *any* two terms of a geometric sequence, you can use that information to find a formula for the nth term of the sequence.

Example 5   Finding a Term of a Geometric Sequence

The 4th term of a geometric sequence is 125, and the 10th term is 125/64. Find the 14th term. (Assume that the terms of the sequence are positive.)

Solution

The 10th term is related to the fourth term by the equation

$$a_{10} = a_4 r^6. \qquad \text{Multiply 4th term by } r^{10-4}.$$

Because $a_{10} = 125/64$ and $a_4 = 125$, you can solve for r as follows.

$$\frac{125}{64} = 125r^6$$

$$\frac{1}{64} = r^6 \quad \Longrightarrow \quad \frac{1}{2} = r$$

You can obtain the 14th term by multiplying the 10th term by r^4.

$$a_{14} = a_{10} r^4 = \frac{125}{64}\left(\frac{1}{2}\right)^4 = \frac{125}{1024}$$

✓ *Checkpoint* Now try Exercise 31.

STUDY TIP

Remember that r is the common ratio of consecutive terms of a geometric sequence. So, in Example 5,

$$a_{10} = a_1 r^9$$

$$= a_1 \cdot r \cdot r \cdot r \cdot r^6$$

$$= a_1 \cdot \frac{a_2}{a_1} \cdot \frac{a_3}{a_2} \cdot \frac{a_4}{a_3} \cdot r^6$$

$$= a_4 r^6.$$

The Sum of a Finite Geometric Sequence

The formula for the sum of a *finite* geometric sequence is as follows. See Appendix B for a proof of this formula.

The Sum of a Finite Geometric Sequence

The sum of the finite geometric sequence

$$a_1, \ a_1r, \ a_1r^2, \ a_1r^3, \ a_1r^4, \ \ldots, \ a_1r^{n-1}$$

with common ratio $r \neq 1$ is given by

$$S_n = \sum_{i=1}^{n} a_1 r^{i-1} = a_1 \left(\frac{1 - r^n}{1 - r} \right).$$

Example 6   Finding the Sum of a Finite Geometric Sequence

Find the sum $\displaystyle\sum_{n=1}^{12} 4(0.3)^n$.

Solution

By writing out a few terms, you have

$$\sum_{n=1}^{12} 4(0.3)^n = 4(0.3)^1 + 4(0.3)^2 + 4(0.3)^3 + \cdots + 4(0.3)^{12}.$$

Now, because $a_1 = 4(0.3)$, $r = 0.3$, and $n = 12$, you can apply the formula for the sum of a finite geometric sequence to obtain

$$\sum_{n=1}^{12} 4(0.3)^n = a_1 \left(\frac{1 - r^n}{1 - r} \right) \qquad \text{Formula for sum of a finite geometric sequence}$$

$$= 4(0.3) \left[\frac{1 - (0.3)^{12}}{1 - 0.3} \right] \qquad \text{Substitute } 4(0.3) \text{ for } a_1, \ 0.3 \text{ for } r, \text{ and } 12 \text{ for } n.$$

$$\approx 1.714. \qquad \text{Use a calculator.}$$

✓ *Checkpoint*   Now try Exercise 49.

TECHNOLOGY TIP

Using the *sum sequence* feature of a graphing utility, you can calculate the sum of the sequence in Example 6 to be 1.7142848, as shown below.

Calculate the sum beginning at $n = 0$. You should obtain a sum of 1.7142848.

When using the formula for the sum of a geometric sequence, be careful to check that the index begins at $i = 1$. If the index begins at $i = 0$, you must adjust the formula for the nth partial sum. For instance, if the index in Example 6 had begun with $n = 0$, the sum would have been

$$\sum_{n=0}^{12} 4(0.3)^n = 4(0.3)^0 + \sum_{n=1}^{12} 4(0.3)^n$$

$$= 4 + \sum_{n=1}^{12} 4(0.3)^n$$

$$\approx 4 + 1.714$$

$$= 5.714.$$

Geometric Series

The sum of the terms of an infinite geometric sequence is called an **infinite geometric series** or simply a **geometric series.**

The formula for the sum of a *finite geometric sequence* can, depending on the value of r, be extended to produce a formula for the sum of an *infinite geometric series*. Specifically, if the common ratio r has the property that $|r| < 1$, it can be shown that r^n becomes arbitrarily close to zero as n increases without bound. Consequently,

$$a_1\left(\frac{1 - r^n}{1 - r}\right) \longrightarrow a_1\left(\frac{1 - 0}{1 - r}\right) \quad \text{as} \quad n \longrightarrow \infty.$$

This result is summarized as follows.

The Sum of an Infinite Geometric Series

If $|r| < 1$, then the infinite geometric series

$$a_1 + a_1r + a_1r^2 + a_1r^3 + \cdots + a_1r^{n-1} + \cdots$$

has the sum

$$S = \sum_{i=0}^{\infty} a_1r^i = \frac{a_1}{1 - r}.$$

Note that if $|r| \geq 1$, the series does not have a sum.

Example 7 Finding the Sum of an Infinite Geometric Series

Use a graphing utility to find the first six partial sums of the series. Then find the sum of the series.

$$\sum_{n=1}^{\infty} 4(0.6)^{n-1}$$

Solution

You can use the *cumulative sum* feature to find the first six partial sums of the series, as shown in Figure 9.14. By scrolling to the right, you can determine that the first six partial sums are as follows.

4, 6.4, 7.84, 8.704, 9.2224, 9.53344

Use the formula for the sum of an infinite geometric series to find the sum.

$$\sum_{n=1}^{\infty} 4(0.6)^{n-1} = 4(1) + 4(0.6) + 4(0.6)^2 + 4(0.6)^3 + \cdots + 4(0.6)^{n-1} + \cdots$$

$$= \frac{4}{1 - (0.6)} = 10 \qquad \frac{a_1}{1 - r}$$

✓ *Checkpoint* Now try Exercise 67.

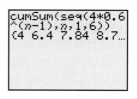

Figure 9.14

Example 8   Finding the Sum of an Infinite Geometric Series

Find the sum $3 + 0.3 + 0.03 + 0.003 + \cdots$.

Solution

$$3 + 0.3 + 0.03 + 0.003 + \cdots = 3 + 3(0.1) + 3(0.1)^2 + 3(0.1)^3 + \cdots$$

$$= \frac{3}{1 - (0.1)} \qquad \frac{a_1}{1 - r}$$

$$= \frac{10}{3}$$

$$\approx 3.33$$

☑ *Checkpoint* Now try Exercise 69.

Exploration

Notice in Example 7 that when using a graphing utility to find the sum of a series, you cannot enter ∞ as the upper limit of summation. Can you still find the sum using a graphing utility? If so, which partial sum will result in 10, the exact sum of the series?

Application

Example 9   Increasing Annuity

A deposit of $50 is made on the first day of each month in a savings account that pays 6% compounded monthly. What is the balance at the end of 2 years? (This type of savings plan is called an **increasing annuity.**)

Solution

The first deposit will gain interest for 24 months, and its balance will be

$$A_{24} = 50\left(1 + \frac{0.06}{12}\right)^{24} = 50(1.005)^{24}.$$

The second deposit will gain interest for 23 months, and its balance will be

$$A_{23} = 50\left(1 + \frac{0.06}{12}\right)^{23} = 50(1.005)^{23}.$$

The last deposit will gain interest for only 1 month, and its balance will be

$$A_1 = 50\left(1 + \frac{0.06}{12}\right)^1 = 50(1.005).$$

The total balance in the annuity will be the sum of the balances of the 24 deposits. Using the formula for the sum of a finite geometric sequence, with $A_1 = 50(1.005)$ and $r = 1.005$, you have

$$S_n = a_1\left(\frac{1 - r^n}{1 - r}\right) \qquad \text{Formula for sum of a finite geometric sequence}$$

$$S_{24} = 50(1.005)\left[\frac{1 - (1.005)^{24}}{1 - 1.005}\right] \qquad \text{Substitute } 50(1.005) \text{ for } a_1, 1.005 \text{ for } r, \text{ and } 24 \text{ for } n.$$

$$\approx \$1277.96. \qquad \text{Simplify.}$$

☑ *Checkpoint* Now try Exercise 85.

STUDY TIP

Recall from Section 4.1 that the compound interest formula is

$$A = P\left(1 + \frac{r}{n}\right)^{nt}.$$

So, in Example 9, $50 is the principal, 0.06 is the interest rate, 12 is the number of compoundings per year, and 2 is the time in years. If you substitute these values, you obtain

$$A = 50\left(1 + \frac{0.06}{12}\right)^{12(2)}$$

$$= 50\left(1 + \frac{0.06}{12}\right)^{24}$$

9.3	**Exercises**

Vocabulary Check

Fill in the blanks.

1. A sequence is called a _____ sequence if the ratios between consecutive terms are the same. This ratio is called the _____ ratio.

2. The nth term of a geometric sequence has the form _____ .

3. The formula for the sum of a finite geometric sequence is given by _____ .

4. The sum of the terms of an infinite geometric sequence is called a _____ .

5. The formula for the sum of an infinite geometric series is given by _____ .

In Exercises 1–10, determine whether or not the sequence is geometric. If it is, find the common ratio.

1. 5, 15, 45, 135, . . .

2. 3, 12, 48, 192, . . .

3. 6, 18, 30, 42, . . .

4. 1, -2, 4, -8, . . .

5. 1, $-\frac{1}{2}$, $\frac{1}{4}$, $-\frac{1}{8}$, . . .

6. 5, 1, 0.2, 0.04, . . .

7. $\frac{1}{8}$, $\frac{1}{4}$, $\frac{1}{2}$, 1, . . .

8. 9, -6, 4, $-\frac{8}{3}$, . . .

9. 1, $\frac{1}{2}$, $\frac{1}{3}$, $\frac{1}{4}$, . . .

10. $\frac{1}{5}$, $\frac{2}{7}$, $\frac{3}{9}$, $\frac{4}{11}$, . . .

In Exercises 11–18, write the first five terms of the geometric sequence.

11. $a_1 = 6$, $r = 3$

12. $a_1 = 4$, $r = 2$

13. $a_1 = 1$, $r = \frac{1}{2}$

14. $a_1 = 2$, $r = \frac{1}{3}$

15. $a_1 = 5$, $r = -\frac{1}{10}$

16. $a_1 = 6$, $r = -\frac{1}{4}$

17. $a_1 = 1$, $r = e$

18. $a_1 = 4$, $r = \sqrt{3}$

In Exercises 19–24, write the first five terms of the geometric sequence. Find the common ratio and write the nth term of the sequence as a function of n.

19. $a_1 = 64$, $a_{k+1} = \frac{1}{2}a_k$

20. $a_1 = 81$, $a_{k+1} = \frac{1}{3}a_k$

21. $a_1 = 9$, $a_{k+1} = 2a_k$

22. $a_1 = 5$, $a_{k+1} = -3a_k$

23. $a_1 = 6$, $a_{k+1} = -\frac{3}{2}a_k$

24. $a_1 = 30$, $a_{k+1} = -\frac{2}{3}a_k$

In Exercises 25–32, find the nth term of the geometric sequence. Use the *table* feature of a graphing utility to verify your answer numerically.

25. $a_1 = 4$, $r = \frac{1}{2}$, $n = 10$

26. $a_1 = 5$, $r = \frac{3}{2}$, $n = 8$

27. $a_1 = 6$, $r = -\frac{1}{3}$, $n = 12$

28. $a_1 = 8$, $r = -\frac{3}{4}$, $n = 9$

29. $a_1 = 500$, $r = 1.02$, $n = 14$

30. $a_1 = 1000$, $r = 1.005$, $n = 11$

31. $a_2 = -18$, $a_5 = \frac{2}{3}$, $n = 6$

32. $a_3 = \frac{16}{3}$, $a_5 = \frac{64}{27}$, $n = 7$

In Exercises 33–36, find the indicated nth term of the geometric sequence.

33. 9th term: 7, 21, 63, . . .

34. 7th term: 3, 36, 432, . . .

35. 10th term: 5, 30, 180, . . .

36. 22nd term: 4, 8, 16, . . .

In Exercises 37–40, use a graphing utility to graph the first 10 terms of the sequence.

37. $a_n = 12(-0.75)^{n-1}$

38. $a_n = 20(1.25)^{n-1}$

39. $a_n = 2(1.3)^{n-1}$

40. $a_n = 10(-1.2)^{n-1}$

In Exercises 41 and 42, find the first four terms of the sequence of partial sums of the geometric series. In a sequence of partial sums, the term S_n is the sum of the first n terms of the sequence. For instance, S_2 is the sum of the first two terms.

41. 8, -4, 2, -1, $\frac{1}{2}$, . . .

42. 8, 12, 18, 27, $\frac{81}{2}$, . . .

In Exercises 43 and 44, use a graphing utility to create a table showing the sequence of partial sums for the first 10 terms of the series.

43. $\displaystyle\sum_{n=1}^{\infty} 16\left(\tfrac{1}{2}\right)^{n-1}$

44. $\displaystyle\sum_{n=1}^{\infty} 4(0.2)^{n-1}$

In Exercises 45–54, find the sum. Use a graphing utility to verify your result.

45. $\displaystyle\sum_{n=1}^{9} 2^{n-1}$

46. $\displaystyle\sum_{n=1}^{9} (-2)^{n-1}$

47. $\displaystyle\sum_{i=1}^{7} 64\left(-\tfrac{1}{2}\right)^{i-1}$

48. $\displaystyle\sum_{i=1}^{6} 32\left(\tfrac{1}{4}\right)^{i-1}$

49. $\displaystyle\sum_{n=0}^{20} 3\left(\tfrac{3}{2}\right)^{n}$

50. $\displaystyle\sum_{n=0}^{15} 2\left(\tfrac{4}{3}\right)^{n}$

51. $\displaystyle\sum_{i=1}^{10} 8\left(-\tfrac{1}{4}\right)^{i-1}$

52. $\displaystyle\sum_{i=1}^{10} 5\left(-\tfrac{1}{3}\right)^{i-1}$

53. $\displaystyle\sum_{n=0}^{5} 300(1.06)^{n}$

54. $\displaystyle\sum_{n=0}^{6} 500(1.04)^{n}$

In Exercises 55–58, use summation notation to write the sum.

55. $5 + 15 + 45 + \cdots + 3645$

56. $7 + 14 + 28 + \cdots + 896$

57. $2 - \tfrac{1}{2} + \tfrac{1}{8} - \cdots + \tfrac{1}{2048}$

58. $15 - 3 + \tfrac{3}{5} - \cdots - \tfrac{3}{625}$

In Exercises 59–72, find the sum of the infinite geometric series, if possible. If not possible, explain why.

59. $\displaystyle\sum_{n=0}^{\infty} \left(\tfrac{1}{2}\right)^{n}$

60. $\displaystyle\sum_{n=0}^{\infty} 2\left(\tfrac{2}{3}\right)^{n}$

61. $\displaystyle\sum_{n=0}^{\infty} \left(-\tfrac{1}{2}\right)^{n}$

62. $\displaystyle\sum_{n=0}^{\infty} 2\left(-\tfrac{2}{3}\right)^{n}$

63. $\displaystyle\sum_{n=1}^{\infty} 2\left(\tfrac{7}{3}\right)^{n-1}$

64. $\displaystyle\sum_{n=1}^{\infty} \tfrac{1}{2}(4)^{n}$

65. $\displaystyle\sum_{n=0}^{\infty} (0.4)^{n}$

66. $\displaystyle\sum_{n=0}^{\infty} 4(0.5)^{n}$

67. $\displaystyle\sum_{n=0}^{\infty} -3(0.9)^{n}$

68. $\displaystyle\sum_{n=0}^{\infty} -10(0.2)^{n}$

69. $8 + 6 + \tfrac{9}{2} + \tfrac{27}{8} + \cdots$

70. $9 + 6 + 4 + \tfrac{8}{3} + \cdots$

71. $3 - 1 + \tfrac{1}{3} - \tfrac{1}{9} + \cdots$

72. $-6 + 5 - \tfrac{25}{6} + \tfrac{125}{36} - \cdots$

In Exercises 73–76, find the rational number representation of the repeating decimal.

73. $0.\overline{36}$

74. $0.\overline{297}$

75. $0.3\overline{18}$

76. $1.3\overline{8}$

77. **Compound Interest**   A principal of $1000 is invested at 3% interest. Find the amount after 10 years if the interest is compounded (a) annually, (b) semiannually, (c) quarterly, (d) monthly, and (e) daily.

78. **Compound Interest**   A principal of $2500 is invested at 4% interest. Find the amount after 20 years if the interest is compounded (a) annually, (b) semi-annually, (c) quarterly, (d) monthly, and (e) daily.

79. **Annuity**   A deposit of $100 is made at the beginning of each month in an account that pays 6% interest, compounded monthly. The balance A in the account at the end of 5 years is given by

$$A = 100\left(1 + \frac{0.06}{12}\right)^{1} + \cdots + 100\left(1 + \frac{0.06}{12}\right)^{60}.$$

Find A.

80. **Annuity**   A deposit of $50 is made at the beginning of each month in an account that pays 8% interest, compounded monthly. The balance A in the account at the end of 5 years is given by

$$A = 50\left(1 + \frac{0.08}{12}\right)^{1} + \cdots + 50\left(1 + \frac{0.08}{12}\right)^{60}.$$

Find A.

81. **Annuity**   A deposit of P dollars is made at the beginning of each month in an account earning an annual interest rate r, compounded monthly. The balance A after t years is given by

$$A = P\left(1 + \frac{r}{12}\right) + P\left(1 + \frac{r}{12}\right)^{2} + \cdots$$
$$+ P\left(1 + \frac{r}{12}\right)^{12t}.$$

Show that the balance is given by

$$A = P\left[\left(1 + \frac{r}{12}\right)^{12t} - 1\right]\left(1 + \frac{12}{r}\right).$$

82. **Annuity**   A deposit of P dollars is made at the beginning of each month in an account earning an annual interest rate r, compounded continuously. The balance A after t years is given by

$$A = Pe^{r/12} + Pe^{2r/12} + \cdots + Pe^{12tr/12}.$$

Show that the balance is given by

$$A = \frac{Pe^{r/12}(e^{rt} - 1)}{e^{r/12} - 1}.$$

Annuities In Exercises 83–86, consider making monthly deposits of P dollars in a savings account earning an annual interest rate r. Use the results of Exercises 81 and 82 to find the balance A after t years if the interest is compounded (a) monthly and (b) continuously.

83. $P = \$50$, $r = 7\%$, $t = 20$ years

84. $P = \$75$, $r = 4\%$, $t = 25$ years

85. $P = \$100$, $r = 5\%$, $t = 40$ years

86. $P = \$20$, $r = 6\%$, $t = 50$ years

87. ***Geometry*** The sides of a square are 16 inches in length. A new square is formed by connecting the midpoints of the sides of the original square, and two of the resulting triangles are shaded (see figure). If this process is repeated five more times, determine the total area of the shaded region.

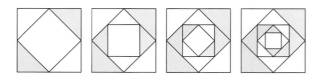

88. ***Geometry*** The sides of a square are 27 inches in length. New squares are formed by dividing the original square into nine squares. The center square is then shaded (see figure). If this process is repeated three more times, determine the total area of the shaded region.

89. ***Data Analysis*** The table shows the population a_n (in thousands) of New Zealand from 1995 to 2001. (Source: U.S. Census Bureau)

Year	Population, a_n
1995	3566
1996	3621
1997	3676
1998	3726
1999	3774
2000	3820
2001	3864

(a) The data in the table can be approximated by the sequence

$$a_n = 3343(1.013)^n, \quad n = 5, 6, \ldots, 11$$

where n represents the year, with $n = 5$ corresponding to 1995. Using this sequence, describe the rate at which the population of New Zealand is growing.

(b) Use the sequence in part (a) to predict the population of New Zealand in 2010.

(c) Use the sequence in part (a) to determine in what year the population of New Zealand will reach 4.1 million.

90. ***Data Analysis*** The table shows the revenue a_n (in billions of dollars) of AT&T Wireless Services from 1997 to 2002. (Source: AT&T Wireless Services)

Year	Revenue, a_n
1997	4.7
1998	5.4
1999	7.6
2000	10.4
2001	13.6
2002	15.6

(a) The data in the table can be approximated by the sequence

$$a_n = 0.737(1.296)^n, \quad n = 7, 8, \ldots, 12$$

where n represents the year, with $n = 7$ corresponding to 1997. Using this sequence, describe the rate at which the revenues of AT&T Wireless Service are growing.

(b) Use the sequence in part (a) and the formula for the sum of a finite geometric sequence to approximate the total revenues earned during this six-year period.

91. ***Salary*** You go to work for a company that pays $\$0.01$ the first day, $\$0.02$ the second day, $\$0.04$ the third day, and so on. If the daily wage keeps doubling, what will your total income be after working (a) 29 days? (b) 30 days? (c) 31 days?

∫ **92. *Distance*** A ball is dropped from a height of 16 feet. Each time it drops h feet, it rebounds $0.81h$ feet.

(a) Find the total vertical distance traveled by the ball.

(b) The ball takes the following times (in seconds) for each fall.

$s_1 = -16t^2 + 16$, $s_1 = 0$ if $t = 1$

$s_2 = -16t^2 + 16(0.81)$, $s_2 = 0$ if $t = 0.9$

$s_3 = -16t^2 + 16(0.81)^2$, $s_3 = 0$ if $t = (0.9)^2$

$s_4 = -16t^2 + 16(0.81)^3$, $s_4 = 0$ if $t = (0.9)^3$

⋮   ⋮

$s_n = -16t^2 + 16(0.81)^{n-1}$, $s_n = 0$ if $t = (0.9)^{n-1}$

Beginning with s_2, the ball takes the same amount of time to bounce up as it does to fall, and so the total time elapsed before it comes to rest is

$$t = 1 + 2\sum_{n=1}^{\infty}(0.9)^n.$$

Find this total time.

Synthesis

True or False? **In Exercises 93 and 94, determine whether the statement is true or false. Justify your answer.**

93. A sequence is geometric if the ratios of consecutive differences of consecutive terms are the same.

94. You can find the nth term of a geometric sequence by multiplying its common ratio by the first term of the sequence raised to the $(n-1)$th power.

In Exercises 95 and 96, write the first five terms of the geometric sequence.

95. $a_1 = 3, r = \dfrac{x}{2}$

96. $a_1 = \dfrac{1}{2}, r = 7x$

In Exercises 97 and 98, find the nth term of the geometric sequence.

97. $a_1 = 100, r = e^x, n = 9$

98. $a_1 = 4, r = \dfrac{4x}{3}, n = 6$

99. *Graphical Reasoning* Use a graphing utility to graph each function. Identify the horizontal asymptote of the graph and determine its relationship to the sum.

(a) $f(x) = 6\left[\dfrac{1-(0.5)^x}{1-(0.5)}\right]$, $\displaystyle\sum_{n=0}^{\infty}6\left(\dfrac{1}{2}\right)^n$

(b) $f(x) = 2\left[\dfrac{1-(0.8)^x}{1-(0.8)}\right]$, $\displaystyle\sum_{n=0}^{\infty}2\left(\dfrac{4}{5}\right)^n$

100. *Writing* Write a brief paragraph explaining why the terms of a geometric sequence decrease in magnitude when $-1 < r < 1$.

101. *Writing* Write a brief paragraph explaining how to use the first two terms of a geometric sequence to find the nth term.

102. *Exploration* You will need a piece of string or yarn, a pair of scissors, and a tape measure. Measure out any length of string at least 5 feet long. Double over the string and cut it in half. Take one of the resulting halves, double it over, and cut it in half. Continue this process until you are no longer able to cut a length of string in half. How many cuts were you able to make? Construct a sequence of the resulting string lengths after each cut, starting with the original length of the string. Find a formula for the nth term of this sequence. How many cuts could you theoretically make? Write a short paragraph discussing why you were not able to make that many cuts.

Review

103. *Average Speed* A truck traveled at an average speed of 50 miles per hour on a 200-mile trip. On the return trip, the average speed was 42 miles per hour. Find the average speed for the round trip.

104. *Work Rate* Your friend can mow a lawn in 4 hours and you can mow it in 6 hours. How long will it take both of you to mow the lawn working together?

In Exercises 105–108, find the determinant of the matrix.

105. $\begin{bmatrix} 4 & -1 \\ 6 & 2 \end{bmatrix}$

106. $\begin{bmatrix} 1 & 3 \\ -2 & 5 \end{bmatrix}$

107. $\begin{bmatrix} -1 & 3 & 4 \\ -2 & 8 & 0 \\ 2 & 5 & -1 \end{bmatrix}$

108. $\begin{bmatrix} -1 & 0 & 4 \\ -4 & 3 & 5 \\ 0 & 2 & -3 \end{bmatrix}$

9.4 Mathematical Induction

Introduction

In this section you will study a form of mathematical proof called **mathematical induction.** It is important that you clearly see the logical need for it, so let's take a closer look at a problem discussed in Example 5(a) on page 710.

$$S_1 = 1 = 1^2$$

$$S_2 = 1 + 3 = 2^2$$

$$S_3 = 1 + 3 + 5 = 3^2$$

$$S_4 = 1 + 3 + 5 + 7 = 4^2$$

$$S_5 = 1 + 3 + 5 + 7 + 9 = 5^2$$

Judging from the pattern formed by these first five sums, it appears that the sum of the first n odd integers is

$$S_n = 1 + 3 + 5 + 7 + 9 + \cdots + (2n - 1) = n^2.$$

Although this particular formula is valid, it is important for you to see that recognizing a pattern and then simply *jumping to the conclusio*n that the pattern must be true for all values of n is *not* a logically valid method of proof. There are many examples in which a pattern appears to be developing for small values of n but then fails at some point. One of the most famous cases of this is the conjecture by the French mathematician Pierre de Fermat (1601–1665), who speculated that all numbers of the form

$$F_n = 2^{2^n} + 1, \quad n = 0, 1, 2, \ldots$$

are prime. For $n = 0, 1, 2, 3,$ and 4, the conjecture is true.

$$F_0 = 3$$

$$F_1 = 5$$

$$F_2 = 17$$

$$F_3 = 257$$

$$F_4 = 65{,}537$$

The size of the next *Fermat number* ($F_5 = 4{,}294{,}967{,}297$) is so great that it was difficult for Fermat to determine whether or not it was prime. However, another well-known mathematician, Leonhard Euler (1707–1783), later found a factorization

$$F_5 = 4{,}294{,}967{,}297$$

$$= 641(6{,}700{,}417)$$

which proved that F_5 is not prime and therefore Fermat's conjecture was false.

Just because a rule, pattern, or formula seems to work for several values of n, you cannot simply decide that it is valid for *all* values of n without going through a *legitimate proof.* Mathematical induction is one method of proof.

What you should learn

- Use mathematical induction to prove statements involving a positive integer n.
- Find the sums of powers of integers.
- Find finite differences of sequences.

Why you should learn it

Finite differences can be used to determine what type of model can be used to represent a sequence. For instance, in Exercise 49 on page 733, you will use finite differences to find a model that represents the average sales price of a new mobile home in the southern region of the United States.

Photodisc/Getty Images

The Principle of Mathematical Induction

Let P_n be a statement involving the positive integer n. If

1. P_1 is true, and
2. the truth of P_k implies the truth of P_{k+1} for every positive k

then P_n must be true for all positive integers n.

STUDY TIP

It is important to recognize that *both* parts of the Principle of Mathematical Induction are necessary.

To apply the Principle of Mathematical Induction, you need to be able to determine the statement P_{k+1} for a given statement P_k. To detemine P_{k+1}, substitute $k + 1$ for k in the statement P_k.

Example 1 A Preliminary Example

Find P_{k+1} for each P_k.

a. $P_k : S_k = \dfrac{k^2(k+1)^2}{4}$

b. $P_k : S_k = 1 + 5 + 9 + \cdot \cdot \cdot + [4(k-1) - 3] + (4k - 3)$

c. $P_k : k + 3 < 5k^2$

d. $P_k : 3^k \geq 2k + 1$

Solution

a. $P_{k+1} : S_{k+1} = \dfrac{(k+1)^2(k+1+1)^2}{4}$ Replace k by $k + 1$.

$= \dfrac{(k+1)^2(k+2)^2}{4}$ Simplify.

b. $P_{k+1} : S_{k+1} = 1 + 5 + 9 + \cdot \cdot \cdot + \{4[(k+1) - 1] - 3\} + [4(k+1) - 3]$

$= 1 + 5 + 9 + \cdot \cdot \cdot + (4k - 3) + (4k + 1)$

c. $P_{k+1} : (k+1) + 3 < 5(k+1)^2$

$k + 4 < 5(k^2 + 2k + 1)$

d. $P_{k+1} : 3^{k+1} \geq 2(k+1) + 1$

$3^{k+1} \geq 2k + 3$

✓ *Checkpoint* Now try Exercise 3.

A well-known illustration used to explain why the Principle of Mathematical Induction works is the unending line of dominoes represented by Figure 9.15. If the line actually contains infinitely many dominoes, it is clear that you could not knock down the entire line by knocking down only *one domino* at a time. However, suppose it were true that each domino would knock down the next one as it fell. Then you could knock them all down simply by pushing the first one and starting a chain reaction. Mathematical induction works in the same way. If the truth of P_k implies the truth of P_{k+1} and if P_1 is true, the chain reaction proceeds as follows: P_1 implies P_2, P_2 implies P_3, P_3 implies P_4, and so on.

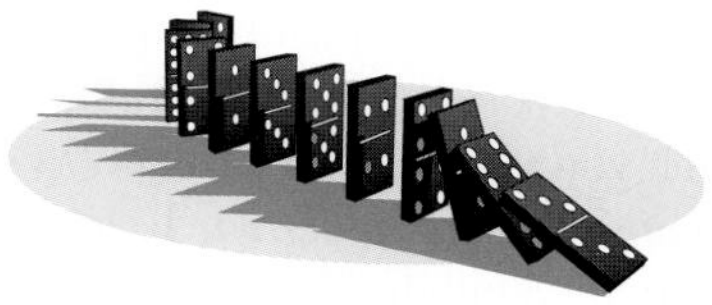

Figure 9.15

When using mathematical induction to prove a *summation* formula (such as the one in Example 2), it is helpful to think of S_{k+1} as

$$S_{k+1} = S_k + a_{k+1}$$

where a_{k+1} is the $(k + 1)$th term of the original sum.

Example 2 Using Mathematical Induction

Use mathematical induction to prove the following formula.

$$S_n = 1 + 3 + 5 + 7 + \cdots + (2n - 1) = n^2$$

Solution

Mathematical induction consists of two distinct parts. First, you must show that the formula is true when $n = 1$.

1. When $n = 1$, the formula is valid because

$$S_1 = 1 = 1^2.$$

The second part of mathematical induction has two steps. The first step is to assume that the formula is valid for *some* integer k. The second step is to use this assumption to prove that the formula is valid for the next integer, $k + 1$.

2. Assuming that the formula

$$S_k = 1 + 3 + 5 + 7 + \cdots + (2k - 1) = k^2$$

is true, you must show that the formula $S_{k+1} = (k + 1)^2$ is true.

$$
\begin{aligned}
S_{k+1} &= 1 + 3 + 5 + 7 + \cdots + (2k - 1) + [2(k + 1) - 1] \\
&= [1 + 3 + 5 + 7 + \cdots + (2k - 1)] + (2k + 2 - 1) \\
&= S_k + (2k + 1) \qquad \text{Group terms to form } S_k. \\
&= k^2 + 2k + 1 \qquad \text{Replace } S_k \text{ by } k^2. \\
&= (k + 1)^2
\end{aligned}
$$

Combining the results of parts (1) and (2), you can conclude by mathematical induction that the formula is valid for all positive integer values of n.

✓ *Checkpoint* Now try Exercise 7.

It occasionally happens that a statement involving natural numbers is *not* true for the first $k - 1$ positive integers but *is* true for all values of $n \geq k$. In these instances, you use a slight variation of the Principle of Mathematical Induction in which you verify P_k rather than P_1. This variation is called the *extended principle of mathematical induction.* To see the validity of this principle, note from Figure 9.15 that all but the first $k - 1$ dominoes can be knocked down by knocking over the kth domino. This suggests that you can prove a statement P_n to be true for $n \geq k$ by showing that P_k is true and that P_k implies P_{k+1}. In Exercises 23–28 in this section, you are asked to apply this extension of mathematical induction.

Example 3   Using Mathematical Induction

Use mathematical induction to prove the formula

$$S_n = 1^2 + 2^2 + 3^2 + 4^2 + \cdots + n^2 = \frac{n(n + 1)(2n + 1)}{6}$$

for all integers $n \geq 1$.

Solution

1. When $n = 1$, the formula is valid because

$$S_1 = 1^2 = \frac{1(1 + 1)(2 \cdot 1 + 1)}{6} = \frac{1(2)(3)}{6}.$$

2. Assuming that

$$S_k = 1^2 + 2^2 + 3^2 + 4^2 + \cdots + k^2 = \frac{k(k + 1)(2k + 1)}{6}$$

you must show that

$$S_{k+1} = \frac{(k + 1)(k + 1 + 1)[2(k + 1) + 1]}{6} = \frac{(k + 1)(k + 2)(2k + 3)}{6}.$$

To do this, write the following.

$$
\begin{aligned}
S_{k+1} &= S_k + a_{k+1} \\
&= (1^2 + 2^2 + 3^2 + 4^2 + \cdots + k^2) + (k + 1)^2 \\
&= \frac{k(k + 1)(2k + 1)}{6} + (k + 1)^2 \qquad \text{By assumption} \\
&= \frac{k(k + 1)(2k + 1) + 6(k + 1)^2}{6} \\
&= \frac{(k + 1)[k(2k + 1) + 6(k + 1)]}{6} \\
&= \frac{(k + 1)(2k^2 + 7k + 6)}{6} \\
&= \frac{(k + 1)(k + 2)(2k + 3)}{6}
\end{aligned}
$$

Combining the results of parts (1) and (2), you can conclude by mathematical induction that the formula is valid for *all* integers $n \geq 1$.

✓ *Checkpoint* Now try Exercise 13.

> **STUDY TIP**
>
> Remember that when adding rational expressions, you must first find the least common denominator (LCD). In Example 3, the LCD is 6.

When proving a formula by mathematical induction, the only statement that you *need* to verify is P_1. As a check, it is a good idea to try verifying some of the other statements. For instance, in Example 3, try verifying P_2 and P_3.

Sums of Powers of Integers

The formula in Example 3 is one of a collection of useful summation formulas. This and other formulas dealing with the sums of various powers of the first n positive integers are summarized below.

Sums of Powers of Integers

1. $\displaystyle\sum_{i=1}^{n} i = 1 + 2 + 3 + 4 + \cdots + n = \frac{n(n+1)}{2}$

2. $\displaystyle\sum_{i=1}^{n} i^2 = 1^2 + 2^2 + 3^2 + 4^2 + \cdots + n^2 = \frac{n(n+1)(2n+1)}{6}$

3. $\displaystyle\sum_{i=1}^{n} i^3 = 1^3 + 2^3 + 3^3 + 4^3 + \cdots + n^3 = \frac{n^2(n+1)^2}{4}$

4. $\displaystyle\sum_{i=1}^{n} i^4 = 1^4 + 2^4 + 3^4 + 4^4 + \cdots + n^4 = \frac{n(n+1)(2n+1)(3n^2+3n-1)}{30}$

5. $\displaystyle\sum_{i=1}^{n} i^5 = 1^5 + 2^5 + 3^5 + 4^5 + \cdots + n^5 = \frac{n^2(n+1)^2(2n^2+2n-1)}{12}$

Each of these formulas for sums can be proven by mathematical induction. (See Exercises 13–16 in this section.)

Example 4   Proving an Inequality by Mathematical Induction

Prove that $n < 2^n$ for all positive integers n.

Solution

1. For $n = 1$ and $n = 2$, the formula is true because

$$1 < 2^1 \text{ and } 2 < 2^2.$$

2. Assuming that

$$k < 2^k$$

you need to show that $k + 1 < 2^{k+1}$. Note first that

$$2^{k+1} = 2(2^k) > 2(k) = 2k. \qquad \text{By assumption}$$

Because $2k = k + k > k + 1$ for all $k > 1$, it follows that

$$2^{k+1} > 2k > k + 1$$

or

$$k + 1 < 2^{k+1}.$$

Combining the results of parts (1) and (2), you can conclude by mathematical induction that $n < 2^n$ for all integers $n \geq 1$.

✓ *Checkpoint* Now try Exercise 23.

Finite Differences

The **first differences** of a sequence are found by subtracting consecutive terms. The **second differences** are found by subtracting consecutive first differences. The first and second differences of the sequence 3, 5, 8, 12, 17, 23, . . . are as follows.

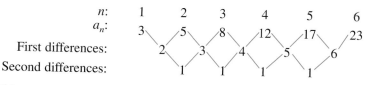

For this sequence, the second differences are all the same. When this happens, and the second differences are nonzero, the sequence has a perfect *quadratic* model. If the first differences are all the same nonzero number, the sequence has a *linear* model—that is, it is arithmetic.

Example 5   Finding a Quadratic Model

Find the quadratic model for the sequence 3, 5, 8, 12, 17, 23, ...

Solution

You know from the second differences shown above that the model is quadratic and has the form

$$a_n = an^2 + bn + c.$$

By substituting 1, 2, and 3 for n, you can obtain a system of three linear equations in three variables.

$a_1 = a(1)^2 + b(1) + c = 3$ Substitute 1 for n.

$a_2 = a(2)^2 + b(2) + c = 5$ Substitute 2 for n.

$a_3 = a(3)^2 + b(3) + c = 8$ Substitute 3 for n.

You now have a system of three equations in a, b, and c.

$$\begin{cases} a + b + c = 3 & \text{Equation 1} \\ 4a + 2b + c = 5 & \text{Equation 2} \\ 9a + 3b + c = 8 & \text{Equation 3} \end{cases}$$

Solving this system of equations using techniques discussed in Chapter 8, you can find the solution to be $a = \frac{1}{2}$, $b = \frac{1}{2}$, and $c = 2$. So, the quadratic model is

$$a_n = \tfrac{1}{2}n^2 + \tfrac{1}{2}n + 2.$$

Check the values of a_1, a_2, and a_3 as follows.

Check

$a_1 = \frac{1}{2}(1)^2 + \frac{1}{2}(1) + 2 = 3$ Solution checks. ✓

$a_2 = \frac{1}{2}(2)^2 + \frac{1}{2}(2) + 2 = 5$ Solution checks. ✓

$a_3 = \frac{1}{2}(3)^2 + \frac{1}{2}(3) + 2 = 8$ Solution checks. ✓

✓ *Checkpoint* Now try Exercise 45.

9.4   Exercises

Vocabulary Check

Fill in the blanks.

1. The first step in proving a formula by _____ is to show that the formula is true when $n = 1$.
2. The _____ differences of a sequence are found by subtracting consecutive terms.
3. A sequence is an _____ sequence if the first differences are all the same nonzero number.
4. If the _____ differences of a sequence are all the same nonzero number, then the sequence has a perfect quadratic model.

In Exercises 1–6, find P_{k+1} for the given P_k.

1. $P_k = \dfrac{5}{k(k+1)}$

2. $P_k = \dfrac{4}{(k+2)(k+3)}$

3. $P_k = \dfrac{3(2k+1)}{k-1}$

4. $P_k = \dfrac{k}{2}(5k-3)$

5. $P_k = 1 + 6 + 11 + \cdots + [5(k-1)-4] + (5k-4)$

6. $P_k = 7 + 13 + 19 + \cdots + [6(k-1)+1] + (6k+1)$

In Exercises 7–18, use mathematical induction to prove the formula for every positive integer n.

7. $2 + 4 + 6 + 8 + \cdots + 2n = n(n+1)$

8. $3 + 11 + 19 + 27 + \cdots + (8n-5) = n(4n-1)$

9. $3 + 8 + 13 + 18 + \cdots + (5n-2) = \dfrac{n}{2}(5n+1)$

10. $1 + 4 + 7 + 10 + \cdots + (3n-2) = \dfrac{n}{2}(3n-1)$

11. $1 + 2 + 2^2 + 2^3 + \cdots + 2^{n-1} = 2^n - 1$

12. $2(1 + 3 + 3^2 + 3^3 + \cdots + 3^{n-1}) = 3^n - 1$

13. $1 + 2 + 3 + 4 + \cdots + n = \dfrac{n(n+1)}{2}$

14. $1^3 + 2^3 + 3^3 + 4^3 + \cdots + n^3 = \dfrac{n^2(n+1)^2}{4}$

15. $\displaystyle\sum_{i=1}^{n} i^4 = \dfrac{n(n+1)(2n+1)(3n^2+3n-1)}{30}$

16. $\displaystyle\sum_{i=1}^{n} i^5 = \dfrac{n^2(n+1)^2(2n^2+2n-1)}{12}$

17. $\displaystyle\sum_{i=1}^{n} i(i+1) = \dfrac{n(n+1)(n+2)}{3}$

18. $\displaystyle\sum_{i=1}^{n} \dfrac{1}{(2i-1)(2i+1)} = \dfrac{n}{2n+1}$

In Exercises 19–22, find the sum using the formulas for the sums of powers of integers.

19. $\displaystyle\sum_{n=1}^{10} n^3$

20. $\displaystyle\sum_{n=1}^{5} n^4$

21. $\displaystyle\sum_{n=1}^{6} (n^2 - n)$

22. $\displaystyle\sum_{n=1}^{20} (n^3 - n)$

In Exercises 23–28, prove the inequality for the indicated integer values of n.

23. $n! > 2^n, \quad n \geq 4$

24. $\left(\dfrac{4}{3}\right)^n > n, \quad n \geq 7$

25. $\dfrac{1}{\sqrt{1}} + \dfrac{1}{\sqrt{2}} + \dfrac{1}{\sqrt{3}} + \cdots + \dfrac{1}{\sqrt{n}} > \sqrt{n}, \quad n \geq 2$

26. $\left(\dfrac{x}{y}\right)^{n+1} < \left(\dfrac{x}{y}\right)^{n}, \quad n \geq 1$ and $0 < x < y$

27. $(1 + a)^n \geq na, \quad n \geq 1$ and $a > 1$

28. $3^n > n\,2^n, \quad n \geq 1$

In Exercises 29–36, use mathematical induction to prove the property for all positive integers n.

29. $(ab)^n = a^n b^n$

30. $\left(\dfrac{a}{b}\right)^n = \dfrac{a^n}{b^n}$

31. If $x_1 \neq 0, x_2 \neq 0, \ldots, x_n \neq 0$, then
$(x_1 x_2 x_3 \cdots x_n)^{-1} = x_1^{-1} x_2^{-1} x_3^{-1} \cdots x_n^{-1}$.

32. If $x_1 > 0, x_2 > 0, \ldots, x_n > 0$, then
$\ln(x_1 x_2 \cdots x_n) = \ln x_1 + \ln x_2 + \cdots + \ln x_n$.

33. Generalized Distributive Law:
$x(y_1 + y_2 + \cdots + y_n) = xy_1 + xy_2 + \cdots + xy_n$

34. $(a + bi)^n$ and $(a - bi)^n$ are complex conjugates for all $n \geq 1$.

35. A factor of $(n^3 + 3n^2 + 2n)$ is 3.

36. A factor of $(n^3 + 5n + 6)$ is 3.

In Exercises 37–44, write the first five terms of the sequence beginning with the given term. Then calculate the first and second differences of the sequence. Does the sequence have a linear model, a quadratic model, or neither?

37. $a_1 = 0$
$a_n = a_{n-1} + 3$

38. $a_1 = 2$
$a_n = n - a_{n-1}$

39. $a_1 = 3$
$a_n = a_{n-1} - n$

40. $a_2 = -3$
$a_n = -2a_{n-1}$

41. $a_0 = 0$
$a_n = a_{n-1} + n$

42. $a_0 = 2$
$a_n = (a_{n-1})^2$

43. $a_1 = 2$
$a_n = a_{n-1} + 2$

44. $a_1 = 0$
$a_n = a_{n-1} + 2n$

In Exercises 45–48, find a quadratic model for the sequence with the indicated terms.

45. $a_0 = 3, \ a_1 = 3, \ a_4 = 15$

46. $a_0 = 7, \ a_1 = 6, \ a_3 = 10$

47. $a_0 = -3, \ a_2 = 1, \ a_4 = 9$

48. $a_0 = 3, \ a_2 = 0, \ a_6 = 36$

49. *Data Analysis*   The table shows the average sales price a_n (in thousands of dollars) of a new mobile home in the southern region of the United States from 1995 to 2001.   (Source: U.S. Census Bureau)

Year	Average sales price, a_n
1995	33.3
1996	35.5
1997	38.0
1998	40.1
1999	41.9
2000	44.2
2001	46.1

(a) Find the first differences of the data shown in the table.

(b) Use your results from part (a) to determine whether a linear model can be used to approximate the data. If so, use the *regression* feature of a graphing utility to find the model. Let n represent the year, with $n = 5$ corresponding to 1995.

(c) Use the model in part (b) and the *table* feature of a graphing utility to create a table of values for the model from 1995 to 2001. How closely does the model represent the original data?

(d) Use the model in part (b) to estimate the average sales price in 2006.

Synthesis

50. *Writing*   In your own words, explain what is meant by a proof by mathematical induction.

True or False?   In Exercises 51–53, determine whether the statement is true or false. Justify your answer.

51. If the statement P_k is true and P_k implies P_{k+1}, then P_1 is also true.

52. If a sequence is arithmetic, then the first differences of the sequence are all zero.

53. A sequence with n terms has $n - 1$ second differences.

54. *Think About It*   What conclusion can be drawn from the given information about the sequence of statements P_n?

(a) P_3 is true and P_k implies P_{k+1}.

(b) $P_1, P_2, P_3, \ldots, P_{50}$ are all true.

(c) $P_1, P_2,$ and P_3 are all true, but the truth of P_k does not imply that P_{k+1} is true.

(d) P_2 is true and P_{2k} implies P_{2k+2}.

Review

In Exercises 55–58, find the product.

55. $(2x^2 - 1)^2$

56. $(2x - y)^2$

57. $(5 - 4x)^3$

58. $(2x - 4y)^3$

In Exercises 59–62, use synthetic division to divide.

59. $(x^3 + x^2 - 10x + 8) \div (x + 4)$

60. $(x^3 - 4x^2 - 29x - 24) \div (x - 8)$

61. $(4x^3 + 11x^2 - 43x + 10) \div (x + 5)$

62. $(6x^3 - 35x^2 - 8x + 12) \div (x - 6)$

In Exercises 63–66, simplify the expression.

63. $3\sqrt{-27} - \sqrt{-12}$

64. $\sqrt[3]{125} + 4\sqrt[3]{-8} - 2\sqrt[3]{-54}$

65. $10\left(\sqrt[3]{64} - 2\sqrt[3]{-16}\right)$ **66.** $\left(-5 + \sqrt{-9}\right)^2$

9.5   The Binomial Theorem

Binomial Coefficients

Recall that a *binomial* is a polynomial that has two terms. In this section, you will study a formula that provides a quick method of raising a binomial to a power. To begin, look at the expansion of

$$(x + y)^n$$

for several values of n.

$$(x + y)^0 = 1$$
$$(x + y)^1 = x + y$$
$$(x + y)^2 = x^2 + 2xy + y^2$$
$$(x + y)^3 = x^3 + 3x^2y + 3xy^2 + y^3$$
$$(x + y)^4 = x^4 + 4x^3y + 6x^2y^2 + 4xy^3 + y^4$$
$$(x + y)^5 = x^5 + 5x^4y + 10x^3y^2 + 10x^2y^3 + 5xy^4 + y^5$$

There are several observations you can make about these expansions.

1. In each expansion, there are $n + 1$ terms.
2. In each expansion, x and y have symmetric roles. The powers of x decrease by 1 in successive terms, whereas the powers of y increase by 1.
3. The sum of the powers of each term is n. For instance, in the expansion of $(x + y)^5$, the sum of the powers of each term is 5.

$$4 + 1 = 5 \quad 3 + 2 = 5$$

$$(x + y)^5 = x^5 + 5\overbrace{x^4y^1} + 10\overbrace{x^3y^2} + 10x^2y^3 + 5x^1y^4 + y^5$$

4. The coefficients increase and then decrease in a symmetric pattern.

The coefficients of a binomial expansion are called **binomial coefficients.** To find them, you can use the **Binomial Theorem.** See Appendix B for a proof of this theorem.

What you should learn

- Use the Binomial Theorem to calculate binomial coefficients.
- Use Pascal's Triangle to calculate binomial coefficients.
- Use binomial coefficients to write binomial expansions.

Why you should learn it

You can use binomial coefficients to predict future behavior. For instance, in Exercise 86 on page 740, you are asked to use binomial coefficients to find the probability that a baseball player gets three hits during the next 10 times at bat.

Jonathan Daniel/Getty Images

The Binomial Theorem

In the expansion of $(x + y)^n$

$$(x + y)^n = x^n + nx^{n-1}y + \cdots + {}_nC_r\, x^{n-r}y^r + \cdots + nxy^{n-1} + y^n$$

the coefficient of $x^{n-r}y^r$ is

$${}_nC_r = \frac{n!}{(n-r)!r!}.$$

The symbol $\binom{n}{r}$ is often used in place of ${}_nC_r$ to denote binomial coefficients.

Example 1   Finding Binomial Coefficients

Find each binomial coefficient.

a. $_8C_2$ **b.** $\binom{10}{3}$ **c.** $_7C_0$ **d.** $\binom{8}{8}$

Solution

a. $_8C_2 = \dfrac{8!}{6! \cdot 2!} = \dfrac{(8 \cdot 7) \cdot 6!}{6! \cdot 2!} = \dfrac{8 \cdot 7}{2 \cdot 1} = 28$

b. $\binom{10}{3} = \dfrac{10!}{7! \cdot 3!} = \dfrac{(10 \cdot 9 \cdot 8) \cdot 7!}{7! \cdot 3!} = \dfrac{10 \cdot 9 \cdot 8}{3 \cdot 2 \cdot 1} = 120$

c. $_7C_0 = \dfrac{7!}{7! \cdot 0!} = 1$

d. $\binom{8}{8} = \dfrac{8!}{0! \cdot 8!} = 1$

✓ *Checkpoint* Now try Exercise 5.

TECHNOLOGY TIP

Most graphing utilities are programmed to evaluate $_nC_r$. The figure below shows how one graphing utility evaluates the binomial coefficient in Example 1(a). For instructions on how to use the $_nC_r$ feature, see Appendix A; for specific keystrokes, go to the text website at *college.hmco.com*.

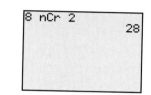

When $r \neq 0$ and $r \neq n$, as in parts (a) and (b) of Example 1, there is a simple pattern for evaluating binomial coefficients that works because there will always be factorial terms that divide out from the expression.

$$_8C_2 = \overset{\text{2 factors}}{\dfrac{8 \cdot 7}{\underset{\text{2 factorial}}{2 \cdot 1}}} \quad \text{and} \quad \binom{10}{3} = \overset{\text{3 factors}}{\dfrac{10 \cdot 9 \cdot 8}{\underset{\text{3 factorial}}{3 \cdot 2 \cdot 1}}}$$

Example 2   Finding Binomial Coefficients

Find each binomial coefficient using the pattern shown above.

a. $_7C_3$ **b.** $_7C_4$ **c.** $_{12}C_1$ **d.** $_{12}C_{11}$

Solution

a. $_7C_3 = \dfrac{7 \cdot 6 \cdot 5}{3 \cdot 2 \cdot 1} = 35$ **b.** $_7C_4 = \dfrac{7 \cdot 6 \cdot 5 \cdot 4}{4 \cdot 3 \cdot 2 \cdot 1} = 35$

c. $_{12}C_1 = \dfrac{12}{1} = 12$

d. $_{12}C_{11} = \dfrac{12!}{1! \cdot 11!} = \dfrac{(12) \cdot 11!}{1! \cdot 11!} = \dfrac{12}{1} = 12$

✓ *Checkpoint* Now try Exercise 7.

It is not a coincidence that the results in parts (a) and (b) of Example 2 are the same and that the results in parts (c) and (d) are the same. In general, it is true that

$$_nC_r = {_nC_{n-r}}.$$

Exploration

Find each pair of binomial coefficients.

a. $_7C_0, \ _7C_7$ **d.** $_7C_1, \ _7C_6$

b. $_8C_0, \ _8C_8$ **e.** $_8C_1, \ _8C_7$

c. $_{10}C_0, \ _{10}C_{10}$ **f.** $_{10}C_1, \ _{10}C_9$

What do you observe about the pairs in (a), (b), and (c)? What do you observe about the pairs in (d), (e), and (f)? Write two conjectures from your observations. Develop a convincing argument for your two conjectures.

Pascal's Triangle

There is a convenient way to remember the pattern for binomial coefficients. By arranging the coefficients in a triangular pattern, you obtain the following array, which is called **Pascal's Triangle.** This triangle is named after the famous French mathematician Blaise Pascal (1623–1662).

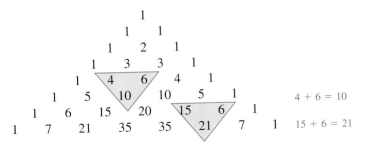

$$4 + 6 = 10$$
$$15 + 6 = 21$$

The first and last number in each row of Pascal's Triangle is 1. Every other number in each row is formed by adding the two numbers immediately above the number. Pascal noticed that the numbers in this triangle are precisely the same numbers as the coefficients of binomial expansions, as follows.

$$(x + y)^0 = 1 \qquad \text{0th row}$$
$$(x + y)^1 = 1x + 1y \qquad \text{1st row}$$
$$(x + y)^2 = 1x^2 + 2xy + 1y^2 \qquad \text{2nd row}$$
$$(x + y)^3 = 1x^3 + 3x^2y + 3xy^2 + 1y^3 \qquad \text{3rd row}$$
$$(x + y)^4 = 1x^4 + 4x^3y + 6x^2y^2 + 4xy^3 + 1y^4 \qquad \vdots$$
$$(x + y)^5 = 1x^5 + 5x^4y + 10x^3y^2 + 10x^2y^3 + 5xy^4 + 1y^5$$
$$(x + y)^6 = 1x^6 + 6x^5y + 15x^4y^2 + 20x^3y^3 + 15x^2y^4 + 6xy^5 + 1y^6$$
$$(x + y)^7 = 1x^7 + 7x^6y + 21x^5y^2 + 35x^4y^3 + 35x^3y^4 + 21x^2y^5 + 7xy^6 + 1y^7$$

The top row of Pascal's Triangle is called the *zeroth row* because it corresponds to the binomial expansion $(x + y)^0 = 1$. Similarly, the next row is called the *first row* because it corresponds to the binomial expansion $(x + y)^1 = 1(x) + 1(y)$. In general, the *nth row* of Pascal's Triangle gives the coefficients of $(x + y)^n$.

Exploration

Complete the table and describe the result.

n	r	$_nC_r$	$_nC_{n-r}$
9	5		
7	1		
12	4		
6	0		
10	7		

What characteristics of Pascal's Triangle are illustrated by the table?

Example 3   Using Pascal's Triangle

Use the seventh row of Pascal's Triangle to find the binomial coefficients.

$$_8C_0, \, _8C_1, \, _8C_2, \, _8C_3, \, _8C_4, \, _8C_5, \, _8C_6, \, _8C_7, \, _8C_8$$

Solution

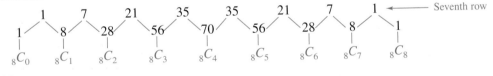

✓ *Checkpoint* Now try Exercise 17.

Binomial Expansions

As mentioned at the beginning of this section, when you write out the coefficients for a binomial that is raised to a power, you are **expanding a binomial.** The formulas for binomial coefficients give you an easy way to expand binomials, as demonstrated in the next four examples.

Example 4   Expanding a Binomial

Write the expansion for the expression $(x + 1)^3$.

Solution

The binomial coefficients from the third row of Pascal's Triangle are

   1, 3, 3, 1.

Therefore, the expansion is as follows.

$$(x + 1)^3 = (1)x^3 + (3)x^2(1) + (3)x(1^2) + (1)(1^3)$$

$$= x^3 + 3x^2 + 3x + 1$$

✓ *Checkpoint* Now try Exercise 21.

To expand binomials representing *differences*, rather than sums, you alternate signs. Here are two examples.

$$(x - 1)^3 = x^3 - 3x^2 + 3x - 1$$

$$(x - 1)^4 = x^4 - 4x^3 + 6x^2 - 4x + 1$$

Example 5   Expanding Binomial Expressions

Write the expansion for each expression.

a. $(2x - 3)^4$

b. $(x - 2y)^4$

Solution

The binomial coefficients from the fourth row of Pascal's Triangle are

   1, 4, 6, 4, 1.

Therefore, the expansions are as follows.

a. $(2x - 3)^4 = (1)(2x)^4 - (4)(2x)^3(3) + (6)(2x)^2(3^2) - (4)(2x)(3^3) + (1)(3^4)$

$$= 16x^4 - 96x^3 + 216x^2 - 216x + 81$$

b. $(x - 2y)^4 = (1)x^4 - (4)x^3(2y) + (6)x^2(2y)^2 - (4)x(2y)^3 + (1)(2y)^4$

$$= x^4 - 8x^3y + 24x^2y^2 - 32xy^3 + 16y^4$$

✓ *Checkpoint* Now try Exercise 29.

TECHNOLOGY TIP

You can use a graphing utility to check the expansion in Example 5(a) by graphing the original binomial expression and the expansion in the same viewing window. The graphs should coincide, as shown below.

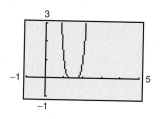

Example 6   Expanding a Binomial

Write the expansion for the expression $(x^2 + 4)^3$.

Solution

Use the third row of Pascal's Triangle, as follows.

$$(x^2 + 4)^3 = (1)(x^2)^3 + (3)(x^2)^2(4) + (3)x^2(4^2) + (1)(4^3)$$

$$= x^6 + 12x^4 + 48x^2 + 64$$

✓ *Checkpoint* Now try Exercise 35.

Sometimes you will need to find a specific term in a binomial expansion. Instead of writing out the entire expansion, you can use the fact that, from the Binomial Theorem, the $(r + 1)$st term is

$$_nC_r x^{n-r} y^r.$$

For example, if you wanted to find the third term of the expression in Example 6, you could use the above formula with $n = 3$ and $r = 2$ to obtain

$$_3C_2(x^2)^{3-2} \cdot 4^2 = 3(x^2) \cdot 16$$

$$= 48x^2.$$

Example 7   Finding a Term or Coefficient in a Binomial Expansion

a. Find the sixth term of $(a + 2b)^8$.

b. Find the coefficient of the term a^6b^5 in the expansion of $(2a - 5b)^{11}$.

Solution

a. To find the sixth term in this binomial expansion, use $n = 8$ and $r = 5$ [the formula is for the $(r + 1)$st term, so r is one less than the number of the term that you are looking for] to get

$$_8C_5 a^{8-5}(2b)^5 = 56 \cdot a^3 \cdot (2b)^5$$

$$= 56(2^5)a^3b^5$$

$$= 1792a^3b^5.$$

b. In this case, $n = 11$, $r = 5$, $x = 2a$, and $y = -5b$. Substitute these values to obtain

$$_nC_r x^{n-r} y^r = {}_{11}C_5(2a)^6(-5b)^5$$

$$= (462)(64a^6)(-3125b^5)$$

$$= -92,400,000.$$

So, the coefficient is $-92,400,000$.

✓ *Checkpoint* Now try Exercise 47.

9.5 Exercises

Vocabulary Check

Fill in the blanks.

1. The coefficients of a binomial expansion are called _____ .
2. To find binomial coefficients you can use the _____ or _____ .
3. The notation used to denote a binomial coefficient is _____ or _____ .
4. When you write out the coefficients for a binomial that is raised to a power, you are _____ a _____ .

In Exercises 1–10, find the binomial coefficient.

1. $_7C_5$

2. $_9C_6$

3. $\binom{12}{0}$

4. $\binom{20}{20}$

5. $_{20}C_{15}$

6. $_{12}C_3$

7. $_{14}C_1$

8. $_{18}C_{17}$

9. $\binom{100}{98}$

10. $\binom{10}{7}$

In Exercises 11–16, use a graphing utility to find $_nC_r$.

11. $_{32}C_{28}$

12. $_{17}C_4$

13. $_{22}C_9$

14. $_{52}C_{47}$

15. $_{41}C_{36}$

16. $_{34}C_4$

In Exercises 17–20, use Pascal's Triangle to find the binomial coefficient.

17. $_7C_4$

18. $_6C_3$

19. $_8C_5$

20. $_5C_2$

In Exercises 21–42, use the Binomial Theorem to expand and simplify the expression.

21. $(x + 2)^4$

22. $(x + 1)^6$

23. $(a + 3)^3$

24. $(a + 2)^4$

25. $(y - 2)^4$

26. $(y - 2)^5$

27. $(x + y)^5$

28. $(x + y)^6$

29. $(3r + 2s)^6$

30. $(4x + 3y)^4$

31. $(x - y)^5$

32. $(2x - y)^5$

33. $(1 - 4x)^3$

34. $(5 - 2y)^3$

35. $(x^2 + y^2)^4$

36. $(x^2 + y^2)^6$

37. $\left(\dfrac{1}{x} + y\right)^5$

38. $\left(\dfrac{1}{x} + 2y\right)^6$

39. $2(x - 3)^4 + 5(x - 3)^2$

40. $3(x + 1)^5 + 4(x + 1)^3$

41. $-3(x - 2)^3 - 4(x + 1)^6$

42. $5(x + 2)^5 - 2(x - 1)^2$

In Exercises 43–46, expand the binomial by using Pascal's Triangle to determine the coefficients.

43. $(3t - s)^5$

44. $(3 - 2z)^4$

45. $(x + 2y)^5$

46. $(3y + 2)^5$

In Exercises 47–54, find the specified nth term in the expansion of the binomial.

47. $(x + 8)^{10}$, $n = 4$

48. $(x - 5)^6$, $n = 7$

49. $(x - 6y)^5$, $n = 3$

50. $(x - 10z)^7$, $n = 4$

51. $(4x + 3y)^9$, $n = 8$

52. $(5a + 6b)^5$, $n = 5$

53. $(10x - 3y)^{12}$, $n = 9$

54. $(7x + 2y)^{15}$, $n = 8$

In Exercises 55–62, find the coefficient a of the term in the expansion of the binomial.

Binomial	Term
55. $(x + 3)^{12}$	ax^4
56. $(x + 4)^{12}$	ax^5
57. $(x - 2y)^{10}$	ax^8y^2
58. $(4x - y)^{10}$	ax^2y^8
59. $(3x - 2y)^9$	ax^6y^3
60. $(2x - 3y)^8$	ax^4y^4
61. $(x^2 + y)^{10}$	ax^8y^6
62. $(z^2 - 1)^{12}$	az^6

In Exercises 63–66, use the Binomial Theorem to expand and simplify the expression.

63. $\left(\sqrt{x} + 5\right)^4$

64. $\left(4\sqrt{t} - 1\right)^3$

65. $(x^{2/3} - y^{1/3})^3$

66. $(u^{3/5} + v^{1/5})^5$

In Exercises 67–70, expand the expression in the difference quotient and simplify.

$$\frac{f(x + h) - f(x)}{h}$$

67. $f(x) = x^3$

68. $f(x) = x^4$

69. $f(x) = \sqrt{x}$

70. $f(x) = \dfrac{1}{x}$

In Exercises 71–76, use the Binomial Theorem to expand the complex number. Simplify your result.

71. $(1 + i)^4$

72. $(4 - i)^5$

73. $(2 - 3i)^6$

74. $\left(5 + \sqrt{-9}\right)^3$

75. $\left(-\dfrac{1}{2} + \dfrac{\sqrt{3}}{2}i\right)^3$

76. $\left(5 - \sqrt{3}i\right)^4$

Approximation   In Exercises 77–80, use the Binomial Theorem to approximate the quantity accurate to three decimal places. For example, in Exercise 77, use the expansion

$$(1.02)^8 = (1 + 0.02)^8 = 1 + 8(0.02) + 28(0.02)^2 + \cdots .$$

77. $(1.02)^8$

78. $(2.005)^{10}$

79. $(2.99)^{12}$

80. $(1.98)^9$

Graphical Reasoning   In Exercises 81 and 82, use a graphing utility to graph f and g in the same viewing window. What is the relationship between the two graphs? Use the Binomial Theorem to write the polynomial function g in standard form.

81. $f(x) = x^3 - 4x, \quad g(x) = f(x + 3)$

82. $f(x) = -x^4 + 4x^2 - 1, \quad g(x) = f(x - 5)$

Graphical Reasoning   In Exercises 83 and 84, use a graphing utility to graph the functions in the given order and in the same viewing window. Compare the graphs. Which two functions have identical graphs and why?

83. (a) $f(x) = (1 - x)^3$

    (b) $g(x) = 1 - 3x$

    (c) $h(x) = 1 - 3x + 3x^2$

    (d) $p(x) = 1 - 3x + 3x^2 - x^3$

84. (a) $f(x) = \left(1 - \frac{1}{2}x\right)^4$

    (b) $g(x) = 1 - 2x + \frac{3}{2}x^2$

    (c) $h(x) = 1 - 2x + \frac{3}{2}x^2 - \frac{1}{2}x^3$

    (d) $p(x) = 1 - 2x + \frac{3}{2}x^2 - \frac{1}{2}x^3 + \frac{1}{16}x^4$

Probability   In Exercises 85–88, consider n independent trials of an experiment in which each trial has two possible outcomes, success or failure. The probability of a success on each trial is p and the probability of a failure is $q = 1 - p$. In this context, the term $_nC_k \, p^k q^{n-k}$ in the expansion of $(p + q)^n$ gives the probability of k successes in the n trials of the experiment.

85. A fair coin is tossed seven times. To find the probability of obtaining four heads, evaluate the term

$$_7C_4\left(\tfrac{1}{2}\right)^4\left(\tfrac{1}{2}\right)^3$$

in the expansion of $\left(\frac{1}{2} + \frac{1}{2}\right)^7$.

86. The probability of a baseball player getting a hit during any given time at bat is $\frac{1}{4}$. To find the probability that the player gets three hits during the next 10 times at bat, evaluate the term

$$_{10}C_3\left(\tfrac{1}{4}\right)^3\left(\tfrac{3}{4}\right)^7$$

in the expansion of $\left(\frac{1}{4} + \frac{3}{4}\right)^{10}$.

87. The probability of a sales representative making a sale with any one customer is $\frac{1}{3}$. The sales representative makes eight contacts a day. To find the probability of making four sales, evaluate the term

$$_8C_4\left(\tfrac{1}{3}\right)^4\left(\tfrac{2}{3}\right)^4$$

in the expansion of $\left(\frac{1}{3} + \frac{2}{3}\right)^8$.

88. To find the probability that the sales representative in Exercise 87 makes four sales if the probability of a sale with any one customer is $\frac{1}{2}$, evaluate the term

$$_8C_4\left(\tfrac{1}{2}\right)^4\left(\tfrac{1}{2}\right)^4$$

in the expansion of $\left(\frac{1}{2} + \frac{1}{2}\right)^8$.

89. *Life Insurance*   The average amount of life insurance per household f (in thousands of dollars) from 1985 through 2000 can be approximated by

$$f(t) = 0.018t^2 + 5.15t + 41.6, \quad 5 \le t \le 20$$

where t represents the year, with $t = 5$ corresponding to 1985 (see figure). (Source: American Council of Life Insurance)

(a) You want to adjust the model so that $t = 5$ corresponds to 1995 rather than 1985. To do this, you shift the graph of f 10 units *to the left* and obtain $g(t) = f(t + 10)$. Write $g(t)$ in standard form.

(b) Use a graphing utility to graph f and g in the same viewing window.

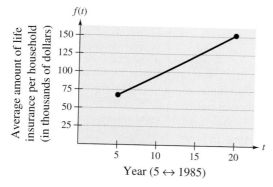

Figure for 89

90. *Education*   The average tuition, room, and board costs f (in dollars) for undergraduates from 1980 through 2001 can be approximated by the model

$$f(t) = 3.65t^2 + 308.7t + 2846, \quad 0 \le t \le 21$$

where t represents the year, with $t = 0$ corresponding to 1980 (see figure).   (Source: U.S. Department of Education)

(a) You want to adjust the model so that $t = 0$ corresponds to 1990 rather than 1980. To do this, you shift the graph of f 10 units to the left and obtain $g(t) = f(t + 10)$. Write $g(t)$ in standard form.

(b) Use a graphing utility to graph f and g in the same viewing window.

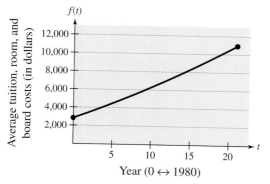

Synthesis

True or False?   In Exercises 91 and 92, determine whether the statement is true or false. Justify your answer.

91. One of the terms in the expansion of $(x - 2y)^{12}$ is $7920x^4y^8$.

92. The x^{10}-term and the x^{14}-term in the expansion of $(x^2 + 3)^{12}$ have identical coefficients.

93. *Writing*   In your own words, explain how to form the rows of Pascal's Triangle.

94. Form rows 8–10 of Pascal's Triangle.

95. *Think About It*   How do the expansions of $(x + y)^n$ and $(x - y)^n$ differ?

96. *Error Analysis*   You are a math instructor and receive the following solutions from one of your students on a quiz. Find the error(s) in each solution and write a short paragraph discussing ways that your student could avoid the error(s) in the future.

(a) Find the second term in the expansion of $(2x - 3y)^5$.

$$5(2x)^4(3y)^2 = 720x^4y^2$$

(b) Find the fourth term in the expansion of $\left(\frac{1}{2}x + 7y\right)^6$.

$$_6C_4\left(\tfrac{1}{2}x\right)^2(7y)^4 = 9003.75x^2y^4$$

Proof   In Exercises 97–100, prove the property for all integers r and n, where $0 \le r \le n$.

97. $_nC_r = {_nC_{n-r}}$

98. $_nC_0 - {_nC_1} + {_nC_2} - \cdots \pm {_nC_n} = 0$

99. $_{n+1}C_r = {_nC_r} + {_nC_{r-1}}$

100. The sum of the numbers in the nth row of Pascal's Triangle is 2^n.

Review

In Exercises 101–104, describe the relationship between the graphs of f and g.

101. $g(x) = f(x) + 8$ **102.** $g(x) = f(x - 3)$

103. $g(x) = f(-x)$ **104.** $g(x) = -f(x)$

In Exercises 105 and 106, find the inverse of the matrix.

105. $\begin{bmatrix} -6 & 5 \\ -5 & 4 \end{bmatrix}$ **106.** $\begin{bmatrix} 1.2 & -2.3 \\ -2 & 4 \end{bmatrix}$

9.6   Counting Principles

Simple Counting Problems

The last two sections of this chapter present a brief introduction to some of the basic counting principles and their application to probability. In the next section, you will see that much of probability has to do with counting the number of ways an event can occur.

What you should learn

- Solve simple counting problems.
- Use the Fundamental Counting Principle to solve more complicated counting problems.
- Use permutations to solve counting problems.
- Use combinations to solve counting problems.

Why you should learn it

You can use counting principles to solve counting problems that occur in real life. For instance, in Exercises 17 and 18 on page 749, you are asked to use counting principles to determine the number of possible ways of forming license plate numbers.

Tony Freeman/PhotoEdit

Example 1   Selecting Pairs of Numbers at Random

Eight pieces of paper are numbered from 1 to 8 and placed in a box. One piece of paper is drawn from the box, its number is written down, and the piece of paper is *returned to the box*. Then, a second piece of paper is drawn from the box, and its number is written down. Finally, the two numbers are added together. In how many different ways can a sum of 12 be obtained?

Solution

To solve this problem, count the number of different ways that a sum of 12 can be obtained using two numbers from 1 to 8.

First number   4 5 6 7 8

Second number   8 7 6 5 4

From this list, you can see that a sum of 12 can occur in five different ways.

 Checkpoint   Now try Exercise 7.

Example 2   Selecting Pairs of Numbers at Random

Eight pieces of paper are numbered from 1 to 8 and placed in a box. Two pieces of paper are drawn from the box *at the same time*, and the numbers on the pieces of paper are written down and totaled. In how many different ways can a sum of 12 be obtained?

Solution

To solve this problem, count the number of different ways that a sum of 12 can be obtained using two *different* numbers from 1 to 8.

First number   4 5 7 8

Second number   8 7 5 4

So, a sum of 12 can be obtained in four different ways.

 Checkpoint   Now try Exercise 8.

STUDY TIP

The difference between the counting problems in Examples 1 and 2 can be described by saying that the random selection in Example 1 occurs **with replacement,** whereas the random selection in Example 2 occurs **without replacement,** which eliminates the possibility of choosing two 6's.

The Fundamental Counting Principle

Examples 1 and 2 describe simple counting problems in which you can *list* each possible way that an event can occur. When it is possible, this is always the best way to solve a counting problem. However, some events can occur in so many different ways that it is not feasible to write out the entire list. In such cases, you must rely on formulas and counting principles. The most important of these is the **Fundamental Counting Principle.**

> **Fundamental Counting Principle**
>
> Let E_1 and E_2 be two events. The first event E_1 can occur in m_1 different ways. After E_1 has occurred, E_2 can occur in m_2 different ways. The number of ways that the two events can occur is $m_1 \cdot m_2$.

Example 3   Using the Fundamental Counting Principle

How many different pairs of letters from the English alphabet are possible?

Solution

There are two events in this situation. The first event is the choice of the first letter, and the second event is the choice of the second letter. Because the English alphabet contains 26 letters, it follows that the number of two-letter pairs is

$$26 \cdot 26 = 676.$$

 Checkpoint Now try Exercise 9.

Example 4   Using the Fundamental Counting Principle

Telephone numbers in the United States currently have 10 digits. The first three are the *area code* and the next seven are the *local telephone number*. How many different telephone numbers are possible within each area code? (Note that at this time, a local telephone number cannot begin with 0 or 1.)

Solution

Because the first digit cannot be 0 or 1, there are only eight choices for the first digit. For each of the other six digits, there are 10 choices.

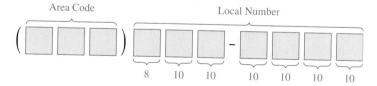

So, the number of local telephone numbers that are possible within each area code is $8 \cdot 10 \cdot 10 \cdot 10 \cdot 10 \cdot 10 \cdot 10 = 8{,}000{,}000$.

 Checkpoint Now try Exercise 21.

Permutations

One important application of the Fundamental Counting Principle is in determining the number of ways that n elements can be arranged (in order). An ordering of n elements is called a **permutation** of the elements.

Definition of Permutation

A **permutation** of n different elements is an ordering of the elements such that one element is first, one is second, one is third, and so on.

Example 5 Finding the Number of Permutations of n Elements

How many permutations are possible of the letters A, B, C, D, E, and F?

Solution

Consider the following reasoning.

First position:	Any of the *six* letters
Second position:	Any of the remaining *five* letters
Third position:	Any of the remaining *four* letters
Fourth position:	Any of the remaining *three* letters
Fifth position:	Any of the remaining *two* letters
Sixth position:	The *one* remaining letter

So, the number of choices for the six positions are as follows.

Permutations of six letters

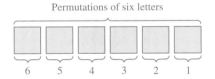

The total number of permutations of the six letters is

$$6! = 6 \cdot 5 \cdot 4 \cdot 3 \cdot 2 \cdot 1 = 720.$$

✓ *Checkpoint* Now try Exercise 39.

Number of Permutations of n Elements

The number of permutations of n elements is given by

$$n \cdot (n - 1) \cdots 4 \cdot 3 \cdot 2 \cdot 1 = n!.$$

In other words, there are $n!$ different ways that n elements can be ordered.

It is useful, on occasion, to order a *subset* of a collection of elements rather than the entire collection. For example, you might want to choose and order r elements out of a collection of n elements. Such an ordering is called a **permutation of n elements taken r at a time.**

Example 6   Counting Horse Race Finishes

Eight horses are running in a race. In how many different ways can these horses come in first, second, and third? (Assume that there are no ties.)

Solution

Here are the different possibilities.

Win (first position):	*Eight* choices
Place (second position):	*Seven* choices
Show (third position):	*Six* choices

The numbers of choices for the three positions are as follows.

Different orders of horses

8   7   6

So, using the Fundamental Counting Principle, you can determine that there are

$8 \cdot 7 \cdot 6 = 336$

different ways in which the eight horses can come in first, second, and third.

✓ *Checkpoint* Now try Exercise 41.

Permutations of n Elements Taken r at a Time

The number of **permutations of n elements taken r at a time** is given by

$$_nP_r = \frac{n!}{(n-r)!}$$

$$= n(n-1)(n-2) \cdots (n-r+1).$$

Using this formula, you can rework Example 6 to find that the number of permutations of eight horses taken three at a time is

$$_8P_3 = \frac{8!}{5!}$$

$$= \frac{8 \cdot 7 \cdot 6 \cdot 5!}{5!}$$

$$= 336$$

which is the same answer obtained in the example.

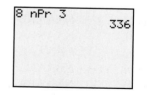

TECHNOLOGY TIP   Most graphing utilities are programmed to evaluate $_nP_r$. Figure 9.16 shows how one graphing utility evaluates the permutation $_8P_3$. For instructions on how to use the $_nP_r$ feature, see Appendix A; for specific keystrokes, go to the text website at *college.hmco.com*.

Figure 9.16

Remember that for permutations, order is important. So, if you are looking at the possible permutations of the letters A, B, C, and D taken three at a time, the permutations (A, B, D) and (B, A, D) would be counted as different because the *order* of the elements is different.

Suppose, however, that you are asked to find the possible permutations of the letters A, A, B, and C. The total number of permutations of the four letters would be $_4P_4 = 4!$. However, not all of these arrangements would be *distinguishable* because there are two A's in the list. To find the number of distinguishable permutations, you can use the following formula.

Distinguishable Permutations

Suppose a set of n objects has n_1 of one kind of object, n_2 of a second kind, n_3 of a third kind, and so on, with

$$n = n_1 + n_2 + n_3 + \cdots + n_k.$$

The number of **distinguishable permutations** of the n objects is given by

$$\frac{n!}{n_1! \cdot n_2! \cdot n_3! \cdots n_k!}.$$

Example 7   Distinguishable Permutations

In how many distinguishable ways can the letters in BANANA be written?

Solution

This word has six letters, of which three are A's, two are N's, and one is a B. So, the number of distinguishable ways in which the letters can be written is

$$\frac{6!}{3! \cdot 2! \cdot 1!} = \frac{6 \cdot 5 \cdot 4 \cdot 3!}{3! \cdot 2!} = 60.$$

The 60 different arrangements are as follows.

AAABNN	AAANBN	AAANNB	AABANN
AABNAN	AABNNA	AANABN	AANANB
AANBAN	AANBNA	AANNAB	AANNBA
ABAANN	ABANAN	ABANNA	ABNAAN
ABNANA	ABNNAA	ANAABN	ANAANB
ANABAN	ANABNA	ANANAB	ANANBA
ANBAAN	ANBANA	ANBNAA	ANNAAB
ANNABA	ANNBAA	BAAANN	BAANAN
BAANNA	BANAAN	BANANA	BANNAA
BNAAAN	BNAANA	BNANAA	BNNAAA
NAAABN	NAAANB	NAABAN	NAABNA
NAANAB	NAANBA	NABAAN	NABANA
NABNAA	NANAAB	NANABA	NANBAA
NBAAAN	NBAANA	NBANAA	NBNAAA
NNAAAB	NNAABA	NNABAA	NNBAAA

✓ *Checkpoint* Now try Exercise 45.

Combinations

When you count the number of possible permutations of a set of elements, order is important. As a final topic in this section, you will look at a method for selecting subsets of a larger set in which order *is not* important. Such subsets are called **combinations of *n* elements taken *r* at a time.** For instance, the combinations

{A, B, C}   and   {B, A, C}

are equivalent because both sets contain the same three elements, and the order in which the elements are listed is not important. So, you would count only one of the two sets. A common example of a combination is a card game in which the player is free to reorder the cards after they have been dealt.

Example 8   Combinations of *n* Elements Taken *r* at a Time

In how many different ways can three letters be chosen from the letters A, B, C, D, and E? (The order of the three letters is not important.)

Solution

The following subsets represent the different combinations of three letters that can be chosen from five letters.

{A, B, C}   {A, B, D}

{A, B, E}   {A, C, D}

{A, C, E}   {A, D, E}

{B, C, D}   {B, C, E}

{B, D, E}   {C, D, E}

From this list, you can conclude that there are 10 different ways in which three letters can be chosen from five letters.

 Checkpoint Now try Exercise 49.

Combination of *n* Elements Taken *r* at a Time

The number of **combinations of *n* elements taken *r* at a time** is given by

$$_nC_r = \frac{n!}{(n-r)!r!}.$$

Note that the formula for $_nC_r$ is the same one given for binomial coefficients. To see how this formula is used, solve the counting problem in Example 8. In that problem, you are asked to find the number of combinations of five elements taken three at a time. So, $n = 5$, $r = 3$, and the number of combinations is

$$_5C_3 = \frac{5!}{2!3!} = \frac{5 \cdot \overset{2}{\cancel{4}} \cdot 3!}{2 \cdot 1 \cdot 3!} = 10$$

which is the same answer obtained in Example 8.

Example 9   Counting Card Hands

A standard poker hand consists of five cards dealt from a deck of 52. How many different poker hands are possible? (After the cards are dealt, the player may reorder them, so order is not important.)

Solution

You can find the number of different poker hands by using the formula for the number of combinations of 52 elements taken five at a time, as follows.

$$_{52}C_5 = \frac{52!}{47!5!}$$

$$= \frac{52 \cdot 51 \cdot 50 \cdot 49 \cdot 48 \cdot 47!}{5 \cdot 4 \cdot 3 \cdot 2 \cdot 1 \cdot 47!}$$

$$= 2{,}598{,}960$$

 Checkpoint   Now try Exercise 55.

Example 10   Forming a Team

You are forming a 12-member swim team from 10 girls and 15 boys. The team must consist of five girls and seven boys. How many different 12-member teams are possible?

Solution

There are $_{10}C_5$ ways of choosing five girls. There are $_{15}C_7$ ways of choosing seven boys. By the Fundamental Counting Principle, there are $_{10}C_5 \cdot {}_{15}C_7$ ways of choosing five girls and seven boys.

$$_{10}C_5 \cdot {}_{15}C_7 = \frac{10!}{5! \cdot 5!} \cdot \frac{15!}{8! \cdot 7!}$$

$$= 252 \cdot 6435$$

$$= 1{,}621{,}620$$

So, there are 1,621,620 12-member swim teams possible. You can verify this by using the $_nC_r$ feature of a graphing utility, as shown in Figure 9.17.

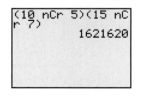

Figure 9.17

 Checkpoint   Now try Exercise 59.

9.6   Exercises

Vocabulary Check

Fill in the blanks.

1. The _____ states that if there are m_1 ways for one event to occur and m_2 ways for a second event to occur, then there are $m_1 \cdot m_2$ ways for both events to occur.

2. An ordering of n elements is called a _____ of the elements.

3. The number of permutations of n elements taken r at a time is given by the formula _____ .

4. The number of _____ of n objects is given by $\dfrac{n!}{n_1! \cdot n_2! \cdot n_3! \cdot \ \cdots \ \cdot n_k!}$.

5. When selecting subsets of a larger set in which order is not important, you are finding the number of _____ of n elements taken r at a time.

Random Selection   In Exercises 1–8, determine the number of ways in which a computer can randomly generate one or more such integers from 1 through 12.

1. An odd integer

2. An even integer

3. A prime integer

4. An integer that is greater than 6

5. An integer that is divisible by 4

6. An integer that is divisible by 3

7. Two integers whose sum is 8

8. Two distinct integers whose sum is 8

9. *Consumer Awareness*   A customer can choose one of four amplifiers, one of six compact disc players, and one of five speaker models for an entertainment system. Determine the number of possible system configurations.

10. *Consumer Awareness*   A customer in a computer store can choose one of four monitors, one of two keyboards, and one of three computers. If all the choices are compatible, determine the number of possible system configurations.

11. *Job Applicants*   A college needs two additional faculty members: a chemist and a statistician. In how many ways can these positions be filled if there are three applicants for the chemistry position and eight applicants for the statistics position?

12. *Course Schedule*   A college student is preparing a course schedule for the next semester. The student must select one of two mathematics courses, one of three science courses, and one of five courses from the social sciences and humanities. How many schedules are possible?

13. *True-False Exam*   In how many ways can a 10-question true-false exam be answered? (Assume that no questions are omitted.)

14. *True-False Exam*   In how many ways can a six-question true-false exam be answered? (Assume that no questions are omitted.)

15. *Recreation*   Four people are lining up for a ride on a toboggan, but only two of the four are willing to take the first position. With that constraint, in how many ways can the four people be seated on the toboggan?

16. *Travel*   Four people are taking a long trip in a four-seat car. Three of the people agree to share the driving. In how many different arrangements can the four people sit?

17. *License Plate*   In the state of Colorado the automobile license plates consist of a three-digit number followed by three letters. How many distinct license plates can be formed?

18. *License Plate*   In the state of Ohio the automobile license plates consist of two letters followed by a two-digit number, followed by two letters. How many distinct license plates can be formed?

19. *Three-Digit Numbers*   How many three-digit numbers can be formed under each condition?

    (a) The leading digit cannot be zero.

    (b) The leading digit cannot be zero and no repetition of digits is allowed.

(c) The leading digit cannot be zero and the number must be a multiple of 5.

(d) The number is at least 400.

20. **Four-Digit Numbers** How many four-digit numbers can be formed under each condition?

(a) The leading digit cannot be zero.

(b) The leading digit cannot be zero and no repetition of digits is allowed.

(c) The leading digit cannot be zero and the number must be less than 5000.

(d) The leading digit cannot be zero and the number must be even.

21. **Telephone Numbers** In 2003, the state of Nevada had two area codes. Using the information about telephone numbers given in Example 4, how many telephone numbers could Nevada's phone system have accommodated?

22. **Telephone Numbers** In 2003, the state of Kansas had four area codes. Using the information about telephone numbers given in Example 4, how many telephone numbers could Kansas's phone system have accommodated?

23. **Entertainment** Three couples have reserved seats in a row for a concert. In how many different ways can they be seated if

(a) there are no seating restrictions?

(b) the two members of each couple wish to sit together?

24. **Single File** In how many orders can five girls and three boys walk through a doorway single file if

(a) there are no restrictions?

(b) the girls walk through before the boys?

In Exercises 25–30, evaluate $_nP_r$ using the formula from this section.

25. $_4P_4$

26. $_5P_5$

27. $_8P_3$

28. $_{20}P_2$

29. $_5P_4$

30. $_7P_4$

In Exercises 31 and 32, solve for n.

31. $14 \cdot {}_nP_3 = {}_{n+2}P_4$

32. $_nP_5 = 18 \cdot {}_{n-2}P_4$

In Exercises 33–38, evaluate using a graphing utility.

33. $_{20}P_6$

34. $_{100}P_5$

35. $_{120}P_4$

36. $_{10}P_8$

37. $_{20}C_4$

38. $_{10}C_7$

39. **Posing for a Photograph** In how many ways can five children line up in a row?

40. **Riding in a Car** In how many ways can four people sit in a four-passenger car?

41. **Choosing Officers** From a pool of 12 candidates, the offices of president, vice-president, secretary, and treasurer will be filled. In how many ways can the offices be filled?

42. **Manufacturing** Four processes are involved in assembling a product, and they can be performed in any order. The management wants to test each order to determine which is the least time consuming. How many different orders will have to be tested?

In Exercises 43–46, find the number of distinguishable permutations of the group of letters.

43. A, A, G, E, E, E, M

44. B, B, B, T, T, T, T, T

45. A, L, G, E, B, R, A

46. M, I, S, S, I, S, S, I, P, P, I

47. Use the letters A, B, C, and D.

(a) Write all permutations of the letters.

(b) Write all permutations of the letters if the letters B and C must remain between the letters A and D.

48. Use the letters A, B, C, D, E, and F.

(a) Write all possible selections of two letters that can be formed from the letters. (The order of the two letters is not important.)

(b) Write all possible selections of three letters that can be formed from the letters. (The order of the three letters is not important.)

49. **Forming an Experimental Group** In order to conduct an experiment, four students are randomly selected from a class of 20. How many different groups of four students are possible?

50. **Exam Questions** You can answer any 12 questions from a total of 14 questions on an exam. In how many different ways can you select the questions?

51. **Lottery** In Maryland's Lotto game a player chooses six distinct numbers from 1 to 49. In how many ways can a player select the six numbers?

52. **Lottery** In Connecticut's Cash 5 game a player chooses five distinct numbers from 1 to 35. In how many ways can a player select the five numbers?

53. *Geometry*   Three points that are not collinear determine three lines. How many lines are determined by nine points, no three of which are collinear?

54. *Defective Units*   A shipment of 25 television sets contains three defective units. In how many ways can a vending company purchase four of these units and receive (a) all good units, (b) two good units, and (c) at least two good units?

55. *Poker Hand*   You are dealt five cards from an ordinary deck of 52 playing cards. In how many ways can you get a full house? (A full house consists of three of one kind and two of another. For example, 8-8-8-5-5 and K-K-K-10-10 are full houses.)

56. *Card Hand*   Five cards are chosen from a standard deck of 52 cards. How many five-card combinations contain two jacks and three aces?

57. *Job Applicants*   A clothing manufacturer interviews 12 people for four openings in the human resources department of the company. Five of the 12 people are women. If all 12 are qualified, in how many ways can the employer fill the four positions if (a) the selection is random and (b) exactly two women are selected?

58. *Job Applicants*   A law office interviews paralegals for 10 openings. There are 13 paralegals with two years of experience and 20 paralegals with one year of experience. How many combinations of seven paralegals with two years of experience and three paralegals with one year of experience are possible?

59. *Forming a Committee*   A six-member research committee is to be formed having one administrator, three faculty members, and two students. There are seven administrators, 12 faculty members, and 20 students in contention for the committee. How many six-member committees are possible?

60. *Interpersonal Relationships*   The number of possible interpersonal relationships increases dramatically as the size of a group increases. Determine the number of different two-person relationships that are possible in a group of people of size (a) 3, (b) 8, (c) 12, and (d) 20.

Geometry   **In Exercises 61–64, find the number of diagonals of the polygon. (A line segment connecting any two nonadjacent vertices is called a *diagonal* of a polygon.)**

61. Pentagon

62. Hexagon

63. Octagon

64. Decagon

Synthesis

True or False?   **In Exercises 65 and 66, determine whether the statement is true or false. Justify your answer.**

65. The number of pairs of letters that can be formed from any of the first 13 letters in the alphabet (A–M), where repetitions are allowed, is an example of a permutation.

66. The number of permutations of n elements can be derived by using the Fundamental Counting Principle.

67. *Think About It*   Can your calculator evaluate $_{100}P_{80}$? If not, explain why.

68. *Writing*   Explain in your own words the meaning of $_nP_r$.

69. What is the relationship between $_nC_r$ and $_nC_{n-r}$?

70. Without calculating the numbers, determine which of the following is greater. Explain.

(a) The number of combinations of 10 elements taken six at a time

(b) The number of permutations of 10 elements taken six at a time

Proof   **In Exercises 71–74, prove the identity.**

71. $_nP_{n-1} = {_nP_n}$

72. $_nC_n = {_nC_0}$

73. $_nC_{n-1} = {_nC_1}$

74. $_nC_r = \dfrac{_nP_r}{r!}$

Review

In Exercises 75–78, solve the equation. Round your answer to three decimal places, if necessary.

75. $\sqrt{x-3} = x-6$

76. $\dfrac{4}{t} + \dfrac{3}{2t} = 1$

77. $\log_2(x-3) = 5$

78. $e^{x/3} = 16$

In Exercises 79–82, use Cramer's Rule to solve the system of equations.

79. $\begin{cases} -5x + 3y = -14 \\ 7x - 2y = 2 \end{cases}$

80. $\begin{cases} 8x + y = 35 \\ 6x + 2y = 10 \end{cases}$

81. $\begin{cases} -3x - 4y = -1 \\ 9x + 5y = -4 \end{cases}$

82. $\begin{cases} 10x - 11y = -74 \\ -8x - 4y = 8 \end{cases}$

9.7   Probability

The Probability of an Event

Any happening whose result is uncertain is called an **experiment.** The possible results of the experiment are **outcomes,** the set of all possible outcomes of the experiment is the **sample space** of the experiment, and any subcollection of a sample space is an **event.**

For instance, when a six-sided die is tossed, the sample space can be represented by the numbers 1 through 6. For this experiment, each of the outcomes is *equally likely*.

To describe a sample space in such a way that each outcome is equally likely, you must sometimes distinguish between or among various outcomes in ways that appear artificial. Example 1 illustrates such a situation.

Tony Freeman/PhotoEdit

Example 1   Finding the Sample Space

Find the sample space for each of the following.

a. One coin is tossed.

b. Two coins are tossed.

c. Three coins are tossed.

Solution

a. Because the coin will land either heads up (denoted by H) or tails up (denoted by T), the sample space is

$$S = \{H, T\}.$$

b. Because either coin can land heads up or tails up, the possible outcomes are as follows.

   $HH =$ heads up on both coins

   $HT =$ heads up on first coin and tails up on second coin

   $TH =$ tails up on first coin and heads up on second coin

   $TT =$ tails up on both coins

So, the sample space is

$$S = \{HH, HT, TH, TT\}.$$

Note that this list distinguishes between the two cases HT and TH, even though these two outcomes appear to be similar.

c. Following the notation of part (b), the sample space is

$$S = \{HHH, HHT, HTH, HTT, THH, THT, TTH, TTT\}.$$

Note that this list distinguishes between the cases HHT, HTH, and THH, and between the cases HTT, THT, and TTH.

✓ *Checkpoint*   Now try Exercise 1.

To calculate the probability of an event, count the number of outcomes in the event and in the sample space. The *number of outcomes* in event E is denoted by $n(E)$, and the number of outcomes in the sample space S is denoted by $n(S)$. The probability that event E will occur is given by $n(E)/n(S)$.

The Probability of an Event

If an event E has $n(E)$ equally likely outcomes and its sample space S has $n(S)$ equally likely outcomes, the **probability** of event E is given by

$$P(E) = \frac{n(E)}{n(S)}.$$

Because the number of outcomes in an event must be less than or equal to the number of outcomes in the sample space, the probability of an event must be a number from 0 to 1, inclusive. That is,

$$0 \le P(E) \le 1,$$

as indicated in Figure 9.18. If $P(E) = 0$, event E *cannot occur*, and E is called an **impossible event**. If $P(E) = 1$, event E *must occur*, and E is called a **certain event.**

Exploration

Toss two coins 40 times and write down the number of heads that occur on each toss (0, 1, or 2). How many times did two heads occur? How many times would you expect two heads to occur if you did the experiment 1000 times?

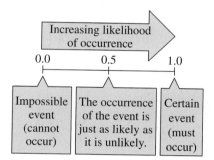

Figure 9.18

Example 2   Finding the Probability of an Event

a. Two coins are tossed. What is the probability that both land heads up?

b. A card is drawn from a standard deck of playing cards. What is the probability that it is an ace?

Solution

a. Following the procedure in Example 1(b), let

$$E = \{HH\}$$

and

$$S = \{HH, HT, TH, TT\}.$$

The probability of getting two heads is

$$P(E) = \frac{n(E)}{n(S)} = \frac{1}{4}.$$

b. Because there are 52 cards in a standard deck of playing cards and there are four aces (one of each suit), the probability of drawing an ace is

$$P(E) = \frac{n(E)}{n(S)}$$

$$= \frac{4}{52} = \frac{1}{13}.$$

✓ *Checkpoint* Now try Exercise 7.

STUDY TIP

You can write a probability as a fraction, a decimal, or a percent. For instance, in Example 2(a), the probability of getting two heads can be written as $\frac{1}{4}$, 0.25, or 25%.

Example 3 Finding the Probability of an Event

Two six-sided dice are tossed. What is the probability that a total of 7 is rolled? (See Figure 9.19.)

Solution

Because there are six possible outcomes on each die, you can use the Fundamental Counting Principle to conclude that there are $6 \cdot 6 = 36$ different outcomes when two dice are tossed. To find the probability of rolling a total of 7, you must first count the number of ways this can occur.

Figure 9.19

First die	1	2	3	4	5	6
Second die	6	5	4	3	2	1

So, a total of 7 can be rolled in six ways, which means that the probability of rolling a 7 is

$$P(E) = \frac{n(E)}{n(S)} = \frac{6}{36} = \frac{1}{6}.$$

✓ *Checkpoint* Now try Exercise 15.

You could have written out each sample space in Examples 2 and 3 and simply counted the outcomes in the desired events. For larger sample spaces, however, using the counting principles discussed in Section 9.6 should save you time.

Example 4 Finding the Probability of an Event

Twelve-sided dice, as shown in Figure 9.20, can be constructed (in the shape of regular dodecahedrons) such that each of the numbers from 1 to 6 appears twice on each die. Prove that these dice can be used in any game requiring ordinary six-sided dice without changing the probabilities of different outcomes.

Solution

For an ordinary six-sided die, each of the numbers 1, 2, 3, 4, 5, and 6 occurs only once, so the probability of any particular number coming up is

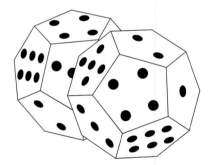

Figure 9.20

$$P(E) = \frac{n(E)}{n(S)} = \frac{1}{6}.$$

For a 12-sided die, each number occurs twice, so the probability of any particular number coming up is

$$P(E) = \frac{n(E)}{n(S)} = \frac{2}{12} = \frac{1}{6}.$$

✓ *Checkpoint* Now try Exercise 17.

Example 5   The Probability of Winning a Lottery

In Delaware's Lotto game, a player chooses six different numbers from 1 to 38. If these six numbers match the six numbers drawn (in any order) by the lottery commission, the player wins (or shares) the top prize. What is the probability of winning the top prize if the player buys one ticket?

Solution

To find the number of elements in the sample space, use the formula for the number of combinations of 38 elements taken six at a time.

$$n(S) = {}_{38}C_6$$

$$= \frac{38 \cdot 37 \cdot 36 \cdot 35 \cdot 34 \cdot 33}{6 \cdot 5 \cdot 4 \cdot 3 \cdot 2 \cdot 1} = 2{,}760{,}681$$

If a person buys only one ticket, the probability of winning is

$$P(E) = \frac{n(E)}{n(S)} = \frac{1}{2{,}760{,}681}.$$

✓ **Checkpoint**   Now try Exercise 19.

Example 6   Random Selection

The numbers of colleges and universities in various regions of the United States in 2001 are shown in Figure 9.21. One institution is selected at random. What is the probability that the institution is in one of the three southern regions?   (Source: U.S. National Center for Education Statistics)

Solution

From the figure, the total number of colleges and universities is 4178. Because there are $383 + 274 + 687 = 1344$ colleges and universities in the three southern regions, the probability that the institution is in one of these regions is

$$P(E) = \frac{n(E)}{n(S)} = \frac{1344}{4178} \approx 0.322.$$

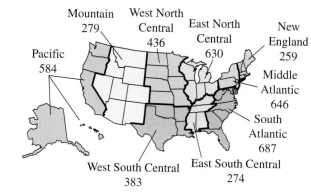

Figure 9.21

✓ **Checkpoint**   Now try Exercise 31.

Mutually Exclusive Events

Two events A and B (from the same sample space) are **mutually exclusive** if A and B have no outcomes in common. In the terminology of sets, the intersection of A and B is the empty set, which is expressed as

$$P(A \cap B) = 0.$$

For instance, if two dice are tossed, the event A of rolling a total of 6 and the event B of rolling a total of 9 are mutually exclusive. To find the probability that one or the other of two mutually exclusive events will occur, you can *add* their individual probabilities.

Probability of the Union of Two Events

If A and B are events in the same sample space, the probability of A *or* B occurring is given by

$$P(A \cup B) = P(A) + P(B) - P(A \cap B).$$

If A and B are mutually exclusive, then

$$P(A \cup B) = P(A) + P(B).$$

Example 7   The Probability of a Union

One card is selected from a standard deck of 52 playing cards. What is the probability that the card is either a heart or a face card?

Solution

Because the deck has 13 hearts, the probability of selecting a heart (event A) is

$$P(A) = \frac{13}{52}.$$

Similarly, because the deck has 12 face cards, the probability of selecting a face card (event B) is

$$P(B) = \frac{12}{52}.$$

Because three of the cards are hearts and face cards (see Figure 9.22), it follows that

$$P(A \cap B) = \frac{3}{52}.$$

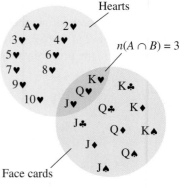

Figure 9.22

Finally, applying the formula for the probability of the union of two events, you can conclude that the probability of selecting a heart or a face card is

$$P(A \cup B) = P(A) + P(B) - P(A \cap B)$$

$$= \frac{13}{52} + \frac{12}{52} - \frac{3}{52} = \frac{22}{52} \approx 0.423.$$

 Checkpoint Now try Exercise 43.

Example 8   Probability of Mutually Exclusive Events

The personnel department of a company has compiled data on the number of employees who have been with the company for various periods of time. The results are shown in the table.

Years of service	Number of employees
0–4	157
5–9	89
10–14	74
15–19	63
20–24	42
25–29	38
30–34	37
35–39	21
40–44	8

If an employee is chosen at random, what is the probability that the employee has (a) 4 or fewer years of service and (b) 9 or fewer years of service?

Solution

a. To begin, add the number of employees and find that the total is 529. Next, let event A represent choosing an employee with 0 to 4 years of service. Then the probability of choosing an employee who has 4 or fewer years of service is

$$P(A) = \frac{157}{529} \approx 0.297.$$

b. Let event B represent choosing an employee with 5 to 9 years of service. Then

$$P(B) = \frac{89}{529}.$$

Because event A from part (a) and event B have no outcomes in common, you can conclude that these two events are mutually exclusive and that

$$P(A \cup B) = P(A) + P(B)$$

$$= \frac{157}{529} + \frac{89}{529}$$

$$= \frac{246}{529}$$

$$\approx 0.465.$$

So, the probability of choosing an employee who has 9 or fewer years of service is about 0.465.

✓ *Checkpoint* Now try Exercise 45.

Independent Events

Two events are **independent** if the occurrence of one has no effect on the occurrence of the other. For instance, rolling a total of 12 with two six-sided dice has no effect on the outcome of future rolls of the dice. To find the probability that two independent events will occur, *multiply* the probabilities of each.

Probability of Independent Events

If A and B are **independent events,** the probability that both A and B will occur is given by

$$P(A \text{ and } B) = P(A) \cdot P(B).$$

Example 9   Probability of Independent Events

A random number generator on a computer selects three integers from 1 to 20. What is the probability that all three numbers are less than or equal to 5?

Solution
The probability of selecting a number from 1 to 5 is

$$P(A) = \frac{5}{20} = \frac{1}{4}.$$

So, the probability that all three numbers are less than or equal to 5 is

$$P(A) \cdot P(A) \cdot P(A) = \left(\frac{1}{4}\right)\left(\frac{1}{4}\right)\left(\frac{1}{4}\right) = \frac{1}{64}.$$

 Checkpoint Now try Exercise 46.

Example 10   Probability of Independent Events

In 2001, approximately 65% of the population of the United States was 25 years old or older. In a survey, 10 people were chosen at random from the population. What is the probability that all 10 were 25 years old or older?   (Source: U.S. Census Bureau)

Solution
Let A represent choosing a person who was 25 years old or older. The probability of choosing a person who was 25 years old or older is 0.65, the probability of choosing a second person who was 25 years old or older is 0.65, and so on. Because these events are independent, you can conclude that the probability that all 10 people were 25 years old or older is

$$[P(A)]^{10} = (0.65)^{10} \approx 0.0135.$$

 Checkpoint Now try Exercise 47.

The Complement of an Event

The **complement of an event** A is the collection of all outcomes in the sample space that are *not* in A. The complement of event A is denoted by A'. Because $P(A \text{ or } A') = 1$ and because A and A' are mutually exclusive, it follows that $P(A) + P(A') = 1$. So, the probability of A' is given by

$$P(A') = 1 - P(A).$$

For instance, if the probability of *winning* a game is

$$P(A) = \frac{1}{4}$$

then the probability of *losing* the game is

$$P(A') = 1 - \frac{1}{4}$$

$$= \frac{3}{4}.$$

Exploration

You are in a class with 22 other people. What is the probability that at least two out of the 23 people will have a birthday on the same day of the year? What if you know the probability of everyone having the same birthday? Do you think this information would help you to find the answer?

Probability of a Complement

Let A be an event and let A' be its complement. If the probability of A is $P(A)$, then the probability of the complement is given by

$$P(A') = 1 - P(A).$$

Example 11   Finding the Probability of a Complement

A manufacturer has determined that a machine averages one faulty unit for every 1000 it produces. What is the probability that an order of 200 units will have one or more faulty units?

Solution

To solve this problem as stated, you would need to find the probabilities of having exactly one faulty unit, exactly two faulty units, exactly three faulty units, and so on. However, using complements, you can simply find the probability that all units are perfect and then subtract this value from 1. Because the probability that any given unit is perfect is 999/1000, the probability that all 200 units are perfect is

$$P(A) = \left(\frac{999}{1000} \right)^{200}$$

$$\approx 0.8186.$$

So, the probability that at least one unit is faulty is

$$P(A') = 1 - P(A)$$

$$\approx 0.1814.$$

✓ *Checkpoint* Now try Exercise 49.

9.7   Exercises

Vocabulary Check

In Exercises 1–7, fill in the blanks.

1. An _____ is an event whose result is uncertain, and the possible results of the event are called _____ .

2. The set of all possible outcomes of an experiment is called the _____ .

3. To determine the _____ of an event, you can use the formula $P(E) = \dfrac{n(E)}{n(S)}$, where $n(E)$ is the number of outcomes in the event and $n(S)$ is the number of outcomes in the sample space.

4. If $P(E) = 0$, then E is an _____ event, and if $P(E) = 1$, then E is a _____ event.

5. If two events from the same sample space have no outcomes in common, then the two events are _____ .

6. If the occurrence of one event has no effect on the occurrence of a second event, then the events are _____ .

7. The _____ of an event A is the collection of all outcomes in the sample space that are not in A.

8. Match the probability formula with the correct probability name.
   - (a) Probability of the union of two events
   - (b) Probability of mutually exclusive events
   - (c) Probability of independent events
   - (d) Probability of a complement
   - (i) $P(A \cup B) = P(A) + P(B)$
   - (ii) $P(A') = 1 - P(A)$
   - (iii) $P(A \cup B) = P(A) + P(B) - P(A \cap B)$
   - (iv) $P(A \text{ and } B) = P(A) \cdot P(B)$

In Exercises 1–6, determine the sample space for the experiment.

1. A coin and a six-sided die are tossed.

2. A six-sided die is tossed twice and the sum of the results is recorded.

3. A taste tester has to rank three varieties of orange juice, A, B, and C, according to preference.

4. Two marbles are selected (without replacement) from a sack containing two red marbles, two blue marbles, and one yellow marble. The color of each marble is recorded.

5. Two county supervisors are selected from five supervisors, A, B, C, D, and E, to study a recycling plan.

6. A sales representative makes presentations about a product in three homes per day. In each home there may be a sale (denote by S) or there may be no sale (denote by F).

Tossing a Coin   **In Exercises 7–10, find the probability for the experiment of tossing a coin three times. Use the sample space $S = \{HHH, HHT, HTH, HTT, THH, THT, TTH, TTT\}$.**

7. The probability of getting exactly two tails

8. The probability of getting a head on the first toss

9. The probability of getting at least one head

10. The probability of getting at least two heads

Drawing a Card   **In Exercises 11–14, find the probability for the experiment of selecting one card from a standard deck of 52 playing cards.**

11. The card is a face card.

12. The card is not a face card.

13. The card is a red face card.

14. The card is an 8 or lower. (Aces are low.)

Tossing a Die   **In Exercises 15–18, find the probability for the experiment of tossing a six-sided die twice.**

15. The sum is 5.

16. The sum is at least 8.

17. The sum is less than 11.

18. The sum is odd or prime.

Drawing Marbles **In Exercises 19–22, find the probability for the experiment of drawing two marbles (without replacement) from a bag containing one green, two yellow, and three red marbles.**

19. Both marbles are red.

20. Both marbles are yellow.

21. Neither marble is yellow.

22. The marbles are of different colors.

In Exercises 23–26, you are given the probability that an event *will* happen. Find the probability that the event *will not* happen.

23. $P(E) = 0.8$

24. $P(E) = 0.29$

25. $P(E) = \frac{1}{3}$

26. $P(E) = \frac{5}{6}$

In Exercises 27–30, you are given the probability that an event *will not* happen. Find the probability that the event *will* happen.

27. $P(E') = 0.12$

28. $P(E') = 0.84$

29. $P(E') = \frac{13}{20}$

30. $P(E') = \frac{61}{100}$

31. *Graphical Reasoning* In 2001 there were approximately 6.7 million unemployed workers in the United States. The circle graph shows the age profile of these unemployed workers. (Source: U.S. Bureau of Labor Statistics)

(a) Estimate the number of unemployed workers in the age group 16–19.

(b) What is the probability that a person selected at random from the population of unemployed workers is in the 25–44 age group?

(c) What is the probability that a person selected at random from the population of unemployed workers is in the 45–64 age group?

(d) What is the probability that a person selected at random from the population of unemployed workers is 45 or over?

Ages of Unemployed Workers

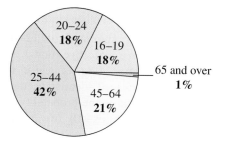

32. *Graphical Reasoning* The circle graph shows the number of children of the 42 U.S. presidents. (Source: Time Almanac 2003)

(a) Determine the number of presidents who had no children.

(b) Determine the number of presidents who had four children.

(c) What is the probability that a president selected at random had five or more children?

(d) What is the probability that a president selected at random had three children?

Children of U.S. Presidents

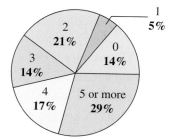

33. *Data Analysis* One hundred college students were interviewed to determine their political-party affiliations and whether they favored a balanced-budget amendment to the Constitution. The results of the study are listed in the table, where *D* represents Democrat and *R* represents Republican.

	D	R	Total
Favor	23	32	55
Oppose	25	9	34
Unsure	7	4	11
Total	55	45	100

A person is selected at random from the sample. Find the probability that the described person is selected.

(a) A person who doesn't favor the amendment

(b) A Republican

(c) A Democrat who favors the amendment

34. *Data Analysis*   A study of the effectiveness of a flu vaccine was conducted with a sample of 500 people. Some participants in the study were given no vaccine, some were given one injection, and some were given two injections. The results of the study are given in the table.

	Flu	No flu	Total
No vaccine	7	149	156
One injection	2	52	54
Two injections	13	277	290
Total	22	478	500

A person is selected at random from the sample. Find each probability.

(a) The person had two injections.

(b) The person did not get the flu.

(c) The person got the flu and had one injection.

35. *Alumni Association*   A college sends a survey to selected members of the class of 2004. Of the 1254 people who graduated that year, 672 are women, of whom 124 went on to graduate school. Of the 582 male graduates, 198 went on to graduate school. An alumni member is selected at random. What is the probability that the person is (a) female, (b) male, and (c) female and did not attend graduate school?

36. *Education*   In a high school graduating class of 128 students, 52 are on the honor roll. Of these, 48 are going on to college; of the other 76 students, 56 are going on to college. A student is selected at random from the class. What is the probability that the person chosen is (a) going to college, (b) not going to college, and (c) not going to college and on the honor roll?

37. *Election*   Taylor, Moore, and Perez are candidates for public office. It is estimated that Moore and Perez have about the same probability of winning, and Taylor is believed to be twice as likely to win as either of the others. Find the probability of each candidate's winning the election.

38. *Payroll Error*   The employees of a company work in six departments: 31 are in sales, 54 are in research, 42 are in marketing, 20 are in engineering, 47 are in finance, and 58 are in production. One employee's paycheck is lost. What is the probability that the employee works in the research department?

In Exercises 39–46, the sample spaces are large and you should use the counting principles discussed in Section 9.6.

39. *Preparing for a Test*   A class is given a list of 20 study problems from which 10 will be chosen as part of an upcoming exam. A given student knows how to solve 15 of the problems. Find the probability that the student will be able to answer (a) all 10 questions on the exam, (b) exactly 8 questions on the exam, and (c) at least 9 questions on the exam.

40. *Payroll Mix-Up*   Five paychecks and envelopes are addressed to five different people. The paychecks are randomly inserted into the envelopes. What is the probability that (a) exactly one paycheck is inserted in the correct envelope and (b) at least one paycheck is inserted in the correct envelope?

41. *Game Show*   On a game show you are given five digits to arrange in the proper order to form the price of a car. If you are correct, you win the car. What is the probability of winning, given the following conditions?

(a) You guess the position of each digit.

(b) You know the first digit and guess the others.

42. *Card Game*   The deck of a card game is made up of 108 cards. Twenty-five each are red, yellow, blue, and green, and eight are wild cards. Each player is randomly dealt a seven-card hand.

(a) What is the probability that a hand will contain exactly two wild cards?

(b) What is the probability that a hand will contain two wild cards, two red cards, and three blue cards?

43. *Drawing a Card*   One card is selected at random from a standard deck of 52 playing cards. Find the probability that (a) the card is an even-numbered card, (b) the card is a heart or a diamond, and (c) the card is a nine or a face card.

44. *Poker Hand*   Five cards are drawn from an ordinary deck of 52 playing cards. What is the probability of getting a full house? (A full house consists of three of one kind and two of another kind.)

45. *Defective Units*   A shipment of 12 microwave ovens contains three defective units. A vending company has ordered four of these units, and because all are packaged identically, the selection will be random. What is the probability that (a) all four units are good, (b) exactly two units are good, and (c) at least two units are good?

46. *Random Number Generator* Two integers from 1 through 40 are chosen by a random number generator. What is the probability that (a) the numbers are both even, (b) one number is even and one is odd, (c) both numbers are less than 30, and (d) the same number is chosen twice?

47. *Consumerism* Suppose that the methods used by shoppers to pay for merchandise are as shown in the circle graph. Two shoppers are chosen at random. What is the probability that both shoppers paid for their purchases only in cash?

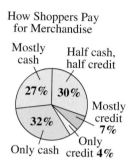

How Shoppers Pay
for Merchandise

Mostly cash   Half cash, half credit

27%   30%

Mostly credit
7%

32%

Only credit 4%

Only cash

48. *Flexible Work Hours* In a survey, people were asked if they would prefer to work flexible hours— even if it meant slower career advancement—so they could spend more time with their families. The results of the survey are shown in the circle graph. Three people from the survey are chosen at random. What is the probability that all three people would prefer flexible work hours?

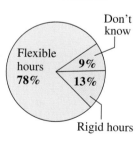

Flexible Work Hours

Don't know

Flexible hours
78%   9%

13%

Rigid hours

49. *Backup System* A space vehicle has an independent backup system for one of its communication networks. The probability that either system will function satisfactorily for the duration of a flight is 0.985. What is the probability that during a given flight (a) both systems function satisfactorily, (b) at least one system functions satisfactorily, and (c) both systems fail?

50. *Backup Vehicle* A fire company keeps two rescue vehicles to serve the community. Because of the demand on the vehicles and the chance of mechanical failure, the probability that a specific vehicle is available when needed is 90%. The availability of one vehicle is *independent* of the other. Find the probability that (a) both vehicles are available at a given time, (b) neither vehicle is available at a given time, and (c) at least one vehicle is available at a given time.

51. *Making a Sale* A sales representative makes sales on approximately one-fifth of all calls. On a given day, the representative contacts six potential clients. What is the probability that a sale will be made with (a) all six contacts, (b) none of the contacts, and (c) at least one contact?

52. *A Boy or a Girl?* Assume that the probability of the birth of a child of a particular sex is 50%. In a family with four children, what is the probability that (a) all the children are boys, (b) all the children are the same sex, and (c) there is at least one boy?

53. *Geometry* You and a friend agree to meet at your favorite fast-food restaurant between 5:00 and 6:00 P.M. The one who arrives first will wait 15 minutes for the other, after which the first person will leave (see figure). What is the probability that the two of you will actually meet, assuming that your arrival times are random within the hour?

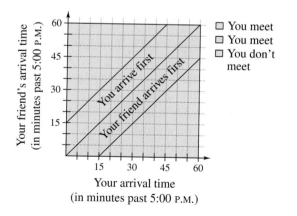

Your friend's arrival time (in minutes past 5:00 P.M.)

You arrive first
Your friend arrives first

□ You meet
□ You meet
□ You don't meet

Your arrival time
(in minutes past 5:00 P.M.)

54. *Estimating π* A coin of diameter d is dropped onto a paper that contains a grid of squares d units on a side (see figure).

(a) Find the probability that the coin covers a vertex of one of the squares on the grid.

(b) Perform the experiment 100 times and use the results to approximate π.

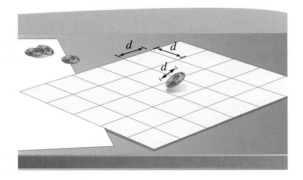

Synthesis

True or False? **In Exercises 55 and 56, determine whether the statement is true or false. Justify your answer.**

55. If the probability of an outcome in a sample space is 1, then the probability of the other outcomes in the sample space is 0.

56. Rolling a number less than 3 on a normal six-sided die has a probability of $\frac{1}{3}$. The complement of this event is to roll a number greater than 3, and its probability is $\frac{1}{2}$.

57. *Pattern Recognition and Exploration* Consider a group of n people.

(a) Explain why the following pattern gives the probability that the n people have distinct birthdays.

$$n = 2: \quad \frac{365}{365} \cdot \frac{364}{365} = \frac{365 \cdot 364}{365^2}$$

$$n = 3: \quad \frac{365}{365} \cdot \frac{364}{365} \cdot \frac{363}{365} = \frac{365 \cdot 364 \cdot 363}{365^3}$$

(b) Use the pattern in part (a) to write an expression for the probability that four people ($n = 4$) have distinct birthdays.

(c) Let P_n be the probability that the n people have distinct birthdays. Verify that this probability can be obtained recursively by

$$P_1 = 1 \quad \text{and} \quad P_n = \frac{365 - (n-1)}{365}P_{n-1}.$$

(d) Explain why $Q_n = 1 - P_n$ gives the probability that at least two people in a group of n people have the same birthday.

(e) Use the results of parts (c) and (d) to complete the table.

n	10	15	20	23	30	40	50
P_n							
Q_n							

(f) How many people must be in a group so that the probability of at least two of them having the same birthday is greater than $\frac{1}{2}$? Explain.

58. *Think About It* The weather forecast indicates that the probability of rain is 40%. Explain what this means.

Review

In Exercises 59–62, solve the rational equation.

59. $\dfrac{2}{x-5} = 4$

60. $\dfrac{3}{2x+3} - 4 = \dfrac{-1}{2x+3}$

61. $\dfrac{3}{x-2} + \dfrac{x}{x+2} = 1$

62. $\dfrac{2}{x} - \dfrac{5}{x-2} = \dfrac{-13}{x^2-2x}$

In Exercises 63–66, solve the equation algebraically. Round your result to three decimal places.

63. $e^x + 7 = 35$

64. $200e^{-x} = 75$

65. $4\ln 6x = 16$

66. $5\ln 2x - 4 = 11$

In Exercises 67–70, evaluate $_nP_r$. Verify your result using a graphing utility.

67. $_5P_3$

68. $_{10}P_4$

69. $_{11}P_8$

70. $_9P_2$

In Exercises 71–74, evaluate $_nC_r$. Verify your result using a graphing utility.

71. $_6C_2$

72. $_9C_5$

73. $_{11}C_8$

74. $_{16}C_{13}$

9 Chapter Summary

What did you learn?

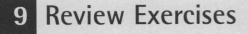

9.1 In Exercises 1–4, write the first five terms of the sequence. (Assume n begins with 1.)

1. $a_n = 2 + \dfrac{6}{n}$ **2.** $a_n = \dfrac{(-1)^n}{2n-1}$

3. $a_n = \dfrac{72}{n!}$ **4.** $a_n = n(n-1)$

In Exercises 5–8, use a graphing utility to graph the first 10 terms of the sequence (Assume n begins with 1.)

5. $a_n = \frac{3}{2}n$ **6.** $a_n = \dfrac{3n}{n+2}$

7. $a_n = 4(0.4)^{n-1}$ **8.** $a_n = -8(0.5)^{n-1}$

In Exercises 9–12, simplify the factorial expression.

9. $\dfrac{18!}{20!}$ **10.** $\dfrac{10!}{8!}$

11. $\dfrac{2! \cdot 5!}{6!}$ **12.** $\dfrac{9! \cdot 6!}{6! \cdot 8!}$

In Exercises 13–20, find the sum.

13. $\displaystyle\sum_{i=1}^{6} 5$ **14.** $\displaystyle\sum_{k=2}^{5} 4k$

15. $\displaystyle\sum_{j=1}^{4} \dfrac{6}{j^2}$ **16.** $\displaystyle\sum_{i=1}^{8} \dfrac{i}{i+1}$

17. $\displaystyle\sum_{k=1}^{10} 2k^3$ **18.** $\displaystyle\sum_{j=0}^{4} (j^2 + 1)$

19. $\displaystyle\sum_{n=0}^{10} (n^2 + 3)$ **20.** $\displaystyle\sum_{n=1}^{100} \left(\dfrac{1}{n} - \dfrac{1}{n+1}\right)$

In Exercises 21–24, use sigma notation to write the sum. Then use a graphing utility to find the sum.

21. $\dfrac{1}{2(1)} + \dfrac{1}{2(2)} + \dfrac{1}{2(3)} + \cdots + \dfrac{1}{2(20)}$

22. $2(1^2) + 2(2^2) + 2(3^2) + \cdots + 2(9^2)$

23. $\dfrac{1}{2} + \dfrac{2}{3} + \dfrac{3}{4} + \cdots + \dfrac{9}{10}$

24. $1 - \dfrac{1}{3} + \dfrac{1}{9} - \dfrac{1}{27} + \cdots$

In Exercises 25–28, find (a) the fourth partial sum and (b) the sum of the infinite series.

25. $\displaystyle\sum_{k=1}^{\infty} \dfrac{5}{10^k}$ **26.** $\displaystyle\sum_{k=1}^{\infty} 8\left(\tfrac{1}{10}\right)^k$

27. $\displaystyle\sum_{k=1}^{\infty} 2\left(\tfrac{1}{100}\right)^k$ **28.** $\displaystyle\sum_{k=1}^{\infty} 7\left(\tfrac{1}{10}\right)^k$

29. *Compound Interest*   A deposit of $2500 is made in an account that earns 8% interest compounded quarterly. The balance in the account after n quarters is given by

$$a_n = 2500\left(1 + \dfrac{0.08}{4}\right)^n, \quad n = 1, 2, 3, \ldots$$

(a) Compute the first eight terms of this sequence.

(b) Find the balance in this account after 10 years by computing the 40th term of the sequence.

30. *Revenue*   The revenue a_n (in billions of dollars) for United Parcel Service, Inc. from 1997 to 2002 can be approximated by the model

$$a_n = -0.251n^2 + 6.58n - 11.5,$$
$$n = 7, 8, \ldots, 12$$

where n is the year, with $n = 7$ corresponding to 1997. Find the terms of this finite sequence and use a graphing utility to construct a bar graph that represents the sequence.   (Source: United Parcel Service, Inc.)

9.2 In Exercises 31–34, determine whether or not the sequence is arithmetic. If it is, find the common difference.

31. $5, 3, 1, -1, -3, \ldots$ **32.** $0, 1, 3, 6, 10, \ldots$

33. $\frac{1}{2}, 1, \frac{3}{2}, 2, \frac{5}{2}, \ldots$ **34.** $\frac{9}{9}, \frac{8}{9}, \frac{7}{9}, \frac{6}{9}, \frac{5}{9}, \ldots$

In Exercises 35–38, write the first five terms of the arithmetic sequence.

35. $a_1 = 3, \ d = 4$ **36.** $a_1 = 8, \ d = -2$

37. $a_4 = 10, \ a_{10} = 28$ **38.** $a_2 = 14, \ a_6 = 22$

In Exercises 39–42, write the first five terms of the arithmetic sequence. Find the common difference and write the nth term of the sequence as a function of n.

39. $a_1 = 35, \quad a_{k+1} = a_k - 3$

40. $a_1 = 15,\quad a_{k+1} = a_k + \frac{5}{2}$

41. $a_1 = 9,\quad a_{k+1} = a_k + 7$

42. $a_1 = 100,\quad a_{k+1} = a_k - 5$

In Exercises 43 and 44, find a formula for a_n for the arithmetic sequence and find the sum of the first 20 terms of the sequence.

43. $a_1 = 100,\quad d = -3$ **44.** $a_1 = 10,\quad a_3 = 28$

In Exercises 45–48, find the partial sum. Use a graphing utility to verify your result.

45. $\displaystyle\sum_{j=1}^{10} (2j - 3)$ **46.** $\displaystyle\sum_{j=1}^{8} (20 - 3j)$

47. $\displaystyle\sum_{k=1}^{11} \left(\tfrac{2}{3}k + 4\right)$ **48.** $\displaystyle\sum_{k=1}^{25} \left(\frac{3k + 1}{4}\right)$

49. Find the sum of the first 100 positive multiples of 5.

50. Find the sum of the integers from 20 to 80 (inclusive).

51. *Job Offer* The starting salary for an accountant is $34,000 with a guaranteed salary increase of $2250 per year for the first 4 years of employment. Determine (a) the salary during the fifth year and (b) the total compensation through 5 full years of employment.

52. *Baling Hay* In his first trip baling hay around a field, a farmer makes 123 bales. In his second trip he makes 11 fewer bales. Because each trip is shorter than the preceding trip, the farmer estimates that the same pattern will continue. Estimate the total number of bales made if there are another six trips around the field.

9.3 In Exercises 53–56, determine whether or not the sequence is geometric. If it is, find the common ratio.

53. 5, 10, 20, 40, . . .   **54.** 54, −18, 6, −2, . . .

55. $\frac{1}{2}, \frac{2}{3}, \frac{3}{4}, \frac{4}{5}, \ldots$ **56.** $\frac{1}{3}, -\frac{2}{3}, \frac{4}{3}, -\frac{8}{3}, \ldots$

In Exercises 57–60, write the first five terms of the geometric sequence.

57. $a_1 = 4,\ r = -\frac{1}{4}$ **58.** $a_1 = 2,\ r = 2$

59. $a_1 = 9,\ a_3 = 4$ **60.** $a_1 = 2,\ a_3 = 12$

In Exercises 61–64, write the first five terms of the geometric sequence. Find the common ratio and write the nth term of the sequence as a function of n.

61. $a_1 = 120,\quad a_{k+1} = \frac{1}{3}a_k$

62. $a_1 = 200,\quad a_{k+1} = 0.1a_k$

63. $a_1 = 25,\quad a_{k+1} = -\frac{3}{5}a_k$

64. $a_1 = 18,\quad a_{k+1} = \frac{5}{3}a_k$

In Exercises 65–68, write an expression for the nth term of the geometric sequence and find the sum of the first 20 terms of the sequence.

65. $a_1 = 16,\quad a_2 = -8$ **66.** $a_3 = 6,\quad a_4 = 1$

67. $a_1 = 100,\quad r = 1.05$ **68.** $a_1 = 5,\quad r = 0.2$

In Exercises 69–76, find the sum. Use a graphing utility to verify your result.

69. $\displaystyle\sum_{i=1}^{7} 2^{i-1}$ **70.** $\displaystyle\sum_{i=1}^{5} 3^{i-1}$

71. $\displaystyle\sum_{n=1}^{7} (-4)^{n-1}$ **72.** $\displaystyle\sum_{n=1}^{4} 12\left(-\tfrac{1}{2}\right)^{n-1}$

73. $\displaystyle\sum_{n=0}^{4} 250(1.02)^n$ **74.** $\displaystyle\sum_{n=0}^{5} 400(1.08)^n$

75. $\displaystyle\sum_{i=1}^{10} 10\left(\tfrac{3}{5}\right)^{i-1}$ **76.** $\displaystyle\sum_{i=1}^{15} 20(0.2)^{i-1}$

In Exercises 77–80, find the sum of the infinite geometric series.

77. $\displaystyle\sum_{i=1}^{\infty} \left(\tfrac{7}{8}\right)^{i-1}$ **78.** $\displaystyle\sum_{i=1}^{\infty} \left(\tfrac{1}{3}\right)^{i-1}$

79. $\displaystyle\sum_{k=1}^{\infty} 4\left(\tfrac{2}{3}\right)^{k-1}$ **80.** $\displaystyle\sum_{k=1}^{\infty} 1.3\left(\tfrac{1}{10}\right)^{k-1}$

81. *Depreciation* A company buys a fleet of six vans for $120,000. During the next 5 years, the fleet will depreciate at a rate of 30% per year. (That is, at the end of each year, the depreciated value is 70% of the value at the beginning of the year.)

(a) Find the formula for the nth term of a geometric sequence that gives the value of the fleet t full years after it was purchased.

(b) Find the depreciated value of the fleet at the end of 5 full years.

82. *Annuity* A deposit of $75 is made at the beginning of each month in an account that pays 4% interest, compounded monthly. The balance A in the account at the end of 4 years is given by

$$A = 75\left(1 + \frac{0.04}{12}\right)^1 + \cdots + 75\left(1 + \frac{0.04}{12}\right)^{48}.$$

Find A.

9.4 In Exercises 83–86, use mathematical induction to prove the formula for every positive integer n.

83. $2 + 7 + \cdots + (5n - 3) = \dfrac{n}{2}(5n - 1)$

84. $1 + \dfrac{3}{2} + 2 + \dfrac{5}{2} + \cdots + \dfrac{1}{2}(n + 1) = \dfrac{n}{4}(n + 3)$

85. $\displaystyle\sum_{i=0}^{n-1} ar^i = \dfrac{a(1 - r^n)}{1 - r}$

86. $\displaystyle\sum_{k=0}^{n-1} (a + kd) = \dfrac{n}{2}[2a + (n - 1)d]$

In Exercises 87–90, find the sum using the formulas for the sums of powers of integers.

87. $\displaystyle\sum_{n=1}^{30} n$

88. $\displaystyle\sum_{n=1}^{10} n^2$

89. $\displaystyle\sum_{n=1}^{7} (n^4 - n)$

90. $\displaystyle\sum_{n=1}^{6} (n^5 - n^2)$

In Exercises 91–94, write the first five terms of the sequence beginning with a_1. Then calculate the first and second differences of the sequence. Does the sequence have a linear model, a quadratic model, or neither?

91. $a_1 = 5$
$a_n = a_{n-1} + 5$

92. $a_1 = -3$
$a_n = a_{n-1} - 2n$

93. $a_1 = 16$
$a_n = a_{n-1} - 1$

94. $a_1 = 1$
$a_n = n - a_{n-1}$

9.5 In Exercises 95–98, find the binomial coefficient. Use a graphing utility to verify your result.

95. $_{10}C_8$

96. $_{12}C_5$

97. $\dbinom{9}{4}$

98. $\dbinom{14}{12}$

In Exercises 99–102, use Pascal's Triangle to find the binomial coefficient.

99. $_6C_3$

100. $_9C_7$

101. $\dbinom{8}{4}$

102. $\dbinom{10}{5}$

In Exercises 103–108, use the Binomial Theorem to expand and simplify the expression. Simplify your answer. $\left(\text{Remember that } i = \sqrt{-1}.\right)$

103. $(x + 5)^4$

104. $(y - 3)^3$

105. $(a - 4b)^5$

106. $(3x + y)^7$

107. $(7 + 2i)^4$

108. $(4 - 5i)^3$

9.6

109. *Numbers in a Hat*   Slips of paper numbered 1 through 14 are placed in a hat. In how many ways can two numbers be drawn so that the sum of the numbers is 12? Assume the random selection is without replacement.

110. *Aircraft Boarding*   Eight people are boarding an aircraft. Two have tickets for first class and board before those in economy class. In how many ways can the eight people board the aircraft?

111. *Course Schedule*   A college student is preparing a course schedule for the next semester. The student must select one of four mathematics courses, one of six biology courses, and one of two art courses. How many schedules are possible?

112. *Telephone Numbers*   The same three-digit prefix is used for all of the telephone numbers in a small town. How many different telephone numbers are possible by changing only the last four digits?

In Exercises 113 and 114, find the number of distinguishable permutations of the group of letters.

113. C, A, L, C, U, L, U, S

114. I, N, V, E, R, T, E, B, R, A, T, E

115. *Sports*   There are 10 bicyclists entered in a race. In how many different orders could the ten bicyclists finish?

116. *Sports*   From a pool of 7 juniors and 11 seniors, four co-captains will be chosen for the football team. How many different combinations are possible if two juniors and two seniors are to be chosen?

117. *Exam Questions*   A student can answer any 15 questions from a total of 20 questions on an exam. In how many different ways can the student select the questions?

118. *Lottery*   In the Lotto Texas game, a player chooses six distinct numbers from 1 to 54. In how many ways can a player select the six numbers?

9.7

119. *Apparel*   A man has five pairs of socks (no two pairs are the same color). He randomly selects two socks from a drawer. What is the probability that he gets a matched pair?

120. *Bookshelf Order*   A child returns a five-volume set of books to a bookshelf. The child is not able to read, and so cannot distinguish one volume from another. What is the probability that the books are shelved in the correct order?

121. *Data Analysis*   A sample of college students, faculty members, and administrators were asked whether they favored a proposed increase in the annual activity fee to enhance student life on campus. The results of the study are shown in the table.

	Favor	Oppose	Total
Students	237	163	400
Faculty	37	38	75
Admin.	18	7	25
Total	292	208	500

A person is selected at random from the sample. Find each probability.

(a) The person is not in favor of the proposal.

(b) The person is a student.

(c) The person is a faculty member and is in favor of the proposal.

122. *Tossing a Die*   A six-sided die is rolled six times. What is the probability that each side appears exactly once?

123. *Poker Hand*   Five cards are drawn from an ordinary deck of 52 playing cards. Find the probability of getting two pairs. (For example, the hand could be A-A-5-5-Q or 4-4-7-7-K.)

124. *Drawing a Card*   You randomly select a card from a 52-card deck. What is the probability that the card is *not* a club?

Synthesis

True or False?   **In Exercises 125 and 126, determine whether the statement is true or false. Justify your answer.**

125. $\dfrac{(n+2)!}{n!} = (n+2)(n+1)$

126. $\displaystyle\sum_{k=1}^{8} 3k = 3\sum_{k=1}^{8} k$

127. *Writing*   In your own words, explain what makes a sequence (a) arithmetic and (b) geometric.

128. *Think About It*   How do the two sequences differ?

(a) $a_n = \dfrac{(-1)^n}{n}$ (b) $a_n = \dfrac{(-1)^{n+1}}{n}$

129. *Graphical Reasoning*   The graphs of two sequences are shown below. Identify each sequence as arithmetic or geometric. Explain your reasoning.

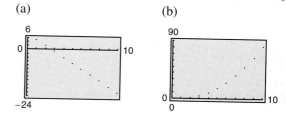

130. *Population Growth*   Consider an idealized population with the characteristic that each member of the population produces one offspring at the end of every time period. If each member has a life span of three time periods and the population begins with 10 newborn members, then the following table shows the population during the first five time periods.

Age Bracket	Time Period				
	1	2	3	4	5
0–1	10	10	20	40	70
1–2		10	10	20	40
2–3			10	10	20
Total	10	20	40	70	130

The sequence for the total population has the property that

$$S_n = S_{n-1} + S_{n-2} + S_{n-3}, \quad n > 3.$$

Find the total population during the next five time periods.

131. *Writing*   Explain what a recursive formula is.

132. *Writing*   Explain why the terms of a geometric sequence of positive terms decrease when $0 < r < 1$.

133. *Think About It*   How do the expansions of $(x - y)^n$ and $(-x + y)^n$ differ?

134. The probability of an event must be a real number in what interval? Is the interval open or closed?

9 | Chapter Test

Take this test as you would take a test in class. After you are finished, check your work against the answers given in the back of the book.

In Exercises 1 and 2, write the first five terms of the sequence.

1. $a_n = \left(-\frac{2}{3}\right)^{n-1}$ (Begin with $n = 1$.)

2. $a_1 = 12$ and $a_{k+1} = a_k + 4$

3. Simplify $\dfrac{11! \cdot 4!}{4! \cdot 7!}$.

In Exercises 4 and 5, find a formula for the nth term of the sequence.

4. Arithmetic: $a_1 = 5000, \quad d = -100$

5. Geometric: $a_1 = 4, \quad a_{k+1} = \frac{1}{2}a_k$

6. Use sigma notation to write $\dfrac{2}{3(1) + 1} + \dfrac{2}{3(2) + 1} + \cdots + \dfrac{2}{3(12) + 1}$.

In Exercises 7–9, find the sum.

7. $\displaystyle\sum_{n=1}^{7} (8n - 5)$ 8. $\displaystyle\sum_{n=1}^{8} 24\left(\frac{1}{6}\right)^{n-1}$ 9. $\displaystyle\sum_{n=1}^{\infty} 5\left(\frac{1}{10}\right)^{n-1}$

10. Use mathematical induction to prove the formula

$$3 + 6 + 9 + \cdots + 3n = \frac{3n(n + 1)}{2}.$$

11. Use the Binomial Theorem to expand and simplify $(2a - 5b)^4$.

12. Find the coefficient of the term x^3y^5 in the expansion of $(3x + 2y)^8$.

In Exercises 13–16, evaluate the expression.

13. $_9C_3$ 14. $_{20}C_3$ 15. $_9P_2$ 16. $_{70}P_3$

17. How many distinct license plates can be issued consisting of one letter followed by a three-digit number?

18. Four students are randomly selected from a class of 25 to answer questions from a reading assignment. In how many ways can the four be selected?

19. A card is drawn from a standard deck of 52 playing cards. Find the probability that it is a red face card.

20. Two spark plugs require replacement in a four-cylinder engine. The mechanic randomly removes two plugs. Find the probability that they are the two defective plugs.

21. Two integers from 1 to 60 are chosen by a random number generator. What is the probability that (a) both numbers are odd, (b) both numbers are less than 12, and (c) the same number is chosen twice?

22. A weather forecast indicates that the probability of snow is 75%. What is the probability that it will not snow?

The cables for suspension bridges are parabolic in shape. You can use the techniques illustrated in this chapter to determine the height of the cables above the roadway of the Golden Gate Bridge.

David W. Hamilton/Getty Images

10 Topics in Analytic Geometry

What You Should Learn

In this chapter, you will learn how to:

■ Recognize, graph, and write equations of conics with vertex or center at the origin.

■ Recognize, graph, and write equations of conics that have been shifted vertically and/or horizontally in the plane.

■ Evaluate sets of parametric equations for given values of the parameter and graph curves that are represented by sets of parametric equations.

■ Rewrite sets of parametric equations as single rectangular equations and find sets of parametric equations for graphs.

■ Plot points in the polar coordinate system and convert equations from rectangular to polar form and vice versa.

■ Graph polar equations.

■ Write conics in terms of eccentricity and write equations of conics in polar form.

10.1 Conics

Introduction

Conic sections were discovered during the classical Greek period, 600 to 300 B.C. The early Greek studies were largely concerned with the geometric properties of conics. It was not until the early 17th century that the broad applicability of conics became apparent and played a prominent role in the early development of calculus.

A **conic section** (or simply **conic**) is the intersection of a plane and a double-napped cone. Notice in Figure 10.1 that in the formation of the four basic conics, the intersecting plane does not pass through the vertex of the cone. When the plane does pass through the vertex, the resulting figure is a **degenerate conic,** as shown in Figure 10.2.

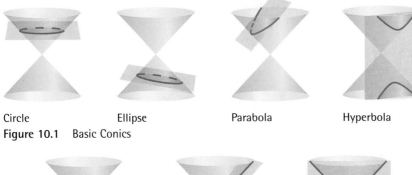

Circle   Ellipse   Parabola   Hyperbola

Figure 10.1   Basic Conics

Photodisc/Getty Images

Point   Line   Two intersecting lines

Figure 10.2   Degenerate Conics

There are several ways to approach the study of conics. You could begin by defining conics in terms of the intersections of planes and cones, as the Greeks did, or you could define them algebraically, in terms of the general second-degree equation

$$Ax^2 + Bxy + Cy^2 + Dx + Ey + F = 0.$$

However, you will study a third approach, in which each of the conics is defined as a *locus* (collection) of points satisfying a certain geometric property. For example, in Section P.5, you saw how the definition of a circle as *the collection of all points (x, y) that are equidistant from a fixed point (h, k)* led easily to the standard form of the equation of a circle

$$(x - h)^2 + (y - k)^2 = r^2. \qquad \text{Equation of circle}$$

From the equation above, the center of the circle is (h, k) and the radius is r. To review circles and their graphs, see Sections P.5 and 1.1.

Parabolas

In Section 3.1, you learned that the graph of the quadratic function

$$f(x) = ax^2 + bx + c$$

is a parabola that opens upward or downward. The following definition of a parabola is more general in the sense that it is independent of the orientation of the parabola.

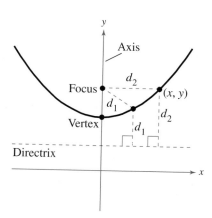

> ### Definition of a Parabola
>
> A **parabola** is the set of all points (x, y) in a plane that are equidistant from a fixed line, the **directrix,** and a fixed point, the **focus,** not on the line. (See Figure 10.3.) The midpoint between the focus and the directrix is the **vertex,** and the line passing through the focus and the vertex is the **axis** of the parabola.

Figure 10.3

> ### Standard Equation of a Parabola (Vertex at Origin)
>
> The **standard form of the equation of a parabola** with vertex at $(0, 0)$ and directrix $y = -p$ is given by
>
> $$x^2 = 4py, \qquad p \neq 0. \qquad \text{Vertical axis}$$
>
> For directrix $x = -p$, the equation is given by
>
> $$y^2 = 4px, \qquad p \neq 0. \qquad \text{Horizontal axis}$$
>
> The focus is on the axis p units (directed distance) from the vertex.

See Appendix B for a proof of the standard form of the equation of a parabola.

Notice that a parabola can have a vertical or a horizontal axis and that a parabola is symmetric with respect to its axis. Examples of each are shown in Figure 10.4. If you know the equation of the axis, it is easy to sketch the graph of a parabola by hand.

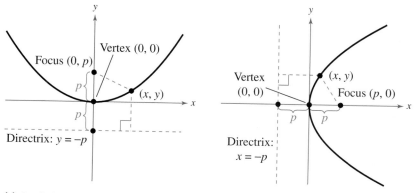

(a) Parabola with vertical axis: $x^2 = 4py$ (b) Parabola with horizontal axis: $y^2 = 4px$

Figure 10.4

Example 1   Finding the Focus of a Parabola

Find the focus of the parabola whose equation is $y = -2x^2$.

Solution

Because the squared term in the equation involves x, you know that the axis is vertical, and the equation is of the form $x^2 = 4py$. You can write the original equation in this form as follows.

$$x^2 = -\tfrac{1}{2}y \quad \Longrightarrow \quad x^2 = 4\left(-\tfrac{1}{8}\right)y \qquad \text{Write in standard form.}$$

So, $p = -\tfrac{1}{8}$. Because p is negative, the parabola opens downward (see Figure 10.5), and the focus of the parabola is

$$(0, p) = \left(0, -\tfrac{1}{8}\right). \qquad \text{Focus}$$

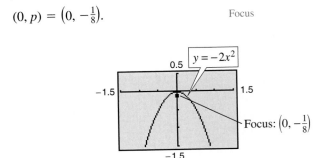

Figure 10.5

✓ *Checkpoint* Now try Exercise 5.

Example 2   A Parabola with a Horizontal Axis

Find the standard form of the equation of the parabola with vertex at the origin and focus at $(2, 0)$.

Solution

The axis of the parabola is horizontal, passing through $(0, 0)$ and $(2, 0)$, as shown in Figure 10.6. So, the standard form is $y^2 = 4px$. Because the focus is $p = 2$ units from the vertex, the equation is

$$y^2 = 4(2)x \quad \Longrightarrow \quad y^2 = 8x.$$

The equation $y^2 = 8x$ does not define y as a function of x. So, to use a graphing utility to graph $y^2 = 8x$, you need to break the graph into two equations, $y_1 = 2\sqrt{2x}$ and $y_2 = -2\sqrt{2x}$, each of which is a function of x.

✓ *Checkpoint* Now try Exercise 11.

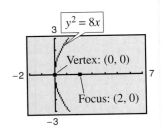

Figure 10.6

Parabolas occur in a wide variety of applications. For instance, a parabolic reflector can be formed by revolving a parabola about its axis. The resulting surface has the property that all incoming rays parallel to the axis are reflected through the focus of the parabola. This is the principle behind the construction of the parabolic mirrors used in reflecting telescopes. Conversely, the light rays emanating from the focus of a parabolic reflector used in a flashlight are all parallel to one another, as shown in Figure 10.7.

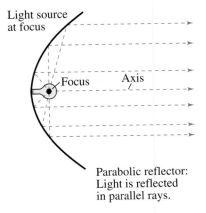

Parabolic reflector:
Light is reflected in parallel rays.

Figure 10.7

Ellipses

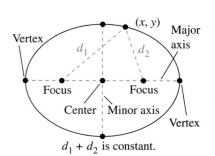

Figure 10.8

Definition of an Ellipse

An **ellipse** is the set of all points (x, y) in a plane, the sum of whose distances from two distinct fixed points **(foci)** is constant. (See Figure 10.8.)

The line through the foci intersects the ellipse at two points called **vertices.** The chord joining the vertices is the **major axis**, and its midpoint is the **center** of the ellipse. The chord perpendicular to the major axis at the center is the **minor axis**. (See Figure 10.8).

You can visualize the definition of an ellipse by imagining two thumbtacks placed at the foci, as shown in Figure 10.9. If the ends of a fixed length of string are fastened to the thumbtacks and the string is drawn taut with a pencil, the path traced by the pencil will be an ellipse.

The standard form of the equation of an ellipse takes one of two forms, depending on whether the major axis is horizontal or vertical.

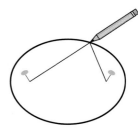

Figure 10.9

Standard Equation of an Ellipse (Center at Origin)

The **standard form of the equation of an ellipse** with center at the origin and major and minor axes of lengths $2a$ and $2b$, respectively (where $0 < b < a$), is given by

$$\frac{x^2}{a^2} + \frac{y^2}{b^2} = 1 \qquad \text{or} \qquad \frac{x^2}{b^2} + \frac{y^2}{a^2} = 1.$$

The vertices and foci lie on the major axis, a and c units, respectively, from the center, as shown in Figure 10.10. Moreover, a, b, and c are related by the equation $c^2 = a^2 - b^2$.

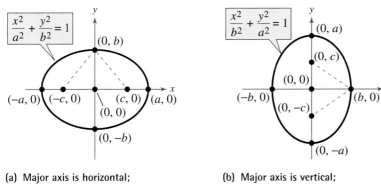

(a) Major axis is horizontal; minor axis is vertical.

(b) Major axis is vertical; minor axis is horizontal.

Figure 10.10

In Figure 10.10, note that because the sum of the distances from a point on the ellipse to each focus is constant, $c^2 = a^2 - b^2$ as follows.

$$2\sqrt{b^2 + c^2} = (a + c) + (a - c)$$

$$\sqrt{b^2 + c^2} = a$$

$$c^2 = a^2 - b^2$$

Exploration

An ellipse can be drawn using two thumbtacks placed at the foci of the ellipse, a string of fixed length (greater than the distance between the tacks), and a pencil, as shown in Figure 10.9. Try doing this. Vary the length of the string and the distance between the thumbtacks. Explain how to obtain ellipses that are almost circular. Explain how to obtain ellipses that are long and narrow.

Example 3 Finding the Standard Equation of an Ellipse

Find the standard form of the equation of the ellipse shown in Figure 10.11.

Solution

From Figure 10.11, the foci occur at $(-2, 0)$ and $(2, 0)$. So, the center of the ellipse is $(0, 0)$, the major axis is horizontal, and the ellipse has an equation of the form

$$\frac{x^2}{a^2} + \frac{y^2}{b^2} = 1. \qquad \text{Standard form}$$

Also from Figure 10.11, the length of the major axis is $2a = 6$. So, $a = 3$. Moreover, the distance from the center to either focus is $c = 2$. Finally,

$$b^2 = a^2 - c^2 = 3^2 - 2^2 = 9 - 4 = 5.$$

Substituting $a^2 = 3^2$ and $b^2 = \left(\sqrt{5}\right)^2$ yields the equation in standard form.

$$\frac{x^2}{3^2} + \frac{y^2}{\left(\sqrt{5}\right)^2} = 1$$

☑ *Checkpoint* Now try Exercise 45.

Figure 10.11

> **TECHNOLOGY SUPPORT**
>
> For instructions on how to use the *zoom* and *trace* features, see Appendix A; for specific keystrokes, go to the text website at *college.hmco.com*.

Example 4 Sketching an Ellipse

Sketch the ellipse given by $4x^2 + y^2 = 36$, and identify the vertices.

Algebraic Solution

$$\frac{4x^2}{36} + \frac{y^2}{36} = \frac{36}{36} \qquad \begin{array}{l}\text{Divide each side of}\\ \text{original equation by 36.}\end{array}$$

$$\frac{x^2}{3^2} + \frac{y^2}{6^2} = 1 \qquad \text{Write in standard form.}$$

Because the denominator of the y^2-term is larger than the denominator of the x^2-term, you can conclude that the major axis is vertical. Moreover, because $a = 6$, the vertices are $(0, -6)$ and $(0, 6)$. Finally, because $b = 3$, the endpoints of the minor axis are $(-3, 0)$ and $(3, 0)$, as shown in Figure 10.12.

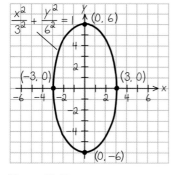

Figure 10.12

☑ *Checkpoint* Now try Exercise 33.

Graphical Solution

Solve the equation of the ellipse for y as follows.

$$4x^2 + y^2 = 36$$

$$y^2 = 36 - 4x^2$$

$$y = \pm\sqrt{36 - 4x^2}$$

Then use a graphing utility to graph $y_1 = \sqrt{36 - 4x^2}$ and $y_2 = -\sqrt{36 - 4x^2}$ in the same viewing window. Be sure to use a square setting. From the graph in Figure 10.13, you can see that the major axis is vertical. You can use the *zoom* and *trace* features to approximate the vertices to be $(0, 6)$ and $(0, -6)$.

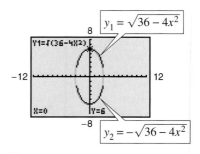

Figure 10.13

TECHNOLOGY TIP   Note that in the graphical solution of Example 4, the ellipse was graphed using two separate equations. When a graphing utility is used to graph conics, it may not connect the two equations. This is because some graphing utilities are limited in their resolution. So, in this text, a blue curve is placed behind the graphing utility's display to indicate where the graph should appear.

Hyperbolas

The definition of a **hyperbola** is similar to that of an ellipse. The difference is that for an ellipse, the *sum* of the distances between the foci and a point on the ellipse is constant, whereas for a hyperbola the *difference* of the distances between the foci and a point on the hyperbola is constant.

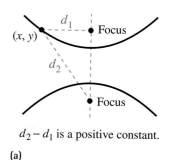

$d_2 - d_1$ is a positive constant.

(a)

> ### Definition of a Hyperbola
>
> A **hyperbola** is the set of all points (x, y) in a plane, the difference of whose distances from two distinct fixed points (**foci**) is a positive constant. [See Figure 10.14(a).]

The graph of a hyperbola has two disconnected parts (**branches**). The line through the two foci intersects the hyperbola at two points (**vertices**). The line segment connecting the vertices is the **transverse axis,** and the midpoint of the transverse axis is the **center** of the hyperbola. [See Figure 10.14(b).]

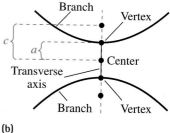

(b)

Figure 10.14

> ### Standard Equation of a Hyperbola (Center at Origin)
>
> The **standard form of the equation of a hyperbola** with center at the origin (where $a \neq 0$ and $b \neq 0$) is given by
>
> $$\frac{x^2}{a^2} - \frac{y^2}{b^2} = 1 \qquad \text{or} \qquad \frac{y^2}{a^2} - \frac{x^2}{b^2} = 1.$$
>
> The vertices and foci are a and c units from the center, respectively. Moreover, a, b, and c are related by the equation $b^2 = c^2 - a^2$. (See Figure 10.15.)

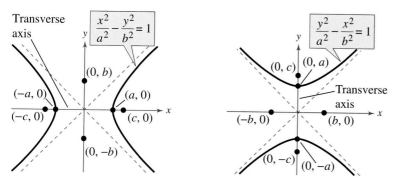

Figure 10.15

Example 5 Finding the Standard Equation of a Hyperbola

Find the standard form of the equation of the hyperbola with foci at $(-3, 0)$ and $(3, 0)$ and vertices at $(-2, 0)$ and $(2, 0)$, as shown in Figure 10.16.

Solution

From the graph, you can determine that $c = 3$ because the foci are three units from the center. Moreover, $a = 2$ because the vertices are two units from the center. So, it follows that

$$b^2 = c^2 - a^2$$

$$= 3^2 - 2^2$$

$$= 9 - 4$$

$$= 5.$$

Because the transverse axis is horizontal, the standard form of the equation is

$$\frac{x^2}{a^2} - \frac{y^2}{b^2} = 1.$$

Finally, substitute $a^2 = 2^2$ and $b^2 = \left(\sqrt{5}\right)^2$ to obtain

$$\frac{x^2}{2^2} - \frac{y^2}{\left(\sqrt{5}\right)^2} = 1. \qquad \text{Write in standard form.}$$

✓ *Checkpoint* Now try Exercise 61.

Figure 10.16

TECHNOLOGY TIP

To use a graphing utility to graph the hyperbola in Example 5, first solve for y^2 to obtain $y^2 = 5(x^2 - 4)/4$. Then enter the positive and negative square roots of the right-hand side of the equation as

$$y_1 = \sqrt{5(x^2 - 4)}/2$$

$$y_2 = -\sqrt{5(x^2 - 4)}/2.$$

An important aid in sketching the graph of a hyperbola is the determination of its *asymptotes*, as shown in Figure 10.17. Each hyperbola has two asymptotes that intersect at the center of the hyperbola. Furthermore, the asymptotes pass through the corners of a rectangle of dimensions $2a$ by $2b$. The line segment of length $2b$, joining $(0, b)$ and $(0, -b)$ [or $(-b, 0)$ and $(b, 0)$], is the **conjugate axis** of the hyperbola.

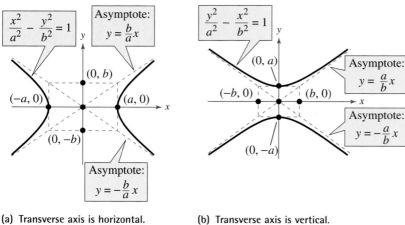

(a) Transverse axis is horizontal.
   Conjugate axis is vertical.

(b) Transverse axis is vertical.
   Conjugate axis is horizontal.

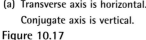
Figure 10.17

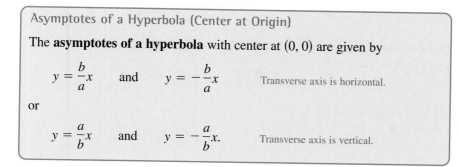

Asymptotes of a Hyperbola (Center at Origin)

The **asymptotes of a hyperbola** with center at $(0, 0)$ are given by

$$y = \frac{b}{a}x \quad \text{and} \quad y = -\frac{b}{a}x \qquad \text{Transverse axis is horizontal.}$$

or

$$y = \frac{a}{b}x \quad \text{and} \quad y = -\frac{a}{b}x. \qquad \text{Transverse axis is vertical.}$$

Example 6   Sketching the Graph of a Hyperbola

Sketch the hyperbola whose equation is $4x^2 - y^2 = 16$.

Algebraic Solution

$$4x^2 - y^2 = 16 \qquad \text{Write original equation.}$$

$$\frac{4x^2}{16} - \frac{y^2}{16} = \frac{16}{16} \qquad \text{Divide each side by 16.}$$

$$\frac{x^2}{2^2} - \frac{y^2}{4^2} = 1 \qquad \text{Write in standard form.}$$

Because the x^2-term is positive, you can conclude that the transverse axis is horizontal and that the vertices occur at $(-2, 0)$ and $(2, 0)$. Moreover, the endpoints of the conjugate axis occur at $(0, -4)$ and $(0, 4)$, and you can sketch the rectangle shown in Figure 10.18. Finally, by drawing the asymptotes through the corners of this rectangle, you can complete the sketch shown in Figure 10.19. Note that the equations of the asymptotes are $y = 2x$ and $y = -2x$.

Graphical Solution

Solve the equation of the hyperbola for y as follows.

$$4x^2 - y^2 = 16$$

$$4x^2 - 16 = y^2$$

$$\pm\sqrt{4x^2 - 16} = y$$

Then use a graphing utility to graph $y_1 = \sqrt{4x^2 - 16}$ and $y_2 = -\sqrt{4x^2 - 16}$ in the same viewing window. Be sure to use a square setting. From the graph in Figure 10.20, you can see that the transverse axis is horizontal. You can use the *zoom* and *trace* features to approximate the vertices to be $(-2, 0)$ and $(2, 0)$.

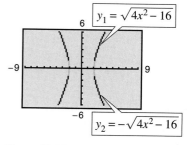

Figure 10.20

Figure 10.18   **Figure 10.19**

✓ *Checkpoint* Now try Exercise 55.

Example 7   Finding the Standard Equation of a Hyperbola

Find the standard form of the equation of the hyperbola that has vertices at $(0, -3)$ and $(0, 3)$ and asymptotes $y = -2x$ and $y = 2x$, as shown in Figure 10.21.

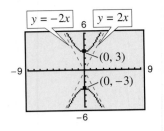

Solution

Because the transverse axis is vertical, the asymptotes are of the form

$$y = \frac{a}{b}x \quad \text{and} \quad y = -\frac{a}{b}x.$$

Using the fact that $y = 2x$ and $y = -2x$, you can determine that

$$\frac{a}{b} = 2.$$

Because $a = 3$, you can determine that $b = \frac{3}{2}$. Finally, you can conclude that the hyperbola has the following equation.

$$\frac{y^2}{3^2} - \frac{x^2}{(3/2)^2} = 1 \qquad \text{Write in standard form.}$$

☑ *Checkpoint* Now try Exercise 63.

Figure 10.21

Example 8   Identifying Conics

Identify each conic by writing its equation in standard form.

a. $6x^2 + y^2 - 36 = 0$ **b.** $4y^2 - 12x^2 - 48 = 0$

Solution

a. $6x^2 + y^2 - 36 = 0$ Write original equation.

$\quad\quad 6x^2 + y^2 = 36$ Add 36 to each side.

$$\frac{x^2}{6} + \frac{y^2}{36} = 1 \qquad \text{Divide each side by 36.}$$

$$\frac{x^2}{\left(\sqrt{6}\right)^2} + \frac{y^2}{6^2} = 1 \qquad \text{Write in standard form.}$$

From the standard form, you can see that the equation represents an ellipse.

b. $4y^2 - 12x^2 - 48 = 0$ Write original equation.

$\quad\quad 4y^2 - 12x^2 = 48$ Add 48 to each side.

$$\frac{y^2}{12} - \frac{x^2}{4} = 1 \qquad \text{Divide each side by 48.}$$

$$\frac{y^2}{\left(\sqrt{12}\right)^2} - \frac{x^2}{2^2} = 1 \qquad \text{Write in standard form.}$$

From the standard form, you can see that the equation represents a hyperbola.

☑ *Checkpoint* Now try Exercise 77.

10.1 Exercises

Vocabulary Check

Fill in the blanks.

1. A _____ is the intersection of a plane and a double-napped cone.

2. A _____ is the set of all points (x, y) in a plane that are equidistant from a fixed lined, called the _____ , and a fixed point, called the _____ , not on the line.

3. The _____ of a parabola is the midpoint between the focus and the directrix.

4. The line that passes through the focus and vertex of a parabola is called the _____ of the parabola.

5. An _____ is the set of all points (x, y) in a plane, the sum of whose distances from two distinct fixed points is constant.

6. The chord joining the vertices of an ellipse is called the _____ , and its midpoint is the _____ of the ellipse.

7. The chord perpendicular to the major axis at the center of an ellipse is called the _____ of the ellipse.

8. A _____ is the set of all points (x, y) in a plane, the difference of whose distances from two distinct fixed points is a positive constant.

9. The graph of a hyperbola has two disconnected parts called _____ .

10. The line segment connecting the vertices of a hyperbola is called the _____ , and the midpoint of the line segment is the _____ of the hyperbola.

In Exercises 1–4, find the standard form of the equation of the circle with the center at the origin and satisfying the given conditions.

1. Radius: 6

2. Radius: 1

3. Diameter: $\frac{10}{7}$

4. Diameter: $2\sqrt{7}$

In Exercises 5–10, find the vertex and focus of the parabola and sketch its graph. Use a graphing utility to verify your graph.

5. $y = \frac{1}{2}x^2$

6. $y = -4x^2$

7. $y^2 = -6x$

8. $y^2 = 3x$

9. $x^2 + 8y = 0$

10. $x + y^2 = 0$

In Exercises 11–20, find the standard form of the equation of the parabola with vertex at the origin.

11. Focus: $\left(0, -\frac{3}{2}\right)$

12. Focus: $\left(\frac{5}{2}, 0\right)$

13. Focus: $(-2, 0)$

14. Focus: $(0, -2)$

15. Directrix: $y = -1$

16. Directrix: $y = 2$

17. Directrix: $x = 3$

18. Directrix: $x = -2$

19. Horizontal axis and passes through the point $(4, 6)$

20. Vertical axis and passes through the point $(-2, -2)$

In Exercises 21–24, find the standard form of the equation of the parabola and determine the coordinates of the focus.

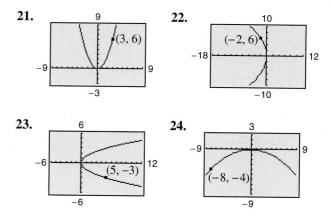

21. (3, 6)

22. (−2, 6)

23. (5, −3)

24. (−8, −4)

In Exercises 25 and 26, use a graphing utility to graph the parabola and the line (called one of its *tangent lines*). Identify the point of intersection (called the *point of tangency*).

Parabola	Tangent Line
25. $y^2 - 8x = 0$	$x - y + 2 = 0$
26. $x^2 + 12y = 0$	$x + y - 3 = 0$

In Exercises 27–34, find the center and vertices of the ellipse and sketch its graph. Use a graphing utility to verify your graph.

27. $\dfrac{x^2}{25} + \dfrac{y^2}{16} = 1$

28. $\dfrac{x^2}{144} + \dfrac{y^2}{169} = 1$

29. $\dfrac{x^2}{\frac{25}{9}} + \dfrac{y^2}{\frac{16}{9}} = 1$

30. $\dfrac{x^2}{4} + \dfrac{y^2}{\frac{1}{4}} = 1$

31. $\dfrac{x^2}{9} + \dfrac{y^2}{5} = 1$

32. $\dfrac{x^2}{28} + \dfrac{y^2}{64} = 1$

33. $4x^2 + y^2 = 1$

34. $4x^2 + 9y^2 = 36$

In Exercises 35–38, use a graphing utility to graph the ellipse. (*Hint:* Use two equations.)

35. $5x^2 + 3y^2 = 15$

36. $x^2 + 16y^2 = 16$

37. $6x^2 + y^2 = 36$

38. $4x^2 + 25y^2 = 100$

In Exercises 39–48, find the standard form of the equation of the ellipse with center at the origin.

39.
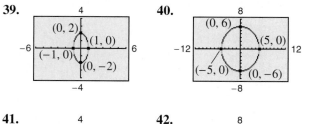

40.

41.
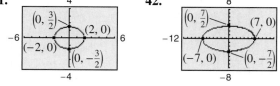

42.

43. Vertices: $(\pm 5, 0)$; foci: $(\pm 2, 0)$

44. Vertices: $(0, \pm 8)$; foci: $(0, \pm 4)$

45. Foci: $(\pm 5, 0)$; major axis of length 12

46. Foci: $(\pm 2, 0)$; major axis of length 8

47. Vertices: $(0, \pm 5)$; passes through the point $(4, 2)$

48. Major axis vertical; passes through the points $(0, 4)$ and $(2, 0)$

In Exercises 49–56, find the center, vertices, and foci of the hyperbola and sketch its graph, using asymptotes as sketching aids. Use a graphing utility to verify your graph.

49. $x^2 - y^2 = 1$

50. $\dfrac{x^2}{9} - \dfrac{y^2}{16} = 1$

51. $\dfrac{y^2}{1} - \dfrac{x^2}{9} = 1$

52. $\dfrac{y^2}{4} - \dfrac{x^2}{1} = 1$

53. $\dfrac{y^2}{25} - \dfrac{x^2}{144} = 1$

54. $\dfrac{x^2}{36} - \dfrac{y^2}{4} = 1$

55. $4y^2 - x^2 = 1$

56. $4y^2 - 9x^2 = 36$

In Exercises 57–60, use a graphing utility to graph the hyperbola and its asymptotes.

57. $2x^2 - 3y^2 = 6$

58. $3y^2 - 5x^2 = 15$

59. $4y^2 - 6x^2 = 12$

60. $8x^2 - 3y^2 = 24$

In Exercises 61–68, find the standard form of the equation of the hyperbola with center at the origin.

61. Vertices: $(0, \pm 2)$; foci: $(0, \pm 4)$

62. Vertices: $(\pm 3, 0)$; foci: $(\pm 5, 0)$

63. Vertices: $(\pm 1, 0)$; asymptotes: $y = \pm 3x$

64. Vertices: $(0, \pm 3)$; asymptotes: $y = \pm 3x$

65. Foci: $(0, \pm 8)$; asymptotes: $y = \pm 4x$

66. Foci: $(\pm 10, 0)$; asymptotes: $y = \pm \frac{3}{4}x$

67.
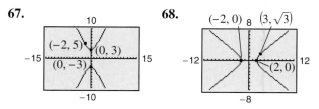

68.

In Exercises 69–76, match the equation with its graph. [The graphs are labeled (a), (b), (c), (d), (e), (f), (g), and (h).]

(a)
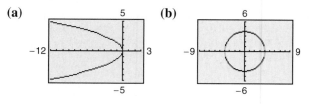

(b)

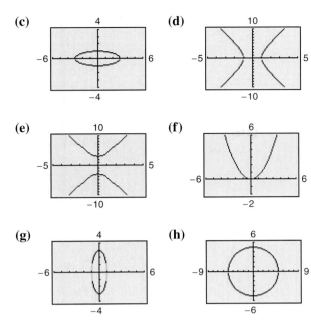

(c) **(d)** **(e)** **(f)** **(g)** **(h)**

69. $x^2 = 2y$ **70.** $y^2 = -2x$

71. $9x^2 + y^2 = 9$ **72.** $x^2 + 9y^2 = 9$

73. $9x^2 - y^2 = 9$ **74.** $y^2 - 9x^2 = 9$

75. $x^2 + y^2 = 16$ **76.** $x^2 + y^2 = 25$

In Exercises 77–82, identify the conic by writing the equation in standard form.

77. $4x^2 + 4y^2 - 16 = 0$ **78.** $4y^2 - 5x^2 + 20 = 0$

79. $3y^2 - 6x = 0$ **80.** $2x^2 + 4y^2 - 12 = 0$

81. $4x^2 + y^2 - 16 = 0$ **82.** $2x^2 - 12y = 0$

83. *Satellite Antenna*   Write an equation for a cross section of the parabolic television dish antenna shown in the figure.

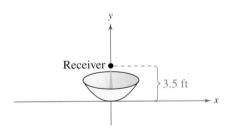

84. *Suspension Bridge*   Each cable of the Golden Gate Bridge is suspended (in the shape of a parabola) between two towers that are 1280 meters apart. The top of each tower is 152 meters above the roadway. The cables touch the roadway midway between the towers.

(a) Draw a sketch of the bridge. Locate the origin of a rectangular coordinate system at the center of the roadway. Label the coordinates of the known points.

(b) Write an equation that models the cables.

(c) Complete the table by finding the height y of the suspension cables over the roadway at a distance of x meters from the center of the bridge.

x	0	200	400	500	600
y					

Table for 84

85. *Beam Deflection*   A simply supported beam is 64 feet long and has a load at the center (see figure). The deflection (bending) of the beam at its center is 1 inch. The shape of the deflected beam is parabolic.

(a) Find an equation of the parabola. (Assume that the origin is at the center of the beam.)

(b) How far from the center of the beam is the deflection equal to $\frac{1}{2}$ inch?

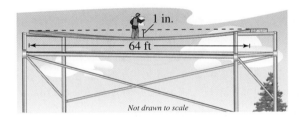

86. *Architecture*   A fireplace arch is to be constructed in the shape of a semiellipse. The opening is to have a height of 2 feet at the center and a width of 6 feet along the base (see figure). The contractor draws the outline of the ellipse on the wall by the method discussed on page 775. Give the required positions of the tacks and the length of the string.

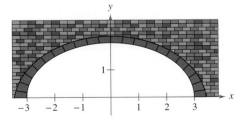

87. *Architecture*   A semielliptical arch over a tunnel for a road through a mountain has a major axis of 110 feet and a height at the center of 40 feet.

(a) Draw a rectangular coordinate system on a sketch of the tunnel with the center of the road entering the tunnel at the origin. Identify the coordinates of the known points.

(b) Find an equation of the semielliptical arch over the tunnel.

(c) Determine the height of the arch 5 feet from the edge of the tunnel.

88. *Geometry*   A line segment through a focus of an ellipse with endpoints on the ellipse and perpendicular to the major axis is called a **latus rectum** of the ellipse. Therefore, an ellipse has two latera recta. Knowing the length of the latera recta is helpful in sketching an ellipse because this information yields other points on the curve (see figure). Show that the length of each latus rectum is $2b^2/a$.

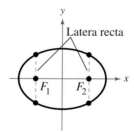

In Exercises 89–92, sketch the graph of the ellipse using the latera recta (see Exercise 88).

89. $\dfrac{x^2}{4} + \dfrac{y^2}{1} = 1$ **90.** $\dfrac{x^2}{9} + \dfrac{y^2}{16} = 1$

91. $9x^2 + 4y^2 = 36$ **92.** $5x^2 + 3y^2 = 15$

93. *Navigation*   Long distance radio navigation for aircraft and ships uses synchronized pulses transmitted by widely separated transmitting stations. These pulses travel at the speed of light (186,000 miles per second). The difference in the times of arrival of these pulses at an aircraft or ship is constant on a hyperbola having the transmitting stations as foci. Assume that two stations 300 miles apart are positioned on a rectangular coordinate system at coordinates $(-150, 0)$ and $(150, 0)$ and that a ship is traveling on a path with coordinates $(x, 75)$, as shown in the figure. Find the x-coordinate of the position of

the ship when the time difference between the pulses from the transmitting stations is 1000 microseconds (0.001 second).

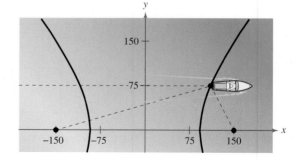

Figure for 93

94. *Optics*   A hyperbolic mirror (used in some telescopes) has the property that a light ray directed at the focus will be reflected to the other focus. The focus of a hyperbolic mirror (see figure) has coordinates $(24, 0)$. Find the vertex of the mirror if its mount at the top edge of the mirror has coordinates $(24, 24)$.

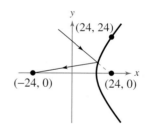

Synthesis

True or False?   **In Exercises 95–97, determine whether the statement is true or false. Justify your answer.**

95. The equation $9x^2 - 16y^2 = 144$ represents an ellipse.

96. The major axis of the ellipse given by $y^2 + 16x^2 = 64$ is vertical.

97. It is possible for a parabola to intersect its directrix.

98. *Exploration*   Consider the parabola $x^2 = 4py$.

(a) Use a graphing utility to graph the parabola for $p = 1$, $p = 2$, $p = 3$, and $p = 4$. Describe the effect on the graph as p increases.

(b) Locate the focus of each parabola in part (a).

(c) For each parabola in part (a), find the length of the chord passing through the focus parallel to the directrix. How can the length of this chord be determined directly from $x^2 = 4py$?

(d) Explain how the result of part (c) can be used as a sketching aid when graphing parabolas.

99. Exploration   Let (x_1, y_1) be the coordinates of a point on the parabola $x^2 = 4py$. The equation of the line that just touches the parabola at the point (x_1, y_1), called a *tangent line*, is given by

$$y - y_1 = \frac{x_1}{2p}(x - x_1).$$

(a) What is the slope of the tangent line?

(b) For each parabola in Exercise 98, find the equations of the tangent lines at the endpoints of the chord. Use a graphing utility to graph the parabola and tangent lines.

100. Exploration   Consider the ellipse

$$\frac{x^2}{a^2} + \frac{y^2}{b^2} = 1, \quad a + b = 20.$$

(a) The area of the ellipse is given by $A = \pi ab$. Write the area of the ellipse as a function of a.

(b) Find the equation of an ellipse with an area of 264 square centimeters.

(c) Complete the table using your equation from part (a) and make a conjecture about the shape of the ellipse with maximum area.

a	8	9	10	11	12	13
A						

(d) Use a graphing utility to graph the area function to support your conjecture in part (c).

Think About It   In Exercises 101 and 102, which part of the graph of the ellipse $4x^2 + 9y^2 = 36$ is represented by the equation? (Do not graph.)

101. $x = -\frac{3}{2}\sqrt{4 - y^2}$ **102.** $y = \frac{2}{3}\sqrt{9 - x^2}$

Think About It   In Exercises 103 and 104, which part of the graph of the hyperbola $4x^2 - 9y^2 = 36$ is represented by the equation? (Do not graph.)

103. $y = -\frac{2}{3}\sqrt{x^2 - 9}$ **104.** $x = \frac{3}{2}\sqrt{y^2 + 4}$

105. Think About It   Is the graph of $x^2 + 4y^4 = 4$ an ellipse? Explain.

106. Think About It   The graph of $x^2 - y^2 = 0$ is a degenerate conic. Sketch this graph and identify the degenerate conic.

107. Writing   Write a paragraph discussing the change in the shape and orientation of the graph of the ellipse

$$\frac{x^2}{a^2} + \frac{y^2}{16} = 1$$

as a increases from 1 to 8.

108. Writing   At the beginning of this section, you learned that each type of conic section can be formed by the intersection of a plane and a double-napped cone. Write a short paragraph describing examples of physical situations in which hyperbolas are formed.

109. Use the definition of an ellipse to derive the standard form of the equation of an ellipse.

110. Use the definition of a hyperbola to derive the standard form of the equation of a hyperbola.

Review

In Exercises 111–114, factor the expression completely.

111. $12x^2 + 7x - 10$ **112.** $25x^3 - 60x^2 + 36x$

113. $12z^4 + 17z^3 + 5z^2$ **114.** $x^3 + 3x^2 - 4x - 12$

In Exercises 115–118, find a polynomial with real coefficients that has the given zeros.

115. $0, 3, 4$ **116.** $-6, 1$

117. $-3, 1 + \sqrt{2}, 1 - \sqrt{2}$

118. $3, 2 + i, 2 - i$

119. Find all the zeros of $f(x) = 2x^3 - 3x^2 + 50x - 75$ if one of the zeros is $x = \frac{3}{2}$.

120. List the possible rational zeros of the function

$$g(x) = 6x^4 + 7x^3 - 29x^2 - 28x + 20.$$

121. Aircraft Boarding   In how many different ways can 11 people board an airplane?

122. Exam Questions   A student can answer any 15 questions from a total of 18 questions on an exam. In how many different ways can the student select the questions?

10.2 Translations of Conics

Vertical and Horizontal Shifts of Conics

In Section 10.1 you looked at conic sections whose graphs were in *standard position* (centered at the origin). In this section you will study the equations of conic sections that have been shifted vertically or horizontally in the plane.

What you should learn

● Recognize equations of conics that have been shifted vertically and/or horizontally in the plane.

● Write and graph equations of conics that have been shifted vertically and/or horizontally in the plane.

Why you should learn it

In some real-life applications, it is convenient to use a conic whose center or vertex is not the origin. For instance, Exercise 77 on page 793 shows how an ellipse can be used to model the equation of a satellite's orbit around Earth. In this application, it is convenient to have one of the ellipse's foci at the origin.

eStock Photo

Standard Forms of Equations of Conics

Circle: Center $= (h, k)$; radius $= r$

$$(x - h)^2 + (y - k)^2 = r^2$$

Ellipse: Center $= (h, k)$

Major axis length $= 2a$; minor axis length $= 2b$

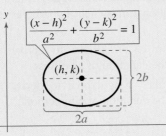
$$\frac{(x - h)^2}{a^2} + \frac{(y - k)^2}{b^2} = 1$$

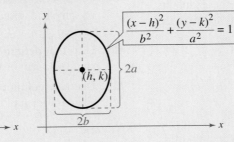
$$\frac{(x - h)^2}{b^2} + \frac{(y - k)^2}{a^2} = 1$$

Hyperbola: Center $= (h, k)$

Transverse axis length $= 2a$; conjugate axis length $= 2b$

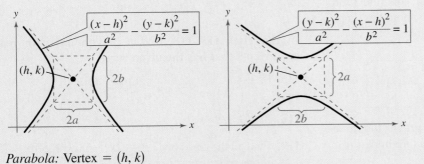
$$\frac{(x - h)^2}{a^2} - \frac{(y - k)^2}{b^2} = 1$$

$$\frac{(y - k)^2}{a^2} - \frac{(x - h)^2}{b^2} = 1$$

Parabola: Vertex $= (h, k)$

Directed distance from vertex to focus $= p$

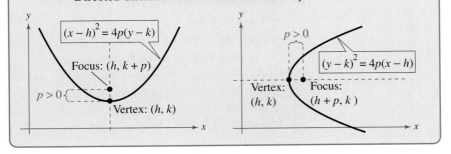
$$(x - h)^2 = 4p(y - k)$$
Focus: $(h, k + p)$

$p > 0$

Vertex: (h, k)

$$(y - k)^2 = 4p(x - h)$$
$p > 0$

Vertex: (h, k)

Focus: $(h + p, k)$

Example 1   Translations of Conic Sections

Describe the translation of the graph of each conic.

a. $(x - 1)^2 + (y + 2)^2 = 3^2$ **b.** $\dfrac{(x - 2)^2}{3^2} + \dfrac{(y - 1)^2}{2^2} = 1$

c. $\dfrac{(x - 3)^2}{1^2} - \dfrac{(y - 2)^2}{3^2} = 1$ **d.** $(x + 2)^2 = 4(-1)(y - 3)$

Solution

a. The graph of $(x - 1)^2 + (y + 2)^2 = 3^2$ is a circle whose center is the point $(1, -2)$ and whose radius is 3, as shown in Figure 10.22. Note that the graph of the circle has been shifted one unit to the right and two units downward from standard position.

b. The graph of

$$\dfrac{(x - 2)^2}{3^2} + \dfrac{(y - 1)^2}{2^2} = 1$$

is an ellipse whose center is the point $(2, 1)$. The major axis of the ellipse is horizontal and of length $2(3) = 6$, and the minor axis of the ellipse is vertical and of length $2(2) = 4$, as shown in Figure 10.23. Note that the graph of the ellipse has been shifted two units to the right and one unit upward from standard position.

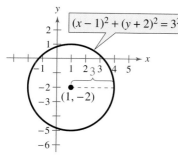

Figure 10.22

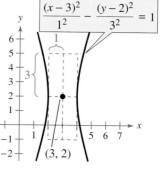

Figure 10.23

Figure 10.24

c. The graph of

$$\dfrac{(x - 3)^2}{1^2} - \dfrac{(y - 2)^2}{3^2} = 1$$

is a hyperbola whose center is the point $(3, 2)$. The transverse axis is horizontal and of length $2(1) = 2$, and the conjugate axis is vertical and of length $2(3) = 6$, as shown in Figure 10.24. Note that the graph of the hyperbola has been shifted three units to the right and two units upward from standard position.

d. The graph of $(x + 2)^2 = 4(-1)(y - 3)$ is a parabola whose vertex is the point $(-2, 3)$. The axis of the parabola is vertical. The focus is one unit above or below the vertex and, because $p = -1$, it follows that the focus lies *below* the vertex and the parabola opens downward, as shown in Figure 10.25. Note that the graph of the parabola has been shifted two units to the left and three units upward from standard position.

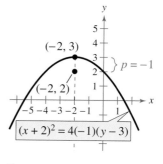

Figure 10.25

✓ *Checkpoint* Now try Exercise 3.

Equations of Conics in Standard Form

STUDY TIP

For a review of completing the square, refer to Section 2.4.

Example 2   Finding the Standard Form of a Parabola

Find the vertex and focus of the parabola $x^2 - 2x + 4y - 3 = 0$.

Solution

$x^2 - 2x + 4y - 3 = 0$	Write original equation.
$x^2 - 2x + 1 = -4y + 3 + 1$	Group terms and add 1 to each side.
$(x - 1)^2 = -4y + 4$	Write in completed square form.
$(x - 1)^2 = 4(-1)(y - 1)$	Write in standard form.

From this standard form, it follows that $h = 1, k = 1$, and $p = -1$. Because the axis is vertical and p is negative, the parabola opens downward. The vertex is $(h, k) = (1, 1)$, and the focus is $(h, k + p) = (1, 0)$. See Figure 10.26.

✓ *Checkpoint* Now try Exercise 23.

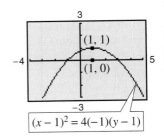

Figure 10.26

Example 3   Sketching an Ellipse

Sketch the graph of the ellipse $x^2 + 4y^2 + 6x - 8y + 9 = 0$.

Algebraic Solution

$x^2 + 4y^2 + 6x - 8y + 9 = 0$	Write original equation.
$(x^2 + 6x + \quad) + 4(y^2 - 2y + \quad) = -9$	Group terms and factor 4 out of y-terms.
$(x^2 + 6x + 9) + 4(y^2 - 2y + 1) = 4$	Add 9 and $4(1) = 4$ to each side.
$(x + 3)^2 + 4(y - 1)^2 = 4$	Write in completed square form.
$\dfrac{(x + 3)^2}{2^2} + \dfrac{(y - 1)^2}{1^2} = 1$	Write in standard form.

From this standard form, it follows that the center is $(h, k) = (-3, 1)$. Because $a^2 = 2^2$ and $b^2 = 1^2$, the endpoints of the major axis lie two units to the right and left of the center and the endpoints of the minor axis lie one unit up and down from the center. The ellipse is shown in Figure 10.27.

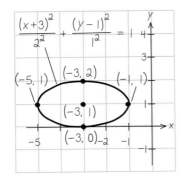
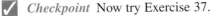

Figure 10.27

✓ *Checkpoint* Now try Exercise 37.

Graphical Solution

Write the completed square form of the ellipse as shown in the *Algebraic Solution*. Then solve the equation for y.

$$(y - 1)^2 = 1 - \frac{(x + 3)^2}{4}$$

$$y_1 = 1 + \sqrt{1 - \frac{(x + 3)^2}{4}}$$

$$y_2 = 1 - \sqrt{1 - \frac{(x + 3)^2}{4}}$$

Then use a graphing utility to graph y_1 and y_2 in the same viewing window, as shown in Figure 10.28. Use the *zoom* and *trace* features to approximate the endpoints of the major and minor axes.

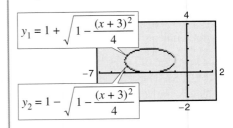

Figure 10.28

Example 4   Sketching a Hyperbola

Sketch the graph of the hyperbola given by the equation

$$y^2 - 4x^2 + 4y + 24x - 41 = 0.$$

Solution

$$y^2 - 4x^2 + 4y + 24x - 41 = 0 \qquad \text{Write original equation.}$$

$$(y^2 + 4y + \boxed{}) - (4x^2 - 24x + \boxed{}) = 41 \qquad \text{Group terms.}$$

$$(y^2 + 4y + \boxed{}) - 4(x^2 - 6x + \boxed{}) = 41 \qquad \text{Factor 4 out of } x\text{-terms.}$$

$$(y^2 + 4y + 4) - 4(x^2 - 6x + 9) = 41 + 4 - 4(9) \qquad \begin{array}{l}\text{Add 4 to and subtract} \\ 4(9) = 36 \text{ from each} \\ \text{side.}\end{array}$$

$$(y + 2)^2 - 4(x - 3)^2 = 9 \qquad \begin{array}{l}\text{Write in completed} \\ \text{square form.}\end{array}$$

$$\frac{(y + 2)^2}{9} - \frac{4(x - 3)^2}{9} = 1 \qquad \text{Divide each side by 9.}$$

$$\frac{(y + 2)^2}{9} - \frac{(x - 3)^2}{\dfrac{9}{4}} = 1 \qquad \text{Rewrite 4 as } \frac{1}{\frac{1}{4}}.$$

$$\frac{(y + 2)^2}{3^2} - \frac{(x - 3)^2}{\left(\dfrac{3}{2}\right)^2} = 1 \qquad \text{Write in standard form.}$$

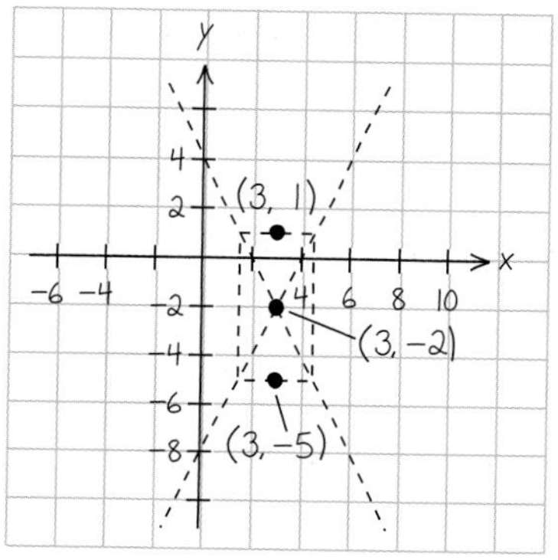

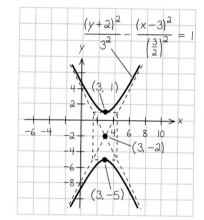

Figure 10.29

Figure 10.30

From this standard form, it follows that the transverse axis is vertical and the center lies at $(h, k) = (3, -2)$. Because the denominator of the y-term is $a^2 = 3^2$, you know that the vertices occur three units above and below the center.

$$(3, 1) \qquad \text{and} \qquad (3, -5) \qquad\qquad \text{Vertices}$$

To sketch the hyperbola, draw a rectangle whose top and bottom pass through the vertices. Because the denominator of the x-term is $b^2 = \left(\frac{3}{2}\right)^2$, locate the sides of the rectangle $\frac{3}{2}$ units to the right and left of the center. Sketch the asymptotes by drawing lines through the opposite corners of the rectangle, as shown in Figure 10.29. Using these asymptotes, you can complete the graph of the hyperbola, as shown in Figure 10.30.

✓ *Checkpoint* Now try Exercise 55.

To find the foci in Example 4, first find c.

$$c^2 = a^2 + b^2 = 9 + \frac{9}{4} = \frac{45}{4} \quad \Longrightarrow \quad c = \frac{3\sqrt{5}}{2}$$

Because the transverse axis is vertical, the foci lie c units above and below the center.

$$\left(3, -2 + \tfrac{3}{2}\sqrt{5}\right) \qquad \text{and} \qquad \left(3, -2 - \tfrac{3}{2}\sqrt{5}\right) \qquad\qquad \text{Foci}$$

Example 5   Finding the Standard Equation of an Ellipse

Find the standard form of the equation of the ellipse whose vertices are $(2, -2)$ and $(2, 4)$. The length of the minor axis of the ellipse is 4, as shown in Figure 10.31.

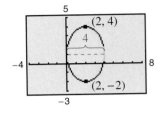

Figure 10.31

Solution

The center of the ellipse lies at the midpoint of its vertices. So, the center is

$(h, k) = (2, 1).$ Center

Because the vertices lie on a vertical line and are six units apart, it follows that the major axis is vertical and has a length of $2a = 6$. So, $a = 3$. Moreover, because the minor axis has a length of 4, it follows that $2b = 4$, which implies that $b = 2$. Therefore, the standard form of the ellipse is as follows.

$$\frac{(x - h)^2}{b^2} + \frac{(y - k)^2}{a^2} = 1 \qquad \text{Major axis is vertical.}$$

$$\frac{(x - 2)^2}{2^2} + \frac{(y - 1)^2}{3^2} = 1 \qquad \text{Write in standard form.}$$

✓ *Checkpoint* Now try Exercise 43.

An interesting application of conics involves the orbits of comets in our solar system. Of the 610 comets identified prior to 1970, 245 have elliptical orbits, 295 have parabolic orbits, and 70 have hyperbolic orbits. For example, Halley's comet has an elliptical orbit, and the reappearance of this comet can be predicted to occur every 76 years. The center of the sun is a focus of each of these orbits, and each orbit has a vertex at the point where the comet is closest to the sun, as shown in Figure 10.32.

If p is the distance between the vertex and the focus in meters, and v is the speed of the comet at the vertex in meters per second, then the type of orbit is determined as follows.

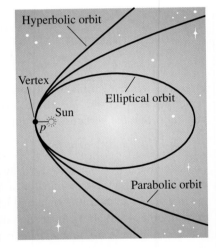

Figure 10.32

1. Ellipse: $v < \sqrt{\dfrac{2GM}{p}}$

2. Parabola: $v = \sqrt{\dfrac{2GM}{p}}$

3. Hyperbola: $v > \sqrt{\dfrac{2GM}{p}}$

In each of these equations, $M = 1.989 \times 10^{30}$ kilograms (the mass of the sun) and $G \approx 6.67 \times 10^{-11}$ cubic meters per kilogram-second squared (the universal gravitational constant).

10.2   Exercises

Vocabulary Check

In Exercises 1–7, match the description of the conic with its standard equation. [The equations are labeled (a), (b), (c), (d), (e), (f), and (g).]

(a) $\dfrac{(x-h)^2}{a^2} - \dfrac{(y-k)^2}{b^2} = 1$

(b) $(x-h)^2 = 4p(y-k)$

(c) $\dfrac{(x-h)^2}{a^2} + \dfrac{(y-k)^2}{b^2} = 1$

(d) $\dfrac{(x-h)^2}{b^2} + \dfrac{(y-k)^2}{a^2} = 1$

(e) $(x-h)^2 + (y-k)^2 = r^2$

(f) $(y-k)^2 = 4p(x-h)$

(g) $\dfrac{(y-k)^2}{a^2} - \dfrac{(x-h)^2}{b^2} = 1$

1. Circle

2. Ellipse with vertical major axis

3. Parabola with vertical axis

4. Hyperbola with horizontal transverse axis

5. Ellipse with horizontal major axis

6. Parabola with horizontal axis

7. Hyperbola with vertical transverse axis

In Exercises 1–6, describe the translation of the graph of the conic.

1. $(x+2)^2 + (y-1)^2 = 4$

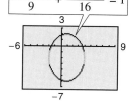

2. $(y-1)^2 = 4(2)(x+2)$

3. $\dfrac{(y+3)^2}{4} - (x-1)^2 = 1$

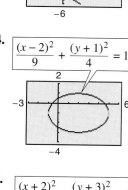

4. $\dfrac{(x-2)^2}{9} + \dfrac{(y+1)^2}{4} = 1$

5. $\dfrac{(x-1)^2}{9} + \dfrac{(y+2)^2}{16} = 1$

6. $\dfrac{(x+2)^2}{4} - \dfrac{(y+3)^2}{9} = 1$

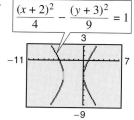

In Exercises 7–12, identify the center and radius of the circle.

7. $x^2 + y^2 = 49$

8. $x^2 + y^2 = 1$

9. $(x+2)^2 + (y-7)^2 = 16$

10. $(x+9)^2 + (y+1)^2 = 36$

11. $(x-1)^2 + y^2 = 15$

12. $x^2 + (y+12)^2 = 24$

In Exercises 13–16, write the equation of the circle in standard form. Then identify its center and radius.

13. $x^2 + y^2 - 2x + 6y + 9 = 0$

14. $x^2 + y^2 - 10x - 6y + 25 = 0$

15. $4x^2 + 4y^2 + 12x - 24y + 41 = 0$

16. $9x^2 + 9y^2 + 54x - 36y + 17 = 0$

In Exercises 17–24, find the vertex, focus, and directrix of the parabola, and sketch its graph. Use a graphing utility to verify your graph.

17. $(x+1)^2 + 8(y+2) = 0$

18. $(x+3) + (y-2)^2 = 0$

19. $\left(y - \frac{1}{2}\right)^2 = 2(x-5)$

20. $\left(x - \frac{1}{2}\right)^2 = 4(y+3)$

21. $y = \frac{1}{4}(x^2 - 2x + 5)$

22. $4x - y^2 - 2y - 33 = 0$

23. $y^2 + 6y + 8x + 25 = 0$

24. $y^2 - 4y - 4x = 0$

In Exercises 25–32, find the standard form of the equation of the parabola.

25. **26.**

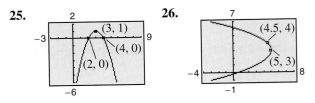

27. Vertex: $(3, 2)$; focus: $(1, 2)$

28. Vertex: $(-1, 2)$; focus: $(-1, 0)$

29. Vertex: $(0, 4)$; directrix: $y = 2$

30. Vertex: $(-2, 1)$; directrix: $x = 1$

31. Focus: $(2, 2)$; directrix: $x = -2$

32. Focus: $(0, 0)$; directrix: $y = 4$

In Exercises 33–40, find the center, foci, and vertices of the ellipse, and sketch its graph. Use a graphing utility to verify your answer.

33. $\dfrac{(x-1)^2}{9} + \dfrac{(y-3)^2}{25} = 1$

34. $\dfrac{(x-6)^2}{4} + \dfrac{(y+7)^2}{16} = 1$

35. $(x+2)^2 + \dfrac{(y-4)^2}{\frac{1}{4}} = 1$

36. $\dfrac{(x-3)^2}{\frac{25}{9}} + (y-8)^2 = 1$

37. $9x^2 + 4y^2 + 36x - 24y + 36 = 0$

38. $9x^2 + 4y^2 - 36x + 8y + 31 = 0$

39. $16x^2 + 25y^2 - 32x + 50y + 16 = 0$

40. $9x^2 + 25y^2 - 36x - 50y + 61 = 0$

In Exercises 41–50, find the standard form of the equation of the ellipse.

41. **42.**

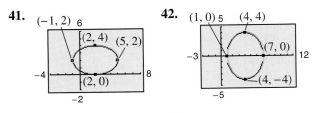

43. Vertices: $(0, 2)$, $(4, 2)$; minor axis of length 2

44. Foci: $(0, 0)$, $(4, 0)$; major axis of length 8

45. Foci: $(0, 0)$, $(0, 8)$; major axis of length 16

46. Center: $(2, -1)$; vertex: $(2, \frac{1}{2})$; minor axis of length 2

47. Vertices: $(3, 1)$, $(3, 9)$; minor axis of length 6

48. Center: $(3, 2)$; $a = 3c$; foci: $(1, 2)$, $(5, 2)$

49. Center: $(0, 4)$; $a = 2c$; vertices: $(-4, 4)$, $(4, 4)$

50. Vertices: $(5, 0)$, $(5, 12)$; endpoints of minor axis: $(0, 6)$, $(10, 6)$

In Exercises 51–60, find the center, vertices, and foci of the hyperbola, and sketch its graph, using asymptotes as sketching aids. Use a graphing utility to verify your graph.

51. $\dfrac{(x+1)^2}{4} - \dfrac{(y-2)^2}{1} = 1$

52. $\dfrac{(x-1)^2}{144} - \dfrac{(y+4)^2}{25} = 1$

53. $(y+6)^2 - (x-2)^2 = 1$

54. $\dfrac{(y+1)^2}{\frac{1}{4}} - \dfrac{(x+3)^2}{\frac{1}{9}} = 1$

55. $9x^2 - y^2 - 36x - 6y + 18 = 0$

56. $x^2 - 9y^2 + 36y - 72 = 0$

57. $x^2 - 9y^2 + 2x - 54y - 80 = 0$

58. $16y^2 - x^2 + 2x + 64y + 63 = 0$

59. $9y^2 - 4x^2 + 8x + 18y + 41 = 0$

60. $11y^2 - 3x^2 + 12x + 44y + 48 = 0$

In Exercises 61–68, find the standard form of the equation of the hyperbola.

61. **62.**

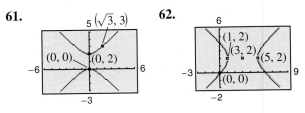

63. Vertices: $(2, 0)$, $(6, 0)$; foci: $(0, 0)$, $(8, 0)$

64. Vertices: $(2, 3)$, $(2, -3)$; foci: $(2, 5)$, $(2, -5)$

65. Vertices: $(4, 1)$, $(4, 9)$; foci: $(4, 0)$, $(4, 10)$

66. Vertices: $(-2, 1)$, $(2, 1)$; foci: $(-3, 1)$, $(3, 1)$

67. Vertices: $(2, 3)$, $(2, -3)$; passes through the point $(0, 5)$

68. Vertices: $(3, 0)$, $(3, 4)$; asymptotes: $y = \frac{2}{3}x$, $y = 4 - \frac{2}{3}x$

In Exercises 69–76, identify the conic by writing the equation in standard form. Then sketch its graph.

69. $x^2 + y^2 - 6x + 4y + 9 = 0$

70. $x^2 + 4y^2 - 6x + 16y + 21 = 0$

71. $4x^2 - y^2 - 4x - 3 = 0$

72. $y^2 - 4y - 4x = 0$

73. $4x^2 + 3y^2 + 8x - 24y + 51 = 0$

74. $4y^2 - 2x^2 - 4y - 8x - 15 = 0$

75. $25x^2 - 10x - 200y - 119 = 0$

76. $4x^2 + 4y^2 - 16y + 15 = 0$

77. *Satellite Orbit* A satellite in a 100-mile-high circular orbit around Earth has a velocity of approximately 17,500 miles per hour. If this velocity is multiplied by $\sqrt{2}$, the satellite will have the minimum velocity necessary to escape Earth's gravity, and it will follow a parabolic path with the center of Earth as the focus.

(a) Find the escape velocity of the satellite.

(b) Find an equation of its path (assume the radius of Earth is 4000 miles).

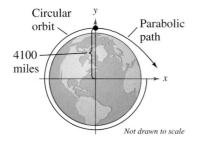

Not drawn to scale

78. *Projectile Motion* A cargo plane is flying at an altitude of 30,000 feet and a speed of 540 miles per hour (792 feet per second). How many feet will a supply crate dropped from the plane travel horizontally before it hits the ground if the path of the crate is modeled by

$$x^2 = -39{,}204(y - 30{,}000)?$$

79. *Path of a Projectile* The path of a softball is modeled by

$$-12.5(y - 7.125) = (x - 6.25)^2.$$

The coordinates x and y are measured in feet, with $x = 0$ corresponding to the position from which the ball was thrown.

(a) Use a graphing utility to graph the trajectory of the softball.

(b) Use the *trace* feature of the graphing utility to approximate the highest point the ball reaches and the distance the ball travels.

80. *Revenue* The revenue R generated by the sale of x 32-inch televisions is modeled by

$$R = 375x - \tfrac{3}{2}x^2.$$

(a) Use a graphing utility to graph the function.

(b) Use the *trace* feature of the graphing utility to approximate *graphically* the sales that will maximize the revenue.

(c) Use the *table* feature of the graphing utility to approximate *numerically* the sales that will maximize the revenue.

(d) Find the coordinates of the vertex to find *algebraically* the sales that will maximize the revenue.

(e) Compare the results of parts (b), (c), and (d). What did you learn as a result of using all three approaches?

In Exercises 81–84, e is called the *eccentricity* of an ellipse and is defined by $e = c/a$. It measures the flatness of the ellipse.

81. Find an equation of the ellipse with vertices $(\pm 5, 0)$ and eccentricity $e = \tfrac{3}{5}$.

82. Find an equation of the ellipse with vertices $(0, \pm 8)$ and eccentricity $e = \tfrac{1}{2}$.

83. *Planetary Motion* The planet Pluto moves in an elliptical orbit with the sun at one of the foci. The length of half of the major axis is 3.67×10^9 miles and the eccentricity is 0.249. Find the smallest distance (*perihelion*) and the greatest distance (*aphelion*) of Pluto from the center of the sun.

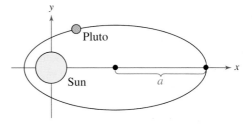

84. *Satellite Orbit*   The first artificial satellite to orbit Earth was Sputnik I (launched by the former Soviet Union in 1957). Its highest point above Earth's surface was 947 kilometers, and its lowest point was 228 kilometers. The center of Earth is the focus of the elliptical orbit, and the radius of Earth is 6378 kilometers. Find the eccentricity of the orbit.

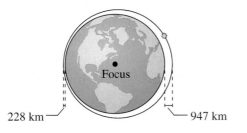

228 km   Focus   947 km

85. *Sports*   In Australia, football by *Australian Rules* (or rugby) is played on elliptical fields. The fields can be a maximum of 170 yards wide and a maximum of 200 yards long. Let the center of a field of maximum size be represented by the point $(0, 85)$. Find the standard form of the equation of the ellipse that represents this field.   (Source: Australian Football League)

86. *Astronomy*   Halley's comet has an elliptical orbit with the sun at one focus. The eccentricity of the orbit is approximately 0.97. The length of the major axis of the orbit is approximately 35.88 astronomical units. (An astronomical unit is about 93 million miles.) Find the standard form of the equation of the orbit. Place the center of the orbit at the origin and place the major axis on the x-axis.

Synthesis

True or False?   In Exercises 87 and 88, determine whether the statement is true or false. Justify your answer.

87. The conic represented by the equation $3x^2 + 2y^2 - 18x - 16y + 58 = 0$ is an ellipse.

88. The graphs of $x^2 + 10y - 10x + 5 = 0$ and $x^2 + 16y^2 + 10x - 32y - 23 = 0$ do not intersect.

Think About It   In Exercises 89 and 90, change the equation so that its graph matches the description.

89. $(y - 3)^2 = 6(x + 1)$; upper half of parabola

90. $(y + 1)^2 = 2(x - 2)$; lower half of parabola

Think About It   In Exercises 91 and 92, change the equation so that its graph matches the description.

91. $\dfrac{(x - 3)^2}{9} + \dfrac{y^2}{4} = 1$; right half of ellipse

92. $\dfrac{(x + 1)^2}{16} + \dfrac{(y - 2)^2}{25} = 1$; bottom half of ellipse

93. *Exploration*   Consider the ellipse $\dfrac{x^2}{a^2} + \dfrac{y^2}{b^2} = 1$.

(a) Show that the equation of the ellipse can be written as

$$\frac{(x - h)^2}{a^2} + \frac{(y - k)^2}{a^2(1 - e^2)} = 1$$

where e is the eccentricity.

(b) Use a graphing utility to graph the ellipse

$$\frac{(x - 2)^2}{4} + \frac{(y - 3)^2}{4(1 - e^2)} = 1$$

for $e = 0.95, 0.75, 0.5, 0.25,$ and 0.

(c) Make a conjecture about the change in the shape of the ellipse as e approaches 0.

94. *Writing*   Use your school's library, the Internet, or some other reference source to research information about the orbits of comets in our solar system. What can you find about the orbits of comets that have been identified since 1970? Write a summary of your results. Identify your source(s).

Review

In Exercises 95–98, use sigma notation to write the sum. Then use a graphing utility to find the sum.

95. $\dfrac{1}{6(1)} + \dfrac{1}{6(2)} + \dfrac{1}{6(3)} + \cdots + \dfrac{1}{6(9)}$

96. $\dfrac{7}{2 + 1} + \dfrac{7}{2 + 2} + \dfrac{7}{2 + 3} + \cdots + \dfrac{7}{2 + 12}$

97. $1 - \dfrac{1}{4} + \dfrac{1}{16} - \cdots + \dfrac{1}{65536}$

98. $\dfrac{1}{9} + \dfrac{4}{27} + \dfrac{7}{81} + \dfrac{10}{243}$

In Exercises 99–102, use the Binomial Theorem to expand and simplify the expression.

99. $(x - 4)^4$ 100. $(x - 3)^6$

101. $(3x + 1)^5$ 102. $(x^2 - 2)^5$

10.3 Parametric Equations

Plane Curves

Up to this point, you have been representing a graph by a single equation involving *two* variables such as x and y. In this section, you will study situations in which it is useful to introduce a *third* variable to represent a curve in the plane.

To see the usefulness of this procedure, consider the path of an object that is propelled into the air at an angle of $45°$. If the initial velocity of the object is 48 feet per second, it can be shown that the object follows the parabolic path

$$y = -\frac{x^2}{72} + x \qquad \text{Rectangular equation}$$

as shown in Figure 10.33. However, this equation does not tell the whole story. Although it does tell you *where* the object has been, it doesn't tell you *when* the object was at a given point (x, y) on the path. To determine this time, you can introduce a third variable t, called a **parameter.** It is possible to write both x and y as functions of t to obtain the **parametric equations**

$$x = 24\sqrt{2}\,t \qquad \text{Parametric equation for } x$$

$$y = -16t^2 + 24\sqrt{2}\,t. \qquad \text{Parametric equation for } y$$

From this set of equations you can determine that at time $t = 0$, the object is at the point $(0, 0)$. Similarly, at time $t = 1$, the object is at the point $\left(24\sqrt{2},\ 24\sqrt{2} - 16\right)$, and so on.

What you should learn

- Evaluate sets of parametric equations for given values of the parameter.
- Graph curves that are represented by sets of parametric equations.
- Rewrite sets of parametric equations as single rectangular equations by eliminating the parameter.
- Find sets of parametric equations for graphs.

Why you should learn it

Parametric equations are useful for modeling the path of an object. For instance, in Exercise 59 on page 802, a set of parametric equations is used to model the path of a baseball.

SuperStock

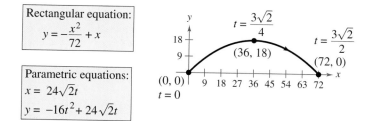

Curvilinear motion: two variables for position, one variable for time
Figure 10.33

For this particular motion problem, x and y are continuous functions of t, and the resulting path is a **plane curve.** (Recall that a *continuous function* is one whose graph can be traced without lifting the pencil from the paper.)

Definition of a Plane Curve

If f and g are continuous functions of t on an interval I, the set of ordered pairs $(f(t), g(t))$ is a **plane curve** C. The equations given by

$$x = f(t) \qquad \text{and} \qquad y = g(t)$$

are **parametric equations** for C, and t is the **parameter.**

STUDY TIP

It is not necessary for the parameter in a set of parametric equations to represent time. The parameter is often used to represent an *angle*, as shown in Example 4.

Sketching a Plane Curve

One way to sketch a curve represented by a pair of parametric equations is to plot points in the xy-plane. Each set of coordinates (x, y) is determined from a value chosen for the parameter t. By plotting the resulting points in the order of *increasing* values of t, you trace the curve in a specific direction. This is called the **orientation** of the curve.

Example 1   Sketching a Plane Curve

Sketch the curve given by the parametric equations

$$x = t^2 - 4 \quad \text{and} \quad y = \frac{t}{2}, \quad -2 \le t \le 3.$$

Describe the orientation of the curve.

Solution

Using values of t in the interval, the parametric equations yield the points (x, y) shown in the table.

t	-2	-1	0	1	2	3
x	0	-3	-4	-3	0	5
y	-1	$-\frac{1}{2}$	0	$\frac{1}{2}$	1	$\frac{3}{2}$

By plotting these points in the order of increasing t, you obtain the curve shown in Figure 10.34. The arrows on the curve indicate its orientation as t increases from -2 to 3. So, if a particle were moving on this curve, it would start at $(0, -1)$ and then move along the curve to the point $\left(5, \frac{3}{2}\right)$.

✓ *Checkpoint* Now try Exercises 7(a) and (b).

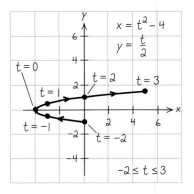

Figure 10.34

Note that the graph shown in Figure 10.34 does not define y as a function of x. This points out one benefit of parametric equations—they can be used to represent graphs that are more general than graphs of functions.

Two different sets of parametric equations can have the same graph. For example, the set of parametric equations

$$x = 4t^2 - 4 \quad \text{and} \quad y = t, \quad -1 \le t \le \frac{3}{2}$$

has the same graph as the set given in Example 1. However, by comparing the values of t in Figures 10.34 and 10.35, you can see that this second graph is traced out more *rapidly* (considering t as time) than the first graph. So, in applications, different parametric representations can be used to represent various *speeds* at which objects travel along a given path.

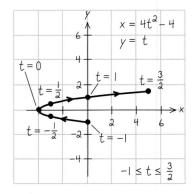

Figure 10.35

TECHNOLOGY TIP Most graphing utilities have a *parametric* mode. So, another way to display a curve represented by a pair of parametric equations is to use a graphing utility, as shown in Example 2. For instructions on how to use the *parametric* mode, see Appendix A; for specifice keystrokes, go to the text website at *college.hmco.com*.

Example 2   Using a Graphing Utility in Parametric Mode

Use a graphing utility to graph the curves represented by the parametric equations. Using the graph and the Vertical Line Test, for which curve is y a function of x?

a. $x = t^2, y = t^3$ **b.** $x = t, y = t^3$ **c.** $x = t^2, y = t$

Solution

Begin by setting the graphing utility to *parametric* mode. When choosing a viewing window, you must set not only minimum and maximum values of x and y, but also minimum and maximum values of t.

a. Enter the parametric equations for x and y, as shown in Figure 10.36. Use the viewing window shown in Figure 10.37. The curve is shown in Figure 10.38. From the graph, you can see that y *is not* a function of x.

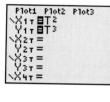
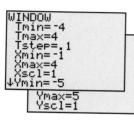

Figure 10.36   Figure 10.37   Figure 10.38

b. Enter the parametric equations for x and y, as shown in Figure 10.39. Use the viewing window shown in Figure 10.40. The curve is shown in Figure 10.41. From the graph, you can see that y *is* a function of x.

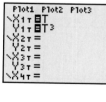
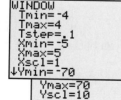
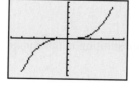

Figure 10.39   Figure 10.40   Figure 10.41

c. Enter the parametric equations for x and y, as shown in Figure 10.42. Use the viewing window shown in Figure 10.43. The curve is shown in Figure 10.44. From the graph, you can see that y *is not* a function of x.

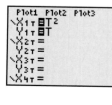
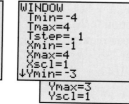
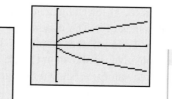

Figure 10.42   Figure 10.43   Figure 10.44

 Checkpoint   Now try Exercise 7(c).

TECHNOLOGY TIP

Notice in Example 2 that in order to set the viewing windows of parametric graphs, you have to scroll down to enter the Ymax and Yscl values.

Eliminating the Parameter

Many curves that are represented by sets of parametric equations have graphs that can also be represented by rectangular equations (in x and y). The process of finding the rectangular equation is called **eliminating the parameter.**

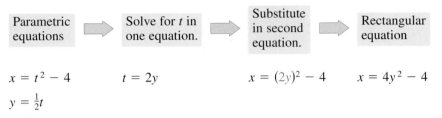

| Parametric equations | → | Solve for t in one equation. | → | Substitute in second equation. | → | Rectangular equation |

$x = t^2 - 4$ $t = 2y$ $x = (2y)^2 - 4$ $x = 4y^2 - 4$

$y = \frac{1}{2}t$

Now you can recognize that the equation $x = 4y^2 - 4$ represents a parabola with a horizontal axis and vertex at $(-4, 0)$.

When converting equations from parametric to rectangular form, you may need to alter the domain of the rectangular equation so that its graph matches the graph of the parametric equations. This situation is demonstrated in Example 3.

Example 3   Eliminating the Parameter

Identify the curve represented by the equations

$$x = \frac{1}{\sqrt{t + 1}} \quad \text{and} \quad y = \frac{t}{t + 1}.$$

Solution

Solving for t in the equation for x produces

$$x^2 = \frac{1}{t + 1} \quad \text{or} \quad \frac{1}{x^2} = t + 1$$

which implies that $t = (1/x^2) - 1$. Substituting in the equation for y, you obtain

$$y = \frac{t}{t + 1}$$

$$= \frac{\left(\frac{1}{x^2}\right) - 1}{\left(\frac{1}{x^2}\right) - 1 + 1} = \frac{\dfrac{1 - x^2}{x^2}}{\left(\dfrac{1}{x^2}\right)} \cdot \frac{x^2}{x^2} = 1 - x^2.$$

From the rectangular equation, you can recognize that the curve is a parabola that opens downward and has its vertex at $(0, 1)$, as shown in Figure 10.45. The rectangular equation is defined for all values of x. The parametric equation for x, however, is defined only when $t > -1$. From the graph of the parametric equations, you can see that x is always positive, as shown in Figure 10.46. So, you should restrict the domain of x to positive values, as shown in Figure 10.47.

✓ *Checkpoint* Now try Exercise 7(d).

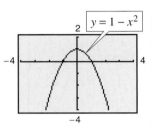

Figure 10.45

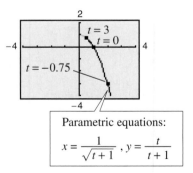

Figure 10.46

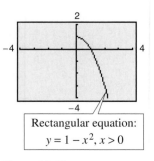

Figure 10.47

Example 4   Eliminating the Parameter

Sketch the curve represented by $x = 3 \cos \theta$ and $y = 4 \sin \theta$, $0 \le \theta \le 2\pi$, by eliminating the parameter.

Solution

Begin by solving for $\cos \theta$ and $\sin \theta$ in the equations.

$$\cos \theta = \frac{x}{3} \quad \text{and} \quad \sin \theta = \frac{y}{4} \qquad \text{Solve for } \cos \theta \text{ and } \sin \theta.$$

Use the identity $\sin^2 \theta + \cos^2 \theta = 1$ to form an equation involving only x and y.

$$\cos^2 \theta + \sin^2 \theta = 1 \qquad \text{Pythagorean identity}$$

$$\left(\frac{x}{3}\right)^2 + \left(\frac{y}{4}\right)^2 = 1 \qquad \text{Substitute } \frac{x}{3} \text{ for } \cos \theta \text{ and } \frac{y}{4} \text{ for } \sin \theta.$$

$$\frac{x^2}{9} + \frac{y^2}{16} = 1 \qquad \text{Rectangular equation}$$

From this rectangular equation, you can see that the graph is an ellipse centered at $(0, 0)$, with vertices $(0, 4)$ and $(0, -4)$, and minor axis of length $2b = 6$, as shown in Figure 10.48. Note that the elliptic curve is traced out *counterclockwise*.

✓ *Checkpoint* Now try Exercise 19.

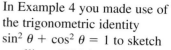

> ### Exploration
>
> In Example 4 you made use of the trigonometric identity $\sin^2 \theta + \cos^2 \theta = 1$ to sketch an ellipse. Which trigonometric identity would you use to obtain the graph of a hyperbola? Sketch the curve represented by $x = 3 \sec \theta$ and $y = 4 \tan \theta$, $0 \le \theta \le 2\pi$, by eliminating the parameter.

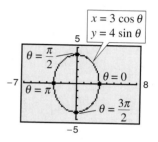

Figure 10.48

Finding Parametric Equations for a Graph

How can you determine a set of parametric equations for a given graph or a given physical description? From the discussion following Example 1, you know that such a representation is not unique. This is further demonstrated in Example 5.

Example 5   Finding Parametric Equations for a Given Graph

Find a set of parametric equations to represent the graph of $y = 1 - x^2$ using the parameters (a) $t = x$ and (b) $t = 1 - x$.

Solution

a. Letting $t = x$, you obtain the following parametric equations.

$$x = t \qquad \text{Parametric equation for } x$$

$$y = 1 - t^2 \qquad \text{Parametric equation for } y$$

The graph of these equations is shown in Figure 10.49.

b. Letting $t = 1 - x$, you obtain the following parametric equations.

$$x = 1 - t \qquad \text{Parametric equation for } x$$

$$y = 1 - (1 - t)^2 = 2t - t^2 \qquad \text{Parametric equation for } y$$

The graph of these equations is shown in Figure 10.50. Note that the graphs in Figures 10.49 and 10.50 have opposite orientations.

✓ *Checkpoint* Now try Exercise 43.

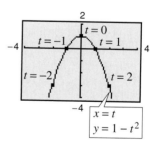

Figure 10.49

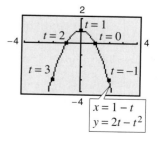

Figure 10.50

10.3 Exercises

Vocabulary Check

Fill in the blanks.

1. If f and g are continuous functions of t on an interval I, the set of ordered pairs $(f(t), g(t))$ is a _____ C. The equations given by $x = f(t)$ and $y = g(t)$ are _____ for C, and t is the _____ .

2. The _____ of a curve is the direction in which the curve is traced out for increasing values of the parameter.

3. The process of converting a set of parametric equations to rectangular form is called _____ the _____ .

In Exercises 1–6, match the set of parametric equations with its graph. [The graphs are labeled (a), (b), (c), (d), (e), and (f).]

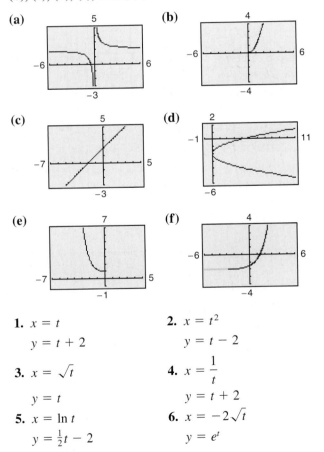

1. $x = t$
   $y = t + 2$

2. $x = t^2$
   $y = t - 2$

3. $x = \sqrt{t}$
   $y = t$

4. $x = \dfrac{1}{t}$
   $y = t + 2$

5. $x = \ln t$
   $y = \frac{1}{2}t - 2$

6. $x = -2\sqrt{t}$
   $y = e^t$

7. Consider the parametric equations $x = \sqrt{t}$ and $y = 2 - t$.

   (a) Create a table of x- and y-values using $t = 0, 1, 2, 3,$ and 4.

   (b) Plot the points (x, y) generated in part (a) and sketch a graph of the parametric equations.

   (c) Use a graphing utility to graph the curve represented by the parametric equations.

   (d) Find the rectangular equation by eliminating the parameter. Sketch its graph. How does the graph differ from those in parts (b) and (c)?

8. Consider the parametric equations $x = 4\cos^2\theta$ and $y = 2\sin\theta$.

   (a) Create a table of x- and y-values using $\theta = -\pi/2, -\pi/4, 0, \pi/4,$ and $\pi/2$.

   (b) Plot the points (x, y) generated in part (a) and sketch a graph of the parametric equations.

   (c) Use a graphing utility to graph the curve represented by the parametric equations.

   (d) Find the rectangular equation by eliminating the parameter. Sketch its graph. How does the graph differ from those in parts (b) and (c)?

In Exercises 9–24, sketch the curve represented by the parametric equations (indicate the orientation of the curve). Use a graphing utility to confirm your result. Then eliminate the parameter and write the corresponding rectangular equation whose graph represents the curve. Adjust the domain of the resulting rectangular equation, if necessary.

9. $x = t$
   $y = -4t$

10. $x = t$
    $y = \frac{1}{2}t$

11. $x = 3t - 3$
    $y = 2t + 1$

12. $x = 3 - 2t$
    $y = 2 + 3t$

13. $x = \frac{1}{4}t$
    $y = t^2$

14. $x = t$
    $y = t^3$

15. $x = t + 2$
    $y = t^2$

16. $x = \sqrt{t}$
    $y = 1 - t$

17. $x = 2t$
    $y = |t - 2|$

18. $x = |t - 1|$
    $y = t + 2$

19. $x = 3 \cos \theta$
    $y = 3 \sin \theta$

20. $x = \cos \theta$
    $y = 3 \sin \theta$

21. $x = e^{-t}$
    $y = e^{3t}$

22. $x = e^{2t}$
    $y = e^t$

23. $x = t^3$
    $y = 3 \ln t$

24. $x = \ln 2t$
    $y = 2t^2$

In Exercises 25–30, use a graphing utility to graph the curve represented by the parametric equations.

25. $x = 4 + 2 \cos \theta$
    $y = -1 + \sin \theta$

26. $x = 4 + 2 \cos \theta$
    $y = -1 + 2 \sin \theta$

27. $x = 4 \sec \theta$
    $y = 3 \tan \theta$

28. $x = \sec \theta$
    $y = \tan \theta$

29. $x = t/2$
    $y = \ln(t^2 + 1)$

30. $x = 10 - 0.01e^t$
    $y = 0.4t^2$

In Exercises 31 and 32, determine how the plane curves differ from each other.

31. (a) $x = t$
    $y = 2t + 1$

(b) $x = \cos \theta$
    $y = 2 \cos \theta + 1$

(c) $x = e^{-t}$
    $y = 2e^{-t} + 1$

(d) $x = e^t$
    $y = 2e^t + 1$

32. (a) $x = 2\sqrt{t}$
    $y = 4 - \sqrt{t}$

(b) $x = 2\sqrt[3]{t}$
    $y = 4 - \sqrt[3]{t}$

(c) $x = 2(t + 1)$
    $y = 3 - t$

(d) $x = -2t^2$
    $y = 4 + t^2$

In Exercises 33–36, eliminate the parameter and obtain the standard form of the rectangular equation.

33. Line through (x_1, y_1) and (x_2, y_2):
    $x = x_1 + t(x_2 - x_1)$
    $y = y_1 + t(y_2 - y_1)$

34. Circle: $x = h + r \cos \theta$
    $y = k + r \sin \theta$

35. Ellipse: $x = h + a \cos \theta$
    $y = k + b \sin \theta$

36. Hyperbola: $x = h + a \sec \theta$
    $y = k + b \tan \theta$

In Exercises 37–42, use the results of Exercises 33–36 to find a set of parametric equations for the line or conic.

37. Line: Passes through $(0, 0)$ and $(5, -2)$

38. Line: Passes through $(1, 4)$ and $(5, -2)$

39. Circle: Center: $(2, 1)$; Radius: 4

40. Circle: Center: $(-3, 1)$; Radius: 3

41. Ellipse: Vertices: $(\pm 5, 0)$; Foci: $(\pm 4, 0)$

42. Hyperbola: Vertices: $(0, \pm 1)$; Foci: $(0, \pm 2)$

In Exercises 43–50, find two different sets of parametric equations for the given rectangular equation.

43. $y = 4x - 3$

44. $y = 5 - 7x$

45. $y = \dfrac{1}{x}$

46. $y = \dfrac{1}{2x}$

47. $y = x^2 + 4$

48. $y = 6x^2 - 5$

49. $y = x^3 + 2x$

50. $y = 1 - 8x^3$

In Exercises 51–54, use a graphing utility to graph the curve represented by the parametric equations.

51. Cycloid: $x = 2(\theta - \sin \theta)$, $y = 2(1 - \cos \theta)$

52. Prolate cycloid: $x = 2\theta - 4 \sin \theta$, $y = 2 - 4 \cos \theta$

53. Witch of Agnesi: $x = 2 \cot \theta$, $y = 2 \sin^2 \theta$

54. Folium of Descartes: $x = \dfrac{3t}{1 + t^3}$, $y = \dfrac{3t^2}{1 + t^3}$

In Exercises 55–58, match the parametric equations with the correct graph. [The graphs are labeled (a), (b), (c), and (d).]

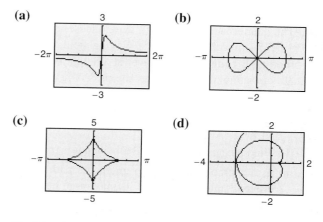

55. Lissajous curve: $x = 2 \cos \theta$, $y = \sin 2\theta$

56. Evolute of ellipse: $x = 2 \cos^3 \theta$, $y = 4 \sin^3 \theta$

57. Involute of circle: $x = \frac{1}{2}(\cos\theta + \theta\sin\theta)$

$\qquad\qquad\qquad\quad y = \frac{1}{2}(\sin\theta - \theta\cos\theta)$

58. Serpentine curve: $x = \frac{1}{2}\cot\theta, y = 4\sin\theta\cos\theta$

Projectile Motion **In Exercises 59 and 60, consider a projectile launched at a height of h feet above the ground at an angle of θ with the horizontal. The initial velocity is v_0 feet per second and the path of the projectile is modeled by the parametric equations**

$$x = (v_0\cos\theta)t \quad \text{and} \quad y = h + (v_0\sin\theta)t - 16t^2.$$

59. The center-field fence in a ballpark is 10 feet high and 400 feet from home plate. A baseball is hit 3 feet above the ground. It leaves the bat at an angle of θ degrees with the horizontal at a speed of 100 miles per hour (see figure).

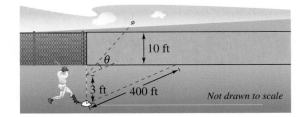

(a) Write a set of parametric equations for the path of the baseball.

(b) Use a graphing utility to graph the path of the baseball for $\theta = 15°$. Is the hit a home run?

(c) Use a graphing utility to graph the path of the baseball for $\theta = 23°$. Is the hit a home run?

(d) Find the minimum angle required for the hit to be a home run.

60. The quarterback of a football team releases a pass at a height of 7 feet above the playing field, and the football is caught by a receiver at a height of 4 feet, 30 yards directly downfield. The pass is released at an angle of $35°$ with the horizontal.

(a) Write a set of parametric equations for the path of the football.

(b) Find the speed of the football when it is released.

(c) Use a graphing utility to graph the path of the football and approximate its maximum height.

(d) Find the time the receiver has to position himself after the quarterback releases the football.

Synthesis

True or False? **In Exercises 61 and 62, determine whether the statement is true or false. Justify your answer.**

61. The two sets of parametric equations $x = t$, $y = t^2 + 1$ and $x = 3t$, $y = 9t^2 + 1$ correspond to the same rectangular equation.

62. The graph of the parametric equations $x = t^2$ and $y = t^2$ is the line $y = x$.

63. As θ increases, the ellipse given by the parametric equations $x = \cos\theta$ and $y = 2\sin\theta$ is traced out *counterclockwise*. Find a parametric representation for which the same ellipse is traced out *clockwise*.

64. ***Think About It*** The graph of the parametric equations $x = t^3$ and $y = t - 1$ is shown below. Would the graph change for the equations $x = (-t)^3$ and $y = -t - 1$? If so, how would it change?

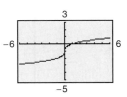

65. ***Writing*** Write a short paragraph explaining why parametric equations are useful.

66. ***Writing*** Explain the process of sketching a plane curve given by parametric equations. What is meant by the orientation of the curve?

Review

In Exercises 67–70, find all solutions of the equation.

67. $5x^2 + 8 = 0$

68. $x^2 - 6x + 4 = 0$

69. $4x^2 + 4x - 11 = 0$

70. $x^4 - 18x^2 + 18 = 0$

In Exercises 71–74, find the sum. Use a graphing utility to verify your result.

71. $\displaystyle\sum_{n=1}^{50} 8n$

72. $\displaystyle\sum_{n=1}^{200} (n - 8)$

73. $\displaystyle\sum_{n=1}^{40}\left(300 - \frac{1}{2}n\right)$

74. $\displaystyle\sum_{n=1}^{70}\frac{7 - 5n}{12}$

10.4 Polar Coordinates

Introduction

What you should learn

● Plot points and find multiple representations of points in the polar coordinate system.

● Convert points from rectangular to polar form and vice versa.

● Convert equations from rectangular to polar form and vice versa.

Why you should learn it

Polar coordinates offer a different mathematical perspective on graphing. For instance, in Exercises 5–12 on page 807, you see that a polar coordinate can be written in more than one way.

So far, you have been representing graphs of equations as collections of points (x, y) on the rectangular coordinate system, where x and y represent the directed distances from the coordinate axes to the point (x, y). In this section, you will study a second coordinate system called the **polar coordinate system.**

To form the polar coordinate system in the plane, fix a point O, called the **pole** (or **origin**), and construct from O an initial ray called the **polar axis,** as shown in Figure 10.51. Then each point P in the plane can be assigned **polar coordinates** (r, θ) as follows.

1. $r = directed\ distance$ from O to P

2. $\theta = directed\ angle$, counterclockwise from the polar axis to segment $\overline{OP}$

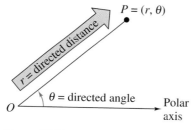

Figure 10.51

Example 1   Plotting Points in the Polar Coordinate System

a. The point $(r, \theta) = (2, \pi/3)$ lies two units from the pole on the terminal side of the angle $\theta = \pi/3$, as shown in Figure 10.52.

b. The point $(r, \theta) = (3, -\pi/6)$ lies three units from the pole on the terminal side of the angle $\theta = -\pi/6$, as shown in Figure 10.53.

c. The point $(r, \theta) = (3, 11\pi/6)$ coincides with the point $(3, -\pi/6)$, as shown in Figure 10.54.

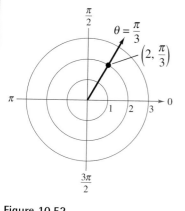

Figure 10.52

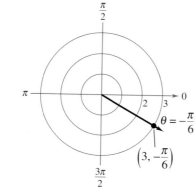

Figure 10.53

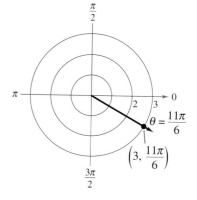

Figure 10.54

✓ **Checkpoint** Now try Exercise 5.

In rectangular coordinates, each point (x, y) has a unique representation. This is not true for polar coordinates. For instance, the coordinates (r, θ) and $(r, \theta + 2\pi)$ represent the same point, as illustrated in Example 1. Another way to obtain multiple representations of a point is to use negative values for r. Because r is a *directed distance*, the coordinates (r, θ) and $(-r, \theta + \pi)$ represent the same point. In general, the point (r, θ) can be represented as

$$(r, \theta) = (r, \theta \pm 2n\pi) \qquad \text{or} \qquad (r, \theta) = (-r, \theta \pm (2n + 1)\pi)$$

where n is any integer. Moreover, the pole is represented by $(0, \theta)$, where θ is any angle.

Example 2 Multiple Representations of Points

Plot the point $(3, -3\pi/4)$ and find three additional polar representations of this point, using $-2\pi < \theta < 2\pi$.

Solution

The point is shown in Figure 10.55. Three other representations are as follows.

$$\left(3, -\frac{3\pi}{4} + 2\pi\right) = \left(3, \frac{5\pi}{4}\right) \qquad \text{Add } 2\pi \text{ to } \theta.$$

$$\left(-3, -\frac{3\pi}{4} - \pi\right) = \left(-3, -\frac{7\pi}{4}\right) \qquad \text{Replace } r \text{ by } -r; \text{ subtract } \pi \text{ from } \theta.$$

$$\left(-3, -\frac{3\pi}{4} + \pi\right) = \left(-3, \frac{\pi}{4}\right) \qquad \text{Replace } r \text{ by } -r; \text{ add } \pi \text{ to } \theta.$$

✓ *Checkpoint* Now try Exercise 7.

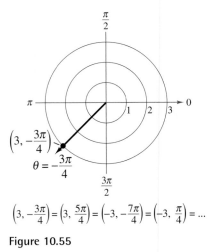

$$\left(3, -\frac{3\pi}{4}\right) = \left(3, \frac{5\pi}{4}\right) = \left(-3, -\frac{7\pi}{4}\right) = \left(-3, \frac{\pi}{4}\right) = \cdots$$

Figure 10.55

Coordinate Conversion

To establish the relationship between polar and rectangular coordinates, let the polar axis coincide with the positive x-axis and the pole with the origin, as shown in Figure 10.56. Because (x, y) lies on a circle of radius r, it follows that $r^2 = x^2 + y^2$. Moreover, for $r > 0$, the definitions of the trigonometric functions imply that

$$\tan \theta = \frac{y}{x}, \qquad \cos \theta = \frac{x}{r}, \qquad \text{and} \qquad \sin \theta = \frac{y}{r}.$$

You can show that the same relationships hold for $r < 0$.

Figure 10.56

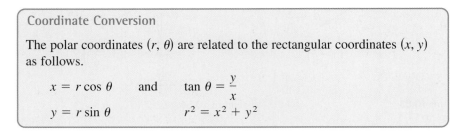

Coordinate Conversion

The polar coordinates (r, θ) are related to the rectangular coordinates (x, y) as follows.

$$x = r \cos \theta \qquad \text{and} \qquad \tan \theta = \frac{y}{x}$$

$$y = r \sin \theta \qquad \qquad r^2 = x^2 + y^2$$

Example 3   Polar-to-Rectangular Conversion

Convert each point to rectangular coordinates.

a. $(2, \pi)$ **b.** $\left(\sqrt{3}, \dfrac{\pi}{6}\right)$

Solution

a. For the point $(r, \theta) = (2, \pi)$, you have the following.

$$x = r \cos \theta = 2 \cos \pi = -2$$
$$y = r \sin \theta = 2 \sin \pi = 0$$

The rectangular coordinates are $(x, y) = (-2, 0)$. (See Figure 10.57.)

b. For the point $(r, \theta) = \left(\sqrt{3}, \pi/6\right)$, you have the following.

$$x = \sqrt{3} \cos \frac{\pi}{6} = \sqrt{3}\left(\frac{\sqrt{3}}{2}\right) = \frac{3}{2}$$

$$y = \sqrt{3} \sin \frac{\pi}{6} = \sqrt{3}\left(\frac{1}{2}\right) = \frac{\sqrt{3}}{2}$$

The rectangular coordinates are $(x, y) = \left(3/2, \sqrt{3}/2\right)$. (See Figure 10.57.)

✓ *Checkpoint* Now try Exercise 13.

Example 4   Rectangular-to-Polar Conversion

Convert each point to polar coordinates.

a. $(-1, 1)$ **b.** $(0, 2)$

Solution

a. For the second-quadrant point $(x, y) = (-1, 1)$, you have

$$\tan \theta = \frac{y}{x} = \frac{1}{-1} = -1$$

$$\theta = \frac{3\pi}{4}.$$

Because θ lies in the same quadrant as (x, y), use positive r.

$$r = \sqrt{x^2 + y^2} = \sqrt{(-1)^2 + (1)^2} = \sqrt{2}$$

So, *one* set of polar coordinates is $(r, \theta) = \left(\sqrt{2}, 3\pi/4\right)$, as shown in Figure 10.58.

b. Because the point $(x, y) = (0, 2)$ lies on the positive y-axis, choose

$$\theta = \pi/2 \quad \text{and} \quad r = 2.$$

This implies that one set of polar coordinates is $(r, \theta) = (2, \pi/2)$, as shown in Figure 10.59.

✓ *Checkpoint* Now try Exercise 27.

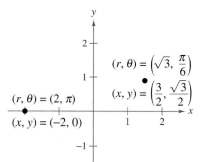

Figure 10.57

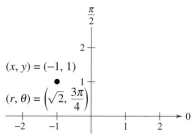

Figure 10.58

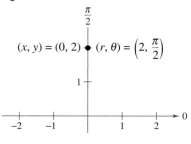

Figure 10.59

Equation Conversion

By comparing Examples 3 and 4, you see that point conversion from the polar to the rectangular system is straightforward, whereas point conversion from the rectangular to the polar system is more involved. For equations, the opposite is true. To convert a rectangular equation to polar form, you simply replace x by $r \cos \theta$ and y by $r \sin \theta$. For instance, the rectangular equation $y = x^2$ can be written in polar form as follows.

$$y = x^2 \qquad \text{Rectangular equation}$$

$$r \sin \theta = (r \cos \theta)^2 \qquad \text{Polar equation}$$

$$r = \sec \theta \tan \theta \qquad \text{Simplest form}$$

On the other hand, converting a polar equation to rectangular form requires considerable ingenuity.

Example 5 demonstrates several polar-to-rectangular conversions that enable you to sketch the graphs of some polar equations.

Example 5 Converting Polar Equations to Rectangular Form

Describe the graph of each polar equation and find the corresponding rectangular equation.

a. $r = 2$ **b.** $\theta = \dfrac{\pi}{3}$ **c.** $r = \sec \theta$

Solution

a. The graph of the polar equation $r = 2$ consists of all points that are two units from the pole. In other words, this graph is a circle centered at the origin with a radius of 2, as shown in Figure 10.60. You can confirm this by converting to rectangular form, using the relationship $r^2 = x^2 + y^2$.

$$r = 2 \quad \Longrightarrow \quad r^2 = 2^2 \quad \Longrightarrow \quad x^2 + y^2 = 2^2$$

Polar equation   Rectangular equation

b. The graph of the polar equation $\theta = \pi/3$ consists of all points on the line that make an angle of $\pi/3$ with the positive x-axis, as shown in Figure 10.61. To convert to rectangular form, you make use of the relationship $\tan \theta = y/x$.

$$\theta = \frac{\pi}{3} \quad \Longrightarrow \quad \tan \theta = \sqrt{3} \quad \Longrightarrow \quad y = \sqrt{3}x$$

Polar equation   Rectangular equation

c. The graph of the polar equation $r = \sec \theta$ is not evident by simple inspection, so you convert to rectangular form by using the relationship $r \cos \theta = x$.

$$r = \sec \theta \quad \Longrightarrow \quad r \cos \theta = 1 \quad \Longrightarrow \quad x = 1$$

Polar equation   Rectangular equation

Now you see that the graph is a vertical line, as shown in Figure 10.62.

✓ *Checkpoint* Now try Exercise 75.

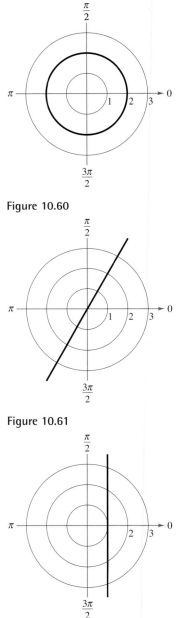

Figure 10.60

Figure 10.61

Figure 10.62

10.4 Exercises

Vocabulary Check

Fill in the blanks.

1. The origin of the polar coordinate system is called the _____ .

2. For the point (r, θ), r is the _____ from O to P and θ is the _____ counterclockwise from the polar axis to segment $\overline{OP}$.

3. To graph the point (r, θ), you use the _____ coordinate system.

In Exercises 1–4, a point in polar coordinates is given. Find the corresponding rectangular coordinates for the point.

1. $\left(4, \dfrac{\pi}{2}\right)$
2. $\left(4, \dfrac{3\pi}{2}\right)$

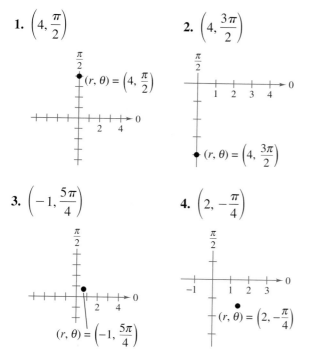

3. $\left(-1, \dfrac{5\pi}{4}\right)$
4. $\left(2, -\dfrac{\pi}{4}\right)$

In Exercises 5–12, plot the point given in polar coordinates and find three additional polar representations of the point, using $-2\pi < \theta < 2\pi$.

5. $\left(3, \dfrac{5\pi}{6}\right)$
6. $\left(2, \dfrac{3\pi}{4}\right)$

7. $\left(-1, -\dfrac{\pi}{3}\right)$
8. $\left(-3, -\dfrac{7\pi}{6}\right)$

9. $\left(\sqrt{3}, \dfrac{5\pi}{6}\right)$
10. $\left(5\sqrt{2}, -\dfrac{11\pi}{6}\right)$

11. $\left(\dfrac{3}{2}, -\dfrac{3\pi}{2}\right)$
12. $\left(0, -\dfrac{\pi}{4}\right)$

In Exercises 13–20, plot the point given in polar coordinates and find the corresponding rectangular coordinates for the point.

13. $\left(4, -\dfrac{\pi}{3}\right)$
14. $\left(2, \dfrac{7\pi}{6}\right)$

15. $\left(-1, -\dfrac{3\pi}{4}\right)$
16. $\left(-3, -\dfrac{2\pi}{3}\right)$

17. $\left(0, -\dfrac{7\pi}{6}\right)$
18. $\left(0, \dfrac{5\pi}{4}\right)$

19. $\left(\sqrt{2}, 2.36\right)$
20. $(-3, -1.57)$

In Exercises 21–24, use a graphing utility to find the rectangular coordinates for the point given in polar coordinates.

21. $\left(2, \dfrac{3\pi}{4}\right)$
22. $\left(-2, \dfrac{7\pi}{6}\right)$

23. $(-4.5, 1.3)$
24. $(8.25, 3.5)$

In Exercises 25–32, plot the point given in rectangular coordinates and find *two* sets of polar coordinates for the point for $0 \le \theta < 2\pi$.

25. $(-7, 0)$
26. $(0, -5)$

27. $(1, 1)$
28. $(-3, -3)$

29. $\left(-\sqrt{3}, -\sqrt{3}\right)$
30. $\left(\sqrt{3}, -1\right)$

31. $(6, 9)$
32. $(5, 12)$

In Exercises 33–38, use a graphing utility to find one set of polar coordinates for the point given in rectangular coordinates.

33. $(3, -2)$
34. $(-4, 1)$

35. $\left(\sqrt{3}, 2\right)$
36. $\left(3\sqrt{2}, 3\sqrt{2}\right)$

37. $\left(\dfrac{5}{2}, \dfrac{4}{3}\right)$
38. $\left(\dfrac{11}{4}, -\dfrac{5}{8}\right)$

In Exercises 39–56, convert the rectangular equation to polar form. Assume $a > 0$.

39. $x^2 + y^2 = 81$

40. $x^2 + y^2 = 16$

41. $y = 4$

42. $y = x$

43. $x = 8$

44. $x = a$

45. $3x - 6y + 2 = 0$

46. $4x + 7y - 2 = 0$

47. $xy = 4$

48. $2xy = 1$

49. $(x^2 + y^2)^2 = 9(x^2 - y^2)$

50. $y^2 - 8x - 16 = 0$

51. $x^2 + y^2 - 6x = 0$

52. $x^2 + y^2 - 8y = 0$

53. $x^2 + y^2 - 2ax = 0$

54. $x^2 + y^2 - 2ay = 0$

55. $y^2 = x^3$

56. $x^2 = y^3$

In Exercises 57–72, convert the polar equation to rectangular form.

57. $r = 6 \sin \theta$

58. $r = 2 \cos \theta$

59. $\theta = \dfrac{4\pi}{3}$

60. $\theta = \dfrac{5\pi}{3}$

61. $r = 4$

62. $r = 10$

63. $r = -3 \csc \theta$

64. $r = 2 \sec \theta$

65. $r^2 = \cos \theta$

66. $r^2 = \sin 2\theta$

67. $r = 2 \sin 3\theta$

68. $r = 3 \cos 2\theta$

69. $r = \dfrac{1}{1 - \cos \theta}$

70. $r = \dfrac{2}{1 + \sin \theta}$

71. $r = \dfrac{6}{2 - 3 \sin \theta}$

72. $r = \dfrac{6}{2 \cos \theta - 3 \sin \theta}$

In Exercises 73–78, describe the graph of the polar equation and find the corresponding rectangular equation. Sketch its graph.

73. $r = 7$

74. $r = 8$

75. $\theta = \dfrac{\pi}{4}$

76. $\theta = \dfrac{7\pi}{6}$

77. $r = 3 \sec \theta$

78. $r = 2 \csc \theta$

Synthesis

True or False?   In Exercises 79 and 80, determine whether the statement is true or false. Justify your answer.

79. If (r_1, θ_1) and (r_2, θ_2) represent the same point in the polar coordinate system, then $|r_1| = |r_2|$.

80. If (r, θ_1) and (r, θ_2) represent the same point in the polar coordinate system, then $\theta_1 = \theta_2 + 2\pi n$ for some integer n.

81. *Think About It*

(a) Show that the distance between the points (r_1, θ_1) and (r_2, θ_2) is

$$\sqrt{r_1^2 + r_2^2 - 2r_1 r_2 \cos(\theta_1 - \theta_2)}.$$

(b) Describe the positions of the points relative to each other for $\theta_1 = \theta_2$. Simplify the Distance Formula for this case. Is the simplification what you expected? Explain.

(c) Simplify the Distance Formula for $\theta_1 - \theta_2 = 90°$. Is the simplification what you expected? Explain.

(d) Choose two points on the polar coordinate system and find the distance between them. Then choose different polar representations of the same two points and apply the Distance Formula again. Discuss the result.

82. *Writing*   Write a short paragraph explaining the differences between the rectangular coordinate system and the polar coordinate system.

Review

In Exercises 83–88, use the Law of Sines or the Law of Cosines to solve the triangle.

83. $a = 13, b = 19, c = 25$

84. $A = 24°, a = 10, b = 6$

85. $A = 56°, C = 38°, c = 12$

86. $B = 71°, a = 21, c = 29$

87. $C = 35°, a = 8, b = 4$

88. $B = 64°, b = 52, c = 44$

In Exercises 89–92, use any method to solve the system of equations.

89. $\begin{cases} 5x - 7y = -11 \\ -3x + y = -3 \end{cases}$

90. $\begin{cases} 3x + 5y = 10 \\ 4x - 2y = -5 \end{cases}$

91. $\begin{cases} 3a - 2b + c = 0 \\ 2a + b - 3c = 0 \\ a - 3b + 9c = 0 \end{cases}$

92. $\begin{cases} 5u + 7v + 9w = 15 \\ u - 2v - 3w = 7 \\ 8u - 2v + w = 0 \end{cases}$

10.5 Graphs of Polar Equations

Introduction

In previous chapters you sketched graphs on rectangular coordinate systems. You began with the basic point-plotting method. Then you used sketching aids such as a graphing utility, symmetry, intercepts, asymptotes, periods, and shifts to further investigate the nature of the graph. This section approaches curve sketching on the polar coordinate system similarly.

Example 1   Graphing a Polar Equation by Point Plotting

Sketch the graph of the polar equation $r = 4 \sin \theta$ by hand.

Solution

The sine function is periodic, so you can get a full range of r-values by considering values of θ in the interval $0 \le \theta \le 2\pi$, as shown in the table.

θ	0	$\dfrac{\pi}{6}$	$\dfrac{\pi}{3}$	$\dfrac{\pi}{2}$	$\dfrac{2\pi}{3}$	$\dfrac{5\pi}{6}$	π	$\dfrac{7\pi}{6}$	$\dfrac{3\pi}{2}$	$\dfrac{11\pi}{6}$	2π
r	0	2	$2\sqrt{3}$	4	$2\sqrt{3}$	2	0	-2	-4	-2	0

By plotting these points as shown in Figure 10.63, it appears that the graph is a circle of radius 2 whose center is the point $(x, y) = (0, 2)$.

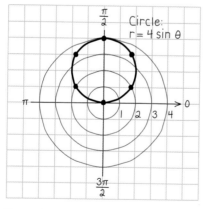

Figure 10.63

✓ *Checkpoint* Now try Exercise 23.

You can confirm the graph found in Example 1 in three ways.

1. *Convert to Rectangular Form*   Multiply each side of the polar equation by r and convert the result to rectangular form.

2. *Use a Polar Coordinate Mode*   Set your graphing utility to *polar* mode and graph the polar equation. (Use $0 \le \theta \le \pi$, $-6 \le x \le 6$, and $-4 \le y \le 4$.)

3. *Use a Parametric Mode*   Set your graphing utility to *parametric* mode and graph $x = (4 \sin t) \cos t$ and $y = (4 \sin t) \sin t$.

Most graphing utilities have a *polar* graphing mode. If yours doesn't, you can rewrite the polar equation $r = f(\theta)$ in parametric form, using t as a parameter, as follows.

$$x = f(t) \cos t \qquad \text{and} \qquad y = f(t) \sin t$$

Symmetry

In Figure 10.63, note that as θ increases from 0 to 2π the graph is traced out twice. Moreover, note that the graph is *symmetric with respect to the line* $\theta = \pi/2$. Had you known about this symmetry and retracing ahead of time, you could have used fewer points. The three important types of symmetry to consider in polar curve sketching are shown in Figure 10.64.

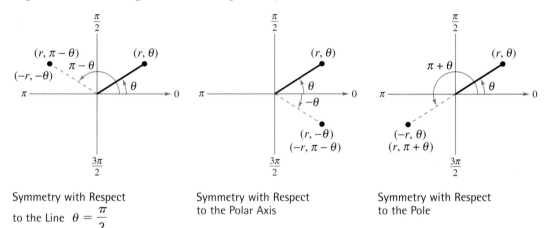

Symmetry with Respect to the Line $\theta = \dfrac{\pi}{2}$

Symmetry with Respect to the Polar Axis

Symmetry with Respect to the Pole

Figure 10.64

Testing for Symmetry in Polar Coordinates

The graph of a polar equation is symmetric with respect to the following if the given substitution yields an equivalent equation.

1. **The line $\theta = \dfrac{\pi}{2}$:**   Replace (r, θ) by $(r, \pi - \theta)$ or $(-r, -\theta)$.

2. **The polar axis:**   Replace (r, θ) by $(r, -\theta)$ or $(-r, \pi - \theta)$.

3. **The pole:**   Replace (r, θ) by $(r, \pi + \theta)$ or $(-r, \theta)$.

You can determine the symmetry of the graph of $r = 2 \sin \theta$ as follows.

1. Replace (r, θ) by $(-r, -\theta)$:

$$-r = 2 \sin(-\theta) \quad \Longrightarrow \quad r = -2 \sin(-\theta) = 2 \sin \theta$$

2. Replace (r, θ) by $(r, -\theta)$: $r = 2 \sin(-\theta) = -2 \sin \theta$

3. Replace (r, θ) by $(-r, \theta)$: $-r = 2 \sin \theta \quad \Longrightarrow \quad r = -2 \sin \theta$

So, the graph of $r = 2 \sin \theta$ is symmetric with respect to the line $\theta = \pi/2$.

STUDY TIP

Recall from Section 5.3 that the sine function is odd. That is, $\sin(-\theta) = -\sin \theta$.

Example 2   Using Symmetry to Sketch a Polar Graph

Use symmetry to sketch the graph of $r = 3 + 2 \cos \theta$ by hand.

Solution

Replacing (r, θ) by $(r, -\theta)$ produces

$$r = 3 + 2 \cos(-\theta) = 3 + 2 \cos \theta. \qquad \cos(-u) = \cos u$$

So, by using the even trigonometric identity, you can conclude that the curve is symmetric with respect to the polar axis. Plotting the points in the table and using polar axis symmetry, you obtain the graph shown in Figure 10.65. This graph is called a **limaçon**.

θ	0	$\dfrac{\pi}{6}$	$\dfrac{\pi}{3}$	$\dfrac{\pi}{2}$	$\dfrac{2\pi}{3}$	$\dfrac{5\pi}{6}$	π
r	5	$3 + \sqrt{3}$	4	3	2	$3 - \sqrt{3}$	1

Use a graphing utility to confirm this graph.

☑ *Checkpoint* Now try Exercise 27.

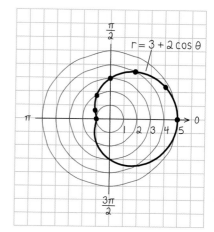

Figure 10.65

The three tests for symmetry in polar coordinates on page 810 are sufficient to guarantee symmetry, but they are not necessary. For instance, Figure 10.66 shows the graph of

$$r = \theta + 2\pi. \qquad \text{Spiral of Archimedes}$$

From the figure, you can see that the graph is symmetric with respect to the line $\theta = \pi/2$. Yet the tests on page 810 fail to indicate symmetry because neither of the following replacements yields an equivalent equation.

Original Equation	*Replacement*	*New Equation*
$r = \theta + 2\pi$	(r, θ) by $(-r, -\theta)$	$-r = -\theta + 2\pi$
$r = \theta + 2\pi$	(r, θ) by $(r, \pi - \theta)$	$r = -\theta + 3\pi$

The equations discussed in Examples 1 and 2 are of the form

$$r = 4 \sin \theta = f(\sin \theta)$$

and

$$r = 3 + 2 \cos \theta = g(\cos \theta).$$

The graph of the first equation is symmetric with respect to the line $\theta = \pi/2$, and the graph of the second equation is symmetric with respect to the polar axis. This observation can be generalized to yield the following *quick tests for symmetry*.

1. The graph of $r = f(\sin \theta)$ is symmetric with respect to the line $\theta = \dfrac{\pi}{2}$.

2. The graph of $r = g(\cos \theta)$ is symmetric with respect to the polar axis.

TECHNOLOGY TIP

The *table* feature of a graphing utility is very useful in constructing tables of values for polar equations. Set your graphing utility to *polar* mode and enter the polar equation in Example 2. You can verify the table of values in Example 2 by starting the table at $\theta = 0$ and incrementing the value of θ by $\pi/6$. For instructions on how to use the *table* feature and *polar* mode, see Appendix A; for specific keystrokes, go to the text website at *college.hmco.com*.

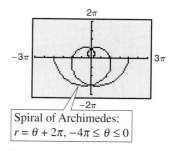

Spiral of Archimedes:
$r = \theta + 2\pi, -4\pi \le \theta \le 0$

Figure 10.66

Zeros and Maximum r-Values

Two additional aids to sketching graphs of polar equations involve knowing the θ-values for which $|r|$ is maximum and knowing the θ-values for which $r = 0$. In Example 1, the maximum value of $|r|$ for $r = 4 \sin \theta$ is $|r| = 4$, and this occurs when $\theta = \pi/2$ (see Figure 10.63). Moreover, $r = 0$ when $\theta = 0$.

Example 3   Finding Maximum r-Values of a Polar Graph

Find the maximum value of r for the graph of $r = 1 - 2 \cos \theta$.

Graphical Solution

Because the polar equation is of the form

$$r = 1 - 2 \cos \theta = g(\cos \theta)$$

you know the graph is symmetric with respect to the polar axis. You can confirm this by graphing the polar equation. Set your graphing utility to *polar* mode and enter the equation, as shown in Figure 10.67. (In the graph, θ varies from 0 to 2π.) To find the maximum r-value for the graph, use your graphing utility's *trace* feature and you should find that the graph has a maximum r-value of 3, as shown in Figure 10.68. This value of r occurs when $\theta = \pi$. In the graph, note that the point $(3, \pi)$ is farthest from the pole.

Numerical Solution

To approximate the maximum value of r for the graph of $r = 1 - 2 \cos \theta$, use the *table* feature of a graphing utility to create a table that begins at $\theta = 0$ and increments by $\pi/12$, as shown in Figure 10.69. From the table, the maximum value of r appears to be 3 when $\theta = 3.1416 \approx \pi$. By creating a second table that begins at $\theta = \pi/2$ and increments by $\pi/24$, as shown in Figure 10.70, the maximum value of r still appears to be 3 when $\theta = 3.1416 \approx \pi$.

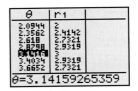

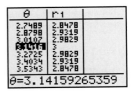

Figure 10.69

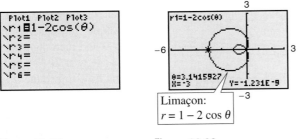

Limaçon:
$r = 1 - 2 \cos \theta$

Figure 10.67   Figure 10.68

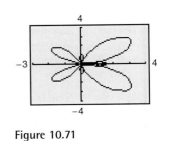

Figure 10.70

Note how the negative r-values determine the *inner loop* of the graph in Figure 10.68. This type of graph is a limaçon.

✓ *Checkpoint* Now try Exercise 15.

Exploration

The graph of the polar equation $r = e^{\cos \theta} - 2 \cos 4\theta + \sin^5(\theta/12)$ is called *the butterfly curve*, as shown in Figure 10.71.

a. The graph at the right was produced using $0 \le \theta \le 2\pi$. Does this show the entire graph? Explain your reasoning.

b. Use the *trace* feature of your graphing utility to approximate the maximum r-value of the graph. Does this value change if you use $0 \le \theta \le 4\pi$ instead of $0 \le \theta \le 2\pi$? Explain.

Figure 10.71

Some curves reach their zeros and maximum r-values at more than one point, as shown in Example 4.

Example 4   Analyzing a Polar Graph

Analyze the graph of $r = 2 \cos 3\theta$.

Solution

Symmetry: With respect to the polar axis

Maximum value of $|r|$: $|r| = 2$ when $3\theta = 0, \pi, 2\pi, 3\pi$

          or $\theta = 0, \pi/3, 2\pi/3, \pi$

Zeros of r: $r = 0$ when $3\theta = \pi/2, 3\pi/2, 5\pi/2$

          or $\theta = \pi/6, \pi/2, 5\pi/6$

θ	0	$\dfrac{\pi}{12}$	$\dfrac{\pi}{6}$	$\dfrac{\pi}{4}$	$\dfrac{\pi}{3}$	$\dfrac{5\pi}{12}$	$\dfrac{\pi}{2}$
r	2	$\sqrt{2}$	0	$-\sqrt{2}$	-2	$-\sqrt{2}$	0

By plotting these points and using the specified symmetry, zeros, and maximum values, you can obtain the graph shown in Figure 10.72. This graph is called a **rose curve,** and each loop on the graph is called a *petal*. Note how the entire curve is generated as θ increases from 0 to π.

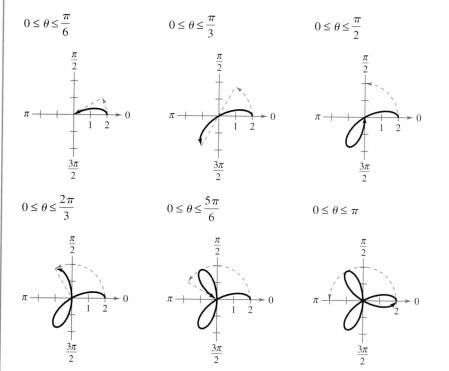

Exploration

Notice that the rose curve in Example 4 has three petals. How many petals do the rose curves $r = 2 \cos 4\theta$ and $r = 2 \sin 3\theta$ have? Determine the numbers of petals for the curves $r = 2 \cos n\theta$ and $r = 2 \sin n\theta$, where n is a positive integer.

Figure 10.72

✓ *Checkpoint* Now try Exercise 31.

Special Polar Graphs

Several important types of graphs have equations that are simpler in polar form than in rectangular form. For example, the circle

$$r = 4 \sin \theta$$

in Example 1 has the more complicated rectangular equation

$$x^2 + (y - 2)^2 = 4.$$

Several other types of graphs that have simple polar equations are shown below.

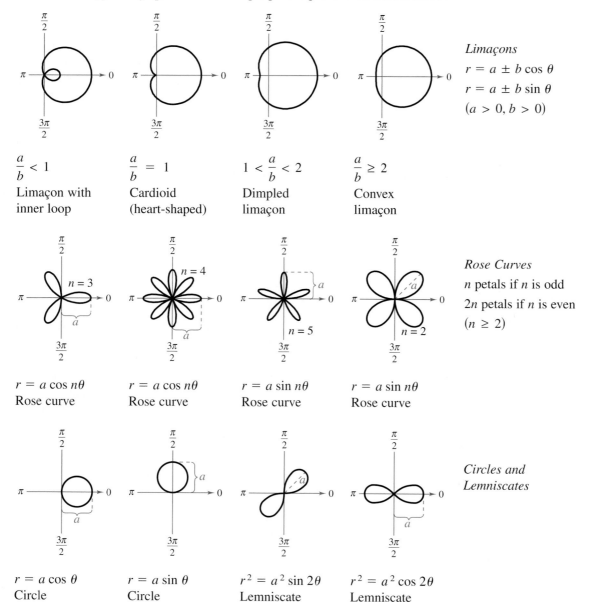

Limaçons
$r = a \pm b \cos \theta$
$r = a \pm b \sin \theta$
$(a > 0, b > 0)$

$\dfrac{a}{b} < 1$

Limaçon with
inner loop

$\dfrac{a}{b} = 1$

Cardioid
(heart-shaped)

$1 < \dfrac{a}{b} < 2$

Dimpled
limaçon

$\dfrac{a}{b} \geq 2$

Convex
limaçon

Rose Curves
n petals if n is odd
$2n$ petals if n is even
$(n \geq 2)$

$n = 3$

$r = a \cos n\theta$
Rose curve

$n = 4$

$r = a \cos n\theta$
Rose curve

$n = 5$

$r = a \sin n\theta$
Rose curve

$n = 2$

$r = a \sin n\theta$
Rose curve

Circles and Lemniscates

$r = a \cos \theta$
Circle

$r = a \sin \theta$
Circle

$r^2 = a^2 \sin 2\theta$
Lemniscate

$r^2 = a^2 \cos 2\theta$
Lemniscate

Example 5   Analyzing a Rose Curve

Analyze the graph of $r = 3 \cos 2\theta$.

Solution

Type of curve: Rose curve with $2n = 4$ petals

Symmetry: With respect to the polar axis, the line $\theta = \pi/2$, and
                          the pole

Maximum value of $|r|$: $|r| = 3$ when $\theta = 0, \pi/2, \pi, 3\pi/2$

Zeros of r: $r = 0$ when $\theta = \pi/4, 3\pi/4$

Using a graphing utility, enter the equation as shown in Figure 10.73 (with
$0 \le \theta \le 2\pi$). You should obtain the graph shown in Figure 10.74.

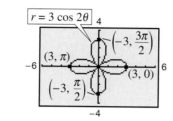

Figure 10.73   Figure 10.74

✓ *Checkpoint* Now try Exercise 35.

Example 6   Analyzing a Lemniscate

Analyze the graph of $r^2 = 9 \sin 2\theta$.

Solution

Type of curve: Lemniscate

Symmetry: With respect to the pole

Maximum value of $|r|$: $|r| = 3$ when $\theta = \pi/4$

Zeros of r: $r = 0$ when $\theta = 0, \pi/2$

Using a graphing utility, enter the equation as shown in Figure 10.75 (with
$0 \le \theta \le 2\pi$). You should obtain the graph shown in Figure 10.76.

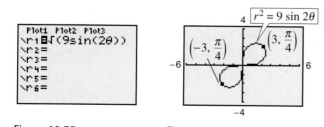

Figure 10.75   Figure 10.76

✓ *Checkpoint* Now try Exercise 43.

10.5 | Exercises

Vocabulary Check

Fill in the blanks.

1. The graph of $r = f(\sin \theta)$ is symmetric with respect to the line _____ .
2. The graph of $r = g(\cos \theta)$ is symmetric with respect to the _____ .
3. The equation $r = 2 + \cos \theta$ represents a _____ .
4. The equation $r = 2 \cos \theta$ represents a _____ .
5. The equation $r^2 = 4 \sin 2\theta$ represents a _____ .
6. The equation $r = 1 + \sin \theta$ represents a _____ .

In Exercises 1–6, identify the type of polar graph.

1.

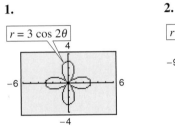

$r = 3 \cos 2\theta$

2.

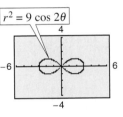

$r = 5 - 5 \sin \theta$

3.

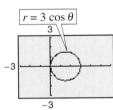

$r = 3 \cos \theta$

4.

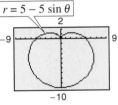

$r^2 = 9 \cos 2\theta$

5.

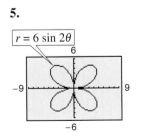

$r = 6 \sin 2\theta$

6.

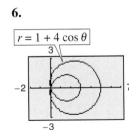

$r = 1 + 4 \cos \theta$

In Exercises 7–14, test for symmetry with respect to $\theta = \pi/2$, the polar axis, and the pole.

7. $r = 10 + 4 \cos \theta$

8. $r = 16 \cos 3\theta$

9. $r = \dfrac{4}{1 + \sin \theta}$

10. $r = \dfrac{2}{1 - \cos \theta}$

11. $r = 6 \sin \theta$

12. $r = 2 \csc \theta \cos \theta$

13. $r^2 = 25 \sin 2\theta$

14. $r^2 = 36 \cos 4\theta$

In Exercises 15–18, find the maximum value of $|r|$ and any zeros of r. Verify your answers numerically.

15. $r = 10 - 10 \sin \theta$

16. $r = 6 + 12 \cos \theta$

17. $r = 4 \cos 3\theta$

18. $r = \sin 2\theta$

In Exercises 19–34, sketch the graph of the polar equation. Use a graphing utility to verify your graph.

19. $r = 4$

20. $r = 2$

21. $\theta = \dfrac{\pi}{4}$

22. $\theta = -\dfrac{5\pi}{3}$

23. $r = 3 \sin \theta$

24. $r = 2 \cos \theta$

25. $r = 3(1 - \cos \theta)$

26. $r = 4(1 + \sin \theta)$

27. $r = 3 - 4 \cos \theta$

28. $r = 1 - 2 \sin \theta$

29. $r = 3 + 6 \sin \theta$

30. $r = 4 + 5 \cos \theta$

31. $r = 5 \cos 3\theta$

32. $r = -\sin 5\theta$

33. $r = 7 \sin 2\theta$

34. $r = 3 \cos 5\theta$

In Exercises 35–48, use a graphing utility to graph the polar equation. Describe your viewing window.

35. $r = 6 \cos 2\theta$

36. $r = \cos 2\theta$

37. $r = 2(3 - \sin \theta)$

38. $r = 6 - 4 \sin \theta$

39. $r = 3 - 6 \cos \theta$

40. $r = 3 - 2 \sin \theta$

41. $r = \dfrac{3}{\sin \theta - 2 \cos \theta}$

42. $r = \dfrac{6}{2 \sin \theta - 3 \cos \theta}$

43. $r^2 = 4 \cos 2\theta$

44. $r^2 = 9 \sin \theta$

45. $r = 4 \sin \theta \cos^2 \theta$

46. $r = 2 \cos(3\theta - 2)$

47. $r = 2 \csc \theta + 5$

48. $r = 4 - \sec \theta$

In Exercises 49–54, use a graphing utility to graph the polar equation. Find an interval for θ for which the graph is traced *only once*.

49. $r = 3 - 2 \cos \theta$

50. $r = 2(1 - 2 \sin \theta)$

51. $r = 2 \cos\left(\dfrac{3\theta}{2}\right)$

52. $r = 3 \sin\left(\dfrac{5\theta}{2}\right)$

53. $r^2 = \sin 2\theta$

54. $r^2 = \dfrac{1}{\theta}$

In Exercises 55–58, use a graphing utility to graph the polar equation and show that the indicated line is an asymptote of the graph.

Name of Graph	Polar Equation	Asymptote
55. Conchoid	$r = 2 - \sec \theta$	$x = -1$
56. Conchoid	$r = 2 + \csc \theta$	$y = 1$
57. Hyperbolic spiral	$r = \dfrac{2}{\theta}$	$y = 2$
58. Strophoid	$r = 2 \cos 2\theta \sec \theta$	$x = -2$

Synthesis

True or False? **In Exercises 59 and 60, determine whether the statement is true or false. Justify your answer.**

59. The graph of $r = 10 \sin 5\theta$ is a rose curve with 5 petals.

60. A rose curve will always have symmetry with respect to the line $\theta = \pi/2$.

61. ***Writing*** Use a graphing utility to graph the polar equation

$$r = \cos 5\theta + n \cos \theta, \quad 0 \le \theta < \pi$$

for the integers $n = -5$ to $n = 5$. As you graph these equations, you should see the graph change shape from a heart to a bell. Write a short paragraph explaining what values of n produce the heart portion of the curve and what values of n produce the bell.

62. The graph of $r = f(\theta)$ is rotated about the pole through an angle ϕ. Show that the equation of the rotated graph is $r = f(\theta - \phi)$.

63. Consider the graph of $r = f(\sin \theta)$.

(a) Show that if the graph is rotated counterclockwise $\pi/2$ radians about the pole, the equation of the rotated graph is $r = f(-\cos \theta)$.

(b) Show that if the graph is rotated counterclockwise π radians about the pole, the equation of the rotated graph is $r = f(-\sin \theta)$.

(c) Show that if the graph is rotated counterclockwise $3\pi/2$ radians about the pole, the equation of the rotated graph is $r = f(\cos \theta)$.

In Exercises 64–66, use the results of Exercises 62 and 63.

64. Write an equation for the limaçon $r = 2 - \sin \theta$ after it has been rotated through the given angle.

(a) $\dfrac{\pi}{4}$ (b) $\dfrac{\pi}{2}$ (c) π (d) $\dfrac{3\pi}{2}$

65. Write an equation for the rose curve $r = 2 \sin 2\theta$ after it has been rotated through the given angle.

(a) $\dfrac{\pi}{6}$ (b) $\dfrac{\pi}{2}$ (c) $\dfrac{2\pi}{3}$ (d) π

66. Sketch the graph of each equation.

(a) $r = 1 - \sin \theta$ (b) $r = 1 - \sin\left(\theta - \dfrac{\pi}{4}\right)$

67. ***Exploration*** Use a graphing utility to graph the polar equation $r = 2 + k \cos \theta$ for $k = 0$, $k = 1$, $k = 2$, and $k = 3$. Identify each graph.

68. ***Exploration*** Consider the polar equation $r = 3 \sin k\theta$.

(a) Use a graphing utility to graph the equation for $k = 1.5$. Find the interval for θ for which the graph is traced only once.

(b) Use a graphing utility to graph the equation for $k = 2.5$. Find the interval for θ for which the graph is traced only once.

(c) Is it possible to find an interval for θ for which the graph is traced only once for any rational number k? Explain.

Review

In Exercises 69–74, sketch the graph of the function. (Include two full periods.)

69. $f(x) = \frac{1}{2} \sin 2x$

70. $f(x) = \cos\left(x - \dfrac{3\pi}{4}\right)$

71. $g(x) = \tan \dfrac{x}{3}$

72. $g(x) = 1 + \csc x$

73. $h(x) = \sec\left(x + \dfrac{\pi}{4}\right)$

74. $h(x) = 2 \cot x$

10.6 Polar Equations of Conics

What you should learn

- Define conics in terms of eccentricities.
- Write and graph equations of conics in polar form.
- Use equations of conics in polar form to model real-life problems.

Why you should learn it

The orbits of planets and satellites can be modeled by polar equations. For instance, in Exercise 45 on page 823, you will use a polar equation to model the orbits of Neptune and Pluto.

Alternative Definition of Conics

In Section 10.1, you learned that the rectangular equations of ellipses and hyperbolas take simple forms when the origin lies at the *center*. As it happens, there are many important applications of conics in which it is more convenient to use one of the *foci* as the origin. In this section you will learn that polar equations of conics take simple forms if one of the foci lies at the pole.

To begin, consider the following alternative definition of a conic that uses the concept of eccentricity (a measure of the flatness of the conic).

Alternative Definition of a Conic

The locus of a point in the plane which moves so that its distance from a fixed point (**focus**) is in a constant ratio to its distance from a fixed line (**directrix**) is a **conic**. The constant ratio is the **eccentricity** of the conic and is denoted by e. Moreover, the conic is an **ellipse** if $e < 1$, a **parabola** if $e = 1$, and a **hyperbola** if $e > 1$. (See Figure 10.77.)

Kevin Kelley/Getty Images

In Figure 10.77, note that for each type of conic, the focus is at the pole.

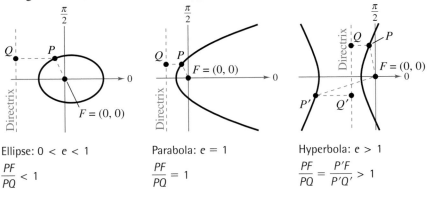

Ellipse: $0 < e < 1$
$$\frac{PF}{PQ} < 1$$

Parabola: $e = 1$
$$\frac{PF}{PQ} = 1$$

Hyperbola: $e > 1$
$$\frac{PF}{PQ} = \frac{P'F}{P'Q'} > 1$$

Figure 10.77

Polar Equations of Conics

The benefit of locating a focus of a conic at the pole is that the equation of the conic becomes simpler. See Appendix B for a proof of the polar equations of conics.

Polar Equations of Conics

The graph of a polar equation of the form

$$\textbf{1. } r = \frac{ep}{1 \pm e \cos \theta} \quad \text{or} \quad \textbf{2. } r = \frac{ep}{1 \pm e \sin \theta}$$

is a conic, where $e > 0$ is the eccentricity and $|p|$ is the distance between the focus (pole) and the directrix.

Equations of the form

$$r = \frac{ep}{1 \pm e \cos \theta} \qquad \text{Vertical directrix}$$

correspond to conics with vertical directrices and equations of the form

$$r = \frac{ep}{1 \pm e \sin \theta} \qquad \text{Horizontal directrix}$$

correspond to conics with horizontal directrices. Moreover, the converse is also true—that is, any conic with a focus at the pole and having a horizontal or vertical directrix can be represented by one of the given equations.

Example 1   Identifying a Conic from Its Equation

Identify the type of conic represented by the equation $r = \dfrac{15}{3 - 2 \cos \theta}$.

Algebraic Solution

To identify the type of conic, rewrite the equation in the form $r = ep/(1 \pm e \cos \theta)$.

$$r = \frac{15}{3 - 2 \cos \theta}$$

$$= \frac{5}{1 - (2/3) \cos \theta} \qquad \begin{array}{l}\text{Divide numerator} \\ \text{and denominator by 3.}\end{array}$$

Because $e = \frac{2}{3} < 1$, you can conclude that the graph is an ellipse.

✓ *Checkpoint* Now try Exercise 9.

Graphical Solution

Use a graphing utility in *polar* mode to graph $r = \dfrac{15}{3 - 2 \cos \theta}$. Be sure to use a square setting. From the graph in Figure 10.78, you can see that the conic appears to be an ellipse.

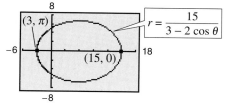

Figure 10.78

For the ellipse in Figure 10.78, the major axis is horizontal and the vertices lie at $(r, \theta) = (15, 0)$ and $(r, \theta) = (3, \pi)$. So, the length of the *major* axis is $2a = 18$. To find the length of the *minor* axis, you can use the equations $e = c/a$ and $b^2 = a^2 - c^2$ to conclude that

$$b^2 = a^2 - c^2$$

$$= a^2 - (ea)^2$$

$$= a^2(1 - e^2). \qquad \text{Ellipse}$$

Because $e = \frac{2}{3}$, you have $b^2 = 9^2 \left[1 - \left(\frac{2}{3} \right)^2 \right] = 45$, which implies that $b = \sqrt{45} = 3\sqrt{5}$. So, the length of the minor axis is $2b = 6\sqrt{5}$. A similar analysis for hyperbolas yields

$$b^2 = c^2 - a^2$$

$$= (ea)^2 - a^2$$

$$= a^2(e^2 - 1). \qquad \text{Hyperbola}$$

STUDY TIP

Because

$$r = \frac{ep}{1 \pm e \cos \theta}$$

is of the form $r = g(\cos \theta)$, you know that the graph of r is symmetric with respect to the polar axis. Similarly,

$$r = \frac{ep}{1 \pm e \sin \theta}$$

is of the form $r = f(\sin \theta)$, so the graph of r is symmetric with respect to the line $\theta = \pi/2$.

Example 2 Analyzing the Graph of a Polar Equation

Analyze the graph of the polar equation

$$r = \frac{32}{3 + 5 \sin \theta}.$$

Solution

Dividing the numerator and denominator by 3 produces

$$r = \frac{32/3}{1 + (5/3) \sin \theta}.$$

Because $e = \frac{5}{3} > 1$, the graph is a hyperbola. The transverse axis of the hyperbola lies on the line $\theta = \pi/2$ and the vertices occur at $(r, \theta) = (4, \pi/2)$ and $(r, \theta) = (-16, 3\pi/2)$. Because the length of the transverse axis is 12, you can see that $a = 6$. To find b, write

$$b^2 = a^2(e^2 - 1) = 6^2 \left[\left(\frac{5}{3} \right)^2 - 1 \right] = 64.$$

So, $b = 8$. The asymptotes are $y = 10 \pm \frac{3}{4}x$, as shown in Figure 10.79.

✓ *Checkpoint* Now try Exercise 21.

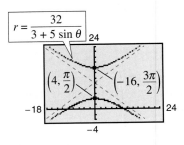

Figure 10.79

In the next example, you are asked to find a polar equation for a specified conic. To do this, let p be the distance between the pole and the directrix.

1. Horizontal directrix above the pole: $r = \dfrac{ep}{1 + e \sin \theta}$

2. Horizontal directrix below the pole: $r = \dfrac{ep}{1 - e \sin \theta}$

3. Vertical directrix to the right of the pole: $r = \dfrac{ep}{1 + e \cos \theta}$

4. Vertical directrix to the left of the pole: $r = \dfrac{ep}{1 - e \cos \theta}$

Example 3 Finding the Polar Equation of a Conic

Find the polar equation of the parabola whose focus is the pole and whose directrix is the line $y = 3$.

Solution

From Figure 10.80, you can see that the directrix is horizontal and above the pole. Moreover, because the eccentricity of a parabola is $e = 1$ and the distance between the pole and the directrix is $p = 3$, you have the equation

$$r = \frac{ep}{1 + e \sin \theta} = \frac{3}{1 + \sin \theta}.$$

✓ *Checkpoint* Now try Exercise 27.

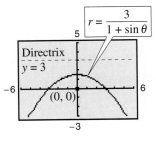

Figure 10.80

Application

Kepler's Laws (listed below), named after the German astronomer Johannes Kepler (1571–1630), can be used to describe the orbits of the planets about the sun.

1. Each planet moves in an elliptical orbit with the sun as a focus.
2. A ray from the sun to the planet sweeps out equal areas of the ellipse in equal times.
3. The square of the period (the time it takes for a planet to orbit the sun) is proportional to the cube of the mean distance between the planet and the sun.

Although Kepler simply stated these laws on the basis of observation, they were later validated by Isaac Newton (1642–1727). In fact, Newton was able to show that each law can be deduced from a set of universal laws of motion and gravitation that govern the movement of all heavenly bodies, including comets and satellites. This is illustrated in the next example, which involves the comet named after the English mathematician and physicist Edmund Halley (1656–1742).

   If you use Earth as a reference with a period of 1 year and a distance of 1 astronomical unit (an *astronomical unit* is defined as the mean distance between Earth and the sun, or about 93 million miles), the proportionality constant in Kepler's third law is 1. For example, because Mars has a mean distance to the sun of $d \approx 1.524$ astronomical units, its period P is given by $d^3 = P^2$. So, the period of Mars is $P \approx 1.88$ years.

Example 4   Halley's Comet

Halley's comet has an elliptical orbit with an eccentricity of $e \approx 0.967$. The length of the major axis of the orbit is approximately 35.88 astronomical units. Find a polar equation for the orbit. How close does Halley's comet come to the sun?

Solution

Using a vertical axis, as shown in Figure 10.81, choose an equation of the form $r = ep/(1 + e \sin \theta)$. Because the vertices of the ellipse occur at $\theta = \pi/2$ and $\theta = 3\pi/2$, you can determine the length of the major axis to be the sum of the r-values of the vertices. That is,

$$2a = \frac{0.967p}{1 + 0.967} + \frac{0.967p}{1 - 0.967} \approx 29.79p \approx 35.88.$$

So, $p \approx 1.204$ and $ep \approx (0.967)(1.204) \approx 1.164$. Using this value of ep in the equation, you have

$$r = \frac{1.164}{1 + 0.967 \sin \theta}$$

where r is measured in astronomical units. To find the closest point to the sun (the focus), substitute $\theta = \pi/2$ into this equation to obtain

$$r = \frac{1.164}{1 + 0.967 \sin(\pi/2)} \approx 0.59 \text{ astronomical units} \approx 55{,}000{,}000 \text{ miles}.$$

 Checkpoint Now try Exercise 41.

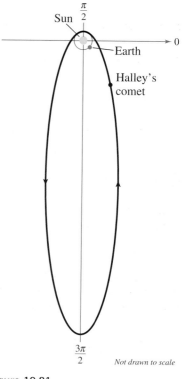

Figure 10.81

10.6   Exercises

Vocabulary Check

In Exercises 1 and 2, fill in the blanks.

1. The locus of a point in the plane which moves so that its distance from a fixed point (focus) is in a constant ratio to its distance from a fixed line (directrix) is a _____ .

2. The constant ratio is the _____ of the conic and is denoted by _____ .

3. Match the conic with its eccentricity.

(a) $e < 1$ (i) ellipse

(b) $e = 1$ (ii) hyperbola

(c) $e > 1$ (iii) parabola

Graphical Reasoning   **In Exercises 1–4, use a graphing utility to graph the polar equation when (a) $e = 1$, (b) $e = 0.5$, and (c) $e = 1.5$. Identify the conic for each equation.**

1. $r = \dfrac{2e}{1 + e \cos \theta}$

2. $r = \dfrac{2e}{1 - e \cos \theta}$

3. $r = \dfrac{2e}{1 - e \sin \theta}$

4. $r = \dfrac{2e}{1 + e \sin \theta}$

In Exercises 5–8, match the polar equation with its graph. [The graphs are labeled (a), (b), (c), and (d).]

(a)

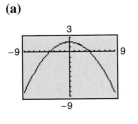

(b)

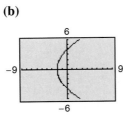

(c)

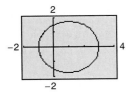

(d)

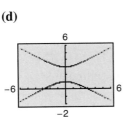

5. $r = \dfrac{4}{1 - \cos \theta}$

6. $r = \dfrac{3}{2 - \cos \theta}$

7. $r = \dfrac{3}{1 + 2 \sin \theta}$

8. $r = \dfrac{4}{1 + \sin \theta}$

In Exercises 9–18, identify the conic represented by the equation algebraically. Use a graphing utility to confirm your result.

9. $r = \dfrac{3}{1 - \cos \theta}$

10. $r = \dfrac{2}{1 + \sin \theta}$

11. $r = \dfrac{4}{4 - \cos \theta}$

12. $r = \dfrac{7}{7 + \sin \theta}$

13. $r = \dfrac{8}{4 + 3 \sin \theta}$

14. $r = \dfrac{6}{3 - 2 \cos \theta}$

15. $r = \dfrac{6}{2 + \sin \theta}$

16. $r = \dfrac{5}{-1 + 2 \cos \theta}$

17. $r = \dfrac{3}{4 - 8 \cos \theta}$

18. $r = \dfrac{10}{3 + 9 \sin \theta}$

In Exercises 19–22, use a graphing utility to graph the polar equation. Identify the graph.

19. $r = \dfrac{-5}{1 - \sin \theta}$

20. $r = \dfrac{-1}{2 + 4 \sin \theta}$

21. $r = \dfrac{14}{14 + 17 \sin \theta}$

22. $r = \dfrac{12}{2 - \cos \theta}$

In Exercises 23–26, use a graphing utility to graph the rotated conic.

23. $r = \dfrac{3}{1 - \cos(\theta - \pi/4)}$ (See Exercise 9.)

24. $r = \dfrac{7}{7 + \sin(\theta - \pi/3)}$ (See Exercise 12.)

25. $r = \dfrac{8}{4 + 3\sin(\theta + \pi/6)}$ (See Exercise 13.)

26. $r = \dfrac{5}{-1 + 2\cos(\theta + 2\pi/3)}$ (See Exercise 16.)

In Exercises 27–38, find a polar equation of the conic with its focus at the pole.

	Conic	Eccentricity	Directrix
27.	Parabola	$e = 1$	$x = -1$
28.	Parabola	$e = 1$	$y = -4$
29.	Ellipse	$e = \frac{1}{2}$	$y = 1$
30.	Ellipse	$e = \frac{3}{4}$	$y = -4$
31.	Hyperbola	$e = 2$	$x = 1$
32.	Hyperbola	$e = \frac{3}{2}$	$x = -1$

	Conic	Vertex or Vertices
33.	Parabola	$\left(1, -\dfrac{\pi}{2}\right)$
34.	Parabola	$(8, 0)$
35.	Ellipse	$(2, 0), (10, \pi)$
36.	Ellipse	$\left(2, \dfrac{\pi}{2}\right), \left(4, \dfrac{3\pi}{2}\right)$
37.	Hyperbola	$\left(1, \dfrac{3\pi}{2}\right), \left(9, \dfrac{3\pi}{2}\right)$
38.	Hyperbola	$\left(4, \dfrac{\pi}{2}\right), \left(-1, \dfrac{3\pi}{2}\right)$

39. *Planetary Motion*   The planets travel in elliptical orbits with the sun at one focus. Assume that the focus is at the pole, the major axis lies on the polar axis, and the length of the major axis is $2a$ (see figure). Show that the polar equation of the orbit of a planet is

$$r = \frac{(1 - e^2)a}{1 - e\cos\theta}$$

where e is the eccentricity.

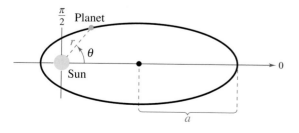

40. *Planetary Motion*   Use the result of Exercise 39 to show that the minimum distance (*perihelion*) from the sun to a planet is

$$r = a(1 - e)$$

and that the maximum distance (*aphelion*) is

$$r = a(1 + e).$$

Planetary Motion   **In Exercises 41–44, use the results of Exercises 39 and 40 to find the polar equation of the orbit of the planet and the perihelion and aphelion distances.**

41.	Earth	$a = 92.956 \times 10^6$ miles
		$e = 0.0167$
42.	Mercury	$a = 35.983 \times 10^6$ miles
		$e = 0.2056$
43.	Jupiter	$a = 77.841 \times 10^7$ kilometers
		$e = 0.0484$
44.	Saturn	$a = 142.673 \times 10^7$ kilometers
		$e = 0.0542$

45. *Planetary Motion*   Use the results of Exercises 39 and 40, where for the planet Neptune, $a = 4.498 \times 10^9$ kilometers and $e = 0.0086$ and for the planet Pluto, $a = 5.906 \times 10^9$ kilometers and $e = 0.2488$.

(a) Find the polar equation of the orbit of each planet.

(b) Find the perihelion and aphelion distances for each planet.

(c) Use a graphing utility to graph both Neptune's and Pluto's equation of orbit in the same viewing window.

(d) Do the orbits of the two planets intersect? Will the two planets ever collide? Why or why not?

(e) Is Pluto ever closer to the sun than Neptune? Why is Pluto called the ninth planet and Neptune the eighth planet?

46. *Explorer 18*   On November 27, 1963, the United States launched *Explorer 18*. Its low and high points above the surface of Earth were 119 miles and 122,800 miles, respectively (see figure). The center of Earth is at one focus of the orbit.

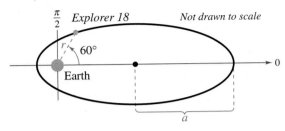

(a) Find the polar equation for the orbit (assume the radius of Earth is 4000 miles).

(b) Find the distance between the surface of Earth and the satellite when $\theta = 60°$.

(c) Find the distance between the surface of Earth and the satellite when $\theta = 30°$.

Synthesis

True or False? **In Exercises 47 and 48, determine whether the statement is true or false. Justify your answer.**

47. The graph of $r = 4/(-3 - 3 \sin \theta)$ has a horizontal directrix above the pole.

48. The conic represented by the following equation is an ellipse.

$$r^2 = \frac{16}{9 - 4 \cos\left(\theta + \dfrac{\pi}{4}\right)}$$

49. Show that the polar equation for the ellipse

$$\frac{x^2}{a^2} + \frac{y^2}{b^2} = 1 \quad \text{is} \quad r^2 = \frac{b^2}{1 - e^2 \cos^2 \theta}.$$

50. Show that the polar equation for the hyperbola

$$\frac{x^2}{a^2} - \frac{y^2}{b^2} = 1 \quad \text{is} \quad r^2 = \frac{-b^2}{1 - e^2 \cos^2 \theta}.$$

In Exercises 51 and 52, use the results of Exercises 49 and 50 to write the polar form of the equation of the conic.

51. $\dfrac{x^2}{169} + \dfrac{y^2}{144} = 1$

52. $\dfrac{x^2}{9} - \dfrac{y^2}{16} = 1$

53. *Exploration*   Consider the polar equation

$$r = \frac{4}{1 - 0.4 \cos \theta}.$$

(a) Identify the conic without graphing the equation.

(b) Without graphing the following polar equations, describe how each differs from the given polar equation.

$$r = \frac{4}{1 + 0.4 \cos \theta}, \quad r = \frac{4}{1 - 0.4 \sin \theta}$$

(c) Use a graphing utility to verify your results in part (b).

54. *Exploration*   The equation

$$r = \frac{ep}{1 \pm e \sin \theta}$$

is the equation of an ellipse with $e < 1$. What happens to the lengths of both the major axis and the minor axis when the value of e remains fixed and the value of p changes? Use an example to explain your reasoning.

55. *Writing*   In your own words, define the term *eccentricity* and explain how it can be used to classify conics.

56. What conic does the polar equation given by $r = a \sin \theta + b \cos \theta$ represent?

Review

In Exercises 57–60, solve the equation.

57. $4\sqrt{3} \tan \theta - 3 = 1$

58. $6 \cos x - 2 = 1$

59. $12 \sin^2 \theta = 9$

60. $9 \csc^2 x - 10 = 2$

In Exercises 61–64, find the value of the trigonometric function given that u and v are in Quadrant IV and $\sin u = -\frac{3}{5}$ and $\cos v = 1/\sqrt{2}$.

61. $\cos(u + v)$ **62.** $\sin(u + v)$

63. $\sin(u - v)$ **64.** $\cos(u - v)$

In Exercises 65–68, evaluate the expression. Do not use a calculator.

65. $_{12}C_9$ **66.** $_{18}C_{16}$

67. $_{10}P_3$ **68.** $_{29}P_2$

10 Chapter Summary

What did you learn?

	Review Exercises
Section 10.1	
☐ Recognize the four basic conics: circles, parabolas, ellipses, and hyperbolas.	1–6
☐ Recognize, graph, and write equations of parabolas (vertex at origin).	7–16
☐ Recognize, graph, and write equations of ellipses (center at origin).	17–26
☐ Recognize, graph, and write equations of hyperbolas (center at origin).	27–34
Section 10.2	
☐ Recognize equations of conics that have been shifted vertically and/or horizontally in the plane.	35–42
☐ Write and graph equations of conics that have been shifted vertically and/or horizontally in the plane.	43–56
Section 10.3	
☐ Evaluate sets of parametric equations for given values of the parameter.	57, 58
☐ Graph curves that are represented by sets of parametric equations.	59–70
☐ Rewrite sets of parametric equations as single rectangular equations by eliminating the parameter.	59–64
☐ Find sets of parametric equations for graphs.	71–74
Section 10.4	
☐ Plot points and find multiple representations of points in the polar coordinate system.	75–78
☐ Convert points from rectangular to polar form and vice versa.	79–86
☐ Convert equations from rectangular to polar form and vice versa.	87–98
Section 10.5	
☐ Graph polar equations by point plotting.	99–104
☐ Use symmetry as a sketching aid.	105–112
☐ Use zeros and maximum r-values as sketching aids.	105–112
☐ Recognize special polar graphs.	105–112
Section 10.6	
☐ Define conics in terms of eccentricities.	113–118
☐ Write and graph equations of conics in polar form.	119–122
☐ Use equations of conics in polar form to model real-life problems.	123, 124

10 Review Exercises

10.1 In Exercises 1–6, match the equation with its graph. [The graphs are labeled (a), (b), (c), (d), (e), and (f).]

(a)

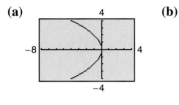

(b)

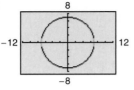

(c)

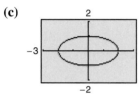

(d)

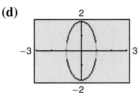

(e)

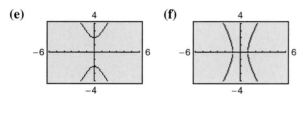

(f)

1. $4x^2 + y^2 = 4$ 2. $y^2 = -4x$
3. $x^2 + 4y^2 = 4$ 4. $x^2 + y^2 = 49$
5. $4x^2 - y^2 = 4$ 6. $y^2 - 4x^2 = 4$

In Exercises 7–10, find the standard form of the equation of the parabola.

7. Vertex: $(0, 0)$; passes through: $(1, 2)$; vertical axis

8. Vertex: $(0, 0)$; passes through: $(4, -2)$; vertical axis

9. Vertex: $(0, 0)$; focus: $(-6, 0)$

10. Vertex: $(0, 0)$; focus: $(0, 3)$

In Exercises 11–14, find the vertex and focus of the parabola and sketch its graph. Use a graphing utility to verify your graph.

11. $4x - y^2 = 0$ 12. $y = -\frac{1}{8}x^2$
13. $\frac{1}{2}y^2 + 18x = 0$ 14. $\frac{1}{4}y - 8x^2 = 0$

15. **Satellite Antenna** A cross section of a large parabolic antenna (see figure) is modeled by

$$y = \frac{x^2}{200}, \quad -100 \le x \le 100.$$

The receiving and transmitting equipment is positioned at the focus. Find the coordinates of the focus.

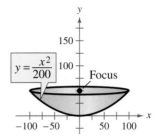

16. **Suspension Bridge** Each cable of a suspension bridge is suspended (in the shape of a parabola) between two towers (see figure). An equation that models the cables is given by $y = \frac{1}{180}x^2$, where x and y are measured in meters. Find the coordinates of the focus.

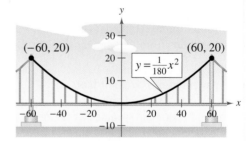

In Exercises 17–20, find the standard form of the equation of the ellipse with center at the origin.

17. Vertices: $(\pm 5, 0)$; foci: $(\pm 4, 0)$
18. Vertices: $(0, \pm 10)$; foci: $(0, \pm 8)$
19. Vertices: $(0, \pm 6)$; passes through $(2, 2)$
20. Vertices: $(\pm 7, 0)$; passes through $(0, \pm 6)$

In Exercises 21–24, find the center and vertices of the ellipse and sketch its graph. Use a graphing utility to verify your graph.

21. $\dfrac{x^2}{4} + \dfrac{y^2}{16} = 1$ 22. $\dfrac{x^2}{9} + \dfrac{y^2}{8} = 1$

23. $6x^2 + 4y^2 = 36$ **24.** $3x^2 + 8y^2 = 48$

25. *Architecture* A semielliptical archway is to be formed over the entrance to an estate. The arch is to be set on pillars that are 10 feet apart and is to have a height (atop the pillars) of 4 feet. Where should the foci be in order to sketch the arch?

26. *Wading Pool* You are building a wading pool that is in the shape of an ellipse. Your plans give an equation for the elliptical shape of the pool measured in feet as

$$\frac{x^2}{324} + \frac{y^2}{196} = 1.$$

Find the longest distance across the pool, the shortest distance, and the distance between the foci.

In Exercises 27–30, find the standard form of the equation of the hyperbola with center at the origin.

27. Vertices: $(\pm 1, 0)$;   asymptotes: $y = \pm 2x$

28. Vertices: $(0, \pm 2)$;   asymptotes: $y = \pm \dfrac{2}{\sqrt{5}}x$

29. Vertices: $(0, \pm 1)$;   foci: $(0, \pm 3)$

30. Vertices: $(\pm 4, 0)$;   foci: $(\pm 6, 0)$

In Exercises 31–34, find the center, vertices, and foci of the hyperbola and sketch its graph. Use a graphing utility to verify your graph.

31. $\dfrac{y^2}{9} - \dfrac{x^2}{64} = 1$ **32.** $\dfrac{x^2}{49} - \dfrac{y^2}{36} = 1$

33. $5y^2 - 4x^2 = 20$ **34.** $x^2 - y^2 = \dfrac{9}{4}$

10.2 **In Exercises 35–42, identify the conic by writing its equation in standard form. Then sketch its graph and describe the translation.**

35. $x^2 - 6x + 2y + 9 = 0$

36. $y^2 - 12y - 8x + 20 = 0$

37. $x^2 + 9y^2 + 10x - 18y + 25 = 0$

38. $16x^2 + 16y^2 - 16x + 24y - 3 = 0$

39. $4x^2 - 4y^2 - 4x + 8y - 11 = 0$

40. $x^2 - 9y^2 + 10x + 18y + 7 = 0$

41. $4x^2 + y^2 - 16x + 15 = 0$

42. $9x^2 - y^2 - 72x + 8y + 119 = 0$

In Exercises 43–46, find the standard form of the equation of the parabola.

43. Vertex: $(4, 2)$;   focus: $(4, 0)$

44. Vertex: $(2, 0)$;   focus: $(0, 0)$

45. Vertex: $(0, 2)$;   directrix: $x = -3$

46. Vertex: $(2, 2)$;   directrix: $y = 0$

In Exercises 47–50, find the standard form of the equation of the ellipse.

47. Vertices: $(-3, 0)$, $(7, 0)$;   foci: $(0, 0)$, $(4, 0)$

48. Vertices: $(2, 0)$, $(2, 4)$;   foci: $(2, 1)$, $(2, 3)$

49. Vertices: $(0, 1)$, $(4, 1)$;
endpoints of the minor axis: $(2, 0)$, $(2, 2)$

50. Vertices: $(-4, -1)$, $(-4, 11)$;
endpoints of the minor axis: $(-6, 5)$, $(-2, 5)$

In Exercises 51–54, find the standard form of the equation of the hyperbola.

51. Vertices: $(-10, 3)$, $(6, 3)$;   foci: $(-12, 3)$, $(8, 3)$

52. Vertices: $(2, 2)$, $(-2, 2)$;   foci: $(4, 2)$, $(-4, 2)$

53. Foci: $(0, 0)$, $(8, 0)$;   asymptotes: $y = \pm 2(x - 4)$

54. Foci: $(3, \pm 2)$;   asymptotes: $y = \pm 2(x - 3)$

55. *Architecture* A church window (see figure) is bounded on top by a parabola and below by the arc of a circle.

(a) Find equations for the parabola and circle.

(b) Use a graphing utility to complete the table showing the vertical distance d between the circle and the parabola for the given values of x.

x	0	1	2	3	4
d					

56. *Astronomy* The comet Encke has an elliptical orbit with the sun at one focus. Encke's orbit ranges from 0.34 to 4.08 astronomical units from the sun. Find the standard equation of the orbit. Place the center of the orbit at the origin and place the major axis on the x-axis.

10.3 In Exercises 57 and 58, complete the table for each set of parametric equations. Plot the points (x, y) and sketch a graph of the parametric equations.

57. $x = 3t - 2$ and $y = 7 - 4t$

t	-2	-1	0	1	2	3
x						
y						

58. $x = \dfrac{1}{5}t$ and $y = \dfrac{4}{t - 1}$

t	-1	0	2	3	4	5
x						
y						

In Exercises 59–64, sketch the curve represented by the parametric equations (indicate the orientation of the curve). Then eliminate the parameter and write the corresponding rectangular equation whose graph represents the curve. Adjust the domain of the resulting rectangular equation, if necessary.

59. $x = 5t - 1$
  $y = 2t + 5$

60. $x = 4t + 1$
  $y = 8 - 3t$

61. $x = t^2 + 2$
  $y = 4t^2 - 3$

62. $x = \dfrac{4}{t}$
  $y = t^2 - 1$

63. $x = \dfrac{1}{t}$
  $y = t$

64. $x = t$
  $y = \dfrac{1}{t}$

In Exercises 65–70, use a graphing utility to graph the curve represented by the parametric equations.

65. $x = \sqrt[3]{t}$
  $y = t$

66. $x = t$
  $y = \sqrt[3]{t}$

67. $x = t^2$
  $y = \sqrt{t}$

68. $x = t + 4$
  $y = t^2$

69. $x = 6 \cos \theta$
  $y = 6 \sin \theta$

70. $x = 3 + 3 \cos \theta$
  $y = 2 + 5 \sin \theta$

In Exercises 71–74, find two different sets of parametric equations for the given rectangular equation.

71. $y = 6x + 2$

72. $y = 10 - x$

73. $y = x^2 + 2$

74. $y = 2x^3 + 5x$

10.4 In Exercises 75–78, plot the point given in polar coordinates and find three additional polar representations of the point, using $-2\pi < \theta < 2\pi$.

75. $\left(1, \dfrac{\pi}{4}\right)$

76. $\left(-5, -\dfrac{\pi}{3}\right)$

77. $\left(\sqrt{5}, -\dfrac{4\pi}{3}\right)$

78. $\left(\sqrt{10}, \dfrac{3\pi}{4}\right)$

In Exercises 79–82, plot the point given in polar coordinates and find the corresponding rectangular coordinates for the point.

79. $\left(5, -\dfrac{7\pi}{6}\right)$

80. $\left(-4, \dfrac{2\pi}{3}\right)$

81. $\left(3, \dfrac{3\pi}{4}\right)$

82. $\left(0, \dfrac{\pi}{2}\right)$

In Exercises 83–86, plot the point given in rectangular coordinates and find two sets of polar coordinates for the point for $0 \le \theta < 2\pi$.

83. $(0, -9)$

84. $(-3, 4)$

85. $(5, -5)$

86. $\left(-3, -\sqrt{3}\right)$

In Exercises 87–92, convert the rectangular equation to polar form.

87. $x^2 + y^2 = 9$

88. $x^2 + y^2 = 20$

89. $x^2 + y^2 - 4x = 0$

90. $x^2 + y^2 - 6y = 0$

91. $xy = 5$

92. $xy = -2$

In Exercises 93–98, convert the polar equation to rectangular form.

93. $r = 5$

94. $r = 12$

95. $r = 3 \cos \theta$

96. $r = 8 \sin \theta$

97. $r^2 = \cos 2\theta$

98. $r^2 = \sin \theta$

10.5 In Exercises 99–104, sketch the graph of the polar equation.

99. $r = 5$

100. $r = 3$

101. $\theta = \dfrac{\pi}{2}$

102. $\theta = -\dfrac{5\pi}{6}$

103. $r = 5 \cos \theta$

104. $r = 2 \sin \theta$

In Exercises 105–112, identify and then sketch the graph of the polar equation. Identify any symmetry, maximum *r*-values, and zeros of *r*. Use a graphing utility to verify your graph.

105. $r = 5 + 4 \cos \theta$

106. $r = 1 + 4 \sin \theta$

107. $r = 3 - 5 \sin \theta$

108. $r = 2 - 6 \cos \theta$

109. $r = -3 \cos 2\theta$

110. $r = \cos 5\theta$

111. $r^2 = 5 \sin 2\theta$

112. $r^2 = \cos 2\theta$

10.6 In Exercises 113–118, identify the conic represented by the equation algebraically. Then use a graphing utility to graph the polar equation.

113. $r = \dfrac{2}{1 - \sin \theta}$

114. $r = \dfrac{1}{1 + 2 \sin \theta}$

115. $r = \dfrac{4}{5 - 3 \cos \theta}$

116. $r = \dfrac{6}{-1 + 4 \cos \theta}$

117. $r = \dfrac{5}{6 + 2 \sin \theta}$

118. $r = \dfrac{3}{4 - 4 \cos \theta}$

In Exercises 119–122, find a polar equation of the conic with its focus at the pole.

Conic	Vertex or Vertices
119. Parabola	Vertex: $(2, \pi)$
120. Parabola	Vertex: $(2, \pi/2)$
121. Ellipse	Vertices: $(5, 0), (1, \pi)$
122. Hyperbola	Vertices: $(1, 0), (7, 0)$

123. **Planetary Motion** The planet Mars has an elliptical orbit with an eccentricity of $e \approx 0.093$. The length of the major axis of the orbit is approximately 3.05 astronomical units. Find a polar equation for the orbit and its perihelion and aphelion distances.

124. **Astronomy** An asteroid takes a parabolic path with Earth as its focus. It is about 6,000,000 miles from Earth at its closest approach. Write the polar equation of the path of the asteroid with its vertex at $\theta = -\pi/2$. Find the distance between the asteroid and Earth when $\theta = -\pi/3$.

True or False? In Exercises 125 and 126, determine whether the statement is true or false. Justify your answer.

125. The graph of $\frac{1}{4}x^2 - y^4 = 1$ represents the equation of a hyperbola.

126. There is only one set of parametric equations that represents the line $y = 3 - 2x$.

Writing In Exercises 127 and 128, an equation and four variations are given. In your own words, describe how the graph of each of the variations differs from the graph of the original equation.

127. $y^2 = 8x$

(a) $(y - 2)^2 = 8x$

(b) $y^2 = 8(x + 1)$

(c) $y^2 = -8x$

(d) $y^2 = 4x$

128. $\dfrac{x^2}{4} + \dfrac{y^2}{9} = 1$

(a) $\dfrac{x^2}{9} + \dfrac{y^2}{4} = 1$

(b) $\dfrac{x^2}{4} + \dfrac{y^2}{4} = 1$

(c) $\dfrac{x^2}{4} + \dfrac{y^2}{25} = 1$

(d) $\dfrac{(x - 3)^2}{4} + \dfrac{y^2}{9} = 1$

129. Consider an ellipse whose major axis is horizontal and 10 units in length. The number *b* in the standard form of the equation of the ellipse must be less than what real number? Describe the change in the shape of the ellipse as *b* approaches this number.

130. The graph of the parametric equations $x = 2 \sec t$ and $y = 3 \tan t$ is shown in the figure. Would the graph change for the equations $x = 2 \sec(-t)$ and $y = 3 \tan(-t)$? If so, how would it change?

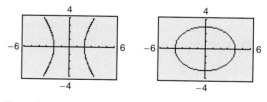

Figure for 130   Figure for 131

131. The path of a moving object is modeled by the parametric equations $x = 4 \cos t$ and $y = 3 \sin t$, where *t* is time (see figure). How would the path change for each of the following?

(a) $x = 4 \cos 2t$, $y = 3 \sin 2t$

(b) $x = 5 \cos t$, $y = 3 \sin t$

10 | Chapter Test

Take this test as you would take a test in class. After you are finished, check your work against the answers given in the back of the book.

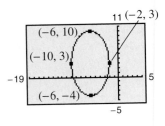

In Exercises 1–3, graph the conic and identify any vertices and foci.

1. $y^2 - 8x = 0$ **2.** $y^2 - 4x + 4 = 0$ **3.** $x^2 - 4y^2 - 4x = 0$

4. Find the standard form of the equation of the parabola with focus $(8, -2)$ and directrix $x = 4$, and sketch the parabola.

5. Find the standard form of the equation of the ellipse shown at the right.

Figure for 5

6. Find the standard form of the equation of the hyperbola with vertices $(0, \pm 3)$ and asymptotes $y = \pm \frac{3}{2}x$.

7. Use a graphing utility to graph the conic $x^2 - \dfrac{y^2}{4} = 1$. Describe your viewing window.

In Exercises 8–10, sketch the curve represented by the parametric equations. Then eliminate the parameter and write the corresponding rectangular equation whose graph represents the curve.

8. $x = t^2 - 6$
$y = \dfrac{1}{2}t - 1$

9. $x = \sqrt{t^2 + 2}$
$y = \dfrac{t}{4}$

10. $x = 2 + 3 \cos \theta$
$y = 2 \sin \theta$

In Exercises 11–13, find two different sets of parametric equations for the given rectangular equation.

11. $y = 7x + 6$ **12.** $y = x^2 + 10$ **13.** $y = \frac{1}{4}x - 5$

14. Convert the polar coordinate $\left(-14, \dfrac{5\pi}{3}\right)$ to rectangular form.

15. Convert the rectangular coordinate $(2, -2)$ to polar form and find two additional representations of this point.

16. Convert the rectangular equation $x^2 + y^2 - 12y = 0$ to polar form.

In Exercises 17–20, identify the conic represented by the polar equation algebraically. Then use a graphing utility to graph the polar equation.

17. $r = 2 + 3 \sin \theta$ **18.** $r = 8 \cos 3\theta$

19. $r = \dfrac{4}{2 + \cos \theta}$ **20.** $r = \dfrac{8}{4 + 6 \sin \theta}$

21. Find a polar equation of an ellipse with its focus at the pole, an eccentricity of $e = \frac{1}{4}$, and directrix at $y = 4$.

22. Find a polar equation of a hyperbola with its focus at the pole, an eccentricity of $e = \frac{5}{4}$, and directrix $y = 2$.

8–10 Cumulative Test

Take this test to review the material from earlier chapters. After you are finished, check your work against the answers given in the back of the book.

In Exercises 1–4, use any method to solve the system of equations.

1. $\begin{cases} -x - 3y = 5 \\ 4x + 2y = 10 \end{cases}$

2. $\begin{cases} 2x - y^2 = 0 \\ x - y = 4 \end{cases}$

3. $\begin{cases} 2x - 3y + z = 13 \\ -4x + y - 2z = -6 \\ x - 3y + 3z = 12 \end{cases}$

4. $\begin{cases} x - 4y + 3z = 5 \\ 5x + 2y - z = 1 \\ -2x - 8y = 30 \end{cases}$

In Exercises 5–8, perform the matrix operations given

$$A = \begin{bmatrix} -3 & 0 & -4 \\ 2 & 4 & 5 \\ -4 & 8 & 1 \end{bmatrix} \quad \text{and} \quad B = \begin{bmatrix} -1 & 5 & 2 \\ 6 & -3 & 3 \\ 0 & 4 & -2 \end{bmatrix}.$$

5. $3A - 2B$ **6.** $5A + 3B$ **7.** AB **8.** BA

9. Find (a) the inverse of A (if it exists), and (b) the determinant of A.

$$A = \begin{bmatrix} 1 & 2 & -1 \\ 3 & 7 & -10 \\ -5 & -7 & -15 \end{bmatrix}$$

10. Use a determinant to find the area of the triangle with vertices $(0, 0)$, $(6, 2)$, and $(8, 10)$.

11. Write the first five terms of each sequence a_n. (Assume n begins with 1.)

(a) $a_n = \dfrac{(-1)^{n+1}}{2n + 3}$

(b) $a_n = 3(2)^{n-1}$

In Exercises 12–15, find the sum. Use a graphing utility to verify your result.

12. $\sum\limits_{k=1}^{6} (7k - 2)$ **13.** $\sum\limits_{k=1}^{4} \dfrac{2}{k^2 + 4}$ **14.** $\sum\limits_{n=0}^{10} 9\left(\dfrac{3}{4}\right)^n$ **15.** $\sum\limits_{n=1}^{\infty} 8(0.9)^{n-1}$

16. Use mathematical induction to prove the formula

$$3 + 7 + 11 + 15 + \cdots + (4n - 1) = n(2n + 1).$$

In Exercises 17–20, use the Binomial Theorem to expand and simplify the expression.

17. $(x + 3)^4$ **18.** $(2x + y^2)^5$ **19.** $(x - 2y)^6$ **20.** $(3a - 4b)^8$

In Exercises 21 and 22, find the number of distinguishable permutations of the group of letters.

21. B, A, S, K, E, T, B, A, L, L **22.** A, N, T, A, R, C, T, I, C, A

In Exercises 23–26, identify the conic and sketch its graph.

23. $\dfrac{(y + 3)^2}{36} - \dfrac{(x - 5)^2}{121} = 1$

24. $\dfrac{(x - 2)^2}{4} + \dfrac{(y + 1)^2}{9} = 1$

25. $y^2 - x^2 = 16$

26. $x^2 + y^2 - 2x - 4y + 1 = 0$

In Exercises 27–29, find the standard form of the equation of the conic.

27.   **28.**   **29.**

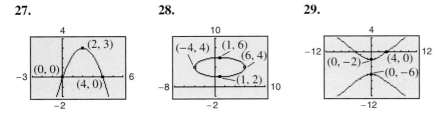

In Exercises 30–32, (a) sketch the curve represented by the parametric equations, (b) use a graphing utility to verify your graph, and (c) eliminate the parameter and write the corresponding rectangular equation whose graph represents the curve. Adjust the domain of the resulting rectangular equation, if necessary.

30. $x = 2t + 1$
  $y = t^2$

31. $x = \cos \theta$
  $y = 2 \sin^2 \theta$

32. $x = 4 \ln t$
  $y = \frac{1}{2}t^2$

In Exercises 33–36, plot the point given in polar coordinates and find three additional polar representations for $-2\pi < \theta < 2\pi$.

33. $\left(8, \dfrac{5\pi}{6}\right)$ **34.** $\left(5, -\dfrac{3\pi}{4}\right)$ **35.** $\left(-2, \dfrac{5\pi}{4}\right)$ **36.** $\left(-3, -\dfrac{11\pi}{6}\right)$

37. Convert the rectangular equation $-8x - 3y + 5 = 0$ to polar form.

38. Convert the polar equation $r = \dfrac{2}{4 - 5 \cos \theta}$ to rectangular form.

In Exercises 39–41, identify the graph represented by the polar equation algebraically. Then use a graphing utility to graph the polar equation.

39. $r = -\dfrac{\pi}{6}$ **40.** $r = 3 - 2 \sin \theta$ **41.** $r = 2 + 5 \cos \theta$

42. The salary for the first year of a job is \$28,000. During the next 14 years, the salary increases by 5% each year. Determine the total compensation over the 15-year period.

43. On a game show, the digits 3, 4, and 5 must be arranged in the proper order to form the price of an appliance. If they are arranged correctly, the contestant wins the appliance. What is the probability of winning if the contestant knows that the price is at least \$400?

44. A parabolic archway is 16 meters high at the vertex. At a height of 14 meters, the width of the archway is 12 meters, as shown in the figure at the right. How wide is the archway at ground level?

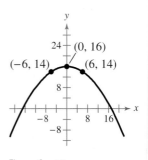

Figure for 44

Appendix A   Technology Support

Introduction

Graphing utilities such as graphing calculators and computers with graphing software are very valuable tools for visualizing mathematical principles, verifying solutions to equations, exploring mathematical ideas, and developing mathematical models. Although graphing utilities are extremely helpful in learning mathematics, their use does not mean that learning algebra is any less important. In fact, the combination of knowledge of mathematics and the use of graphing utilities enables you to explore mathematics more easily and to a greater depth. If you are using a graphing utility in this course, it is up to you to learn its capabilities and to practice using this tool to enhance your mathematical learning.

In this text, there are many opportunities to use a graphing utility, some of which are described below.

Uses of a Graphing Utility

1. Check or validate answers to problems obtained using algebraic methods.

2. Discover and explore algebraic properties, rules, and concepts.

3. Graph functions, and approximate solutions to equations involving functions.

4. Efficiently perform complicated mathematical procedures such as those found in many real-life applications.

5. Find mathematical models for sets of data.

In this appendix, the features of graphing utilities are discussed from a generic perspective and are listed in alphabetical order. To learn how to use the features of a specific graphing utility, consult your user's manual or the website for this text found at *college.hmco.com*. Additional keystroke guides are available for most graphing utilities, and your college library may have a videotape on how to use your graphing utility.

Many graphing utilities are designed to act as "function graphers." In this course, functions and their graphs are studied in detail. You may recall from previous courses that a function can be thought of as a rule that describes the relationship between two variables. These rules are frequently written in terms of x and y. For example, the equation

$$y = 3x + 5$$

represents y as a function of x.

Many graphing utilities have an *equation editor* that requires that an equation be written in "$y =$" form in order to be entered, as shown in Figure A.1. (You should note that your *equation editor* screen may not look like the screen shown in Figure A.1.)

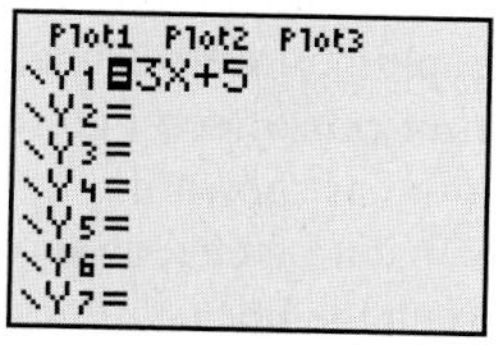

Figure A.1

Cumulative Sum Feature

The *cumulative sum* feature finds partial sums of a series. For example, to find the first four partial sums of the series

$$\sum_{k=1}^{4} 2(0.1)^k$$

choose the *cumulative sum* feature, which is found in the *operations* menu of the *list* feature (see Figure A.2). To use this feature, you will also have to use the *sequence* feature (see Figure A.2 and page A15). You must enter an expression for the sequence, a variable, the lower limit of summation, and the upper limit of summation, as shown in Figure A.3. After pressing [ENTER], you can see that the first four partial sums are 0.2, 0.22, 0.222, and 0.2222. You may have to scroll to the right in order to see all the partial sums.

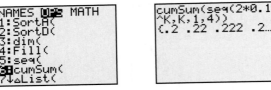

Figure A.2   Figure A.3

Determinant Feature

The *determinant* feature evaluates the determinant of a square matrix. For example, to evaluate the determinant of the matrix shown at the right, enter the 3×3 matrix into the graphing utility using the *matrix editor*, as shown in Figure A.4. Then choose the *determinant* feature from the *math* menu of the *matrix* feature, as shown in Figure A.5. Once you choose the matrix name, A, press [ENTER] and you should obtain a determinant of -50, as shown in Figure A.6.

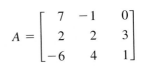

$$A = \begin{bmatrix} 7 & -1 & 0 \\ 2 & 2 & 3 \\ -6 & 4 & 1 \end{bmatrix}$$

Figure A.4   Figure A.5   Figure A.6

Draw Inverse Feature

The *draw inverse* feature graphs the inverse function of a one-to-one function. For instance, to graph the inverse function of $f(x) = x^3 + 4$, first enter the function into the *equation editor* (see Figure A.7) and graph the function (using a square viewing window), as shown in Figure A.8. Then choose the *draw inverse* feature from the *draw* feature menu, as shown in Figure A.9. You must enter the function you want to graph the inverse function of, as shown in Figure A.10. Finally, press [ENTER] to obtain the inverse function of $f(x) = x^3 + 4$, as shown in Figure A.11. This feature can only be used when the graphing utility is in *function* mode.

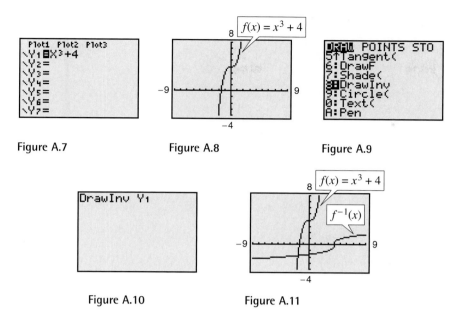

Figure A.7   Figure A.8   Figure A.9

Figure A.10   Figure A.11

Elementary Row Operations Features

Most graphing utilities can perform elementary row operations on matrices.

Row Swap Feature

The *row swap* feature interchanges two rows of a matrix. To interchange rows 1 and 3 of the matrix shown at the right, first enter the matrix into the graphing utility using the *matrix editor*, as shown in Figure A.12. Then choose the *row swap* feature from the *math* menu of the *matrix* feature, as shown in Figure A.13. When using this feature, you must enter the name of the matrix and the two rows that are to be interchanged. After pressing ENTER, you should obtain the matrix shown in Figure A.14. Because the resulting matrix will be used to demonstrate the other elementary row operation features, use the *store* feature to copy the resulting matrix to [A], as shown in Figure A.15.

$$A = \begin{bmatrix} -1 & -2 & 1 & 2 \\ 2 & -4 & 6 & -2 \\ 1 & 3 & -3 & 0 \end{bmatrix}$$

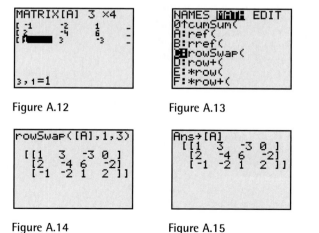

Figure A.12   Figure A.13

Figure A.14   Figure A.15

TECHNOLOGY TIP

The *store* feature of a graphing utility is used to store a value in a variable or to copy one matrix to another matrix. For instance, as shown at the left, after performing a row operation on a matrix, you can copy the answer to another matrix (see Figure A.15). You can then perform another row operation on the copied matrix. If you want to continue performing row operations to obtain a matrix in row-echelon form or reduced row-echelon form, you must copy the resulting matrix to a new matrix before each operation.

Row Addition and Row Multiplication and Addition Features

The *row addition* and *row multiplication and addition* features add a multiple of a row of a matrix to another row of the same matrix. To add row 1 to row 3 of the matrix stored in [A], choose the *row addition* feature from the *math* menu of the *matrix* feature, as shown in Figure A.16. When using this feature, you must enter the name of the matrix and the two rows that are to be added. After pressing [ENTER], you should obtain the matrix shown in Figure A.17. Copy the resulting matrix to [A].

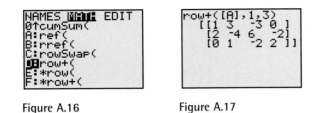

Figure A.16   Figure A.17

To add -2 times row 1 to row 2 of the matrix stored in [A], choose the *row multiplication and addition* feature from the *math* menu of the *matrix* feature, as shown in Figure A.18. When using this feature, you must enter the constant, the name of the matrix, the row the constant is multiplied by, and the row to be added to. After pressing [ENTER], you should obtain the matrix shown in Figure A.19. Copy the resulting matrix to [A].

Figure A.18   Figure A.19

Row Multiplication Feature

The *row multiplication* feature multiplies a row of a matrix by a nonzero constant. To multiply row 2 of the matrix stored in [A] by $-\frac{1}{10}$, choose the *row multiplication* feature from the *math* menu of the *matrix* feature, as shown in Figure A.20. When using this feature, you must enter the constant, the name of the matrix, and the row to be multiplied. After pressing [ENTER], you should obtain the matrix shown in Figure A.21.

Figure A.20   Figure A.21

Intersect Feature

The *intersect* feature finds the point(s) of intersection of two graphs. The *intersect* feature is found in the *calculate* menu (see Figure A.22). To find the point(s) of intersection of the graphs of $y_1 = -x + 2$ and $y_2 = x + 4$, first enter the equations in the *equation editor*, as shown in Figure A.23. Then graph the equations, as shown in Figure A.24. Next, use the *intersect* feature to find the point of intersection. Trace the cursor along the graph of y_1 near the intersection and press ENTER (see Figure A.25). Then trace the cursor along the graph of y_2 near the intersection and press ENTER (see Figure A.26). Marks are then placed on the graph at these points (see Figure A.27). Finally, move the cursor near the point of intersection and press ENTER. From Figure A.28, you can see that the coordinates of the point of intersection are displayed in the bottom of the window. So, the point of intersection is $(-1, 3)$.

Figure A.22

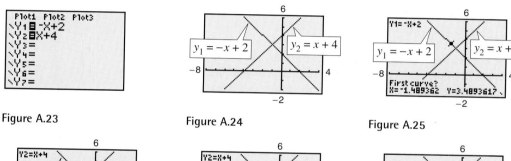

Figure A.23

Figure A.24

Figure A.25

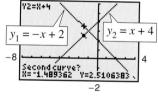

Figure A.26

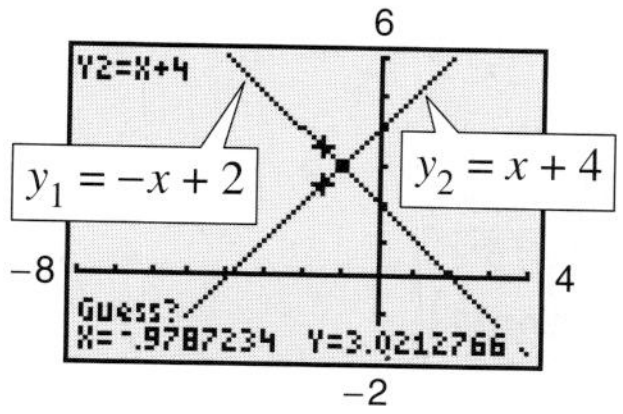

Figure A.27

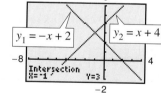

Figure A.28

List Editor

Most graphing utilities can hold data in lists. The *list editor* can be used to create tables and to hold statistical data. The *list editor* can be found in the *edit* menu of the *statistics* feature, as shown in Figure A.29. To enter the numbers 1 through 10 into a list, first choose a list (L_1) and then begin entering the data into each row, as shown in Figure A.30.

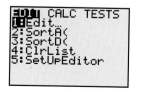

Figure A.29

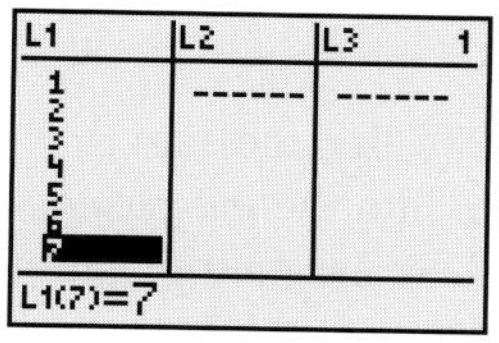

Figure A.30

You can also attach a formula to a list. For instance, you can multiply each of the data values in L_1 by 3. First, display the *list editor* and move the cursor to the top line. Then move the cursor onto the list to which you want to attach the formula (L_2). Finally, enter the formula $3* L_1$ (see Figure A.31) and then press ENTER. You should obtain the list shown in Figure A.32.

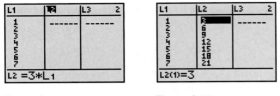

Figure A.31   Figure A.32

Matrix Feature

The *matrix* feature of a graphing utility has many uses, such as evaluating a determinant and performing row operations.

Matrix Editor

You can define, display, and edit matrices using the *matrix editor*. The *matrix editor* can be found in the *edit* menu of the *matrix* feature. For instance, to enter the matrix shown at the right, first choose the matrix name [A], as shown in Figure A.33. Then enter the dimension of the matrix (in this case, the dimension is 2×3) and enter the entries of the matrix, as shown in Figure A.34. To display the matrix on the home screen, choose the *name* menu of the *matrix* feature and select the matrix [A] (see Figure A.35), then press ENTER. The matrix A should now appear on the home screen, as shown in Figure A.36.

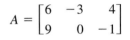

$$A = \begin{bmatrix} 6 & -3 & 4 \\ 9 & 0 & -1 \end{bmatrix}$$

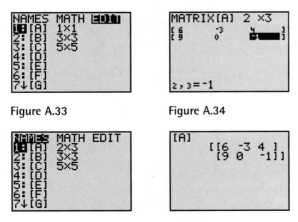

Figure A.33   Figure A.34

Figure A.35   Figure A.36

Matrix Operations

Most graphing utilities can perform matrix operations. To find the sum $A + B$ of the matrices shown at the right, first enter the matrices into the *matrix editor* as [A] and [B]. Then find the sum as shown in Figure A.37. Scalar multiplication can be performed in a similar manner. For example, you can evaluate $7A$, where A is the matrix at the right, as shown in Figure A.38. To find the product AB of the matrices A and B at the right, first be sure that the product is defined. Because the number of columns of A (2 columns) equals the number of rows of B (2 rows), you can find the product AB, as shown in Figure A.39.

$$A = \begin{bmatrix} -3 & 5 \\ 0 & 4 \end{bmatrix}$$

$$B = \begin{bmatrix} 7 & -2 \\ -1 & 2 \end{bmatrix}$$

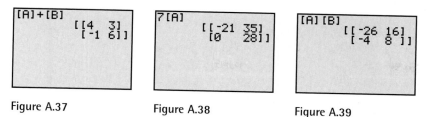

Figure A.37   Figure A.38   Figure A.39

Inverse Matrix

Some graphing utilities may not have an *inverse matrix* feature. However, you can find the inverse of a square matrix by using the inverse key $\boxed{x^{-1}}$. To find the inverse of the matrix shown at the right, enter the matrix in the *matrix editor* as [A]. Then find the inverse as shown in Figure A.40.

$$A = \begin{bmatrix} 1 & -2 & 1 \\ -1 & 3 & 0 \\ 2 & 4 & 5 \end{bmatrix}$$

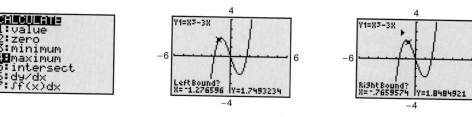

Figure A.40

Maximum and Minimum Features

The *maximum* and *minimum* features find relative extrema of a function. For instance, the graph of $y = x^3 - 3x$ is shown in Figure A.41. From the figure, the graph appears to have a relative maximum at $x = -1$ and a relative minimum at $x = 1$. To find the exact values of the relative extrema, you can use the *maximum* and *minimum* features found in the *calculate* menu (see Figure A.42). First, to find the relative maximum, choose the *maximum* feature and trace the cursor along the graph to a point left of the maximum and press $\boxed{\text{ENTER}}$ (see Figure A.43). Then trace the cursor along the graph to a point right of the maximum and press $\boxed{\text{ENTER}}$ (see Figure A.44). Note the two arrows near the top of the display marking the left and right bounds, as shown in Figure A.45. Next, trace the cursor along the graph between the two bounds and as close to the maximum as you can (see Figure A.45) and press $\boxed{\text{ENTER}}$. From Figure A.46, you can see that the coordinates of the maximum point are displayed in the bottom of the window. So, the relative maximum is $(-1, 2)$.

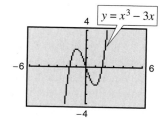

Figure A.41

Figure A.42   Figure A.43   Figure A.44

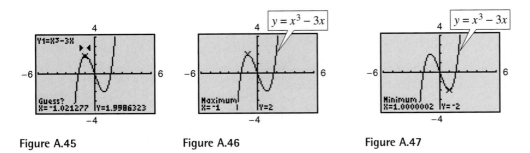

Figure A.45   Figure A.46   Figure A.47

You can find the relative minimum in a similar manner. From Figure A.47, you can see that the relative minimum is $(1, -2)$.

Mean and Median Features

In real-life applications, you often encounter large data sets and want to calculate statistical values. The *mean* and *median* features calculate the mean and median of a data set. For instance, in a survey, 100 people were asked how much money (in dollars) per week they withdraw from an automatic teller machine (ATM). The results are shown in the table below. The frequency represents the number of responses.

Amount	10	20	30	40	50	60	70	80	90	100
Frequency	3	8	10	19	24	13	13	7	2	1

To find the mean and median of the data set, first enter the data in the *list editor*, as shown in Figure A.48. Enter the amount in L_1 and the frequency in L_2. Then choose the *mean* feature from the *math* menu of the *list* feature, as shown in Figure A.49. When using this feature, you must enter a list and a frequency list (if applicable). In this case, the list is L_1 and the frequency list is L_2. After pressing ENTER, you should obtain a mean of \$49.80, as shown in Figure A.50. You can follow the same steps (except choose the *median* feature) to find the median of the data. You should obtain a median of \$50, as shown in Figure A.51.

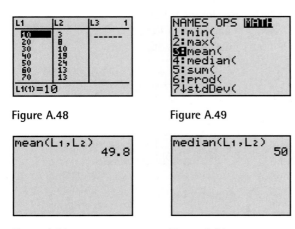

Figure A.48   Figure A.49

Figure A.50   Figure A.51

Mode Settings

Mode settings of a graphing utility control how the utility displays and interprets numbers and graphs. The default mode settings are shown in Figure A.52.

Figure A.52

Radian and Degree Modes

The trigonometric functions can be applied to angles measured in either radians or degrees. When your graphing utility is in *radian* mode, it interprets angle values as radians and displays answers in radians. When your graphing utility is in *degree* mode, it interprets angle values as degrees and displays answers in degrees. For instance, to calculate $\sin(\pi/6)$, make sure the calculator is in *radian* mode. You should obtain an answer of 0.5, as shown in Figure A.53. To calculate $\sin 45°$, make sure the calculator is in *degree* mode, as shown in Figure A.54. You should obtain an approximate answer of 0.7071, as shown in Figure A.55. If you did not change the mode of the calculator before evaluating $\sin 45°$, you would obtain an answer of approximately 0.8509, which is the sine of 45 radians.

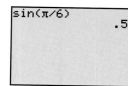

Figure A.53

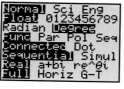

Figure A.54

Figure A.55

Function, Parametric, Polar, and Sequence Modes

Most graphing utilities can graph using four different modes.

Function Mode   The *function* mode is used to graph standard algebraic and trigonometric functions. For instance, to graph $y = 2x^2$, use the *function* mode, as shown in Figure A.52. Then enter the equation in the *equation editor*, as shown in Figure A.56. Using a standard viewing window (see Figure A.57), you obtain the graph shown in Figure A.58.

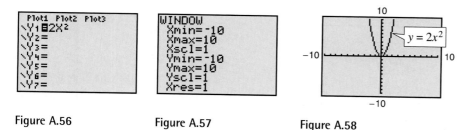

Figure A.56   Figure A.57   Figure A.58

Parametric Mode   To graph parametric equations such as $x = t + 1$ and $y = t^2$, use the *parametric* mode, as shown in Figure A.59. Then enter the equations in the *equation editor*, as shown in Figure A.60. Using the viewing window shown in Figure A.61, you obtain the graph shown in Figure A.62.

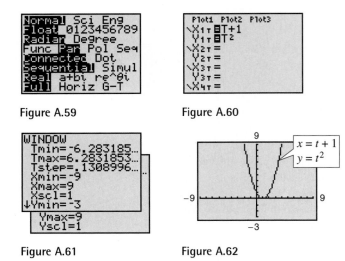

Figure A.59

Figure A.60

Figure A.61

Figure A.62

Polar Mode   To graph polar equations of the form $r = f(\theta)$, you can use the *polar* mode of a graphing utility. For instance, to graph the polar equation $r = 2 \cos \theta$, use the *polar* mode (and *radian* mode), as shown in Figure A.63. Then enter the equation in the *equation editor*, as shown in Figure A.64. Using the viewing window shown in Figure A.65, you obtain the graph shown in Figure A.66.

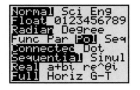

Figure A.63

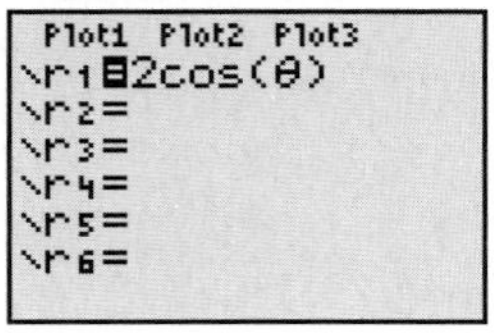

Figure A.64

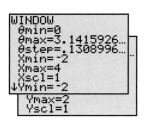

Figure A.65

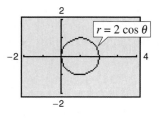

Figure A.66

Sequence Mode   To graph the first five terms of a sequence such as $a_n = 4n - 5$, use the *sequence* mode, as shown in Figure A.67. Then enter the sequence in the *equation editor*, as shown in Figure A.68 (assume that n begins with 1). Using the viewing window shown in Figure A.69, you obtain the graph shown in Figure A.70.

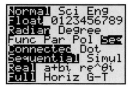

Figure A.67

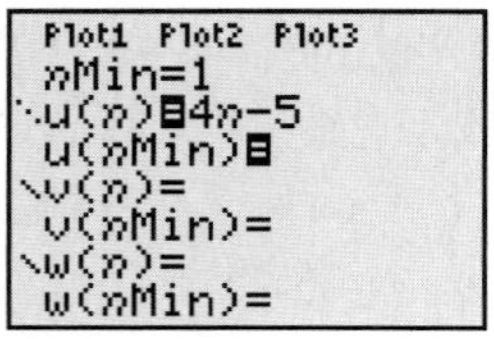

Figure A.68

TECHNOLOGY TIP

Note that when using the different graphing modes of a graphing utility, the utility uses different variables. When the utility is in *function* mode, it uses the variables x and y. In *parametric* mode, the utility uses the variables x, y, and t. In *polar* mode, the utility uses the variables r and θ. In *sequence* mode, the utility uses the variables u (instead of a) and n.

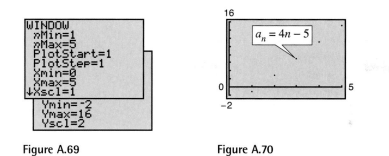

Figure A.69   Figure A.70

Connected and Dot Modes

Graphing utilities use the point-plotting method to graph functions. When a graphing utility is in *connected* mode, the utility connects the points that are plotted. When the utility is in *dot* mode, it does not connect the points that are plotted. For example, the graph of $y = x^3$ in *connected* mode is shown in Figure A.71. To graph this function using *dot* mode, first change the mode to *dot* mode (see Figure A.72) and then graph the equation, as shown in Figure A.73. As you can see from Figure A.73, the graph is a collection of dots.

Figure A.71   Figure A.72   Figure A.73

A problem arises when using the *connected* mode of some graphing utilities. Graphs with vertical asymptotes, such as rational functions and tangent functions, appear to be connected. For instance, the graph of

$$y = \frac{1}{x + 3}$$

is shown in Figure A.74. Notice how the two portions of the graph appear to be connected with a vertical line at $x = -3$. From your study of rational functions, you know that the graph has a vertical asymptote at $x = -3$ and therefore is undefined when $x = -3$. When using a graphing utility to graph rational functions and other functions that have vertical asymptotes, you should use the *dot* mode to eliminate extraneous vertical lines. Because the *dot* mode of a graphing utility displays graphs as a collection of dots rather than as a smooth curve, in this text, a blue or light red curve is placed behind the graphing utility's display to indicate where the graph should appear, as shown in Figure A.75.

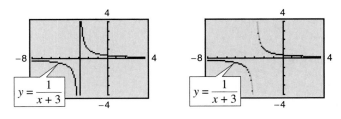

Figure A.74   Figure A.75

$_nC_r$ Feature

The $_nC_r$ feature calculates binomial coefficients and the number of combinations of n elements taken r at a time. For example, to find the number of combinations of eight elements taken five at a time, enter 8 (the n-value) on the home screen and choose the $_nC_r$ feature from the *probability* menu of the *math* feature (see Figure A.76). Next, enter 5 (the r-value) on the home screen and press ENTER. You should obtain 56, as shown in Figure A.77.

Figure A.76　　　　　　　Figure A.77

$_nP_r$ Feature

The $_nP_r$ feature calculates the number of permutations of n elements taken r at a time. For example, to find the number of permutations of six elements taken four at a time, enter 6 (the n-value) on the home screen and choose the $_nP_r$ feature from the *probability* menu of the *math* feature (see Figure A.78). Next enter 4 (the r-value) on the home screen and press ENTER. You should obtain 360, as shown in Figure A.79.

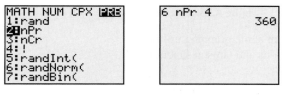

Figure A.78　　　　　　　Figure A.79

One-Variable Statistics Feature

Graphing utilities are useful when calculating statistical values for a set of data. The *one-variable statistics* feature analyzes data with one measured variable. This feature outputs the mean of the data, the sum of the data, the sum of the data squared, the sample standard deviation of the data, the population standard deviation of the data, the number of data points, the minimum data value, the maximum data value, the first quartile of the data, the median of the data, and the third quartile of the data. Consider the following data, which shows the hourly earnings (in dollars) for 12 retail sales associates.

$$5.95, \ 8.15, \ 6.35, \ 7.05, \ 6.80, \ 6.10, \ 7.15, \ 8.20, \ 6.50, \ 7.50, \ 7.95, \ 9.25$$

You can use the *one-variable statistics* feature to determine the mean and standard deviation of the data. First, enter the data in the *list editor*, as shown in Figure A.80. Then choose the *one-variable statistics* feature from the *calculate* menu of the *statistics* feature, as shown in Figure A.81. When using this feature, you must enter a list. In this case, the list is L_1. From Figure A.82, you can see that the mean of the data is $\bar{x} \approx 7.25$ and the standard deviation of the data is $\sigma x \approx 0.95$.

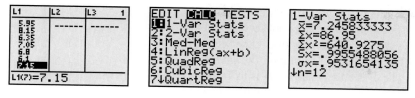

Figure A.80   Figure A.81   Figure A.82

Regression Feature

Throughout the text, you are asked to use the *regression* feature of a graphing utility to find models for sets of data. Most graphing utilities have built-in regression programs for the following.

Regression	*Form of Model*
Linear	$y = ax + b$ or $y = a + bx$
Quadratic	$y = ax^2 + bx + c$
Cubic	$y = ax^3 + bx^2 + cx + d$
Quartic	$y = ax^4 + bx^3 + cx^2 + dx + e$
Logarithmic	$y = a + b \ln(x)$
Exponential	$y = ab^x$
Power	$y = ax^b$
Logistic	$y = \dfrac{c}{1 + ae^{-bx}}$
Sine	$y = a \sin(bx + c) + d$

For example, you can find a linear model for the number y of television sets (in millions) in U.S. households for the years 1996 through 2003, shown in the table. (Source: Nielsen Media Research)

Year	Number, y
1996	222.8
1997	228.7
1998	235.0
1999	240.3
2000	245.0
2001	248.2
2002	254.4
2003	260.2

First, let x represent the year, with $x = 6$ corresponding to 1996. Then enter the data in the *list editor*, as shown in Figure A.83. Note that L_1 contains the years and L_2 contains the numbers of television sets that correspond to the years. Now choose the *linear regression* feature from the *calculate* menu of the *statistics* feature, as shown in Figure A.84. From Figure A.85, you can see that a linear model for the data is given by $y = 5.17x + 192.7$.

When you use the *regression* feature of a graphing utility, you will notice that the program may also output an "*r*-value." (For some calculators, make sure you select the *diagnostics on* feature before you use the *regression* feature. Otherwise, the calculator will not output an *r*-value.) The *r*-value or *correlation coefficient* measures how well the model fits the data. The closer the value of $|r|$ is to 1, the better the fit. For the data above, $r \approx 0.998$, which implies that the model is a good fit for the data.

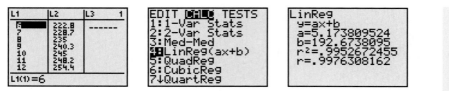

Figure A.83   Figure A.84   Figure A.85

Row-Echelon and Reduced Row-Echelon Features

Some graphing utilities have features that can automatically transform a matrix to row-echelon form and reduced row-echelon form. These features can be used to check your solutions to systems of equations.

Row-Echelon Feature

Consider the system of equations and the corresponding augmented matrix shown below.

Linear System	*Augmented Matrix*

$$\begin{cases} 2x + 5y - 3z = 4 \\ 4x + y = 2 \\ -x + 3y - 2z = -1 \end{cases} \qquad \begin{bmatrix} 2 & 5 & -3 & \vdots & 4 \\ 4 & 1 & 0 & \vdots & 2 \\ -1 & 3 & -2 & \vdots & -1 \end{bmatrix}$$

You can use the *row-echelon* feature of a graphing utility to write the augmented matrix in row-echelon form. First, enter the matrix into the graphing utility using the *matrix editor*, as shown in Figure A.86. Next, choose the *row-echelon* feature from the *math* menu of the *matrix* feature, as shown in Figure A.87. When using this feature, you must enter the name of the matrix. In this case, the matrix is [A]. You should obtain the matrix shown in Figure A.88. You may have to scroll to the right in order to see all the entries of the matrix.

Figure A.86   Figure A.87   Figure A.88

Reduced Row-Echelon Feature

To write the augmented matrix in reduced row-echelon form, follow the same steps used to write a matrix in row-echelon form except choose the *reduced row-echelon form* feature, as shown in Figure A.89. You should obtain the matrix shown in Figure A.90. From Figure A.90, you can conclude that the solution to the system is $x = 3$, $y = -10$, and $z = -16$.

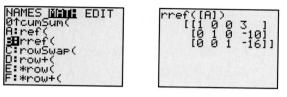

Figure A.89   Figure A.90

Sequence Feature

The *sequence* feature is used to display the terms of sequences. For instance, to determine the first five terms of the arithmetic sequence

$$a_n = 3n + 5 \qquad \text{Assume } n \text{ begins with 1.}$$

set the graphing utility to *sequence* mode. Then choose the *sequence* feature from the *operations* menu of the *list* feature, as shown in Figure A.91. When using this feature, you must enter the sequence, the variable (in this case n), the beginning value (in this case 1), and the end value (in this case 5). The first five terms of the sequence are 8, 11, 14, 17, and 20, as shown in Figure A.92. You may have to scroll to the right in order to see all the terms of the sequence.

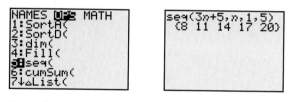

Figure A.91   Figure A.92

Shade Feature

Most graphing utilities have a *shade* feature that can be used to graph inequalities. For instance, to graph the inequality $y \leq 2x - 3$, first enter the equation $y = 2x - 3$ into the *equation editor*, as shown in Figure A.93. Next, using a standard viewing window (see Figure A.94), graph the equation, as shown in Figure A.95.

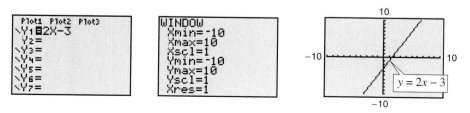

Figure A.93   Figure A.94   Figure A.95

Because the inequality sign is $\leq$, you want to shade the region below the line $y = 2x - 3$. Choose the *shade* feature from the *draw* feature menu, as shown in Figure A.96. You must enter a lower function and an upper function. In this case, the lower function is -10 (this is the least y-value in the viewing window) and the upper function is Y_1 ($y = 2x - 3$), as shown in Figure A.97. Then press [ENTER] to obtain the graph shown in Figure A.98.

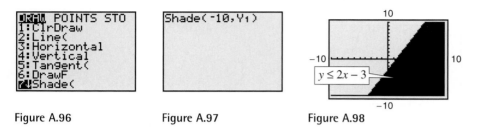

Figure A.96   Figure A.97   Figure A.98

If you wanted to graph the inequality $y \geq 2x - 3$ (using a standard viewing window), you would enter the lower function as Y_1 ($y = 2x - 3$) and the upper function as 10 (the greatest y-value in the viewing window).

Sum Feature

The *sum* feature finds the sum of a list of data. For instance, the data below represents a student's quiz scores on 10 quizzes throughout an algebra course.

     22, 23, 19, 24, 20, 15, 25, 21, 18, 24

To find the total quiz points the student earned, enter the data in the *list editor*, as shown in Figure A.99. To find the sum, choose the *sum* feature from the *math* menu of the *list* feature, as shown in Figure A.100. You must enter a list. In this case the list is L_1. You should obtain a sum of 211, as shown in Figure A.101.

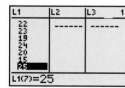  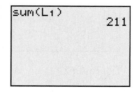

Figure A.99   Figure A.100   Figure A.101

Sum Sequence Feature

The *sum* feature and the *sequence* feature can be combined to find the sum of sequences and series. For example, to find the sum

$$\sum_{k=0}^{10} 5^{k+1}$$

first choose the *sum* feature from the *math* menu of the *list* feature, as shown in Figure A.102. Then choose the *sequence* feature from the *operations* menu of the *list* feature, as shown in Figure A.103. You must enter an expression for the sequence, a variable, the lower limit of summation, and the upper limit of summation. After pressing ⬚ENTER⬚, you should obtain the sum 61,035,155, as shown in Figure A.104.

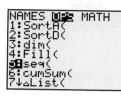

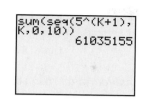

Figure A.102   Figure A.103   Figure A.104

Table Feature

Most graphing utilities are capable of displaying a table of values with x-values and one or more corresponding y-values. These tables can be used to check solutions of an equation and to generate ordered pairs to assist in graphing an equation by hand.

To use the *table* feature, enter an equation into the *equation editor*. The table may have a setup screen, which allows you to select the starting x-value and the table step or x-increment. You may then have the option of automatically generating values for x and y or building your own table using the *ask* mode (see Figure A.105).

Figure A.105

For example, enter the equation

$$y = \frac{3x}{x + 2}$$

into the *equation editor*, as shown in Figure A.106. In the table setup screen, set the table to start at $x = -4$ and set the table step to 1, as shown in Figure A.107. When you view the table, notice that the first x-value is -4 and that each value after it increases by 1. Also notice that the Y_1 column gives the resulting y-value for each x-value, as shown in Figure A.108. The table shows that the y-value when $x = -2$ is ERROR. This means that the equation is undefined when $x = -2$.

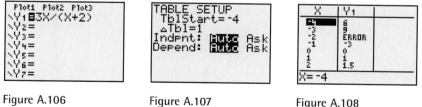

Figure A.106   Figure A.107   Figure A.108

With the same equation in the *equation editor*, set the independent variable in the table to *ask* mode, as shown in Figure A.109. In this mode, you do not need to set the starting x-value or the table step because you are entering any value you choose for x. You may enter any real value for x—integers, fractions, decimals, irrational numbers, and so forth. If you enter $x = 1 + \sqrt{3}$, the graphing utility may rewrite the number as a decimal approximation, as shown in Figure A.110. You can continue to build your own table by entering additional x-values in order to generate y-values, as shown in Figure A.111.

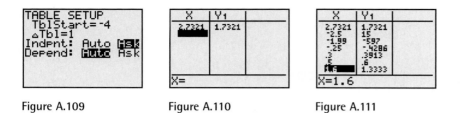

Figure A.109   Figure A.110   Figure A.111

If you have several equations in the *equation editor*, the table may generate *y*-values for each equation.

Tangent Feature

Some graphing utilities have the capability of drawing a tangent line to a graph at a given point. For instance, consider the equation

$$y = -x^3 + x + 2.$$

To draw the line tangent to the point $(1, 2)$, enter the equation into the *equation editor*, as shown in Figure A.112. Using the viewing window shown in Figure A.113, graph the equation, as shown in Figure A.114. Next, choose the *tangent* feature from the *draw* feature menu, as shown in Figure A.115. You can either move the cursor to select a point or you can enter the *x*-value at which you want the tangent line to be drawn. Because you want the tangent line to the point $(1, 2)$, enter 1 (see Figure A.116) and then press $\boxed{\text{ENTER}}$. The *x*-value you entered and the equation of the tangent line are displayed at the bottom of the window, as shown in Figure A.117.

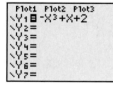

Figure A.112

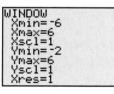

Figure A.113

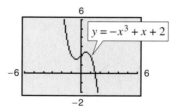

Figure A.114

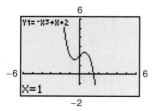

Figure A.115

Figure A.116

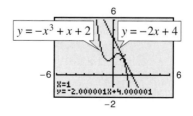

Figure A.117

Trace Feature

For instructions on how to use the *trace* feature, see the Zoom and Trace Features description on page A22.

Value Feature

The *value* feature finds the value of a function y for a given x-value. To find the value of a function such as $f(x) = 0.5x^2 - 1.5x$ at $x = 1.8$, first enter the function into the *equation editor* (see Figure A.118) and then graph the function (using a standard viewing window), as shown in Figure A.119. Next, choose the *value* feature from the *calculate* menu, as shown in Figure A.120. You will see "X= " displayed at the bottom of the window. Enter the x-value, in this case $x = 1.8$, as shown in Figure A.121. When entering an x-value, be sure it is between the Xmin and Xmax values you entered for the viewing window. Then press $\boxed{\text{ENTER}}$. From Figure A.122, you can see that when $x = 1.8$, $y = -1.08$.

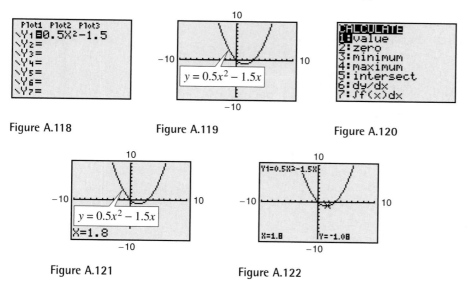

Figure A.118   Figure A.119   Figure A.120

Figure A.121   Figure A.122

Viewing Window

A viewing window for a graph is a rectangular portion of the coordinate plane. A viewing window is determined by the following six values (see Figure A.123).

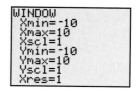

Figure A.123

Xmin = the smallest value of x

Xmax = the largest value of x

Xscl = the number of units per tick mark on the x-axis

Ymin = the smallest value of y

Ymax = the largest value of y

Yscl = the number of units per tick mark on the y-axis

When you enter these six values into a graphing utility, you are setting the viewing window. On some graphing utilities there is a seventh value on the viewing window labeled Xres. This sets the pixel resolution (1 through 8). For instance, when Xres = 1, functions are evaluated and graphed at each pixel on the *x*-axis. Some graphing utilities have a standard viewing window, as shown in Figure A.124. To initialize the standard viewing window quickly, choose the *standard viewing window* feature from the *zoom* feature menu (see page A22), as shown in Figure A.125.

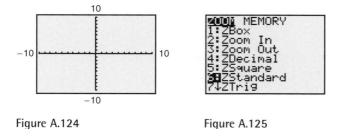

Figure A.124   Figure A.125

By choosing different viewing windows for a graph, it is possible to obtain different impressions of the graph's shape. For instance, Figure A.126 shows four different viewing windows for the graph of

$$y = 0.1x^4 - x^3 + 2x^2.$$

Of these, the view shown in part (a) is the most complete.

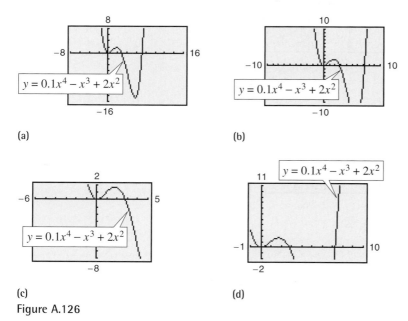

Figure A.126

On most graphing utilities, the display screen is two-thirds as high as it is wide. On such screens, you can obtain a graph with a true geometric perspective by using a square setting—one in which

$$\frac{\text{Ymax} - \text{Ymin}}{\text{Xmax} - \text{Xmin}} = \frac{2}{3}.$$

One such setting is shown in Figure A.127. Notice that the *x* and *y* tick marks are equally spaced on a square setting, but not on a standard setting (see Figure A.124). To initialize the square viewing window quickly, choose the *square viewing window* feature from the *zoom* feature menu (see page A22), as shown in Figure A.128.

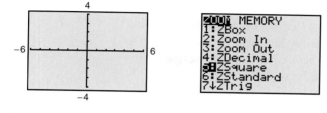

Figure A.127   Figure A.128

To see how the viewing window affects the geometric perspective, graph the semicircles $y_1 = \sqrt{9 - x^2}$ and $y_2 = -\sqrt{9 - x^2}$ using a standard viewing window, as shown in Figure A.129. Notice how the circle appears elliptical rather than circular. Now graph y_1 and y_2 using a square viewing window, as shown in Figure A.130. Notice how the circle appears circular. (Note that when you graph the two semicircles, your graphing utility may not connect them. This is because some graphing utilities are limited in their resolution. So, in this text, a blue or light red curve is placed behind the graphing utility's display to indicate where the graph should appear.)

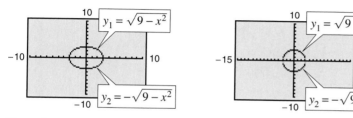

Figure A.129   Figure A.130

Zero or Root Feature

The *zero* or *root* feature finds the real zeros of the various types of functions studied in this text. To find the zeros of a function such as

$$f(x) = 2x^3 - 4x$$

first enter the function into the *equation editor*, as shown in Figure A.131. Now graph the equation (using a standard viewing window), as shown in Figure A.132. From the graph you can see that the graph of the function crosses the x-axis three times, so the function has three zeros.

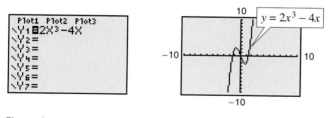

Figure A.131   Figure A.132

To find these zeros, choose the *zero* feature found in the *calculate* menu (see Figure A.133). Next, trace the cursor along the graph to a point left of one of the zeros and press ENTER (see Figure A.134). Then trace the cursor along the graph to a point right of the zero and press ENTER (see Figure A.135). Note the two arrows near the top of the display marking the left and right bounds, as shown in Figure A.136. Now trace the cursor along the graph between the two bounds and as close to the zero as you can (see Figure A.137) and press ENTER . From Figure A.138, you can see that one zero of the function is $x \approx -1.414214$.

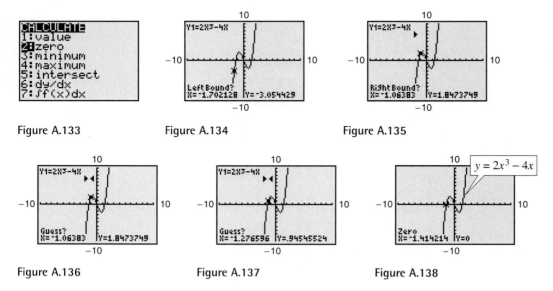

Figure A.133   Figure A.134   Figure A.135

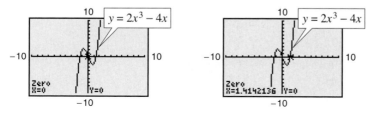

Figure A.136   Figure A.137   Figure A.138

Repeat this process to determine that the other two zeros of the function are $x = 0$ (see Figure A.139) and $x \approx 1.414214$ (see Figure A.140).

Figure A.139   Figure A.140

Zoom and Trace Features

The *zoom* feature enables you to quickly adjust the viewing window of a graph (see Figure A.141). For example, the *zoom box* feature allows you to create a new viewing window by drawing a box around any part of the graph.

Figure A.141

The *trace* feature moves from point to point along a graph. For instance, enter the equation $y = 2x^3 - 3x + 2$ into the *equation editor* (see Figure A.142) and graph the equation, as shown in Figure A.143. To activate the *trace* feature, press $\boxed{\text{TRACE}}$; then use the arrow keys to move the cursor along the graph. As you trace the graph, the coordinates of each point are displayed, as shown in Figure A.144.

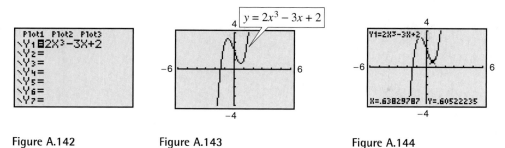

Figure A.142   Figure A.143   Figure A.144

The *trace* feature combined with the *zoom* feature enables you to obtain better and better approximations of desired points on a graph. For instance, you can use the *zoom* feature to approximate the *x*-intercept of the graph of $y = 2x^3 - 3x + 2$. From the viewing window shown in Figure A.143, the graph appears to have only one *x*-intercept. This intercept lies between -2 and -1. To zoom in on the *x*-intercept, choose the *zoom-in* feature from the *zoom* feature menu, as shown in Figure A.145. Next, trace the cursor to the point you want to zoom in on, in this case the *x*-intercept (see Figure A.146). Then press $\boxed{\text{ENTER}}$. You should obtain the graph shown in Figure A.147. Now, using the *trace* feature, you can approximate the *x*-intercept to be $x \approx -1.468085$, as shown in Figure A.148. Use the *zoom-in* feature again to obtain the graph shown in Figure A.149. Using the *trace* feature, you can approximate the *x*-intercept to be $x \approx -1.476064$, as shown in Figure A.150.

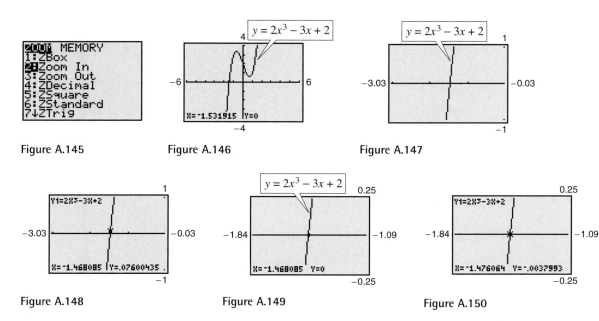

Figure A.145   Figure A.146   Figure A.147

Figure A.148   Figure A.149   Figure A.150

Here are some suggestions for using the *zoom* feature.

1. With each successive zoom-in, adjust the scale so that the viewing window shows at least one tick mark on each side of the *x*-intercept.

2. The error in your approximation will be less than the distance between two scale marks.

3. The *trace* feature can usually be used to add one more decimal place of accuracy without changing the viewing window.

You can adjust the scale in Figure A.150 to obtain a better approximation of the *x*-intercept. Using the suggestions above, change the viewing window settings so that the viewing window shows at least one tick mark on each side of the *x*-intercept, as shown in Figure A.151. From Figure A.151, you can determine that the error in your approximation will be less than 0.001 (the Xscl value). Then, using the *trace* feature, you can improve the approximation, as shown in Figure A.152. To three decimal places, the *x*-intercept is $x \approx -1.476$.

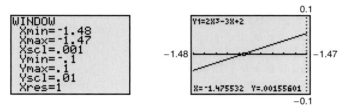

Figure A.151   Figure A.152

Appendix B Proofs of Selected Theorems

Section P.5, page 51

> **The Midpoint Formula**
>
> The midpoint of the line segment joining the points (x_1, y_1) and (x_2, y_2) is given by the Midpoint Formula
>
> $$\text{Midpoint} = \left(\frac{x_1 + x_2}{2}, \frac{y_1 + y_2}{2} \right).$$

Proof

Using the figure, you must show that

$$d_1 = d_2 \quad \text{and} \quad d_1 + d_2 = d_3.$$

By the Distance Formula, you obtain

$$d_1 = \sqrt{\left(\frac{x_1 + x_2}{2} - x_1 \right)^2 + \left(\frac{y_1 + y_2}{2} - y_1 \right)^2}.$$

Now, to simplify the expressions within the parentheses, you must find the least common denominator. The least common denominator is 2. Because both expressions have a denominator of 2^2, factor $\frac{1}{4}$ out of the expressions and then simplify the radical as follows.

$$d_1 = \tfrac{1}{2}\sqrt{(x_2 - x_1)^2 + (y_2 - y_1)^2}$$

To find d_2 and d_3, use the same procedure as above to obtain

$$d_2 = \sqrt{\left(x_2 - \frac{x_1 + x_2}{2} \right)^2 + \left(y_2 - \frac{y_1 + y_2}{2} \right)^2} = \frac{1}{2}\sqrt{(x_2 - x_1)^2 + (y_2 - y_1)^2}$$

$$d_3 = \sqrt{(x_2 - x_1)^2 + (y_2 - y_1)^2}.$$

So, it follows that $d_1 = d_2$ and $d_1 + d_2 = d_3$.

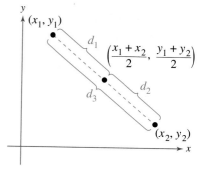

Midpoint Formula

Section 3.3, page 268

> **The Remainder Theorem**
>
> If a polynomial $f(x)$ is divided by $x - k$, the remainder is
>
> $$r = f(k).$$

Proof

From the Division Algorithm, you have

$$f(x) = (x - k)q(x) + r(x)$$

and because either $r(x) = 0$ or the degree of $r(x)$ is less than the degree of $x - k$, you know that $r(x)$ must be a constant. That is, $r(x) = r$. Now, by evaluating $f(x)$ at $x = k$, you have

$$f(k) = (k - k)q(k) + r = (0)q(k) + r = r.$$

Section 3.3, page 268

The Factor Theorem

A polynomial $f(x)$ has a factor $(x - k)$ if and only if $f(k) = 0$.

Proof

Using the Division Algorithm with the factor $(x - k)$, you have

$$f(x) = (x - k)q(x) + r(x).$$

By the Remainder Theorem, $r(x) = r = f(k)$, and you have

$$f(x) = (x - k)q(x) + f(k)$$

where $q(x)$ is a polynomial of lesser degree than $f(x)$. If $f(k) = 0$, then

$$f(x) = (x - k)q(x)$$

and you see that $(x - k)$ is a factor of $f(x)$. Conversely, if $(x - k)$ is a factor of $f(x)$, division of $f(x)$ by $(x - k)$ yields a remainder of 0. So, by the Remainder Theorem, you have $f(k) = 0$.

Section 3.4, page 279

Linear Factorization Theorem

If $f(x)$ is a polynomial of degree n, where $n > 0$, then f has precisely n linear factors

$$f(x) = a_n(x - c_1)(x - c_2) \cdots (x - c_n)$$

where $c_1, c_2, \ldots, c_n$ are complex numbers.

Proof

Using the Fundamental Theorem of Algebra, you know that f must have at least one zero, c_1. Consequently, $(x - c_1)$ is a factor of $f(x)$, and you have

$$f(x) = (x - c_1)f_1(x).$$

If the degree of $f_1(x)$ is greater than zero, you again apply the Fundamental Theorem to conclude that f_1 must have a zero c_2, which implies that

$$f(x) = (x - c_1)(x - c_2)f_2(x).$$

It is clear that the degree of $f_1(x)$ is $n - 1$, that the degree of $f_2(x)$ is $n - 2$, and that you can repeatedly apply the Fundamental Theorem n times until you obtain

$$f(x) = a_n(x - c_1)(x - c_2) \cdots (x - c_n)$$

where a_n is the leading coefficient of the polynomial $f(x)$.

Section 3.4, page 281

> **Factors of a Polynomial**
>
> Every polynomial of degree $n > 0$ with real coefficients can be written as the product of linear and quadratic factors with real coefficients, where the quadratic factors have no real zeros.

Proof

To begin, use the Linear Factorization Theorem to conclude that $f(x)$ can be completely factored in the form

$$f(x) = d(x - c_1)(x - c_2)(x - c_3) \cdots (x - c_n).$$

If each c_i is real, there is nothing more to prove. If any c_i is complex ($c_i = a + bi$, $b \neq 0$), then, because the coefficients of $f(x)$ are real, you know that the conjugate $c_j = a - bi$ is also a zero. By multiplying the corresponding factors, you obtain

$$(x - c_i)(x - c_j) = [x - (a + bi)][x - (a - bi)]$$
$$= x^2 - 2ax + (a^2 + b^2)$$

where each coefficient is real.

Section 4.3, page 344

> **Properties of Logarithms**
>
> Let a be a positive number such that $a \neq 1$, and let n be a real number. If u and v are positive real numbers, the following properties are true.
>
> | *Logarithm with Base a* | *Natural Logarithm* |
> |---|---|
> | **1.** $\log_a(uv) = \log_a u + \log_a v$ | **1.** $\ln(uv) = \ln u + \ln v$ |
> | **2.** $\log_a \dfrac{u}{v} = \log_a u - \log_a v$ | **2.** $\ln \dfrac{u}{v} = \ln u - \ln v$ |
> | **3.** $\log_a u^n = n \log_a u$ | **3.** $\ln u^n = n \ln u$ |

Proof

Each of the above three properties of logarithms can be proved by using properties of exponential functions. To prove Property 1, let

$$x = \log_a u \quad \text{and} \quad y = \log_a v.$$

The corresponding exponential forms of these two equations are

$$a^x = u \quad \text{and} \quad a^y = v.$$

Multiplying u and v produces $uv = a^x a^y = a^{x+y}$. The corresponding logarithmic form of $uv = a^{x+y}$ is $\log_a(uv) = x + y$. So, $\log_a(uv) = \log_a u + \log_a v$. The other two properties can be proved in a similar manner.

Section 6.4, page 508

> ### Sum and Difference Formulas
>
> $$\sin(u + v) = \sin u \cos v + \cos u \sin v$$
>
> $$\tan(u + v) = \frac{\tan u + \tan v}{1 - \tan u \tan v}$$
>
> $$\sin(u - v) = \sin u \cos v - \cos u \sin v$$
>
> $$\cos(u + v) = \cos u \cos v - \sin u \sin v$$
>
> $$\tan(u - v) = \frac{\tan u - \tan v}{1 + \tan u \tan v}$$
>
> $$\cos(u - v) = \cos u \cos v + \sin u \sin v$$

Proof

You can use the figures at the right for the proofs of the formulas for $\cos(u \pm v)$. In the top figure, let A be the point $(1, 0)$ and then use u and v to locate the points $B = (x_1, y_1)$, $C = (x_2, y_2)$, and $D = (x_3, y_3)$ on the unit circle. So, $x_i^2 + y_i^2 = 1$ for $i = 1, 2,$ and 3. For convenience, assume that $0 < v < u < 2\pi$. In the bottom figure, note that arcs AC and BD have the same length. So, line segments AC and BD are also equal in length, which implies that

$$\sqrt{(x_2 - 1)^2 + (y_2 - 0)^2} = \sqrt{(x_3 - x_1)^2 + (y_3 - y_1)^2}$$

$$x_2^2 - 2x_2 + 1 + y_2^2 = x_3^2 - 2x_1x_3 + x_1^2 + y_3^2 - 2y_1y_3 + y_1^2$$

$$(x_2^2 + y_2^2) + 1 - 2x_2 = (x_3^2 + y_3^2) + (x_1^2 + y_1^2) - 2x_1x_3 - 2y_1y_3$$

$$1 + 1 - 2x_2 = 1 + 1 - 2x_1x_3 - 2y_1y_3$$

$$x_2 = x_3x_1 + y_3y_1.$$

Finally, by substituting the values $x_2 = \cos(u - v)$, $x_3 = \cos u$, $x_1 = \cos v$, $y_3 = \sin u$, and $y_1 = \sin v$, you obtain

$$\cos(u - v) = \cos u \cos v + \sin u \sin v.$$

The formula for $\cos(u + v)$ can be established by considering $u + v = u - (-v)$ and using the formula just derived to obtain

$$\cos(u + v) = \cos[u - (-v)]$$

$$= \cos u \cos(-v) + \sin u \sin(-v)$$

$$= \cos u \cos v - \sin u \sin v.$$

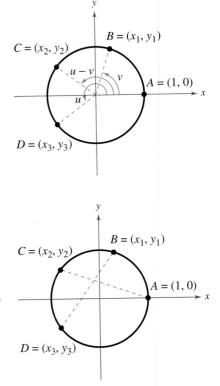

Section 6.5, page 515

> ### Double-Angle Formulas
>
> $$\sin 2u = 2 \sin u \cos u$$
>
> $$\cos 2u = \cos^2 u - \sin^2 u$$
>
> $$\tan 2u = \frac{2 \tan u}{1 - \tan^2 u}$$
>
> $$= 2 \cos^2 u - 1$$
>
> $$= 1 - 2 \sin^2 u$$

Proof

To prove all three formulas, let $v = u$ in the corresponding sum formulas.

$$\sin 2u = \sin(u + u)$$
$$= \sin u \cos u + \cos u \sin u = 2 \sin u \cos u$$

$$\cos 2u = \cos(u + u)$$
$$= \cos u \cos u - \sin u \sin u = \cos^2 u - \sin^2 u$$

$$\tan 2u = \tan(u + u)$$
$$= \frac{\tan u + \tan u}{1 - \tan u \tan u} = \frac{2 \tan u}{1 - \tan^2 u}$$

Section 6.5, page 517

Power-Reducing Formulas

$$\sin^2 u = \frac{1 - \cos 2u}{2} \qquad \cos^2 u = \frac{1 + \cos 2u}{2} \qquad \tan^2 u = \frac{1 - \cos 2u}{1 + \cos 2u}$$

Proof

The first two formulas can be verified by solving for $\sin^2 u$ and $\cos^2 u$, respectively, in the double-angle formulas

$$\cos 2u = 1 - 2 \sin^2 u \qquad \text{and} \qquad \cos 2u = 2 \cos^2 u - 1.$$

The third formula can be verified using the fact that

$$\tan^2 u = \frac{\sin^2 u}{\cos^2 u}.$$

Section 6.5, page 520

Sum-to-Product Formulas

$$\sin u + \sin v = 2 \sin\left(\frac{u + v}{2}\right) \cos\left(\frac{u - v}{2}\right)$$

$$\sin u - \sin v = 2 \cos\left(\frac{u + v}{2}\right) \sin\left(\frac{u - v}{2}\right)$$

$$\cos u + \cos v = 2 \cos\left(\frac{u + v}{2}\right) \cos\left(\frac{u - v}{2}\right)$$

$$\cos u - \cos v = -2 \sin\left(\frac{u + v}{2}\right) \sin\left(\frac{u - v}{2}\right)$$

Proof

To prove the first formula, let $x = u + v$ and $y = u - v$. Then substitute $u = (x + y)/2$ and $v = (x - y)/2$ in the product-to-sum formula.

$$\sin u \cos v = \frac{1}{2}[\sin(u + v) + \sin(u - v)]$$

$$\sin\left(\frac{x + y}{2}\right)\cos\left(\frac{x - y}{2}\right) = \frac{1}{2}(\sin x + \sin y)$$

$$2\sin\left(\frac{x + y}{2}\right)\cos\left(\frac{x - y}{2}\right) = \sin x + \sin y$$

Section 7.1, page 532

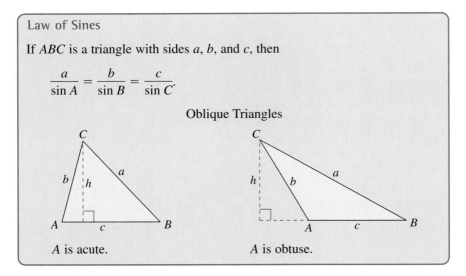

Law of Sines

If ABC is a triangle with sides a, b, and c, then

$$\frac{a}{\sin A} = \frac{b}{\sin B} = \frac{c}{\sin C}.$$

Oblique Triangles

A is acute. A is obtuse.

Proof

Let h be the altitude of either triangle found in the figure above. Then you have

$$\sin A = \frac{h}{b} \quad \text{or} \quad h = b \sin A \quad \text{and} \quad \sin B = \frac{h}{a} \quad \text{or} \quad h = a \sin B.$$

Equating these two values of h, you have

$$a \sin B = b \sin A \quad \text{or} \quad \frac{a}{\sin A} = \frac{b}{\sin B}.$$

Note that $\sin A \neq 0$ and $\sin B \neq 0$ because no angle of a triangle can have a measure of $0°$ or $180°$. In a similar manner, by constructing an altitude from vertex B to side AC (extended in the obtuse triangle), as shown at the right, you have

$$\sin A = \frac{h}{c} \quad \text{or} \quad h = c \sin A \quad \text{and} \quad \sin C = \frac{h}{a} \quad \text{or}$$

$$h = a \sin C.$$

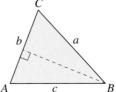

A is acute.

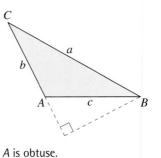

A is obtuse.

Equating these two values of h, you have

$$a \sin C = c \sin A \qquad \text{or} \qquad \frac{a}{\sin A} = \frac{c}{\sin C}.$$

By the Transitive Property of Equality you know that

$$\frac{a}{\sin A} = \frac{b}{\sin B} = \frac{c}{\sin C}.$$

So, the Law of Sines is established.

Section 7.2, page 541

> ### Law of Cosines
>
> | *Standard Form* | *Alternative Form* |
> |---|---|
> | $a^2 = b^2 + c^2 - 2bc \cos A$ | $\cos A = \dfrac{b^2 + c^2 - a^2}{2bc}$ |
> | $b^2 = a^2 + c^2 - 2ac \cos B$ | $\cos B = \dfrac{a^2 + c^2 - b^2}{2ac}$ |
> | $c^2 = a^2 + b^2 - 2ab \cos C$ | $\cos C = \dfrac{a^2 + b^2 - c^2}{2ab}$ |

Proof

Consider a triangle that has three acute angles, as shown in the figure. Note that vertex B has coordinates $(c, 0)$. Furthermore, C has coordinates (x, y), where $x = b \cos A$ and $y = b \sin A$. Because a is the distance from vertex C to vertex B, it follows that

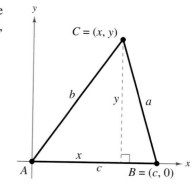

$$a = \sqrt{(x - c)^2 + (y - 0)^2} \qquad \text{Distance Formula}$$

$$a^2 = (b \cos A - c)^2 + (b \sin A)^2 \qquad \begin{array}{l}\text{Square each side and}\\\text{substitute for } x \text{ and } y.\end{array}$$

$$a^2 = b^2 \cos^2 A - 2bc \cos A + c^2 + b^2 \sin^2 A \qquad \text{Expand.}$$

$$a^2 = b^2(\sin^2 A + \cos^2 A) + c^2 - 2ab \cos A \qquad \text{Factor out } b^2.$$

$$a^2 = b^2 + c^2 - 2bc \cos A. \qquad \sin^2 A + \cos^2 A = 1$$

A similar argument can be used for a triangle having an obtuse angle.

Section 7.2, page 544

> ### Heron's Area Formula
>
> Given any triangle with sides of lengths a, b, and c, the area of the triangle is given by
>
> $$\text{Area} = \sqrt{s(s - a)(s - b)(s - c)}$$
>
> where $s = \dfrac{a + b + c}{2}$.

Proof

From Section 7.1, you know that

$$\text{Area} = \frac{1}{2}bc \sin A$$ Formula for the area of an oblique triangle

$$= \sqrt{\frac{1}{4}b^2c^2 \sin^2 A}$$ Square the right side. Then take the square root of the right side.

$$= \sqrt{\frac{1}{4}b^2c^2(1 - \cos^2 A)}$$ Pythagorean Identity

$$= \sqrt{\left[\frac{1}{2}bc(1 + \cos A)\right]\left[\frac{1}{2}bc(1 - \cos A)\right]}.$$ Factor.

Using the Law of Cosines, you can show that

$$\frac{1}{2}bc(1 + \cos A) = \frac{a + b + c}{2} \cdot \frac{-a + b + c}{2}$$

and

$$\frac{1}{2}bc(1 - \cos A) = \frac{a - b + c}{2} \cdot \frac{a + b - c}{2}.$$

Letting $s = (a + b + c)/2$, these two equations can be rewritten as

$$\frac{1}{2}bc(1 + \cos A) = s(s - a)$$

and

$$\frac{1}{2}bc(1 - \cos A) = (s - b)(s - c).$$

So, by substituting into the last formula for area, you can conclude that

$$\text{Area} = \sqrt{s(s - a)(s - b)(s - c)}.$$

Section 7.4, page 562

> ### Properties of the Dot Product
>
> Let $\mathbf{u}$, $\mathbf{v}$, and $\mathbf{w}$ be vectors in the plane or in space and let c be a scalar.
>
> 1. $\mathbf{u} \cdot \mathbf{v} = \mathbf{v} \cdot \mathbf{u}$
> 2. $\mathbf{0} \cdot \mathbf{v} = 0$
> 3. $\mathbf{u} \cdot (\mathbf{v} + \mathbf{w}) = \mathbf{u} \cdot \mathbf{v} + \mathbf{u} \cdot \mathbf{w}$
> 4. $\mathbf{v} \cdot \mathbf{v} = \|\mathbf{v}\|^2$
> 5. $c(\mathbf{u} \cdot \mathbf{v}) = c\mathbf{u} \cdot \mathbf{v} = \mathbf{u} \cdot c\mathbf{v}$

Proof

Let $\mathbf{u} = \langle u_1, u_2 \rangle$, $\mathbf{v} = \langle v_1, v_2 \rangle$, $\mathbf{w} = \langle w_1, w_2 \rangle$, $\mathbf{0} = \langle 0, 0 \rangle$, and let c be a scalar.

1. $\mathbf{u} \cdot \mathbf{v} = u_1 v_1 + u_2 v_2 = v_1 u_1 + v_2 u_2 = \mathbf{v} \cdot \mathbf{u}$

2. $\mathbf{0} \cdot \mathbf{v} = 0 \cdot v_1 + 0 \cdot v_2 = 0$

3. $\mathbf{u} \cdot (\mathbf{v} + \mathbf{w}) = \mathbf{u} \cdot \langle v_1 + w_1, v_2 + w_2 \rangle$
$$= u_1(v_1 + w_1) + u_2(v_2 + w_2)$$
$$= u_1 v_1 + u_1 w_1 + u_2 v_2 + u_2 w_2$$
$$= (u_1 v_1 + u_2 v_2) + (u_1 w_1 + u_2 w_2)$$
$$= \mathbf{u} \cdot \mathbf{v} + \mathbf{u} \cdot \mathbf{w}$$

4. $\mathbf{v} \cdot \mathbf{v} = v_1^2 + v_2^2$
$$= \left(\sqrt{v_1^2 + v_2^2} \right)^2$$
$$= \|\mathbf{v}\|^2$$

5. $c(\mathbf{u} \cdot \mathbf{v}) = c(\langle u_1, u_2 \rangle \cdot \langle v_1, v_2 \rangle)$
$$= c(u_1 v_1 + u_2 v_2)$$
$$= (cu_1)v_1 + (cu_2)v_2$$
$$= \langle cu_1, cu_2 \rangle \cdot \langle v_1, v_2 \rangle = c\mathbf{u} \cdot \mathbf{v}$$
$$= u_1(cv_1) + u_2(cv_2)$$
$$= \langle u_1, u_2 \rangle \cdot \langle cv_1, cv_2 \rangle = \mathbf{u} \cdot c\mathbf{v}$$

Section 7.4, page 563

> ### Angle Between Two Vectors
> If θ is the angle between two nonzero vectors $\mathbf{u}$ and $\mathbf{v}$, then
> $$\cos \theta = \frac{\mathbf{u} \cdot \mathbf{v}}{\|\mathbf{u}\| \, \|\mathbf{v}\|}.$$

Proof

Consider the triangle determined by vectors $\mathbf{u}$, $\mathbf{v}$, and $\mathbf{v} - \mathbf{u}$, as shown in the figure. By the Law of Cosines, you can write

$$\|\mathbf{v} - \mathbf{u}\|^2 = \|\mathbf{u}\|^2 + \|\mathbf{v}\|^2 - 2\|\mathbf{u}\| \, \|\mathbf{v}\| \cos \theta$$

$$(\mathbf{v} - \mathbf{u}) \cdot (\mathbf{v} - \mathbf{u}) = \|\mathbf{u}\|^2 + \|\mathbf{v}\|^2 - 2\|\mathbf{u}\| \, \|\mathbf{v}\| \cos \theta$$

$$(\mathbf{v} - \mathbf{u}) \cdot \mathbf{v} - (\mathbf{v} - \mathbf{u}) \cdot \mathbf{u} = \|\mathbf{u}\|^2 + \|\mathbf{v}\|^2 - 2\|\mathbf{u}\| \, \|\mathbf{v}\| \cos \theta$$

$$\mathbf{v} \cdot \mathbf{v} - \mathbf{u} \cdot \mathbf{v} - \mathbf{v} \cdot \mathbf{u} + \mathbf{u} \cdot \mathbf{u} = \|\mathbf{u}\|^2 + \|\mathbf{v}\|^2 - 2\|\mathbf{u}\| \, \|\mathbf{v}\| \cos \theta$$

$$\|\mathbf{v}\|^2 - 2\mathbf{u} \cdot \mathbf{v} + \|\mathbf{u}\|^2 = \|\mathbf{u}\|^2 + \|\mathbf{v}\|^2 - 2\|\mathbf{u}\| \, \|\mathbf{v}\| \cos \theta$$

$$-2\mathbf{u} \cdot \mathbf{v} = -2\|\mathbf{u}\| \, \|\mathbf{v}\| \cos \theta$$

$$\cos \theta = \frac{\mathbf{u} \cdot \mathbf{v}}{\|\mathbf{u}\| \, \|\mathbf{v}\|}.$$

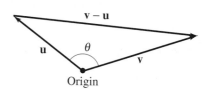

$\mathbf{v} - \mathbf{u}$

$\mathbf{u}$ θ $\mathbf{v}$

Origin

Section 9.1, page 701

> ## Properties of Sums
>
> **1.** $\displaystyle\sum_{i=1}^{n} c = cn,$ c is a constant.   **2.** $\displaystyle\sum_{i=1}^{n} ca_i = c\sum_{i=1}^{n} a_i,$ c is a constant.
>
> **3.** $\displaystyle\sum_{i=1}^{n} (a_i + b_i) = \sum_{i=1}^{n} a_i + \sum_{i=1}^{n} b_i$
>
> **4.** $\displaystyle\sum_{i=1}^{n} (a_i - b_i) = \sum_{i=1}^{n} a_i - \sum_{i=1}^{n} b_i$

Proof

Each of these properties follows directly from the properties of real numbers. For example, note the use of the Distributive Property in the proof of Property 2.

$$\sum_{i=1}^{n} ca_i = ca_1 + ca_2 + ca_3 + \cdots + ca_n$$

$$= c(a_1 + a_2 + a_3 + \cdots + a_n) = c\sum_{i=1}^{n} a_i$$

Section 9.2, page 710

> ## The Sum of a Finite Arithmetic Sequence
>
> The sum of a finite arithmetic sequence with n terms is given by
>
> $$S_n = \frac{n}{2}(a_1 + a_n).$$

Proof

Begin by generating the terms of the arithmetic sequence in two ways. In the first way, repeatedly add d to the first term to obtain

$$S_n = a_1 + a_2 + a_3 + \cdots + a_{n-2} + a_{n-1} + a_n$$
$$= a_1 + [a_1 + d] + [a_1 + 2d] + \cdots + [a_1 + (n-1)d].$$

In the second way, repeatedly subtract d from the nth term to obtain

$$S_n = a_n + a_{n-1} + a_{n-2} + \cdots + a_3 + a_2 + a_1$$
$$= a_n + [a_n - d] + [a_n - 2d] + \cdots + [a_n - (n-1)d].$$

If you add these two versions of S_n, the multiples of d subtract out and you obtain

$$2S_n = (a_1 + a_n) + (a_1 + a_n) + (a_1 + a_n) + \cdots + (a_1 + a_n) \qquad n \text{ terms}$$
$$2S_n = n(a_1 + a_n)$$

$$S_n = \frac{n}{2}(a_1 + a_n).$$

Section 9.3, page 719

> **The Sum of a Finite Geometric Sequence**
>
> The sum of the geometric sequence
>
> $$a_1, \ a_1r, \ a_1r^2, \ a_1r^3, \ a_1r^4, \ldots, a_1r^{n-1}$$
>
> with common ratio $r \neq 1$ is given by $S_n = \sum_{i=1}^{n} a_1 r^{i-1} = a_1 \left(\dfrac{1 - r^n}{1 - r} \right)$.

Proof

Begin by writing out the nth partial sum.

$$S_n = a_1 + a_1r + a_1r^2 + \cdots + a_1r^{n-2} + a_1r^{n-1}$$

Multiplication by r yields

$$rS_n = a_1r + a_1r^2 + a_1r^3 + \cdots + a_1r^{n-1} + a_1r^n.$$

Subtracting the second equation from the first yields

$$S_n - rS_n = a_1 - a_1r^n.$$

So, $S_n(1 - r) = a_1(1 - r^n)$, and, because $r \neq 1$, you have

$$S_n = a_1 \left(\frac{1 - r^n}{1 - r} \right).$$

Section 9.5, page 734

> **The Binomial Theorem**
>
> In the expansion of $(x + y)^n$
>
> $$(x + y)^n = x^n + nx^{n-1}y + \cdots + {}_nC_r\, x^{n-r}y^r + \cdots + nxy^{n-1} + y^n$$
>
> the coefficient of $x^{n-r}y^r$ is ${}_nC_r = \dfrac{n!}{(n - r)!r!}$.

Proof

The Binomial Theorem can be proved quite nicely using mathematical induction. The steps are straightforward but look complicated, so only an outline of the proof is presented.

1. If $n = 1$, you have

    $$(x + y)^1 = x^1 + y^1 = {}_1C_0x + {}_1C_1y$$

    and the formula is valid.

2. Assuming that the formula is true for $n = k$, the coefficient of $x^{k-r}y^r$ is

    $$_kC_r = \frac{k!}{(k - r)!r!} = \frac{k(k - 1)(k - 2) \cdots (k - r + 1)}{r!}.$$

To show that the formula is true for $n = k + 1$, look at the coefficient of $x^{k+1-r}y^r$ in the expansion of

$$(x + y)^{k+1} = (x + y)^k(x + y).$$

From the right-hand side, you can determine that the term involving $x^{k+1-r}y^r$ is the sum of two products.

$$(_kC_r x^{k-r}y^r)(x) + (_kC_{r-1}x^{k+1-r}y^{r-1})(y)$$

$$= \left[\frac{k!}{(k-r)!r!} + \frac{k!}{(k+1-r)!(r-1)!} \right]x^{k+1-r}y^r$$

$$= \left[\frac{(k+1-r)k!}{(k+1-r)!r!} + \frac{k!r}{(k+1-r)!r!} \right]x^{k+1-r}y^r$$

$$= \left[\frac{k!(k+1-r+r)}{(k+1-r)!r!} \right]x^{k+1-r}y^r$$

$$= \left[\frac{(k+1)!}{(k+1-r)!r!} \right]x^{k+1-r}y^r$$

$$= {}_{k+1}C_r x^{k+1-r}y^r$$

So, by mathematical induction, the Binomial Theorem is valid for all positive integers n.

Section 10.1, page 773

Standard Equation of a Parabola (Vertex at Origin)

The standard form of the equation of a parabola with vertex at $(0, 0)$ and directrix $y = -p$ is

$$x^2 = 4py, \qquad p \neq 0. \qquad \text{Vertical axis}$$

For directrix $x = -p$, the equation is

$$y^2 = 4px, \qquad p \neq 0. \qquad \text{Horizontal axis}$$

The focus is on the axis p units (directed distance) from the vertex.

Proof

Because the two cases are similar, a proof will be given for the first case only. Suppose the directrix $(y = -p)$ is parallel to the x-axis. In the figure, you assume that $p > 0$, and because p is the directed distance from the vertex to the focus, the focus must lie above the vertex. Because the point (x, y) is equidistant from $(0, p)$ and $y = -p$, you can apply the Distance Formula to obtain

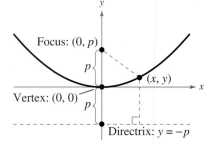

$$\sqrt{(x - 0)^2 + (y - p)^2} = y + p$$

$$x^2 + (y - p)^2 = (y + p)^2$$

$$x^2 + y^2 - 2py + p^2 = y^2 + 2py + p^2$$

$$x^2 = 4py.$$

Section 10.6, page 818

> **Polar Equations of Conics**
>
> The graph of a polar equation of the form
>
> $$\textbf{1. } r = \frac{ep}{1 \pm e \cos \theta} \quad \text{or} \quad \textbf{2. } r = \frac{ep}{1 \pm e \sin \theta}$$
>
> is a conic, where $e > 0$ is the eccentricity and $|p|$ is the distance between the focus (pole) and the directrix.

Proof

A proof for $r = ep/(1 + e \cos \theta)$ with $p > 0$ is listed here. The proofs of the other cases are similar. In the figure, consider a vertical directrix, p units to the right of the focus $F = (0, 0)$. If $P = (r, \theta)$ is a point on the graph of

$$r = \frac{ep}{1 + e \cos \theta}$$

the distance between P and the directrix is

$$
\begin{aligned}
PQ &= |p - x| \\
&= |p - r \cos \theta| \\
&= \left| p - \left(\frac{ep}{1 + e \cos\theta} \right) \cos \theta \right| \\
&= \left| p \left(1 - \frac{e \cos\theta}{1 + e \cos \theta} \right) \right| \\
&= \left| p \left(\frac{1 + e \cos\theta - e \cos \theta}{1 + e \cos \theta} \right) \right| \\
&= \left| \frac{p}{1 + e \cos \theta} \right| \\
&= \left| \frac{r}{e} \right|.
\end{aligned}
$$

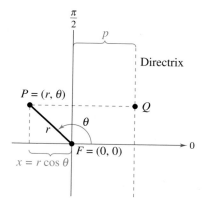

Moreover, because the distance between P and the pole is simply $PF = |r|$, the ratio of PF to PQ is

$$
\begin{aligned}
\frac{PF}{PQ} &= \frac{|r|}{|r/e|} \\
&= |e| \\
&= e
\end{aligned}
$$

and by definition, the graph of the equation must be a conic.

Appendix C   Concepts in Statistics

C.1   Measures of Central Tendency and Dispersion

Mean, Median, and Mode

In many real-life situations, it is helpful to describe data by a single number that is most representative of the entire collection of numbers. Such a number is called a **measure of central tendency.** The most commonly used measures are as follows.

1. The **mean,** or **average,** of n numbers is the sum of the numbers divided by n.

2. The **median** of n numbers is the middle number when the numbers are written in numerical order. If n is even, the median is the average of the two middle numbers.

3. The **mode** of n numbers is the number that occurs most frequently. If two numbers tie for most frequent occurrence, the collection has two modes and is called **bimodal.**

What you should learn

● Find and interpret the mean, median, and mode of a set of data.

● Determine the measure of central tendency that best represents a set of data.

● Find the standard deviation of a set of data.

● Use box-and-whisker plots.

Why you should learn it

Measures of central tendency and dispersion provide a convenient way to describe and compare sets of data. For instance, in Exercise 32 on page A46, the mean and standard deviation are used to analyze the price of gold for the years 1982 through 2001.

Example 1   Comparing Measures of Central Tendency

On an interview for a job, the interviewer tells you that the average annual income of the company's 25 employees is $60,849. The actual annual incomes of the 25 employees are shown below. What are the mean, median, and mode of the incomes?

$17,305,	$478,320,	$45,678,	$18,980,	$17,408,
$25,676,	$28,906,	$12,500,	$24,540,	$33,450,
$12,500,	$33,855,	$37,450,	$20,432,	$28,956,
$34,983,	$36,540,	$250,921,	$36,853,	$16,430,
$32,654,	$98,213,	$48,980,	$94,024,	$35,671

Solution

The mean of the incomes is

$$\text{Mean} = \frac{17{,}305 + 478{,}320 + 45{,}678 + 18{,}980 + \cdots + 35{,}671}{25}$$

$$= \frac{1{,}521{,}225}{25} = \$60{,}849.$$

To find the median, order the incomes as follows.

$12,500,	$12,500,	$16,430,	$17,305,	$17,408,
$18,980,	$20,432,	$24,540,	$25,676,	$28,906,
$28,956,	$32,654,	$33,450,	$33,855,	$34,983,
$35,671,	$36,540,	$36,853,	$37,450,	$45,678,
$48,980,	$94,024,	$98,213,	$250,921,	$478,320

From this list, you can see that the median income is $33,450. You can also see that $12,500 is the only income that occurs more than once. So, the mode is $12,500.

 Checkpoint Now try Exercise 1.

In Example 1, was the interviewer telling you the truth about the annual incomes? Technically, the person was telling the truth because the average is (generally) defined to be the mean. However, of the three measures of central tendency—*mean:* $60,849, *median:* $33,450, *mode:* $12,500—it seems clear that the median is most representative. The mean is inflated by the two highest salaries.

Choosing a Measure of Central Tendency

Which of the three measures of central tendency is most representative of a particular data set? The answer is that it depends on the distribution of the data *and* the way in which you plan to use the data.

For instance, in Example 1, the mean salary of $60,849 does not seem very representative to a potential employee. To a city income tax collector who wants to estimate 1% of the total income of the 25 employees, however, the mean is precisely the right measure.

Example 2   Choosing a Measure of Central Tendency

Which measure of central tendency is most representative of the data given in each frequency distribution?

a.

Number	1	2	3	4	5	6	7	8	9
Frequency	7	20	15	11	8	3	2	0	15

b.

Number	1	2	3	4	5	6	7	8	9
Frequency	9	8	7	6	5	6	7	8	9

c.

Number	1	2	3	4	5	6	7	8	9
Frequency	6	1	2	3	5	5	4	3	0

Solution

a. For this data, the mean is 4.23, the median is 3, and the mode is 2. Of these, the median or mode is probably the most representative measure.

b. For this data, the mean and median are each 5 and the modes are 1 and 9 (the distribution is bimodal). Of these, the mean or median is the most representative measure.

c. For this data, the mean is 4.59, the median is 5, and the mode is 1. Of these, the mean or median is the most representative measure.

 Checkpoint Now try Exercise 13.

Variance and Standard Deviation

Very different sets of numbers can have the same mean. You will now study two **measures of dispersion,** which give you an idea of how much the numbers in a set differ from the mean of the set. These two measures are called the *variance* of the set and the *standard deviation* of the set.

> **Definition of Variance and Standard Deviation**
>
> Consider a set of numbers $\{x_1, x_2, \ldots, x_n\}$ with a mean of $\bar{x}$. The **variance** of the set is
> $$v = \frac{(x_1 - \bar{x})^2 + (x_2 - \bar{x})^2 + \cdots + (x_n - \bar{x})^2}{n}$$
> and the **standard deviation** of the set is $\sigma = \sqrt{v}$ (σ is the lowercase Greek letter *sigma*).

The standard deviation of a set is a measure of how much a typical number in the set differs from the mean. The greater the standard deviation, the more the numbers in the set vary from the mean. For instance, each of the following sets has a mean of 5.

$$\{5, 5, 5, 5\}, \qquad \{4, 4, 6, 6\}, \qquad \text{and} \qquad \{3, 3, 7, 7\}$$

The standard deviations of the sets are 0, 1, and 2.

$$\sigma_1 = \sqrt{\frac{(5-5)^2 + (5-5)^2 + (5-5)^2 + (5-5)^2}{4}} = 0$$

$$\sigma_2 = \sqrt{\frac{(4-5)^2 + (4-5)^2 + (6-5)^2 + (6-5)^2}{4}} = 1$$

$$\sigma_3 = \sqrt{\frac{(3-5)^2 + (3-5)^2 + (7-5)^2 + (7-5)^2}{4}} = 2$$

Example 3   Estimations of Standard Deviation

Consider the three frequency distributions represented by the bar graphs in Figure C.1. Which set has the smallest standard deviation? Which has the largest?

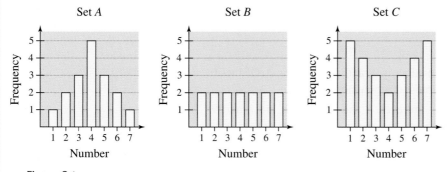

Figure C.1

Solution

Of the three sets, the numbers in set A are grouped most closely to the center and the numbers in set C are the most dispersed. So, set A has the smallest standard deviation and set C has the largest standard deviation.

 Checkpoint Now try Exercise 15.

Example 4   Find Standard Deviation

Find the standard deviation of each set shown in Example 3.

Solution

Because of the symmetry of each bar graph, you can conclude that each has a mean of $\bar{x} = 4$. The standard deviation of set A is

$$\sigma = \sqrt{\frac{(-3)^2 + 2(-2)^2 + 3(-1)^2 + 5(0)^2 + 3(1)^2 + 2(2)^2 + (3)^2}{17}}$$

$$\approx 1.53.$$

The standard deviation of set B is

$$\sigma = \sqrt{\frac{2(-3)^2 + 2(-2)^2 + 2(-1)^2 + 2(0)^2 + 2(1)^2 + 2(2)^2 + 2(3)^2}{14}}$$

$$= 2.$$

The standard deviation of set C is

$$\sigma = \sqrt{\frac{5(-3)^2 + 4(-2)^2 + 3(-1)^2 + 2(0)^2 + 3(1)^2 + 4(2)^2 + 5(3)^2}{26}}$$

$$\approx 2.22.$$

These values confirm the results of Example 3. That is, set A has the smallest standard deviation and set C has the largest.

✓ *Checkpoint* Now try Exercise 21.

The following alternative formula provides a more efficient way to compute the standard deviation.

Alternative Formula for Standard Deviation

The standard deviation of $\{x_1, x_2, \ldots, x_n\}$ is given by

$$\sigma = \sqrt{\frac{x_1^2 + x_2^2 + \cdots + x_n^2}{n} - \bar{x}^2}.$$

Because of lengthy computations, this formula is difficult to verify. Conceptually, however, the process is straightforward. It consists of showing that the expressions

$$\sqrt{\frac{(x_1 - \bar{x})^2 + (x_2 - \bar{x})^2 + \cdots + (x_n - \bar{x})^2}{n}}$$

and

$$\sqrt{\frac{x_1^2 + x_2^2 + \cdots + x_n^2}{n} - \bar{x}^2}$$

are equivalent. Try verifying this equivalence for the set $\{x_1, x_2, x_3\}$ with $\bar{x} = (x_1 + x_2 + x_3)/3$.

TECHNOLOGY TIP

Calculating the standard deviation of a large data set can become time consuming. Most graphing utilities have *statistical* features that can be used to find different statistical values of data sets. Enter the data from set A of Example 3 into the *list editor* of a graphing utility. Then use the *one-variable statistics* feature to verify the solution to Example 4 as shown below.

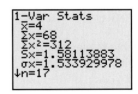

In the figure above, the standard deviation is represented as σx, which is about 1.53. For instructions on how to use the *one-variable statistics* feature, see Appendix A; for specific keystrokes, go to the text website at *college.hmco.com*.

Example 5   Using the Alternative Formula

Use the alternative formula for standard deviation to find the standard deviation of the following set of numbers.

5, 6, 6, 7, 7, 8, 8, 8, 9, 10

Solution

Begin by finding the mean of the set, which is 7.4. So, the standard deviation is

$$\sigma = \sqrt{\frac{5^2 + 2(6^2) + 2(7^2) + 3(8^2) + 9^2 + 10^2}{10} - (7.4)^2}$$

$$= \sqrt{\frac{568}{10} - 54.76} = \sqrt{2.04} \approx 1.43.$$

You can use the *one-variable statistics* feature of a graphing utility to check this result.

 Checkpoint Now try Exercise 27.

A well-known theorem in statistics, called *Chebychev's Theorem*, states that at least

$$1 - \frac{1}{k^2}$$

of the numbers in a distribution must lie within k standard deviations of the mean. So, at least 75% of the numbers in a collection must lie within two standard deviations of the mean, and at least 88.9% of the numbers must lie within three standard deviations of the mean. For most distributions, these percentages are low. For instance, in all three distributions shown in Example 3, 100% of the numbers lie within two standard deviations of the mean.

Example 6   Describing a Distribution

The table at the right shows the number of outpatient visits to hospitals (in millions) in each state and the District of Columbia in 2000. Find the mean and standard deviation of the numbers. What percent of the numbers lie within two standard deviations of the mean?   (Source: Health Forum)

Solution

Begin by entering the numbers into a graphing utility. Then use the *one-variable statistics* feature to obtain $\bar{x} \approx 10.24$ and $\sigma = 10.52$. The interval that contains all numbers that lie within two standard deviations of the mean is

$$[10.24 - 2(10.52), 10.24 + 2(10.52)] \quad \text{or} \quad [-10.80, 31.28].$$

From the table you can see that all but three of the numbers (96%) lie in this interval—all but the numbers that correspond to the numbers of outpatient visits to hospitals in California, New York, and Pennsylvania.

 Checkpoint Now try Exercise 32.

AK	1	MT	3
AL	8	NC	12
AR	4	ND	2
AZ	5	NE	3
CA	45	NH	3
CO	7	NJ	16
CT	7	NM	3
DC	1	NV	2
DE	2	NY	46
FL	22	OH	27
GA	11	OK	5
HI	3	OR	7
IA	9	PA	32
ID	2	RI	2
IL	25	SC	8
IN	14	SD	2
KS	5	TN	10
KY	9	TX	29
LA	10	UT	5
MA	17	VA	10
MD	6	VT	1
ME	3	WA	10
MI	25	WI	11
MN	7	WV	5
MO	15	WY	1
MS	4		

Box-and-Whisker Plots

Standard deviation is the measure of dispersion that is associated with the mean. **Quartiles** measure dispersion associated with the median.

> **Definition of Quartiles**
>
> Consider an ordered set of numbers whose median is m. The **lower quartile** is the median of the numbers that occur on or before m. The **upper quartile** is the median of the numbers that occur on or after m.

Example 7   Finding Quartiles of a Set

Find the lower and upper quartiles of the following set.

34, 14, 24, 16, 12, 18, 20, 24, 16, 26, 13, 27

Solution

Begin by ordering the set.

12, 13, 14,   16, 16, 18,   20, 24, 24,   26, 27, 34

1st 25%   2nd 25%   3rd 25%   4th 25%

The median of the entire set is 19. The median of the six numbers that are less than 19 is 15. So, the lower quartile is 15. The median of the six numbers that are greater than 19 is 25. So, the upper quartile is 25.

✓ *Checkpoint* Now try Exercise 35(a).

Quartiles are represented graphically by a **box-and-whisker plot,** as shown in Figure C.2. In the plot, notice that five numbers are listed: the smallest number, the lower quartile, the median, the upper quartile, and the largest number. Also notice that the numbers are spaced proportionally, as though they were on a real number line.

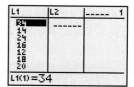

Figure C.3

12   15   19   25   34

Figure C.2

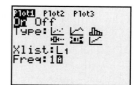

Figure C.4

TECHNOLOGY TIP You can use a graphing utility to graph the box-and-whisker plot in Figure C.2. First enter the data into the graphing utility's *list editor,* as shown in Figure C.3. Then use the *statistical plotting* feature to set up the box-and-whisker plot, as shown in Figure C.4. Finally, display the box-and-whisker plot (using the *ZoomStat* feature), as shown in Figure C.5.

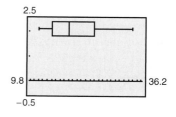

Figure C.5

The next example shows how to find quartiles when the number of elements in a set is not divisible by 4.

Example 8 Sketching Box-and-Whisker Plots

Sketch a box-and-whisker plot for each data set.

a. 82, 82, 83, 85, 87, 89, 90, 94, 95, 95, 96, 98, 99

b. 11, 13, 13, 15, 17, 17, 20, 24, 24, 27

Solution

a. This set has 13 numbers. The median is 90 (the seventh number). The lower quartile is 84 (the median of the first six numbers). The upper quartile is 95.5 (the median of the last six numbers). See Figure C.6.

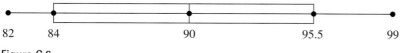

82   84   90   95.5   99

Figure C.6

b. This set has 10 numbers. The median is 17 (the average of the fifth and sixth numbers). The lower quartile is 13 (the median of the first five numbers). The upper quartile is 24 (the median of the last five numbers). See Figure C.7.

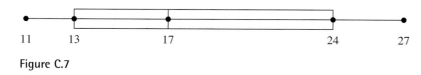

11   13   17   24   27

Figure C.7

✓ *Checkpoint* Now try Exercise 37(b).

C.1 Exercises

Vocabulary Check

Fill in the blanks.

1. A single number that is the most representative of a data set is called a _____ of _____ .
2. If two numbers are tied for the most frequent occurrence, the collection has two _____ and is called _____ .
3. Two measures of dispersion are called the _____ and the _____ of a data set.
4. _____ measure dispersion associated with the median.

In Exercises 1–6, find the mean, median, and mode of the set of measurements.

1. 5, 12, 7, 14, 8, 9, 7
2. 30, 37, 32, 39, 33, 34, 32
3. 5, 12, 7, 24, 8, 9, 7
4. 20, 37, 32, 39, 33, 34, 32
5. 5, 12, 7, 14, 9, 7
6. 30, 37, 32, 39, 34, 32

7. Reasoning

(a) Compare your answers in Exercises 1 and 3 with those in Exercises 2 and 4. Which of the measures of central tendency is sensitive to extreme measurements? Explain your reasoning.

(b) Add 6 to each measurement in Exercise 1 and calculate the mean, median, and mode of the revised measurements. How are the measures of central tendency changed?

(c) If a constant k is added to each measurement in a set of data, how will the measures of central tendency change?

8. Consumer Awareness A person had the following monthly bills for electricity. What are the mean and median of the collection of bills?

January	$67.92	February	$59.84
March	$52.00	April	$52.50
May	$57.99	June	$65.35
July	$81.76	August	$74.98
September	$87.82	October	$83.18
November	$65.35	December	$57.00

9. Car Rental A car rental company kept the following record of the numbers of miles a rental car was driven. What are the mean, median, and mode of this data?

Monday	410	Tuesday	260
Wednesday	320	Thursday	320
Friday	460	Saturday	150

10. Families A study was done on families having six children. The table shows the numbers of families in the study with the indicated numbers of girls. Determine the mean, median, and mode of the data.

Number of girls	0	1	2	3	4	5	6
Frequency	1	24	45	54	50	19	7

11. Think About It Construct a collection of numbers that has the following properties. If this is not possible, explain why.

Mean = 6, median = 4, mode = 4

12. Think About It Construct a collection of numbers that has the following properties. If this is not possible, explain why.

mean = 6, median = 6, mode = 4

13. Test Scores An English professor records the following scores for a 100-point exam.

99, 64, 80, 77, 59, 72, 87, 79, 92, 88, 90, 42, 20, 89, 42, 100, 98, 84, 78, 91

Which measure of central tendency best describes these test scores?

14. Shoe Sales A salesman sold eight pairs of men's brown dress shoes. The sizes of the eight pairs were as follows: $10\frac{1}{2}$, 8, 12, $10\frac{1}{2}$, 10, $9\frac{1}{2}$, 11, and $10\frac{1}{2}$. Which measure (or measures) of central tendency best describes the typical shoe size for this data?

In Exercises 15 and 16, line plots of data sets are given. Determine the mean and standard deviation of each set.

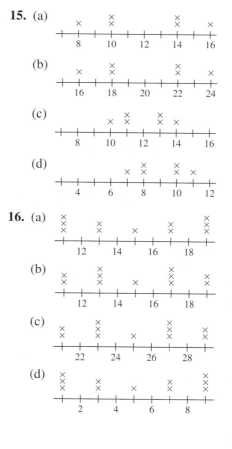

15. (a)

(b)

(c)

(d)

16. (a)

(b)

(c)

(d)

In Exercises 17–24, find the mean ($\bar{x}$), variance (v), and standard deviation (σ) of the set.

17. 4, 10, 8, 2
18. 3, 15, 6, 9, 2
19. 0, 1, 1, 2, 2, 2, 3, 3, 4
20. 2, 2, 2, 2, 2, 2
21. 1, 2, 3, 4, 5, 6, 7
22. 1, 1, 1, 5, 5, 5
23. 49, 62, 40, 29, 32, 70
24. 1.5, 0.4, 2.1, 0.7, 0.8

In Exercises 25–28, use the alternative formula to find the standard deviation of the set.

25. 2, 4, 6, 6, 13, 5
26. 246, 336, 473, 167, 219, 359
27. 8.1, 6.9, 3.7, 4.2, 6.1
28. 9.0, 7.5, 3.3, 7.4, 6.0

29. *Reasoning*   Without calculating the standard deviation, explain why the set {4, 4, 20, 20} has a standard deviation of 8.

30. *Reasoning*   If the standard deviation of a set of numbers is 0, what does this imply about the set?

31. *Test Scores*   An instructor adds five points to each student's exam score. Will this change the mean or standard deviation of the exam scores? Explain.

32. *Price of Gold*   The following data represents the average prices of gold (in dollars per fine ounce) for the years 1982 to 2001. Use a computer or graphing utility to find the mean, variance, and standard deviation of the data. What percent of the data lies within two standard deviations of the mean?   (Source: U.S. Bureau of Mines and U.S. Geological Survey)

376,	424,	361,	318,	368,
478,	438,	383,	385,	363,
345,	361,	385,	386,	389,
332,	295,	280,	280,	272

33. *Think About It*   The histograms represent the test scores of two classes of a college course in mathematics. Which histogram has the smaller standard deviation?

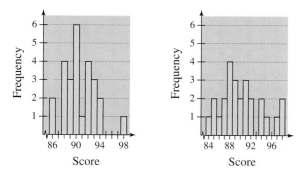

34. *Test Scores*   The scores on a mathematics exam given to 600 science and engineering students at a college had a mean and standard deviation of 235 and 28, respectively. Use Chebychev's Theorem to determine the intervals containing at least $\frac{3}{4}$ and at least $\frac{8}{9}$ of the scores. How would the intervals change if the standard deviation were 16?

In Exercises 35–38, (a) find the lower and upper quartiles of the data and (b) sketch a box-and-whisker plot for the data without using a graphing utility.

35. 23, 15, 14, 23, 13, 14, 13, 20, 12
36. 11, 10, 11, 14, 17, 16, 14, 11, 8, 14, 20
37. 46, 48, 48, 50, 52, 47, 51, 47, 49, 53
38. 25, 20, 22, 28, 24, 28, 25, 19, 27, 29, 28, 21

In Exercises 39–42, use a graphing utility to create a box-and-whisker plot for the data.

39. 19, 12, 14, 9, 14, 15, 17, 13, 19, 11, 10, 19
40. 9, 5, 5, 5, 6, 5, 4, 12, 7, 10, 7, 11, 8, 9, 9
41. 20.1, 43.4, 34.9, 23.9, 33.5, 24.1, 22.5, 42.4, 25.7, 17.4, 23.8, 33.3, 17.3, 36.4, 21.8
42. 78.4, 76.3, 107.5, 78.5, 93.2, 90.3, 77.8, 37.1, 97.1, 75.5, 58.8, 65.6

43. *Product Lifetime*   A company has redesigned a product in an attempt to increase the lifetime of the product. The two sets of data list the lifetimes (in months) of 20 units with the original design and 20 units with the new design. Create a box-and-whisker plot for each set of data, and then comment on the differences between the plots.

Original Design

15.1	78.3	56.3	68.9	30.6
27.2	12.5	42.7	72.7	20.2
53.0	13.5	11.0	18.4	85.2
10.8	38.3	85.1	10.0	12.6

New Design

55.8	71.5	25.6	19.0	23.1
37.2	60.0	35.3	18.9	80.5
46.7	31.1	67.9	23.5	99.5
54.0	23.2	45.5	24.8	87.8

C.2 Least Squares Regression

In many of the examples and exercises in this text, you have been asked to use the *regression* feature of a graphing utility to find mathematical models for sets of data. The *regression* feature of a graphing utility uses the **method of least squares** to find a mathematical model for a set of data. As a measure of how well a model fits a set of data points

$$\{(x_1, y_1), (x_2, y_2), (x_3, y_3), \ldots, (x_n, y_n)\}$$

you can add the squares of the differences between the actual y-values and the values given by the model to obtain the **sum of the squared differences.** For instance, the table shows the heights x (in feet) and the diameters y (in inches) of eight trees. The table also shows the values of a linear model $y^* = 0.54x - 29.5$ for each x-value. The sum of squared differences for the model is 51.7.

x	70	72	75	76	85	78	77	80
y	8.3	10.5	11.0	11.4	12.9	14.0	16.3	18.0
y^*	8.3	9.38	11.0	11.54	16.4	12.62	12.08	13.7
$(y - y^*)^2$	0	1.2544	0	0.0196	12.25	1.9044	17.8084	18.49

The model that has the *least* sum of squared differences is the **least squares regression** line for the data. The least squares regression line for the data in the table is $y \approx 0.43x - 20.3$. The sum of squared differences is 43.3.

To find the least squares regression line $y = ax + b$ for the points $\{(x_1, y_1), (x_2, y_2), (x_3, y_3), \ldots, (x_n, y_n)\}$ algebraically, you need to solve the following system for a and b.

$$\begin{cases} nb + \left(\displaystyle\sum_{i=1}^{n} x_i\right)a = \displaystyle\sum_{i=1}^{n} y_i \\ \left(\displaystyle\sum_{i=1}^{n} x_i\right)b + \left(\displaystyle\sum_{i=1}^{n} x_i^2\right)a = \displaystyle\sum_{i=1}^{n} x_i y_i \end{cases}$$

In the system,

$$\sum_{i=1}^{n} x_i = x_1 + x_2 + \cdots + x_n$$

$$\sum_{i=1}^{n} y_i = y_1 + y_2 + \cdots + y_n$$

$$\sum_{i=1}^{n} x_i^2 = x_1^2 + x_2^2 + \cdots + x_n^2$$

$$\sum_{i=1}^{n} x_i y_i = x_1 y_1 + x_2 y_2 + \cdots + x_n y_n.$$

TECHNOLOGY TIP   Recall from Section 2.6 that when you use the *regression* feature of a graphing utility, the program may output a correlation coefficient, r. When $|r|$ is close to 1, the model is a good fit for the data.

What you should learn

- Use the sum of squared differences to determine a least squares regression line.
- Find a least squares regression line for a set of data.
- Find a least squares regression parabola for a set of data.

Why you should learn it

The method of least squares provides a way of creating a mathematical model for a set of data, which can then be analyzed.

TECHNOLOGY SUPPORT

For instructions on how to use the *regression* feature, see Appendix A; for specific keystrokes, go to the text website at *college.hmco.com*.

Example 1   Finding a Least Squares Regression Line

Find the least squares regression line for $(-3, 0)$, $(-1, 1)$, $(0, 2)$, and $(2, 3)$.

Solution

Begin by constructing a table, as shown below.

x	y	xy	x^2
-3	0	0	9
-1	1	-1	1
0	2	0	0
2	3	6	4
$\sum_{i=1}^{n} x_i = -2$	$\sum_{i=1}^{n} y_i = 6$	$\sum_{i=1}^{n} x_i y_i = 5$	$\sum_{i=1}^{n} x_i^2 = 14$

Applying the system for the least squares regression line with $n = 4$ produces

$$\begin{cases} nb + \left(\sum_{i=1}^{n} x_i \right) a = \sum_{i=1}^{n} y_i \\ \left(\sum_{i=1}^{n} x_i \right) b + \left(\sum_{i=1}^{n} x_i^2 \right) a = \sum_{i=1}^{n} x_i y_i \end{cases}$$

$$\Rightarrow \quad \begin{cases} 4b - 2a = 6 \\ -2b + 14a = 5 \end{cases}.$$

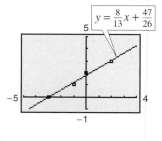

$$y = \frac{8}{13}x + \frac{47}{26}$$

Solving this system of equations produces $a = \frac{8}{13}$ and $b = \frac{47}{26}$. So, the least squares regression line is $y = \frac{8}{13}x + \frac{47}{26}$, as shown in Figure C.8.

Figure C.8

The least squares regression parabola $y = ax^2 + bx + c$ for the points

$$\{(x_1, y_1), (x_2, y_2), (x_3, y_3), \ldots, (x_n, y_n)\}$$

is obtained in a similar manner by solving the following system of three equations in three unknowns for a, b, and c.

$$\begin{cases} nc + \left(\sum_{i=1}^{n} x_i \right) b + \left(\sum_{i=1}^{n} x_i^2 \right) a = \sum_{i=1}^{n} y_i \\ \left(\sum_{i=1}^{n} x_i \right) c + \left(\sum_{i=1}^{n} x_i^2 \right) b + \left(\sum_{i=1}^{n} x_i^3 \right) a = \sum_{i=1}^{n} x_i y_i \\ \left(\sum_{i=1}^{n} x_i^2 \right) c + \left(\sum_{i=1}^{n} x_i^3 \right) b + \left(\sum_{i=1}^{n} x_i^4 \right) a = \sum_{i=1}^{n} x_i^2 y_i \end{cases}$$

C.2   Exercises

In Exercises 1–4, find the least squares regression line for the points. Verify your answer with a graphing utility.

1. $(-4, 1)$, $(-3, 3)$, $(-2, 4)$, $(-1, 6)$

2. $(0, -1)$, $(2, 0)$, $(4, 3)$, $(6, 5)$

3. $(-3, 1)$, $(-1, 2)$, $(1, 2)$, $(4, 3)$

4. $(0, -1)$, $(2, 1)$, $(3, 2)$, $(5, 3)$

Appendix D | Solving Linear Equations and Inequalities

Linear Equations

What you should learn

● Solve linear equations in one variable.

● Solve linear inequalities in one variable.

Why you should learn it

The method of solving linear equations is used to determine the intercepts of the graph of a linear function. The method of solving linear inequalities is used to determine the domains of different functions.

A *linear equation* in one variable x is an equation that can be written in the standard form $ax + b = 0$, where a and b are real numbers with $a \neq 0$.

A linear equation has exactly one solution. To see this, consider the following steps. (Remember that $a \neq 0$.)

$$ax + b = 0 \qquad \text{Original equation}$$

$$ax = -b \qquad \text{Subtract } b \text{ from each side.}$$

$$x = -\frac{b}{a} \qquad \text{Divide each side by } a.$$

To solve a linear equation in x, isolate x on one side of the equation by creating a sequence of *equivalent* (and usually simpler) equations, each having the same solution(s) as the original equation. The operations that yield equivalent equations come from the Substitution Principle and the Properties of Equality studied in Chapter P.

Generating Equivalent Equations

An equation can be transformed into an *equivalent equation* by one or more of the following steps.

	Original Equation	*Equivalent Equation*
1. Remove symbols of grouping, combine like terms, or simplify fractions on one or both sides of the equation.	$2x - x = 4$	$x = 4$
2. Add (or subtract) the same quantity to (from) *each* side of the equation.	$x + 1 = 6$	$x = 5$
3. Multiply (or divide) *each* side of the equation by the same *nonzero* quantity.	$2x = 6$	$x = 3$
4. Interchange the two sides of the equation.	$2 = x$	$x = 2$

After solving an equation, you should check each solution in the original equation. For example, you can check the solution to the equation in step 2 above as follows.

$$x + 1 = 6 \qquad \text{Write original equation.}$$

$$5 + 1 \stackrel{?}{=} 6 \qquad \text{Substitute 5 for } x.$$

$$6 = 6 \qquad \text{Solution checks. } \checkmark$$

Example 1   Solving Linear Equations

a. $3x - 6 = 0$ Original equation

$3x - 6 + 6 = 0 + 6$ Add 6 to each side.

$3x = 6$ Simplify.

$x = 2$ Divide each side by 3.

b. $4(2x + 3) = 6$ Original equation

$8x + 12 = 6$ Distributive Property

$8x + 12 - 12 = 6 - 12$ Subtract 12 from each side.

$8x = -6$ Simplify.

$\dfrac{8x}{8} = \dfrac{-6}{8}$ Divide each side by 8.

$x = -\dfrac{3}{4}$ Simplify.

✓ *Checkpoint* Now try Exercise 15.

Linear Inequalities

Solving a linear inequality in one variable is much like solving a linear equation in one variable. To solve the inequality, you isolate the variable on one side using transformations that produce *equivalent inequalities*, which have the same solution(s) as the original inequality.

Generating Equivalent Inequalities

An inequality can be transformed into an *equivalent inequality* by one or more of the following steps.

	Original Inequality	*Equivalent Inequality*
1. Remove symbols of grouping, combine like terms, or simplify fractions on one or both sides of the inequality.	$4x + x \geq 2$	$5x \geq 2$
2. Add (or subtract) the same number to (from) *each* side of the inequality.	$x - 3 < 5$	$x < 8$
3. Multiply (or divide) each side of the inequality by the same *positive* number.	$\frac{1}{2}x > 3$	$x > 6$
4. Multiply (or divide) each side of the inequality by the same *negative* number and *reverse* the inequality symbol.	$-2x \leq 6$	$x \geq -3$

Example 2   Solving Linear Inequalities

a. $x + 5 \geq 3$ Original inequality

$x + 5 - 5 \geq 3 - 5$ Subtract 5 from each side.

$x \geq -2$ Simplify.

The solution is all real numbers greater than or equal to -2, which is denoted by $[-2, \infty)$. Check several numbers that are greater than or equal to -2 in the original inequality.

b. $-4.2m > \ 6.3$ Original inequality

$\dfrac{-4.2m}{-4.2} < \dfrac{6.3}{-4.2}$ Divide each side by -4.2 and reverse inequality symbol.

$m < -1.5$ Simplify.

The solution is all real numbers less than -1.5, which is denoted by $(-\infty, -1.5)$. Check several numbers that are less than -1.5 in the original inequality.

> **STUDY TIP**
>
> Remember that when you multiply or divide by a negative number, you *must reverse* the inequality symbol, as shown in Example 2(b).

✓ *Checkpoint* Now try Exercise 29.

D Exercises

Vocabulary Check

Fill in the blanks.

1. A _____ equation in one variable x is an equation that can be written in the standard form $ax + b = 0$.

2. To solve a linear inequality, isolate the variable on one side using transformations that produce _____ .

In Exercises 1–22, solve the equation and check your solution.

1. $x + 11 = 15$

2. $x + 3 = 9$

3. $x - 2 = 5$

4. $x - 5 = 1$

5. $3x = 12$

6. $2x = 6$

7. $\dfrac{x}{5} = 4$

8. $\dfrac{x}{4} = 5$

9. $8x + 7 = 39$

10. $12x - 5 = 43$

11. $24 - 7x = 3$

12. $13 + 6x = 61$

13. $8x - 5 = 3x + 20$

14. $7x + 3 = 3x - 17$

15. $-2(x + 5) = 10$

16. $4(3 - x) = 9$

17. $2x + 3 = 2x - 2$

18. $8(x - 2) = 4(2x - 4)$

19. $\frac{3}{2}(x + 5) - \frac{1}{4}(x + 24) = 0$

20. $\frac{3}{2}x + \frac{1}{4}(x - 2) = 10$

21. $0.25x + 0.75(10 - x) = 3$

22. $0.60x + 0.40(100 - x) = 50$

In Exercises 23–44, solve the inequality and check your solution.

23. $x + 6 < 8$

24. $3 + x > -10$

25. $-x - 8 > -17$

26. $-3 + x < 19$

27. $6 + x \leq -8$

28. $x - 10 \geq -6$

29. $\frac{4}{5}x > 8$

30. $\frac{2}{3}x < -4$

31. $-\frac{3}{4}x > -3$

32. $-\frac{1}{6}x < -2$

33. $4x < 12$

34. $10x > -40$

35. $-11x \leq -22$

36. $-7x \geq 21$

37. $x - 3(x + 1) \geq 7$

38. $2(4x - 5) - 3x \leq -15$

39. $7x - 12 < 4x + 6$

40. $11 - 6x \leq 2x + 7$

41. $\frac{3}{4}x - 6 \leq x - 7$

42. $3 + \frac{2}{7}x > x - 2$

43. $3.6x + 11 \geq -3.4$

44. $15.6 - 1.3x < -5.2$

Appendix E   Systems of Inequalities

E.1   Solving Systems of Inequalities

The Graph of an Inequality

The statements $3x - 2y < 6$ and $2x^2 + 3y^2 \geq 6$ are inequalities in two variables. An ordered pair (a, b) is a **solution of an inequality** in x and y if the inequality is true when a and b are substituted for x and y, respectively. The **graph of an inequality** is the collection of all solutions of the inequality. To sketch the graph of an inequality, begin by sketching the graph of the *corresponding equation*. The graph of the equation will normally separate the plane into two or more regions. In each such region, one of the following must be true.

1. *All* points in the region are solutions of the inequality.

2. *No* point in the region is a solution of the inequality.

So, you can determine whether the points in an entire region satisfy the inequality by simply testing *one* point in the region.

> ### Sketching the Graph of an Inequality in Two Variables
>
> **1.** Replace the inequality sign with an equal sign and sketch the graph of the corresponding equation. Use a dashed line for $<$ or $>$ and a solid line for $\leq$ or $\geq$. (A dashed line means all points on the line or curve *are not* solutions of the inequality. A solid line means all points on the line or curve *are* solutions of the inequality.)
>
> **2.** Test one point in each of the regions formed by the graph in Step 1. If the point satisfies the inequality, shade the entire region to denote that every point in the region satisfies the inequality.

Example 1   Sketching the Graph of an Inequality

Sketch the graph of $y \geq x^2 - 1$ by hand.

Solution

Begin by graphing the corresponding *equation* $y = x^2 - 1$, which is a parabola, as shown in Figure E.1. By testing a point *above* the parabola $(0, 0)$ and a point *below* the parabola $(0, -2)$, you can see that $(0, 0)$ satisfies the inequality because $0 \geq 0^2 - 1$ and that $(0, -2)$ does not satisfy the inequality because $-2 \not\geq 0^2 - 1$. So, the points that satisfy the inequality are those lying above and those lying on the parabola.

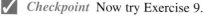 *Checkpoint* Now try Exercise 9.

What you should learn

- Sketch graphs of inequalities in two variables.
- Solve systems of inequalities.
- Use systems of inequalities in two variables to model and solve real-life problems.

Why you should learn it

Systems of inequalities in two variables can be used to model and solve real-life problems. For instance, Exercise 71 on page A61 shows how to use a system of inequalities to analyze the compositions of dietary supplements.

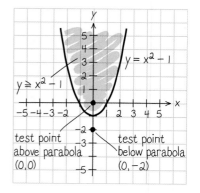

Figure E.1

The inequality in Example 1 is a nonlinear inequality in two variables. Most of the following examples involve **linear inequalities** such as $ax + by < c$ (a and b are not both zero). The graph of a linear inequality is a half-plane lying on one side of the line $ax + by = c$.

Example 2 Sketching the Graphs of Linear Inequalities

Sketch the graph of each linear inequality.

a. $x > -2$ **b.** $y \le 3$

Solution

a. The graph of the corresponding equation $x = -2$ is a vertical line. The points that satisfy the inequality $x > -2$ are those lying to the right of (but not on) this line, as shown in Figure E.2.

b. The graph of the corresponding equation $y = 3$ is a horizontal line. The points that satisfy the inequality $y \le 3$ are those lying below (or on) this line, as shown in Figure E.3.

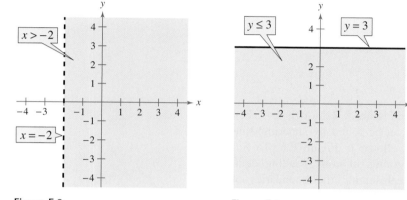

Figure E.2   Figure E.3

✓ *Checkpoint* Now try Exercise 13.

TECHNOLOGY TIP

A graphing utility can be used to graph an inequality. For instance, to graph $y \ge x - 2$, enter $y = x - 2$ and use the *shade* feature of the graphing utility to shade the correct part of the graph. You should obtain the graph shown below.

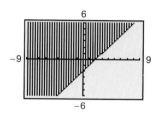

For instructions on how to use the *shade* feature, see Appendix A; for specific keystrokes, go to the text website at *college.hmco.com*.

Example 3 Sketching the Graph of a Linear Inequality

Sketch the graph of $x - y < 2$.

Solution

The graph of the corresponding equation $x - y = 2$ is a line, as shown in Figure E.4. Because the origin $(0, 0)$ satisfies the inequality, the graph consists of the half-plane lying above the line. (Try checking a point below the line. Regardless of which point below the line you choose, you will see that it does not satisfy the inequality.)

✓ *Checkpoint* Now try Exercise 15.

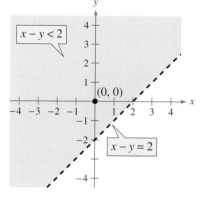

Figure E.4

To graph a linear inequality, it can help to write the inequality in slope-intercept form. For instance, by writing $x - y < 2$ in Example 3 in the form

$$y > x - 2$$

you can see that the solution points lie *above* the line $y = x - 2$ (or $x - y = 2$), as shown in Figure E.4.

Systems of Inequalities

Many practical problems in business, science, and engineering involve systems of linear inequalities. A **solution of a system of inequalities** in x and y is a point (x, y) that satisfies each inequality in the system.

To sketch the graph of a system of inequalities in two variables, first sketch the graph of each individual inequality (on the same coordinate system) and then find the region that is *common* to every graph in the system. For systems of *linear* inequalities, it is helpful to find the vertices of the solution region.

Example 4   Solving a System of Inequalities

Sketch the graph (and label the vertices) of the solution set of the system.

$$\begin{cases} x - y < 2 & \text{Inequality 1} \\ x > -2 & \text{Inequality 2} \\ y \le 3 & \text{Inequality 3} \end{cases}$$

Solution

The graphs of these inequalities are shown in Figures E.2 through E.4. The triangular region common to all three graphs can be found by superimposing the graphs on the same coordinate system, as shown in Figure E.5. To find the vertices of the region, solve the three systems of corresponding equations obtained by taking pairs of equations representing the boundaries of the individual regions and solving these pairs of equations.

Vertex A: $(-2, -4)$ *Vertex B:* $(5, 3)$ *Vertex C:* $(-2, 3)$

$$\begin{cases} x - y = 2 \\ x = -2 \end{cases} \qquad \begin{cases} x - y = 2 \\ y = 3 \end{cases} \qquad \begin{cases} x = -2 \\ y = 3 \end{cases}$$

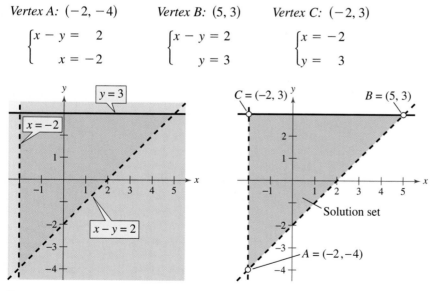

Figure E.5

> **STUDY TIP**
>
> Using different colored pencils to shade the solution of each inequality in a system makes identifying the solution of the system of inequalities easier. The region common to every graph in the system is where all shaded regions overlap. This region represents the solution set of the system.

Note in Figure E.5 that the vertices of the region are represented by open dots. This means that the vertices *are not* solutions of the system of inequalities.

✓ *Checkpoint* Now try Exercise 39.

For the triangular region shown in Figure E.5, each point of intersection of a pair of boundary lines corresponds to a vertex. With more complicated regions, two border lines can sometimes intersect at a point that is not a vertex of the region, as shown in Figure E.6. To keep track of which points of intersection are actually vertices of the region, you should sketch the region and refer to your sketch as you find each point of intersection.

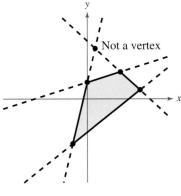

Figure E.6

Example 5   Solving a System of Inequalities

Sketch the region containing all points that satisfy the system of inequalities.

$$\begin{cases} x^2 - y \le 1 & \text{Inequality 1} \\ -x + y \le 1 & \text{Inequality 2} \end{cases}$$

Solution

As shown in Figure E.7, the points that satisfy the inequality $x^2 - y \le 1$ are the points lying above (or on) the parabola given by

$y = x^2 - 1.$ Parabola

The points that satisfy the inequality $-x + y \le 1$ are the points lying below (or on) the line given by

$y = x + 1.$ Line

To find the points of intersection of the parabola and the line, solve the system of corresponding equations.

$$\begin{cases} x^2 - y = 1 \\ -x + y = 1 \end{cases}$$

Using the method of substitution, you can find the solutions to be $(-1, 0)$ and $(2, 3)$. So, the region containing all points that satisfy the system is indicated by the purple shaded region in Figure E.7.

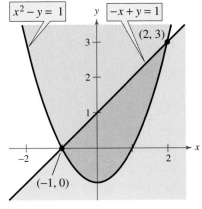

Figure E.7

✓ **Checkpoint** Now try Exercise 47.

When solving a system of inequalities, you should be aware that the system might have no solution, or it might be represented by an unbounded region in the plane. These two possibilities are shown in Examples 6 and 7.

Example 6   A System with No Solution

Sketch the solution set of the system of inequalities.

$$\begin{cases} x + y > 3 & \text{Inequality 1} \\ x + y < -1 & \text{Inequality 2} \end{cases}$$

Solution

From the way the system is written, it is clear that the system has no solution, because the quantity $(x + y)$ cannot be both less than -1 and greater than 3. Graphically, the inequality $x + y > 3$ is represented by the half-plane lying above the line $x + y = 3$, and the inequality $x + y < -1$ is represented by the half-plane lying below the line $x + y = -1$, as shown in Figure E.8. These two half-planes have no points in common. So the system of inequalities has no solution.

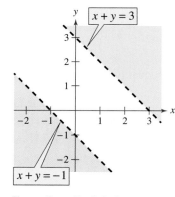

Figure E.8   No Solution

 Checkpoint Now try Exercise 43.

Example 7   An Unbounded Solution Set

Sketch the solution set of the system of inequalities.

$$\begin{cases} x + y < 3 & \text{Inequality 1} \\ x + 2y > 3 & \text{Inequality 2} \end{cases}$$

Solution

The graph of the inequality $x + y < 3$ is the half-plane that lies below the line $x + y = 3$, as shown in Figure E.9. The graph of the inequality $x + 2y > 3$ is the half-plane that lies above the line $x + 2y = 3$. The intersection of these two half-planes is an *infinite wedge* that has a vertex at $(3, 0)$. This unbounded region represents the solution set.

 Checkpoint Now try Exercise 45.

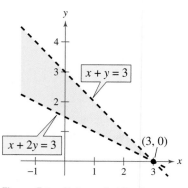

Figure E.9   Unbounded Region

Applications

The next example discusses two concepts that economists call *consumer surplus* and *producer surplus*. As shown in Figure E.10, the *point of equilibrium* is defined by the price p and the number of units x that satisfy both the demand and supply equations. Consumer surplus is defined as the area of the region that lies *below* the demand curve, *above* the horizontal line passing through the equilibrium point, and to the right of the p-axis. Similarly, the producer surplus is defined as the area of the region that lies *above* the supply curve, *below* the horizontal line passing through the equilibrium point, and to the right of the p-axis. The consumer surplus is a measure of the amount that consumers would have been willing to pay *above what they actually paid*, whereas the producer surplus is a measure of the amount that producers would have been willing to receive *below what they actually received*.

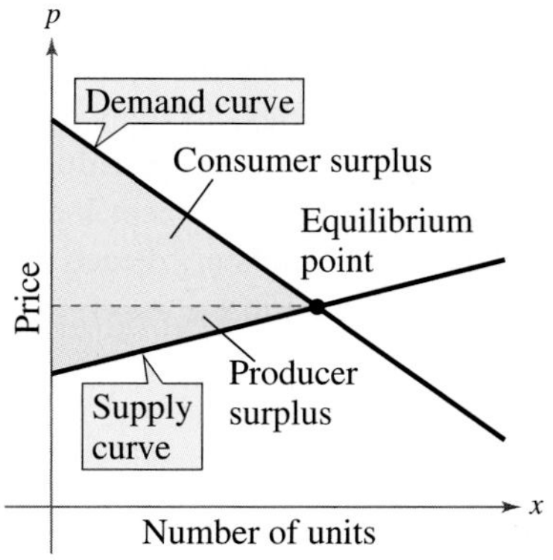

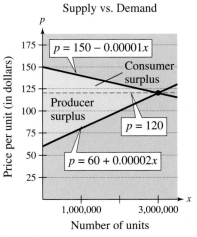

Figure E.10

Example 8   Consumer Surplus and Producer Surplus

The demand and supply functions for a new type of calculator are given by

$$\begin{cases} p = 150 - 0.00001x & \text{Demand equation} \\ p = 60 + 0.00002x & \text{Supply equation} \end{cases}$$

where p is the price (in dollars) and x represents the number of units. Find the consumer surplus and producer surplus for these two equations.

Solution

Begin by finding the point of equilibrium by setting the two equations equal to each other and solving for x.

$$60 + 0.00002x = 150 - 0.00001x \qquad \text{Set equations equal to each other.}$$

$$0.00003x = 90 \qquad \text{Combine like terms.}$$

$$x = 3{,}000{,}000 \qquad \text{Solve for } x.$$

So, the solution is $x = 3{,}000{,}000$, which corresponds to an equilibrium price of $p = \$120$. So, the consumer surplus and producer surplus are the areas of the following triangular regions.

Consumer Surplus	*Producer Surplus*
$\begin{cases} p \le 150 - 0.00001x \\ p \ge 120 \\ x \ge 0 \end{cases}$	$\begin{cases} p \ge 60 + 0.00002x \\ p \le 120 \\ x \ge 0 \end{cases}$

In Figure E.11, you can see that the consumer and producer surpluses are defined as the areas of the shaded triangles.

$$\text{Consumer surplus} = \tfrac{1}{2}(\text{base})(\text{height}) = \tfrac{1}{2}(3{,}000{,}000)(30) = \$45{,}000{,}000$$

$$\text{Producer surplus} = \tfrac{1}{2}(\text{base})(\text{height}) = \tfrac{1}{2}(3{,}000{,}000)(60) = \$90{,}000{,}000$$

✓ *Checkpoint* Now try Exercise 65.

Supply vs. Demand

Figure E.11

Example 9 Nutrition

The minimum daily requirements from the liquid portion of a diet are 300 calories, 36 units of vitamin A, and 90 units of vitamin C. A cup of dietary drink X provides 60 calories, 12 units of vitamin A, and 10 units of vitamin C. A cup of dietary drink Y provides 60 calories, 6 units of vitamin A, and 30 units of vitamin C. Set up a system of linear inequalities that describes how many cups of each drink should be consumed each day to meet the minimum daily requirements for calories and vitamins.

Solution

Begin by letting x and y represent the following.

x = number of cups of dietary drink X

y = number of cups of dietary drink Y

To meet the minimum daily requirements, the following inequalities must be satisfied.

$$\begin{cases} 60x + 60y \geq 300 & \text{Calories} \\ 12x + 6y \geq 36 & \text{Vitamin A} \\ 10x + 30y \geq 90 & \text{Vitamin C} \\ x \geq 0 \\ y \geq 0 \end{cases}$$

The last two inequalities are included because x and y cannot be negative. The graph of this system of inequalities is shown in Figure E.12. (More is said about this application in Example 6 of Section E.2.)

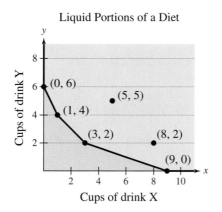

Liquid Portions of a Diet

Figure E.12

From the graph, you can see that two solutions (other than the vertices) that will meet the minimum daily requirements for calories and vitamins are $(5, 5)$ and $(8, 2)$. There are many other solutions.

 Checkpoint Now try Exercise 71.

E.1 Exercises

Vocabulary Check

Fill in the blanks.

1. An ordered pair (a, b) is a _____ of an inequality in x and y if the inequality is true when a and b are substituted for x and y, respectively.

2. The _____ of an inequality is the collection of all solutions of the inequality.

3. The graph of a _____ inequality is a half-plane lying on one side of the line $ax + by = c$.

4. The _____ of _____ is defined by the price p and the number of units x that satisfy both the demand and supply equations.

In Exercises 1–8, match the inequality with its graph. [The graphs are labeled (a), (b), (c), (d), (e), (f), (g), and (h).]

1. $x < 2$
2. $y \geq 3$
3. $2x + 3y \geq 6$
4. $2x - y \leq -2$
5. $x^2 + y^2 < 9$
6. $(x - 2)^2 + (y - 3)^2 > 9$
7. $xy > 1$
8. $y \leq 1 - x^2$

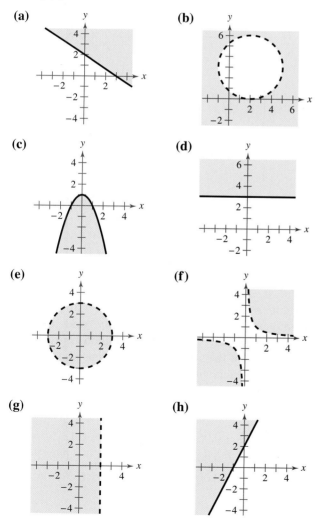

(a)   **(b)**

(c)   **(d)**

(e)   **(f)**

(g)   **(h)**

In Exercises 9–20, sketch the graph of the inequality.

9. $y < 2 - x^2$
10. $y^2 - x < 0$
11. $x \geq 4$
12. $x \leq -5$
13. $y \geq -1$
14. $y \leq 3$
15. $2y - x \geq 4$
16. $5x + 3y \geq -15$
17. $y > 3x^2 + 1$
18. $4x + y^2 > 1$
19. $(x + 1)^2 + y^2 < 9$
20. $(x - 1)^2 + (y - 4)^2 > 9$

In Exercises 21–32, use a graphing utility to graph the inequality. Use the *shade* feature to shade the region representing the solution.

21. $y \geq \frac{2}{3}x - 1$
22. $y \leq 6 - \frac{3}{2}x$
23. $y < -3.8x + 1.1$
24. $y \geq -20.74 + 2.66x$
25. $x^2 + 5y - 10 \leq 0$
26. $2x^2 - y - 3 > 0$
27. $y \leq \dfrac{1}{1 + x^2}$
28. $y > \dfrac{-10}{x^2 + x + 4}$
29. $y < \ln x$
30. $y \geq 4 - \ln(x + 5)$
31. $y > 3^{-x-4}$
32. $y \leq 2^{2x-1} - 3$

In Exercises 33–36, write an inequality for the shaded region shown in the graph.

33.   **34.**

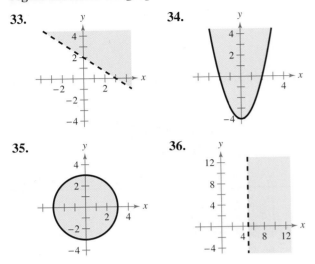

35.   **36.**

In Exercises 37 and 38, determine whether each ordered pair is a solution of the system of inequalities.

37. $\begin{cases} -2x + 5y \geq 3 \\ \quad\quad y < 4 \\ -4x + 2y < 7 \end{cases}$ (a) $(0, 2)$ (b) $(-6, 4)$
    (c) $(-8, -2)$ (d) $(-3, 2)$

38. $\begin{cases} x^2 + y^2 \geq 36 \\ -3x + y \leq 10 \\ \frac{2}{3}x - y \geq 5 \end{cases}$ (a) $(-1, 7)$ (b) $(-5, 1)$
    (c) $(6, 0)$ (d) $(4, -8)$

In Exercises 39–56, sketch the graph of the solution of the system of inequalities.

39. $\begin{cases} x + y \leq 1 \\ -x + y \leq 1 \\ \quad\quad y \geq 0 \end{cases}$ **40.** $\begin{cases} 3x + 2y < 6 \\ x \quad\quad > 0 \\ \quad\quad y > 0 \end{cases}$

41. $\begin{cases} -3x + 2y < 6 \\ x - 4y > -2 \\ 2x + y < 3 \end{cases}$ **42.** $\begin{cases} x - 7y > -36 \\ 5x + 2y > 5 \\ 6x - 5y > 6 \end{cases}$

43. $\begin{cases} 3x + y \leq y^2 \\ x - y > 0 \end{cases}$ **44.** $\begin{cases} y^2 - 3x \geq 9 \\ x + y \geq -3 \end{cases}$

45. $\begin{cases} 2x + y < 2 \\ x + 3y > 2 \end{cases}$ **46.** $\begin{cases} x - 2y < -6 \\ 2x - 4y > -9 \end{cases}$

47. $\begin{cases} x < y^2 \\ x > y + 2 \end{cases}$ **48.** $\begin{cases} x - y^2 > 0 \\ x - y < 2 \end{cases}$

49. $\begin{cases} x^2 + y^2 \leq 9 \\ x^2 + y^2 \geq 1 \end{cases}$ **50.** $\begin{cases} x^2 + y^2 \leq 25 \\ 4x - 3y \leq 0 \end{cases}$

51. $\begin{cases} y \leq \sqrt{3x} + 1 \\ y \geq x^2 + 1 \end{cases}$ **52.** $\begin{cases} y < -x^2 + 2x + 3 \\ y > x^2 - 4x + 3 \end{cases}$

53. $\begin{cases} y < x^3 - 2x + 1 \\ y > -2x \\ x \leq 1 \end{cases}$ **54.** $\begin{cases} y \geq x^4 - 2x^2 + 1 \\ y \leq 1 - x^2 \end{cases}$

55. $\begin{cases} x^2 y \geq 1 \\ 0 < x \leq 4 \\ y \leq 4 \end{cases}$ **56.** $\begin{cases} y \leq e^{-x^2/2} \\ y \geq 0 \\ -2 \leq x \leq 2 \end{cases}$

In Exercises 57–64, find a set of inequalities to describe the region.

57.   **58.**

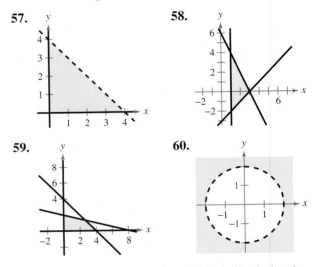

59.   **60.**

61. Rectangle: Vertices at $(2, 1)$, $(5, 1)$, $(5, 7)$, $(2, 7)$

62. Parallelogram: Vertices at $(0, 0)$, $(4, 0)$, $(1, 4)$, $(5, 4)$

63. Triangle: Vertices at $(0, 0)$, $(5, 0)$, $(2, 3)$

64. Triangle: Vertices at $(-1, 0)$, $(1, 0)$, $(0, 1)$

Supply and Demand In Exercises 65–68, graph the system representing the consumer surplus and producer surplus for the supply and demand equations. Be sure to shade the region representing the solution of the system. Find the consumer surplus and the producer surplus.

	Demand	*Supply*
65.	$p = 50 - 0.5x$	$p = 0.125x$
66.	$p = 100 - 0.05x$	$p = 25 + 0.1x$
67.	$p = 300 - 0.0002x$	$p = 225 + 0.0005x$
68.	$p = 140 - 0.00002x$	$p = 80 + 0.00001x$

In Exercises 69–72, (a) find a system of inequalities that models the problem and (b) graph the system, shading the region that represents the solution of the system.

69. *Investment Analysis*   A person plans to invest some or all of $30,000 in two different interest-bearing accounts. Each account is to contain at least $7500, and one account should have at least twice the amount that is in the other account.

70. *Ticket Sales*   For a summer concert event, one type of ticket costs $20 and another costs $35. The promoter of the concert must sell at least 20,000 tickets, including at least 10,000 of the $20 tickets and at least 5000 of the $35 tickets, and the gross receipts must total at least $300,000 in order for the concert to be held.

71. *Nutrition*   A dietitian is asked to design a special dietary supplement using two different foods. The minimum daily requirements of the new supplement are 280 units of calcium, 160 units of iron, and 180 units of vitamin B. Each ounce of food X contains 20 units of calcium, 15 units of iron, and 10 units of vitamin B. Each ounce of food Y contains 10 units of calcium, 10 units of iron, and 20 units of vitamin B.

72. *Inventory*   A store sells two models of computers. Because of the demand, the store stocks at least twice as many units of model A as units of model B. The costs to the store for models A and B are $800 and $1200, respectively. The management does not want more than $20,000 in computer inventory at any one time, and it wants at least four model A computers and two model B computers in inventory at all times.

73. *Construction*   You plan an exercise facility that has an indoor running track with an exercise floor inside the track (see figure). The track must be at least 125 meters long, and the exercise floor must have an area of at least 500 square meters.

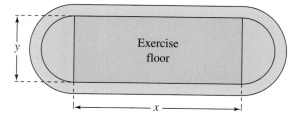

Exercise floor

(a) Find a system of inequalities describing the requirements of the facility.

(b) Sketch the graph of the system in part (a).

74. *Graphical Reasoning*   Two concentric circles have radii of x and y meters, where $y > x$ (see figure). The area between the boundaries of the circles must be at least 10 square meters.

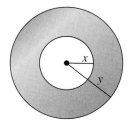

(a) Find an inequality describing the constraints on the circles.

(b) Graph the inequality in part (a).

(c) Identify the graph of the line $y = x$ in relation to the boundary of the inequality. Explain its meaning in the context of the problem.

Synthesis

True or False?   **In Exercises 75 and 76, determine whether the statement is true or false. Justify your answer.**

75. The area of the figure defined by the system below is 99 square units.
$$\begin{cases} x \geq -3 \\ x \leq 6 \\ y \leq 5 \\ y \geq -6 \end{cases}$$

76. The graph below shows the solution of the system
$$\begin{cases} y \leq 6 \\ -4x - 9y > 6. \\ 3x + y^2 \geq 2 \end{cases}$$

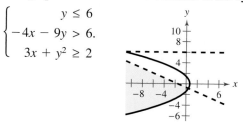

77. *Think About It*   After graphing the boundary of an inequality in x and y, how do you decide on which side of the boundary the solution set of the inequality lies?

78. *Writing*   Describe the difference between the solution set of a system of equations and the solution set of a system of inequalities.

E.2 Linear Programming

Linear Programming: A Graphical Approach

Many applications in business and economics involve a process called **optimization,** in which you are asked to find the minimum or maximum value of a quantity. In this section you will study an optimization strategy called **linear programming.**

A two-dimensional linear programming problem consists of a linear **objective function** and a system of linear inequalities called **constraints.** The objective function gives the quantity that is to be maximized (or minimized), and the constraints determine the set of **feasible solutions.** For example, suppose you are asked to maximize the value of

$$z = ax + by \qquad \text{Objective function}$$

subject to a set of constraints that determines the region in Figure E.13. Because every point in the shaded region satisfies each constraint, it is not clear how you should find the point that yields a maximum value of z. Fortunately, it can be shown that if there is an optimal solution, it must occur at one of the vertices. So, *you can find the maximum value of z by testing z at each of the vertices.*

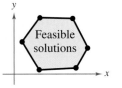

Figure E.13

Optimal Solution of a Linear Programming Problem

If a linear programming problem has a solution, it must occur at a vertex of the set of feasible solutions. If there is more than one solution, at least one of them must occur at such a vertex. In either case, the value of the objective function is unique.

Here are some guidelines for solving a linear programming problem in two variables in which an objective function is to be maximized *or* minimized.

Solving a Linear Programming Problem

1. Sketch the region corresponding to the system of constraints. (The points inside or on the boundary of the region are *feasible solutions*.)

2. Find the vertices of the region.

3. Test the objective function at each of the vertices and select the values of the variables that optimize the objective function. For a bounded region, both a minimum and a maximum value will exist. (For an unbounded region, *if* an optimal solution exists, it will occur at a vertex.)

Example 1   Solving a Linear Programming Problem

Find the maximum value of

$$z = 3x + 2y \qquad \text{Objective function}$$

subject to the following constraints.

$$\left.\begin{array}{r} x \ge 0 \\ y \ge 0 \\ x + 2y \le 4 \\ x - y \le 1 \end{array}\right\} \qquad \text{Constraints}$$

Solution

The constraints form the region shown in Figure E.14. At the four vertices of this region, the objective function has the following values.

At $(0, 0)$: $z = 3(0) + 2(0) = 0$

At $(1, 0)$: $z = 3(1) + 2(0) = 3$

At $(2, 1)$: $z = 3(2) + 2(1) = 8$ Maximum value of z

At $(0, 2)$: $z = 3(0) + 2(2) = 4$

So, the maximum value of z is 8, and this value occurs when $x = 2$ and $y = 1$.

✓ *Checkpoint* Now try Exercise 13.

In Example 1, try testing some of the *interior* points in the region. You will see that the corresponding values of z are less than 8. Here are some examples.

At $(1, 1)$: $z = 3(1) + 2(1) = 5$

At $\left(1, \frac{1}{2}\right)$: $z = 3(1) + 2\left(\frac{1}{2}\right) = 4$

At $\left(\frac{1}{2}, \frac{3}{2}\right)$: $z = 3\left(\frac{1}{2}\right) + 2\left(\frac{3}{2}\right) = \frac{9}{2}$

To see why the maximum value of the objective function in Example 1 must occur at a vertex, consider writing the objective function in the form

$$y = -\frac{3}{2}x + \frac{z}{2} \qquad \text{Family of lines}$$

where $z/2$ is the y-intercept of the objective function. This equation represents a family of lines, each of slope $-\frac{3}{2}$. Of these infinitely many lines, you want the one that has the largest z-value while still intersecting the region determined by the constraints. In other words, of all the lines with a slope of $-\frac{3}{2}$, you want the one that has the largest y-intercept *and* intersects the given region, as shown in Figure E.15. It should be clear that such a line will pass through one (or more) of the vertices of the region.

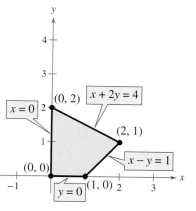

Figure E.14

STUDY TIP

Remember that a vertex of a region can be found using a system of linear equations. The system will consist of the equations of the lines passing through the vertex.

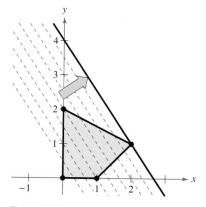

Figure E.15

The next example shows that the same basic procedure can be used to solve a problem in which the objective function is to be *minimized*.

Example 2   Solving a Linear Programming Problem

Find the minimum value of

$$z = 5x + 7y \qquad \text{Objective function}$$

where $x \geq 0$ and $y \geq 0$, subject to the following constraints.

$$\left.\begin{array}{r} 2x + 3y \geq 6 \\ 3x - y \leq 15 \\ -x + y \leq 4 \\ 2x + 5y \leq 27 \end{array}\right\} \qquad \text{Constraints}$$

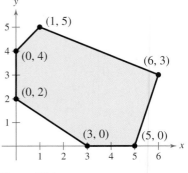

Figure E.16

Solution

The region bounded by the constraints is shown in Figure E.16. By testing the objective function at each vertex, you obtain the following.

At $(0, 2)$: $z = 5(0) + 7(2) = 14$ Minimum value of z
At $(0, 4)$: $z = 5(0) + 7(4) = 28$
At $(1, 5)$: $z = 5(1) + 7(5) = 40$
At $(6, 3)$: $z = 5(6) + 7(3) = 51$
At $(5, 0)$: $z = 5(5) + 7(0) = 25$
At $(3, 0)$: $z = 5(3) + 7(0) = 15$

So, the minimum value of z is 14, and this value occurs when $x = 0$ and $y = 2$.

✓ *Checkpoint* Now try Exercise 15.

Example 3   Solving a Linear Programming Problem

Find the maximum value of

$$z = 5x + 7y \qquad \text{Objective function}$$

where $x \geq 0$ and $y \geq 0$, subject to the following constraints.

$$\left.\begin{array}{r} 2x + 3y \geq 6 \\ 3x - y \leq 15 \\ -x + y \leq 4 \\ 2x + 5y \leq 27 \end{array}\right\} \qquad \text{Constraints}$$

Solution

This linear programming problem is identical to that given in Example 2 above, *except* that the objective function is *maximized* instead of minimized. Using the values of z at the vertices shown above, you can conclude that the maximum value of z is 51, and that this value occurs when $x = 6$ and $y = 3$.

✓ *Checkpoint* Now try Exercise 17.

It is possible for the maximum (or minimum) value in a linear programming problem to occur at *two* different vertices. For instance, at the vertices of the region shown in Figure E.17, the objective function

$$z = 2x + 2y \qquad \text{Objective function}$$

has the following values.

At $(0, 0)$: $z = 2(0) + 2(0) = 0$

At $(0, 4)$: $z = 2(0) + 2(4) = 8$

At $(2, 4)$: $z = 2(2) + 2(4) = 12$ Maximum value of z

At $(5, 1)$: $z = 2(5) + 2(1) = 12$ Maximum value of z

At $(5, 0)$: $z = 2(5) + 2(0) = 10$

In this case, you can conclude that the objective function has a maximum value (of 12) not only at the vertices $(2, 4)$ and $(5, 1)$, but also at *any point on the line segment connecting these two vertices*, as shown in Figure E.17. Note that by rewriting the objective function as

$$y = -x + \frac{1}{2}z$$

you can see that its graph has the same slope as the line through the vertices $(2, 4)$ and $(5, 1)$.

Some linear programming problems have no optimal solutions. This can occur if the region determined by the constraints is *unbounded*.

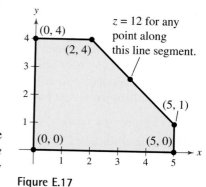

Figure E.17

Example 4   An Unbounded Region

Find the maximum value of

$$z = 4x + 2y \qquad \text{Objective function}$$

where $x \geq 0$ and $y \geq 0$, subject to the following constraints.

$$\left. \begin{array}{r} x + 2y \geq 4 \\ 3x + y \geq 7 \\ -x + 2y \leq 7 \end{array} \right\} \qquad \text{Constraints}$$

Solution

The region determined by the constraints is shown in Figure E.18. For this unbounded region, there is no maximum value of z. To see this, note that the point $(x, 0)$ lies in the region for all values of $x \geq 4$. By choosing large values of x, you can obtain values of $z = 4(x) + 2(0) = 4x$ that are as large as you want. So, there is no maximum value of z. For the vertices of the region, the objective function has the following values. So, there *is* a minimum value of z, $z = 10$, which occurs at the vertex $(2, 1)$.

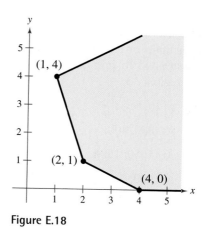

Figure E.18

At $(1, 4)$: $z = 4(1) + 2(4) = 12$

At $(2, 1)$: $z = 4(2) + 2(1) = 10$ Minimum value of z

At $(4, 0)$: $z = 4(4) + 2(0) = 16$

✓ *Checkpoint* Now try Exercise 27.

Applications

Example 5 shows how linear programming can be used to find the maximum profit in a business application.

Example 5   Optimizing Profit

A manufacturer wants to maximize the profit for selling two types of boxed chocolates. A box of chocolate covered creams yields a profit of $1.50 per box and a box of chocolate covered cherries yields a profit of $2.00 per box. Market tests and available resources have indicated the following constraints.

1. The combined production level should not exceed 1200 boxes per month.

2. The demand for a box of chocolate covered cherries is no more than half the demand for a box of chocolate covered creams.

3. The production level of a box of chocolate covered creams is less than or equal to 600 boxes plus three times the production level of a box of chocolate covered cherries.

Solution

Let x be the number of boxes of chocolate covered creams and y be the number of boxes of chocolate covered cherries. The objective function (for the combined profit) is given by

$P = 1.5x + 2y.$ Objective function

The three constraints translate into the following linear inequalities.

1. $x + y \le 1200$ $x + y \le 1200$

2. $y \le \frac{1}{2}x$ $-x + 2y \le 0$

3. $x \le 3y + 600$ $x - 3y \le 600$

Because neither x nor y can be negative, you also have the two additional constraints of $x \ge 0$ and $y \ge 0$. Figure E.19 shows the region determined by the constraints. To find the maximum profit, test the value of P at the vertices of the region.

At $(0, 0)$: $P = 1.5(0) + 2(0) = 0$

At $(800, 400)$: $P = 1.5(800) + 2(400) = 2000$ Maximum profit

At $(1050, 150)$: $P = 1.5(1050) + 2(150) = 1875$

At $(600, 0)$: $P = 1.5(600) + 2(0) = 900$

So, the maximum profit is $2000, and it occurs when the monthly production consists of 800 boxes of chocolate covered creams and 400 boxes of chocolate covered cherries.

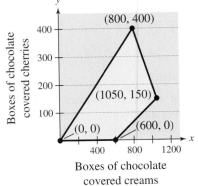

Figure E.19

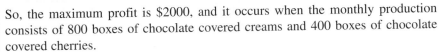

✓ *Checkpoint* Now try Exercise 31.

In Example 5, suppose the manufacturer improves the production of chocolate covered creams so that a profit of $2.50 per box is obtained. The maximum
profit can now be found using the objective function $P = 2.5x + 2y$. By testing
the values of P at the vertices of the region, you find that the maximum profit is
now $2925, which occurs when $x = 1050$ and $y = 150$.

Example 6   Optimizing Cost

The minimum daily requirements from the liquid portion of a diet are 300 calories, 36 units of vitamin A, and 90 units of vitamin C. A cup of dietary drink X
costs $0.12 and provides 60 calories, 12 units of vitamin A, and 10 units of vita-
min C. A cup of dietary drink Y costs $0.15 and provides 60 calories, 6 units of
vitamin A, and 30 units of vitamin C. How many cups of each drink should be
consumed each day to minimize the cost and still meet the daily requirements?

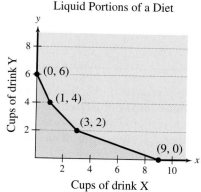

Liquid Portions of a Diet

Figure E.20

Solution

As in Example 9 on page A58, let x be the number of cups of dietary drink X and
let y be the number of cups of dietary drink Y.

For Calories:	$60x + 60y \geq 300$	
For Vitamin A:	$12x + 6y \geq 36$	
For Vitamin C:	$10x + 30y \geq 90$	Constraints
	$x \geq 0$	
	$y \geq 0$	

The cost C is given by

$$C = 0.12x + 0.15y.$$ Objective function

The graph of the region determined by the constraints is shown in Figure E.20.
To determine the minimum cost, test C at each vertex of the region.

At $(0, 6)$: $C = 0.12(0) + 0.15(6) = 0.90$

At $(1, 4)$: $C = 0.12(1) + 0.15(4) = 0.72$

At $(3, 2)$: $C = 0.12(3) + 0.15(2) = 0.66$ Minimum value of C

At $(9, 0)$: $C = 0.12(9) + 0.15(0) = 1.08$

So, the minimum cost is $0.66 per day, and this cost occurs when three cups of
drink X and two cups of drink Y are consumed each day.

✓ *Checkpoint* Now try Exercise 33.

E.2 Exercises

Vocabulary Check

Fill in the blanks.

1. In the process called _____ , you are asked to find the minimum or maximum value of a quantity.

2. The _____ of a linear programming problem gives the quantity that is to be maximized or minimized.

3. The _____ of a linear programming problem determine the set of _____ .

In Exercises 1–12, find the minimum and maximum values of the objective function and where they occur, subject to the indicated constraints. (For each exercise, the graph of the region determined by the constraints is provided.)

1. Objective function:

$z = 3x + 5y$

Constraints:

$x \geq 0$

$y \geq 0$

$x + y \leq 6$

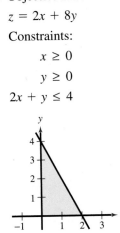

3. Objective function:

$z = 10x + 7y$

Constraints:

See Exercise 1.

5. Objective function:

$z = 3x + 2y$

Constraints:

$x \geq 0$

$y \geq 0$

$x + 3y \leq 15$

$4x + y \leq 16$

2. Objective function:

$z = 2x + 8y$

Constraints:

$x \geq 0$

$y \geq 0$

$2x + y \leq 4$

4. Objective function:

$z = 7x + 3y$

Constraints:

See Exercise 2.

6. Objective function:

$z = 4x + 3y$

Constraints:

$x \geq 0$

$2x + 3y \geq 6$

$3x - 2y \leq 9$

$x + 5y \leq 20$

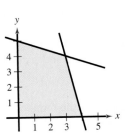

Figure for 5

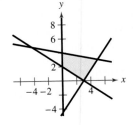

Figure for 6

7. Objective function:

$z = 5x + 0.5y$

Constraints:

See Exercise 5.

9. Objective function:

$z = 10x + 7y$

Constraints:

$0 \leq x \leq 60$

$0 \leq y \leq 45$

$5x + 6y \leq 420$

8. Objective function:

$z = x + 6y$

Constraints:

See Exercise 6.

10. Objective function:

$z = 50x + 35y$

Constraints:

$x \geq 0$

$y \geq 0$

$8x + 9y \leq 7200$

$8x + 9y \geq 5400$

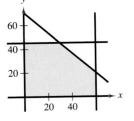

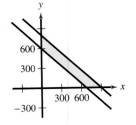

11. Objective function:

$z = 25x + 30y$

Constraints:

See Exercise 9.

12. Objective function:

$z = 15x + 20y$

Constraints:

See Exercise 10.

In Exercises 13–22, sketch the region determined by the constraints. Then find the minimum and maximum values of the objective function and where they occur, subject to the indicated constraints.

13. Objective function:

$z = 6x + 10y$

Constraints:

$$x \geq 0$$
$$y \geq 0$$
$$2x + 5y \leq 10$$

14. Objective function:

$z = 7x + 8y$

Constraints:

$$x \geq 0$$
$$y \geq 0$$
$$x + \tfrac{1}{2}y \leq 4$$

15. Objective function:

$z = 3x + 4y$

Constraints:

$$x \geq 0$$
$$y \geq 0$$
$$2x + 5y \leq 50$$
$$4x + y \leq 28$$

16. Objective function:

$z = 4x + 5y$

Constraints:

$$x \geq 0$$
$$y \geq 0$$
$$2x + 2y \leq 10$$
$$x + 2y \leq 6$$

17. Objective function:

$z = 4x + y$

Constraints:

$$x \geq 0$$
$$y \geq 0$$
$$x + 2y \leq 40$$
$$2x + 3y \geq 72$$

18. Objective function:

$z = x$

Constraints:

$$x \geq 0$$
$$y \geq 0$$
$$2x + 3y \leq 60$$
$$2x + y \leq 28$$
$$4x + y \leq 48$$

19. Objective function:

$z = x + 4y$

Constraints:
See Exercise 17.

20. Objective function:

$z = y$

Constraints:
See Exercise 18.

21. Objective function:

$z = 2x + 3y$

Constraints:
See Exercise 17.

22. Objective function:

$z = 3x + 2y$

Constraints:
See Exercise 18.

Exploration In Exercises 23–26, perform the following.

(a) Graph the region bounded by the following constraints.

$$3x + y \leq 15$$
$$4x + 3y \leq 30$$
$$x \geq 0$$
$$y \geq 0$$

(b) Graph the objective function for the given maximum value of z on the same set of coordinate axes as the graph of the constraints.

(c) Use the graph to determine the feasible point or points that yield the maximum. Explain how you arrived at your answer.

Objective Function	*Maximum*
23. $z = 2x + y$	$z = 12$
24. $z = 5x + y$	$z = 25$
25. $z = x + y$	$z = 10$
26. $z = 3x + y$	$z = 15$

In Exercises 27–30, the linear programming problem has an unusual characteristic. Sketch a graph of the solution region for the problem and describe the unusual characteristic. The objective function is to be maximized in each case.

27. Objective function:

$z = x + y$

Constraints:

$$x \geq 0$$
$$y \geq 0$$
$$-x + y \leq 1$$
$$-x + 2y \leq 4$$

28. Objective function:

$z = 2.5x + y$

Constraints:

$$x \geq 0$$
$$y \geq 0$$
$$3x + 5y \leq 15$$
$$5x + 2y \leq 10$$

29. Objective function:

$z = x + y$

Constraints:

$$x \geq 0$$
$$y \geq 0$$
$$-x + y \leq 0$$
$$-3x + y \geq 3$$

30. Objective function:

$z = -x + 2y$

Constraints:

$$x \geq 0$$
$$y \geq 0$$
$$x \leq 10$$
$$x + y \leq 7$$

31. ***Optimizing Revenue*** An accounting firm has 800 hours of staff time and 96 hours of reviewing time available each week. The firm charges $2000 for an audit and $300 for a tax return. Each audit requires 100 hours of staff time and 8 hours of review time. Each tax return requires 12.5 hours of staff time and 2 hours of review time. What numbers of audits and tax returns will yield the maximum revenue? What is the maximum revenue?

32. *Optimizing Profit* A manufacturer produces two models of snowboards. The amounts of time (in hours) required for assembling, painting, and packaging the two models are as follows.

	Model A	Model B
Assembling	2.5	3
Painting	2	1
Packaging	0.75	1.25

The total amounts of time available for assembling, painting, and packaging are 4000 hours, 2500 hours, and 1500 hours, respectively. The profits per unit are $50 for model A and $52 for model B. How many of each model should be produced to maximize profit? What is the maximum profit?

33. *Optimizing Cost* A farming cooperative mixes two brands of cattle feed. Brand X costs $25 per bag and contains two units of nutritional element A, two units of element B, and two units of element C. Brand Y costs $20 per bag and contains one unit of nutritional element A, nine units of element B, and 3 units of element C. The minimum requirements for nutrients A, B, and C are 12 units, 36 units, and 24 units, respectively. Find the number of bags of each brand that should be mixed to produce a mixture having a minimum cost per bag. What is the minimum cost?

34. *Optimizing Cost* A pet supply company mixes two brands of dry dog food. Brand X costs $15 per bag and contains eight units of nutritional element A, one unit of nutritional element B, and two units of nutritional element C. Brand Y costs $30 per bag and contains two units of nutritional element A, one unit of nutritional element B, and seven units of nutritional element C. Each bag of mixed dog food must contain at least 16 units, 5 units, and 20 units of nutritional elements A, B, and C, respectively. Find the numbers of bags of brands X and Y that should be mixed to produce a mixture meeting the minimum nutritional requirements and having a minimum cost per bag. What is the minimum cost?

Synthesis

True or False? In Exercises 35 and 36, determine whether the statement is true or false. Justify your answer.

35. If an objective function has a maximum value at the adjacent vertices $(4, 7)$ and $(8, 3)$, you can conclude that it also has a maximum value at the points $(4.5, 6.5)$ and $(7.8, 3.2)$.

36. When solving a linear programming problem, if the objective function has a maximum value at two adjacent vertices, you can assume that there are an infinite number of points that will produce the maximum value.

Think About It In Exercises 37–40, find an objective function that has a maximum or minimum value at the indicated vertex of the constraint region shown below. (There are many correct answers.)

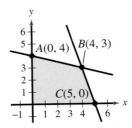

37. The maximum occurs at vertex A.

38. The maximum occurs at vertex B.

39. The maximum occurs at vertex C.

40. The minimum occurs at vertex C.

In Exercises 41 and 42, determine values of t such that the objective function has a maximum value at each indicated vertex.

41. Objective function:

$z = 3x + ty$

Constraints:

$x \geq 0$

$y \geq 0$

$x + 3y \leq 15$

$4x + y \leq 16$

(a) $(0, 5)$

(b) $(3, 4)$

42. Objective function:

$z = 3x + ty$

Constraints:

$x \geq 0$

$y \geq 0$

$x + 2y \leq 4$

$x - y \leq 1$

(a) $(2, 1)$

(b) $(0, 2)$

Answers to Odd-Numbered Exercises and Tests

Chapter P

Section P.1   (page 9)

> **Vocabulary Check**   (page 9)
>
> **1.** rational   **2.** Irrational   **3.** absolute value
>
> **4.** composite   **5.** prime   **6.** variables, constants
>
> **7.** terms   **8.** coefficient   **9.** Zero-Factor Property

1. (a) 5, 1   (b) 5, 0, 1   (c) $-9, 5, 0, 1, -4, -1$

  (d) $-9, -\frac{7}{2}, 5, \frac{2}{3}, 0, 1, -4, -1$ (e) $\sqrt{2}$

3. (a) 1, 20   (b) 1, 20   (c) $-13, 1, -10, 20$

  (d) $2.01, 0.666\ldots, -13, 1, -10, 20$

  (e) $0.010110111\ldots$

5. (a) $\frac{6}{3}, 3$ (b) $\frac{6}{3}, 3$ (c) $\frac{6}{3}, -2, 3, -3$

  (d) $-\frac{1}{3}, \frac{6}{3}, -7.5, -2, 3, -3$ (e) $-\pi, \frac{1}{2}\sqrt{2}$

7. 0.625   **9.** $0.\overline{123}$ **11.** $-9.\overline{09}$ **13.** $\frac{23}{5}$ **15.** $\frac{13}{2}$

17. $-1 < 2.5$

19.

$-4 > -8$

21.

$\frac{3}{2} < 7$

23.

$\frac{5}{6} > \frac{2}{3}$

25. (a) $x \le 5$ is the set of all real numbers less than or equal to 5.

  (b)   (c) Unbounded

27. (a) $x < 0$ is the set of all negative real numbers.

  (b)   (c) Unbounded

29. (a) $-2 < x < 2$ is the set of all real numbers greater than -2 and less than 2.

  (b)   (c) Bounded

31. (a) $-1 \le x < 0$ is the set of all negative real numbers greater than or equal to -1.

  (b)   (c) Bounded

33. $x < 0; (-\infty, 0)$ **35.** $y \ge 0; [0, \infty)$

37. $-1 \le p < 9; [-1, 9)$

39. The set of all real numbers greater than -6

41. The set of all real numbers less than or equal to 2

43. 10   **45.** -9

47. 1 for $x > -2$; undefined for $x = -2$; -1 for $x < -2$

49. $|-3| > -|-3|$ **51.** $-5 = -|5|$

53. $-|-2| = -|2|$ **55.** 51   **57.** $\frac{5}{2}$ **59.** $\frac{128}{75}$

61. $|x - 5| \le 3$ **63.** $|y - 0| \ge 6$ **65.** 179 miles

67. $|\$113{,}356 - \$112{,}700| = \$656 > \$500$

  $0.05(\$112{,}700) = \$5635$

  Because the actual expenses differ from the budget by more than \$500, there is failure to meet the "budget variance test."

69. $|\$37{,}335 - \$37{,}640| = \$305 < \$500$

  $0.05(\$37{,}640) = \$1882$

  Because the difference between the actual expenses and the budget is less than \$500 and less than 5% of the budgeted amount, there is compliance with the "budget variance test."

71. Receipts $= \$92.5$ billion,

  $|\text{Receipts} - \text{Expenditures}| = \$0.3$ billion

  There was a surplus of \$0.3 billion.

73. Receipts $= \$517.1$ billion,

  $|\text{Receipts} - \text{Expenditures}| = \$73.8$ billion

  There was a deficit of \$73.8 billion.

75. Receipts $= \$2025.2$ billion,

  $|\text{Receipts} - \text{Expenditures}| = \$236.4$ billion

  There was a surplus of \$236.4 billion.

77. Terms: $7x, 4$; coefficient: 7

79. Terms: $\sqrt{3}x^2, -8x, -11$; coefficients: $\sqrt{3}, -8$

81. Terms: $4x^3, \frac{x}{2}, -5$; coefficients: $4, \frac{1}{2}$

83. (a) -10 (b) -6 **85.** (a) -10 (b) 0

87. Commutative Property of Addition

89. Multiplicative Inverse Property

91. Distributive Property

93. Associative Property of Addition

95. $\dfrac{1}{2}$ **97.** $\dfrac{3}{8}$ **99.** $\dfrac{11x}{12}$ **101.** $\dfrac{96}{x}$ **103.** $-\dfrac{7}{5}$

105. -36 **107.** 1.56   **109.** 13.33

111. (a)

n	1	0.5	0.01	0.0001	0.000001
$5/n$	5	10	500	50,000	5,000,000

(b) $5/n$ approaches ∞ as n approaches 0.

113. False. A contradiction can be shown using the numbers $a = 2$ and $b = 1$. $2 > 1$, but $\frac{1}{2} \not> \frac{1}{1}$.

115. (a) $-A$ is negative.   (b) $B - A$ is negative.

117. (a) No. If u is negative while v is positive, or vice versa, the expressions will not be equal.

(b) $|u + v| \le |u| + |v|$

119. Answers will vary. Sample answer: Natural numbers are the integers from 1 to infinity. Whole numbers are integers from 0 to infinity. A rational number can be expressed as the ratio of two integers; an irrational number cannot.

Section P.2   (page 21)

Vocabulary Check   (page 21)

1. exponent, base   **2.** scientific notation

3. square root   **4.** principal nth root

5. index, radicand   **6.** simplest form

7. conjugates   **8.** rationalizing   **9.** power, index

1. (a) 48   (b) 81   **3.** (a) 729   (b) -9

5. (a) 243   (b) $-\frac{3}{4}$ **7.** (a) $\frac{5}{6}$ (b) 4   **9.** $\frac{7}{4}$

11. -54 **13.** 1   **15.** (a) $-125z^3$ (b) $5x^6$

17. (a) $\dfrac{7}{x}$ (b) $\dfrac{4}{3}(x + y)^2$, $x + y \ne 0$

19. (a) $\dfrac{x^2}{y^2}$, $x \ne 0$ (b) $\dfrac{b^5}{a^5}$, $b \ne 0$

21. -1600 **23.** 2.125   **25.** 5.73×10^7

27. 8.99×10^{-5} **29.** 564,000,000

31. 0.0000000000000000016022   **33.** 5×10^4 or 50,000

35. (a) 4.907×10^{17} (b) 1.479

37. (a) 67,082.039   (b) 39.791   **39.** 11   **41.** 3

43. -125 **45.** $\frac{1}{8}$ **47.** -4 **49.** -7.225

51. 21.316   **53.** 14.499

55. (a) 3   (b) $2x\sqrt[5]{3}$

57. (a) $3y^2\sqrt{6x}$ (b) $2\sqrt[3]{4a^2/b^2}$

59. (a) $34\sqrt{2}$ (b) $22\sqrt{2}$

61. (a) $13\sqrt{x + 1}$ (b) $18\sqrt{5x}$

63. $\sqrt{5} + \sqrt{3} > \sqrt{5 + 3}$ **65.** $5 > \sqrt{3^2 + 2^2}$

67. $\dfrac{\sqrt{3}}{3}$ **69.** $\dfrac{\sqrt{14} + 2}{2}$ **71.** $\dfrac{2}{\sqrt{2}}$

73. $\dfrac{2}{3(\sqrt{5} - \sqrt{3})}$ **75.** $64^{1/3} = 4$ **77.** $\sqrt[5]{32} = 2$

79. $(-216)^{1/3} = -6$ **81.** $8^{3/4} = 27$ **83.** $\dfrac{2}{x}$

85. $\dfrac{1}{x^3}$, $x > 0$ **87.** (a) $\sqrt{3}$ (b) $\sqrt[3]{(x + 1)^2}$

89. (a) $2\sqrt[4]{2}$ (b) $\sqrt[8]{2x}$ **91.** $T = \dfrac{\pi}{2} \approx 1.57$ seconds

93. 0.026 inches   **95.** True. $x^{k+1}/x = x^k x/x = x^k$ $(x \ne 0)$.

97. $1 = \dfrac{a^n}{a^n} = a^{n-n} = a^0$

99. When any positive integer is squared, the units digit is 0, 1, 4, 5, 6, or 9. Therefore, $\sqrt{5233}$ is not an integer.

Section P.3   (page 32)

Vocabulary Check   (page 32)

1. n, a_n **2.** zero polynomial   **3.** monomial

4. First, Outer, Inner, Last   **5.** prime

6. perfect square trinomial

1. d   **2.** e   **3.** b   **4.** a   **5.** f   **6.** c

7. Answers will vary, but first term is $-2x^3$.

9. Answers will vary, but first term has form $-ax^4$, $a > 0$.

11. $4x^2 + 3x + 2$

Degree: 2; leading coefficient: 4

13. $x^7 + 1$

Degree: 7; leading coefficient: 1

15. $-2x^5 + 6x^4 - x + 1$

Degree: 5; leading coefficient: -2

17. Polynomial: $-2x^3 + 7x + 10$ **19.** Not a polynomial

21. $-2x - 10$ **23.** $3x^3 - 2x + 2$

25. $8.1x^3 + 29.7x^2 + 11$ **27.** $3x^3 - 6x^2 + 3x$

29. $-15z^2 + 5z$ **31.** $-4x^4 + 4x$ **33.** $-7.5x^3 - 15x$

35. $-\frac{1}{4}x^2 - 6x$ **37.** $x^2 + 7x + 12$ **39.** $6x^2 - 7x - 5$

41. $4x^2 - 20xy + 25y^2$ **43.** $x^2 - 100$ **45.** $x^2 - 4y^2$

47. $4r^4 - 25$ **49.** $x^3 + 3x^2 + 3x + 1$

51. $8x^3 - 12x^2y + 6xy^2 - y^3$ **53.** $\frac{1}{4}x^2 - 5x + 25$

55. $\frac{1}{16}x^2 - 9$ **57.** $5.76x^2 + 14.4x + 9$

59. $-3x^4 - x^3 - 12x^2 - 19x - 5$

61. $m^2 - 6m + 9 - n^2$

63. $x^2 + 2xy + y^2 - 6y - 6x + 9$ **65.** $2x^2 + 2x$

67. $u^4 - 16$ **69.** $2(x + 4)$ **71.** $2x(x^2 - 3)$

73. $(x - 5)(3x + 8)$ **75.** $(x + 8)(x - 8)$

77. $2(4y + 3)(4y - 3)$ **79.** $\left(2x - \frac{1}{3}\right)\left(2x + \frac{1}{3}\right)$

81. $[(x - 1) - 2][(x - 1) + 2] = (x - 3)(x + 1)$

83. $(x - 2)^2$ **85.** $\left(x + \frac{1}{2}\right)^2$ **87.** $(2t + 1)^2$

89. $\left(3t + \frac{1}{4}\right)^2$ **91.** $(x - 2)(x^2 + 2x + 4)$

93. $(y + 6)(y^2 - 6y + 36)$ **95.** $\left(x - \frac{2}{3}\right)\left(x^2 + \frac{2}{3}x + \frac{4}{9}\right)$

97. $(2x - 1)(4x^2 + 2x + 1)$ **99.** $\left(\frac{1}{2}x + 1\right)\left(\frac{1}{4}x^2 - \frac{1}{2}x + 1\right)$

101. $(x - 1)(x + 2)$ **103.** $(s - 2)(s - 3)$

105. $-(y - 4)(y + 5)$ **107.** $(3x - 2)(x - 1)$

109. $(2x + 1)(x - 1)$ **111.** $(5x + 1)(x + 5)$

113. $-(5u - 2)(u + 3)$ **115.** $(x - 1)(x^2 + 2)$

117. $(3x + 2)(2x - 1)$ **119.** $x(x + 4)(x - 4)$

121. $x^2(x - 1)$ **123.** $(x - 1)^2$ **125.** $(2x - 1)^2$

127. $-2x(x - 2)(x + 1)$ **129.** $(9x + 1)(x + 1)$

131. $\frac{1}{96}(3x + 2)(4x - 3)$ **133.** $(3x + 1)(x^2 + 5)$

135. $(u + 2)(3 - u^2)$ **137.** $-z(z + 10)$

139. $(x + 1)^2(x - 1)^2$ **141.** $2(t - 2)(t^2 + 2t + 4)$

143. $2(2x - 1)(4x - 1)$ **145.** $-(x + 1)(x - 3)(x + 9)$

147. $7(x^2 + 1)(3x^2 - 1)$ **149.** $-2x(x - 5)^3(x + 5)$

151. (a) $500r^2 + 1000r + 500$

(b)
r	$2\frac{1}{2}\%$	3%	4%
$500(1 + r)^2$	525.31	530.45	540.80

r	$4\frac{1}{2}\%$	5%
$500(1 + r)^2$	546.01	551.25

(c) Amount increases with increasing r.

153. $V = x(15 - 2x)\left(\dfrac{45 - 3x}{2}\right)$

$= \dfrac{3}{2}x(x - 15)(2x - 15)$

x (cm)	3	5	7
V (cu cm)	486	375	84

155. (a) $T(x) = 0.0475x^2 + 1.099x + 0.23$

(b)
x (mi/hr)	30	40	55
T (ft)	75.95	120.19	204.36

(c) Stopping distance increases as speed increases.

157. b   **158.** c   **159.** a   **160.** d

161.

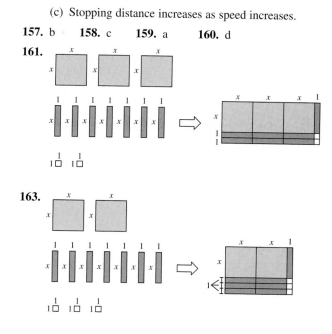

163.

165. $4\pi(r + 1)$ **167.** $4(6 - x)(6 + x)$

169. $-14, 14, -2, 2$ **171.** $-51, 51, -15, 15, -27, 27$

173. $2, -3$ (Answers will vary.)

175. $3, -8$ (Answers will vary.)

177. (a) $V = \pi h(R - r)(R + r)$

(b) $V = 2\pi\left[\left(\dfrac{R + r}{2}\right)(R - r)\right]h$

179. False. $(x^2 - 1)(x^2 + 1)$ becomes a fourth-degree polynomial.

181. False. Counterexample:

$x^2 + 2^2 \neq (x + 2)^2$ if $x = 3$.

183. n

185. A polynomial is in factored form if it is written as a product, not as a sum.

187. $9x^2 - 9x - 54 = 9(x^2 - x - 6)$

$= 9(x + 2)(x - 3)$

189. No. $(-3x^2 + 2x + 1) + (3x^2 - x - 2)$

$= x - 1$, which is a first-degree polynomial.

Section P.4   (page 44)

1. All real numbers   **3.** All nonnegative real numbers

5. All real numbers such that $x \neq 3$

7. All real numbers such that $x \geq -7$

9. $3x,\ x \neq 0$ **11.** $\dfrac{3x}{2},\ x \neq 0$ **13.** $\dfrac{3y}{y+1},\ x \neq 0$

15. $-\dfrac{4y}{5},\ y \neq \dfrac{1}{2}$ **17.** $-\dfrac{1}{2},\ x \neq 5$

19. $y - 4,\ y \neq -4$ **21.** $\dfrac{x(x+3)}{x-2},\ x \neq -2$

23. $\dfrac{y-4}{y+6},\ y \neq 3$ **25.** $-(x^2+1),\ x \neq 2$ **27.** $z - 2$

29.

x	0	1	2	3	4	5	6
$\dfrac{x^2 - 2x - 3}{x - 3}$	1	2	3	Undef.	5	6	7
$x + 1$	1	2	3	4	5	6	7

The expressions are equivalent except at $x = 3$.

31. Only common factors of the numerator and denominator can be canceled. In this case, factors of terms were incorrectly canceled.

33. $\dfrac{\pi}{4}$ **35.** $\dfrac{1}{5(x-2)},\ x \neq 1$ **37.** $\dfrac{r+1}{r},\ r \neq 1$

39. $\dfrac{t-3}{(t+3)(t-2)},\ t \neq -2$ **41.** $\dfrac{3}{2},\ x \neq -y$

43. $\dfrac{x+5}{x-1}$ **45.** $-\dfrac{2x^2 - 5x - 18}{(2x+1)(x+3)}$ **47.** $-\dfrac{2}{x-2}$

49. $-\dfrac{x^2+3}{(x+1)(x-2)(x-3)}$ **51.** $-\dfrac{x^2 - 2x + 2}{x(x^2+1)}$

53. $\dfrac{1}{2},\ x \neq 2$ **55.** $x(x+1),\ x \neq -1, 0$

57. $-\dfrac{2x+h}{x^2(x+h)^2},\ h \neq 0$ **59.** $\dfrac{2x-1}{2x},\ x > 0$

61. $x^{-2}(x^7 - 2) = \dfrac{x^7 - 2}{x^2}$ **63.** $-\dfrac{1}{(x^2+1)^5}$

65. $\dfrac{2x^3 - 2x^2 - 5}{(x-1)^{1/2}}$ **67.** $\dfrac{2x^2 - 1}{x^{5/2}}$ **69.** $\dfrac{1}{\sqrt{x+2} + \sqrt{x}}$

71. (a) $\dfrac{1}{16}$ minute   (b) $\dfrac{x}{16}$ minute(s)   (c) $\dfrac{60}{16} = \dfrac{15}{4}$ minutes

73. $\dfrac{x}{2(2x+1)}$

75. (a)

t	0	2	4	6	8	10
T	75	55.9	48.3	45	43.3	42.3

t	12	14	16	18	20	22
T	41.7	41.3	41.1	40.9	40.7	40.6

(b) 40

77. False. The domain of the left-hand side is $x^n \neq 1$.

79. Completely factor the numerator and denominator to determine if they have any common factors.

Section P.5   (page 54)

1. A: $(2, 6)$; B: $(-6, -2)$; C: $(4, -4)$; D: $(-3, 2)$

3.   **5.**

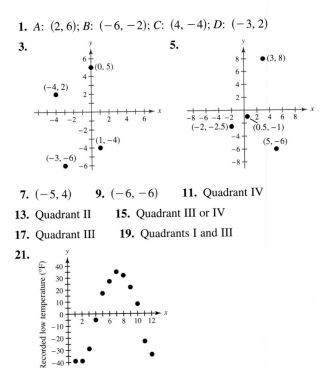

7. $(-5, 4)$ **9.** $(-6, -6)$ **11.** Quadrant IV

13. Quadrant II   **15.** Quadrant III or IV

17. Quadrant III   **19.** Quadrants I and III

21.

23. 8   **25.** 5   **27.** 13   **29.** $\dfrac{\sqrt{277}}{6}$

31. $\sqrt{71.78}$ **33.** (a) $4, 3, 5$ (b) $4^2 + 3^2 = 5^2$

35. (a) $10, 3, \sqrt{109}$ (b) $10^2 + 3^2 = \left(\sqrt{109}\right)^2$

37. $\left(\sqrt{5}\right)^2 + \left(\sqrt{45}\right)^2 = \left(\sqrt{50}\right)^2$

39. Opposite sides have equal lengths of $2\sqrt{5}$ and $\sqrt{85}$.

41. (a)

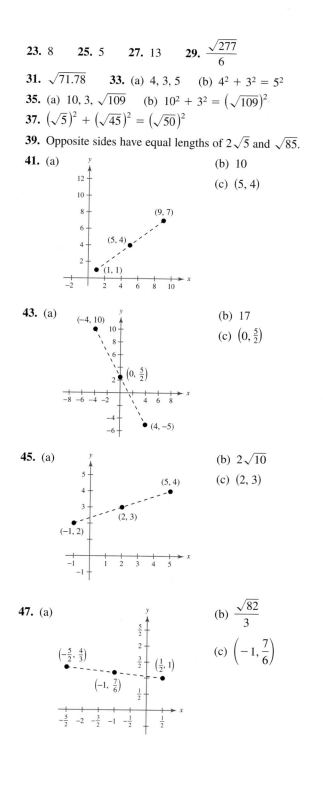

(b) 10

(c) $(5, 4)$

43. (a)

(b) 17

(c) $\left(0, \dfrac{5}{2}\right)$

45. (a)

(b) $2\sqrt{10}$

(c) $(2, 3)$

47. (a)

(b) $\dfrac{\sqrt{82}}{3}$

(c) $\left(-1, \dfrac{7}{6}\right)$

49. (a)

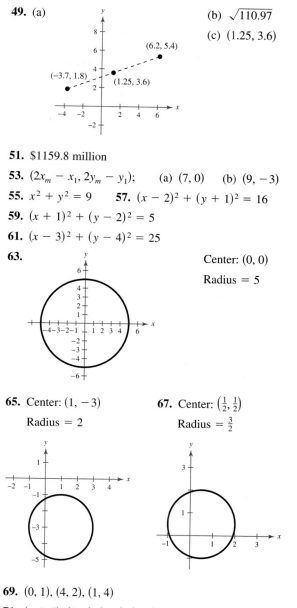

(b) $\sqrt{110.97}$

(c) $(1.25, 3.6)$

51. \$1159.8 million

53. $(2x_m - x_1, 2y_m - y_1)$;   (a) $(7, 0)$ (b) $(9, -3)$

55. $x^2 + y^2 = 9$ **57.** $(x - 2)^2 + (y + 1)^2 = 16$

59. $(x + 1)^2 + (y - 2)^2 = 5$

61. $(x - 3)^2 + (y - 4)^2 = 25$

63.

Center: $(0, 0)$

Radius $= 5$

65. Center: $(1, -3)$ **67.** Center: $\left(\dfrac{1}{2}, \dfrac{1}{2}\right)$

Radius $= 2$ Radius $= \dfrac{3}{2}$

69. $(0, 1), (4, 2), (1, 4)$

71. $(-1, 5), (2, 8), (4, 5), (1, 2)$ **73.** 65

75. (a) Answers will vary. Sample answer: The number of artists elected each year seems to be nearly steady except for the first few years. Estimate: between 6 and 8 new members in 2005.

(b) The Rock and Roll Hall of Fame was opened in 1986.

77. $5\sqrt{74} \approx 43$ yards   **79.** False; 15 times

81. False. It could be a rhombus.

83. No. The scales depend on the magnitudes of the quantities measured.

Section P.6   (page 63)

Vocabulary Check   (page 63)

1. Statistics   **2.** Line plots   **3.** histogram

4. frequency distribution   **5.** bar graph

6. Line graphs

1. (a) $1.709   (b) 0.19

3.

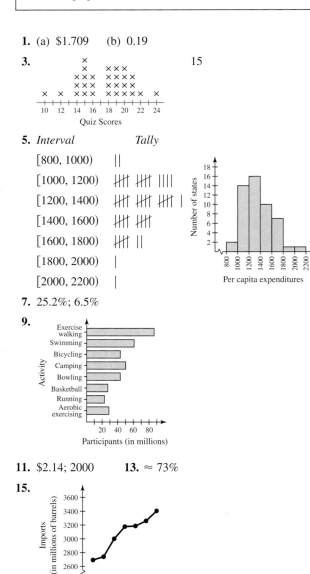

15

Quiz Scores

5.

Interval	Tally
[800, 1000)	||
[1000, 1200)	ʜʜ ʜʜ ||||
[1200, 1400)	ʜʜ ʜʜ ʜʜ |
[1400, 1600)	ʜʜ ʜʜ
[1600, 1800)	ʜʜ ||
[1800, 2000)	|
[2000, 2200)	|

7. 25.2%; 6.5%

9.

Participants (in millions)

11. $2.14; 2000   **13.** $\approx$ 73%

15.

Answers will vary. Sample answer: The graph shows crude oil imports have risen from 1995 to 2001.

17.

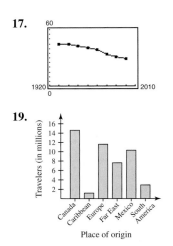

19.

Place of origin

21. A bar graph is similar to a histogram, except that the bars can be either horizontal or vertical and the labels of the bars are not necessarily numbers. Another difference between a bar graph and a histogram is that the bars in a bar graph are usually separated by spaces.

23. Answers will vary.

Review Exercises   (page 68)

1. (a) 11   (b) 11   (c) 11, -14

(d) 11, $-14, -\frac{8}{9}, \frac{5}{2}, 0.4$ (e) $\sqrt{6}$

3. (a) $0.8\overline{3}$ (b) 0.875

$\frac{5}{6} < \frac{7}{8}$

5. The set consists of all real numbers less than or equal to 7.

7. 122   **9.** $|x - 8| \geq 3$ **11.** $|y + 30| < 5$

13. (a) -13 (b) 27   **15.** (a) -18 (b) -12

17. Associative Property of Addition

19. Commutative Property of Multiplication   **21.** $\frac{14}{9}$

23. $\frac{1}{24}$ **25.** $\frac{47x}{60}$ **27.** (a) $-8z^3$ (b) $3a^3b^2$

29. (a) $\frac{3u^5}{v^4}$ (b) m^{-2} **31.** 4.38×10^{10}

33. 483,600,000   **35.** 78   **37.** $2x^2$ **39.** $\frac{3}{4}$

41. $\frac{x}{3}\sqrt[3]{2}$ **43.** $2\sqrt{2}$ **45.** $3\sqrt{3x}$ **47.** $\sqrt{2x}(2x + 1)$

49. 288 square inches. Square, because the width and height are equal.

51. $2 + \sqrt{3}$ **53.** $\frac{5}{2\sqrt{5}}$ **55.** 729   **57.** $6x^{9/10}$

59. $-2x^5 - x^4 + 3x^3 + 15x^2 + 5$; degree: 5; leading coefficient: -2

61. $-3x^2 - 7x + 1$ **63.** $2x^3 - x^2 + 3x - 9$

65. $x^5 - 2x^4 + x^3 - x^2 + 2x - 1$ **67.** $y^6 + y^4 - y^3 - y$

69. $x^2 - 64$ **71.** $x^3 - 12x^2 + 48x - 64$

73. $m^2 - 8m - n^2 + 16$

75. $(x + 3)(x + 5) = 5(x + 3) + x(x + 3)$

Distributive Property

77. $7(x + 5)$ **79.** $x(x^2 - 1) = x(x - 1)(x + 1)$

81. $2x(x^2 + 9x - 2)$

83. (a)

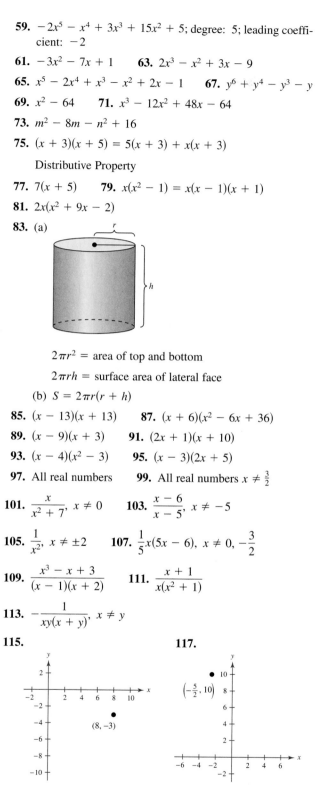

$2\pi r^2 =$ area of top and bottom

$2\pi rh =$ surface area of lateral face

(b) $S = 2\pi r(r + h)$

85. $(x - 13)(x + 13)$ **87.** $(x + 6)(x^2 - 6x + 36)$

89. $(x - 9)(x + 3)$ **91.** $(2x + 1)(x + 10)$

93. $(x - 4)(x^2 - 3)$ **95.** $(x - 3)(2x + 5)$

97. All real numbers **99.** All real numbers $x \neq \frac{3}{2}$

101. $\dfrac{x}{x^2 + 7}$, $x \neq 0$ **103.** $\dfrac{x - 6}{x - 5}$, $x \neq -5$

105. $\dfrac{1}{x^2}$, $x \neq \pm 2$ **107.** $\dfrac{1}{5}x(5x - 6)$, $x \neq 0, -\dfrac{3}{2}$

109. $\dfrac{x^3 - x + 3}{(x - 1)(x + 2)}$ **111.** $\dfrac{x + 1}{x(x^2 + 1)}$

113. $-\dfrac{1}{xy(x + y)}$, $x \neq y$

115.

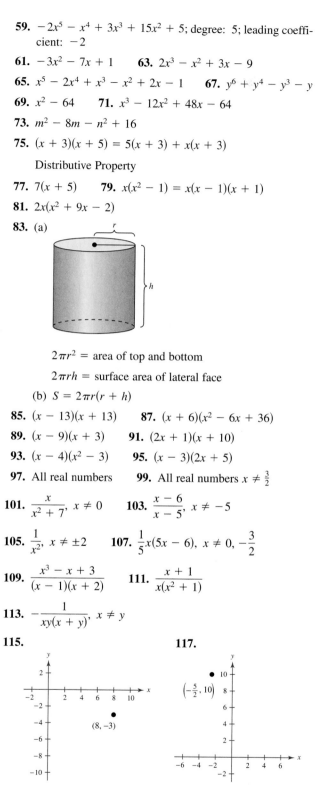

Quadrant IV

117.

$\left(-\frac{5}{2}, 10\right)$

Quadrant II

119. Quadrant IV

121.

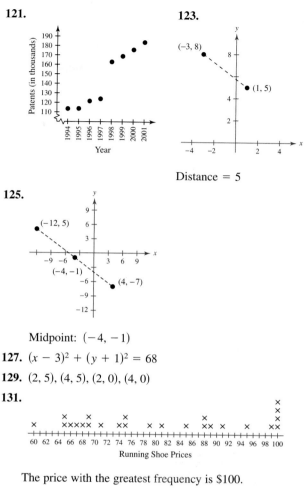

Year

123.

$(-3, 8)$

$(1, 5)$

Distance = 5

125.

$(-12, 5)$

$(-4, -1)$

$(4, -7)$

Midpoint: $(-4, -1)$

127. $(x - 3)^2 + (y + 1)^2 = 68$

129. $(2, 5), (4, 5), (2, 0), (4, 0)$

131.

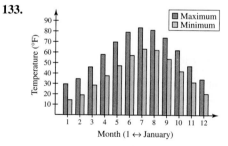

Running Shoe Prices

The price with the greatest frequency is \$100.

133.

Month (1 ↔ January)

135. False. $\dfrac{x^3 - 1}{x - 1}$ is undefined at $x = 1$.

137. You must raise 2 to the fourth power as well.

$(2x)^4 = 2^4x^4 = 16x^4$

139. Radicals cannot be combined unless the index and the radicand are the same.

Chapter Test (page 72)

1. $-\frac{10}{3} > -|-4|$ **2.** 56

3. Additive Identity Property

4. (a) -18 (b) $\frac{4}{27}$ (c) $-\frac{8}{343}$ (d) $\frac{8}{729}$

5. (a) 25 (b) 6 (c) 1.8×10^{5} (d) 2.7×10^{13}

6. (a) $12z^{8}$ (b) $\dfrac{1}{(u-2)^{7}}$ (c) $\dfrac{3x^{2}}{y^{2}}$

7. (a) $15z\sqrt{2z}$ (b) $-10\sqrt{y}$ (c) $\dfrac{2}{v}\sqrt[3]{\dfrac{2}{v^{2}}}$

8. $-2x^{5} - x^{4} + 3x^{3} + 3$; degree: 5; leading coefficient: -2

9. $2x^{2} - 3x - 5$ **10.** $x^{2} - 5$ **11.** $8,\ x \neq 3$

12. $\dfrac{x-1}{2x},\ x \neq \pm 1$ **13.** $x^{2}(2x+1)(x-2)$

14. $(x-2)(x+2)^{2}$ **15.** $(2x-3)(4x^{2}+6x+9)$

16. (a) $4\sqrt[3]{4}$ (b) $-3\left(1+\sqrt{3}\right)$

17. $\dfrac{5}{6}\sqrt{3}x^{2}$

18.

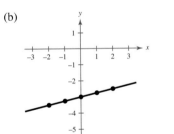

19.

Midpoint: $\left(2, \frac{5}{2}\right)$; Distance: $\sqrt{89}$

Chapter 1

Section 1.1   (page 82)

Vocabulary Check   (page 82)

1. solution point   **2.** graph   **3.** intercepts

1. (a) Yes   (b) Yes   **3.** (a) No   (b) Yes

5. (a) No   (b) Yes

7.

x	-1	0	1	$\frac{3}{2}$	2
y	5	3	1	0	-1
Solution point	$(-1, 5)$	$(0, 3)$	$(1, 1)$	$\left(\frac{3}{2}, 0\right)$	$(2, -1)$

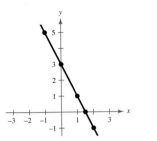

9.

x	-1	0	1	2	3
y	3	0	-1	0	3
Solution point	$(-1, 3)$	$(0, 0)$	$(1, -1)$	$(2, 0)$	$(3, 3)$

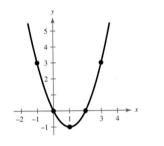

11. (a)

x	-2	-1	0	1	2
y	$-\frac{7}{2}$	$-\frac{13}{4}$	-3	$-\frac{11}{4}$	$-\frac{5}{2}$

(b)

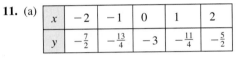

(c)

x	-2	-1	0	1	2
y	$-\frac{5}{2}$	$-\frac{11}{4}$	-3	$-\frac{13}{4}$	$-\frac{7}{2}$

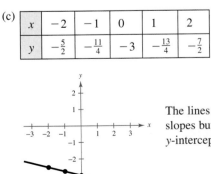

The lines have opposite slopes but the same y-intercept.

13. d   **14.** c   **15.** f   **16.** e   **17.** a   **18.** b

19.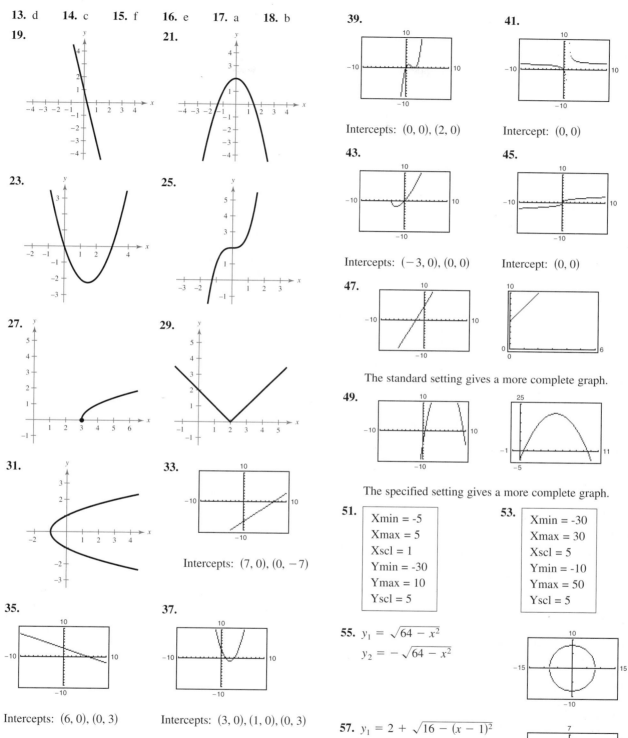

21.

23.

25.

27.

29.

31.

33.

Intercepts: $(7, 0), (0, -7)$

35.

37.

Intercepts: $(6, 0), (0, 3)$

Intercepts: $(3, 0), (1, 0), (0, 3)$

39.

Intercepts: $(0, 0), (2, 0)$

41.

Intercept: $(0, 0)$

43.

Intercepts: $(-3, 0), (0, 0)$

45.

Intercept: $(0, 0)$

47.

The standard setting gives a more complete graph.

49.

The specified setting gives a more complete graph.

51.
```
Xmin = -5
Xmax = 5
Xscl = 1
Ymin = -30
Ymax = 10
Yscl = 5
```

53.
```
Xmin = -30
Xmax = 30
Xscl = 5
Ymin = -10
Ymax = 50
Yscl = 5
```

55. $y_1 = \sqrt{64 - x^2}$
$y_2 = -\sqrt{64 - x^2}$

57. $y_1 = 2 + \sqrt{16 - (x - 1)^2}$
$y_2 = 2 - \sqrt{16 - (x - 1)^2}$

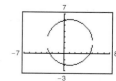

59. The graphs are identical. Distributive Property

61. The graphs are identical. Associative Property of Multiplication

63.

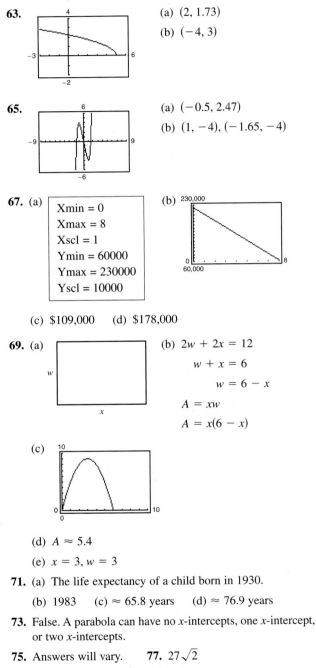

(a) $(2, 1.73)$

(b) $(-4, 3)$

65.

(a) $(-0.5, 2.47)$

(b) $(1, -4), (-1.65, -4)$

67. (a)

Xmin = 0
Xmax = 8
Xscl = 1
Ymin = 60000
Ymax = 230000
Yscl = 10000

(b) 230,000

(c) $109,000   (d) $178,000

69. (a)

(b) $2w + 2x = 12$
$w + x = 6$
$w = 6 - x$
$A = xw$
$A = x(6 - x)$

(c)

(d) $A \approx 5.4$

(e) $x = 3, w = 3$

71. (a) The life expectancy of a child born in 1930.

(b) 1983   (c) ≈ 65.8 years   (d) ≈ 76.9 years

73. False. A parabola can have no x-intercepts, one x-intercept, or two x-intercepts.

75. Answers will vary.   **77.** $27\sqrt{2}$

79. $7^7 = 823,543$ **81.** $2x^2 + 8x + 11$

Section 1.2   (page 94)

Vocabulary Check   (page 94)

1. (a) iii   (b) i   (c) v   (d) ii   (e) iv

2. slope   **3.** parallel   **4.** perpendicular

5. linear extrapolation

1. (a) L_2 (b) L_3 (c) L_1

3.

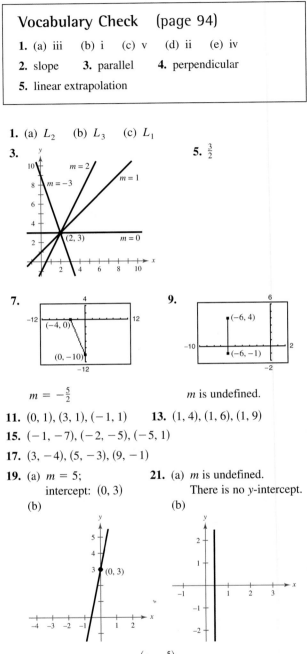

5. $\frac{3}{2}$

7.

9.

$m = -\frac{5}{2}$

m is undefined.

11. $(0, 1), (3, 1), (-1, 1)$ **13.** $(1, 4), (1, 6), (1, 9)$

15. $(-1, -7), (-2, -5), (-5, 1)$

17. $(3, -4), (5, -3), (9, -1)$

19. (a) $m = 5$; intercept: $(0, 3)$

(b)

21. (a) m is undefined. There is no y-intercept.

(b)

23. (a) $m = 0$; intercept: $\left(0, -\frac{5}{3}\right)$

(b)

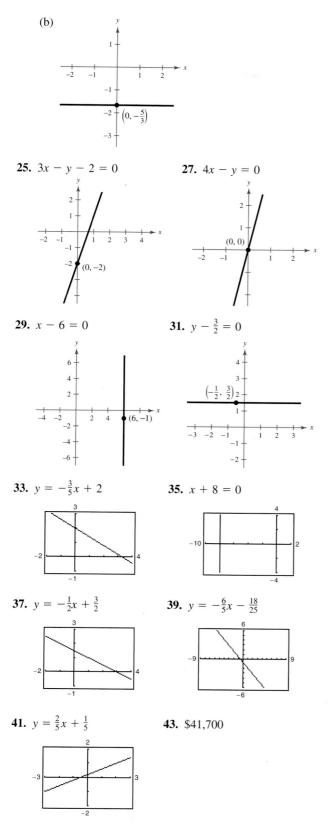

25. $3x - y - 2 = 0$

27. $4x - y = 0$

29. $x - 6 = 0$

31. $y - \frac{3}{2} = 0$

33. $y = -\frac{3}{5}x + 2$

35. $x + 8 = 0$

37. $y = -\frac{1}{2}x + \frac{3}{2}$

39. $y = -\frac{6}{5}x - \frac{18}{25}$

41. $y = \frac{2}{5}x + \frac{1}{5}$

43. \$41,700

45. $\frac{1}{2}$; $(0, -2)$; a line that rises from the left to the right

47. Undefined; none; a vertical line at $x = -6$

49.

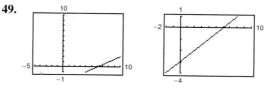

The second setting gives a more complete graph, with a better view of the intercepts.

51. Perpendicular   **53.** Parallel

55. (a) $y = 2x - 3$ (b) $y = -\frac{1}{2}x + 2$

57. (a) $y = -\frac{3}{4}x + \frac{3}{8}$ (b) $y = \frac{4}{3}x + \frac{127}{2}$

59. (a) $x = 3$ (b) $y = -2$

61.

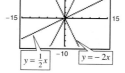

$y = \frac{1}{2}x$ and $y = -2x$ are perpendicular.

63.

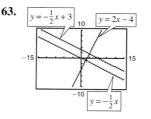

$y = -\frac{1}{2}x$ and $y = -\frac{1}{2}x + 3$ are parallel. Both are perpendicular to $y = 2x - 4$.

65. (a) Sales increase of \$135

(b) No sales increase

(c) Sales decrease of \$40

67. (a) Increase: 1998–1999; Decrease: 2001–2002

(b) $y = -0.05x + 0.68$

(c) There is a decrease of \$0.05 per year.

(d) Using the equation from part (b),
$y = -0.05(16) + 0.68 = -0.12.$
Answers will vary.

69. 12 feet   **71.** $V = 125t + 2040$

73. $V = -2000t + 28,400$

75. b; slope $= -10$; the amount owed decreases by \$10 per week.

76. c; slope $= 1.50$; the hourly wage increases by 1.50 per unit produced.

77. a; slope $= 0.35$; expenses increase by $0.35 per mile.

78. d; slope $= -100$; the value depreciates $100 per year.

79. $F = \frac{9}{5}C + 32$

81. (a) $V = -175t + 875$

(b)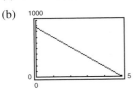

t	0	1	2	3	4	5
V	875	700	525	350	175	0

83. (a) $C = 16.75t + 36,500$ (b) $R = 27t$

(c) $P = 10.25t - 36,500$ (d) $t \approx 3561$ hours

85. (a) Increase of ≈ 639 students per year

(b) $71,531$; $79,838$; $81,755$

(c) $y = 639t + 75,365$; $m = 639$; the slope determines the average increase in enrollment.

87. False. The slopes $\left(\frac{2}{7} \text{ and } -\frac{11}{7}\right)$ are not equal.

89.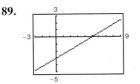

x- and *y*-intercepts

91. $3x + 2y - 6 = 0$ **93.** $12x + 3y + 2 = 0$

95. The line with slope -3 is steeper than the line with slope $\frac{5}{2}$.

97. Use the slope formula to show that $\overline{AB}$ is perpendicular to $\overline{AC}$.

99. Yes. $x + 20$ **101.** No   **103.** No

105. $(x - 9)(x + 3)$ **107.** $(2x - 5)(x + 8)$

Section 1.3 (page 107)

Vocabulary Check (page 107)

1. domain, range, function

2. independent, dependent   **3.** piecewise defined

4. implied domain   **5.** difference quotient

1. Yes; each element of the domain is assigned to exactly one element of the range.

3. No; the National League, an element in the domain, is assigned to three items in the range, the Cubs, the Pirates, and the Dodgers; the American League, an element in the domain, is also assigned to three items in the range, the Orioles, the Yankees, and the Twins.

5. Yes. Each input value is matched with one output value.

7. No. The same input value is matched with two different output values.

9. (a) Function

(b) Not a function because the element 1 in A corresponds to two elements, -2 and 1, in B.

(c) Function   (d) Not a function because the element 2 in A corresponds to no element in B.

11. Each is a function. To each year there corresponds one and only one circulation.

13. Not a function   **15.** Function   **17.** Function

19. Not a function   **21.** Function   **23.** Not a function

25. (a) $\frac{1}{5}$ (b) 1   (c) $\frac{1}{4t + 1}$ (d) $\frac{1}{x + c + 1}$

27. (a) -1 (b) -9 (c) $2x - 5$

29. (a) 0   (b) -0.75 (c) $x^2 + 2x$

31. (a) 1   (b) 2.5   (c) $3 - 2|x|$

33. (a) $-\frac{1}{9}$ (b) Undefined   (c) $\frac{1}{y^2 + 6y}$

35. (a) 1   (b) -1 (c) $1, x \neq 0$

37. (a) -1 (b) 2   (c) 6

39.

t	-5	-4	-3	-2	-1
$h(t)$	1	$\frac{1}{2}$	0	$\frac{1}{2}$	1

41.

x	-2	-1	0	1	2
$f(x)$	5	$\frac{9}{2}$	4	1	0

43. 5   **45.** $\frac{4}{3}$ **47.** $2, -1$ **49.** All real numbers x

51. All real numbers t except $t = 0$ **53.** All real numbers x

55. All real numbers x except $x = 0, -2$

57. All real numbers y such that $y > 10$

59.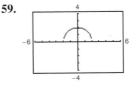

Domain: $[-2, 2]$; range: $[0, 2]$

61.

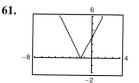

Domain: $(-\infty, \infty)$; range: $[0, \infty)$

63. $\{(-2, 4), (-1, 1), (0, 0), (1, 1), (2, 4)\}$

65. $\{(-2, 4), (-1, 3), (0, 2), (1, 3), (2, 4)\}$

67. $A = \dfrac{C^2}{4\pi}$

69. (a) \$3375

(b)

Yes, it is a function.

(c) $P(x) = \begin{cases} 30x, & x \le 100 \\ 45x - 0.15x^2, & x > 100 \end{cases}$

71. $A = \dfrac{x^2}{2(x - 2)}, \; x > 2$

73. (a) $V = x^2 y$

$= x^2(108 - 4x)$

$= 108x^2 - 4x^3$

(b) $0 < x < 27$

(c)

(d) $x = 18$ in., $y = 36$ in.

75. $7 \le x \le 12, 1 \le x \le 6$; Answers will vary.

77. 4.63; \$4630 in monthly revenue in November.

79.

t	0	1	2	3	4	5
$n(t)$	575	650.3	707.2	745.7	765.8	791

t	6	7	8	9	10
$n(t)$	817.8	844.6	871.4	898.2	925

81. (a)

y	5	10	20
$F(y)$	26,474	149,760	847,170

y	30	40
$F(y)$	2,334,527	4,792,320

Each time the depth is doubled, the force increases by more than 2 times.

(b)

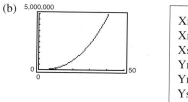

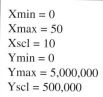

Xmin = 0
Xmax = 50
Xscl = 10
Ymin = 0
Ymax = 5,000,000
Yscl = 500,000

(c) Depth ≈ 21.37 feet

Use the *trace* and *zoom* features on a graphing utility.

83. $2, c \ne 0$ **85.** $3 + h, h \ne 0$ **87.** $-\dfrac{1}{t}, t \ne 1$

89. False. The range is $[-1, \infty)$.   **91.** $r(x) = \dfrac{32}{x}; c = 32$

93. The domain is the set of input values of a function. The range is the set of output values.

95. $\dfrac{12x + 20}{x + 2}$ **97.** $\dfrac{(x + 6)(x + 10)}{5(x + 3)}, x \ne 0, \dfrac{1}{2}$

Section 1.4   (page 121)

Vocabulary Check   (page 121)

1. ordered pairs   **2.** Vertical Line Test

3. decreasing   **4.** minimum

5. greatest integer   **6.** even

1. Domain: $(-\infty, \infty)$; range: $(-\infty, 1]$; 1

3. Domain: $[-4, 4]$; range: $[0, 4]$; 4

5.

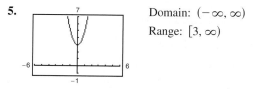

Domain: $(-\infty, \infty)$

Range: $[3, \infty)$

7. 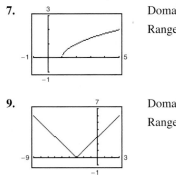 Domain: $[1, \infty)$

Range: $[0, \infty)$

9. 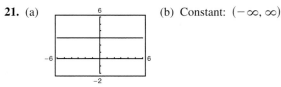 Domain: $(-\infty, \infty)$

Range: $[0, \infty)$

11. Function. Graph the given function over the window shown in the figure.

13. Not a function. Solve for y and graph the two resulting functions.

15. Function. Solve for y and graph the resulting function.

17. Increasing on $(-\infty, \infty)$

19. Increasing on $(-\infty, 0), (2, \infty)$

Decreasing on $(0, 2)$

21. (a) 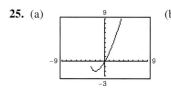 (b) Constant: $(-\infty, \infty)$

23. (a) 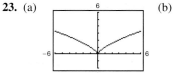 (b) Decreasing on $(-\infty, 0)$

Increasing on $(0, \infty)$

25. (a) (b) Increasing on $(-2, \infty)$

Decreasing on $(-3, -2)$

27. (a) (b) Decreasing on $(-\infty, -1)$

Constant on $(-1, 1)$;

Increasing on $(1, \infty)$

29. 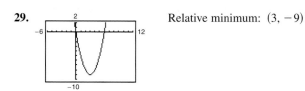 Relative minimum: $(3, -9)$

31. 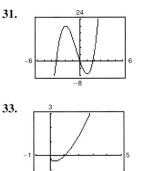 Relative minimum: $(1, -7)$

Relative maximum: $(-2, 20)$

33.  Minimum: $(0.33, -0.38)$

35. (a) Answers will vary.

(b) Relative minimum at $(2, -9)$

(c) Answers will vary.

37. (a) Answers will vary.

(b) Relative minimum at $(1.63, -8.71)$

Relative maximum at $(-1.63, 8.71)$

(c) Answers will vary.

39. (a) Answers will vary.   (b) Relative minimum at $(4, 0)$

(c) Answers will vary.

41. 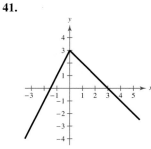   **43.**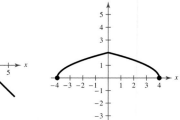

45. 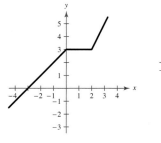   **47.**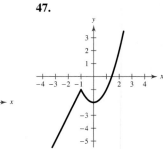

49. Neither even nor odd   **51.** Odd function

53. Odd function   **55.** Even function

57. (a) $\left(\frac{3}{2}, 4\right)$ (b) $\left(\frac{3}{2}, -4\right)$

59. (a) $(-4, 9)$ (b) $(-4, -9)$

61. (a) $(-x, -y)$ (b) $(-x, y)$

63. Even function

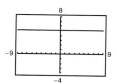

65. Neither even nor odd

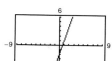

(b)

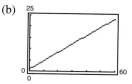

$7.89

67. Even function

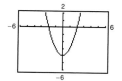

69. Neither even nor odd

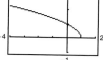

83. $h = -x^2 + 4x - 3, \ 1 \le x \le 3$

85. $L = \frac{1}{2}y^2, \ 0 \le y \le 4$

87. (a)

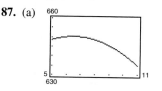

71. Neither even nor odd

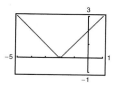

(b) Increasing from 1995 to 1996; decreasing from 1996 to 2001

(c) $\approx 650{,}200$

89. False. Counterexample: $f(x) = \sqrt{1 + x^2}$

73. $(-\infty, 4]$

75. $(-\infty, -3], [3, \infty)$

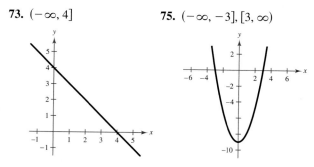

91. If $y = a_{2n+1}x^{2n+1} + a_{2n-1}x^{2n-1} + \cdots + a_3x^3 + a_1x$, each exponent is odd. Then

$$f(-x) = -a_{2n+1}x^{2n+1} - a_{2n-1}x^{2n-1} - \cdots - a_3x^3 - a_1x,$$

which is equal to $-f(x)$. Therefore, by definition, the original function is odd.

93. (a) Even. g is a reflection in the x-axis.

(b) Even. g is a reflection in the y-axis.

(c) Even. g is a vertical shift downward.

(d) Neither even nor odd. g is shifted to the right and reflected in the x-axis.

77.

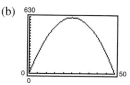

Domain: $(-\infty, \infty)$

Range: $[0, 2)$

Sawtooth pattern

95. No. x is not a function of y because horizontal lines can be drawn to intersect the graph twice. Therefore, each y-value corresponds to two distinct x-values when $-5 < y < 5$.

79. (a) Answers will vary.

(b)

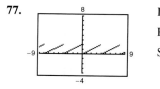

97. Terms: $-2x^2, 8x$; coefficients: $-2, 8$

99. Terms: $\dfrac{x}{3}, -5x^2, x^3$; coefficients: $\dfrac{1}{3}, -5, 1$

101. (a) $d = 4\sqrt{5}$ (b) Midpoint: $(2, 5)$

103. (a) $d = \sqrt{41}$ (b) Midpoint: $\left(\frac{1}{2}, \frac{3}{2}\right)$

105. (a) 29   (b) -6 (c) $5x - 16$

(c) 625 square meters; 25×25 meters

107. (a) 0   (b) 36   (c) $6\sqrt{3}$ **109.** $h + 4, h \ne 0$

81. (a) C_2 is the appropriate model. The cost of the first minute is $1.05 and the cost increases $0.38 when the next minute begins, and so on.

Section 1.5   (page 131)

Vocabulary Check   (page 131)

1. quadratic function   2. absolute value function

3. rigid transformations   4. $-f(x), f(-x)$

5. $c > 1, 0 < c < 1$

6. (a) ii   (b) iv   (c) iii   (d) i

1.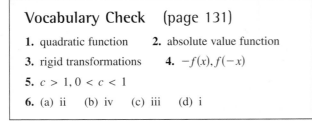

3.

5.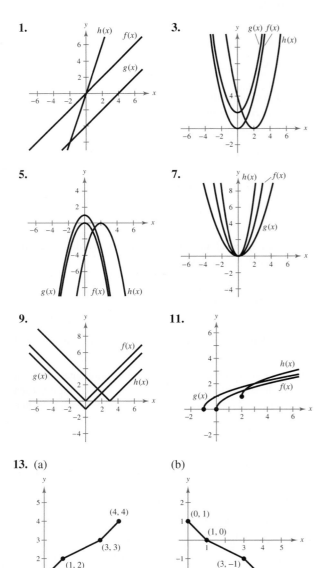

7.

9.

11.

13. (a)

(b)

(c)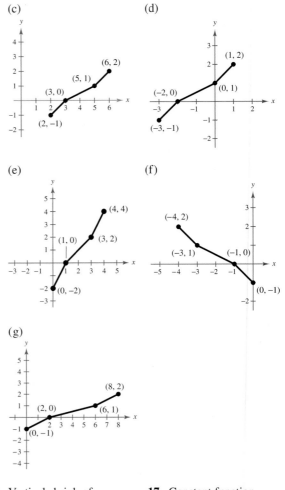

(d)

(e)

(f)

(g)

15. Vertical shrink of $y = x$ **17.** Constant function

$y = \frac{1}{2}x$ $y = 7$

19. Reflection in the x-axis and a vertical shift of $y = \sqrt{x}$

$y = 1 - \sqrt{x}$

21. Horizontal shift of $y = |x|$ **23.** Vertical shift of $y = x^2$

$y = |x + 2|$ $y = x^2 - 1$

25. Reflection in the x-axis of $y = x^3$, followed by a vertical shift

$y = 1 - x^3$

27. Reflection in the x-axis and vertical shift one unit downward

29. Horizontal shift two units to the right

31. Horizontal shrink

33. Horizontal shift five units to the left

35. Reflection in the x-axis   **37.** Vertical stretch

39. Reflection in the x-axis and vertical shift four units upward

41. Horizontal shift two units to the left and vertical shrink

43. Horizontal stretch and vertical shift two units upward

45.

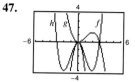

g is a horizontal shift and h is a vertical shrink.

47.

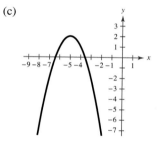

g is a vertical shrink and a reflection in the x-axis and h is a reflection in the y-axis.

49. $g(x) = -(x^3 - 3x^2) + 1$

51. (a) $f(x) = x^2$

(b) Horizontal shift five units to the left, reflection in the x-axis, and vertical shift two units upward

(c)

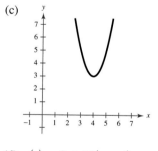

(d) $g(x) = 2 - f(x + 5)$

53. (a) $f(x) = x^2$

(b) Horizontal shift four units to the right, vertical stretch, and vertical shift three units upward

(c)

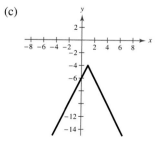

(d) $g(x) = 3 + 2f(x - 4)$

55. (a) $f(x) = x^3$

(b) Horizontal shift two units to the right and vertical stretch

(c)

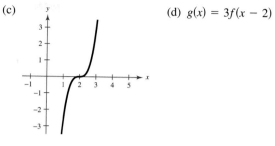

(d) $g(x) = 3f(x - 2)$

57. (a) $f(x) = x^3$

(b) Horizontal shift one unit to the right and vertical shift two units upward

(c)

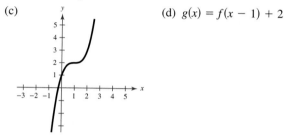

(d) $g(x) = f(x - 1) + 2$

59. (a) $f(x) = |x|$

(b) Horizontal shift four units to the left and vertical shift eight units upward

(c)

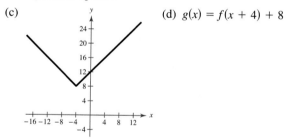

(d) $g(x) = f(x + 4) + 8$

61. (a) $f(x) = |x|$

(b) Horizontal shift one unit to the right, reflection in the x-axis, vertical stretch, and vertical shift four units downward

(c)

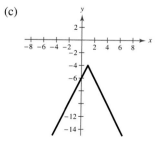

(d) $g(x) = -2f(x - 1) - 4$

63. (a) $f(x) = \sqrt{x}$

(b) Horizontal shift three units to the left, reflection in the x-axis, vertical shrink, and vertical shift one unit downward

(c)

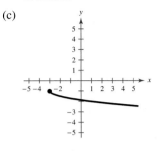

(d) $g(x) = -\frac{1}{2}f(x + 3) - 1$

65. (a)

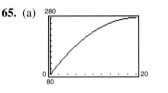

(b) $P(x) = 55 + 20x - 0.5x^2$; vertical shift

(c) $P(x) = 80 + \dfrac{1}{5}x - \dfrac{x^2}{20,000}$; horizontal stretch

67. (a) Vertical shrink and vertical shift

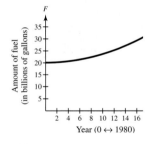

(b) $G(t) = 0.036t^2 + 0.72t + 23.7$, $-10 \le t \le 10$

69. True. The absolute value function is an even function.

71. (a)   (b)

   (c)   (d)

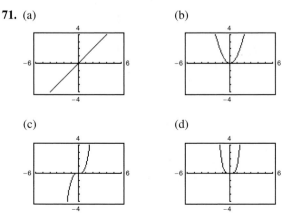

(e)   (f)

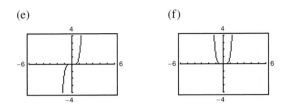

All the graphs pass through the origin. The graphs of the odd powers of x are symmetric with respect to the origin and the graphs of the even powers are symmetric with respect to the y-axis. As the powers increase, the graphs become flatter in the interval $-1 < x < 1$.

73. Neither   **75.** All real numbers x except $x = 9$

77. All real numbers x such that $-10 \le x \le 10$

Section 1.6   (page 141)

1.   **3.**

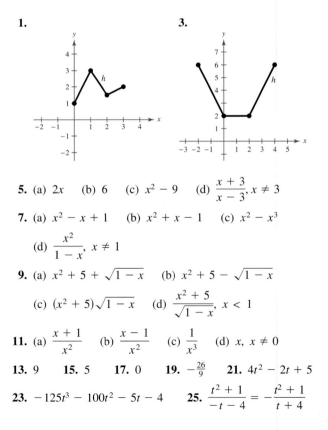

5. (a) $2x$ (b) 6 (c) $x^2 - 9$ (d) $\dfrac{x + 3}{x - 3}$, $x \ne 3$

7. (a) $x^2 - x + 1$ (b) $x^2 + x - 1$ (c) $x^2 - x^3$

(d) $\dfrac{x^2}{1 - x}$, $x \ne 1$

9. (a) $x^2 + 5 + \sqrt{1 - x}$ (b) $x^2 + 5 - \sqrt{1 - x}$

(c) $(x^2 + 5)\sqrt{1 - x}$ (d) $\dfrac{x^2 + 5}{\sqrt{1 - x}}$, $x < 1$

11. (a) $\dfrac{x + 1}{x^2}$ (b) $\dfrac{x - 1}{x^2}$ (c) $\dfrac{1}{x^3}$ (d) x, $x \ne 0$

13. 9 **15.** 5 **17.** 0 **19.** $-\frac{26}{9}$ **21.** $4t^2 - 2t + 5$

23. $-125t^3 - 100t^2 - 5t - 4$ **25.** $\dfrac{t^2 + 1}{-t - 4} = -\dfrac{t^2 + 1}{t + 4}$

27.

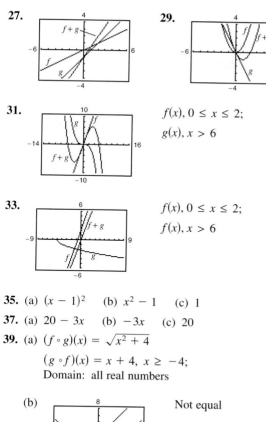

29.

31. $f(x), 0 \le x \le 2;$

$g(x), x > 6$

33. $f(x), 0 \le x \le 2;$

$f(x), x > 6$

35. (a) $(x - 1)^2$ (b) $x^2 - 1$ (c) 1

37. (a) $20 - 3x$ (b) $-3x$ (c) 20

39. (a) $(f \circ g)(x) = \sqrt{x^2 + 4}$

$(g \circ f)(x) = x + 4, \ x \ge -4;$

Domain: all real numbers

(b)

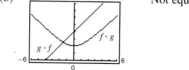

Not equal

41. (a) $(f \circ g)(x) = x - \frac{8}{3}; \ (g \circ f)(x) = x - 8;$

Domain: all real numbers

(b)

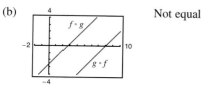

Not equal

43. (a) $(f \circ g)(x) = x^4; \ (g \circ f)(x) = x^4;$

Domain: all real numbers

(b)

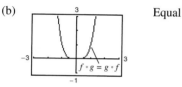

Equal

45. (a) $(f \circ g)(x) = 24 - 5x; \ (g \circ f)(x) = -5x$

(b) $24 - 5x \ne -5x$

(c)

x	0	1	2	3
$g(x)$	4	3	2	1
$(f \circ g)(x)$	24	19	14	9

x	0	1	2	3
$f(x)$	4	9	14	19
$(g \circ f)(x)$	0	-5	-10	-15

47. (a) $(f \circ g)(x) = \sqrt{x^2 + 1}; \ (g \circ f)(x) = x + 1, \ x \ge -6$

(b) $x + 1 \ne \sqrt{x^2 + 1}$

(c)

x	0	1	2	3
$g(x)$	-5	-4	-1	4
$(f \circ g)(x)$	1	$\sqrt{2}$	$\sqrt{5}$	$\sqrt{10}$

x	0	1	2	3
$f(x)$	$\sqrt{6}$	$\sqrt{7}$	$\sqrt{8}$	3
$(g \circ f)(x)$	1	2	3	4

49. (a) $(f \circ g)(x) = |2x + 2|; \ (g \circ f)(x) = 2|x + 3| - 1$

(b) $(f \circ g)(x) = \begin{cases} 2x + 2, & x \ge -1 \\ -2x - 2, & x < -1 \end{cases}$

$(g \circ f)(x) = \begin{cases} 2x + 5, & x \ge -3 \\ -2x - 7, & x < -3 \end{cases}$

$(f \circ g)(x) \ne (g \circ f)(x)$

(c)

x	0	-1	-3	-5
$g(x)$	-1	-3	-7	-11
$(f \circ g)(x)$	2	0	4	8

x	0	-1	-3	-5
$f(x)$	3	2	0	2
$(g \circ f)(x)$	5	3	-1	3

51. (a) 3 (b) 0 **53.** (a) 0 (b) 4

55. $f(x) = x^2, \ g(x) = 2x + 1$

57. $f(x) = \sqrt[3]{x}, \ g(x) = x^2 - 4$

59. $f(x) = \dfrac{1}{x}, \ g(x) = x + 2$

61. $f(x) = x^2 + 2x, \ g(x) = x + 4$

63. (a) All real numbers x such that $x \geq -4$

(b) All real numbers

(c) All real numbers

65. (a) All real numbers

(b) All real numbers x such that $x \geq 0$

(c) All real numbers x such that $x \geq 0$

67. (a) All real numbers x except $x = 0$

(b) All real numbers

(c) All real numbers x except $x = -3$

69. (a) All real numbers   (b) All real numbers

(c) All real numbers

71. (a) All real numbers

(b) All real numbers x except $x = \pm 2$

(c) All real numbers x except $x = \pm 2$

73. (a) $T = \frac{3}{4}x + \frac{1}{15}x^2$

(b)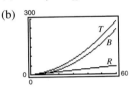

(c) B. For example, $B(60) = 240$, whereas $R(60)$ is only 45.

75.

Year	1994	1995	1996	1997
y_1	138.7	147.7	156.6	165.5
y_2	314.9	326.7	342.2	361.4
y_3	40.3	45.0	49.0	52.3

Year	1998	1999	2000
y_1	174.4	183.4	192.3
y_2	384.5	411.3	441.9
y_3	54.9	56.7	57.8

77. $(A \circ r)(t) = 0.36\pi t^2$

$(A \circ r)(t)$ represents the area of the circle at time t.

79. (a) $(C \circ x)(t) = 3000t + 750$

$(C \circ x)(t)$ represents the cost after t production hours.

(b) 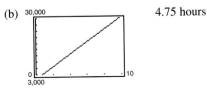   4.75 hours

81. $g(f(x))$ represents three percent of an amount over $500,000.

83. False. $(f \circ g)(x) = 6x + 1 \neq 6x + 6 = (g \circ f)(x)$

85. To prove that the product of two odd functions f and g is an even function, show that $(fg)(-x) = (fg)(x)$.

$(fg)(-x) = f(-x)g(-x)$

$= [-f(x)][-g(x)] = f(x)g(x) = (fg)(x)$

To prove that the product of two even functions f and g is an even function, show that $(fg)(-x) = (fg)(x)$.

$(fg)(-x) = f(-x)g(-x) = f(x)g(x) = (fg)(x)$

87. Prove $g(-x) = g(x)$.

$g(-x) = \frac{1}{2}[f(-x) + f(x)] = \frac{1}{2}[f(x) + f(-x)] = g(x)$

Prove $h(-x) = -h(x)$.

$h(-x) = \frac{1}{2}[f(-x) - f(x)] = -\frac{1}{2}[f(x) - f(-x)]$

$= -h(x)$

89. $(0, -5), (1, -5), (2, -7)$ (Answers will vary.)

91. $\left(0, 2\sqrt{6}\right), \left(1, \sqrt{23}\right), \left(2, 2\sqrt{5}\right)$ (Answers will vary.)

93. $10x - y + 38 = 0$ **95.** $30x + 11y - 34 = 0$

97.   **99.**

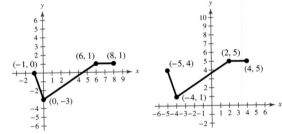

101.

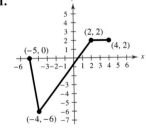

Section 1.7   (page 152)

1. $f^{-1}(x) = \dfrac{x}{6}$ **3.** $f^{-1}(x) = x - 7$

5. $f^{-1}(x) = \frac{1}{2}(x - 1)$ **7.** $f^{-1}(x) = x^3$

9. (a) $f(g(x)) = f\left(-\dfrac{2x + 6}{7}\right)$

$$= -\frac{7}{2}\left(-\frac{2x + 6}{7}\right) - 3 = x$$

$$g(f(x)) = g\left(-\frac{7}{2}x - 3\right)$$

$$= -\frac{2\left(-\frac{7}{2}x - 3\right) + 6}{7} = x$$

(b)

x	0	2	-2	6
$f(x)$	-3	-10	4	-24

x	-3	-10	4	-24
$g(x)$	0	2	-2	6

11. (a) $f(g(x)) = f\left(\sqrt[3]{x - 5}\right) = \left(\sqrt[3]{x - 5}\right)^3 + 5 = x$

$g(f(x)) = g(x^3 + 5) = \sqrt[3]{(x^3 + 5) - 5} = x$

(b)

x	0	1	-1	-2	4
$f(x)$	5	6	4	-3	69

x	5	6	4	-3	69
$g(x)$	0	1	-1	-2	4

13. (a) $f(g(x)) = f(8 + x^2)$

$$= -\sqrt{(8 + x^2) - 8}$$

$$= -\sqrt{x^2} = -(-x) = x, \ x \le 0$$

$$g(f(x)) = g\left(-\sqrt{x - 8}\right)$$

$$= 8 + \left(-\sqrt{x - 8}\right)^2$$

$$= 8 + (x - 8) = x, \ x \ge 8$$

(b)

x	8	9	12	15
$f(x)$	0	-1	-2	$-\sqrt{7}$

x	0	-1	-2	$-\sqrt{7}$
$g(x)$	8	9	12	15

15. $f(g(x)) = f\left(\sqrt[3]{x}\right) = \left(\sqrt[3]{x}\right)^3 = x$

$g(f(x)) = g(x^3) = \sqrt[3]{x^3} = x$

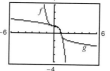

Reflections in the line $y = x$

17. $f(g(x)) = f(x^2 + 4), \ \ x \ge 0$

$$= \sqrt{(x^2 + 4) - 4} = x$$

$$g(f(x)) = g\left(\sqrt{x - 4}\right)$$

$$= \left(\sqrt{x - 4}\right)^2 + 4 = x$$

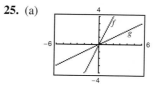

Reflections in the line $y = x$

19. $f(g(x)) = f\left(\sqrt[3]{1 - x}\right) = 1 - \left(\sqrt[3]{1 - x}\right)^3 = x$

$g(f(x)) = g(1 - x^3) = \sqrt[3]{1 - (1 - x^3)} = x$

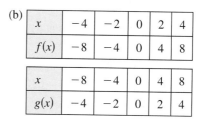

Reflections in the line $y = x$

21. c   **22.** b   **23.** a   **24.** d

25. (a)

(b)

x	-4	-2	0	2	4
$f(x)$	-8	-4	0	4	8

x	-8	-4	0	4	8
$g(x)$	-4	-2	0	2	4

27. (a)

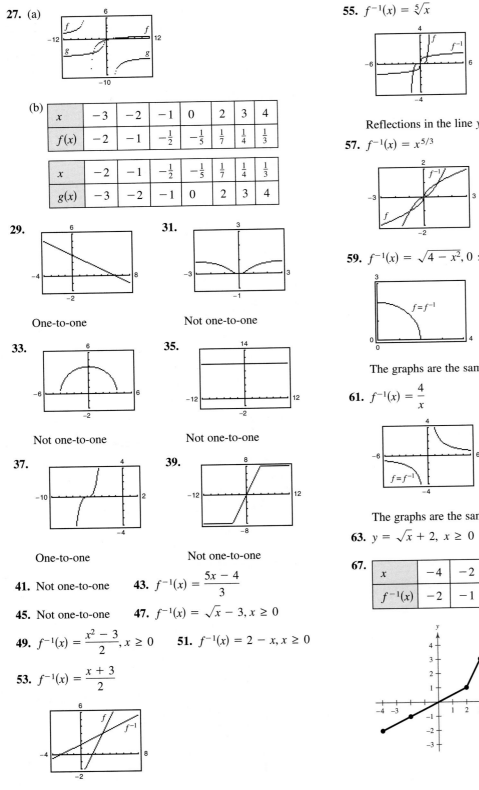

(b)

x	-3	-2	-1	0	2	3	4
$f(x)$	-2	-1	$-\frac{1}{2}$	$-\frac{1}{5}$	$\frac{1}{7}$	$\frac{1}{4}$	$\frac{1}{3}$

x	-2	-1	$-\frac{1}{2}$	$-\frac{1}{5}$	$\frac{1}{7}$	$\frac{1}{4}$	$\frac{1}{3}$
$g(x)$	-3	-2	-1	0	2	3	4

29.

One-to-one

31.

Not one-to-one

33.

Not one-to-one

35.

Not one-to-one

37.

One-to-one

39.

Not one-to-one

41. Not one-to-one

43. $f^{-1}(x) = \dfrac{5x - 4}{3}$

45. Not one-to-one

47. $f^{-1}(x) = \sqrt{x} - 3, x \geq 0$

49. $f^{-1}(x) = \dfrac{x^2 - 3}{2}, x \geq 0$

51. $f^{-1}(x) = 2 - x, x \geq 0$

53. $f^{-1}(x) = \dfrac{x + 3}{2}$

Reflections in the line $y = x$

55. $f^{-1}(x) = \sqrt[5]{x}$

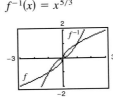

Reflections in the line $y = x$

57. $f^{-1}(x) = x^{5/3}$

Reflections in the line $y = x$

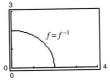

59. $f^{-1}(x) = \sqrt{4 - x^2}, 0 \leq x \leq 2$

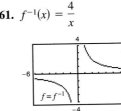

The graphs are the same.

61. $f^{-1}(x) = \dfrac{4}{x}$

The graphs are the same.

63. $y = \sqrt{x} + 2, x \geq 0$ **65.** $y = x - 2, x \geq 0$

67.

x	-4	-2	2	3
$f^{-1}(x)$	-2	-1	1	3

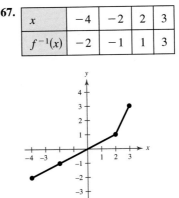

69. (a) and (b)

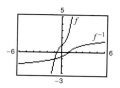

(c) Inverse function because it satisfies the Vertical Line Test

71. (a) and (b)

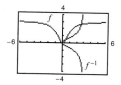

(c) Not an inverse function because the inverse relation does not satisfy the Vertical Line Test

73. 32   **75.** 600   **77.** $2\sqrt[3]{x + 3}$

79. $\dfrac{x + 1}{2}$ **81.** $\dfrac{x + 1}{2}$

83. (a) Yes

(b) $f^{-1}(t)$ represents the year new car sales totaled $\$t$ billion.

(c) 10 or 2000

(d) No. The inverse is not a function because f is not one-to-one.

85. False. For example, $y = x^2$ is even, but does not have an inverse.

87. Answers will vary.   **89.** $9x$, $x \neq 0$

91. $-(x + 6)$, $x \neq 6$ **93.** Function

95. Not a function   **97.** Function

Review Exercises   (page 156)

1.

x	-2	0	2	3	4
y	3	2	1	$\frac{1}{2}$	0
Solution point	$(-2, 3)$	$(0, 2)$	$(2, 1)$	$\left(3, \frac{1}{2}\right)$	$(4, 0)$

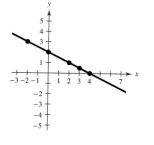

3.

x	-2	-1	0	1	2
y	0	3	4	3	0
Solution point	$(-2, 0)$	$(-1, 3)$	$(0, 4)$	$(1, 3)$	$(2, 0)$

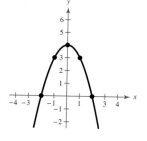

5.

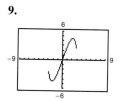

Intercepts: $(-1, 0)$, $\left(0, \frac{1}{4}\right)$

7.

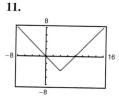

Intercepts: $(0, 0)$, $(\pm 2\sqrt{2}, 0)$

9.

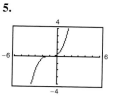

Intercepts: $(0, 0)$, $(\pm 3, 0)$

11.

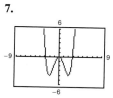

Intercepts: $(0, 0)$, $(8, 0)$

13.

```
Xmin = -20
Xmax = 50
Xscl = 10
Ymin = -2
Ymax = 1
Yscl = 0.5
```

15. (a)

```
Xmin = 0
Xmax = 6
Xscl = 1
Ymin = 7000
Ymax = 14000
Yscl = 1000
```

(b)

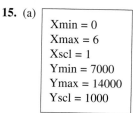

(c) 4

17.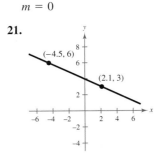

$m = 0$

19.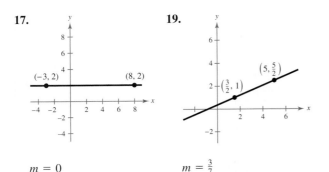

$m = \frac{3}{7}$

21.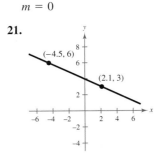

$m = -\frac{5}{11}$

23. $x - 4y - 6 = 0$; $(6, 0), (10, 1), (-2, -2)$

25. $3x - 2y - 10 = 0$; $(4, 1), (2, -2), (-2, 8)$

27. $5x + 5y + 24 = 0$; $\left(-5, \frac{1}{5}\right), \left(-4, -\frac{4}{5}\right), \left(-6, \frac{6}{5}\right)$

29. $y = 6$; $(0, 6), (1, 6), (-1, 6)$

31. $x = 10$; $(10, 1), (10, 3), (10, -2)$

33. $y = -1$

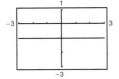

35. $y = \frac{2}{7}x + \frac{2}{7}$

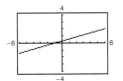

37. $V = 850t + 8250$ **39.** \$210,000

41. (a) $5x - 4y - 23 = 0$ **43.** (a) $x = -6$

    (b) $4x + 5y - 2 = 0$ (b) $y = 2$

45. (a) Not a function because element 20 in A corresponds to two elements, 4 and 6, in B.

    (b) Function

(c) Function

(d) Not a function because 30 in A corresponds to no element in B.

47. Not a function   **49.** Function

51. (a) 5   (b) 17   (c) $t^4 + 1$ (d) $-x^2 - 1$

53. (a) -3 (b) -1 (c) 2   (d) 6

55. All real numbers   **57.** $[-5, 5]$

59. All real numbers s except $s = 3$

61. (a) $C = 5.35x + 16{,}000$ (b) $P = 2.85x - 16{,}000$

63. $2h + 4x + 3$, $h \neq 0$

65.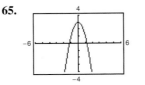

Domain: $(-\infty, \infty)$
Range: $(-\infty, 3]$

67.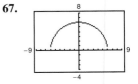

Domain: $[-6, 6]$
Range: $[0, 6]$

69. (a)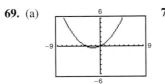

71. (a)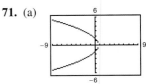

(b) Function

(b) Not a function

73. (a)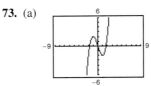

(b) Increasing on $(-\infty, -1), (1, \infty)$

    Decreasing on $(-1, 1)$

75. (a)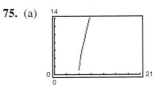

(b) Increasing on $(6, \infty)$

77. Relative maximum: $(0, 16)$

    Relative minimums: $(-2, 0), (2, 0)$

79. Relative maximum: $(3, 27)$

81.

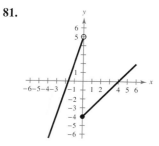

83. Even

85. Constant function $f(x) = C$; vertical shift 2 units downward; $g(x) = -2$

87. Cubic function $f(x) = x^3$; reflection in the x-axis and vertical shift two units downward; $g(x) = -x^3 - 2$

89. (a) Quadratic function

    (b) Vertical shift six units downward

    (c)

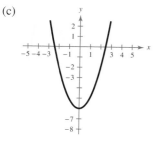

    (d) $h(x) = f(x) - 6$

91. (a) Cubic function

    (b) Horizontal shift one unit to the right, and vertical shift seven units upward

    (c)

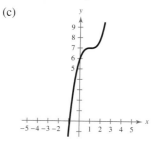

    (d) $h(x) = f(x - 1) + 7$

93. (a) Square root function

    (b) Vertical shift five units downward

(c)

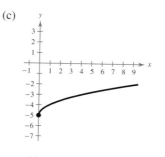

    (d) $h(x) = f(x) - 5$

95. (a) Quadratic function

    (b) Reflection in the x-axis and vertical shift three units downward

    (c)

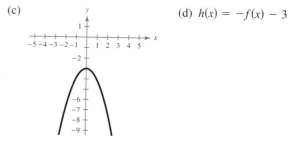

    (d) $h(x) = -f(x) - 3$

97. (a) Quadratic function

    (b) Vertical stretch, reflection in the x-axis, and vertical shift three units upward

    (c)

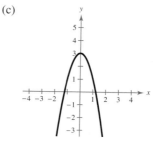

    (d) $h(x) = -2f(x) + 3$

99. (a) Absolute value function

    (b) Vertical shrink, reflection in the x-axis, and vertical shift nine units upward

    (c)

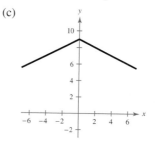

    (d) $h(x) = -\frac{1}{2}f(x) + 9$

101. -7 **103.** -42 **105.** 5

107. 23   **109.** -97

111.

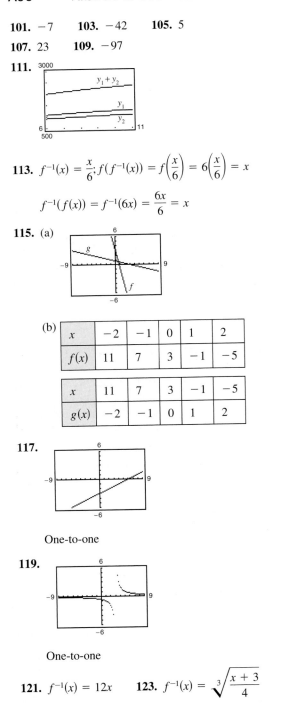

113. $f^{-1}(x) = \dfrac{x}{6}; f(f^{-1}(x)) = f\left(\dfrac{x}{6}\right) = 6\left(\dfrac{x}{6}\right) = x$

$f^{-1}(f(x)) = f^{-1}(6x) = \dfrac{6x}{6} = x$

115. (a)

(b)

x	-2	-1	0	1	2
$f(x)$	11	7	3	-1	-5

x	11	7	3	-1	-5
$g(x)$	-2	-1	0	1	2

117.

One-to-one

119.

One-to-one

121. $f^{-1}(x) = 12x$ **123.** $f^{-1}(x) = \sqrt[3]{\dfrac{x+3}{4}}$

125. $f^{-1}(x) = x^2 - 10, \ x \geq 0$

127. False. The point $(-1, 28)$ does not lie on the graph of the function $g(x) = -(x-6)^2 - 3$.

129. False. For example, $f(x) = 4 - x = f^{-1}(x)$

Chapter Test   (page 160)

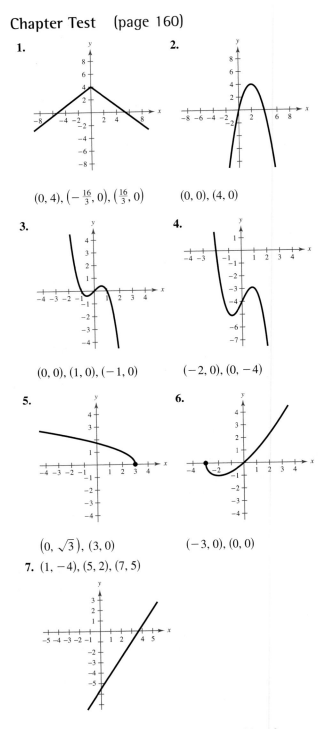

1.

$(0, 4), \left(-\dfrac{16}{3}, 0\right), \left(\dfrac{16}{3}, 0\right)$

2.

$(0, 0), (4, 0)$

3.

$(0, 0), (1, 0), (-1, 0)$

4.

$(-2, 0), (0, -4)$

5.

$\left(0, \sqrt{3}\right), (3, 0)$

6.

$(-3, 0), (0, 0)$

7. $(1, -4), (5, 2), (7, 5)$

8. (a) $5x + 2y - 8 = 0$ (b) $2x - 5y + 20 = 0$

9. No. To some x there corresponds more than one value of y.

10. (a) -9 (b) 1   (c) $|t - 4| - 15$

11. $(-\infty, 3]$ **12.** $C = 5.60x + 24,000$

$P = 93.9x - 24,000$

13. Increasing: $(-2, 0)$, $(2, \infty)$

Decreasing: $(-\infty, -2)$, $(0, 2)$

14. Increasing: $(-2, 2)$

Constant: $(-\infty, -2)$, $(2, \infty)$

15.
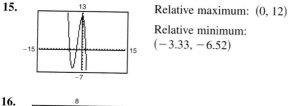

Relative maximum: $(0, 12)$

Relative minimum: $(-3.33, -6.52)$

16.
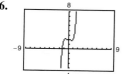

Relative minimum: $(0.77, 1.81)$

Relative maximum: $(-0.77, 2.19)$

17. (a) $f(x) = x^3$

(b) Horizontal shift five units to the right, reflection in the x-axis, vertical stretch, and vertical shift three units upward

(c)
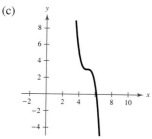

18. (a) $f(x) = \sqrt{x}$

(b) Reflection in the y-axis and a horizontal shift seven units to the left

(c)
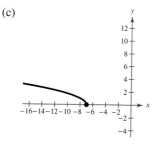

19. (a) $f(x) = |x|$

(b) Reflection in the y-axis (no effect), vertical stretch, and vertical shift seven units downward

(c)
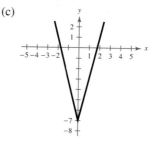

20. (a) $x^2 - \sqrt{2-x}$, $(-\infty, 2]$ (b) $\dfrac{x^2}{\sqrt{2-x}}$, $(-\infty, 2)$

(c) $2 - x$, $(-\infty, 2]$ (d) $\sqrt{2 - x^2}$, $\left[-\sqrt{2}, \sqrt{2}\right]$

21. $f^{-1}(x) = \sqrt[3]{x - 8}$ **22.** No inverse

23. $f^{-1}(x) = \left(\tfrac{8}{3}x\right)^{2/3}$, $x \geq 0$

Chapter 2
Section 2.1   (page 168)

Vocabulary Check   (page 168)

1. equation   **2.** solve   **3.** identity, conditional

4. $ax + b = 0$ **5.** extraneous

6. Mathematical modeling   **7.** formulas

1. (a) Yes   (b) No   (c) No   (d) No

3. (a) Yes   (b) No   (c) No   (d) No

5. (a) No   (b) No   (c) No   (d) Yes

7. Identity   **9.** Identity   **11.** Conditional   **13.** $-\dfrac{96}{23}$

15. -10 **17.** $-\dfrac{6}{5}$ **19.** 10 **21.** 4 **23.** 5

25. $\dfrac{11}{6}$ **27.** $\dfrac{5}{3}$ **29.** No solution   **31.** $h = \dfrac{2A}{b}$

33. $P = A\left(1 + \dfrac{r}{n}\right)^{-nt}$ **35.** $r = \dfrac{S - a}{S - L}$ **37.** $b = \dfrac{3V}{4\pi a^2}$

39. 61.2 inches

41. (a)
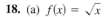

(b) $l = 1.5w$; $P = 5w$

(c) 5 meters × 7.5 meters

43. (a) Test average $= \dfrac{\text{test 1} + \text{test 2} + \text{test 3} + \text{test 4}}{4}$

(b) 97

45. 3 hours   **47.** ≈ 46.3 miles per hour

49. ≈ 8.33 minutes

51. (a)   (b) 91.4 feet

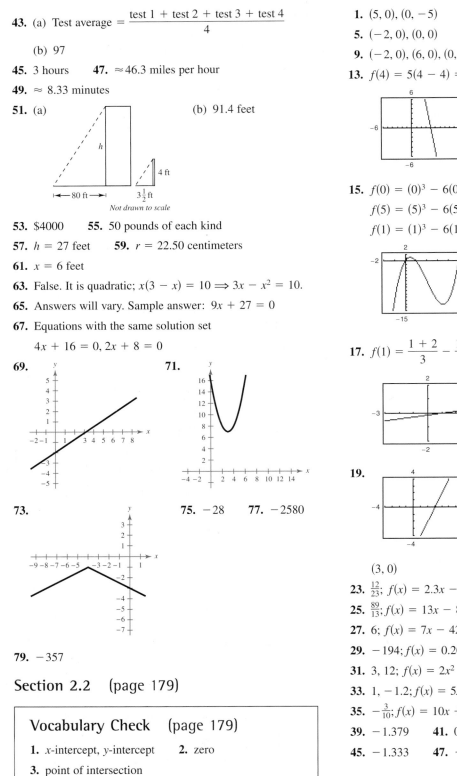

80 ft $3\frac{1}{2}$ ft   4 ft h

Not drawn to scale

53. \$4000   **55.** 50 pounds of each kind

57. $h = 27$ feet   **59.** $r = 22.50$ centimeters

61. $x = 6$ feet

63. False. It is quadratic; $x(3 - x) = 10 \Rightarrow 3x - x^2 = 10.$

65. Answers will vary. Sample answer: $9x + 27 = 0$

67. Equations with the same solution set

$4x + 16 = 0,\ 2x + 8 = 0$

69.   **71.**

73.

79. -357

Section 2.2   (page 179)

Vocabulary Check   (page 179)

1. x-intercept, y-intercept   **2.** zero

3. point of intersection

1. $(5, 0), (0, -5)$ **3.** $(-2, 0), (1, 0), (0, -2)$

5. $(-2, 0), (0, 0)$ **7.** No intercepts

9. $(-2, 0), (6, 0), (0, -2)$ **11.** $(1, 0), \left(0, \frac{1}{2}\right)$

13. $f(4) = 5(4 - 4) = 0$

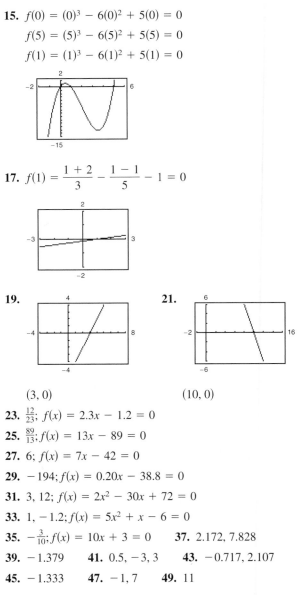

15. $f(0) = (0)^3 - 6(0)^2 + 5(0) = 0$

$f(5) = (5)^3 - 6(5)^2 + 5(5) = 0$

$f(1) = (1)^3 - 6(1)^2 + 5(1) = 0$

17. $f(1) = \dfrac{1 + 2}{3} - \dfrac{1 - 1}{5} - 1 = 0$

19.   **21.**

$(3, 0)$ $(10, 0)$

23. $\frac{12}{23}; f(x) = 2.3x - 1.2 = 0$

25. $\frac{89}{13}; f(x) = 13x - 89 = 0$

27. $6; f(x) = 7x - 42 = 0$

29. $-194; f(x) = 0.20x - 38.8 = 0$

31. $3, 12; f(x) = 2x^2 - 30x + 72 = 0$

33. $1, -1.2; f(x) = 5x^2 + x - 6 = 0$

35. $-\frac{3}{10}; f(x) = 10x + 3 = 0$ **37.** $2.172, 7.828$

39. -1.379 **41.** $0.5, -3, 3$ **43.** $-0.717, 2.107$

45. -1.333 **47.** $-1, 7$ **49.** 11

51. (a)

x	-1	0	1	2	3	4
$3.2x - 5.8$	-9	-5.8	-2.6	0.6	3.8	7

(b) $1 < x < 2$; Answers will vary.

(c)

x	1.5	1.6	1.7
$3.2x - 5.8$	-1	-0.68	-0.36

x	1.8	1.9	2
$3.2x - 5.8$	-0.04	0.28	0.6

(d) $1.8 < x < 1.9$

To improve accuracy, evaluate the expression in this interval and determine where the sign changes.

(e) $x = 1.8125$

53. $(1, 1)$ **55.** $(2, 2)$ **57.** $(-1, 3), (2, 6)$ **59.** $(4, 1)$

61. $(1.449, 1.899), (-3.449, -7.899)$

63. $(0, 0), (-2, 8), (2, 8)$

65. (a) 6.46

(b) $\frac{1.73}{0.27} \approx 6.41$. The second method decreases the accuracy.

67. (a) $t(x) = \dfrac{x}{63} + \dfrac{280 - x}{54}$

(b)
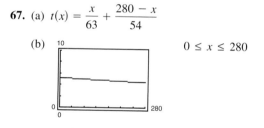
$0 \le x \le 280$

(c) 164.5 miles

69. (a) $A = 0.33(55 - x) + x$

(b)
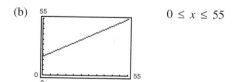
$0 \le x \le 55$

(c) 22.2 gallons

71. (a) $A(x) = 12x$

(b)
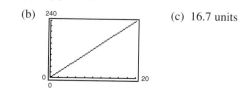
(c) 16.7 units

73. (a) 5200 cubic feet   (b) $y = \frac{1}{8}x$

(c) Answers will vary.

(d)

(e)

d	3	5	7	9
V	720	2000	3600	5200

(f) 8.5 feet   (g) 38,896 gallons

75. 1993; To answer the question algebraically, solve the equation $0.91t + 56.4 = 60$. To answer the question graphically, graph the model $y_1 = 0.91x + 56.4$ and the horizontal line $y_2 = 60$ in the same viewing window and determine where the lines intersect.

77. True   **79.** False. The lines could be identical.

81. $\dfrac{4\sqrt{3}}{5}$ **83.** $\dfrac{3(8 - \sqrt{11})}{53}$

85. $3x^2 + 13x - 30$ **87.** $4x^2 - 81$

Section 2.3   (page 189)

Vocabulary Check   (page 189)

1. (a) ii   (b) iii   (c) i   **2.** $\sqrt{-1}, -1$

3. complex, $a + bi$ **4.** real, imaginary

5. Mandelbrot Set

1. $a = -9, b = 4$ **3.** $a = 6, b = 5$ **5.** $4 + 5i$

7. 7   **9.** $-1 - 5i$ **11.** -75 **13.** $0.3i$

15. $11 - i$ **17.** $7 - 3\sqrt{2}i$ **19.** $-14 + 20i$

21. $\frac{19}{6} + \frac{37}{6}i$ **23.** $-4.2 + 7.5i$ **25.** $-2\sqrt{3}$

27. -10 **29.** $5 + i$ **31.** $-20 + 32i$ **33.** 24

35. $80i$ **37.** $4 - 3i; 25$ **39.** $-6 + \sqrt{5}i; 41$

41. $-\sqrt{20}i; 20$ **43.** $3 + \sqrt{-2}; 11$

45. $-6i$ **47.** $\frac{8}{41} + \frac{10}{41}i$ **49.** $\frac{3}{5} + \frac{4}{5}i$

51. $-\frac{40}{1681} - \frac{9}{1681}i$ **53.** $-\frac{1}{2} - \frac{5}{2}i$ **55.** $\frac{62}{949} + \frac{297}{949}i$

57. $-1 + 6i$ **59.** $-375\sqrt{3}i$ **61.** i

63. (a) 8   (b) 8   (c) 8; Answers will vary.

65. $4 + 3i$

67.

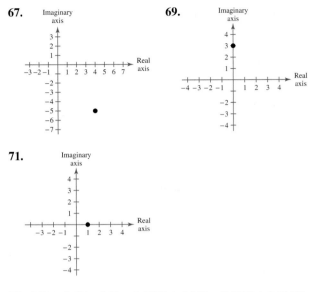

69.

71.

73. $0.5i$, $-0.25 + 0.5i$, $-0.1875 + 0.25i$, $-0.0273 + 0.4063i$, $-0.1643 + 0.4778i$, $-0.2013 + 0.3430i$; Yes, bounded

75. $1, 2, 5, 26, 677, 458{,}330$; No, unbounded

77. $3.12 - 0.97i$

79. False. Any real number is equal to its conjugate.

81. $16x^2 - 25$ **83.** $3x^2 + \frac{23}{2}x - 2$

Section 2.4   (page 205)

Vocabulary Check   (page 205)

1. quadratic equation

2. factoring, extracting square roots, completing the square, quadratic formula

3. discriminant

4. position, $-16t^2 + v_0t + s_0$, initial velocity, initial height

1. $2x^2 + 5x - 3 = 0$ **3.** $3x^2 - 60x - 10 = 0$

5. $0, -\frac{1}{2}$ **7.** $4, -2$ **9.** $3, -\frac{1}{2}$ **11.** $2, -6$

13. $-a - b, -a + b$ **15.** ± 7

17. $16, 8$ **19.** $\frac{1}{2} \pm \sqrt{3}$; $2.23, -1.23$

21. 2 **23.** $-8, 4$ **25.** $-3 \pm \sqrt{7}$ **27.** $1 \pm \dfrac{\sqrt{6}}{3}$

29. $2 \pm 2\sqrt{3}$ **31.** $-\dfrac{5}{4} \pm \dfrac{\sqrt{89}}{4}$

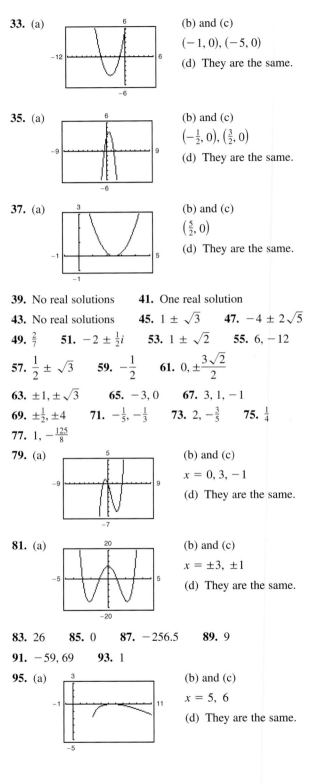

33. (a)

(b) and (c)

$(-1, 0), (-5, 0)$

(d) They are the same.

35. (a)

(b) and (c)

$\left(-\frac{1}{2}, 0\right), \left(\frac{3}{2}, 0\right)$

(d) They are the same.

37. (a)

(b) and (c)

$\left(\frac{5}{2}, 0\right)$

(d) They are the same.

39. No real solutions   **41.** One real solution

43. No real solutions   **45.** $1 \pm \sqrt{3}$ **47.** $-4 \pm 2\sqrt{5}$

49. $\frac{2}{7}$ **51.** $-2 \pm \frac{1}{2}i$ **53.** $1 \pm \sqrt{2}$ **55.** $6, -12$

57. $\frac{1}{2} \pm \sqrt{3}$ **59.** $-\frac{1}{2}$ **61.** $0, \pm\dfrac{3\sqrt{2}}{2}$

63. $\pm 1, \pm \sqrt{3}$ **65.** $-3, 0$ **67.** $3, 1, -1$

69. $\pm\frac{1}{2}, \pm 4$ **71.** $-\frac{1}{5}, -\frac{1}{3}$ **73.** $2, -\frac{3}{5}$ **75.** $\frac{1}{4}$

77. $1, -\frac{125}{8}$

79. (a)

(b) and (c)

$x = 0, 3, -1$

(d) They are the same.

81. (a)

(b) and (c)

$x = \pm 3, \pm 1$

(d) They are the same.

83. 26 **85.** 0 **87.** -256.5 **89.** 9

91. $-59, 69$ **93.** 1

95. (a)

(b) and (c)

$x = 5, 6$

(d) They are the same.

97. (a)

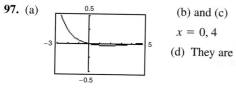

(b) and (c)

$x = 0, 4$

(d) They are the same.

99. $4, -5$ **101.** $\dfrac{-3 \pm \sqrt{21}}{6}$ **103.** $2, -\dfrac{3}{2}$

105. $3, -2$ **107.** $\sqrt{3}, -3$

109. (a)

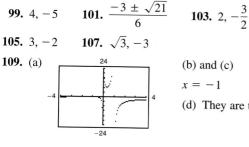

(b) and (c)

$x = -1$

(d) They are the same.

111. (a)

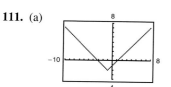

(b) and (c)

$x = 1, -3$

(d) They are the same.

113. $x^2 + x - 30 = 0$ **115.** $x^3 - 4x^2 - 2x + 8 = 0$

117. $x^4 - 3x^2 - 4 = 0$ **119.** $x = 6 \text{ or } -4$

121. (a)

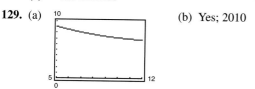

(b) $1632 = w^2 + 14w$

(c) Width: 34 feet; length: 48 feet

123. 14 centimeters $\times$ 14 centimeters

125. (a) $s = -16t^2 + 1815$

(b)

t	0	2	4	6
s	1815	1751	1559	1239

t	8	10	12
s	791	215	-489

(c) $[10, 12]$; 10 seconds; ≈ 10.65 seconds

127. (a) $s = -16t^2 + 45t + 5.5$ (b) 24 feet

(c) ≈ 2.8 seconds

129. (a)

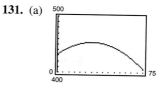

(b) Yes; 2010

131. (a)

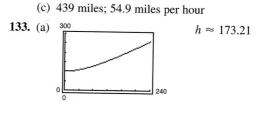

(b) 453 miles; 15 hours

(c) 439 miles; 54.9 miles per hour

133. (a)

$h \approx 173.21$

(b)

h	160	165	170	175	180	185
d	188.7	192.9	197.2	201.6	205.9	210.3

(c) 173.2

(d) Solving graphically or numerically yields an approximate solution. An exact solution is obtained by solving algebraically.

135. Eastbound plane: ≈ 550 miles per hour

Northbound plane: ≈ 600 miles per hour

137. False. Both solutions are complex.

139. False. For example, $|x| = x^2 + x + 3$ has two extraneous solutions.

141. (a) $0, -\dfrac{b}{a}$ (b) $0, 1$ **143.** $x^2(x - 3)(x^2 + 3x + 9)$

145. $(x + 5)(x - \sqrt{2})(x + \sqrt{2})$ **147.** Function

149. Not a function   **151.** Function

Section 2.5   (page 219)

Vocabulary Check   (page 219)

1. negative   **2.** double   **3.** $-a \le x \le a$

4. $x \le -a, \ x \ge a$ **5.** zeros, undefined values

1. d   **2.** a   **3.** c   **4.** b

5. (a) Yes   (b) No   (c) Yes   (d) No

7. (a) No   (b) Yes   (c) Yes   (d) No

9. $x > -4$ **11.** $x < -\dfrac{1}{2}$

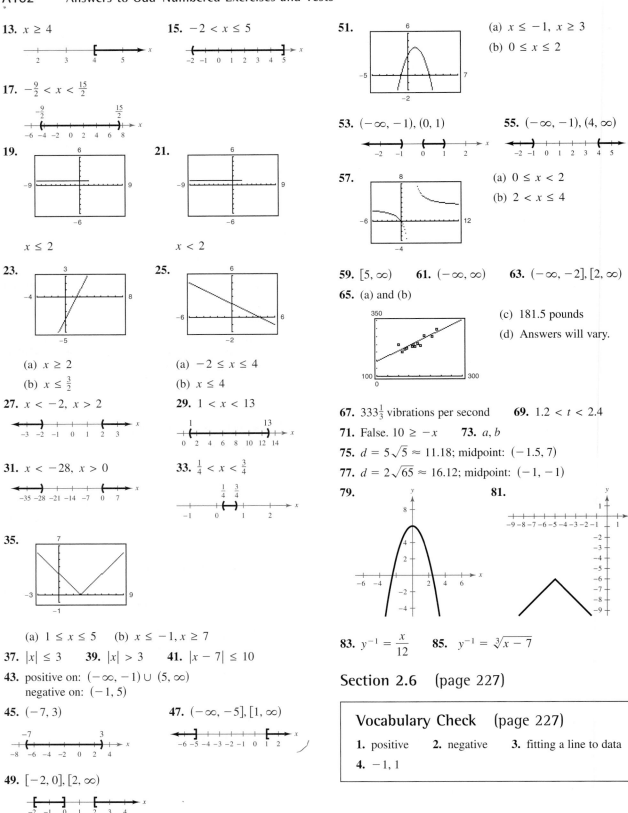

13. $x \geq 4$

15. $-2 < x \leq 5$

17. $-\frac{9}{2} < x < \frac{15}{2}$

19.

21.

$x \leq 2$

$x < 2$

23.

25.

(a) $x \geq 2$

(b) $x \leq \frac{3}{2}$

(a) $-2 \leq x \leq 4$

(b) $x \leq 4$

27. $x < -2, \ x > 2$

29. $1 < x < 13$

31. $x < -28, \ x > 0$

33. $\frac{1}{4} < x < \frac{3}{4}$

35.

(a) $1 \leq x \leq 5$ (b) $x \leq -1, x \geq 7$

37. $|x| \leq 3$ **39.** $|x| > 3$ **41.** $|x - 7| \leq 10$

43. positive on: $(-\infty, -1) \cup (5, \infty)$
negative on: $(-1, 5)$

45. $(-7, 3)$

47. $(-\infty, -5], [1, \infty)$

49. $[-2, 0], [2, \infty)$

51.

(a) $x \leq -1, \ x \geq 3$

(b) $0 \leq x \leq 2$

53. $(-\infty, -1), (0, 1)$

55. $(-\infty, -1), (4, \infty)$

57.

(a) $0 \leq x < 2$

(b) $2 < x \leq 4$

59. $[5, \infty)$ **61.** $(-\infty, \infty)$ **63.** $(-\infty, -2], [2, \infty)$

65. (a) and (b)

(c) 181.5 pounds

(d) Answers will vary.

67. $333\frac{1}{3}$ vibrations per second   **69.** $1.2 < t < 2.4$

71. False. $10 \geq -x$ **73.** a, b

75. $d = 5\sqrt{5} \approx 11.18$; midpoint: $(-1.5, 7)$

77. $d = 2\sqrt{65} \approx 16.12$; midpoint: $(-1, -1)$

79.

81.

83. $y^{-1} = \dfrac{x}{12}$ **85.** $y^{-1} = \sqrt[3]{x - 7}$

Section 2.6   (page 227)

Vocabulary Check   (page 227)

1. positive   **2.** negative   **3.** fitting a line to data

4. $-1, 1$

1. (a)

(b) Yes. The monthly sales increase as the experience of the sales representatives increases.

3. Negative correlation   **5.** No correlation

7. (a)

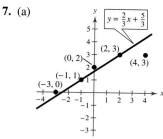

(b) $y = 0.46x + 1.6$

(c)

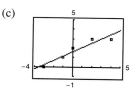

(d) The models appear valid.

9. (a)

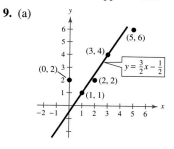

(b) $y = 0.95x + 0.92$

(c)

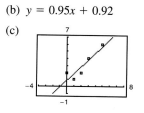

(d) The models appear valid.

11. (a)

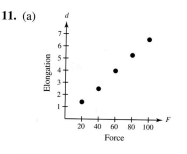

(b) $d = 0.07F - 0.3$

(c) $d = 0.066F$; This model fits the data better.

(d) 3.63 centimeters

13. (a) $S = 0.2t - 0.14$

(b)

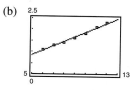

(c) The slope represents the average annual increase in salaries (in millions of dollars).

(d) $3.1 million

15. (a) $S = 0.183t + 7.013$

(b)

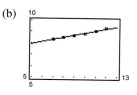

(c) The slope represents the average annual increase in spending.

(d) $10.31

(e)

Year	1997	1998	1999
Actual, S	8.30	8.50	8.65
Model, S	8.29	8.48	8.66

Year	2000	2001	2002
Actual, S	8.80	9.00	9.25
Model, S	8.84	9.03	9.21

The model fits well.

17. (a) $y = -0.024x + 5.06$

(b) The winning times decrease as time in years increases.

(c)

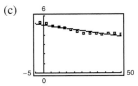

(d) The model is a reasonably close fit.

(e) Answers will vary.

19. True. To have positive correlation, the y-values tend to increase as x increases.

21. Answers will vary.   **23.** $P \le 2$ **25.** $-3 \le z \le 10$

27. $5x - 10, x \ne -2$ **29.** (a) 10   (b) $2w^2 + 5w + 7$

31. (a) 5   (b) 1   **33.** $-\frac{3}{5}$ **35.** $-\frac{1}{4}, \frac{3}{2}$

37. $\dfrac{7 \pm \sqrt{17}}{4}$

Review Exercises   (page 232)

1. (a) No   (b) No   (c) No   (d) Yes

3. $x = 9$ **5.** $x = \frac{1}{2}$ **7.** $x = \frac{7}{3}$

9. September: \$325,000; October: \$364,000

11. $2\frac{6}{7}$ liters

13. (a)   (b) $h = \frac{64}{3}$ meters

h

8 m   2 m   75 cm

15. $-3.5°C$ **17.** $(-3, 0), (0, 3)$ **19.** $(1, 0), (8, 0), (0, 8)$

21. $(0, -7)$ **23.** $x = 2.2$ **25.** $x = -1.301$

27. $x = 0.338, 1.307$ **29.** $(1, -2)$

31. $(4.5, -3.125), (-3, 2.5)$ **33.** $6 + 5i$ **35.** $2 + 7i$

37. $3 + 7i$ **39.** $40 + 65i$ **41.** $-4 - 46i$

43. -80 **45.** $1 - 6i$ **47.** $\frac{17}{26} + \frac{7}{26}i$

49.   **51.**

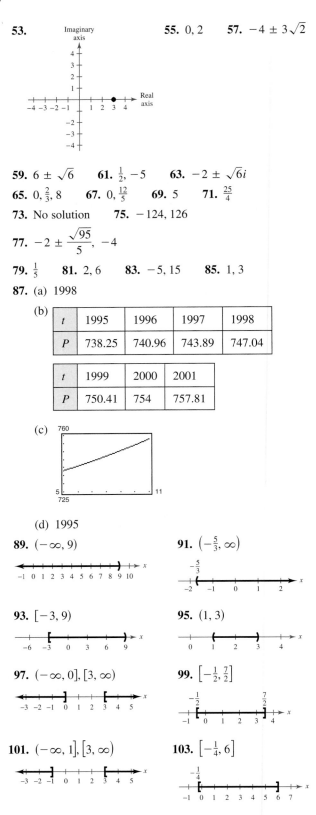

53.   **55.** $0, 2$ **57.** $-4 \pm 3\sqrt{2}$

59. $6 \pm \sqrt{6}$ **61.** $\frac{1}{2}, -5$ **63.** $-2 \pm \sqrt{6}i$

65. $0, \frac{2}{3}, 8$ **67.** $0, \frac{12}{5}$ **69.** 5   **71.** $\frac{25}{4}$

73. No solution   **75.** $-124, 126$

77. $-2 \pm \dfrac{\sqrt{95}}{5}, -4$

79. $\frac{1}{5}$ **81.** $2, 6$ **83.** $-5, 15$ **85.** $1, 3$

87. (a) 1998

(b)

t	1995	1996	1997	1998
P	738.25	740.96	743.89	747.04

t	1999	2000	2001
P	750.41	754	757.81

(c)

(d) 1995

89. $(-\infty, 9)$ **91.** $\left(-\frac{5}{3}, \infty\right)$

93. $[-3, 9)$ **95.** $(1, 3)$

97. $(-\infty, 0], [3, \infty)$ **99.** $\left[-\frac{1}{2}, \frac{7}{2}\right]$

101. $(-\infty, 1], [3, \infty)$ **103.** $\left[-\frac{1}{4}, 6\right]$

105. $[-4, 0], [4, \infty)$

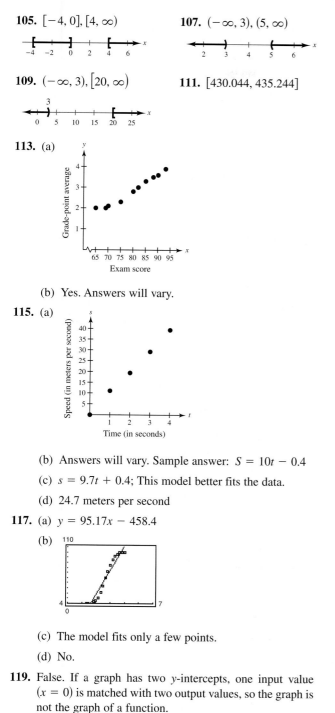

107. $(-\infty, 3), (5, \infty)$

109. $(-\infty, 3), [20, \infty)$

111. $[430.044, 435.244]$

113. (a)

(b) Yes. Answers will vary.

115. (a)

(b) Answers will vary. Sample answer: $S = 10t - 0.4$

(c) $s = 9.7t + 0.4$; This model better fits the data.

(d) 24.7 meters per second

117. (a) $y = 95.17x - 458.4$

(b)

(c) The model fits only a few points.

(d) No.

119. False. If a graph has two y-intercepts, one input value $(x = 0)$ is matched with two output values, so the graph is not the graph of a function.

121. False. For example, the slope of the regression line for $(1, 4), (2, 3), (3, 2),$ and $(4, 1)$ is -1.

123. The real zeros of a function are the values of x at which the graph of the function crosses the x-axis (the x-intercepts). They are also the values of x that satisfy the equation $f(x) = 0$.

125. $\sqrt{-6} \cdot \sqrt{-6} = 6i^2$ or -6, not 6.

127. (a) 1   (b) i (c) -1 (d) $-i$

Chapter Test   (page 236)

1. $x = 3$ **2.** $x = \frac{2}{15}$ **3.** $-9 - 18i$

4. $6 + (2\sqrt{5} + \sqrt{14})i$ **5.** $13 + 4i$ **6.** $-17 + 14i$

7. $\frac{43}{37} + \frac{38}{37}i$ **8.** $1 + 2i$

9.

No x-intercepts

No real zeros

10.

No x-intercepts

No real zeros

11.

One x-intercept: $(0, 0)$

One real zero: 0

12.

One x-intercept: $(0, 0)$

One real zero: 0

13. 1, 9   **14.** $-6 \pm \sqrt{38}$ **15.** $\pm\frac{9}{2}$

16. $-3, \frac{1}{5}$ **17.** $\pm 2, \frac{4}{3}$ **18.** 2   **19.** $\pm\sqrt{58}$

20. $-\frac{5}{2}, \frac{11}{4}$

21. $\left(\frac{7}{6}, \frac{17}{8}\right)$ **22.** $(3, 13)$

23. $\left(-7, -\frac{2}{3}\right)$

24. $L = 14.8t + 92$; 2004

Cumulative Test for Chapters P–2
(page 237)

1. $\dfrac{7x^3}{16y^5}$, $x \neq 0$ 2. $9\sqrt{15}$ 3. $2x^2y\sqrt{7y}$

4. $7x - 10$ 5. $x^3 - x^2 - 5x + 6$ 6. $\dfrac{x - 1}{(x + 1)(x + 3)}$

7. $(3 + x)(7 - x)$ 8. $x(1 + x)(1 - 6x)$

9. $2(3 - 2x)(9 + 6x + 4x^2)$

10. Midpoint: $(1.5, -2)$; $d = 2\sqrt{61} \approx 15.62$

11. $\left(x + \frac{1}{2}\right)^2 + (y + 8)^2 = \frac{25}{16}$

12.
13.
14.

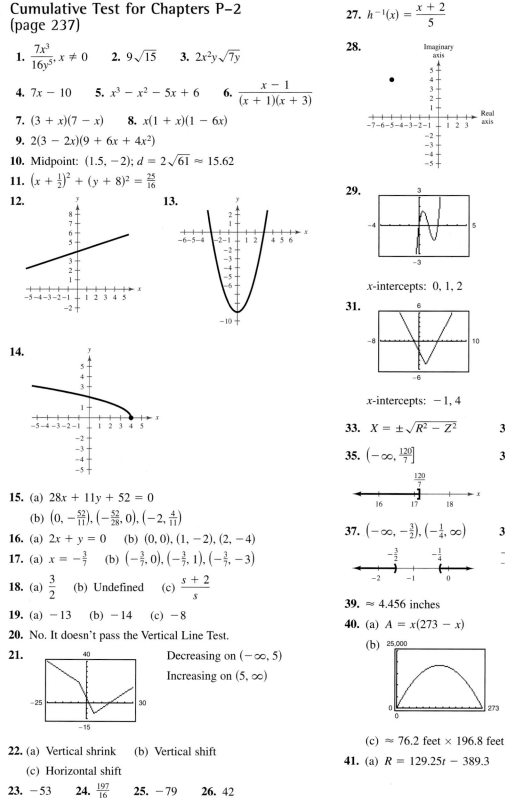

15. (a) $28x + 11y + 52 = 0$

   (b) $\left(0, -\frac{52}{11}\right), \left(-\frac{52}{28}, 0\right), \left(-2, \frac{4}{11}\right)$

16. (a) $2x + y = 0$ (b) $(0, 0), (1, -2), (2, -4)$

17. (a) $x = -\frac{3}{7}$ (b) $\left(-\frac{3}{7}, 0\right), \left(-\frac{3}{7}, 1\right), \left(-\frac{3}{7}, -3\right)$

18. (a) $\dfrac{3}{2}$ (b) Undefined   (c) $\dfrac{s + 2}{s}$

19. (a) -13 (b) -14 (c) -8

20. No. It doesn't pass the Vertical Line Test.

21.   Decreasing on $(-\infty, 5)$
       Increasing on $(5, \infty)$

22. (a) Vertical shrink   (b) Vertical shift

   (c) Horizontal shift

23. -53 24. $\frac{197}{16}$ 25. -79 26. 42

27. $h^{-1}(x) = \dfrac{x + 2}{5}$

28.

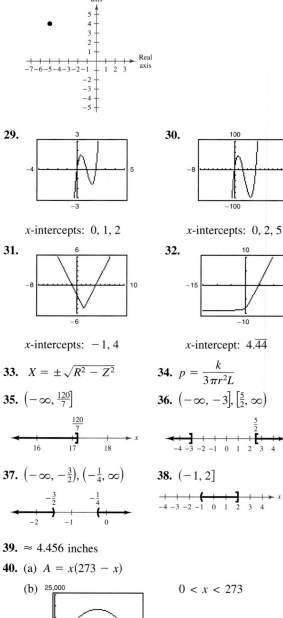

29.   x-intercepts: $0, 1, 2$

30.   x-intercepts: $0, 2, 5$

31.   x-intercepts: $-1, 4$

32.   x-intercept: $4.\overline{44}$

33. $X = \pm\sqrt{R^2 - Z^2}$ 34. $p = \dfrac{k}{3\pi r^2 L}$

35. $\left(-\infty, \frac{120}{7}\right]$

36. $\left(-\infty, -3\right], \left[\frac{5}{2}, \infty\right)$

37. $\left(-\infty, -\frac{3}{2}\right), \left(-\frac{1}{4}, \infty\right)$

38. $(-1, 2]$

39. ≈ 4.456 inches

40. (a) $A = x(273 - x)$

   (b) $0 < x < 273$

   (c) ≈ 76.2 feet $\times$ 196.8 feet

41. (a) $R = 129.25t - 389.3$

(b)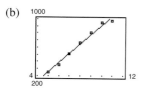

(c) As time increases, revenues increase.

(d) $1,808,000,000

(e)

Year	1995	1996	1997	1998
Actual, R	253.4	360.1	508.8	669.8
Model, R	257.0	386.2	515.5	644.7

Year	1999	2000	2001
Actual, R	805.3	944.7	971.2
Model, R	774.0	903.2	1032.5

The model is a good fit.

Chapter 3

Section 3.1   (page 247)

1. g   2. c   3. b   4. h   5. f   6. a

7. e   8. d

9.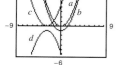

(a) Vertical shrink

(b) Vertical shrink and vertical shift one unit downward

(c) Vertical shrink and a horizontal shift three units to the left

(d) Vertical shrink, reflection in the x-axis, a horizontal shift three units to the left, and a vertical shift one unit downward

11.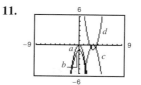

(a) Vertical stretch and a reflection in the x-axis

(b) Vertical stretch, a reflection in the x-axis, and vertical shift one unit down

(c) Vertical stretch, reflection in the x-axis, and horizontal shift three units to the right

(d) Vertical stretch, horizontal shift three units to the right, and vertical shift one unit downward

13. Vertex: $(0, 25)$
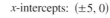
x-intercepts: $(\pm 5, 0)$

15. Vertex: $(0, -4)$

x-intercepts: $(\pm 2\sqrt{2}, 0)$

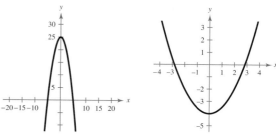

17. Vertex: $(-4, -3)$
x-intercepts:
$$(\pm\sqrt{3} - 4, 0)$$

19. Vertex: $(4, 0)$
x-intercept: $(4, 0)$

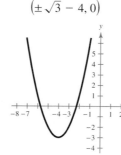

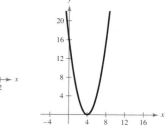

21. Vertex: $\left(\frac{1}{2}, 1\right)$
x-intercept: None

23. Vertex: $(1, 6)$
x-intercepts:
$$\left(1 \pm \sqrt{6}, 0\right)$$

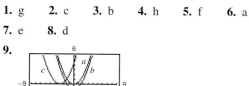

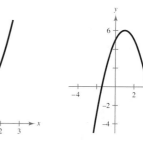

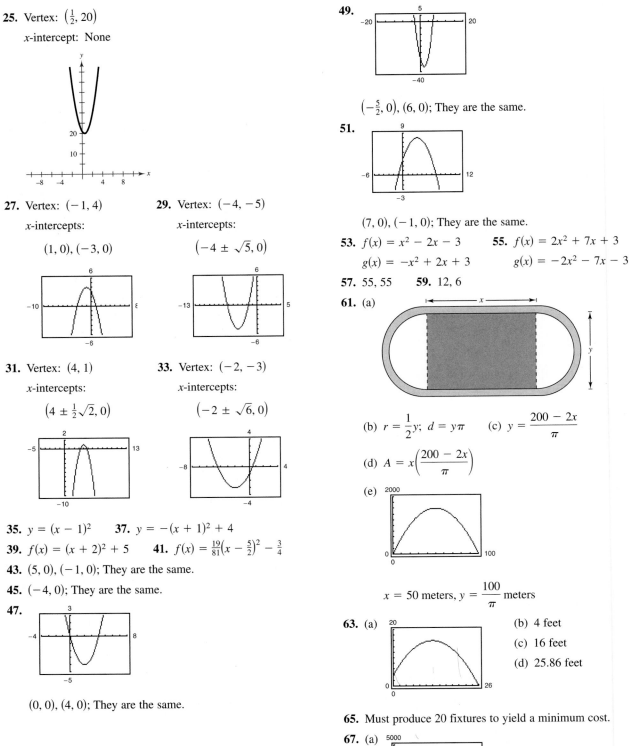

25. Vertex: $\left(\frac{1}{2}, 20\right)$

x-intercept: None

27. Vertex: $(-1, 4)$

x-intercepts:

$(1, 0), (-3, 0)$

29. Vertex: $(-4, -5)$

x-intercepts:

$\left(-4 \pm \sqrt{5}, 0\right)$

31. Vertex: $(4, 1)$

x-intercepts:

$\left(4 \pm \frac{1}{2}\sqrt{2}, 0\right)$

33. Vertex: $(-2, -3)$

x-intercepts:

$\left(-2 \pm \sqrt{6}, 0\right)$

35. $y = (x - 1)^2$ **37.** $y = -(x + 1)^2 + 4$

39. $f(x) = (x + 2)^2 + 5$ **41.** $f(x) = \frac{19}{81}\left(x - \frac{5}{2}\right)^2 - \frac{3}{4}$

43. $(5, 0), (-1, 0)$; They are the same.

45. $(-4, 0)$; They are the same.

47.

$(0, 0), (4, 0)$; They are the same.

49.

$\left(-\frac{5}{2}, 0\right), (6, 0)$; They are the same.

51.

$(7, 0), (-1, 0)$; They are the same.

53. $f(x) = x^2 - 2x - 3$ **55.** $f(x) = 2x^2 + 7x + 3$

$g(x) = -x^2 + 2x + 3$ $g(x) = -2x^2 - 7x - 3$

57. 55, 55   **59.** 12, 6

61. (a)

(b) $r = \frac{1}{2}y$; $d = y\pi$ (c) $y = \dfrac{200 - 2x}{\pi}$

(d) $A = x\left(\dfrac{200 - 2x}{\pi}\right)$

(e)

$x = 50$ meters, $y = \dfrac{100}{\pi}$ meters

63. (a)   (b) 4 feet

(c) 16 feet

(d) 25.86 feet

65. Must produce 20 fixtures to yield a minimum cost.

67. (a)

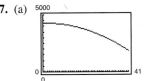

(b) 4276 cigarettes per person; Answers will vary.

(c) ≈ 8564 cigarettes per smoker per year;

    ≈ 23.5 cigarettes per smoker per day

69. True. The vertex is $(0, -1)$ and the parabola opens down.

71. Model (a). The profits are positive and rising.

73. $(1.2, 6.8)$ **75.** $(2, 5), (-3, 0)$ **77.** $-5 - 3i$

79. $19 - 25i$

Section 3.2   (page 260)

Vocabulary Check   (page 260)

1. continuous   **2.** Leading Coefficient Test

3. $n, n - 1$, relative extrema

4. solution, $(x - a)$, x-intercept   **5.** touches, crosses

6. Intermediate Value

---

1. f   **2.** h   **3.** c   **4.** a   **5.** e   **6.** d

7. g   **8.** b

9. (a)   (b)

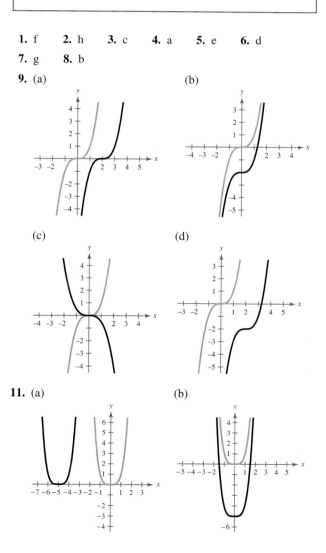

(c)   (d)

11. (a)   (b)

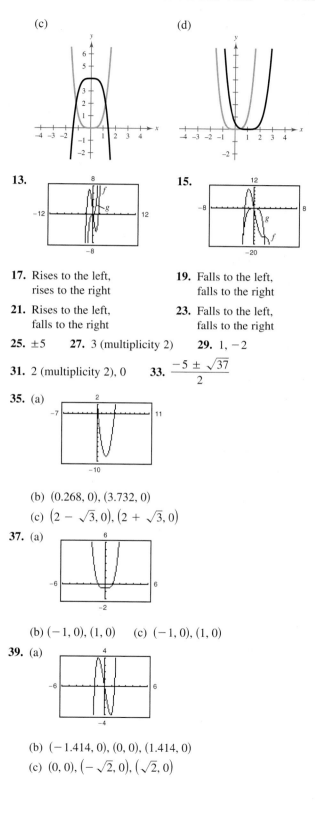

(c)   (d)

13.   **15.**

17. Rises to the left,   **19.** Falls to the left,
    rises to the right   falls to the right

21. Rises to the left,   **23.** Falls to the left,
    falls to the right   falls to the right

25. ± 5 **27.** 3 (multiplicity 2)   **29.** $1, -2$

31. 2 (multiplicity 2), 0   **33.** $\dfrac{-5 \pm \sqrt{37}}{2}$

35. (a)

(b) $(0.268, 0), (3.732, 0)$

(c) $\left(2 - \sqrt{3}, 0\right), \left(2 + \sqrt{3}, 0\right)$

37. (a)

(b) $(-1, 0), (1, 0)$ (c) $(-1, 0), (1, 0)$

39. (a)

(b) $(-1.414, 0), (0, 0), (1.414, 0)$

(c) $(0, 0), \left(-\sqrt{2}, 0\right), \left(\sqrt{2}, 0\right)$

41. (a)

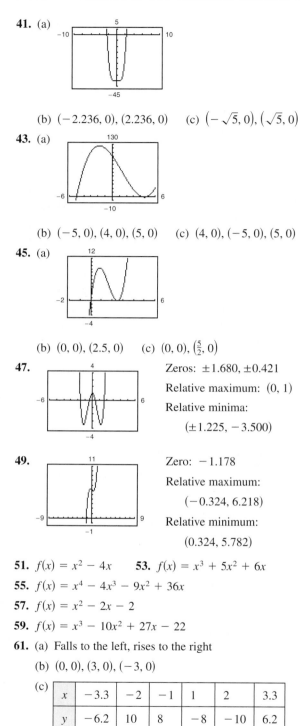

(b) $(-2.236, 0), (2.236, 0)$ (c) $\left(-\sqrt{5}, 0\right), \left(\sqrt{5}, 0\right)$

43. (a)

(b) $(-5, 0), (4, 0), (5, 0)$ (c) $(4, 0), (-5, 0), (5, 0)$

45. (a)

(b) $(0, 0), (2.5, 0)$ (c) $(0, 0), \left(\frac{5}{2}, 0\right)$

47.

Zeros: $\pm 1.680, \pm 0.421$

Relative maximum: $(0, 1)$

Relative minima:

$(\pm 1.225, -3.500)$

49.

Zero: -1.178

Relative maximum:

$(-0.324, 6.218)$

Relative minimum:

$(0.324, 5.782)$

51. $f(x) = x^2 - 4x$ **53.** $f(x) = x^3 + 5x^2 + 6x$

55. $f(x) = x^4 - 4x^3 - 9x^2 + 36x$

57. $f(x) = x^2 - 2x - 2$

59. $f(x) = x^3 - 10x^2 + 27x - 22$

61. (a) Falls to the left, rises to the right

(b) $(0, 0), (3, 0), (-3, 0)$

(c)

x	-3.3	-2	-1	1	2	3.3
y	-6.2	10	8	-8	-10	6.2

(d)

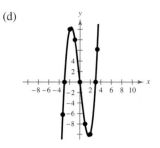

63. (a) Rises to the left and right

(b) No zeros

(c)

t	-3	-2	-1	0	1	2	3
y	7.5	5.8	4.5	3.8	3.5	3.8	4.5

(d)

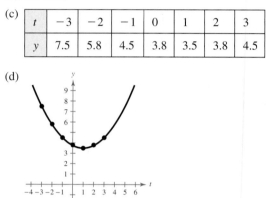

65. (a) Falls to the left, rises to the right

(b) $(0, 0), (3, 0)$

(c)

x	-2.5	-1	1	2	4
y	-34.4	-4	-2	-4	16

(d)

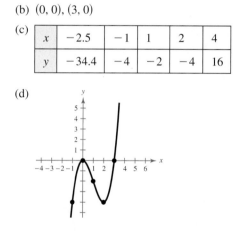

67. (a) Rises to the left, falls to the right

(b) $(0, 0), (-5, 0)$

(c)

x	-6	-4	-3	-2	1	2
y	36	-16	-18	-12	-6	-28

(d)

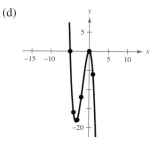

69. (a) Falls to the left, rises to the right

(b) $(0, 0), (4, 0)$

(c)

x	-2	-1	1	2	3	5
y	-24	-5	-3	-8	-9	25

(d)

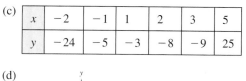

71. (a) Falls to the left and right

(b) $(-2, 0), (2, 0)$

(c)

t	-4	-3	-1	0	1	3
y	-36	-6.3	-2.3	-4	-2.3	-6.3

(d)

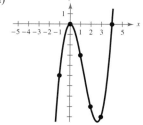

73. (a)

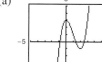

(b) $-0.879, 1.347, 2.532$

$(-1, 0), (1, 2), (2, 3)$

75. (a)

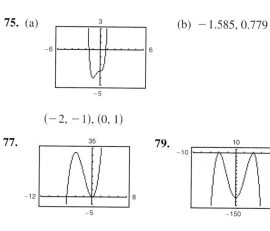

(b) $-1.585, 0.779$

$(-2, -1), (0, 1)$

77.

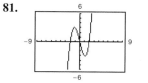

Two x-intercepts

79.

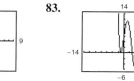

Symmetric to the y-axis
Two x-intercepts

81.

Symmetric to the origin
Three x-intercepts

83.

Three x-intercepts

85. (a) Answers will vary.

(b) Domain: $0 < x < 18$

(c)

Height, x	Volume, V
1	$1[36 - 2(1)]^2 = 1156$
2	$2[36 - 2(2)]^2 = 2048$
3	$3[36 - 2(3)]^2 = 2700$
4	$4[36 - 2(4)]^2 = 3136$
5	$5[36 - 2(5)]^2 = 3380$
6	$6[36 - 2(6)]^2 = 3456$
7	$7[36 - 2(7)]^2 = 3388$

$5 < x < 7$

(d)

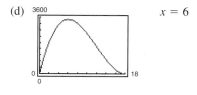

$x = 6$

87. $(200, 160)$

89.

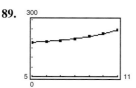

The model is a good fit.

91. Northeast: \$505,920; South: \$81,085; Answers will vary.

93. False. It can have at most five turning points.

95. b; Answers will vary; No; Answers will vary.

96. d; Answers will vary; No; Answers will vary.

97. a; Answers will vary; No; Answers will vary.

98. c; Answers will vary; No; Answers will vary.

99. 69   **101.** $-\frac{1408}{49} \approx -28.73$ **103.** 109

105. $x > -8$ **107.** $-26 \le x < 7$

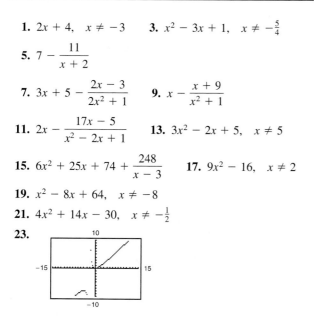

Section 3.3   (page 275)

Vocabulary Check   (page 275)

1. $f(x)$ is the dividend, $d(x)$ is the divisor, $q(x)$ is the quotient, and $r(x)$ is the remainder.

2. improper, proper   **3.** synthetic division

4. Rational Zero   **5.** Descartes's Rule, Signs

6. Remainder Theorem

7. upper bound, lower bound

---

1. $2x + 4, \quad x \ne -3$ **3.** $x^2 - 3x + 1, \quad x \ne -\frac{5}{4}$

5. $7 - \dfrac{11}{x + 2}$

7. $3x + 5 - \dfrac{2x - 3}{2x^2 + 1}$ **9.** $x - \dfrac{x + 9}{x^2 + 1}$

11. $2x - \dfrac{17x - 5}{x^2 - 2x + 1}$ **13.** $3x^2 - 2x + 5, \quad x \ne 5$

15. $6x^2 + 25x + 74 + \dfrac{248}{x - 3}$ **17.** $9x^2 - 16, \quad x \ne 2$

19. $x^2 - 8x + 64, \quad x \ne -8$

21. $4x^2 + 14x - 30, \quad x \ne -\frac{1}{2}$

23.

25. $f(x) = (x - 4)(x^2 + 3x - 2) + 3, \; f(4) = 3$

27. $f(x) = \left(x - \sqrt{2}\right)\left[x^2 + \left(3 + \sqrt{2}\right)x + 3\sqrt{2}\right] - 8,$
$\quad f\left(\sqrt{2}\right) = -8$

29. $f(x) = \left(x - 1 + \sqrt{3}\right)\left[4x^2 - \left(2 + 4\sqrt{3}\right)x - \left(2 + 2\sqrt{3}\right)\right],$
$\quad f\left(1 - \sqrt{3}\right) = 0$

31. (a) 1   (b) 4   (c) 4   (d) 1954

33. (a) 97   (b) $-\frac{5}{3}$ (c) 17   (d) -199

35. $(x - 2)(x + 3)(x - 1)$ **37.** $(2x - 1)(x - 5)(x - 2)$
    Zeros: $2, -3, 1$ Zeros: $\frac{1}{2}, 5, 2$

39. (a) Answers will vary.   (b) $(2x - 1)$
    (c) $(x + 2)(x - 1)(2x - 1)$ (d) $-2, 1, \frac{1}{2}$

41. (a) Answers will vary.   (b) $(x - 1), (x - 2)$
    (c) $(x - 5)(x + 4)(x - 1)(x - 2)$ (d) $-4, 1, 2, 5$

43. (a) Answers will vary.   (b) $(x + 7)$
    (c) $(2x + 1)(3x - 2)(x + 7)$ (d) $-7, -\frac{1}{2}, \frac{2}{3}$

45. $\pm 1, \pm 3$

47. $\pm 1, \pm 3, \pm 5, \pm 9, \pm 15, \pm 45, \pm \frac{1}{2}, \pm \frac{3}{2}, \pm \frac{5}{2}, \pm \frac{9}{2}, \pm \frac{15}{2}, \pm \frac{45}{2}$

49. $-1, 2$ **51.** $-6, \frac{1}{2}, 1$

53. (a) $-2, 0.268, 3.732$ (b) -2
    (c) $h(t) = (t + 2)\left(t - 2 + \sqrt{3}\right)\left(t - 2 - \sqrt{3}\right)$

55. (a) $0, 3, 4, -1.414, 1.414$ (b) $0, 3, 4$
    (c) $h(x) = x(x - 3)(x - 4)\left(x + \sqrt{2}\right)\left(x - \sqrt{2}\right)$

57. 4, 2 or 0 positive real zeros, no negative real zeros

59. 2 or 0 positive real zeros, 1 negative real zero

61. (a) 1 positive real zero, 2 or 0 negative real zeros
    (b) $\pm 1, \pm 2, \pm 4$
    (c)

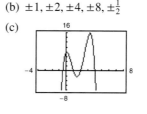

    (d) $-2, -1, 2$

63. (a) 3 or 1 positive real zeros, 1 negative real zero
    (b) $\pm 1, \pm 2, \pm 4, \pm 8, \pm \frac{1}{2}$
    (c)

    (d) $-\frac{1}{2}, 1, 2, 4$

65. (a) 2 or 0 positive real zeros, 1 negative real zero

(b) $\pm 1, \pm 3, \pm\frac{1}{2}, \pm\frac{3}{2}, \pm\frac{1}{4}, \pm\frac{3}{4}, \pm\frac{1}{8}, \pm\frac{3}{8}, \pm\frac{1}{16}, \pm\frac{3}{16}, \pm\frac{1}{32}, \pm\frac{3}{32}$

(c)

(d) $-\frac{1}{8}, \frac{3}{4}, 1$

67 and 69. Answers will vary.   **71.** $\pm 2, \pm\frac{3}{2}$ **73.** $\pm 1, \frac{1}{4}$

75. d   **76.** a   **77.** b   **78.** c

79. (a)

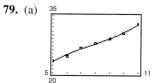

The model is a good fit.

(b)

Year	1995	1996	1997	1998
Actual, R	23.07	24.41	26.48	27.81
Model, R	22.92	24.81	26.30	27.60

Year	1999	2000	2001
Actual, R	28.92	30.37	32.87
Model, R	28.96	30.60	32.77

The model is a close fit.

(c) $R(18) = 82.10$. No; the model will turn sharply upward.

81. (a) Answers will vary.

(b)

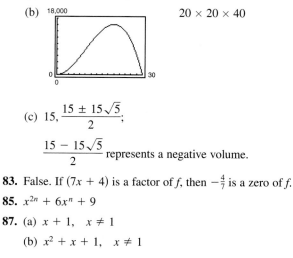

$20 \times 20 \times 40$

(c) $15, \dfrac{15 \pm 15\sqrt{5}}{2}$;

$\dfrac{15 - 15\sqrt{5}}{2}$ represents a negative volume.

83. False. If $(7x + 4)$ is a factor of f, then $-\frac{4}{7}$ is a zero of f.

85. $x^{2n} + 6x^n + 9$

87. (a) $x + 1, \quad x \neq 1$

(b) $x^2 + x + 1, \quad x \neq 1$

(c) $x^3 + x^2 + x + 1, \quad x \neq 1$

$\dfrac{x^n - 1}{x - 1} = x^{n-1} + x^{n-2} + \cdots + x^2 + x + 1, \quad x \neq 1$

89. $\pm\dfrac{5}{3}$ **91.** $\dfrac{-3 \pm \sqrt{3}}{2}$ **93.** $f(x) = x^2 + 12x$

95. $f(x) = x^4 - 6x^3 + 3x^2 + 10x$

Section 3.4   (page 284)

1. $-3, 0, 0$ **3.** $-9, -2i, 2i$

5. Zeros: $4, -i, i$. One real zero; they are the same.

7. Zeros: $\sqrt{2}i, \sqrt{2}i, -\sqrt{2}i, -\sqrt{2}i$. No real zeros; they are the same.

9. $2 \pm \sqrt{3}$

$\left(x - 2 - \sqrt{3}\right)\left(x - 2 + \sqrt{3}\right)$

11. $6 \pm \sqrt{10}$

$\left(x - 6 - \sqrt{10}\right)\left(x - 6 + \sqrt{10}\right)$

13. $\pm 5i$

$(x + 5i)(x - 5i)$

15. $\pm 3, \pm 3i$

$(x + 3)(x - 3)(x + 3i)(x - 3i)$

17. $\dfrac{1 \pm \sqrt{223}i}{2}$

$\left(z - \dfrac{1 - \sqrt{223}i}{2}\right)\left(z - \dfrac{1 + \sqrt{223}i}{2}\right)$

19. $-5, 4 \pm 3i$

$(t + 5)(t - 4 + 3i)(t - 4 - 3i)$

21. $1 \pm \sqrt{5}i, -\frac{1}{5}$

$(5x + 1)\left(x - 1 + \sqrt{5}i\right)\left(x - 1 - \sqrt{5}i\right)$

23. $\pm i, \pm 3i$

$(x + i)(x - i)(x + 3i)(x - 3i)$

25. $2, 2, \pm 2i$

$(x - 2)^2(x + 2i)(x - 2i)$

27. (a) $7 \pm \sqrt{3}$

(b) $\left(x - 7 - \sqrt{3}\right)\left(x - 7 + \sqrt{3}\right)$

(c) $\left(7 \pm \sqrt{3}, 0\right)$

29. (a) $-7 \pm \sqrt{5}$

(b) $\left(x + 7 - \sqrt{5}\right)\left(x + 7 + \sqrt{5}\right)$

(c) $\left(-7 \pm \sqrt{5}, 0\right)$

31. (a) $-6, 3 \pm 4i$

(b) $(x + 6)(x - 3 - 4i)(x - 3 + 4i)$

(c) $(-6, 0)$

33. (a) $\pm 4i, \pm 3i$

(b) $(x + 4i)(x - 4i)(x + 3i)(x - 3i)$

(c) None

35. $x^3 - 3x^2 + x - 3$ **37.** $x^3 + 6x^2 + x - 34$

39. $x^4 + 8x^3 + 9x^2 - 10x + 100$

41. (a) $(x^2 + 1)(x^2 - 7)$

(b) $(x^2 + 1)\left(x + \sqrt{7}\right)\left(x - \sqrt{7}\right)$

(c) $(x + i)(x - i)\left(x + \sqrt{7}\right)\left(x - \sqrt{7}\right)$

43. (a) $(x^2 - 6)(x^2 - 2x + 3)$

(b) $\left(x + \sqrt{6}\right)\left(x - \sqrt{6}\right)(x^2 - 2x + 3)$

(c) $\left(x + \sqrt{6}\right)\left(x - \sqrt{6}\right)\left(x - 1 - \sqrt{2}i\right)\left(x - 1 + \sqrt{2}i\right)$

45. $-\frac{3}{2}, \pm 5i$ **47.** $-3, 5 \pm 2i$

49. $-\frac{2}{3}, 1 \pm \sqrt{3}i$ **51.** $\frac{3}{4}, \frac{1}{2}\left(1 \pm \sqrt{5}i\right)$

53. (a) $1.000, 2.000$ (b) $-3 \pm \sqrt{2}i$

55. (a) 0.750 (b) $\frac{1}{2} \pm \frac{\sqrt{5}}{2}i$

57. No. Setting $h = 64$ and solving the resulting equation yields imaginary roots.

59. False. A polynomial can only have an even number of complex zeros, so one of the zeros of a third-degree polynomial must be real.

61. (a) $k = 4$ (b) $k < 0$

63.

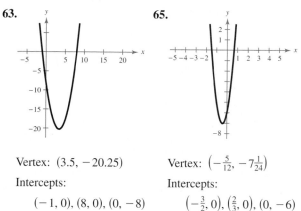

Vertex: $(3.5, -20.25)$

Intercepts:

$(-1, 0), (8, 0), (0, -8)$

65.

Vertex: $\left(-\frac{5}{12}, -7\frac{1}{24}\right)$

Intercepts:

$\left(-\frac{3}{2}, 0\right), \left(\frac{2}{3}, 0\right), (0, -6)$

Section 3.5   (page 292)

Vocabulary Check   (page 292)

1. rational functions   **2.** vertical asymptote

3. horizontal asymptote

1. (a)

x	$f(x)$	x	$f(x)$
0.5	-2	1.5	2
0.9	-10	1.1	10
0.99	-100	1.01	100
0.999	-1000	1.001	1000

x	$f(x)$	x	$f(x)$
5	0.25	-5	$-0.\overline{16}$
10	$0.\overline{1}$	-10	$-0.\overline{09}$
100	$0.\overline{01}$	-100	$-0.\overline{0099}$
1000	$0.\overline{001}$	-1000	$-0.\overline{000999}$

(b) Vertical asymptote: $x = 1$

Horizontal asymptote: $y = 0$

(c) Domain: all real numbers x except $x = 1$

3. (a)

x	$f(x)$	x	$f(x)$
0.5	3	1.5	9
0.9	27	1.1	33
0.99	297	1.01	303
0.999	2997	1.001	3003

x	$f(x)$	x	$f(x)$
5	3.75	-5	-2.5
10	$3.\overline{33}$	-10	-2.727
100	$3.\overline{03}$	-100	-2.97
1000	$3.\overline{003}$	-1000	-2.997

(b) Vertical asymptote: $x = 1$

Horizontal asymptotes: $y = \pm 3$

(c) Domain: all real numbers x except $x = 1$

5. (a)

x	$f(x)$
0.5	-1
0.9	-12.79
0.99	-147.8
0.999	-1498

x	$f(x)$
1.5	5.4
1.1	17.29
1.01	152.3
1.001	1502.3

x	$f(x)$
5	3.125
10	$3.\overline{03}$
100	$3.\overline{0003}$
1000	3

x	$f(x)$
-5	3.125
-10	$3.\overline{03}$
-100	$3.\overline{0003}$
-1000	3.000003

(b) Vertical asymptotes: $x = \pm 1$

Horizontal asymptote: $y = 3$

(c) Domain: all real numbers x except $x = \pm 1$

7. a   **8.** d   **9.** c   **10.** e

11. b   **12.** f

13. (a) Domain: all real numbers x except $x = 0$

(b) Vertical asymptote: $x = 0$

Horizontal asymptote: $y = 0$

15. (a) Domain: all real numbers x except $x = 2$

(b) Vertical asymptote: $x = 2$

Horizontal asymptote: $y = -1$

17. (a) Domain: all real numbers x except $x = 0, \frac{1}{2}$

(b) Vertical asymptote: $x = \frac{1}{2}$

Horizontal asymptote: $y = \frac{1}{2}$

19. (a) Domain: all real numbers

(b) Vertical asymptote: none

Horizontal asymptote: $y = 3$

21. (a) Domain: all real numbers x except $x = 0$

(b) Vertical asymptote: $x = 0$

Horizontal asymptotes: $y = \pm 1$

23. (a) Domain of f: all real numbers x except $x = -2$

Domain of g: all real numbers

(b) Vertical asymptote: None

(c)

x	-4	-3	-2.5	-2	-1.5	-1	0
$f(x)$	-6	-5	-4.5	Undef.	-3.5	-3	-2
$g(x)$	-6	-5	-4.5	-4	-3.5	-3	-2

(d) Values differ only where f is undefined.

25. (a) Domain of f: all real numbers x except $x = 0, 3$; domain of g: all real numbers x except $x = 0$

(b) Vertical asymptote: $x = 0$

(c)

x	-1	-0.5	0	0.5	2	3	4
$f(x)$	-1	-2	Undef.	2	$\frac{1}{2}$	Undef.	$\frac{1}{4}$
$g(x)$	-1	-2	Undef.	2	$\frac{1}{2}$	$\frac{1}{3}$	$\frac{1}{4}$

(d) Values differ only where f is undefined and g is defined.

27. 4; less than; greater than

29. 2; greater than; less than   **31.** ± 2 **33.** 7

35. (a) \$28.33 million   (b) \$170 million

(c) \$765 million

(d)

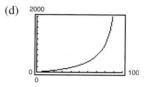

Answers will vary.

(e) No. The function is undefined at the 100% level.

37. (a) $y = \dfrac{1}{0.445 - 0.007x}$

(b)

Age, x	16	32	44	50	60
Near point, y	3.0	4.5	7.3	10.5	40

(c) No; the function is negative for $x = 70$.

39. (a)

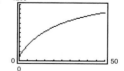

(b) 333 deer, 500 deer, 800 deer

(c) 1500. Because the degrees of the numerator and the denominator are equal, the limiting size is the ratio of the leading coefficients, $60/0.04 = 1500$.

41. False. The degree of the denominator gives the maximum possible number of vertical asymptotes, and the degree is finite.

43. $f(x) = \dfrac{1}{x^2 + x - 2}$ **45.** $f(x) = \dfrac{2x^2}{1 + x^2}$

47. $x - y - 1 = 0$ **49.** $3x - y + 1 = 0$

51. $x + 9 + \dfrac{42}{x - 4}$ **53.** $2x - 9 + \dfrac{34}{x + 5}$

Section 3.6 (page 301)

1.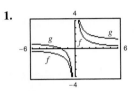

Vertical shift

3.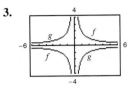

Reflection in the x-axis

5.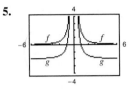

Vertical shift

7.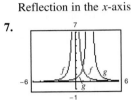

Horizontal shift

9.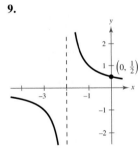

11.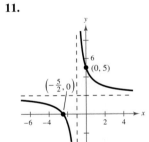

13.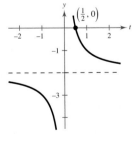

15.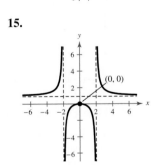

17.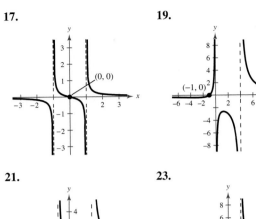

19.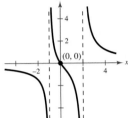

21.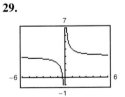

23.

There is a hole at $x = -3$.

25.

There is a hole at $x = -1$.

27.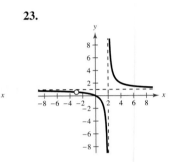

Domain: $(-\infty, 1), (1, \infty)$

Vertical asymptote: $x = 1$

Horizontal asymptote:
$y = -1$

29.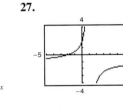

Domain: $(-\infty, 0), (0, \infty)$

Vertical asymptote: $t = 0$

Horizontal asymptote: $y = 3$

31.

Domain: $(-\infty, \infty)$

Horizontal asymptote: $y = 0$

33.

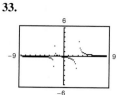

Domain: $(-\infty, -2)$,
$\qquad (-2, 3), (3, \infty)$

Vertical asymptote:
$\qquad x = -2, x = 3$

Horizontal asymptote: $y = 0$

35.

Domain: $(-\infty, 0), (0, \infty)$

Vertical asymptote: $x = 0$

Horizontal asymptote:
$\qquad y = 0$

37.

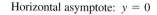

There are two horizontal asymptotes, $y = 6$ and $y = -6$.

39.

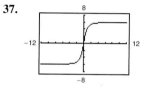

There are two horizontal asymptotes, $y = 4$ and $y = -4$, and one vertical asymptote, $x = -1$.

41.

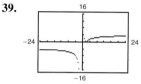

The graph crosses the horizontal asymptote, $y = 4$.

43.

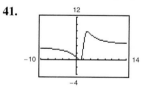

45.

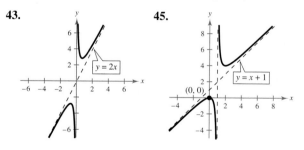

47.

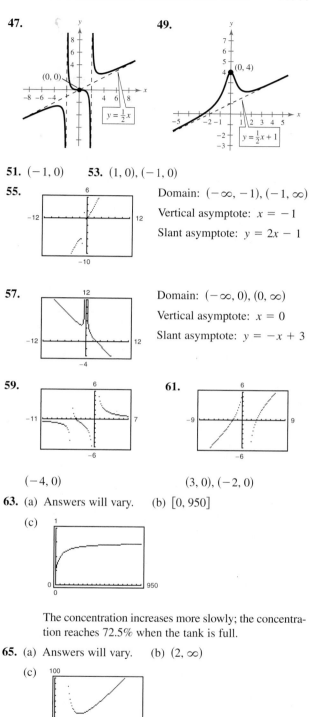

49.

51. $(-1, 0)$ **53.** $(1, 0), (-1, 0)$

55.

Domain: $(-\infty, -1), (-1, \infty)$

Vertical asymptote: $x = -1$

Slant asymptote: $y = 2x - 1$

57.

Domain: $(-\infty, 0), (0, \infty)$

Vertical asymptote: $x = 0$

Slant asymptote: $y = -x + 3$

59.

$(-4, 0)$

61.

$(3, 0), (-2, 0)$

63. (a) Answers will vary.   (b) $[0, 950]$

(c)

The concentration increases more slowly; the concentration reaches 72.5% when the tank is full.

65. (a) Answers will vary.   (b) $(2, \infty)$

(c)

5.9 inches $\times$ 11.8 inches

67.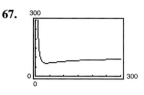

$x \approx 40$

69. (a) $C = 0$. The chemical will eventually dissipate.

(b) 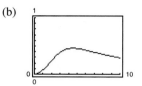 $t \approx 4.5$ hours

(c) Before ≈ 2.6 hours and after ≈ 8.3 hours

71. (a) $y = 0.44t + 1.8$ (b) $y = \dfrac{1}{-0.016t + 0.32}$

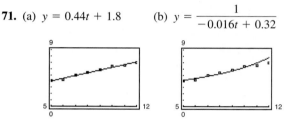

(c)

t	5	6	7	8	9
Linear model	4.0	4.4	4.9	5.3	5.8
Rational model	4.2	4.5	4.8	5.2	5.7

t	10	11	12
Linear model	6.2	6.6	7.1
Rational model	6.3	6.9	7.8

The linear model more closely represents the actual data.

73. False. A graph with a vertical asymptote is not continuous.

75.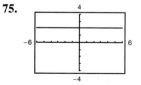

The denominator is a factor of the numerator.

77. $f(x) = \dfrac{x^2 - x - 6}{x - 2}$ **79.** $\dfrac{512}{x^3}$ **81.** $\dfrac{x^2}{5y^2}, x \neq 0$

83. 3

85.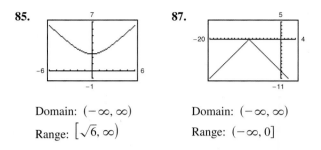

Domain: $(-\infty, \infty)$

Range: $\left[\sqrt{6}, \infty\right)$

87.

Domain: $(-\infty, \infty)$

Range: $(-\infty, 0]$

Section 3.7 (page 309)

Vocabulary Check   (page 309)

1. linear   **2.** quadratic

---

1. Quadratic   **3.** Linear   **5.** Neither

7. (a) 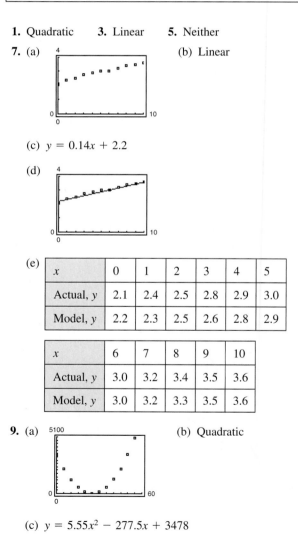   (b) Linear

(c) $y = 0.14x + 2.2$

(d)

(e)

x	0	1	2	3	4	5
Actual, y	2.1	2.4	2.5	2.8	2.9	3.0
Model, y	2.2	2.3	2.5	2.6	2.8	2.9

x	6	7	8	9	10
Actual, y	3.0	3.2	3.4	3.5	3.6
Model, y	3.0	3.2	3.3	3.5	3.6

9. (a)   (b) Quadratic

(c) $y = 5.55x^2 - 277.5x + 3478$

(d)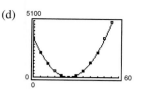

(e)

x	0	5	10	15	20
Actual, y	3480	2235	1250	565	150
Model, y	3478	2229	1258	564	148

x	25	30	35	40
Actual, y	12	145	575	1275
Model, y	9	148	564	1258

x	45	50	55
Actual, y	2225	3500	5010
Model, y	2229	3478	5004

11. (a) 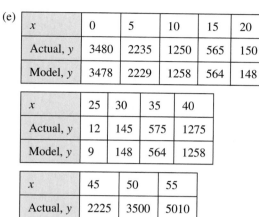   (b) Quadratic

(c) $y = -0.120x^2 + 0.21x + 7.5$

(d)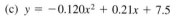

(e)

x	-5	-4	-3	-2	-1	0
Actual, y	3.8	4.7	5.5	6.2	7.1	7.9
Model, y	3.5	4.7	5.8	6.6	7.2	7.5

x	1	2	3	4	5	6	7
Actual, y	8.1	7.7	6.9	6.0	5.6	4.4	3.2
Model, y	7.6	7.4	7.1	6.4	5.6	4.4	3.1

13. (a)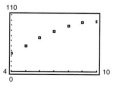

(b) $y = -1.30t^2 + 29.0t - 61$

(c) 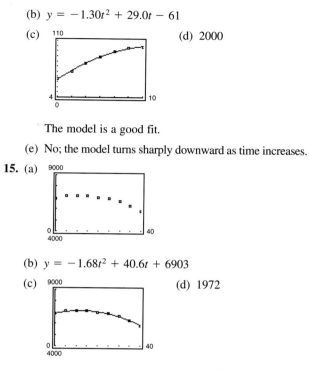   (d) 2000

The model is a good fit.

(e) No; the model turns sharply downward as time increases.

15. (a)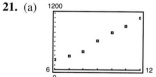

(b) $y = -1.68t^2 + 40.6t + 6903$

(c) 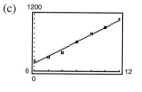   (d) 1972

The model is a good fit.

(e) No; the model continues to decrease as time increases.

17. (a) $y = 2.48x + 1.1$; $y = 0.071x^2 + 1.69x + 2.7$

(b) 0.990, 0.995

(c) Quadratic

19. (a) $y = -0.89x + 0.2$

$y = 0.001x^2 - 0.89x + 0.2$

(b) 0.9998, 0.9999

(c) Quadratic

21. (a)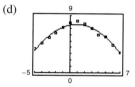

(b) $S = 155.01t - 774.2$

(c)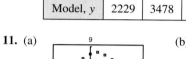

(d) $S = 6.660t^2 + 35.14t - 261.4$

(e)

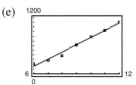

(f) Linear; $1861.1 million

23. True. The Leading Coefficient Test guarantees that a parabola with a negative leading coefficient will have a maximum as its vertex.

25. (a) $(f \circ g)(x) = 2x^2 + 5$

(b) $(g \circ f)(x) = 4(x^2 - x + 1)$

27. (a) $(f \circ g)(x) = x$

(b) $(g \circ f)(x) = x$

29. $f^{-1}(x) = \dfrac{x - 5}{2}$ **31.** $f^{-1}(x) = \sqrt{x - 5}$

33.

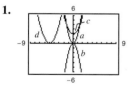

35.

Review Exercises   (page 313)

1.

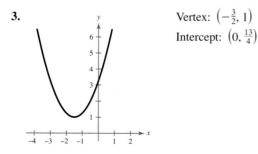

(a) Vertical stretch

(b) Vertical stretch and reflection in the x-axis

(c) Vertical shift

(d) Horizontal shift

3.

Vertex: $\left(-\dfrac{3}{2}, 1\right)$

Intercept: $\left(0, \dfrac{13}{4}\right)$

5.

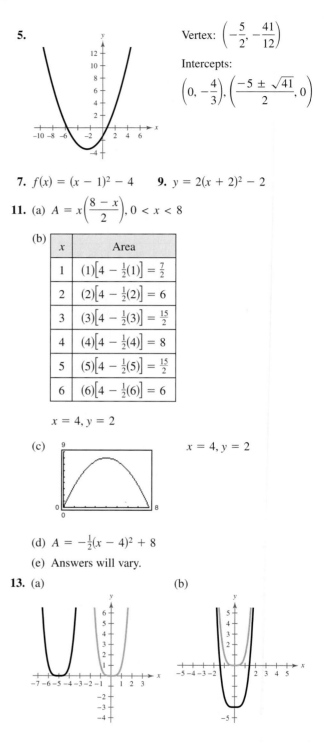

Vertex: $\left(-\dfrac{5}{2}, -\dfrac{41}{12}\right)$

Intercepts:

$\left(0, -\dfrac{4}{3}\right), \left(\dfrac{-5 \pm \sqrt{41}}{2}, 0\right)$

7. $f(x) = (x - 1)^2 - 4$ **9.** $y = 2(x + 2)^2 - 2$

11. (a) $A = x\left(\dfrac{8 - x}{2}\right), 0 < x < 8$

(b)

x	Area
1	$(1)\left[4 - \frac{1}{2}(1)\right] = \frac{7}{2}$
2	$(2)\left[4 - \frac{1}{2}(2)\right] = 6$
3	$(3)\left[4 - \frac{1}{2}(3)\right] = \frac{15}{2}$
4	$(4)\left[4 - \frac{1}{2}(4)\right] = 8$
5	$(5)\left[4 - \frac{1}{2}(5)\right] = \frac{15}{2}$
6	$(6)\left[4 - \frac{1}{2}(6)\right] = 6$

$x = 4, y = 2$

(c)

$x = 4, y = 2$

(d) $A = -\dfrac{1}{2}(x - 4)^2 + 8$

(e) Answers will vary.

13. (a)   (b)

(c)

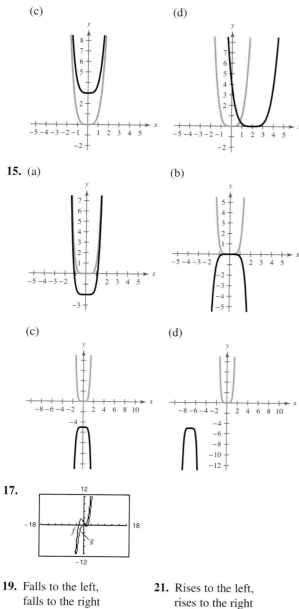

(d)

15. (a)

(b)

(c)

(d)

17.

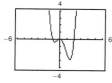

19. Falls to the left, falls to the right

21. Rises to the left, rises to the right

23. (a)

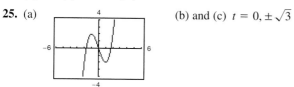

(b) and (c) $x = -1, 0, 2$

25. (a)

(b) and (c) $t = 0, \pm\sqrt{3}$

27. (a)

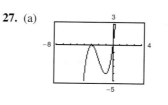

(b) and (c) $x = -3, 0$

29. (a) $(-3, -2), (-1, 0), (0, 1)$

(b) $(-2.247, 0), (-0.555, 0), (0.802, 0)$

31. (a) $(-3, -2), (2, 3)$ (b) $(-2.570, 0), (2.570, 0)$

33.

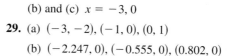

35. $8x + 5 + \dfrac{2}{3x - 2}$

37. $x^2 - 2, \ x \neq \pm 1$ **39.** $5x + 2, \ x \neq \dfrac{3 \pm \sqrt{5}}{2}$

41. $3x^2 + 5x + 8 + \dfrac{10}{2x^2 - 1}$

43. $\dfrac{1}{4}x^3 - \dfrac{9}{2}x^2 + 9x - 18 + \dfrac{36}{x + 2}$

45. $6x^3 - 27x, \ x \neq \dfrac{2}{3}$ **47.** $3x^2 + 2x + 20 + \dfrac{58}{x - 4}$

49. (a) -421 (b) -9

51. (a) Answers will vary.

(b) $(x + 1)(x + 7)$

(c) $f(x) = (x - 4)(x + 1)(x + 7)$

(d) $x = 4, -1, -7$

53. (a) Answers will vary.

(b) $(x + 1)(x - 4)$

(c) $f(x) = (x + 2)(x - 3)(x + 1)(x - 4)$

(d) $x = -2, 3, -1, 4$

55. $\pm 1, \pm 3, \pm\frac{3}{2}, \pm\frac{3}{4}, \pm\frac{1}{2}, \pm\frac{1}{4}$

57. $\frac{5}{6}, \pm 2i$ **59.** $-1, \frac{3}{2}, 3, \frac{2}{3}$

61. 2 or 0 positive real zeros

1 negative real zero

63. Answers will vary.

65. $x = 0, 2, 2$ **67.** $x = -4, 6, \pm 2i$

69. Zeros: $2, -\frac{3}{2}, 1 \pm i$;

$(x - 2)(2x + 3)[x - (1 - i)][x - (1 + i)]$

71. Zeros: $4, \dfrac{3 \pm \sqrt{15}i}{2}$;

$$(x - 4)\left[x - \left(\dfrac{3 + \sqrt{15}i}{2}\right)\right]\left[x - \dfrac{3 - \sqrt{15}i}{2}\right]$$

73. (a) $2, 1 \pm i$

(b) $(x - 2)(x - 1 - i)(x - 1 + i)$ (c) $(2, 0)$

75. (a) $-4, -1 \pm \sqrt{2}i$

(b) $(x + 4)(x + 1 + \sqrt{2}i)(x + 1 - \sqrt{2}i)$

(c) $(-4, 0)$

77. (a) $\pm 3i, \pm 5i$

(b) $(x - 3i)(x + 3i)(x - 5i)(x + 5i)$ (c) None

79. $f(x) = x^4 + 4x^3 + 29x^2 + 100x + 100$

81. $f(x) = x^4 + 9x^3 + 48x^2 + 78x - 136$

83. (a) $(x^2 + 4)(x^2 - 2)$

(b) $(x^2 + 4)(x - \sqrt{2})(x + \sqrt{2})$

(c) $(x + 2i)(x - 2i)(x - \sqrt{2})(x + \sqrt{2})$

85. (a) $(x^2 + 9)(x^2 - 2x - 1)$

(b) $(x^2 + 9)(x - 1 + \sqrt{2})(x - 1 - \sqrt{2})$

(c) $(x + 3i)(x - 3i)(x - 1 + \sqrt{2})(x - 1 - \sqrt{2})$

87. (a) Domain: all real numbers x except $x = 1$

(b) Vertical asymptote: $x = 1$

Horizontal asymptote: $y = -1$

89. (a) Domain: all real numbers x except $x = 6, -3$

(b) Vertical asymptotes: $x = 6, x = -3$

Horizontal asymptote: $y = 0$

91. (a) Domain: all real numbers x except $x = 7$

(b) Vertical asymptote: $x = 7$

Horizontal asymptote: $y = -1$

93. (a) Domain: all real numbers x except $x = \pm\dfrac{\sqrt{6}}{2}$

(b) Vertical asymptotes: $x = \pm\dfrac{\sqrt{6}}{2}$

Horizontal asymptote: $y = 2$

95. (a) Domain: all real numbers x except $x = 5, -3$

(b) Vertical asymptotes: $x = -3$

Horizontal asymptote: $y = 0$

97. (a) Domain: all real numbers

(b) Vertical asymptote: none

Horizontal asymptotes: $y = \pm 1$

99. (a) \$176 million; \$528 million; \$1584 million

(b)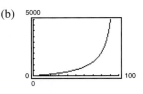

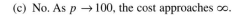

Answers will vary.

(c) No. As $p \to 100$, the cost approaches ∞.

101.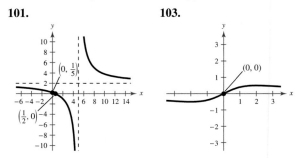

103.

105.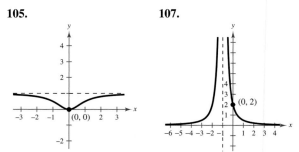

107.

109.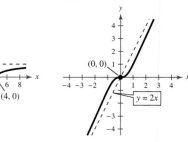

111.

There is a hole at $x = -4$.

113.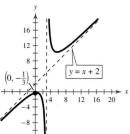

115. (a)

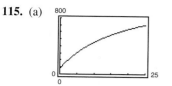

(b) 304,000; 453,333; 702,222

(c) 1,200,000, because N has a horizontal asymptote at $y = 1,200,000$.

117. Quadratic   **119.** Linear

121. (a)

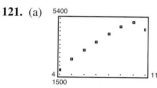

(b) $R = -68.707t^2 + 1485.31t - 3126.7$

(c)

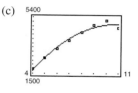

Answers will vary.

(d) 2000

(e) No; the model begins to decrease rapidly over time.

123. False. For the graph of a rational function to have a slant asymptote, the degree of its numerator must be exactly one more than the degree of its denominator.

125. The divisor is a factor of the dividend.

Chapter Test   (page 318)

1. (a) Reflection in the x-axis followed by a vertical shift six units up

(b) Horizontal shift of $\frac{3}{2}$ units to the right

2. Vertex: $(-2, -1)$

Intercepts: $(0, 3), (-3, 0), (-1, 0)$

3. $y = (x - 3)^2 - 6$

4. (a) 50 feet

(b) 5; Changing the constant term results in a vertical shift of the graph and therefore changes the maximum height.

5. $3x + \dfrac{x - 1}{x^2 + 1}$ **6.** $2x^3 + 4x^2 + 3x + 6 + \dfrac{9}{x - 2}$

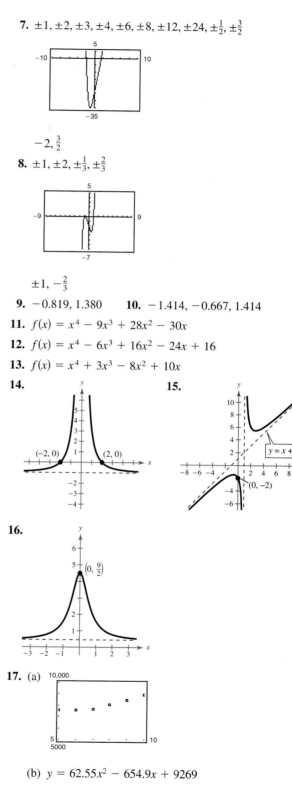

7. $\pm 1, \pm 2, \pm 3, \pm 4, \pm 6, \pm 8, \pm 12, \pm 24, \pm\frac{1}{2}, \pm\frac{3}{2}$

$-2, \frac{3}{2}$

8. $\pm 1, \pm 2, \pm\frac{1}{3}, \pm\frac{2}{3}$

$\pm 1, -\frac{2}{3}$

9. $-0.819, 1.380$ **10.** $-1.414, -0.667, 1.414$

11. $f(x) = x^4 - 9x^3 + 28x^2 - 30x$

12. $f(x) = x^4 - 6x^3 + 16x^2 - 24x + 16$

13. $f(x) = x^4 + 3x^3 - 8x^2 + 10x$

14.

15.

16.

17. (a)

(b) $y = 62.55x^2 - 654.9x + 9269$

(c)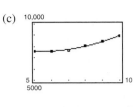

10,000

5000   5   10

The model is a good fit.

(d) 2006

Chapter 4

Section 4.1   (page 329)

Vocabulary Check   (page 329)

1. algebraic   **2.** transcendental

3. natural exponential, natural

4. $A = P\left(1 + \dfrac{r}{n}\right)^{nt}$ **5.** $A = Pe^{rt}$

1. 4112.033   **3.** 0.006   **5.** 18,297.851

7.

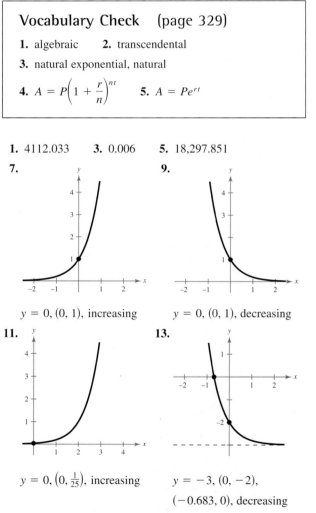

$y = 0, (0, 1)$, increasing

9.

$y = 0, (0, 1)$, decreasing

11.

$y = 0, \left(0, \frac{1}{25}\right)$, increasing

13.

$y = -3, (0, -2),$
$(-0.683, 0)$, decreasing

15. d   **16.** a   **17.** c   **18.** b

19. Right shift of five units

21. Left shift of four units and reflection in the x-axis

23. 9897.129   **25.** 54.164   **27.** 1.516

29.

x	-1	0	1	2	3
$f(x)$	0.4	1	2.5	6.3	15.6

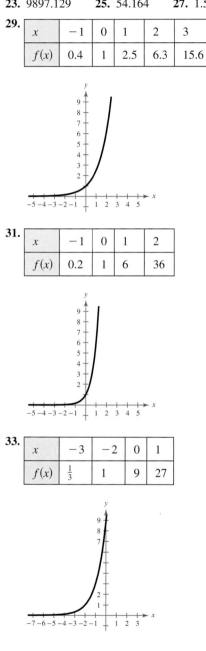

31.

x	-1	0	1	2
$f(x)$	0.2	1	6	36

33.

x	-3	-2	0	1
$f(x)$	$\frac{1}{3}$	1	9	27

35.

x	-7	-6	-5	-4	-3
$f(x)$	0.1	0.4	1.1	3	8

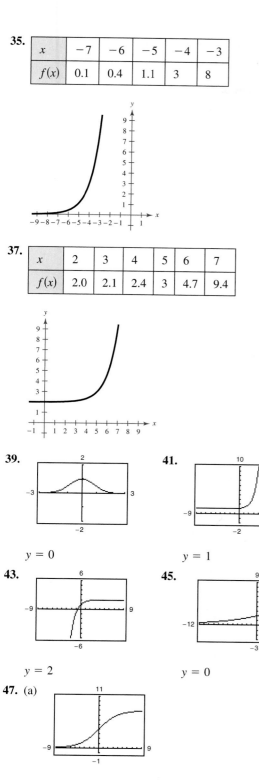

37.

x	2	3	4	5	6	7
$f(x)$	2.0	2.1	2.4	3	4.7	9.4

39.

$y = 0$

41.

$y = 1$

43.

$y = 2$

45.

$y = 0$

47. (a)

(b)

x	-30	-20	-10	0
$f(x)$	0.0000024	0.00036	0.054	4

x	10	20	30
$f(x)$	7.95	7.9996	7.999998

$y = 0, y = 8$

49. (a)

(b)

x	-20	-10	0	3	3.4	3.46
$f(x)$	-3.03	-3.22	-6	-34	-230	-2617

x	3.47	4	5	7	10	15	25
$f(x)$	3516	27	8	2.9	1.1	0.3	0.04

$y = -3, y = 0, x \approx 3.46$

51. (a)

(b) Decreasing on $(-\infty, 0), (2, \infty)$

Increasing on $(0, 2)$

(c) Relative minimum at $(0, 0)$

Relative maximum at $(2, 0.541)$

53. (a)

(b) Increasing on $(-\infty, 1.443)$

Decreasing on $(1.443, \infty)$

(c) Relative maximum at $(1.443, 4.246)$

55.

n	1	2	4	12
A	3200.21	3205.09	3207.57	3209.23

n	365	Continuous
A	3210.04	3210.06

57.

n	1	2	4	12
A	5477.81	5520.10	5541.79	5556.46

n	365	Continuous
A	5563.61	5563.85

59.

t	1	10	20
A	12,489.73	17,901.90	26,706.49

t	30	40	50
A	39,841.40	59,436.39	88,668.67

61.

t	1	10	20
A	12,427.44	17,028.81	24,165.03

t	30	40	50
A	34,291.81	48,662.40	69,055.23

63. (a)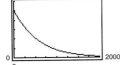

(b) $421.12

(c) $350.13

(d)

x	100	200	300	400
p	849.53	717.64	603.25	504.94

x	500	600	700
p	421.12	350.13	290.35

65. (a) 25 grams   (b) 16.30 grams

(c)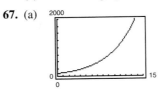

(d) Never. The graph has a horizontal asymptote at $Q = 0$.

67. (a)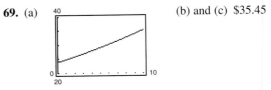

(b) and (c) $P(0) = 100$; $P(5) \approx 300$; $P(10) \approx 900$

69. (a) 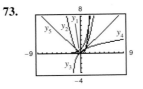   (b) and (c) $35.45

71. True. $f(x) = 1^x$ is not an exponential function because the definition of an exponential function is $f(x) = a^x$, $a > 0$, $a \neq 1$.

73.

(a) $y_1 = e^x$

(b) The exponential function increases at a faster rate.

(c) It usually implies rapid growth.

75.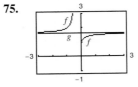

$f(x)$ approaches $g(x) = 1.6487$.

77. $f^{-1}(x) = \dfrac{x + 7}{5}$ **79.** $f^{-1}(x) = x^3 - 8$

81.

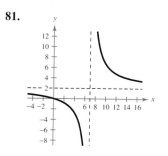

Section 4.2 (page 339)

1. $4^3 = 64$ **3.** $7^{-2} = \frac{1}{49}$ **5.** $32^{2/5} = 4$ **7.** $e^0 = 1$

9. $\log_5 125 = 3$ **11.** $\log_{81} 3 = \frac{1}{4}$ **13.** $\log_6 \frac{1}{36} = -2$

15. $\ln 20.0855 \ldots = 3$ **17.** 4 **19.** -2

21. 2.538 **23.** 7.022 **25.** 9 **27.** 2 **29.** $\frac{1}{10}$

31. **33.**

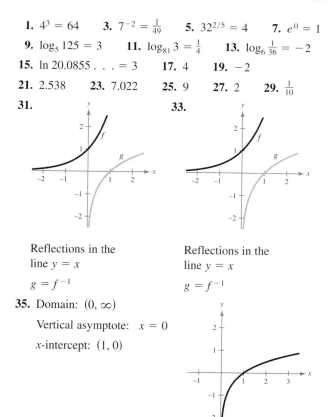

Reflections in the
line $y = x$

$g = f^{-1}$

Reflections in the
line $y = x$

$g = f^{-1}$

35. Domain: $(0, \infty)$

Vertical asymptote: $x = 0$

x-intercept: $(1, 0)$

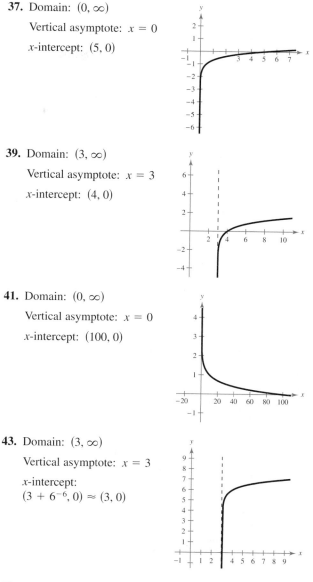

37. Domain: $(0, \infty)$

Vertical asymptote: $x = 0$

x-intercept: $(5, 0)$

39. Domain: $(3, \infty)$

Vertical asymptote: $x = 3$

x-intercept: $(4, 0)$

41. Domain: $(0, \infty)$

Vertical asymptote: $x = 0$

x-intercept: $(100, 0)$

43. Domain: $(3, \infty)$

Vertical asymptote: $x = 3$

x-intercept:
$(3 + 6^{-6}, 0) \approx (3, 0)$

45. b **46.** c **47.** d **48.** a

49. Because $g(x) = -f(x)$, the graph of g can be obtained by reflecting the graph of f in the x-axis.

51. Because $g(x) = 4 - f(x)$, the graph of g can be obtained by reflecting the graph of f in the x-axis and then shifting the graph of f 4 units upward.

53. 1.869 **55.** 0.693 **57.** 2 **59.** 1.8

61. Domain: $(1, \infty)$

Vertical asymptote: $x = 1$

x-intercept: $(2, 0)$

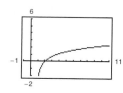

63. Domain: $(-\infty, 0)$

Vertical asymptote: $x = 0$

x-intercept: $(-1, 0)$

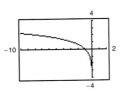

65. (a)

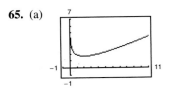

(b) Domain: $(0, \infty)$

(c) Decreasing on $(0, 2)$; increasing on $(2, \infty)$

(d) Relative minimum: $(2, 1.693)$

67. (a)

(b) Domain: $(0, \infty)$

(c) Decreasing on $(0, 0.37)$; increasing on $(0.37, \infty)$

(d) Relative minimum: $(0.37, -1.47)$

69. (a) 80   (b) 68.1   (c) 62.3

71. (a)

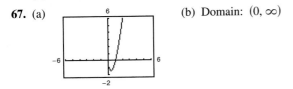

K	1	2	4	6	8	10	12
t	0	12.6	25.2	32.6	37.8	41.9	45.2

Answers will vary.

(b)

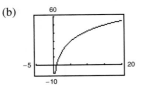

73. (a) 120 decibels   (b) 100 decibels

(c) No. Answers will vary.

75.

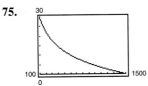

17.66 cubic feet per minute

77. False. Reflect $g(x)$ in the line $y = x$.

79. $\log_a x$ is the inverse of a^x only if $0 < a < 1$ and $a > 1$, so $\log_a x$ is defined only for $0 < a < 1$ and $a > 1$.

81. (a) False   (b) True   (c) True   (d) False

83. (a)

x	1	5	10	10^2
$f(x)$	0	0.322	0.230	0.046

x	10^4	10^6
$f(x)$	0.00092	0.0000138

(b) 0

(c)

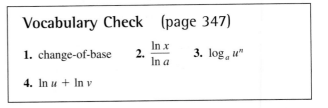

85. $(x + 3)(x - 1)$ **87.** $(4x + 3)(3x - 1)$

89. $(4x + 5)(4x - 5)$ **91.** $x(2x - 9)(x + 5)$

93. 15   **95.** 4300   **97.** $x = 2.75$ **99.** $x = 27.\overline{6}$

101. Vertical asymptote: $x = -8$

Horizontal asymptote: $y = 0$

103. Vertical asymptotes: $x = \frac{5}{2}, x = -3$

Horizontal asymptote: $y = 0$

105. Vertical asymptote: $x = 6$

Horizontal asymptote: $y = 1$

Section 4.3   (page 347)

Vocabulary Check   (page 347)

1. change-of-base   **2.** $\dfrac{\ln x}{\ln a}$ **3.** $\log_a u^n$

4. $\ln u + \ln v$

1. (a) $\dfrac{\log_{10} x}{\log_{10} 5}$ (b) $\dfrac{\ln x}{\ln 5}$ **3.** (a) $\dfrac{\log_{10} x}{\log_{10} \frac{1}{5}}$ (b) $\dfrac{\ln x}{\ln \frac{1}{5}}$

5. (a) $\dfrac{\log_{10} \frac{3}{10}}{\log_{10} a}$ (b) $\dfrac{\ln \frac{3}{10}}{\ln a}$

7. (a) $\dfrac{\log_{10} x}{\log_{10} 2.6}$ (b) $\dfrac{\ln x}{\ln 2.6}$

9. 1.771   **11.** -2 **13.** -0.102 **15.** 2.691

17. $\frac{3}{2}$ **19.** $6 + \ln 5$

21. $\log_5 \frac{1}{250} = \log_5 1 - \log_5 250$

$= 0 - \log_5 (125 \cdot 2)$

$= -(\log_5 125 + \log_5 2)$

$= -3 - \log_5 2$

23. $\log_{10} 5 + \log_{10} x$ **25.** $\log_{10} 5 - \log_{10} x$

27. $4 \log_8 x$ **29.** $\frac{1}{2} \ln z$ **31.** $\ln x + \ln y + \ln z$

33. $2 \ln a + \frac{1}{2} \ln(a - 1)$ **35.** $\frac{1}{3} \ln x - \frac{1}{3} \ln y$

37. $\ln(x + 1) + \ln(x - 1) - 3 \ln x, \ x > 1$

39. $4 \ln x + \frac{1}{2} \ln y - 5 \ln z$

41. $2 \log_b x - 2 \log_b y - 3 \log_b z$

43. (a)

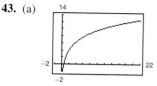

(b)

x	1	2	3	4	5	6
y_1	1.6094	3.8712	5.2417	6.2383	7.0255	7.6779
y_2	1.6094	3.8712	5.2417	6.2383	7.0255	7.6779

x	7	8	9	10	11
y_1	8.2356	8.7232	9.1566	9.5468	9.9017
y_2	8.2356	8.7232	9.1566	9.5468	9.9017

(c) $y_1 = y_2$ for positive values of x.

45. $\ln 4x$ **47.** $\log_4 \dfrac{z}{y}$ **49.** $\log_2(x + 3)^2$

51. $\log_3 \sqrt[3]{7x}$ **53.** $\ln \dfrac{x}{(x + 1)^3}$ **55.** $\ln \dfrac{x - 2}{x + 2}$

57. $\ln \dfrac{x}{(x^2 - 4)^2}$ **59.** $\ln \sqrt[3]{\dfrac{x(x + 3)^2}{x^2 - 1}}$

61. $\ln \dfrac{\sqrt[3]{y(y + 4)^2}}{y - 1}$

63. (a)

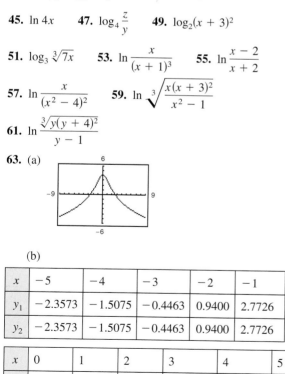

(b)

x	-5	-4	-3	-2	-1
y_1	-2.3573	-1.5075	-0.4463	0.9400	2.7726
y_2	-2.3573	-1.5075	-0.4463	0.9400	2.7726

x	0	1	2	3	4	5
y_1	4.1589	2.7726	0.9400	-0.4463	-1.5075	-2.3573
y_2	4.1589	2.7726	0.9400	-0.4463	-1.5075	-2.3573

(c) $y_1 = y_2$

65. (a)

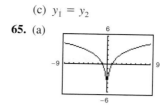

(b)

x	-5	-4	-3	-2	-1	0
y_1	3.2189	2.7726	2.1972	1.3863	0	Error
y_2	Error	Error	Error	Error	Error	Error

x	1	2	3	4	5
y_1	0	1.3863	2.1972	2.7726	3.2189
y_2	0	1.3863	2.1972	2.7726	3.2189

(c) No. The domains differ.

67. 2   **69.** 6.8   **71.** -4 is not in the domain of $\log_2 x$.

73. 2   **75.** -4 **77.** 8   **79.** $-\frac{1}{2}$

81. (a) $120 + 10 \log_{10} I$

(b) and (c)

I	10^{-4}	10^{-6}	10^{-8}	10^{-10}	10^{-12}	10^{-14}
β	80	60	40	20	0	-20

83. (a)

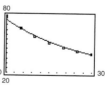

(b) $T = 54.438(0.964)^t + 21$

(c) $\ln(T - 21) = -0.037t + 3.997$

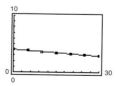

$T = e^{(-0.037t + 3.997)} + 21$

(d) $\dfrac{1}{T - 21} = 0.0012t + 0.0162$

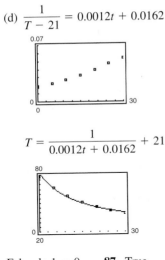

$$T = \dfrac{1}{0.0012t + 0.0162} + 21$$

85. False. $\ln 1 = 0$ **87.** True

89. False. $f(\sqrt{x}) = \frac{1}{2}f(x)$ **91.** True

93. $\log_b u^n = \log_b(\underbrace{u \times u \times u \times \cdots \times u})$

n u's multiplied together

$= \underbrace{\log_b u + \log_b u + \cdots + \log_b u}$

n terms

$= n \log_b u$

95. $f(x) = \dfrac{\log_{10} x}{\log_{10} 2}$ **97.** $f(x) = \dfrac{\log_{10} \sqrt{x}}{\log_{10} 3}$

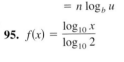

99. $f(x) = \dfrac{\log_{10}(x/3)}{\log_{10} 5}$

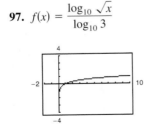

101. 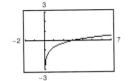 $f(x) = h(x)$

The graphs are identical because for each positive value of x, Property 2 of logarithms holds.

103. $\dfrac{3x^4}{2y^3}$ **105.** $1, x \neq 0, y \neq 0$ **107.** $3 \pm \sqrt{7}$

109. $\pm 4, \pm \sqrt{3}$ **111.** $\pm 2, 6$

Section 4.4   (page 357)

1. (a) Yes   (b) No

3. (a) No   (b) Yes   (c) Yes, approximate

5. (a) Yes, approximate   (b) No   (c) Yes

7. (a) Yes   (b) Yes, approximate   (c) No

9.   **11.**

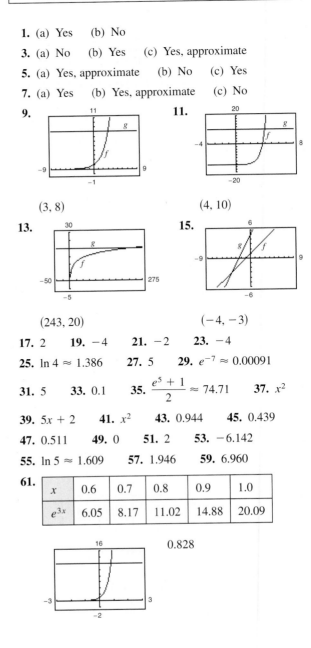

$(3, 8)$ $(4, 10)$

13.   **15.**

$(243, 20)$ $(-4, -3)$

17. 2   **19.** -4 **21.** -2 **23.** -4

25. $\ln 4 \approx 1.386$ **27.** 5   **29.** $e^{-7} \approx 0.00091$

31. 5   **33.** 0.1   **35.** $\dfrac{e^5 + 1}{2} \approx 74.71$ **37.** x^2

39. $5x + 2$ **41.** x^2 **43.** 0.944   **45.** 0.439

47. 0.511   **49.** 0   **51.** 2   **53.** -6.142

55. $\ln 5 \approx 1.609$ **57.** 1.946   **59.** 6.960

61.

x	0.6	0.7	0.8	0.9	1.0
e^{3x}	6.05	8.17	11.02	14.88	20.09

0.828

63.

x	5	6	7	8	9
$20(100 - e^{x/2})$	1756	1598	1338	908	200

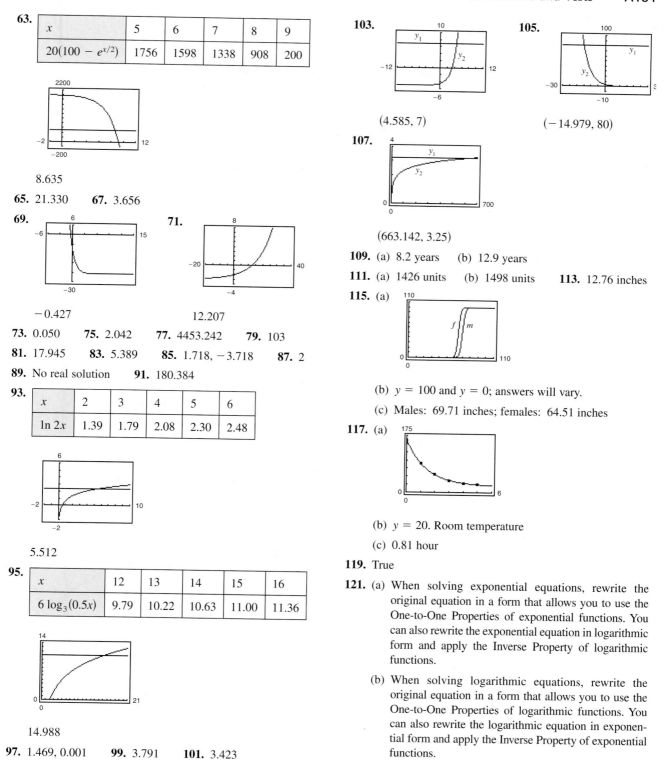

8.635

65. 21.330   **67.** 3.656

69.

−0.427

71.

12.207

73. 0.050   **75.** 2.042   **77.** 4453.242   **79.** 103

81. 17.945   **83.** 5.389   **85.** 1.718, −3.718   **87.** 2

89. No real solution   **91.** 180.384

93.

x	2	3	4	5	6
$\ln 2x$	1.39	1.79	2.08	2.30	2.48

5.512

95.

x	12	13	14	15	16
$6\log_3(0.5x)$	9.79	10.22	10.63	11.00	11.36

14.988

97. 1.469, 0.001   **99.** 3.791   **101.** 3.423

103.

(4.585, 7)

105.

(−14.979, 80)

107.

(663.142, 3.25)

109. (a) 8.2 years   (b) 12.9 years

111. (a) 1426 units   (b) 1498 units   **113.** 12.76 inches

115. (a)

(b) $y = 100$ and $y = 0$; answers will vary.

(c) Males: 69.71 inches; females: 64.51 inches

117. (a)

(b) $y = 20$. Room temperature

(c) 0.81 hour

119. True

121. (a) When solving exponential equations, rewrite the original equation in a form that allows you to use the One-to-One Properties of exponential functions. You can also rewrite the exponential equation in logarithmic form and apply the Inverse Property of logarithmic functions.

(b) When solving logarithmic equations, rewrite the original equation in a form that allows you to use the One-to-One Properties of logarithmic functions. You can also rewrite the logarithmic equation in exponential form and apply the Inverse Property of exponential functions.

123. Yes. Doubling time: $t = \dfrac{\ln 2}{r}$

Quadrupling time: $t = \dfrac{\ln 4}{r} = 2\left(\dfrac{\ln 2}{r}\right)$

125.

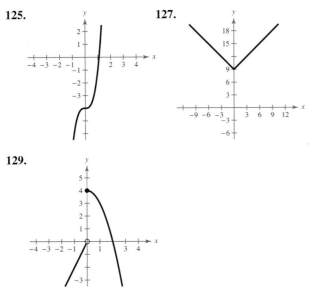

127.

129.

Section 4.5   (page 368)

Vocabulary Check   (page 368)

1. $y = ae^{bx}, b > 0$

2. $y = a + b \ln x, y = a + b \log_{10} x$

3. logistic growth   **4.** bell-shaped curve

5. sigmoidal

1. c   **2.** e   **3.** b   **4.** a   **5.** d   **6.** f

Initial Investment	Annual % Rate	Time to Double	Amount After 10 Years
7. $1000	3.5%	19.8 yr	$1419.07
9. $750	8.94%	7.75 yr	$1833.67
11. $500	9.5%	7.30 yr	$1292.85
13. $6376.28	4.5%	15.4 yr	$10,000.00

15.

r	2%	4%	6%	8%	10%	12%
t	54.93	27.47	18.31	13.73	10.99	9.16

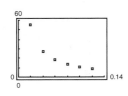

17.

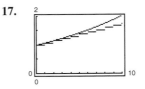

Continuous compounding

Isotope	Half-Life (years)	Initial Quantity	Amount After 1000 Years
19. ^{226}Ra	1600	10 g	6.48 g
21. ^{14}C	5730	3 g	2.66 g

23. $y = e^{0.768x}$ **25.** $y = 4e^{-0.2773x}$

27. (a) Australia: $y = 19.2e^{0.00848t}$; 24.8 million

Canada: $y = 31.3e^{0.00915t}$; 41.2 million

Philippines: $y = 81.2e^{0.0187t}$; 142.3 million

South Africa: $y = 43.4e^{-0.0054t}$; 36.9 million

Turkey: $y = 65.7e^{0.01095t}$; 91.2 million

(b) b; Population changes at a faster rate for a greater magnitude of b.

(c) b; b is positive when the population is increasing and negative when the population is decreasing.

29. 3.15 hours   **31.** 95.8%

33. (a) $V = -7000t + 32,000$ (b) $V = 32,000e^{-0.2877t}$

(c)

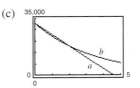

Exponential

(d) 1 year: Linear: $25,000; exponential: $24,000
3 years: Linear: $11,000; exponential: $13,499

(e) Value decreases by $7000 each year

35. (a) $S(t) = 100(1 - e^{-0.1625t})$

(b)

(c) 55,625 units

37. (a)

(b) 100

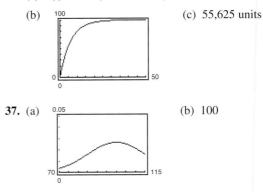

39. (a)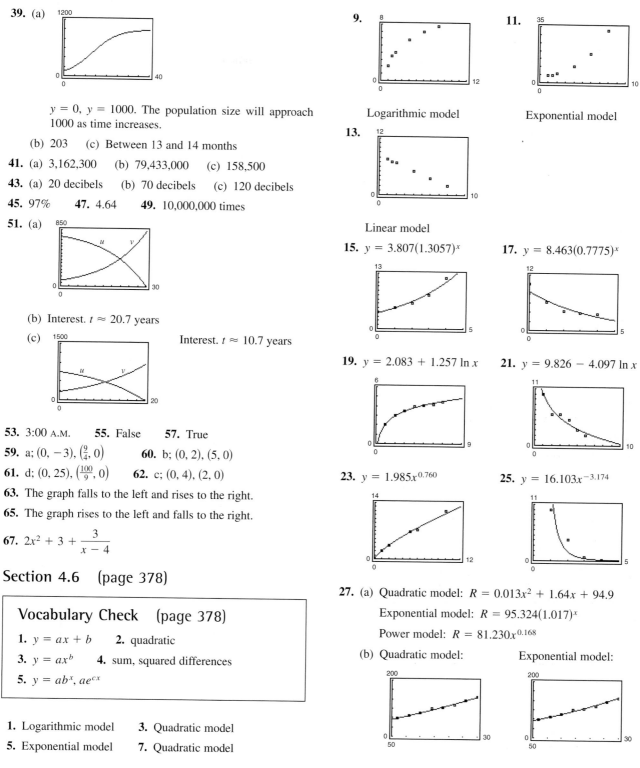

$y = 0$, $y = 1000$. The population size will approach 1000 as time increases.

(b) 203   (c) Between 13 and 14 months

41. (a) 3,162,300   (b) 79,433,000   (c) 158,500

43. (a) 20 decibels   (b) 70 decibels   (c) 120 decibels

45. 97%   **47.** 4.64   **49.** 10,000,000 times

51. (a)

(b) Interest. $t \approx 20.7$ years

(c)

Interest. $t \approx 10.7$ years

53. 3:00 A.M.   **55.** False   **57.** True

59. a; $(0, -3)$, $\left(\frac{9}{4}, 0\right)$ **60.** b; $(0, 2)$, $(5, 0)$

61. d; $(0, 25)$, $\left(\frac{100}{9}, 0\right)$ **62.** c; $(0, 4)$, $(2, 0)$

63. The graph falls to the left and rises to the right.

65. The graph rises to the left and falls to the right.

67. $2x^2 + 3 + \dfrac{3}{x - 4}$

Section 4.6   (page 378)

Vocabulary Check   (page 378)

1. $y = ax + b$ **2.** quadratic

3. $y = ax^b$ **4.** sum, squared differences

5. $y = ab^x$, ae^{cx}

1. Logarithmic model   **3.** Quadratic model

5. Exponential model   **7.** Quadratic model

9.

Logarithmic model

11.

Exponential model

13.

Linear model

15. $y = 3.807(1.3057)^x$ **17.** $y = 8.463(0.7775)^x$

19. $y = 2.083 + 1.257 \ln x$ **21.** $y = 9.826 - 4.097 \ln x$

23. $y = 1.985x^{0.760}$ **25.** $y = 16.103x^{-3.174}$

27. (a) Quadratic model: $R = 0.013x^2 + 1.64x + 94.9$

Exponential model: $R = 95.324(1.017)^x$

Power model: $R = 81.230x^{0.168}$

(b) Quadratic model: Exponential model:

Power model:

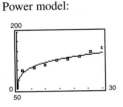

(c) Quadratic   (d) 165.7 million

29. (a) $y = 3.1x + 251$

(b) $y = (251.5)(1.01)^x$

(c) No. The linear model is a better fit than the exponential model because the r-value of the linear model has an absolute value closer to one.

(d) Linear model: 306.8 million

Exponential model: 300.8 million

31. (a) Linear model: $T = -1.24t + 73.0$;

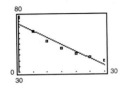

The data does not appear to be linear. Answers may vary.

(b) Quadratic model: $T = 0.034t^2 - 2.26t + 77.3$;

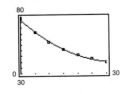

The data appears to be quadratic. When $t = 60$, the temperature of the water should decrease, not increase according to the model.

(c) Exponential model: $T = 54.438(0.964)^t + 21$

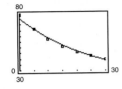

(d) Answers will vary.

33. (a) $S = 925.73(1.15)^x$

(b)

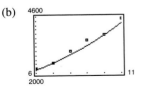

(c) The model represents the data well.

(d) \$9,962,000,000

35. (a) Linear model: $y = 15.79x + 47.9$

Logarithmic model: $y = -97.5 + 131.92 \ln x$

Quadratic model: $y = -1.968x^2 + 49.24x - 88.6$

Exponential model: $y = 83.94(1.09)^x$

Power model: $y = 36.51x^{0.7525}$

(b) Linear model: Logarithmic model:

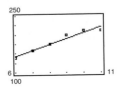 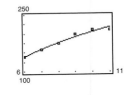

Quadratic model: Exponential model:

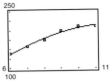 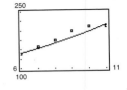

Power model:

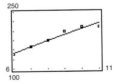

By comparing the graphs, the quadratic model best fits the data.

(c) Linear: 217.2

Logarithmic: 120.91

Quadratic: 72.7

Exponential: 523.2

Power: 189.0

By comparing the sums of the squared differences, the quadratic model best fits the data.

(d) Linear: 0.9526

Logarithmic: 0.9736

Quadratic: 0.9841

Exponential: 0.9402

Power: 0.9697

By comparing the r^2-values, the quadratic model best fits the data.

(e) The quadratic model best fits the data.

37. True

39. Slope: $-\frac{2}{5}$; y-intercept: $(0, 2)$

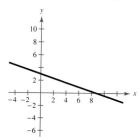

41. Slope: $-\frac{12}{35}$; y-intercept: $(0, 3)$

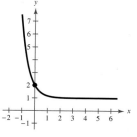

Review Exercises   (page 383)

1. 10.3254   **3.** 0.0001   **5.** c   **6.** d

7. b   **8.** a

9.

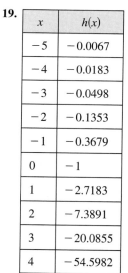

Horizontal asymptote: x-axis

y-intercept: $(0, 1)$

Increasing on $(-\infty, \infty)$

11.

Horizontal asymptote: $y = 1$

y-intercept: $(0, 2)$

Decreasing on $(-\infty, \infty)$

13. 2980.958   **15.** 0.122

17.

x	$h(x)$
-5	0.0025
-4	0.0067
-3	0.0183
-2	0.0498
-1	0.1353
0	0.3679
1	1
2	2.7183
3	7.3891
4	20.0855

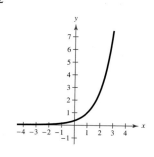

19.

x	$h(x)$
-5	-0.0067
-4	-0.0183
-3	-0.0498
-2	-0.1353
-1	-0.3679
0	-1
1	-2.7183
2	-7.3891
3	-20.0855
4	-54.5982

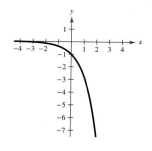

21.

x	$f(x)$
-5	48.7300
-4	29.5562
-3	17.9268
-2	10.8731
-1	6.5949
0	4
1	2.4261
2	1.4715
3	0.8925
4	0.5413
5	0.3283

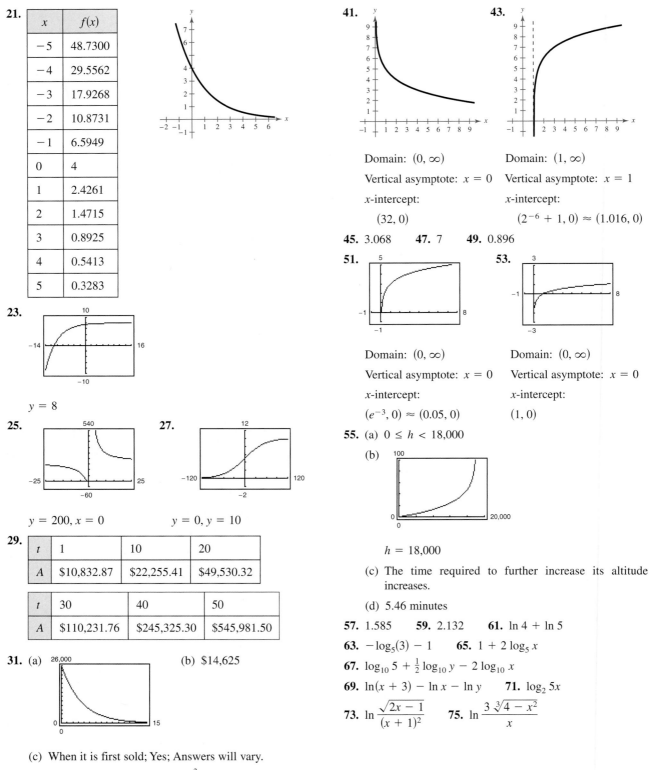

23.

$y = 8$

25.   **27.**

$y = 200, x = 0$ $y = 0, y = 10$

29.

t	1	10	20
A	\$10,832.87	\$22,255.41	\$49,530.32

t	30	40	50
A	\$110,231.76	\$245,325.30	\$545,981.50

31. (a)   (b) \$14,625

(c) When it is first sold; Yes; Answers will vary.

33. $\log_4 64 = 3$ **35.** $\log_{25} 125 = \frac{3}{2}$ **37.** 3   **39.** -1

41.   **43.**

Domain: $(0, \infty)$ Domain: $(1, \infty)$

Vertical asymptote: $x = 0$ Vertical asymptote: $x = 1$

x-intercept: x-intercept:

$\quad (32, 0)$ $\quad (2^{-6} + 1, 0) \approx (1.016, 0)$

45. 3.068   **47.** 7   **49.** 0.896

51.   **53.**

Domain: $(0, \infty)$ Domain: $(0, \infty)$

Vertical asymptote: $x = 0$ Vertical asymptote: $x = 0$

x-intercept: x-intercept:

$\quad (e^{-3}, 0) \approx (0.05, 0)$ $\quad (1, 0)$

55. (a) $0 \le h < 18,000$

(b)

$\quad h = 18,000$

(c) The time required to further increase its altitude increases.

(d) 5.46 minutes

57. 1.585   **59.** 2.132   **61.** $\ln 4 + \ln 5$

63. $-\log_5(3) - 1$ **65.** $1 + 2 \log_5 x$

67. $\log_{10} 5 + \frac{1}{2} \log_{10} y - 2 \log_{10} x$

69. $\ln(x + 3) - \ln x - \ln y$ **71.** $\log_2 5x$

73. $\ln \dfrac{\sqrt{2x - 1}}{(x + 1)^2}$ **75.** $\ln \dfrac{3\sqrt[3]{4 - x^2}}{x}$

77. (a)

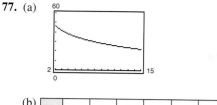

(b)

h	4	6	8	10	12	14
s	38	33	30	27	25	23

(c) The decrease in productivity starts to level off.

79. 3   **81.** −3   **83.** 2401   **85.** $e^4 \approx 54.598$

87. $\ln 12 \approx 2.485$ **89.** $-\dfrac{\ln 44}{5} \approx -0.757$

91. $\dfrac{\ln 22}{\ln 2} \approx 4.459$ **93.** $\log_5 17 \approx 1.760$

95. $\ln 5 \approx 1.609$, $\ln 2 \approx 0.693$

97. $\frac{1}{3}e^{8.2} \approx 1213.650$ **99.** $\frac{1}{4}e^{15/2} \approx 452.011$

101. $3e^2 \approx 22.167$ **103.** $e^4 - 1 \approx 53.598$

105. No solution   **107.** $\frac{9}{10}$ **109.** ≈ 15.2 years

111. e   **112.** b   **113.** f   **114.** d   **115.** a

116. c   **117.** $y = 2e^{0.1014x}$

119. $y = \frac{1}{2}e^{0.4605x}$ **121.** $k \approx 0.0259$; 606,000

123. (a) 5.78%   (b) $10,595.03

125. (a) 7.7 weeks   (b) 13.3 weeks

127. Logistic model   **129.** Logarithmic model

131. (a) Quadratic model: $m = 0.03x^2 + 0.1x + 10$

Exponential model: $m = 8.73(1.05)^x$

Power model: $m = 3.466x^{0.647}$

(b) Quadratic model: Exponential model:

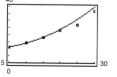

Power model:

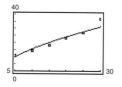

(c) Exponential model

(d) 53,000 screens

133. (a) $P = \dfrac{9999.887}{1 + 19.0e^{-0.2x}}$

(b)

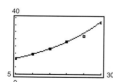

(c) The model fits the data well.

(d) 10,000 fish

135. True   **137.** False. $\ln(xy) = \ln x + \ln y$

139. False. $x > 0$

141. (a)

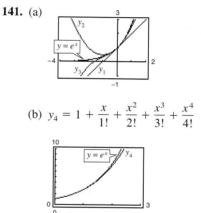

(b) $y_4 = 1 + \dfrac{x}{1!} + \dfrac{x^2}{2!} + \dfrac{x^3}{3!} + \dfrac{x^4}{4!}$

The graph of y_4 is close to the graph of $y = e^x$ near the point $(0, 1)$. As n becomes larger, the polynomials are better approximations of $y = e^x$.

Chapter Test   (page 388)

1. 1123.690   **2.** 687.291   **3.** 0.497   **4.** 22.198

5.

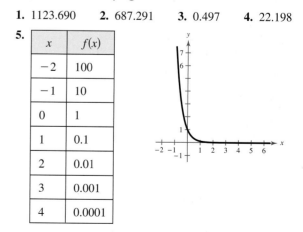

x	$f(x)$
−2	100
−1	10
0	1
1	0.1
2	0.01
3	0.001
4	0.0001

6.

x	$f(x)$
-2.0	-0.00077
-1.5	-0.00189
-1.0	-0.00463
-0.5	-0.01134
0	-0.02778
0.5	-0.06804
1.0	-0.16667
1.5	-0.40825
2	-1
2.5	-2.44949
3	-6
3.5	-14.69694
4	-36

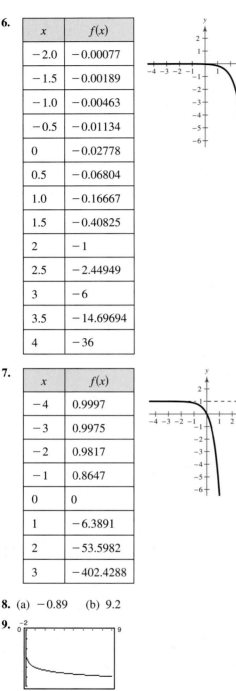

7.

x	$f(x)$
-4	0.9997
-3	0.9975
-2	0.9817
-1	0.8647
0	0
1	-6.3891
2	-53.5982
3	-402.4288

8. (a) -0.89 (b) 9.2

9.

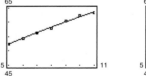

Domain: $(0, \infty)$

Vertical asymptote: $x = 0$

x-intercept: $(1 \times 10^{-6}, 0) \approx (0, 0)$

10.

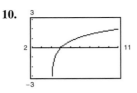

Domain: $(4, \infty)$

Vertical asymptote: $x = 4$

x-intercept: $(5, 0)$

11.

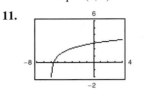

Domain: $(-6, \infty)$

Vertical asymptote: $x = -6$

x-intercept: $(e^{-1} - 6, 0) \approx (-5.632, 0)$

12. 1.945 **13.** 0.115 **14.** 1.328

15. $\log_2 3 + 4 \log_2 a$ **16.** $\ln 5 + \frac{1}{2} \ln x - \ln 6$

17. $\log_3 13y$ **18.** $\ln\left(\dfrac{x^4}{y^4}\right)$ **19.** 1.321

20. $\dfrac{800}{501} \approx 1.597$ **21.** 54.96%

22. (a) Quadratic model: $R = -0.031x^2 + 2.37x + 41.4$

Exponential model: $R = 44.863(1.0328)^x$

Power model: $R = 35.06298x^{0.2466}$

(b) Quadratic model: Exponential model:

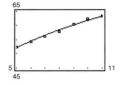   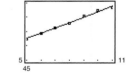

Power model:

(c) Quadratic model   (d) $\$72,700,000,000$

Cumulative Test for Chapters 3–4
(page 389)

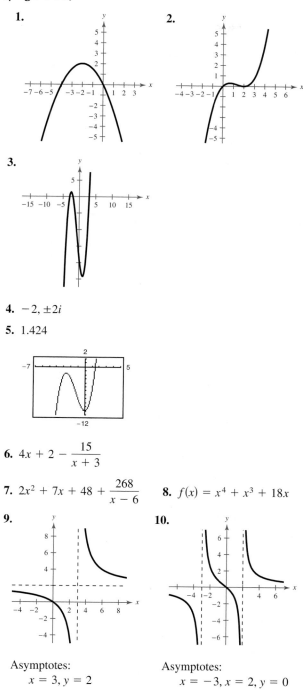

1.

2.

3.

4. $-2, \pm 2i$

5. 1.424

6. $4x + 2 - \dfrac{15}{x + 3}$

7. $2x^2 + 7x + 48 + \dfrac{268}{x - 6}$

8. $f(x) = x^4 + x^3 + 18x$

9.

Asymptotes:
$x = 3, y = 2$

10.

Asymptotes:
$x = -3, x = 2, y = 0$

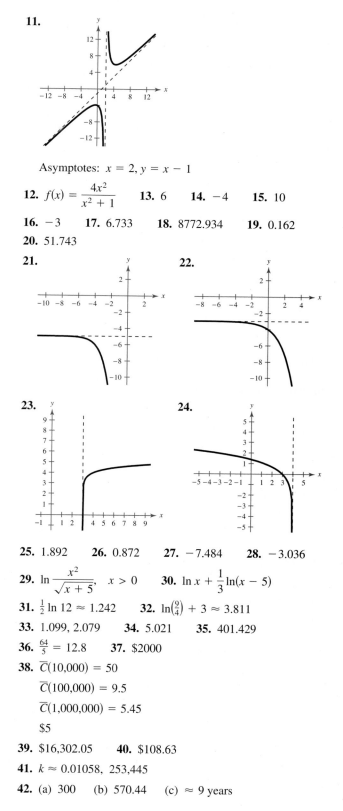

11.

Asymptotes: $x = 2, y = x - 1$

12. $f(x) = \dfrac{4x^2}{x^2 + 1}$ **13.** 6   **14.** -4 **15.** 10

16. -3 **17.** 6.733   **18.** 8772.934   **19.** 0.162

20. 51.743

21.

22.

23.

24.

25. 1.892   **26.** 0.872   **27.** -7.484 **28.** -3.036

29. $\ln \dfrac{x^2}{\sqrt{x + 5}}, \quad x > 0$ **30.** $\ln x + \dfrac{1}{3}\ln(x - 5)$

31. $\frac{1}{2}\ln 12 \approx 1.242$ **32.** $\ln\left(\frac{9}{4}\right) + 3 \approx 3.811$

33. 1.099, 2.079   **34.** 5.021   **35.** 401.429

36. $\frac{64}{5} = 12.8$ **37.** \$2000

38. $\overline{C}(10,000) = 50$

$\overline{C}(100,000) = 9.5$

$\overline{C}(1,000,000) = 5.45$

\$5

39. \$16,302.05   **40.** \$108.63

41. $k \approx 0.01058$, 253,445

42. (a) 300   (b) 570.44   (c) ≈ 9 years

43. (a) Quadratic model: $y = 0.248x^2 - 0.28x + 50.3$

Exponential model: $y = 42.62(1.05)^x$

Power model: $y = 32.45x^{0.3329}$

(b) Quadratic model: Exponential model:

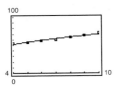

Power model:

(c) Quadratic model

(d) 109,000

Chapter 5

Section 5.1   (page 399)

Vocabulary Check   (page 399)

1. Trigonometry   **2.** angle

3. standard position   **4.** coterminal   **5.** degree

6. complementary   **7.** supplementary

8. radian   **9.** linear   **10.** angular

1. $210°$ **3.** $-45°$

5. (a) Quadrant II   (b) Quadrant IV

7. (a) Quadrant III   (b) Quadrant I

9. (a)   (b)

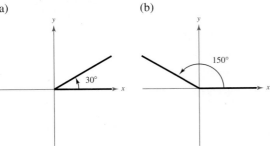

11. (a)   (b)

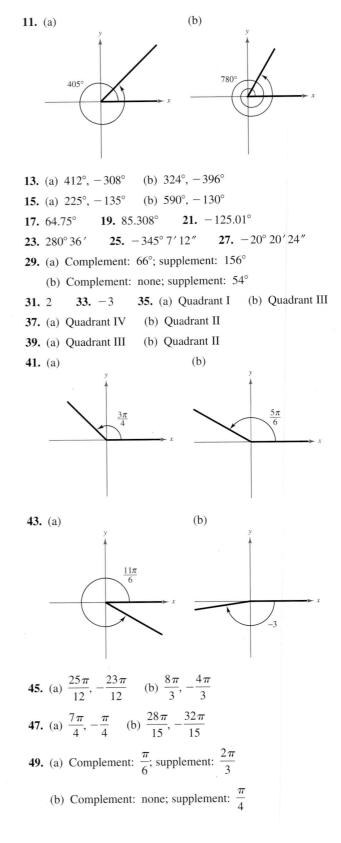

13. (a) $412°, -308°$ (b) $324°, -396°$

15. (a) $225°, -135°$ (b) $590°, -130°$

17. $64.75°$ **19.** $85.308°$ **21.** $-125.01°$

23. $280°36'$ **25.** $-345°7'12''$ **27.** $-20°20'24''$

29. (a) Complement: $66°$; supplement: $156°$

   (b) Complement: none; supplement: $54°$

31. 2 **33.** -3 **35.** (a) Quadrant I   (b) Quadrant III

37. (a) Quadrant IV   (b) Quadrant II

39. (a) Quadrant III   (b) Quadrant II

41. (a)   (b)

43. (a)   (b)

45. (a) $\dfrac{25\pi}{12}, -\dfrac{23\pi}{12}$ (b) $\dfrac{8\pi}{3}, -\dfrac{4\pi}{3}$

47. (a) $\dfrac{7\pi}{4}, -\dfrac{\pi}{4}$ (b) $\dfrac{28\pi}{15}, -\dfrac{32\pi}{15}$

49. (a) Complement: $\dfrac{\pi}{6}$; supplement: $\dfrac{2\pi}{3}$

   (b) Complement: none; supplement: $\dfrac{\pi}{4}$

51. (a) $\dfrac{\pi}{6}$ (b) $\dfrac{5\pi}{6}$ **53.** (a) $-\dfrac{\pi}{9}$ (b) $-\dfrac{4\pi}{3}$

55. (a) $270°$ (b) $-210°$ **57.** (a) $420°$ (b) $-39°$

59. 2.199 **61.** -3.776 **63.** -0.014 **65.** $25.714°$

67. $292.5°$ **69.** $-114.592°$ **71.** $\frac{6}{5}$ rad   **73.** $4\frac{4}{7}$ rad

75. $\frac{8}{29}$ rad   **77.** $2\frac{12}{29}$ rad

79. 15π inches ≈ 47.12 inches

81. 2 meters   **83.** 1141.81 miles   **85.** $4°\,2'\,33.02''$

87. $\frac{5}{12}$ rad $\approx 23.87°$

89. (a) $540° \approx 9.42$ rad   (b) $900° \approx 15.71$ rad

 (c) $1260° \approx 21.99$ rad

91. 20.16π inches per second

93. False. A radian is larger: 1 rad $\approx 57.3°$.

95. (a) An angle is in standard position when the origin is the vertex and the initial side coincides with the positive x-axis.

 (b) A negative angle is generated by a clockwise rotation.

 (c) Angles that have the same initial and terminal sides are called coterminal angles.

 (d) An obtuse angle is between $90°$ and $180°$.

97. $\dfrac{50\pi}{3}$ square meters

99. (a) $A = 0.4r^2$, $r > 0$; $s = 0.8r$, $r > 0$

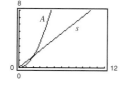

 The area function changes more rapidly for $r > 1$ because it is quadratic, whereas the arc length function is linear.

 (b) $A = 50\theta$, $0 < \theta < 2\pi$; $s = 10\theta$, $0 < \theta < 2\pi$

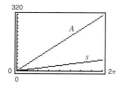

101. $y = 4.54x + 16.4$ **103.** $x = -7, -4$

105. $x = -5, 0, 2$

Section 5.2   (page 410)

Vocabulary Check   (page 410)

1. (a) iii (b) vi (c) ii (d) v (e) i (f) iv

2. hypotenuse, opposite, adjacent

3. elevation, depression

1. $\sin \theta = \dfrac{3}{5}$

 $\cos \theta = \dfrac{4}{5}$

 $\tan \theta = \dfrac{3}{4}$

 $\csc \theta = \dfrac{5}{3}$

 $\sec \theta = \dfrac{5}{4}$

 $\cot \theta = \dfrac{4}{3}$

3. $\sin \theta = \dfrac{8}{17}$

 $\cos \theta = \dfrac{15}{17}$

 $\tan \theta = \dfrac{8}{15}$

 $\csc \theta = \dfrac{17}{8}$

 $\sec \theta = \dfrac{17}{15}$

 $\cot \theta = \dfrac{15}{8}$

5. $\sin \theta = \dfrac{1}{3}$

 $\cos \theta = \dfrac{2\sqrt{2}}{3}$

 $\tan \theta = \dfrac{\sqrt{2}}{4}$

 $\csc \theta = 3$

 $\sec \theta = \dfrac{3\sqrt{2}}{4}$

 $\cot \theta = 2\sqrt{2}$

The triangles are similar and corresponding sides are proportional.

7. $\sin \theta = \dfrac{3}{5}$

 $\cos \theta = \dfrac{4}{5}$

 $\tan \theta = \dfrac{3}{4}$

 $\csc \theta = \dfrac{5}{3}$

 $\sec \theta = \dfrac{5}{4}$

 $\cot \theta = \dfrac{4}{3}$

The triangles are similar and corresponding sides are proportional.

9. $\cos \theta = \dfrac{\sqrt{5}}{3}$

 $\tan \theta = \dfrac{2\sqrt{5}}{5}$

 $\csc \theta = \dfrac{3}{2}$

 $\sec \theta = \dfrac{3\sqrt{5}}{5}$

 $\cot \theta = \dfrac{\sqrt{5}}{2}$

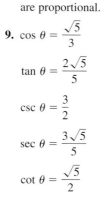

11. $\sin\theta = \dfrac{\sqrt{15}}{4}$

$\cos\theta = \dfrac{1}{4}$

$\tan\theta = \sqrt{15}$

$\csc\theta = \dfrac{4\sqrt{15}}{15}$

$\cot\theta = \dfrac{\sqrt{15}}{15}$

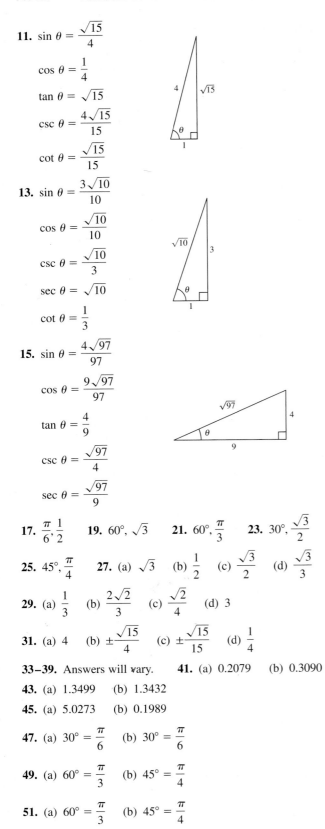

13. $\sin\theta = \dfrac{3\sqrt{10}}{10}$

$\cos\theta = \dfrac{\sqrt{10}}{10}$

$\csc\theta = \dfrac{\sqrt{10}}{3}$

$\sec\theta = \sqrt{10}$

$\cot\theta = \dfrac{1}{3}$

15. $\sin\theta = \dfrac{4\sqrt{97}}{97}$

$\cos\theta = \dfrac{9\sqrt{97}}{97}$

$\tan\theta = \dfrac{4}{9}$

$\csc\theta = \dfrac{\sqrt{97}}{4}$

$\sec\theta = \dfrac{\sqrt{97}}{9}$

17. $\dfrac{\pi}{6}, \dfrac{1}{2}$ **19.** $60°, \sqrt{3}$ **21.** $60°, \dfrac{\pi}{3}$ **23.** $30°, \dfrac{\sqrt{3}}{2}$

25. $45°, \dfrac{\pi}{4}$ **27.** (a) $\sqrt{3}$ (b) $\dfrac{1}{2}$ (c) $\dfrac{\sqrt{3}}{2}$ (d) $\dfrac{\sqrt{3}}{3}$

29. (a) $\dfrac{1}{3}$ (b) $\dfrac{2\sqrt{2}}{3}$ (c) $\dfrac{\sqrt{2}}{4}$ (d) 3

31. (a) 4 (b) $\pm\dfrac{\sqrt{15}}{4}$ (c) $\pm\dfrac{\sqrt{15}}{15}$ (d) $\dfrac{1}{4}$

33–39. Answers will vary.   **41.** (a) 0.2079 (b) 0.3090

43. (a) 1.3499 (b) 1.3432

45. (a) 5.0273 (b) 0.1989

47. (a) $30° = \dfrac{\pi}{6}$ (b) $30° = \dfrac{\pi}{6}$

49. (a) $60° = \dfrac{\pi}{3}$ (b) $45° = \dfrac{\pi}{4}$

51. (a) $60° = \dfrac{\pi}{3}$ (b) $45° = \dfrac{\pi}{4}$

53. $35\sqrt{3}$ **55.** $\dfrac{38\sqrt{3}}{3}$

57. (a) 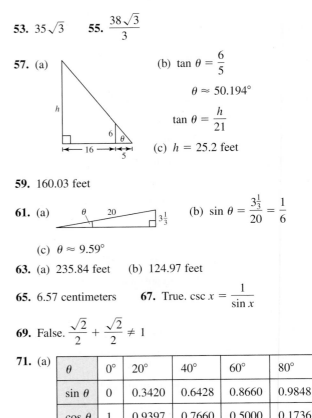   (b) $\tan\theta = \dfrac{6}{5}$

$\theta \approx 50.194°$

$\tan\theta = \dfrac{h}{21}$

(c) $h = 25.2$ feet

59. 160.03 feet

61. (a) 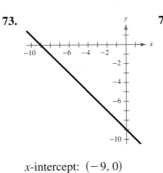   (b) $\sin\theta = \dfrac{3\frac{1}{3}}{20} = \dfrac{1}{6}$

(c) $\theta \approx 9.59°$

63. (a) 235.84 feet   (b) 124.97 feet

65. 6.57 centimeters   **67.** True. $\csc x = \dfrac{1}{\sin x}$

69. False. $\dfrac{\sqrt{2}}{2} + \dfrac{\sqrt{2}}{2} \ne 1$

71. (a)

θ	0°	20°	40°	60°	80°
$\sin\theta$	0	0.3420	0.6428	0.8660	0.9848
$\cos\theta$	1	0.9397	0.7660	0.5000	0.1736
$\tan\theta$	0	0.3640	0.8391	1.7321	5.6713

(b) Sine: increasing

Cosine: decreasing

Tangent: increasing

(c)

θ	0°	20°	40°	60°	80°
$\tan\theta = \dfrac{\sin\theta}{\cos\theta}$	0	0.3640	0.8391	1.7321	5.6713

73.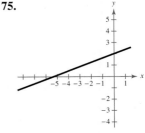

x-intercept: $(-9, 0)$

y-intercept: $(0, -9)$

75.

x-intercept: $\left(-\dfrac{16}{3}, 0\right)$

y-intercept: $(0, 2)$

77. 18.661   **79.** 17.420

Section 5.3   (page 423)

1. (a) $\sin \theta = \frac{3}{5}$

$\cos \theta = \frac{4}{5}$

$\tan \theta = \frac{3}{4}$

$\csc \theta = \frac{5}{3}$

$\sec \theta = \frac{5}{4}$

$\cot \theta = \frac{4}{3}$

(b) $\sin \theta = -\frac{15}{17}$

$\cos \theta = -\frac{8}{17}$

$\tan \theta = \frac{15}{8}$

$\csc \theta = -\frac{17}{15}$

$\sec \theta = -\frac{17}{8}$

$\cot \theta = \frac{8}{15}$

3. (a) $\sin \theta = -\frac{1}{2}$

$\cos \theta = -\frac{\sqrt{3}}{2}$

$\tan \theta = \frac{\sqrt{3}}{3}$

$\csc \theta = -2$

$\sec \theta = -\frac{2\sqrt{3}}{3}$

$\cot \theta = \sqrt{3}$

(b) $\sin \theta = \frac{\sqrt{17}}{17}$

$\cos \theta = -\frac{4\sqrt{17}}{17}$

$\tan \theta = -\frac{1}{4}$

$\csc \theta = \sqrt{17}$

$\sec \theta = -\frac{\sqrt{17}}{4}$

$\cot \theta = -4$

5. $\sin \theta = \frac{24}{25}$

$\cos \theta = \frac{7}{25}$

$\tan \theta = \frac{24}{7}$

$\csc \theta = \frac{25}{24}$

$\sec \theta = \frac{25}{7}$

$\cot \theta = \frac{7}{24}$

7. $\sin \theta = -\frac{12}{13}$

$\cos \theta = \frac{5}{13}$

$\tan \theta = -\frac{12}{5}$

$\csc \theta = -\frac{13}{12}$

$\sec \theta = \frac{13}{5}$

$\cot \theta = -\frac{5}{12}$

9. $\sin \theta = \frac{5\sqrt{29}}{29}$

$\cos \theta = -\frac{2\sqrt{29}}{29}$

$\tan \theta = -\frac{5}{2}$

$\csc \theta = \frac{\sqrt{29}}{5}$

$\sec \theta = -\frac{\sqrt{29}}{2}$

$\cot \theta = -\frac{2}{5}$

11. $\sin \theta = \frac{9\sqrt{85}}{85}$

$\cos \theta = -\frac{2\sqrt{85}}{85}$

$\tan \theta = -\frac{9}{2}$

$\csc \theta = \frac{\sqrt{85}}{9}$

$\sec \theta = -\frac{\sqrt{85}}{2}$

$\cot \theta = -\frac{2}{9}$

13. Quadrant III

15. Quadrant IV

17. $\sin \theta = \frac{3}{5}$

$\cos \theta = -\frac{4}{5}$

$\tan \theta = -\frac{3}{4}$

$\csc \theta = \frac{5}{3}$

$\sec \theta = -\frac{5}{4}$

$\cot \theta = -\frac{4}{3}$

19. $\sin \theta = -\frac{15}{17}$

$\cos \theta = \frac{8}{17}$

$\tan \theta = -\frac{15}{8}$

$\csc \theta = -\frac{17}{15}$

$\sec \theta = \frac{17}{8}$

$\cot \theta = -\frac{8}{15}$

21. $\sin \theta = \frac{\sqrt{3}}{2}$

$\cos \theta = -\frac{1}{2}$

$\tan \theta = -\sqrt{3}$

$\csc \theta = \frac{2\sqrt{3}}{3}$

$\sec \theta = -2$

$\cot \theta = -\frac{\sqrt{3}}{3}$

23. $\sin \theta = 0$

$\cos \theta = -1$

$\tan \theta = 0$

$\csc \theta$ is undefined.

$\sec \theta = -1$

$\cot \theta$ is undefined.

25. $\sin \theta = \frac{\sqrt{2}}{2}$

$\cos \theta = -\frac{\sqrt{2}}{2}$

$\tan \theta = -1$

$\csc \theta = \sqrt{2}$

$\sec \theta = -\sqrt{2}$

$\cot \theta = -1$

27. $\sin \theta = -\frac{2\sqrt{5}}{5}$

$\cos \theta = -\frac{\sqrt{5}}{5}$

$\tan \theta = 2$

$\csc \theta = -\frac{\sqrt{5}}{2}$

$\sec \theta = -\sqrt{5}$

$\cot \theta = \frac{1}{2}$

29. -1 **31.** 0   **33.** 1   **35.** Undefined

37. $\theta' = 60°$

39. $\theta' = 45°$

41. $\theta' = \frac{\pi}{3}$

43. $\theta' = \frac{\pi}{6}$

45. $\theta' = 28°$

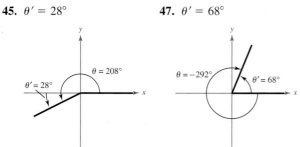

47. $\theta' = 68°$

79. 5.7588 **81.** 0.6052 **83.** −0.2369

85. 0.8391 **87.** −2.9238

89. (a) $30° = \dfrac{\pi}{6}, 150° = \dfrac{5\pi}{6}$ (b) $210° = \dfrac{7\pi}{6}, 330° = \dfrac{11\pi}{6}$

91. (a) $60° = \dfrac{\pi}{3}, 120° = \dfrac{2\pi}{3}$ (b) $135° = \dfrac{3\pi}{4}, 315° = \dfrac{7\pi}{4}$

93. (a) $150° = \dfrac{5\pi}{6}, 210° = \dfrac{7\pi}{6}$ (b) $60° = \dfrac{\pi}{3}, 300° = \dfrac{5\pi}{3}$

49. $\theta' = \dfrac{\pi}{5}$

51. $\theta' = 3.5 - \pi \approx 0.358$

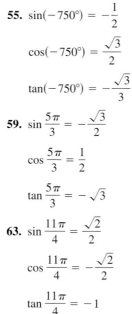

95. $\left(\dfrac{\sqrt{2}}{2}, \dfrac{\sqrt{2}}{2}\right)$

$\sin\dfrac{\pi}{4} = \dfrac{\sqrt{2}}{2}$

$\cos\dfrac{\pi}{4} = \dfrac{\sqrt{2}}{2}$

$\tan\dfrac{\pi}{4} = 1$

97. $\left(-\dfrac{\sqrt{3}}{2}, \dfrac{1}{2}\right)$

$\sin\dfrac{5\pi}{6} = \dfrac{1}{2}$

$\cos\dfrac{5\pi}{6} = -\dfrac{\sqrt{3}}{2}$

$\tan\dfrac{5\pi}{6} = -\dfrac{\sqrt{3}}{3}$

99. $\left(-\dfrac{1}{2}, -\dfrac{\sqrt{3}}{2}\right)$

$\sin\dfrac{4\pi}{3} = -\dfrac{\sqrt{3}}{2}$

$\cos\dfrac{4\pi}{3} = -\dfrac{1}{2}$

$\tan\dfrac{4\pi}{3} = \sqrt{3}$

101. $(0, -1)$

$\sin\dfrac{3\pi}{2} = -1$

$\cos\dfrac{3\pi}{2} = 0$

$\tan\dfrac{3\pi}{2}$ is undefined.

53. $\sin 225° = -\dfrac{\sqrt{2}}{2}$

$\cos 225° = -\dfrac{\sqrt{2}}{2}$

$\tan 225° = 1$

55. $\sin(-750°) = -\dfrac{1}{2}$

$\cos(-750°) = \dfrac{\sqrt{3}}{2}$

$\tan(-750°) = -\dfrac{\sqrt{3}}{3}$

103. (a) −1 (b) −0.4

105. (a) 0.25, 2.89 (b) 1.82, 4.46

107. (a) 29°F (b) 70°F (c) 31.75°F

109. (a) 2 centimeters

(b)

t	0.50	1.02	1.54	2.07	2.59
y	−1.20	0.71	−0.42	0.25	−0.15

57. $\sin(-240°) = \dfrac{\sqrt{3}}{2}$

$\cos(-240°) = -\dfrac{1}{2}$

$\tan(-240°) = -\sqrt{3}$

59. $\sin\dfrac{5\pi}{3} = -\dfrac{\sqrt{3}}{2}$

$\cos\dfrac{5\pi}{3} = \dfrac{1}{2}$

$\tan\dfrac{5\pi}{3} = -\sqrt{3}$

(c) Friction within the system damps the oscillations and is modeled by the factor e^{-t}.

(d) 0.26 second, 0.79 second

111. 0.79 amperes **113.** True. $\theta' = 180° - \theta$

115. True

117. (a)

61. $\sin\left(-\dfrac{7\pi}{6}\right) = \dfrac{1}{2}$

$\cos\left(-\dfrac{7\pi}{6}\right) = -\dfrac{\sqrt{3}}{2}$

$\tan\left(-\dfrac{7\pi}{6}\right) = -\dfrac{\sqrt{3}}{3}$

63. $\sin\dfrac{11\pi}{4} = \dfrac{\sqrt{2}}{2}$

$\cos\dfrac{11\pi}{4} = -\dfrac{\sqrt{2}}{2}$

$\tan\dfrac{11\pi}{4} = -1$

θ	0°	20°	40°	60°	80°
$\sin\theta$	0	0.3420	0.6428	0.8660	0.9848
$\sin(180° - \theta)$	0	0.3420	0.6428	0.8660	0.9848

65. $\sin\left(-\dfrac{17\pi}{6}\right) = -\dfrac{1}{2}$

$\cos\left(-\dfrac{17\pi}{6}\right) = -\dfrac{\sqrt{3}}{2}$

$\tan\left(-\dfrac{17\pi}{6}\right) = \dfrac{\sqrt{3}}{3}$

67. $\dfrac{4}{5}$ **69.** $-\dfrac{\sqrt{13}}{2}$

(b) $\sin\theta = \sin(180° - \theta)$

71. $\dfrac{8}{5}$ **73.** 0.1736 **75.** 2.1445 **77.** −0.3420

119. (a)

Function	sin x	cos x	tan x
Domain	$(-\infty, \infty)$	$(-\infty, \infty)$	All reals except $\frac{\pi}{2} + n\pi$
Range	$[-1, 1]$	$[-1, 1]$	$(-\infty, \infty)$
Even/Odd	Odd	Even	Odd
Period	2π	2π	π
Zeros	$n\pi$	$\frac{\pi}{2} + n\pi$	$n\pi$

Function	csc x	sec x	cot x
Domain	All reals except $n\pi$	All reals except $\frac{\pi}{2} + n\pi$	All reals except $n\pi$
Range	$(-\infty, -1) \cup (1, \infty)$	$(-\infty, -1) \cup (1, \infty)$	$(-\infty, \infty)$
Even/Odd	Odd	Even	Odd
Period	2π	2π	π
Zeros	None	None	$\frac{\pi}{2} + n\pi$

Patterns and conclusions may vary.

121.

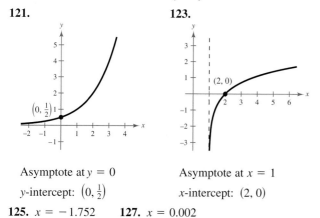

Asymptote at $y = 0$

y-intercept: $\left(0, \frac{1}{2}\right)$

123.

Asymptote at $x = 1$

x-intercept: $(2, 0)$

125. $x = -1.752$ **127.** $x = 0.002$

Section 5.4   (page 434)

1. Period: π **3.** Period: 4π **5.** Period: 2
Amplitude: 3   Amplitude: $\frac{5}{2}$ Amplitude: $\frac{2}{3}$

7. Period: 2π **9.** Period: $\frac{\pi}{5}$ **11.** Period: 3π
Amplitude: 2   Amplitude: 3   Amplitude: $\frac{1}{4}$

13. Period: $\frac{1}{2}$ **15.** g is a shift of f π units to the right.
Amplitude: $\frac{1}{3}$

17. g is a reflection of f in the x-axis.

19. g is a reflection of f in the x-axis and has five times the amplitude of f.

21. g is a shift of f five units upward.

23. g has twice the amplitude of f.

25. g is a horizontal shift of f π units to the right.

27.

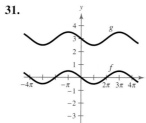

29.

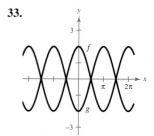

31.

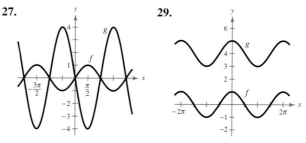

33.

35.

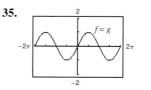

37.

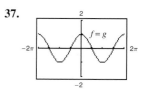

39.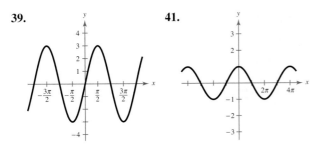

41.

43.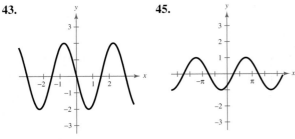

45.

47.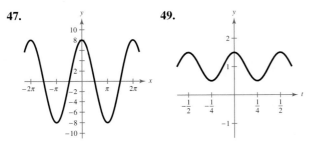

49.

51.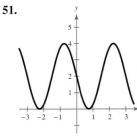

53.

55.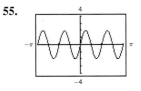

Amplitude: 2

Period: $\dfrac{\pi}{2}$

57.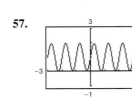

Amplitude: 1

Period: 1

59.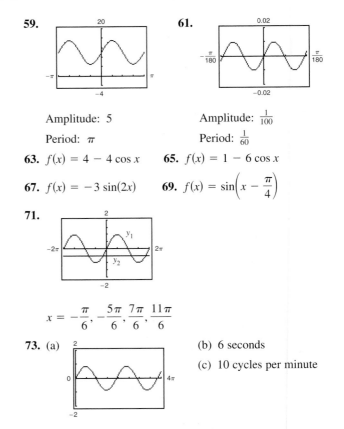

Amplitude: 5

Period: π

61.

Amplitude: $\dfrac{1}{100}$

Period: $\dfrac{1}{60}$

63. $f(x) = 4 - 4 \cos x$

65. $f(x) = 1 - 6 \cos x$

67. $f(x) = -3 \sin(2x)$

69. $f(x) = \sin\!\left(x - \dfrac{\pi}{4}\right)$

71.

$$x = -\dfrac{\pi}{6},\ -\dfrac{5\pi}{6},\ \dfrac{7\pi}{6},\ \dfrac{11\pi}{6}$$

73. (a)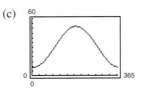

(b) 6 seconds

(c) 10 cycles per minute

(d) The period of the model would change because the time for a respiratory cycle would decrease.

75. (a)

(b) Minimum height: 5 feet

Maximum height: 55 feet

77. (a) 365 days. The cycle is 1 year.

(b) 30.3 gallons per day; the average is the constant term of the model.

(c)

Consumption exceeds 40 gallons per day from the beginning of May through part of September.

79. (a)

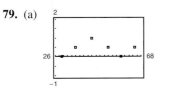

(b) $y = 0.508 \sin(0.216x - 1.616) + 0.539$

(c)

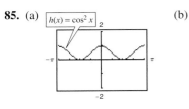

The model is a good fit.

(d) 29.09

(e) 100%

81. True. $2\pi \cdot \dfrac{10}{3} = \dfrac{20\pi}{3}$ **83.** True

85. (a)

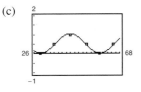

Even

(b)

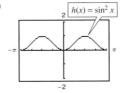

Even

(c)

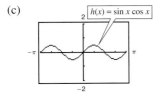

Odd

87.

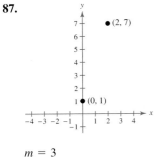

$m = 3$

89. 487.014°

Section 5.5   (page 445)

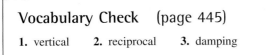

Vocabulary Check   (page 445)

1. vertical   **2.** reciprocal   **3.** damping

1. c   **2.** d   **3.** e   **4.** f   **5.** a   **6.** b

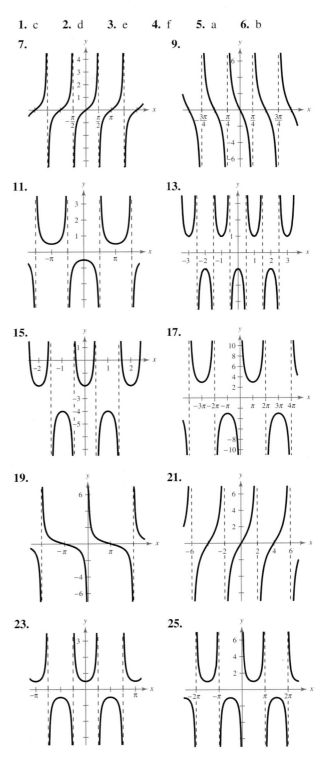

27.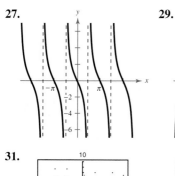

29.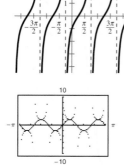

31.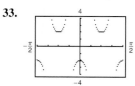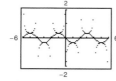

Answers will vary.

33.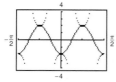

Answers will vary.

35.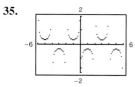

Answers will vary.

37. $-\dfrac{7\pi}{4}, -\dfrac{3\pi}{4}, \dfrac{\pi}{4}, \dfrac{5\pi}{4}$ **39.** $-\dfrac{4\pi}{3}, -\dfrac{2\pi}{3}, \dfrac{2\pi}{3}, \dfrac{4\pi}{3}$

41. Even   **43.** Odd

45.    **47.**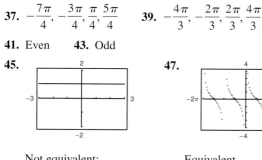

Not equivalent;   Equivalent
y_1 is undefined at $x = 0$.

49. d; as x approaches 0, $f(x)$ approaches 0.

50. a; as x approaches 0, $f(x)$ approaches 0.

51. b; as x approaches 0, $g(x)$ approaches 0.

52. c; as x approaches 0, $g(x)$ approaches 0.

53.   **55.**

57. 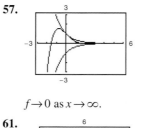   **59.**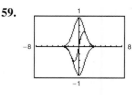

$f \to 0$ as $x \to \infty$. $g \to 0$ as $x \to \infty$.

61.

As $x \to 0$ from the left, $f \to -\infty$

As $x \to 0$ from the right, $f \to \infty$

63.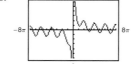

Approaches 1

65. 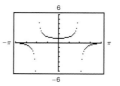   **67.** $d = 5 \cot x$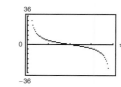

Approaches 1

69. As the predator population increases, the number of prey decreases. When the number of prey is small, the number of predators decreases.

71. (a)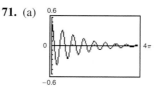

(b) Not periodic and damped; goes to 0 as t increases.

73. (a) Yes. To each t there corresponds one and only one value of y.

(b) 1.3 oscillations per second

(c) $y = 12(0.221)^t \cos(8.2t)$

(d) $y = 12e^{-1.5t} \cos(8.2t)$

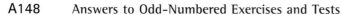
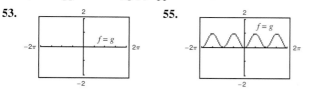
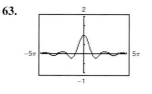

(e)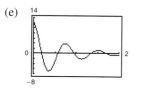

Answers will vary.

75. True

77. $f(x)$ approaches ∞ as x approaches $\pi/2$ from the left.
$f(x)$ approaches $-\infty$ as x approaches $\pi/2$ from the right.

79. (a)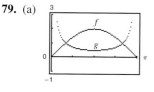

(b) $\dfrac{\pi}{6} < x < \dfrac{5\pi}{6}$

(c) Sine approaches 0 and cosecant approaches ∞ because the cosecant is the reciprocal of the sine.

81. Distributive Property   **83.** Additive Identity Property

85. Not one-to-one

87. One-to-one. $f^{-1}(x) = \dfrac{x^2 + 14}{3}$, $x \geq 0$

Section 5.6   (page 456)

Vocabulary Check   (page 456)

1. $\sin^{-1} x$, $-1 \leq x \leq 1$

2. $y = \arccos x$, $0 \leq y \leq \pi$

3. $y = \tan^{-1} x$, $-\infty < x < \infty$, $-\dfrac{\pi}{2} < y < \dfrac{\pi}{2}$

---

1. (a) $\dfrac{\pi}{6}$ (b) 0   **3.** (a) $\dfrac{\pi}{6}$ (b) $-\dfrac{\pi}{4}$

5. (a) $-\dfrac{\pi}{3}$ (b) $\dfrac{\pi}{3}$ **7.** (a) $\dfrac{\pi}{3}$ (b) $-\dfrac{\pi}{6}$

9. (a)

x	-1	-0.8	-0.6	-0.4	-0.2
y	3.1416	2.4981	2.2143	1.9823	1.7722

x	0	0.2	0.4	0.6	0.8	1
y	1.5708	1.3694	1.1593	0.9273	0.6435	0

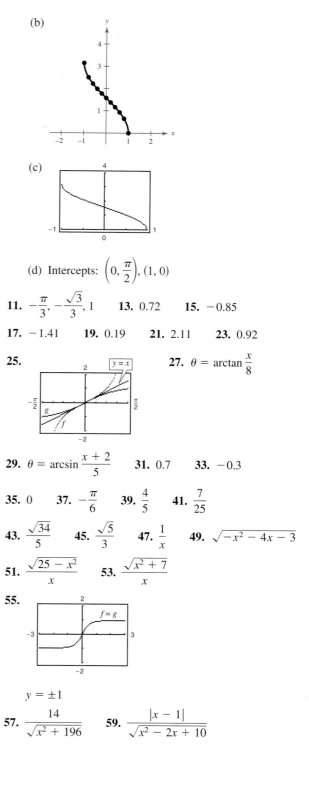

(b)

(c)

(d) Intercepts: $\left(0, \dfrac{\pi}{2}\right)$, $(1, 0)$

11. $-\dfrac{\pi}{3}$, $-\dfrac{\sqrt{3}}{3}$, 1   **13.** 0.72   **15.** -0.85

17. -1.41 **19.** 0.19   **21.** 2.11   **23.** 0.92

25.   **27.** $\theta = \arctan \dfrac{x}{8}$

29. $\theta = \arcsin \dfrac{x + 2}{5}$ **31.** 0.7   **33.** -0.3

35. 0   **37.** $-\dfrac{\pi}{6}$ **39.** $\dfrac{4}{5}$ **41.** $\dfrac{7}{25}$

43. $\dfrac{\sqrt{34}}{5}$ **45.** $\dfrac{\sqrt{5}}{3}$ **47.** $\dfrac{1}{x}$ **49.** $\sqrt{-x^2 - 4x - 3}$

51. $\dfrac{\sqrt{25 - x^2}}{x}$ **53.** $\dfrac{\sqrt{x^2 + 7}}{x}$

55.

$y = \pm 1$

57. $\dfrac{14}{\sqrt{x^2 + 196}}$ **59.** $\dfrac{|x - 1|}{\sqrt{x^2 - 2x + 10}}$

61.

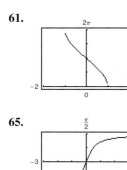

63.

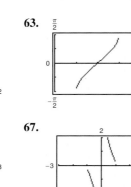

65.

67.

69. $3\sqrt{2}\sin\left(2t + \dfrac{\pi}{4}\right)$

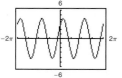

The two forms are equivalent.

71. (a) $\theta = \arcsin\dfrac{10}{s}$ (b) 0.19 rad, 0.39 rad

73. (a) $\theta = \arctan\dfrac{s}{750}$ (b) 0.49 rad, 1.13 rad

75. (a) $\theta = \arctan\dfrac{6}{x}$ (b) 0.54 rad, 1.11 rad

77. False. $5\pi/6$ is not in the range of the arcsine.

79.

81.

83–85. Answers will vary.   **87.** $\dfrac{\sqrt{2}}{2}$ **89.** $\dfrac{\sqrt{3}}{3}$

91. $\cos\theta = \dfrac{\sqrt{11}}{6}$

$\tan\theta = \dfrac{5\sqrt{11}}{11}$

$\csc\theta = \dfrac{6}{5}$

$\sec\theta = \dfrac{6\sqrt{11}}{11}$

$\cot\theta = \dfrac{\sqrt{11}}{5}$

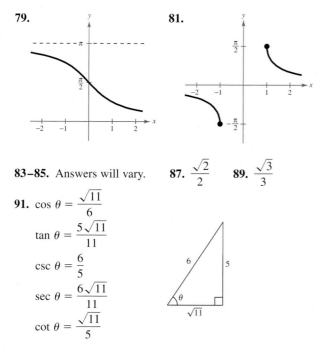

93. $\cos\theta = \dfrac{\sqrt{7}}{4}$

$\tan\theta = \dfrac{3\sqrt{7}}{7}$

$\csc\theta = \dfrac{4}{3}$

$\sec\theta = \dfrac{4\sqrt{7}}{7}$

$\cot\theta = \dfrac{\sqrt{7}}{3}$

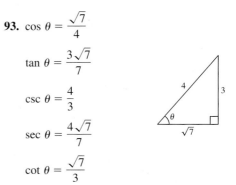

Section 5.7   (page 466)

1. $B = 70°$ **3.** $a \approx 8.26$ **5.** $c \approx 17.09$
$a \approx 3.64$ $c \approx 25.38$ $A \approx 20.6°$
$c \approx 10.64$ $A = 19°$ $A \approx 69.4°$

7. $a \approx 45.25$ **9.** $a \approx 91.34$
$A \approx 70.5°$ $b \approx 420.70$
$B \approx 19.5°$ $B = 77°45'$

11. 2.56 inches   **13.** 6.30 feet

15. (a)   (b) $L = 60\cot\theta$

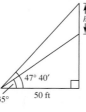

(c)

θ	10°	20°	30°	40°	50°
L	340	165	104	72	50

(d) No. Cotangent is not a linear function.

17. 19.7 feet

19. (a)

(b) $h = 50(\tan 47°40' - \tan 35°)$ (c) 19.9 feet

21. ≈ 2090 feet   **23.** $\approx 38.3°$ **25.** 75.97°

27. ≈ 5099feet   **29.** ≈ 0.66 mile

31. About 105 nautical miles south and 58 nautical miles west

33. (a) N 58° E   (b) ≈ 68.82 meters   **35.** N 56.31° W

37. ≈1933.3 feet   **39.** 17,054 feet ≈ 3.23 miles

41. 78.69°   **43.** ≈ 35.3°   **45.** $y = \sqrt{3}\,r$

47. 29.4 inches   **49.** $a ≈ 12.2, b ≈ 7$

51. $y = 8 \sin \pi t$ **53.** $d = 3 \cos\left(\dfrac{4\pi}{3}t\right)$

55. (a) 4   (b) 4   (c) 4   (d) $\frac{1}{16}$

57. (a) $\frac{1}{16}$ (b) 70   (c) 0   (d) $\frac{1}{140}$ **59.** $\omega = 528\pi$

61. (a)

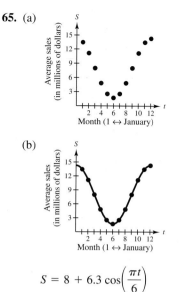

(b) $\dfrac{\pi}{8}$ second

(c) $\dfrac{\pi}{32}$ second

63. (a)

Base 1	Base 2	Altitude	Area
8	8 + 16 cos 10°	8 sin 10°	22.1
8	8 + 16 cos 20°	8 sin 20°	42.5
8	8 + 16 cos 30°	8 sin 30°	59.7
8	8 + 16 cos 40°	8 sin 40°	72.7
8	8 + 16 cos 50°	8 sin 50°	80.5
8	8 + 16 cos 60°	8 sin 60°	83.1
8	8 + 16 cos 70°	8 sin 70°	80.7

(b)

Base 1	Base 2	Altitude	Area
8	8 + 16 cos 56°	8 sin 56°	82.73
8	8 + 16 cos 58°	8 sin 58°	83.04
8	8 + 16 cos 59°	8 sin 59°	83.11
8	8 + 16 cos 60°	8 sin 60°	83.14
8	8 + 16 cos 61°	8 sin 61°	83.11
8	8 + 16 cos 62°	8 sin 62°	83.04

83.14 square feet

(c) $A = 64(1 + \cos \theta)(\sin \theta)$

(d)

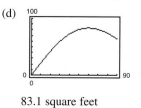

83.1 square feet

65. (a)

Average sales (in millions of dollars) vs Month (1 ↔ January)

(b)

Average sales (in millions of dollars) vs Month (1 ↔ January)

$$S = 8 + 6.3 \cos\left(\dfrac{\pi t}{6}\right)$$

The model is a good fit.

(c) 12 months. Yes; sales of outerwear is seasonal.

(d) Maximum displacement of 6.3 million dollars from the average sales of 8 million dollars

67. False. $a = \dfrac{22.56}{\tan(48.1°)}$ **69.** $4x - y + 6 = 0$

71. $4x + 5y - 22 = 0$ **73.** All real numbers

75. All real numbers

77. All real numbers x except $x = 0, x = 2$

79. All real numbers   **81.** $x = 1.994$

83. $x = \pm 4.851$

Review Exercises   (page 473)

1. $-320°$

3. (a)

(b) Quadrant I

(c) $400°, -320°$

5. (a)

(b) Quadrant III

(c) $250°, -470°$

7. Complement: 85°; supplement: 175°

9. Complement: none; supplement: 9°   **11.** 135.279°

13. 5.381°   **15.** 135° 17′ 24″   **17.** −85° 21′ 36″

19. (a)

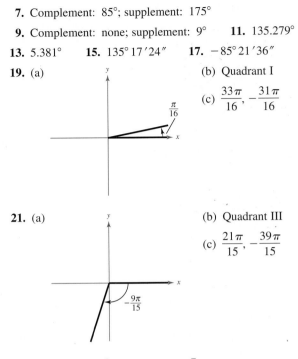

(b) Quadrant I

(c) $\dfrac{33\pi}{16}, -\dfrac{31\pi}{16}$

21. (a)

(b) Quadrant III

(c) $\dfrac{21\pi}{15}, -\dfrac{39\pi}{15}$

23. Complement: $\dfrac{3\pi}{8}$; supplement: $\dfrac{7\pi}{8}$

25. Complement: $\dfrac{\pi}{5}$; supplement: $\dfrac{7\pi}{10}$

27. 8.3776 rad   **29.** −0.5760 rad   **31.** 128.57°

33. −200.54°   **35.** 2.083 rad   **37.** 48.171 meters

39. 100π centimeters per second

41. $\sin\theta = \dfrac{5\sqrt{61}}{61}$

$\cos\theta = \dfrac{6\sqrt{61}}{61}$

$\tan\theta = \dfrac{5}{6}$

$\csc\theta = \dfrac{\sqrt{61}}{5}$

$\sec\theta = \dfrac{\sqrt{61}}{6}$

$\cot\theta = \dfrac{6}{5}$

43. $\sin\theta = \dfrac{\sqrt{65}}{9}$

$\cos\theta = \dfrac{4}{9}$

$\tan\theta = \dfrac{\sqrt{65}}{4}$

$\csc\theta = \dfrac{9\sqrt{65}}{65}$

$\sec\theta = \dfrac{9}{4}$

$\cot\theta = \dfrac{4\sqrt{65}}{65}$

45. Answers will vary.   **47.** (a) 0.1045   (b) 0.1045

49. (a) 0.7071   (b) 1.4142   **51.** ≈ 235 feet

53. $\sin\theta = \dfrac{4}{5}$

$\cos\theta = \dfrac{3}{5}$

$\tan\theta = \dfrac{4}{3}$

$\csc\theta = \dfrac{5}{4}$

$\sec\theta = \dfrac{5}{3}$

$\cot\theta = \dfrac{3}{4}$

55. $\sin\theta = \dfrac{2\sqrt{53}}{53}$

$\cos\theta = -\dfrac{7\sqrt{53}}{53}$

$\tan\theta = -\dfrac{2}{7}$

$\csc\theta = \dfrac{\sqrt{53}}{2}$

$\sec\theta = -\dfrac{\sqrt{53}}{7}$

$\cot\theta = -\dfrac{7}{2}$

57. $\sin\theta = \dfrac{15\sqrt{241}}{241}$

$\cos\theta = \dfrac{4\sqrt{241}}{241}$

$\tan\theta = \dfrac{15}{4}$

$\csc\theta = \dfrac{\sqrt{241}}{15}$

$\sec\theta = \dfrac{\sqrt{241}}{4}$

$\cot\theta = \dfrac{4}{15}$

59. $\sin\theta = -\dfrac{\sqrt{11}}{6}$

$\cos\theta = \dfrac{5}{6}$

$\tan\theta = -\dfrac{\sqrt{11}}{5}$

$\csc\theta = -\dfrac{6\sqrt{11}}{11}$

$\cot\theta = -\dfrac{5\sqrt{11}}{11}$

61. $\cos\theta = -\dfrac{\sqrt{55}}{8}$

$\tan\theta = -\dfrac{3\sqrt{55}}{55}$

$\csc\theta = \dfrac{8}{3}$

$\sec\theta = -\dfrac{8\sqrt{55}}{55}$

$\cot\theta = -\dfrac{\sqrt{55}}{3}$

63. $\theta' = 84°$

$\theta = 264°$

$\theta' = 84°$

65. $\theta' = \dfrac{\pi}{5}$

$\theta' = \dfrac{\pi}{5}$

$\theta = -\dfrac{6\pi}{5}$

67. $\sin 240° = -\dfrac{\sqrt{3}}{2}$

$\cos 240° = -\dfrac{1}{2}$

$\tan 240° = \sqrt{3}$

69. $\sin(-210°) = \dfrac{1}{2}$

$\cos(-210°) = -\dfrac{\sqrt{3}}{2}$

$\tan(-210°) = -\dfrac{\sqrt{3}}{3}$

71. $\sin\left(-\dfrac{9\pi}{4}\right) = -\dfrac{\sqrt{2}}{2}$

$\cos\left(-\dfrac{9\pi}{4}\right) = \dfrac{\sqrt{2}}{2}$

$\tan\left(-\dfrac{9\pi}{4}\right) = -1$

73. $\sin\dfrac{\pi}{2} = 1$

$\cos\dfrac{\pi}{2} = 0$

$\tan\dfrac{\pi}{2}$ is undefined.

75. $\left(-\dfrac{1}{2}, \dfrac{\sqrt{3}}{2}\right)$

$\sin\dfrac{2\pi}{3} = \dfrac{\sqrt{3}}{2}$

$\cos\dfrac{2\pi}{3} = -\dfrac{1}{2}$

$\tan\dfrac{2\pi}{3} = -\sqrt{3}$

77. $\left(-\dfrac{\sqrt{3}}{2}, -\dfrac{1}{2}\right)$

$\sin\dfrac{7\pi}{6} = -\dfrac{1}{2}$

$\cos\dfrac{7\pi}{6} = -\dfrac{\sqrt{3}}{2}$

$\tan\dfrac{7\pi}{6} = \dfrac{\sqrt{3}}{3}$

79.

81.

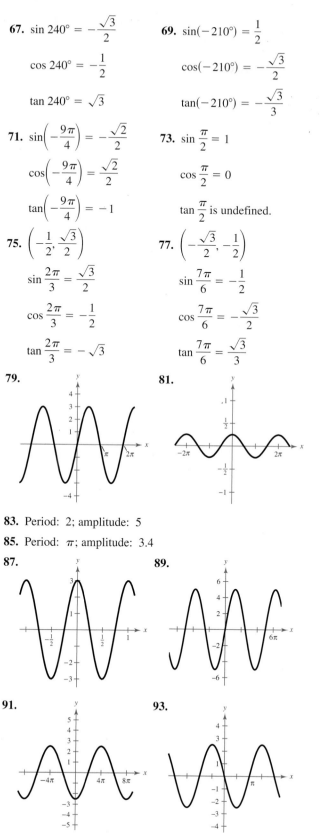

83. Period: 2; amplitude: 5

85. Period: π; amplitude: 3.4

87.

89.

91.

93.

95.

97.

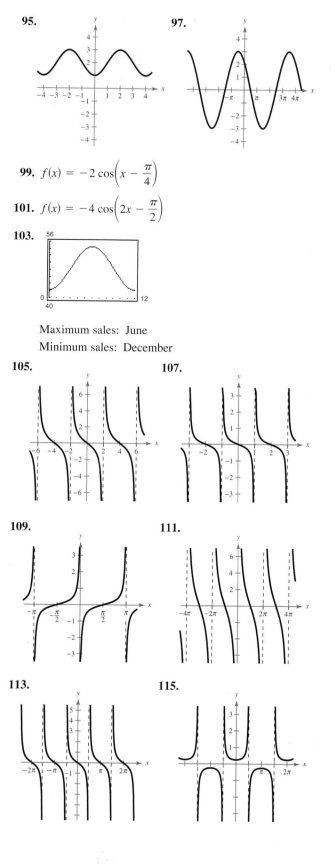

99. $f(x) = -2\cos\left(x - \dfrac{\pi}{4}\right)$

101. $f(x) = -4\cos\left(2x - \dfrac{\pi}{2}\right)$

103.

Maximum sales: June

Minimum sales: December

105.

107.

109.

111.

113.

115.

117. **119.**

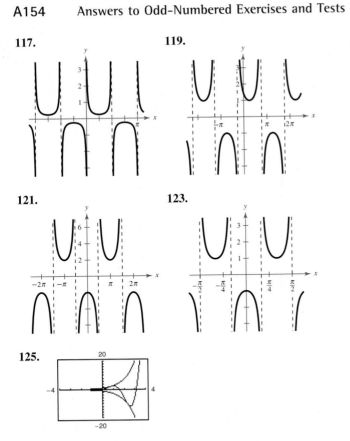

121. **123.**

125.

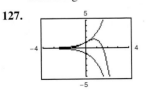

As $x \to \infty$, f cycles between $-\infty$ and ∞, but always increasing in absolute value.

127.

As $x \to \infty$, f cycles between $-\infty$ and ∞, but always increasing in absolute value.

129. (a) $\dfrac{\pi}{2}$ (b) Does not exist   **131.** (a) $\dfrac{\pi}{4}$ (b) $\dfrac{5\pi}{6}$

133. 1.14   **135.** -1.22 **137.** -1.49

139. 0.68   **141.** $\theta = \arcsin\left(\dfrac{x+3}{16}\right)$

143. $\dfrac{\sqrt{-x^2 + 2x}}{-x^2 + 2x}$ **145.** $\dfrac{2\sqrt{4 - 2x^2}}{4 - x^2}$ **147.** $\theta \approx 57.3°$

149. 6.8°, 2516 feet   **151.** 49,987 feet $\approx$ 9.47 miles

153. $y = 3 \cos\left(\dfrac{2\pi t}{15}\right)$

155. False. y is a function but is not one-to-one on $30° \le \theta \le 150°$.

157. As θ increases from 0° to 90°, x will decrease from 12 to 0 centimeters and y will increase from 0 to 12 centimeters. Therefore,

$\sin \theta = \dfrac{y}{12}$ will increase from 0 to 1 and

$\cos \theta = \dfrac{x}{12}$ will decrease from 1 to 0. So,

$\tan \theta = \dfrac{y}{x}$ will increase without bound.

When $\theta = 90°$, the tangent will be undefined.

159. (a)

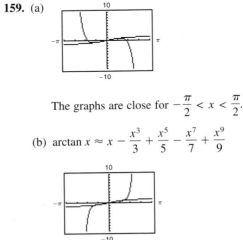

The graphs are close for $-\dfrac{\pi}{2} < x < \dfrac{\pi}{2}$.

(b) $\arctan x \approx x - \dfrac{x^3}{3} + \dfrac{x^5}{5} - \dfrac{x^7}{7} + \dfrac{x^9}{9}$

The accuracy of the approximation increases as additional terms are added.

Chapter Test   (page 478)

1. (a)

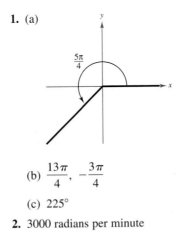

(b) $\dfrac{13\pi}{4}$, $-\dfrac{3\pi}{4}$

(c) 225°

2. 3000 radians per minute

3. $\sin \theta = \dfrac{4\sqrt{17}}{17}$

$\cos \theta = -\dfrac{\sqrt{17}}{17}$

$\tan \theta = -4$

$\csc \theta = \dfrac{\sqrt{17}}{4}$

$\sec \theta = -\sqrt{17}$

$\cot \theta = -\dfrac{1}{4}$

4. $\sin \theta = \pm\dfrac{6\sqrt{61}}{61}$

$\cos \theta = \pm\dfrac{5\sqrt{61}}{61}$

$\csc \theta = \pm\dfrac{\sqrt{61}}{6}$

$\sec \theta = \pm\dfrac{\sqrt{61}}{5}$

$\cot \theta = \dfrac{5}{6}$

5. $\theta' = 75°$

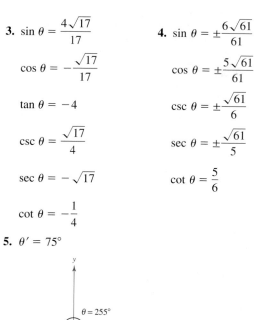

6. Quadrant III   **7.** $150°, 210°$ **8.** $1.33, 1.81$

9. $\sin \theta = \dfrac{4}{5}$

$\tan \theta = -\dfrac{4}{3}$

$\csc \theta = \dfrac{5}{4}$

$\sec \theta = -\dfrac{5}{3}$

$\cot \theta = -\dfrac{3}{4}$

10.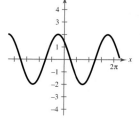

11.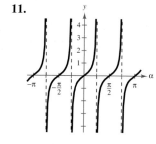

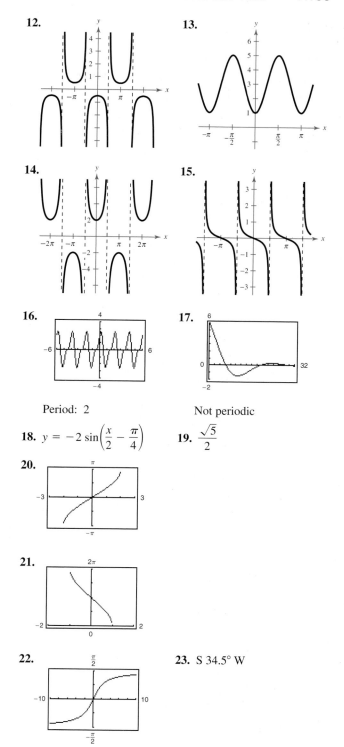

12.

13.

14.

15.

16.

Period: 2

17.

Not periodic

18. $y = -2 \sin\left(\dfrac{x}{2} - \dfrac{\pi}{4}\right)$ **19.** $\dfrac{\sqrt{5}}{2}$

20.

21.

22.   **23.** S $34.5°$ W

Chapter 6

Section 6.1   (page 485)

Vocabulary Check   (page 485)

1. $\sec u$ **2.** $\tan u$ **3.** $\cot u$ **4.** $\csc u$

5. $\cot^2 u$ **6.** $\sec^2 u$ **7.** $\sin u$ **8.** $\csc u$

9. $-\sin u$ **10.** $\cos u$

1. $\tan x = \sqrt{3}$

$\csc x = \dfrac{2\sqrt{3}}{3}$

$\sec x = 2$

$\cot x = \dfrac{\sqrt{3}}{3}$

3. $\cos \theta = \dfrac{\sqrt{2}}{2}$

$\tan \theta = -1$

$\csc \theta = -\sqrt{2}$

$\cot \theta = -1$

5. $\sin x = -\dfrac{7}{25}$

$\cos x = -\dfrac{24}{25}$

$\csc x = -\dfrac{25}{7}$

$\cot x = \dfrac{24}{7}$

7. $\cos \phi = -\dfrac{12}{13}$

$\tan \phi = -\dfrac{5}{12}$

$\csc \phi = \dfrac{13}{5}$

$\cot \phi = -\dfrac{12}{5}$

9. $\sin x = \dfrac{2}{3}$

$\cos x = -\dfrac{\sqrt{5}}{3}$

$\csc x = \dfrac{3}{2}$

$\sec x = -\dfrac{3\sqrt{5}}{5}$

$\cot x = -\dfrac{\sqrt{5}}{2}$

11. $\sin \theta = -\dfrac{2\sqrt{5}}{5}$

$\cos \theta = -\dfrac{\sqrt{5}}{5}$

$\csc \theta = -\dfrac{\sqrt{5}}{2}$

$\sec \theta = -\sqrt{5}$

$\cot \theta = \dfrac{1}{2}$

13. $\sin \theta = 0$

$\cos \theta = -1$

$\tan \theta = 0$

$\sec \theta = -1$

$\cot \theta$ is undefined

15. d   **16.** a   **17.** b   **18.** f   **19.** e

20. c   **21.** b   **22.** c   **23.** f   **24.** a

25. e   **26.** d   **27.** $\cos x$ **29.** $\cos^2 \phi$

31. $\cos x$ **33.** 1   **35.** $\cot x$ **37.** $1 + \sin y$

39–49. Answers will vary.   **51.** $\cos^2 x$ **53.** $\cos x + 2$

55. $\sec^4 x$ **57.** $\sin^2 x - \cos^2 x$ **59.** $(\csc x - 1)\cot^2 x$

61. $1 + 2\sin x \cos x$ **63.** $\tan^2 x$ **65.** $2 \csc^2 x$

67. $-\cot x$ **69.** $1 + \cos y$ **71.** $3(\sec x + \tan x)$

73.

x	0.2	0.4	0.6	0.8
y_1	0.1987	0.3894	0.5646	0.7174
y_2	0.1987	0.3894	0.5646	0.7174

x	1.0	1.2	1.4
y_1	0.8415	0.9320	0.9854
y_2	0.8415	0.9320	0.9854

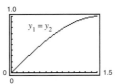

75.

x	0.2	0.4	0.6	0.8
y_1	1.2230	1.5085	1.8958	2.4650
y_2	1.2230	1.5085	1.8958	2.4650

x	1.0	1.2	1.4
y_1	3.4082	5.3319	11.6814
y_2	3.4082	5.3319	11.6814

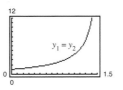

77. $\csc x$ **79.** $\tan x$ **81.** $5 \cos \theta$ **83.** $3 \tan \theta$

85. $0 \le \theta \le \pi$ **87.** $0 \le \theta < \dfrac{\pi}{2}, \dfrac{3\pi}{2} < \theta < 2\pi$

89. $\ln|\cot \theta|$ **91.** $\ln|(\cos x)(1 + \sin x)|$

93. (a) $\csc^2(132°) - \cot^2(132°) \approx 1.8107 - 0.8107 = 1$

(b) $\csc^2\left(\dfrac{2\pi}{7}\right) - \cot^2\left(\dfrac{2\pi}{7}\right) \approx 1.63596 - 0.63596 = 1$

95. (a) $\cos(90° - 80°) = \cos(10°) \approx 0.9848;$

$\sin(80°) \approx 0.9848$

(b) $\cos\left(\dfrac{\pi}{2} - 0.8\right) \approx 0.7174;\ \sin(0.8) \approx 0.7174$

97. Answers will vary.   **99.** True, for all $\theta \ne n\pi$.

101. 1, 1   **103.** $\infty, 0$

105. $\cos \theta = \pm \sqrt{1 - \sin^2 \theta}$

$\tan \theta = \pm \dfrac{\sin \theta}{\sqrt{1 - \sin^2 \theta}}$

$\csc \theta = \dfrac{1}{\sin \theta}$

$\sec \theta = \pm \dfrac{1}{\sqrt{1 - \sin^2 \theta}}$

$\cot \theta = \pm \dfrac{\sqrt{1 - \sin^2 \theta}}{\sin \theta}$

The sign depends on the choice of θ.

107. Answers will vary. Sample answer:

$\sin \theta = \dfrac{o}{h}$

$\cos \theta = \dfrac{a}{h}$

$o^2 + a^2 = h^2$

$\dfrac{o^2}{h^2} + \dfrac{a^2}{h^2} = 1$

$\left(\dfrac{o}{h}\right)^2 + \left(\dfrac{a}{h}\right)^2 = 1$

$\sin^2 \theta + \cos^2 \theta = 1$

109. 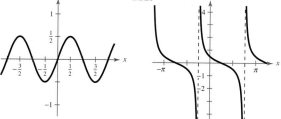   **111.**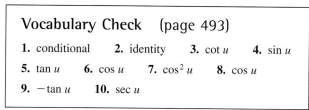

Section 6.2   (page 493)

Vocabulary Check   (page 493)

1. conditional   **2.** identity   **3.** $\cot u$ **4.** $\sin u$

5. $\tan u$ **6.** $\cos u$ **7.** $\cos^2 u$ **8.** $\cos u$

9. $-\tan u$ **10.** $\sec u$

1–9. Answers will vary.

11.

x	0.2	0.4	0.6	0.8
y_1	4.8348	2.1785	1.2064	0.6767
y_2	4.8348	2.1785	1.2064	0.6767

x	1.0	1.2	1.4
y_1	0.3469	0.1409	0.0293
y_2	0.3469	0.1409	0.0293

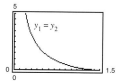

13.

x	0.2	0.4	0.6	0.8
y_1	4.8348	2.1785	1.2064	0.6767
y_2	4.8348	2.1785	1.2064	0.6767

x	1.0	1.2	1.4
y_1	0.3469	0.1409	0.0293
y_2	0.3469	0.1409	0.0293

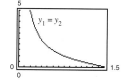

15.

x	0.2	0.4	0.6	0.8
y_1	5.0335	2.5679	1.7710	1.3940
y_2	5.0335	2.5679	1.7710	1.3940

x	1.0	1.2	1.4
y_1	1.1884	1.0729	1.0148
y_2	1.1884	1.0729	1.0148

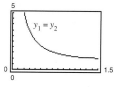

17.

x	0.2	0.4	0.6	0.8
y_1	5.1359	2.7880	2.1458	2.0009
y_2	5.1359	2.7880	2.1458	2.0009

x	1.0	1.2	1.4
y_1	2.1995	2.9609	5.9704
y_2	2.1995	2.9609	5.9704

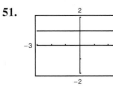

19. $\cot(-x) = -\cot(x)$, so

$(1 + \tan x)[1 + \cot(-x)] = \tan x - \cot x.$

21–49. Answers will vary.

51.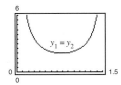

$y = 1$

53.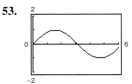

$y = \sin x$

55 and 57. Answers will vary.   **59.** 1   **61.** 2

63 and 65. Answers will vary.   **67.** $\mu = \tan \theta,\ W \neq 0$

69. True   **71.** False. $\sin^2\left(\dfrac{\pi}{4}\right) + \cos^2\left(\dfrac{\pi}{4}\right) \neq 1 + \tan^2\left(\dfrac{\pi}{4}\right)$

73. $\sqrt{\tan^2 x} = |\tan x|;\ \dfrac{3\pi}{4}$ **75.** Answers will vary.

77. Answers will vary.

Sample answer: $y = x^3 - x^2 + 64x - 64$

79. Answers will vary.

Sample answer: $y = x^3 - 16x^2 + 85x - 148$

81. 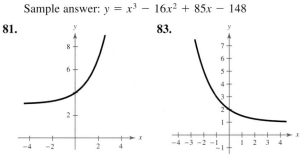   **83.**

85. Quadrant II   **87.** Quadrant IV

89. $A = 10°$
$b \approx 90.74$
$c \approx 92.14$

91. $A = 60.26°$
$B \approx 29.74°$
$c \approx 16.12$

Section 6.3   (page 504)

Vocabulary Check   (page 504)

1. general   **2.** quadratic   **3.** extraneous

1–5. Answers will vary.   **7.** $\dfrac{2\pi}{3} + 2n\pi, \dfrac{4\pi}{3} + 2n\pi$

9. $\dfrac{\pi}{6} + 2n\pi, \dfrac{11\pi}{6} + 2n\pi$

11. $\dfrac{\pi}{3} + n\pi, \dfrac{2\pi}{3} + n\pi$ **13.** $\dfrac{\pi}{3} + n\pi, \dfrac{2\pi}{3} + n\pi$

15. $\dfrac{\pi}{3} + n\pi, \dfrac{2\pi}{3} + n\pi$ **17.** $\dfrac{2\pi}{3}, \dfrac{5\pi}{3}$

19. $\dfrac{\pi}{4}, \dfrac{3\pi}{4}, \dfrac{5\pi}{4}, \dfrac{7\pi}{4}$ **21.** $0, \dfrac{\pi}{6}, \dfrac{5\pi}{6}, \pi, \dfrac{7\pi}{6}, \dfrac{11\pi}{6}$

23. $\dfrac{\pi}{3}, \pi, \dfrac{5\pi}{3}$ **25.** No solution   **27.** $\dfrac{\pi}{3}, \dfrac{5\pi}{3}$

29. $2.0344, 5.1760, \dfrac{\pi}{4}, \dfrac{5\pi}{4}$ **31.** $3.6652, 4.7124, 5.7596$

33. $0.8614, 5.4218$ **35.** 1.5708 **37.** $0.5236, 2.6180$

39. $\dfrac{2\pi}{3} + n\pi, \dfrac{5\pi}{6} + n\pi$ **41.** $\dfrac{\pi}{8} + \dfrac{n\pi}{4}$

43. $\dfrac{n\pi}{3}, \dfrac{\pi}{4} + n\pi$ **45.** $\dfrac{\pi}{2} + 4n\pi, \dfrac{7\pi}{2} + 4n\pi$ **47.** $-1, 3$

49. ± 2 **51.** $1.1071, 4.2487$ **53.** $0.8603, 3.4256$

55. $0, 2.6779, 3.1416, 5.8195$

57. $0.3398, 0.8481, 2.2935, 2.8018$ **59.** $-1.154, 0.534$

61. 1.110

63. (a)

Maximum: $(0.785, 1.41)$
Minimum: $(3.93, -1.41)$

(b) $\dfrac{\pi}{4}, \dfrac{5\pi}{4}$

65. 1

67. (a) All real numbers x except $x = 0$

(b) y-axis symmetry; horizontal asymptote: $y = 1$

(c) Oscillates (d) Infinite number of solutions

(e) Yes. 0.6366

69. 0.04 second, 0.43 second, 0.83 second

71. May, June, July **73.** $37°, 53°$

75. (a)

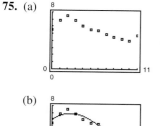

(b)

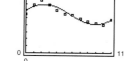

The model fits the data.

(c) The constant term; 5.51%

(d) Approximately 13 years

(e) 2007

77. False. Reasons will vary.

79. False. The domain of the sine function does not include 3.4.

81. 2.164 rad **83.** -0.007 rad

85. 24.249 **87.** 2290.4 feet

Section 6.4 (page 512)

Vocabulary Check (page 512)

1. $\sin u \cos v - \cos u \sin v$

2. $\cos u \cos v - \sin u \sin v$ **3.** $\dfrac{\tan u + \tan v}{1 - \tan u \tan v}$

4. $\sin u \cos v + \cos u \sin v$

5. $\cos u \cos v + \sin u \sin v$ **6.** $\dfrac{\tan u - \tan v}{1 + \tan u \tan v}$

---

1. (a) $-\dfrac{1}{2}$ (b) $-\dfrac{3}{2}$ **3.** (a) 0 (b) $\dfrac{1 + \sqrt{3}}{2}$

5. (a) $\dfrac{-\sqrt{2} - \sqrt{6}}{4}$ (b) $\dfrac{-\sqrt{2} - \sqrt{3}}{2}$

7. $\sin 105° = \dfrac{\sqrt{6} + \sqrt{2}}{4}$

$\cos 105° = \dfrac{\sqrt{2} - \sqrt{6}}{4}$

$\tan 105° = -2 - \sqrt{3}$

9. $\sin 195° = \dfrac{\sqrt{2} - \sqrt{6}}{4}$

$\cos 195° = \dfrac{-\sqrt{2} - \sqrt{6}}{4}$

$\tan 195° = 2 - \sqrt{3}$

11. $\sin \dfrac{11\pi}{12} = \dfrac{\sqrt{6} - \sqrt{2}}{4}$

$\cos \dfrac{11\pi}{12} = \dfrac{-\sqrt{6} - \sqrt{2}}{4}$

$\tan \dfrac{11\pi}{12} = -2 + \sqrt{3}$

13. $\sin\left(-\dfrac{\pi}{12}\right) = \dfrac{\sqrt{2} - \sqrt{6}}{4}$

$\cos\left(-\dfrac{\pi}{12}\right) = \dfrac{\sqrt{6} + \sqrt{2}}{4}$

$\tan\left(-\dfrac{\pi}{12}\right) = -2 + \sqrt{3}$

15. $\sin 75° = \dfrac{\sqrt{2} + \sqrt{6}}{4}$

$\cos 75° = \dfrac{\sqrt{6} - \sqrt{2}}{4}$

$\tan 75° = \sqrt{3} + 2$

17. $\sin(-225°) = \dfrac{\sqrt{2}}{2}$

$\cos(-225°) = -\dfrac{\sqrt{2}}{2}$

$\tan(-225°) = -1$

19. $\sin \dfrac{13\pi}{12} = \dfrac{-\sqrt{6} + \sqrt{2}}{4}$

$\cos \dfrac{13\pi}{12} = \dfrac{-\sqrt{6} - \sqrt{2}}{4}$

$\tan \dfrac{13\pi}{12} = 2 - \sqrt{3}$

21. $\sin\left(-\dfrac{7\pi}{12}\right) = -\dfrac{\sqrt{6} + \sqrt{2}}{4}$

$\cos\left(-\dfrac{7\pi}{12}\right) = \dfrac{\sqrt{2} - \sqrt{6}}{4}$

$\tan\left(-\dfrac{7\pi}{12}\right) = 2 + \sqrt{3}$

23. $\cos 70°$ **25.** $\tan 239°$

27. $\sin 2.3$ **29.** $\cos \dfrac{12\pi}{35}$

31.

x	0.2	0.4	0.6	0.8
y_1	0.6621	0.7978	0.9017	0.9696
y_2	0.6621	0.7978	0.9017	0.9696

x	1.0	1.2	1.4
y_1	0.9989	0.9883	0.9384
y_2	0.9989	0.9883	0.9384

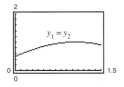

33.

x	0.2	0.4	0.6	0.8
y_1	0.9605	0.8484	0.6812	0.4854
y_2	0.9605	0.8484	0.6812	0.4854

x	1.0	1.2	1.4
y_1	0.2919	0.1313	0.0289
y_2	0.2919	0.1313	0.0289

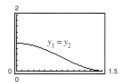

35. $-\frac{63}{65}$ **37.** $-\frac{63}{16}$ **39.** $\frac{3}{5}$ **41.** $\frac{44}{125}$ **43.** 1

45. $\dfrac{2x^2 - \sqrt{1 - x^2}}{\sqrt{4x^2 + 1}}$ **47–53.** Answers will vary.

55. $\dfrac{\pi}{2}$ **57.** $0, \dfrac{\pi}{3}, \pi, \dfrac{5\pi}{3}$ **59.** 0.7854, 5.4978

61. 0, 3.1416   **63.** Answers will vary.

65. False. $\cos(u \pm v) = \cos u \cos v \mp \sin u \sin v$

67 and 69. Answers will vary.

71. (a) $\sqrt{2} \sin\left(\theta + \dfrac{\pi}{4}\right)$ (b) $\sqrt{2} \cos\left(\theta - \dfrac{\pi}{4}\right)$

73. (a) $13 \sin(3\theta + 0.3948)$ (b) $13 \cos(3\theta - 1.1760)$

75. $2 \cos \theta$ **77.** Answers will vary.

79. $u + v = w$. Answers will vary.   **81.** $(0, 19), (38, 0)$

83. $(0, 4), (2, 0), (7, 0)$ **85.** $\dfrac{\pi}{6}$ **87.** $\dfrac{\pi}{2}$

Section 6.5   (page 522)

1. $\frac{3}{5}$ **3.** $\frac{7}{25}$ **5.** $\frac{24}{7}$ **7.** $\frac{25}{24}$

9. $0, 1.0472, 3.1416, 5.2360; 0, \dfrac{\pi}{3}, \pi, \dfrac{5\pi}{3}$

11. $0.2618, 1.3090, 3.4034, 4.4506; \dfrac{\pi}{12}, \dfrac{5\pi}{12}, \dfrac{13\pi}{12}, \dfrac{17\pi}{12}$

13. $0, 2.0944, 4.1888; 0, \dfrac{2\pi}{3}, \dfrac{4\pi}{3}$

15. $0, 1.5708, 3.1416, 4.7124; 0, \dfrac{\pi}{2}, \pi, \dfrac{3\pi}{2}$

17. $\sin 2u = \frac{24}{25}$ **19.** $\sin 2u = \frac{4}{5}$
$\cos 2u = \frac{7}{25}$ $\cos 2u = \frac{3}{5}$
$\tan 2u = \frac{24}{7}$ $\tan 2u = \frac{4}{3}$

21. $\sin 2u = -\dfrac{4\sqrt{21}}{25}$

$\cos 2u = -\dfrac{17}{25}$

$\tan 2u = \dfrac{4\sqrt{21}}{17}$

23. $4 \sin 2x$ **25.** $6 \cos 2x$

27. $\frac{1}{8}(3 + 4 \cos 2x + \cos 4x)$ **29.** $\frac{1}{8}(1 - \cos 4x)$

31. $\frac{1}{32}(2 + \cos 2x - 2 \cos 4x - \cos 6x)$

33. $\dfrac{4\sqrt{17}}{17}$ **35.** $\dfrac{1}{4}$ **37.** $\sqrt{17}$ **39.** $\dfrac{8}{17}$

41. $\sin 15° = \dfrac{\sqrt{2 - \sqrt{3}}}{2}$ **43.** $\sin 112° 30' = \dfrac{\sqrt{2 + \sqrt{2}}}{2}$

$\cos 15° = \dfrac{\sqrt{2 + \sqrt{3}}}{2}$ $\cos 112° 30' = -\dfrac{\sqrt{2 - \sqrt{2}}}{2}$

$\tan 15° = 2 - \sqrt{3}$ $\tan 112° 30' = -1 - \sqrt{2}$

45. $\sin\dfrac{\pi}{8} = \dfrac{\sqrt{2-\sqrt{2}}}{2}$

$\cos\dfrac{\pi}{8} = \dfrac{\sqrt{2+\sqrt{2}}}{2}$

$\tan\dfrac{\pi}{8} = \sqrt{2}-1$

47. $\sin\dfrac{3\pi}{8} = \dfrac{\sqrt{2+\sqrt{2}}}{2}$

$\cos\dfrac{3\pi}{8} = \dfrac{\sqrt{2-\sqrt{2}}}{2}$

$\tan\dfrac{3\pi}{8} = \sqrt{2}+1$

49. $\sin\dfrac{u}{2} = \dfrac{5\sqrt{26}}{26}$

$\cos\dfrac{u}{2} = \dfrac{\sqrt{26}}{26}$

$\tan\dfrac{u}{2} = 5$

51. $\sin\dfrac{u}{2} = \sqrt{\dfrac{89-5\sqrt{89}}{178}}$

$\cos\dfrac{u}{2} = -\sqrt{\dfrac{89+5\sqrt{89}}{178}}$

$\tan\dfrac{u}{2} = \dfrac{5-\sqrt{89}}{8}$

53. $\sin\dfrac{u}{2} = \dfrac{3\sqrt{10}}{10}$

$\cos\dfrac{u}{2} = -\dfrac{\sqrt{10}}{10}$

$\tan\dfrac{u}{2} = -3$

55. $|\sin 3x|$ **57.** $-|\tan 4x|$ **59.** $\dfrac{\pi}{3}, \dfrac{5\pi}{3}$

61. $\dfrac{\pi}{3}, \pi, \dfrac{5\pi}{3}$ **63.** $3\sin\dfrac{2\pi}{3}$ **65.** $\dfrac{1}{2}(\sin 8\theta + \sin 2\theta)$

67. $\dfrac{5}{2}(\cos 8\beta + \cos 2\beta)$ **69.** $2\cos 3\theta \sin 2\theta$

71. $2\cos\alpha \sin\beta$ **73.** $-2\sin\theta\sin\dfrac{\pi}{2} = -2\sin\theta$

75. $\dfrac{\sqrt{2}}{2}$ **77.** $\dfrac{\sqrt{6}}{2}$ **79.** $0, \dfrac{\pi}{4}, \dfrac{\pi}{2}, \dfrac{3\pi}{4}, \pi, \dfrac{5\pi}{4}, \dfrac{3\pi}{2}, \dfrac{7\pi}{4}$

81. $\dfrac{\pi}{6}, \dfrac{5\pi}{6}$ **83.** $\dfrac{25}{169}$ **85.** $\dfrac{4}{13}$

87–97. Answers will vary.

99.
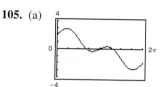

101.
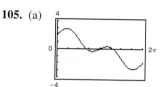

103. (a)

Maximum: (3.1416, 3)

(b) π

105. (a)

Minimum: $(5.5839, -2.8642)$

Maximum: $(0.6993, 2.8642)$

(b) $0.6993, 2.6078, 3.6754, 5.5839$

107. $2x\sqrt{1-x^2}$ **109.** $1-2x^2$ **111.** $\dfrac{1-x^2}{1+x^2}$

113. $r = \dfrac{1}{16}v_0^2\sin\theta\cos\theta$

115. (a) $\theta = \pi$ (c) 760 mph; 3420 mph

(b) $\theta = 0.4482$ (d) $\cos\theta = 1 - \dfrac{2}{m^2}$

117. False. $\sin\dfrac{x}{2} = -\sqrt{\dfrac{1-\cos x}{2}}$ for $\pi \le \dfrac{x}{2} \le 2\pi$.

119. (a)

(b) $y = \sin 2x$

(c) Answers will vary.

121. Answers will vary.

123. (a)

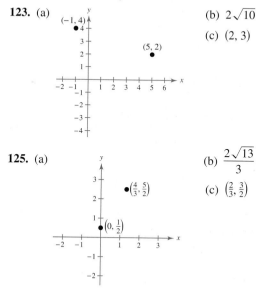

(b) $2\sqrt{10}$

(c) $(2, 3)$

125. (a)

(b) $\dfrac{2\sqrt{13}}{3}$

(c) $\left(\dfrac{2}{3}, \dfrac{3}{2}\right)$

127. (a) Complement: $35°$; supplement: $125°$

(b) Complement: none; supplement: $18°$

129. (a) Complement: $\dfrac{4\pi}{9}$; supplement: $\dfrac{17\pi}{18}$

(b) Complement: $\dfrac{\pi}{20}$; supplement: $\dfrac{11\pi}{20}$

131. 0.467 rad

133.   **135.**

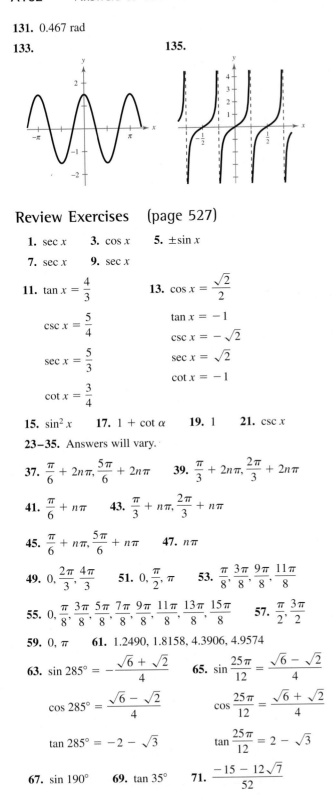

Review Exercises   (page 527)

1. $\sec x$ **3.** $\cos x$ **5.** $\pm\sin x$

7. $\sec x$ **9.** $\sec x$

11. $\tan x = \dfrac{4}{3}$ **13.** $\cos x = \dfrac{\sqrt{2}}{2}$

$\csc x = \dfrac{5}{4}$ $\tan x = -1$

                               $\csc x = -\sqrt{2}$

$\sec x = \dfrac{5}{3}$ $\sec x = \sqrt{2}$

                               $\cot x = -1$

$\cot x = \dfrac{3}{4}$

15. $\sin^2 x$ **17.** $1 + \cot\alpha$ **19.** 1   **21.** $\csc x$

23–35. Answers will vary.

37. $\dfrac{\pi}{6} + 2n\pi, \dfrac{5\pi}{6} + 2n\pi$ **39.** $\dfrac{\pi}{3} + 2n\pi, \dfrac{2\pi}{3} + 2n\pi$

41. $\dfrac{\pi}{6} + n\pi$ **43.** $\dfrac{\pi}{3} + n\pi, \dfrac{2\pi}{3} + n\pi$

45. $\dfrac{\pi}{6} + n\pi, \dfrac{5\pi}{6} + n\pi$ **47.** $n\pi$

49. $0, \dfrac{2\pi}{3}, \dfrac{4\pi}{3}$ **51.** $0, \dfrac{\pi}{2}, \pi$ **53.** $\dfrac{\pi}{8}, \dfrac{3\pi}{8}, \dfrac{9\pi}{8}, \dfrac{11\pi}{8}$

55. $0, \dfrac{\pi}{8}, \dfrac{3\pi}{8}, \dfrac{5\pi}{8}, \dfrac{7\pi}{8}, \dfrac{9\pi}{8}, \dfrac{11\pi}{8}, \dfrac{13\pi}{8}, \dfrac{15\pi}{8}$ **57.** $\dfrac{\pi}{2}, \dfrac{3\pi}{2}$

59. $0, \pi$ **61.** 1.2490, 1.8158, 4.3906, 4.9574

63. $\sin 285° = -\dfrac{\sqrt{6} + \sqrt{2}}{4}$ **65.** $\sin \dfrac{25\pi}{12} = \dfrac{\sqrt{6} - \sqrt{2}}{4}$

$\cos 285° = \dfrac{\sqrt{6} - \sqrt{2}}{4}$ $\cos \dfrac{25\pi}{12} = \dfrac{\sqrt{6} + \sqrt{2}}{4}$

$\tan 285° = -2 - \sqrt{3}$ $\tan \dfrac{25\pi}{12} = 2 - \sqrt{3}$

67. $\sin 190°$ **69.** $\tan 35°$ **71.** $\dfrac{-15 - 12\sqrt{7}}{52}$

73. $\dfrac{507\sqrt{7} - 960}{1121}$ **75.** $\dfrac{5\sqrt{7} - 36}{52}$

77–81. Answers will vary.   **83.** $\dfrac{\pi}{6}, \dfrac{11\pi}{6}$

85. $\sin 2u = \dfrac{20\sqrt{6}}{49}$ **87.** $\sin 2u = -\dfrac{36}{85}$

$\cos 2u = -\dfrac{1}{49}$ $\cos 2u = \dfrac{77}{85}$

$\tan 2u = -20\sqrt{6}$ $\tan 2u = -\dfrac{36}{77}$

89 and 91. Answers will vary.   **93.** $15°, 75°$

95. $\dfrac{1}{32}(10 - 15\cos 2x + 6\cos 4x - \cos 6x)$

97. $\dfrac{1}{8}(3 + 4\cos 4x + \cos 8x)$

99. $\sin 105° = \dfrac{\sqrt{2 + \sqrt{3}}}{2}$

$\cos 105° = -\dfrac{\sqrt{2 - \sqrt{3}}}{2}$

$\tan 105° = -2 - \sqrt{3}$

101. $\sin \dfrac{7\pi}{8} = \dfrac{\sqrt{2 - \sqrt{2}}}{2}$ **103.** $\sin \dfrac{u}{2} = \dfrac{\sqrt{10}}{10}$

$\cos \dfrac{7\pi}{8} = -\dfrac{\sqrt{2 + \sqrt{2}}}{2}$ $\cos \dfrac{u}{2} = \dfrac{3\sqrt{10}}{10}$

$\tan \dfrac{7\pi}{8} = 1 - \sqrt{2}$ $\tan \dfrac{u}{2} = \dfrac{1}{3}$

105. $\sin \dfrac{u}{2} = \dfrac{3\sqrt{14}}{14}$

$\cos \dfrac{u}{2} = \dfrac{\sqrt{70}}{14}$

$\tan \dfrac{u}{2} = \dfrac{3\sqrt{5}}{5}$

107. $-|\cos 5x|$ **109.** $V = \sin \dfrac{\theta}{2} \cos \dfrac{\theta}{2}$ cubic meters

111. $3 \sin \dfrac{\pi}{2} = 3$ **113.** $\dfrac{1}{2}(\cos\alpha - \cos 5\alpha)$

115. $2 \cos \dfrac{5\theta}{2} \cos \dfrac{\theta}{2}$ **117.** $2 \cos x \sin \dfrac{\pi}{4} = \sqrt{2} \cos x$

119. $y = \dfrac{1}{2}\sqrt{10} \sin\left(8t - \arctan \dfrac{1}{3}\right)$ **121.** $\dfrac{1}{2}\sqrt{10}$

123. False. $\cos \dfrac{\theta}{2} > 0$ **125.** True

127. Answers will vary.   **129.** $y_3 = y_2 + 1$

Chapter Test   (page 530)

1. $\sin\theta = -\dfrac{6\sqrt{61}}{61}$ **2.** 1   **3.** 1

$\cos\theta = -\dfrac{5\sqrt{61}}{61}$

$\csc\theta = -\dfrac{\sqrt{61}}{6}$

$\sec\theta = -\dfrac{\sqrt{61}}{5}$

$\cot\theta = \dfrac{5}{6}$

4. $\csc\theta\sec\theta$ **5.** $\dfrac{\pi}{2} < \theta \le \pi, \dfrac{3\pi}{2} < \theta < 2\pi$

6.

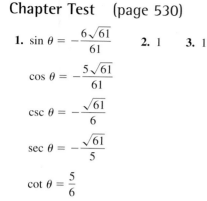

7–12. Answers will vary.   **13.** $-2 - \sqrt{3}$

14. $\dfrac{1}{16}\left[\dfrac{10 - 15\cos 2x + 6\cos 4x - \cos 6x}{1 + \cos 2x}\right]$

15. $\tan 2\theta$ **16.** $3(\cos 2\theta - \cos 10\theta)$

17. $2\cos 4\theta\cos\theta$ **18.** $0, \dfrac{3\pi}{4}, \pi, \dfrac{7\pi}{4}$

19. $\dfrac{\pi}{6}, \dfrac{\pi}{2}, \dfrac{5\pi}{6}, \dfrac{3\pi}{2}$ **20.** $\dfrac{\pi}{6}, \dfrac{5\pi}{6}, \dfrac{7\pi}{6}, \dfrac{11\pi}{6}$

21. $\dfrac{\pi}{6}, \dfrac{5\pi}{6}, \dfrac{3\pi}{2}$ **22.** 1.306

23. $\sin 2u = \dfrac{4}{5}$ **24.** 76.52°

$\cos 2u = -\dfrac{3}{5}$

$\tan 2u = -\dfrac{4}{3}$

Chapter 7

Section 7.1   (page 538)

Vocabulary Check   (page 538)

1. oblique   **2.** $\dfrac{b}{\sin B}$

3. $\frac{1}{2}bc\sin A; \frac{1}{2}ab\sin C; \frac{1}{2}ac\sin B$

1. $C = 105°$, $b \approx 16.97$, $c \approx 23.18$

3. $C = 110°$, $b \approx 22.44$, $c \approx 24.35$

5. $B \approx 21.6°$, $C \approx 122.4°$, $c \approx 11.49$

7. $B = 60.9°$, $b \approx 19.32$, $c \approx 6.36$

9. $B = 42°4'$, $a \approx 22.05$, $b \approx 14.88$

11. $A \approx 10°11'$, $C \approx 154°19'$, $c \approx 11.03$

13. $B \approx 18°13'$, $C \approx 51°32'$, $c \approx 40.06$

15. $B \approx 48.74°$, $C \approx 21.26°$, $c \approx 48.23$

17. No solution

19. Two solutions

$B \approx 72.2°$, $C \approx 49.8°$, $c \approx 10.27$

$B \approx 107.8°$, $C \approx 14.2°$, $c \approx 3.3$

21. ≈ 28.2 square units   **23.** ≈ 1782.3 square units

25. 2878.4 square units

27. (a)

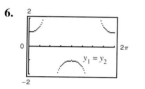

(b) $\dfrac{16}{\sin 70°} = \dfrac{h}{\sin 34°}$ (c) ≈ 9.5 meters

29. 240°

31. (a)

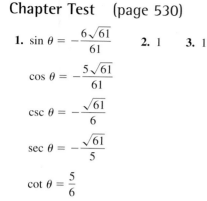

(b) 4385.71 feet

(c) 3061.80 feet

33. ≈ 15.53 kilometers; ≈ 42.43 kilometers   **35.** $\approx 16.1°$

37. (a) $\alpha \approx 5.36°$

(b) $\beta = \sin^{-1}\left(\dfrac{d\sin\theta}{58.36}\right)$

(c) $d = \sin(84.64 - \theta)\left[\dfrac{58.36}{\sin\theta}\right]$

(d)

θ	10°	20°	30°	40°	50°	60°
d	324.1	154.2	95.2	63.8	43.3	28.1

39. False. The triangle can't be solved if only three angles are known.

41. Yes, the Law of Sines can be used to solve a right triangle provided that at least one side is given.

43. $\tan \theta = -\frac{12}{5}$; $\csc \theta = -\frac{13}{12}$; $\sec \theta = \frac{13}{5}$; $\cot \theta = -\frac{5}{12}$

45. $\sin \theta = \frac{\sqrt{122}}{122}$; $\cos \theta = -\frac{11\sqrt{122}}{122}$; $\tan \theta = -\frac{1}{11}$;

$\csc \theta = \sqrt{122}$

47. $3(\sin 11\theta + \sin 5\theta)$ **49.** $\frac{3}{2}\left(\sin \frac{11\pi}{6} + \sin \frac{3\pi}{2}\right)$

Section 7.2 (page 545)

1. $A \approx 26.4°$, $B \approx 36.3°$, $C \approx 117.3°$

3. $B \approx 29.4°$, $C \approx 100.6°$, $a \approx 23.38$

5. $A \approx 36.9°$, $B \approx 53.1°$, $C = 90°$

7. $A \approx 103.5°$, $B \approx 38.2°$, $C \approx 38.2°$

9. $b \approx 8.58$, $A \approx 154°14'$, $C \approx 17°31'$

11. $b \approx 9.94$, $A \approx 37.1°$, $C \approx 67.6°$

	a	b	c	d	θ	ϕ
13.	4	8	11.64	4.96	30°	150°
15.	10	14	20	13.86	68.2°	111.8°
17.	15	16.96	25	20	77.2°	102.8°

19. ≈ 19.81 square units   **21.** ≈ 15.52 square units

23. ≈ 35.19 square units

25.

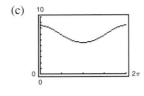

N
W ← → E
S
Rosemont
648 mi
32°
Centerville
75°
810 mi
Franklin

1357.8 miles, 236°

27. 483.4 meters   **29.** 43.3 miles

31. (a) N 58.4° W   (b) S 81.5° W

33. $PQ \approx 9.4$, $QS \approx 5.0$, $RS \approx 12.8$

35. $\approx 18,618$ square feet

37. (a) $49 = 2.25 + x^2 - 3x \cos \theta$

(b) $x = \frac{1}{2}\left(3 \cos \theta + \sqrt{9 \cos^2\theta + 187}\right)$

(c)

10

0   2π
0

(d) 6 inches

39. False. A triangle cannot be formed with sides of lengths 10 feet, 16 feet, and 5 feet.

41. False. $s = \dfrac{a + b + c}{2}$ **43.** $-\dfrac{\pi}{2}$ **45.** $\dfrac{\pi}{3}$

47. $-2 \sin\left(\dfrac{7\pi}{12}\right) \sin\left(\dfrac{\pi}{4}\right)$

Section 7.3 (page 557)

1. $\|\mathbf{u}\| = \|\mathbf{v}\| = \sqrt{17}$, $\text{slope}_\mathbf{u} = \text{slope}_\mathbf{v} = \frac{1}{4}$

$\mathbf{u}$ and $\mathbf{v}$ have the same magnitude and direction, so they are equal.

3. $\langle 4, 3 \rangle$, $\|\mathbf{v}\| = 5$ **5.** $\langle -3, 2 \rangle$, $\|\mathbf{v}\| = \sqrt{13} \approx 3.61$

7. $\langle 0, 5 \rangle$, $\|\mathbf{v}\| = 5$ **9.** $\left\langle -\frac{9}{2}, -\frac{5}{2} \right\rangle$, $\|\mathbf{v}\| = \frac{\sqrt{106}}{2} \approx 5.15$

11. $\langle 8, 6 \rangle$, $\|\mathbf{v}\| = 10$

13.

y

v

x

−v

15.

y

u + v

v

u

x

17.

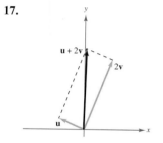

19. (a) $\langle 11, 3 \rangle$ (b) $\langle -3, 1 \rangle$ (c) $\langle -13, 1 \rangle$ (d) $\langle 23, 9 \rangle$

21. (a) $\langle -4, -4 \rangle$ (b) $\langle -8, -12 \rangle$ (c) $\langle -18, -28 \rangle$

　　(d) $\langle -22, -28 \rangle$

23. (a) $3\mathbf{i} - 2\mathbf{j}$ (b) $-\mathbf{i} + 4\mathbf{j}$ (c) $-4\mathbf{i} + 11\mathbf{j}$

　　(d) $6\mathbf{i} + \mathbf{j}$

25. $\langle 1, 0 \rangle$ **27.** $\left\langle -\dfrac{\sqrt{2}}{2}, \dfrac{\sqrt{2}}{2} \right\rangle$ **29.** $\left\langle -\dfrac{24}{25}, -\dfrac{7}{25} \right\rangle$

31. $\dfrac{4}{5}\mathbf{i} - \dfrac{3}{5}\mathbf{j}$ **33.** $\mathbf{j}$ **35.** $\dfrac{40\sqrt{61}}{61}\mathbf{i} + \dfrac{48\sqrt{61}}{61}\mathbf{j}$

37. $\dfrac{21}{5}\mathbf{i} + \dfrac{28}{5}\mathbf{j}$ **39.** $-8\mathbf{i}$ **41.** $7\mathbf{i} + 4\mathbf{j}$ **43.** $3\mathbf{i} + 8\mathbf{j}$

45. $\mathbf{v} = \left\langle 3, -\dfrac{3}{2} \right\rangle$ **47.** $\mathbf{v} = \langle 4, 3 \rangle$

49. $\mathbf{v} = \left\langle \dfrac{7}{2}, -\dfrac{1}{2} \right\rangle$ **51.** $\|\mathbf{v}\| = 5, \theta = 30°$

53. $\|\mathbf{v}\| = 6\sqrt{2}, \theta = 315°$ **55.** $\|\mathbf{v}\| = \sqrt{29}, \theta = 111.8°$

57. $\mathbf{v} = \langle 3, 0 \rangle$ **59.** $\mathbf{v} = \left\langle -\dfrac{3\sqrt{6}}{2}, \dfrac{3\sqrt{2}}{2} \right\rangle$

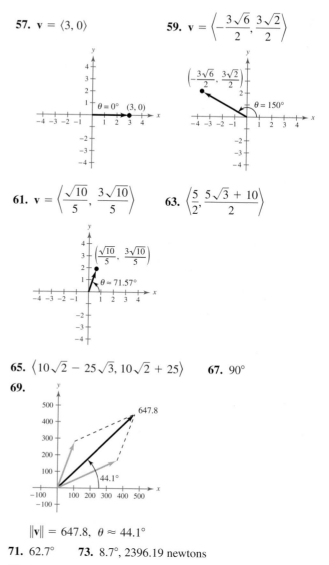

61. $\mathbf{v} = \left\langle \dfrac{\sqrt{10}}{5}, \dfrac{3\sqrt{10}}{5} \right\rangle$ **63.** $\left\langle \dfrac{5}{2}, \dfrac{5\sqrt{3} + 10}{2} \right\rangle$

65. $\langle 10\sqrt{2} - 25\sqrt{3}, 10\sqrt{2} + 25 \rangle$ **67.** $90°$

69.

$\|\mathbf{v}\| = 647.8, \theta \approx 44.1°$

71. $62.7°$ **73.** $8.7°$, 2396.19 newtons

75. Horizontal component: $70 \cos 40° \approx 53.62$ feet per second

　　Vertical component: $70 \sin 40° \approx 45.00$ feet per second

77. $T_{AC} \approx 3611.1$ pounds, $T_{BC} \approx 2169.5$ pounds

79. (a) $T = 3000 \sec \theta$; Domain: $0° \le \theta < 90°$

(b)

θ	10°	20°	30°
T	3046.3	3192.5	3464.1

θ	40°	50°	60°
T	3916.2	4667.2	6000.0

(c)

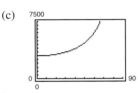

(d) The component in the direction of the motion of the barge decreases.

81. N 26.7° E, 130.35 kilometers per hour

83. (a) 12.1°, 357.85 newtons

(b) $M = 10\sqrt{660 \cos\theta + 709}$,

$\alpha = \arctan\dfrac{15\sin\theta}{15\cos\theta + 22}$

(c)

θ	0°	30°	60°	90°
M	370.0	357.9	322.3	266.3
α	0°	12.1°	23.8°	34.3°

θ	120°	150°	180°
M	194.7	117.2	70.0
α	41.9°	39.8°	0°

(d)

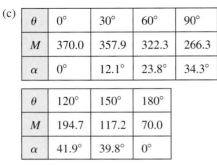

(e) For increasing θ, the two vectors tend to work against each other, resulting in a decrease in the magnitude of the resultant.

85. True   **87.** True. $a = b = 0$

89. (a) 0°   (b) 180°

(c) No. The magnitude is equal to the sum when the angle between the vectors is 0°.

91. Answers will vary.   **93.** $\langle 1, 3\rangle$ or $\langle -1, -3\rangle$

95. $12x^3 y^7$, $x \neq 0$, $y \neq 0$ **97.** $48xy^2$, $x \neq 0$

99. 7.14×10^5 **101.** $7\cos\theta$ **103.** $10\csc\theta$

105. $\dfrac{\pi}{2} + n\pi$, $\pi + 2n\pi$ **107.** $\dfrac{\pi}{3} + 2n\pi$, $\dfrac{5\pi}{3} + 2n\pi$

Section 7.4   (page 569)

Vocabulary Check   (page 569)

1. dot product   **2.** $\dfrac{\mathbf{u}\cdot\mathbf{v}}{\|\mathbf{u}\|\,\|\mathbf{v}\|}$ **3.** orthogonal

4. $\left(\dfrac{\mathbf{u}\cdot\mathbf{v}}{\|\mathbf{v}\|^2}\right)\mathbf{v}$ **5.** $\|\mathrm{proj}_{\overrightarrow{PQ}}\,\mathbf{F}\|\,\|\overrightarrow{PQ}\|$, $\mathbf{F}\cdot\overrightarrow{PQ}$

1. 0   **3.** 14   **5.** 8, scalar   **7.** $\langle 2, -8\rangle$, vector

9. 4, scalar   **11.** 13   **13.** $5\sqrt{41}$ **15.** 4

17. 90°   **19.** 70.56°   **21.** 90°   **23.** $\dfrac{5\pi}{12}$

25.   **27.**

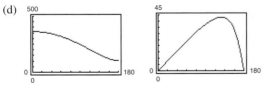

91.33°   90°

29. 26.6°, 63.4°, 90°   **31.** -20 **33.** Parallel

35. Neither   **37.** Orthogonal

39. $\dfrac{16}{17}\langle 4, 1\rangle$, $\mathbf{u} = \left\langle\dfrac{64}{17}, \dfrac{16}{17}\right\rangle + \left\langle -\dfrac{13}{17}, \dfrac{52}{17}\right\rangle$

41. $\dfrac{45}{229}\langle 2, 15\rangle$, $\mathbf{u} = \left\langle\dfrac{90}{229}, \dfrac{675}{229}\right\rangle + \left\langle -\dfrac{90}{229}, \dfrac{12}{229}\right\rangle$ **43.** $\mathbf{u}$

45. 0   **47.** $\langle 3, -1\rangle$, $\langle -3, 1\rangle$ **49.** $-\dfrac{3}{4}\mathbf{i} - \dfrac{1}{2}\mathbf{j}$, $\dfrac{3}{4}\mathbf{i} + \dfrac{1}{2}\mathbf{j}$

51. 32   **53.** \$37,289; total revenue

55. (a) Force $= 30{,}000 \sin d°$

(b)

d	0°	1°	2°	3°	4°	5°
Force	0	523.6	1047.0	1570.1	2092.7	2614.7

d	6°	7°	8°	9°	10°
Force	3135.9	3656.1	4175.2	4693.0	5209.4

(c) 29,885.8 pounds

57. (a) $W = 15{,}691\dfrac{\sqrt{3}}{2}d$

(b)

d	0	200	400	800
Work	0	2,717,760.9	5,435,521.8	10,871,043.7

59. 725.0 foot-pounds

61. True. The zero vector is orthogonal to every vector.

63. Orthogonal. $\mathbf{u} \cdot \mathbf{v} = 0$

65. (a) $\mathbf{u}$ and $\mathbf{v}$ are parallel.   (b) $\mathbf{u}$ and $\mathbf{v}$ are orthogonal.

67. Answers will vary.

69. g is a horizontal shift four units to the right.

71. g is a vertical shift six units upward.

73. $-1 + 2i$ **75.** $15 + 12i$ **77.** 10   **79.** $\frac{47}{26} - \frac{27}{26}i$

81.   **83.**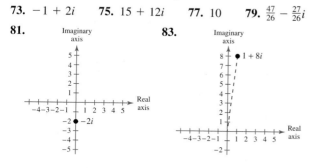

85. 8 people

Section 7.5   (page 580)

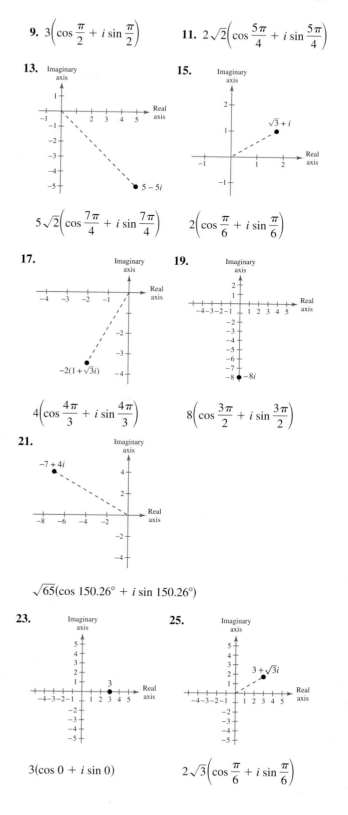

9. $3\left(\cos \dfrac{\pi}{2} + i \sin \dfrac{\pi}{2}\right)$ **11.** $2\sqrt{2}\left(\cos \dfrac{5\pi}{4} + i \sin \dfrac{5\pi}{4}\right)$

13.   **15.**

$5\sqrt{2}\left(\cos \dfrac{7\pi}{4} + i \sin \dfrac{7\pi}{4}\right)$ $2\left(\cos \dfrac{\pi}{6} + i \sin \dfrac{\pi}{6}\right)$

17.   **19.**

$4\left(\cos \dfrac{4\pi}{3} + i \sin \dfrac{4\pi}{3}\right)$ $8\left(\cos \dfrac{3\pi}{2} + i \sin \dfrac{3\pi}{2}\right)$

21.

$\sqrt{65}(\cos 150.26° + i \sin 150.26°)$

23.   **25.**

$3(\cos 0 + i \sin 0)$ $2\sqrt{3}\left(\cos \dfrac{\pi}{6} + i \sin \dfrac{\pi}{6}\right)$

1. 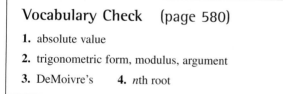   **3.**

6   4

5. 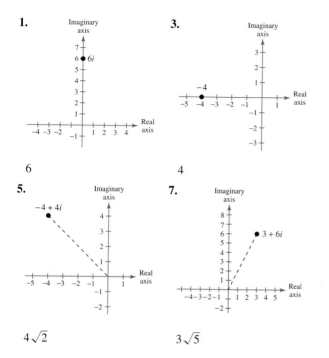   **7.**

$4\sqrt{2}$ $3\sqrt{5}$

27.

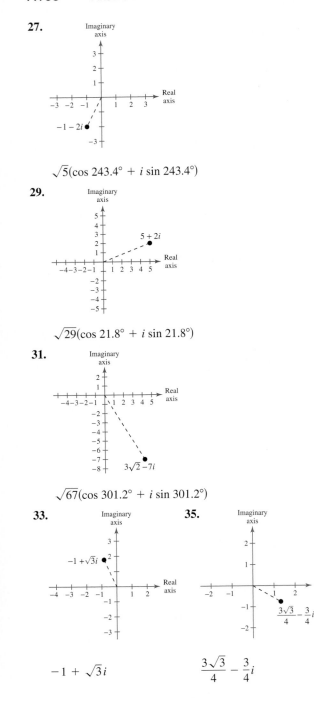

$\sqrt{5}(\cos 243.4° + i \sin 243.4°)$

29.

Imaginary axis

$\sqrt{29}(\cos 21.8° + i \sin 21.8°)$

31.

Imaginary axis

$\sqrt{67}(\cos 301.2° + i \sin 301.2°)$

33.   **35.**

Imaginary axis

$-1 + \sqrt{3}i$ $\dfrac{3\sqrt{3}}{4} - \dfrac{3}{4}i$

37.

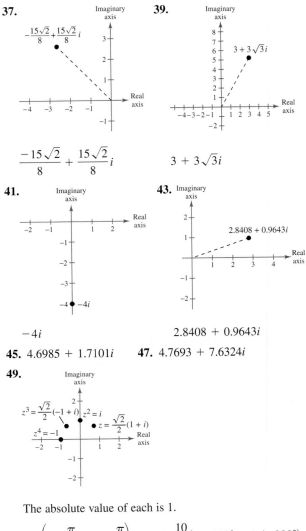

$\dfrac{-15\sqrt{2}}{8} + \dfrac{15\sqrt{2}}{8}i$

39.

Imaginary axis

$3 + 3\sqrt{3}i$

41.

Imaginary axis

$-4i$

43.

Imaginary axis

$2.8408 + 0.9643i$

45. $4.6985 + 1.7101i$ **47.** $4.7693 + 7.6324i$

49.

Imaginary axis

The absolute value of each is 1.

51. $12\left(\cos \dfrac{\pi}{2} + i \sin \dfrac{\pi}{2}\right)$ **53.** $\dfrac{10}{9}(\cos 200° + i \sin 200°)$

55. $\dfrac{11}{50}(\cos 130° + i \sin 130°)$ **57.** $\cos 30° + i \sin 30°$

59. $\dfrac{1}{2}(\cos 80° + i \sin 80°)$

61. $6[\cos(-48°) + i \sin(-48°)]$

63. (a) $2\sqrt{2}\left(\cos \dfrac{\pi}{4} + i \sin \dfrac{\pi}{4}\right)$

$\sqrt{2}\left[\cos\left(-\dfrac{\pi}{4}\right) + i \sin\left(-\dfrac{\pi}{4}\right)\right]$

(b) $4(\cos 0 + i \sin 0) = 4$

(c) 4

65. (a) $2\left[\cos\left(-\dfrac{\pi}{2}\right) + i\sin\left(-\dfrac{\pi}{2}\right)\right]$

$\sqrt{2}\left(\cos\dfrac{\pi}{4} + i\sin\dfrac{\pi}{4}\right)$

(b) $2\sqrt{2}\left[\cos\left(-\dfrac{\pi}{4}\right) + i\sin\left(-\dfrac{\pi}{4}\right)\right] = 2 - 2i$

(c) $2 - 2i$

67. (a) $5(\cos 0 + i\sin 0)$

$\sqrt{13}(\cos 56.31° + i\sin 56.31°)$

(b) $\dfrac{5\sqrt{13}}{13}[\cos(-56.31°) + i\sin(-56.31°)]$

$\approx 0.7692 - 1.1538i$

(c) $\dfrac{10}{13} - \dfrac{15}{13}i \approx 0.7692 - 1.1538i$

69.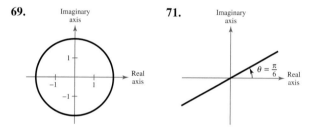

71.

73. $-2 + 2i$ **75.** $-32i$ **77.** $-32\sqrt{3} + 32i$

79. $\dfrac{125}{2} + \dfrac{125\sqrt{3}}{2}i$ **81.** i

83. $140.0189 + 1014.3819i$ **85.** $-597.00 - 122.00i$

87. $\dfrac{81}{2} + \dfrac{81\sqrt{3}}{2}i$

89. $\sqrt[6]{1}\left[\cos\dfrac{0 + 2\pi(4)}{6} + i\sin\dfrac{0 + 2\pi(4)}{6}\right] = -\dfrac{1}{2}(1 + \sqrt{3}i)$

91. (a) $2(\cos 30° + i\sin 30°)$ (b) and (c) $8i$

$2(\cos 150° + i\sin 150°)$

$2(\cos 270° + i\sin 270°)$

93. (a) $\sqrt{5}(\cos 60° + i\sin 60°)$

$\sqrt{5}(\cos 240° + i\sin 240°)$

(b)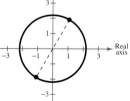

(c) $\dfrac{\sqrt{5}}{2} + \dfrac{\sqrt{15}}{2}i,\ -\dfrac{\sqrt{5}}{2} - \dfrac{\sqrt{15}}{2}i$

95. (a) $2\left(\cos\dfrac{\pi}{3} + i\sin\dfrac{\pi}{3}\right)$

$2\left(\cos\dfrac{5\pi}{6} + i\sin\dfrac{5\pi}{6}\right)$

$2\left(\cos\dfrac{4\pi}{3} + i\sin\dfrac{4\pi}{3}\right)$

$2\left(\cos\dfrac{11\pi}{6} + i\sin\dfrac{11\pi}{6}\right)$

(b)

(c) $1 + \sqrt{3}i,\ -\sqrt{3} + i,\ -1 - \sqrt{3}i,\ \sqrt{3} - i$

97. (a) $3\left(\cos\dfrac{\pi}{2} + i\sin\dfrac{\pi}{2}\right)$

$3\left(\cos\dfrac{7\pi}{6} + i\sin\dfrac{7\pi}{6}\right)$

$3\left(\cos\dfrac{11\pi}{6} + i\sin\dfrac{11\pi}{6}\right)$

(b)

(c) $3i,\ -\dfrac{3\sqrt{3}}{2} - \dfrac{3}{2}i,\ \dfrac{3\sqrt{3}}{2} - \dfrac{3}{2}i$

99. (a) $5\left(\cos\dfrac{4\pi}{9} + i\sin\dfrac{4\pi}{9}\right)$

$5\left(\cos\dfrac{10\pi}{9} + i\sin\dfrac{10\pi}{9}\right)$

$5\left(\cos\dfrac{16\pi}{9} + i\sin\dfrac{16\pi}{9}\right)$

(b)

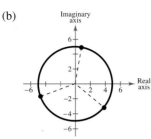

(c) $0.8682 + 4.9240i,\ -4.6985 - 1.7101i,$
$3.8302 - 3.2139i$

101. (a) $4\left(\cos\dfrac{\pi}{6} + i\sin\dfrac{\pi}{6}\right)$

$4\left(\cos\dfrac{5\pi}{6} + i\sin\dfrac{5\pi}{6}\right)$

$4\left(\cos\dfrac{3\pi}{2} + i\sin\dfrac{3\pi}{2}\right)$

(b)

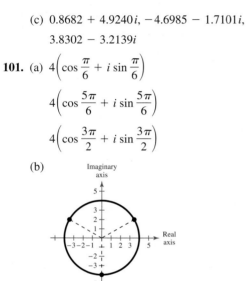

(c) $2\sqrt{3} + 2i,\ -2\sqrt{3} + 2i,\ -4i$

103. (a) $\cos 0 + i\sin 0$

$\cos\dfrac{2\pi}{5} + i\sin\dfrac{2\pi}{5}$

$\cos\dfrac{4\pi}{5} + i\sin\dfrac{4\pi}{5}$

$\cos\dfrac{6\pi}{5} + i\sin\dfrac{6\pi}{5}$

$\cos\dfrac{8\pi}{5} + i\sin\dfrac{8\pi}{5}$

(b)

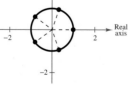

(c) $1, 0.3090 + 0.9511i,\ -0.8090 + 0.5878i,$
$-0.8090 - 0.5878i,\ 0.3090 - 0.9511i$

105. (a) $5\left(\cos\dfrac{\pi}{3} + i\sin\dfrac{\pi}{3}\right)$

$5(\cos\pi + i\sin\pi)$

$5\left(\cos\dfrac{5\pi}{3} + i\sin\dfrac{5\pi}{3}\right)$

(b)

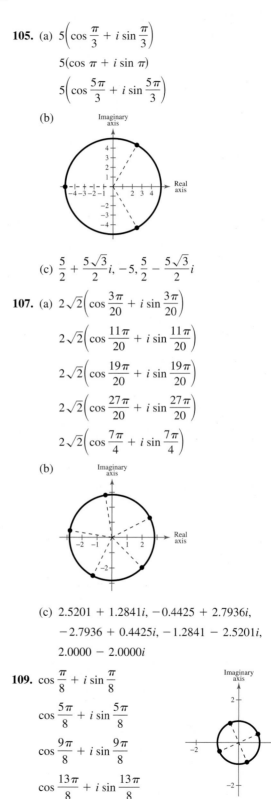

(c) $\dfrac{5}{2} + \dfrac{5\sqrt{3}}{2}i,\ -5,\ \dfrac{5}{2} - \dfrac{5\sqrt{3}}{2}i$

107. (a) $2\sqrt{2}\left(\cos\dfrac{3\pi}{20} + i\sin\dfrac{3\pi}{20}\right)$

$2\sqrt{2}\left(\cos\dfrac{11\pi}{20} + i\sin\dfrac{11\pi}{20}\right)$

$2\sqrt{2}\left(\cos\dfrac{19\pi}{20} + i\sin\dfrac{19\pi}{20}\right)$

$2\sqrt{2}\left(\cos\dfrac{27\pi}{20} + i\sin\dfrac{27\pi}{20}\right)$

$2\sqrt{2}\left(\cos\dfrac{7\pi}{4} + i\sin\dfrac{7\pi}{4}\right)$

(b)

(c) $2.5201 + 1.2841i,\ -0.4425 + 2.7936i,$
$-2.7936 + 0.4425i,\ -1.2841 - 2.5201i,$
$2.0000 - 2.0000i$

109. $\cos\dfrac{\pi}{8} + i\sin\dfrac{\pi}{8}$

$\cos\dfrac{5\pi}{8} + i\sin\dfrac{5\pi}{8}$

$\cos\dfrac{9\pi}{8} + i\sin\dfrac{9\pi}{8}$

$\cos\dfrac{13\pi}{8} + i\sin\dfrac{13\pi}{8}$

111. $3\left(\cos\dfrac{\pi}{5} + i\sin\dfrac{\pi}{5}\right)$

$3\left(\cos\dfrac{3\pi}{5} + i\sin\dfrac{3\pi}{5}\right)$

$3(\cos\pi + i\sin\pi)$

$3\left(\cos\dfrac{7\pi}{5} + i\sin\dfrac{7\pi}{5}\right)$

$3\left(\cos\dfrac{9\pi}{5} + i\sin\dfrac{9\pi}{5}\right)$

113. $2\left(\cos\dfrac{3\pi}{8} + i\sin\dfrac{3\pi}{8}\right)$

$2\left(\cos\dfrac{7\pi}{8} + i\sin\dfrac{7\pi}{8}\right)$

$2\left(\cos\dfrac{11\pi}{8} + i\sin\dfrac{11\pi}{8}\right)$

$2\left(\cos\dfrac{15\pi}{8} + i\sin\dfrac{15\pi}{8}\right)$

115. $\sqrt[6]{2}\left(\cos\dfrac{7\pi}{12} + i\sin\dfrac{7\pi}{12}\right)$

$\sqrt[6]{2}\left(\cos\dfrac{5\pi}{4} + i\sin\dfrac{5\pi}{4}\right)$

$\sqrt[6]{2}\left(\cos\dfrac{23\pi}{12} + i\sin\dfrac{23\pi}{12}\right)$

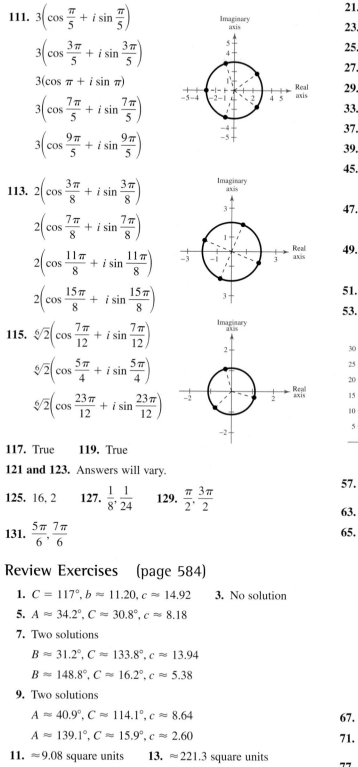

117. True   **119.** True

121 and 123. Answers will vary.

125. 16, 2   **127.** $\dfrac{1}{8}, \dfrac{1}{24}$ **129.** $\dfrac{\pi}{2}, \dfrac{3\pi}{2}$

131. $\dfrac{5\pi}{6}, \dfrac{7\pi}{6}$

Review Exercises   (page 584)

1. $C = 117°, b \approx 11.20, c \approx 14.92$ **3.** No solution

5. $A \approx 34.2°, C \approx 30.8°, c \approx 8.18$

7. Two solutions

$B \approx 31.2°, C \approx 133.8°, c \approx 13.94$

$B \approx 148.8°, C \approx 16.2°, c \approx 5.38$

9. Two solutions

$A \approx 40.9°, C \approx 114.1°, c \approx 8.64$

$A \approx 139.1°, C \approx 15.9°, c \approx 2.60$

11. ≈ 9.08 square units   **13.** ≈ 221.3 square units

15. ≈ 15.3 meters   **17.** ≈ 31.0 feet

19. $A \approx 15.3°, B \approx 20.6°, C \approx 144.1°$

21. $A \approx 41.7°, B \approx 93.3°, c \approx 6.37$

23. $A = 35°, C = 35°, b \approx 6.55$

25. $b \approx 29.09, A \approx 9.9°, C \approx 20.1°$

27. $A \approx 57.8°, B \approx 35.5°, C \approx 86.7°$

29. ≈ 4.3 feet, ≈ 12.6 feet   **31.** ≈ 1135.5 miles

33. ≈ 9.8 square units   **35.** ≈ 511.7 square units

37. $\|\mathbf{u}\| = \|\mathbf{v}\| = \sqrt{61}$, slope$_\mathbf{u}$ = slope$_\mathbf{v}$ = $\dfrac{5}{6}$

39. $\langle 7, -5\rangle$ **41.** $\langle 7, -7\rangle$ **43.** $\langle -4, 4\sqrt{3}\rangle$

45. (a) $\langle -4, 3\rangle$ (b) $\langle 2, -9\rangle$ (c) $\langle -3, -9\rangle$

(d) $\langle -11, -3\rangle$

47. (a) $\langle -1, 6\rangle$ (b) $\langle -9, -2\rangle$ (c) $\langle -15, 6\rangle$

(d) $\langle -17, 18\rangle$

49. (a) $7\mathbf{i} + 2\mathbf{j}$ (b) $-3\mathbf{i} - 4\mathbf{j}$ (c) $6\mathbf{i} - 3\mathbf{j}$

(d) $20\mathbf{i} + \mathbf{j}$

51. (a) $3\mathbf{i} + 6\mathbf{j}$ (b) $5\mathbf{i} - 6\mathbf{j}$ (c) $12\mathbf{i}$ (d) $18\mathbf{i} + 12\mathbf{j}$

53. $\langle 30, 9\rangle$ **55.** $\langle 74, -5\rangle$

57. $\langle 0, -1\rangle$ **59.** $\dfrac{1}{\sqrt{29}}\langle 5, -2\rangle$ **61.** $9\mathbf{i} - 8\mathbf{j}$

63. $10\sqrt{2}[(\cos 135°)\mathbf{i} + (\sin 135°)\mathbf{j}]$

65.

Magnitude: 32.62

Direction: 44.67°

67. 60°, 76.8 pounds   **69.** 180 pounds

71. 422.3 miles per hour, 130.4°   **73.** -20 **75.** 7

77. 25   **79.** -40 **81.** 160.5°   **83.** $\dfrac{11\pi}{12}$

85.

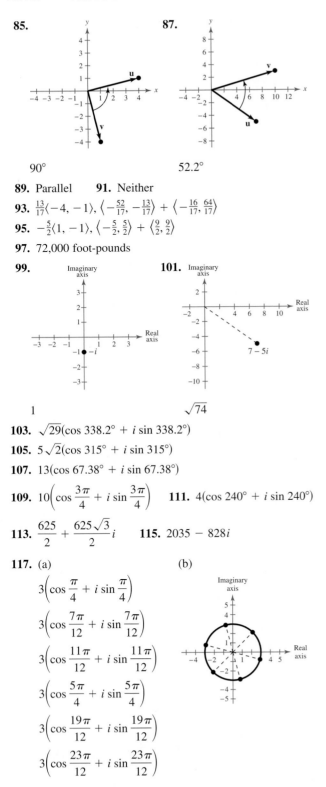

90°

87.

52.2°

(c)

$$\frac{3\sqrt{2}}{2} + \frac{3\sqrt{2}}{2}i, \; -0.7765 + 2.898i, \; -2.898 + 0.7765i,$$

$$-\frac{3\sqrt{2}}{2} - \frac{3\sqrt{2}}{2}i, \; 0.7765 - 2.898i, \; 2.898 - 0.7765i$$

119. (a) $2(\cos 0 + i \sin 0)$ (b)

$$2\left(\cos \frac{2\pi}{3} + i \sin \frac{2\pi}{3}\right)$$

$$2\left(\cos \frac{4\pi}{3} + i \sin \frac{4\pi}{3}\right)$$

(c) $2, \; -1 + \sqrt{3}i, \; -1 - \sqrt{3}i$

121. $4\left(\cos \frac{\pi}{4} + i \sin \frac{\pi}{4}\right) = 2\sqrt{2} + 2\sqrt{2}i$

$$4\left(\cos \frac{3\pi}{4} + i \sin \frac{3\pi}{4}\right) = -2\sqrt{2} + 2\sqrt{2}i$$

$$4\left(\cos \frac{5\pi}{4} + i \sin \frac{5\pi}{4}\right) = -2\sqrt{2} - 2\sqrt{2}i$$

$$4\left(\cos \frac{7\pi}{4} + i \sin \frac{7\pi}{4}\right) = 2\sqrt{2} - 2\sqrt{2}i$$

89. Parallel   **91.** Neither

93. $\frac{13}{17}\langle -4, -1 \rangle, \; \left\langle -\frac{52}{17}, -\frac{13}{17} \right\rangle + \left\langle -\frac{16}{17}, \frac{64}{17} \right\rangle$

95. $-\frac{5}{2}\langle 1, -1 \rangle, \; \left\langle -\frac{5}{2}, \frac{5}{2} \right\rangle + \left\langle \frac{9}{2}, \frac{9}{2} \right\rangle$

97. 72,000 foot-pounds

99.   **101.**

1   $\sqrt{74}$

103. $\sqrt{29}(\cos 338.2° + i \sin 338.2°)$

105. $5\sqrt{2}(\cos 315° + i \sin 315°)$

107. $13(\cos 67.38° + i \sin 67.38°)$

109. $10\left(\cos \frac{3\pi}{4} + i \sin \frac{3\pi}{4}\right)$ **111.** $4(\cos 240° + i \sin 240°)$

113. $\frac{625}{2} + \frac{625\sqrt{3}}{2}i$ **115.** $2035 - 828i$

117. (a)

$$3\left(\cos \frac{\pi}{4} + i \sin \frac{\pi}{4}\right)$$

$$3\left(\cos \frac{7\pi}{12} + i \sin \frac{7\pi}{12}\right)$$

$$3\left(\cos \frac{11\pi}{12} + i \sin \frac{11\pi}{12}\right)$$

$$3\left(\cos \frac{5\pi}{4} + i \sin \frac{5\pi}{4}\right)$$

$$3\left(\cos \frac{19\pi}{12} + i \sin \frac{19\pi}{12}\right)$$

$$3\left(\cos \frac{23\pi}{12} + i \sin \frac{23\pi}{12}\right)$$

(b)

123. $2\left(\cos \frac{\pi}{2} + i \sin \frac{\pi}{2}\right) = 2i$

$$2\left(\cos \frac{7\pi}{6} + i \sin \frac{7\pi}{6}\right) = -\sqrt{3} - i$$

$$2\left(\cos \frac{11\pi}{6} + i \sin \frac{11\pi}{6}\right) = \sqrt{3} - i$$

125. True   **127.** Direction and magnitude

129. $z_1 z_2 = -4, \dfrac{z_1}{z_2} = -\dfrac{1}{4} z_1{}^2$

Chapter Test (page 588)

1. $C = 46°, a \approx 14.71, b \approx 24.78$

2. $A \approx 25.2°, B \approx 48.2°, C \approx 106.6°$

3. $B \approx 45.9°, C \approx 99.1°, a \approx 6.39$

4. Two solutions

$\qquad B \approx 52.1°, C \approx 102.9°, c \approx 34.59$

$\qquad B \approx 127.9°, C \approx 27.1°, c \approx 16.16$

5. No solution **6.** $B \approx 14.8°, C \approx 15.2°, c \approx 4.93$

7. ≈ 675 feet **8.** ≈ 2337 square meters

9. $\mathbf{w} = \langle 12, 13 \rangle, \|\mathbf{w}\| \approx 17.69$

10. (a) $\langle -4, 4 \rangle$ (b) $\langle 6, -16 \rangle$ (c) $\langle 2, -24 \rangle$

11. (a) $\langle 0, 7 \rangle$ (b) $\langle -5, -18 \rangle$ (c) $\langle -11, -20 \rangle$

12. (a) $\langle 13, 17 \rangle$ (b) $\langle -17, -28 \rangle$ (c) $\langle -1, -14 \rangle$

13. (a) $\langle 0, -1 \rangle$ (b) $\langle 5, 9 \rangle$ (c) $\langle 11, 17 \rangle$

14. $\dfrac{\sqrt{65}}{65}\langle 7, 4 \rangle$ **15.** $\left\langle \dfrac{18\sqrt{34}}{17}, -\dfrac{30\sqrt{34}}{17} \right\rangle$

16. $\theta \approx 14.9°, \approx 250.2$ pounds **17.** 3

18. $\approx 105.9°$ **19.** No, because $\mathbf{u} \cdot \mathbf{v} = 24$, not 0.

20. $\left\langle \frac{185}{26}, \frac{37}{26} \right\rangle, \mathbf{u} = \left\langle \frac{185}{26}, \frac{37}{26} \right\rangle + \left\langle -\frac{29}{26}, \frac{145}{26} \right\rangle$

21. $z = 2\sqrt{2}\left(\cos \dfrac{3\pi}{4} + i \sin \dfrac{3\pi}{4} \right)$ **22.** $-50 - 50\sqrt{3}\,i$

23. $-\dfrac{6561}{2} + \dfrac{6561\sqrt{3}}{2}i$ **24.** $5832i$

25. $4\left(\cos \dfrac{\pi}{12} + i \sin \dfrac{\pi}{12} \right) \approx 3.8637 + 1.0353i$

$\qquad 4\left(\cos \dfrac{7\pi}{12} + i \sin \dfrac{7\pi}{12} \right) \approx -1.0353 + 3.8637i$

$\qquad 4\left(\cos \dfrac{13\pi}{12} + i \sin \dfrac{13\pi}{12} \right) \approx -3.8637 - 1.0353i$

$\qquad 4\left(\cos \dfrac{19\pi}{12} + i \sin \dfrac{19\pi}{12} \right) \approx 1.0353 - 3.8637i$

26. $5\left(\cos \dfrac{\pi}{8} + i \sin \dfrac{\pi}{8} \right)$

$\qquad 5\left(\cos \dfrac{5\pi}{8} + i \sin \dfrac{5\pi}{8} \right)$

$\qquad 5\left(\cos \dfrac{9\pi}{8} + i \sin \dfrac{9\pi}{8} \right)$

$\qquad 5\left(\cos \dfrac{13\pi}{8} + i \sin \dfrac{13\pi}{8} \right)$

Cumulative Test for Chapters 5–7 (page 589)

1. (a)

(b) $240°$

(c) $-\dfrac{2\pi}{3}$

(d) $60°$

(e) $\sin(-120°) = -\dfrac{\sqrt{3}}{2}$

$\qquad \cos(-120°) = -\dfrac{1}{2}$

$\qquad \tan(-120°) = \sqrt{3}$

$\qquad \csc(-120°) = -\dfrac{2\sqrt{3}}{3}$

$\qquad \sec(-120°) = -2$

$\qquad \cot(-120°) = \dfrac{\sqrt{3}}{3}$

2. $134.6°$ **3.** $-\dfrac{3}{5}$

4.

5.

6.

7. $a = -3, b = \pi, c = 0$

8. 6.7 **9.** $\dfrac{3}{4}$ **10.** $\sqrt{1 - 4x^2}$ **11.** $2 \tan \theta$

12–14. Answers will vary. **15.** $\dfrac{3\pi}{2} + 2n\pi$

16. $\dfrac{\pi}{6} + n\pi, \dfrac{5\pi}{6} + n\pi$ **17.** $1.7646, 4.5186$

18.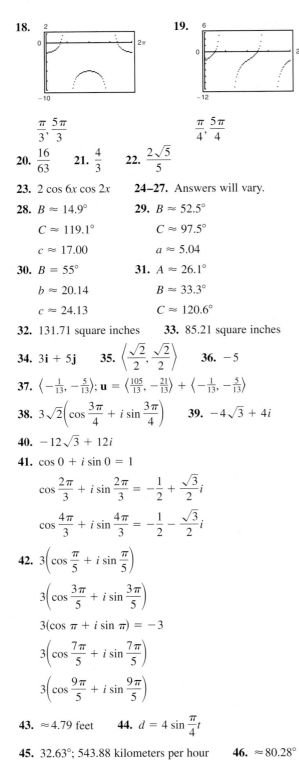

19.

$\dfrac{\pi}{3}, \dfrac{5\pi}{3}$

$\dfrac{\pi}{4}, \dfrac{5\pi}{4}$

20. $\dfrac{16}{63}$ **21.** $\dfrac{4}{3}$ **22.** $\dfrac{2\sqrt{5}}{5}$

23. $2\cos 6x \cos 2x$ **24–27.** Answers will vary.

28. $B \approx 14.9°$
$C \approx 119.1°$
$c \approx 17.00$

29. $B \approx 52.5°$
$C \approx 97.5°$
$a \approx 5.04$

30. $B = 55°$
$b \approx 20.14$
$c \approx 24.13$

31. $A \approx 26.1°$
$B \approx 33.3°$
$C \approx 120.6°$

32. 131.71 square inches   **33.** 85.21 square inches

34. $3\mathbf{i} + 5\mathbf{j}$ **35.** $\left\langle \dfrac{\sqrt{2}}{2}, \dfrac{\sqrt{2}}{2} \right\rangle$ **36.** -5

37. $\left\langle -\dfrac{1}{13}, -\dfrac{5}{13} \right\rangle$; $\mathbf{u} = \left\langle \dfrac{105}{13}, -\dfrac{21}{13} \right\rangle + \left\langle -\dfrac{1}{13}, -\dfrac{5}{13} \right\rangle$

38. $3\sqrt{2}\left(\cos \dfrac{3\pi}{4} + i \sin \dfrac{3\pi}{4} \right)$ **39.** $-4\sqrt{3} + 4i$

40. $-12\sqrt{3} + 12i$

41. $\cos 0 + i \sin 0 = 1$

$\cos \dfrac{2\pi}{3} + i \sin \dfrac{2\pi}{3} = -\dfrac{1}{2} + \dfrac{\sqrt{3}}{2}i$

$\cos \dfrac{4\pi}{3} + i \sin \dfrac{4\pi}{3} = -\dfrac{1}{2} - \dfrac{\sqrt{3}}{2}i$

42. $3\left(\cos \dfrac{\pi}{5} + i \sin \dfrac{\pi}{5} \right)$

$3\left(\cos \dfrac{3\pi}{5} + i \sin \dfrac{3\pi}{5} \right)$

$3(\cos \pi + i \sin \pi) = -3$

$3\left(\cos \dfrac{7\pi}{5} + i \sin \dfrac{7\pi}{5} \right)$

$3\left(\cos \dfrac{9\pi}{5} + i \sin \dfrac{9\pi}{5} \right)$

43. ≈ 4.79 feet   **44.** $d = 4 \sin \dfrac{\pi}{4}t$

45. $32.63°$; 543.88 kilometers per hour   **46.** $\approx 80.28°$

Chapter 8

Section 8.1   (page 599)

1. (a) No   (b) No   (c) No   (d) Yes

3. (a) No   (b) Yes   (c) No   (d) No

5. $(2, 2)$ **7.** $(2, 6), (-1, 3)$

9. $(0, 2), \left(\sqrt{3}, 2 - 3\sqrt{3} \right), \left(-\sqrt{3}, 2 + 3\sqrt{3} \right)$

11. $(0, 0), (2, -4)$ **13.** $(4, 4)$ **15.** $(5, 5)$ **17.** $\left(\dfrac{1}{2}, 3 \right)$

19. $(1, 1)$ **21.** $\left(\dfrac{20}{3}, \dfrac{40}{3} \right)$ **23.** No solution

25. No solution   **27.** $(0, 0), (1, 1), (-1, -1)$

29. $(4, 3)$ **31.** $\left(\dfrac{5}{2}, \dfrac{3}{2} \right)$ **33.** $(2, 2), (4, 0)$

35. $(1, 4), (4, 7)$

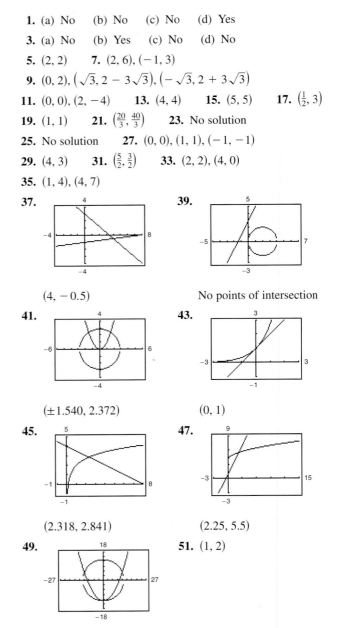

37.

$(4, -0.5)$

39.

No points of intersection

41.

$(\pm 1.540, 2.372)$

43.

$(0, 1)$

45.

$(2.318, 2.841)$

47.

$(2.25, 5.5)$

49.

$(0, -13), (\pm 12, 5)$

51. $(1, 2)$

53. $(-2, 0), \left(\frac{29}{10}, \frac{21}{10}\right)$ **55.** $(0.25, 1.5)$ **57.** $(0.287, 1.751)$

59. $(0, 1), (1, 0)$ **61.** $\left(-4, -\frac{1}{4}\right), \left(\frac{1}{2}, 2\right)$

63.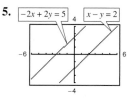

192 units; $1,910,400

65.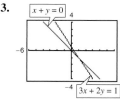

3133 units; $10,308

67. (a) $C = 35.45x + 16,000$

$R = 55.95x$

(b)

781 units

69. Sales greater than $11,667

71. (a) $\begin{cases} x + y = 20,000 \\ 0.065x + 0.085y = 1600 \end{cases}$

(b)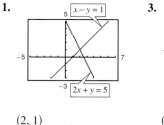

More invested at 6.5% means less invested at 8.5% and less interest.

(c) $5000

73. (a) $F_{\text{VCR}} = -107.86x^2 + 1583.6x - 3235$

$F_{\text{DVD}} = 21.14x^2 + 143.8x - 1939$

(b)

(c) and (d) 2000

(e) They are the same.

75. 6 meters × 9 meters   **77.** 8 miles × 12 miles

79. False. You could solve for either variable.

81. For a linear system the result will be a contradictory equation such as $0 = N$, where N is a nonzero real number. For a nonlinear system there may be an equation with imaginary roots.

83. (Answers are not unique.)

(a) $y = x + 1$ (b) $y = 0$ (c) $y = -2$

85. $2x + 7y - 45 = 0$ **87.** $y - 3 = 0$

89. $30x - 17y - 18 = 0$

91. Domain: All real numbers x except $x = 6$

Asymptotes: $y = 0, x = 6$

93. Domain: All real numbers x except $x = \pm 4$

Asymptotes: $y = 1, x = \pm 4$

Section 8.2   (page 609)

Vocabulary Check   (page 609)

1. method, elimination   **2.** equivalent

3. consistent, inconsistent

1. (2, 1)

3. (1, −1)

5. Inconsistent

7. $\left(\frac{5}{2}, \frac{3}{4}\right)$ **9.** $(3, 4)$ **11.** $(4, -1)$ **13.** $\left(\frac{12}{7}, \frac{18}{7}\right)$

15. Inconsistent   **17.** b   **18.** a   **19.** c   **20.** d

21. $\left(-\frac{6}{35}, \frac{43}{35}\right)$ **23.** Inconsistent

25. All points on the line $6x + 8y - 1 = 0$ **27.** $(5, -2)$

29. All points on the line $5x - 6y - 3 = 0$ **31.** $(101, 96)$

33. Consistent; $(5, 2)$

35. Inconsistent

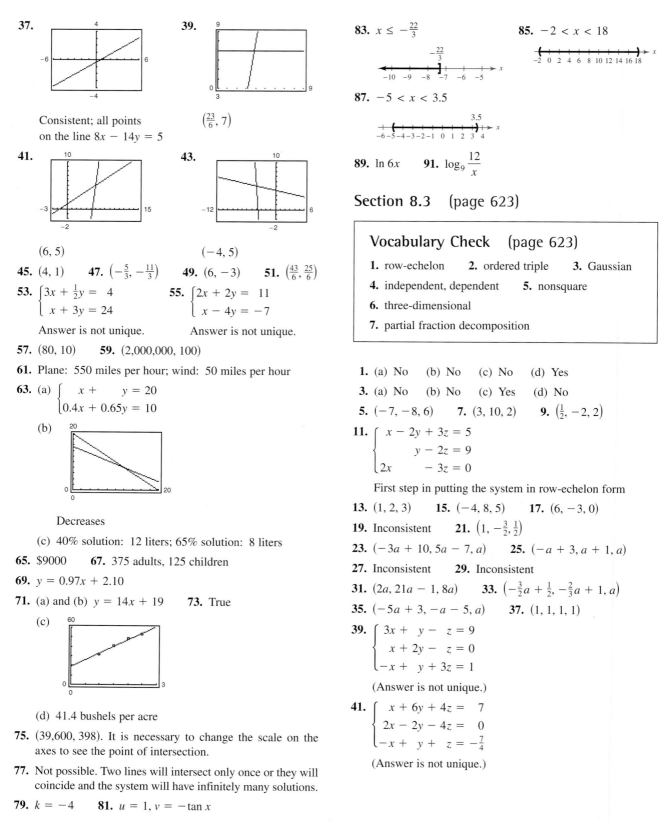

37.

Consistent; all points
on the line $8x - 14y = 5$

39.

$\left(\frac{23}{6}, 7\right)$

41.

$(6, 5)$

43.

$(-4, 5)$

45. $(4, 1)$ **47.** $\left(-\frac{5}{3}, -\frac{11}{3}\right)$ **49.** $(6, -3)$ **51.** $\left(\frac{43}{6}, \frac{25}{6}\right)$

53. $\begin{cases} 3x + \frac{1}{2}y = 4 \\ x + 3y = 24 \end{cases}$ **55.** $\begin{cases} 2x + 2y = 11 \\ x - 4y = -7 \end{cases}$

Answer is not unique.   Answer is not unique.

57. $(80, 10)$ **59.** $(2,000,000, 100)$

61. Plane: 550 miles per hour; wind: 50 miles per hour

63. (a) $\begin{cases} x + y = 20 \\ 0.4x + 0.65y = 10 \end{cases}$

(b)

Decreases

(c) 40% solution: 12 liters; 65% solution: 8 liters

65. \$9000   **67.** 375 adults, 125 children

69. $y = 0.97x + 2.10$

71. (a) and (b) $y = 14x + 19$ **73.** True

(c)

(d) 41.4 bushels per acre

75. $(39,600, 398)$. It is necessary to change the scale on the axes to see the point of intersection.

77. Not possible. Two lines will intersect only once or they will coincide and the system will have infinitely many solutions.

79. $k = -4$ **81.** $u = 1, v = -\tan x$

83. $x \leq -\frac{22}{3}$

85. $-2 < x < 18$

87. $-5 < x < 3.5$

89. $\ln 6x$ **91.** $\log_9 \dfrac{12}{x}$

Section 8.3   (page 623)

Vocabulary Check   (page 623)

1. row-echelon   2. ordered triple   3. Gaussian

4. independent, dependent   5. nonsquare

6. three-dimensional

7. partial fraction decomposition

1. (a) No   (b) No   (c) No   (d) Yes

3. (a) No   (b) No   (c) Yes   (d) No

5. $(-7, -8, 6)$ **7.** $(3, 10, 2)$ **9.** $\left(\frac{1}{2}, -2, 2\right)$

11. $\begin{cases} x - 2y + 3z = 5 \\ y - 2z = 9 \\ 2x - 3z = 0 \end{cases}$

First step in putting the system in row-echelon form

13. $(1, 2, 3)$ **15.** $(-4, 8, 5)$ **17.** $(6, -3, 0)$

19. Inconsistent   **21.** $\left(1, -\frac{3}{2}, \frac{1}{2}\right)$

23. $(-3a + 10, 5a - 7, a)$ **25.** $(-a + 3, a + 1, a)$

27. Inconsistent   **29.** Inconsistent

31. $(2a, 21a - 1, 8a)$ **33.** $\left(-\frac{3}{2}a + \frac{1}{2}, -\frac{2}{3}a + 1, a\right)$

35. $(-5a + 3, -a - 5, a)$ **37.** $(1, 1, 1, 1)$

39. $\begin{cases} 3x + y - z = 9 \\ x + 2y - z = 0 \\ -x + y + 3z = 1 \end{cases}$

(Answer is not unique.)

41. $\begin{cases} x + 6y + 4z = 7 \\ 2x - 2y - 4z = 0 \\ -x + y + z = -\frac{7}{4} \end{cases}$

(Answer is not unique.)

43.

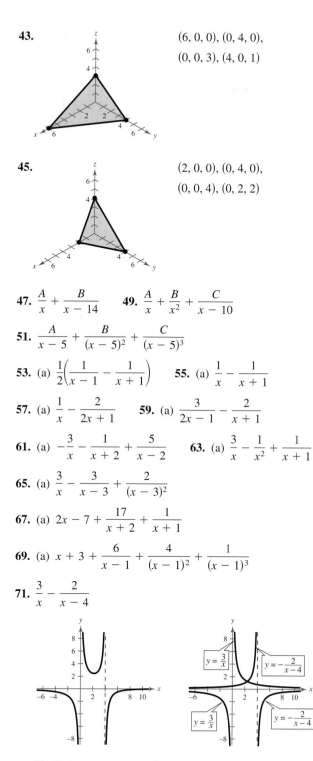

$(6, 0, 0), (0, 4, 0),$
$(0, 0, 3), (4, 0, 1)$

45.

$(2, 0, 0), (0, 4, 0),$
$(0, 0, 4), (0, 2, 2)$

47. $\dfrac{A}{x} + \dfrac{B}{x-14}$ **49.** $\dfrac{A}{x} + \dfrac{B}{x^2} + \dfrac{C}{x-10}$

51. $\dfrac{A}{x-5} + \dfrac{B}{(x-5)^2} + \dfrac{C}{(x-5)^3}$

53. (a) $\dfrac{1}{2}\left(\dfrac{1}{x-1} - \dfrac{1}{x+1}\right)$ **55.** (a) $\dfrac{1}{x} - \dfrac{1}{x+1}$

57. (a) $\dfrac{1}{x} - \dfrac{2}{2x+1}$ **59.** (a) $\dfrac{3}{2x-1} - \dfrac{2}{x+1}$

61. (a) $-\dfrac{3}{x} - \dfrac{1}{x+2} + \dfrac{5}{x-2}$ **63.** (a) $\dfrac{3}{x} - \dfrac{1}{x^2} + \dfrac{1}{x+1}$

65. (a) $\dfrac{3}{x} - \dfrac{3}{x-3} + \dfrac{2}{(x-3)^2}$

67. (a) $2x - 7 + \dfrac{17}{x+2} + \dfrac{1}{x+1}$

69. (a) $x + 3 + \dfrac{6}{x-1} + \dfrac{4}{(x-1)^2} + \dfrac{1}{(x-1)^3}$

71. $\dfrac{3}{x} - \dfrac{2}{x-4}$

Vertical asymptotes are the same.

73. $s = -16t^2 + 144$ **75.** $s = -16t^2 - 32t + 500$

77. $y = \frac{1}{2}x^2 - 2x$ **79.** $y = x^2 - 6x + 8$

81. $x^2 + y^2 - 4x = 0$ **83.** $x^2 + y^2 + 6x - 8y = 0$

85. $\$366{,}666.67$ at 8%, $\$316{,}666.67$ at 9%, and $\$91{,}666.67$ at 10%

87. $\$156{,}250 + 0.75s$ in certificates of deposit

$\$125{,}000$ in municipal bonds

$\$218{,}750 - 1.75s$ in blue-chip stocks

s in growth stocks

89. 20 liters of spray X, 18 liters of spray Y, and 16 liters of spray Z

91. Use four medium trucks or two large trucks, one medium truck, and two small trucks. (Other answers are possible.)

93. $I_1 = 1, I_2 = 2, I_3 = 1$ **95.** $y = -\frac{5}{24}x^2 - \frac{3}{10}x + \frac{41}{6}$

97. $y = x^2 - x$

99. (a) $y = 0.165x^2 - 6.55x + 103$

(b)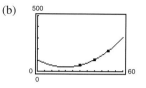

(c) 453 feet

101. (a) $\dfrac{2000}{7-4x} - \dfrac{2000}{11-7x}, \ 0 \le x \le 1$

(b)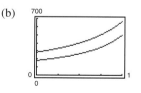

103. False. The leading coefficients are not all 1.

105. False. $\dfrac{A}{x+10} + \dfrac{B}{x-10} + \dfrac{C}{(x-10)^2}$

107. $\dfrac{1}{2a}\left(\dfrac{1}{a+x} + \dfrac{1}{a-x}\right)$ **109.** $\dfrac{1}{a}\left(\dfrac{1}{y} + \dfrac{1}{a-y}\right)$

Answers will vary.   Answers will vary.

111. No. There are two arithmetic errors. They are the constant in the second equation and the coefficient of z in the third equation.

113. $x = 5, y = 5, \lambda = -5$

115. $x = \dfrac{\sqrt{2}}{2}, y = \dfrac{1}{2}, \lambda = 1$

$x = -\dfrac{\sqrt{2}}{2}, y = \dfrac{1}{2}, \lambda = 1$

$x = 0, y = 0, \lambda = 0$

117.

119.

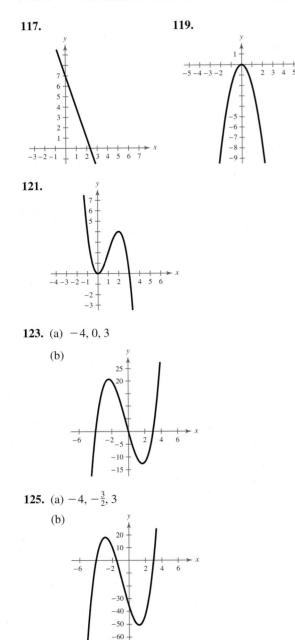

121.

123. (a) $-4, 0, 3$

(b)

125. (a) $-4, -\frac{3}{2}, 3$

(b)

127.

x	-6	-5	-4	-3	0
y	11	-1	-4	-4.75	-4.996

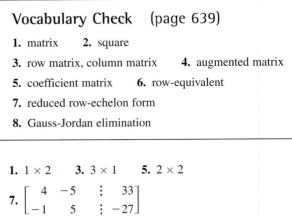

129.

x	-3	-1	0	1	2	3
y	-2.9	-2.6	-2	-0.7	2.5	9.9

Section 8.4   (page 639)

Vocabulary Check   (page 639)

1. matrix   **2.** square

3. row matrix, column matrix   **4.** augmented matrix

5. coefficient matrix   **6.** row-equivalent

7. reduced row-echelon form

8. Gauss-Jordan elimination

1. 1×2 **3.** 3×1 **5.** 2×2

7. $\begin{bmatrix} 4 & -5 & \vdots & 33 \\ -1 & 5 & \vdots & -27 \end{bmatrix}$

9. $\begin{bmatrix} 1 & 10 & -2 & \vdots & 2 \\ 5 & -3 & 4 & \vdots & 0 \\ 2 & 1 & 0 & \vdots & 6 \end{bmatrix}$

11. $\begin{cases} x + 2y = 7 \\ 2x - 3y = 4 \end{cases}$ **13.** $\begin{cases} 9x + 12y + 3z = 0 \\ -2x + 18y + 5z = 10 \\ x + 7y - 8z = -4 \end{cases}$

15. $\begin{bmatrix} 1 & 4 & 3 \\ 0 & 2 & -1 \end{bmatrix}$

17. $\begin{bmatrix} 1 & 1 & 4 & -1 \\ 0 & 5 & -2 & 6 \\ 0 & 3 & 20 & 4 \end{bmatrix}, \begin{bmatrix} 1 & 1 & 4 & -1 \\ 0 & 1 & -\frac{2}{5} & \frac{6}{5} \\ 0 & 3 & 20 & 4 \end{bmatrix}$

19. Add 5 times R_2 to R_1.   **21.** Interchange R_1 and R_2.

23. Reduced row-echelon form

25. Not in row-echelon form

27. (a) $\begin{bmatrix} 1 & 2 & 3 \\ 0 & -5 & -10 \\ 3 & 1 & -1 \end{bmatrix}$ (b) $\begin{bmatrix} 1 & 2 & 3 \\ 0 & -5 & -10 \\ 0 & -5 & -10 \end{bmatrix}$

(c) $\begin{bmatrix} 1 & 2 & 3 \\ 0 & -5 & -10 \\ 0 & 0 & 0 \end{bmatrix}$ (d) $\begin{bmatrix} 1 & 2 & 3 \\ 0 & 1 & 2 \\ 0 & 0 & 0 \end{bmatrix}$

(e) $\begin{bmatrix} 1 & 0 & -1 \\ 0 & 1 & 2 \\ 0 & 0 & 0 \end{bmatrix}$

The matrix is in reduced row-echelon form.

29.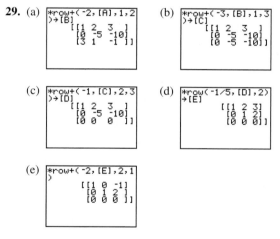

The matrix is in reduced row-echelon form.

31. $\begin{bmatrix} 1 & 1 & 0 & 5 \\ 0 & 1 & 2 & 0 \\ 0 & 0 & 1 & -1 \end{bmatrix}$ **33.** $\begin{bmatrix} 1 & -1 & -1 & 1 \\ 0 & 1 & 6 & 3 \\ 0 & 0 & 0 & 0 \end{bmatrix}$

35. $\begin{bmatrix} 1 & 0 & 0 \\ 0 & 1 & 0 \\ 0 & 0 & 1 \end{bmatrix}$ **37.** $\begin{bmatrix} 1 & 0 & 3 & 16 \\ 0 & 1 & 2 & 12 \end{bmatrix}$

39. $\begin{cases} x - 2y = 4 \\ \quad\;\; y = -3 \end{cases}$ **41.** $\begin{cases} x - y + 2z = 4 \\ \quad\;\; y - z = 2 \\ \qquad\quad z = -2 \end{cases}$

$(-2, -3)$

$(8, 0, -2)$

43. $(7, -5)$ **45.** $(-4, -8, 2)$ **47.** $(3, 2)$

49. Inconsistent **51.** $\left(\frac{3}{2}, -\frac{1}{4}\right)$ **53.** Inconsistent

55. $(4, -3, 2)$ **57.** $(2a + 1, 3a + 2, a)$ **59.** $(7, -3, 4)$

61. $(0, 2 - 4a, a)$ **63.** $(1, 0, 4, -2)$ **65.** $(-2a, a, a)$

67. Yes; $(-1, 1, -3)$ **69.** No **71.** $y = x^2 + 2x + 5$

73. $150{,}000 at 7\%, \$750{,}000 at 8\%, and \$600{,}000 at 10\%$

75. $I_1 = \frac{13}{10}, I_2 = \frac{11}{5}, I_3 = \frac{9}{10}$

77. (a) $y = -3.1t^2 + 57.1t - 218.9$

(b)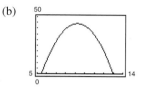

(c) $\$19.90$

(d) $-\$59.90$; No. The average price is negative.

79. (a) $x_1 = s, x_2 = t, x_3 = 600 - s,$

$x_4 = s - t, x_5 = 500 - t, x_6 = s, x_7 = t$

(b) $x_1 = 0, x_2 = 0, x_3 = 600, x_4 = 0, x_5 = 500,$

$x_6 = 0, x_7 = 0$

(c) $x_1 = 0, x_2 = -500, x_3 = 600, x_4 = 500,$

$x_5 = 1000, x_6 = 0, x_7 = -500$

81. False. It is a 2×4 matrix.

83. $\begin{cases} x + y + 7z = -1 \\ x + 2y + 11z = 0 \\ 2x + y + 10z = -3 \end{cases}$

(Answer is not unique.)

85. Gauss-Jordan elimination was not performed on the last column of the matrix. Answers will vary.

87. **89.**

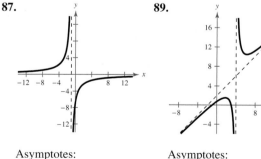

Asymptotes: Asymptotes:

$x = -1, y = 0$ $x = 4, y = x + 2$

91.

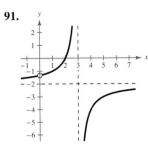

Asymptotes: $x = 3$, $y = -2$

There is a hole at $x = 0$.

Section 8.5   (page 654)

1. $x = -4$, $y = 22$ **3.** $x = -1$, $y = 4$, $z = 6$

5. (a) $\begin{bmatrix} 3 & -2 \\ 1 & 7 \end{bmatrix}$ (b) $\begin{bmatrix} -1 & 0 \\ 3 & -9 \end{bmatrix}$

(c) $\begin{bmatrix} 3 & -3 \\ 6 & -3 \end{bmatrix}$ (d) $\begin{bmatrix} -1 & -1 \\ 8 & -19 \end{bmatrix}$

7. (a) $\begin{bmatrix} 9 & 5 \\ 1 & -2 \\ -3 & 15 \end{bmatrix}$ (b) $\begin{bmatrix} 7 & -7 \\ 3 & 8 \\ -5 & -5 \end{bmatrix}$

(c) $\begin{bmatrix} 24 & -3 \\ 6 & 9 \\ -12 & 15 \end{bmatrix}$ (d) $\begin{bmatrix} 22 & -15 \\ 8 & 19 \\ -14 & -5 \end{bmatrix}$

9. (a) $\begin{bmatrix} 5 & 5 & -2 & 4 & 4 \\ -5 & 10 & 0 & -4 & -7 \end{bmatrix}$

(b) $\begin{bmatrix} 3 & 5 & 0 & 2 & 4 \\ 7 & -6 & -4 & 2 & 7 \end{bmatrix}$

(c) $\begin{bmatrix} 12 & 15 & -3 & 9 & 12 \\ 3 & 6 & -6 & -3 & 0 \end{bmatrix}$

(d) $\begin{bmatrix} 10 & 15 & -1 & 7 & 12 \\ 15 & -10 & -10 & 3 & 14 \end{bmatrix}$

11. (a) Not possible   (b) Not possible

(c) $\begin{bmatrix} 18 & 0 & 9 \\ -3 & -12 & 0 \end{bmatrix}$ (d) Not possible

13. $\begin{bmatrix} -8 & -7 \\ 15 & -1 \end{bmatrix}$ **15.** $\begin{bmatrix} -24 & -4 & 12 \\ -12 & 32 & 12 \end{bmatrix}$

17. $\begin{bmatrix} -1 & 5 \\ 1 & -2 \end{bmatrix}$ **19.** $\begin{bmatrix} -14.646 & 21.306 \\ -41.546 & -69.137 \\ 78.117 & -32.064 \end{bmatrix}$

21. $\begin{bmatrix} 14 & 4 \\ -52 & 0 \end{bmatrix}$ **23.** $\begin{bmatrix} -6 & -9 \\ -1 & 0 \\ 17 & -10 \end{bmatrix}$

25. $\begin{bmatrix} 3 & 3 \\ -0.5 & 0 \\ -6.5 & 5.5 \end{bmatrix}$ **27.** Not possible   **29.** $\begin{bmatrix} -2 & 51 \\ -8 & 33 \\ 0 & 27 \end{bmatrix}$

31. $\begin{bmatrix} 1 & 0 & 0 \\ 0 & 1 & 0 \\ 0 & 0 & \frac{7}{2} \end{bmatrix}$ **33.** $\begin{bmatrix} -15 & -5 & -25 & -45 \\ -18 & -6 & -30 & -54 \end{bmatrix}$

35. (a) $\begin{bmatrix} 0 & 15 \\ 8 & 11 \end{bmatrix}$ (b) $\begin{bmatrix} -3 & 2 \\ 39 & 14 \end{bmatrix}$ (c) $\begin{bmatrix} 11 & 6 \\ 15 & 14 \end{bmatrix}$

37. (a) $\begin{bmatrix} 0 & -10 \\ 10 & 0 \end{bmatrix}$ (b) $\begin{bmatrix} 0 & -10 \\ 10 & 0 \end{bmatrix}$ (c) $\begin{bmatrix} 8 & -6 \\ 6 & 8 \end{bmatrix}$

39. (a) $\begin{bmatrix} 7 & 7 & 14 \\ 8 & 8 & 16 \\ -1 & -1 & -2 \end{bmatrix}$ (b) $[13]$ (c) Not possible

41. $\begin{bmatrix} 70 & -17 & 73 \\ 32 & 11 & 6 \\ 16 & -38 & 70 \end{bmatrix}$ **43.** $\begin{bmatrix} 151 & 25 & 48 \\ 516 & 279 & 387 \\ 47 & -20 & 87 \end{bmatrix}$

45. Not possible   **47.** $\begin{bmatrix} 5 & 8 \\ -4 & -16 \end{bmatrix}$ **49.** $\begin{bmatrix} -4 & 10 \\ 3 & 14 \end{bmatrix}$

51. (a) No   (b) Yes   (c) No   (d) No

53. (a) No   (b) Yes   (c) No   (d) No

55. (a) $\begin{bmatrix} -1 & 1 \\ -2 & 1 \end{bmatrix} \begin{bmatrix} x_1 \\ x_2 \end{bmatrix} = \begin{bmatrix} 4 \\ 0 \end{bmatrix}$ (b) $\begin{bmatrix} 4 \\ 8 \end{bmatrix}$

57. (a) $\begin{bmatrix} -2 & -3 \\ 6 & 1 \end{bmatrix} \begin{bmatrix} x_1 \\ x_2 \end{bmatrix} = \begin{bmatrix} -4 \\ -36 \end{bmatrix}$ (b) $\begin{bmatrix} -7 \\ 6 \end{bmatrix}$

59. (a) $\begin{bmatrix} 1 & -2 & 3 \\ -1 & 3 & -1 \\ 2 & -5 & 5 \end{bmatrix} \begin{bmatrix} x_1 \\ x_2 \\ x_3 \end{bmatrix} = \begin{bmatrix} 9 \\ -6 \\ 17 \end{bmatrix}$ (b) $\begin{bmatrix} 1 \\ -1 \\ 2 \end{bmatrix}$

61. (a) $\begin{bmatrix} 1 & -5 & 2 \\ -3 & 1 & -1 \\ 0 & -2 & 5 \end{bmatrix} \begin{bmatrix} x_1 \\ x_2 \\ x_3 \end{bmatrix} = \begin{bmatrix} -20 \\ 8 \\ -16 \end{bmatrix}$ (b) $\begin{bmatrix} -1 \\ 3 \\ -2 \end{bmatrix}$

63. $\begin{bmatrix} -4 & 0 \\ 8 & 2 \end{bmatrix}$ **65.** $\begin{bmatrix} 84 & 60 & 30 \\ 42 & 120 & 84 \end{bmatrix}$

67. $[\$1037.50 \quad \$1400.00 \quad \$1012.50]$

The entries represent the total profit made at each outlet.

69. $\begin{bmatrix} \$15{,}770 & \$18{,}300 \\ \$26{,}500 & \$29{,}250 \\ \$21{,}260 & \$24{,}150 \end{bmatrix}$

The entries are the total wholesale and retail prices of the inventory at each outlet.

71. $\begin{bmatrix} 0.40 & 0.15 & 0.15 \\ 0.28 & 0.53 & 0.17 \\ 0.32 & 0.32 & 0.68 \end{bmatrix}$

P^2 represents the changes in party affiliations after two elections.

73. True. To add two matrices, you add corresponding entries.

75. Not possible   **77.** Not possible   **79.** 2×2

81. 2×3 **83.** $AC = BC = \begin{bmatrix} 2 & 3 \\ 2 & 3 \end{bmatrix}$

85. (a) $A^2 = \begin{bmatrix} -1 & 0 \\ 0 & -1 \end{bmatrix}, A^3 = \begin{bmatrix} -i & 0 \\ 0 & -i \end{bmatrix}, A^4 = \begin{bmatrix} 1 & 0 \\ 0 & 1 \end{bmatrix}$

The entries on the main diagonal are i^2 in A^2, i^3 in A^3, and i^4 in A^4.

(b) $B^2 = \begin{bmatrix} 1 & 0 \\ 0 & 1 \end{bmatrix}$

B^2 is the identity matrix.

87. (a) $A = \begin{bmatrix} 0 & 2 \\ 0 & 0 \end{bmatrix}, B = \begin{bmatrix} 0 & 2 & 3 \\ 0 & 0 & 4 \\ 0 & 0 & 0 \end{bmatrix}$

(Answers are not unique.)

(b) A^2 and B^3 are zero matrices.

(c) $A = \begin{bmatrix} 0 & 2 & 3 & 4 \\ 0 & 0 & 5 & 6 \\ 0 & 0 & 0 & 7 \\ 0 & 0 & 0 & 0 \end{bmatrix}$

A^4 is the zero matrix.

(d) A^n is the zero matrix.

89. $\ln\left(\dfrac{64}{\sqrt[3]{x^2 + 3}} \right)$ **91.** $\ln\left(\dfrac{\sqrt{x}(x + 5)}{\sqrt{x - 8}} \right)$

Section 8.6   (page 665)

1–11. Answers will vary.

13. $\begin{bmatrix} \frac{1}{2} & 0 \\ 0 & \frac{1}{3} \end{bmatrix}$ **15.** $\begin{bmatrix} -3 & 2 \\ -2 & 1 \end{bmatrix}$ **17.** $\begin{bmatrix} 1 & -1 \\ 2 & -1 \end{bmatrix}$

19. Does not exist   **21.** $\begin{bmatrix} 1 & 1 & -1 \\ -3 & 2 & -1 \\ 3 & -3 & 2 \end{bmatrix}$

23. Does not exist   **25.** Does not exist

27. $\begin{bmatrix} -12 & -5 & -9 \\ -4 & -2 & -4 \\ -8 & -4 & -6 \end{bmatrix}$ **29.** $\frac{5}{11}\begin{bmatrix} 0 & -4 & 2 \\ -22 & 11 & 11 \\ 22 & -6 & -8 \end{bmatrix}$

31. $\begin{bmatrix} 1 & 0 & 1 & 0 \\ 0 & 1 & 0 & 1 \\ 2 & 0 & 1 & 0 \\ 0 & 1 & 0 & 2 \end{bmatrix}$ **33.** $\begin{bmatrix} \frac{1}{4} & \frac{1}{8} \\ -\frac{1}{4} & -\frac{5}{8} \end{bmatrix}$

35. $\frac{1}{59}\begin{bmatrix} 16 & 15 \\ -4 & 70 \end{bmatrix}$ **37.** $(5, 0)$ **39.** $(-8, -6)$

41. $(3, 8, -11)$ **43.** $(2, 1, 0, 0)$ **45.** $(2, -2)$

47. Not possible because A is not invertible

49. $(-4, -8)$ **51.** $(-1, 3, 2)$

53. Not possible because A is not invertible

55. $(5, 0, -2, 3)$

57. \$10,000 in AAA-rated bonds, \$5000 in A-rated bonds, and \$10,000 in B-rated bonds

59. \$20,000 in AAA-rated bonds, \$15,000 in A-rated bonds, and \$30,000 in B-rated bonds

61. (a) $I_1 = -3$ amperes, $I_2 = 8$ amperes, $I_3 = 5$ amperes

(b) $I_1 = \frac{5}{7}$ ampere, $I_2 = \frac{10}{7}$ amperes, $I_3 = \frac{15}{7}$ amperes

63. True   **65.** Answers will vary.

67. $\dfrac{9}{2x + 6}, \quad x \neq 0$ **69.** $\dfrac{x^2 + 2x - 13}{x(x - 2)}, \quad x \neq \pm 3$

71. $\ln 3 \approx 1.099$ **73.** $\dfrac{e^{12/7}}{3} \approx 1.851$

Section 8.7   (page 673)

1. 4   **3.** 16   **5.** 28   **7.** -24 **9.** 0

11. -9 **13.** -0.002

15. (a) $M_{11} = -5, M_{12} = 2, M_{21} = 4, M_{22} = 3$

(b) $C_{11} = -5, C_{12} = -2, C_{21} = -4, C_{22} = 3$

17. (a) $M_{11} = 30, M_{12} = 12, M_{13} = 11, M_{21} = -36,$

$M_{22} = 26, M_{23} = 7, M_{31} = -4, M_{32} = -42,$

$M_{33} = 12$

(b) $C_{11} = 30, C_{12} = -12, C_{13} = 11, C_{21} = 36, C_{22} = 26,$

$C_{23} = -7, C_{31} = -4, C_{32} = 42, C_{33} = 12$

19. (a) -75 (b) -75 **21.** (a) 170   (b) 170

23. -58 **25.** -30 **27.** -168 **29.** 412

31. -60 **33.** -168 **35.** -336 **37.** 410

39. (a) -3 (b) -2 (c) $\begin{bmatrix} -2 & 0 \\ 0 & -3 \end{bmatrix}$ (d) 6

41. (a) 2   (b) -6 (c) $\begin{bmatrix} 1 & 4 & 3 \\ -1 & 0 & 3 \\ 0 & 2 & 0 \end{bmatrix}$ (d) -12

43. (a) -25 (b) -220

(c) $\begin{bmatrix} -7 & -16 & -1 & -28 \\ -4 & -14 & -11 & 8 \\ 13 & 4 & 4 & -4 \\ -2 & 3 & 2 & 2 \end{bmatrix}$ (d) 5500

45–49. Answers will vary.   **51.** $-4, -1$

53. $8uv - 1$ **55.** e^{5x} **57.** $1 - \ln x$ **59.** True

61. Answers will vary.

Sample answer: $A = \begin{bmatrix} 1 & 3 \\ -2 & 4 \end{bmatrix}$, $B = \begin{bmatrix} -4 & 0 \\ 3 & 5 \end{bmatrix}$

$|A + B| = -30$, $|A| + |B| = -10$

63. (a) Columns 2 and 3 are interchanged.

(b) Rows 1 and 3 are interchanged.

65. (a) 5 is factored from the first row.

(b) 4 and 3 are factored from the second and third columns, respectively.

67. $(x - 2)(x - 1)$ **69.** $(2y - 3)^2$ **71.** $(2, -4)$

Section 8.8   (page 684)

Vocabulary Check   (page 684)

1. collinear   **2.** Cramer's Rule   **3.** cryptogram

4. uncoded, coded

1. 14   **3.** $\frac{5}{2}$ **5.** $\frac{33}{8}$ **7.** $x = \frac{16}{5}, 0$

9. Collinear   **11.** Not collinear   **13.** $x = -3$

15. $(-3, -2)$ **17.** Not possible   **19.** $\left(\frac{32}{7}, \frac{30}{7}\right)$

21. $(-1, 3, 2)$ **23.** $\left(0, -\frac{1}{2}, \frac{1}{2}\right)$ **25.** 250 square miles

27. Uncoded: $[3 \quad 1 \quad 12], [12 \quad 0 \quad 13], [5 \quad 0 \quad 20],$
$[15 \quad 13 \quad 15], [18 \quad 18 \quad 15], [23 \quad 0 \quad 0]$

Encoded: $-68 \ 21 \ 35 \ -66 \ 14 \ 39 \ -115 \ 35 \ 60$
$-62 \ 15 \ 32 \ -54 \ 12 \ 27 \ 23 \ -23 \ 0$

29. $38 \ 63 \ 51 \ -1 \ -14 \ -32 \ 58 \ 119 \ 133 \ 44 \ 88 \ 95$

31. HAPPY NEW YEAR   **33.** TEST ON FRIDAY

35. True   **37.** Answers will vary.

39. $x + 4y - 19 = 0$ **41.** $2x - 7y - 27 = 0$

43. 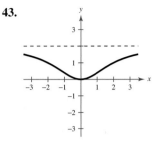   Asymptote: $y = 2$

Review Exercises   (page 688)

1. $(1, 1)$ **3.** $(5, 4)$ **5.** $(0, 0), (2, 8), (-2, 8)$

7. $\left(2, -\frac{1}{2}\right)$ **9.** $(0, 0), (-3, 3)$

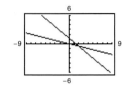

11. $(4, 4)$

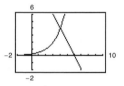

13. 4762 units   **15.** 96 meters × 144 meters   **17.** $\left(\frac{5}{2}, 3\right)$

19. $\left(-\frac{1}{2}, \frac{4}{5}\right)$ **21.** $(0, 0)$ **23.** $\left(\frac{14}{5} + \frac{8}{5}a, a\right)$

25. 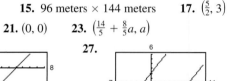   **27.**

Consistent; $(1.6, -2.4)$ Inconsistent

29.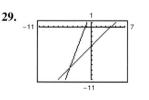

Consistent; $(-4.6, -8.6)$

31. $\left(\frac{500,000}{7}, \frac{159}{7}\right)$

33. 218.75 miles per hour; 193.75 miles per hour

35. $(2, -4, -5)$ **37.** $\left(\frac{38}{17}, \frac{40}{17}, -\frac{63}{17}\right)$

39. $(3a + 4, 2a + 5, a)$

41. $(a - 4, a - 3, a)$

43.

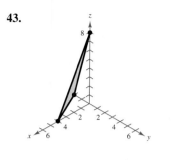

$(0, 0, 8), (0, -2, 0), (4, 0, 0), (1, -1, 2)$

45. (a) $\dfrac{3}{x + 2} - \dfrac{4}{x + 4}$ **47.** (a) $\dfrac{1}{2}\left(\dfrac{3}{x - 1} - \dfrac{x - 3}{x^2 + 1}\right)$

49. $y = 2x^2 + x - 5$

51. Spray X: 10 gallons; Spray Y: 5 gallons;

   Spray Z: 12 gallons

53. 3×1 **55.** 1×1

57. $\begin{bmatrix} 3 & -10 & \vdots & 15 \\ 5 & 4 & \vdots & 22 \end{bmatrix}$ **59.** $\begin{bmatrix} 8 & -7 & 4 & \vdots & 12 \\ 3 & -5 & 2 & \vdots & 20 \\ 5 & 3 & -3 & \vdots & 26 \end{bmatrix}$

61. $\begin{cases} 5x + y + 7z = -9 \\ 4x + 2y = 10 \\ 9x + 4y + 2z = 3 \end{cases}$

63. $\begin{bmatrix} 1 & 0 & 0 \\ 0 & 1 & 0 \\ 0 & 0 & 1 \end{bmatrix}$ **65.** $\begin{bmatrix} 1 & 0 & 3 & -2 \\ 0 & 1 & 4 & -3 \end{bmatrix}$

67. $\begin{bmatrix} 1 & 0 & 1 \\ 0 & 1 & 1 \\ 0 & 0 & 0 \end{bmatrix}$ **69.** $(10, -12)$ **71.** $(-0.2, 0.7)$

73. $\left(\tfrac{1}{2}, -\tfrac{1}{3}, 1\right)$ **75.** $\left(-2a - 5, \tfrac{5}{3}a + \tfrac{2}{3}, a\right)$

77. $(2, -3, 3)$ **79.** $(2, 3, -1)$

81. $(3, 0, -4)$ **83.** $(2, 6, -10, -3)$

85. $x = 12, y = -7$ **87.** $x = 1, y = 11$

89. (a) $\begin{bmatrix} 17 & -17 \\ 13 & 2 \end{bmatrix}$ (b) $\begin{bmatrix} -3 & 23 \\ -15 & 8 \end{bmatrix}$

   (c) $\begin{bmatrix} 28 & 12 \\ -4 & 20 \end{bmatrix}$ (d) $\begin{bmatrix} 37 & -57 \\ 41 & -4 \end{bmatrix}$

91. (a) $\begin{bmatrix} 6 & 5 & 8 \\ 1 & 7 & 8 \\ 5 & 1 & 4 \end{bmatrix}$ (b) $\begin{bmatrix} 6 & -5 & 6 \\ 9 & -9 & -4 \\ 1 & 3 & 2 \end{bmatrix}$

   (c) $\begin{bmatrix} 24 & 0 & 28 \\ 20 & -4 & 8 \\ 12 & 8 & 12 \end{bmatrix}$ (d) $\begin{bmatrix} 6 & 15 & 10 \\ -7 & 23 & 20 \\ 9 & -1 & 6 \end{bmatrix}$

93. $\begin{bmatrix} -13 & -8 & 18 \\ 0 & 11 & -19 \end{bmatrix}$ **95.** $\begin{bmatrix} 2 & 1 & 12 \\ -13 & 1 & -17 \\ -30 & -54 & 40 \end{bmatrix}$

97. $\begin{bmatrix} 48 & -18 & -3 \\ 15 & 51 & 33 \end{bmatrix}$ **99.** $\begin{bmatrix} -14 & -4 \\ 7 & -17 \\ -17 & -2 \end{bmatrix}$

101. $\dfrac{1}{3}\begin{bmatrix} 9 & 2 \\ -4 & 11 \\ 10 & 0 \end{bmatrix}$ **103.** $\begin{bmatrix} 14 & -2 & 8 \\ 14 & -10 & 40 \\ 36 & -12 & 48 \end{bmatrix}$

105. $\begin{bmatrix} 11 & -6 \\ 18 & 39 \end{bmatrix}$ **107.** $\begin{bmatrix} 14 & -22 & 22 \\ 19 & -41 & 80 \\ 42 & -66 & 66 \end{bmatrix}$

109. $\begin{bmatrix} 1 & 17 \\ 12 & 36 \end{bmatrix}$ **111.** $\begin{bmatrix} 96 & 84 & 108 & 48 \\ 60 & 36 & 96 & 24 \\ 108 & 72 & 120 & 60 \end{bmatrix}$

113. Answers will vary.

115. $\begin{bmatrix} 4 & -5 \\ 5 & -6 \end{bmatrix}$ **117.** $\begin{bmatrix} 13 & 6 & -4 \\ -12 & -5 & 3 \\ 5 & 2 & -1 \end{bmatrix}$

119. $\begin{bmatrix} \tfrac{1}{5} & \tfrac{1}{5} \\ \tfrac{1}{10} & -\tfrac{1}{15} \end{bmatrix}$ **121.** $\begin{bmatrix} \tfrac{1}{2} & -1 & -\tfrac{1}{2} \\ \tfrac{1}{2} & -\tfrac{2}{3} & -\tfrac{5}{6} \\ 0 & \tfrac{2}{3} & \tfrac{1}{3} \end{bmatrix}$

123. $\begin{bmatrix} 1 & -1 \\ 4 & -\tfrac{7}{2} \end{bmatrix}$ **125.** Does not exist   **127.** $(36, 11)$

129. $(2, -1, -2)$ **131.** $(6, 1, -1)$ **133.** $(-3, 1)$

135. $(1, 1, -2)$ **137.** -42 **139.** 550

141. (a) $M_{11} = 4, M_{12} = 7, M_{21} = -1, M_{22} = 2$

   (b) $C_{11} = 4, C_{12} = -7, C_{21} = 1, C_{22} = 2$

143. (a) $M_{11} = 30, M_{12} = -12, M_{13} = -21, M_{21} = 20,$

     $M_{22} = 19, M_{23} = 22, M_{31} = 5, M_{32} = -2,$

     $M_{33} = 19$

   (b) $C_{11} = 30, C_{12} = 12, C_{13} = -21, C_{21} = -20,$

     $C_{22} = 19, C_{23} = -22, C_{31} = 5, C_{32} = 2, C_{33} = 19$

145. 130 **147.** -3 **149.** 279 **151.** -96

153. 16 **155.** 1.75 **157.** Not collinear   **159.** $(1, 2)$

161. $(4, 7)$ **163.** $(-1, 4, 5)$ **165.** $(0, -2.4, -2.6)$

167. Uncoded: $[12 \ 15 \ 15], [11 \ 0 \ 15], [21 \ 20 \ 0],$
       $[2 \ 5 \ 12], [15 \ 23 \ 0]$

   Encoded: $-21 \ 6 \ 0 \ -68 \ 8 \ 45 \ 102 \ -42$
       $-60 \ -53 \ 20 \ 21 \ 99 \ -30 \ -69$

169. SEE YOU FRIDAY

171. False, the solution may have irrational numbers.

173. Elementary row operations correspond to the operations performed on a system of equations.

Chapter Test   (page 694)

1. $(4, -2)$ **2.** $(0, -1), (1, 0), (2, 1)$

3. $(8, 5), (2, -1)$ **4.** $\left(\frac{28}{9}, -\frac{31}{9}\right)$ **5.** $(-1, 5, 2)$

6. $\left(-\frac{3}{5}a, \frac{4}{5}a, a\right)$ **7.** $y = -\frac{1}{2}x^2 + x + 6$

8. $\dfrac{5}{x - 1} + \dfrac{3}{(x - 1)^2}$ **9.** $(-2a + 1.5, 2a + 1, a)$

10. $(5, 2, -6)$

11. (a) $\begin{bmatrix} 1 & 0 & 4 \\ -7 & -6 & -1 \\ 0 & 4 & 0 \end{bmatrix}$

(b) $\begin{bmatrix} 15 & 12 & 12 \\ -12 & -12 & 0 \\ 3 & 6 & 0 \end{bmatrix}$

(c) $\begin{bmatrix} 7 & 4 & 12 \\ -18 & -16 & -2 \\ 1 & 10 & 0 \end{bmatrix}$

(d) $\begin{bmatrix} 36 & 20 & 4 \\ -28 & -24 & -4 \\ 10 & 8 & 2 \end{bmatrix}$

12. $A^{-1} = \begin{bmatrix} \frac{1}{2} & \frac{2}{5} \\ 1 & \frac{3}{5} \end{bmatrix}$; $(13, 22)$

13. 67 **14.** -2 **15.** 7 **16.** $\left(\frac{3}{4}, -\frac{1}{2}\right)$

17. $x_1 = 700 - s - t, x_2 = 300 - s - t,$

$x_3 = s, x_4 = 100 - t, x_5 = t$

Chapter 9

Section 9.1   (page 703)

1. $7, 9, 11, 13, 15$ **3.** $2, 4, 8, 16, 32$

5. $-\frac{1}{2}, \frac{1}{4}, -\frac{1}{8}, \frac{1}{16}, -\frac{1}{32}$ **7.** $2, \frac{3}{2}, \frac{4}{3}, \frac{5}{4}, \frac{6}{5}$ **9.** $\frac{1}{2}, \frac{2}{5}, \frac{3}{10}, \frac{4}{17}, \frac{5}{26}$

11. $0, 1, 0, \frac{1}{2}, 0$ **13.** $\frac{1}{2}, \frac{3}{4}, \frac{7}{8}, \frac{15}{16}, \frac{31}{32}$

15. $1, \dfrac{1}{2^{3/2}}, \dfrac{1}{3^{3/2}}, \dfrac{1}{4^{3/2}} = \dfrac{1}{8}, \dfrac{1}{5^{3/2}}$ **17.** $-1, \dfrac{1}{4}, -\dfrac{1}{9}, \dfrac{1}{16}, -\dfrac{1}{25}$

19. $3, 15, 35, 63, 99$ **21.** -73 **23.** $\frac{100}{101}$

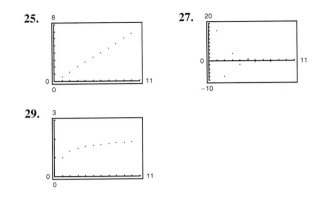

25.   **27.**

29.

31. $9, 15, 21, 27, 33, 39, 45, 51, 57, 63$

33. $3, \frac{5}{2}, \frac{7}{3}, \frac{9}{4}, \frac{11}{5}, \frac{13}{6}, \frac{15}{7}, \frac{17}{8}, \frac{19}{9}, \frac{21}{10}$ **35.** c   **36.** b

37. d   **38.** a   **39.** $a_n = 3n - 2$ **41.** $a_n = n^2 - 1$

43. $a_n = \dfrac{n + 1}{n + 2}$ **45.** $a_n = \dfrac{(-1)^{n+1}}{2^n}$

47. $a_n = 1 + \dfrac{1}{n}$ **49.** $a_n = \dfrac{1}{n!}$

51. $a_n = (-1)^n + 2(1)^n = (-1)^n + 2$

53. $28, 24, 20, 16, 12$ **55.** $3, 4, 6, 10, 18$

57. $6, 8, 10, 12, 14$; $a_n = 2n + 4$

59. $81, 27, 9, 3, 1$; $a_n = \dfrac{243}{3^n}$ **61.** $1, 1, \dfrac{1}{2}, \dfrac{1}{6}, \dfrac{1}{24}$

63. $1, \dfrac{1}{3}, \dfrac{2}{5}, \dfrac{6}{7}, \dfrac{8}{3}$ **65.** $1, \dfrac{1}{2}, \dfrac{1}{24}, \dfrac{1}{720}, \dfrac{1}{40,320}$ **67.** $\dfrac{1}{12}$

69. 495 **71.** $n + 1$ **73.** $\dfrac{1}{2n(2n + 1)}$ **75.** 35

77. 40 **79.** 30 **81.** $\frac{9}{5}$ **83.** 238 **85.** 30

87. 81 **89.** $\dfrac{47}{60}$ **91.** $\displaystyle\sum_{i=1}^{9} \dfrac{1}{3i} \approx 0.94299$

93. $\displaystyle\sum_{i=1}^{8} \left[2\left(\dfrac{i}{8}\right) + 3\right] = 33$ **95.** $\displaystyle\sum_{i=1}^{6} (-1)^{i+1} 3i = -546$

97. $\displaystyle\sum_{i=1}^{20} \dfrac{(-1)^{i+1}}{i^2} \approx 0.821$ **99.** $\displaystyle\sum_{i=1}^{5} \dfrac{2^i - 1}{2^{i+1}} \approx 2.0156$

101. $\frac{75}{16}$ **103.** $-\frac{3}{2}$ **105.** $\frac{2}{3}$ **107.** $\frac{1}{9}$

109. (a) $A_1 = \$5037.50, A_2 = \$5075.28,$

$A_3 = \$5113.35, A_4 = \$5151.70,$

$A_5 = \$5190.33, A_6 = \$5229.26,$

$A_7 = \$5268.48, A_8 = \$5307.99$

(b) $\$6741.74$

111. (a) $a_3 = 719,630, a_4 = 732,320,$

$a_5 = 747,750, a_6 = 765,920,$

$a_7 = 786,830, a_8 = 810,480,$

$a_9 = 836,870, a_{10} = 866,000, a_{11} = 897,870$

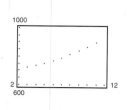

(b) Enrollment will increase.

113. $3,491,000,000; The sums are approximately the same.

115. True

117. 1, 1, 2, 3, 5, 8, 13, 21, 34, 55, 89, 144;

$1, 2, \frac{3}{2}, \frac{5}{3}, \frac{8}{5}, \frac{13}{8}, \frac{21}{13}, \frac{34}{21}, \frac{55}{34}, \frac{89}{55}$

119. $x, \dfrac{x^2}{2}, \dfrac{x^3}{6}, \dfrac{x^4}{24}, \dfrac{x^5}{120}$

121. $-\dfrac{x^2}{2}, \dfrac{x^4}{24}, -\dfrac{x^6}{720}, \dfrac{x^8}{40,320}, -\dfrac{x^{10}}{3,628,800}$

123. (a) $\begin{bmatrix} 8 & 1 \\ -3 & 7 \end{bmatrix}$ (b) $\begin{bmatrix} -22 & -7 \\ 3 & -18 \end{bmatrix}$

(c) $\begin{bmatrix} 18 & 9 \\ 18 & 0 \end{bmatrix}$ (d) $\begin{bmatrix} 0 & 6 \\ 27 & 18 \end{bmatrix}$

125. (a) $\begin{bmatrix} -3 & -7 & 4 \\ 4 & 4 & 1 \\ 1 & 4 & 3 \end{bmatrix}$ (b) $\begin{bmatrix} 8 & 17 & -14 \\ -12 & -13 & -9 \\ -3 & -15 & -10 \end{bmatrix}$

(c) $\begin{bmatrix} -2 & 7 & -16 \\ 4 & 42 & 45 \\ 1 & 23 & 48 \end{bmatrix}$ (d) $\begin{bmatrix} 16 & 31 & 42 \\ 10 & 47 & 31 \\ 13 & 22 & 25 \end{bmatrix}$

Section 9.2   (page 713)

Vocabulary Check   (page 713)

1. arithmetic, common   **2.** $a_n = dn + c$

3. nth partial sum

1. Arithmetic sequence, $d = -2$

3. Arithmetic sequence, $d = -\frac{1}{2}$

5. Arithmetic sequence, $d = 8$

7. Arithmetic sequence, $d = 0.6$

9. 21, 34, 47, 60, 73

Arithmetic sequence, $d = 13$

11. $\frac{1}{2}, \frac{1}{3}, \frac{1}{4}, \frac{1}{5}, \frac{1}{6}$

Not an arithmetic sequence

13. 143, 136, 129, 122, 115

Arithmetic sequence, $d = -7$

15. $1, 4, \frac{7}{3}, \frac{7}{2}, \frac{13}{5}$

Not an arithmetic sequence

17. $a_n = -2 + 3n$ **19.** $a_n = 108 - 8n$

21. $a_n = \frac{13}{2} - \frac{5}{2}n$ **23.** $a_n = \frac{10}{3}n + \frac{5}{3}$

25. $a_n = 103 - 3n$ **27.** 5, 11, 17, 23, 29

29. $-10, -22, -34, -46, -58$ **31.** $-2, 2, 6, 10, 14$

33. 22.45, 20.725, 19, 17.275, 15.55

35. 15, 19, 23, 27, 31; $d = 4$; $a_n = 11 + 4n$

37. $\frac{7}{2}, \frac{13}{4}, 3, \frac{11}{4}, \frac{5}{2}$; $d = -\frac{1}{4}$; $a_n = -\frac{1}{4}n + \frac{15}{4}$

39. 59   **41.** 18.6

43.   **45.**

47. $-1, 3, 7, 11, 15, 19, 23, 27, 31, 35$

49. 19.25, 18.5, 17.75, 17, 16.25, 15.5, 14.75, 14, 13.25, 12.5

51. 1.505, 1.51, 1.515, 1.52, 1.525, 1.53, 1.535, 1.54, 1.545, 1.55

53. 620   **55.** 41   **57.** 4000   **59.** 10,000

61. 1275   **63.** 25,250   **65.** 355   **67.** 129,250

69. 520   **71.** 2725   **73.** 10,120

75. (a) $40,000   (b) $217,500

77. 405 bricks   **79.** 585 seats   **81.** $150,000

83. True. Given a_1 and a_2, you know $d = a_2 - a_1$. Hence, $a_n = a_1 + (n - 1)d$.

85. $x, 3x, 5x, 7x, 9x, 11x, 13x, 15x, 17x, 19x$ **87.** 4

89. (a) 4, 9, 16, 25, 36

(b) The sum of the first n positive odd integers is n^2; 49

(c) $\dfrac{n}{2}[1 + (2n - 1)] = n^2$

91. $(1, 5, -1)$ **93.** 15 square units

Section 9.3   (page 722)

Vocabulary Check   (page 722)

1. geometric, common   **2.** $a_n = a_1 r^{n-1}$

3. $S_n = \displaystyle\sum_{i=1}^{n} a_1 r^{i-1} = a_1\left(\dfrac{1 - r^n}{1 - r}\right)$

4. geometric series   **5.** $S = \displaystyle\sum_{i=0}^{\infty} a_1 r^i = \dfrac{a_1}{1 - r}$

1. Geometric sequence, $r = 3$

3. Not a geometric sequence

5. Geometric sequence, $r = -\frac{1}{2}$

7. Geometric sequence, $r = 2$

9. Not a geometric sequence   **11.** 6, 18, 54, 162, 486

13. $1, \frac{1}{2}, \frac{1}{4}, \frac{1}{8}, \frac{1}{16}$ **15.** $5, -\frac{1}{2}, \frac{1}{20}, -\frac{1}{200}, \frac{1}{2000}$

17. $1, e, e^2, e^3, e^4$

19. 64, 32, 16, 8, 4; $r = \frac{1}{2}$;   **21.** 9, 18, 36, 72, 144; $r = 2$;

$a_n = 128\left(\frac{1}{2}\right)^n$ $a_n = 9(2)^{n-1}$

23. $6, -9, \frac{27}{2}, -\frac{81}{4}, \frac{243}{8}; r = -\frac{3}{2}$;

$a_n = 6\left(-\frac{3}{2}\right)^{n-1}$

25. $\left(\frac{1}{2}\right)^7$ **27.** $-\frac{2}{3^{10}}$ **29.** $500(1.02)^{13}$ **31.** $-\frac{2}{9}$

33. 45,927   **35.** 50,388,480

37. 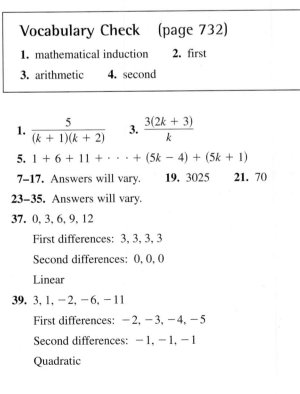   **39.**

41. 8, 4, 6, 5

43.

n	S_n
1	16
2	24
3	28
4	30
5	31
6	31.5
7	31.75
8	31.875
9	31.9375
10	31.96875

45. 511   **47.** 43

49. 29,921.31   **51.** 6.4   **53.** 2092.60

55. $\sum_{n=1}^{7} 5(3)^{n-1}$ **57.** $\sum_{n=1}^{7} 2\left(-\frac{1}{4}\right)^{n-1}$ **59.** 2   **61.** $\frac{2}{3}$

63. Series does not have a finite sum.   **65.** $\frac{5}{3}$

67. -30 **69.** 32   **71.** $\frac{9}{4}$ **73.** $\frac{4}{11}$ **75.** $\frac{7}{22}$

77. (a) \$1343.92   (b) \$1346.86   (c) \$1348.35

(d) \$1349.35   (e) \$1349.84

79. \$7011.89   **81.** Answers will vary.

83. (a) \$26,198.27   (b) \$26,263.88

85. (a) \$153,237.86   (b) \$153,657.02

87. 126 square inches

89. (a) 1.3%   (b) 4,328,000 people   (c) 2005

91. (a) \$5,368,709.11   (b) \$10,737,418.23

(c) \$21,474,836.47

93. False. Any arithmetic sequence can be used as a counterexample.

95. $3, \frac{3x}{2}, \frac{3x^2}{4}, \frac{3x^3}{8}, \frac{3x^4}{16}$ **97.** $100e^{8x}$

99. (a)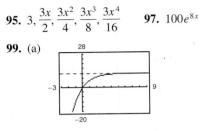

Horizontal asymptote: $y = 12$

Corresponds to the sum of the series

(b)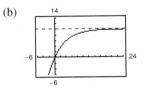

Horizontal asymptote: $y = 10$

Corresponds to the sum of the series

101. Divide the second term by the first to obtain the common ratio. The nth term is the first term times the common ratio raised to the $n - 1$ power.

103. 45.65 miles per hour   **105.** 14   **107.** -102

Section 9.4   (page 732)

Vocabulary Check   (page 732)

1. mathematical induction   **2.** first

3. arithmetic   **4.** second

1. $\dfrac{5}{(k+1)(k+2)}$ **3.** $\dfrac{3(2k+3)}{k}$

5. $1 + 6 + 11 + \cdots + (5k - 4) + (5k + 1)$

7–17. Answers will vary.   **19.** 3025   **21.** 70

23–35. Answers will vary.

37. 0, 3, 6, 9, 12

First differences: 3, 3, 3, 3

Second differences: 0, 0, 0

Linear

39. $3, 1, -2, -6, -11$

First differences: $-2, -3, -4, -5$

Second differences: $-1, -1, -1$

Quadratic

41. 0, 1, 3, 6, 10

    First differences: 1, 2, 3, 4

    Second differences: 1, 1, 1

    Quadratic

43. 2, 4, 6, 8, 10

    First differences: 2, 2, 2, 2

    Second differences: 0, 0, 0

    Linear

45. $a_n = n^2 - n + 3$ **47.** $a_n = \frac{1}{2}n^2 + n - 3$

49. (a) First differences: 2.2, 2.5, 2.1, 1.8, 2.3, 1.9

    (b) Yes; $a_n = 2.13n + 22.8$

    (c)

Year	Original data	Data using model
1995	33.3	33.5
1996	35.5	35.6
1997	38.0	37.7
1998	40.1	39.8
1999	41.9	42.0
2000	44.2	44.1
2001	46.1	46.2

    The model represents the data well.

    (d) $56,900

51. False. Not necessarily

53. False. It has $n - 2$ second differences.

55. $4x^4 - 4x^2 + 1$ **57.** $-64x^3 + 240x^2 - 300x + 125$

59. $x^2 - 3x + 2$ **61.** $4x^2 - 9x + 2$

63. $7\sqrt{3}i$ **65.** $40(1 + \sqrt[3]{2})$

Section 9.5 (page 739)

Vocabulary Check (page 739)

1. binomial coefficients

2. Binomial Theorem, Pascal's Triangle   **3.** $_nC_r$

4. expanding, binomial

---

1. 21   **3.** 1   **5.** 15,504   **7.** 14   **9.** 4950

11. 35,960   **13.** 497,420   **15.** 749,398   **17.** 35

19. 56   **21.** $x^4 + 8x^3 + 24x^2 + 32x + 16$

23. $a^3 + 9a^2 + 27a + 27$

25. $y^4 - 8y^3 + 24y^2 - 32y + 16$

27. $x^5 + 5x^4y + 10x^3y^2 + 10x^2y^3 + 5xy^4 + y^5$

29. $729r^6 + 2916r^5s + 4860r^4s^2 + 4320r^3s^3 + 2160r^2s^4$
    $+ 576rs^5 + 64s^6$

31. $x^5 - 5x^4y + 10x^3y^2 - 10x^2y^3 + 5xy^4 - y^5$

33. $1 - 12x + 48x^2 - 64x^3$

35. $x^8 + 4x^6y^2 + 6x^4y^4 + 4x^2y^6 + y^8$

37. $\dfrac{1}{x^5} + \dfrac{5y}{x^4} + \dfrac{10y^2}{x^3} + \dfrac{10y^3}{x^2} + \dfrac{5y^4}{x} + y^5$

39. $2x^4 - 24x^3 + 113x^2 - 246x + 207$

41. $-4x^6 - 24x^5 - 60x^4 - 83x^3 - 42x^2 - 60x + 20$

43. $243t^5 - 405t^4s + 270t^3s^2 - 90t^2s^3 + 15ts^4 - s^5$

45. $x^5 + 10x^4y + 40x^3y^2 + 80x^2y^3 + 80xy^4 + 32y^5$

47. $61,440x^7$ **49.** $360x^3y^2$ **51.** $1,259,712x^2y^7$

53. $32,476,950,000x^4y^8$ **55.** 3,247,695   **57.** 180

59. $-489,888$ **61.** 210

63. $x^2 + 20x^{3/2} + 150x + 500x^{1/2} + 625$

65. $x^2 - 3x^{4/3}y^{1/3} + 3x^{2/3}y^{2/3} - y$

67. $3x^2 + 3xh + h^2,\ h \neq 0$

69. $\dfrac{\sqrt{x + h} - \sqrt{x}}{h} = \dfrac{1}{\sqrt{x + h} + \sqrt{x}},\ h \neq 0$

71. -4 **73.** $2035 + 828i$ **75.** 1   **77.** 1.172

79. 510,568.785

81.

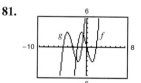

    g is shifted three units to the left.

    $g(x) = x^3 + 9x^2 + 23x + 15$

83.

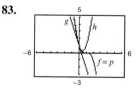

    85. 0.273   **87.** 0.171

    $p(x)$ is the expansion of $f(x)$.

89. (a) $g(t) = 0.018t^2 + 5.51t + 94.9,\ -5 \leq t \leq 10$

    (b)

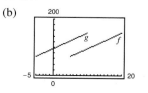

91. False. The correct term is $126{,}720x^4y^8$.

93. The first and last numbers in each row are 1. Every other number in each row is formed by adding the two numbers immediately above the number.

95. Alternating terms of $(x - y)^n$ are negative.

97 and 99. Answers will vary.

101. $g(x)$ is shifted eight units up from $f(x)$.

103. $g(x)$ is the reflection of $f(x)$ in the y-axis.

105. $\begin{bmatrix} 4 & -5 \\ 5 & -6 \end{bmatrix}$

Section 9.6 (page 749)

Vocabulary Check (page 749)

1. Fundamental Counting Principle **2.** permutation

3. $_nP_r = \dfrac{n!}{(n-r)!}$ **4.** distinguishable permutations

5. combinations

1. 6 **3.** 5 **5.** 3 **7.** 7 **9.** 120 **11.** 24

13. 1024 **15.** 12 **17.** 17,576,000

19. (a) 900 (b) 648 (c) 180 (d) 600

21. 16,000,000 **23.** (a) 720 (b) 48 **25.** 24

27. 336 **29.** 120 **31.** $n = 5$ or $n = 6$

33. 27,907,200 **35.** 197,149,680 **37.** 4845

39. 120 **41.** 11,880 **43.** 420 **45.** 2520

47. (a) ABCD, ABDC, ACBD, ACDB, ADBC, ADCB,
BACD, BADC, CABD, CADB, DABC, DACB,
BCAD, BDAC, CBAD, CDAB, DBAC, DCAB,
BCDA, BDCA, CBDA, CDBA, DBCA, DCBA

 (b) ABCD, ACBD, DBCA, DCBA

49. 4845 **51.** 13,983,816 **53.** 36 **55.** 3744

57. (a) 495 (b) 210 **59.** 292,600 **61.** 5

63. 20 **65.** False. This is an example of a combination.

67. For some calculators the answer is too large.

69. They are equal. **71 and 73.** Answers will vary.

75. 8.303 **77.** 35 **79.** $(-2, -8)$ **81.** $(-1, 1)$

Section 9.7 (page 760)

Vocabulary Check (page 760)

1. experiment, outcomes **2.** sample space

3. probability **4.** impossible, certain

5. mutually exclusive **6.** independent

7. complement **8.** (a) iii (b) i (c) iv (d) ii

1. $\{(H, 1), (H, 2), (H, 3), (H, 4), (H, 5), (H, 6),$
$(T, 1), (T, 2), (T, 3), (T, 4), (T, 5), (T, 6)\}$

3. $\{ABC, ACB, BAC, BCA, CAB, CBA\}$

5. $\{(A, B), (A, C), (A, D), (A, E), (B, C), (B, D),$
$(B, E), (C, D), (C, E), (D, E)\}$

7. $\frac{3}{8}$ **9.** $\frac{7}{8}$ **11.** $\frac{3}{13}$ **13.** $\frac{3}{26}$ **15.** $\frac{1}{9}$ **17.** $\frac{11}{12}$

19. $\frac{1}{5}$ **21.** $\frac{2}{5}$ **23.** 0.2 **25.** $\frac{2}{3}$ **27.** 0.88 **29.** $\frac{7}{20}$

31. (a) 1.2 million (b) 0.42 (c) 0.21 (d) Over 0.22

33. (a) 0.34 (b) 0.45 (c) 0.23

35. (a) $\frac{672}{1254}$ (b) $\frac{582}{1254}$ (c) $\frac{548}{1254}$

37. $P(\{\text{Taylor wins}\}) = \frac{1}{2}$
$P(\{\text{Moore wins}\}) = P(\{\text{Perez wins}\}) = \frac{1}{4}$

39. (a) $\frac{21}{1292} \approx 0.016$ (b) $\frac{225}{646} \approx 0.348$ (c) $\frac{49}{323} \approx 0.152$

41. (a) $\frac{1}{120}$ (b) $\frac{1}{24}$ **43.** (a) $\frac{5}{13}$ (b) $\frac{1}{2}$ (c) $\frac{4}{13}$

45. (a) $\frac{14}{55}$ (b) $\frac{12}{55}$ (c) $\frac{54}{55}$ **47.** 0.1024

49. (a) 0.9702 (b) 0.9998 (c) 0.0002

51. (a) $\dfrac{1}{15{,}625}$ (b) $\dfrac{4096}{15{,}625}$ (c) $\dfrac{11{,}529}{15{,}625}$

53. $\frac{7}{16}$ **55.** True

57. (a) As you consider successive people with distinct birthdays, the probabilities must decrease to take into account the birth dates already used. Because the birth dates of people are independent events, multiply the respective probabilities of distinct birthdays.

 (b) $\frac{365}{365} \cdot \frac{364}{365} \cdot \frac{363}{365} \cdot \frac{362}{365}$

 (c) Answers will vary.

 (d) Q_n is the probability that the birthdays are *not* distinct, which is equivalent to at least two people having the same birthday.

 (e)

n	10	15	20	23	30	40	50
P_n	0.88	0.75	0.59	0.49	0.29	0.11	0.03
Q_n	0.12	0.25	0.41	0.51	0.71	0.89	0.97

 (f) 23

59. $x = \frac{11}{2}$ **61.** $x = -10$ **63.** $\ln 28 \approx 3.332$

65. $x = \frac{1}{6}e^4 \approx 9.100$ **67.** 60   **69.** 6,652,800

71. 15   **73.** 165

Review Exercises   (page 766)

1. $8, 5, 4, \frac{7}{2}, \frac{16}{5}$ **3.** $72, 36, 12, 3, \frac{3}{5}$

5.   **7.**

9. $\frac{1}{380}$ **11.** $\frac{1}{3}$ **13.** 30   **15.** $\frac{205}{24}$ **17.** 6050

19. 418   **21.** $\sum\limits_{k=1}^{20} \frac{1}{2k}$; 1.799   **23.** $\sum\limits_{k=1}^{9} \frac{k}{k+1}$; 7.071

25. (a) $\frac{1111}{2000}$ (b) $\frac{5}{9}$ **27.** (a) $\frac{2,020,202}{100,000,000}$ (b) $\frac{2}{99}$

29. (a) $a_1 = \$2550.00$, $a_2 = \$2601.00$, $a_3 = \$2653.02$,

$a_4 = \$2706.08$, $a_5 = \$2760.20$, $a_6 = \$2815.41$,

$a_7 = \$2871.71$, $a_8 = \$2929.15$

(b) $5520.10

31. Arithmetic sequence, $d = -2$

33. Arithmetic sequence, $d = \frac{1}{2}$

35. 3, 7, 11, 15, 19   **37.** 1, 4, 7, 10, 13

39. 35, 32, 29, 26, 23; $d = -3$;

$a_n = 38 - 3n$

41. 9, 16, 23, 30, 37; $d = 7$;

$a_n = 2 + 7n$

43. $a_n = 103 - 3n$; 1430   **45.** 80   **47.** 88

49. 25,250   **51.** (a) $43,000   (b) $192,500

53. Geometric sequence, $r = 2$

55. Not a geometric sequence

57. $4, -1, \frac{1}{4}, -\frac{1}{16}, \frac{1}{64}$ **59.** $9, 6, 4, \frac{8}{3}, \frac{16}{9}$ or $9, -6, 4, -\frac{8}{3}, \frac{16}{9}$

61. $120, 40, \frac{40}{3}, \frac{40}{9}, \frac{40}{27}$; $r = \frac{1}{3}$;

$a_n = 120\left(\frac{1}{3}\right)^{n-1}$

63. $25, -15, 9, -\frac{27}{5}, \frac{81}{25}$; $r = -\frac{3}{5}$;

$a_n = 25\left(-\frac{3}{5}\right)^{n-1}$

65. $a_n = 16\left(-\frac{1}{2}\right)^{n-1}$; 10.67

67. $a_n = 100(1.05)^{n-1}$; 3306.60   **69.** 127   **71.** 3277

73. 1301.01   **75.** 24.85   **77.** 8   **79.** 12

81. (a) $a_t = 120,000(0.7)^t$ (b) $20,168.40

83 and 85. Answers will vary.   **87.** 465   **89.** 4648

91. 5, 10, 15, 20, 25

First differences: 5, 5, 5, 5

Second differences: 0, 0, 0

Linear model

93. 16, 15, 14, 13, 12

First differences: $-1, -1, -1, -1$

Second differences: 0, 0, 0

Linear model

95. 45   **97.** 126   **99.** 20   **101.** 70

103. $x^4 + 20x^3 + 150x^2 + 500x + 625$

105. $a^5 - 20a^4b + 160a^3b^2 - 640a^2b^3 + 1280ab^4 - 1024b^5$

107. $1241 + 2520i$ **109.** 10   **111.** 48   **113.** 5040

115. 3,628,800   **117.** 15,504   **119.** $\frac{1}{9}$

121. (a) 0.416   (b) 0.8   (c) 0.074   **123.** 0.0475

125. True. $\dfrac{(n+2)!}{n!} = \dfrac{(n+2)(n+1)n!}{n!} = (n+2)(n+1)$

127. (a) Each term is obtained by adding the same constant (common difference) to the preceding term.

(b) Each term is obtained by multiplying the same constant (common ratio) by the preceding term.

129. (a) Arithmetic. There is a constant difference between consecutive terms.

(b) Geometric. Each term is a constant multiple of the preceding term. In this case the common ratio is greater than 1.

131. Each term of the sequence is defined using a previous term or terms.

133. If n is even, the expressions are the same. If n is odd, the expressions are negatives of each other.

Chapter Test   (page 770)

1. $1, -\frac{2}{3}, \frac{4}{9}, -\frac{8}{27}, \frac{16}{81}$ **2.** 12, 16, 20, 24, 28   **3.** 7920

4. $a_n = 5100 - 100n$ **5.** $a_n = 4\left(\frac{1}{2}\right)^{n-1}$

6. $\sum\limits_{n=1}^{12} \dfrac{2}{3n+1}$ **7.** 189   **8.** 28.80   **9.** $\frac{50}{9}$

10. Answers will vary.

11. $16a^4 - 160a^3b + 600a^2b^2 - 1000ab^3 + 625b^4$

12. 48,384   **13.** 84   **14.** 1140   **15.** 72

16. 328,440   **17.** 26,000   **18.** 12,650   **19.** $\frac{3}{26}$

20. $\frac{1}{6}$ **21.** (a) $\frac{1}{4}$ (b) $\frac{121}{3600}$ (c) $\frac{1}{60}$ **22.** 0.25

Chapter 10

Section 10.1   (page 781)

1. $x^2 + y^2 = 36$ **3.** $x^2 + y^2 = \frac{25}{49}$

5. Vertex: $(0, 0)$ **7.** Vertex: $(0, 0)$

Focus: $\left(0, \frac{1}{2}\right)$ Focus: $\left(-\frac{3}{2}, 0\right)$

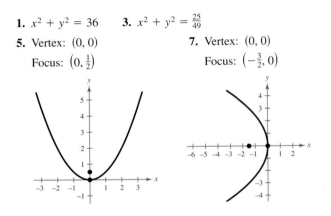

9. Vertex: $(0, 0)$

Focus: $(0, -2)$

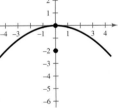

11. $x^2 = -6y$ **13.** $y^2 = -8x$ **15.** $x^2 = 4y$

17. $y^2 = -12x$ **19.** $y^2 = 9x$

21. $y = \frac{2}{3}x^2$; focus: $\left(0, \frac{3}{8}\right)$ **23.** $x = \frac{5}{9}y^2$; focus: $\left(\frac{9}{20}, 0\right)$

25.

$(2, 4)$

27. Center: $(0, 0)$ **29.** Center: $(0, 0)$

Vertices: $(\pm 5, 0)$ Vertices: $\left(\pm\frac{5}{3}, 0\right)$

31. Center: $(0, 0)$ **33.** Center: $(0, 0)$

Vertices: $(\pm 3, 0)$ Vertices: $(0, \pm 1)$

35.   **37.**

39. $\frac{x^2}{1} + \frac{y^2}{4} = 1$ **41.** $\frac{x^2}{4} + \frac{y^2}{\frac{9}{4}} = 1$ **43.** $\frac{x^2}{25} + \frac{y^2}{21} = 1$

45. $\frac{x^2}{36} + \frac{y^2}{11} = 1$ **47.** $\frac{x^2}{\frac{400}{21}} + \frac{y^2}{25} = 1$

49. Center: $(0, 0)$

Vertices: $(\pm 1, 0)$

Foci: $\left(\pm\sqrt{2}, 0\right)$

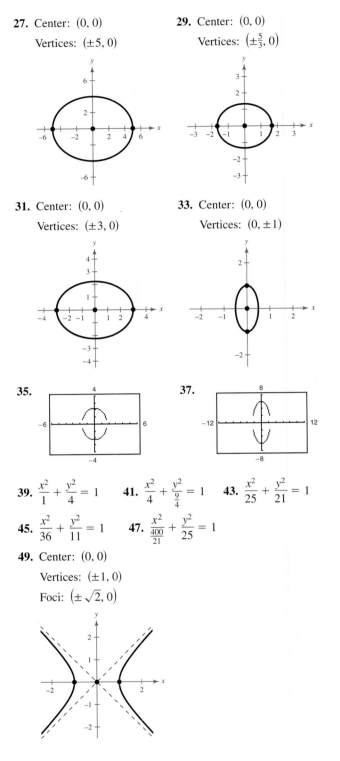

51. Center: $(0, 0)$
Vertices: $(0, \pm1)$
Foci: $\left(0, \pm\sqrt{10}\right)$

53. Center: $(0, 0)$
Vertices: $(0, \pm5)$
Foci: $(0, \pm13)$

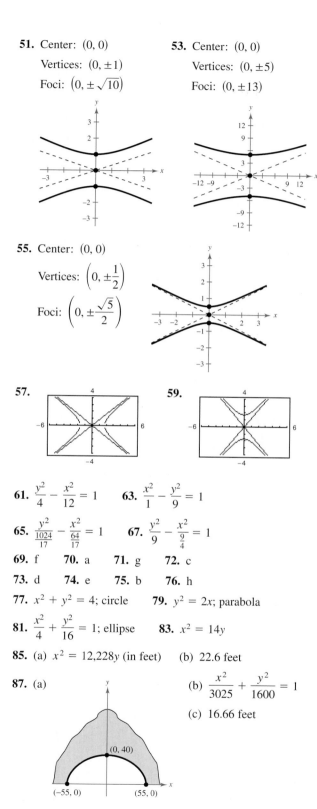

55. Center: $(0, 0)$
Vertices: $\left(0, \pm\dfrac{1}{2}\right)$
Foci: $\left(0, \pm\dfrac{\sqrt{5}}{2}\right)$

57.

59.

61. $\dfrac{y^2}{4} - \dfrac{x^2}{12} = 1$ **63.** $\dfrac{x^2}{1} - \dfrac{y^2}{9} = 1$

65. $\dfrac{y^2}{\frac{1024}{17}} - \dfrac{x^2}{\frac{64}{17}} = 1$ **67.** $\dfrac{y^2}{9} - \dfrac{x^2}{\frac{9}{4}} = 1$

69. f   **70.** a   **71.** g   **72.** c

73. d   **74.** e   **75.** b   **76.** h

77. $x^2 + y^2 = 4$; circle   **79.** $y^2 = 2x$; parabola

81. $\dfrac{x^2}{4} + \dfrac{y^2}{16} = 1$; ellipse   **83.** $x^2 = 14y$

85. (a) $x^2 = 12{,}228y$ (in feet)   (b) 22.6 feet

87. (a)

(b) $\dfrac{x^2}{3025} + \dfrac{y^2}{1600} = 1$

(c) 16.66 feet

89.

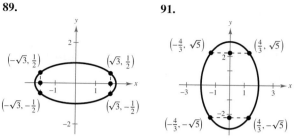

91.

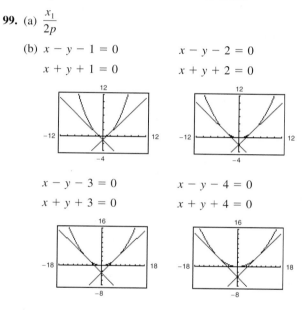

93. $x \approx 110.3$ miles

95. False. The equation represents a hyperbola.

97. False. If the graph intersected the directrix, there would exist points nearer the directrix than the focus.

99. (a) $\dfrac{x_1}{2p}$

(b) $x - y - 1 = 0$ $x - y - 2 = 0$
$x + y + 1 = 0$ $x + y + 2 = 0$

$x - y - 3 = 0$ $x - y - 4 = 0$
$x + y + 3 = 0$ $x + y + 4 = 0$

101. Left half   **103.** Bottom half

105. No. Only second-degree equations can be ellipses.

107. The shape continuously changes from an ellipse with a vertical major axis of length 8 and a minor axis of length 2 to a circle with a diameter of 8 and then to an ellipse with a horizontal major axis of length 16 and a minor axis of length 8.

109. Answers will vary.   **111.** $(3x - 2)(4x + 5)$

113. $z^2(12z + 5)(z + 1)$

115. Answers will vary.   **117.** Answers will vary.
Sample answer: Sample answer:
$f(x) = x^3 - 7x^2 + 12x$ $f(x) = x^3 + x^2 - 7x - 3$

119. $\frac{3}{2}, \pm5i$ **121.** 39,916,800

Section 10.2   (page 791)

Vocabulary Check   (page 791)

1. e   **2.** d   **3.** b   **4.** a   **5.** c

6. f   **7.** g

1. The graph is a circle whose center is the point $(-2, 1)$ and whose radius is 2.

3. The graph is a hyperbola whose center is the point $(1, -3)$. The transverse axis is vertical and of length 4, and the conjugate axis is horizontal and of length 2.

5. The graph is an ellipse whose center is the point $(1, -2)$. The major axis of the ellipse is vertical and of length 8. The minor axis of the ellipse is horizontal and of length 6.

7. Center: $(0, 0)$
   Radius: 7

9. Center: $(-2, 7)$
   Radius: 4

11. Center: $(1, 0)$
   Radius: $\sqrt{15}$

13. $(x - 1)^2 + (y + 3)^2 = 1$
   Center: $(1, -3)$
   Radius: 1

15. $\left(x + \frac{3}{2}\right)^2 + (y - 3)^2 = 1$
   Center: $\left(-\frac{3}{2}, 3\right)$
   Radius: 1

17. Vertex: $(-1, -2)$
   Focus: $(-1, -4)$
   Directrix: $y = 0$

19. Vertex: $\left(5, \frac{1}{2}\right)$
   Focus: $\left(\frac{11}{2}, \frac{1}{2}\right)$
   Directrix: $x = \frac{9}{2}$

21. Vertex: $(1, 1)$
   Focus: $(1, 2)$
   Directrix: $y = 0$

23. Vertex: $(-2, -3)$
   Focus: $(-4, -3)$
   Directrix: $x = 0$

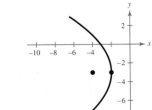

25. $(x - 3)^2 = -(y - 1)$ **27.** $(y - 2)^2 = -8(x - 3)$

29. $x^2 = 8(y - 4)$ **31.** $(y - 2)^2 = 8x$

33. Center: $(1, 3)$
   Vertices: $(1, -2), (1, 8)$
   Foci: $(1, -1), (1, 7)$

35. Center: $(-2, 4)$
   Vertices: $(-1, 4), (-3, 4)$
   Foci: $\left(-2 \pm \frac{\sqrt{3}}{2}, 4\right)$

37. Center: $(-2, 3)$
   Vertices: $(-2, 6)$,
      $(-2, 0)$
   Foci: $\left(-2, 3 \pm \sqrt{5}\right)$

39. Center: $(1, -1)$
   Vertices: $\left(\frac{9}{4}, -1\right)$,
      $\left(-\frac{1}{4}, -1\right)$
   Foci: $\left(\frac{7}{4}, -1\right), \left(\frac{1}{4}, -1\right)$

41. $\dfrac{(x - 2)^2}{9} + \dfrac{(y - 2)^2}{4} = 1$

43. $\dfrac{(x - 2)^2}{4} + \dfrac{(y - 2)^2}{1} = 1$ **45.** $\dfrac{x^2}{48} + \dfrac{(y - 4)^2}{64} = 1$

47. $\dfrac{(x - 3)^2}{9} + \dfrac{(y - 5)^2}{16} = 1$ **49.** $\dfrac{x^2}{16} + \dfrac{(y - 4)^2}{12} = 1$

51. Center: $(-1, 2)$
   Vertices: $(-3, 2), (1, 2)$
   Foci: $\left(-1 \pm \sqrt{5}, 2\right)$

53. Center: $(2, -6)$
   Vertices: $(2, -5), (2, -7)$
   Foci: $\left(2, -6 \pm \sqrt{2}\right)$

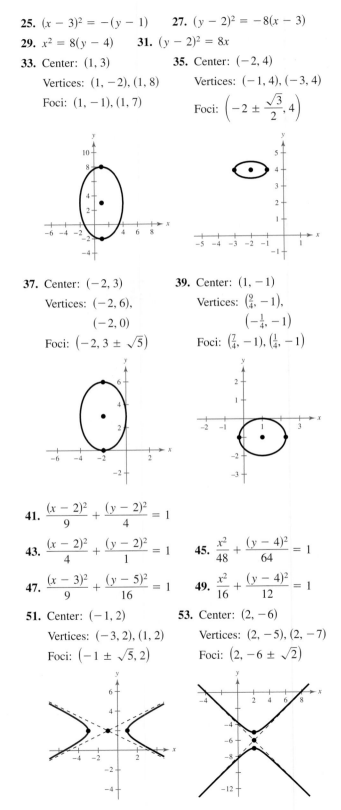

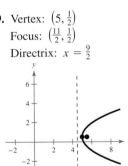

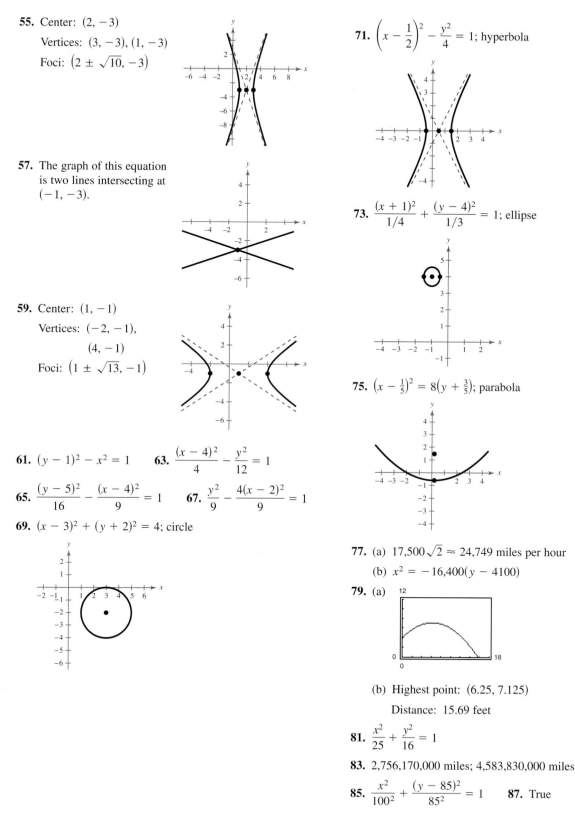

55. Center: $(2, -3)$
Vertices: $(3, -3), (1, -3)$
Foci: $\left(2 \pm \sqrt{10}, -3\right)$

57. The graph of this equation is two lines intersecting at $(-1, -3)$.

59. Center: $(1, -1)$
Vertices: $(-2, -1)$, $(4, -1)$
Foci: $\left(1 \pm \sqrt{13}, -1\right)$

61. $(y - 1)^2 - x^2 = 1$

63. $\dfrac{(x - 4)^2}{4} - \dfrac{y^2}{12} = 1$

65. $\dfrac{(y - 5)^2}{16} - \dfrac{(x - 4)^2}{9} = 1$

67. $\dfrac{y^2}{9} - \dfrac{4(x - 2)^2}{9} = 1$

69. $(x - 3)^2 + (y + 2)^2 = 4$; circle

71. $\left(x - \dfrac{1}{2}\right)^2 - \dfrac{y^2}{4} = 1$; hyperbola

73. $\dfrac{(x + 1)^2}{1/4} + \dfrac{(y - 4)^2}{1/3} = 1$; ellipse

75. $\left(x - \tfrac{1}{5}\right)^2 = 8\left(y + \tfrac{3}{5}\right)$; parabola

77. (a) $17{,}500\sqrt{2} \approx 24{,}749$ miles per hour
(b) $x^2 = -16{,}400(y - 4100)$

79. (a)

(b) Highest point: $(6.25, 7.125)$
Distance: 15.69 feet

81. $\dfrac{x^2}{25} + \dfrac{y^2}{16} = 1$

83. $2{,}756{,}170{,}000$ miles; $4{,}583{,}830{,}000$ miles

85. $\dfrac{x^2}{100^2} + \dfrac{(y - 85)^2}{85^2} = 1$ **87.** True

89. $y = \sqrt{6(x + 1)} + 3$ **91.** $x = \frac{3}{2}(2 + \sqrt{4 - y^2})$

93. (a) Answers will vary.

(b) $e = 0.95$ $e = 0.75$

(b)

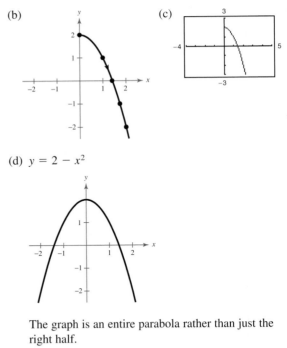

(c)

$e = 0.5$ $e = 0.25$

$e = 0$

(d) $y = 2 - x^2$

The graph is an entire parabola rather than just the right half.

(c) The ellipse becomes more circular.

95. $\sum_{n=1}^{9} \frac{1}{6n} = 0.4715$ **97.** $\sum_{n=0}^{8} \left(-\frac{1}{4}\right)^n = 0.80$

99. $x^4 - 16x^3 + 96x^2 - 256x + 256$

101. $243x^5 + 405x^4 + 270x^3 + 90x^2 + 15x + 1$

Section 10.3   (page 800)

Vocabulary Check   (page 800)

1. plane curve, parametric equations, parameter

2. orientation   **3.** eliminating, parameter

1. c   **2.** d   **3.** b   **4.** a   **5.** f   **6.** e

7. (a)

t	0	1	2	3	4
x	0	1	1.414	1.732	2
y	2	1	0	-1	-2

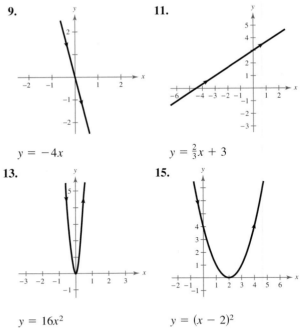

9.

$y = -4x$

11.

$y = \frac{2}{3}x + 3$

13.

$y = 16x^2$

15.

$y = (x - 2)^2$

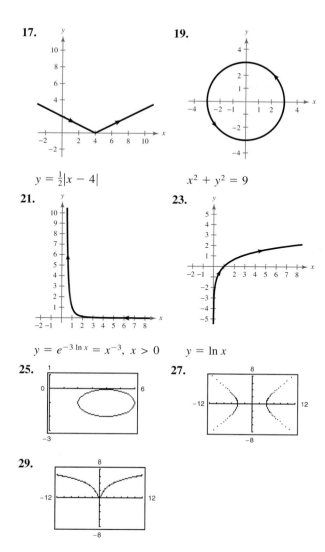

17.

$y = \frac{1}{2}|x - 4|$

19.

$x^2 + y^2 = 9$

21.

$y = e^{-3 \ln x} = x^{-3}, \; x > 0$

23.

$y = \ln x$

25.

27.

29.

31. Each curve represents a portion of the line $y = 2x + 1$.

Domain	Orientation
(a) $(-\infty, \infty)$	Left to right
(b) $[-1, 1]$	Depends on θ
(c) $(0, \infty)$	Right to left
(d) $(0, \infty)$	Left to right

33. $y - y_1 = \dfrac{y_2 - y_1}{x_2 - x_1}(x - x_1)$

35. $\dfrac{(x - h)^2}{a^2} + \dfrac{(y - k)^2}{b^2} = 1$

37. $x = 5t$
$y = -2t$

39. $x = 2 + 4\cos\theta$
$y = 1 + 4\sin\theta$

41. $x = 5\cos\theta$
$y = 3\sin\theta$

43. Answers will vary. Sample answers:
$x = t, \; y = 4t - 3$
$x = \frac{1}{4}t + \frac{3}{4}, \; y = t$

45. Answers will vary. Sample answers:
$x = t, \; y = \dfrac{1}{t}$
$x = t^3, \; y = \dfrac{1}{t^3}$

47. Answers will vary. Sample answers:
$x = t, \; y = t^2 + 4$
$x = t^3, \; y = t^6 + 4$

49. Answers will vary. Sample answers:
$x = t, \; y = t^3 + 2t$
$x = \frac{1}{2}t, \; y = \frac{1}{8}t^3 + t$

51.

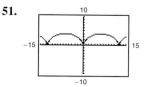

53.

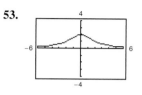

55. b   **56.** c   **57.** d   **58.** a

59. (a) $x = (146.67 \cos\theta)t$
$y = 3 + (146.67 \sin\theta)t - 16t^2$

(b) $x = 141.7t$
$y = 3 + 38.0t - 16t^2$

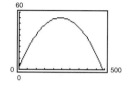

No

(c) $x = 135.0t$
$y = 3 + 57.3t - 16t^2$

Yes

(d) About $19.4°$

61. True. Both sets of parametric equations correspond to $y = x^2 + 1$.

63. $x = \cos \theta$ or $x = -\cos \theta$

$y = -2 \sin \theta$ $y = 2 \sin \theta$

65. Answers will vary.   **67.** $x = \pm \dfrac{2\sqrt{10}}{5} i$

69. $x = -\frac{1}{2} \pm \sqrt{3}$ **71.** 10,200   **73.** 11,590

Section 10.4   (page 807)

Vocabulary Check   (page 807)

1. pole   **2.** directed distance, directed angle

3. polar

---

1. $(0, 4)$ **3.** $\left(\dfrac{\sqrt{2}}{2}, \dfrac{\sqrt{2}}{2} \right)$

5.

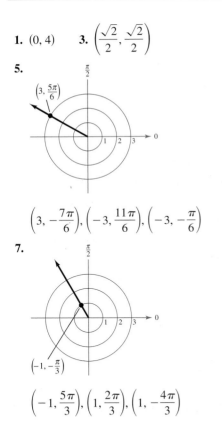

$\left(3, -\dfrac{7\pi}{6} \right), \left(-3, \dfrac{11\pi}{6} \right), \left(-3, -\dfrac{\pi}{6} \right)$

7.

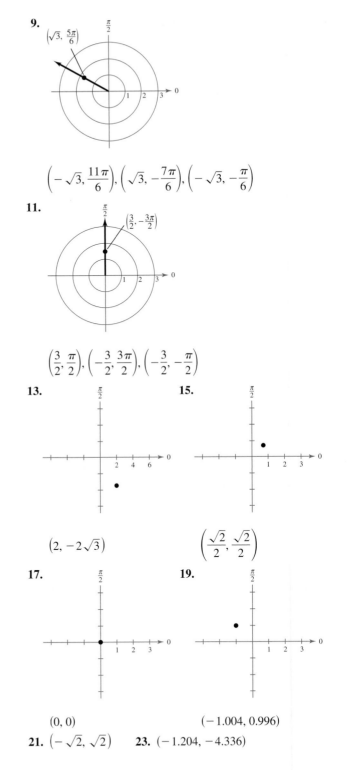

9.

$\left(-\sqrt{3}, \dfrac{11\pi}{6} \right), \left(\sqrt{3}, -\dfrac{7\pi}{6} \right), \left(-\sqrt{3}, -\dfrac{\pi}{6} \right)$

11.

$\left(\dfrac{3}{2}, \dfrac{\pi}{2} \right), \left(-\dfrac{3}{2}, \dfrac{3\pi}{2} \right), \left(-\dfrac{3}{2}, -\dfrac{\pi}{2} \right)$

13.   **15.**

$(2, -2\sqrt{3})$ $\left(\dfrac{\sqrt{2}}{2}, \dfrac{\sqrt{2}}{2} \right)$

17.   **19.**

$(0, 0)$ $(-1.004, 0.996)$

21. $\left(-\sqrt{2}, \sqrt{2} \right)$ **23.** $(-1.204, -4.336)$

25.

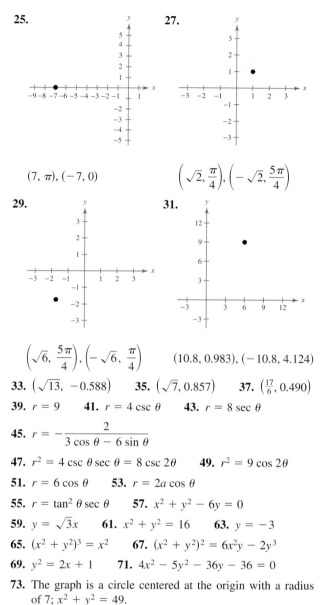

$(7, \pi), (-7, 0)$

27.

$\left(\sqrt{2}, \dfrac{\pi}{4}\right), \left(-\sqrt{2}, \dfrac{5\pi}{4}\right)$

29.

31.

$\left(\sqrt{6}, \dfrac{5\pi}{4}\right), \left(-\sqrt{6}, \dfrac{\pi}{4}\right)$ $(10.8, 0.983), (-10.8, 4.124)$

33. $\left(\sqrt{13}, -0.588\right)$ **35.** $\left(\sqrt{7}, 0.857\right)$ **37.** $\left(\frac{17}{6}, 0.490\right)$

39. $r = 9$ **41.** $r = 4 \csc \theta$ **43.** $r = 8 \sec \theta$

45. $r = -\dfrac{2}{3 \cos \theta - 6 \sin \theta}$

47. $r^2 = 4 \csc \theta \sec \theta = 8 \csc 2\theta$ **49.** $r^2 = 9 \cos 2\theta$

51. $r = 6 \cos \theta$ **53.** $r = 2a \cos \theta$

55. $r = \tan^2 \theta \sec \theta$ **57.** $x^2 + y^2 - 6y = 0$

59. $y = \sqrt{3}x$ **61.** $x^2 + y^2 = 16$ **63.** $y = -3$

65. $(x^2 + y^2)^3 = x^2$ **67.** $(x^2 + y^2)^2 = 6x^2y - 2y^3$

69. $y^2 = 2x + 1$ **71.** $4x^2 - 5y^2 - 36y - 36 = 0$

73. The graph is a circle centered at the origin with a radius of 7; $x^2 + y^2 = 49$.

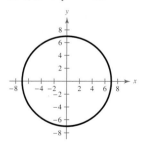

75. The graph consists of all points on the line that make an angle of $\pi/4$ with the positive x-axis; $x - y = 0$.

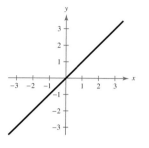

77. The graph is a vertical line through $(3, 0)$; $x - 3 = 0$.

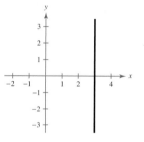

79. True. Because r is a directed distance, (r, θ) can be represented by $(-r, \theta \pm (2n + 1)\pi)$, so $|r| = |-r|$.

81. (a) Answers will vary.

(b) The points lie on a line.
$$d = \sqrt{r_1^2 + r_2^2 - 2r_1 r_2} = |r_1 - r_2|$$

(c) $d = \sqrt{r_1^2 + r_2^2}$ (Pythagorean Theorem)

Answers will vary.

(d) Answers will vary. The distance formulas should give the same results.

83. $A \approx 30.7°$ **85.** $a \approx 16.16$ **87.** $A \approx 119.1°$
$B \approx 48.2°$ $b \approx 19.44$ $B \approx 25.9°$
$C \approx 101.1°$ $B = 86°$ $c \approx 5.25$

89. $(2, 3)$ **91.** $(0, 0, 0)$

Section 10.5   (page 816)

Vocabulary Check   (page 816)

1. $\theta = \dfrac{\pi}{2}$ **2.** polar axis   **3.** convex limaçon

4. circle   **5.** lemniscate   **6.** cardioid

1. Rose curve   **3.** Circle   **5.** Rose curve

7. Polar axis   **9.** $\theta = \dfrac{\pi}{2}$ **11.** $\theta = \dfrac{\pi}{2}$ **13.** Pole

15. Maximum: $|r| = 20$ when $\theta = \dfrac{3\pi}{2}$

Zero: $r = 0$ when $\theta = \dfrac{\pi}{2}$

17. Maximum: $|r| = 4$ when $\theta = 0, \dfrac{\pi}{3}, \dfrac{2\pi}{3}, \pi$

Zeros: $r = 0$ when $\theta = \dfrac{\pi}{6}, \dfrac{\pi}{2}, \dfrac{5\pi}{6}$

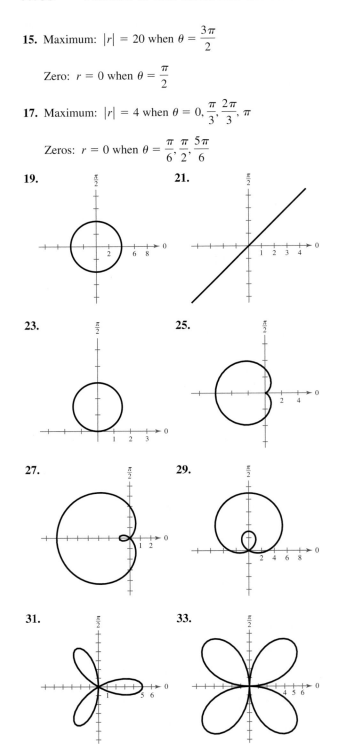

19.

21.

23.

25.

27.

29.

31.

33.

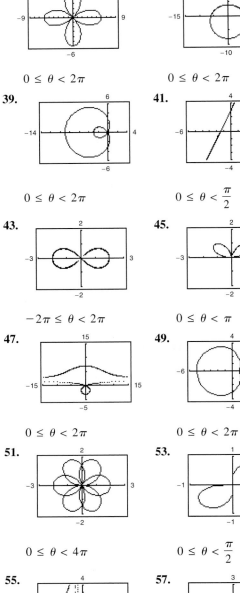

35.

37.

$0 \le \theta < 2\pi$

$0 \le \theta < 2\pi$

39.

41.

$0 \le \theta < 2\pi$

$0 \le \theta < \dfrac{\pi}{2}$

43.

45.

$-2\pi \le \theta < 2\pi$

$0 \le \theta < \pi$

47.

49.

$0 \le \theta < 2\pi$

$0 \le \theta < 2\pi$

51.

53.

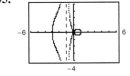

$0 \le \theta < 4\pi$

$0 \le \theta < \dfrac{\pi}{2}$

55.

57.

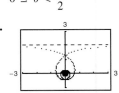

59. True

61. $n = -5$ $n = -4$

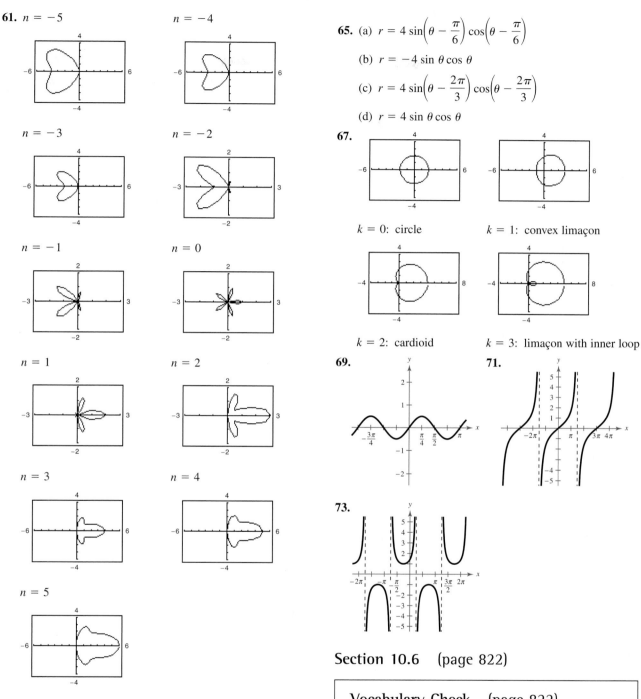

$n = -3$ $n = -2$

$n = -1$ $n = 0$

$n = 1$ $n = 2$

$n = 3$ $n = 4$

$n = 5$

Negative values of n produce the heart portion, positive values of n produce the bell portion.

63. (a), (b), and (c) Answers will vary.

65. (a) $r = 4 \sin\left(\theta - \dfrac{\pi}{6}\right) \cos\left(\theta - \dfrac{\pi}{6}\right)$

(b) $r = -4 \sin\theta \cos\theta$

(c) $r = 4 \sin\left(\theta - \dfrac{2\pi}{3}\right) \cos\left(\theta - \dfrac{2\pi}{3}\right)$

(d) $r = 4 \sin\theta \cos\theta$

67.

$k = 0$: circle $k = 1$: convex limaçon

$k = 2$: cardioid $k = 3$: limaçon with inner loop

69.   **71.**

73.

Section 10.6   (page 822)

Vocabulary Check   (page 822)

1. conic   **2.** eccentricity, e

3. (a) i   (b) iii   (c) ii

1. (a) parabola **3.** (a) parabola
    (b) ellipse   (b) ellipse
    (c) hyperbola   (c) hyperbola

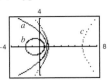 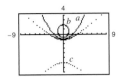

5. b **6.** c **7.** d **8.** a

9. Parabola **11.** Ellipse **13.** Ellipse

15. Ellipse **17.** Hyperbola

19. 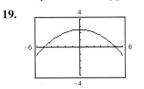 **21.**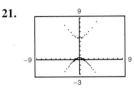

Parabola   Hyperbola

23. **25.**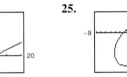

27. $r = \dfrac{1}{1 - \cos\theta}$ **29.** $r = \dfrac{1}{2 + \sin\theta}$

31. $r = \dfrac{2}{1 + 2\cos\theta}$ **33.** $r = \dfrac{2}{1 - \sin\theta}$

35. $r = \dfrac{10}{3 + 2\cos\theta}$ **37.** $r = \dfrac{9}{4 - 5\sin\theta}$

39. Answers will vary.

41. $r = \dfrac{9.2930 \times 10^7}{1 - 0.0167\cos\theta}$

Perihelion: 9.1404×10^7 miles

Aphelion: 9.4508×10^7 miles

43. $r = \dfrac{7.7659 \times 10^8}{1 - 0.0484\cos\theta}$

Perihelion: 7.4073×10^8 kilometers

Aphelion: 8.1609×10^8 kilometers

45. (a) $r_{\text{Neptune}} = \dfrac{4.4977 \times 10^9}{1 - 0.0086\cos\theta}$

$r_{\text{Pluto}} = \dfrac{5.5404 \times 10^9}{1 - 0.2488\cos\theta}$

(b) Neptune: Perihelion $= 4.4593 \times 10^9$ kilometers

Aphelion $= 4.5367 \times 10^9$ kilometers

Pluto: Perihelion $= 4.4366 \times 10^9$ kilometers

Aphelion $= 7.3754 \times 10^9$ kilometers

(c)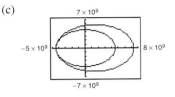

(d) No. Answers will vary.

(e) Answers will vary.

47. False. The equation can be rewritten as

$r = \dfrac{-4/3}{1 + \sin\theta}.$

Because ep is negative, you know that e is positive and p is negative and that p represents the distance between the pole and the directrix, so the directrix has to be below the pole.

49. Answers will vary. **51.** $r^2 = \dfrac{24{,}336}{169 - 25\cos^2\theta}$

53. (a) Ellipse

(b) $r = \dfrac{4}{1 + 0.4\cos\theta}$ is reflected about the line $\theta = \dfrac{\pi}{2}$.

$r = \dfrac{4}{1 - 0.4\sin\theta}$ is rotated $90°$ counterclockwise.

55. Answers will vary. **57.** $\dfrac{\pi}{6} + n\pi$

59. $\dfrac{\pi}{3} + n\pi, \dfrac{2\pi}{3} + n\pi$

61. $\dfrac{\sqrt{2}}{10}$ **63.** $\dfrac{\sqrt{2}}{10}$ **65.** 220 **67.** 720

Review Exercises (page 826)

1. d **2.** a **3.** c **4.** b **5.** f **6.** e

7. $x^2 = \frac{1}{2}y$ **9.** $y^2 = -24x$

11. Vertex: $(0, 0)$ **13.** Vertex: $(0, 0)$

Focus: $(1, 0)$ Focus: $(-9, 0)$

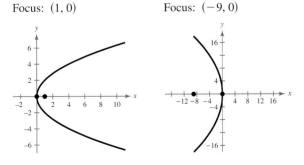

15. $(0, 50)$ **17.** $\dfrac{x^2}{25} + \dfrac{y^2}{9} = 1$ **19.** $\dfrac{2x^2}{9} + \dfrac{y^2}{36} = 1$

21. Center: $(0, 0)$ **23.** Center: $(0, 0)$
   Vertices: $(0, \pm 4)$ Vertices: $(0, \pm 3)$

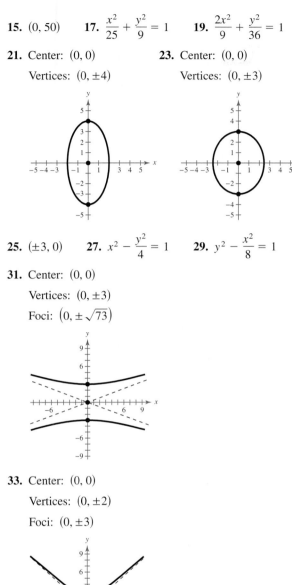

25. $(\pm 3, 0)$ **27.** $x^2 - \dfrac{y^2}{4} = 1$ **29.** $y^2 - \dfrac{x^2}{8} = 1$

31. Center: $(0, 0)$
   Vertices: $(0, \pm 3)$
   Foci: $\left(0, \pm \sqrt{73}\right)$

33. Center: $(0, 0)$
   Vertices: $(0, \pm 2)$
   Foci: $(0, \pm 3)$

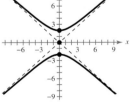

35. $(x - 3)^2 = -2y$; parabola

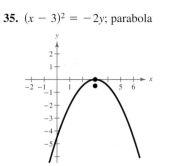

The parabola is shifted three units to the right.

37. $\dfrac{(x + 5)^2}{9} + \dfrac{(y - 1)^2}{1} = 1$; ellipse

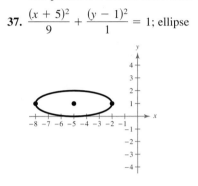

The ellipse is shifted five units to the left and one unit upward.

39. $\dfrac{\left(x - \frac{1}{2}\right)^2}{2} - \dfrac{(y - 1)^2}{2} = 1$; hyperbola

The hyperbola is shifted $\frac{1}{2}$ unit to the right and one unit upward.

41. $\dfrac{(x - 2)^2}{\frac{1}{4}} + y^2 = 1$; ellipse

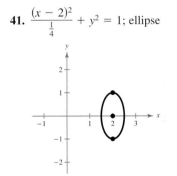

The ellipse is shifted two units to the right.

43. $(x - 4)^2 = -8(y - 2)$

45. $(y - 2)^2 = 12x$

47. $\dfrac{(x - 2)^2}{25} + \dfrac{y^2}{21} = 1$　　**49.** $\dfrac{(x - 2)^2}{4} + (y - 1)^2 = 1$

51. $\dfrac{(x + 2)^2}{64} - \dfrac{(y - 3)^2}{36} = 1$　　**53.** $\dfrac{5(x - 4)^2}{16} - \dfrac{5y^2}{64} = 1$

55. (a) $y = 4 - \frac{1}{4}x^2$

$x^2 + \left(y + 4\sqrt{3}\right)^2 = 64$

(b)

x	0	1	2	3	4
d	2.928	2.741	2.182	1.262	0

57.

t	-2	-1	0	1	2	3
x	-8	-5	-2	1	4	7
y	15	11	7	3	-1	-5

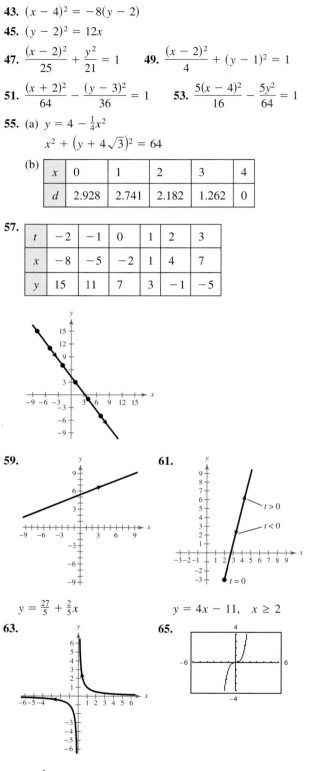

59.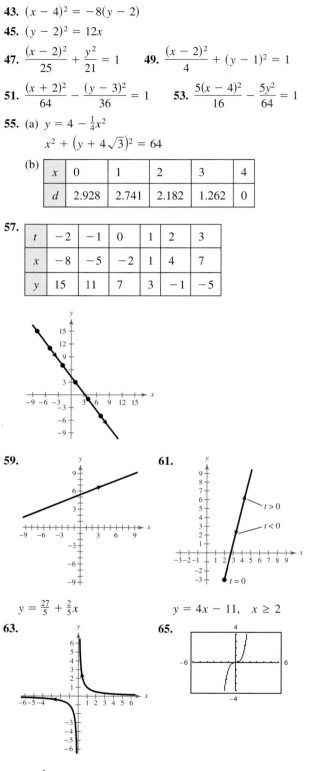

$y = \frac{27}{5} + \frac{2}{5}x$

61.

$y = 4x - 11, \quad x \geq 2$

63.

$y = \dfrac{1}{x}$

65.

67.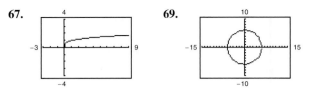

69.

71. Answers will vary. Sample answers:

$x = t,\ y = 6t + 2;\ x = 3t,\ y = 18t + 2$

73. Answers will vary. Sample answers:

$x = t,\ y = t^2 + 2;\ x = -t,\ y = t^2 + 2$

75.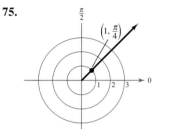

$\left(1, -\dfrac{7\pi}{4}\right), \left(-1, \dfrac{5\pi}{4}\right), \left(-1, -\dfrac{3\pi}{4}\right)$

77.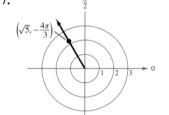

$\left(\sqrt{5}, \dfrac{2\pi}{3}\right), \left(-\sqrt{5}, -\dfrac{\pi}{3}\right), \left(-\sqrt{5}, \dfrac{5\pi}{3}\right)$

79.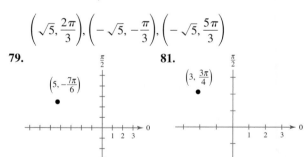

$\left(-\dfrac{5\sqrt{3}}{2}, \dfrac{5}{2}\right)$

81.

$\left(-\dfrac{3\sqrt{2}}{2}, \dfrac{3\sqrt{2}}{2}\right)$

83.

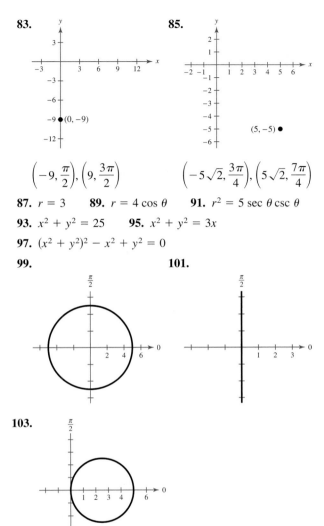

85.

$\left(-9, \dfrac{\pi}{2}\right), \left(9, \dfrac{3\pi}{2}\right)$ $\left(-5\sqrt{2}, \dfrac{3\pi}{4}\right), \left(5\sqrt{2}, \dfrac{7\pi}{4}\right)$

87. $r = 3$ **89.** $r = 4\cos\theta$ **91.** $r^2 = 5\sec\theta\csc\theta$

93. $x^2 + y^2 = 25$ **95.** $x^2 + y^2 = 3x$

97. $(x^2 + y^2)^2 - x^2 + y^2 = 0$

99.   **101.**

103.

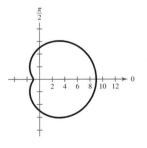

105. Dimpled limaçon

Symmetry: Polar axis
Maximum: $|r| = 9$ when $\theta = 0, 2\pi$
Zeros of r: None

107. Limaçon with a loop

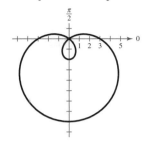

Symmetry: The line $\theta = \dfrac{\pi}{2}$

Maximum: $|r| = 8$ when $\theta = \dfrac{3\pi}{2}$

Zeros: $r = 0$ when $\theta = 0.644, 2.498$

109. Rose curve

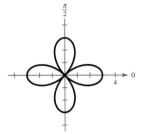

Symmetry: Pole, polar axis, and the line $\theta = \dfrac{\pi}{2}$

Maximum: $|r| = 3$ when $\theta = 0, \dfrac{\pi}{2}, \pi, \dfrac{3\pi}{2}, 2\pi$

Zeros of r: $r = 0$ when $\theta = \dfrac{\pi}{4}, \dfrac{3\pi}{4}, \dfrac{5\pi}{4}, \dfrac{7\pi}{4}$

111. Lemniscate

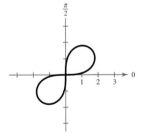

Symmetry: With respect to the pole

Maximum: $|r| = \sqrt{5}$ when $\theta = \dfrac{\pi}{4}, \dfrac{5\pi}{4}$

Zeros: $r = 0$ when $\theta = 0, \dfrac{\pi}{2}, \pi, \dfrac{3\pi}{2}$

113. Parabola   **115.** Ellipse

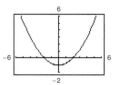

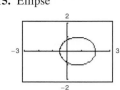

117. Ellipse

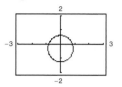

119. $r = \dfrac{4}{1 - \cos \theta}$ **121.** $r = \dfrac{5}{3 - 2 \cos \theta}$

123. $r = \dfrac{1.512}{1 - 0.093 \cos \theta}$

Perihelion: 1.3833 astronomical units

Aphelion: 1.6670 astronomical units

125. False. The equation of a hyperbola is a second-degree equation.

127. (a) Vertical translation

    (b) Horizontal translation

    (c) Reflection in the y-axis

    (d) Parabola opens more slowly.

129. 5; The ellipse becomes more circular and approaches a circle of radius 5.

131. (a) The speed would double.

    (b) The elliptical orbit would be flatter. The length of the major axis is greater.

Chapter Test   (page 830)

1.

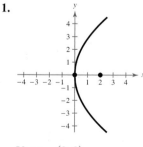

Vertex: $(0, 0)$

Focus: $(2, 0)$

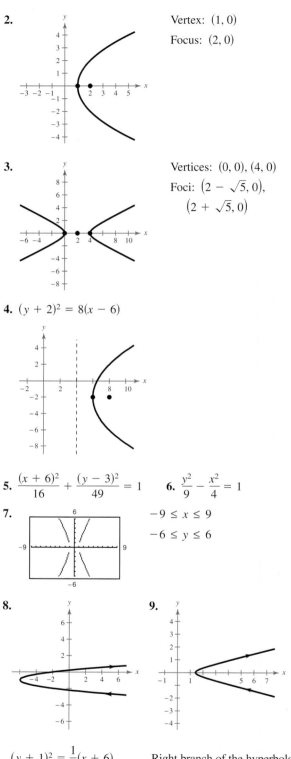

2. Vertex: $(1, 0)$

Focus: $(2, 0)$

3. Vertices: $(0, 0), (4, 0)$

Foci: $\left(2 - \sqrt{5}, 0\right),$

$\left(2 + \sqrt{5}, 0\right)$

4. $(y + 2)^2 = 8(x - 6)$

5. $\dfrac{(x + 6)^2}{16} + \dfrac{(y - 3)^2}{49} = 1$ **6.** $\dfrac{y^2}{9} - \dfrac{x^2}{4} = 1$

7. $-9 \le x \le 9$

$-6 \le y \le 6$

8.   **9.**

$(y + 1)^2 = \dfrac{1}{4}(x + 6)$

Right branch of the hyperbola

$\dfrac{x^2}{2} - \dfrac{y^2}{0.125} = 1$

10.

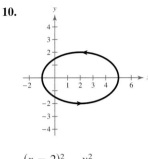

$$\frac{(x-2)^2}{9} + \frac{y^2}{4} = 1$$

11. Answers will vary. Sample answers:

$x = t, y = 7t + 6; x = 2t, y = 14t + 6$

12. Answers will vary. Sample answers:

$x = t, y = t^2 + 10; x = 2t, y = 4t^2 + 10$

13. Answers will vary. Sample answers:

$x = t, y = \frac{1}{4}t - 5; x = 2t, y = \frac{1}{2}t - 5$

14. $(x, y) = \left(-7, 7\sqrt{3}\right)$

15. $\left(2\sqrt{2}, \frac{7\pi}{4}\right), \left(2\sqrt{2}, -\frac{\pi}{4}\right), \left(-2\sqrt{2}, \frac{3\pi}{4}\right)$

16. $r = 12 \sin \theta$

17. Limaçon

18. Rose curve

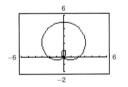

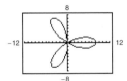

19. Ellipse

20. Hyperbola

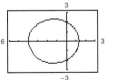

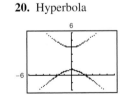

21. $r = \dfrac{4}{4 + \sin \theta}$ **22.** $r = \dfrac{10}{4 + 5 \sin \theta}$

Cumulative Test for Chapters 8–10
(page 831)

1. $(4, -3)$ **2.** $(8, 4), (2, -2)$ **3.** $(0.6, -4, -0.2)$

4. $(1, -4, -4)$

5. $\begin{bmatrix} -7 & -10 & -16 \\ -6 & 18 & 9 \\ -12 & 16 & 7 \end{bmatrix}$ **6.** $\begin{bmatrix} -18 & 15 & -14 \\ 28 & 11 & 34 \\ -20 & 52 & -1 \end{bmatrix}$

7. $\begin{bmatrix} 3 & -31 & 2 \\ 22 & 18 & 6 \\ 52 & -40 & 14 \end{bmatrix}$ **8.** $\begin{bmatrix} 5 & 36 & 31 \\ -36 & 12 & -36 \\ 16 & 0 & 18 \end{bmatrix}$

9. (a) $\begin{bmatrix} -175 & 37 & -13 \\ 95 & -20 & 7 \\ 14 & -3 & 1 \end{bmatrix}$

(b) 1

10. 22 square units

11. (a) $\frac{1}{5}, -\frac{1}{7}, \frac{1}{9}, -\frac{1}{11}, \frac{1}{13}$ (b) 3, 6, 12, 24, 48

12. 135   **13.** $\frac{47}{52} \approx 0.9038$ **14.** ≈ 34.4795 **15.** 80

16. Answers will vary.

17. $x^4 + 12x^3 + 54x^2 + 108x + 81$

18. $32x^5 + 80x^4y^2 + 80x^3y^4 + 40x^2y^6 + 10xy^8 + y^{10}$

19. $x^6 - 12x^5y + 60x^4y^2 - 160x^3y^3 + 240x^2y^4$
   $- 192xy^5 + 64y^6$

20. $6561a^8 - 69,984a^7b + 326,592a^6b^2 - 870,912a^5b^3$
   $+ 1,451,520a^4b^4 - 1,548,288a^3b^5 + 1,032,192a^2b^6$
   $- 393,216ab^7 + 65,536b^8$

21. 453,600   **22.** 151,200

23. Hyperbola

24. Ellipse

25. Hyperbola

26. Circle

27. $(x - 2)^2 = -\frac{4}{3}(y - 3)$ **28.** $\dfrac{(x-1)^2}{25} + \dfrac{(y-4)^2}{4} = 1$

29. $\dfrac{(y+4)^2}{4} - \dfrac{x^2}{16/3} = 1$

30. (a)

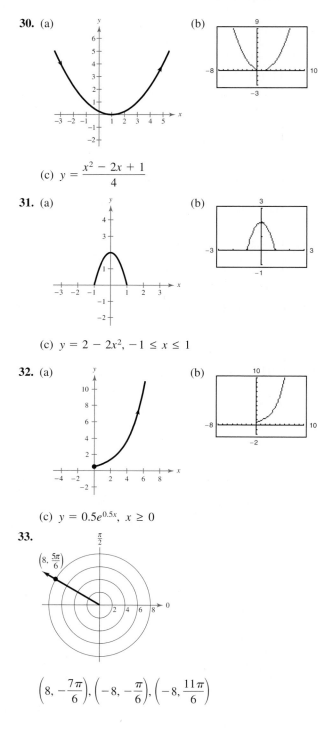

(b)

(c) $y = \dfrac{x^2 - 2x + 1}{4}$

31. (a)

(b)

(c) $y = 2 - 2x^2, \ -1 \le x \le 1$

32. (a)

(b)

(c) $y = 0.5e^{0.5x}, \ x \ge 0$

33.

$\left(8, -\dfrac{7\pi}{6}\right), \left(-8, -\dfrac{\pi}{6}\right), \left(-8, \dfrac{11\pi}{6}\right)$

34.

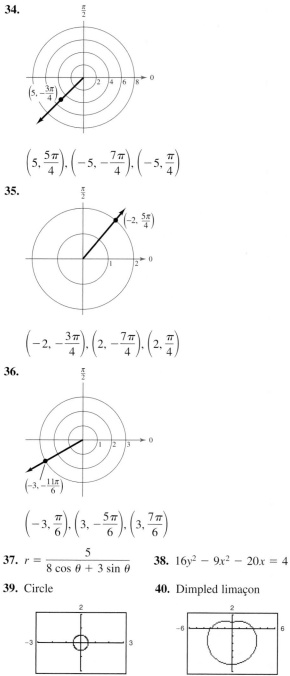

$\left(5, \dfrac{5\pi}{4}\right), \left(-5, -\dfrac{7\pi}{4}\right), \left(-5, \dfrac{\pi}{4}\right)$

35.

$\left(-2, -\dfrac{3\pi}{4}\right), \left(2, -\dfrac{7\pi}{4}\right), \left(2, \dfrac{\pi}{4}\right)$

36.

$\left(-3, \dfrac{\pi}{6}\right), \left(3, -\dfrac{5\pi}{6}\right), \left(3, \dfrac{7\pi}{6}\right)$

37. $r = \dfrac{5}{8 \cos \theta + 3 \sin \theta}$

38. $16y^2 - 9x^2 - 20x = 4$

39. Circle

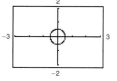

40. Dimpled limaçon

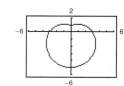

41. Limaçon

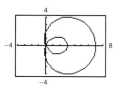

42. $604,199.78 \approx \$604,000$ **43.** $\frac{1}{4}$

44. $24\sqrt{2}$ meters ≈ 33.9 meters

Appendices

Appendix C.1   (page A44)

> ### Vocabulary Check   (page A44)
>
> **1.** measure, central tendency   **2.** modes, bimodal
>
> **3.** variance, standard deviation   **4.** Quartiles

1. Mean: 8.86; median: 8; mode: 7

3. Mean: 10.29; median: 8; mode: 7

5. Mean: 9; median: 8; mode: 7

7. (a) The mean is sensitive to extreme values.

 (b) Mean: 14.86; median: 14; mode: 13

  Each is increased by 6.

 (c) Each will increase by k.

9. Mean: 320; median: 320; mode: 320

11. One possibility: $\{4, 4, 10\}$

13. The median gives the most representative description.

15. (a) $\bar{x} = 12$; $\sigma = 2.83$ (b) $\bar{x} = 20$; $\sigma = 2.83$

 (c) $\bar{x} = 12$; $\sigma = 1.41$ (d) $\bar{x} = 9$; $\sigma = 1.41$

17. $\bar{x} = 6$, $v = 10$, $\sigma = 3.16$

19. $\bar{x} = 2$, $v = \frac{4}{3}$, $\sigma = 1.15$ **21.** $\bar{x} = 4$, $v = 4$, $\sigma = 2$

23. $\bar{x} = 47$, $v = 226$, $\sigma = 15.03$ **25.** 3.42   **27.** 1.65

29. $\bar{x} = 12$ and $|x_i - 12| = 8$ for all x_i.

31. The mean will increase by 5, but the standard deviation will not change.

33. First histogram

35. (a) Upper quartile: 21.5

  Lower quartile: 13

 (b)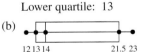
  12 13 14   21.5 23

37. (a) Upper quartile: 51

  Lower quartile: 47

 (b)
  46 47   48.5   51   53

39.
 9
  11.5 14   18 19

41.
 17.3   43.4
  24.1 34.9
 21.8

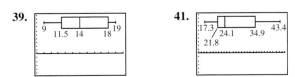

43.

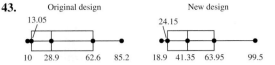

Original design   New design

13.05   24.15

10 28.9   62.6 85.2   18.9 41.35 63.95   99.5

From the plots, you can see that the lifetimes of the sample units made by the new design are greater than the lifetimes of the sample units made by the original design. (The median lifetime increased by more than 12 months.)

Appendix C.2   (page A48)

1. $y = 1.6x + 7.5$ **3.** $y = 0.262x + 1.93$

Appendix D   (page A51)

> ### Vocabulary Check   (page A51)
>
> **1.** linear   **2.** equivalent equations

1. 4   **3.** 7   **5.** 4   **7.** 20   **9.** 4   **11.** 3

13. 5   **15.** -10 **17.** No solution   **19.** $-\frac{6}{5}$

21. 9   **23.** $x < 2$ **25.** $x < 9$ **27.** $x \le -14$

29. $x > 10$ **31.** $x < 4$ **33.** $x < 3$ **35.** $x \ge 2$

37. $x \le -5$ **39.** $x < 6$ **41.** $x \ge 4$ **43.** $x \ge -4$

Appendix E.1   (page A59)

> ### Vocabulary Check   (page A59)
>
> **1.** solution   **2.** graph   **3.** linear
>
> **4.** point, equilibrium

1. g   **2.** d   **3.** a   **4.** h   **5.** e   **6.** b

7. f   **8.** c

9.   **11.**

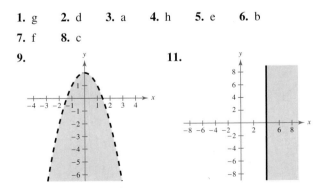

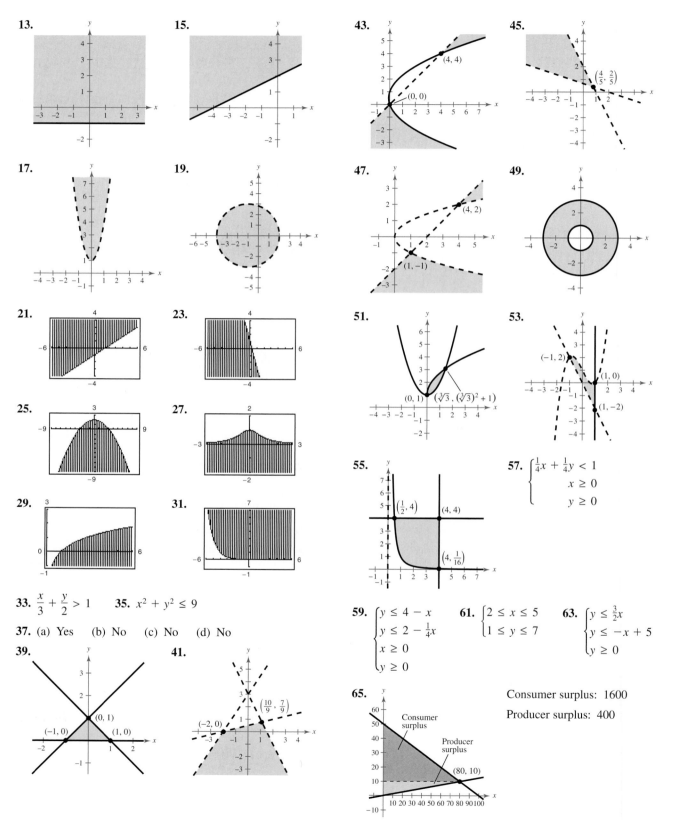

13. **15.**

17. **19.**

21. **23.**

25. **27.**

29. **31.**

33. $\dfrac{x}{3} + \dfrac{y}{2} > 1$ **35.** $x^2 + y^2 \le 9$

37. (a) Yes   (b) No   (c) No   (d) No

39. **41.**

43. **45.**

47. **49.**

51. **53.**

55. **57.** $\begin{cases} \frac{1}{4}x + \frac{1}{4}y < 1 \\ \quad\quad x \ge 0 \\ \quad\quad y \ge 0 \end{cases}$

59. $\begin{cases} y \le 4 - x \\ y \le 2 - \frac{1}{4}x \\ x \ge 0 \\ y \ge 0 \end{cases}$ **61.** $\begin{cases} 2 \le x \le 5 \\ 1 \le y \le 7 \end{cases}$ **63.** $\begin{cases} y \le \frac{3}{2}x \\ y \le -x + 5 \\ y \ge 0 \end{cases}$

65. Consumer surplus: 1600

Producer surplus: 400

67.

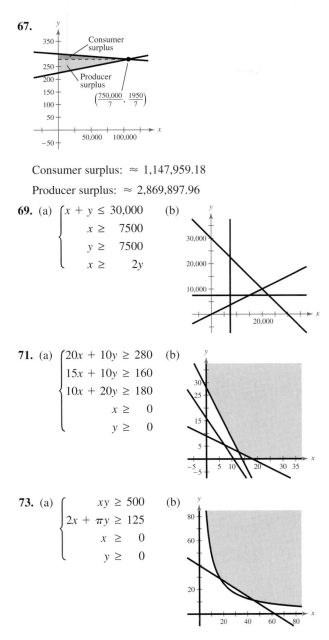

Consumer surplus: $\approx 1,147,959.18$

Producer surplus: $\approx 2,869,897.96$

69. (a) $\begin{cases} x + y \le 30,000 \\ \quad x \ge \quad 7500 \\ \quad y \ge \quad 7500 \\ \quad x \ge \quad 2y \end{cases}$ (b)

71. (a) $\begin{cases} 20x + 10y \ge 280 \\ 15x + 10y \ge 160 \\ 10x + 20y \ge 180 \\ \qquad x \ge \quad 0 \\ \qquad y \ge \quad 0 \end{cases}$ (b)

73. (a) $\begin{cases} \quad xy \ge 500 \\ 2x + \pi y \ge 125 \\ \quad x \ge \quad 0 \\ \quad y \ge \quad 0 \end{cases}$ (b)

75. True   **77.** Test a point on either side.

Appendix E.2   (page A68)

Vocabulary Check   (page A68)

1. optimization   **2.** objective function

3. constraints, feasible solutions

1. Minimum at $(0, 0)$: 0

Maximum at $(0, 6)$: 30

3. Minimum at $(0, 0)$: 0

Maximum at $(6, 0)$: 60

5. Minimum at $(0, 0)$: 0

Maximum at $(3, 4)$: 17

7. Minimum at $(0, 0)$: 0

Maximum at $(4, 0)$: 20

9. Minimum at $(0, 0)$: 0

Maximum at $(60, 20)$: 740

11. Minimum at $(0, 0)$: 0

Maximum at any point on the line segment connecting $(60, 20)$ and $(30, 45)$: 2100

13.

15.

Minimum at $(0, 0)$: 0

Maximum at $(5, 0)$: 30

Minimum at $(0, 0)$: 0

Maximum at $(5, 8)$: 47

17.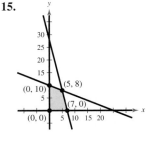

19.

Minimum at $(24, 8)$: 104

Maximum at $(40, 0)$: 160

Minimum at $(36, 0)$: 36

Maximum at $(24, 8)$: 56

21.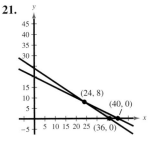

Minimum at any point on the line segment connecting $(24, 8)$ and $(36, 0)$: 72

Maximum at $(40, 0)$: 80

23. (a) and (b)   (c) $(3, 6)$

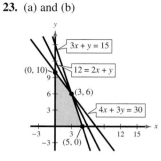

25. (a) and (b)   (c) $(0, 10)$

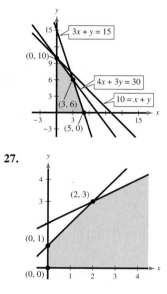

27.

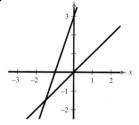

The constraints do not form a closed set of points. Therefore, $z = x + y$ is unbounded.

29.

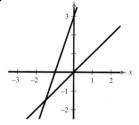

The feasible set is empty.

31. Four audits, 32 tax returns

Maximum revenue: $17,600

33. Three bags of Brand X, six bags of Brand Y

Minimum cost: $195

35. True   **37.** $z = x + 5y$ (Answer is not unique.)

39. $z = 4x + y$ (Answer is not unique.)

41. (a) $t > 9$ (b) $\frac{3}{4} < t < 9$

Index of Applications

Index

COMMON FORMULAS

Temperature

$$F = \frac{9}{5}C + 32$$

F = degrees Fahrenheit
C = degrees Celsius

Distance

$$d = rt$$

d = distance traveled
t = time
r = rate

Simple Interest

$$I = Prt$$

I = interest
P = principal
r = annual interest rate
t = time in years

Compound Interest

$$A = P\left(1 + \frac{r}{n}\right)^{nt}$$

A = balance
P = principal
r = annual interest rate
n = compoundings per year
t = time in years

Coordinate Plane: Midpoint Formula

$$\left(\frac{x_1 + x_2}{2}, \frac{y_1 + y_2}{2}\right)$$

midpoint of line segment
joining (x_1, y_1) and (x_2, y_2)

Coordinate Plane: Distance Formula

$$d = \sqrt{(x_2 - x_1)^2 + (y_2 - y_1)^2}$$

d = distance
between points
(x_1, y_1) and (x_2, y_2)

Quadratic Formula

If $p(x) = ax^2 + bx + c$, $a \neq 0$ and $b^2 - 4ac \geq 0$, then the real zeros of p are

$$x = \frac{-b \pm \sqrt{b^2 - 4ac}}{2a}.$$

CONVERSIONS

Length and Area

1 foot = 12 inches
1 mile = 5280 feet
1 kilometer = 1000 meters
1 kilometer ≈ 0.621 mile
1 meter ≈ 3.281 feet
1 foot ≈ 0.305 meter

1 yard = 3 feet
1 mile = 1760 yards
1 meter = 100 centimeters
1 mile ≈ 1.609 kilometers
1 meter ≈ 39.370 inches
1 foot ≈ 30.480 centimeters

1 meter = 1000 millimeters
1 centimeter ≈ 0.394 inch
1 inch ≈ 2.540 centimeters
1 acre = 4840 square yards
1 square mile = 640 acres

Volume

1 gallon = 4 quarts
1 gallon = 231 cubic inches
1 liter = 1000 milliliters
1 liter ≈ 1.057 quarts
1 gallon ≈ 3.785 liters

1 quart = 2 pints
1 gallon ≈ 0.134 cubic foot
1 liter = 100 centiliters
1 liter ≈ 0.264 gallon
1 quart ≈ 0.946 liter

1 pint = 16 fluid ounces
1 cubic foot ≈ 7.48 gallons

Weight and Mass on Earth

1 ton = 2000 pounds
1 kilogram ≈ 2.205 pounds

1 pound = 16 ounces
1 pound ≈ 0.454 kilogram

1 kilogram = 1000 grams
1 gram ≈ 0.035 ounce

FORMULAS FROM GEOMETRY

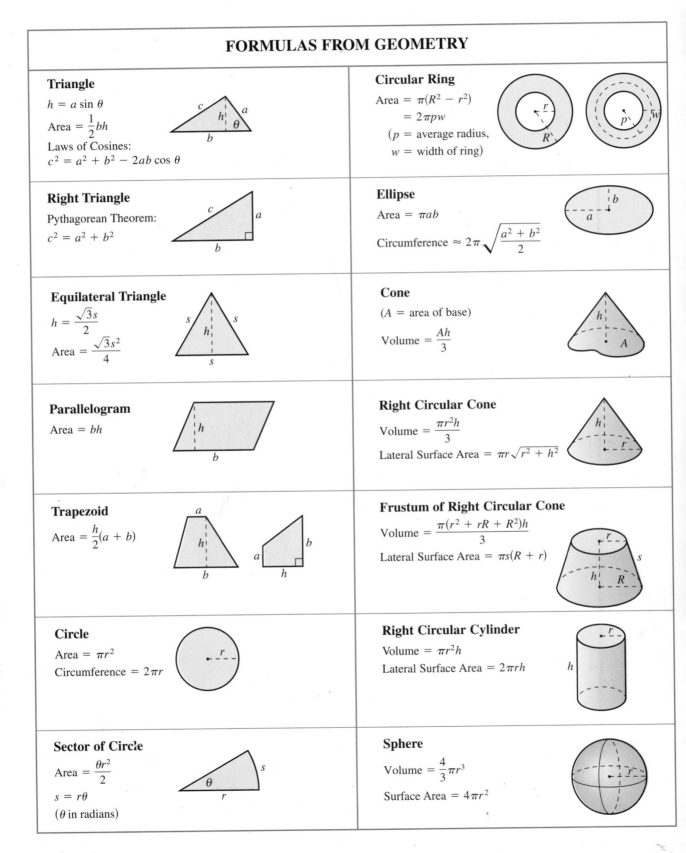

Triangle

$h = a \sin \theta$

$\text{Area} = \dfrac{1}{2}bh$

Laws of Cosines:

$c^2 = a^2 + b^2 - 2ab \cos \theta$

Right Triangle

Pythagorean Theorem:

$c^2 = a^2 + b^2$

Equilateral Triangle

$h = \dfrac{\sqrt{3}s}{2}$

$\text{Area} = \dfrac{\sqrt{3}s^2}{4}$

Parallelogram

$\text{Area} = bh$

Trapezoid

$\text{Area} = \dfrac{h}{2}(a + b)$

Circle

$\text{Area} = \pi r^2$

$\text{Circumference} = 2\pi r$

Sector of Circle

$\text{Area} = \dfrac{\theta r^2}{2}$

$s = r\theta$

(θ in radians)

Circular Ring

$\text{Area} = \pi(R^2 - r^2)$

$\quad\quad = 2\pi pw$

(p = average radius,

w = width of ring)

Ellipse

$\text{Area} = \pi ab$

$\text{Circumference} \approx 2\pi\sqrt{\dfrac{a^2 + b^2}{2}}$

Cone

(A = area of base)

$\text{Volume} = \dfrac{Ah}{3}$

Right Circular Cone

$\text{Volume} = \dfrac{\pi r^2 h}{3}$

$\text{Lateral Surface Area} = \pi r\sqrt{r^2 + h^2}$

Frustum of Right Circular Cone

$\text{Volume} = \dfrac{\pi(r^2 + rR + R^2)h}{3}$

$\text{Lateral Surface Area} = \pi s(R + r)$

Right Circular Cylinder

$\text{Volume} = \pi r^2 h$

$\text{Lateral Surface Area} = 2\pi rh$

Sphere

$\text{Volume} = \dfrac{4}{3}\pi r^3$

$\text{Surface Area} = 4\pi r^2$